eleventh edition

A Problem Solving Approach to

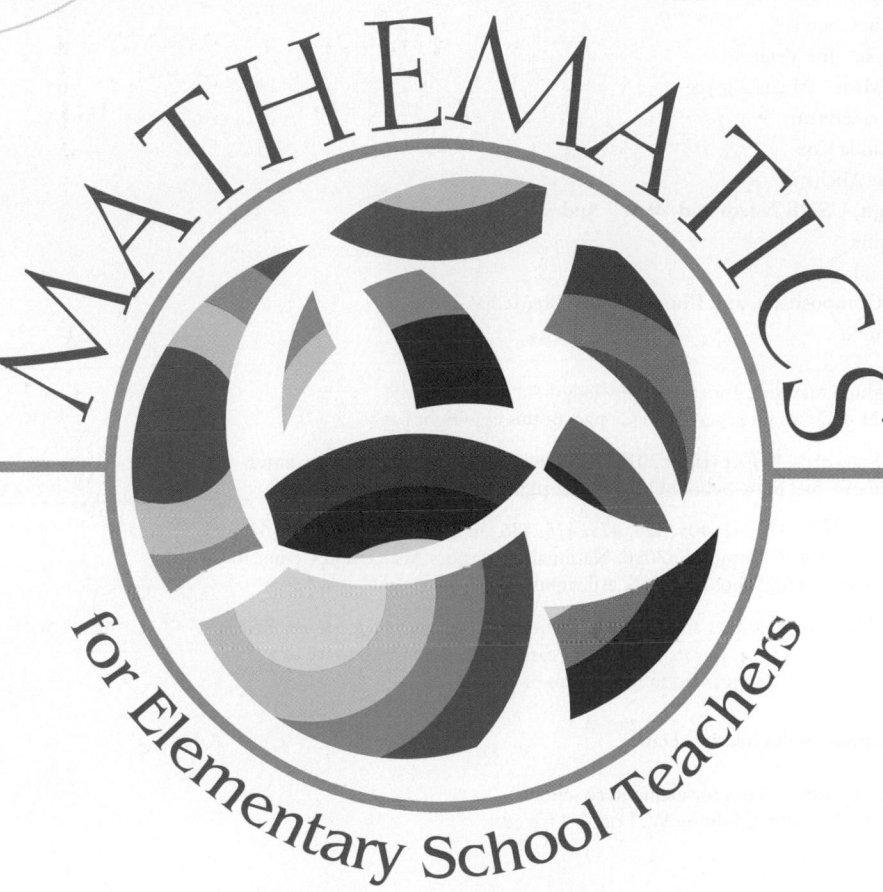

MATHEMATICS
for Elementary School Teachers

RICK BILLSTEIN
University of Montana

SHLOMO LIBESKIND
University of Oregon

JOHNNY W. LOTT
University of Montana

PEARSON

Boston   Columbus   Indianapolis   New York   San Francisco   Upper Saddle River
Amsterdam   Cape Town   Dubai   London   Madrid   Milan   Munich   Paris   Montreal   Toronto
Delhi   Mexico City   Sao Paulo   Sydney   Hong Kong   Seoul   Singapore   Taipei   Tokyo

Editor in Chief Anne Kelly
Acquisitions Editor Marnie Greenhut
Senior Content Editor Rachel Reeve
Assistant Editor Elle Driska
Senior Managing Editor Karen Wernholm
Senior Production Project Manager Patty Bergin
Digital Assets Manager Marianne Groth
Supplements Production Coordinator Kerri Consalvo
Manager, Multimedia Production Christine Stavrou
Project Supervisor, Math XL Bob Carroll
Executive Marketing Manager Roxanne McCarley
Marketing Assistant Caitlin Crain
Senior Technology Specialist Joe Vetere
Rights and Permissions Advisor Michael Joyce
Image Manager Rachel Youdelman
Procurement Specialist Linda Cox
Senior Media Buyer Ginny Michaud
Associate Director of Design, USHE North and West Andrea Nix
Senior Designer Beth Paquin
Text Design Susan Raymond
Production Coordination, Composition, and Illustrations PreMediaGlobal
Cover Design Susan Raymond

For permission to use copyrighted material, grateful acknowledgment is made to the copyright holders on pages 921–922, which is hereby made part of this copyright page.

p. 176: Common Core State Standards: © Copyright 2010. National Governors Association Center for Best Practices and Council of Chief State School Officers. All rights reserved.

pp. 190, 223, 245, 263, 278, 286, 297, 317, 342, 405, 425, 473, 475, 486, 509, 543, 574, 585, 591, 630, 651, 704, 770–771, 852, 857, 891, 904: © Copyright 2010. National Governors Association Center for Best Practices and Council of Chief State School Officers. All rights reserved.

Many of the designations used by manufacturers and sellers to distinguish their products are claimed as trademarks. Where those designations appear in this book, and Pearson Education was aware of a trademark claim, the designations have been printed in initial caps or all caps.

Library of Congress Cataloging-in-Publication Data
Billstein, Rick.
  A problem solving approach to mathematics for elementary school
teachers / Rick Billstein, Shlomo Libeskind, Johnny W. Lott.—11th ed.
    p. cm.
  ISBN 978-0-321-75666-4
  1. Mathematics—Study and teaching (Elementary) 2. Problem
solving—Study and teaching (Elementary) I. Libeskind, Shlomo.
II. Lott, Johnny W., 1944– III. Title.
  QA135.6.B55 2012
  372.7'044—dc23

                          2011042972

1 2 3 4 5 6 7 8 9 10—CRK—15 14 12 11

ISBN-13: 978-0-321-75666-4
ISBN-10: 0-321-75666-5

Contents

Preface

The eleventh edition of *A Problem Solving Approach to Mathematics for Elementary School Teachers* is designed to prepare outstanding future elementary and middle school teachers. This edition continues to be heavily concept- and skill-based, with an emphasis on active and collaborative learning. The content has been revised and updated to better prepare students to become teachers in their own classrooms.

National Standards for Mathematics

- **Common Core State Standards for Mathematics** The National Governors Association spearheaded the effort to develop the *Common Core Standards (2010)*; they have been adopted by a majority of states and are used in this text to highlight concepts. The complete text of the *Common Core Standards* is found at www.corestandards.org.
- **Curriculum Focal Points** The National Council of Teachers of Mathematics published *Curriculum Focal Points for Pre-kindergarten through Grade 8 Mathematics* (2006) that describes the essential mathematical concepts and skills to which the mathematics in each chapter relates. *Focal Points* are referred to and set off throughout the text.
- **Principles and Standards** We focus on the National Council of Teachers of Mathematics (NCTM) publication, *Principles and Standards of School Mathematics* (2000) (hereafter referred to as *PSSM*).
- The complete text of both the NCTM *Principles and Standards* and *Focal Points* can be found online at www.nctm.org.

Our Goals

- To present appropriate mathematics in an intellectually honest and mathematically correct manner.
- To use problem solving as an integral part of mathematics.
- To approach mathematics in a sequence that instills confidence AND challenges students.
- To provide opportunities for alternate forms of teaching and learning.
- To provide communication and technology problems to develop writing skills that allow students to practice reasoning and explanation through mathematical exposition.
- To provide core mathematics for prospective elementary and middle school teachers in a way that challenges them to determine why mathematics is done as it is.
- To provide core mathematics that allows instructors to use methods integrated with content.
- To assist prospective teachers with connecting mathematics, its ideas, and its applications.
- To assist future teachers in becoming familiar with the content and philosophy of the national standards listed above.

The eleventh edition gives instructors a variety of approaches to teaching, encourages discussion and collaboration among future teachers and with their instructors, and aids the integration of projects into the curriculum. Most importantly, it promotes discovery and active learning.

New to This Edition

- The entire text has been streamlined to help users focus on the most important concepts.
- Algebraic thinking is still at the forefront of mathematics in the early grades and must be in the thinking of prospective teachers. We build on previous knowledge in early chapters of the book and have an intensive emphasis on algebra in Chapter 8, "Real Numbers and Algebraic Thinking," that allows more in-depth coverage. This is especially relevant for future middle school teachers.
- Many users of the previous editions requested that the concepts of integers and number theory be in separate chapters, so those are now Chapter 4, "Number Theory," and Chapter 5, "Integers."
- Because proportional reasoning is so closely connected to rational numbers, these topics are the basis of Chapter 6, "Rational Numbers and Proportional Reasoning."
- All chapters have been revised and reorganized to build on the mathematics in the *Common Core Standards*.

ix

- *Integrating Mathematics and Pedagogy* (IMAP) video references have been added to the Annotated Instructor's Edition and allow future teachers to see elementary and middle school students working out numerical concepts. These videos provide an opportunity for valuable classroom discussion of the mathematics and knowledge of student understanding needed to teach concepts. The IMAP videos are available in MyMathLab.
- Questions pertaining to some of the IMAP videos are now assignable in MyMathLab.
- The assessments are still in two forms, A and B, organized so that exercises in Assessment A (with answers in the text) have parallel exercises in Assessment B. Added to the Connections section are additional technology problems using the *Geometer's Sketchpad* and *GeoGebra*. The Annotated Instructor's Edition of the text contains answers to Assessments A and B and Mathematical Connections.

Content Highlights

Chapter 1 An Introduction to Problem Solving
This chapter has been reorganized and shortened to make it friendlier. Much of the detail work on series has been moved to later chapters to allow students to gain a knowledge of problem-solving techniques with less algebraic manipulation at this stage.

Chapter 2 Numeration Systems and Sets
This chapter includes a historical development of numeration systems to explore how various cultures handled concepts and computations. Different number bases are used to help students understand the base-ten system and to address issues students have when learning the Hindu-Arabic system. Set theory and set operations with properties are introduced as a basis for learning whole-number concepts.

Chapter 3 Whole Numbers and Their Operations
This chapter models addition and subtraction of whole numbers. It emphasizes the *missing-addend model*, the definition of subtraction in terms of addition, and discusses various algorithms for addition and subtraction including those in different bases. Models for multiplication and division of whole numbers, properties of these operations with emphasis on the distributive property of multiplication over addition, and various algorithms are covered in depth. Mental mathematics and estimation with whole numbers feature prominently.

Chapter 4 Number Theory
In the eleventh edition, a separate chapter on number theory does not depend on *integers*, which are introduced in Chapter 5. Concepts of divisibility with divisibility tests are discovered. Prime numbers, prime factorization, greatest common divisor and least common multiple as well as the Euclidean Algorithm are explored with many new exercises added. A module on Clock Arithmetic is available online.*

Chapter 5 Integers
This much-slimmed-down chapter concentrates only on integers, their operations, and properties.

Chapter 6 Rational Numbers and Proportional Reasoning
This chapter has been revised to follow many recommendations in the *Common Core Standards*. Videos showing elementary students learning fraction concepts are now included so that future teachers can observe what happens when elementary students absorb what is taught and how they work with those concepts. Proportional reasoning, one of the most important concepts taught in middle school mathematics, is covered in great depth in its natural setting.

Chapter 7 Decimals: Rational Numbers and Percent
This chapter has been reorganized to focus on decimal representation of rational numbers. Discussion of percent includes the computing of simple and compound interest as well as estimation involving percents.

Chapter 8 Real Numbers and Algebraic Thinking
With an introduction to real numbers in the opening sections, the chapter combines knowledge of real numbers with algebraic skills to give a review of algebra needed to teach in grades K through 8. This includes work in the coordinate plane and with spreadsheets. A module on Using Real Numbers in Equations is available online.*

Chapter 9 Probability
Content has been reordered with many new exercises and student book pages added. *Common Core Standards* have been addressed with content designed to accompany these standards. The use of permutations and combinations in probability is included in this chapter.

---

* Online modules are available at www.pearsonhighered.com/mathstatsresources

Chapter 10 Data Analysis/Statistics: An Introduction
Chapter 10 opens with Designing Experiments/Collecting Data, a section based on *Guidelines for Assessment and Instruction in Statistics Education (GAISE) Report: A preK–12 Curriculum Framework* (2005) by the American Statistical Association. This section, aligned with the *Common Core Standards*, focuses on designing studies and surveys. In the following sections, data, graphs, examples, and assessment exercises have been updated and new material added.

Chapter 11 Introductory Geometry
This chapter has been reorganized to allow students to explore some of the ramifications of different definitions in mathematics used in schools. Linear measure is introduced to emphasize its importance in the curriculum. Also symmetries are now introduced as an early concept that could be used to form geometrical definitions. The Networks module is now offered online.*

Chapter 12 Congruence and Similarity with Constructions
Congruence and constructions sections have been expanded to allow more exploration of circles and quadrilaterals. The concept of similarity is used to reintroduce slope of a line and its properties. Many new exercises have been added. A module on Trigonometric Ratios via Similarity is available online.*

Chapter 13 Congruence and Similarity with Transformations
Because of the prominence of motion geometry in the *Common Core Standards*, this chapter appears earlier among the geometry sections. It focuses on connections among transformations and dilations in congruence and similarity.

Chapter 14 Area, Pythagorean Theorem, and Volume
Chapter 14 continues a reorganization of the geometry chapters. Geometry in three dimensions is included with the topics of area, the Pythagorean theorem, and volume. Many topics have been shifted and new material added, for example, the subsection *Comparing Measurements of Similar Figures*. Assessment sets and examples have been updated.

Technology Usage
Virtually all mathematics standards have included the use of technology as a tool for learning mathematics, yet the manner and type of usage in classrooms is as varied as the classrooms and teachers themselves. We strongly support the use of technology as a learning tool and have since the inception of this book. In this edition, online modules discuss the use of technology. These modules are designed for a brief introduction to the use of spreadsheets and graphing calculators as indicated but it is expected that many instructors using the text will naturally incorporate those tools in their teaching. Moreover, parallel modules on the use of *Geometer's Sketchpad* and *GeoGebra* are available. The *GeoGebra* module is new to this edition.

References to the online geometry module problems and lab activities are included in the Mathematical Connections section of the assessments under the heading GSP/GeoGebra Activities. It is noted that there are more problems and activities in the online modules than are listed in the text. This is purposefully done to allow instructors to use it in the manner that is most pedagogically and mathematically desirable for their courses.

Features
Active Learning
- *Now Try This* problems appear throughout each chapter, and are intended to help students become actively involved in their learning, to facilitate the development and improvement of their critical thinking and problem-solving skills, and to stimulate both in-class and out-of-class discussion. Answers are in both the Annotated Instructor's Edition and Student Edition.
- An *e-Manipulatives CD* is packaged in the text to help students investigate, explore, and practice new concepts. The CD is intended to be used in conjunction with the main text, and is referenced in the margins throughout the book. The e-manipulatives are also available within MyMathLab.
- References to the *Activity Manual* are found throughout the Annotated Instructor's Edition as a guide to more fully integrate activities into the course.
- *Brain Teasers* provide a different avenue for problem solving. They are solved in both the Annotated Instructor's Edition and Student Edition and may be assigned or used by the instructor to challenge students.

Pedagogical Tools

- *Problem-solving strategies* are highlighted in italics, and problem-solving examples help show students how to put these strategies to work.
- *Historical Notes* add context and humanize the mathematics.
- **New!** *Chapter Summaries* at the end of each chapter are organized in a chart format along with page numbers referencing the content to help students review the chapter.
- *Chapter Reviews* at the end of each chapter allow students to test themselves in preparation for exams.

Assessment

- The *exercise sets* are organized into Assessments A, B, and Mathematical Connections. Assessment A exercises have answers in the text so that students can check their work. Assessment B contains exercises similar to those in Assessment A, but answers are not given in the student text. Mathematical Connections is divided into the following categories of problems: Communication, Open-Ended, Cooperative Learning, Questions from the Classroom, TIMSS, NAEP, GSP/GeoGebra (where appropriate), and Review. Answers to Mathematical Connections questions (except the GeoGebra and GSP exercises which are answered online in the appendices) are in the Annotated Instructor's Edition while the Student Edition includes odd answers only.
- Relevant and realistic problems are more accessible and appealing to students of diverse backgrounds.

Professional Development

- Relevant quotations from the *Common Core Standards*, *PSSM*, and *Focal Points* are incorporated throughout the text to provide a context for the content.
- *Questions from the Classroom* present questions as they might be posed by K through 8 students. They appear at the end of each section as part of Mathematical Connections.
- *School Book Pages* are included to show how the mathematics is actually introduced in the K through 8 classroom and are referenced throughout the text. Students are asked to complete many of the activities so they can see what is expected in elementary schools. An icon is included next to the text discussion that is related to the School Book Page displayed.
- A separate section entitled *GSP/GeoGebra* appears when appropriate as a part of Mathematical Connections in the assessment section.

Student and Instructor Resources

For the Student

E-Manipulatives CD

- Available packaged in the text and within MyMathLab.
- 21 Flash-based manipulatives investigate, explore, and practice new concepts, helping students develop a conceptual understanding of key ideas.
- Helps explore the way elementary students would use manipulatives in the classroom.
- References to these e-manipulatives are in the margin of the main text.

Activities Manual

Mathematics Activities for Elementary School Teachers: A Problem Solving Approach, by Dan Dolan, *Project to Increase Mastery of Mathematics and Science, Wesleyan University*; Jim Williamson, *University of Montana*; and Mari Muri, *Project to Increase Mastery of Mathematics and Science, Wesleyan University*
- ISBN 0-321-75878-1 | 978-0-321-75878-1
- Provides hands-on, manipulative-based activities keyed to the text that involve future elementary school teachers discovering concepts, solving problems, and exploring mathematical ideas.
- **New!** Colorful, perforated paper manipulatives in a convenient storage pouch.
- Activities can also be adapted for use with elementary students at a later time.
- References to these activities are in the margin of the Annotated Instructor's Edition.

Student's Solutions Manual by David Yopp, *Montana State University*
- ISBN 0-321-78332-8 | 978-0321-78332-5
- Provides detailed, worked-out solutions to all of the **Assessment A** problems and **Chapter Review** exercises.

Video Resources

- Digitized videos for student use at home or on campus though MyMathLab. Ideal for distance learning and supplemental instruction. They include optional English and Spanish subtitles.
- Includes examples from the text and supports visualization and problem solving.

Connecting Mathematics for Elementary Teachers: How Children Learn Mathematics by David Feikes, *Purdue University North Central*; Keith Schwingendorf, *Purdue University North Central*; and Jeff Gregg, *Purdue University Calumet*
- ISBN 0-321-54266-5 | 978-0-321-54266-3
- Understanding mathematics and how children think about mathematics can help you become a better teacher. Provides general descriptions of children's learning and shows how children approach mathematics differently than adults.

When Will I Ever Teach This? An Activities Manual for Mathematics for Elementary Teachers by Sharon E. Taylor, *Georgia Southern University* and Susie Lanier, *Georgia Southern University*
- ISBN 0-321-23717-X | 978-0-321-23717-0
- The best way to demonstrate to students the need to learn certain topics is to show pages from a real K through 8 textbook. This allows students to see when and where a topic occurs in the curriculum and also to see how it is presented in a text.

For the Instructor

Annotated Instructor's Edition
- ISBN 0-321-75667-3 | 978-0-321-75667-1
- This special edition includes answers to the text exercises on the page where they occur and includes answers to the Preliminary Problems, Now Try This activities, Mathematical Connections questions and Brain Teasers.
- Margin notes referencing the *Integrating Mathematics and Pedagogy* (IMAP) videos and *Activities Manual* are included.

Online Supplements
The following instructor material is available for download from Pearson's online catalog at www.pearsonhighered. com or within MyMathLab.

Instructor's Solutions Manual by David Yopp, *Montana State University*
- Provides detailed, worked-out solutions to all of the **Assessment A & B** exercises and **Chapter Review** exercises.
- **New!** Provides detailed solutions to all the *Mathematical Connection* problems.

Instructor's Testing Manual
- Comprehensive worksheets contain two forms of chapter assessments with answers for each.

Insider's Guide
- Includes resources to assist instructors with course preparation.
- Provides helpful teaching tips correlated to the sections of the text, as well as general teaching advice and tips on using manipulatives.

Instructor's Guide for
Mathematics Activities for Elementary School Teachers: A Problem Solving Approach, by Dan Dolan, *Project to Increase Mastery of Mathematics and Science, Wesleyan University*; Jim Williamson, *University of Montana*; and Mari Muri, *Project to Increase Mastery of Mathematics and Science, Wesleyan University*
- Contains answers for all activities, as well as additional teaching suggestions for some activities.

Instructor's Guide for
Connecting Mathematics for Elementary Teachers: How Children Learn Mathematics by David Feikes, *Purdue University North Central*; Keith Schwingendorf, *Purdue University North Central*; and Jeff Gregg, *Purdue University Calumet*
- Contains tips and teaching suggestions on how to incorporate the book into your course and syllabus.

PowerPoint® Lecture Slides
- Provides section-by-section coverage of key topics and concepts along with examples.

TestGen®
- TestGen® enables instructors to build, edit, print, and administer tests using a computerized bank of questions developed to cover all the objectives of the text.
- TestGen is algorithmically based, allowing instructors to create multiple but equivalent versions of the same question or test with the click of a button. Instructors can also modify test bank questions or add new questions.

Online Learning

MyMathLab® Online Course (access code required)
MyMathLab delivers **proven results** in helping individual students succeed.

- MyMathLab has a consistently positive impact on the quality of learning in higher education math instruction. MyMathLab can be successfully implemented in any environment—lab-based, hybrid, fully online, traditional—and demonstrates the quantifiable difference that integrated usage has on student retention, subsequent success, and overall achievement.
- MyMathLab's comprehensive online gradebook automatically tracks students' results on tests, quizzes, homework, and in the study plan. You can use the gradebook to quickly intervene if your students have trouble, or to provide positive feedback on a job well done. The data within MyMathLab is easily exported to a variety of spreadsheet programs, such as Microsoft Excel. Instructors can determine which points of data to export, and then analyze the results to determine success.

MyMathLab provides **engaging experiences** that personalize, stimulate, and measure learning for each student.

- **Tutorial Exercises**: The homework and practice exercises in MyMathLab and MyStatLab are correlated to the exercises in the textbook, and they regenerate algorithmically to give students unlimited opportunity for practice and mastery. The software offers immediate, helpful feedback when students enter incorrect answers.
- **Multimedia Learning Aids**: Exercises include guided solutions, sample problems, animations, videos, and eText clips for extra help at point-of-use.
- **Expert Tutoring**: Although many students describe the whole of MyMathLab as "like having personal tutor," students using MyMathLab and MyStatLab do have access to live tutoring from Pearson, from qualified math and statistics instructors who provide tutoring sessions for students via MyMathLab and MyStatLab.

And, MyMathLab comes from a **trusted partner** with educational expertise and an eye on the future.

- Knowing that you are using a Pearson product means knowing that you are using quality content. That means that our eTexts are accurate, that our assessment tools work, and that our questions are error-free. And whether you are just getting started with MyMathLab, or have a question along the way, we're here to help you learn about our technologies and how to incorporate them into your course.

New to the MyMathLab course

- A new type of problem using Integrated Mathematics and Pedagogy (IMAP) videos to test students' understanding of concepts and content in the context of children's reasoning processes.
- A new type of problem using the eManipulatives where students interact with the eManipulative and then answer assigned questions.
- An IMAP implementation guide and correlation providing additional material on how to integrate the IMAP videos into your course.
- e-Manipulatives and IMAP videos integrated into the multimedia textbook.
- The Image Resource Library contains all art from the text, for instructors to use in their own presentations and handouts.

To learn more about how MyMathLab combines proven learning applications with powerful assessment, visit www.mymathlab.com or contact your Pearson representative.

MyMathLab® Ready to Go Course (access code required)
These new Ready to Go courses provide students with all the same great MyMathLab features that you're used to, but make it easier for instructors to get started. Each course included pre-assigned homeworks and quizzes to make creating your course even simpler. Ask Your Pearson representative about the details for this particular course or to see a copy of this course.

MathXL® Online Course (access code required)
MathXL® is the homework and assessment engine that runs MyMathLab. (MyMathLab is MathXL plus a learning management system.) With MathXL, instructors can:

- Create, edit, and assign online homework and tests using algorithmically generated exercises correlated at the objective level to the textbook.
- Create and assign their own online exercises and import TestGen tests for added flexibility.
- Maintain records of all student work tracked in MathXL's online gradebook.

With MathXL, students can:

- Take chapter tests in MathXL and receive personalized study plans and/or personalized homework assignments based on their test results.
- Use the study plan and/or the homework to link directly to tutorial exercises for the objectives they need to study.
- Access supplemental animations and video clips directly from selected exercises.

MathXL is available to qualified adopters. For more information, visit the website at www.mathxl.com, or contact a Pearson representative.

Acknowledgments

For past editions of this book, many noted and illustrious mathematics educators and mathematicians have served as reviewers. To honor the work of the past as well as to honor the reviewers of this edition, we list all but place asterisks by this edition's reviewers.

Leon J. Ablon
Paul Ache
G.L, Alexanderson
Haldon Anderson
Bernadette Antkoviak
*Renee Austin
Richard Avery
Sue H. Baker
Jane Barnard
Joann Becker
Cindy Bernlohr
James Bierden
Jackie Blagg
Jim Boone
Sue Boren
Barbara Britton
Beverly R. Broomell
Anne Brown
Jane Buerger
Maurice Burke
David Bush
Laura Cameron
*Karen Cannon
Louis J. Chatterley
Phyllis Chinn
Donald J. Dessart
Ronald Dettmers
Jackie Dewar
Nicole Duvernoy
Amy Edwards
Lauri Edwards
Margaret Ehringer
Rita Eisele
Albert Filano
Marjorie Fitting
Michael Flom
*Pari L. Ford
*Marie Franzosa
Martha Gady
Edward A. Gallo
Dwight Galster
*Melinda Gann
Sandy Geiger

Glenadine Gibb
Don Gilmore
Diane Ginsbach
Elizabeth Gray
*Jerrold Grossman
Alice Guckin
Jennifer Hegeman
Joan Henn
Boyd Henry
Linda Hintzman
Alan Hoffer
E. John Hornsby, Jr.
Patricia A. Jaberg
Judith E. Jacobs
Donald James
Thomas R. Jay
Jeff Johannes
Jerry Johnson
Wilburn C. Jones
Robert Kalin
Sarah Kennedy
Steven D. Kerr
Leland Knauf
Margret F. Kothmann
Kathryn E. Lenz
Hester Lewellen
Ralph A. Liguori
Richard Little
Susan B. Lloyd
Don Loftsgaarden
Sharon Louvier
*Carol A. Lucas
Stanley Lukawecki
Lou Ann Martin
Judith Merlau
Barbara Moses
Cynthia Naples
Charles Nelson
Glenn Nelson
Kathy Nickell
Bethany Noblitt
Dale Oliver
Mark Oursland

Linda Padilla
Dennis Parker
Clyde Paul
Keith Peck
Barbara Pence
Glen L. Pfeifer
Debra Pharo
Jack Porter
Edward Rathnell
Sandra Rucker
Jennifer Rutherford
Helen R. Santiz
*Sharon Saxton
Sherry Scarborough
Jane Schielack
Barbara Shabell
M. Geralda Shaefer
Nancy Shell
Wade H. Sherard
Gwen Shufelt
Julie Sliva
Ron Smit
Joe K. Smith
William Sparks
Virginia Strawderman
Mary M. Sullivan
Viji Sundar
Sharon Taylor
Jo Temple
C. Ralph Verno
Hubert Voltz
John Wagner
Edward Wallace
Virginia Warfield
Lettie Watford
Mark F. Weiner
Grayson Wheatley
*Bill D. Whitmire
*Teri Willard
Jim Williamson
Ken Yoder
Jerry L. Young
Deborah Zopf

Dedication

To my students who for the past 46 years have provided me with enjoyment and challenge. Each new edition of this book reflects experiences learned in the classroom.
—Rick Billstein

To my co-authors, Rick and Shlomo, with whom I have grown mathematically and pedagogically in developing this book.
—Johnny W. Lott

To my dear cousin Rony Eviattar, and to the memory of her parents Chaim Libeskind and Alta Libeskind-Edelsztein
—Shlomo Libeskind

An Introduction to Problem Solving

Preliminary Problem

A person removing 2 eggs at a time from a basket has 1 left; the person removing them 3 at a time has 2 left; and the person removing them 5 at a time has 3 left. What is the fewest number of eggs possible in the basket?

If needed, see Hint on page 46.

Problem solving has long been recognized as one of the hallmarks of mathematics. George Pólya (1887–1985), one of the great mathematicians and teachers of the twentieth century, pointed out that "solving a problem means finding a way out of difficulty, a way around an obstacle, attaining an aim which was not immediately attainable" (Pólya 1981, p. ix). In *Principles and Standards for School Mathematics (PSSM)*, (NCTM 2000), we find the following:

> Problem solving means engaging in a task for which the solution method is not known in advance. In order to find a solution, students must draw on their knowledge, and through this process, they will often develop new mathematical understandings. Solving problems is not only a goal of learning mathematics but also a major means of doing so. (p. 52)

Common Core State Standards for Mathematics (CCSSM) (2010) states: "Mathematically proficient students start by explaining to themselves the meaning of a problem and looking for its solution." (p. 6)

Students learn mathematics as a result of solving problems. *Exercises* are routine practice for skill building and serve a purpose in learning mathematics, but problem solving must be a focus of school mathematics. A reasonable amount of tension and discomfort improves problem-solving performance. Mathematical experience often determines whether situations are *problems* or *exercises*.

Worthwhile, interesting problems must be a part of elementary students' mathematical experience. Otherwise, students may wind up with attitudes similar to Luann's in the cartoon.

Luann copyright © 2011 GEC Inc./Distributed by United Feature Syndicate, Inc.

Mathematical problem solving may occur when:

1. Students are presented with a situation that they understand but do not know how to proceed directly to a solution.
2. Students are interested in finding the solution and attempt to do so.
3. Students are required to use mathematical ideas to solve the problem.

In this text, we present many opportunities to solve problems. Each chapter opens with a problem that can be solved by using the concepts developed in the chapter. We give a hint for the solution to the problem at the end of each chapter. Throughout the text, some problems are solved using a four-step process and others solved using different formats.

Working with other students to solve problems can enhance problem-solving ability and communication skills. We encourage *cooperative learning* and working in groups whenever possible. To encourage group work and help identify when cooperative learning could be useful, we identify activities and problems where group discussions might lead to strategies for solving the problem and learning mathematics.

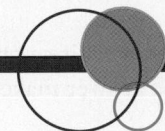

1-1 Mathematics and Problem Solving

If problems are approached in only one way, a mind-set may be formed. For example, consider the following: Spell the word *spot* three times out loud. "S-P-O-T! S-P-O-T! S-P-O-T!" Now answer the question "What do we do when we come to a green light?" Write an answer.

If we answer "Stop," we may be guilty of having formed a mind-set. We do not stop at a *green* light.

Consider the following problem: "A shepherd had 36 sheep. All but 10 died. How many lived?" If we answer "10," we are ready to try some problems. If not, we did not understand the question. *Understanding the problem* is the first step in the four-step problem-solving process developed by George Pólya. Using the four-step process does not guarantee a solution to a problem, but it does provide a systematic means of approaching it.

Four-Step Problem-Solving Process

1. **Understanding the problem**
   a. Can the problem be stated differently?
   b. What is to be found or what is needed?
   c. What are the unknowns?
   d. What information is obtained from the problem?
   e. What information, if any, is missing or not needed?

2. **Devising a plan**
   The following list of strategies, although not exhaustive, is very useful:
   a. Look for a pattern.
   b. Examine related problems and determine whether the same techniques applied to them can be applied to the current problem.
   c. Examine a simpler or special case of the problem to gain insight into the solution of the original problem.
   d. Make a table or list.
   e. Identify a subgoal.
   f. Make a diagram.
   g. Use guess and check.
   h. Work backward.
   i. Write an equation.

3. **Carrying out the plan**
   a. Implement the strategy or strategies in step 2 and perform any necessary actions or computations.
   b. Attend to precision in language and mathematics used.
   c. Check each step of the plan along the way. This may be intuitive checking or a formal proof of each step.
   d. Keep an accurate record of all work.

Historical Note

George Pólya (1887–1985) was born in Hungary and received his Ph.D. from the University of Budapest. He moved to the United States in 1940, and after a brief stay at Brown University, joined the faculty at Stanford University. In addition to being a preeminent mathematician, he focused on mathematics education. At Stanford, he published 10 books, including *How To Solve It* (1945), which has been translated into 23 languages. ●

4. **Looking back**
   a. Check the results in the original problem. (In some cases, this will require a proof.)
   b. Interpret the solution in terms of the original problem. Does the answer make sense? Is it reasonable? Does it answer the question that was asked?
   c. Determine whether there is another method of finding the solution.
   d. If possible, determine other related or more general problems for which the techniques will work.

The role Pólya's problem-solving process plays in the teaching of elementary mathematics is discussed in *PSSM* in the following:

> An obvious question is, How should these strategies be taught? Should they receive explicit attention, and how should they be integrated with the mathematics curriculum? As with any other component of the mathematical tool kit, strategies must receive instructional attention if students are expected to learn them. (p. 54)

Strategies for Problem Solving

We next provide a variety of problems with different contexts to provide experience in problem solving. Frequently, a variety of problem-solving strategies is necessary to solve these and other problems. These strategies are used to discover or construct the means to achieve a solution. For each strategy described, we give an example that can be solved with that strategy. Often, problems can be solved in more than one way. There is no one best strategy to use.

In many of the examples, we use the set of **natural numbers**, 1, 2, 3, ... Note that the three dots, an *ellipsis*, are used to represent missing terms.

Strategy: Look for a Pattern

Problem Solving   **Gauss's Problem**

As a student, Carl Gauss and his class were asked to find the sum of the first 100 natural numbers. The teacher expected to keep the class occupied for some time, but Gauss gave the answer almost immediately. How might he have done it?

Understanding the Problem The natural numbers are 1, 2, 3, 4, ... Thus, the problem is to find the sum $1 + 2 + 3 + 4 + \ldots + 100$.

Devising a Plan The strategy *look for a pattern* is useful here. One story about young Gauss reports that he listed the sum, and wrote the same sum backwards as in Figure 1-1. If $S = 1 + 2 + 3 + 4 + 5 + \ldots + 98 + 99 + 100$, then Gauss could have computed as follows using an identified pattern.

$$
\begin{array}{rcccccccccc}
S = & 1 + & 2 + & 3 + & 4 + & 5 + & \ldots + & 98 + & 99 + & 100 \\
+\ S = & 100 + & 99 + & 98 + & 97 + & 96 + & \ldots + & 3 + & 2 + & 1 \\
\hline
2S = & 101 + & 101 + & 101 + & 101 + & 101 + & \ldots + & 101 + & 101 + & 101
\end{array}
$$

Figure 1-1

To discover the original sum from the last equation, Gauss could have divided the sum, $2S$, in Figure 1-1 by 2.

Historical Note

Carl Gauss (1777–1855), one of the greatest mathematicians of all time, was born to humble parents in Brunswick, Germany. He was an infant prodigy who later made contributions in many areas of science as well as mathematics. After Gauss's death, the King of Hanover ordered a commemorative medal prepared in his honor. On the medal was an inscription referring to Gauss as the "Prince of Mathematics."

Carrying Out the Plan There are 100 sums of 101. Thus, $2S = 100 \cdot 101$ and $S = \dfrac{100 \cdot 101}{2}$, or 5050.

Looking Back Note that the sum in each pair $(1, 100), (2, 99), (3, 98), \ldots, (100, 1)$ is always 101 and there are 100 pairs with this sum. This technique can be used to solve a more general problem of finding the sum of the first n natural numbers $1 + 2 + 3 + 4 + 5 + 6 + \ldots + n$. We use the same plan as before and notice the relationship in Figure 1-2. There are n sums of $n + 1$ for a total of $n(n + 1)$. Therefore, $2S = n(n + 1)$ and $S = \dfrac{n(n + 1)}{2}$.

$$
\begin{array}{rccccccc}
S = & 1 & + & 2 & + & 3 & + & 4 & + \ldots + & n \\
+\ S = & n & + & (n-1) & + & (n-2) & + & (n-3) & + \ldots + & 1 \\
\hline
2S = & (n+1) & + & (n+1) & + & (n+1) & + & (n+1) & + \ldots + & (n+1)
\end{array}
$$

Figure 1-2

A different strategy for finding a sum of consecutive natural numbers involves the strategy of *making a diagram* and thinking of the sum geometrically as a stack of blocks. This alternative method is explored in exercise 2 of Assessment 1-1A.

 NOW TRY THIS 1-1

One cut through a log produces two pieces, two cuts in the same direction as the first produce three pieces, and three similar cuts produce four pieces. How many pieces are produced by ten such cuts? Assume the cuts are made in the same manner as the first three cuts. How many pieces are produced by n such cuts?

Strategy: Examine a Related Problem

Problem Solving **Sums of Even Natural Numbers**

Find the sum of the even natural numbers less than or equal to 100. Find that sum and generalize the result.

Understanding the Problem Even natural numbers are 2, 4, 6, 8, 10, ... The problem is to find the sum of these numbers: $2 + 4 + 6 + 8 + \ldots + 100$.

Devising a Plan Recognizing that the sum can be *related to Gauss's original problem* helps us devise a plan. Consider the following:

$$2 + 4 + 6 + 8 + \ldots + 100 = 2 \cdot 1 + 2 \cdot 2 + 2 \cdot 3 + 2 \cdot 4 + \ldots + 2 \cdot 50$$
$$= 2(1 + 2 + 3 + 4 + \ldots + 50)$$

Thus, we can use Gauss's method to find the sum of the first 50 natural numbers and then double that.

Carrying Out the Plan We carry out the plan as follows:

$$2 + 4 + 6 + 8 + \ldots + 100 = 2(1 + 2 + 3 + 4 + \ldots + 50)$$
$$= 2[50(50 + 1)/2]$$
$$= 2550$$

Thus, the sum is 2550.

Looking Back A different way to approach this problem is to realize that there are 25 sums of 102, as shown in Figure 1-3. (Why are there 25 sums to consider?)

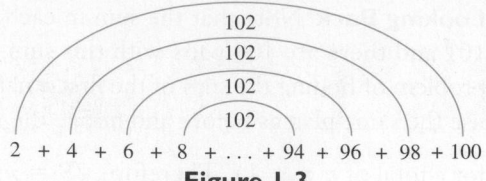

2 + 4 + 6 + 8 + ... + 94 + 96 + 98 + 100

Figure 1-3

Thus, the sum is $25 \cdot 102$, or 2550.

 NOW TRY THIS 1-2

  a. Find the sum of the odd natural numbers less than 100.
  b. Find the sum of consecutive natural numbers shown: $25 + 26 + 27 + ... + 120$.

Strategy: Examine a Simpler Case

 One strategy for solving a complex problem is to *examine a simpler case* of the problem and then consider other parts of the original problem. An example is shown on the grade 7 student page on page 7, where 3^{50} means $\underbrace{3 \cdot 3 \cdot 3 \cdot ... \cdot 3}_{\text{Fifty 3s}}$. (Note that two lines after the table 4, 3^{49} should be 3^{48}.)

 NOW TRY THIS 1-3

Each of 16 people in a round-robin handball tournament played each other person exactly once. How many games were played?

Strategy: Make a Table

 An often-used strategy in problem solving is *making a table*. A table can be used to look for patterns that emerge in the problem, which in turn can lead to a solution. An example of this strategy is shown on the grade 6 student page on page 8.

 NOW TRY THIS 1-4

Molly and Karly started a new job the same day. After they start work, Molly is to visit the home office every 15 days and Karly is to visit the home office every 18 days. How many days will it be before they both visit the home office the same day?

Strategy: Identify a Subgoal

In attempting to devise a plan for solving a problem, a solution to a somewhat easier or more familiar related problem could make it easier. In such a case, finding the solution to the easier problem may become a *subgoal*. The magic square problem on page 8 shows an example of this.

School Book Page SOLVING A SIMPLER PROBLEM

Work a Simpler Problem

When to Use This Strategy If a problem seems to have many steps, you may be able to *Work a Simpler Problem* first. The result may give you a clue about the solution of the original problem.

When you simplify 3^{50}, what number is in the ones place?

Understand

You know that 3^{50} is a large number to calculate. You need to find the number in the ones place.

Plan

It is not easy to simplify 3^{50} with paper and pencil. Simplify easier expressions such as $3^2, 3^3$, and 3^4, to see what number is in the ones place.

Carry Out

The table shows the values of the first 10 powers of 3.

Power	Value
3^1	3
3^2	9
3^3	27
3^4	81
3^5	243
3^6	729
3^7	2,187
3^8	6,561
3^9	19,683
3^{10}	59,049

Notice that the ones digits in the value column repeat in the pattern 3, 9, 7, and 1. Every fourth power of 3 has a ones digit of 1. Since 4 is divisible by 4, 3^{48} has a ones digit of 1. Then the ones digit of 3^{49} is 3 and the ones digit of 3^{50} is 9.

Check

When the exponent of 3 is 2, 6, or 10, the ones digit is 9. The numbers 2, 6, and 10 are divisible by 2 but not by 4. Since 50 is divisible by 2 but not by 4, the ones digit of 3^{50} is 9.

xliv   Problem Solving Handbook

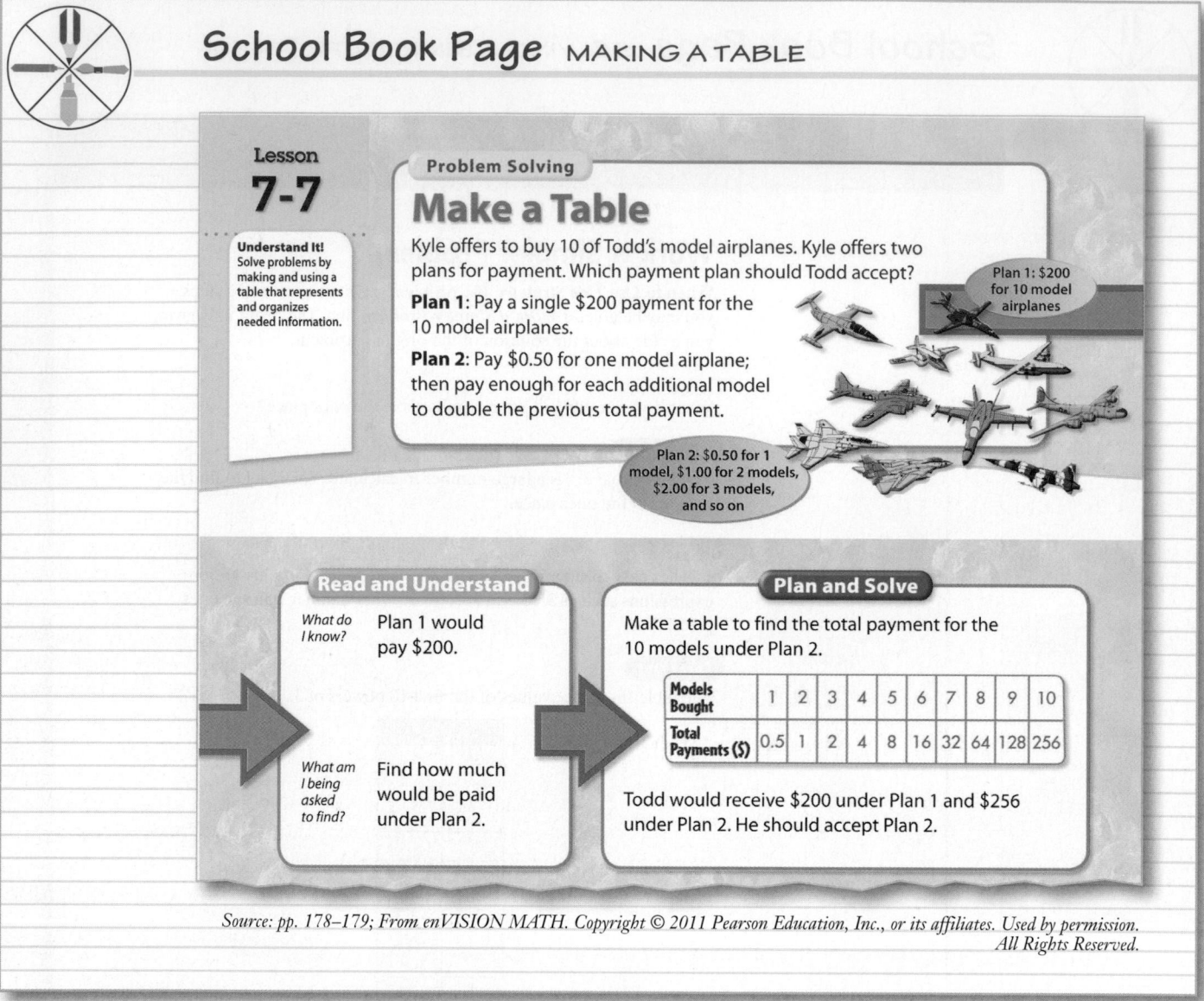

School Book Page MAKING A TABLE

Lesson
7-7

Understand It!
Solve problems by making and using a table that represents and organizes needed information.

Problem Solving

Make a Table

Kyle offers to buy 10 of Todd's model airplanes. Kyle offers two plans for payment. Which payment plan should Todd accept?

Plan 1: Pay a single $200 payment for the 10 model airplanes.

Plan 2: Pay $0.50 for one model airplane; then pay enough for each additional model to double the previous total payment.

Plan 1: $200 for 10 model airplanes

Plan 2: $0.50 for 1 model, $1.00 for 2 models, $2.00 for 3 models, and so on

Read and Understand

What do I know? Plan 1 would pay $200.

What am I being asked to find? Find how much would be paid under Plan 2.

Plan and Solve

Make a table to find the total payment for the 10 models under Plan 2.

Models Bought	1	2	3	4	5	6	7	8	9	10
Total Payments ($)	0.5	1	2	4	8	16	32	64	128	256

Todd would receive $200 under Plan 1 and $256 under Plan 2. He should accept Plan 2.

Figure 1-4

Problem Solving **A Magic Square**

Arrange the numbers 1 through 9 into a square subdivided into nine smaller squares like the one shown in Figure 1-4 so that the sum of every row, column, and main diagonal is the same. The result is a *magic square*.

Understanding the Problem Each of the nine numbers 1, 2, 3, . . ., 9 must be placed in the small squares, a different number in each square, so that the sums of the numbers in each row, in each column, and in each of the two major diagonals are the same.

Devising a Plan If we knew the fixed sum of the numbers in each row, column, and diagonal, we would have a better idea of which numbers can appear together in a single row, column, or diagonal. Thus *the subgoal* is to find that fixed sum. The sum of the nine numbers, $1 + 2 + 3 + \ldots + 9$, equals 3 times the sum in one row. (Why?) Consequently, the fixed sum can be found using the process developed by Gauss. We have $(1 + 2 + 3 + \ldots + 9)/3 = \left(\dfrac{9 \cdot 10}{2}\right)/3$, or 15, so the sum in each row, column, and diagonal must be 15. Next, we need to decide what numbers could occupy the various squares. The number in the center space will appear in four sums, each adding to 15 (two diagonals, the second row, and the second column). Each number in the corners will appear in three sums of 15. (Why?) If we write 15 as a sum of three different numbers 1 through 9 in all possible ways, we could then count how many sums contain each of the numbers 1 through 9. The numbers that appear in at least four sums are candidates for placement in the center square, whereas the numbers that appear in at least three sums are candidates for the corner squares. Thus the new *subgoal* is to write 15 in as many ways as possible as a sum of three different numbers from 1, 2, 3, . . ., 9.

Carrying Out the Plan The sums of 15 can be written systematically as follows:

$$9 + 5 + 1$$
$$9 + 4 + 2$$
$$8 + 6 + 1$$
$$8 + 5 + 2$$
$$8 + 4 + 3$$
$$7 + 6 + 2$$
$$7 + 5 + 3$$
$$6 + 5 + 4$$

Note that the order of the numbers in sums like $9 + 5 + 1$ is irrelevant because the order in which additions are done does not matter. In the list, 1 appears in only two sums, 2 in three sums, 3 in two sums, and so on. Table 1-1 summarizes this information.

Table 1-1

Number	1	2	3	4	5	6	7	8	9
Number of sums containing the number	2	3	2	3	4	3	2	3	2

The only number that appears in four sums is 5; hence, 5 must be in the center of the square. (Why?) Because 2, 4, 6, and 8 appear 3 times each, they must go in the corners. Suppose we choose 2 for the upper left corner. Then 8 must be in the lower right corner. This is shown in Figure 1-5(a). Now we could place 6 in the lower left corner or upper right corner. If we choose the upper right corner, we obtain the result in Figure 1-5(b). The magic square can now be completed, as shown in Figure 1-5(c).

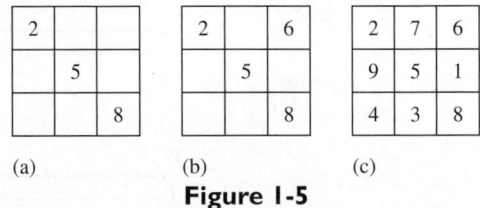

(a)   (b)   (c)

Figure 1-5

Looking Back We have seen that 5 was the only number among the given numbers that could appear in the center. However, we had various choices for a corner, and so it seems that the magic square we found is not the only one possible. Can you find all the others?

Another way to see that 5 could be in the center square is to consider the sums $1 + 9, 2 + 8, 3 + 7, 4 + 6$, as shown in Figure 1-6. We could add 5 to each to obtain 15.

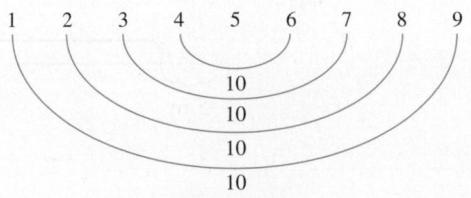

Figure 1-6

● NOW TRY THIS 1-5

Five friends decided to give a party and split the costs equally. Al spent $4.75 on invitations, Betty spent $12.00 for drinks and $5.25 on vegetables, Carl spent $24.00 for pizza, Dani spent $6.00 on paper plates and napkins, and Ellen spent $13.00 on decorations. Determine who owes money to whom and how the money can be paid.

Strategy: Make a Diagram

In the following problem, *making a diagram* helps us to understand the problem and work toward a solution.

Problem Solving   50-m Race Problem

Bill and Jim ran a 50-m race three times. The speed of the runners did not vary. In the first race, Jim was at the 45-m mark when Bill crossed the finish line.

  a. In the second race, Jim started 5 m ahead of Bill, who lined up at the starting line. Who won?
  b. In the third race, Jim started at the starting line and Bill started 5 m behind. Who won?

Understanding the Problem When Bill and Jim ran a 50-m race, Bill won by 5 m; that is, whenever Bill covered 50 m, at the same time Jim covered only 45 m. If Bill started at the starting line and Jim had a 5-m head start or Jim started at the starting line and Bill started 5 m behind, we are to determine who would win in each case.

Devising a Plan A strategy to determine the winner under each condition is to *make a diagram*. A diagram for the first 50-m race is given in Figure 1-7(a). In this case, Bill won by 5 m. In the second race, Jim had a 5-m head start and hence when Bill ran 50 m to the finish line, Jim ran only 45 m. Because Jim is 45 m from the finish line, he reached the finish line at the same time as Bill did. This is shown in Figure 1-7(b). In the third race, because Bill started 5 m behind, we use Figure 1-7(a) but move Bill back 5 m, as shown in Figure 1-7(c). From the diagram we determine the results in each case.

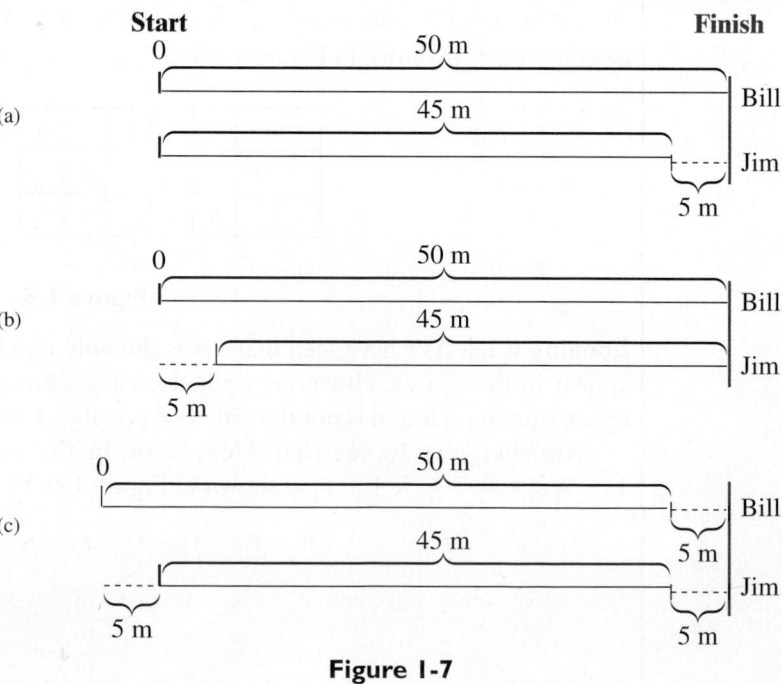

Figure 1-7

Carrying Out the Plan From Figure 1-7(b) we see that if Jim had a 5-m head start, then the race ends in a tie. If Bill started 5 m behind Jim, then at 45 m they would be tied. Because Bill is faster than Jim, Bill would cover the last 5 m faster than Jim and win the race.

Looking Back The diagrams show the solution makes sense and is appropriate. Other problems can be investigated involving racing and handicaps. For example, if Bill and Jim run on a 50-m oval track, how many laps will it take for Bill to lead Jim by one full lap? (Assume the same speeds as earlier.)

 NOW TRY THIS 1-6

An elevator stopped at the middle floor of a building. It then moved up 4 floors, stopped, moved down 6 floors, stopped, and then moved up 10 floors and stopped. The elevator was now 3 floors from the top floor. How many floors does the building have?

Strategy: Use Guess and Check

In the strategy of *guess and check*, we first guess at a solution using as reasonable a guess as possible. Then we check to see whether the guess is correct. If not, the next step is to learn as much as possible about the solution based on the guess before making a next guess. This strategy can be regarded as a form of trial and error, whereby the information about the error helps us choose what to try next. The guess-and-check strategy is often used when a student does not know how to solve the problem more efficiently or if the student does not yet have the tools to solve the problem in a faster way. Frank Lester (1975)* suggested that students in grades 1–3 rely primarily on a *guess-and-check* strategy when faced with a mathematical problem. In grades 6–12 this tendency decreases. Older students benefit more from the observed "errors" after a guess when formulating a new "trial."

 The grade 7 student page (page 12) gives an example of this strategy identified as "*systematic guess and check*."

 NOW TRY THIS 1-7

A cryptarithm is a collection of words in which each unique letter represents a unique digit. Find the digits that can be substituted in the following:

$$
\begin{array}{r}
S\,U\,N \\
+\,F\,U\,N \\
\hline
S\,W\,I\,M
\end{array}
$$

Strategy: Work Backward

 In some problems, it is easier to start with the result and to *work backward*. This is demonstrated as a test-taking strategy on the student page on page 13. Note that choice A can be eliminated with mental math and is not discussed.

*Lester, F. "Developmental Aspects of Children's Ability to Understand Mathematical Proof." *Journal for Research in Mathematics Education* 6 (1975): 14–25.

School Book Page SYSTEMATIC GUESS AND CHECK

Systematic Guess and Check

When to Use This Strategy The strategy *Systematic Guess and Check* works well when you can start by making a reasonable estimate of the answer.

Construction A group of students is building a sailboat. The students have 48 ft² of material to make a sail. They design the sail in the shape of a right triangle as shown below. Find the length of the base and the height.

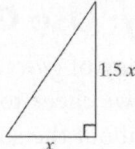

Understand

The diagram shows that the height is 1.5 times the length of the base.

Plan

Test possible dimensions of the triangle formed by the boom (base), the mast (height), and the sail. Check to see if they produce the desired area. Organize your results in a table.

Carry Out

Boom	Mast	Area	Conclusion
6	9	$\frac{1}{2} \cdot 6 \cdot 9 = 27$	Too low
10	15	$\frac{1}{2} \cdot 10 \cdot 15 = 75$	Too high
8	12	$\frac{1}{2} \cdot 8 \cdot 12 = 48$	✔

Check

A triangle with a base length of 8 ft and a height of 12 ft has an area of 48 ft².

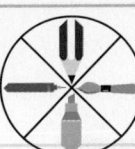

School Book Page WORK BACKWARD

Working Backward

The problem-solving strategy *Work Backward* is useful when taking multiple-choice tests. Work backward by testing each choice in the original problem. You will eliminate incorrect answers. Eventually you will find the correct answer.

EXAMPLE

A fruit stand is selling 8 bananas for $1.00. At this rate, how much will 20 bananas cost?

Ⓐ $1.50   Ⓑ $2.00   Ⓒ $2.50   Ⓓ $3.00

Use mental math to test the choices that are easy to use.

$2.00 is twice $1.00. Twice 8 is only 16, so choice B is not the answer.

$3.00 is three times $1.00. Three times 8 is 24, so choice D is not the correct answer.

Since 20 is between 16 and 24, the cost must be between $2.00 and $3.00. The correct answer is choice C.

Source: p. 353; From MIDDLE GRADES MATH (Course 1) Copyright © 2010 Pearson Education, Inc., or its affiliates. Used by permission. All Rights Reserved.

 NOW TRY THIS 1-8

When Linda added all her test scores and divided by 11 (the number of tests), she found her average to be 80. Her teacher tells her she can drop her single low score of 50. What is her new average?

Strategy: Write an Equation

Even though algebraic thinking is involved in the strategy *writing an equation* and may evoke thoughts of traditional algebra, a closer look reveals that algebraic thinking starts very early in students' school lives. For example, finding the missing addend in a problem like

$$\begin{array}{r} 14 \\ -\square \\ \hline 3 \end{array}$$

could be thought of algebraically as $14 - \square = 3$, or as $3 + \square = 14$. In a traditional algebra course, this might be seen as $14 - x = 3$ or $3 + x = 14$ with 11 as a solution. We use such algebraic thinking long before formal algebra is taught. And the strategy of *writing an equation* is used in this text before more formal approaches to algebra are seen in Chapter 8 *Real Numbers and Algebra*.

 A student example of *writing an equation* to solve a problem is seen on the grade 6 student page (p. 14).

School Book Page WRITE AN EQUATION

Write an Equation

When to Use This Strategy To *write an equation* is one way of organizing the information needed to solve a problem.

Discount A bicycle is on sale for $139.93. This is 30% off the regular price. What is the regular price of the bicycle?

Understand

The sale price of the bicycle, $139.93, is 30% off the regular price. You need to find the regular price.

Plan

Translate the words into an equation. You will pay 100% − 30% = 70% of the regular price.

Carry Out

The percent you pay times the regular price equals the sale price.

Words   percent you pay times regular price equals sale price

Equation   Let r = the regular price.

$$70\% \qquad \times \qquad r \qquad = \qquad \$139.93$$

$0.7r = 139.93$ ← Write 70% as a decimal: 0.7.

$0.7r \div 0.7 = 139.93 \div 0.7$ ← Divide each side by 0.7 to find r.

$r = \$199.90$ ← Simplify.

The regular price of the bicycle is $199.90.

Check

The regular price is about $200. The sale price is about 70% of $200, or $140. This is close to the sale price.

xlviii   **Problem Solving Handbook**

 **Assessment 1-1A**

1. Use the approach in Gauss's Problem to find the following sums (do not use formulas):
   a. $1 + 2 + 3 + 4 + \ldots + 99$
   b. $1 + 3 + 5 + 7 + \ldots + 1001$
2. Use the ideas behind the drawings in (a) and (b) to find the solution to Gauss's problem. Explain your reasoning.

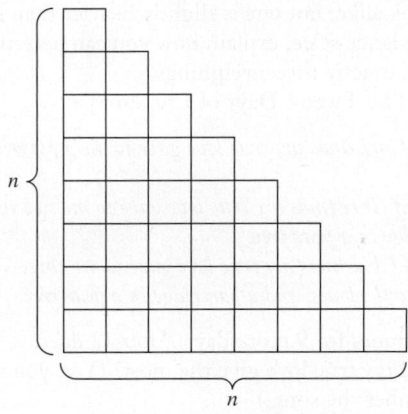

(a)

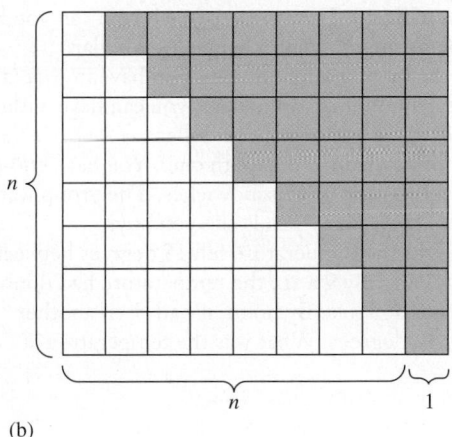

(b)

3. Find the sum of $36 + 37 + 38 + 39 + \ldots + 146 + 147$.
4. Cookies are sold singly or in packages of 2 or 6. With this packaging, how many ways can you buy a dozen cookies?
5. A nursery rhyme states:

   As I was going to St. Ives
   I met a man with seven wives.
   Every wife had seven sacks,
   Every sack had seven cats,
   Every cat had seven kits.
   Kits, cats, sacks, and wives,
   How many were going to St. Ives?

   Explain how many are going to St. Ives.

6. How many triangles are in the following figure?

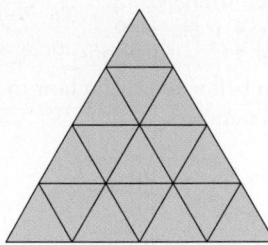

7. Without computing each sum, find which is greater, O or E, and by how much.

   $$O = 1 + 3 + 5 + 7 + \ldots + 97$$
   $$E = 2 + 4 + 6 + 8 + \ldots + 98$$

8. Alabama, Bubba, Cory, and Dandy are in a horse race. Bubba is the slowest, Cory is faster than Alabama but slower than Dandy. Name the finishing order of the horses.
9. How many ways can you make change for a $50 bill using $5, $10, and $20 bills?
10. The following is a magic square (all rows, columns, and diagonals sum to the same number). Find the values of each letter.

17	a	7
12	22	b
c	d	27

11. Frankie and Johnny began reading a novel on the same day. Frankie reads 8 pages a day and Johnny reads 5 pages a day. If Frankie is on page 72, what page is Johnny on?
12. The 14 digits of a credit card are written in the boxes shown. If the sum of any three consecutive digits is 20, what is the value of A?

A	7										7	4

13. Three closed boxes (A, B, and C) of fruit arrive as a gift from a friend. Each box is mislabeled. How could you choose only one fruit from one box to decide how the boxes should be labeled?

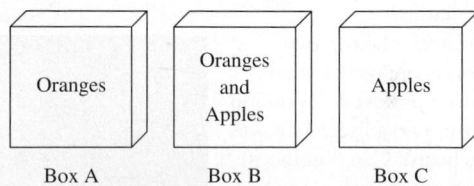

14. A compass and a ruler together cost $4. The compass costs 90¢ more than the ruler. How much does the compass cost?
15. Kathy stood on the middle rung of a ladder. She climbed up three rungs, moved down five rungs, and then climbed up seven rungs. Then she climbed up the remaining six rungs to the top of the ladder. How many rungs are there in the whole ladder?

Assessment 1-1B

1. Use the approach in Gauss's Problem to find the following sums (do not use formulas):
   a. $1 + 2 + 3 + 4 + \ldots + 49$
   b. $1 + 3 + 5 + 7 + \ldots + 2009$
2. Use the diagram below to explain how to find the sum of the first 100 natural numbers.

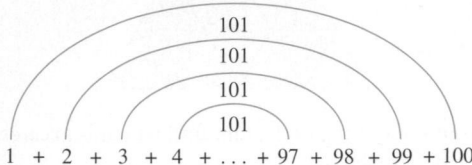

101
101
101
101
$1 + 2 + 3 + 4 + \ldots + 97 + 98 + 99 + 100$

3. Find the sum of $58 + 59 + 60 + 61 + \ldots + 203$.
4. Eve Merriam* entitled her children's book *12 Ways to Get to 11* (1993). Using only addition and natural numbers, describe 12 ways that one can arrive at the sum of 11.
5. Explain why in a drawer containing only two different colors of socks one must draw only three socks to find a matching pair.
6. How many squares are in the following figure?

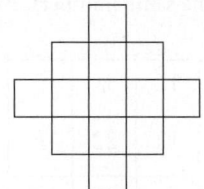

7. If $E = 2 + 4 + 6 + 8 + \ldots + 98$ and $P = 1 + 3 + 5 + 7 + \ldots + 99$, which is greater, E or P, and by how much?
8. The sign says that you are leaving Missoula, Butte is 120 mi away, and Bozeman is 200 mi away. There is a rest stop halfway between Butte and Bozeman. How far is the rest stop from Missoula if both Butte and Bozeman are in the same direction?

9. Marc goes to the store with exactly $1.00 in change. He has at least one of each coin less than a half-dollar coin, but he does not have a half-dollar coin.
   a. What is the least number of coins he could have?
   b. What is the greatest number of coins he could have?
10. Find a 3-by-3 magic square using the numbers 3, 5, 7, 9, 11, 13, 15, 17, and 19.
11. Eight marbles look alike, but one is slightly heavier than the others. Using a balance scale, explain how you can determine the heavier one in exactly three weighings.
12. Recall the song "The Twelve Days of Christmas":

    On the first day of Christmas my true love gave to me a partridge in a pear tree.
    On the second day of Christmas my true love gave to me two turtle doves and a partridge in a pear tree.
    On the third day of Christmas my true love gave to me three French hens, two turtle doves, and a partridge in a pear tree.

    This pattern continues for 9 more days. After 12 days,
    a. which gifts did my true love give the most? (Yes, you will have to remember the song.)
    b. how many total gifts did my true love give to me?
13. a. Suppose you have quarters, dimes, and pennies with a total value of $1.19. How many of each coin can you have without being able to make change for a dollar?
    b. Tell why the combination of coins you have in part (a) is the greatest amount of money that you can have without being able to make change for a dollar.
14. Suppose you buy lunch for the math club. You have enough money to buy 20 salads or 15 sandwiches. The group wants 12 sandwiches. How many salads can you buy?
15. One winter night the temperature fell 15 degrees between midnight and 5 A.M. By 9 A.M., the temperature had doubled from what it was at 5 A.M. By noon, it had risen another 10 degrees to 32 degrees. What was the temperature at midnight?

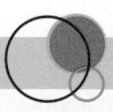

Mathematical Connections 1-1

Communication

1. Why is teaching problem solving an important part of mathematics?
2. In the checkerboard, two squares on opposite corners have been removed. A domino can cover two adjacent squares on the board. Can dominoes be arranged in such a way that all the remaining squares on the board can be coverd with no dominoes overlapping or hanging off the board? If not, why not? (Hint: Each domino must cover one black and

one red square. Compare this with the number of each color of squares on the board.)
3. a. If eight people shake hands with one another exactly once, how many handshakes take place?
   b. Compare strategies for working the problem. How are they the same? How are they different?
   c. Find as many ways as possible to do the problem.
   d. Generalize the solution for n people.

Open-Ended

4. Choose a problem-solving strategy and make up a problem that would use this strategy. Write the solution using Pólya's four-step approach.
5. The distance around the world is approximately 40,000 km. Approximately how many people of average size would it take to stretch around the world if they were holding hands?

*Merriam, E. *12 Ways to Get to 11*. New York: Aladdin Paperbacks, 1993.

Cooperative Learning

6. Work in pairs on the following version of a game called NIM. A calculator is needed for each pair.
   a. Player 1 presses $\boxed{1}$ and $\boxed{+}$ or $\boxed{2}$ and $\boxed{+}$. Player 2 does the same. The players take turns until the target number of 21 is reached. The first player to make the display read 21 is the winner. Determine a strategy for deciding who always wins.
   b. Try a game of NIM using the digits 1, 2, 3, and 4, with a target number of 104. The first player to reach 104 wins. What is the winning strategy?
   c. Try a game of NIM using the digits 3, 5, and 7, with a target number of 73. The first player to exceed 73 loses. What is the winning strategy?
   d. Now play Reverse NIM with the keys $\boxed{1}$ and $\boxed{2}$. Instead of $\boxed{+}$, use $\boxed{-}$. Put 21 on the display. Let the target number be 0. Determine a strategy for winning Reverse NIM.
   e. Try Reverse NIM using the digits 1, 2, and 3 and starting with 24 on the display. The target number is 0. What is the winning strategy?
   f. Try Reverse NIM using the digits 3, 5, and 7 and starting with 73 on the display. The first player to display a negative number loses. What is the winning strategy?

Questions from the Classroom

7. John asks why the last step of Pólya's four-step problem-solving process, *looking back*, is necessary since he has already given the answer. What could you tell him?
8. A student asks why he just can't make "random guesses" rather than "intelligent guesses" when using the guess-and-check problem-solving strategy. How do you respond?
9. Rob says that it is possible to create a magic square with the numbers 1, 3, 4, 5, 6, 7, 8, 9, and 10. How do you respond?

Trends in Mathematics and Science Study (TIMSS) Question

4	11	6
9		5
8	3	10

The rule for the table is that numbers in each row and column must add up to the same number. What number goes in the center of the table?
   a. 1   b. 2
   c. 7   d. 12

TIMSS, Grade 4, 2003

Joe knows that a pen costs 1 zed more than a pencil.
His friend bought 2 pens and 3 pencils for 17 zeds.
How many zeds will Joe need to buy 1 pen and 2 pencils?

TIMSS, Grade 8, 2007

National Assessment of Educational Progress (NAEP) Question

There will be 58 people at a breakfast and each person will eat 2 eggs. There are 12 eggs in each carton. How many cartons of eggs will be needed for the breakfast?
   a. 9   b. 10
   c. 72   d. 116

NAEP, Grade 4, 2007

BRAIN TEASER Ten women are fishing all in a row in a boat. One seat in the center of the boat is empty. The five women in the front of the boat want to change seats with the five women in the back of the boat. A person can move from her seat to the next empty seat or she can step over one person without capsizing the boat. What is the minimum number of moves needed for the five women in front to change places with the five in back?

BRAIN TEASER Place a half-dollar, a quarter, and a nickel in position A as shown in Figure 1-8. Try to move these coins, one at a time, to position C. At no time may a larger coin be placed on a smaller coin. Coins may be placed in position B. How many moves does it take to get them to position C? Now add a penny to the pile and see how many moves are required. This is a simple case of the famous Tower of Hanoi problem, in which ancient Brahman priests were required to move a pile of 64 disks of decreasing size, after which the world would end. How long would it take at a rate of one move per second?

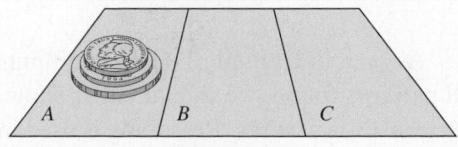

A   B   C

Figure 1-8

1-2 Explorations with Patterns

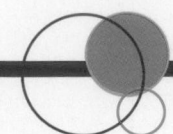

E-Manipulative Activity

Additional practice with patterns not involving numbers can be found in the *Patterns* activity on the E-Manipulatives disk. The activity involves completing simple and complex patterns of symbols and colored blocks.

Mathematics has been described as the study of patterns. Patterns are everywhere—in wallpaper, tiles, traffic, and even television schedules. Police investigators study case files to find the *modus operandi*, or pattern of operation, when a series of crimes is committed. Scientists look for patterns in order to isolate variables so that they can reach valid conclusions in their research.

Non-numerical patterns abound. For young children, a pattern could appear as shown in Now Try This 1-9.

 NOW TRY THIS 1-9

a. Find three more terms to continue a pattern:

$$\bigcirc, \triangle, \triangle, \bigcirc, \triangle, \triangle, \bigcirc, \underline{\quad}, \underline{\quad}, \underline{\quad}$$

b. Describe in words the pattern found in part (a).

Patterns can be surprising. Consider Example 1-1.

 **EXAMPLE 1-1**

a. Describe any patterns seen in the following:

$$1 + 0 \cdot 9 = 1$$
$$2 + 1 \cdot 9 = 11$$
$$3 + 12 \cdot 9 = 111$$
$$4 + 123 \cdot 9 = 1111$$
$$5 + 1234 \cdot 9 = 11111$$

b. Do the patterns continue? Why or why not?

Solution

a. There are several possible patterns. For example, the numbers on the far left are natural numbers. The pattern starts with 1 and continues to the next greater natural number in each successive line. The numbers "in the middle" are products of two numbers, the second of which is 9; the left-most number in the first product is 0; after that the left-most number in each product is formed using natural numbers and including an additional natural number in each successive line. The numbers after the "=" sign are formed using 1s and include an additional 1 in each successive line.

b. The pattern in the complete equation appears to continue for a number of cases, but it does not continue in general; for example,

$$13 + 123456789101112 \cdot 9 = 1,111,111,101,910,021$$

This pattern breaks down when the pattern of digits in the number being multiplied by 9 contains previously used digits.

As seen in Example 1-1, determining a pattern on the basis of a few cases is not reliable. For all patterns found, we should either show the pattern does not hold in general or justify that the pattern always works. Reasoning is used in both cases.

Reasoning

Some books list various types of reasoning as a problem-solving strategy. However, we think that reasoning underlies problem solving. *PSSM* states that students at all grade levels should be enabled to

- recognize reasoning and proof as fundamental aspects of mathematics
- make and evaluate mathematical conjectures
- develop and evaluate mathematical arguments and proofs
- select and use various types of reasoning and methods of proof. (p. 402)

For students to recognize reasoning and proof as fundamental aspects of mathematics, it is necessary that they use both reasoning and proof in their studies. However, it must be recognized that the level of use depends on the level of the students and their understanding of mathematics. For example, from very early ages, students use inductive reasoning to look for regularities in patterns based on a very few cases and to develop **conjectures** statements or conclusions that have not been proven. **Inductive reasoning** is the method of making generalizations based on observations and patterns. Such reasoning may or may not be accurate, and conjectures based on inductive reasoning may or may not be true. The validity, or truth, of conjectures in mathematics relies on **deductive reasoning**—the use of mathematical axioms, theorems, definitions, undefined terms assumed to be true, and logic for proof.

Throughout mathematics, there is a fine interweaving of inductive reasoning and conjecturing to develop conclusions thought to be true. Deductive reasoning is required to prove those conclusions. In this book we show how inductive reasoning may lead to false conclusions or false conjectures. We show how deductive reasoning is used to prove true conjectures.

Inductive and Deductive Reasoning

Scientists make observations and propose general laws based on patterns. Statisticians use patterns when they form conclusions based on collected data. This process of *inductive reasoning* may lead to new discoveries; its weakness is that conclusions are drawn only from the collected evidence. If not all cases have been checked, another case may prove the conclusion false. For example, considering only that $0^2 = 0$ and that $1^2 = 1$, we might conjecture that *every number squared is equal to itself*. When we find an example ($2^2 = 4$) that contradicts the conjecture, that **counterexample** proves the conjecture false. Students frequently experience difficulty with the concept of a counterexample. Sometimes finding a counterexample is difficult, but not finding one immediately does not make a conjecture true.

Next, consider a pattern that does work and helps solve a problem. How can you find the sum of three consecutive natural numbers without performing the addition? Three examples are given below.

$$14 + 15 + 16 \qquad (\mathbf{45})$$
$$19 + 20 + 21 \qquad (\mathbf{60})$$
$$99 + 100 + 101 \qquad (\mathbf{300})$$

After studying the sums, a pattern of multiplying the middle number by 3 emerges. The pattern suggests other mathematical questions to consider. For example,

1. Does this work for any three consecutive natural numbers?
2. How can we find the sum of any odd number of consecutive natural numbers?
3. What happens if there is an even number of consecutive natural numbers?

To answer question (1), we give a proof showing that the sum of three consecutive natural numbers is equal to 3 times the middle number. This proof is an example of *deductive reasoning*.

Proof

Let n be the first of three consecutive natural numbers. Then the three numbers are n, $n + 1$, and $n + 2$. The sum of these three numbers is $n + (n + 1) + (n + 2) = 3n + 3 = 3(n + 1)$. Therefore, the sum of the three consecutive natural numbers is 3 times the middle number.

The Danger of Making Conjectures Based on a Few Cases

In grade 5, *PSSM*, we find the following:

> Students should move toward reasoning that depends on relationships and properties. Students need to be challenged with questions such as, What if I gave you twenty more problems like this to do—would they all work the same way? How do you know? (p. 190)

The following discussion illustrates the danger of making a conjecture based on a few cases. In Figure 1-9, we choose points on a circle and connect them to form distinct, nonoverlapping regions. In this figure, 2 points determine 2 regions, 3 points determine 4 regions, and 4 points determine 8 regions. What is the maximum number of regions that would be determined by 10 points?

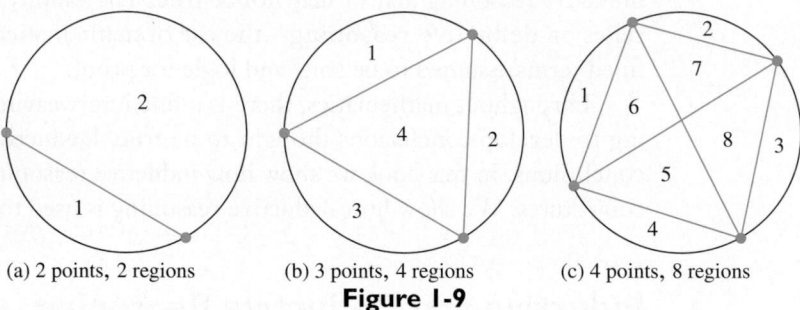

(a) 2 points, 2 regions   (b) 3 points, 4 regions   (c) 4 points, 8 regions

Figure 1-9

The data from Figure 1-9 are recorded in Table 1-2. It appears that each time the number of points increases by 1, the number of regions doubles. If this were true, then for 5 points we would have 2 times the number of regions with 4 points, or $2 \cdot 8 = 16 = 2^4$, and so on. If we base our conjecture on this pattern, we might believe that for 10 points, we would have 2^9, or 512 regions. (Why?)

Table 1-2

Number of points	2	3	4	5	6	...	10
Maximum number of regions	2	4	8				?

An initial check for this conjecture is to see whether we obtain 16 regions for 5 points. We obtain a diagram similar to that in Figure 1-10, and our guess of 16 regions is confirmed. For 6 points, the pattern predicts that the number of regions will be 32. Choose the points so that they are neither symmetrically arranged nor equally spaced and count the regions carefully. You should obtain 31 regions and not 32 regions as predicted. No matter how the points are located on the circle, the guess of 32 regions is not correct. The counterexample tells us that the doubling pattern is not correct; note that it does not tell us whether or not there are 512 regions with 10 points, but only that the pattern is not what we conjectured. Thus, in the context of counting the number of regions of a circle, the pattern is incorrect.

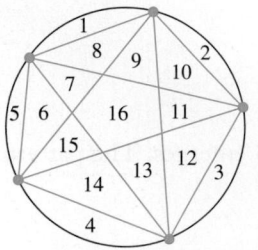

Figure 1-10

🔵 NOW TRY THIS 1-10

A *prime number* is a natural number with exactly two distinct positive numbers, 1 and the number itself, that divide it with 0 remainder; for example, 2, 3, 5, 7, 11, 13 are primes. One day Amy makes a *conjecture* that the formula, $y = x^2 + x + 11$ will produce only prime numbers if she substitutes the natural numbers, 1, 2, 3, 4, 5, . . . for x. She shows her work so far in Table 1-3 for $x = 1, 2, 3, 4$.

Table 1-3

x	1	2	3	4
y	13	17	23	31

a. What type of reasoning is Amy using?
b. Try the next several natural numbers and see whether they seem to work.
c. Can you find a counterexample to show that Amy's conjecture is false?

Arithmetic Sequences

A **sequence** is an ordered arrangement of numbers, figures, or objects. A sequence has items or terms identified as *1st*, *2nd*, *3rd*, and so on. Often, sequences can be classified by their properties. For example, what property do the following first three sequences have that the fourth does not?

a. 1, 2, 3, 4, 5, 6, . . .
b. 0, 5, 10, 15, 20, 25, . . .
c. 2, 6, 10, 14, 18, 22, . . .
d. 1, 11, 111, 1111, 11111, 111111, . . .

In each of the first three sequences, each term—starting from the second term—is obtained from the preceding term by adding a fixed number, the **common difference** or **difference**. In part (a) the difference is 1, in part (b) the difference is 5, and in part (c) the difference is 4. Sequences such as the first three are arithmetic sequences. An **arithmetic sequence** is one in which each successive term from the second term on is obtained from the previous term by the addition or subtraction of a fixed number. The sequence in part (d) is not arithmetic because there is no single fixed number that can be added to or subtract from the previous term to obtain the next term.

Arithmetic sequences can be generated from objects, as shown in Example 1-2.

EXAMPLE 1-2

Find a numerical pattern in the number of matchsticks required to continue the sequence shown in Figure 1-11.

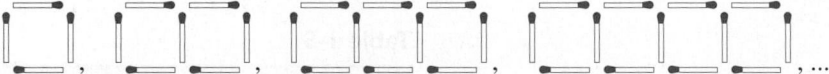

Figure 1-11

Solution   Assume the matchsticks are arranged so that each figure has one more square on the right than the preceding figure. Note that the addition of a square to an arrangement requires the addition of three matchsticks each time. Thus, with this assumption, the numerical pattern obtained is 4, 7, 10, 13, 16, 19, . . ., an arithmetic sequence starting at 4 and having a difference of 3.

An informal description of an arithmetic sequence is one that can be described as an "add d" pattern, where d is the common difference. In Example 1-2, $d = 3$. In the language of children, the pattern in Example 1-2 is "add 3." This is an example of a **recursive pattern**. In a recursive pattern, after one or more consecutive terms are given to start, each successive term of the sequence is obtained from the previous term(s). For example, 3, 6, 9, . . . is another "add 3" sequence starting with 3.

A recursive pattern is typically used in a spreadsheet, as seen in Table 1-4 where the index column tracks the order of the terms. The headers for the columns are A, B, etc. The first entry in the A column (in the A1 cell) is 4; and to find the term in the A2 cell, we use the number in the A1 cell and add 3. The pattern is continued using the *Fill Down* command. In spreadsheet language, the formula $= A1 + 3$ finds any term after the first by adding 3 to the previous term. A formula based on a recursive pattern is a **recursive formula**. (For more explicit directions on using a spreadsheet, see the Technology Manual, which can be found online at www.pearson-highered.com/Billstein11einfo.)

Table 1-4

	A	B
1	4	
2	7	
3	10	
4	13	
5	16	
6	19	
7	22	
8	25	
9	28	
10		
11		
12		
13		

Index Column →

If we want to find the number of matchsticks in the 100th figure in Example 1-2, we can use the spreadsheet or we can find an explicit formula or a general rule for finding the number of matchsticks when given the number of the term. The problem-solving strategy of *making a table* is again helpful here.

The spreadsheet in Table 1-4 provides an easy way to *make a table*. The index column gives the numbers of the terms and column A gives the terms of the sequence. If we are building such a table without a spreadsheet, it might look like Table 1-5. Notice that each term is a sum of 4 and a certain number of 3s. We see that the number of 3s is 1 less than the number of the term. This pattern should continue, since the first term is $4 + 0 \cdot 3$ and each time we increase the number of the term by 1, we add one *more* 3. Thus, it seems that the 100th term is $4 + (100 - 1)3$; and, in general, the **nth term** is $4 + (n - 1)3$. Note that $4 + (n - 1)3$ could be written as $3n + 1$.

Table 1-5

Number of Term	Term
1	4
2	$7 = 4 + 3 = 4 + 1 \cdot 3$
3	$10 = (4 + 1 \cdot 3) + 3 = 4 + 2 \cdot 3$
4	$13 = (4 + 2 \cdot 3) + 3 = 4 + 3 \cdot 3$
.	.
.	.
.	.
n	$4 + (n - 1)3 = 3n + 1$

Still a different approach to finding the number of matchsticks in the 100th term of Figure 1-11 might be as follows: If the matchstick figure has 100 squares, we could find the total number of matchsticks by adding the number of horizontal and vertical sticks. There are $2 \cdot 100$ placed horizontally. (Why?) Notice that in the first figure, there are 2 matchsticks placed vertically; in the second, 3; and in the third, 4. In the 100th figure, there should be $100 + 1$ vertical matchsticks.

Altogether there will be $2 \cdot 100 + (100 + 1)$, or 301, matchsticks in the 100th figure. Similarly, in the nth figure, there would be $2n$ horizontal and $n + 1$ vertical matchsticks, for a total of $3n + 1$. This discussion is summarized in Table 1-6.

Table 1-6

Number of Term	Number of Matchsticks Horizontally	Number of Matchsticks Vertically	Total
1	2	2	4
2	4	3	7
3	6	4	10
4	8	5	13
⋮	⋮	⋮	⋮
100	200	101	301
⋮	⋮	⋮	⋮
n	$2n$	$n + 1$	$2n + (n + 1) = 3n + 1$

If we are given the value of the term, we can use the formula for the nth term in Table 1-6 to *work backward* to find the number of the term. For example, given the term 1798, we can write an equation: $3n + 1 = 1798$. Therefore, $3n = 1797$ and $n = 599$. Consequently, 1798 is the 599th term. We could obtain the same answer by solving $4 + (n - 1)3 = 1798$ for n.

In the matchstick problem, we found the nth term of a sequence. If the nth term of a sequence is given, we can find any term of the sequence, as shown in Example 1-3.

**EXAMPLE 1-3**

Find the first four terms of a sequence, the nth term of which is given by the following, and determine whether the sequence seems to be arithmetic:

a. $4n + 3$ **b.** $n^2 - 1$

Solution

a.

Number of Term	Term
1	$4 \cdot 1 + 3 = 7$
2	$4 \cdot 2 + 3 = 11$
3	$4 \cdot 3 + 3 = 15$
4	$4 \cdot 4 + 3 = 19$

Hence, the first four terms of the sequence are 7, 11, 15, 19. This sequence seems arithmetic with difference 4.

b.

Number of Term	Term
1	$1^2 - 1 = 0$
2	$2^2 - 1 = 3$
3	$3^2 - 1 = 8$
4	$4^2 - 1 = 15$

Thus, the first four terms of the sequence are 0, 3, 8, 15. This sequence is not arithmetic because it has no common difference.

We generalize our work with arithmetic sequences in Chapter 8.

EXAMPLE 1-4

The diagrams in Figure 1-12 show the molecular structure of alkanes, a class of hydrocarbons. C represents a carbon atom and H a hydrogen atom. A connecting segment shows a chemical bond.

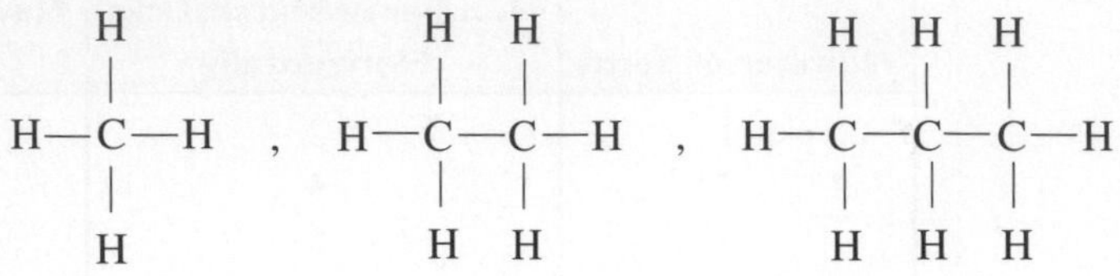

```
        H                   H   H                 H   H   H
        |                   |   |                 |   |   |
   H — C — H    ,      H — C — C — H    ,    H — C — C — C — H
        |                   |   |                 |   |   |
        H                   H   H                 H   H   H
```

methane (C_1H_4)   ethane (C_2H_6)   propane (C_3H_8)

Figure 1-12

a. Hectane is an alkane with 100 carbon atoms. How many hydrogen atoms does it have?

b. Write a general rule for alkanes C_nH_m showing the relationship between m and n.

Solution

a. To determine the relationship between the number of carbon and hydrogen atoms, we first study the drawing of the alkanes and disregard the extreme left and right hydrogen atoms in each. With this restriction, we see that for every carbon atom, there are two hydrogen atoms. Therefore, there are twice as many hydrogen atoms as carbon atoms plus the two hydrogen atoms at the extremes. For example, when there are 3 carbon atoms, there are $(2 \cdot 3) + 2$, or 8, hydrogen atoms. This notion is summarized in Table 1-7. If we extend the table for 4 carbon atoms, we get $(2 \cdot 4) + 2$, or 10, hydrogen atoms. For 100 carbon atoms, there are $(2 \cdot 100) + 2$, or 202, hydrogen atoms.

Table 1-7

No. of Carbon Atoms	No. of Hydrogen Atoms
1	4
2	6
3	8
.	.
.	.
.	.
100	202
.	.
.	.
.	.
n	m

b. In general, for n carbon atoms there would be n hydrogen atoms attached above, n attached below, and 2 attached on the sides. Hence, the total number of hydrogen atoms m would be $2n + 2$. It follows that the number of hydrogen atoms is $m = 2n + 2$.

EXAMPLE 1-5

A theater is set up so that there are 20 seats in the first row, and 4 additional seats in each consecutive row to the back of the theater where there are 144 seats. How many rows are there in the theater?

Solution
Two strategies lend themselves to this problem. One is to *build a table* and to consider the entries as seen in Table 1-8.

Table 1-8

Row Number	Number of Seats
1	20
2	$20 + 4$
3	$20 + 2 \cdot 4$
4	$20 + 3 \cdot 4$
5	$20 + 4 \cdot 4$
.	.
.	.
.	.
n	$20 + (n-1)4$

Observe that in Table 1-8 when we write the number of seats as 20 plus the number of additional 4 seats in consecutive rows, the number of 4s added is one less than the number of the row. We know that in the last row there are 144 seats. Thus, we have the following:

$144 = 20 + (n-1)4$, but 20 added to 124 is 144 so $124 = (n-1)4$.

Now $4 \cdot 31$ is 124 giving us $31 = n - 1$.

Therefore, $n = 32$, and there are 32 rows in the theater.

A different way to solve the problem is to use a spreadsheet as seen in Table 1-9, where the number of the row is seen in the index column and the entry in cell A1 indicates 20 seats in that row. Filling down the A column using the recursive formula = A1 + 4, we find 144 seats in row 32. Thus, there are 32 rows in the theater.

Table 1-9

	A	B
1	20	
2	24	
3	28	
4	32	
5	36	
6	40	
7	44	
8	48	
9	52	
10	56	
11	60	
12	64	
13	68	
14	72	
15	76	
16	80	
17	84	
18	88	

Spreadsheet continued.

19	92	
20	96	
21	100	
22	104	
23	108	
24	112	
25	116	
26	120	
27	124	
28	128	
29	132	
30	136	
31	140	
32	144	
33	148	
34	152	
35	156	
36	160	
37	164	

Fibonacci Sequence

Dan Brown's popular book *The Da Vinci Code* brought renewed interest to one of the most famous sequences of all time, the **Fibonacci sequence**. The Fibonacci sequence is hinted at in the following Foxtrot cartoon.

The Fibonacci series in the cartoon is actually a sequence with 0 as a starting term. More typically, the sequence is seen as follows:

$$1, 1, 2, 3, 5, 8, 13, 21, 34, 55, 89, 144, \ldots$$

This sequence is not *arithmetic* as there is no fixed difference, d. The first two terms of the Fibonacci sequence are 1, 1 and each subsequent term is the sum of the previous two.

● NOW TRY THIS 1-11

Starting with the first two terms, the seeds, as 1, 1,

   a. Add the first three Fibonacci numbers.
   b. Add the first four Fibonacci numbers.
   c. Add the first five Fibonacci numbers.
   d. Add the first six Fibonacci numbers.
   e. Add the first seven Fibonacci numbers.
   f. What pattern is there in the sums in parts (a)–(e) and any of the remaining numbers in the Fibonacci sequence?

Geometric Sequences

A child has 2 biological parents, 4 grandparents, 8 great grandparents, 16 great-great grandparents, and so on. The number of generational ancestors form the **geometric sequence** 2, 4, 8, 16, 32, ... Each successive term of a geometric sequence is obtained from its predecessor by multiplying by a fixed nonzero number, the **ratio**. In this example, both the first term and

○– Historical Note

Leonardo de Pisa was born around 1170. His real family name was Bonaccio but he preferred the nickname Fibonacci, derived from *filius Bonacci*, meaning "son of Bonacci." In his book *Liber Abaci* (1202) he described the now-famous rabbit problem, whose solution, the sequence 1, 1, 2, 3, 5, 8, 13, 21, . . . , became known as the *Fibonacci sequence* with terms called *Fibonacci numbers*.

the ratio are 2. (The ratio is 2 because each person has two parents.) To find the nth term examine the pattern in Table 1-10.

Table 1-10

Number of Term	Term
1	$2 = 2^1$
2	$4 = 2 \cdot 2 = 2^2$
3	$8 = (2 \cdot 2) \cdot 2 = 2^3$
4	$16 = (2 \cdot 2 \cdot 2) \cdot 2 = 2^4$
5	$32 = (2 \cdot 2 \cdot 2 \cdot 2) \cdot 2 = 2^5$
⋮	⋮

In Table 1-10, when the given term is written as a power of 2, the number of the term is the **exponent**. Following this pattern, the 10th term is 2^{10}, or 1024, the 100th term is 2^{100}, and the nth term is 2^n. Thus, the number of ancestors in the nth previous generation is 2^n. The notation used in Table 1-10 can be generalized as follows.

Definition of a^n

$$\text{If } n \text{ is a natural number, then } a^n = \overbrace{a \cdot a \cdot a \cdot \ldots \cdot a}^{n \text{ factors}}.$$

If $n = 0$ and $a \neq 0$, then $a^0 = 1$.

Geometric sequences play an important role in everyday life. For example, suppose we have $1000 in a bank that pays 5% interest annually. (Note that 5% = 0.05.) If no money is added or taken out, then at the end of the first year we have all of the money we started with plus 5% more.

Year 1: $\$1000 + 0.05(\$1000) = \$1000(1 + 0.05) = \$1000(1.05) = \$1050$

If no money is added or taken out, then at the end of the second year we would have 5% more money than the previous year.

Year 2: $\$1050 + 0.05(\$1050) = \$1050(1 + 0.05) = \$1050(1.05) = \$1102.50$

The amount of money in the account after any number of years can be found by noting that every dollar invested for one year becomes $1 + 0.05 \cdot 1$, or 1.05 dollars. Therefore, the amount in each year is obtained by multiplying the amount from the previous year by 1.05. The amounts in the bank after each year form a geometric sequence because the amount in each year (starting from year 2) is obtained by multiplying the amount in the previous year by the same number, 1.05. This is summarized in Table 1-11. ($\approx$ means approximately equal to.)

Table 1-11

Number of Term (Year)	Term (Amount at the Beginning of Each Year)
1	$1000
2	$\$1000(1.05)^1 = \$1050.00$
3	$\$1000(1.05)^2 = \$1102.50$
4	$\$1000(1.05)^3 \approx \$1157.63$
⋮	⋮
n	$\$1000(1.05)^{n-1}$

NOW TRY THIS 1-12

a. Two bacteria are in a dish. The number of bacteria triples every hour. Following this pattern, find the number of bacteria in the dish after 10 hours and after n hours.

b. Suppose that instead of increasing geometrically as in part (a), the number of bacteria increases arithmetically by 3 each hour. Compare the growth after 10 hours and after n hours. Comment on the difference in growth of a geometric sequence versus an arithmetic sequence.

Other Sequences

Figurate numbers, based on geometrical patterns, provide examples of sequences that are neither arithmetic nor geometric. Such numbers can be represented by dots arranged in the shape of certain geometric figures. The number 1 is the beginning of most patterns involving figurate numbers. The arrays in Figure 1-13 represents the first four terms of the sequence of **triangular numbers**.

1 dot   3 dots   6 dots   10 dots

Figure 1-13

The triangular numbers can be written numerically as $1, 3, 6, 10, 15, \ldots$. The sequence $1, 3, 6, 10, 15, \ldots$ is not an arithmetic sequence because there is no *common difference*, as Figure 1-14 shows. It is not a geometric sequence because there is no common ratio. It is not a Fibonacci sequence.

(First difference)

Figure 1-14

However, the sequence of differences, $2, 3, 4, 5, \ldots$, appears to form an arithmetic sequence with difference 1, as Figure 1-15 shows. The next successive terms for the original sequence are shown in color in Figure 1-15.

```
               1   3   6   10   15   21   28
                \ / \ / \ / \  / \  / \  /
(First difference)  2   3   4   5    6    7
                     \ / \ / \  / \  / \  /
(Second difference)   1   1   1    1    1
```

Figure 1-15

Table 1-12 suggests a pattern for finding the next terms and the nth term for the triangular numbers. The second term is obtained from the first term by adding 2; the third term is obtained from the second term by adding 3; and so on.

Table 1-12

Number of Term	Term
1	1
2	$3 = 1 + 2$
3	$6 = 1 + 2 + 3$
4	$10 = 1 + 2 + 3 + 4$
5	$15 = 1 + 2 + 3 + 4 + 5$
.	.
.	.
.	.
10	$55 = 1 + 2 + 3 + 4 + 5 + 6 + 7 + 8 + 9 + 10$

In general, because the *n*th triangular number has *n* dots in the *n*th row, it is equal to the sum of the dots in the previous triangular number (the $(n - 1)$st one) plus the *n* dots in the *n*th row. Following this pattern, the 10th term is $1 + 2 + 3 + 4 + 5 + 6 + 7 + 8 + 9 + 10$, or 55, and the *n*th term is $1 + 2 + 3 + 4 + 5 + \ldots + (n - 1) + n$. This problem is similar to Gauss's Problem in Section 1-1. Because of the work done in Section 1-1, we know that this sum can be expressed as

$$\frac{n(n + 1)}{2}.$$

Next consider the first four *square numbers* in Figure 1-16. These square numbers, 1, 4, 9, 16 can be written as $1^2, 2^2, 3^2, 4^2$. Continuing, the number of dots in the 10th array would be 10^2, the number of dots in the 100th array would be 100^2, and the number of dots in the *n*th array would be n^2. The sequence of square numbers is neither arithmetic nor geometric. Investigate whether the sequence of first differences is an arithmetic sequence and tell why.

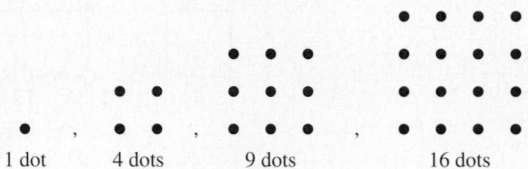

1 dot   4 dots   9 dots   16 dots

Figure 1-16

EXAMPLE 1-6

Use differences to find a pattern. Then assuming that the pattern discovered continues, find the seventh term in each of the following sequences:

a. 5, 6, 14, 29, 51, 80, . . .   **b.** 2, 3, 9, 23, 48, 87, . . .

Solution

a. Figure 1-17 shows the sequence of first differences.

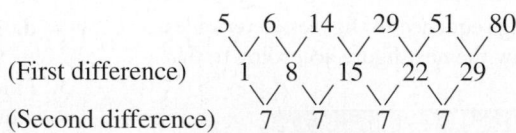

Figure 1-17

To discover a pattern for the original sequence, we try to find a pattern for the sequence of differences 1, 8, 15, 22, 29, ... This sequence is an arithmetic sequence with fixed difference 7 as seen in Figure 1-18.

              5   6   14   29   51   80
(First difference)   1   8   15   22   29
(Second difference)   7   7   7   7

Figure 1-18

Thus, the sixth term in the first difference row is $29 + 7$, or 36, and the seventh term in the original sequence is $80 + 36$, or 116. What number follows 116?

b. Because the second difference is not a fixed number, we go on to the third difference as in Figure 1-19.

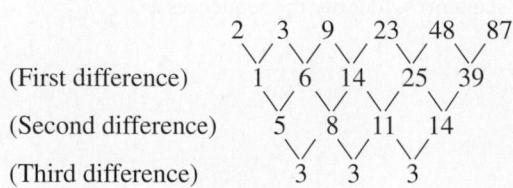

              2   3   9   23   48   87
(First difference)   1   6   14   25   39
(Second difference)   5   8   11   14
(Third difference)   3   3   3

Figure 1-19

The third difference is a fixed number; therefore, the second difference is an arithmetic sequence. The fifth term in the second-difference sequence is $14 + 3$, or 17; the sixth term in the first-difference sequence is $39 + 17$, or 56; and the seventh term in the original sequence is $87 + 56$, or 143.

NOW TRY THIS 1-13

Figure 1-20 shows the first three figures of arrays of sticks with the number of sticks written below the figures.

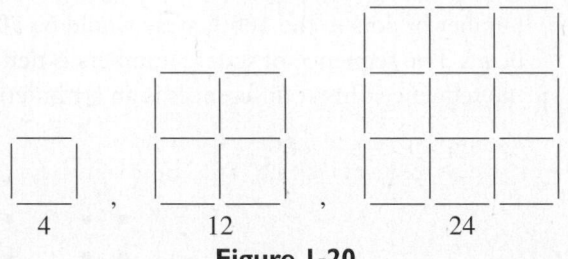

Figure 1-20

a. Draw the next array of sticks.
b. Build a table showing the term number and the number of sticks for $n = 1, 2, 3, 4$.
c. Use differences to predict the number of sticks for $n = 5, 6, 7$.
d. Is finding differences the best way to determine how many sticks there are in the 100th term? Tell how you would find that term.

When asked to find a pattern for a given sequence, we first look for some easily recognizable pattern and determine whether the sequence is arithmetic or geometric. If a pattern is unclear, taking successive differences may help. *It is possible that none of the methods described reveal a pattern.*

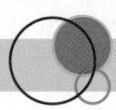

Assessment 1-2 A

1. For each of the following sequences of figures, determine a possible pattern and draw the next figure according to that pattern:
   a. ![figures], . . .
   b. △, ◿◺, ◿▽◺, ◿▽▽◺, . . .
   c. ![figures], . . .

2. In each of the following, list three terms that continue the arithmetic or geometric sequences. Identify the sequences as arithmetic or geometric.
   a. 1, 3, 5, 7, 9
   b. 0, 50, 100, 150, 200
   c. 3, 6, 12, 24, 48

   d. 10, 100, 1,000, 10,000, 100,000
   e. 9, 13, 17, 21, 25, 29

3. Find the 100th term and the nth term for each of the sequences in exercise 2.

4. Use a traditional clock face to determine the next three terms in the following sequence:

   $$1, 6, 11, 4, 9, \ldots$$

5. The pattern $1, 8, 27, 64, 125, \ldots$ is a cubic pattern named because $1 = 1 \cdot 1 \cdot 1$ or 1^3, $8 = 2 \cdot 2 \cdot 2$ or 2^3, and so on.
   a. What is the least 4-digit number greater than 1000 in this pattern?
   b. What is the greatest 3-digit number in this pattern?
   c. If this pattern was produced in a normal spreadsheet, what is the number in cell A14?

6. The first windmill has 5 matchstick squares, the second has 9, and the third has 13, as shown. How many matchstick squares are in **(a)** the 10th windmill? **(b)** the *n*th windmill? **(c)** How many matchsticks will it take to build the *n*th windmill?

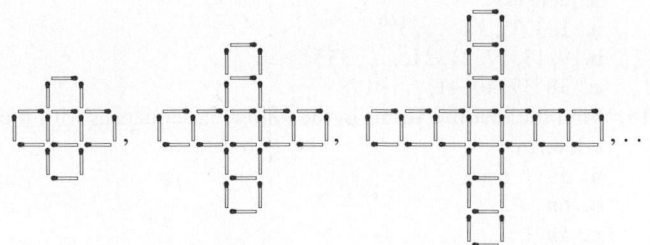

7. In the following sequence, the figures are made of cubes that are glued together. If the exposed surface needs to be painted, how many squares will be painted in **(a)** the 10th figure? **(b)** the *n*th figure?

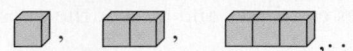

8. The school population for a certain school is predicted to increase by 50 students per year for the next 10 years. If the current enrollment is 700 students, what will the enrollment be after 10 years?

9. Joe's annual income has been increasing each year by the same dollar amount. The first year his income was $24,000, and the ninth year his income was $31,680. In which year was his income $45,120?

10. The first difference of a sequence is the arithmetic sequence 2, 4, 6, 8, 10, ... Find the first six terms of the original sequence in each of the following cases:
    a. The first term of the original sequence is 3.
    b. The sum of the first two terms of the original sequence is 10.
    c. The fifth term of the original sequence is 35.

11. List the next three terms to continue a pattern in each of the following. (Finding differences may be helpful.)
    a. 5, 6, 14, 32, 64, 115, 191
    b. 0, 2, 6, 12, 20, 30, 42

12. How many terms are there in each of the following sequences?
    a. 51, 52, 53, 54, . . ., 151
    b. 1, 2, 2^2, 2^3, . . ., 2^{60}

---

    c. 10, 20, 30, 40, . . ., 2000
    d. 1, 2, 4, 8, 16, 32, . . ., 1024

13. Find the first five terms in sequences with the following *n*th terms.
    a. $n^2 + 2$
    b. $5n - 1$
    c. $10^n - 1$
    d. $3n + 2$

14. Find a counterexample for each of the following:
    a. If *n* is a natural number, then $(n + 5)/5 = n + 1$.
    b. If *n* is a natural number, then $(n + 4)^2 = n + 16$.

15. Assume that the following patterns are built of square tile and the pattern continues. Answer the questions that follow.

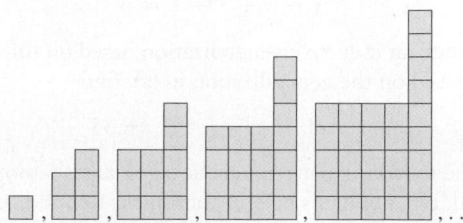

    a. How many square tiles are there in the sixth figure?
    b. How many square tiles are in the *n*th figure?
    c. Is there a figure that has exactly 1259 square tiles? If so, which one? (*Hint*: To determine if there is a figure in the sequence containing exactly 1259 square tiles, first think about the greatest square number less than 1259.)

16. Consider the sequences given in the table below. Find the least number, *n*, such that the *n*th term of the geometric sequence is greater than the corresponding term in the arithmetic sequence.

Number of term	1	2	3	4	5	6	. . .	*n*
Arithmetic	300	500	700	900	1100	1300	. . .	
Geometric	2	4	8	16	32	64	. . .	

17. Start out with a piece of paper. Next, cut the piece of paper into five pieces. Take any one of the pieces and cut it into five pieces, and so on.
    a. What sequence can be obtained through this process?
    b. What is the total number of pieces obtained after *n* cuts?

Assessment 1-2 B

1. In each of the following, determine a possible pattern and draw the next figure according to that pattern if the sequence continues.

    a.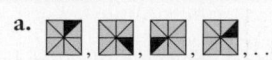

    b.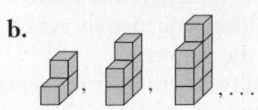

    c.

2. In each of the following, list three terms that continue the arithmetic or geometric sequences. Identify the sequences as arithmetic or geometric.
    a. 8, 11, 14, 17, 20, . . .
    b. 5, 15, 45, 135, 405, . . .

c. 2, 7, 12, 17, 22, . . .
d. 1, 1, 1, 1, 1, . . .
e. 2, 10, 50, 250, 1250, . . .
3. Find the 100th term and the nth term for each of the sequences in exercise 2.
4. Use a traditional clock face to determine the next three terms in the following sequence:

$$1, 9, 5, 1, \ldots$$

5. Observe the following pattern:

$$1 + 3 = 2^2,$$
$$1 + 3 + 5 = 3^2,$$
$$1 + 3 + 5 + 7 = 4^2$$

   a. State an inductive generalization based on this pattern.
   b. Based on the generalization in (a), find

$$1 + 3 + 5 + 7 + \ldots + 35$$

6. In the following pattern, one hexagon takes 6 toothpicks to build, two hexagons take 11 toothpicks to build, and so on. How many toothpicks would it take to build (a) 10 hexagons? (b) n hexagons?

7. Each successive figure below is made of small triangles like the first one in the sequence. Conjecture the number of small triangles needed to make (a) the 100th figure and (b) the nth figure.

$\triangle$, $\triangle\triangle$, $\triangle\triangle\triangle$, . . .

8. A tank contains 15,360 L of water. At the end of each subsequent day, half of the water is removed and not replaced. How much water is left in the tank after 10 days?
9. The Washington Middle School schedule is an arithmetic sequence. Each period is the same length and includes a 4th period lunch. The first three periods begin at 8:10 A.M., 9:00 A.M., and 9:50 A.M., respectively. At what time does the eighth period begin?
10. The first difference of a sequence is 3, 6, 9, 12, 15, . . . Find the first six terms of the original sequence in each of the following cases:
   a. The first term of the original sequence is 3.
   b. The sum of the first two terms of the original sequence is 7.
   c. The fifth term of the original sequence is 34.

11. List the next three terms to continue a pattern in each of the following. (Finding differences may be helpful.)
   a. 3, 8, 15, 24, 35, 48, . . .
   b. 1, 7, 18, 37, 67, 111, . . .
12. How many terms are there in each of the following sequences?
   a. 1, 3, 3^2, 3^3, . . . , 3^{60}
   b. 9, 13, 17, 21, 25, . . . , 353
   c. 38, 39, 40, 41, . . . , 198
13. Find the first five terms in the following sequences with the nth term.
   a. $5n + 6$
   b. $6n - 2$
   c. $5n + 1$
   d. $n^2 - 1$
14. Find a counterexample for each of the following:
   a. If n is a natural number, then $(3 + n)/3 = n$.
   b. If n is a natural number, then $(n - 2)^2 = n^2 - 2^2$.
15. Assume the following pattern with terms built of square tile figures continues and answer the questions that follow.

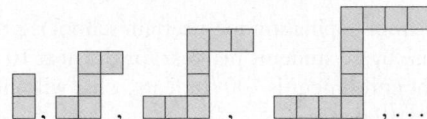

   a. How many square tiles are there in the sixth figure?
   b. How many square tiles are in the nth figure?
   c. Is there a figure that has exactly 449 square tiles? If so, which one?
16. Consider the sequences given in the table below. Find the least number, n, such that the nth term of the geometric sequence is greater than the corresponding term in the arithmetic sequence.

Number of term	1	2	3	4	5	6	. . .	n
Arithmetic	200	500	800	1100	1400	1700	. . .	
Geometric	1	3	9	27	81	243	. . .	

17. Female bees are born from fertilized eggs, and male bees are born from unfertilized eggs. This means that a male bee has only a mother, whereas a female bee has a mother and a father. If the ancestry of a male bee is traced 10 generations including the generation of the male bee, how many bees are there in all 10 generations? (Hint: The Fibonacci sequence might be helpful.)

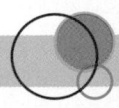

Mathematical Connections 1-2

Communication

1. a. If a fixed number is added to each term of an arithmetic sequence, is the resulting sequence an arithmetic sequence? Justify the answer.

   b. If each term of an arithmetic sequence is multiplied by a fixed number, will the resulting sequence always be an arithmetic sequence? Justify the answer.
   c. If the corresponding terms of two arithmetic sequences are added, is the resulting sequence arithmetic?

2. A student says she read that Thomas Robert Malthus (1766–1834), a renowned British economist and demographer, claimed that the increase of population will take place, if unchecked, in a geometric sequence, whereas the supply of food will increase in only an arithmetic sequence. This theory implies that population increases faster than food production. The student is wondering why. How do you respond?

Open-Ended

3. Patterns can be used to count the number of dots on the Chinese checkerboard; two patterns are shown here. Determine several other patterns to count the dots.

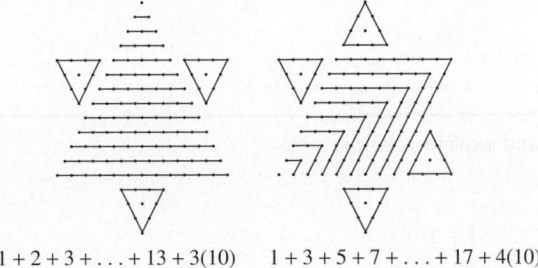

$1 + 2 + 3 + \ldots + 13 + 3(10) \qquad 1 + 3 + 5 + 7 + \ldots + 17 + 4(10)$

4. Make up a pattern involving figurate numbers and find a formula for the 100th term. Describe the pattern and how to find the 100th term.

5. A sequence that follows the same pattern as the Fibonacci sequence but in which the first two terms are any numbers except 1s is called a *Fibonacci type sequence*. Choose a few such sequences and answer the questions in Now Try This 1-11. Do these sequences behave in the same way?

Cooperative Learning

6. The following pattern is called *Pascal's triangle*. It was named for the mathematician Blaise Pascal (1623–1662).

```
                1
              1   1
            1   2   1
          1   3   3   1
        1   4   6   4   1
      1   5  10  10   5   1
    1   6  15  20  15   6   1
  1   7  21  35  35  21   7   1
```

   a. Have each person in the group find four different patterns in the triangle and then share them with the rest of the group.
   b. Add the numbers in each horizontal row. Discuss the pattern that occurs.
   c. Use part (b) to find the sum in the 16th row.
   d. What is the sum of the numbers in the *n*th row?

7. If the following pattern continued indefinitely, the resulting figure would be called the *Sierpinski triangle*, or *Sierpinski gasket*.

In a group, determine each of the following. Discuss different counting strategies.
   a. How many black triangles would be in the fifth figure?
   b. How many white triangles would be in the fifth figure?
   c. If the pattern is continued for *n* figures, how many black triangles will there be?

Questions from the Classroom

8. Joey said that 4, 24, 44, and 64 all have remainder 0 when divided by 4, so all numbers that end in 4 must have 0 remainder when divided by 4. How do you respond?

9. Al and Betty were asked to extend the sequence 2, 4, 8, ... Al said his answer of 2, 4, 8, 16, 32, 64, . . . was the correct one. Betty said Al was wrong and it should be 2, 4, 8, 14, 22, 32, 44, ... What do you tell these students?

10. A student claims the sequence 6, 6, 6, 6, 6, . . . never changes, so it is neither arithmetic nor geometric. How do you respond?

Review Problems

11. In a baseball league consisting of 10 teams, each team plays each of the other teams twice. How may games will be played?

12. How many ways can you make change for 40¢ using only nickels, dimes, and quarters?

13. Tents hold 2, 3, 5, 6, or 12 people. What combinations of tents are possible to sleep 26 people if all tents are fully occupied and only one 12-person tent is used?

Trends in Mathematics and Science Study (TIMSS) Questions

The numbers in the sequence 7, 11, 15, 19, 23, . . . increase by four. The numbers in the sequence 1, 10, 19, 28, 37, . . . increase by nine. The number 19 is in both sequences. If the two sequences are continued, what is the next number that is in BOTH the first and the second sequences?

TIMSS, Grade 8, 2003

Matchsticks are arranged as shown in the figures.

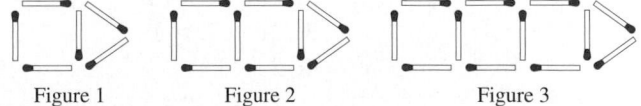

| Figure 1 | Figure 2 | Figure 3 |

If the pattern is continued, how many matchsticks would be used to make Figure 10?
   a. 30   b. 33   c. 36
   d. 39   e. 42

TIMSS, Grade 8, 2003

The three figures below are divided into small congruent triangles.

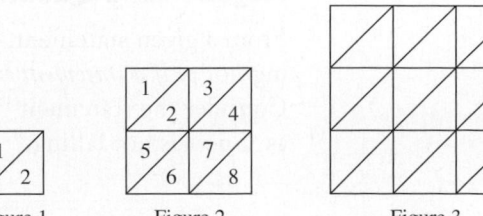

| Figure 1 | Figure 2 | Figure 3 |

a. Complete the table below. First, fill in how many small triangles make up Figure 3. Then, find the number of small triangles that would be needed for the fourth figure if the sequence of figures is extended.

Figure	Numbers of Small Triangles
1	2
2	8
3	
4	

b. The sequence of figures is extended to the seventh figure. How many small triangles would be needed for Figure 7?

c. The sequence of figures is extended to the 50th figure. Explain a way to find the number of small triangles in the 50th figure that does not involve drawing it and counting the number of triangles.

TIMSS, Grade 8, 2003

BRAIN TEASER

Find the next row in the following pattern and explain your pattern:

$$1$$
$$1 \quad 1$$
$$2 \quad 1$$
$$1 \quad 2 \quad 1 \quad 1$$
$$1 \quad 1 \quad 1 \quad 2 \quad 2 \quad 1$$

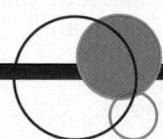

1-3 Reasoning and Logic: An Introduction

Logic is a tool used in mathematical thinking and problem solving. It is essential for reasoning and, although it cannot be taught in a single unit, we present a few "basics" of logic in this section. In logic, a **statement** *is a sentence that is either true or false, but not both.* The following expressions are not statements because their truth values cannot be determined:

1. She has blue eyes.
2. $x + 7 = 18$
3. $2y + 7 > 1$
4. $2 + 3$

5. How are you?
6. Look out!
7. Lincoln was the best president.

Expressions (**1**), (**2**), and (**3**) become statements if, for (**1**), "she" is identified, and for (**2**) and (**3**), values are assigned to x and y, respectively. However, an expression involving *he* or *she* or x or y may already be a statement. For example, "If he is over 210 cm tall, then he is over 2 m tall" and "$2(x + y) = 2x + 2y$" are both statements because they are true no matter who *he* is or what the numerical values of x and y are.

Negation and Quantifiers

From a given statement, p, it is possible to create the **negation** of p denoted by $\sim p$ and meaning *not p. If a statement is true, its negation is false, and if the statement is false, its negation is true.* Consider the statement "Snow is falling." The negation of this statement may be stated simply as "Snow is not falling."

EXAMPLE 1-7

Negate each of the following statements:

  a. $2 + 3 = 5$
  b. A hexagon has six sides.

Solution

  a. $2 + 3 \neq 5$
  b. A hexagon does not have six sides.

At a given time, sentences such as "The shirt I have on is blue" and "The shirt I have on is green" are statements. However, they are not negations of each other. A statement and its negation must have opposite truth values in all possible cases. If the shirt I have on is actually red, then both of the statements are false and, hence, cannot be negations of each other. However, at a given time the statements "The shirt I have on is blue" and "The shirt I have on is not blue" are negations of each other because they have opposite truth values no matter what color the shirt really is.

Some statements involve **quantifiers** and are more complicated to negate. Quantifiers include words such as *all*, *some*, *every*, and *there exists*.

- The quantifiers *all*, *every*, and *no* refer to each and every element in a set and are called *universal quantifiers*.
- The quantifiers *some* and *there exists at least one* refer to one or more, or possibly all, of the elements in a set and are called *existential quantifiers*.
- *All*, *every*, and *each* have the same mathematical meaning. Similarly, *some* and *there exists at least one* have the same meaning.

Assume the following is true: "Some professors at Paxson University have blue eyes." This means that at least one professor at Paxson University has blue eyes. It does not rule out the possibilities that *all* the Paxson professors have blue eyes or that some of the Paxson professors do not have blue eyes. Because the negation of a true statement is false, neither "Some professors at Paxson University do not have blue eyes" nor "All professors at Paxson have blue eyes" are negations of the original statement. One possible negation of the original statement is "No professors at Paxson University have blue eyes."

To discover if one statement is a negation of another, we use arguments similar to the preceding one to determine whether they have opposite truth values in all possible cases.

General forms of quantified statements with their negations follow.

Statement	Negation
Some *a* are *b*.	No *a* is *b*.
Some *a* are not *b*.	All *a* are *b*.
All *a* are *b*.	Some *a* are not *b*.
No *a* is *b*.	Some *a* are *b*.

EXAMPLE 1-8

Negate each of the following regardless of its truth value:

  a. All students like hamburgers.
  b. Some people like mathematics.
  c. There exists a natural number n such that $3n = 6$.
  d. For all natural numbers n, $3n = 3n$.

Solution

a. Some students do not like hamburgers.
b. No people like mathematics.
c. For all natural numbers n, $3n \neq 6$.
d. There exists a natural number n such that $3n \neq 3n$.

Truth Tables and Compound Statements

To investigate the truth of statements, consider the following puzzle by a foremost writer of logic puzzles, Raymond Smullyan. One of Smullyan's books, *The Lady or the Tiger?*, has a puzzle about a prisoner who must make a choice between two rooms. Each room has a sign on the door and the prisoner knows that exactly *one* of the signs is true. The signs are shown in Figure 1-21.

IN THIS ROOM THERE IS A LADY AND IN THE OTHER ROOM THERE IS A TIGER.	IN ONE OF THESE ROOMS THERE IS A LADY AND IN ONE OF THESE ROOMS THERE IS A TIGER.
Door Sign for Room 1	Door Sign for Room 2

Figure 1-21

With the information on the signs in Figure1-21, the prisoner can choose the correct room. Discuss Smullyan's puzzle and try to find a solution before reading on.

Consider that if the sign on Room 1 is true, then the sign on Room 2 must be true. Since this cannot happen, the sign on Room 2 must be true, making the sign on Room 1 false. Because the sign on Room 1 is false, the lady can't be in Room 1 and must be in Room 2.

A symbolic system can help in the study of logic. **Truth tables** are often used to show all possible true-false patterns for statements. Table 1-13 summarizes the truth tables for p and $\sim p$.

Table 1-13

Statement p	Negation $\sim p$
T	F
F	T

From two given statements, it is possible to create a new, **compound statement** by using a connective such as *and*. A compound statement may be formed by combining two or more statements. For example, "Snow is falling" and "The ski run is open" together with *and* give "Snow is falling and the ski run is open." Other compound statements can be obtained by using the connective *or*. For example, "Snow is falling or the ski run is open." The symbols $\wedge$ and $\vee$ are used to represent the connectives *and* and *or*, respectively. For example, if p represents "Snow is falling" and q represents "The ski run is open," then "Snow is falling and the ski run is open" is denoted by $p \wedge q$. Similarly, "Snow is falling or the ski run is open" is denoted by $p \vee q$.

Table 1-14

p	q	Conjunction $p \wedge q$
T	T	T
T	F	F
F	T	F
F	F	F

The truth value of any compound statement, such as $p \wedge q$, is defined using the truth value of each of the simple statements. Because each of the statements p and q may be either true or false, there are four distinct possibilities for the truth value of $p \wedge q$, as shown in Table 1-14. The compound statement is the **conjunction** of p and q and is defined to be true if, and only if, both p and q are true. Otherwise, it is false.

The compound statement $p \vee q$ (*p or q*) is a **disjunction**. In everyday language, *or* is not always interpreted in the same way. In logic, we use an *inclusive or*. The statement "I will go to a movie or I will read a book" means I will either go to a movie, or read a book, or do both. Hence, in logic, *p or q*, symbolized $p \vee q$, is defined to be false if both p and q are false and true in all other cases. This is summarized in Table 1-15.

Table 1-15

p	q	Disjunction $p \vee q$
T	T	T
T	F	T
F	T	T
F	F	F

EXAMPLE 1-9

Classify each of the following as true or false:

p: $2 + 3 = 5$ q: $2 \cdot 3 = 6$ r: $5 + 3 = 9$

a. $p \wedge q$ **c.** $\sim p \vee r$ **e.** $\sim (p \wedge q)$
b. $q \vee r$ **d.** $\sim p \wedge \sim q$ **f.** $(p \wedge q) \vee \sim r$

Solution
  a. p is true and q is true, so $p \wedge q$ is true.
  b. q is true and r is false, so $q \vee r$ is true.
  c. $\sim p$ is false and r is false, so $\sim p \vee r$ is false.
  d. $\sim p$ is false and $\sim q$ is false, so $\sim p \wedge \sim q$ is false.
  e. $p \wedge q$ is true so $\sim (p \wedge q)$ is false.
  f. $p \wedge q$ is true and $\sim r$ is true, so $(p \wedge q) \vee \sim r$ is true.

Truth tables are used not only to summarize the truth values of compound statements; they also are used to determine if two statements are **logically equivalent**. *Two statements are logically equivalent if, and only if, they have the same truth values in every possible situation.* If p and q are logically equivalent, we write $p \equiv q$.

EXAMPLE 1-10

Show that $\sim (p \wedge q) \equiv \sim p \vee \sim q$ [First of De Morgan's Laws].

Solution   Two statements are logically equivalent if they have the same truth values. Truth tables for these statements are given in Table 1-16.

Table 1-16

p	q	$p \wedge q$	$\sim (p \wedge q)$	$\sim p$	$\sim q$	$\sim p \vee \sim q$
T	T	T	F	F	F	F
T	F	F	T	F	T	T
F	T	F	T	T	F	T
F	F	F	T	T	T	T

Because the two statements have the same truth values, we know that $\sim (p \wedge q) \equiv \sim p \vee \sim q$.

Example 1-10 shows that $\sim (p \wedge q) \equiv \sim p \vee \sim q$. In the same way we can show that $\sim (p \vee q) \equiv \sim p \wedge \sim q$. We call these equivalencies **De Morgan's Laws** and state them as Theorem 1-1.

Historical Note

George Boole (1815–1864), born in Lincoln, England, is called "the father of logic." As a professor at Queens College in Ireland, he used symbols to represent concepts and developed a system of algebraic manipulations to accompany the symbols. His work, a marriage of logic and mathematics, known as Boolean algebra, has applications in computer science.

> **Theorem 1-1:   De Morgan's Laws**
> **a.** $\sim(p \land q) \equiv \sim p \lor \sim q$
> **b.** $\sim(p \lor q) \equiv \sim p \land \sim q$

NOW TRY THIS 1-14

Use truth tables to confirm the second De Morgan Law.

Conditionals and Biconditionals

Statements expressed in the form "if p, then q" are **conditionals**, or **implications**, and are denoted by $p \rightarrow q$. Such statements also can be read "p implies q." The "if" part of a conditional is the **hypothesis** of the implication and the "then" part is the **conclusion**. Many types of statements can be put in "if-then" form. An example follows:

Statement:	All equilateral triangles have acute angles.
If-then form:	If a triangle is equilateral, then it has acute angles.

Hypothesis   Conclusion

An implication may also be thought of as a promise. Suppose Betty makes the promise "If I get a raise, then I will take you to dinner." If Betty keeps her promise, the implication is true; if Betty breaks her promise, the implication is false. Consider the four possibilities in Table 1-17.

Table 1-17

Case	p	q	Translation of Symbols	Result
(1)	T	T	Betty gets the raise; she takes you to dinner.	Promise Kept
(2)	T	F	Betty gets the raise; she does not take you to dinner.	Promise Broken
(3)	F	T	Betty does not get the raise; she takes you to dinner.	Promise Kept
(4)	F	F	Betty does not get the raise; she does not take you to dinner.	Promise Kept

Table 1-18

p	q	Implication $p \rightarrow q$
T	T	T
T	F	F
F	T	T
F	F	T

The only case in which Betty breaks her promise is when she gets her raise and fails to take you to dinner, case (2). If she does not get the raise, she can either take you to dinner or not without breaking her promise. The definition of implication is summarized in Table 1-18. Observe that the only case for which the implication is false is when p is true and q is false.

An implication can be worded in several equivalent ways, as follows:

1. If the sun is shining, then the swimming pool is open. (If p, then q.)
2. If the sun is shining, the swimming pool is open. (If p, q.)
3. The swimming pool is open if the sun is shining. (q if p.)
4. The sun is shining implies the swimming pool is open. (p implies q.)
5. The sun is shining only if the pool is open. (p only if q.)
6. The sun's shining is a sufficient condition for the swimming pool to be open. (p is a sufficient condition for q.)
7. The swimming pool's being open is a necessary condition for the sun to be shining. (q is a necessary condition for p.)

A statement in the form $p \rightarrow q$ has three related implication statements, as follows:

Statement:	If p, then q.	$p \rightarrow q$
Converse:	If q, then p.	$q \rightarrow p$
Inverse:	If not p, then not q.	$\sim p \rightarrow \sim q$
Contrapositive:	If not q, then not p.	$\sim q \rightarrow \sim p$

EXAMPLE 1-11

Write the converse, the inverse, and the contrapositive for the following statement:

If I am in San Francisco, then I am in California.

Solution   *Converse:*   If I am in California, then I am in San Francisco.
             Inverse: If I am not in San Francisco, then I am not in California.
             Contrapositive: If I am not in California, then I am not in San Francisco.

Example 1-11 illustrates that if an implication is true, its converse and inverse are not necessarily true. However, the contrapositive is true. We check these observations on the following true statement: *If a number is a natural number, the number is not 0.* The natural numbers are 1, 2, 3, 4, 5, 6, ... We check the truth of the converse, inverse, and contrapositive.

Inverse: *If a number is not a natural number, then it is 0.* This is false, since ⁻6 is not a natural number but it also is not 0.

Converse: *If a number is not 0, then it is a natural number.* This is false, since ⁻6 is not 0 but neither is it a natural number.

Contrapositive: *If a number is 0, then it is not a natural number.* This is true because the natural numbers are 1, 2, 3, 4, 5, 6 . . . .

The contrapositive of the last statement is the original statement. Hence, the preceding discussion suggests that a statement and its contrapositive are logically equivalent. It follows that a statement and its contrapositive cannot have opposite truth values. We summarize this in the following theorem.

> **Theorem 1-2:   Equivalence of a statement and its contrapositive**
> The implication, $p \rightarrow q$, and its contrapositive, $\sim q \rightarrow \sim p$, are logically equivalent; that is,
> $p \rightarrow q \equiv \sim q \rightarrow \sim p$.

EXAMPLE 1-12

Use truth tables to prove that $p \rightarrow q \equiv \sim q \rightarrow \sim p$.

Solution   Truth tables for these statements are given in Table 1-19.

Table 1-19

p	q	$p \rightarrow q$	$\sim q$	$\sim p$	$\sim q \rightarrow \sim p$
T	T	T	F	F	T
T	F	F	T	F	F
F	T	T	F	T	T
F	F	T	T	T	T

Because the two statements have the same truth values, $p \rightarrow q \equiv \sim q \rightarrow \sim p$.

NOW TRY THIS 1-15

The implication "If Betty gets a raise (p), then she will take you to dinner (q)" motivated the truth table for $p \rightarrow q$. The only case where Betty broke her promise is when she gets her raise and then fails to take you to dinner: that is, $p \wedge \sim q$. Therefore, $p \wedge \sim q$ is a candidate for the negation of $p \rightarrow q$. To investigate whether $p \wedge \sim q$ is the negation of $p \rightarrow q$, use truth tables to determine whether $\sim (p \rightarrow q) \equiv p \wedge \sim q$.

Connecting a statement and its converse with the connective *and* gives $(p \rightarrow q) \wedge (q \rightarrow p)$. This compound statement can be written as $p \leftrightarrow q$ and usually is read "**p if, and only if, q.**" The statement "*p* if, and only if, *q*" is a **biconditional**.

 NOW TRY THIS 1-16

Build a truth table to determine when a biconditional is true.

Valid Reasoning

In problem solving, the reasoning is said to be **valid** if the conclusion follows unavoidably from the hypotheses. Consider the following examples:

Hypotheses: All dogs are animals.
                Goofy is a dog.

Conclusion: Goofy is an animal.

The statement "All dogs are animals" can be pictured with the **Euler diagram** in Figure 1-22(a).

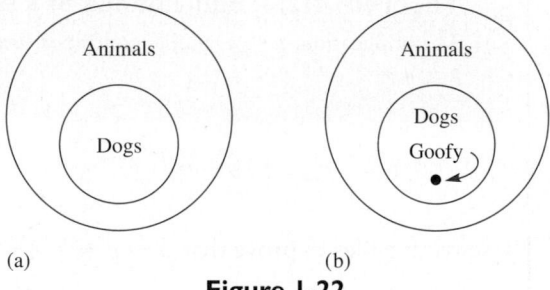

(a)   (b)

Figure 1-22

The information "Goofy is a dog" implies that Goofy now also belongs to the circle containing dogs, as pictured in Figure 1-22(b). Goofy must also belong to the circle containing animals. Thus, the reasoning is valid because it is impossible to draw a picture that satisfies the hypotheses and contradicts the conclusion.

Consider the following argument:

Hypotheses: All elementary schoolteachers are mathematically literate.
                Some mathematically literate people are not children.

Conclusion: No elementary schoolteacher is a child.

Let E be the set of elementary schoolteachers, M be the set of mathematically literate people, and C be the set of children. Then the statement "All elementary schoolteachers are mathematically literate" can be pictured as in Figure 1-23(a). The statement "Some mathematically literate people are not children" can be pictured in several ways. Three of these are illustrated in Figure 1-23(b) through (d). The argument appears to be true with the drawings in Figure 1-23(b) and (c).

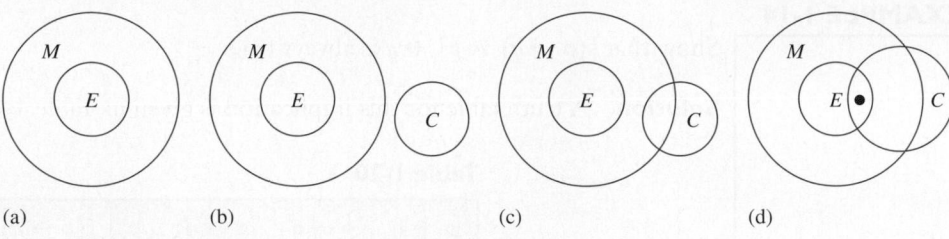

Figure 1-23

However, according to Figure 1-23(d), it is possible that some elementary schoolteachers are children, as noted by the dot placed in the figure and yet the given statements are satisfied. Therefore, the conclusion that "No elementary schoolteacher is a child" does not follow from the given hypotheses. Hence, the reasoning is not valid.

If even one picture can be drawn to satisfy the hypotheses of an argument and contradict the conclusion, the argument is not valid. However, *to show that an argument is valid, all possible pictures must show that there are no contradictions.* There must be no way to satisfy the hypotheses and contradict the conclusion if the argument is valid.

EXAMPLE 1-13

Determine whether the following argument is valid:

Hypotheses: In Washington, D.C., all lobbyists wear suits.
No one in Washington, D.C., over 6 ft tall wears a suit.

Conclusion: Persons over 6 ft tall are not lobbyists in Washington, D.C.

Solution   If L represents the lobbyists in Washington, D.C., and S the people who wear suits, the first hypothesis is pictured as shown in Figure 1-24 (a). If W represents the people in Washington, D.C., over 6 ft tall, the second hypothesis is pictured in Figure 1-24 (b). Because people over 6 ft tall are outside the circle representing suit wearers and lobbyists are in the circle S, the conclusion is valid and no person over 6 ft tall is a lobbyist in Washington, D.C.

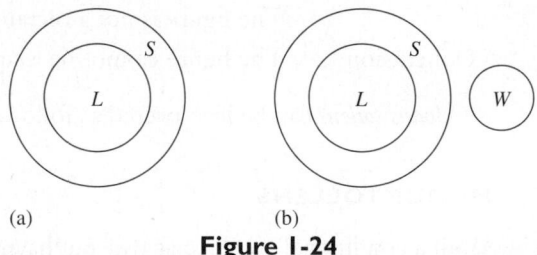

(a)   (b)
Figure 1-24

A different method for determining whether an argument is valid uses **direct reasoning** and a form of argument called the **law of detachment** (or *modus ponens*). For example, consider the following:

Hypotheses: If the sun is shining, then we shall go to the beach.
The sun is shining.

Conclusion: We shall go to the beach.

In general, the law of detachment (or *modus ponens*) is stated as follows:

LAW OF DETACHMENT (*MODUS PONENS*)

If the statement "if p, then q" is true, and p is true, then q must be true.

EXAMPLE 1-14

Show that $[(p \rightarrow q) \wedge p] \rightarrow q$ is always true.

Solution   A truth table for this implication is given in Table 1-20.

Table 1-20

p	q	$p \rightarrow q$	$(p \rightarrow q) \wedge p$	$[(p \rightarrow q) \wedge p] \rightarrow q$
T	T	T	T	T
T	F	F	F	T
F	T	T	F	T
F	F	T	F	T

The statement $[(p \rightarrow q) \wedge p] \rightarrow q$ is a **tautology;** that is, a statement that is true all the time.

EXAMPLE 1-15

Determine whether the following argument is valid if the hypotheses are true and x is a natural number:

Hypotheses: If $x > 2$, then $x^2 > 4$.
                     $x > 2$
Conclusion: $x^2 > 4$

Solution   Using the law of detachment, *modus ponens*, we see that the conclusion is valid.

A different type of reasoning, **indirect reasoning**, uses a form of argument called *modus tollens*. For example, consider the following:

Hypotheses: If a figure is a square, then it is a rectangle.
                        The figure is not a rectangle.
Conclusion: The figure cannot be a square.

Modus tollens can be interpreted as follows:

MODUS TOLLENS

With a conditional accepted as true but having a false conclusion, the hypothesis must be false. Symbolically if $p \rightarrow q$ is true, and q is not true, then p is not true.

Modus tollens follows from the fact that an implication and its contrapositive are logically equivalent.

EXAMPLE 1-16

Determine conclusions for each of the following pairs of true statements:

 a. If a person lives in Boston, then the person lives in Massachusetts. Jessica does not live in Massachusetts.
 b. If $x = 3$, then $2x \neq 7$. And we know that $2x = 7$.

Solution
 a. Jessica does not live in Boston (*modus tollens*).
 b. $x \neq 3$ (*modus tollens*)

The final reasoning argument we consider here involves the **chain rule** (*transitivity*). Consider the following:

Hypotheses: If I save, I will retire early.
                      If I retire early, I will play golf.

Conclusion: If I save, I will play golf.

In general, the chain rule is stated as follows:

CHAIN RULE (TRANSITIVITY)

If "if p, then q" and "if q, then r" are true, then "if p, then r" is true.

People often make invalid conclusions based on advertising or other information. Assume, for example, the statement "Healthy people eat Super-Bran cereal" is true. Are the following conclusions valid?

If a person eats Super-Bran cereal, then the person is healthy.

If a person is not healthy, the person does not eat Super-Bran cereal.

If the original statement is denoted by $p \rightarrow q$, where p is "a person is healthy" and q is "a person eats Super-Bran cereal," then the first conclusion is the converse of $p \rightarrow q$; that is, $q \rightarrow p$, and the second conclusion is the inverse of $p \rightarrow q$; that is, $\sim p \rightarrow \sim q$. Hence, neither is valid.

EXAMPLE 1-17

Determine valid conclusions for the following true statements:

   a. If a triangle is equilateral, then it is isosceles. If a triangle is isosceles, it has at least two congruent sides.
   b. If a number is a whole number, then the number is an integer. If a number is an integer, then the number is a rational number. If a number is a rational number, then the number is a real number.

Solution
   a. If a triangle is equilateral, then it has at least two congruent sides.
   b. If a number is a whole number, then it is a real number.

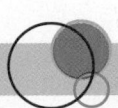

Assessment 1-3A

1. Determine which of the following are statements and then classify each statement as true or false:
   a. $2 + 4 = 8$
   b. Los Angeles is a state in the United States.
   c. What time is it?
   d. $3 \cdot 2 = 6$
   e. This statement is false.
2. Use quantifiers to make each of the following true, where n is a natural number:
   a. $n + 8 = 11$ **b.** $n^2 = 4$
   c. $n + 3 = 3 + n$ **d.** $5n + 4n = 9n$

3. Use quantifiers to make each equation in exercise 2 false.
4. Write the negation of each of the following statements:
   a. This book has 500 pages.
   b. $3 \cdot 5 = 15$
   c. All dogs have four legs.
   d. Some rectangles are squares.
   e. Not all rectangles are squares.
   f. No dogs have fleas.
5. Identify the following as true or false:
   a. For some natural numbers n, $n < 6$ and $n > 3$.
   b. For all natural numbers n, $n > 0$ or $n < 5$.

6. Complete each of the following truth tables:

a.
p	$\sim p$	$\sim(\sim p)$
T		
F		

b.
p	$\sim p$	$p \vee \sim p$	$p \wedge \sim p$
T			
F			

   c. Based on part (a), is p logically equivalent to $\sim(\sim p)$?

   d. Based on part (b), is $p \vee \sim p$ logically equivalent to $p \wedge \sim p$?

7. If q stands for "This course is easy" and r stands for "Lazy students do not study," write each of the following in symbolic form:

   a. This course is easy, and lazy students do not study.

   b. Lazy students do not study, or this course is not easy.

   c. It is false that both this course is easy and lazy students do not study.

   d. This course is not easy.

8. If p is false and q is true, find the truth values for each of the following:

   a. $p \wedge q$ b. $\sim p$

   c. $\sim(\sim p)$ d. $p \wedge \sim q$

   e. $\sim(\sim p \wedge q)$

9. Find the truth value for each statement in exercise 8 if p is false and q is false.

10. For each of the following, is the pair of statements logically equivalent?

   a. $\sim(p \vee q)$ and $\sim p \vee \sim q$

   b. $\sim(p \wedge q)$ and $\sim p \wedge \sim q$

11. Complete the following truth table:

p	q	$\sim p$	$\sim p \wedge q$
T	T		
T	F		
F	T		
F	F		

12. Write each of the following in symbolic form if p is the statement "It is raining" and q is the statement "The grass is wet."

   a. If it is raining, then the grass is wet.

   b. If it is not raining, then the grass is wet.

   c. If it is raining, then the grass is not wet.

   d. The grass is wet if it is raining.

   e. The grass is not wet implies that it is not raining.

   f. The grass is wet if, and only if, it is raining.

13. For each of the following implications, state the converse, inverse, and contrapositive:

   a. If $x = 5$, then $2x = 10$.

   b. If you do not like this book, then you do not like mathematics.

   c. If you do not use Ultra Brush toothpaste, then you have cavities.

   d. If you are good at logic, then your grades are high.

14. Write a statement logically equivalent to the statement "If a number is a multiple of 8, then it is a multiple of 4."

15. Investigate the validity of each of the following arguments:

   a. All squares are quadrilaterals.
      All quadrilaterals are polygons.
      Conclusion: All squares are polygons.

   b. All teachers are intelligent.
      Some teachers are rich.
      Conclusion: Some intelligent people are rich.

   c. If a student is a freshman, then the student takes mathematics.
      Jane is a sophomore.
      Conclusion: Jane does not take mathematics.

16. For each of the following, form a conclusion that follows logically from the given statements:

   a. Some freshmen like mathematics.
      All people who like mathematics are intelligent.

   b. If I study for the final, then I will pass the final.
      If I pass the final, then I will pass the course.
      If I pass the course, then I will look for a teaching job.

   c. Every equilateral triangle is isosceles.
      There exist triangles that are equilateral.

17. Write the following in if-then form:

   a. Every figure that is a square is a rectangle.

   b. All integers are rational numbers.

   c. Polygons with exactly three sides are triangles.

18. Use De Morgan's Laws to write a negation of each of the following:

   a. $3 \cdot 2 = 6$ and $1 + 1 \neq 3$.

   b. You can pay me now or you can pay me later.

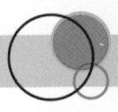

Assessment 1-3B

1. Determine which of the following are statements and then classify each statement as true or false:

   a. Shut the window.

   b. He is in town.

   c. $2 \cdot 2 = 2 + 2$

   d. $2 + 3 = 8$

   e. Stay put!

2. Use quantifiers to make each of the following true, where n is a natural number:

   a. $n + 0 = n$

   b. $n + 1 = n + 2$

   c. $3(n + 2) = 12$

   d. $n^3 = 8$

3. Use quantifiers to make each equation in exercise 2 false.
4. Write the negation of each of the following statements:
   a. $6 < 8$
   b. Some cats do not have nine lives.
   c. All squares are rectangles.
   d. Not all numbers are positive.
   e. Some people have blond hair.
5. Identify the following as true or false:
   a. For some natural numbers n, $n > 5$ and $n > 2$.
   b. For all natural numbers n, $n > 5$ or $n < 5$.
6. a. If you know that p is true, what can you conclude about the truth value of $p \lor q$, even if you don't know the truth value of q?
   b. If you know that p is false, what can you conclude about the truth value of $p \rightarrow q$, even if you don't know the truth value of q?
7. If q stands for "You said goodbye" and r stands for "I said hello," write each of the following in symbolic form:
   a. You said goodbye and I said hello.
   b. You said goodbye and I did not say hello.
   c. I did not say hello or you did not say goodbye.
   d. It is false that both you said goodbye and I said hello.
8. If p is false and q is true, find the truth values for each of the following:
   a. $p \lor q$ b. $\sim q$
   c. $\sim p \lor q$ d. $\sim (p \lor q)$
   e. $\sim q \land \sim p$
9. Find the truth value for each statement in exercise 8 if p is false and q is false.
10. For each of the following, is the pair of statements logically equivalent?
    a. $\sim (p \lor q)$ and $\sim p \land \sim q$
    b. $\sim (p \land q)$ and $\sim p \lor \sim q$
11. Complete the following truth table:

p	q	$\sim q$	$p \lor \sim q$
T	T		
T	F		
F	T		
F	F		

12. Write each of the following in symbolic form if p is the statement "You build it" and q is the statement "They will come:"
    a. If you build it, they will come.
    b. If you do not build it, then they will come.
    c. If you build it, they will not come.
    d. They will come if you build it.
    e. If you do not build it, then they will not come.
    f. If they will not come, then you do not build it.
13. For each of the following implications, state the converse, inverse, and contrapositive:
    a. If $x = 3$, then $x^2 = 9$.
    b. If it snows, then classes are canceled.
14. Iris makes the true statement "If it rains, then I am going to the movies." Does it follow logically that if it does not rain, then Iris does not go to the movies?
15. Investigate the validity of each of the following arguments:
    a. All women are mortal.
       Hypatia was a woman.
       Conclusion: Hypatia was mortal.
    b. All rainy days are cloudy.
       Today is not cloudy.
       Conclusion: Today is not rainy.
    c. Some students like skiing.
       Al is a student.
       Conclusion: Al likes skiing.
16. For each of the following, form a conclusion that follows logically from the given statements:
    a. All college students are poor.
       Helen is a college student.
    b. All engineers need mathematics.
       Ron does not need mathematics.
    c. All bicycles have tires.
       All tires use rubber.
17. Write each of the following in if-then form:
    a. All natural numbers are real numbers.
    b. Every circle is a closed figure.
18. Use De Morgan's Laws to write a negation of each of the following:
    a. $3 + 5 \neq 9$ and $3 \cdot 5 = 15$.
    b. I am going or she is going.

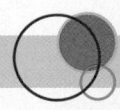

Mathematical Connections 1-3

Communication

1. Explain why commands and questions are not statements.
2. Explain how to write the negation of a quantified statement in the form "Some As are Bs." Give an example.
3. a. Describe under what conditions a disjunction is true.
   b. Describe under what conditions an implication is true.
4. What does the use of an "inclusive" *or* mean?

5. Describe Dr. No as completely as possible.

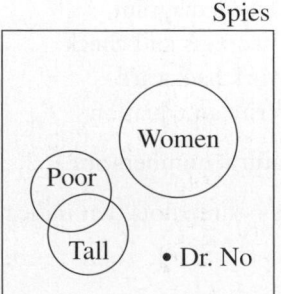

6. Consider the nursery rhyme:

For want of a nail, the shoe was lost.
For want of a shoe, the horse was lost.
For want of a horse, the rider was lost.
For want of a rider, the battle was lost.
For want of a battle, the war was lost.
Therefore, for want of a nail, the war was lost.

  a. Write each line as an *if-then* statement.
  b. Does the conclusion (the *therefore* statement) follow logically. Why?

7. In an e-mail address line, a comma or a semicolon is used to separate addresses. From an e-mail sender's standpoint, explain the logical meaning of the punctuation mark.

Open-Ended

8. Give two examples from mathematics for each of the following:
  a. A statement and its converse are true.
  b. A statement is true, but its converse is false.

  c. An "if, and only if," true statement.
  d. An "if, and only if," false statement.

Cooperative Learning

9. Discuss the paradox arising from the following:
  a. This textbook is 2000 pages long.
  b. The author of this textbook is Dante.
  c. The statements (a), (b), and (c) are all false.

Questions from the Classroom

10. A student says that she does not see how a compound statement consisting of two simple sentences that are false can be true. How do you respond?

11. A student says that if the hypothesis is false, an argument cannot be valid. How do you respond?

 **Hint for Solving the Preliminary Problem**

The strategy of *looking for a pattern* might be useful here. For example, we could make three lists based on the number of eggs being removed and left over in each case and then search the lists until matching numbers are found in all lists. The least of these should be the minimum number of eggs.

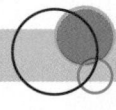 ## Chapter I Summary

1-2 Explorations with Patterns

Reasoning, both **inductive** and **deductive**, is used in problem solving.	19–20
Inductive reasoning—a method of making generalizations based on observations and patterns	19
Conjecture—an statement thought to be true but not yet proved	19
Counterexample—an example contradicting a conjecture	19
Sequence—a ordered arrangment of terms that may be numbers, figures, or objects	21
• **Arithmetic sequence**—a sequence with each successive term from the second on obtained from the previous term by the addition or subtraction of a fixed number, the difference	21
• **Geometric sequence**—a sequence with each successive term from the second on obtained from the previous term by the multiplication of a fixed nonzero number, the ratio	26–27
• **Fibonacci sequence**—1, 1, 2, 3, 5, 8, 13, . . .	27
• **Fibonacci-type sequence**—$a, b, a + b, a + 2b, 2a + 3b, \ldots$	33
• **Figurate numbers**—sequences of numbers formed by counting the dots used to form geometric figures	28
• **Recursive sequence**—a sequence in which one or more consecutive terms are given to start and each successive term is obtained from previous terms	21–22
Finding **common differences**—a technique used to discover patterns	28
Exponentiation—If n is a natural number, then $a^n = \underbrace{a \cdot a \cdot a \cdot \ldots \cdot a}_{n \text{ factors}}$. If $n = 0$ and $a \neq 0$, then $a^0 = 1$.	27

1-3 Reasoning and Logic: An Introduction

Statement—a sentence that is either true or false but not both	34
Negation of a statement, p—a statement, not p, or $\sim p$ having the opposite truth value of p	34
Universal quantifiers—words such as *all* and *every*	35
Existential quantifiers—words or phrases such as *some* and *there exists at least one*	35
Compound statement—a statement formed from combinations of simple statements	36
Conjunction—a compound statement formed using "and" (in symbols: $p \wedge q$) and defined to be true if, and only if both p and q are true	36
Disjunction—a compound statement formed using "or" (in symbols: $p \vee q$) and defined to be true if, and only if, either p, or q, or both are true	36
Logically equivalent statements—statements with the same truth value	37
Conditional statement or implication—a statement in the form "if p, then q" (in symbols $p \rightarrow q$) and defined to be true unless p is true and q is false	38
Theorem: (Equivalence of a statement and its contrapositive) The implication, $p \rightarrow q$, and its Contrapositive, $\sim q \rightarrow \sim p$, are logically equivalent	39
Biconditional statement—$p \rightarrow q$ and $q \rightarrow p$ (in symbols $p \leftrightarrow q$) and referred to as "p if, and only if, q"	40
Other forms of statements—**contrapositive, inverse,** and **converse**	38–39

Laws to determine the validity of arguments—**law of detachment** (*modus ponens*), *modus tollens*, and the **chain rule**	41–43
• **Law of Detachment** (*modus ponens*): If the statement "if p, then q" is true, then q must be true.	41
• **Modus tollens**: If $p \rightarrow q$ is true, and q is not true, then p is not true.	42
• **Chain Rule** (**Transitivity**): If "if p, then q" and "if q, then r" are true, then "if p, then r" is true.	43
Tautology—a statement that is always true	42
Theorem: De Morgan's Laws	
a. $\sim(p \wedge q) \equiv \sim p \vee \sim q$	
b. $\sim(p \vee q) \equiv \sim p \wedge \sim q$	38

Chapter 1 Review

1. If today is Sunday, July 4, and next year is not a leap year, what day of the week will July 4 be on next year?

2. A coded message was written as 19-5-3-18-5-20 3-15-4-5-19. What did the message say?

3. A nursery rhyme states:

    A diller, a dollar, a ten o'clock scholar!
    What makes you come so soon?
    You used to come at ten o'clock,
    But now you come at noon.

    Explain whether the rhyme makes sense mathematically.

4. List three more terms that complete a pattern in each of the following; explain your reasoning, and tell whether each sequence is arithmetic or geometric, or neither.
    a. 0, 1, 3, 6, 10, ____, ____, ____,
    b. 52, 47, 42, 37, ____, ____, ____,
    c. 6400, 3200, 1600, 800, ____, ____, ____,
    d. 1, 2, 3, 5, 8, 13, ____, ____, ____,
    e. 2, 5, 8, 11, 14, ____, ____, ____,
    f. 1, 4, 16, 64, ____, ____, ____,
    g. 0, 4, 8, 12, ____, ____, ____,
    h. 1, 8, 27, 64, ____, ____, ____,

5. Find a possible nth term in each of the following and explain your reasoning.
    a. 5, 8, 11, 14, . . .
    b. 3, 9, 27, 81, 243, . . .
    c. 0, 7, 26, 63, . . .

6. Find the first five terms of the sequences whose nth term is given as follows:
    a. $3n - 2$
    b. $n^2 + n$
    c. $4n - 1$

7. Find the following sums:
    a. $2 + 4 + 6 + 8 + 10 + \ldots + 200$
    b. $51 + 52 + 53 + 54 + \ldots + 151$

8. Produce a counterexample, if possible, to disprove each of the following:
    a. If two odd numbers are added, then the sum is odd.
    b. If a number is odd, then it ends in a 1 or a 3.
    c. If two even numbers are added, then the sum is even.

9. Complete the following magic square; that is, complete the square so that the sum in each row, column, and diagonal is the same.

16	3	2	13
	10		
9		7	12
4		14	

10. How many people can be seated at 12 square tables lined up end to end if each table individually holds four persons?

11. A shirt and a tie sell for $9.50. The shirt costs $5.50 more than the tie. What is the cost of the tie?

12. If fence posts are to be placed in a row 5 m apart, how many posts are needed for 100 m of fence?

13. If a complete rotation of a car tire moves a car forward 6 ft, how many rotations of the tire occur before the tire goes off its 50,000 mi warranty?

14. The members of Mrs. Grant's class are standing in a circle; they are evenly spaced and are numbered in order. The student with number 7 is standing directly

across from the student with number 17. How many students are in the class?

15. A carpenter has three large boxes. Inside each large box are two medium-sized boxes. Inside each medium-sized box are five small boxes. How many boxes are there altogether?

16. Use differences to find the next term in the following sequence:

$$5, 15, 37, 77, 141, \underline{\hspace{1cm}},$$

17. An ant farm can hold 100,000 ants. If the farm held 1500 ants on the first day, 3000 ants on the second day, 6000 ants on the third day, and so on forming a geometric sequence, in how many days will the farm be full?

18. Toma's team entered a mathematics contest where teams of students compete by answering questions that are worth either 3 points or 5 points. No partial credit was given. Toma's team scored 44 points on 12 questions. How many 5-point questions did the team answer correctly?

19. Three pieces of wood are needed for a project. They are to be cut from a 90-cm-long piece of wood. The longest piece is to be 3 times as long as the middle-sized piece and the shortest piece is to be 10 cm shorter than the middle-sized piece. How long are the pieces?

20. How many four-digit numbers have the same digits as 1993?

21. Al, Betty, Carl, and Dan were each born in a different season. Al was born in February; Betty was born in the fall; Carl was born in the spring. Determine the season in which each was born.

22. We have two containers, one of which holds 7 cups and the other holds 4 cups. How can we measure exactly 5 cups of water if we have an unlimited amount of water with which to start?

23. The following geometric arrays suggest a sequence of numbers: 2, 6, 12, 20, . . .

   a. Find the next three terms
   b. Find the 100th term
   c. Find the nth term

24. Each side of each pentagon below is 1 unit long.

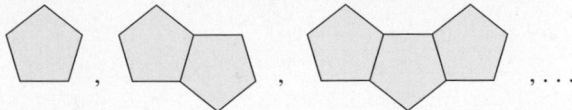

   a. Draw a possible next figure in the sequence.
   b. What is the perimeter (distance around) of each of the first four figures?
   c. What is the perimeter of the 100th figure?
   d. What is the perimeter of the nth figure?

25. Explain the difference between the following two statements: (i) All students passed the final. (ii) Some students passed the final.

26. Which of the following are statements?
   a. The moon is inhabited.
   b. $3 + 5 = 8$
   c. $n + 7 = 15$
   d. Some women have Ph.D.'s in mathematics.

27. Negate each of the following:
   a. Some women smoke.
   b. $3 + 5 = 8$
   c. Beethoven wrote only classical music.

28. Write the converse, inverse, and contrapositive of the following: If we have a rock concert, someone will faint.

29. Use truth tables to show that $p \rightarrow \sim q \equiv q \rightarrow \sim p$.

30. Construct truth tables for each of the following:
   a. $(p \wedge \sim q) \vee (p \wedge q)$
   b. $[(p \vee q) \wedge \sim p] \rightarrow q$

31. Find valid conclusions for the following hypotheses:
   a. All Americans love Mom and apple pie.
      Joe Czernyu is an American.
   b. Steel eventually rusts.
      The Statue of Liberty has a steel structure.
   c. Albertina passed Math 100 or Albertina dropped out.
      Albertina did not drop out.

32. Write the following argument symbolically and then determine its validity:
   If you are fair-skinned, you will sunburn.
   If you sunburn, you will not go to the dance.
   If you do not go to the dance, your parents will want to know why you didn't go to the dance.
   Your parents do not want to know why you didn't go to the dance.
   Conclusion: You are not fair-skinned.

33. State whether the conclusion is valid and tell why.
   If Bob scores at least 80 on the final, he will pass the course.
   Bob did not pass the course.
   Conclusion: Bob did not score at least 80 on the final.

Numeration Systems and Sets

Preliminary Problem

John has applied for the registrar's job at a small college. He submitted the following report to the hiring committee on a survey he did of 100 students: 45 take mathematics; 40 take chemistry; 47 take physics; 20 take mathematics and chemistry; 15 take chemistry and physics; 10 take mathematics and physics; 8 take all three of these subjects; and 10 students take none of these three subjects. Do you think John should be hired on the basis of his report? Explain why.

If needed, see Hint on page 93.

In the NCTM document *Curriculum Focal Points for Prekindergarten through Grade 8 Mathematics: A Quest for Coherence*, the Council suggested specific topics that must be taught in grades pre-K through 8. In that document, as early as pre-K we find the following:

> Children develop an understanding of the meanings of whole numbers and recognize the number of objects in small groups without counting—the first and most basic mathematical algorithm. They understand that number words refer to quantity. They use one-to-one correspondence to solve problems by matching sets and comparing number amounts and in counting objects to 10 and beyond. (p. 11)

In this chapter, we introduce several counting systems. Next, we discuss set theory and the ways in which it adds structure to our number system.

2-1 Numeration Systems

In this section, we introduce various number systems and compare them to the system of numbers used today in the United States. Comparing our current system with ancient systems helps to develop a clearer appreciation of our system. The commonly used base-ten system relies on 10 digits—0, 1, 2, 3, 4, 5, 6, 7, 8, and 9. The written symbols for the digits, such as 2 or 5, are **numerals**. Different cultures developed different numerals over the years to represent numbers. Table 2-1 shows other representations along with how they relate to the digits 0 through 9 and the number 10.

Table 2-1

	0	1	2	3	4	5	6	7	8	9	10
Babylonian		▼	▼▼	▼▼▼	▼▼▼▼	▼▼▼/▼▼	▼▼▼/▼▼▼	▼▼▼▼/▼▼▼	▼▼▼▼/▼▼▼▼	▼▼▼▼▼/▼▼▼▼	◄
Egyptian		I	II	III	IIII	II/II (5)	III/II	III/III	IIII/III	III/III/III	∩
Mayan	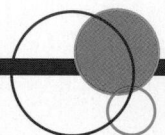	•	••	•••	••••	—	⎯•⎯	⎯••⎯	⎯•••⎯	⎯••••⎯	══
Greek		α	β	γ	δ	∈	φ	ζ	η	υ	ι
Roman		I	II	III	IV	V	VI	VII	VIII	IX	X
Hindu	0	1	7	3	8	4	6	1	8	9	
Arabic	•	١	٢	٣	٤	٥	٦	٧	٨	٩	
Hindu-Arabic	0	1	2	3	4	5	6	7	8	9	10

Table 2-1 shows rudiments of different sets of numbers. A **numeration system** is a collection of properties and symbols agreed upon to represent numbers systematically. Through the study of various numeration systems, we explore the evolution of our current system, the Hindu-Arabic system.

Hindu-Arabic Numeration System

The **Hindu-Arabic numeration system** that we use today was developed by the Hindus and transported to Europe by the Arabs—hence, the name *Hindu-Arabic*. The Hindu-Arabic system relies on the following properties:

1. All numerals are constructed from the 10 digits—0, 1, 2, 3, 4, 5, 6, 7, 8, and 9.
2. Place value is based on powers of 10, the number base of the system.

Because the Hindu-Arabic system is based on powers of 10, the system is a base ten, or a decimal, system. **Place value** assigns a value to a digit depending on its placement in a numeral. To find the value of a digit in this system, we multiply the place value of the digit by its **face value**, where the face value is a digit. For example, in the numeral 5984, the 5 has place value "thousands," the 9 has place value "hundreds," the 8 has place value "tens," and the 4 has place value "ones" or "units," as seen in Figure 2-1. The values of the respective digits are $5 \cdot 1000$ or 5000, $9 \cdot 100$ or 900, $8 \cdot 10$ or 80, and $4 \cdot 1$ or 4.

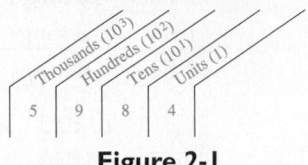

Figure 2-1

We could write 5984 in **expanded form** as $5 \cdot 10^3 + 9 \cdot 10^2 + 8 \cdot 10 + 4 \cdot 1$. In the expanded form of 5984, exponents are used. For example, 1000, or $10 \cdot 10 \cdot 10$, is written as 10^3. In this case, 10 is a **factor** of the product. In general, we have the following:

Definition of a^n

If a is any number and n is any natural number, then

$$a^n = \underbrace{a \cdot a \cdot a \cdot \ldots \cdot a}_{n \text{ factors}}.$$

If $n = 0$, $a \neq 0$, then $a^0 = 1$.

A set of base-ten blocks (or Dienes blocks), shown in Figure 2-2, can be used to better understand place value. The blocks consist of *units*, *longs*, *flats*, and *blocks*, representing 1, 10, 100, and 1000, respectively.

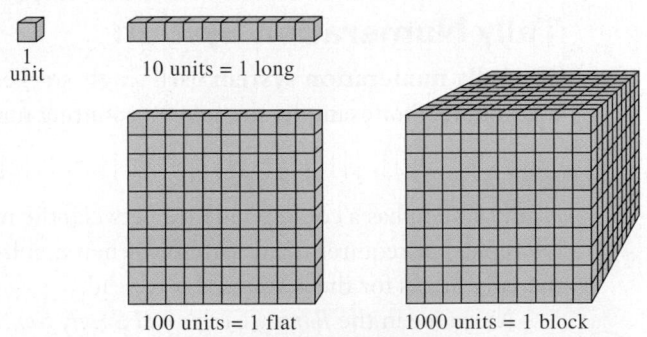

Figure 2-2

$1 \text{ long} \rightarrow 10^1 = 1 \text{ row of 10 units}$

$1 \text{ flat} \rightarrow 10^2 = 1 \text{ row of 10 longs, or 100 units}$

$1 \text{ block} \rightarrow 10^3 = 1 \text{ row of 10 flats, or 100 longs, or 1000 units}$

Students trade blocks by regrouping. That is, they take a set of base-ten blocks representing a number and trade them until they have the fewest possible pieces representing the same number. For example, suppose you have 58 units and want to trade them for other base-ten blocks. You start trading the units for as many longs as possible. Five sets of 10 units each can be traded for 5 longs. Thus, 58 units can be traded so that you now have 5 longs and 8 units. In terms of numbers, this is analogous to rewriting 58 as $5 \cdot 10 + 8$.

EXAMPLE 2-1

What is the fewest number of pieces you can receive in a fair exchange for 11 flats, 17 longs, and 16 units?

Solution

11 flats	17 longs	~~16 units~~	(16 units = 1 long and 6 units)
	1 long	6 units	(Trade)
11 flats	18 longs	6 units	(After the first trade)

11 flats	~~18 longs~~	6 units	(18 longs = 1 flat and 8 longs)
1 flat	8 longs		(Trade)
12 flats	8 longs	6 units	(After the second trade)

	~~12 flats~~	8 longs	6 units	(12 flats = 1 block and 2 flats)
1 block	2 flats			(Trade)
1 block	2 flats	8 longs	6 units	(After the third trade)

Therefore, the fewest number of pieces is $1 + 2 + 8 + 6 = 17$. Notice that as a result of the trading we obtain 1 block, 2 flats, 8 longs, and 6 units that can be written as $11 \cdot 100 + 17 \cdot 10 + 16$ or $1 \cdot 10^3 + 2 \cdot 10^2 + 8 \cdot 10 + 6$, which implies that there are 1286 units.

NOW TRY THIS 2-1

a. Use trading with base-ten blocks (as shown in Figure 2-2) to re-write 3 blocks, 12 flats, 11 longs, and 17 units with the fewest number of blocks. Write a Hindu-Arabic number to represent this fewest number of blocks.
b. Write 3282 in expanded form.

Next, we discuss other numeration systems. The study of such systems provides a historical perspective on the development of numeration systems and helps us better understand our own system.

Tally Numeration System

The **tally numeration system** used single strokes, or tally marks, to represent each object that was counted; for example, the first 10 counting numbers are

|, ||, |||, ||||, |||||, ||||||, |||||||, ||||||||, |||||||||, ||||||||||

A tally system has a correspondence between the marks and the items being counted. The system is simple, but requires many symbols, when numbers are great. Also as numbers become greater, the tally marks for them are harder to read.

As we see in the *Barney Google and Snuffy Smith* cartoon, the tally system can be improved by *grouping*. We see that the tallies are grouped into fives by placing a diagonal across four tallies to make a group of five. Grouping makes it easier to read the numeral.

Historical Note

The invention of the Hindu-Arabic numeration system is considered one of the most important developments in mathematics. The system was introduced in India and then transmitted by the Arabs to North Africa and Spain and then to the rest of Europe. Historians trace the use of zero as a placeholder to the fourth century BCE (Before the Common Era). Arab mathematicians extended the decimal system to include fractions. The Italian mathematician Fibonacci, also known as Leonardo de Pisa (1170–1250), studied in Algeria and brought back with him the new numeration system, which he described and used in a book he published in 1202.

Barney Google copyright ©2005 King Features Syndicate

Egyptian Numeration System

The Egyptian numeration system, dating to about 3400 BCE, used tally marks for the first nine numerals. The Egyptians improved on the system based only on tally marks by developing a *grouping system* to represent certain sets of numbers. This makes the numbers easier to record. For example, the Egyptians used a heel bone symbol, ∩, to stand for a grouping of 10 tally marks.

$$||||||||| \to \cap$$

Table 2-2 shows other numerals that the Egyptians used in their system. Some of the symbols from the Karnak temple in Luxor are depicted in Figure 2-3.

Table 2-2

Egyptian Numeral	Description	Hindu-Arabic Equivalent
|	Vertical staff	1
∩	Heel bone	10
୨	Scroll	100
𐦀	Lotus flower	1000
∅	Pointing finger	10,000
◠	Polliwog or burbot	100,000
𓀀	Astonished man	1,000,000

Figure 2-3

In its simplest form, the Egyptian system involved an **additive property**; that is, the value of a number was the sum of the face values of the numerals. The Egyptians customarily wrote the numerals in decreasing order from left to right, as in ꙮ999∩∩II. The number can be converted to base ten as shown below:

ꙮ	represents	100,000
999	represents	300 (100 + 100 + 100)
∩∩	represents	20 (10 + 10)
II	represents	2 (1 + 1)
ꙮ999∩∩II	represents	100,322

NOW TRY THIS 2-2

a. Use the Egyptian system to represent 1,312,322.

b. Use the Hindu-Arabic system to represent ꙮꙮ𓏲𓏲𓏲∩∩IIII.

c. What disadvantages do you see in the Egyptian system compared to the Hindu-Arabic system?

Babylonian Numeration System

The Babylonian numeration system was developed at about the same time as the Egyptian system. The symbols in Table 2-3 were made using a stylus either vertically or horizontally on clay tablets.

Table 2-3

Babylonian Numeral	Hindu-Arabic Equivalent
▼	1
<	10

The clay tablets were heated and dried to preserve a permanent record. Babylonian symbols on a clay tablet are pictured in Figure 2-4.

Figure 2-4

The Babylonian numerals 1 through 59 were similar to the Egyptian numerals, but the vertical staff and the heel bone were replaced by the symbols shown in Table 2-3. For example, << **VVV** represented 23.

The Babylonian numeration system used a *place-value system*. Numbers greater than 59 were represented by repeated groupings of 60, much as we use groupings of 10 today. For example, **VV** << represents $2 \cdot 60 + 20$, or 140. The space indicates that **VV** represents $2 \cdot 60$ rather than 2. Numerals immediately to the left of a second space have a value $60 \cdot 60$ or 60^2 times their face value, and so on.

<< **V**	represents	$20 \cdot 60 + 1$, or 1201
<**V** <**V** **V**	represents	$11 \cdot 60 \cdot 60 + 11 \cdot 60 + 1$, or $11 \cdot 60^2 + 11 \cdot 60 + 1$, or 40,261
V <**V** <**V** **V**	represents	$1 \cdot 60 \cdot 60 \cdot 60 + 11 \cdot 60 \cdot 60 + 11 \cdot 60 + 1$, or $1 \cdot 60^3 + 11 \cdot 60^2 + 11 \cdot 60 + 1$, or 256,261

The initial Babylonian system was inadequate by today's standards. For example, the symbol **VV** could have represented 2 or $2 \cdot 60$. Later, the Babylonians introduced the symbol ⚊ as a placeholder for missing position values. Using this symbol, < <<**V** represented $10 \cdot 60 + 21$ and < ⚊ <<**V** represented $10 \cdot 60^2 + 0 \cdot 60 + 21$. In this sense, ⚊ represented 0.

NOW TRY THIS 2-3

a. Use the Babylonian system to represent 12,321.
b. Use the Hindu-Arabic system to represent **VV** <**V** **V**.
c. What advantages does the Hindu-Arabic system have over the Babylonian system?

Mayan Numeration System

In the early development of numeration systems, people frequently used parts of their bodies to count. Fingers could be matched to objects to stand for one, two, three, four, or five objects. Two hands could then stand for a set of ten objects. In warmer climates where people went barefoot, people may have used their toes as well as their fingers for counting. The Mayans introduced an attribute that was not present in the Egyptian or early Babylonian systems, namely, a symbol for zero. The Mayan system used only three symbols, which Table 2-4 shows, and based their system primarily on 20 with vertical groupings.

Table 2-4

Mayan Numeral	Hindu-Arabic Equivalent
•	1
‒	5
 👁	0

The symbols for the first eleven numers (beginning with zero) in the Mayan system are shown in Table 2-1. Notice the groupings of five, where each horizontal bar represents a group of five. Thus, the symbol for 19 was ≣, or three 5s and four 1s. The symbol for 20 was 👁, which represents one group of 20 plus zero 1s. In Figure 2-5(a), we have $2 \cdot 5 + 3 \cdot 1$ (or 13) groups of 20 plus $2 \cdot 5 + 1 \cdot 1$ (or 11), for a total of 271. In Figure 2-5(b), we have $3 \cdot 5 + 1 \cdot 1$ (or 16) groups of 20 and zero 1s, for a total of 320.

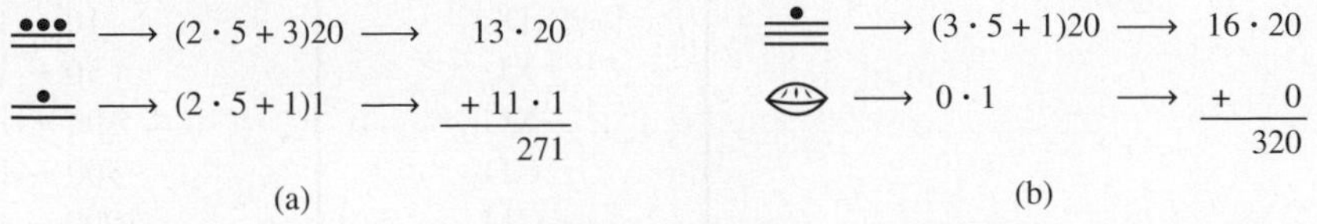

••• ‒	⟶	$(2 \cdot 5 + 3)20$	⟶	$13 \cdot 20$	
• ‒‒	⟶	$(2 \cdot 5 + 1)1$	⟶	$\dfrac{+ 11 \cdot 1}{271}$	
(a)					

• ‒‒‒	⟶	$(3 \cdot 5 + 1)20$	⟶	$16 \cdot 20$	
👁	⟶	$0 \cdot 1$	⟶	$\dfrac{+ \quad 0}{320}$	
(b)					

Figure 2-5

In a true base-twenty system, the place value of the symbols in the third position vertically from the bottom should be 20^2, or 400. However, the Mayans used $20 \cdot 18$, or 360, instead of 400. (The number 360 is an approximation of the length of a calendar year, which consisted of 18 months of 20 days each, plus 5 "unlucky" days.) Thus, instead of place values of 1, 20, 20^2, 20^3, 20^4, and so on, the Mayans used 1, 20, $20 \cdot 18$, $20^2 \cdot 18$, $20^3 \cdot 18$, and so on. For example, in Figure 2-6(a), we have $5 + 1$ (or 6) groups of 360, plus $2 \cdot 5 + 2$ (or 12) groups of 20, plus $5 + 4$ (or 9) groups of 1, for a total of 2409. In Figure 2-6(b), we have $2 \cdot 5$ (or 10) groups of 360, plus 0 groups of 20, plus two 1s, for a total of 3602. Spacing is important in the Mayan system. For example, if two horizontal bars are placed close together, as in $\equiv$, the symbols represent $5 + 5 = 10$. If the bars are spaced apart, as in $\overline{\underline{}}$, then the value is $5 \cdot 20 + 5 \cdot 1 = 105$.

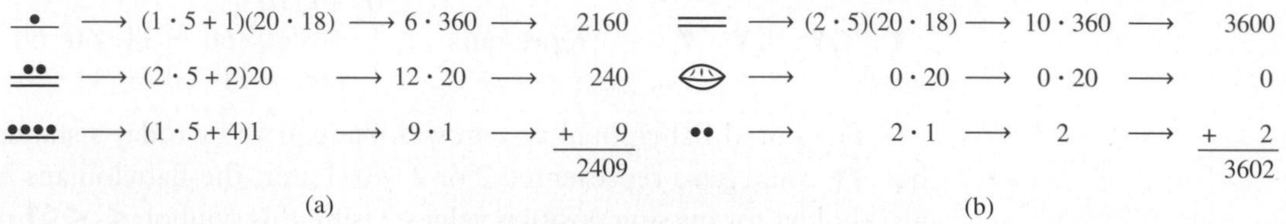

$$\underline{\bullet} \longrightarrow (1 \cdot 5 + 1)(20 \cdot 18) \longrightarrow 6 \cdot 360 \longrightarrow 2160 \qquad \equiv \longrightarrow (2 \cdot 5)(20 \cdot 18) \longrightarrow 10 \cdot 360 \longrightarrow 3600$$

$$\underline{\overset{\bullet\bullet}{}} \longrightarrow (2 \cdot 5 + 2)20 \longrightarrow 12 \cdot 20 \longrightarrow 240 \qquad \ominus \longrightarrow \qquad 0 \cdot 20 \longrightarrow 0 \cdot 20 \longrightarrow 0$$

$$\underset{\bullet\bullet\bullet\bullet}{} \longrightarrow (1 \cdot 5 + 4)1 \longrightarrow 9 \cdot 1 \longrightarrow \underset{\overline{2409}}{+\ 9} \qquad \bullet\bullet \longrightarrow \qquad 2 \cdot 1 \longrightarrow 2 \longrightarrow \underset{\overline{3602}}{+\ 2}$$

(a)   (b)

Figure 2-6

Roman Numeration System

The Roman numeration system was used in Europe in its early form from the third century BCE. It remains in use today, as seen on cornerstones, on the opening pages of books, and on the faces of some clocks. The Roman system uses only the symbols shown in Table 2-5.

Table 2-5

Roman Numeral	Hindu-Arabic Equivalent
I	1
V	5
X	10
L	50
C	100
D	500
M	1000

Roman numerals can be combined by using an additive property. For example, MDCLXVI represents $1000 + 500 + 100 + 50 + 10 + 5 + 1 = 1666$, CCCXXVIII represents 328, and VI represents 6. Romans wrote their symbols in decreasing order.

To avoid repeating a symbol more than three times, as in IIII, a **subtractive property** was introduced in the Middle Ages. For example, I is less than V, so if it immediately is to the left of V, it is subtracted. Thus, IV has a value of $5 - 1$, or 4, and XC represents $100 - 10$, or 90. Some extensions of the subtractive property could lead to ambiguous results. For example, IXC could be 91 or 89. By custom, 91 is written XCI and 89 is written LXXXIX. In general, only one smaller number symbol can be to the left of a larger number symbol and the pair must be one of those listed in Table 2-6.

Table 2-6

Roman Numeral	Hindu-Arabic Equivalent
IV	$5 - 1$, or 4
IX	$10 - 1$, or 9
XL	$50 - 10$, or 40
XC	$100 - 10$, or 90
CD	$500 - 100$, or 400
CM	$1000 - 100$, or 900

In the Middle Ages, a bar was placed over a Roman numeral to multiply it by 1000. The use of bars is based on a **multiplicative property**. For example, $\overline{V}$ represents $5 \cdot 1000$, or 5000, and $\overline{CDX}$ represents $410 \cdot 1000$, or 410,000. To indicate even greater numbers, more bars appear. For example, $\overline{\overline{V}}$ represents $(5 \cdot 1000)1000$, or 5,000,000; $\overline{\overline{CXI}}$ represents $111 \cdot 1000^3$, or 111,000,000,000; and $\overline{CXI}$ represents $110 \cdot 1000 + 1$, or 110,001.

Several properties might be used to represent some numbers, for example:

$$\overline{DCLIX} = \underbrace{\underbrace{(500 \cdot 1000)}_{\text{Multiplicative}} + \underbrace{(100 + 50)}_{\text{Additive}} + \underbrace{(10 - 1)}_{\text{Subtractive}}}_{\text{Additive}} = 500,159$$

NOW TRY THIS 2-4

a. Write CCXLIX as a Hindu-Arabic numeral.
b. Use Roman numerals to represent each of the following
   (i) 1634   (ii) 5280   (iii) 88
c. Use Mayan numerals to represent each of the following
   (i) 684   (ii) 164

Other Number-base Systems

To better understand our system and to investigate some of the problems that students might have when learning it, we investigate similar systems that have different number bases.

Base Five

The Luo peoples of Kenya used a *quinary*, or base-five, system. A system of this type can be modeled by counting with only one hand. The digits available for counting are 0, 1, 2, 3, and 4. In the "one-hand system," or base-five system, we count 1, 2, 3, 4, 10, where *10 represents one hand and no fingers*. Counting in base five proceeds as shown in Figure 2-7. We write the small "five" below the numeral as a reminder that the numeral is written in base five. **If no base is written, a numeral is assumed to be in base ten**. Also note that 1, 2, 3, 4 are the same, and have the same meaning, in both base five and base ten.

One-Hand System	Base-Five Symbol	Base-Five Blocks
0 fingers	0_{five}	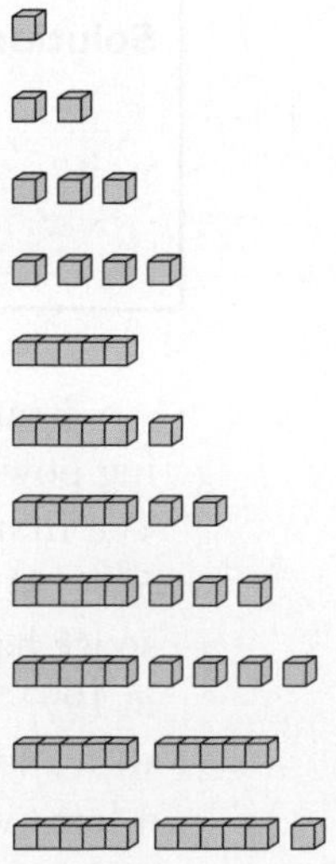
1 finger	1_{five}	
2 fingers	2_{five}	
3 fingers	3_{five}	
4 fingers	4_{five}	
1 hand and 0 fingers	10_{five}	
1 hand and 1 finger	11_{five}	
1 hand and 2 fingers	12_{five}	
1 hand and 3 fingers	13_{five}	
1 hand and 4 fingers	14_{five}	
2 hands and 0 fingers	20_{five}	
2 hands and 1 finger	21_{five}	

Figure 2-7

Counting in base five is similar to counting in base ten. Because we have only five digits (0_{five}, 1_{five}, 2_{five}, 3_{five}, and 4_{five}), 4_{five} plays the role of 9 in base ten. Figure 2-8 shows how we can find the number that comes after 34_{five} by using base five blocks. We see that if we add one more unit block to 34_{five} and perform a trade as shown in Figure 2-8, we obtain 40_{five}. Note that this is read "four zero base five" and not "forty base five."

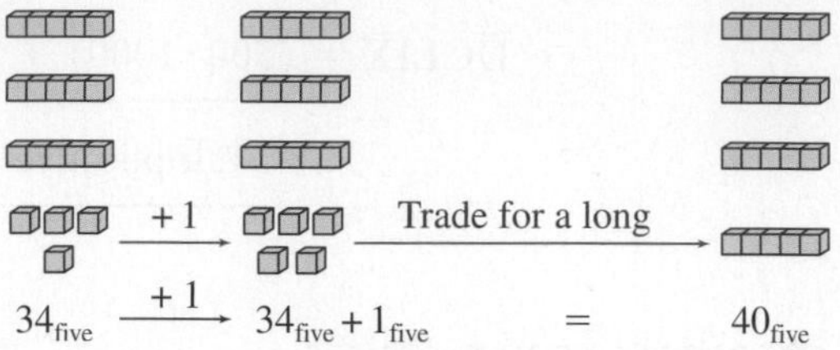

$$34_{\text{five}} \xrightarrow{+1} 34_{\text{five}} + 1_{\text{five}} = 40_{\text{five}}$$

Figure 2-8

What number follows 44_{five}? There are no more two-digit numerals in the system after 44_{five}. In base-ten, the same situation occurs at 99. We use 100 to represent ten 10s, or one 100. In the base-five system, we need a symbol to represent five 5s. To continue the analogy with base ten, we use 100_{five} to represent one group of five 5s or 5^2, zero groups of five, and zero units. The name for 100_{five} is read "one zero zero base five." The number 100 means $1 \cdot 10^2 + 0 \cdot 10^1 + 0$, whereas the numeral 100_{five} means $(1 \cdot 10^2 + 0 \cdot 10^1 + 0)_{\text{five}}$, or $1 \cdot 5^2 + 0 \cdot 5^1 + 0$, or 25.

Examples of base-five numerals along with their base-five block representations and conversions to base ten are given in Figure 2-9. Multibase blocks will be used throughout the text to illustrate various concepts.

Base-Five Numeral	Base-Five Blocks	Base-Ten Numeral
14_{five}	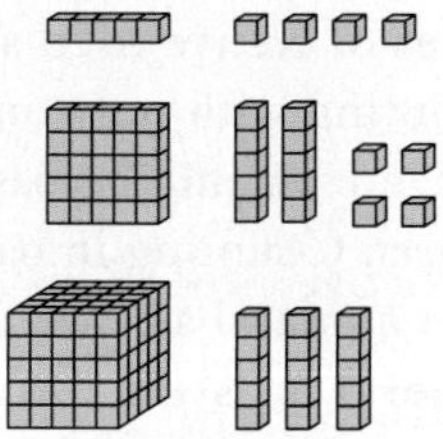	$1 \cdot 5 + 4 = 9$
124_{five}		$1 \cdot 5^2 + 2 \cdot 5 + 4 = 39$
1030_{five}		$1 \cdot 5^3 + 0 \cdot 5^2 + 3 \cdot 5 + 0 \cdot 1 = 140$

Figure 2-9

EXAMPLE 2-2

Convert 11244_{five} to base ten.

Solution
$$
\begin{aligned}
11244_{\text{five}} &= 1 \cdot 5^4 + 1 \cdot 5^3 + 2 \cdot 5^2 + 4 \cdot 5^1 + 4 \cdot 1 \\
&= 1 \cdot 625 + 1 \cdot 125 + 2 \cdot 25 + 4 \cdot 5 + 4 \cdot 1 \\
&= 625 + 125 + 50 + 20 + 4 \\
&= 824
\end{aligned}
$$

Example 2-2 suggests a method for changing a base-five numeral to a base-ten numeral using powers of 5. To convert 824 to base five, we consider how to write 824 using powers of 5. We first determine the greatest power of 5 less than or equal to 824. Because $5^4 = 625$ and $5^5 = 3125$, the greatest power is 5^4. How many 5^4s are contained in 824? There is only one and so we have $824 = 1 \cdot 5^4 + 199$. Next we find the greatest power of 5 in 199. How many 5^3s are in 199? There is only one giving us $824 = 1 \cdot 5^4 + 1 \cdot 5^3 + 74$. Similarly, there are two 5^2s in 74 with 24 left over, giving $824 = 1 \cdot 5^4 + 1 \cdot 5^3 + 2 \cdot 5^2 + 24$. Because there are four 5s in 24 with 4 left over, we have $824 = 1 \cdot 5^4 + 1 \cdot 5^3 + 2 \cdot 5^2 + 4 \cdot 5^1 + 4 \cdot 1$. Therefore, $824 = 11244_{\text{five}}$. A shorthand method for illustrating this conversion follows.

$5^4 = 625 \rightarrow 625\overline{\smash{\big)}\,824}\;\underline{1}$ How many groups of 625 in 824?
-625

$5^3 = 125 \rightarrow 125\overline{\smash{\big)}\,199}\;\underline{1}$ How many groups of 125 in 199?
-125

$5^2 = 25 \rightarrow 25\overline{\smash{\big)}\,74}\;\underline{2}$ How many groups of 25 in 74?
-50

$5^1 = 5 \rightarrow 5\overline{\smash{\big)}\,24}\;\underline{4}$ How many groups of 5 in 24?
-20

$5^0 = 1 \rightarrow 1\overline{\smash{\big)}\,4}\;\underline{4}$ How many 1s in 4?
-4
$\overline{0}$

Thus, $824 = 11244_{\text{five}}$.

 NOW TRY THIS 2-5

A different method of converting 824 to base five is shown using successive divisions by 5. The quotient in each case is placed below the dividend and the remainder is placed on the right, on the same line with the quotient. The answer is read from bottom to top, that is, as 11244_{five}. Use this method to convert 728 to base five.

$$
\begin{array}{r|l}
5 & 824 \\
\hline
5 & 164 \quad 4 \\
\hline
5 & 32 \quad 4 \\
\hline
5 & 6 \quad 2 \\
\hline
& 1 \quad 1
\end{array}
$$

Base Two

Historians tell of early tribes that used base two. Some aboriginal tribes still count "one, two, two and one, two twos, two twos and one, ..." Because base two has only two digits, it is called the **binary system**. Base two is especially important because of its use in computers. One of the two digits is represented by the presence of an electrical signal and the other by the absence of an electrical signal. Although base two works well for some purposes, it is inefficient for everyday use because multidigit numbers are reached very rapidly in counting in this system.

Conversions from base two to base ten, and vice versa, can be accomplished in a manner similar to that used for base five conversions.

EXAMPLE 2-3

   a. Convert 10111_{two} to base ten.
   b. Convert 27 to base two.

Solution
   a. $10111_{\text{two}} = 1 \cdot 2^4 + 0 \cdot 2^3 + 1 \cdot 2^2 + 1 \cdot 2^1 + 1$
   $\phantom{10111_{\text{two}}} = 16 + 0 + 4 + 2 + 1$
   $\phantom{10111_{\text{two}}} = 23$

b.

16)	27	1	How many groups of 16 in 27?
−16			
8)	11	1	How many groups of 8 in 11?
−8			
4)	3	0	How many groups of 4 in 3?
−0			
2)	3	1	How many groups of 2 in 3?
−2			
1)	1	1	How many 1s in 1?
−1			
0			

Alternative Solution:

2	27	
2	13	1
2	6	1
2	3	0
	1	1

Thus, 27 is equivalent to 11011_{two}.

Base Twelve

Another commonly used number-base system is the base twelve, or the duodecimal ("dozens"), system. Eggs are bought by the dozen, and pencils are bought by the *gross* (a dozen dozen). In base twelve, there are 12 digits, just as there are 10 digits in base ten, 5 digits in base five, and 2 digits in base two. In base twelve, new symbols are needed to represent the following groups of x's:

$$\underbrace{xxxxxxxxxx}_{10\ x\text{'s}} \quad \text{and} \quad \underbrace{xxxxxxxxxxx}_{11\ x\text{'s}}$$

The new symbols chosen are T and E, respectively, so that the base twelve digits are $(0, 1, 2, 3, 4, 5, 6, 7, 8, 9, T, E)_{twelve}$. Thus, in base twelve we count $(1, 2, 3, 4, 5, 6, 7, 8, 9, T, E, 10, 11, 12, \ldots, 17, 18, 19, 1T, 1E, 20, 21, 22, \ldots, 28, 29, 2T, 2E, 30, \ldots)_{twelve}$.

EXAMPLE 2-4

a. Convert $E2T_{twelve}$ to base ten.
b. Convert 1277 to base twelve.

Solution

a.
$$E2T_{twelve} = 11 \cdot 12^2 + 2 \cdot 12^1 + 10 \cdot 1$$
$$= 11 \cdot 144 + 24 + 10$$
$$= 1584 + 24 + 10$$
$$= 1618$$

b.

144)	1277	8	How many groups of 144 in 1277?
−1152			
12)	125	T	How many groups of 12 in 125?
−120			
1)	5	5	How many 1s in 5?
−5			
0			

Thus, $1277 = 8T5_{twelve}$.

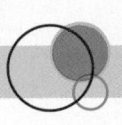

EXAMPLE 2-5

Rob used base twelve to write the following:

$$g36_{twelve} = 1050_{ten}$$

What is the value of g?

Solution   We could write the following equations:

$$g \cdot 12^2 + 3 \cdot 12 + 6 \cdot 1 = 1050$$
$$144g + 36 + 6 = 1050$$
$$144g + 42 = 1050$$
$$144g = 1008$$
$$g = 7$$

Check $736_{twelve} = 7 \cdot 12^2 + 3 \cdot 12 + 6 \cdot 1 = 1050$

Assessment 2-1A

1. For each of the following, tell which numeral represents the greater number and why:
   a. $\overline{\text{MCDXXIV}}$ and $\overline{\overline{\text{MCDXXIV}}}$
   b. 4632 and 46,032
   c. $<$▼▼ and $<$ ▼▼
   d. 999∩∩|| and 𝄃∩|
   e. ≝ and ☺

2. For each of the following, write both the succeeding and the preceding numerals (one more and one less):
   a. MCMXLIX
   b. $<<$ $<$▼
   c. 𝄃99
   d. ⚎

3. If the cornerstone represents when a building was built and it reads MCMXXII, when was this building built?

4. Write each of the following in Roman symbols:
   a. 121
   b. 42

5. Complete the following table, which compares symbols for numbers in different numeration systems:

	Hindu-Arabic	Babylonian	Egyptian	Roman	Mayan
a.	72				
b.		$<$ ▼▼			
c.					

6. For each of the following base ten numerals, give the place value of the underlined digit:
   a. 827,367
   b. 8,421,000

7. Rewrite each of the following as a base ten numeral:
   a. $3 \cdot 10^6 + 4 \cdot 10^3 + 5$
   b. $2 \cdot 10^4 + 1$

8. A certain three-digit natural number has the following properties: The hundreds digit is greater than 7; the tens digit is an odd number; and the sum of the three digits is 10. What could the number be?

9. Study the following counting frame. In the frame, the value of each dot is represented by the number in the box below the dot. For example, the following figure represents the number 154:

..	...	..
64	8	1

What numbers are represented in the frames in (a) and (b)?

a.

...	..	.
25	5	1

b.

.		.	.
8	4	2	1

10. Write the base-four numeral for the base-four representation shown.

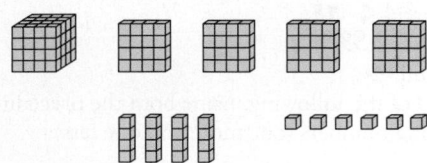

11. Write the first 15 counting numerals for each of the following bases:
    a. base two
    b. base four

12. How many different digits are needed for base twenty?

13. Write 2032_{four} in expanded notation and then convert it to base ten.

14. Determine the greatest three-digit number in each of the following bases:
    a. Base two
    b. Base twelve
15. Find the numeral preceding and succeeding each of the following:
    a. $EE0_{twelve}$
    b. 100000_{two}
    c. 555_{six}
16. What, if anything, is wrong with the following numerals?
    a. 204_{four}
    b. 607_{five}
17. What is the fewest number of base four blocks needed to represent 214?
18. Draw base-five blocks to represent 231_{five}.
19. An introduction to base five is especially suitable for early learning in elementary school, as children can think of making change using quarters, nickels, and pennies. Use only these coins to answer the following:
    a. What is the fewest number of quarters, nickels, and pennies you can receive in a fair exchange for two quarters, nine nickels, and eight pennies?
    b. How could you use the approach in (a) to write 73 in base five?
20. Without converting to base ten, tell which is the lesser for each of the following pairs.
    a. 3030_{four} or 3100_{four}
    b. $EOTE_{twelve}$ or $EOET_{twelve}$
21. Recall that with base-ten blocks, 1 long = 10 units, 1 flat = 10 longs, and 1 block = 10 flats. Make all possible exchanges to obtain the fewest number of pieces and write the corresponding numeral in the given base in the following.
    a. Ten flats in base ten
    b. Twenty flats in base twelve

22. Convert each of the following base ten numerals to a numeral in the indicated bases.
    a. 456 in base five
    b. 1782 in base twelve
    c. 32 in base two
23. Write each of the following numerals in base ten:
    a. 432_{five}
    b. 101101_{two}
    c. $92E_{twelve}$
24. You are asked to distribute $900 in prize money. The dollar amounts for the prizes are $625, $125, $25, $5, and $1. How should this $900 be distributed in order to give the fewest number of prizes?
25. Convert each of the following:
    a. 58 days to weeks and days
    b. 29 hours to days and hours
26. For each of the following, find b if possible. If not possible, tell why.
    a. $b2_{seven} = 44_{ten}$
    b. $5b2_{twelve} = 734_{ten}$
27. Write the following in the indicated base without multiplying out the powers:
    a. $3 \cdot 5^4 + 3 \cdot 5^2$ in base five
    b. $2 \cdot 12^5 + 8 \cdot 12^3 + 12$ in base twelve
28. In a game called WIPEOUT, we are to "wipe out" digits from a calculator's display without changing any of the other digits. "Wipeout" in this case means to replace the chosen digit(s) with a 0. For example, if the initial number is 54,321 and we are to wipe out the 4, we could subtract 4000 to obtain 50,321. Complete the following two problems and then try other numbers or challenge another person to wipe out a number from the number you have placed on the screen:
    a. Wipe out the 2s from 32,420.
    b. Wipe out the 5 from 67,357.

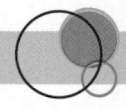

Assessment 2-1B

1. For each of the following, tell which numeral represents the greater number and why:
   a. $\overline{\text{MDCXXIV}}$ and $\overline{\text{MCDXXIV}}$
   b. 3456 and 30,456
   c. $< \blacktriangledown$ and $< \blacktriangledown\blacktriangledown$
   d. 99∩I and 999
   e. ⸪ and ☺
2. For each of the following, name both the preceding and the succeeding numbers (one more and one less):
   a. $\overline{\text{MI}}$
   b. CMXCIX
   c. $< \,<\blacktriangledown$
   d. 𐡁9
   e. •••
3. On the United States one dollar bill, the number MDCCLXXVI is written on the base of the pyramid. What year does this represent?
4. Write each of the following in Roman symbols:
   a. 89
   b. 5202

5. Complete the following table, which compares symbols for numbers in different numeration systems:

	Hindu-Arabic	Babylonian	Egyptian	Roman	Mayan
a.	78				
b.		$< \blacktriangledown$			
c.			𐡁9∩I		

6. For each of the following base ten numerals, give the place value of the underlined digit:
   a. 9<u>7</u>, 998
   b. 8<u>1</u>0, 485
7. Rewrite each of the following as a base ten numeral:
   a. $3 \cdot 10^3 + 5 \cdot 10^2 + 6 \cdot 10$
   b. $9 \cdot 10^6 + 9 \cdot 10 + 9$
8. A two-digit number has the property that the units digit is 4 less than the tens digit and the tens digit is twice the units digit. What is the number?

9. On a counting frame, the following number is represented. What might the number be? Explain your reasoning.

·	··	··
27	9	1

10. Write the base-three numeral for the base-three representation shown.

11. Write the first 10 counting numerals for each of the following bases:
    a. Base three
    b. Base eight
12. How many different digits are needed for base eighteen?
13. Write 2022_{three} in expanded form and then convert it to base ten.
14. Determine the greatest three-digit numeral in each of the following bases:
    a. Base three
    b. Base twelve
15. Find the numeral preceding and succeeding each of the following:
    a. 100_{seven}
    b. 10000_{two}
    c. 101_{two}
16. What, if anything, is wrong with the following numerals?
    a. 306_{four}
    b. 1023_{two}
17. What is the fewest number of base-three blocks needed to represent 79?
18. Draw base-two blocks to represent 1001_{two}.
19. Using a number system based on dozen and gross, how would you describe the representation for 277?

20. Without converting to base ten, tell which is the lesser for each of the following pairs and explain why?
    a. $EET9E_{twelve}$ or $E0T9E_{twelve}$
    b. 1011011_{two} or 101011_{two}
    c. 50555_{six} or 51000_{six}
21. What is the fewest number of multibase blocks that can be used to write the corresponding numeral in the given base?
    a. 10 longs in base four
    b. 10 longs in base three
22. Convert each of the following base-ten numerals to a numeral in the indicated base:
    a. 234 in base four
    b. 1876 in base twelve
    c. 303 in base three
    d. 22 in base two
23. Write each of the following numerals in base ten:
    a. 432_{six}
    b. 11011_{two}
    c. $E29_{twelve}$
24. *Who Wants the Money*, a game show, distributes prizes that are powers of 2. What is the minimum number of prizes that could be distributed from $900?
25. A coffee shop sold 1 cup, 1 pint, and 1 quart of coffee. Express the number of cups sold in base two.
26. For each of the following, find b, if possible. If not possible, tell why.
    a. $b3_{four} = 31_{ten}$
    b. $1b2_{twelve} = 1534_{six}$
27. Using only the number keys on a calculator, fill the display to show the greatest four-digit number if each key can be used only once.

Mathematical Connections 2-1

Communication

1. Ben claims that zero is the same as nothing. Explain how you as a teacher would respond to Ben's statement.
2. What are the major drawbacks to each of the following systems?
   a. Egyptian
   b. Babylonian
   c. Roman
3. a. Why are large numbers in the United States written with commas separating groups of three digits?
   b. Find examples from other countries that do not use commas to separate groups of three digits.
4. In the Roman numeral system explain (a) when you add values, (b) when you subtract values, and (c) when you multiply values. Give an example to illustrate each case.

Open-Ended

5. An inspector of weights and measures uses a special set of weights to check the accuracy of scales. Various weights are placed on a scale to check accuracy of any amount from 1 oz through 15 oz. What is the fewest number of weights the inspector needs? What weights are needed to check the accuracy of scales from 1 oz through 15 oz? From 1 oz through 31 oz?

Cooperative Learning

6. a. Create a numeration system with unique symbols and write a paragraph explaining the properties of the system.
   b. Complete the following table using the system:

Hindu-Arabic Numeral	Your System Numeral
1	
5	
10	
50	
100	
5000	
10,000	
115,280	

Questions from the Classroom

7. A student claims that the Roman system is a base-ten system since it has symbols for 10, 100, and 1000. How do you respond?

8. When using Roman numerals, a student asks whether it is correct to write $\overline{\text{II}}$, as well as MI for 1001. How do you reply?

Trends in Mathematics and Science Study (TIMSS) Questions

Which digit is in the hundreds place in 2345?
   a. 2
   b. 3
   c. 4
   d. 5

TIMSS, Grade 4, 2003

Which number equals 3 ones + 2 tens + 4 hundreds.
   a. 432
   b. 423
   c. 324
   d. 234

TIMSS, Grade 4, 2007

National Assessment of Educational Progress (NAEP) Question

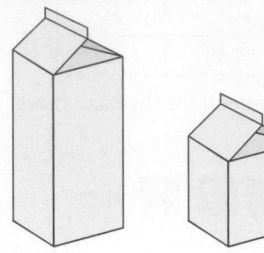

1 quart = 2 pints

Mr. Harper bought 6 pints of milk. How many quarts of milk is this equal to?
   a. 3
   b. 4
   c. 6
   d. 12

NAEP, Grade 4, 2007

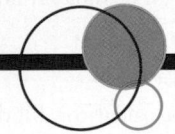

2-2 Describing Sets

In the years from 1871 through 1884, Georg Cantor created *set theory*, which had a profound effect on research and mathematics teaching. Sets, and relations between sets, form a basis for teaching children the *concept* of a whole number, {0, 1, 2, 3, . . .}, and the concept of "less than" as well as addition, subtraction, and multiplication of whole numbers. Understanding whole numbers and operations on whole numbers can be enhanced by the notion of sets. In this section we introduce set notation, relations between sets, set operations, and their properties.

The Language of Sets

A **set** is understood to be any collection of objects. Individual objects in a set are **elements**, or **members**, of the set. For example, each letter is an element of the set of letters in the English language. Capital letters are generally used to name sets. The elements of the set are listed inside a pair of braces, { }. The set A of lowercase letters of the English alphabet can be written in set notation as follows:

$$A = \{a, b, c, d, e, f, g, h, i, j, k, l, m, n, o, p, q, r, s, t, u, v, w, x, y, z\}$$

The order in which the elements are written makes no difference, and *each element is listed only once*. Thus, $\{b, o, k\}$ and $\{k, o, b\}$ are considered to be the same set.

 Historical Note

Georg Cantor (1845–1918) pursued a career in mathematics and obtained his doctorate in Berlin at age 22. Most of his academic career was spent at the University of Halle. His hope of becoming a professor at the University of Berlin did not materialize because his work gained little recognition during his lifetime. However, after his death Cantor's work in set theory was praised as an astonishing product of mathematical thought, one of the most beautiful realizations of human activity.

We show that an element belongs to a set by using the symbol $\in$. For example, $b \in A$. If an element does not belong to a set, we use the symbol $\notin$. For example, $3 \notin A$. In mathematics, the same letter, one lowercase and the other uppercase, cannot be freely interchanged. For example, in the set $A = \{a, b, c\}$ we have $b \in A$ but $B \notin A$.

A set must be **well defined**; that is, if we are given a set and some particular object, then we must be able to tell whether the object does or does not belong to the set. For example, the set of all citizens of Pasadena, California, who ate rice on January 1, 2011, is well defined. We personally may not know if a particular resident of Pasadena ate rice or not, but that resident either belongs or does not belong to the set. On the other hand, the set of all tall people is not well defined because there is no clear meaning of "tall."

We may use sets to define mathematical terms. For example, the set N of *natural numbers* is defined by the following:

$$N = \{1, 2, 3, 4, \ldots\}$$

An *ellipsis* (three dots) indicates that the sequence continues in the same manner.

Two common methods of describing sets are the **listing** or **roster method** and **set-builder notation**, as seen in these examples:

Listing or roster method: $C = \{1, 2, 3, 4\}$

Set-builder notation: $C = \{x \mid x \in N, x < 5\}$

The latter notation is read as follows:

C	$=$	$\{$	x	$\mid$	$x \in N,$	$x < 5\}$
Set C	is	the	all	such	x is an element	x is less
	equal	set	elements	that	of the natural	than 5
	to	of	x		numbers, and	

In the set-builder notation $C = \{x \mid x \in N, x < 5\}$, the comma is a place holder for "and." With this notation both conditions, $x \in N$ and $x < 5$, must be true.

In set-builder notation any lowercase letter can be used to represent a general element. Set-builder notation is useful when the individual elements of a set are not known or they are too numerous to list. For example, the set of decimals between 0 and 1 can be written as

$$D = \{x \mid x \text{ is a decimal between 0 and 1}\}.$$

This is read "D is the set of all elements x such that x is a decimal between 0 and 1." It would be impossible to list all the elements of D. Hence the set-builder notation is indispensable here.

EXAMPLE 2-6

Write the following sets consisting of terms in arithmetic sequences using set-builder notation:

a. $\{2, 4, 6, 8, 10, \ldots\}$ **b.** $\{1, 3, 5, 7, \ldots\}$

Solution

a. $\{x \mid x \text{ is an even natural number}\}$. Or because every even natural number can be written as 2 times some natural number, this set can be written as $\{x \mid x = 2n, n \in N\}$ or, in a somewhat simpler form, as $\{2n \mid n \in N\}$.

b. $\{x \mid x \text{ is an odd natural number}\}$. Or because every odd natural number can be written as some even number minus 1, this set can be written as $\{x \mid x = 2n - 1, n \in N\}$ or $\{2n - 1 \mid n \in N\}$.

EXAMPLE 2-7

Each of the following sets is described in set-builder notation. Write each of the sets by listing its elements.

a. $C = \{2k + 1 \mid k = 3, 4, 5\}$
b. $D = \{x \mid x \text{ is a positive even natural number less than } 8\}$

Solution

a. We substitute $k = 3, 4, 5$ in $2k + 1$ and obtain the corresponding values shown in Table 2-7. Thus, $C = \{7, 9, 11\}$.

Table 2-7

k	$2k + 1$
3	$2 \cdot 3 + 1 = 7$
4	$2 \cdot 4 + 1 = 9$
5	$2 \cdot 5 + 1 = 11$

b. $D = \{2, 4, 6\}$

As noted earlier, the order in which the elements are listed does not matter. If sets A and B are equal, written $A = B$, then every element of A is an element of B, and every element of B is an element of A. If A does not equal B, we write $A \neq B$.

> **Definition of Equal Sets**
> Two sets are **equal** if, and only if, they contain exactly the same elements.

One-to-One Correspondence

One of the most useful concepts in set theory is a **one-to-one correspondence** between two sets. For example, consider the set of people $P = \{\text{Tomas, Dick, Mari}\}$ and the set of swimming lanes $S = \{1, 2, 3\}$. Suppose each person in P is to swim in a lane numbered 1, 2, or 3 so that no two people swim in the same lane. Such a person-lane pairing is a one-to-one correspondence. One way to exhibit a one-to-one correspondence is shown in Figure 2-10 with arrows connecting corresponding elements.

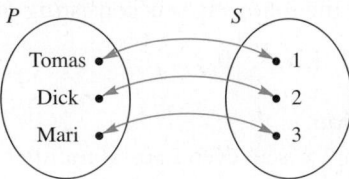

Figure 2-10

Other possible one-to-one correspondences exist between the sets P and S. All six possible one-to-one correspondences between sets P and S can be listed as follows:

1. Tomas $\leftrightarrow$ 1   **2.** Tomas $\leftrightarrow$ 1   **3.** Tomas $\leftrightarrow$ 2
      Dick $\leftrightarrow$ 2   Dick $\leftrightarrow$ 3   Dick $\leftrightarrow$ 1
      Mari $\leftrightarrow$ 3   Mari $\leftrightarrow$ 2   Mari $\leftrightarrow$ 3

4. Tomas ↔ 2   **5.** Tomas ↔ 3   **6.** Tomas ↔ 3
   Dick ↔ 3   Dick ↔ 1   Dick ↔ 2
   Mari ↔ 1   Mari ↔ 2   Mari ↔ 1

Notice that each listing 1–6 above represents a single one-to-one correspondence between the sets P and S. A complete set of one-to-one correspondences between sets P and S can also be listed using a table as in Table 2-8.

Table 2-8

Pairings / Lanes	1	2	3
1.	Tomas	Dick	Mari
2.	Tomas	Mari	Dick
3.	Dick	Tomas	Mari
4.	Dick	Mari	Tomas
5.	Mari	Tomas	Dick
6.	Mari	Dick	Tomas

The general definition of one-to-one correspondence follows.

Definition of One-to-One Correspondence

If the elements of sets P and S can be paired so that for each element of P there is exactly one element of S and for each element of S there is exactly one element of P, then the two sets P and S are said to be in **one-to-one correspondence**.

 NOW TRY THIS 2-6

Consider a set of four people {A, B, C, D} and a set of four swimming lanes {1, 2, 3, 4}.

a. Exhibit all the one-to-one correspondences between the two sets.
b. How many such one-to-one correspondences are there?
c. Find the number of one-to-one correspondences between two sets with five elements each and explain your reasoning.

A tree diagram also lists the possible one-to-one correspondences in Figure 2-11. To read the tree diagram and see the one-to-one correspondence, we follow each branch. The person occupying a specific lane in a correspondence is listed below the lane number. For example, the top branch gives the pairing (Tomas, 1), (Dick, 2), and (Mari, 3).

Observe in Figure 2-11 when assigning a swimmer to lane 1 we have a choice of three people: Tomas, Dick, or Mari. If we put Tomas in lane 1, then he cannot be in lane 2, and hence the second lane must be occupied by either Dick or Mari. In the same way, we see that if Dick is in lane 1, then there are two choices for lane 2: Tomas or Mari. Similarly, if Mari is in lane 1, then again there are two choices for the second lane: Tomas or Dick. Thus, for each of the three ways we can fill the first lane, there are two subsequent ways to fill the second lane, and hence there are $2 + 2 + 2$, or $3 \cdot 2$, or 6 ways to arrange the swimmers in the first two lanes. Notice that for each arrangement of the swimmers in the first two lanes, there remains only one possible swimmer to fill the third lane. For example, if Mari fills the first lane and Dick fills the second, then Tomas must be in the third. Thus, the total number of arrangements for the three swimmers is equal to $3 \cdot 2 \cdot 1$, or 6.

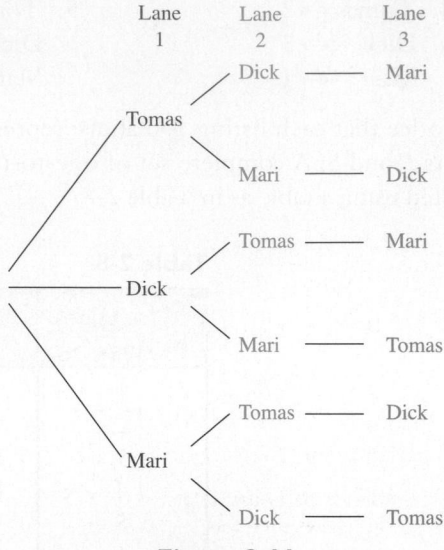

Figure 2-11

Similar reasoning can be used to find how many ice-cream arrangements are possible on a two-scoop cone if 10 flavors are offered. If we count chocolate and vanilla (chocolate on bottom and vanilla on top) different from vanilla and chocolate (vanilla on bottom and chocolate on top) and allow two scoops to be of the same flavor, we can proceed as follows. There are 10 choices for the first scoop, and for each of these 10 choices there are 10 subsequent choices for the second scoop. Thus, the total number of arrangements is $10 \cdot 10$, or 100.

The counting argument used to find the number of possible one-to-one correspondences between the set of swimmers and the set of lanes and the previous problem about ice-cream-scoop arrangements are examples of the Fundamental Counting Principle.

Theorem 2-1:   Fundamental Counting Principle

If event M can occur in m ways and, after it has occurred, event N can occur in n ways, then event M followed by event N can occur in mn ways.

 NOW TRY THIS 2-7

How many one-to-one correspondences are there between two sets with n elements each?

Equivalent Sets

Closely associated with one-to-one correspondences is the concept of **equivalent sets**. For example, suppose a room contains 20 chairs and one student is sitting in each chair with no one standing. There is a one-to-one correspondence between the set of chairs and the set of students in the room. In this case, the set of chairs and the set of students are equivalent sets.

Definition of Equivalent Sets

Two sets A and B are **equivalent**, written $A \sim B$, if, and only if, there exists a one-to-one correspondence between the sets.

The term *equivalent* should not be confused with *equal*. The difference should be made clear by Example 2-8.

EXAMPLE 2-8

Let

$$A = \{p, q, r, s\}, \qquad B = \{a, b, c\}, \qquad C = \{x, y, z\}, \qquad \text{and} \qquad D = \{b, a, c\}.$$

Compare the sets, using the terms *equal* and *equivalent*.

Solution

Each set is both equivalent to and equal to itself.

Sets A and B are not equivalent ($A \nsim B$) and not equal ($A \neq B$).

Sets A and C are not equivalent ($A \nsim C$) and not equal ($A \neq C$).

Sets A and D are not equivalent ($A \nsim D$) and not equal ($A \neq D$).

Sets B and C are equivalent ($B \sim C$) but not equal ($B \neq C$).

Sets B and D are equivalent ($B \sim D$) and equal ($B = D$).

Sets C and D are equivalent ($C \sim D$) but not equal ($C \neq D$).

NOW TRY THIS 2-8

a. If two sets are equivalent, are they necessarily equal? Explain why or why not.

b. If two sets are equal, are they necessarily equivalent? Explain why or why not.

The following *Peanuts* cartoon demonstrates some set theory concepts related to addition, though a child would not be expected to know all these concepts to add 2 and 2.

Peanuts copyright © 1965 and 2010 Peanuts Worldwide LLC. Distributed by United Feature Syndicate, Inc.

Cardinal Numbers

The concept of one-to-one correspondence can be used to consider the notion of two sets having the same number of elements. Without knowing how to count, a child might tell that there are as many fingers on the left hand as on the right hand by matching the fingers on one hand with the fingers on the other hand, as in Figure 2-12. Naturally placing the fingers so that the left thumb touches the right thumb, the left index finger touches the right index finger, and so on, exhibits a one-to-one correspondence between the fingers of the two hands. Similarly, without counting, children realize that if every student in a class sits in a chair and no chairs are empty, there are as many chairs as students, and vice versa.

A one-to-one correspondence between sets helps explain the concept of a number. Consider the five sets $\{a, b\}$, $\{p, q\}$, $\{x, y\}$, $\{b, a\}$, and $\{*, \#\}$; the sets are equivalent to one another and share the property of "twoness"; that is, these sets have the same cardinal number, namely, 2. The **cardinal number** of a set S, denoted $n(S)$, indicates the number of elements in the set S. If $S = \{a, b\}$, the cardinal number of S is 2, and we write $n(S) = 2$. *If two sets, A and B, are equivalent, then A and B have the same cardinal number; that is, $n(A) = n(B)$.*

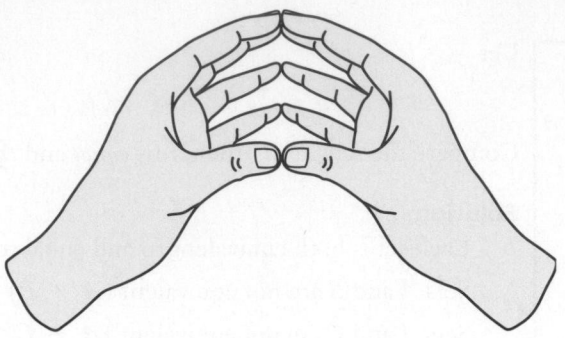

Figure 2-12

The Empty Set

A set that contains no elements has cardinal number 0 and is an **empty**, or **null, set**. The empty set is designated by the symbol $\varnothing$ or { }. Two examples of sets with no elements are the following:

$$C = \{x \mid x \text{ was a state of the United States before 1200 CE}\}$$
$$D = \{x \mid x \text{ is a natural number such that } x^2 = 17\}$$

The empty set is often incorrectly recorded as $\{\varnothing\}$. This set is not empty but contains one element. Likewise, $\{0\}$ does not represent the empty set. Why?

A set is a **finite set** if the cardinal number of the set is zero or a natural number. The set of natural numbers N is an **infinite set**; it is not finite. The set W, containing all the natural numbers and 0, is the set of **whole numbers**: $W = \{0, 1, 2, 3, \ldots\}$. W is an infinite set.

NOW TRY THIS 2-9

Use *reasoning* to explain why there can be no greatest natural number. That is, explain why the set of natural numbers is not a finite set as Dolly implies in the cartoon below.

THE FAMILY CIRCUS®   **By Bil Keane**

11-23
Copyright 1988
Cowles Syndicate, Inc.

"The alphabet ends at 'Z,' but
numbers go on forever."

Family Circus copyright © 1988 and 2011 Bil Keane, Inc.
Distributed by King Features Syndicate

More About Sets

The **universal set**, or the **universe**, denoted U, is the set that contains all elements being considered in a given discussion. Suppose $U = \{x \mid x$ is a person living in California$\}$ and $F = \{x \mid x$ is a female living in California$\}$. The universal set, U, and set F can be represented by a diagram, as in Figure 2-13(a). The universal set is represented by a large rectangle, and F is indicated by the circle inside the rectangle, as shown in Figure 2-13(a). This figure is an example of a **Venn diagram**, named after the Englishman John Venn (1834–1923), who used such diagrams to illustrate ideas in logic. The set of elements in the universe that are not in F, denoted by $\overline{F}$, is the set of males living in California and is the **complement** of F. It is represented by the shaded region in Figure 2-13(b).

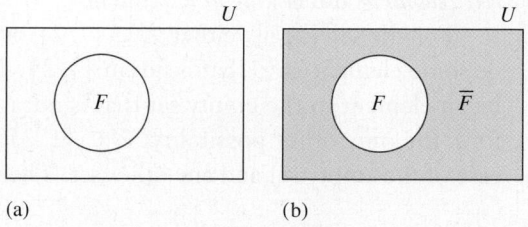

(a)   (b)

Figure 2-13

Definition of Set Complement

The **complement** of a set A, written $\overline{A}$, is the set of all elements in the universal set U that are not in A; that is, $\overline{A} = \{x \mid x \in U$ and $x \notin A\}$.

EXAMPLE 2-9

a. If $U = \{a, b, c, d\}$ and $B = \{c, d\}$, find (i) $\overline{B}$; (ii) $\overline{U}$; (iii) $\overline{\varnothing}$.
b. If $U = \{x \mid x$ is an animal in the zoo$\}$ and $S = \{x \mid x$ is a snake in the zoo$\}$, describe $\overline{S}$.
c. If $U = N$, $E = \{2, 4, 6, 8, \ldots\}$, and $O = \{1, 3, 5, 7, \ldots\}$, find (i) $\overline{E}$; (ii) $\overline{O}$.

Solution

a. (i) $\overline{B} = \{a, b\}$; (ii) $\overline{U} = \varnothing$; (iii) $\overline{\varnothing} = U$
b. Because the individual animals in the zoo are not known, $\overline{S}$ must be described using set-builder notation:

$$\overline{S} = \{x \mid x \text{ is an animal in the zoo and } x \text{ is not a snake}\}$$

c. (i) $\overline{E} = O$; (ii) $\overline{O} = E$

Subsets

Consider the sets $A = \{1, 2, 3, 4, 5, 6\}$ and $B = \{2, 4, 6\}$. All the elements of B are contained in A and we say that B is a **subset** of A. We write $B \subseteq A$. In general, we have the following:

Definition of Subset

For all sets A and B, B is a **subset** of A, written $B \subseteq A$, if, and only if, every element of B is an element of A.

This definition allows B to be equal to A. The definition is written with the phrase "if, and only if," which means "if B is a subset of A, then every element of B is an element of A, and if every element of B is an element of A, then B is a subset of A." *If both $A \subseteq B$ and $B \subseteq A$, then $A = B$.*

When a set A is not a subset of another set B, we write $A \nsubseteq B$. To show that $A \nsubseteq B$, we must find at least one element of A that is not in B. If $A = \{1, 3, 5\}$ and $B = \{1, 2, 3\}$, then A is not a subset of B because 5 is an element of A but not of B. Likewise, $B \nsubseteq A$ because 2 belongs to B but not to A.

The Empty Set as a Subset of Every Set

It is not obvious how the empty set fits the definition of a subset because no elements in the empty set are elements of another set. To investigate this problem, we use the strategies of *indirect reasoning* and *looking at a special case*.

For the set $\{1, 2\}$, either $\varnothing \subseteq \{1, 2\}$ or $\varnothing \nsubseteq \{1, 2\}$. Suppose $\varnothing \nsubseteq \{1, 2\}$. Then there must be some element in $\varnothing$ that is not in $\{1, 2\}$. Because the empty set has no elements, there cannot be an element in the empty set that is not in $\{1, 2\}$. Consequently, $\varnothing \nsubseteq \{1, 2\}$ is false. Therefore, the only other possibility, $\varnothing \subseteq \{1, 2\}$, is true. The same reasoning can be applied in the case of the empty set and any other set. Therefore, *the empty set is a subset of every set.*

Proper Subsets

For sets $A = \{a, b, c\}$ and $B = \{c, b, a\}$, we have $B \subseteq A$, $A \subseteq B$, and $B = A$. If B is a subset of A and B is not equal to A, then B is a **proper subset** of A, written $B \subset A$. This means that every element of B is contained in A and there is at least one element of A that is not in B.

> ### Definition of Proper Subset
> For all sets A and B, B is a **proper subset** of A, written $B \subset A$, if, and only if, $B \subseteq A$ and $B \neq A$; that is, every element of B is an element of A, and there is at least one element of A that is not an element of B.

To indicate a proper subset, sometimes a Venn diagram like the one shown in Figure 2-14 is used, showing a dot (an element) in A that is not in B. An argument similar to the one given above can be used to show that the empty set is a proper subset of every non-empty set.

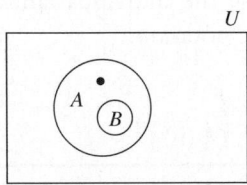

Figure 2-14

EXAMPLE 2-10

Given $A = \{1, 2, 3, 4, 5\}$, $B = \{1, 3\}$, $P = \{x \mid x = 2^n - 1, n \in N\}$:

  a. Identify all the subset relationships that occur among these sets.
  b. Identify all the proper subset relationships that occur among these sets.
  c. If $C = \{2k \mid k \in N\}$ and $D = \{4k \mid k \in N\}$, show that one of the sets is a subset of the other.

Solution

a. Because $2^1 - 1 = 1$, $2^2 - 1 = 3$, $2^3 - 1 = 7$, $2^4 - 1 = 15$, $2^5 - 1 = 31$, and so on, $P = \{1, 3, 7, 15, 31, \ldots\}$. Thus, $B \subseteq P$. Also $B \subseteq A$, $A \subseteq A$, $B \subseteq B$ and $P \subseteq P$.

b. $B \subset A$ and $B \subset P$.

c. $C = \{2 \cdot 1, 2 \cdot 2, 2 \cdot 3, 2 \cdot 4, \ldots\} = \{2, 4, 6, 8, \ldots\}$ and $D = \{4 \cdot 1, 4 \cdot 2, 4 \cdot 3, 4 \cdot 4, \ldots\} = \{4, 8, 12, 16, \ldots\}$. Each element in D appears in C. This is true because $4k = 2(2k)$. Therefore, every element of D is an element of C and $D \subseteq C$.

 NOW TRY THIS 2-10

a. Suppose $A \subset B$. Can we always conclude that $A \subseteq B$?

b. If $A \subseteq B$, does it follow that $A \subset B$?

c. Is the empty set a proper subset of itself? Why?

Subsets and elements of sets are often confused. We say that $2 \in \{1, 2, 3\}$. But because 2 is not a set, we cannot substitute the symbol $\subseteq$ for $\in$. However, $\{2\} \subseteq \{1, 2, 3\}$ and $\{2\} \subset \{1, 2, 3\}$. Notice that $\{2\} \notin \{1, 2, 3\}$.

Inequalities: An Application of Set Concepts

The notion of proper subset and the concept of one-to-one correspondence can be used to define the concept of "less than" among natural numbers. The set $\{a, b, c\}$ has fewer elements than the set $\{w, x, y, z\}$ because when we try to pair the elements of the two sets, as in

$$\{a, b, c\}$$
$$\uparrow \uparrow \uparrow$$
$$\downarrow \downarrow \downarrow$$
$$\{x, y, z, w\}$$

we see that there is at least one element of the second set that is not paired with an element of the first set. The set $\{a, b, c\}$ is equivalent to a proper subset of the set $\{x, y, z, w\}$.

In general, *if A and B are finite sets, A has fewer elements than B if A is equivalent to a proper subset of B.* We say that $n(A)$ is **less than** $n(B)$ and write $n(A) < n(B)$. We say that b is **greater than** a, written $b > a$, if, and only if, $a < b$.

So, if A and B are finite sets and $A \subset B$, then A has fewer elements than B and it is not possible to find a one-to-one correspondence between the sets. Consequently, A and B are not equivalent. However, when both sets are infinite and $A \subset B$, the sets could be equivalent. For example, consider the set N of natural numbers and the set E of even natural numbers. $E \subset N$, but it is still possible to find a one-to-one correspondence between the sets. To do so, we correspond each number in set N to a number in set E that is twice as great. That is, $n \in N$ corresponds to $2n \in E$, as shown next.

$$N = \{1, 2, 3, 4, 5, \ldots, n, \ldots\}$$
$$\uparrow \uparrow \uparrow \uparrow \uparrow \qquad \nwarrow$$
$$\downarrow \downarrow \downarrow \downarrow \downarrow \qquad \searrow$$
$$E = \{2, 4, 6, 8, 10, \ldots, 2n, \ldots\}$$

Notice that in the correspondence, every element of N corresponds to a unique element in E and, conversely, every element of E corresponds to a unique element in N. For example, 11 in N corresponds to $2 \cdot 11$, or 22, in E. And 100 in E corresponds to $100 \div 2$, or 50, in N. Thus, $N \sim E$; that is, N and E are equivalent.

Students sometimes have difficulty with infinite sets and especially with their cardinal numbers, called **transfinite numbers**. As shown, E is a proper subset of N, but because they can be placed in a one-to-one correspondence, they are equivalent and have the same cardinal number. Georg Cantor was the first to introduce the concept of a transfinite number.

Problem Solving   Passing a Senate Measure

A committee of senators consists of Abel, Baro, Carni, and Davis. Suppose each member of the committee has one vote and a simple majority is needed to either pass or reject any measure. A measure that is neither passed nor rejected is considered to be blocked and will be voted on again. Determine the number of ways a measure could be passed or rejected and the number of ways a measure could be blocked.

Understanding the Problem   We are asked to determine how many ways the committee of four could pass or reject a measure and how many ways the committee of four could block a measure. To pass or reject a measure requires a winning coalition, that is, a group of senators who can pass or reject the measure, regardless of what the others do. To block a proposal, there must be a blocking coalition, that is, a group who can prevent any measure from passing but who cannot reject the measure.

Devising a Plan   To solve the problem, we can *make a list* of subsets of the set of senators. Any subset of the set of senators with three or four members will form a winning coalition. Any subset of the set of senators with exactly two members will form a blocking coalition.

Carrying Out the Plan   We list all subsets of the set $S = \{$Abel, Baro, Carni, Davis$\}$ that have at least three elements and all subsets that have exactly two elements. For ease, we identify the members as follows: A—Abel, B—Baro, C—Carni, D—Davis. All the subsets are given next:

$$\varnothing \quad \{A\} \quad \{A, B\} \quad \{A, B, C\} \quad \{A, B, C, D\}$$
$$\{B\} \quad \{A, C\} \quad \{A, B, D\}$$
$$\{C\} \quad \{A, D\} \quad \{A, C, D\}$$
$$\{D\} \quad \{B, C\} \quad \{B, C, D\}$$
$$\{B, D\}$$
$$\{C, D\}$$

There are five subsets with at least three members that can form a winning coalition to pass or reject a measure and six subsets with exactly two members that can block a measure.

Looking Back   Other questions that might be considered include:

1. How many minimal winning coalitions are there? In other words, how many subsets are there of which no proper subset could pass a measure?
2. Devise a method to solve this problem without listing all subsets.
3. In "Carrying Out the Plan," 16 subsets of $\{A, B, C, D\}$ are listed. Use that result to systematically list all the subsets of a committee of five senators. Can you find the number of subsets of the 5-member committee without actually counting the subsets?

NOW TRY THIS 2-11

Suppose a committee of U.S. senators consists of five members.

a. Compare the number of winning coalitions having exactly four members with the number of senators on the committee. What is the reason for the result?
b. Compare the number of winning coalitions having exactly three members with the number of subsets of the committee having exactly two members. What is the reason for the result?

Number of Subsets of a Finite Set

How many subsets can be made from a set containing n elements? To obtain a general formula, we use the strategy of *trying simpler cases* first.

1. If $P = \{a\}$, then P has two subsets, $\varnothing$ and $\{a\}$.
2. If $Q = \{a, b\}$, then Q has four subsets, $\varnothing$, $\{a\}$, $\{b\}$, and $\{a, b\}$.
3. If $R = \{a, b, c\}$, then R has eight subsets, $\varnothing$, $\{a\}$, $\{b\}$, $\{c\}$, $\{a, b\}$, $\{a, c\}$, $\{b, c\}$, and $\{a, b, c\}$.

Methodically list all the subsets of a given set by using a tree diagram. For example, tree diagrams for the subsets of $Q = \{a, b\}$ and $R = \{a, b, c\}$ are given in Figure 2-15(a) and (b) respectively.

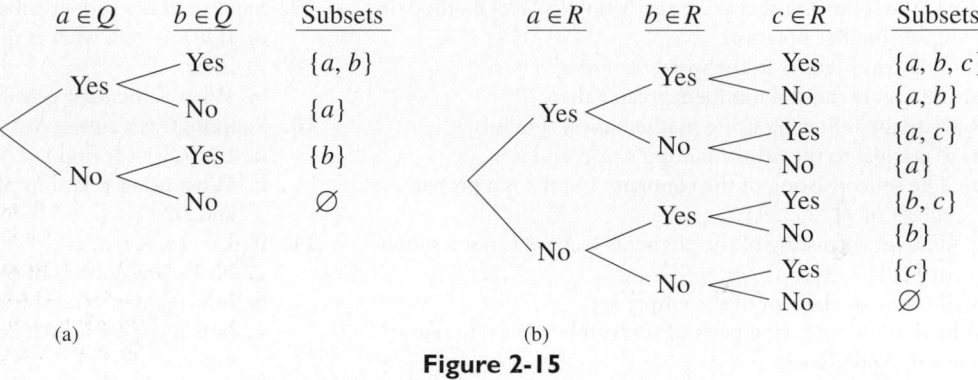

(a)   (b)

Figure 2-15

Using the information from these cases, we *make a table* and *search for a pattern*, as in Table 2-9.

Table 2-9

Number of Elements	Number of Subsets
1	2, or 2^1
2	4, or 2^2
3	8, or 2^3
.	.
.	.
.	.

Table 2-9 suggests that for four elements, there might be 2^4, or 16, subsets. Is this correct? If $S = \{a, b, c, d\}$, then all the subsets of $R = \{a, b, c\}$ are also subsets of S. Eight new subsets are also formed by including the element d in each of the eight subsets of R. The eight new subsets are $\{d\}$, $\{a, d\}$, $\{b, d\}$, $\{c, d\}$, $\{a, b, d\}$, $\{a, c, d\}$, $\{b, c, d\}$, and $\{a, b, c, d\}$. Thus, there are twice as many subsets of set S (with four elements) as there are of set R (with three elements). Consequently, there are $2 \cdot 8$, or 2^4, subsets of a set with four elements. Because including one more element in a finite set doubles the number of possible subsets of the new set, a set with five elements will have $2 \cdot 2^4$, or 2^5, subsets, and so on. In each case, the number of elements and the power of 2 used to obtain the number of subsets are equal. *Therefore, if there are n elements in a set, 2^n subsets can be formed.* If we apply this formula to the empty set—that is, when $n = 0$—then we have $2^0 = 1$. The pattern is meaningful because the empty set has only one subset—itself.

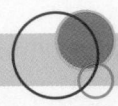

NOW TRY THIS 2-12

a. How many subsets does a set with five elements have?

b. How many proper subsets does a set with four elements have?

c. How many proper subsets does a set with n elements have?

d. How many subsets that include the element a does the set $\{a, b, c, d, e\}$ have?

Assessment 2-2A

1. Write the following sets using the listing (roster) method or using set-builder notation:
   a. The set of letters in the word *mathematics*
   b. The set of natural numbers greater than 20

2. Rewrite the following using mathematical symbols:
   a. P is equal to the set containing a, b, c, and d.
   b. The set consisting of the elements 1 and 2 is a proper subset of $\{1, 2, 3, 4\}$.
   c. The set consisting of the elements 0 and 1 is not a subset of $\{1, 2, 3, 4\}$.
   d. 0 is not an element of the empty set.

3. Which of the following pairs of sets can be placed in one-to-one correspondence?
   a. $\{1, 2, 3, 4, 5\}$ and $\{m, n, o, p, q\}$
   b. $\{a, b, c, d, e, f, \ldots, m\}$ and $\{1, 2, 3, 4, 5, 6, \ldots, 13\}$
   c. $\{x \mid x$ is a letter in the word *mathematics*$\}$ and $\{1, 2, 3, 4, \ldots, 11\}$

4. How many one-to-one correspondences are there between two sets with
   a. 6 elements each?
   b. n elements each?

5. How many one-to-one correspondences are there between the sets $\{x, y, z, u, v\}$ and $\{1, 2, 3, 4, 5\}$ if in each correspondence
   a. x must correspond to 5?
   b. x must correspond to 5 and y to 1?
   c. x, y, and z must correspond to odd numbers?

6. Which of the following represent equal sets?
   $A = \{a, b, c, d\}$ $B = \{x, y, z, w\}$
   $C = \{c, d, a, b\}$ $D = \{x \mid 1 \le x \le 4, \ x \in N\}$
   $E = \varnothing$ $F = \{\varnothing\}$
   $G = \{0\}$ $H = \{\ \}$
   $I = \{x \mid x = 2n + 1, \text{ and } n \in \{0, 1, 2, 3, \ldots\}\}$
   $L = \{x \mid x = 2n - 1, n \in N\}$

7. Find the cardinal number of each of the following sets. Assume the pattern continues in each part:
   a. $\{101, 102, 103, \ldots, 1100\}$
   b. $\{1, 3, 5, \ldots, 1001\}$
   c. $\{1, 2, 4, 8, 16, \ldots, 1024\}$
   d. $\{x \mid x = k^2, k = 1, 2, 3, \ldots, 100\}$
   e. $\{i + j \mid i \in \{1, 2, 3\} \text{ and } j \in \{1, 2, 3\}\}$

8. If U is the set of all college students and A is the set of all college students with a straight-A average, describe $\overline{A}$.

9. Suppose B is a proper subset of C.
   a. If $n(C) = 8$, what is the maximum number of elements in B?
   b. What is the least possible number of elements in B?

10. Suppose C is a subset of D and D is a subset of C.
    a. If $n(C) = 5$, find $n(D)$.
    b. What other relationship exists between sets C and D?

11. If $A = \{a, b, c, d, e\}$,
    a. how many subsets does A have?
    b. how many proper subsets does A have?
    c. how many subsets does A have that include the elements a and e?

12. If a set has 255 proper subsets, how many elements are in the set?

13. Identify all the possible proper subset relationships that occur among the following sets:
    $A = \{3n \mid n \in N\}, B = \{6n \mid n \in N\},$
    $C = \{12n \mid n \in N\}.$

14. Indicate which symbol, $\in$ or $\notin$, makes each of the following statements true:
    a. $0 \underline{\hspace{1cm}} \varnothing$
    b. $\{1\} \underline{\hspace{1cm}} \{1, 2\}$
    c. $1024 \underline{\hspace{1cm}} \{x \mid x = 2^n, n \in N\}$
    d. $3002 \underline{\hspace{1cm}} \{x \mid x = 3n - 1, n \in N\}$

15. Indicate which symbol, $\subseteq$ or $\nsubseteq$, makes each part of problem 14 true.

16. Answer each of the following. If your answer is *no*, tell why.
    a. If $A = B$, can we always conclude that $A \subseteq B$?
    b. If $A \subseteq B$, can we always conclude that $A \subset B$?
    c. If $A \subset B$, can we always conclude that $A \subseteq B$?
    d. If $A \subseteq B$, can we always conclude that $A = B$?

17. Use the definition of *less than* to show each of the following:
    a. $3 < 100$ b. $0 < 3$

18. On a certain senate committee there are seven senators: Abel, Brooke, Cox, Dean, Eggers, Funk, and Gage. Three of these members are to be appointed to a subcommittee. How many possible subcommittees are there?

19. How many two-digit numbers in base ten can be formed if the tens digit cannot be 0 and no digit can be repeated?

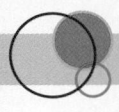

Assessment 2-2B

1. Write the following sets using the listing (roster) method or set-builder notation:
   a. the set of letters in the word *geometry*
   b. the set of natural numbers greater than 7
2. Rewrite the following using mathematical symbols:
   a. Q is equal to the set whose elements are a, b, and c.
   b. The set containing 1 and 3 is equal to the set containing 3 and 1.
   c. The set containing 1 and 3 only is not a proper subset of $\{1, 4, 6, 7\}$
   d. The empty set does not contain 0 as an element.
3. Which of the following pairs of sets can be placed in a one-to-one correspondence?
   a. $\{1, 2, 3, 4\}$ and $\{w, c, y, z\}$
   b. $\{1, 2, 3, \ldots, 25\}$ and $\{a, b, c, d, \ldots, x, y\}$
   c. $\{x \mid x$ is a letter in the word *geometry*$\}$ and $\{1, 2, 3, 4, 5, 6, 7, 8\}$
4. How many one-to-one correspondences exist between two sets with
   a. 8 elements each?
   b. $n - 1$ elements each?
5. How many one-to-one correspondences are there between the sets $\{a, b, c, d\}$ and $\{1, 2, 3, 4\}$ if in each correspondence
   a. b must correspond to 3?
   b. b must correspond to 3 and d to 4?
   c. a and c must correspond to even numbers?
6. Which of the following represent equal sets?
   $A = \{a, b, c\}$ $\quad$ $B = \{x, y\}$
   $C = \{c, a, b\}$ $\quad$ $D = \{x \mid 1 \le x \le 3, x \in N\}$
   $I = \{x \mid x = 2n,$ and $n \in \{0, 1, 2, 3, \ldots\}\}$,
   $K = \{2, 4, 6, 8, 10, 12, \ldots\}$
   $L = \{x \mid x = 2n - 1, n \in N\}$
7. Find the cardinal number of each of the following sets. Assume the pattern continues in each part:
   a. $\{9, 10, 11, \ldots, 99\}$
   b. $\{2, 4, 6, 8, \ldots, 2002\}$
   c. $\{x^2 \mid x = 1, 3, 5, 7, \ldots, 99\}$
   d. $\{x \mid x = x + 1, x \in N\}$
8. If U is the set of all women and G is the set of alumnae of Georgia State University, describe $\overline{G}$.
9. Suppose $A \subseteq B$.
   a. What is the minimum number of elements in set A?

   b. Is it possible for set B to be the empty set? If so, give an example of sets A and B satisfying this. If not, explain why not.
10. If two sets are subsets of each other, what other relationships must they have?
11. If $A = \{1, 2, 3, 4, 5, 6, 7, 8, 9\}$,
    a. how many subsets does A have?
    b. how many proper subsets does A have?
12. If a set has 16 subsets, how many elements are in the set?
13. Identify all possible proper subset relationships that occur among the following sets:
    $A = \{3n + 1 \mid n \in N\}$, $B = \{6n + 1 \mid n \in N\}$,
    $C = \{12n + 1 \mid n \in N\}$
14. Indicate which symbol, $\in$ or $\notin$, makes each of the following statements true:
    a. $\varnothing$ _____ $\varnothing$
    b. $\{2\}$ _____ $\{3, 2, 1\}$
    c. 1022 _____ $\{s \mid s = 2^n - 2, n \in N\}$
    d. 3004 _____ $\{x \mid x = 3n + 1, n \in N\}$
15. Indicate which symbol, $\subseteq$ or $\nsubseteq$, makes each part of problem 14 true.
16. Answer each of the following. If your answer is *no*, tell why.
    a. If $A \subseteq B$, can we always conclude that $A = B$?
    b. If $A \subset B$, can we conclude that $A = B$?
    c. If A and B can be placed in a one-to-one correspondence, must $A = B$?
    d. If A and B can be placed in a one-to-one correspondence, must $A \subseteq B$?
17. Use the definition of *less than* to show each of the following:
    a. $0 < 2$
    b. $99 < 100$
18. How many ways are there to stack an ice-cream cone with 4 scoops if the choices are
    a. vanilla, chocolate, rhubarb, and strawberry and each scoop must be different?
    b. vanilla, chocolate, rhubarb, and strawberry and there are no restrictions on different scoops?
19. How many seven-digit numbers are there when 0 and 1 cannot be the leading number?

Mathematical Connections 2-2

Communication

1. Explain the difference between a well-defined set and one that is not. Give examples.
2. Which of the following sets are not well defined? Explain.
   a. The set of wealthy schoolteachers
   b. The set of great books
   c. The set of natural numbers greater than 100
   d. The set of subsets of $\{1, 2, 3, 4, 5, 6\}$
   e. The set $\{x \mid x \ne x, x \in N\}$

3. Is $\varnothing$ a proper subset of every non-empty set? Explain your reasoning.
4. Explain why $\{\varnothing\}$ has $\varnothing$ as an element and also as a subset.
5. Tell how you would show that $A \nsubseteq B$.
6. Explain why every set is a subset of itself.
7. Define *less than or equal to* in a way similar to the definition of *less than*.

Open-Ended

8. a. Give three examples of sets A and B and a universal set U such that $A \subset B$; find $\overline{A}$ and $\overline{B}$.

b. Based on your observations, conjecture a relationship between $\overline{B}$ and $\overline{A}$.

c. Demonstrate your conjecture in (b) using a Venn diagram.

9. Find an infinite set A such that
a. $\overline{A}$ is finite.
b. $\overline{A}$ is infinite.

10. Describe two sets from real-life situations such that it is clear from using one-to-one correspondence, and not from counting, that one set has fewer elements than the other.

Cooperative Learning

11. Assume the fastest computer can list one subset in approximately 1 microsecond (one-millionth of a second).
a. Use a calculator if necessary to estimate the time in years it would take a computer to list all the subsets of $\{1, 2, 3, \ldots, 64\}$.
b. Estimate the time in years it would take the computer to exhibit all the one-to-one correspondences between the sets $\{1, 2, 3, \ldots, 64\}$ and $\{65, 66, 67, \ldots, 128\}$.

Questions from the Classroom

12. A student argues that $\{\varnothing\}$ is the proper notation for the empty set. What is your response?

13. A student asks if $A \subseteq B$ and $B \subseteq C$, do we know $A \subseteq C$? How do you answer?

14. A student argues that $A = \{1, \{1\}\}$ has only one element. How do you respond?

15. A student states that either $A \subseteq B$ or $B \subseteq A$. Is the student correct?

Review Problems

16. Write 5280 in expanded form.

17. What is the value of MCDX in Hindu-Arabic numerals?

18. Convert each of the following to base ten:
a. $E0T_{\text{twelve}}$
b. 1011_{two}
c. 43_{five}

19. Write $12^4 + 12^2 + 13$ in base twelve.

National Assessment of Educational Progress (NAEP) Question

Four people—A, X, Y, and Z—go to a movie and sit in adjacent seats. If A sits in the aisle seat, list all possible arrangements of the other three people. One of the arrangements is shown below.

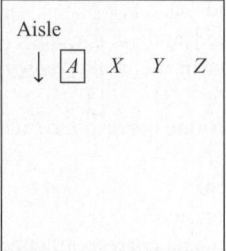

NAEP, Grade 12, 1996

BRAIN TEASER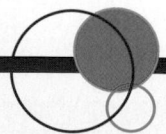

Mr. Gonzales's and Ms. Chan's seventh-grade classes in Paxson Middle School have 24 and 25 students, respectively. Linda, a student in Mr. Gonzales's class, claims that the number of school committees that could be formed to contain at least one student from each class is greater than the number of people in the world. Assuming that a committee can have up to 49 students, find the number of committees and determine whether Linda is right.

2-3 Other Set Operations and Their Properties

Finding the complement of a set is an operation that acts on only one set at a time. In this section, we consider operations that act on two sets at a time.

Set Intersection

Suppose that during the fall quarter, a college wants to mail a survey to all its students who are enrolled in both art and biology classes. To do this, the school officials must identify those students who are taking both classes. If A and B are the set of students taking art courses and the set of students taking biology courses, respectively, during the fall quarter, then the desired set of students includes those common to A and B, or the **intersection** of A and B. The intersection of sets A and B is the shaded region in Figure 2-16. Figure 2-16 depicts the possibility of A and B containing common elements. The intersection might contain no elements. Notice that for all sets A and B, $A \cap B \subseteq A$ and $A \cap B \subseteq B$.

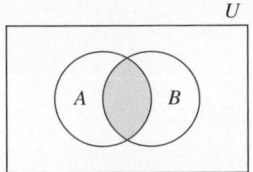

$A \cap B$

Figure 2-16

Definition of Set Intersection

The **intersection** of two sets A and B, written $A \cap B$, is the set of all elements common to both A and B; $A \cap B = \{x \mid x \in A \text{ and } x \in B\}$.

The key word in the definition of *intersection* is *and* (see Chapter 1). In everyday language, as in mathematics, *and* implies that both conditions must be met. In the example, the desired set is the set of those students enrolled in both art and biology.

If sets such as A and B have no elements in common, they are **disjoint sets**. In other words, two sets A and B are disjoint if, and only if, $A \cap B = \varnothing$. For example, if there are no students that are taking both art (A) and biology (B), then the sets are disjoint. Note that the Venn diagram for this situation could be drawn as in Figure 2-17.

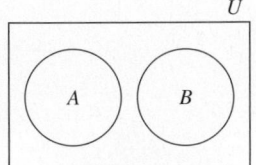

Figure 2-17

EXAMPLE 2-11

Find $A \cap B$ in each of the following:

a. $A = \{1, 2, 3, 4\}$, $B = \{3, 4, 5, 6\}$
b. $A = \{0, 2, 4, 6, \ldots\}$, $B = \{1, 3, 5, 7, \ldots\}$
c. $A = \{2, 4, 6, 8, \ldots\}$, $B = \{1, 2, 3, 4, \ldots\}$

Solution
a. $A \cap B = \{3, 4\}$.
b. $A \cap B = \varnothing$; therefore A and B are disjoint.
c. $A \cap B = A$ because all the elements of A are also in B.

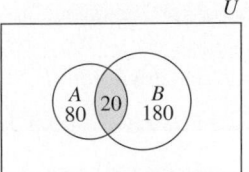

Figure 2-18

If A represents all students enrolled in art classes and B all students enrolled in biology classes, we may use a Venn diagram, taking into account that some students are enrolled in both subjects. If we know that 100 students are enrolled in art and 200 in biology and that 20 of these students are enrolled in both art and biology, then $100 - 20$, or 80, students are enrolled in art but not in biology and $200 - 20$, or 180, are enrolled in biology but not art. We can record this information as in Figure 2-18. Notice that the total number of students in set A is 100 and the total in set B is 200.

Set Union

If A is the set of students taking art courses during the fall quarter and B is the set of students taking biology courses during the fall quarter, then the set of students taking art or biology or both during the fall quarter is the **union** of sets A and B. The union of sets A and B is pictured in Figure 2-19. Notice that for all sets A and B, $A \subseteq A \cup B$, $B \subseteq A \cup B$, and $A \cap B \subseteq A \cup B$.

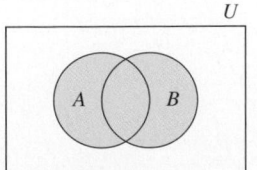

$A \cup B$

Figure 2-19

Definition of Set Union

The **union** of two sets A and B, written $A \cup B$, is the set of all elements in A or in B; $A \cup B = \{x \mid x \in A \text{ or } x \in B\}$.

The key word in the definition of *union* is *or* (see Chapter 1). In mathematics, *or* usually means "one or the other or both." This is known as the *inclusive or*.

EXAMPLE 2-12

Find $A \cup B$ for each of the following:

    a. $A = \{1, 2, 3, 4\}$, $B = \{3, 4, 5, 6\}$
    b. $A = \{0, 2, 4, 6, \ldots\}$, $B = \{1, 3, 5, 7, \ldots\}$
    c. $A = \{2, 4, 6, 8, \ldots\}$, $B = \{1, 2, 3, 4, \ldots\}$

Solution

    a. $A \cup B = \{1, 2, 3, 4, 5, 6\}$.   **b.** $A \cup B = \{0, 1, 2, 3, 4, \ldots\}$.
    c. Because every element of A is already in B, we have $A \cup B = B$.

EXAMPLE 2-13

Find each of the following if $A = \{a, b, c\}$.

    a. $A \cap \emptyset$ **b.** $A \cup \emptyset$ **c.** $\emptyset \cap \emptyset$ **d.** $\emptyset \cup \emptyset$

Solution

    a. $\emptyset$ **b.** A   **c.** $\emptyset$ **d.** $\emptyset$

NOW TRY THIS 2-13

In Figure 2-18, $n(A \cup B) = 80 + 20 + 180 = 280$, but $n(A) + n(B) = 100 + 200 = 300$; hence in general, $n(A \cup B) \neq n(A) + n(B)$. Use the concept of intersection of sets to write a formula for $n(A \cup B)$.

Set Difference

During the fall quarter if A is the set of students taking art classes and B is the set of students taking biology classes, then the set of all students taking biology but not art is called the **complement of A relative to B**, or the **set difference** of B and A.

> **Definition of Relative Complement**
> The **complement of A relative to B**, written $B - A$, is the set of all elements in B that are not in A;
> $B - A = \{x \mid x \in B \text{ and } x \notin A\}$.

Note that $B - A$ is not read as "B minus A." The *minus* sign indicates the subtraction operation on numbers and *set difference* is an operation on sets. A Venn diagram representing $B - A$ is shown in Figure 2-20(a). The shaded region represents all the elements that are in B but not in A. A Venn diagram for $B \cap \overline{A}$ is given in Figure 2-20(b). The shaded region represents all the elements that are in B and in $\overline{A}$. Notice that $B \cap \overline{A} = B - A$ because $B \cap \overline{A}$ is, by definitions of intersection and complement, the set of all elements in B and not in A.

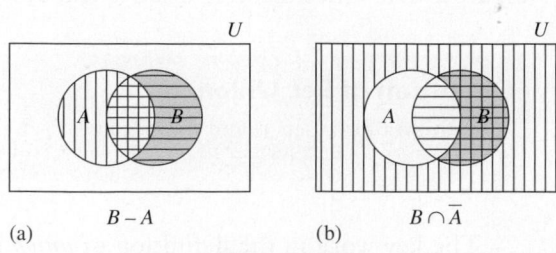

        $B - A$ $B \cap \overline{A}$
  (a)   (b)

Figure 2-20

EXAMPLE 2-14

If $A = \{d, e, f\}$, $B = \{a, b, c, d, e, f\}$, and $C = \{a, b, c\}$, find or answer each of the following:

a. $A - B$
b. $B - A$
c. $B - C$
d. $C - B$
e. To answer parts (a)–(d), does it matter what the universal set is?

Solution

a. $A - B = \varnothing$
b. $B - A = \{a, b, c\}$
c. $B - C = \{d, e, f\}$
d. $C - B = \varnothing$
e. Parts (a)–(d) can be answered independently of the universal set. The definition of set difference relates one set to another, independent of the universal set.

Properties of Set Operations

Because the order of elements in a set is not important, $A \cup B$ is equal to $B \cup A$. It does not matter in which order we write the sets when the union of two sets is involved. Similarly, $A \cap B = B \cap A$. These properties are stated formally next.

> **Theorem 2-2: Commutative Property of Set Intersection and Commutative Property of Set Union**
>
> For all sets A and B, $A \cap B = B \cap A$ is the **commutative property of set intersection**. Similarly, $A \cup B = B \cup A$ is the **commutative property of set union**.

NOW TRY THIS 2-14

Use Venn diagrams and other means to find whether grouping is important when the same operation is involved. For example, is it always true that $A \cap (B \cap C) = (A \cap B) \cap C$? Similar questions should be investigated involving union and set difference.

In answering Now Try This 2-14, the following properties become evident:

> **Theorem 2-3: Associative Property of Set Intersection and Associative Property of Set Union**
>
> For all sets A, B, and C, $A \cap (B \cap C) = (A \cap B) \cap C$ is the **associative property of set intersection**. Similarly, $A \cup (B \cup C) = (A \cup B) \cup C$ is the **associative property of set union**.

EXAMPLE 2-15

Is grouping important when two different set operations are involved? For example, is it true that $A \cap (B \cup C) = (A \cap B) \cup C$?

Solution

To investigate this, let $A = \{a, b, c\}$, $B = \{c, d\}$, and $C = \{d, e, f\}$. Then

$$A \cap (B \cup C) = \{a, b, c\} \cap (\{c, d\} \cup \{d, e, f\})$$
$$= \{a, b, c\} \cap \{c, d, e, f\}$$
$$= \{c\}$$

$$\text{Also, } (A \cap B) \cup C = (\{a, b, c\} \cap \{c, d\}) \cup \{d, e, f\}$$
$$= \{c\} \cup \{d, e, f\}$$
$$= \{c, d, e, f\}$$

In this case, $A \cap (B \cup C) \neq (A \cap B) \cup C$. So we have found a counterexample; that is, an example illustrating that the general statement is not always true.

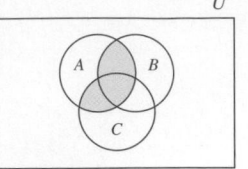

Figure 2-21

To discover an expression that is equal to $A \cap (B \cup C)$, consider the Venn diagram for $A \cap (B \cup C)$ shown by the shaded region in Figure 2-21. In the figure, $A \cap B$ and $A \cap C$ are subsets of the shaded region. The union of $A \cap B$ and $A \cap C$ is the entire shaded region. Thus, $A \cap (B \cup C) = (A \cap B) \cup (A \cap C)$. This property is stated formally next.

Theorem 2-4:   Distributive Property of Set Intersection over Set Union

For all sets A, B, and C, $A \cap (B \cup C) = (A \cap B) \cup (A \cap C)$.

Theorems 2-2, 2-3, and 2-4 can be proved using the definition of set intersection and set union and the rules of logic.

NOW TRY THIS 2-15

In the distributive property of set intersection over set union, if the symbol $\cap$ is replaced by $\cup$ and the symbol $\cup$ is replaced by $\cap$, is the new property true? Explain why. What should this property be called?

EXAMPLE 2-16

Use set notation to describe the shaded portions of the Venn diagrams in Figure 2-22.

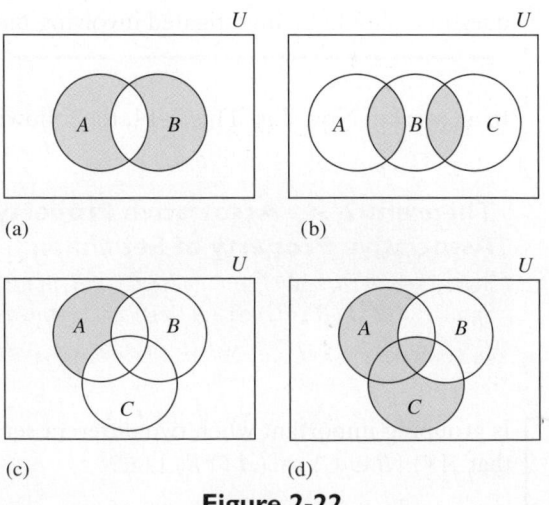

(a)   (b)

(c)   (d)

Figure 2-22

Solution
The solutions can be described in many different, but equivalent, forms. Some possible answers follow:

  a. $(A \cup B) - (A \cap B)$ or $(A \cup B) \cap \overline{A \cap B}$ or $(A - B) \cup (B - A)$
  b. $(A \cap B) \cup (B \cap C)$ or $B \cap (A \cup C)$
  c. $(A - B) - C$ or $A - (B \cup C)$ or $(A - (A \cap B)) - (A \cap C)$
  d. $((A \cup C) - B) \cup (A \cap B \cap C)$ or $(A - (B \cup C)) \cup (C - (A \cup B)) \cup (A \cap C)$

Using Venn Diagrams as a Problem-Solving Tool

Venn diagrams can be used as a problem-solving tool for modeling information, as shown in the following examples.

EXAMPLE 2-17

Suppose M is the set of all students taking mathematics and E is the set of all students taking English. Identify in words and using set notation the students described by each region in Figure 2-23.

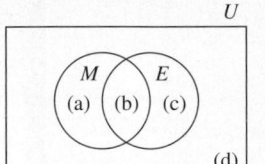

Figure 2-23

Solution
  Region (a) contains all students taking mathematics but not English or $M - E$.

  Region (b) contains all students taking both mathematics and English or $M \cap E$.

  Region (c) contains all students taking English but not mathematics or $E - M$.

  Region (d) contains all students taking neither mathematics nor English or $\overline{M \cup E}$ or $\overline{A} \cap \overline{E}$.

NOW TRY THIS 2-16

The following student page shows the use of Venn diagrams in modeling information. Answer questions 1–5 on the student page.

EXAMPLE 2-18

In a survey that investigated the high school backgrounds of 110 college freshmen, the following information was gathered:

  25 took physics.

  45 took biology.

  48 took mathematics.

  10 took physics and mathematics.

  8 took biology and mathematics.

  6 took physics and biology.

  5 took all three subjects.

  a. How many students took biology but neither physics nor mathematics?
  b. How many took physics, biology, or mathematics?
  c. How many did not take any of the three subjects?

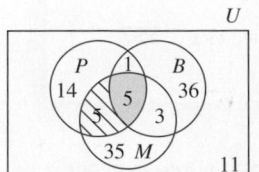

Figure 2-24

Solution
To solve this problem, we *build a model* using sets. Because there are three distinct subjects, we should use three circles. In Figure 2-24, P is the set of students taking physics, B is the set taking biology, and M is the set taking mathematics. The shaded region represents the 5 students who took all three subjects. The lined region represents the students who took physics and mathematics, but who did not take biology.

School Book Page   VENN DIAGRAMS

11-1b   Activity Lab

Venn Diagrams

A Venn diagram shows relationships between sets of items. Each set is represented separately. Items that belong to both sets are represented by the intersection.

EXAMPLE   Using a Venn Diagram

Geography There are 22 states that are all or partly in the eastern time zone, and 15 states that are all or partly in the central time zone. This includes the 5 states that are in both the eastern and central time zones. How many states are in at least one of the two time zones?

Draw a Venn diagram.

There are 17 + 5 + 10, or 32 states in either or both the eastern and central time zones.

Eastern Time Zone States
$22 - 5 = 17$

5

Central Time Zone States
$15 - 5 = 10$

Exercises

Forty students have pets. Thirty-two students have cats or dogs or both, but no birds. Eight students have birds. One student has all three kinds of pets.

cats 15   dogs 14

0

4 birds

1. Copy and complete the Venn diagram.

2. How many students have only dogs and birds? Only dogs and cats?

3. **Reasoning** A new student who has only fish joins the class. How many of the existing sets would this new set overlap?

4. **Geography** There are 10 states that are completely in the central time zone, and 6 states that are completely in the mountain time zone. There are 21 states that are completely in one of the time zones, or in both. Use a Venn diagram to find how many states are in both time zones.

5. **Data Collection** Conduct a survey of your class. Ask students whether they have brothers, sisters, both, or neither. Use a Venn diagram to display the results.

Activity Lab   Venn Diagrams   **537**

Part (a) asks for the number of students in the subset of B that has no element in common with either P or M; that is, $B - (P \cup M)$. Part (b) asks for the number of elements in $P \cup B \cup M$. Finally, part (c) asks for the number of students in $\overline{P \cup B \cup M}$, or $U - (P \cup B \cup M)$. Our strategy is to find the number of students in each of the eight nonoverlapping regions.

One mind set to be aware of in this problem is thinking, for example, that the 25 students who took physics took only physics. That is not necessarily the case.

a. Because a total of 10 students took physics and mathematics and 5 of those also took biology, $10 - 5$, or 5, students took physics and mathematics but not biology. Similarly, because 8 students took biology and mathematics and 5 took all three subjects, $8 - 5$, or 3, took biology and mathematics but not physics. Also $6 - 5$, or 1, student took physics and biology but not mathematics. To find the number of students who took biology but neither physics nor mathematics, we subtract from 45 (the total number who took biology) the number of those that are in the distinct regions that include biology and other subjects; that is, $1 + 5 + 3$, or 9. Because $45 - 9 = 36$, we know that 36 students took biology but neither physics nor mathematics.

b. To find the number of students in all the remaining distinct regions in P, M, or B, we proceed as follows. The number of students who took physics but neither mathematics nor biology is $25 - (1 + 5 + 5)$, or 14. The number of students who took mathematics but neither physics nor biology is $48 - (5 + 5 + 3)$, or 35. Hence the number of students who took mathematics, physics, or biology is $35 + 14 + 36 + 3 + 5 + 5 + 1$, or 99.

c. Because the total number of students is 110, the number of students who did not take any of the three subjects is $110 - 99$, or 11.

E-Manipulative Activity

Additional work with Venn diagrams can be found in the *Venn Diagrams* activity on the E-Manipulatives disk.

Cartesian Products

Another way to produce a set from two given sets is by forming the **Cartesian product**. This formation pairs the elements of one set with the elements of another set in a specific way to create elements in a new set. Suppose a person has three pairs of pants, $P = \{$blue, white, green$\}$, and two shirts, $S = \{$blue, white$\}$. According to the Fundamental Counting Principle, there are $3 \cdot 2$, or 6, possible different pants-and-shirt pairs, as shown in Figure 2-25.

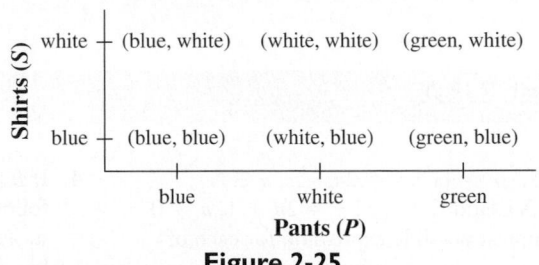

Figure 2-25

The pants-shirt combinations are elements of the set of all possible pairs in which the first member of the pair is an element of set P and the second member is an element of set S. The set of all possible pairs is given in Figure 2-25. Because the first component in each pair represents pants and the second component in each pair represents shirts, the order in which the components are written is important. Thus (white, blue) represents white pants and a blue shirt, whereas (blue, white) represents blue pants and a white shirt. Therefore, the two pairs represent different outfits. Because the order in each pair is important, the pairs are **ordered pairs**. The positions that the ordered pairs occupy within the set of outfits is immaterial. Only the order of the **components** within each pair is significant. The pants-and-shirt pairs suggest the following definition of **equality for ordered pairs**: $(x, y) = (m, n)$ *if, and only if, the first components are equal and the second components are equal.* In symbols, $(x, y) = (m, n)$ if, and only if, $x = m$ and $y = n$. The set of six ordered pairs shown in Figure 2-25, denoted $P \times S$, is an example of a Cartesian product. A formal definition follows.

Definition of Cartesian Product

For any sets A and B, the **Cartesian product** of A and B, written $A \times B$, is the set of all ordered pairs such that the first component of each pair is an element of A and the second component of each pair is an element of B.

$$A \times B = \{(x, y) \mid x \in A, y \in B\}$$

$A \times B$ is commonly read as "A cross B" and should never be read "A times B."

EXAMPLE 2-19

If $A = \{a, b, c\}$ and $B = \{1, 2, 3\}$, find each of the following:

a. $A \times B$
b. $B \times A$
c. $A \times A$

Solution

a. $A \times B = \{(a, 1), (a, 2), (a, 3), (b, 1), (b, 2), (b, 3), (c, 1), (c, 2), (c, 3)\}$
b. $B \times A = \{(1, a), (1, b), (1, c), (2, a), (2, b), (2, c), (3, a), (3, b), (3, c)\}$
c. $A \times A = \{(a, a), (a, b), (a, c), (b, a), (b, b), (b, c), (c, a), (c, b), (c, c)\}$

It is possible to form a Cartesian product involving the empty set. Suppose $A = \{1, 2\}$. Because there are no elements in $\varnothing$, no ordered pairs (x, y) with $x \in A$ and $y \in \varnothing$ are possible, so $A \times \varnothing = \varnothing$. This is true for all sets A. Similarly, $\varnothing \times A = \varnothing$ for all sets A. There is an analogy between the last equation and the multiplication fact that $0 \cdot a = 0$, where a is a whole number. In Chapter 3 we use the concept of Cartesian product to define multiplication of whole numbers.

Assessment 2-3A

1. If $N = \{1, 2, 3, 4, \ldots\}$, $A = \{x \mid x = 2n - 1, n \in N\}$, $B = \{x \mid x = 2n, n \in N\}$, and $C = \{x \mid x = 2n + 1, n = 0$ or $n \in N\}$, find the simplest possible expression for each of the following:
   a. $A \cup C$
   b. $A \cup B$
   c. $A \cap B$
2. Decide whether the following pairs of sets are always equal:
   a. $A \cap B$ and $B \cap A$
   b. $A \cup B$ and $B \cup A$
   c. $A \cup (B \cup C)$ and $(A \cup B) \cup C$
   d. $A \cup A$ and $A \cup \varnothing$
3. Tell whether each of the following is true for all sets A and B. If false, give a counterexample.
   a. $A \cup \varnothing = A$
   b. $A - B = B - A$
   c. $\overline{A \cap B} = \overline{A} \cap \overline{B}$
   d. $(A \cup B) - A = B$
   e. $(A - B) \cup A = (A - B) \cup (B - A)$

4. If $B \subseteq A$, find a simpler expression for each of the following:
   a. $A \cap B$
   b. $A \cup B$
5. For each of the following, shade the portion of the Venn diagram that illustrates the set:
   a. $A \cup B$
   b. $\overline{A \cap B}$
   c. $(A \cap B) \cup (A \cap C)$
   d. $(A \cup B) \cap \overline{C}$
   e. $(A \cap B) \cup C$

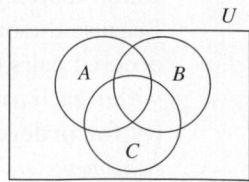

6. If S is a subset of universe U, find each of the following:
   a. $S \cup \overline{S}$
   b. $\overline{U}$
   c. $S \cap \overline{S}$
   d. $\varnothing \cap S$
7. For each of the following conditions, find $A - B$:
   a. $A \cap B = \varnothing$
   b. $B = \varnothing$
   c. $B = U$
8. If for sets A and B we know that $A - B = \varnothing$, is it necessarily true that $A \subseteq B$? Justify your answer.
9. Use set notation to identify each of the following shaded regions:

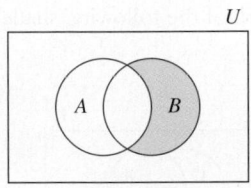 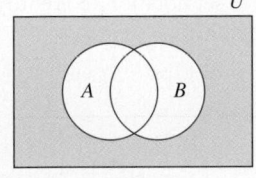

   a.   **b.**

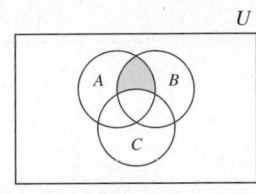

   c.

10. In the following, shade the portion of the Venn diagram that represents the given set:

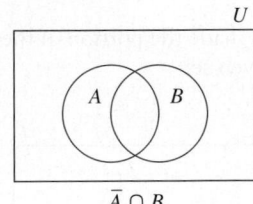

$$\overline{A} \cap B$$

11. Use Venn diagrams to determine if each of the following is true:
    a. $A \cup (B \cap C) = (A \cup B) \cap C$
    b. $A - (B - C) = (A - B) - C$
12. For each of the following pairs of sets, which is a subset of the other? If neither is a subset of the other, explain why.
    a. $A \cap B$ and $A \cap B \cap C$
    b. $A \cup B$ and $A \cup B \cup C$
13. a. If A has three elements and B has two elements, what is the greatest number of elements possible in (i) $A \cup B$? (ii) $A \cap B$? (iii) $B - A$? (iv) $A - B$?
    b. If A has n elements and B has m elements, what is the greatest number of elements possible in (i) $A \cup B$? (ii) $A \cap B$? (iii) $B - A$? (iv) $A - B$?
14. If $n(A) = 4$, $n(B) = 5$, and $n(C) = 6$, what is the greatest and least number of elements possible in
    a. $A \cup B \cup C$?
    b. $A \cap B \cap C$?

15. Given that the universe is the set of all humans, $B = \{x \mid x \text{ is a college basketball player}\}$, and $S = \{x \mid x \text{ is a college student more than 200 cm tall}\}$, describe each of the following in words:
    a. $B \cap S$
    b. $\overline{S}$
    c. $B \cup S$
    d. $\overline{B \cup S}$
    e. $\overline{B} \cap S$
    f. $B \cap \overline{S}$
16. Of the eighth graders at the Paxson School, 7 played basketball, 9 played volleyball, 10 played soccer, 1 played basketball and volleyball only, 1 played basketball and soccer only, 2 played volleyball and soccer only, and 2 played volleyball, basketball, and soccer. How many played one or more of the three sports?
17. In a fraternity with 30 members, 18 take mathematics, 5 take both mathematics and biology, and 8 take neither mathematics nor biology. How many take biology but not mathematics?
18. In Paul's bicycle shop, 40 bicycles were inspected. If 20 needed new tires and 30 needed gear repairs, answer the following:
    a. What is the greatest number of bikes that could have needed both?
    b. What is the least number of bikes that could have needed both?
    c. What is the greatest number of bikes that could have needed neither?
19. The Red Cross looks for three types of antigens in blood tests: A, B, and Rh. When the antigen A or B is present, it is listed, but if both of these antigens are absent, the blood is type O. If the Rh antigen is present, the blood is positive; otherwise, it is negative. If a laboratory technician reports the following results after testing the blood samples of 100 people, how many were classified as O negative? Explain your reasoning.

Number of Samples	Antigen in Blood
40	A
18	B
82	Rh
5	A and B
31	A and Rh
11	B and Rh
4	A, B, and Rh

20. Classify the following as true or false. If false, give a counterexample. Assume that A and B are finite sets.
    a. If $n(A) = n(B)$, then $A = B$.
    b. If $A - B = \varnothing$, then $A = B$.
    c. If $A \subset B$, then $n(A) < n(B)$.
21. Three announcers each try to predict the winners of next Sunday's professional football games. The only team not picked that is playing Sunday is the Giants. The choices for each person were as follows:

Phyllis: Cowboys, Steelers, Vikings, Bills
Paula: Steelers, Packers, Cowboys, Redskins
Rashid: Redskins, Vikings, Jets, Cowboys

If the only teams playing Sunday are those just mentioned, which teams will play which other teams?

22. Let $A = \{x, y\}$ and $B = \{a, b, c\}$. Answer each of the following:
   a. $A \times B$
   b. $B \times A$
   c. Does $A \times B = B \times A$?

23. For each of the following, the Cartesian product $C \times D$ is given by the sets listed. Find C and D.
   a. $\{(a, b), (a, c), (a, d), (a, e)\}$
   b. $\{(1, 1), (1, 2), (1, 3), (2, 1), (2, 2), (2, 3)\}$
   c. $\{(0, 1), (0, 0), (1, 1), (1, 0)\}$

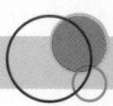

Assessment 2–3B

1. If $W = \{0, 1, 2, 3, \ldots\}$, $A = \{x \mid x = 2n + 1, n \in W\}$, $B = \{x \mid x = 2n, n \in W\}$, and $N = \{1, 2, 3, \ldots\}$, find the simplest possible expression for each of the following:
   a. $W - A$
   b. $A \cap B$
   c. $W \cap N$

2. Decide whether the following pairs of sets are always equal.
   a. $A - B$ and $B - A$
   b. $A - B$ and $A \cap \overline{B}$
   c. $A \cap (B \cap C)$ and $(A \cap B) \cap C$
   d. $B \cup \varnothing$ and $B \cap B$

3. Tell whether each of the following is true for all sets A, B, and C. If false, give a counterexample.
   a. $A - B = A - \varnothing$
   b. $\overline{A \cup B} = \overline{A} \cup \overline{B}$
   c. $A \cap (B \cup C) = (A \cap B) \cup C$
   d. $(A - B) \cap A = A$
   e. $A - (B \cap C) = (A - B) \cap (A - C)$

4. If $X \subseteq Y$, find a simpler expression for each of the following:
   a. $X - Y$
   b. $X \cap \overline{Y}$

5. For each of the following, shade the portion of the Venn diagram that illustrates the set:
   a. $A \cap \overline{C}$
   b. $\overline{A \cup B}$
   c. $(A \cap B) \cup (B \cap C)$
   d. $A \cup (B \cap C)$
   e. $A \cup \overline{B} \cap C$

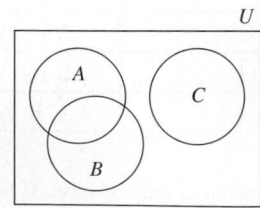

6. If A is a subset of universe U, find each of the following:
   a. $A \cup U$
   b. $U - A$
   c. $A - \varnothing$
   d. $\varnothing \cap A$

7. For each of the following conditions, find $B - A$.
   a. $A = B$
   b. $B \subseteq A$

8. Give two examples of sets A and B for which $B - A = \varnothing$. Show that in each example $B \subseteq A$.

9. Use set notation to identify each of the following shaded regions:

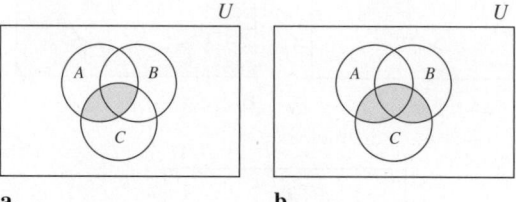

a.   b.

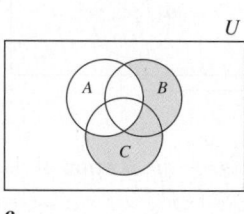

c.

10. In the following, shade the portion of the Venn diagram that represents the given set:

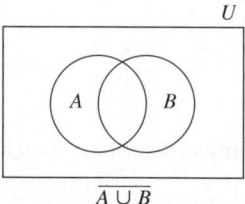

$\overline{A \cup B}$

11. Use Venn diagrams to determine whether each of the following is true:
   a. $A - (B \cap C) = (A - B) \cap (A - C)$
   b. $A - (B \cup C) = (A - B) \cup (A - C)$

12. For each of the following pairs of sets, explain which is a subset of the other. If neither is a subset of the other, explain why.
   a. $A - B$ and $A - (B - C)$
   b. $A \cup B$ and $(A \cup B) - \varnothing$

13. a. If $n(A \cup B) = 22$, $n(A \cap B) = 8$, and $n(B) = 12$, find $n(A)$.
    b. If $n(A) = 8$, $n(B) = 14$, and $n(A \cap B) = 5$, find $n(A \cup B)$.

14. The equation $\overline{A \cup B} = \overline{A} \cap \overline{B}$ and a similar equation for $\overline{A \cap B}$ are referred to as *De Morgan's Laws* in honor of

the famous British mathematician who first discovered them.
   a. Use Venn diagrams to show that $\overline{A \cup B} = \overline{A} \cap \overline{B}$.
   b. Discover an equation similar to the one in part (a) involving $\overline{A \cap B}, \overline{A}$, and $\overline{B}$. Use Venn diagrams to show that the equation holds.
   c. Verify the equations in (a) and (b) for specific sets.

15. Suppose P is the set of all eighth-grade students at the Paxson School, with B the set of all students in the band and C the set of all students in the choir. Identify in words the students described by each region of the following figure:

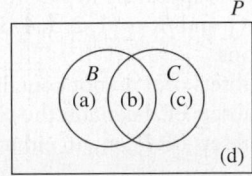

16. Fill in the Venn diagram with the appropriate numbers based on the following information:

$n(A) = 26$ $n(B \cap C) = 12$
$n(B) = 32$ $n(A \cap C) = 8$
$n(C) = 23$ $n(A \cap B \cap C) = 3$
$n(A \cap B) = 10$ $n(U) = 65$

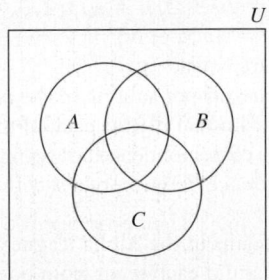

17. Write the letters in the appropriate sections of the following Venn diagram using the following information (Ignore the case of the letters):

Set A contains the letters in the word *Iowa*.
Set B contains the letters in the word *Hawaii*.
Set C contains the letters in the word *Ohio*.

The universal set U contains the letters in the word *Washington*.

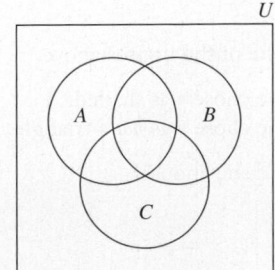

18. Students at Hellgate High School were asked if they were taking courses in algebra (A), biology (B), or chemistry (C). Use the diagram to answer the following questions.
   a. Which region (or regions) represents the students who do not take algebra? (List the letters)
   b. Which region (or regions) represents the students who take biology or chemistry? (List the letters)
   c. A student who took both algebra and biology could be in which region (or regions)? (List the letters)
   d. How would you describe region (a) in sentence form?
   e. How would you describe region (f) in sentence form?
   f. Describe region (f) using set notation.
   g. If we were to combine regions (d) and (g) to form one large region, describe it using set notation.
   h. Describe region (g) using set notation.

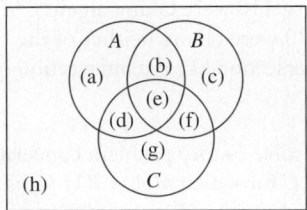

19. A pollster interviewed 500 university seniors who owned credit cards. She reported that 240 owned Goldcard, 290 had Supercard, and 270 had Thriftcard. Of those seniors, the report said that 80 owned only a Goldcard and a Supercard, 70 owned only a Goldcard and a Thriftcard, 60 owned only a Supercard and a Thriftcard, and 50 owned all three cards. When the report was submitted for publication in the local campus newspaper, the editor refused to publish it, claiming the poll was not accurate. Was the editor right? Why or why not?

20. In a survey of 1000 investors it was determined that 600 invested in stocks, 575 in bonds, and 300 in both stocks and bonds.
   a. How many invested only in stocks?
   b. How many invested in stocks or bonds?
   c. How many did not invest in either stocks or bonds?

21. In a survey of 150 students, 90 were taking algebra and 30 were taking biology.
   a. What is the least number of students who could have been taking both courses?
   b. What is the greatest number of students who could be taking both courses?
   c. What is the greatest number of students who could have been taking neither course?

22. Tell whether each of the following is true or false and explain why:
   a. $(2, 5) = (5, 2)$
   b. $(2, 5) = \{2, 5\}$

23. Answer each of the following:
   a. If A has five elements and B has four elements, how many elements are in $A \times B$?
   b. If A has m elements and B has n elements, how many elements are in $A \times B$?
   c. If A has m elements, B has n elements, and C has p elements, how many elements are in $(A \times B) \times C$?

Mathematical Connections 2-3

Communication

1. Answer each of the following and justify your answer:
   a. If $a \in A \cap B$, is it true that $a \in A \cup B$?
   b. If $a \in A \cup B$, is it true that $a \in A \cap B$?
2. Explain how $\overline{A}$ is related to $U - A$.
3. Is the operation of forming Cartesian products commutative? Explain why or why not.
4. If A and B are sets, is it always true that $n(A - B) = n(A) - n(B)$? Explain.
5. If $A \subset B$ and $n(A) = 13$, how many elements could set B have? Explain.
6. Explain whether the following could be true: In a survey of 220 students, 110 were taking algebra, 90 were taking biology, and 20 were taking neither of the two subjects.
7. How is set intersection like the intersection of two highways?

Open-Ended

8. Make up and solve a story problem concerning specific sets A, B, and C for which $n(A \cup B \cup C)$ is known and it is required to find $n(A)$, $n(B)$, and $n(C)$.
9. Describe a real-life situation that can be represented by each of the following:
   a. $A \cap \overline{B}$
   b. $A \cap B \cap C$
   c. $A - (B \cup C)$

Cooperative Learning

10. A set of attribute blocks consists of 32 blocks. Each block is identified by its own shape, size, and color. The four shapes in a set are square, triangle, rhombus, and circle; the four colors are red, yellow, blue, and green; the two sizes are large and small. In addition to the blocks, each set contains a group of 20 cards. Ten cards specify one of the attributes of the blocks (for example, red, large, square). The other 10 cards are negation cards and specify the lack of an attribute (for example, not green, not circle). Many set-type problems can be studied with these blocks. For example, let A be the set of all green blocks and B be the set of all large blocks. As a group; using the set of all blocks as the universal set, describe elements in each set listed here to determine which are equal:
    a. $A \cup B$ and $B \cup A$
    b. $\overline{A \cap B}$ and $\overline{A} \cap \overline{B}$
    c. $\overline{A \cap B}$ and $\overline{A} \cup \overline{B}$
    d. $A - B$ and $A \cap \overline{B}$

Questions from the Classroom

11. A student asks, "If $A = \{a, b, c\}$ and $B = \{b, c, d\}$, why isn't it true that $A \cup B = \{a, b, c, b, c, d\}$?" What is your response?
12. A student claims that the complement bar can be "broken" over the operation of intersections; that is, $\overline{A \cap B} = \overline{A} \cap \overline{B}$. What is your response?
13. A student is asked to find all one-to-one correspondences between two given sets. He finds the Cartesian product of the sets and claims that his answer is correct because it includes all possible pairings between the elements of the sets. How do you respond?
14. A student argues that adding two sets A and B, or $A + B$, and taking the union of two sets, $A \cup B$, is the same thing. How do you respond?

Review Problems

15. In base two, does the number "two" exist? Explain your reasoning.
16. Write 81 in base three.
17. a. Write $\{4, 5, 6, 7, 8, 9\}$ using set-builder notation.
    b. Write $\{x \mid x = 5n, n = 3, 6, \text{or } 9\}$ using the listing method.
18. Find the number of elements in the following sets:
    a. $\{x \mid x \text{ is a letter in } commonsense\}$
    b. The set of letters appearing in the word *committee*
19. If $A = \{1, 2, 3, 4\}$ and $B = \{1, 2, 3, 4, 5\}$, answer the following questions:
    a. How many subsets of A do not contain the element 1?
    b. How many subsets of A contain the element 1?
    c. How many subsets of A contain either the element 1 or 2?
    d. How many subsets of A contain neither the element 1 nor 2?
    e. How many subsets of B contain the element 5 and how many do not?
    f. If all the subsets of A are known, how can all the subsets of B be listed systematically? How many subsets of B are there?
20. Let $A = \{2, 4, 6, 8, 10, \ldots\}$
    $B = \{x \mid x = 2n + 2, n = 0, 1, 2, 3, 4, \ldots\}$
    $C = \{x \mid x = 4n, n \in N\}$
    a. Which sets are equal?
    b. Which sets are proper subsets of the other sets?
21. Give examples from real life for each of the following:
    a. A one-to-one correspondence between two sets
    b. A correspondence between two sets that is not one-to-one
22. If there are six teams in the Alpha league and five teams in the Beta league and if each team from one league plays each team from the other league exactly once, how many games are played?
23. José has four pairs of slacks, five shirts, and three sweaters. From how many combinations can he choose if he chooses a pair of slacks, a shirt, and a sweater each day?

National Assessment of Educational Progress (NAEP) Question

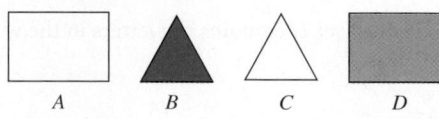

Melissa chose one of the figures above.

- The figure she chose was shaded.
- The figure she chose was *not* a triangle.

Which figure did she choose?
a. A
b. B
c. C
d. D

NAEP, Grade 4, 2007

Hint for Solving the Preliminary Problem

A Venn diagram is a good tool for sorting the data in this problem. Distinguish among the sets of students taking mathematics, chemistry, and physics by representing each set with a circle in a Venn diagram. Sorting the information into the various sections representing each bit of information will help determine whether the information that John reported was indeed accurate.

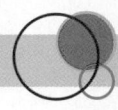

Chapter 2 Summary

2-1 Numeration Systems	**Pages**
A **numeration system** is a collection of properties and symbols agreed upon to represent numbers systematically.	52
Numerals are written symbols such as 2, 5 or 123.	52
The **Hindu-Arabic numeration system** is a base-ten system constructed from the 10 digits—0, 1, 2, 3, 4, 5, 6, 7, 8, 9. It uses place value based on powers of 10, the number base of the system.	52

- **Place value** assigns a value to a digit depending on its placement in a numeral. — 53
- **Face value** is the value of the digit. — 53
- The **value** of a digit is the product of its face value and place value. — 53
- If a is any number and n is any natural number, then

$$a^n = \underbrace{a \cdot a \cdot a \cdot \ldots \cdot a}_{n \text{ factors}}$$

  where a is the **base** and n is the exponent. — 53
- If $a \neq 0$, then $a^0 = 1$.

The **tally numeration system** used single strokes or tally marks to represent numbers.	54
The **Egyptian numeration system** used a *grouping system* to represent certain sets of numbers.	55–56
The **Babylonian numeration system** used a *place-value* system.	56–57
The **Mayan numeration system** based their system primarily on 20 with vertical groupings.	57–58
The **Roman numeration system** used the following properties:	58–59

- **Additive property** — 58
- **Subtractive property** — 58
- **Multiplicative property** — 58

| **Other number base systems** are studied to better understand our base ten system. | 59–63 |

2-2 Describing Sets	
A **set** is understood to be any collection of objects.	66

- An **element** or **member** is any individual object in the set. — 66
- A set must be **well defined**; that is, if we are given a set and some particular object, then we must be able to tell whether the object does or does not belong to the set. — 67

Sets can be specified by either using the **listing (roster) method** or **set-builder notation**.	67
Two sets are **equal** if, and only if, they contain exactly the same elements.	68
If the elements of sets P and S can be paired so that for each element of P there is exactly one element of S and for each element of S there is exactly one element of P, then the two sets P and S are in **one-to-one correspondence**.	68–70
Theorem: Fundamental Counting Principle: If event M can occur in m ways and, after it has occurred, event N can occur in n ways, then event M followed by event N can occur in mn ways.	70
Two sets A and B are **equivalent**, written $A \sim B$, if, and only if, there exists a one-to-one correspondence between the sets.	70
The **cardinal number** of a set S, denoted by $n(S)$, indicates the number of elements in the set S.	71
The **empty set** or **null set** is a set that contains no elements.	72
The **universal set** or **universe**, denoted U, is the set that contains all elements being discussed.	73
The **complement** of a set A, written $\overline{A}$, is the set of all elements in the universal set U that are not in A; that is, $\overline{A} = \{x \,|\, x \in U, x \notin A\}$.	73
For all sets A and B, B is a **subset** of A, written $B \subseteq A$, if, and only if, every element of B is an element of A.	73
Set B is a **proper subset** of set A written $B \subset A$, if, and only if, every element of B is an element of A and there exists at least one element of A that is not in B.	74
If there are n elements in a set, 2^n subsets can be formed.	77

2-3 Other Set Operations and Their Properties

The **intersection** of two sets A and B, written $A \cap B$, is the set of all elements common to both A and B; $A \cap B = \{x \,|\, x \in A, x \in B\}$.	80–81
The **union** of two sets A and B, written $A \cup B$, is the set of all elements in A or in B or both; $A \cup B = \{x \,|\, x \in A \text{ or } x \in B\}$.	81
The **complement of A relative to B**, written $B - A$, is the set of all elements in B that are not in A; $B - A = \{x \,|\, x \in B, x \notin A\}$.	82
Properties of Set Operations • **Theorem: Commutative Properties of Set Intersection and Set Union:** For all sets A and B, $A \cap B = B \cap A$, $A \cup B = B \cup A$.	83
• **Theorem: Associative Properties of Set Intersection and Set Union:** For all sets A, B, and C, $A \cap (B \cap C) = (A \cap B) \cap C$ and $A \cup (B \cup C) = (A \cup B) \cup C$.	83
• **Theorem: Distributive Properties of Set Intersection over Set Union:** For all sets A, B, and C, $A \cap (B \cup C) = (A \cap B) \cup (A \cap C)$.	84
Venn diagrams can be used as a problem-solving tool. Venn diagrams can be used as a problem-solving tool.	85
For any sets A and B, the **Cartesian product** of A and B, written $A \times B$, is the set of all ordered pairs such that the first component of each pair is an element of A and the second component of each pair is an element of B; $A \times B = \{(x, y) \,|\, x \in A \text{ and } y \in B\}$.	87–88

Chapter 2 Review

1. For each of the following base-ten numerals, tell the place value for the underlined digits:
   a. 4<u>3</u>2
   b. <u>3</u>432
   c. 19<u>3</u>24

2. Convert each of the following to a base-ten numeral.
   a. $\overline{\text{CDXLIV}}$
   b. 432_{five}
   c. $ET0_{\text{twelve}}$
   d. 1011_{two}
   e. 4136_{seven}

3. Convert each of the following to a numeral in the indicated system:
   a. 999 to Roman
   b. 86 to Egyptian
   c. 123 to Mayan
   d. 346_{ten} to base five
   e. 27_{ten} to base two

4. Simplify each of the following, if possible. Write your answers in exponential form, a^b.
   a. $3^4 \cdot 3^7 \cdot 3^6$
   b. $2^{10} \cdot 2^{11}$

5. Write the base-three numeral for the base-three blocks shown.

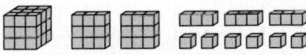

6. What is the fewest number of base-three blocks needed to represent 51?

7. Draw multibase blocks to represent
   a. 123_{four}
   b. 24_{five}

8. a. The first digit from the left (the lead digit) of a base-ten numeral is 4 followed by 10 zeros. What is the place value of 4?
   b. A number in base five has 10 digits. What is the place value of the second digit from the left?
   c. A number in base two has lead digit 1 followed by 30 zeros and units digit 1. What is the place value of the lead digit?

9. Write the following base-ten numerals in the indicated base without performing any multiplications:
   a. $10^{10} + 23$ in base ten
   b. $2^{10} + 1$ in base two
   c. $5^{10} + 1$ in base five
   d. $10^{10} - 1$ in base ten
   e. $2^{10} - 1$ in base two
   f. $12^5 - 1$ in base twelve

10. Give an example of a base other than ten used in a real-life situation. How is it used?

11. Describe the important characteristics of each of the following systems:
    a. Egyptian   b. Babylonian
    c. Roman   d. Hindu-Arabic

12. Write 128 in each of the following bases:
    a. Five
    b. Two
    c. Twelve

13. Write each of the following in the indicated bases without multiplying out the various powers:
    a. $2^{10} + 2^3$ in base two
    b. $11 \cdot 12^5 + 10 \cdot 12^3 + 20$ in base twelve

14. If $123_b = 83$, solve for b.

15. How many proper subsets does $A = \{a, b, c, d\}$ have?

16. List all the subsets of $\{m, a, t, h\}$.

17. Let
    $U = \{u, n, i, v, e, r, s, a, l\}$
    $A = \{r, a, v, e\}$ $C = \{l, i, n, e\}$
    $B = \{a, r, e\}$ $D = \{s, a, l, e\}$
    Find each of the following:
    a. $A \cup B$ b. $C \cap D$
    c. $\overline{D}$
    d. $A \cap \overline{D}$
    e. $\overline{B \cup C}$
    f. $(B \cup C) \cap D$
    g. $(\overline{A} \cup B) \cap (C \cap \overline{D})$
    h. $(C \cap D) \cap A$
    i. $n(B - A)$
    j. $n(\overline{C})$
    k. $n(C \times D)$

18. Indicate the following sets by shading the figure:

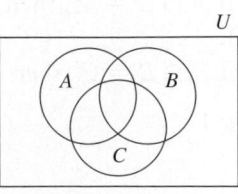 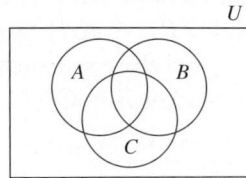

    a. $A \cap (B \cup C)$ b. $\overline{A \cup B} \cap C$

19. Suppose you are playing a word game with seven distinct letters. How many different arrangements of the seven letters can there be?

20. a. Show one possible one-to-one correspondence between sets D and E if $D = \{t, h, e\}$ and $E = \{e, n, d\}$.
    b. How many one-to-one correspondences between sets D and E are possible?

21. Use a Venn diagram to determine whether $A \cap (B \cup C) = (A \cap B) \cup C$ for all sets A, B, and C.

22. According to a student survey, 16 students liked history, 19 liked English, 18 liked mathematics, 8 liked mathematics and English, 5 liked history and English, 7 liked history and mathematics, 3 liked all three subjects, and every student liked at least one of the subjects. Draw a Venn diagram describing this information and answer the following questions:
    a. How many students were in the survey?
    b. How many students liked only mathematics?
    c. How many students liked English and mathematics but not history?

23. Describe, using symbols, the shaded portion in each of the following figures:

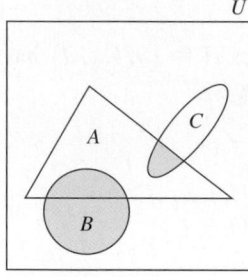
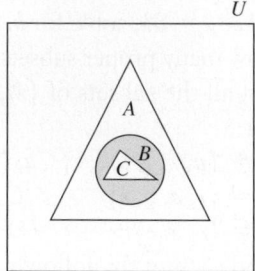

    a.   b.

24. Classify each of the following as true or false. If false, tell why.
    a. For all sets A and B, either $A \subseteq B$ or $B \subseteq A$.
    b. The empty set is a proper subset of every set.
    c. For all sets A and B, if $A \sim B$, then $A = B$.
    d. The set $\{5, 10, 15, 20, \ldots\}$ is a finite set.
    e. No set is equivalent to a proper subset of itself.
    f. If A is an infinite set and $B \subseteq A$, then B also is an infinite set.
    g. For all finite sets A and B, if $A \cap B \neq \emptyset$, then $n(A \cup B) \neq n(A) + n(B)$.
    h. If A and B are sets such that $A \cap B = \emptyset$, then $A = \emptyset$ or $B = \emptyset$.

25. Decide whether each of the following is always true for finite sets A and B.
    a. $n(A \cup B) = n(A - B) + n(B - A) + n(A \cap B)$
    b. $n(A \cup B) = n(A - B) + n(B) = n(B - A) + n(A)$

26. Suppose P and Q are equivalent sets and $n(P) = 17$.
    a. What is the minimum number of elements in $P \cup Q$?
    b. What is the maximum number of elements in $P \cup Q$?
    c. What is the minimum number of elements in $P \cap Q$?
    d. What is the maximum number of elements in $P \cap Q$?

27. Case Eastern Junior College awarded 26 varsity letters in crew, 15 in swimming, and 16 in soccer. If awards went to 46 students and only 2 lettered in all sports, how many students lettered in just two of the three sports?

28. If $A \subseteq B$, which of the following are always true? Give a counterexample if false.
    a. $\overline{A} \subseteq \overline{B}$
    b. $\overline{B} \subseteq \overline{A}$
    c. $A \cup B = B$
    d. $A \cap B = A$
    e. $\overline{A} \cap \overline{B} = \overline{B}$
    f. $\overline{A} \cup \overline{B} = A$

29. If $n(A) = 3$, $n(B) = 4$, and $n(c) = 2$, find $n(A \times B \times C)$.

30. Tell whether each of the following is true or false. If false, give a counterexample.
    a. If $A \cup B = A \cup C$, then $B = C$.
    b. If $A \cap B = A \cap C$, then $B = C$.

31. Using the definitions of less than or greater than, show that each of the following inequalities is true:
    a. $3 < 13$ b. $12 > 9$

32. Heidi has a brown pair and a gray pair of slacks; a brown blouse, a yellow blouse, and a white blouse; and a blue sweater and a white sweater. How many different outfits does she have if each outfit she wears consists of slacks, a blouse, and a sweater?

Whole Numbers and Their Operations

Preliminary Problem

Abby baked 124 cookies and while they were cooling she stepped outside. She saw a group of children run into the kitchen and run back out with cookies in their hands. When she returned to the kitchen she found only 7 cookies left on the cooking racks. If each child took the same number of cookies and the number of children who took cookies was more than 10 and fewer than 30, how many were in the group?

If needed, see Hint on page 168.

In Section 2-2 we saw one-to-one correspondence used to introduce children to the concept of a number. In NCTM's pre-kindergarten *Focal Points*, we find the following:

> Children develop an understanding of the meanings of whole numbers and recognize the number of objects in small groups without counting and by counting—the first and most basic mathematical algorithm. They understand that number words refer to quantity. They use one-to-one correspondence to solve problems by matching sets and comparing number amounts and in counting objects to 10 and beyond. (p. 11)

In the following *Peanuts* cartoon, it seems that Rerun has not learned to associate number words with a collection of objects. He will soon learn that a set of fingers can be put into one-to-one correspondence with sets of objects to be counted. The word *three* will not only be associated with Lucy's three upheld fingers but with other sets of objects.

Peanuts copyright © 1992 and 2010 Peanuts Worldwide LLC. Distributed by United Feature Syndicate, Inc.

In this chapter, we investigate operations and algorithms involving whole numbers. The grade 3 *Common Core Standards* state that students should be able to:

- Fluently add and subtract within 1000 using strategies and algorithms based on place value, properties of operations, and/or the relationship between addition and subtraction. (p. 24)
- Fluently multiply and divide within 100, using strategies such as the relationship between multiplication and division (e.g., knowing that $8 \times 5 = 40$, one knows $40 \div 5 = 8$) or properties of operations. By the end of grade 3, know from memory all products of two one-digit numbers. (p. 23)
- Solve two-step word problems using the four operations. Represent these problems using equations with the letter standing for the unknown quantity. Assess the reasonableness of answers using mental computation and estimation strategies including rounding. (p. 23)

3-1 Addition and Subtraction of Whole Numbers

When zero is included with the set of natural numbers, $N = \{1, 2, 3, 4, 5, \ldots\}$, we have the set of whole numbers, denoted $W = \{0, 1, 2, 3, 4, 5, \ldots\}$. In this section, we provide a variety of models for teaching computational skills involving whole numbers and revisit mathematics for deeper understanding that teachers need.

Addition of Whole Numbers

Children encounter addition in their preschool years by combining sets of objects and wanting to know how many objects are in the combined set. They may count the objects to find the cardinal

○— **Historical Note** ─────────────────────────────────

Historians think that the word *zero* originated from the Hindu word *sūnya*, which means "void." Then *sūnya* was translated into the Arabic *sifr*, which when translated to Latin became *zephirum*, from which the word *zero* was derived.   ●

number of the combined set or may "count on." *Counting on* is an addition strategy where addition is performed by counting on from one of the numbers, for example, 5 + 3 can be computed by starting at 5 and counting 6, 7, 8.

Set Model

A set model is one way to represent addition of whole numbers. Suppose Jane has 4 blocks in one pile and 3 in another. If she combines the two groups, how many objects are there in the combined group? Figure 3-1 shows the solution. The combined set of blocks is the union of the disjoint sets of 4 blocks and 3 blocks. *Note the importance of the sets being disjoint or having no elements in common.* If the sets have common elements, then an incorrect conclusion can be drawn when using the set model.

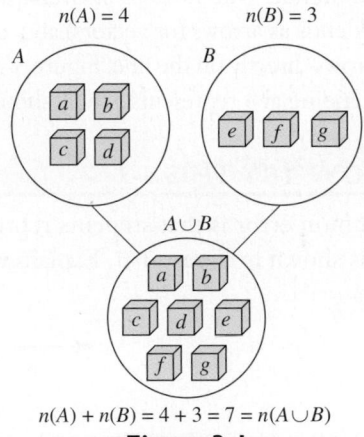

$$n(A) + n(B) = 4 + 3 = 7 = n(A \cup B)$$

Figure 3-1

Using set terminology, we define addition formally.

Definition of Addition of Whole Numbers

Let A and B be two disjoint finite sets. If $n(A) = a$ and $n(B) = b$, then $a + b = n(A \cup B)$.

The numbers a and b in $a + b$ are the **addends** and $a + b$ is the **sum**.

NOW TRY THIS 3-1

If the sets in the preceding definition of addition of whole numbers are not disjoint, explain why the definition is incorrect.

Number-Line (Measurement) Model

The set model for addition is not always the best model for addition. For example, consider the following:

1. Josh has 4 feet of red ribbon and 3 feet of white ribbon. How many feet of ribbon does he have altogether?
2. One day, Gail drank 4 ounces of orange juice in the morning and 3 ounces at lunchtime. If she drank no other orange juice that day, how many ounces of orange juice did she drink for the entire day?

A *number line* can be used to model the preceding and other whole-number additions. Any line marked with two fundamental points, one representing 0 and the other representing 1, can be turned into a number line. The points representing 0 and 1 mark the ends of a *unit segment*. Other points can be marked and labeled, as shown in Figure 3-2. Any two consecutive whole numbers on the number line in Figure 3-2 mark the ends of a segment that has the same length as the unit segment.

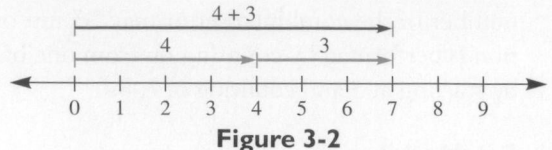

Figure 3-2

Addition problems may be modeled using directed arrows (*vectors*) on the number line. For example, the sum of 4 + 3 is shown in Figure 3-2. Arrows representing the addends, 4 and 3, are combined into one arrow representing the sum 4 + 3.

Students need to understand that the sum represented by any two directed arrows can be found by placing the endpoint of the first directed arrow at 0 and then joining to it the directed arrow for the second number with no gaps or overlaps. The sum of the numbers can then be read. We have depicted the addends as arrows (or vectors) above the number line, but students typically concatenate (connect) the arrows directly on the line. Figure 3-2 poses an inherent problem for students. If an arrow starting at 0 and ending at 3 represents 3, why should an arrow starting at 4 and ending at 7 represent 3?

NOW TRY THIS 3-2

A common error is that students represent 3 as an arrow on the number line sometimes starting at 1, as shown in Figure 3-3. Explain why this is not appropriate.

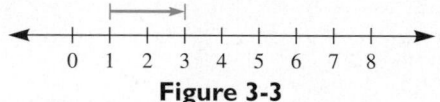

Figure 3-3

Ordering Whole Numbers

In the grade 1 *Focal Points*, the authors state the following:

> Children compare and order whole numbers (at least to 100) to develop an understanding of and solve problems involving the relative sizes of these numbers . . . . They understand the sequential order of the counting numbers and their relative magnitudes and represent numbers on a number line. (p. 13)

In Chapter 2, we used the concept of a set and one-to-one correspondence to define *less than* and *greater than* relations. A horizontal number line can also be used to describe **greater than** and **less than** relations on the set of whole numbers. For example, in Figure 3-2, notice that 4 is to the left of 7 on the number line. We say, "four is less than seven," and we write $4 < 7$. Equivalently we also say "seven is greater than four" and write $7 > 4$. Because 4 is less than 7, there is a natural number that can be added to 4 to get 7, namely, 3. Thus, $4 < 7$ because $4 + 3 = 7$. This discussion can be generalized to form the following definition of *less than*.

Definition of Less Than

For any whole numbers a and b, a is **less than** b, written $a < b$, if, and only if, there exists a natural number k such that $a + k = b$.

Historical Note

The symbol "+" first appeared in a 1417 manuscript and was a short way of writing the Latin word *et*, which means "and." The word *minus* means "less" in Latin. First written as an *m*, it was later shortened to a horizontal bar. In a printed book in 1489 the symbols + and − referred only to surpluses and deficits in business problems.   ●

Sometimes equality is combined with the inequalities, greater than and less than, to give the relations **greater than or equal to** and **less than or equal to**, denoted $\geq$ and $\leq$, respectively. Thus, $a \leq b$ means $a < b$ or $a = b$. The emphasis with respect to these symbols is on the meaning of *or* in logic. Thus $3 \leq 5$, $5 \geq 3$, and $3 \geq 3$ are all true statements.

Whole-Number Addition Properties

Any time two whole numbers are added, a unique whole number is obtained. We say that "the set of whole numbers is *closed under addition*" and has the following property:

> **Theorem 3-1:   Closure Property of Addition of Whole Numbers**
>
> If a and b are whole numbers, then $a + b$ is a whole number.

The closure property implies that the sum of two whole numbers *exists* and that the sum is a *unique whole number*; for example, $5 + 2$ is a unique whole number and that number is 7.

 NOW TRY THIS 3-3

Determine whether each of the following sets is closed under addition:

a. $E = \{2n \mid n \in W\}$
b. $F = \{2n + 1 \mid n \in W\}$
c. $G = \{0, 2\}$

Figure 3-4(a) shows two additions. Pictured above the number line is $3 + 5$ and below the number line is $5 + 3$. The sums are the same. Figure 3-4(b) shows the same sums obtained with colored rods. Both illustrations in Figure 3-4 demonstrate the idea that when two whole numbers are added in either order we obtain the same answer. This property is true in general and is the *commutative property of addition of whole numbers*.

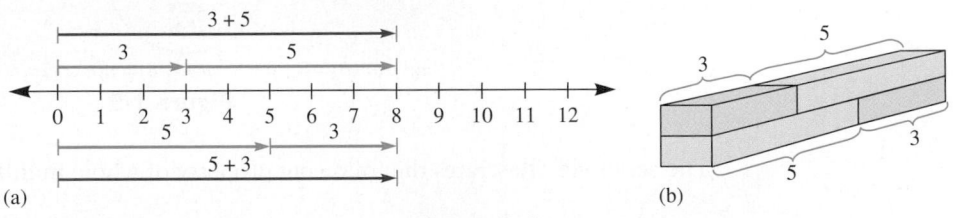

(a)   (b)

Figure 3-4

> **Theorem 3-2:   Commutative Property of Addition of Whole Numbers**
>
> If a and b are any whole numbers, then $a + b = b + a$.

We say that "addition of whole numbers is commutative." The word *commutative* is derived from *commute*, which means "to interchange."

The commutative property of addition of whole numbers is not obvious to many young children. They may be able to find the sum $9 + 2$ and not be able to find the sum $2 + 9$. Using *counting on*, $9 + 2$ can be computed by starting at 9 and then counting on two more as "ten" and "eleven." To compute $2 + 9$ without the commutative property, the *counting on* is more involved. Students need to understand that $2 + 9$ is another name for $9 + 2$.

NOW TRY THIS 3-4

a. Use the set model to show the commutative property for $3 + 5 = 5 + 3$.

b. Use the number line model to show that $4 + 2 = 2 + 4$.

Another property of addition is demonstrated when selecting the order in which to add three or more numbers. For example, we could compute $24 + 8 + 2$ by grouping the 24 and the 8 together: $(24 + 8) + 2 = 32 + 2 = 34$. (The parentheses indicate that the first two numbers are grouped together.) We might also recognize that it is easy to add any number to 10 and compute it as $24 + (8 + 2) = 24 + 10 = 34$. This example illustrates the *associative property of addition of whole numbers*. The word *associative* is derived from the word *associate*, which means "to unite."

> **Theorem 3-3:   Associative Property of Addition of Whole Numbers**
>
> If a, b, and c are whole numbers, then $(a + b) + c = a + (b + c)$.

When several numbers are being added, the parentheses are usually omitted because the grouping does not alter the result.

Another property of addition of whole numbers is seen when one addend is 0. In Figure 3-5, set A has 5 blocks and set B has 0 blocks. The union of sets A and B has only 5 blocks.

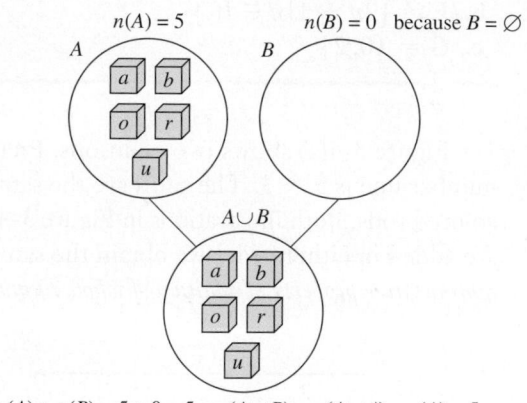

$$n(A) + n(B) = 5 + 0 = 5 = n(A \cup B) = n(A \cup \phi) = n(A) = 5$$

Figure 3-5

The set model illustrates the following property of whole numbers:

> **Theorem 3-4:   Identity Property of Addition of Whole Numbers**
>
> There is a unique whole number 0, the **additive identity**, such that for any whole number a, $a + 0 = a = 0 + a$.

Notice how the commutative, associative, and identity properties are introduced on the grade 3 student page. Work through questions 6–19.

School Book Page

Commutative (Order) Property of Addition: You can add numbers in any order and the sum will be the same.

$$7 + 5 = 5 + 7$$

Identity (Zero) Property of Addition: The sum of zero and any number is that same number.

$$5 + 0 = 5$$

Associative (Grouping) Property of Addition: You can group addends in any way and the sum will be the same.

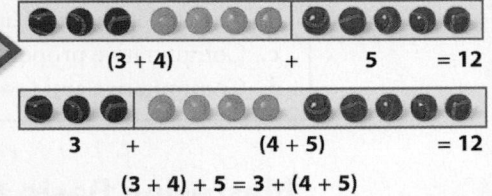

$$(3 + 4) + 5 = 12$$

$$3 + (4 + 5) = 12$$

$$(3 + 4) + 5 = 3 + (4 + 5)$$

Parentheses, (), show what to add first.

Independent Practice

Write each missing number.

6. ▨ + 8 = 8 + 2

7. 19 + ▨ = 19

8. (3 + ▨) + 2 = 2 + 8

9. 4 + (2 + 3) = 4 + ▨

10. 7 + 3 = ▨ + 7

11. ▨ + 25 = 25

12. (3 + ▨) + 6 = 3 + (4 + 6)

13. (6 + 2) + ▨ = 8 + 7

14. (7 + ▨) + 6 = 7 + 6

15. (5 + 6) + 3 = ▨ + (5 + 6)

Problem Solving

16. Reasoning What property of addition is shown in the number sentence 3 + (6 + 5) = (6 + 5) + 3? Explain.

17. Draw objects of 2 different colors to show that 4 + 3 = 3 + 4.

18. A lionfish has 13 spines on its back, 2 near the middle of its underside, and 3 on its underside near its tail. Write two different number sentences to find how many spines a lionfish has in all. What property did you use?

19. Which number sentence matches the picture?

   A 3 + 8 = 11

   B 11 + 0 = 11

   C 11 − 8 = 3

   D 11 − 3 = 8

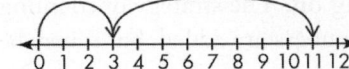

Lesson 2-1   **33**

EXAMPLE 3-1

Which properties are illustrated in each of the following?

a. $5 + 7 = 7 + 5$
b. $1001 + 733$ is a unique whole number.
c. $(3 + 5) + 7 = (5 + 3) + 7$
d. $(8 + 5) + 2 = 2 + (8 + 5) = (2 + 8) + 5$

Solution

a. Commutative property of addition
b. Closure property of addition
c. Commutative property of addition
d. Commutative and associative properties of addition

Mastering Basic Addition Facts

Certain mathematical facts are *basic addition facts*. Basic addition facts are those involving a single digit plus a single digit. In the *Dennis the Menace* cartoon, it seems that Dennis has not mastered basic addition facts.

DENNIS THE MENACE

"MAKE UP YOUR MIND. FIRST YOU TELL ME 3 PLUS 3 IS SIX, AND NOW YOU SAY 4 PLUS 2 IS SIX!"

Dennis the Menace copyright © 1990 and 2011 Hank Ketcham Enterprises. Distributed by North American Syndicate

One method of learning the basic facts is to organize them according to different derived fact strategies. Several strategies are listed below.

1. **Counting on.** The strategy of counting on from one of the addends can be used any time whole numbers are added, but it may be inefficient. It is usually used when one addend is 1, 2, or 3. For example, in the cartoon Dennis could have computed $4 + 2$ by starting at 4 and then counting on 5, 6.

2. **Doubles.** Use of *doubles* such as $3 + 3$ in the cartoon receive special attention. After students master doubles, *doubles* $+ 1$ and *doubles* $+ 2$ can be learned easily. For example, if a student knows $6 + 6 = 12$, then $6 + 7$ is $(6 + 6) + 1$, or 1 more than the double of 6, or 13.

3. **Making 10.** Another strategy is that of *making 10* and then adding any leftover. For example, think of 8 + 5 as shown in Figure 3-6. Notice that we are really using the associative property of addition.

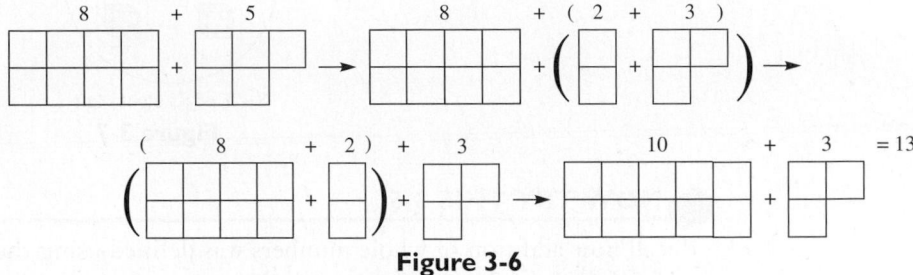

Figure 3-6

4. **Counting back.** The strategy of *counting back* is usually used when one number is 1 or 2 less than 10. For example, because 9 is 1 less than 10, then 9 + 7 is 1 less than 10 + 7. In symbols, this is $9 + 7 = (10 + 7) - 1 = 17 - 1 = 16$.

Many basic facts might be classified under more than one strategy. For example, 9 + 8 can use *making 10* as $9 + 8 = 9 + (1 + 7) = (9 + 1) + 7 = 10 + 7 = 17$ or use a *double* + 1 as $(8 + 8) + 1$.

Subtraction of Whole Numbers

In the grade 1 *Focal Points* we find the following:

| "By comparing a variety of solution strategies children relate addition and subtraction as inverse operations." (p. 13)

In elementary school, operations that "undo" each other are **inverse operations**. Subtraction is the inverse operation for addition. It is sometimes hard for children to understand this inverse relationship, as the cartoon demonstrates.

B.C.

Subtraction of whole numbers models include the *take-away* model, the *missing-addend* model, the *comparison* model, and the *number-line (measurement)* model.

Take-Away Model

To model addition, we imagine a second set of objects as being joined to a first set, but in subtraction, we imagine a second set as being *taken away* from a first set. For example, suppose from a set of 8 blocks 3 are taken away. This is illustrated in Figure 3-7 and the process is recorded as $8 - 3 = 5$.

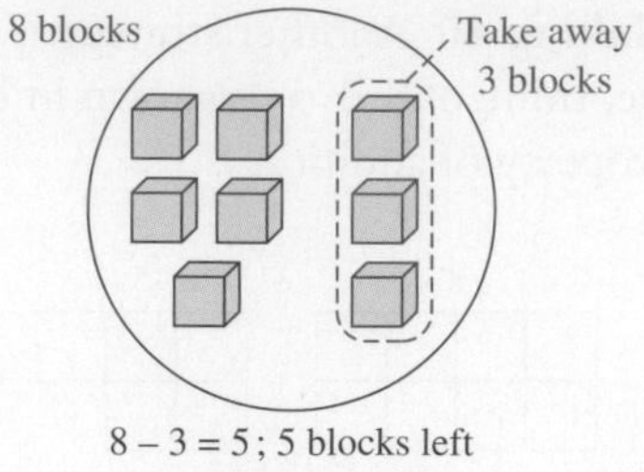

8 blocks   Take away
                3 blocks

$8 - 3 = 5$; 5 blocks left

Figure 3-7

NOW TRY THIS 3-5

Recall how addition of whole numbers was defined using the concept of union of two disjoint sets. Similarly, write a definition of subtraction of whole numbers using the concepts of subsets and set difference.

Missing-Addend Model

A second model for subtraction, the *missing-addend* model, relates subtraction and addition. In Figure 3-8(a) suppose set A has 8 blocks but only 3 are visible. How many blocks must be in the set to have a total of 8? The answer is $8 - 3$, but this can also be thought of as the number of blocks that could be added to 3 blocks in order to get 8 blocks, that is,

$$3 + \boxed{8 - 3} = 8$$

The number $8 - 3$, or 5, is the **missing-addend** in the equation

$$3 + \square = 8$$

We can relate the *missing-addend* approach to a number line. The subtraction $8 - 3$ is illustrated in Figure 3-8(a) using sets and in Figure 3-8(b) using the number line.

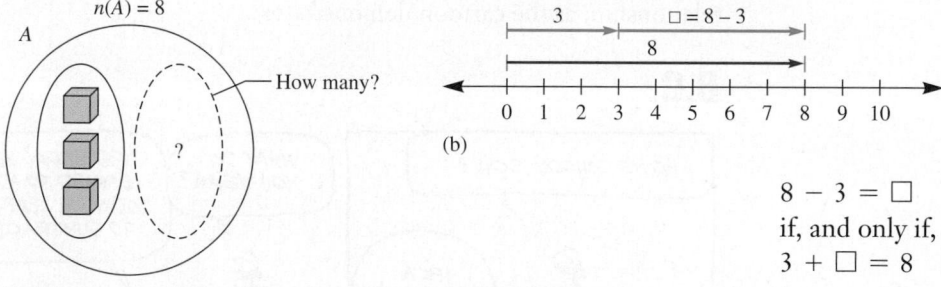

(a)

(b)

$8 - 3 = \square$
if, and only if,
$3 + \square = 8$

Figure 3-8

The missing-addend model gives elementary school students an opportunity to begin algebraic thinking because the concept of the unknown plays a key role in this model.

Cashiers often use the missing-addend model. For example, if the bill for a movie is $8 and you pay $10, the cashier might calculate the change by saying "8 and 2 is 10." This idea can be generalized: For any whole numbers a and b such that $a \geq b$, $a - b$ is the unique whole number such that $b + (a - b) = a$. That is, $a - b$ is the unique solution of the equation $b + \square = a$. The definition can be written as follows:

Definition of Subtraction of Whole Numbers

For any whole numbers a and b such that $a \geq b$, $a - b$ is the unique whole number c such that $b + c = a$.

An alternative but equivalent definition can be given without stipulating that $a \geq b$: "For any whole numbers a and b, $a - b$ is the unique whole number c, if it exists, such that $b + c = a$." Notice that "$a \geq b$" has been deleted and "if it exists" added.

Refer to the partial student page on p. 108 to see how grade 3 students are shown how addition and subtraction are related using a **fact family**.

Comparison Model

Another way to consider subtraction is by using a *comparison* model. Suppose Juan has 8 blocks and Susan has 3 blocks and we want to know how many more blocks Juan has than Susan. We can pair Susan's blocks with some of Juan's blocks, as shown in Figure 3-9, and determine that Juan has 5 more blocks than Susan. We also write this as $8 - 3 = 5$.

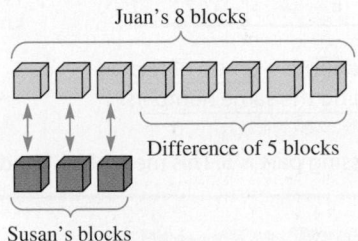

Figure 3-9

Number-Line (Measurement) Model

Subtraction of whole numbers can also be modeled on a number line using directed arrows, as suggested in Figure 3-10, which shows that $5 - 3 = 2$.

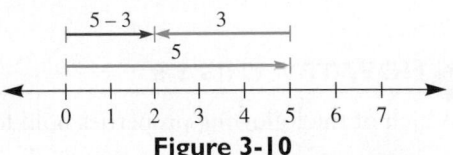

Figure 3-10

In each of the following four problems the answer is 5 but each can be thought of using a different model.

1. **Take-away model.** Al had \$9 and spent \$4. How much did he have left?
2. **Missing-addend model.** Al has read 4 chapters of a 9-chapter book. How many chapters does he have left to read?
3. **Comparison model.** Al has 9 books and Betty has 4 books. How many more books does Al have than Betty?
4. **Number-line model.** Al walked 9 miles from his campsite. Then he turned around and walked 4 miles back and stopped. How far from his campsite was he at that point?

Properties of Subtraction

In an attempt to find $3 - 5$, we use the definition of subtraction: $3 - 5 = c$ if, and only if, $c + 5 = 3$. Since there is no whole number c that satisfies the equation, $3 - 5$ is not meaningful in the set of whole numbers. In general, if $a < b$, then $a - b$ is not meaningful in the set of whole numbers. Therefore, the set of whole numbers is not closed under subtraction.

School Book Page

Another Example Subtract to find a missing addend.

Rick plans on making 13 flags. How many more flags does he need?

The parts and the whole show how addition and subtraction are related.

13 flags in all

8	?

$8 + \blacksquare = 13$

You can write a fact family when you know the parts and the whole.

A fact family is a group of related facts using the same numbers.

$5 + 8 = 13$ $13 - 8 = 5$
$8 + 5 = 13$ $13 - 5 = 8$

The missing part is 5. This means Rick needs to make 5 more flags.

 NOW TRY THIS 3-6

Which of the following properties hold for subtraction of whole numbers? Explain.

a. Closure property
b. Associative property
c. Commutative property
d. Identity property

Using Whole-Number Addition and Subtraction in Equations

Sentences such as $9 + 5 = \square$ and $12 - \Delta = 4$ can be true or false depending on the values of $\square$ and Δ. For example, if $\square = 10$, then $9 + 5 = \square$ is false. If $\Delta = 8$, then $12 - \Delta = 4$ is true. A value that makes the equation true, it is a **solution** to the equation. Thus $8 - 5$ or 3 is the solution to $\square + 5 = 8$.

 NOW TRY THIS 3-7

Find the solution for each of the following where x is a whole number:

a. $x + 8 = 13$ **b.** $15 - x = 8$ **c.** $x > 9$ and $x < 11$
d. $x + b = a$ if $a \geq b$, a and b are whole numbers.

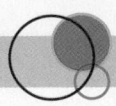

Assessment 3-1A

1. For which of the following is it true that $n(A) + n(B) = n(A \cup B)$?
   a. $A = \{a, b, c\}, B = \{d, e\}$
   b. $A = \{a, b, c\}, B = \{b, c\}$
   c. $A = \{a, b, c\}, B = \varnothing$

2. If $n(A) = 3$, $n(B) = 5$, and $n(A \cup B) = 6$, what do you know about $n(A \cap B)$?

3. Give an example to show why, in the definition of addition, sets A and B must be disjoint.

4. If $n(A) = 3$ and $n(A \cup B) = 6$, answer the following:
   a. What are the possible values for $n(B)$?
   b. If in addition $A \cap B = \varnothing$, what are the possible values of $n(B)$?

5. Which of the following sets are closed under addition? If a set is not closed under addition explain why not.
   a. $B = \{0\}$
   b. $T = \{3n \mid n \in W\}$
   c. $N = \{1, 2, 3, 4, 5, \ldots\}$
   d. $V = \{3, 5, 7\}$
   e. $A = \{x \mid x \in W \text{ and } x > 10\}$
   f. $C = \{0, 1\}$

6. Set A is closed under addition and contains the numbers 2 and 3. List six additional numbers, three even and three odd, that must be in A.

7. Each of the following is an example of one of the properties of whole-number addition. Fill in the blank to make a true statement and identify the property or properties.
   a. $9 + 1 = \underline{} + 9$
   b. $7 + (3 + 5) = (3 + 5) + \underline{}$
   c. $a + \underline{} = a$
   d. $7 + (3 + 5) = (3 + \underline{}) + 5$

8. Each of the following illustrates a property of addition of whole numbers. Identify the property illustrated.
   a. $6 + 3 = 3 + 6$
   b. $(6 + 3) + 5 = 6 + (3 + 5)$
   c. $(6 + 3) + 5 = (3 + 6) + 5$
   d. $5 + 0 = 5 = 0 + 5$
   e. $5 + 0 = 0 + 5$
   f. $(a + c) + d = a + (c + d)$

9. In the definition of *less than*, can the natural number k be replaced by the whole number k? Why or why not?

10. Give a definition of *less than* and *greater than* using the concept of subtraction.

11. Find the next three terms in each of the following arithmetic sequences:
    a. 8, 13, 18, 23, 28, ___, ___, ___
    b. 98, 91, 84, 77, 70, 63, ___, ___, ___

12. If A, B, and C each stand for a different single digit from 1 to 9, answer the following if
$$A + B = C$$
    a. What is the greatest digit that C could be?
    b. What is the greatest digit that A could be?

c. What is the least digit that C could be?
    d. If A, B, and C are even, what number(s) could C be?
    e. If C is 5 more than A, what number(s) could B be?
    f. If A is 3 times B, what number(s) could C be?
    g. If A is odd and A is 5 more than B, what number(s) could C be?

13. A special domino set contains all number pairs from double-0 to double-8, with each number pair occurring only once. For example, the following domino counts as 1-4 and 4-1. How many dominos are in the set?

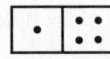

14. a. At a volleyball game, the players stood in a row ordered by height. If Kent is shorter than Mischa, Sally is taller than Mischa, and Vera is taller than Sally, who is the tallest and who is the shortest?
    b. Write possible heights for the players in part (a).

15. Rewrite each of the following subtraction problems as an equivalent addition problem:
    a. $9 - 7 = x$
    b. $x - 6 = 3$
    c. $9 - x = 2$

16. What conditions, if any, must be placed on a, b, and c in each of the following to make sure that the operations can be performed within the set of whole numbers.
    a. $a - b$
    b. $a - (b - c)$

17. Show that each of the following is true. Give a property of addition to justify each step in your argument.
    a. $x + (y + z) = z + (x + y)$
    b. $x + (y + z) = z + (x + z)$

18. Illustrate $8 - 5 = 3$ using each of the following models:
    a. Take-away
    b. Missing addend
    c. Comparison
    d. Number line

19. Find the solution in the set W for each of the following:
    a. $3 + (4 + 7) = (3 + x) + 7$
    b. $8 + 0 = x$
    c. $5 + 8 = 8 + x$
    d. $x + 8 = 12 + 5$
    e. $x + 8 = 5 + (x + 3)$
    f. $x - 2 = 9$
    g. $x - 3 = x + 1$
    h. $0 + x = x + 0$

20. Kelsey has a marbles, Gena has b marbles and Noah has c marbles. If Kelsey has more marbles than Gena and Noah combined, write an expression (using a, b, and c) that shows how many more marbles Kelsey has than Gena and Noah combined.

Assessment 3-1B

1. For which of the following is it true that $n(A) + n(B) = n(A \cup B)$?
   a. $A = \{a, b\}, B = \{d, e\}$
   b. $A = \{a, b, c\}, B = \{b, c, d\}$
   c. $A = \{a\}, B = \varnothing$
2. If $n(A) = 3, n(B) = 5$, and $n(A \cap B) = 1$, what do you know about $n(A \cup B)$?
3. If $n(A) - n(B)$ is defined as $n(A - B)$, what must $A \cap B$ equal?
4. If $n(B) = 4$ and $n(A \cup B) = 6$, answer the following.
   a. What are the possible values of $n(A)$?
   b. If in addition $n(A \cap B) = 0$, what are the possible values of $n(A)$?
5. Explain whether the following given are closed under addition:
   a. $B = \{0, 1\}$
   b. $T = \{0, 4, 8, 12, 16, \ldots\}$
   c. $F = \{5, 6, 7, 8, 9, 10, \ldots\}$
   d. $\{x \mid x \in W \text{ and } x > 100\}$
6. a. Set A is closed under addition and contains the numbers 2, 5, and 8. List six other elements that must be in A.
   b. Set A contains the element 1. What other whole numbers must be in set A for it to be closed under addition?
7. Each of the following is an example of one of the properties of whole-number addition. Fill in the blank to make a true statement and identify the property.
   a. $3 + 4 = \underline{\quad} + 3$
   b. $5 + (4 + 3) = (4 + 3) + \underline{\quad}$
   c. $8 + \underline{\quad} = 8$
   d. $3 + (4 + 5) = (3 + \underline{\quad}) + 5$
   e. $3 + 4$ is a unique $\underline{\quad}$ number.
8. Each of the following is an example of one of the properties for addition of whole numbers. Identify the property illustrated.
   a. $6 + 8 = 8 + 6$
   b. $(6 + 3) + 0 = 6 + 3$
   c. $(6 + 8) + 2 = (8 + 6) + 2$
   d. $(5 + 3) + 2 = 5 + (3 + 2)$
9. Complete the following statement: $a \leq b$ if and only if $b = a + k$ for some $\underline{\qquad\qquad}$ number k.
10. Use the concept of subtraction to define $a \geq b$.
11. Find the next three terms in each of the following arithmetic sequences:
    a. $5, 12, 19, 26, 33, \underline{\quad}, \underline{\quad}, \underline{\quad}$
    b. $63, 59, 55, 51, 47, \underline{\quad}, \underline{\quad}, \underline{\quad}$
12. If A, B, C, and D each stand for a different single digit from 1 to 9, answer each of the following if

$$\begin{array}{r} A \\ + B \\ \hline CD \end{array}$$

a. What is the value of C?
b. Can D be 1? Why?
c. If D is 7, what values can A be?
d. If A is 6 greater than B, what is the value of D?

13. a. A domino set contains all number pairs from double-0 to double-6, with each number pair occurring only once; for example, the following domino counts as 2-4 and 4-2. How many dominoes are in the set?

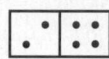

   b. When considering the sum of all dots on a single domino in an ordinary set of dominoes, explain how the commutative property might be important.
14. If $a < b, c > b, d > c$, and $c < e < d$, order the letters from the least to the greatest.
15. Rewrite each of the following subtraction problems as an equivalent addition problem:
    a. $9 - 3 = x$
    b. $x - 5 = 8$
    c. $11 - x = 2$
16. What conditions, if any, must be placed on a, b, and c in each of the following to make sure the operations can be performed within the set of whole numbers?
    a. $b - a$
    b. $b - (a - 3)$
17. Show that each of the following is true. Give a property of addition to justify each step.
    a. $a + (b + c) = c + (a + b)$
    b. $a + (b + c) = (c + b) + a$
18. Illustrate $7 - 3 = 4$ using each of the following models:
    a. Take-away
    b. Missing addend
    c. Comparison
    d. Number line
19. Find the solution in the set W for each of the following:
    a. $12 - x = x + 6$
    b. $(9 - x) - 6 = 1$
    c. $3 + x = x + 3$
    d. $11 - x = 0$
    e. $14 - x = 7 - x$
    f. $x - 3 = 17$
    g. $x + 3 = x - 1$
    h. $0 + x = x - 0$
20. Rob has 11 pencils. Kelly has 5 pencils. Which number sentence shows how many more pencils Rob has than Kelly?
    i. $11 + 5 = 16$
    ii. $16 - 5 = 11$
    iii. $11 - 5 = 6$
    iv. $11 - 6 = 5$

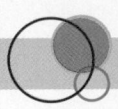 **Mathematical Connections 3-1**

Communication

1. In a survey of 52 students, 22 said they were taking algebra and 30 said they were taking biology. Are there necessarily 52 students taking algebra or biology? Why?
2. To find $9 + 7$, a student says she thinks of $9 + 7$ as $9 + (1 + 6) = (9 + 1) + 6 = 10 + 6 = 16$. What property or properties is she using?
3. Explain in two different ways why if $a > b$ and c is a whole number, then $a + c > b + c$.
4. When subtraction and addition appear in an expression without parentheses, it is agreed that the operations are performed in order of their appearance from left to right. Taking this into account, and assuming that all expressions are meaningful, use an appropriate model for subtraction to explain why each of the following is true:
   a. $a - b - c = a - c - b$
   b. $a - b - c = a - (b + c)$
   c. $(a + b) - c = a + (b - c)$
   d. $a + b - c = a - c + b$
5. Explain whether it is important for elementary students to learn more than one model for the operations of addition and subtraction.
6. Do elementary students still have to learn their basic facts when the calculator is part of the curriculum? Explain.
7. Explain how the model shown can be used to illustrate each of the following addition and subtraction facts:
   a. $9 + 4 = 13$ b. $4 + 9 = 13$
   c. $4 - 13 - 9$ d. $9 = 13 - 4$

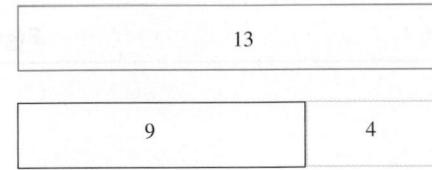

8. How are addition and subtraction related? Explain.
9. Why is 0 not an identity for subtraction? Explain.

Open-Ended

10. Describe any model not in this text that you might use to teach addition to students.
11. Suppose $A \subseteq B$. If $n(A) = a$ and $n(B) = b$, then $b - a$ is defined as $n(B - A)$. Choose two sets A and B and illustrate this definition.
12. a. Create a word problem in which the set model would be more appropriate to show $25 + 8 = 33$.
    b. Create a word problem in which the number-line (measurement) model would be more appropriate to show $25 + 8 = 33$.
    c. Create a word problem or a diagram that explains why $a - (b - c) = a - b + c$ (where all operations result in whole numbers).

Cooperative Learning

13. Use the basic addition fact table for whole numbers shown and discuss with your group each of the following.

+	0	1	2	3	4	5	6	7	8	9
0	0	1	2	3	4	5	6	7	8	9
1	1	2	3	4	5	6	7	8	9	10
2	2	3	4	5	6	7	8	9	10	11
3	3	4	5	6	7	8	9	10	11	12
4	4	5	6	7	8	9	10	11	12	13
5	5	6	7	8	9	10	11	12	13	14
6	6	7	8	9	10	11	12	13	14	15
7	7	8	9	10	11	12	13	14	15	16
8	8	9	10	11	12	13	14	15	16	17
9	9	10	11	12	13	14	15	16	17	18

   a. How does the table illustrate the closure property?
   b. How does the table illustrate the commutative property?
   c. How does the table illustrate the identity property?
   d. How do the addition properties help students learn their basic facts?
14. Suppose that a number system used only four symbols, a, b, c, and d, and the operation Δ; and the system operated as shown in the table. Discuss with your group each of the following:

Δ	a	b	c	d
a	a	b	c	d
b	b	c	d	a
c	c	d	a	b
d	d	a	b	c

   a. Is the system closed? Why?
   b. Is the operation commutative? Why?
   c. Does the operation have an identity? If so, what is it?
   d. Try several examples to investigate the associative property of this operation.
15. Have each person in your group choose a different grade textbook that does whole number operations and report on when and how subtraction of whole numbers is introduced. Compare with different ways subtraction is introduced in this section.

Questions from the Classroom

16. A student says that 0 is the identity for subtraction, because if $a \in W$, then $a - 0 = a$. How do you respond?

17. A student asks why we use subtraction to determine how many more pencils Sam has than Karly if nothing is being taken away. How do you respond?

18. A student claims that the set of whole numbers is closed with respect to subtraction. To show this is true, she shows that $8 - 5 = 3, 5 - 2 = 3, 6 - 1 = 5$, and $12 - 7 = 5$ and says she can show examples like this all day that yield whole numbers when the subtraction is performed. How do you respond?

19. John claims that he can get the same answer to the problem below by adding up (begin with $4 + 7$) or by adding down (begin with $8 + 7$). He wants to know why and if this works all the time. How do you respond?

$$\begin{array}{r} 8 \\ 7 \\ + 4 \\ \hline \end{array}$$

Trends in Mathematics and Science Study (TIMSS) Questions

Ali had 50 apples. He sold some and then had 20 left. Which of these is a number sentence that shows this?

a. $\square - 20 = 50$
b. $20 - \square = 50$
c. $\square - 50 = 20$
d. $50 - \square = 20$

TIMSS, Grade 4, 2007

Which of these has the same value as 342?

a. $3,000 + 400 + 2$
b. $300 + 40 + 2$
c. $30 + 4 + 2$
d. $3 + 4 + 2$

TIMSS, Grade 4, 2003

National Assessment of Educational Progress (NAEP) Questions

$\square - 8 = 21$

What number should be put in the box to make the number sentence above true?

Answer: _____

NAEP, Grade 4, 2009

BRAIN TEASER Use Figure 3-11 to design an *unmagic square*. That is, use each of the digits 1, 2, 3, 4, 5, 6, 7, 8, and 9 exactly once so that every column, row, and diagonal adds to a different sum.

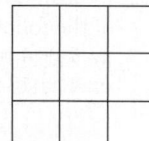

Figure 3-11

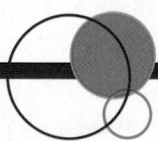

3-2 Algorithms for Whole-Number Addition and Subtraction

In the grade 2 *Focal Points*, we find the following regarding fluency with multidigit addition and subtraction:

> Children develop, discuss, and use efficient, accurate, and generalizable methods to add and subtract multidigit whole numbers. They select and apply appropriate methods to estimate sums and differences or calculate them mentally, depending on the context and numbers involved. They develop fluency with efficient procedures, including standard algorithms, for adding and subtracting whole numbers, understand why the procedures work (on the basis of place value and properties of operations), and use them to solve problems. (p. 14)

The grade 2 *Common Core Standards* point out that students at this level should be able to

- Add and subtract within 1000, using concrete models or drawings and strategies based on place value, properties of operations, and/or the relationship between addition and subtraction; relate the strategy to a written method. (p. 19)
- Understand that in adding or subtracting three-digit numbers, one adds or subtracts hundreds and hundreds, tens and tens, ones and ones, and sometimes it is necessary to compose or decompose tens or hundreds. (p. 19)

The previous section introduced the operations of addition and subtraction of whole numbers, and now, as stated in the *Focal Points* and the *Common Core Standards*, it is time to focus on *computational fluency*—having and using efficient and accurate methods for computing. Standard algorithms are one means to achieve this fluency. An **algorithm** is a step-by-step procedure for solving a problem or performing a computation. The word is derived from the name of the Persian mathematician, astronomer, and geographer al-Khwarizmi (780–850 CE).

This section focuses on developing and understanding algorithms involving addition and subtraction.

Addition Algorithms

In teaching mathematics to young children, we support the transition from concrete to abstract thinking. To help children understand the use of paper-and-pencil algorithms, they should explore addition using manipulatives. If children can touch and move around items such as chips, bean sticks, and an abacus or use base-ten blocks, they often proceed naturally on their own to the creation of algorithms for addition. In what follows, we use base-ten blocks to illustrate the development of an algorithm for whole-number addition.

To add 14 + 23, we start with a concrete model as in Figure 3-12(a), move to the expanded algorithm in Figure 3-12(b), and then to the standard algorithm in Figure 3-12(c).

E-Manipulative Activity

Practice using base-ten blocks to perform additions can be found in the *Adding Blocks* activity.

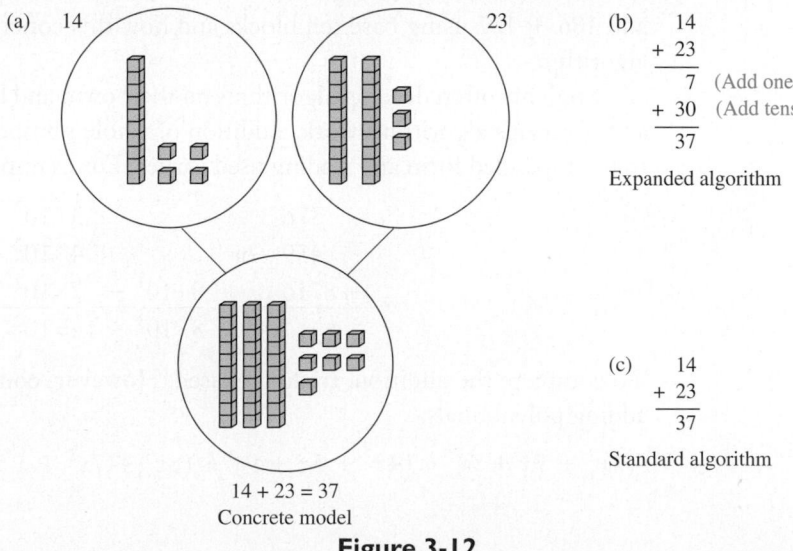

Figure 3-12

A more formal justification for this addition, not usually presented at the elementary school level, is the following:

$$14 + 23 = (1 \cdot 10 + 4) + (2 \cdot 10 + 3) \qquad \text{Expanded form}$$
$$= (1 \cdot 10 + 2 \cdot 10) + (4 + 3) \qquad \text{Commutative and associative properties of addition}$$
$$= (1 + 2)10 + (4 + 3) \qquad \text{Distributive property of multiplication over addition}$$
$$= 3 \cdot 10 + 7 \qquad \text{Single-digit addition facts}$$
$$= 37 \qquad \text{Place value}$$

On the student page (page 115) is an example of adding two-digit numbers with regrouping using base-ten blocks. Kim's way leads to the *expanded algorithm* and Henry's way leads to the *standard algorithm*. Each algorithm is discussed in more detail on page 115. Notice that on the student page, students are asked to estimate their answers before performing an algorithm. This is good practice and leads to the development of number sense and also makes students consider whether their answers are reasonable. Study the student page and answer the *Talk About It* questions.

Once children have mastered regrouping models, they should be ready to use the expanded and standard algorithms. Figure 3-13 shows the computation 37 + 28 using both algorithms. In Figure 3-13(b), notice that when there were more than 10 ones, we regrouped 10 ones as a ten and then added the tens. The words *regroup* or *trade* are now commonly used in the elementary classroom to describe what was called *carrying* in the past.

(a)   37

    + 28

    15 (Add ones)

  + 50 (Add tens)

    65

$$\text{(a)} \quad \begin{array}{r} 37 \\ +\,28 \\ \hline 15 \\ +\,50 \\ \hline 65 \end{array} \quad \begin{array}{l} \\ \\ \text{(Add ones)} \\ \text{(Add tens)} \\ \end{array} \qquad \text{(b)} \quad \begin{array}{r} \overset{1}{3}7 \\ +\,28 \\ \hline 65 \end{array} \quad \text{(Add the ones, regroup, and add the tens)}$$

Expanded algorithm   Standard algorithm

Figure 3-13

Next we add two three digit numbers involving two regroupings. Figure 3-14 shows how to add 186 + 127 using base ten blocks and how this concrete model carries over to the standard algorithm.

Students often develop algorithms on their own, and learning can occur by investigating how and if various algorithms work. Addition of whole numbers using blocks has a natural carryover to the expanded form and trading used earlier. For example, consider the following addition:

$$\begin{array}{r} 376 \\ 459 \quad \text{or} \\ +\,8716 \\ \hline \end{array} \qquad \begin{array}{r} 3 \cdot 10^2 + 7 \cdot 10 + 6 \\ 4 \cdot 10^2 + 5 \cdot 10 + 9 \\ 8 \cdot 10^3 + 7 \cdot 10^2 + 1 \cdot 10 + 6 \\ \hline 8 \cdot 10^3 + 14 \cdot 10^2 + 13 \cdot 10 + 21 \end{array}$$

To complete the addition, trading is used. However, consider an analogous algebra problem of adding polynomials:

$$(3x^2 + 7x + 6) + (4x^2 + 5x + 9) + (8x^3 + 7x^2 + x + 6) \quad \text{or} \qquad \begin{array}{r} 3x^2 + 7x + 6 \\ 4x^2 + 5x + 9 \\ +\,8x^3 + 7x^2 + x + 6 \\ \hline 8x^3 + 14x^2 + 13x + 21 \end{array}$$

Note that if $x = 10$, the addition is the same as given earlier. Also note that knowledge of place value in addition problems aids in algebraic thinking. Next we explore several algorithms that have been used throughout history.

School Book Page ADDING TWO-DIGIT NUMBERS

Lesson 3-1

Key Idea
You can break apart numbers, using place value, to add.

Vocabulary
• regroup

Think It Through
• I should **estimate** so I will know if my answer is reasonable.
• I can **use place-value blocks** to show addition.

Adding Two-Digit Numbers

LEARN

✔ WARM UP
Use mental math.
1. 48 + 20   2. 63 + 11
3. 71 + 8   4. 53 + 5

How do you add two-digit numbers?

Example

Cal counted 46 ladybugs on a log and 78 more on some bushes. How many ladybugs did he count all together?

Find 46 + 78.

Estimate: 46 rounds to 50. 78 rounds to 80.

50 + 80 = 130, so the answer should be about 130.

What You **Think**		What You **Write**

Kim's Way
• Add the ones.
  6 + 8 = 14 ones
• Add the tens.
  4 tens + **7** tens =
  11 tens = **110**
• Find the sum.

11 tens   14 ones

$$\begin{array}{r} 46 \\ +\ 78 \\ \hline 14 \\ 110 \\ \hline 124 \end{array}$$

Henry's Way
• Add the ones.
  6 + 8 = 14 ones
• **Regroup 14** ones into **1** ten **4** ones.
• Add the tens.
  1 ten + **4** tens +
  7 tens = **12** tens
• Find the sum.

14 ones = 1 ten 4 ones

$$\begin{array}{r} 1 \\ 46 \\ +\ 78 \\ \hline 124 \end{array}$$

Cal counted 124 ladybugs all together.

✔ Talk About It

1. Why did Henry write a small 1 above the 4 in the tens place?

2. Why should you estimate when adding two-digit numbers?

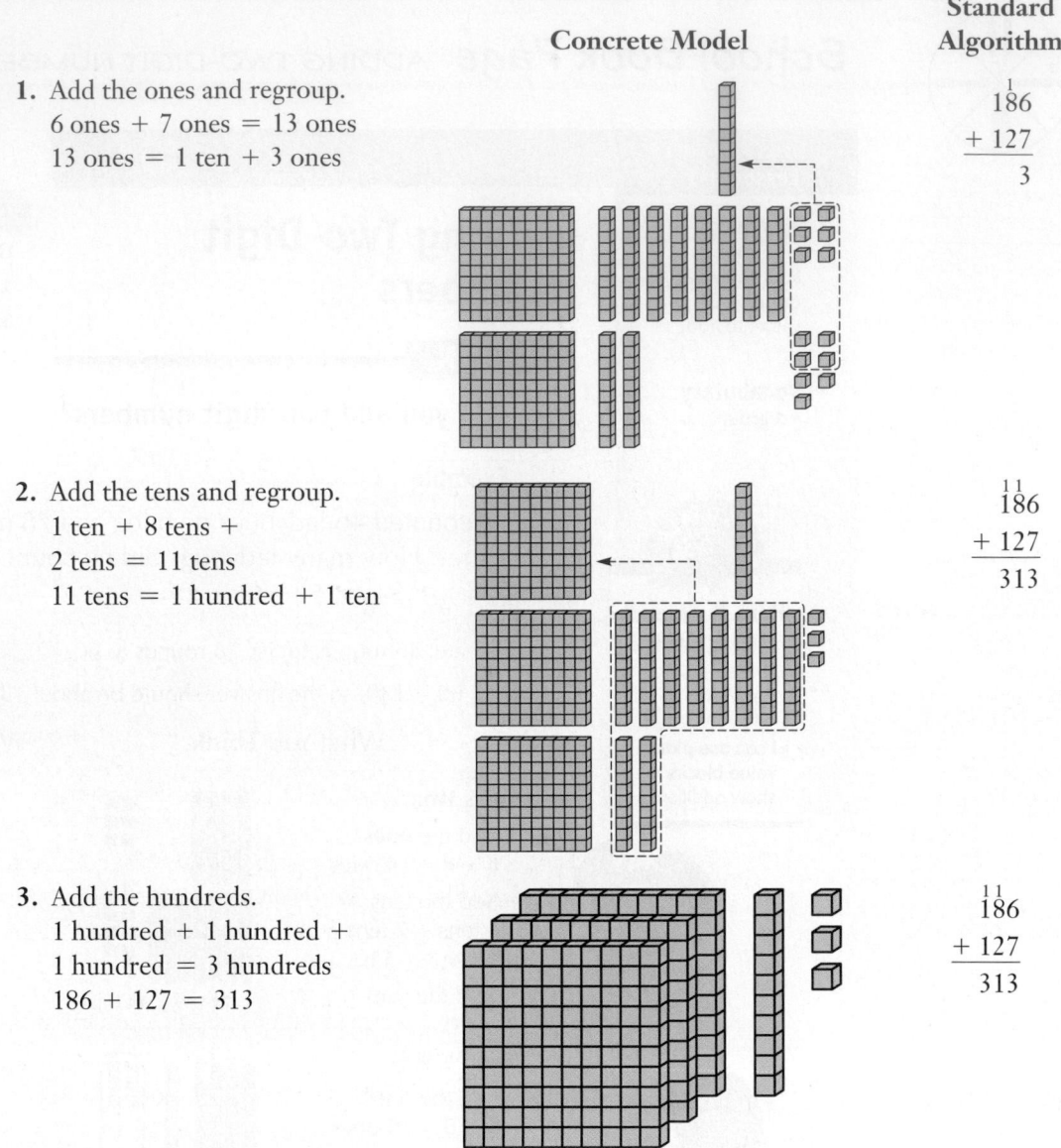

Concrete Model   **Standard Algorithm**

1. Add the ones and regroup.
   6 ones + 7 ones = 13 ones
   13 ones = 1 ten + 3 ones

$$\overset{1}{186}\\+127\\\hline 3$$

2. Add the tens and regroup.
   1 ten + 8 tens +
   2 tens = 11 tens
   11 tens = 1 hundred + 1 ten

$$\overset{11}{186}\\+127\\\hline 313$$

3. Add the hundreds.
   1 hundred + 1 hundred +
   1 hundred = 3 hundreds
   186 + 127 = 313

$$\overset{11}{186}\\+127\\\hline 313$$

Figure 3-14

Left-to-Right Algorithm for Addition

Because children learn to read from left to right, it might seem natural that they may try to add from left to right. When working with base ten blocks, many children in fact do combine the larger pieces first and then move to combining the smaller pieces. This method has the advantage of emphasizing place value. An example of the use of left-to-right algorithm follows:

$$
\begin{array}{rcl}
 & 568 & \quad\quad 568 \\
 & +\,757 & \quad\quad +\,757 \\
(500+700) \rightarrow & 1200 \longrightarrow & 12\cancel{1}\cancel{1}5 \rightarrow 1325 \\
(60+50) \rightarrow & 110 & \quad\quad {}_{3\ 2} \\
(8+7) \rightarrow & \underline{15} & \\
 & 1325 &
\end{array}
$$

Explain why this technique works and try it with 9076 + 4689.

Lattice Algorithm for Addition

This algorithm works through an addition involving two four-digit numbers. For example,

$$
\begin{array}{r}
3\ \ 5\ \ 6\ \ 7 \\
+\ 5\ \ 6\ \ 7\ \ 8 \\
\end{array}
$$

To use the algorithm, add the single-digit numbers by place value on top to the single-digit numbers on the bottom and record the results in a lattice. Then add the sums along the diagonals, as shown. This is very similar to the expanded algorithm introduced earlier.

Scratch Algorithm for Addition

The scratch algorithm for addition is often referred to as a *low-stress algorithm* because it allows students to perform complicated additions by doing a series of additions that involve only two single digits. An example follows:

1. $\begin{array}{r} 87 \\ 6\not{5}_2 \\ +49 \end{array}$ Add the numbers in the units place starting at the top. When the sum is 10 or more, record this sum by scratching a line through the last digit added and writing the number of units next to the scratched digit. For example, since $7 + 5 = 12$, the "scratch" represents 10 and the 2 represents the units.

2. $\begin{array}{r} 87 \\ 6\not{5}_2 \\ +\ 4\not{9}_1 \end{array}$ Continue adding the units, including any new digits written down. When the addition again results in a sum of 10 or more, as with $2 + 9 = 11$, repeat the process described in (1).

3. $\begin{array}{r} ^2 87 \\ 6\not{5}_2 \\ +\ 4\not{9}_1 \\ \hline 1 \end{array}$ When the first column of additions is completed, write the number of units, 1, below the addition line in the proper place value position. Count the number of scratches, 2, and add this number to the second column.

4. $\begin{array}{r} ^2\not{8}_0 7 \\ 6\ \not{5}_2 \\ \not{4}_0\not{9}_1 \\ \hline 2\ \ 0\ \ 1 \end{array}$ Repeat the procedure for each successive column until the last column with non-zero values. At this stage, sum the scratches in that column and place the number to the left of the current value.

Subtraction Algorithms

As with addition, we use base-ten blocks to provide a concrete model for subtraction. Consider how the base ten blocks are used to perform the subtraction $243 - 61$: First we represent 243 with 2 flats, 4 longs, and 3 units, as shown in Figure 3-15.

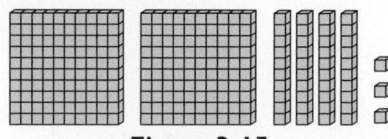

Figure 3-15

To subtract 61 from 243, we try to remove 6 longs and 1 unit from the blocks in Figure 3-15. We can remove 1 unit, as in Figure 3-16.

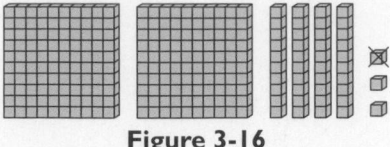

Figure 3-16

To remove 6 longs from Figure 3-16, we have to trade 1 flat for 10 longs, as shown in Figure 3-17.

E-Manipulative Activity

Practice using base-ten blocks to perform subtractions can be found in the *Subtracting Blocks* activity.

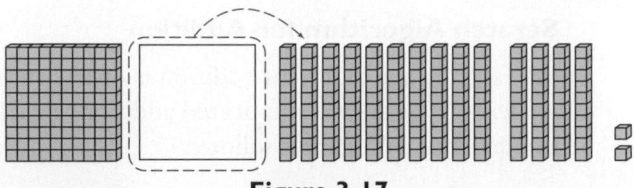

Figure 3-17

Now we can remove, or "take away," 6 longs, leaving 1 flat, 8 longs, and 2 units, or 182, as shown in Figure 3-18.

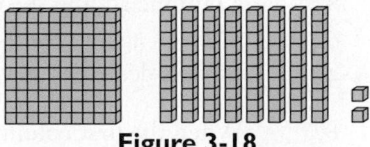

Figure 3-18

 Student work with base ten blocks along with discussions and recorded work lead to the development of the standard algorithm as seen on the student page on page 119. Work through the student page (a)–(h).

 NOW TRY THIS 3-8

Use base-ten blocks and addition to check that $243 - 61 = 182$.

Subtraction of whole numbers using blocks carries over to the expanded form and trading. For example, consider the following subtraction problem done earlier with blocks:

$$
\begin{array}{r}
243 \\
-\ 61
\end{array}
\longrightarrow
\begin{array}{r}
2 \cdot 10^2 + 4 \cdot 10 + 3 \\
-\ (6 \cdot 10 + 1)
\end{array}
\longrightarrow
\begin{array}{r}
1 \cdot 10^2 +\qquad 14 \cdot 10 + 3 \\
-\ (6 \cdot 10 + 1) \\
\hline
1 \cdot 10^2 + (14 - 6)10 + (3 - 1)
\end{array}
$$

$$= 1 \cdot 10^2 + 8 \cdot 10 + 2$$

$$= 182$$

Note that to complete the subtraction, trading was used.

School Book Page MODELS FOR SUBTRACTING THREE-DIGIT NUMBERS

Lesson 3-8

Key Idea
You can use blocks to show regrouping for subtraction.

Materials
• place-value blocks

or  **tools**

Think It Through
I can **use objects** to show a subtraction problem with regrouping.

Models for Subtracting Three-Digit Numbers

LEARN

Activity

How can you subtract with place-value blocks?

Find 255 − 163.

	What You **Show**	What You **Write**

a. Show 255 with place-value blocks.

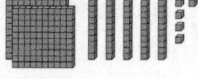

$\begin{array}{r} 2\,5\,5 \\ -1\,6\,3 \\ \hline \end{array}$

b. Subtract the ones. Regroup if needed.

5 > 3. No regrouping is needed.   **5** ones − **3** ones = **2** ones

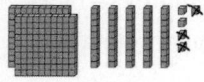

$\begin{array}{r} 2\,5\,5 \\ -1\,6\,3 \\ \hline 2 \end{array}$

c. Subtract the tens. Regroup if needed.

5 tens < 6 tens.

So, regroup 1 hundred for 10 tens.

15 tens − **6** tens = **9** tens

$\begin{array}{r} \overset{1\;15}{2\,\cancel{5}\,5} \\ -1\;6\;3 \\ \hline 9\;2 \end{array}$

d. Subtract the hundreds.

1 hundred − **1** hundred = **0** hundreds

$\begin{array}{r} \overset{1\;15}{\cancel{2}\,\cancel{5}\,5} \\ -1\;6\;3 \\ \hline 9\;2 \end{array}$

e. Find the value of the remaining blocks in Step d: 9 tens 2 ones = 92, so 255 − 163 = 92.

f. In Step b, did you have to regroup to subtract the ones? Explain.

g. In Step c, did you have to regroup to subtract the tens? Explain.

h. Use place-value blocks to subtract.
    243 − 72   145 − 126   223 − 156

Equal-Additions Algorithm

The equal-additions algorithm for subtraction is based on the fact that the difference between two numbers does not change if we add the same amount to both numbers. For example, $93 - 27 = (93 + 3) - (27 + 3)$. Thus, the difference can be computed as $96 - 30 = 66$. Using this approach, subtraction could be performed as follows:

$$
\begin{array}{c}
255 \\
-\,163
\end{array}
\rightarrow
\begin{array}{c}
255 + 7 \\
-(163 + 7)
\end{array}
\rightarrow
\begin{array}{c}
262 \\
-170
\end{array}
\rightarrow
\begin{array}{c}
262 + 30 \\
-(170 + 30)
\end{array}
\rightarrow
\begin{array}{r}
292 \\
-200 \\
\hline
92
\end{array}
$$

NOW TRY THIS 3-9

Jessica claims that a method similar to *equal additions* for subtraction also works for addition. She says that in an addition problem, "you may add the same amount to one number as you subtract from the other. For example, $68 + 29 = (68 - 1) + (29 + 1)$. Thus, the sum can be computed as $67 + 30 = 97$ or as $(68 + 2) + (29 - 2) = 70 + 27 = 97$." (i) Explain why this method is valid and (ii) use it to compute $97 + 69$.

Understanding Addition and Subtraction in Bases Other Than Ten

A look at computation in other bases provides insight into computation in base ten. Use of multibase blocks helps in building an addition table for different bases. Table 3-1 is a base-five addition table.

Table 3-1 Base-Five Addition Table

+	0	1	2	3	4
0	0	1	2	3	4
1	1	2	3	4	10
2	2	3	4	10	11
3	3	4	10	11	12
4	4	10	11	12	13

NOW TRY THIS 3-10

Compute each of the following in base five:

a. $444_{\text{five}} + 1_{\text{five}}$ **b.** $13_{\text{five}} + 44_{\text{five}}$

Using the addition facts in Table 3-1, we develop algorithms for base-five addition similar to those we used for base-ten addition. We show the computation using a concrete model in Figure 3-19(a); in Figure 3-19(b), we use an expanded algorithm; in Figure 3-19(c), we use the standard algorithm.

The subtraction facts for base five can also be derived from the addition facts table by using the definition of subtraction. For example, to find $12_{\text{five}} - 4_{\text{five}}$, recall that $12_{\text{five}} - 4_{\text{five}} = c_{\text{five}}$ if, and only if, $c_{\text{five}} + 4_{\text{five}} = 12_{\text{five}}$. From Table 3-1, we see that $c = 3_{\text{five}}$. An example of subtraction involving regrouping, $32_{\text{five}} - 14_{\text{five}}$, is developed in Figure 3-20.

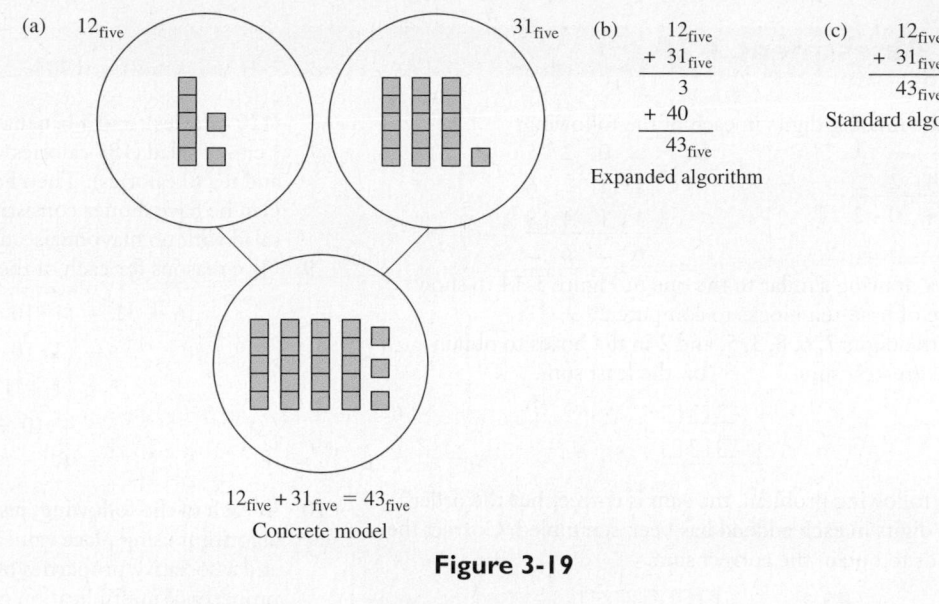

$12_{\text{five}} + 31_{\text{five}} = 43_{\text{five}}$
Concrete model

(b)
$$
\begin{array}{r}
12_{\text{five}} \\
+\ 31_{\text{five}} \\
\hline
3 \\
+\ 40 \\
\hline
43_{\text{five}}
\end{array}
$$
Expanded algorithm

(c)
$$
\begin{array}{r}
12_{\text{five}} \\
+\ 31_{\text{five}} \\
\hline
43_{\text{five}}
\end{array}
$$
Standard algorithm

Figure 3-19

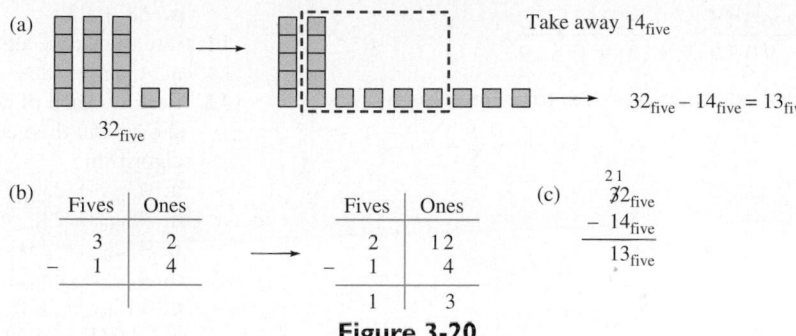

(a) Take away 14_{five}
$32_{\text{five}} - 14_{\text{five}} = 13_{\text{five}}$

32_{five}

(b)

Fives	Ones
3	2
− 1	4

Fives	Ones
2	12
− 1	4
1	3

(c)
$$
\begin{array}{r}
\overset{2\ 1}{\cancel{3}2}_{\text{five}} \\
-\ 14_{\text{five}} \\
\hline
13_{\text{five}}
\end{array}
$$

Figure 3-20

The equal additions algorithm for subtraction and other algorithms that can be used in base-ten computations can also be efficiently used in other bases. For example, consider $432_{\text{five}} - 43_{\text{five}}$:

$$
\begin{array}{r}
432_{\text{five}} \\
-\ 43_{\text{five}}
\end{array}
\rightarrow
\begin{array}{r}
432_{\text{five}} + 2 \\
-\ (43_{\text{five}} + 2)
\end{array}
\rightarrow
\begin{array}{r}
434_{\text{five}} \\
-100_{\text{five}} \\
\hline
334_{\text{five}}
\end{array}
$$

NOW TRY THIS 3-11

a. Build an addition table for base two.

b. Use the addition table from part (a) to perform: **(i)** $1111_{\text{two}} + 111_{\text{two}}$ **(ii)** $1101_{\text{two}} - 111_{\text{two}}$.

BRAIN TEASER

The number on a license plate consists of five digits. When the license plate is looked at upside down, you can still read it, but the value of the upside-down number is 78,633 greater than the real license number. What is the license number?

Assessment 3-2A

1. Find the missing digits in each of the following:

   a.
   ```
       – – 1
     + 4 2 –
     – 4 0 2
   ```
   b.
   ```
       – 0 2 5
       1 1 – 6
     + 3 1 4 8
       6 – 6 –
   ```

2. Make a drawing similar to the one in Figure 3-14 to show the use of base-ten blocks to compute $29 + 37$.

3. Place the digits 7, 6, 8, 3, 5, and 2 in the boxes to obtain
   a. the greatest sum.   **b.** the least sum.
   ```
      □ □ □
    + □ □ □
   ```

4. In the following problem, the sum is correct but the order of the digits in each addend has been scrambled. Correct the addends to obtain the correct sum.
   ```
      2 8 3 4        □ □ □ □
    + 6 3 1 5      + □ □ □ □
      9 0 5 9        9 0 5 9
   ```

5. Use the equal-additions approach to compute each of the following:

   a.
   ```
        93
      – 37
   ```
   b.
   ```
       321
      – 38
   ```

6. Janet worked her addition problems by placing the partial sums as shown here:
   ```
       569
     + 645
        14
        10
        11
      1214
   ```
   a. Use this method to work the following:
   (i)
   ```
      687
    + 549
   ```
   (ii)
   ```
      359
    + 673
   ```
   b. Explain why this algorithm works.

7. Analyze the following computations. Explain what is wrong in each case.

   a.
   ```
       28
      +75
      913
   ```
   b.
   ```
       28
      +75
      121
   ```
   c.
   ```
      305
     -259
      154
   ```
   d.
   ```
       2 10
       3Ø5
     -259
       56
   ```

8. Tom's diet allows only 1500 calories per day. For breakfast, Tom had skim milk (90 calories), a waffle with no syrup (120 calories), and a banana (119 calories). For lunch, he had $\frac{1}{2}$ cup of salad (185 calories) with mayonnaise (110 calories) and tea (0 calories). Then he had pecan pie (570 calories). Can he have dinner consisting of fish (250 calories), $\frac{1}{2}$ cup of salad with no mayonnaise, and tea?

9. Give reasons for each of the following steps:
$$
\begin{aligned}
16 + 31 &= (1 \cdot 10 + 6) + (3 \cdot 10 + 1) \\
&= (1 \cdot 10 + 3 \cdot 10) + (6 + 1) \\
&= (1 + 3)10 + (6 + 1) \\
&= 4 \cdot 10 + 7 \\
&= 47
\end{aligned}
$$

10. In each of the following, justify the standard addition algorithm using place value of the numbers, the commutative and associative properties of addition, and the distributive property of multiplication over addition:
   a. $68 + 23$ **b.** $174 + 285$

11. Use the lattice algorithm to perform each of the following:
   a. $4358 + 3864$ **b.** $4923 + 9897$

12. Perform each of the following operations using the bases shown and then check your answers using a different algorithm:
   a. $43_{five} + 23_{five}$
   b. $43_{five} - 23_{five}$
   c. $432_{five} + 23_{five}$
   d. $42_{five} - 23_{five}$
   e. $110_{two} + 11_{two}$
   f. $10001_{two} - 111_{two}$

13. Construct an addition table for base eight and then perform each of the following:
   a. $573_{eight} - 77_{eight}$ **b.** $765_{eight} - 76_{eight}$

14. Perform each of the following operations:
   a.
   ```
       3 hr 36 min 58 sec
     + 5 hr 56 min 27 sec
   ```
   b.
   ```
       5 hr 36 min 38 sec
     - 3 hr 56 min 58 sec
   ```

15. The following is a supermagic square.

1	15	14	4
12	6	7	9
8	10	11	5
13	3	2	16

   a. Find the sum of each row, the sum of each column, and the sum of each diagonal.
   b. Find the sum of the four numbers in the center.
   c. Find the sum of the four numbers in each corner.
   d. Add 5 to each number in the square. Is the square still a magic square?
   e. Subtract 1 from each number in the square. Is the square still a magic square?

16. Use scratch addition to perform the following:

    a. 432
       976
     + 1418

    b. 32_{five}
        13_{five}
        22_{five}
        43_{five}
        23_{five}
      + 12_{five}

17. Determine what is wrong with the following:

    $$22_{\text{five}}$$
    $$+\ 33_{\text{five}}$$
    $$55_{\text{five}}$$

18. Find the number to put in the blank to make each equation true. Do not convert to base ten.

    a. $3423_{\text{five}} -$ _____ $= 2132_{\text{five}}$
    b. $11011_{\text{two}} +$ _____ $= 100000_{\text{two}}$
    c. $TEE_{\text{twelve}} -$ _____ $= 1$
    d. $1000_{\text{five}} +$ _____ $= 10000_{\text{five}}$

19. The Hawks played the Elks in a basketball game. Based on the following information, complete the scoreboard showing the number of points scored by each team during each quarter and the final score of the game.

	Quarters				
Teams	**1**	**2**	**3**	**4**	**Final Score**
Hawks					
Elks					

    a. The Hawks scored 14 points in the first quarter.
    b. The Hawks were behind by 4 points at the end of the first quarter.

    c. The Elks scored 5 more points in the second quarter than they did in the first quarter.
    d. The Hawks scored 1 fewer points than the Elks in the second quarter.
    e. The Elks outscored the Hawks by 6 points in the fourth quarter.
    f. The Hawks scored a total of 120 points in the game.
    g. The Hawks scored twice as many points in the third quarter as the Elks did in the first quarter.
    h. The Elks scored as many points in the third quarter as the Hawks did in the first two quarters combined.

20. a. Place the numbers 1 through 9 in the following circles so that the sums are the same in each direction:

    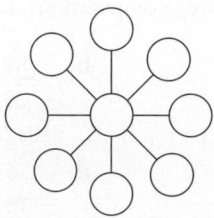

    b. How many different numbers can be placed in the middle to obtain a solution?

21. A palindrome is any number that reads the same backward as forward, for example, 121 and 2332. Try the following: Begin with any multi-digit number. Is it a palindrome? If not, reverse the digits and add this reversed number to the original number. Is the result a palindrome? If not, repeat the procedure until a palindrome is obtained. For example, start with 78. Because 78 is not a palindrome, we add: $78 + 87 = 165$. Because 165 is not a palindrome, we add: $165 + 561 = 726$. Again, 726 is not a palindrome, so we add $726 + 627$ to obtain 1353. Finally, $1353 + 3531$ yields 4884, which is a palindrome.

    Try this method with the following numbers:
    a. 93   b. 588   c. 2003

Assessment 3-2B

1. Find the missing digits in each of the following:

    a.   3 _ _
        − 1 5 9
        _ 2 4

    b. 1 _ _ _ 6
        − 8 3 0 9
        4 9 8 7

2. Make an appropriate drawing like the one in Figure 3-12 to show the use of base ten blocks to compute $46 + 38$.

3. Place the digits 7, 6, 8, 3, 5, and 2 in the boxes to obtain
    a. the greatest difference.
    b. the least positive difference.

    ☐ ☐ ☐
  − ☐ ☐ ☐

4. In the following problem, the sum is correct but the order of the digits in each addend has been scrambled. Correct the addends to obtain the correct sum.

    8 3 5 4   ☐ ☐ ☐ ☐
  + 3 4 5 6   + ☐ ☐ ☐ ☐
    1 1 7 2 9   1 1 7 2 9

5. Use the equal-additions approach to compute each of the following:

    a.   86
        − 38

    b. 5 8 2
        − 4 4

6. Janet computed her addition problems by placing the partial sums as shown here:

$$
\begin{array}{r}
768 \\
+\ 647 \\
\hline
15 \\
10 \\
13 \\
\hline
1415
\end{array}
$$

Use this method to work the following:

a. $\begin{array}{r} 987 \\ +\ 356 \\ \hline \end{array}$ b. $\begin{array}{r} 415 \\ +\ 79 \\ \hline \end{array}$

7. Analyze the following computations. Explain what is wrong in each case.

a. $\begin{array}{r} 135 \\ +\ 47 \\ \hline 172 \end{array}$ b. $\begin{array}{r} 87 \\ +\ 25 \\ \hline 1012 \end{array}$

c. $\begin{array}{r} 57 \\ -\ 38 \\ \hline 21 \end{array}$ d. $\begin{array}{r} 56 \\ -\ 18 \\ \hline 48 \end{array}$

8. George is cooking an elaborate meal for Thanksgiving. He can cook only one thing at a time in his microwave oven. His turkey takes 75 min; the pumpkin pie takes 18 min; rolls take 45 sec; and a cup of coffee takes 30 sec to heat. How much time does he need to cook the meal? When does he need to start in order to complete the cooking at 4 P.M.?

9. Give reasons for each of the following steps:

$$
\begin{aligned}
123 + 45 &= (1 \cdot 10^2 + 2 \cdot 10 + 3) + (4 \cdot 10 + 5) \\
&= 1 \cdot 10^2 + (2 \cdot 10 + 4 \cdot 10) + (3 + 5) \\
&= 1 \cdot 10^2 + (2 + 4)10 + (3 + 5) \\
&= 1 \cdot 10^2 + 6 \cdot 10 + 8 \\
&= 168
\end{aligned}
$$

10. In each of the following justify the standard addition algorithm using place value of the numbers, the commutative and associative properties of addition, and the distributive property of multiplication over addition:

a. $46 + 32$
b. $3214 + 783$

11. Use the lattice algorithm to perform each of the following:

a. $2345 + 8888$
b. $8713 + 4214$

12. Perform each of the following operations using the bases shown:

a. $43_{\text{five}} - 24_{\text{five}}$
b. $143_{\text{five}} + 23_{\text{five}}$
c. $32_{\text{five}} - 23_{\text{five}}$
d. $232_{\text{five}} + 43_{\text{five}}$
e. $110_{\text{two}} + 111_{\text{two}}$
f. $10001_{\text{two}} - 101_{\text{two}}$

13. Construct an addition table for base six, and then perform each of the following subtractions and check your answers by computing an equivalent addition problem.

a. $231_{\text{six}} - 144_{\text{six}}$ b. $342_{\text{six}} - 144_{\text{six}}$

14. Perform each of the following operations (2 c = 1 pt, 2 pt = 1 qt, 4 qt = 1 gal):

a. $\begin{array}{l} 1\ \text{qt}\ 1\ \text{pt}\ 1\ \text{c} \\ +\quad\ 1\ \text{pt}\ 1\ \text{c} \\ \hline \end{array}$

b. $\begin{array}{l} 1\ \text{qt}\qquad 1\text{c} \\ -\quad 1\ \text{pt}\ 1\text{c} \\ \hline \end{array}$

c. $\begin{array}{l} 1\ \text{gal}\ 3\ \text{qt}\ 1\ \text{c} \\ -\qquad\ 4\ \text{qt}\ 2\ \text{c} \\ \hline \end{array}$

15. The following is a supermagic square taken from an engraving called *Melancholia* by Dürer. Notice 1514 in the bottom row, the year the engraving was made.

16	3	2	13
5	10	11	8
9	6	7	12
4	15	14	1

a. Find the sum of each row, the sum of each column, and the sum of each diagonal.
b. Find the sum of the four numbers in the center.
c. Find the sum of the four numbers in each corner.
d. Add 11 to each number in the square. Is the square still a magic square?
e. Subtract 11 from each number in the square. Is the square still a magic square?

16. Use scratch addition to perform the following:

a. $\begin{array}{r} 537 \\ 318 \\ +\ 2345 \\ \hline \end{array}$

b. $\begin{array}{r} 41_{\text{six}} \\ 32_{\text{six}} \\ 22_{\text{six}} \\ 43_{\text{six}} \\ 22_{\text{six}} \\ +\,54_{\text{six}} \\ \hline \end{array}$

17. Determine what is wrong with the following:

$$
\begin{array}{r}
23_{\text{six}} \\
+\ 43_{\text{six}} \\
\hline
66_{\text{six}}
\end{array}
$$

18. Find the number to put in the blank to make each equation true. Do not convert to base ten.

a. $342_{\text{five}} - \underline{\quad\quad} = 213_{\text{five}}$
b. $1101_{\text{two}} - \underline{\quad\quad} = 1011_{\text{two}}$
c. $E08_{\text{twelve}} - \underline{\quad\quad} = 9_{\text{twelve}}$
d. $100_{\text{two}} + \underline{\quad\quad} = 10000_{\text{two}}$

19. The Hawks played the Elks in a basketball game. Based on the following information, complete the scoreboard showing the number of points scored by each team during each quarter and the final score of the game.

	Quarters				
Teams	1	2	3	4	Final Score
Hawks					
Elks					

a. The Hawks scored 15 points in the first quarter.
b. The Hawks were behind by 5 points at the end of the first quarter.

c. The Elks scored 5 more points in the second quarter than they did in the first quarter.

d. The Hawks scored 7 more points than the Elks in the second quarter.

e. The Elks outscored the Hawks by 6 points in the fourth quarter.

f. The Hawks scored a total of 120 points in the game.

g. The Hawks scored twice as many points in the third quarter as the Elks did in the first quarter.

h. The Elks scored as many points in the third quarter as the Hawks did in the first two quarters combined.

20. a. Place the numbers 24 through 32 in the following circles so that the sums are the same in each direction:

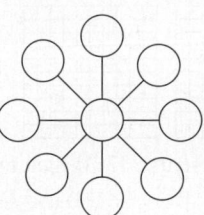

b. How many different numbers can be placed in the middle to obtain a solution?

Mathematical Connections 3-2

Communication

1. Discuss the merit of the following expanded algorithm for addition where we first add the ones, then the tens, then the hundreds, and then the total:

$$
\begin{array}{r}
479 \\
+\ 385 \\
\hline
14 \\
150 \\
+\ 700 \\
\hline
864
\end{array}
$$

2. The following example uses a regrouping approach to subtraction. Discuss the merit of this approach in teaching subtraction.

$$
\begin{array}{r}
843 \\
-\ 568
\end{array}
\rightarrow
\begin{array}{r}
800 + 40 + 3 \\
-(500 + 60 + 8)
\end{array}
\rightarrow
$$

$$
\begin{array}{r}
800 + 30 + 13 \\
-(500 + 60 + 8)
\end{array}
\rightarrow
\begin{array}{r}
700 + 130 + 13 \\
-(500 +\ \ 60 +\ \ 8) \\
\hline
200 +\ \ 70 +\ \ 5\ = 275
\end{array}
$$

3. Explain why the scratch addition algorithm works.

4. Consider the following subtraction algorithm.
   a. Explain why it works.
   b. Use this algorithm to find $787 - 398$.

$$
\begin{aligned}
585 - 277 &= 585 - 200 -\ \ 77 \\
&= 385 -\ \ 77 \\
&= 385 - 100 +\ \ 23 \\
&= 285 +\ \ 23 = 308
\end{aligned}
$$

Open-Ended

5. Search for or develop an algorithm for whole-number addition or subtraction and write a description of your algorithm so that others can understand and use it.

Cooperative Learning

6. In this section you have been exposed to many different algorithms. Discuss in your group whether children should be encouraged to develop and use their own algorithms for whole-number addition and subtraction or whether they should be taught only one algorithm per operation and all students should use only one algorithm.

Questions from the Classroom

7. Tira, a fourth grader, performs addition by adding and subtracting the same number. She added as follows:

$$
\begin{array}{r}
39 \\
+\ 84
\end{array}
\rightarrow
\begin{array}{r}
39 + 1 \\
+\ 84 - 1
\end{array}
\rightarrow
\begin{array}{r}
40 \\
+\ 83 \\
\hline
123
\end{array}
$$

   How would you respond if you were her teacher?

8. Cathy found her own algorithm for subtraction. She subtracted as follows:

$$
\begin{array}{r}
97 \\
-\ 28 \\
\hline
-\ 1 \\
+\ 70 \\
\hline
69
\end{array}
$$

   How would you respond if you were her teacher?

9. To find $68 - 19$, Joe began by finding $6 - 1$ and then $9 - 8$ and wrote 51 as the answer. How would you respond?

10. Jill subtracted $415 - 212$ by writing $4 - 2 = 2$ and $15 - 12 = 3$ and wrote 23 as the answer. How would you help her?

11. Betsy found $518 - 49 = 469$. She was not sure she was correct so she tried to check her answer by adding $518 + 49$. How could you help her?

12. A child is asked to compute $7 + 2 + 3 + 8 + 11$ and writes $7 + 2 = 9 + 3 = 12 + 8 = 20 + 11 = 31$. Noticing that the answer is correct, if you were the teacher how would you react?

13. A student wants to know why if all operations result in whole numbers, then $a - b = (a - c) - (b - c)$. How would you respond?

Review Problems

14. Is the set $\{1, 2, 3\}$ closed under addition? Why?

15. Give an example of the associative property of addition of whole numbers.

16. Find all whole numbers for which:
   a. $20 - x = x$
   b. $20 - x - 6 = 0$
   c. $x + 4 = 3 + x + 1$

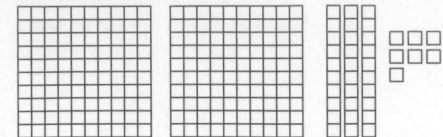
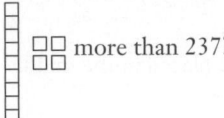
3-3 Multiplication and Division of Whole Numbers

In the grade 3 *Focal Points* we find the following concerning multiplication and division of whole numbers:

> Students understand the meanings of multiplication and division of whole numbers through the use of representations (e.g., equal-sized groups, arrays, area models, and equal "jumps" on number lines for multiplication, and successive subtraction, partitioning, and sharing for division). They use properties of addition and multiplication (e.g., commutativity, associativity, and the distributive property) to multiply whole numbers and apply increasingly sophisticated strategies based on these properties to solve multiplication and division problems involving basic facts. By comparing a variety of solution strategies, students relate multiplication and division as inverse operations. (p. 15)

Further in the grade 3 *Common Core Standards*, the connection is shown between studying whole number operations and the study of algebra.

> • Solve two-step word problems using the four operations. Represent these problems using equations with a letter standing for the unknown quantity. Assess the reasonableness of answers using mental computation and estimation strategies including rounding. (p. 23)
> • Identify arithmetic patterns (including patterns in the addition table or multiplication table), and explain them using properties of operations. For example, observe that 4 times a number is always even, and explain why 4 times a number can be decomposed into two equal addends. (p. 23)

These quotes set the tone and agenda for this section. We discuss representations that can be used to help students understand the meanings of multiplication and division. We develop the distributive property of multiplication over addition along with the relationship of multiplication and division as inverse operations.

Multiplication of Whole Numbers

In this section, we explore the kind of problems that Grampa is having in the *Peanuts* cartoon. Why do you think he would have more trouble with "9 times 8" than "3 times 4"? If multiplication facts are only memorized, they may be forgotten. If students have a conceptual understanding of the basic facts, then all of the basic facts can be determined even if not automatically recalled.

Peanuts copyright © 1987 and 2010 Peanuts Worldwide LLC. Distributed by United Feature Syndicate, Inc.

Repeated-Addition Model Approach

The student page on page 128, shows that if there are 4 groups of 3 brushes, addition can put the groups together. When equal-sized groups are put together we can use multiplication. Think of this as combining 4 sets of 3 objects into a single set. The 4 sets of 3 suggest the following addition:

$$\underbrace{3 + 3 + 3 + 3}_{\text{four 3s}} = 12$$

We write $3 + 3 + 3 + 3$ as $4 \cdot 3$ and say "four times three" or "three multiplied by four." The advantage of the multiplication notation over repeated addition is evident when the number of addends is great: for example, with 25 groups of 3 brushes, we could find the total number of brushes by adding 25 3s or writing this as $25 \cdot 3$.

The *repeated-addition* approach can be illustrated in several ways, including the number line or measurement model and arrays. For example, using colored rods of length 4, the combined length of five rods can be found by joining the rods end-to-end, as in Figure 3-22(a). Figure 3-22(b) shows the process using arrows on a number line.

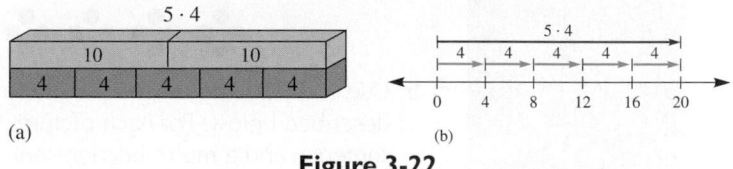

Figure 3-22

The constant feature on a calculator can help relate multiplication to addition. Students can find products on the calculator without using the ⊠ key. For example, if a calculator has the *constant feature*, then 5×3 can be found by starting with and pressing ⊞③⊟ ⊟ ⊟ ⊟ ⊟. Each press of the equal sign will add 3 to the display. (Some calculators work differently.)

Access to only the "repeated-addition" model for multiplication can lead to misunderstanding. In this section, we introduce three other multiplication models: the *array area* and *Cartesian-product* models.

◯ Historical Note

William Oughtred (1574–1660), an English mathematician, was interested in mathematical symbols. He was the first to introduce the "St. Andrew's cross" (×) as the symbol for multiplication. This symbol was not readily adopted because, as Gottfried Wilhelm von Leibniz (1646–1716) objected, it was too easily confused with the letter *x*. Leibniz used the dot (·) for multiplication, which has become common.

School Book Page MULTIPLICATION AS REPEATED ADDITION

Lesson 5-1

Algebra

Key Idea
Multiplying is a quick way of adding equal groups.

Vocabulary
• multiplication
• factor
• product

Materials
• counters
or ⚙ tools

TEST TALK

Think It Through
I can **use objects** to show equal groups.

260

Multiplication as Repeated Addition

✔ WARM UP
1. 2 + 2 + 2
2. 3 + 3 + 3 + 3
3. 5 + 5 + 5 + 5 + 5

LEARN

Activity

How can you find the total?

There are **4 groups of 3** paintbrushes.

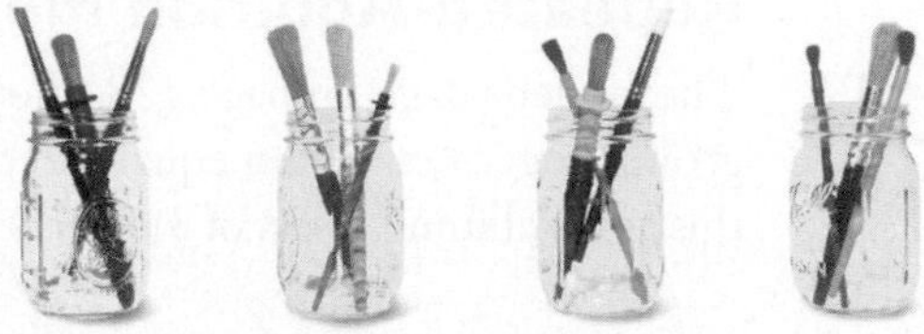

You can use addition to put together groups.

$$3 + 3 + 3 + 3 = 12 \quad \text{Addition sentence}$$

When you put together **equal groups,** you can also use **multiplication.**

What You **Say:**   4 times 3 equals 12

What You **Write:**   4 × 3 = 12   **Multiplication sentence**

                       ↑   ↑   ↑
                    factor factor product

a. Write an addition sentence and a multiplication sentence to show the total number of counters below.

b. Use counters and draw a picture to show the groups described below. For each picture, write an addition sentence and a multiplication sentence to show how many counters in all.

5 groups of 2
4 groups of 5
3 groups of 3

**Take It to the NET
More Examples**
www.scottforesman.com

The Rectangular Array and Area Models

Another representation useful in exploring multiplication of whole numbers is an *array*. An array is suggested when objects are arranged in equal-sized rows, as in Figure 3-23.

Figure 3-23

In Figure 3-24(a), sticks are crossed to create intersection points, thus forming an array of points. The number of intersection points on a single vertical stick is 4 and there are 5 sticks, forming a total of $5 \cdot 4$ points in the array. In Figure 3-24(b), the area model is shown as a 4-by-5 grid. The number of unit squares required to fill in the grid is 20. These models motivate the following definition of multiplication of whole numbers.

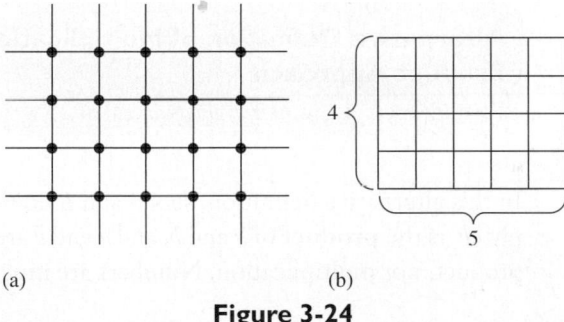

(a)   (b)

Figure 3-24

Definition of Multiplication of Whole Numbers: Repeated Addition Approach

For every whole numbers a and $n \neq 0$,

$$n \cdot a = \underbrace{a + a + a + \ldots + a}_{n \text{ terms}}.$$

If $n = 0$, then $0 \cdot a = 0$.

We typically write $n \cdot a$ as na.

E-Manipulative Activity

Practice using the area model to perform multiplication can be found in the *Multiplication* activity.

Cartesian-Product Model

The *Cartesian-product* model offers another way to discuss multiplication. Suppose you can order a soyburger on light or dark bread with one condiment: mustard, mayonnaise, or horseradish. To show the number of different soyburger orders that a waiter could write for the cook, we use a *tree diagram* model. The ways of writing the order are listed in Figure 3-25, where the bread is chosen from the set $B = \{$light, dark$\}$ and the condiment is chosen from the set $C = \{$mustard, mayonnaise, horseradish$\}$.

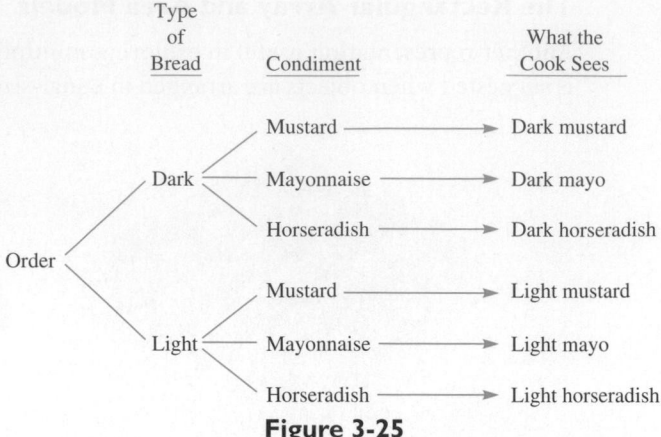

Figure 3-25

Each order can be written as an ordered pair, for example, (dark, mustard). The set of ordered pairs forms the Cartesian product $B \times C$. The Fundamental Counting Principle tells us that the number of ordered pairs in $B \times C$ is $2 \cdot 3$.

The preceding discussion demonstrates how multiplication of whole numbers can be defined in terms of Cartesian products. Thus, an alternative definition of multiplication of whole numbers is as follows:

> **Alternative Definition of Multiplication of Whole Numbers: Cartesian Product Approach**
>
> For finite sets A and B, if $n(A) = a$ and $n(B) = b$, then $a \cdot b = n(A \times B)$.

In this alternative definition, sets A and B do not have to be disjoint. The expression $a \cdot b$, or simply ab, is the **product** of a and b, and a and b are **factors**. Note that $A \times B$ indicates the Cartesian product, not multiplication. Numbers are multiplied, not sets.

◗ NOW TRY THIS 3-12

How would you use the repeated-addition definition of multiplication to explain to a child unfamiliar with the Fundamental Counting Principle that the number of possible outfits consisting of a shirt and pants combination—given 6 shirts and 5 pairs of pants—is $6 \cdot 5$?

The following problems can be solved using the models shown for multiplication. Work through each problem using the suggested model.

1. *Repeated-addition*: One piece of gum costs 5¢; how much do three pieces cost?
2. *Number-line*: If Al walks 5 mph for 3 hr, how far has he walked?
3. *Array*: A sheet of stamps has 4 rows of 5 stamps. How many stamps are there in a sheet?
4. *Area*: If a carpet is 5 ft by 3 ft, what is the area of the carpet?
5. *Cartesian-product*: Al has 5 shirts and 3 pairs of pants; how many different shirt-pants combinations are possible?

Properties of Whole-Number Multiplication

The set of whole numbers is *closed* under multiplication. That is, if we multiply any two whole numbers, the result is a unique whole number. This property is referred to as the *closure property of multiplication of whole numbers*. Multiplication on the set of whole numbers, like addition, has the closure, commutative, and associative properties.

> ### Theorem 3-5: Properties of Multiplication of Whole Numbers
>
> **Closure property of multiplication of whole numbers**   For whole numbers a and b, $a \cdot b$ is a unique whole number.
>
> **Commutative property of multiplication of whole numbers**   For whole numbers a and b, $a \cdot b = b \cdot a$.
>
> **Associative property of multiplication of whole numbers**   For whole numbers a, b, and c, $(a \cdot b) \cdot c = a \cdot (b \cdot c)$.

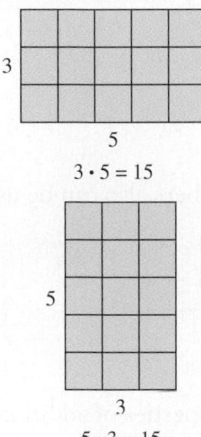

3 ⎡ grid ⎤
5
$3 \cdot 5 = 15$

5
3
$5 \cdot 3 = 15$

Figure 3-26

The *commutative property of multiplication of whole numbers* is illustrated easily by building a 3-by-5 grid and then turning it sideways, as shown in Figure 3-26. We see that the number of 1×1 squares present in either case is 15; that is, $3 \cdot 5 = 15 = 5 \cdot 3$. The commutative property can be verified by first showing that for sets A and B, $n(A \times B) = n(B \times A)$.

The *associative property of multiplication of whole numbers* can be illustrated as follows. Suppose $a = 3$, $b = 5$, and $c = 4$. In Figure 3-27(a), we see a picture of $3(5 \cdot 4)$ blocks. In Figure 3-27(b), we see the same blocks, this time arranged as $4(3 \cdot 5)$. By the commutative property this can be written as $(3 \cdot 5)4$. Because both sets of blocks in Figure 3-27(a) and figure 3-27(b) compress to the set shown in Figure 3-27(c), we see that $3(5 \cdot 4) = (3 \cdot 5)4$. The associative property is useful in computations such as the following:

$$3 \cdot 40 = 3(4 \cdot 10) = (3 \cdot 4)10 = 12 \cdot 10 = 120$$

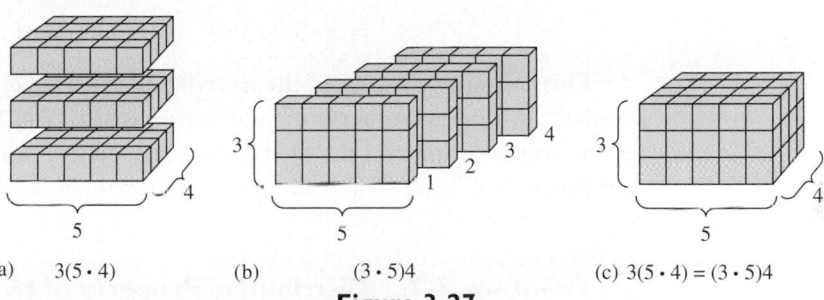

(a) $3(5 \cdot 4)$ (b) $(3 \cdot 5)4$ (c) $3(5 \cdot 4) = (3 \cdot 5)4$

Figure 3-27

The *multiplicative identity for whole numbers* is 1. For example, $3 \cdot 1 = 1 + 1 + 1 = 3$. In general, for any whole number a,

$$a \cdot 1 = \underbrace{1 + 1 + 1 + \ldots + 1}_{a \text{ terms}} = a$$

Thus, $a \cdot 1 = a$, which, along with the commutative property for multiplication implies that $a \cdot 1 = a = 1 \cdot a$. Cartesian products can also be used to show that $a \cdot 1 = a = 1 \cdot a$.

Next, consider multiplication involving 0. For example, $0 \cdot 6$ by definition means we have zero 6s or 0. Also $6 \cdot 0 = 0 + 0 + 0 + 0 + 0 + 0 = 0$. Thus we see that multiplying 0 by 6 or 6 by 0 yields a product of 0. This is an example of the *zero multiplication property*. This property can also be verified by using the definition of multiplication in terms of Cartesian products. In algebra, $3x$ means 3 x's or $x + x + x$. Therefore, $0 \cdot x$ means 0 sets of x, or 0. Thus the following is true.

> ### Theorem 3-6: Multiplication Properties of 1 and 0
>
> **Identity property of multiplication of whole numbers**   There is a unique whole number 1 such that for every whole number a, $a \cdot 1 = a = 1 \cdot a$.
>
> **Zero multiplication property of whole numbers**   For every whole number a, $a \cdot 0 = 0 = 0 \cdot a$.

The Distributive Property of Multiplication over Addition and Subtraction

Next the operations of multiplication and addition combine to form the basis for understanding multiplication algorithms for whole numbers. The area of the large rectangle in Figure 3-28 equals the sum of the areas of the two smaller rectangles and hence $5(3 + 4) = 5 \cdot 3 + 5 \cdot 4$.

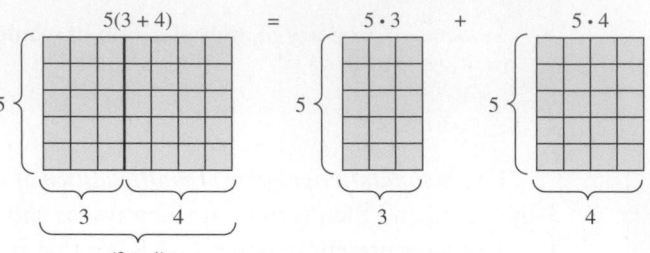

Figure 3-28

Properties of addition and the definition of multiplication of whole numbers also can be used to justify the above result:

$$5(3 + 4) = \underbrace{(3 + 4) + (3 + 4) + (3 + 4) + (3 + 4) + (3 + 4)}_{\text{Five terms}}$$

$$= (3 + 3 + 3 + 3 + 3) + (4 + 4 + 4 + 4 + 4)$$

Commutative and associative properties of addition

$$= 5 \cdot 3 + 5 \cdot 4 \quad \text{Definition of multiplication}$$

This example illustrates the *distributive property of multiplication over addition* for whole numbers. A similar property of subtraction is also true. The distributive property of multiplication over addition and the distributive property of multiplication over subtraction are stated as follows:

Theorem 3-7:   Distributive Property of Multiplication over Addition for Whole Numbers

For all whole numbers a, b, and c,

$$a(b + c) = ab + ac.$$

Theorem 3-8:   Distributive Property of Multiplication over Subtraction for Whole Numbers

For all whole numbers a, b, and c with $b \geq c$,

$$a(b - c) = ab - ac.$$

Because the commutative property of multiplication of whole numbers holds, the distributive property of multiplication over addition can be rewritten as $(b + c)a = ba + ca$.

The distributive property can also be generalized to any finite number of terms. For example, $a(b + c + d) = ab + ac + ad$.

The distributive property can be written as

$$ab + ac = a(b + c)$$

This is commonly referred to as *factoring*. Thus, the factors of $ab + ac$ are a and $(b + c)$.

Students find the distributive property of multiplication over addition useful when doing mental mathematics. For example, $13 \cdot 7 = (10 + 3)7 = 10 \cdot 7 + 3 \cdot 7 = 70 + 21 = 91$. The distributive property of multiplication over addition is important in the study of algebra and in developing algorithms for arithmetic operations. For example, it is used to combine like terms when we work with variables, as in $3x + 5x = (3 + 5)x = 8x$ or $3ab + 2b = (3a + 2)b$.

EXAMPLE 3-2

a. Use an area model to show that $(x + y)(z + w) = xz + xw + yz + yw$.
b. Use the distributive property of multiplication over addition to justify the result in part (a).

Solution

a. Consider the rectangle in Figure 3-29, whose height is $x + y$ and whose length is $z + w$. The area of the entire rectangle is $(x + y)(z + w)$. If we divide the rectangle into smaller rectangles as shown, we notice that the sum of the areas of the four smaller rectangles is $xz + xw + yz + yw$. Because the area of the original rectangle equals the sum of the areas of

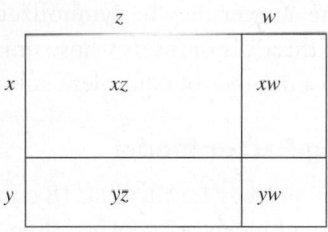

Figure 3-29

the smaller rectangles, the result follows.

b. To apply the distributive property of multiplication over addition, think of $x + y$ as one number and proceed as follows:

$(x + y)(z + w) = (x + y)z + (x + y)w$ The distributive property of multiplication over addition

$= xz + yz + xw + yw$ The distributive property of multiplication over addition

$= xz + xw + yz + yw$ The commutative and associative properties of addition

The properties of whole-number multiplication reduce the 100 basic multiplication facts involving numbers 0–9. For example, 19 facts involve multiplication by 0, and 17 more have a factor of 1. Therefore, knowing the zero multiplication property and the identity multiplication property allows students to know 36 facts. Next, 8 facts are *squares*, such as $5 \cdot 5$, leaving 56 facts. The commutative property cuts this number in half, because if students know $7 \cdot 9$, then they know $9 \cdot 7$. Knowing some multiplication facts students can use the associative and distributive properties to figure out other products. For example, $6 \cdot 5$ can be thought of as $(5 + 1)5 = 5 \cdot 5 + 1 \cdot 5$, or 30.

Division of Whole Numbers

We use three models for division: the *set (partition)* model, the *missing-factor* model, and the *repeated-subtraction* model.

Set (Partition) Model

Suppose we have 18 cookies and want to give an equal number of cookies to each of three friends: Bob, Dean, and Charlie. How many should each person receive? If we draw a picture, we can see that we can divide (or partition) the 18 cookies into three sets, with an equal number of cookies in each set. Figure 3-30 shows that each friend receives 6 cookies.

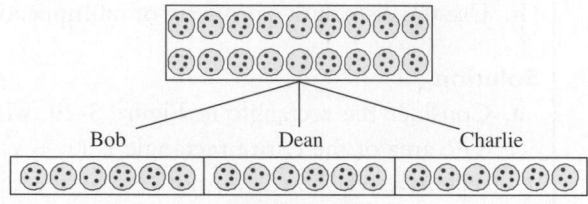

Figure 3-30

The answer may be symbolized as $18 \div 3 = 6$. Thus, $18 \div 3$ is the number of cookies in each of three disjoint sets whose union has 18 cookies. In this approach to division, we partition a set into a number of equivalent subsets.

Missing-Factor Model

Another strategy for dividing 18 cookies among three friends is to use the *missing-factor* model. If each friend receives c cookies, then the three friends receive $3c$, or 18, cookies. Hence, $3c = 18$. Because $3 \cdot 6 = 18$, $c = 6$. We have answered the division computation by using multiplication. This leads to the following definition of division of whole numbers:

> **Definition of Division of Whole Numbers: Missing-Factor Approach**
>
> For any whole numbers a and b, with $b \neq 0$, $a \div b = c$ if, and only if, c is the unique whole number such that $b \cdot c = a$.

The number a is the **dividend**, b is the **divisor**, and c is the **quotient**. Note that $a \div b$ can also be written as $\dfrac{a}{b}$ or $b\overline{)a}$.

Repeated-Subtraction Model Approach

Suppose we have 18 cookies and want to package them in cookie boxes that hold 6 cookies each. How many boxes are needed? We could reason that if one box is filled, then we have $18 - 6$ (or 12) cookies left. If one more box is filled, then there are $12 - 6$ (or 6) cookies left. Finally, we place the last 6 cookies in a third box. This discussion is summarized by writing $18 - 6 - 6 - 6 = 0$. We found by repeated subtraction that $18 \div 6 = 3$. Treating division as repeated subtraction works well if there are no cookies left over. If there are cookies left over, a nonzero remainder arises.

Calculators illustrate the repeated subtraction operation. For example, consider $135 \div 15$. If the calculator has a constant key, $\boxed{\text{K}}$, press $\boxed{1}\boxed{5}\boxed{-}\boxed{\text{K}}\boxed{1}\boxed{3}\boxed{5}\boxed{=}$ ... and then count how many times the $\boxed{=}$ key must be pressed to make the display read 0. Calculators with a different constant feature may require a different sequence of entries. For example, on some calculators, we press ⬚1⬚ ⬚3⬚ ⬚5⬚ ⬚−⬚ ⬚1⬚ ⬚5⬚ ⬚=⬚ and count the number of times the ⬚=⬚ key is pressed to make the display read ⬚0⬚.

The Division Algorithm

Just as the set of whole numbers is not closed under subtraction, it is also not closed under division of whole numbers. For example, to find $27 \div 5$, look for a whole number c such that $5c = 27$.

Table 3-2 shows several products of whole numbers times 5. Since 27 is between 25 and 30, there is no whole number c such that $5c = 27$. Because no whole number c satisfies this equation, $27 \div 5$ has no meaning in the set of whole numbers, and the set of whole numbers is not closed under division.

Table 3-2

$5 \cdot 1$	$5 \cdot 2$	$5 \cdot 3$	$5 \cdot 4$	$5 \cdot 5$	$5 \cdot 6$
5	10	15	20	25	30

Even though the set of whole numbers is not closed under division, practical applications with whole number divisions are common. For example, if 32 apples were to be divided among 6 students, each student would receive 5 apples and 2 apples would remain. The number 2 is the **remainder**. Thus, 32 contains six 5s with a remainder of 2. Observe that the remainder is a whole number less than 5. This operation is illustrated in Figure 3-31. The concept illustrated is the **division algorithm**.

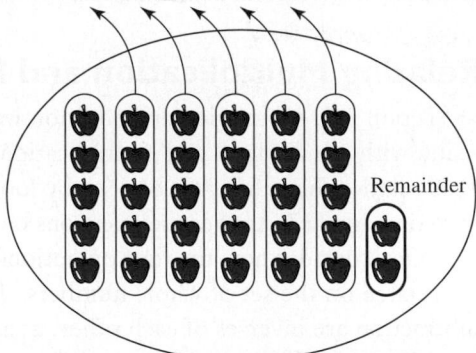

Remainder

$32 = 6 \cdot 5 + 2$ with $0 \leq 2 < 6$, quotient 5, remainder 2

Figure 3-31

The above is frequently written as $32 \div 6 = 5\text{R}2$

Division Algorithm

Given any whole numbers a and b with $b \neq 0$, there exist unique whole numbers q (quotient) and r (remainder) such that

$$a = bq + r \qquad \text{with } 0 \leq r < b.$$

When a is "divided" by b and the remainder is 0, then a is *divisible* by b or b, is a *divisor* of a, or b *divides* a. By the division algorithm, a is divisible by b if $a = bq$ for a unique whole number q. Thus, 63 is divisible by 9 because $63 = 9 \cdot 7$. Notice that 63 is also divisible by 7 and that the remainder is 0.

EXAMPLE 3-3

If 123 is divided by a number and the remainder is 13, what are the possible divisors?

Solution

If 123 is divided by b, then the division algorithm gives:

$$123 = bq + 13 \quad \text{and} \quad b > 13.$$

Using the definition of subtraction, $bq = 123 - 13$, and hence $110 = bq$. Now we are looking for two numbers whose product is 110, where one number is greater than 13. Table 3-3 shows the pairs of whole numbers whose product is 110.

Table 3-3

1	110
2	55
5	22
10	11

The only possible values for b are 110, 55, and 22 because each is greater than 13.

 NOW TRY THIS 3-13

When the marching band was placed in rows of 5, one member was left over. When the members were placed in rows of 6, there was still one member left over. However, when they were placed in rows of 7, nobody was left over. What is the smallest number of members that could have been in the band?

Relating Multiplication and Division as Inverse Operations

 In Section 3-1, subtraction and addition were related as inverse operations. In a similar way, division with remainder 0 and multiplication are related. Division is the inverse of multiplication. This relationship can again be seen by looking at fact families as shown on the grade 3 student page on page 137. Answer the questions on the student page.

Next consider how the four operations of addition, subtraction, multiplication, and division are related on the set of whole numbers. This is shown in Figure 3-32. Note that addition and subtraction are inverses of each other, as are multiplication and division with remainder 0. Also note that multiplication can be viewed as repeated addition, and division can be accomplished using repeated subtraction.

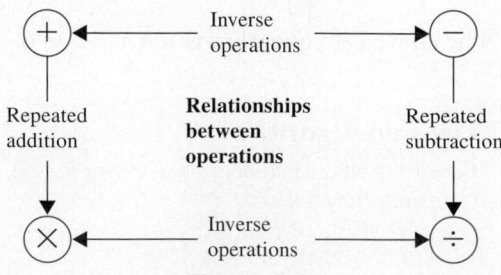

Figure 3-32

In Section 3-1, it was shown that the set of whole numbers is closed under addition and that addition is commutative and associative and has an identity. On the other hand, subtraction did not have these properties. In this section, we have seen that multiplication has some of

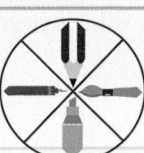

School Book Page RELATING MULTIPLICATION AND DIVISION

Lesson 7-5

Algebra

Key Idea
Fact families show how multiplication and division are connected.

Vocabulary
- array (p. 262)
- fact family (p. 70)
- factor (p. 260)
- product (p. 260)
- dividend
- divisor
- quotient

Think It Through
I can use **what I know** about multiplication to understand division.

Relating Multiplication and Division

✓ **WARM UP**

1. 2×5	2. 5×2
3. 3×4	4. 4×3
5. 7×2	6. 2×7

LEARN

How does an array show division?

In 1818, there were only 20 stars on the United States flag.

There were 4 equal rows of stars.

How many stars were in each row?

The **array** shows:

Multiplication

4 rows of **5** stars = 20 stars

$4 \times 5 = 20$

Division

20 stars in 4 equal rows = **5** stars in each row

$20 \div 4 = 5$

So, there were 5 stars in each row.

How can a fact family help you divide?

A **fact family** shows how multiplication and division are related.

Fact family for 4, 5, and 20:

$$4 \times 5 = 20 \qquad 20 \div 4 = 5$$
$$5 \times 4 = 20 \qquad 20 \div 5 = 4$$

factor × factor = product   dividend ÷ divisor = quotient

✓ **Talk About It**

1. Skip count by 5s to find 4×5. Then start at 20 and skip count by 5s backward to 0. The number of times you count back is the quotient for $20 \div 5$.

2. How can you use the fact $3 \times 6 = 18$ to find $18 \div 3$?

3. **Number Sense** Is $3 \times 5 = 15$ part of the fact family for 3, 4, and 12? Explain.

384

the same properties that hold for addition. Does it follow that division behaves like subtraction? Investigate this in Now Try This 3-14.

 **NOW TRY THIS 3-14**

    a. Provide counterexamples to show that the set of whole numbers is not closed under division and that division is neither commutative nor associative.

    b. Why is 1 not the identity for division?

Division by 0 and 1

Division by 0 and by 1 are frequently misunderstood by students. Before reading on, try to find the values of the following three expressions:

    1. $3 \div 0$ **2.** $0 \div 3$ **3.** $0 \div 0$

Consider the following explanations:

    1. By definition, $3 \div 0 = c$ if c is the unique whole number such that $0 \cdot c = 3$. Since the zero property of multiplication states that $0 \cdot c = 0$ for any whole number c, there is no whole number c such that $0 \cdot c = 3$. Thus, $3 \div 0$ is undefined because there is no answer to the equivalent multiplication problem.

    2. By definition, $0 \div 3 = c$ if c is the unique whole number such that $3 \cdot c = 0$. Because $3 \cdot 0 = 0$, $c = 0$ and $0 \div 3 = 0$. Note that $c = 0$ is the only number that satisfies $3 \cdot c = 0$.

    3. By definition, $0 \div 0 = c$ if c is the unique whole number such that $0 \cdot c = 0$. Notice that for *any* c, $0 \cdot c = 0$. According to the definition of division, c must be unique. Since there is no *unique* number c such that $0 \cdot c = 0$, it follows that $0 \div 0$ is undefined.

Division involving 0 may be summarized as follows. Let n be any non-zero whole number. Then,

    1. $n \div 0$ is undefined;   **2.** $0 \div n = 0$;   **3.** $0 \div 0$ is undefined.

    Recall that $n \cdot 1 = n$ for any whole number n. Thus, by the definition of division, $n \div 1 = n$. For example, $3 \div 1 = 3$, $1 \div 1 = 1$, and $0 \div 1 = 0$.

Order of Operations

Difficulties involving the order of arithmetic operations sometimes arise. For example, many students treat $2 + 3 \cdot 6$ as $(2 + 3)6$, whereas others treat it as $2 + (3 \cdot 6)$. In the first case, the value is 30; in the second case, the value is 20. To avoid confusion, mathematicians agree that when no parentheses are present, multiplications and divisions are performed *before* additions and subtractions. The multiplications and divisions are performed in the order they occur from left to right, and then the additions and subtractions are performed in the order they occur from left to right. Thus, $2 + 3 \cdot 6 = 2 + 18 = 20$. This order of operations is not built into some calculators that display an incorrect answer of 30. The computation $8 - 9 \div 3 \cdot 2 + 3$ is performed as

$$8 - 9 \div 3 \cdot 2 + 3 = 8 - 3 \cdot 2 + 3$$
$$= 8 - 6 + 3$$
$$= 2 + 3$$
$$= 5.$$

Assessment 3-3A

1. For each of the following, find, if possible, the whole numbers that make the equations true:
   a. $3 \cdot \square = 15$
   b. $18 = 6 + 3 \cdot \square$
   c. $\square \cdot (5 + 6) = \square \cdot 5 + \square \cdot 6$

2. Determine whether the following sets are closed under multiplication:
   a. $\{0, 1\}$
   b. $\{2, 4, 6, 8, 10, \ldots\}$
   c. $\{1, 4, 7, 10, 13, \ldots\}$

3. a. If 5 is removed from the set of whole numbers, is the set closed with respect to addition? Explain.
   b. If 5 is removed from the set of whole numbers, is the set closed with respect to multiplication?

4. Rename each of the following using the distributive property of multiplication over addition so that there are no parentheses in the final answer:
   a. $(a + b)(c + d)$
   b. $\square(\Delta + \bigcirc)$
   c. $a(b + c) - ac$

5. Place parentheses, if needed, to make each of the following equations true:
   a. $5 + 6 \cdot 3 = 33$
   b. $8 + 7 - 3 = 12$
   c. $6 + 8 - 2 \div 2 = 13$
   d. $9 + 6 \div 3 = 5$

6. Using the distributive property of multiplication over addition, we can factor as in $x^2 + xy = x(x + y)$. Use the distributive property and other multiplication properties to factor each of the following:
   a. $xy + y^2$
   b. $xy + x$
   c. $a^2b + ab^2$

7. For each of the following, find whole numbers to make the statement true, if possible:
   a. $18 \div 3 = \square$
   b. $\square \div 76 = 0$
   c. $28 \div \square = 7$

8. A sporting goods store has designs for six shirts, four pairs of pants, and three vests. How many different shirt-pants-vest outfits are possible?

9. What multiplication is suggested by the following models?
   a.
   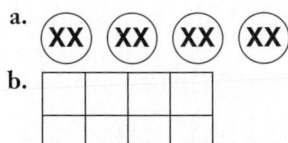
   b.

10. Which property is illustrated in each of the following:
    a. $6(5 \cdot 4) = (6 \cdot 5)4$
    b. $6(5 \cdot 4) = 6(4 \cdot 5)$
    c. $6(5 \cdot 4) = (5 \cdot 4)6$
    d. $1(5 \cdot 4) = 5 \cdot 4$
    e. $(3 + 4) \cdot 0 = 0$
    f. $(3 + 4)(5 + 6) = (3 + 4)5 + (3 + 4)6$

11. Students are overheard making the following statements. What properties justify their statements?
    a. I know that $9 \cdot 7$ is either 63 or 69 and I know they can't both be right.
    b. I know that $9 \cdot 0$ is 0 because I know that any number times 0 is 0.

    c. Any number times 1 is the same as the number we started with, so $9 \cdot 1$ is 9.

12. The product $6 \cdot 14$ can be found by thinking of the problem as $6(10 + 4) = 6 \cdot 10 + 6 \cdot 4 = 60 + 24 = 84$.
    a. What properties are being used?
    b. Use this technique to mentally compute $32 \cdot 12$.

13. Use the distributive property of multiplication over subtraction to compute each of the following:
    a. $9(10 - 2)$
    b. $20(8 - 3)$

14. Show that $(a + b)^2 = a^2 + 2ab + b^2$ using
    a. the distributive property of multiplication over addition and other properties.
    b. an area model.

15. If a and b are whole numbers with $a > b$, use the rectangles in the figure to explain why $(a + b)^2 - (a - b)^2 = 4ab$.

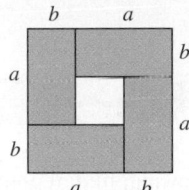

16. Use the property in exercise 14 to compute the following:
    a. 51^2
    b. 52^2
    c. 101^2
    d. 102^2

17. In each of the following, show that the left side of the equation is equal to the right side and give a reason for every step:
    a. $(ab)c = (ca)b$
    b. $(a + b)c = c(b + a)$

18. Factor each of the following:
    a. $xy - y^2$
    b. $47 \cdot 101 - 47$
    c. $ab^2 - ba^2$

19. Rewrite each of the following division problems as a multiplication problem:
    a. $40 \div 8 = 5$
    b. $326 \div 2 = x$

20. Think of a number. Multiply it by 5. Add 5. Divide by 5 and then subtract 1. How does the result compare with your original number? Will this work all the time? Justify your answer.

21. Show that, in general, each of the following is false if a, b, and c are whole numbers:
    a. $(a \div b) \div c = a \div (b \div c)$
    b. $a \div (b + c) = (a \div b) + (a \div c)$

22. Suppose all of the operations result in whole numbers. Explain why $(a + b) \div c = (a \div c) + (b \div c)$.

23. Find the solution for each of the following:
    a. $5x + 2 = 22$
    b. $3x + 7 = x + 13$
    c. $3(x + 4) = 18$
    d. $(x - 5) \div 10 = 9$

24. A new model of car is available in 4 exterior colors and 3 interior colors. Use a tree diagram and specific colors to show how many color schemes are possible for the car.

25. 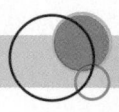 To find $7 \div 5$ on the calculator, press $\boxed{7}\,\boxed{\div}\,\boxed{5}\,\boxed{=}$, which yields 1.4. To find the whole-number remainder, ignore the decimal portion of 1.4, multiply $5 \cdot 1$, and subtract this product from 7. The result is the remainder. Use a calculator to find the whole-number remainder for each of the following divisions:

    a. $28 \div 5$ b. $32 \div 10$
    c. $29 \div 3$ d. $41 \div 7$
    e. $49,382 \div 14$

26. Is it possible to find a whole number less than 100 that when divided by 10 leaves remainder 4 and when divided by 47 leaves remainder 17?

27. Students were divided into 10 teams with 12 on each team. Later, the same students were divided into teams with 8 on each team. How many teams were there then?

28. In each of the following, tell what computation must be done last:

    a. $5(16 - 7) - 18$
    b. $54/(10 - 5 + 4)$

    c. $(14 - 3) + (24 \cdot 2)$
    d. $21,045/345 + 8$

29. Find infinitely many whole numbers that leave remainder 1 upon division by 4.

30. The operation $\odot$ is defined on the set $S = \{a, b, c\}$, as shown in the following table. For example, $a \odot b = b$ and $b \odot a = b$.

$\odot$	a	b	c
a	a	b	c
b	b	c	a
c	c	a	b

    a. Is S closed with respect to $\odot$?
    b. Is $\odot$ commutative on S?
    c. Is there an identity for $\odot$ on S? If yes, what is it?
    d. Try several examples to investigate the associative property for $\odot$ on S.

Assessment 3-3B

1. For each of the following, find, if possible, the whole numbers that make the equations true:

    a. $8 \cdot \square = 24$ b. $28 = 4 + 6 \cdot \square$
    c. $\square \cdot (8 + 6) = \square \cdot 8 + \square \cdot 6$

2. Determine if the following sets are closed under multiplication:

    a. $\{1, 2\}$ b. $\{2k + 1 \mid k \in W\}$
    c. $\{2^{k+1} \mid k \in W\}$

3. a. If 2 is removed from the set of whole numbers, is the set closed with respect to addition? Explain.
   b. If 1 is removed from the set of whole number, is the set closed with respect to multiplication? Explain.

4. Rename each of the following using the distributive property of multiplication over addition so that there are no parentheses in the final answer. Simplify when possible.

    a. $3(x + y + 5)$
    b. $(x + y)(x + y + z)$
    c. $x(y + 1) - x$

5. Place parentheses, if needed, to make each of the following equations true:

    a. $4 + 3 \cdot 2 = 14$
    b. $9 \div 3 + 1 = 4$
    c. $5 + 4 + 9 \div 3 = 6$
    d. $3 + 6 - 2 \div 1 = 7$

6. Using the distributive property of multiplication over addition, we can factor as in $x^2 + xy = x(x + y)$. Use the distributive property and other multiplication properties to factor each of the following:

    a. $47 \cdot 99 + 47$
    b. $(x + 1)y + (x + 1)$
    c. $x^2y + zx^3$

7. For each of the following, find whole numbers to make the statement true, if possible:

    a. $27 \div 9 = \square$
    b. $\square \div 52 = 1$
    c. $13 \div \square = 13$

8. A new car comes in 5 exterior colors and 3 interior colors. How many different looking cars are possible?

9. What multiplication is suggested by the following models?

    a.

    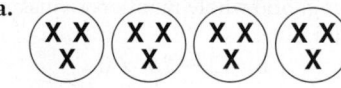

    b. ▦

10. Which property of whole numbers is illustrated in each of the following:

    a. $(5 \cdot 4)0 = 0$
    b. $7(3 \cdot 4) = 7(4 \cdot 3)$
    c. $7(3 \cdot 4) = (3 \cdot 4)7$
    d. $(3 + 4)1 = 3 + 4$
    e. $(3 + 4)5 = 3 \cdot 5 + 4 \cdot 5$
    f. $(1 + 2)(3 + 4) = (1 + 2)3 + (1 + 2)4$

11. Students are overheard making the following statements. What properties justify their statements?

    a. I know if I remember what $7 \cdot 9$ is, then I also know what $9 \cdot 7$ is.
    b. To find $9 \cdot 6$, I just remember that $9 \cdot 5$ is 45 and so $9 \cdot 6$ is just 9 more than 45, or 54.

12. The product $5 \cdot 24$ can be found by thinking of the computation as $5(20 + 4) = 5 \cdot 20 + 5 \cdot 4 = 100 + 20 = 120$.

    a. What property is being used?
    b. Use this technique to mentally compute $8 \cdot 34$.

13. Use the distributive property of multiplication over subtraction to compute each of the following:
    a. $15(10 - 2)$
    b. $30(9 - 2)$
14. Show that if $b > c$, then $a(b - c) = ab - ac$ using an area model suggested by the given figure (express the shaded area in two different ways).

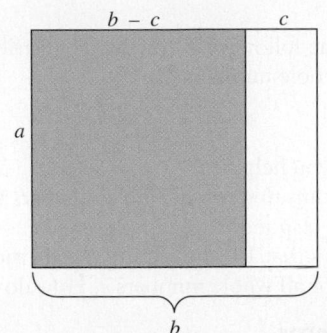

15. Use an area model suggested by the following figure to explain why $(a + b)(a - b) = a^2 - b^2$.

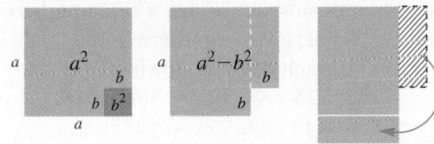

16. Use the formula $(a + b)(a - b) = a^2 - b^2$ to compute the following:
    a. $19 \cdot 21$ b. $25 \cdot 15$
    c. $99 \cdot 101$ d. $101^2 - 99^2$
17. Show that the left-hand side of the equation is equal to the right-hand side and give a reason for every step.
    a. $(ab)c = b(ac)$ b. $a(b + c) = ac + ab$
18. Factor each of the following:
    a. $xy - y$
    b. $(x + 1)y - (x + 1)$
    c. $a^2b^3 - ab^2$
19. Rewrite each of the following division problems as a multiplication problem:
    a. $48 \div x = 16$ b. $x \div 5 = 17$
20. Think of a number. Multiply it by 2. Add 2. Divide by 2. Subtract 1. How does the result compare with your original number? Will this work all the time? Explain your answer.
21. Show that, in general, each of the following is false if a, b, and c are whole numbers:
    a. $a \div b = b \div a$
    b. $a - b = b - a$
22. Suppose all operations result in whole numbers. Explain why $(a - b) \div c = (a \div c) - (b \div c)$.
23. Find the solution for each of the following:
    a. $5x + 8 = 28$ b. $5x + 6 = x + 14$
    c. $5(x + 3) = 35$ d. $(x - 6) \div 3 = 1$
24. String art is formed by connecting evenly spaced nails on the vertical and horizontal axes by segments of string. Connect

the nail farthest from the origin on the vertical axis with the nail closest to the origin on the horizontal axis. Continue until all nails are connected, as shown in the figure that follows. How many intersection points are created with 10 nails on each axis?

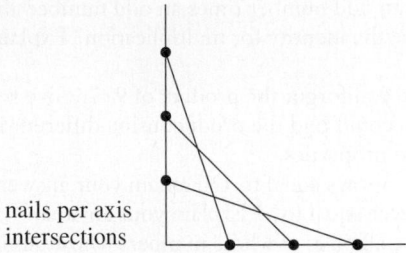

3 nails per axis
3 intersections

25. To find $7 \div 5$ on the calculator, press �7️⃣➗5️⃣🟰, which yields 1.4. To find the whole-number remainder, ignore the decimal portion of 1.4, multiply $5 \cdot 1$, and subtract this product from 7. The result is the remainder. Use a calculator to find the whole-number remainder for each of the following divisions:
    a. $28 \div 8$
    b. $42 \div 10$
    c. $29 \div 13$
    d. $45 \div 7$
    e. $59{,}382 \div 14$
26. Jonah has a large collection of marbles. He notices that if he borrows 5 marbles from a friend, he can arrange the marbles in rows of 13 each. What is the remainder when he divides his original number of marbles by 13?
27. Students were divided into eight teams with nine on each team. Later, the same students were divided into teams with six on each team. How many teams were there then?
28. In each of the following, tell what computation must be done last:
    a. $5 \cdot 6 - 3 \cdot 4 + 2$
    b. $19 - 3 \cdot 4 + 9 \div 3$
    c. $15 - 6 \div 2 \cdot 4$
    d. $5 + (8 - 2)3$
29. Find infinitely many whole numbers that leave remainder 3 upon division by 5.
30. The operation $\odot$ is defined on the set $S = \{a, b, c\}$, as shown in the following table. For example, $a \odot b = b$ and $b \odot a = b$.

$\odot$	a	b	c
a	a	b	c
b	b	a	c
c	c	c	c

    a. Is S closed with respect to $\odot$?
    b. Is $\odot$ commutative on S?
    c. Is there an identity for $\odot$ on S? If yes, what is it?
    d. Try several examples to investigate the associative property for $\odot$ on S.

Mathematical Connections 3-3

Communication

1. Why is an odd number times an odd number always odd?
2. Can 0 be the identity for multiplication? Explain why or why not.
3. Suppose you forgot the product of $9 \cdot 7$. Give several ways that you could find the product using different multiplication facts and properties.
4. Is $x \div x$ always equal to 1? Explain your answer.
5. Is $x \cdot x$ ever equal to x? Explain your answer.
6. Describe all pairs of whole numbers whose sum and product are the same.

Open-Ended

7. Describe a real-life situation that could be represented by the expression $3 + 2 \cdot 6$.

Cooperative Learning

8. Multiplication facts that most children have memorized can be stated in the table that is partially filled:

×	1	2	3	4	5	6	7	8	9
1									
2									
3									
4				16					
5							35		
6									
7									
8									72
9									81

a. Fill out the table of multiplication facts. Find as many patterns as you can. List all the patterns that your group discovered and explain why some of those patterns occur in the table.
b. How can the multiplication table be used to solve division problems?
c. Consider the odd number 35 shown in the multiplication table. Consider all the numbers that surround it. Note that they are all even. Does this happen for all odd numbers in the table? Explain why or why not.

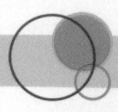

9. Enter a natural number less than 20 on the calculator. If the number is even, divide it by 2; if it is odd, multiply it by 3 and add 1. Next, use the number on the display. Follow the given directions. Repeat the process.
   a. Will the display eventually reach 1?
   b. Which number less than 20 takes the most steps before reaching 1?
   c. Do even or odd numbers reach 1 more quickly?

   d. Investigate what happens with numbers greater than 20.

Questions from the Classroom

10. Suppose a student argued that $0 \div 0 = 1$ because every number divided by itself is 1. How would you help that person?
11. Sue claims the following is true by the distributive law, where a and b are whole numbers:

$$3(ab) = (3a)(3b)$$

How might you help her?
12. A student claims that for all whole numbers $(ab) \div b = a$. How do you respond?
13. A student says that 1 is the identity for division because $a \div 1 = a$ for all whole numbers a. How do you respond?

Review Problems

14. Give an infinite set of even numbers that is not closed under addition.
15. Is the operation of subtraction for whole numbers commutative? If not, give a counterexample.
16. What is wrong in each of the following?

a.	137	b.	35	c.	56	d.	46
	$+56$		$+47$		-29		-17
	183		712		33		39

Trends in Mathematics and Science Study (TIMSS) Questions

In Toshi's class there are twice as many girls as boys. There are 8 boys in the class. What is the total number of boys and girls in the class?
  a. 12   b. 16   c. 20   d. 24

TIMSS, Grade 4, 2007

A piece of rope 204 cm long is cut into 4 equal pieces. Which of these gives the length of each piece in centimeters?
  a. $204 + 4$ b. 204×4 c. $204 - 4$ d. $204 \div 4$

TIMSS, Grade 4, 2007

National Assessment of Educational Progress (NAEP) Question

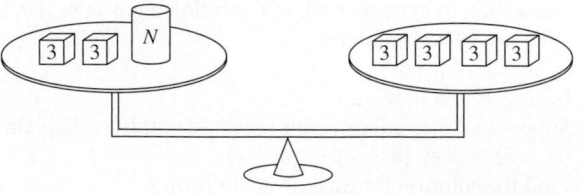

The weights on the scale above are balanced. Each cube weighs 3 pounds. The cylinder weighs N pounds. Which number sentence best describes this situation?
  a. $6 + N = 12$
  b. $6 + N = 4$
  c. $2 + N = 12$
  d. $2 + N = 4$

NAEP, Grade 4, 2007

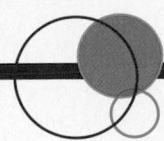

3-4 Algorithms for Whole-Number Multiplication and Division

In the grade 4 *Focal Points*, it states the following with respect to multiplication and division and students' use of algorithms for doing computations:

> They select appropriate methods and apply them accurately to estimate products or calculate them mentally, depending on the context and numbers involved. They develop fluency with efficient procedures, including the standard algorithm, for multiplying whole numbers, understand why the procedures work (on the basis of place value and properties of operations), and use them to solve problems. (p. 16)

In this section, multiplication and division algorithms will be developed using various models. We start with properties of exponents.

Properties of Exponents

In what follows, we introduce some properties of exponents that are useful in this section and in the following chapters.

Definition of a^n

If a, the base, and n, the exponent, are whole numbers and $n \neq 0$, then

$$a^n = \underbrace{a \cdot a \cdot \ldots \cdot a}_{n \text{ factors}} \quad \text{and} \quad a^1 = a$$

When multiplying powers of 10, the definition of exponents is used. For example, $10^2 \cdot 10^1 = (10 \cdot 10)10 = 10^3$, or 10^{2+1}. In general, where a is a whole number and m and n are natural numbers, $a^m \cdot a^n$ is given by the following:

$$a^m \cdot a^n = (\underbrace{a \cdot a \cdot a \cdot \ldots \cdot a}_{m \text{ factors}}) \cdot (\underbrace{a \cdot a \cdot a \cdot \ldots \cdot a}_{n \text{ factors}})$$

$$= \underbrace{a \cdot a \cdot a \cdot \ldots \cdot a}_{m + n \text{ factors}} = a^{m+n}$$

Consequently, $a^m \cdot a^n = a^{m+n}$.

Therefore we have following theorem.

Theorem 3-9

For every whole number a and natural numbers m and n:

$$a^m \cdot a^n = a^{m+n}$$

The above definition and theorem can be used to rewrite an expression such as $(5^2)^3$ using a single exponent:

$$(5^2)^3 = 5^2 \cdot 5^2 \cdot 5^2 = 5^{2+2+2} = 5^{3 \cdot 2} = 5^6$$

This suggest the following theorem:

Theorem 3-10

For every whole number a, and natural numbers m and n:

$$(a^m)^n = a^{mn}$$

Proof

$$(a^m)^n = \underbrace{a^m \cdot a^m \cdot \ldots \cdot a^m}_{n \text{ factors}} = a^{\overbrace{m+m+\ldots+m}^{n \text{ terms}}} = a^{nm}$$

Notice that the preceding theorems involve exponents and multiplication. However, corresponding properties for exponents and addition do not exist, for example $2^5 + 2^3 \neq 2^{5+3}$.

Sometimes it is convenient to write a product such as $2^3 \cdot 5^3$ with a single exponent:

$$2^3 \cdot 5^3 = 2 \cdot 2 \cdot 2 \cdot 5 \cdot 5 \cdot 5 = (2 \cdot 5)(2 \cdot 5)(2 \cdot 5) = (2 \cdot 5)^3$$

In general it can be stated as:

Theorem 3-11

For every whole number a, and natural numbers n:

$$a^n \cdot b^n = (ab)^n$$

The proof of this theorem is similar to the one in the previous example and is left as an exercise. It is often useful to divide exponential expressions with the same base. Consider, writing $2^6 \div 2^2$ with a single exponent. Using the definition of division, $2^6 \div 2^2 = \square$, if, and only if, $\square \cdot 2^2 = 2^6$. Since $\boxed{2^4} \cdot 2^2 = 2^6$, it follows that $2^6 \div 2^2 = 2^4$. This example suggests the following theorem:

Theorem 3-12

If a, m, and n are natural numbers with $m > n$, then

$$a^m \div a^n = a^{m-n}$$

Proof

By definition of division $a^m \div a^n = a^x$ if, and only if, $a^x \cdot a^n = a^m$. We want to find x in terms of m and n. By Theorem 3-9, $a^{x+n} = a^m$. Hence $x + n = m$, so $x = m - n$. Thus $a^m \div a^n = a^{m-n}$.

So far the exponents have been natural numbers. How should a^0 be defined? If the laws of exponents for natural numbers are to hold for the 0 exponent, then $a^0 \cdot a^n = a^{0+n} = a^n$. Because $a^0 \cdot a^n = a^n$ then $a^0 = 1$. However, 0^0 is left undefined.

Definition of a^0 for natural number a

If a is a natural number then

$$a^0 = 1$$

With the above defintion Theorem 3-11 is true for all whole numbers n and Theorem 3-12 is true for all whole numbers m and n with $m \geq n$.

 **EXAMPLE 3-4**

Write each of the following with only one exponent.

   a. $2^6 \cdot 8^5 \cdot 16^3$ **b.** $(9^4 \cdot 36^5) \div 3^{18}$

Solution

   a. $2^6 \cdot 8^5 \cdot 16^3 = 2^6 \cdot (2^3)^5 \cdot (2^4)^3 = 2^6 \cdot 2^{15} \cdot 2^{12}$
   $$= 2^{6+15+12} = 2^{33}$$

   b. $9^4 \cdot 36^5 = (3^2)^4 \cdot (2^2 \cdot 3^2)^5 = 3^{2 \cdot 4} \cdot (2^2)^5 \cdot (3^2)^5$
   $$= 3^8 \cdot 2^{10} \cdot 3^{10}$$
   $$= 3^8 \cdot 3^{10} \cdot 2^{10}$$
   $$= 3^{18} \cdot 2^{10}$$

   Thus $(9^4 \cdot 36^5) \div 3^{18} = (3^{18} \cdot 2^{10}) \div 3^{18}$
   $$= 2^{10}$$

NOW TRY THIS 3-15

Use the fact that $a^m \cdot a^n = a^{m+n}$ along other multiplication properties to explain why the computations in the cartoon are both true.

Hi & Lois copyright © 2005 King Features Syndicate

Multiplication Algorithms

To develop algorithms for multiplying multi-digit whole numbers, the strategy of *examining simpler computations* first is used. Consider $4 \cdot 12$. This computation could be pictured as in Figure 3-33(a) with 4 rows of 12 blocks, or 48 blocks. These blocks in Figure 3-33(a) can also be partitioned to show that $4 \cdot 12 = 4(10 + 2) = 4 \cdot 10 + 4 \cdot 2$. The numbers $4 \cdot 10$ and $4 \cdot 2$ are *partial products*.

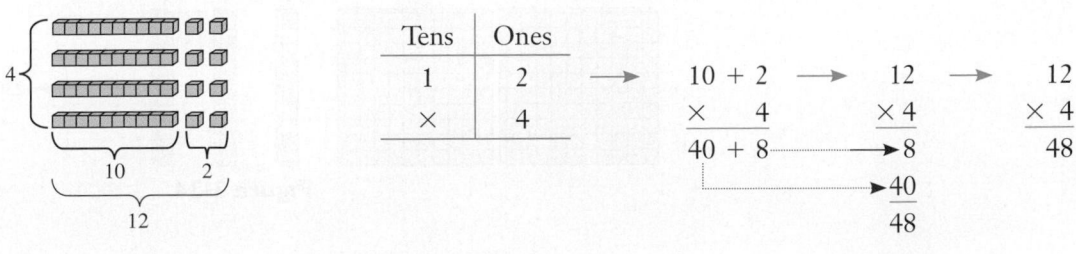

(a)   (b)

Figure 3-33

Figure 3-33(a) illustrates the distributive property of multiplication over addition on the set of whole numbers. The process leading to an algorithm for multiplying $4 \cdot 12$ is seen in Figure 3-33(b). Notice the similarity between the multiplication in Figure 3-33 and the following algebra multiplication:

$$4(x + 2) = 4x + 4 \cdot 2$$
$$= 4x + 8$$

Similarly, notice the analogy between the products

$$23 \cdot 14 = (2 \cdot 10 + 3)(1 \cdot 10 + 4) \quad \text{and} \quad (2x + 3)(1x + 4)$$

The analogy is continued as shown

$$
\begin{array}{r}
2 \cdot 10 + 3 \\
\times\ (1 \cdot 10 + 4) \\
\hline
12 \\
8 \cdot 10 \\
3 \cdot 10 \\
2 \cdot 10^2 \\
\hline
2 \cdot 10^2 + 11 \cdot 10 + 12
\end{array}
\qquad
\begin{array}{r}
2x + 3 \\
\times\ (1x + 4) \\
\hline
8x + 12 \\
2x^2 + 3x \\
\hline
2x^2 + 11x + 12
\end{array}
$$

Multiplication of a three-digit number by a one-digit factor will be explored after discussing multiplication by a power of 10.

Multiplication by 10^n

Next consider multiplication by powers of 10. First, consider what happens when a given number is multiplied by 10, such as $10 \cdot 23$. If we start out with the base-ten block representation of 23, we have 2 longs and 3 units. To multiply by 10, we must replace each piece with a base-ten piece that represents the next higher power of 10. This is shown in Figure 3-34. Notice that the 3 units in 23 when multiplied by 10 become 3 longs or 3 tens. Therefore, after multiplication by 10 there are no units and hence we have 0 in the units place. In general, if we multiply any natural number by 10, we append a 0 to the right of the original number.

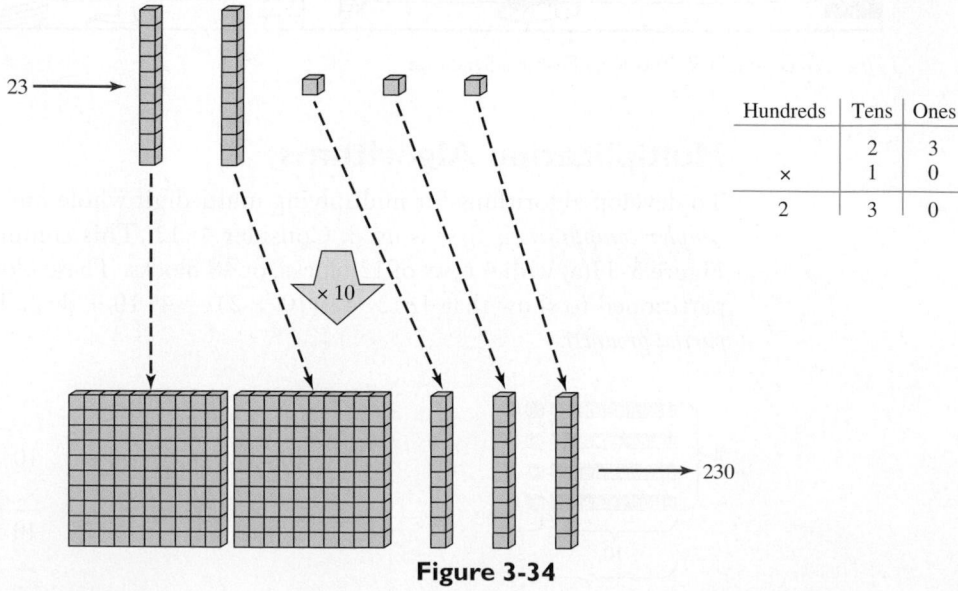

	Hundreds	Tens	Ones
		2	3
$\times$		1	0
	2	3	0

Figure 3-34

The computation $23 \cdot 10$ from Figure 3-34 can be shown as follows:

$$
\begin{aligned}
23 \cdot 10 &= (2 \cdot 10 + 3)10 \\
&= (2 \cdot 10)10 + 3 \cdot 10 \\
&= 2(10 \cdot 10) + 3 \cdot 10 \\
&= 2 \cdot 10^2 + 3 \cdot 10 \\
&= 2 \cdot 10^2 + 3 \cdot 10 + 0 \cdot 1 \\
&= 230
\end{aligned}
$$

To compute products such as $3 \cdot 200$, proceed as follows:

$$
\begin{aligned}
3 \cdot 200 &= 3(2 \cdot 10^2) \\
&= (3 \cdot 2)10^2 \\
&= 6 \cdot 10^2 \\
&= 6 \cdot 10^2 + 0 \cdot 10 + 0 \cdot 1 \\
&= 600
\end{aligned}
$$

Multiplying 6 by 10^2 results in appending two zeros to the right of 6. This idea can be generalized to the statement that *multiplication of any natural number by 10^n where n is a natural number, results in appending n zeros to the right of the number.*

The appending of n zeros to a natural number when multiplying by 10^n can also be explained as follows. First multiply by 10, resulting in appending of one zero (as in $23 \cdot 10 = 230$). Then multiply by another 10, resulting in appending another zero (as in $230 \cdot 10 = 2300$). Since we multiply n times by 10, n zeros are appended to the right of the original natural number.

Multiplication by a power of 10 is helpful in calculating the product of a one-digit number and a three-digit number. In the following example, we assume the previously developed algorithm for multiplying a one-digit number times a two-digit number:

$$
\begin{aligned}
4 \cdot 367 &= 4(3 \cdot 10^2 + 6 \cdot 10 + 7) \\
&= 4(3 \cdot 10^2) + 4(6 \cdot 10) + 4 \cdot 7 \\
&= (4 \cdot 3)10^2 + (4 \cdot 6)10 + 4 \cdot 7 \\
&= 1200 + 240 + 28 \\
&= 1468
\end{aligned}
$$

$$
\begin{array}{r}
367 \\
\times\ 4 \\
\hline
28 \\
240 \\
1200 \\
\hline
1468
\end{array}
$$

NOW TRY THIS 3-16

Use expanded notation and an approach similar to the preceding to calculate $7 \cdot 4589$.

Multiplication with Two-Digit Factors

Consider $14 \cdot 23$. Model this computation by first using base-ten blocks, as shown in Figure 3-35(a), and then showing all the *partial products* and adding, as shown in Figure 3-35(b).

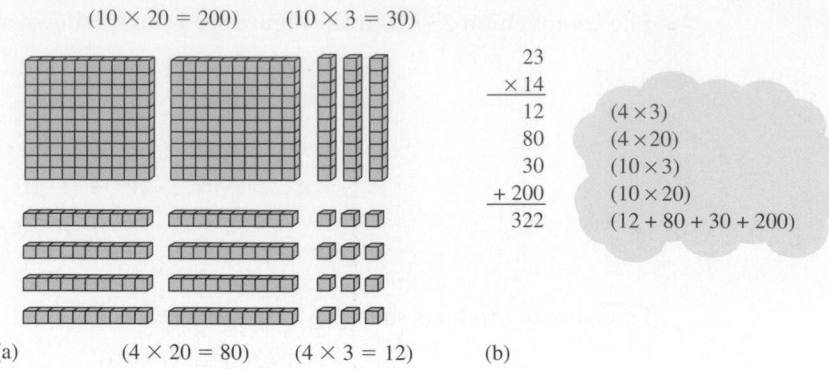

Figure 3-35

This last approach leads to an algorithm for multiplication:

$$\begin{array}{r} 23 \\ \times\ 14 \\ \hline 92 \end{array} \quad (4\cdot23) \quad \text{or} \quad \begin{array}{r} 23 \\ \times\ 14 \\ \hline 92 \end{array}$$

$$\begin{array}{r} \underline{230} \quad (10\cdot23) \\ 322 \end{array} \qquad \begin{array}{r} \underline{23} \\ 322 \end{array}$$

It is not uncommon to see the partial product 230 written without the zero, as 23. The placement of 23 with 3 in the tens column obviates having to write the 0 in the units column. We encourage the inclusion of the zero. This promotes better understanding and helps to avoid errors.

The distributive property of multiplication over addition can be used to explain why the algorithm for multiplication works. Again, consider $14\cdot23$.

$$\begin{aligned} 14\cdot23 &= (10+4)23 \\ &= 10\cdot23 + 4\cdot23 \\ &= 230 + 92 \\ &= 322 \end{aligned}$$

Because algorithms are powerful, there is sometimes a tendency to overapply them or to use paper and pencil for a task that should be done mentally. For example, consider

$$\begin{array}{r} 213 \\ \times\ \ 1000 \\ \hline 000 \\ 000 \\ 000 \\ \underline{213\ \ \ } \\ 213000 \end{array}$$

This application is not wrong but is inefficient. Mental math and estimation are important skills in learning mathematics and should be practiced in addition to paper-and-pencil computations. Children should be encouraged to *estimate* whether their answers are reasonable. For example, in the computation $14\cdot23$, we know that the answer must be between $10\cdot20 = 200$ and $20\cdot30 = 600$ because $10 < 14 < 20$ and $20 < 23 < 30$.

Lattice Multiplication

Lattice multiplication has the advantage of delaying all additions until the single-digit multiplications are complete. Because of this, it is sometimes referred to as a "low-stress algorithm." Students like this algorithm, perhaps because of the structure provided by the lattice. The lattice multiplication algorithm for multiplying 14 and 23 is shown in Figure 3-36. (Determining the reasons why lattice multiplication works is left as an exercise.)

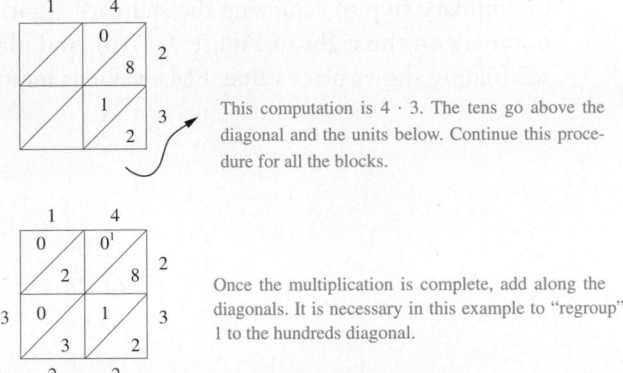

This computation is 4 · 3. The tens go above the diagonal and the units below. Continue this procedure for all the blocks.

Once the multiplication is complete, add along the diagonals. It is necessary in this example to "regroup" 1 to the hundreds diagonal.

Figure 3-36

Division Algorithms

Using Repeated Subtraction to Develop the Standard Division Algorithm

As we have seen in section 3-3 division of whole numbers can be modeled by repeated subtraction. We use this approach in the following question.

> A shopkeeper is packaging juice in cartons that hold 6 bottles each. She has 726 bottles. How many cartons does she need?

We reason that if 1 carton holds 6 bottles, then 10 cartons hold 60 bottles and 100 cartons hold 600 bottles. If 100 cartons are filled, there are $726 - 100 \cdot 6$, or 126, bottles remaining. If 10 more cartons are filled, then $126 - 10 \cdot 6$, or 66, bottles remain. Similarly, if 10 more cartons are filled, $66 - 10 \cdot 6$, or 6, bottles remain. Finally, 1 carton will hold the remaining 6 bottles. The total number of cartons necessary is $100 + 10 + 10 + 1$, or 121. This procedure is summarized in Figure 3-37(a). A more efficient method is shown in Figure 3-37(b).

$$
\begin{array}{rl}
6\overline{)726} & \\
-600 & \text{100 sixes} \\
\hline
126 & \\
-60 & \text{10 sixes} \\
\hline
66 & \\
-60 & \text{10 sixes} \\
\hline
6 & \\
-6 & \text{1 six} \\
\hline
0 & \text{121 sixes}
\end{array}
\qquad
\begin{array}{rl}
6\overline{)726} & \\
-600 & \text{100 sixes} \\
\hline
126 & \\
-120 & \text{20 sixes} \\
\hline
6 & \\
-6 & \text{1 six} \\
\hline
0 & \text{121 sixes}
\end{array}
$$

(a)   (b)

Figure 3-37

Historical Note

Lattice multiplication dates back to tenth-century India. This algorithm was imported to Europe and was popular in the fourteenth and fifteenth centuries. Napier's rods (or bones), developed by John Napier in the early 1600s, were modeled on lattice multiplication. ●

Divisions such as the one in Figure 3-37 are usually shown in elementary school texts in the most efficient form, as in Figure 3-38(b), in which the numbers in color in Figure 3-38(a) are omitted. The technique used in Figure 3-38(a) is often called "scaffolding" and may be used as a preliminary step to achieving the standard algorithm, as in Figure 3-38(b). Scaffolding takes the numbers on the right in Figure 3-37(b), and places them on the top as in Figure 3-38(a). The scaffolding shows place value. Place value is important to understanding the standard algorithm.

$$
\begin{array}{r}
\underline{121} \\
1 \\
20 \\
100 \\
6\overline{)726} \\
-600 \\
\hline
126 \\
-120 \\
\hline
6 \\
-6 \\
\hline
0
\end{array}
\qquad
\begin{array}{r}
121 \\
6\overline{)726} \\
-6 \\
\hline
12 \\
-12 \\
\hline
6 \\
-6 \\
\hline
0
\end{array}
$$

(a)   (b)

Figure 3-38

Using Base-ten Blocks to Develop the Standard Division Algorithm

Students need to see why each move in an algorithm is appropriate rather than just what sequence of moves to make. Next base-ten blocks are used to justify why each move in the standard algorithm is appropriate. In Table 3-4, the base-ten model is on the left with the corresponding steps in the standard algorithm on the right.

Table 3-4

Base-ten Blocks	Algorithm
1. First represent 726 with base-ten blocks. 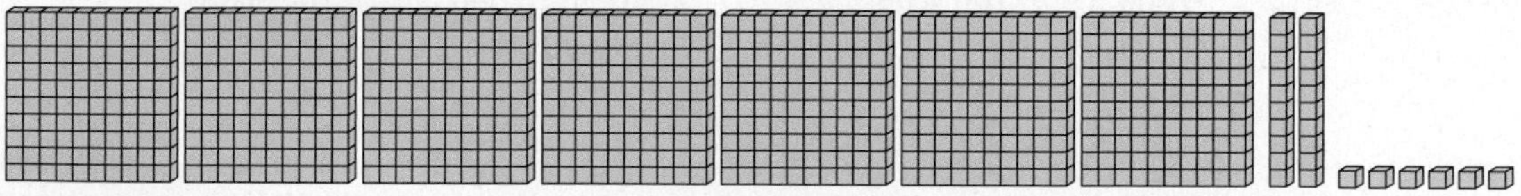	$6\overline{)726}$
2. Next determine how many sets of 6 flats (hundreds) there are in the representation. There is 1 set of 6 flats with 1 flat, 2 longs (tens), and 6 units (ones) left over. 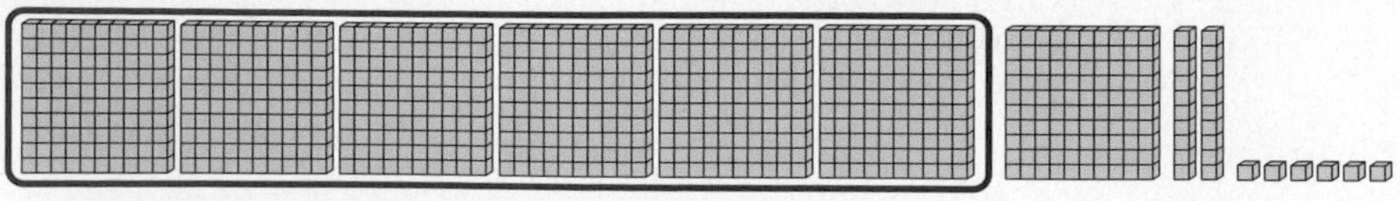	1 set of 6 flats $\begin{array}{r}1\\6\overline{)726}\\-6\\\hline1\end{array}$ 1 flat 2 longs 6 units left over
3. Next, convert the one leftover flat to 10 longs (tens), giving 12 longs (tens) and 6 units (ones). 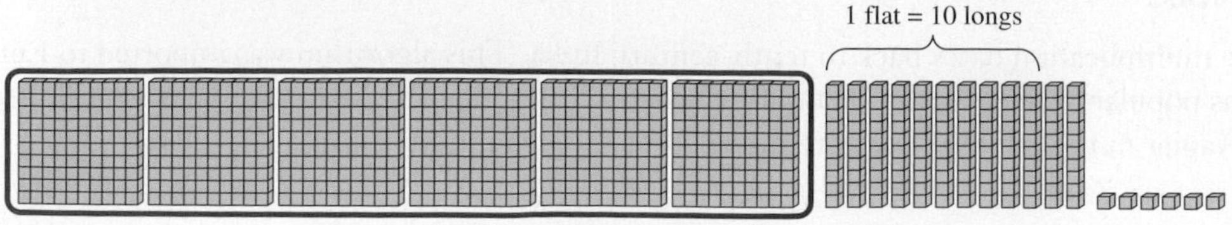 1 flat = 10 longs	1 set of 6 flats $\begin{array}{r}1\\6\overline{)726}\\-6\\\hline12\end{array}$ 12 longs 6 units left over

(continued)

Table 3-4 continued

Base-ten Blocks	Algorithm
4. Then determine how many sets of 6 longs (tens) there are in 12 longs and 6 units, giving 2 sets of 6 longs and 6 units left over.	1 set of 6 flats 2 sets of 6 longs $\begin{array}{r} 12 \\ 6\overline{)726} \\ -\underline{6} \\ 12 \\ -\underline{12} \\ 6 \end{array}$ 6 units left over
5. Finally determine how many sets of 6 units (ones) there are in the 6 remaining units. There is 1 set of 6 units with no units left over (the remainder is 0).	1 set of 6 flats 2 sets of 6 longs 1 set of 6 units $\begin{array}{r} 121 \\ 6\overline{)726} \\ -\underline{6} \\ 12 \\ -\underline{12} \\ 6 \\ -\underline{6} \\ 0 \end{array}$ 0 remainder

Therefore, in the base-ten block representation of 726, there is 1 group of 6 flats (hundreds), 2 groups of 6 longs (tens), and 1 group of 6 units (ones) with none left over. Hence, the quotient is 121 with a remainder of 0. The steps in the algorithm are shown alongside the work with the base-ten blocks.

Short Division

The process used in Table 3-4 is usually referred to as "long" division. Another technique, called "short" division, can be used when the divisor is a one-digit number and most of the work is done mentally. An example of the short division algorithm is given in Figure 3-39.

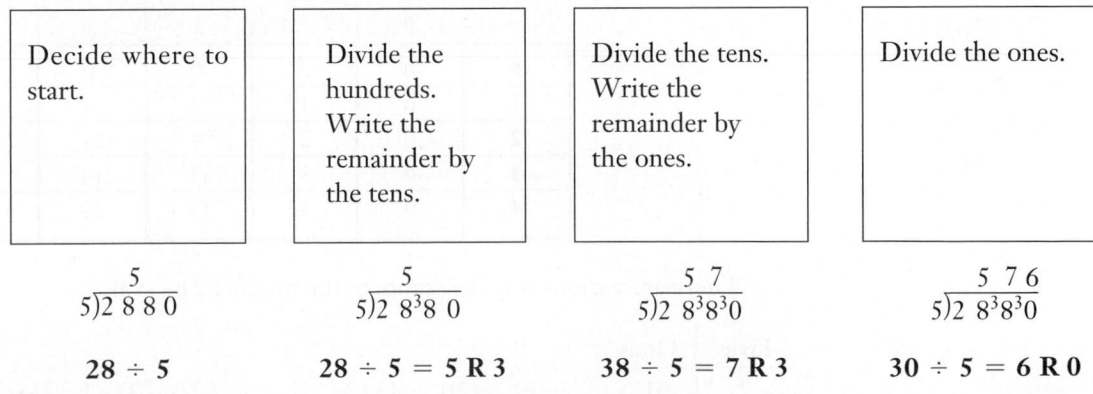

Decide where to start.	Divide the hundreds. Write the remainder by the tens.	Divide the tens. Write the remainder by the ones.	Divide the ones.
$\dfrac{5}{5\overline{)2\ 8\ 8\ 0}}$	$\dfrac{5}{5\overline{)2\ 8^3 8\ 0}}$	$\dfrac{5\ 7}{5\overline{)2\ 8^3 8^3 0}}$	$\dfrac{5\ 7\ 6}{5\overline{)2\ 8^3 8^3 0}}$
$28 \div 5$	$28 \div 5 = 5\ \textbf{R}\ 3$	$38 \div 5 = 7\ \textbf{R}\ 3$	$30 \div 5 = 6\ \textbf{R}\ 0$

Figure 3-39

Division by a Two-digit Divisor

An example of division by a divisor of more than one digit is given next. Consider $32\overline{)2618}$.

1. Estimate the quotient in $32\overline{)2618}$. Because $1 \cdot 32 = 32$, $10 \cdot 32 = 320$, and $100 \cdot 32 = 3200$, the quotient is between 10 and 100.

2. Find the number of tens in the quotient. Because $26 \div 3$ is approximately 8, 26 hundreds divided by 3 tens is approximately 8 tens. Then write the 8 in the tens place, as shown:

$$
\begin{array}{r}
80 \\
32\overline{)2618} \\
-2560 \\
\hline
58
\end{array}
\quad (32 \cdot 80)
$$

3. Find the number of units in the quotient. Because $5 \div 3$ is approximately 1, 5 tens divided by 3 tens is approximately 1. This is shown on the left, with the standard algorithm shown on the right.

$$
\begin{array}{r}
81 \\
\overline{1} \\
80 \\
32\overline{)2618} \\
-2560 \\
\hline
58 \\
-32 \\
\hline
26
\end{array}
\quad (32 \cdot 1) \quad \rightarrow \quad
\begin{array}{r}
81 \text{ R26} \\
32\overline{)2618} \\
-256 \\
\hline
58 \\
-32 \\
\hline
26
\end{array}
$$

Normally grade-school books show the format on the right, which places the remainder beside the quotient.

4. Check: $32 \cdot 81 + 26 = 2618$.

Multiplication and Division in Different Bases

In multiplication, as with addition and subtraction, we identify the basic facts of single-digit multiplication before developing any algorithms. The multiplication facts for base five are given in Table 3-5. These facts can be derived by using repeated addition.

Table 3-5 Base Five Multiplication Table

x	0	1	2	3	4
0	0	0	0	0	0
1	0	1	2	3	4
2	0	2	4	11	13
3	0	3	11	14	22
4	0	4	13	22	31

There are various ways to compute the product $21_{\text{five}} \cdot 3_{\text{five}}$:

Fives	Ones
2	1
×	3

$$
\begin{array}{r}
(20 + 1)_{\text{five}} \\
\times \quad 3_{\text{five}} \\
\hline
(110 + 3)_{\text{five}}
\end{array}
\quad \rightarrow \quad
\begin{array}{r}
21_{\text{five}} \\
\times 3_{\text{five}} \\
\hline
3 \\
110 \\
\hline
113_{\text{five}}
\end{array}
\quad \rightarrow \quad
\begin{array}{r}
21_{\text{five}} \\
\times \ 3_{\text{five}} \\
\hline
113_{\text{five}}
\end{array}
$$

The multiplication of a two-digit number by a two-digit number is developed next:

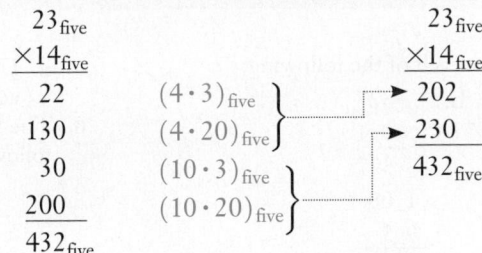

$$
\begin{array}{r}
23_{\text{five}} \\
\times 14_{\text{five}} \\
\hline
22 \\
130 \\
30 \\
200 \\
\hline
432_{\text{five}}
\end{array}
\quad
\begin{array}{l}
(4 \cdot 3)_{\text{five}} \\
(4 \cdot 20)_{\text{five}} \\
(10 \cdot 3)_{\text{five}} \\
(10 \cdot 20)_{\text{five}}
\end{array}
\qquad
\begin{array}{r}
23_{\text{five}} \\
\times 14_{\text{five}} \\
\hline
202 \\
230 \\
\hline
432_{\text{five}}
\end{array}
$$

Lattice multiplication can also be used to multiply numbers in various number bases. This is explored in Assessment 3-4.

Division in different bases can be performed using the multiplication facts and the definition of division. For example, $22_{\text{five}} \div 3_{\text{five}} = c$ if, and only if, $c \cdot 3_{\text{five}} = 22_{\text{five}}$. From Table 3-5, we see that $c = 4_{\text{five}}$. As in base-ten, computing multi-digit divisions efficiently in different bases requires practice. The ideas behind the algorithms for division can be developed by using repeated subtraction. For example, $3241_{\text{five}} \div 43_{\text{five}}$ is computed by means of the repeated-subtraction technique in Figure 3-40(a) and by means of the conventional algorithm in Figure 3-40(b). Thus, $3241_{\text{five}} \div 43_{\text{five}}$ results in quotient 34_{five} and remainder 14_{five}. Using the **division algorithm**, this can be written as $3241_{\text{five}} = 34_{\text{five}} \cdot 43_{\text{five}} + 14_{\text{five}}$.

$$
\begin{array}{r}
43_{\text{five}} \overline{)3241_{\text{five}}} \\
-430 \\
\hline
2311 \\
-430 \\
\hline
1331 \\
-430 \\
\hline
401 \\
-141 \\
\hline
-210 \\
-141 \\
\hline
14
\end{array}
\quad
\begin{array}{l}
(10 \cdot 43)_{\text{five}} \\
\\
(10 \cdot 43)_{\text{five}} \\
\\
(10 \cdot 43)_{\text{five}} \\
\\
(2 \cdot 43)_{\text{five}} \\
\\
(2 \cdot 43)_{\text{five}} \\
(34 \cdot 43)_{\text{five}}
\end{array}
\qquad\qquad
\begin{array}{r}
34_{\text{five}} \ \text{R}14_{\text{five}} \\
43_{\text{five}} \overline{)3241_{\text{five}}} \\
-234 \\
\hline
401 \\
-332 \\
\hline
14_{\text{five}}
\end{array}
$$

(a)   (b)

Figure 3-40

Computations involving base two are demonstrated in Example 3-4.

EXAMPLE 3-5

a. Multiply:

$$
\begin{array}{r}
101_{\text{two}} \\
\times \ 11_{\text{two}}
\end{array}
$$

b. Divide:

$$101_{\text{two}} \overline{)110110_{\text{two}}}$$

Solution

a.
$$
\begin{array}{r}
101_{\text{two}} \\
\times 11_{\text{two}} \\
\hline
101 \\
101 \\
\hline
1111_{\text{two}}
\end{array}
$$

b.
$$
\begin{array}{r}
1010_{\text{two}} \ \text{R}100_{\text{two}} \\
101_{\text{two}} \overline{)110110_{\text{two}}} \\
-101 \\
\hline
111 \\
-101 \\
\hline
100_{\text{two}}
\end{array}
$$

Assessment 3-4A

1. Fill in the missing numbers in each of the following:

   a.
   ```
        4_6
      × 783
      ─────
      1_78
      3408
      _982
      ─────
     3335_8
   ```

   b.
   ```
        327
      × 9_1
      ─────
        327
      1_08
      _9_3
      ──────
     30_ _07
   ```

2. Perform the following multiplications using the lattice multiplication algorithm:

   a. 728
         × 94

   b. 306
         × 24

3. The following chart displays the average daily water use per person for various countries; an approximate population for the countries is also given.

   Compute the daily total water use in each country.

Country/population	Daily water use per person in liters
India/1,200,000,000	140
Brazil/200,000,000	200
Nigeria/160,000,000	40
Japan/130,000,000	375
Australia/23,000,000	500
Kuwait/3,000,000	500
Bahrain/800,000	450

4. Simplify each of the following using properties of exponents. Leave answers as powers.
   a. $5^7 \cdot 5^{12}$
   b. $6^{10} \cdot 6^2 \cdot 6^3$
   c. $10^{296} \cdot 10^{17}$
   d. $2^7 \cdot 10^5 \cdot 5^7$

5. **a.** Which is greater, $2^{80} + 2^{80}$ or 2^{100}? Why?
   b. Which is greatest, 2^{101}, $3 \cdot 2^{100}$, or 2^{102}? Why?

6. The following model illustrates $22 \cdot 13$:

   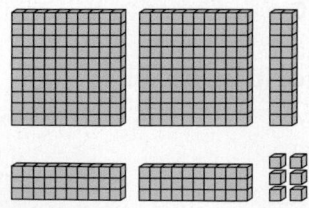

   a. Explain how the partial products are shown in the figure.
   b. Draw a similar model for $15 \cdot 21$.
   c. Draw a similar base five model for the product $43_{\text{five}} \cdot 23_{\text{five}}$. Explain how the model can be used to find the answer in base five.

7. **a.** Compute $110_{\text{two}} \cdot 11_{\text{two}}$.
   b. Use the distributive property of multiplication over addition to explain why multiplication of a natural number in base two by 10_{two} results in annexation of 0 to the number.
   c. Use part (b) to explain why multiplication in base two by 100_{two} results in annexation of 00 to the number.

d. Use the distributive property of multiplication over addition and part (b) to compute $110_{\text{two}} \cdot 11_{\text{two}}$.

8. The Russian peasant algorithm for multiplying $27 \cdot 68$ follows. (Disregard remainders when halving.)

	Halves			Doubles	
		→	27 ×	68	
Halve 27	→	13		136	Double 68
Halve 13	→	6		272	Double 136
Halve 6	→	3		544	Double 272
Halve 3	→	1		1088	Double 544

   In the "Halves" column, choose the odd numbers marked by an arrow. In the "Doubles" column, circle the numbers paired with the odds from the "Halves" column. Add the circled numbers.

   ```
       68
      136
      544
     1088
     ────
     1836  This is the product of 27 · 68.
   ```

   Try this algorithm for $17 \cdot 63$.

9. Answer the following questions based on the activity chart provided:

Activity	Calories Burned per Hour
Playing tennis	462
Snowshoeing	708
Cross-country skiing	444
Playing volleyball	198

   a. How many calories are burned during 3 hr of cross-country skiing?
   b. Jane played tennis for 2 hr while Carolyn played volleyball for 3 hr. Who burned more calories, and how many more?
   c. Lyle went snowshoeing for 3 hr and Maurice went cross-country skiing for 5 hr. Who burned more calories, and how many more?

10. On a 14-day vacation, Glenn increased his caloric intake by 1500 calories per day. He also worked out more than usual by swimming 2 hr a day. Swimming burns 666 calories per hour, and a net gain of 3500 calories adds 1 lb of weight. Did Glenn gain at least 1 lb during his vacation?

11. Dave purchased a $50,000 life insurance policy at the price of $30 for each $1000 of coverage. If he pays the premium quarterly, how much is each installment?

12. Perform each of the following divisions using both the repeated-subtraction and standard algorithms:
    a. $8\overline{)623}$
    b. $36\overline{)298}$
    c. $391\overline{)4001}$

13. Place the digits 4, 5, 7, and 3 in the boxes $\square\overline{)\square\square\square}$ to obtain
    a. the greatest quotient.
    b. the least quotient.

14. Using a calculator, Ralph multiplied by 10 when he should have divided by 10. The display read 300. What should the correct answer be?

15. Consider the following multiplications. Notice that when the digits in the factors are reversed, the products are the same.

$$
\begin{array}{r}
36 \\
\times\,42 \\
\hline
1512
\end{array}
\qquad
\begin{array}{r}
63 \\
\times\,24 \\
\hline
1512
\end{array}
$$

   a. Find other multiplications where this procedure works.
   b. Find a pattern for the numbers that work in this way.

16. Dan has 4520 pennies in three boxes. He says that there are 3 times as many pennies in the first box as in the third and twice as many in the second box as in the first. How much does he have in each box?

17. Gina buys apples from an orchard and then sells them at a country fair in bags of 3 for $1 a bag. She bought 50 boxes of apples, 36 apples in a box, and paid $452. If she sold all but 18 apples, what was her total profit?

18. Discuss possible error patterns in each of the following:

   a.
$$
\begin{array}{r}
35 \\
\times\,26 \\
\hline
90
\end{array}
$$

   b.
$$
\begin{array}{r}
5\,3 \\
5)\overline{2515} \\
-25 \\
\hline
15 \\
-15 \\
\hline
0
\end{array}
$$

19. Give reasons for each of the following steps:

$$
\begin{aligned}
56 \cdot 10 &= (5 \cdot 10 + 6) \cdot 10 \\
&= (5 \cdot 10) \cdot 10 + 6 \cdot 10 \\
&= 5 \cdot (10 \cdot 10) + 6 \cdot 10 \\
&= 5 \cdot 10^2 + 6 \cdot 10 \\
&= 5 \cdot 10^2 + 6 \cdot 10 + 0 \cdot 1 \\
&= 560
\end{aligned}
$$

20. a. Find all whole numbers that leave remainder 3 upon division by 4. Write your answer using set-builder notation.
    b. Write the numbers from part (a) in a sequence starting in increasing order.
    c. What kind of sequence is the one in part (b)? Why?

21. For what possible bases are each of the following computations correct?

   a.
$$
\begin{array}{r}
213 \\
+\,308 \\
\hline
522
\end{array}
$$

   b.
$$
\begin{array}{r}
213 \\
\times\,32 \\
\hline
430 \\
1043 \\
\hline
11300
\end{array}
$$

22. a. Use lattice multiplication to compute $323_{\text{five}} \cdot 42_{\text{five}}$.
    b. Find the least values of a and b such that $32_a = 23_b$.

23. Place the digits 7, 6, 8, and 3 in the boxes to obtain

   a. the greatest product.
   b. the least product.

24. Perform each of these operations using the bases shown:
   a. $32_{\text{five}} \cdot 4_{\text{five}}$
   b. $32_{\text{five}} \div 4_{\text{five}}$
   c. $43_{\text{six}} \cdot 23_{\text{six}}$
   d. $143_{\text{five}} \div 3_{\text{five}}$
   e. $10010_{\text{two}} \div 11_{\text{two}}$
   f. $10110_{\text{two}} \cdot 101_{\text{two}}$

Assessment 3-4B

1. Fill in the missing numbers in the following:

$$
\begin{array}{r}
4\_4 \\
\times\,327 \\
\hline
3\_88 \\
968 \\
452 \\
\hline
1582\_8
\end{array}
$$

2. Perform the following multiplications using the lattice multiplication algorithm:

   a.
$$
\begin{array}{r}
327 \\
\times\,43 \\
\end{array}
$$

   b.
$$
\begin{array}{r}
2618 \\
\times\,137 \\
\end{array}
$$

3. The following chart gives average water usage for various activities for one person for one day:

Use	Average Amount
Taking bath	110 L (liters)
Taking shower	75 L
Flushing toilet	22 L
Washing hands, face	7 L
Getting a drink	1 L
Brushing teeth	1 L
Doing dishes (one meal)	30 L
Cooking (one meal)	18 L

   a. Use the chart to calculate how much water you use each day.

   b. If average American uses approximately 200 L of water per day and there are approximately 310,000,000 people in the United States, on average approximately how much water is used in the United States per day?

4. Simplify each of the following using properties of exponents. Leave answers as powers.

   a. $3^8 \cdot 3^4$ **b.** $5^2 \cdot 5^4 \cdot 5^2$

   c. $6^2 \cdot 2^2 \cdot 3^2$ **d.** $4^8 \cdot 8^4 \cdot 32^5$

5. **a.** Which is greater, $2^{20} + 2^{20}$ or 2^{21}? Why?

   b. Which is greatest, $3^{31}, 9 \cdot 3^{30}$, or 3^{33}? Why?

6. The following model illustrates $13 \cdot 12$:

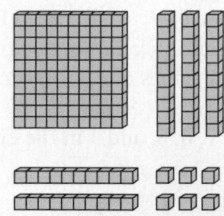

   a. Explain how the partial products are shown in the figure.

   b. Draw a similar model for $12 \cdot 22$.

7. **a.** Compute $14_{\text{five}} \cdot 23_{\text{five}}$.

   b. Use the distributive property of multiplication over addition to explain why multiplication of a natural number in base five by 10_{five} results in annexation of 0 to the number.

   c. Use part (b) to explain why multiplication of a natural number in base five by 100_{five} results in annexation of two 0s to the number.

   d. Use the distributive property of multiplication over addition and part (b) to compute $14_{\text{five}} \cdot 23_{\text{five}}$.

8. Use the Russian peasant algorithm from exercise 8 in Assessment 3-4A to find $31 \cdot 69$.

9. Answer the following questions based on the activity chart:

Activity	Calories Burned per Hour
Playing tennis	462
Snowshoeing	708
Cross-country skiing	444
Playing volleyball	198

   a. How many calories are burned during 4 hr of cross-country skiing?

   b. Jane played tennis for 3 hr while Carolyn played volleyball for 4 hr. Who burned more calories, and how many more?

   c. Lyle went snowshoeing for 4 hr and Maurice went cross-country skiing for 5 hr. Who burned more calories, and how many more?

10. On a 14-day vacation, Glenn increased his calorie intake by 1800 calories per day. He also worked out more than usual by swimming 3 hr a day. Swimming burns 666 calories per hour, and a net gain of 3500 calories adds 1 lb of weight. Did Glenn gain at least 1 lb during his vacation?

11. Sue purchased a $30,000 life-insurance policy at the price of $24 for each $1000 of coverage. If she pays the premium in 12 monthly installments, how much is each installment?

12. Perform each of the following divisions using both the repeated-subtraction and the standard algorithms:

   a. $7)\overline{392}$ **b.** $37)\overline{925}$

   c. $423)\overline{5002}$

13. Place the digits 7, 6, 8, and 3 in the boxes $\square)\overline{\square\,\square\,\square}$ to obtain

   a. the greatest quotient.

   b. the least quotient.

14. Using a calculator, Jody multiplied by 5 when she should have divided by 5. The display read 250. What should the correct answer be?

15. A student wrote the following to his parents:

$$\begin{array}{r} \text{SEND} \\ + \text{ MORE} \\ \hline \text{MONEY} \end{array}$$

If each letter represents a different digit how much money did the student ask for?

16. Debbie has 340 dimes in three boxes. She says that there are 4 times as many dimes in the first box as in the second and 3 times as many in the third box as in the first. How much money in dollars does she have in each box?

17. Xuan saved $5340 in 3 years. If he saved $95 per month in the first year and a fixed amount per month for the next 2 years, how much did he save per month during the last 2 years?

18. Discuss possible error patterns in each of the following:

   a. $\begin{array}{r} 34 \\ \times\ 8 \\ \hline 2432 \end{array}$ **b.** $\begin{array}{r} 34 \\ \times\ 6 \\ \hline 114 \end{array}$

19. Give reasons for each of the following steps:

$$\begin{aligned} 35 \cdot 100 &= (3 \cdot 10 + 5)100 \\ &= (3 \cdot 10 + 5)10^2 \\ &= (3 \cdot 10)10^2 + 5 \cdot 10^2 \\ &= 3(10 \cdot 10^2) + 5 \cdot 10^2 \\ &= 3 \cdot 10^3 + 5 \cdot 10^2 \\ &= 3 \cdot 10^3 + 5 \cdot 10^2 + 0 \cdot 10 + 0 \cdot 1 \\ &= 3500 \end{aligned}$$

20. **a.** Find all the whole numbers that leave remainder 1 upon division by 4. Write your answer using set-builder notation.

   b. Write the numbers from part (a) in a sequence in increasing order.

   c. What kind of sequence is the one in part (b)?

21. For what possible bases are each of the following computations correct?

   a. $\begin{array}{r} 322 \\ -\ 233 \\ \hline 23 \end{array}$ **b.** $\begin{array}{r} 101 \\ 11)\overline{1111} \\ -\ 11 \\ \hline 11 \\ -\ 11 \\ \hline 0 \end{array}$

22. **a.** Use lattice multiplication to compute $423_{\text{five}} \cdot 23_{\text{five}}$.

   b. Find the least values of a and b such that $41_a = 14_b$.

23. Place the digits 7, 6, 8, 3, and 2 in the boxes to obtain

$$\begin{array}{r} \square\square\square \\ \times \quad \square\square \\ \hline \end{array}$$

    a. the greatest product.   **b.** the least product.

24. Perform each of these operations using the bases shown:

    a. $42_{\text{five}} \cdot 3_{\text{five}}$

    b. $22_{\text{five}} \div 4_{\text{five}}$

    c. $32_{\text{five}} \cdot 42_{\text{five}}$

    d. $1313_{\text{five}} \div 23_{\text{five}}$

    e. $101_{\text{two}} \cdot 101_{\text{two}}$

    f. $1001_{\text{two}} \div 11_{\text{two}}$

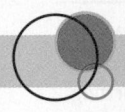

Mathematical Connections 3-4

Communication

1. How would you explain to children how to multiply $345 \cdot 678$, assuming that they know and understand multiplication by a single digit and multiplication by a power of 10?

2. What happens when you multiply any two-digit number by 101? Explain why this happens.

3. Pick a number. Double it. Multiply the result by 3. Add 24. Divide by 6. Subtract your original number. Is the result always the same? Write a convincing argument for what happens.

4. Do you think it is valuable for students to see more than one method of doing computation problems? Why or why not?

5. Tom claims that long division should receive reduced attention in elementary classrooms. Do you agree or disagree? Defend your answer. (Check the web for related research.)

Open-Ended

6. If a student presented a new "algorithm" for computing with whole numbers, describe the process you would recommend to the student to determine whether the algorithm would always work.

Cooperative Learning

7. The traditional sequence for teaching operations in the elementary school is first addition, then subtraction, followed by multiplication, and finally division. Some educators advocate teaching addition followed by multiplication, then subtraction followed by division. Within your group, prepare arguments for teaching the operations in either order listed.

Questions from the Classroom

8. A student asks why should she learn the standard long division algorithm if she can get a correct answer using repeated subtraction. How do you respond?

9. A student divides as follows. How do you help?

$$\begin{array}{r} 15 \\ 6\overline{)36} \\ -6 \\ \hline 30 \\ 30 \end{array}$$

10. A student asks how you can find the quotient and the remainder in a division problem like $593 \div 36$ using a calculator with only 4 arithmetic operations. How do you respond?

11. A student claims that to divide a number with the units digit 0 by 10, she just crosses out the 0 to get the answer. She wants to know if this is always true and why and if the 0 has to be the units digit. How do you respond?

12. A student claims that $m \div n = (mc) \div (nc)$ where m, n, and c are whole numbers, $c \neq 0$, and $m \div n$ is a whole number. She wants to know why. How would you respond, assuming the student does not know anything about fractions?

13. **a.** A student asks if $39 + 41 = 40 + 40$, is it true that $39 \cdot 41 = 40 \cdot 40$. How do you reply?

    b. Another student says that he knows that $39 \cdot 41 \neq 40 \cdot 40$ but he found that $39 \cdot 41 = 40 \cdot 40 - 1$. He also found that $49 \cdot 51 = 50 \cdot 50 - 1$. He wants to know if this pattern continues. How would you respond?

Review Problems

14. Illustrate the identity property of addition for whole numbers.

15. Rename each of the following using the distributive property of multiplication over addition:

    a. $ax + bx + 2x$

    b. $3(a + b) + x(a + b)$

16. At the beginning of a trip, the odometer registered 52,281. At the end of the trip, the odometer registered 59,260. How many miles were traveled on this trip?

17. Write each of the following division problems as a multiplication problem:

    a. $36 \div 4 = 9$

    b. $112 \div 2 = x$

    c. $48 \div x = 6$

    d. $x \div 7 = 17$

Trends in Mathematics and Science Study (TIMSS) Question

Each student needs 8 notebooks for school. How many notebooks are needed for 115 students?

Use the tiles $\boxed{1}$, $\boxed{4}$, and $\boxed{5}$. Write the numbers on the tiles in the boxes below to make the largest answer when you multiply.

$$\begin{array}{r} \square\square \\ \times \quad \square \\ \hline \end{array}$$

$37 \times \blacksquare = 703$.

What is the value of $37 \times \blacksquare + 6$?

TIMSS, Grade 4, 2007

National Assessment of Educational Progress (NAEP) Question

There will be 58 people at a breakfast and each person will eat 2 eggs. There are 12 eggs in each carton. How many cartons of eggs will be needed for the breakfast?

a. 9
b. 10
c. 72
d. 116

NAEP, Grade 4, 2007

 BRAIN TEASER

1. Messages can be coded on paper tape in base two. A hole in the tape represents 1, whereas the absence of a hole represents 0. The value of each hole depends on its position; from left to right, the values are 16, 8, 4, 2, 1 (all powers of 2). Letters of the alphabet may be coded in base two according to their position in the alphabet. For example, G is the seventh letter. Since $7 = 1 \cdot 4 + 1 \cdot 2 + 1$, the holes appear as they do in Figure 3-41:

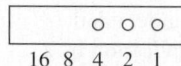

16 8 4 2 1

Figure 3-41

   a. Decode the message in Figure 3-42.

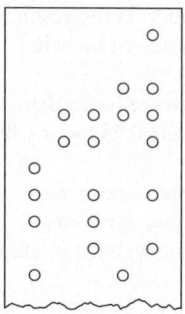

Figure 3-42

   b. Write your name on a tape using base two.
2. Consider the cards in Figure 3-43 which are modeled on base two arithmetic.

Card E		Card D		Card C		Card B		Card A	
16	24	8	24	4	20	2	18	1	17
17	25	9	25	5	21	3	19	3	19
18	26	10	26	6	22	6	22	5	21
19	27	11	27	7	23	7	23	7	23
20	28	12	28	12	28	10	26	9	25
21	29	13	29	13	29	11	27	11	27
22	30	14	30	14	30	14	30	13	29
23	31	15	31	15	31	15	31	15	31

Figure 3-43

   a. Suppose a person under 32 tells you that his age appears on cards E, C, and B. How can you use that information to find the person's age? Explain how the system works in general.
   b. Design card F and redesign cards A to E so that the numbers 1 through 63 appear on the cards.

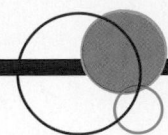

3-5 Mental Mathematics and Estimation for Whole-Number Operations

Focal Points makes the following statements about estimation at the various grade levels. Notice that as the grade level advances, additional operations are included until all four operations are covered.

In the grade 2 *Focal Points*:

> They [students] select and apply appropriate methods to estimate sums and differences or calculate them mentally, depending on the context and numbers involved. (p. 14)

In the grade 4 *Focal Points*:

> They [students] select appropriate methods and apply them accurately to estimate products or calculate them mentally, depending on the context and numbers involved. (p. 16)

In the grade 5 *Focal Points*:

> They [students] select appropriate methods and apply them accurately to estimate quotients or calculate them mentally, depending on the context and numbers involved. (p. 17)

In earlier sections, we focused mainly on paper-and-pencil computational strategies. Now we focus on mental mathematics and computational estimation. **Mental mathematics** is the process of producing an answer to a computation without using computational aids. **Computational estimation** is the process of forming an *approximate* answer to a numerical problem. Facility with estimation strategies helps to determine whether an answer is reasonable. In the *Calvin and Hobbes* cartoon, Calvin is a poor estimator and may very well believe that estimation is neither important nor useful.

Calvin and Hobbes   by Bill Watterson

Calvin & Hobbes copyright © 1990 and 2011 Watterson. Distributed by Universal Uclick. Reprinted with permission. All rights reserved.

Proficiency in mental mathematics can help in everyday estimation skills. It is essential to have these skills even in a time when calculators are readily available. We must be able to judge the reasonableness of calculator answers. Mental mathematics uses a variety of strategies and properties. We consider next several of the most common algorithms for performing operations mentally on whole numbers. Note that the *trading off* algorithm is just the *equal additions* algorithm discussed earlier.

Mental Mathematics: Addition

1. *Adding from the left*

   a.   67 $60 + 30 = 90$ (Add the tens.)
       + 36 $7 + 6 = 13$ (Add the units.)
                  $90 + 13 = 103$ (Add the two sums.)

   b.   36 $30 + 30 = 60$ (Double 30.)
       + 36 $6 + 6 = 12$ (Double 6.)
                  $60 + 12 = 72$ (Add the doubles.)

2. *Breaking up and bridging*

      67        $67 + 30 = 97$ (Add the first number to the tens in the second number.)
     + 36 $97 + 6 = 103$ (Add this sum to the units in the second number.)

3. *Trading off*

   a.   67 $67 + 3 = 70$ (Add 3 to make a multiple of 10.)
       + 36 $36 - 3 = 33$ (Subtract 3 to compensate for the 3 that was added.)
                  $70 + 33 = 103$ (Add the two sums.)

   b.   67 $67 + 30 = 97$ (Add 30 [next multiple of 10 greater than 29].)
       + 29 $97 - 1 = 96$ (Subtract 1 to compensate for the extra 1 that was added.)

4. *Using compatible numbers*
   Compatible numbers are numbers whose sums are easy to calculate mentally.

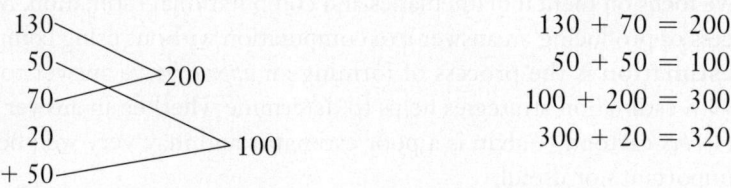

   $130 + 70 = 200$
   $50 + 50 = 100$
   $100 + 200 = 300$
   $300 + 20 = 320$

5. *Making compatible numbers*

      25        $25 + 75 = 100$ ($25 + 75$ adds up to 100.)
     + 79 $100 + 4 = 104$ (Add 4 more units.)

Mental Mathematics: Subtraction

1. *Breaking up and bridging*

      67        $67 - 30 = 37$ (Subtract the tens in the second number from the first number.)
     - 36 $37 - 6 = 31$ (Subtract the units in the second number from the difference.)

2. *Trading off*

      71        $71 + 1 = 72; 39 + 1 = 40$ (Add 1 to both numbers. Perform the
     -39 $72 - 40 = 32$ subtraction, which is easier than the
                                                original problem.)

   Notice that adding 1 to both numbers does not change the answer. (Why?)

3. *Drop the zeros*

      8700 $87 - 5 = 82$ (Notice that there are two zeros in each number. Drop
     - 500 $82 \rightarrow 8200$ these zeros and perform the computation. Then replace
                                        the two zeros to obtain proper place value.)

Another mental mathematics technique for subtraction is called "adding up." This method is based on the *missing addend* approach and is sometimes referred to as the "cashier's algorithm." An example of *adding up* or the *cashier's algorithm* follows.

EXAMPLE 3-5

Noah owed $11 for his groceries. He used a $50 bill to pay. While handing Noah the change, the cashier said, "11, 12, 13, 14, 15, 20, 30, 50." How much change did Noah receive?

Solution Table 3-5 shows what the cashier said and how much money Noah received each time. Since $11 plus $1 is $12, Noah must have received $1 when the cashier said $12. The same reasoning follows for $13, $14, and so on. Thus, the total amount of change that Noah received is given by $1 + $1 + $1 + $1 + $5 + $10 + $20 = $39. In other words, $50 − $11 = $39 because $39 + $11 = $50.

Table 3-5

What the Cashier Said	$11	$12	$13	$14	$15	$20	$30	$50
Amount of Money Noah Received Each Time	0	$1	$1	$1	$1	$5	$10	$20

NOW TRY THIS 3-17

Perform each of the following computations mentally and explain what technique you used to find the answer:

a. 40 + 160 + 29 + 31
b. 3679 − 474
c. 75 + 28
d. 2500 − 700

Mental Mathematics: Multiplication

As with addition and subtraction, mental mathematics is useful for multiplication. For example, consider $8 \cdot 26$. Students may think of this computation in a variety of ways, as shown here.

26 = 20 + 6	26 = 25 + 1	26 = 30 − 4
$8 \cdot 20$ is 160;	$8 \cdot 25$ is 200;	$8 \cdot 30$ is 240;
$8 \cdot 6$ is 48, so	$8 \cdot 1$ is 8 more, so	take off $8 \cdot 4 = 32$.
$8 \cdot 26$ is 160 + 48,	$8 \cdot 26$ is 200 + 8,	So $8 \cdot 26$ is
or 208.	or 208.	240 − 32 = 208.

Next we consider several of the most common strategies for performing mental mathematics using multiplication.

1. *Front-end multiplying*

$$
\begin{array}{r}
64 \\
\times 5 \\
\hline
\end{array}
\qquad
\begin{array}{l}
60 \cdot 5 = 300 \\
4 \cdot 5 = 20 \\
\hline
300 + 20 = 320
\end{array}
$$

(Multiply the number of tens in the first number by 5.)
(Multiply the number of units in the first number by 5.)
(Add the two products.)

2. *Using compatible numbers*

$2 \cdot 9 \cdot 5 \cdot 20 \cdot 5$ Rearrange as $9 \cdot (2 \cdot 5) \cdot (20 \cdot 5) =$
$9 \cdot 10 \cdot 100 = 9000$

3. *Thinking money*

a. $\begin{array}{r} 64 \\ \times\ 5 \\ \hline \end{array}$ Think of the product as 64 nickels, which can be thought of as 32 dimes, which is $32 \times 10 = 320$ cents.

b. $\begin{array}{r} 64 \\ \times 50 \\ \hline \end{array}$ Think of the product as 64 half-dollars, which is 32 dollars, or 3200 cents.

c. $\begin{array}{r} 64 \\ \times 25 \\ \hline \end{array}$ Think of the product as 64 quarters, which is 32 half-dollars, or 16 dollars. Thus we have 1600 cents.

Mental Mathematics: Division

1. *Breaking up the dividend*

$7\overline{)4256}$ $7\overline{)42\,|\,56}$ (Break up the dividend into parts.)

$600\ +\ 8$

$7\overline{)4200\ +\ 56}$ (Divide both parts by 7.)

$600\ +\ 8 = 608$ (Add the answers together.)

2. *Using compatible numbers*

a. $3\overline{)105}$ $105 = 90 + 15$ (Look for numbers that you recognize as divisible by 3 and having a sum of 105.)

$\dfrac{30\ +\ 5}{3\overline{)90\ +\ 15}}.$ Thus $105 \div 3 = 35.$ (Divide both parts and add the answers.)

b. $8\overline{)232}$ $232 = 240 - 8$ (Look for numbers that are easily divisible by 8 and whose difference is 232.)

$\dfrac{30\ -\ 1}{8\overline{)240\ -\ 8}}.$ Thus $232 \div 8 = 29.$ (Divide both parts and take the difference.)

● NOW TRY THIS 3-18

Perform each of the following computations mentally and explain what technique you used to find the answer:

a. $25 \cdot 32 \cdot 4$ b. $123 \cdot 3$ c. $25 \cdot 35$ d. $5075 \div 25$

Computational Estimation

Computational estimation may help determine whether an answer is reasonable or not. This is especially useful when the computation is done on a calculator. Some common estimation strategies for addition are given next.

1. *Front-end with adjustment*

Front-end with adjustment estimation begins by focusing on the lead, or front, digits of the addition. These front, or lead, digits are added and assigned an appropriate place value. At this point we may have an underestimate that needs to be adjusted. The adjustment is made by focusing on the next group of digits. The following example shows how front-end estimation works:

$\begin{array}{r} 4 + 3 + 5 \\ 12\ \text{hundred} \end{array}$ ← $\begin{array}{r} 423 \\ 338 \\ + 561 \\ \hline \end{array}$ $\begin{array}{l} 20 \\ 100 \end{array}$ 120

Steps: **(1) Add front-end digits**
$4 + 3 + 5 = 12$
(2) Place value $= 1200.$
(3) Adjust $61 + 38 \approx 100$ and $20 + 100$ is 120
(4) Adjusted estimate is $1200 + 120 = 1320.$

2. *Grouping to nice numbers*
The strategy used to obtain the adjustment in the preceding example is the *grouping to nice numbers* strategy, which means that numbers that "nicely" fit together are grouped. Another example is given here.

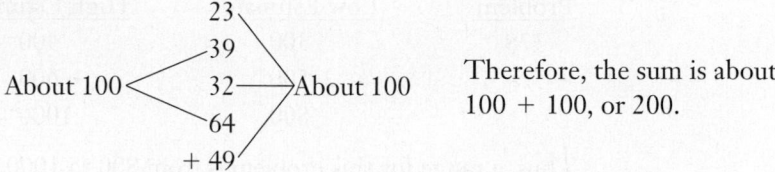

Therefore, the sum is about
100 + 100, or 200.

3. *Clustering*
Clustering is used when a group of numbers cluster around a common value. This strategy is limited to certain kinds of computations. In this example, the numbers seem to cluster around 6000.

$$
\begin{array}{r}
6200 \\
5842 \\
6512 \\
5521 \\
+6319 \\
\end{array}
$$

Estimate the "average"—about 6000.

Multiply the average by the number of values to obtain $5 \cdot 6000 = 30{,}000$.

4. *Rounding*
Rounding is a way of cleaning up numbers so that they are easier to handle. Rounding enables us to find approximate answers to calculations, as follows:

4724	5000	(Round 4724 to 5000)
+3192	+3000	(Round 3192 to 3000)
	8000	(Add the rounded numbers)

1267	1300	(Round 1267 to 1300)
− 510	− 500	(Round 510 to 500)
	800	(Subtract the rounded numbers)

Performing estimations requires a knowledge of place value and rounding techniques. We illustrate a rounding procedure that can be generalized to all rounding situations. For example, suppose we wish to round 4724 to the nearest thousand. We may proceed in four steps (see also Figure 3-44).

a. Determine between which two consecutive thousands the number lies.
b. Determine the midpoint between the thousands.
c. Determine which thousand the number is closer to by observing whether it is greater than or less than the midpoint. (*Not all texts use the same rule for rounding when a number falls at a midpoint.*)
d. If the number to be rounded is greater than or equal to the midpoint, round the given number to the greater thousand; otherwise, round to the lesser thousand. In this case, 4724 rounds to 5000.

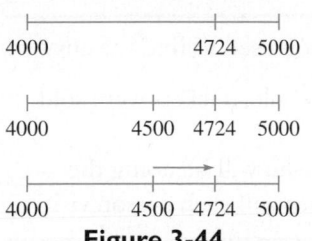

Figure 3-44

5. *Using the range*

It is often useful to know into what *range* an answer falls. The range is determined by finding a low estimate and a high estimate and reporting that the answer falls in this interval. An example follows:

Problem	Low Estimate	High Estimate
378	300	400
+ 524	+ 500	+ 600
	800	1000

Thus, a range for this problem is from 800 to 1000.

 The student page on page 165 shows both the *rounding* and *front-end* estimation strategies applied to a problem.

Estimation: Multiplication and Division

Examples of estimation strategies for multiplication and division are shown next.

1. *Front-end*

$$524$$
$$\times 8$$

$500 \cdot 8 = 4000$ (Start multiplying at the front to obtain a first estimate.)

$20 \cdot 8 = 160$ (Multiply the next important digit by 8.)

$4000 + 160 = 4160$ (Adjust the first estimate by adding the two numbers.)

2. *Compatible numbers*

$5\overline{)4163}$ $5\overline{)4000}$ (Change 4163 to a number close to it that you know is divisible by 5.)

800
$5\overline{)4000}$ (Carry out the division and obtain the first estimate of 800. Various techniques can be used to adjust the first estimate.)

NOW TRY THIS 3-19

Estimate each of the following mentally and explain what technique you used to find the answer:

a. A sold-out concert was held in a theater with a capacity of 4525 people. Tickets were sold for $9 each. Approximately how much money was collected?

b. Fliers are to be delivered to 3625 houses and there are 42 people who will be doing the distribution. If distributed equally, approximately how many houses will each person visit?

Assessment 3-5A

1. Compute each of the following mentally:
   a. $180 + 97 - 23 + 20 - 140 + 26$
   b. $87 - 42 + 70 - 38 + 43$

2. Use compatible numbers to compute each of the following mentally:
   a. $2 \cdot 9 \cdot 5 \cdot 6$ **b.** $8 \cdot 25 \cdot 7 \cdot 4$

3. Compute each of the following mentally and describe your approach.
   a. $475 + 49 + 525$ **b.** $375 - 76$

4. Use breaking up and bridging or front-end multiplying to compute each of the following mentally:
   a. $567 + 38$ **b.** $321 \cdot 3$

5. Use trading off to compute each of the following mentally:
   a. $85 - 49$ **b.** $87 + 33$
   c. $143 - 97$ **d.** $58 + 39$

6. Compute each of the following using the *adding up* (cashier's) algorithm:
   a. $53 - 28$ **b.** $63 - 47$

School Book Page ESTIMATING SUMS AND DIFFERENCES

Lesson 1-9

Key Idea
There is more than one way to estimate sums and differences.

Vocabulary
• front-end estimation
• rounding (p. 26)

Think It Through
I only need an **estimate** because it asks about how many pounds.

Estimating Sums and Differences

How can you estimate sums?

Students at Skyline Elementary collected aluminum cans for recycling. About how many pounds of cans did they collect in all?

✔ WARM UP
Round each number to the place of the underlined digit.

1. 1<u>7</u>.333

2. 5<u>6</u>7,642

3. 38.<u>0</u>45

Recycling Cans				
Grade	3rd	4th	5th	6th
Pounds Collected	398	257	285	318

You can estimate 398 + 257 + 285 + 318 two ways.

Jon used **rounding**.

Kylie used **front-end estimation** and adjusted the estimate.

I'll round each number to the nearest hundred.

$$
\begin{array}{rcr}
398 & \rightarrow & 400 \\
257 & \rightarrow & 300 \\
285 & \rightarrow & 300 \\
+\ 318 & \rightarrow & +\ 300 \\
\hline
 & & 1{,}300
\end{array}
$$

About 1,300 pounds

I'll first add the front-end digits.

$$
\begin{array}{rcr}
398 & \rightarrow & 300 \\
257 & \rightarrow & 200 \\
285 & \rightarrow & 200 \\
+\ 318 & \rightarrow & +\ 300 \\
\hline
 & & 1{,}000
\end{array}
$$

Then I'll adjust to include the remaining numbers.
98 → 100.
85 → 100.
57 + 18 → 100.
Less than 1,300 pounds

7. A car trip took 8 hr of driving at an average of 62 mph. Mentally compute the total number of miles traveled.

8. Compute each of the following mentally. In each case, briefly explain your method.
   a. 86 + 37
   b. 97 + 54
   c. 230 + 60 + 70 + 44 + 40 + 6

9. Round each number to the place value indicated by the digit in bold.
   a. **5**280   b. **1**15,234
   c. 115,**2**34   d. 2,3**2**5

10. Estimate each answer by rounding.
    a. 878 ÷ 29
    b. 25,201 − 19,987
    c. 32 · 28
    d. 2215 + 3023 + 5967 + 975

11. Use front-end estimation with adjustment to estimate each of the following:
    a. 2215 + 3023 + 5987 + 975
    b. 234 + 478 + 987 + 319 + 469

12. a. Would the clustering strategy of estimation be a good one to use in each of the following cases? Why or why not?

    (i)   474   (ii)   483
            1467   475
              64   530
         + 2445   503
                             + 528

    b. Estimate each part of (a) using the following strategies:
       (i) Front-end
       (ii) Grouping to nice numbers
       (iii) Rounding

13. Use the range strategy to estimate each of the following. Explain how you arrived at your estimates.
    a. 22 · 38
    b. 145 + 678
    c. 278 + 36

14. Suppose you had a balance of $3287 in your checking account and you wrote checks for $85, $297, $403, and $523. Estimate your balance and tell what you did.

15. A theater has 38 rows with 23 seats in each row. Estimate the number of seats in the theater and tell how you arrived at your estimate.

16. Without computing, tell which of the following have the same answer. Describe your reasoning.
    a. 44 · 22 and 22 · 11
    b. 22 · 32 and 11 · 64
    c. 13 · 33 and 39 · 11

17. The following is a list of the areas in square miles of Europe's largest countries. Mentally use this information to decide if each of the given statements is true.

    France   211,207
    Spain   194,896
    Sweden   173,731
    Finland   130,119
    Norway   125,181

    a. Sweden is less than 40,000 mi^2 larger than Finland.
    b. France is more than twice the size of Norway.
    c. France is more than 100,000 mi^2 larger than Norway.
    d. Spain is about 21,000 mi^2 larger than Sweden

18. The attendance at a World's Fair for one week follows:

    Monday   72,250
    Tuesday   63,891
    Wednesday   67,490
    Thursday   73,180
    Friday   74,918
    Saturday   68,480

    Estimate the week's attendance and tell what strategy you used.

19. In each of the following, determine whether the estimate given in parentheses is high (higher than the actual answer) or low (lower than the actual answer). Justify your answers without computing the exact values.
    a. 299 · 300 (90,000)
    b. 6001 ÷ 299 (20)
    c. 6000 ÷ 299 (20)
    d. 999 ÷ 99 (10)

20. Use your calculator to calculate 25^2, 35^2, 45^2, and 55^2, and then see if you can find a pattern that will let you find 65^2 and 75^2 mentally.

Assessment 3-5B

1. Compute each of the following mentally:
   a. 160 + 92 − 32 + 40 − 18
   b. 36 + 97 − 80 + 44

2. Use compatible numbers to compute each of the following mentally:
   a. 5 · 11 · 3 · 20
   b. 82 + 37 + 18 + 13

3. Supply reasons for each of the first four steps given here.

$$(525 + 37) + 75 = 525 + (37 + 75)$$
$$= 525 + (75 + 37)$$
$$= (525 + 75) + 37$$
$$= 600 + 37$$
$$= 637$$

4. Use breaking and bridging or front-end multiplying to compute each of the following mentally:
   a. $997 - 32$ **b.** $56 \cdot 30$

5. Use trading off to compute each of the following mentally:
   a. $75 - 38$
   b. $57 + 35$
   c. $137 - 29$
   d. $78 + 49$

6. Compute each of the following using the *adding up* (cashier's) algorithm:
   a. $74 - 63$ **b.** $73 - 57$

7. Compute each of the following mentally. In each case, briefly explain your method.
   a. $81 - 46$ **b.** $98 - 19$
   c. $9700 - 600$

8. A car trip took 7 hr of driving at an average of 69 mph. Mentally compute the total number of miles traveled. Describe your method.

9. Round each number to the place value indicated by the digit in bold.
   a. 3**5**87   **b.** 1**4**8,213
   c. 23,**7**85   **d.** 2,3**5**7

10. Estimate each answer by rounding.
    a. $937 \div 28$
    b. $32,285 - 18,988$
    c. $52 \cdot 48$
    d. $3215 + 3789 + 5987$

11. Use front-end estimation with adjustment to estimate each of the following:
    a. $2345 + 5250 + 4210 + 910$
    b. $345 + 518 + 655 + 270$

12. **a.** Would the clustering strategy of estimation be a good one to use in each of the following cases? Why or why not?

(i)	(ii)
318	2350
2314	1987
57	2036
+ 3489	2103
	+ 1890

    b. Estimate each part of (a) using the following strategies:
       (i) Front-end with adjustment
       (ii) Grouping to nice numbers
       (iii) Rounding

13. Use the range strategy to estimate each of the following. Explain how you arrived at your estimates.
    a. $32 \cdot 47$
    b. $123 + 780$
    c. $482 + 246$

14. Tom estimated $31 \cdot 179$ in the three ways shown.
      (i) $30 \cdot 200 = 6000$
     (ii) $30 \cdot 180 = 5400$
    (iii) $31 \cdot 200 = 6200$
    Without finding the actual product, which estimate do you think is closer to the actual product? Why?

15. About 3540 calories must be burned to lose 1 lb of body weight. Estimate how many calories must be burned to lose 6 lb.

16. Without computing, tell which of the following have the same answer. Describe your reasoning.
    a. $88 \cdot 44$ and $44 \cdot 22$
    b. $93 \cdot 15$ and $31 \cdot 45$
    c. $12 \cdot 18$ and $20 \cdot 17$

17. In each of the following, answer the question using estimation methods if possible. If estimation is not appropriate, explain why not.
    a. Josh has $380 in his checking account. He wants to write checks for $39, $28, $59, and $250. Will he have enough money in his account to cover these checks?
    b. Gila deposited two checks into her account, one for $981 and the other for $1140. Does she have enough money in her account to cover a check for $2000 if we know she has a positive balance to start with?
    c. Alberto and Juan are running for city council. They receive votes from two districts. Alberto receives 3473 votes from one district and 5615 votes from the other district. Juan receives 3463 votes from the first district and 5616 from the second. Who gets elected?
    d. Two rectangular parcels have dimensions 101 ft by 120 ft and 103 ft by 129 ft. Which parcel has greater area? (Recall that the area of a rectangle is length times width.)

18. The attendence at a County Fair for six days follows:

Monday	71,150
Tuesday	64,993
Wednesday	68,490
Thursday	72,980
Friday	84,968
Saturday	69,495

    Estimate the total attendance for the six days.

19. In each of the following, determine if the estimate given in parentheses is high (higher than the actual answer) or low (lower than the actual answer). Justify your answers without computing the exact values.
    a. $398 \cdot 500$ (200,000)
    b. $8001 \div 398$ (20)
    c. $10,000 \div 999$ (10)
    d. $1999 \div 201$ (10)

20. Use your calculator to multiply several two-digit numbers by 99. Then see if you can find a pattern that will let you find the product of any two-digit number and 99 mentally.

Mathematical Connections 3-5

Communication

1. Is the front-end estimate for addition before adjustment always less than the exact sum? Explain why or why not.
2. In the new textbooks, there is an emphasis on mental mathematics and estimation. Explain why these topics are important for today's students.
3. Suppose x and y are positive (greater than 0) whole numbers. If x is greater than y and you estimate $x - y$ by rounding x up and y down, will your estimate always be too high or too low or could it be either? Explain.

Open-Ended

4. Dina calculated each of the following mentally:

$$49 \cdot 51 + 49 \cdot 49 = 4900$$
$$98 \cdot 37 + 2 \cdot 37 = 3700$$
$$99 \cdot 37 + 37 = 3700$$

Explain how she performed the calculations and why her method works. Then give three similar calculations based on the same approach and find the answers mentally.
5. Give several examples from real-world situations where an estimate, rather than an exact answer, is sufficient.
6. a. Give a numerical example of when front-end estimation and rounding can produce the same estimate.
   b. Give an example of when they can produce a different estimate.

Cooperative Learning

7. Prepare a grocery list of 10 items with prices and find the total price using a calculator. Give the list to each member of your group without revealing the total price. Ask them to estimate the total price in one minute. Find who came closest to the total price. Take turns preparing a list.
8. a. Without actually finding the answers determine which is greater: $19,876 \cdot 43$ or $19,875 \cdot 44$. Explain your approach. Compare the explanations with other members in your group.
   b. Come up with three similar pairs of products and determine which is greater. Justify your answers. Compare your answers with the other students in your group.

Questions from the Classroom

9. Molly computed $261 - 48$ by first subtracting 50 from 261 to obtain 211; then, to make up for adding 2 to 48, she subtracted 2 from 211 to obtain an answer of 209. Is her thinking correct? If not, how could you help her?
10. A student asks why he has to learn about any estimation strategy other than rounding. What is your response?
11. In order to finish her homework quickly, an elementary student does her estimation problems by using a calculator to find the exact answers and then rounds them to get her estimate. What do you tell her?

Review Problems

12. Explain why when a number is multiplied by 10 we append a zero to the number.
13. Perform each of the following divisions using both the repeated-subtraction and the standard algorithm.
    a. $18\overline{)623}$ b. $21\overline{)493}$
    c. $97\overline{)1000}$
14. Write each of the answers in exercise 13 in the form $a = bq + r$, where $0 \le r < b$.

National Assessment of Educational Progress (NAEP) Questions

Which of these would be easiest to solve by using mental math?
a. $65.12 - 28.19$
b. 358×2
c. $1,625 \div 3$
d. $100.00 + 10.00$

NAEP, Grade 4, 2007

A loaded trailer truck weighs 26,643 kilograms. When the trailer truck is empty, it weighs 10,547 kilograms. About how much does the load weigh?
a. 14,000 kilograms
b. 16,000 kilograms
c. 18,000 kilograms
d. 36,000 kilograms

NAEP, Grade 4, 2009

BRAIN TEASER The Washington School PTA set up a phone tree in order to reach all of its members. Each person's responsibility, after receiving a call, is to call two other assigned members until all members have been called. Assume that everyone is home and answers the phone and that each phone call takes 30 seconds. If one of the 85 members, the PTA president, makes the first phone call and starts the clock, what is the least amount of time necessary to reach all 85 members of the group?

Hint for Solving the Preliminary Problem

Write $124 - 7$ as a product of two whole numbers each different from 1.

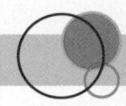

Chapter 3 Summary

3-3 Multiplication and Division of Whole Numbers

3-4 Algorithms for Whole-Number Multiplication and Division

3-5 Mental Mathematics and Estimation for Whole-Number Operations

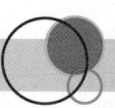

Chapter 3 Review

1. For each of the following where $a, b \in W$, identify the properties of the operation(s) for whole numbers illustrated:
   a. $3(a + b) = 3a + 3b$
   b. $2 + a = a + 2$
   c. $16 \cdot 1 = 1 \cdot 16 = 16$
   d. $6(12 + 3) = 6 \cdot 12 + 6 \cdot 3$
   e. $3(a \cdot 2) = 3(2a)$
   f. $3(2a) = (3 \cdot 2)a$

2. Using the definitions of less than or greater than given in this chapter, prove that each of the following inequalities is true:
   a. $3 < 13$
   b. $12 > 9$

3. For each of the following, find all possible replacements to make the statements true for whole numbers:
   a. $4 \cdot \square - 37 < 27$
   b. $398 = \square \cdot 37 + 28$
   c. $\square \cdot (3 + 4) = \square \cdot 3 + \square \cdot 4$
   d. $42 - \square \geq 16$

4. Use the distributive property of multiplication over addition, other multiplication properties, and addition facts, if possible, to rename each of the following where $a, b, x, y \in W$:
   a. $3a + 7a + 5a$ \qquad b. $3x^2 + 7x^2 - 5x^2$
   c. $x(a + b + y)$ \qquad d. $(x + 5)3 + (x + 5)y$
   e. $3x^2 + x$ \qquad f. $2x^5 + x^3$

5. How many 12-oz cans of juice would it take to give 60 people one 8-oz serving each?

6. Heidi has a brown pair and a gray pair of slacks; a brown blouse, a yellow blouse, and a white blouse; and a blue sweater and a white sweater. How many different outfits does she have if each outfit she wears consists of slacks, a blouse, and a sweater?

7. I am thinking of a whole number. If I divide it by 13, then multiply the answer by 12, then subtract 20, and then add 89, I end up with 93. What was my original number?

8. A ski resort offers a weekend ski package for $80 per person or $6000 for a group of 80 people. Which would be the less expensive option for a group of 80?

9. Josi has a job in which she works 30 hr/wk and gets paid $5/hr. If she works more than 30 hr in a week, she receives $8/hr for each hour over 30 hr. If she worked 38 hr this week, how much did she earn?

10. In a television game show, there are five questions to answer. Each question is worth twice as much as the previous question. If the last question was worth $6400, what was the first question worth?

11. **a.** Think of a number.
    Add 17.
    Double the result.
    Subtract 4.
    Double the result.
    Add 20.
    Divide by 4.
    Subtract 20.
    Your answer will be your original number. Explain how this trick works.
    b. Fill in two more steps that will take you back to your original number.
    Think of a number.
    Add 18.
    Multiply by 4.
    Subtract 7.
    .
    .
    .
    c. Make up a series of instructions such that you will always get back to your original number.

12. Use both the scratch and the traditional algorithms to compute the following:

$$\begin{array}{r} 316 \\ 712 \\ + 91 \\ \hline \end{array}$$

13. Use both the traditional and the lattice multiplication algorithms to compute the following:

$$\begin{array}{r} 613 \\ \times\ 98 \\ \hline \end{array}$$

14. Use both the repeated-subtraction and the conventional algorithms to perform the following:
    a. $912 \overline{)4803}$
    b. $11 \overline{)1011}$
    c. $23_{\text{five}} \overline{)3312_{\text{five}}}$
    d. $11_{\text{two}} \overline{)1011_{\text{two}}}$

15. Use the division algorithm to check your answers in exercise 14.

16. In some calculations a combination of mental math and a calculator is most appropriate. For example, because

$$200 \cdot 97 \cdot 146 \cdot 5 = 97 \cdot 146 (200 \cdot 5)$$
$$= 97 \cdot 146 \cdot 1000$$

we can calculate $97 \cdot 146$ on a calculator and then mentally multiply by 1000. Show how to calculate each of the following using a combination of mental math and a calculator:
    a. $19 \cdot 5 \cdot 194 \cdot 2$
    b. $379 \cdot 4 \cdot 193 \cdot 25$
    c. $8 \cdot 481 \cdot 73 \cdot 125$
    d. $374 \cdot 200 \cdot 893 \cdot 50$

17. Jim was paid $320 a month for 6 mo and $410 a month for 6 mo. What were his total earnings for the year?

18. A soft-drink manufacturer produces 15,600 cans of product each hour. Cans are packed 24 to a case. How many cases could be filled with the cans produced in 4 hr?

19. Apples normally sell for 32¢ each. They go on sale for 3 for 69¢. How much money is saved if you purchase 2 dozen apples while they are on sale?

20. The owner of a bicycle shop reported his inventory of bicycles and tricycles in an unusual way. He said he counted 126 wheels and 108 pedals. How many bikes and how many trikes did he have?

21. Perform each of the following computations:
    a. $\begin{array}{r} 123_{\text{five}} \\ + 34_{\text{five}} \\ \hline \end{array}$ **b.** $\begin{array}{r} 1010_{\text{two}} \\ - 101_{\text{two}} \\ \hline \end{array}$

    c. $\begin{array}{r} 23_{\text{five}} \\ \times 34_{\text{five}} \\ \hline \end{array}$ **d.** $\begin{array}{r} 1001_{\text{two}} \\ \times 101_{\text{two}} \\ \hline \end{array}$

22. Use the distributive property of multiplication over addition to compute $44_{\text{five}} \cdot 34_{\text{five}}$.

23. Use repeated subtraction to find $434_{\text{five}} \div 4_{\text{five}}$ without first converting to base ten. Write the answer using the Division Algorithm.

24. Tell how to use compatible numbers mentally to perform each of the following:

    a. $26 + 37 + 24 - 7$

    b. $4 \cdot 7 \cdot 9 \cdot 25$

25. Compute each of the following mentally. Name the strategy you used to perform your mental math (strategies vary).

    a. $63 \cdot 7$ **b.** $85 - 49$

    c. $(18 \cdot 5)2$ **d.** $2436 \div 6$

26. Estimate the following addition using **(a)** front-end estimation with adjustment and **(b)** rounding.

$$
\begin{array}{r}
543 \\
398 \\
255 \\
408 \\
+\ 998 \\
\end{array}
$$

27. Using clustering, estimate the sum $2345 + 2854 + 2234 + 2203$.

28. In some cases, the distributive property of multiplication over addition or distributive property of multiplication over subtraction can be used to obtain an answer quickly. Use one of the distributive properties to calculate each of the following in as simple a way as possible:

    a. $999 \cdot 47 + 47$

    b. $43 \cdot 59 + 41 \cdot 43$

    c. $1003 \cdot 79 - 3 \cdot 79$

    d. $1001 \cdot 113 - 113$

    e. $101 \cdot 35$

    f. $98 \cdot 35$

29. Recall that addition problems like $3478 + 521$ can be written and computed using expanded notation as shown here, and answer the questions that follow.

$$
\begin{array}{r}
3 \cdot 10^3 + 4 \cdot 10^2 + 7 \cdot 10 + 8 \\
+ \qquad\quad 5 \cdot 10^2 + 2 \cdot 10 + 1 \\
\hline
3 \cdot 10^3 + 9 \cdot 10^2 + 9 \cdot 10 + 9 \\
\end{array}
$$

    a. Write a corresponding addition algebra problem (use x for 10) and find the answer.

    b. Write a subtraction problem and the corresponding algebra problem and find the answer.

    c. Write a multiplication problem and the corresponding algebra problem and compute the answer.

Number Theory

Preliminary Problem

After a game of marbles with his three friends, Jacob said, "If only I had 1 more marble, I would have 4 times as many as Felipe, 5 times as many as Bella, and 7 times as many as Jessie."

What is the least number of marbles Jacob could have had?

If needed, see Hint on page 216.

Number theory started to flourish in the seventeenth century with the work of Pierre de Fermat (1601–1665). Topics in number theory that occur in the elementary school curriculum include factors, multiples, divisibility tests, prime numbers, prime factorizations, greatest common divisors, and least common multiples. The topic of congruences, introduced by Carl Gauss (1777–1855), is also incorporated into the elementary curriculum through clock arithmetic and modular arithmetic. Clock and modular arithmetic give students a look at a mathematical system (available online).

The grade 4 *Common Core Standards* state that students should be able to:

> Find all factor pairs for a whole number in the range 1–100. Recognize that a whole number is a multiple of each of its factors. Determine whether a given whole number in the range 1–100 is a multiple of a given one-digit number. Determine whether a given whole number in the range 1–100 is prime or composite. (p. 29)

Also in grade 6 students should be able to:

> Find the greatest common factor of two whole numbers less than or equal to 100 and the least common multiple of two whole numbers less than or equal to 12. (p. 42)

In addition, *PSSM* states that in grades 6–8:

> All students should use factors, multiples, prime factorization, and relatively prime numbers to solve problems. (p. 214)

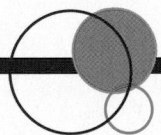

4-1 Divisibility

The concepts of *even* and *odd* for whole numbers are commonly used. For example, during summer water shortages in some parts of the country, houses with even-number addresses can water on even-numbered days of the month and houses with odd-number addresses can water on odd-numbered days. An **even** whole number is a whole number that has remainder 0 when divided by 2. An **odd** whole number is a whole number that leaves remainder 1 when divided by 2. From the Division Algorithm if n is even then $n = 2q$ for some whole number q (the remainder is 0). If n is odd, then $n = 2q + 1$ (why?). In the Division Algorithm if the remainder is 0 when m is divided by n, m is **divisible by** n. The fact that 12 is divisible by 2 can be stated in the following equivalent statements in the left column:

Example	General Statement
12 is divisible by 2.	a is divisible by b.
2 is a divisor of 12.	b is a divisor of a.
12 is a multiple of 2.	a is a multiple of b.
2 is a factor of 12.	b is a factor of a.
2 divides 12.	b divides a.

The statement that "2 divides 12" is written with a vertical segment, as in $2 \mid 12$, where the vertical segment means **divides**. Likewise for $a, b \in W$, "b divides a" is written $b \mid a$. Each statement in the right column above can be written $b \mid a$. We write $5 \nmid 12$ to symbolize that 5 does not divide 12 or that 12 is not divisible by 5. The notation $5 \nmid 12$ also implies that 12 is not a multiple of 5 and 5 is not a factor of 12.

In general, if a is a whole number and b is a non-zero whole number, then a is divisible by b or equivalently b divides a if, and only if, the remainder when a is divided by b is 0. Using the Division Algorithm, this means that there is a unique whole number q (quotient) such that $a = bq$. Thus we have the following definition.

Definition of "Divides"

If a and b are any whole numbers then b divides a, written $b \mid a$, if, and only if, there is a unique whole number q such that $a = bq$.

If $b \mid a$, then b is a **factor**, or a **divisor**, of a, and a is a **multiple** of b.

Do not confuse $b \mid a$ with b divided by a, that is b/a, which is interpreted as $b \div a$. The former, a relation, is either true or false. The latter, an operation on whole numbers, has a numerical value if $a \neq 0$. Note that if a/b is a whole number, then $b \mid a$. Also note that for whole numbers $b \nmid a$ is equivalent to saying that the remainder when a is divided by b is not 0.

EXAMPLE 4-1

Classify each of the following as true or false. Explain your answers.

a. $3 \mid 12$ **b.** $0 \mid 2$ **c.** 0 is even.

d. $8 \nmid 2$ **e.** For all whole numbers a, $1 \mid a$.   **f.** For all non-zero whole numbers a, $a^2 \mid a^5$.

g. $3 \mid 6n$ for all whole numbers n.

h. $3 \nmid (5 \cdot 7 \cdot 9 \cdot 11 + 1)$

i. $0 \mid 0$

Solution

a. $3 \mid 12$ is true because $12 = 4 \cdot 3$.

b. $0 \mid 2$ is false because there is no whole number c such that $2 = c \cdot 0$.

c. 0 is even is true because $0 = 0 \cdot 2$.

d. $8 \nmid 2$ is true because there is no whole number c such that $2 = c \cdot 8$.

e. $1 \mid a$ is true for all whole numbers a because $a = a \cdot 1$.

f. True, because $a^5 = a^2 \cdot a^3$.

g. $3 \mid 6n$ is true. Because $6n = 3 \cdot (2n)$, $6n$ is a multiple of 3 and hence $3 \mid 6n$.

h. True, because $5 \cdot 7 \cdot 9 \cdot 11 + 1 = 3 \cdot (5 \cdot 7 \cdot 3 \cdot 11) + 1$, so $5 \cdot 7 \cdot 9 \cdot 11 + 1$ leaves remainder 1 when divided by 3.

i. $0 \mid 0$ is false because $0 = 0 \cdot q$ for all integers q, so q is not unique.

Suppose we have one pack of gum and we know that the number of pieces, a, is divisible by 5. Then if we had two packs, the total number of pieces of gum is still divisible by 5. The same is true if we had 10 packs, or 100 packs, or in general n packs where n is any whole number. We could record this observation as follows:

$$\text{If } 5 \mid a, \text{ then } 5 \mid na, \text{ where } a \text{ and } n \text{ are whole numbers.}$$

The above statement is generalized in the following theorem.

Historical Note

Pierre de Fermat (1601–1665) was a French lawyer who devoted his leisure time to mathematics—a subject in which he had no formal training. After his death, his son published a new edition of Diophantus's *Arithmetica* with Fermat's notes. One of the notes in the margin asserted that the equation $x^n + y^n = z^n$ has no natural number solutions if n is greater than 2 and commented, "I have found an admirable proof of this, but the margin is too narrow to contain it." Only in 1995, Andrew Wiles, a Princeton University mathematician, proved the theorem.

Theorem 4-1

For any whole numbers a and d, if $d|a$ and n is any whole number, then $d|na$.

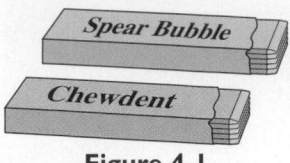

Figure 4-1

The theorem can be stated in an equivalent form:

> *If d is a factor of a (that is, a equals some whole number times d), then d is a factor of any multiple of a.*

Next consider two different brands of chewing gum each having five pieces, as in Figure 4-1. We can divide each pack of gum evenly among five students. In addition, if we opened both packs and put all of the pieces in a bag, we could still divide the number of pieces of gum evenly among the five students. To generalize this notion, if we buy gum in larger packages with a pieces in one package and b pieces in a second package with both a and b divisible by 5, we record the preceding discussion as follows:

$$\text{If } 5|a \text{ and } 5|b, \text{ then } 5|(a + b).$$

If the number, a, of pieces of gum in one package is divisible by 5, but the number, b, of pieces in the other package is not, then the total, $a + b$, cannot be divided evenly among the five students. This is recorded as:

$$\text{If } 5|a \text{ and } 5 \nmid b, \text{ then } 5 \nmid (a + b).$$

● NOW TRY THIS 4-1

If $a, b \in W$, determine the truth of the statement: if $5 \nmid a$ and $5 \nmid b$, then $5 \nmid (a + b)$?

Since subtraction can be defined in terms of addition, results similar to those for addition hold for subtraction. These ideas are generalized in Theorem 4-2.

Theorem 4-2

For any whole numbers a, b, and d:

a. If $d|a$, $d|b$, then $d|(a + b)$.

b. If $d|a$, $d \nmid b$, then $d \nmid (a + b)$.

c. If $d|a$, $d|b$, and $a \geq b$, then $d|(a - b)$.

d. If $d|a$, $d \nmid b$, and $a \geq b$, then $d \nmid (a - b)$.

e. If $d \nmid a$, $d|b$, and $a \geq b$, then $d \nmid (a - b)$.

Theorem 4-2 can be extended. For example, if a, b, c, and d are whole numbers with $d \neq 0$, then,

$$\text{If } d|a, d|b, \text{ and } d|c, \text{ then } d|(a + b + c).$$

The proofs of several theorems in this section are left as exercises, but the proofs of Theorem 4-2(a) and (b) are given as illustrations.

Proof

Theorem 4-2(a) is equivalent to the following:

> If a is a multiple of d and b is a multiple of d, then $a + b$ is a multiple of d.

Notice that "a is a multiple of d" means $a = md$, for some whole number m. Similarly "b is a multiple of d" means $b = nd$, for some whole number n. To show that $a + b$ is a multiple of d, we add as follows:

$$a + b = md + nd$$

Is $md + nd$ a multiple of d? Because $md + nd = (m + n)d$, then $a + b = (m + n)d$. By the closure property of whole number addition, $m + n$ is a whole number. Consequently, $a + b$ is a multiple of d and therefore $d \,|\, (a + b)$.

To prove Theorem 4-2(b), we proceed as follows:

Because $d \,|\, a$, $a = md$ for some $m \in W$. Because $d \nmid b$, when b is divided by d the remainder is not 0; that is, $b = qd + r$, where $q, r \in W$ and $0 < r < d$. Thus

$$a + b = md + qd + r$$
$$= (m + q)d + r$$

Consequently when $a + b$ is divided by d, the quotient is $m + q$ and the remainder is r. Because $r \neq 0$, $d \nmid (a + b)$.

EXAMPLE 4-2

Classify each of the following as true or false, where x, y, and z are whole numbers. If a statement is true, prove it. If a statement is false, provide a counterexample.

a. If $3 \,|\, x$ and $3 \,|\, y$, then $3 \,|\, xy$.   **b.** If $3 \,|\, (x + y)$, then $3 \,|\, x$ and $3 \,|\, y$.
c. If $9 \nmid a$, then $3 \nmid a$.

Solution
a. True; by Theorem 4-1, if $3 \,|\, x$, then, for any whole number y, $3 \,|\, yx$ or $3 \,|\, xy$.
b. False; for example, $3 \,|\, (7 + 2)$, but $3 \nmid 7$ and $3 \nmid 2$.
c. False; for example, $9 \nmid 21$, but $3 \,|\, 21$.

NOW TRY THIS 4-2

If $x, y \in W$, and $3 \,|\, x$, is it true that $3 \,|\, xy$ regardless of whether $3 \,|\, y$ or $3 \nmid y$? Why?

EXAMPLE 4-3

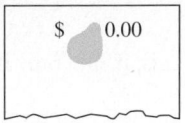

Figure 4-2

Five students found a padlocked money box that had a deposit slip attached to it. The deposit slip was water-spotted, so the currency total appeared as shown in Figure 4-2. One student remarked that if the money listed on the deposit slip was in the box, it could be divided equally among the five students without using coins. How did the student know this?

Solution Because the units digit of the amount of the currency is 0, the solution to the problem is to determine whether all non-zero whole numbers whose units digit is 0 are divisible by 5. Every number that ends in 0 is a multiple of 10; that is, it equals $10k$ for some whole number k. But, $10k = 5 \cdot (2k)$, and hence $5 \,|\, 10k$. Therefore, 5 divides the amount of money in the box, and the student is correct.

Divisibility Rules

As shown in Example 4-3, sometimes it is convenient to know whether one number is divisible by another just by looking at it or by performing a simple test. We showed that if a whole number ends in 0, then the number is divisible by 5. A similar argument can be used to show that if a whole number ends in 5, it is divisible by 5. This is an example of a divisibility rule. Moreover, if the units digit of a whole number is neither 0 nor 5, then the number is not divisible by 5. Elementary school texts frequently state divisibility rules, but such rules have limited practical use except in mental arithmetic.

It is possible to determine whether 1734 is divisible by 17, either by using pencil and paper or a calculator. To check divisibility and avoid decimals, we can use a calculator with an integer division button, $\boxed{\text{INT} \div}$. On such a calculator, integer division can be performed using the following sequence of buttons:

$$\boxed{1}\,\boxed{7}\,\boxed{3}\,\boxed{4}\,\boxed{\text{INT}\div}\,\boxed{1}\,\boxed{7}\,\boxed{=}$$

 to obtain the display $\underset{\underset{\text{Q}}{\smile}}{102}\quad\underset{\underset{\text{R}}{\smile}}{0.}$

This implies $1734/17 = 102$ with a remainder of 0, which, in turn, implies $17 \mid 1734$.

We could reach the conclusion mentally by considering the following:

$$1734 = 1700 + 34$$

Because $17 \mid 1700$ and $17 \mid 34$, by Theorem 4-2(a), we have $17 \mid (1700 + 34)$, or $17 \mid 1734$. Similarly, we could determine mentally that $17 \nmid 1735$.

Divisibility Tests for 2, 5, and 10

To determine mentally whether a given whole number n is divisible by a whole number d, we express n as a sum or difference of two whole numbers, of which at least one is divisible by d. We use a concrete example to get an idea of how this works. Consider the number 1362. This number can be represented as in Figure 4-3. Notice that because 10 and every power of 10 has a factor of two, 2 divides each part of the figure (see dashed segments). Since 2 divides each part of the figure, by the extension of Theorem 4-2, 2 divides the sum of all of the parts and hence $2 \mid 1362$.

Note that if the original number was 1363, then $2 \mid 1362$ and $2 \nmid 1$, so $2 \nmid 1363$. We see that all we have to do is to determine whether the units digit is divisible by 2 in order to determine whether the number is divisible by 2. We can develop a similar test for divisibility by 5 and 10.

> ### Theorem 4-3:   Divisibility Test for 2
> A whole number is divisible by 2 if, and only if, its units digit is divisible by 2; that is, if and only if, the unit digit is even.

> ### Theorem 4-4:   Divisibility Test for 5
> A whole number is divisible by 5 if, and only if, its units digit is divisible by 5; that is, if, and only if, the units digit is 0 or 5.

> ### Theorem 4-5:   Divisibility Test for 10
> A whole number is divisible by 10 if, and only if, its units digit is divisible by 10; that is, if, and only if, the units digit is 0.

Divisibility Tests for 4 and 8

When we consider divisibility rules for 4 and 8, we see that $4 \nmid 10$ and $8 \nmid 10$, so it is not a matter of checking the units digit for divisibility by 4 or 8. However, 4 (which is 2^2) divides 10^2, and 8 (which is 2^3) divides 10^3.

We first develop a divisibility rule for 4. Consider any four-digit whole number n such that $n = a \cdot 10^3 + b \cdot 10^2 + c \cdot 10 + d$. The *subgoal* is to *write the given number as a sum of two numbers*, one of which is as great as possible and divisible by 4. We know that $4 \mid 10^2$ because $10^2 = 4 \cdot 25$ and, consequently, $4 \mid 10^3$. Because $4 \mid 10^2$, then $4 \mid b \cdot 10^2$; and because $4 \mid 10^3$, then

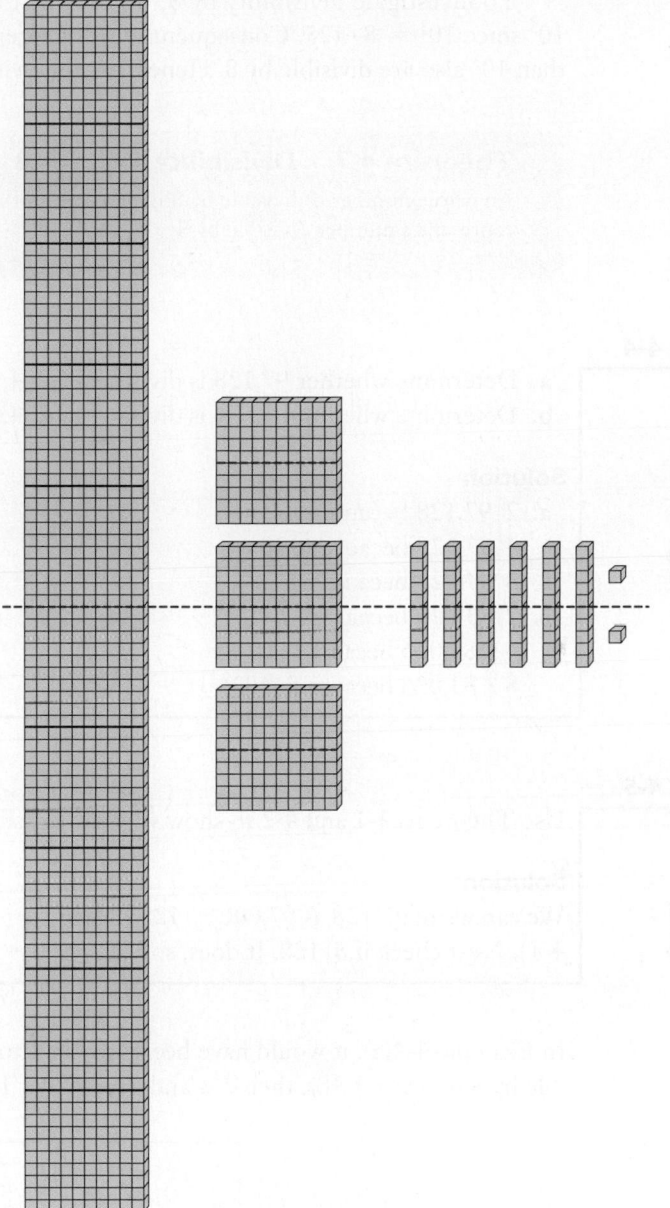

Figure 4-3

$4 | a \cdot 10^3$. Finally, $4 | a \cdot 10^3$ and $4 | b \cdot 10^2$ imply $4 | (a \cdot 10^3 + b \cdot 10^2)$. Now the divisibility of $a \cdot 10^3 + b \cdot 10^2 + c \cdot 10 + d$ by 4 depends on the divisibility of $(c \cdot 10 + d)$ by 4. Notice that $c \cdot 10 + d$ is the number represented by the last two digits in the given number n.

The preceding derivation of divisibility test for 4 can be somewhat shortened by observing that if $abcd$ is a base ten numeral (a, b, c, d are the digits), then $abcd = ab00 + cd = ab \cdot 100 + cd$. Since $4 | 100$, by Theorem 4-1 this implies that $4 | ab \cdot 100$. Thus, if $4 | cd$, then by Theorem 4-2, $4 | ab \cdot 100 + cd$; that is, $4 | abcd$. If, however, $4 \nmid cd$, then Theorem 4-2 implies that $4 \nmid ab \cdot 100 + cd$; that is, $4 \nmid abcd$. We summarize this in the following theorem.

Theorem 4-6: Divisibility Test for 4

A whole number is divisible by 4 if, and only if, the two rightmost digits of the number represent a number divisible by 4.

To investigate divisibility by 8, we note that the least positive power of 10 divisible by 8 is 10^3 since $10^3 = 8 \cdot 125$. Consequently by Theorem 4-1, all whole number powers of 10 greater than 10^3 also are divisible by 8. Hence, the following is a divisibility test for 8.

> **Theorem 4-7:   Divisibility Test for 8**
>
> An whole number is divisible by 8 if, and only if, the three rightmost digits of the whole number represent a number divisible by 8.

EXAMPLE 4-4

a. Determine whether 97,128 is divisible by 2, 4, and 8.
b. Determine whether 83,026 is divisible by 2, 4, and 8.

Solution

a. $2 \mid 97,128$ because $2 \mid 8$
   $4 \mid 97,128$ because $4 \mid 28$
   $8 \mid 97,128$ because $8 \mid 128$
b. $2 \mid 83,026$ because $2 \mid 6$
   $4 \nmid 83,026$ because $4 \nmid 26$
   $8 \nmid 83,026$ because $8 \nmid 026$

EXAMPLE 4-5

Use Theorems 4-1 and 4-2 to show why the divisibility test for 8 works in Example 4-4(a).

Solution

We can write 97,128 as $97,000 + 128$. Because $8 \mid 1000$, then $8 \mid 97 \cdot 1000$ or $8 \mid 97,000$ (Theorem 4-1). Next check if $8 \mid 128$. It does, so $8 \mid (97,000 + 128)$ or $8 \mid 97,128$ (Theorem 4-2).

In Example 4-4(a), it would have been sufficient to check that the given whole number a is divisible by 8 because if $8 \mid a$, then $2 \mid a$ and $4 \mid a$. Why? This relationship is shown in Figure 4-4.

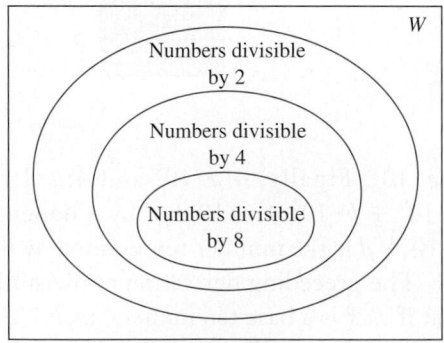

Figure 4-4

Notice that if $8 \nmid a$, we cannot conclude that $4 \nmid a$ or $2 \nmid a$. (Why?)

Divisibility Tests for 3 and 9

Consider a divisibility test for 3. No power of 10 is divisible by 3, but the numbers 9, and 99, and 999, and others of this type are divisible by 3. For example, to determine whether 5721 is divisible by 3, we rewrite the number using 999, 99, and 9, as follows:

$$5721 = 5 \cdot 10^3 + 7 \cdot 10^2 + 2 \cdot 10 + 1$$
$$= 5(999 + 1) + 7(99 + 1) + 2(9 + 1) + 1$$
$$= 5 \cdot 999 + 5 \cdot 1 + 7 \cdot 99 + 7 \cdot 1 + 2 \cdot 9 + 2 \cdot 1 + 1$$
$$= (5 \cdot 999 + 7 \cdot 99 + 2 \cdot 9) + (5 + 7 + 2 + 1)$$

The sum in the first set of parentheses in the last line is divisible by 3, so the divisibility of 5721 by 3 depends on the sum in the second set of parentheses. In this case, $5 + 7 + 2 + 1 = 15$ and $3 \mid 15$, so by Theorem 4-2(a) $3 \mid 5721$. Hence, to test 5721 for divisibility by 3, we test $5 + 7 + 2 + 1$ for divisibility by 3. Notice that $5 + 7 + 2 + 1$ is the sum of the digits of 5721. This suggests the following test for divisibility by 3.

> ### Theorem 4-8: Divisibility Test for 3
>
> A whole number is divisible by 3 if, and only if, the sum of its digits is divisible by 3.

We use a similar argument to generalize the test for divisibility by 3 on any whole number and in particular for any four-digit whole number $n = a \cdot 10^3 + b \cdot 10^2 + c \cdot 10 + d$. Even though $a \cdot 10^3 + b \cdot 10^2 + c \cdot 10 + d$ is not necessarily divisible by 3, the number $a \cdot 999 + b \cdot 99 + c \cdot 9$ *is* divisible by 3. We have the following:

$$a \cdot 10^3 + b \cdot 10^2 + c \cdot 10 + d = a \cdot 1000 + b \cdot 100 + c \cdot 10 + d$$
$$= a \cdot (999 + 1) + b \cdot (99 + 1) + c \cdot (9 + 1) + d$$
$$= (a \cdot 999 + b \cdot 99 + c \cdot 9) + (a \cdot 1 + b \cdot 1 + c \cdot 1 + d)$$
$$= (a \cdot 999 + b \cdot 99 + c \cdot 9) + (a + b + c + d)$$

Because $3 \mid 9, 3 \mid 99$, and $3 \mid 999$, it follows that $3 \mid (a \cdot 999 + b \cdot 99 + c \cdot 9)$. If $3 \mid (a + b + c + d)$, then $3 \mid [(a \cdot 999 + b \cdot 99 + c \cdot 9) + (a + b + c + d)]$; that is, $3 \mid n$. If, on the other hand, if $3 \nmid (a + b + c + d)$, it follows from Theorem 4-2(b) that $3 \nmid n$.

Since $9 \mid 9, 9 \mid 99, 9 \mid 999$, and so on, a test similar to that for divisibility by 3 applies to divisibility by 9. (Why?)

> ### Theorem 4-9: Divisibility Test for 9
>
> A whole number is divisible by 9 if, and only if, the sum of the digits of the whole number is divisible by 9.

EXAMPLE 4-6

Use divisibility tests to determine whether each of the following numbers is divisible by 3 and divisible by 9:

  a. 1002   **b.** 14,238

Solution

a. Because $1 + 0 + 0 + 2 = 3$ and $3 \mid 3$, it follows that $3 \mid 1002$. Because $9 \nmid 3$, it follows that $9 \nmid 1002$.

b. Because $1 + 4 + 2 + 3 + 8 = 18$ and $3 \mid 18$, it follows that $3 \mid 14{,}238$. Because $9 \mid 18$, it follows that $9 \mid 14{,}238$.

EXAMPLE 4-7

The store manager has an invoice for 72 calculators. The first and last digits on the invoice are illegible. The manager can read

$$\$\blacksquare67.9\blacksquare$$

What are the missing digits, and what is the cost of each calculator?

Solution

Let the missing digits be x and y so that the base ten notation for the number is $x67.9y$ dollars, or $x679y$ cents. Because there were 72 calculators sold, the number on the invoice must be divisible by 72. Because the number is divisible by 72 and $72 = 8 \cdot 9$, it must be divisible by 8 and 9, which are factors of 72. For the number on the invoice to be divisible by 8, the three-digit number $79y$ must be divisible by 8. Because $79y$ must be divisible by 8, it is an even number. Therefore, $79y$ must be either 790, 792, 794, 796, or 798. Only the number 792 is divisible by 8, so the last digit, y, on the invoice must be 2.

Because the number on the invoice must be divisible by 9, we know that 9 must divide $x + 6 + 7 + 9 + 2$, or $(x + 24)$. Since 3 is the only single digit that will make $(x + 24)$ divisible by 9, then x must be 3. Therefore, the number on the invoice must be \$367.92. The calculators must cost \$367.92/72, or \$5.11, each.

Divisibility Tests for 11 and 6

The divisibility test for 7 is usually harder to use than actually performing the division, so we omit the test. We state the divisibility test for 11 but omit the proof.

> **Theorem 4-10:   Divisibility Test for 11**
>
> A whole number is divisible by 11 if, and only if, the sum of the digits in the places that are even powers of 10 minus the sum of the digits in the places that are odd powers of 10 is divisible by 11. (If the sums are different subtract the smaller from the greater.)

For example, to test whether 8,471,986 is divisible by 11, we check whether 11 divides the difference $(6 + 9 + 7 + 8) - (8 + 1 + 4)$, or 17. Because $11 \nmid 17$, it follows from the divisibility test for 11 that $11 \nmid 8{,}471{,}986$. A number like 2772 is divisible by 11 because $(2 + 7) - (7 + 2) = 9 - 9 = 0$ and 0 is divisible by 11.

The divisibility test for 6 is related to the divisibility tests for 2 and 3. In Section 4-2, we will be able to show that if $2 \mid n$ and $3 \mid n$, then $(2 \cdot 3) \mid n$, and in general: if a and b have no factors in common, then if $a \mid n$ and $b \mid n$, we can conclude that $ab \mid n$. Consequently, the following divisibility test is true.

> **Theorem 4-11:   Divisibility Test for 6**
>
> A whole number is divisible by 6 if, and only if, the whole number is divisible by both 2 and 3.

Divisibility tests for other numbers are explored in Assessments 4-1A and 4-1B.

 EXAMPLE 4-8

The number 57,729,364,583 has too many digits for most calculator displays. Determine whether it is divisible by each of the following:

a. 2 **b.** 3 **c.** 5 **d.** 6 **e.** 8 **f.** 9 **g.** 10 **h.** 11

Solution
a. No; the units digit, 3, is not divisible by 2.
b. No; the sum of the digits is 59, which is not divisible by 3.
c. No; the units digit is neither 0 nor 5.
d. No; the number is not divisible by 2, (see Theorem 4-11).
e. No; the number formed by the last three digits, 583, is not divisible by 8.
f. No; the sum of the digits is 59, which is not divisible by 9.
g. No; the units digit is not 0.
h. Yes; $(3 + 5 + 6 + 9 + 7 + 5) - (8 + 4 + 3 + 2 + 7) = 35 - 24 = 11$ and 11 is divisible by 11.

NOW TRY THIS 4-3

Fill in the following blanks so that the number 12,506,5_ _. is divisible by 9. List all possibilities.

Problem Solving **A Mistake in the Inventory**

A class from Washington School visited a neighborhood cannery warehouse. The warehouse manager told the class that there were 11,368 cans of juice in the inventory and that the cans were packed in boxes of 6 or 24, depending on the size of the can. One of the students, Sam, thought for a moment and announced that there was a mistake in the inventory. Is Sam's statement correct? Why or why not?

Understanding the Problem The problem is to determine whether the manager's inventory of 11,368 cans was correct. To solve the problem, assume there are no partial boxes of cans; that is, a box must contain exactly 6 or exactly 24 cans of juice.

Devising a Plan We know that the boxes contain either 6 cans or 24 cans, but we do not know how many boxes of each type there are. One strategy for solving this problem is to *find an equation* that involves the total number of cans in all the boxes.

The total number of cans, 11,368, equals the number of cans in all the 6-can boxes plus the number of cans in all the 24-can boxes. If there are n boxes containing 6 cans each, there are $6n$ cans altogether in those boxes. Similarly, if there are m boxes with 24 cans each, those boxes

Historical Note

A twentieth century mathematician who did research in number theory was American Julia Robinson (1919–1985). Robinson was the first woman mathematician to be elected to the National Academy of Sciences and the first woman president of the American Mathematical Society.

contain a total of $24m$ cans. Because the total was reported to be 11,368 cans, we have the equation $6n + 24m = 11,368$. Sam claimed that $6n + 24m \neq 11,368$.

One way to show that $6n + 24m \neq 11,368$ is to show that $6n + 24m$ and 11,368 do not have the same divisors. Both $6n$ and $24m$ are divisible by 6. This implies that $6n + 24m$ must be divisible by 6. If 11,368 is not divisible by 6, then Sam is correct.

Carrying Out the Plan The divisibility test for 6 states that a number is divisible by 6 if, and only if, the number is divisible by both 2 and 3. Because 11,368 ends in 8, it is divisible by 2. However, because $3 \nmid 1 + 1 + 3 + 6 + 8$, $3 \nmid 11,368$. Thus $6 \nmid 11,368$ and therefore Sam is correct.

Looking Back A somewhat shorter way to solve the problem is to notice that because the cans are packed in boxes of 6 or 24, then the number of cans in each box and hence in all the boxes is divisible by 6. However the reported total is not divisible by 6.

Suppose 11,368 had been divisible by 6. Would that have implied that the manager was correct? The answer is no; it would have implied only that we would have to change the approach to the problem.

As a further Looking Back activity, suppose that, given different data, the manager is correct. Can we determine values for m and n? If a computer is available, a spreadsheet can be written to determine all possible whole number values of m and n.

BRAIN TEASER

The following is an argument to show that an ant weighs as much as an elephant. What is wrong?

Let e be the weight of the elephant and a the weight of the ant. Let $e - a = d$. Consequently, $e = a + d$. Multiply each side of $e = a + d$ by $e - a$. Then simplify.

$$e(e - a) = (a + d)(e - a)$$
$$e^2 - ea = ae + de - a^2 - da$$
$$e^2 - ea - de = ae - a^2 - da$$
$$e(e - a - d) = a(e - a - d)$$
$$e = a$$

Thus, the weight of the elephant equals the weight of the ant.

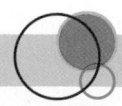

Assessment 4-1A

1. Classify each of the following as true or false. If false, tell why.
   a. 6 is a factor of 30.
   b. 6 is a divisor of 30.
   c. $6 \mid 30$.
   d. 30 is divisible by 6.
   e. 30 is a multiple of 6.
   f. 6 is a multiple of 30.

2. Using divisibility tests, solve each of the following:
   a. There are 1379 children signed up to play in a baseball league. If exactly 9 players are to be placed on each team, will any team be short of players?
   b. A forester has 43,682 seedlings to be planted. Can these be planted in an equal number of rows with 11 seedlings in each row?
   c. There are 261 students to be assigned to 9 teachers so that each teacher has the same number of students. Is this possible?

3. Without using a calculator, test each of the following numbers for divisibility by 2, 3, 4, 5, 6, 8, 9, 10, and 11:
   a. 746,988
   b. 81,342
   c. 15,810

4. Answer each of the following without actually performing the division. Explain how you did it in each case.
   a. Is 34,015 divisible by 17?
   b. Is 34,051 divisible by 17?
   c. Is 19,031 divisible by 19?
   d. Is $2 \cdot 3 \cdot 5 \cdot 7$ divisible by 5?
   e. Is $(2 \cdot 3 \cdot 5 \cdot 7) + 1$ divisible by 5?

5. Justify each of the given statements, assuming that a, b, and c are whole numbers. If a statement cannot be justified by one of the theorems in this section, answer "none."

a. $4 \mid 20$ implies $4 \mid 113 \cdot 20$.
b. $4 \mid 100$ and $4 \nmid 13$ imply $4 \nmid (100 + 13)$.
c. $4 \mid 100$ and $4 \nmid 13$ imply $4 \nmid 1300$.
d. $3 \mid (a + b)$ and $3 \nmid c$ imply $3 \nmid (a + b + c)$.
e. $3 \mid a$ implies $3 \mid a^2$.

6. Classify each of the following as true or false. Justify your answers.
   a. If $b \mid a$, then $(b + c) \mid (a + c)$.
   b. If $b \mid a$, then $b^2 \mid a^3$.
   c. If $b \mid a$, then $b \mid (a + b)$.

7. Justify each of the following:
   a. $7 \mid 210$
   b. $19 \mid (1900 + 38)$
   c. $6 \mid 2^3 \cdot 3^2 \cdot 17^4$
   d. $7 \nmid (4200 + 22)$

8. Classify each of the following as true or false:
   a. If every digit of a number is divisible by 3, the number itself is divisible by 3.
   b. If a number is divisible by 3, then every digit of the number is divisible by 3.
   c. A number is divisible by 3 if, and only if, every digit of the number is a multiple of 3.

9. Fill each of the following blanks with the greatest digit that makes the statement true:
   a. $3 \mid 74\_$
   b. $9 \mid 83\_45$
   c. $11 \mid 6\_55$

10. Place all possible digits in the square, so that the number

    $$527,4\square2$$

    is divisible by
    a. 2   b. 3
    c. 4   d. 9
    e. 11

11. Without using a calculator, classify each of the following as true or false. Justify your answers.
    a. $13 \mid 390,026$ b. $13 \nmid 260,033$
    c. $31 \mid 93^{11}$ d. $23 \nmid 690,068$

12. The bookstore marked some notepads down from $2.00 but still kept the price over $1.00. It sold all of them. The total amount of money from the sale of the pads was $31.45. How many notepads were sold?

13. A group of people ordered pencils. The bill was $2.09. If the original price of each was 12¢ and the price has risen, how much does each cost?

14. To find the remainder when a number is divided by 3 you could use long division. Alternatively, you could find the sum of the digits and then find the remainder when that sum is divided by 3.
    a. Find the remainder when 7,242,815 is divided by 3 using long division and then using the sum of the digits.
    b. Explain why the sum of the digits approach is valid.
    c. Will the sum of the digits approach work to find the remainder when a number is divided by 9? Justify your answer.

15. A test for checking computations is called *casting out nines*. Consider the sum $193 + 24 + 786 = 1003$. The remainders when 193, 24, and 786 are divided by 9 are 4, 6, and 3, respectively. The sum of the remainders, 13, has a remainder of 4 when divided by 9, as does 1003. Checking the remainders in this manner provides a quasi-check for the computation. Find the following sums and use casting out nines to check the sums:
    a. $12,343 + 4546 + 56$
    b. $987 + 456 + 8765$
    c. $10,034 + 3004 + 400 + 20$
    d. Try the check on the subtraction $1003 - 46$.
    e. Try the check on the multiplication $345 \cdot 56$.
    f. Would it make sense to try the check on division? Why or why not?

16. A palindrome is a number that reads the same forward as backward.
    a. Are all five-digit palindromes divisible by 11? Why or why not?
    b. Are all six-digit palindromes divisible by 11? Why or why not?

17. If 21 divides n, what other whole numbers divide n? Why?

18. The numbers x and y are divisible by 5.
    a. Is the sum of x and y divisible by 5? Why?
    b. Is the difference of x and y divisible by 5? Why?
    c. Is the product of x and y divisible by 5? Why?

19. Classify each of the following as true or false, assuming that a, b, c, and d are whole numbers. If a statement is false, give a counterexample.
    a. If $d \mid (a + b)$, then $d \mid a$ and $d \mid b$.
    b. If $d \mid (a + b)$, then $d \mid a$ or $d \mid b$.
    c. If $d \mid ab$, then $d \mid a$ or $d \mid b$.
    d. If $ab \mid c$, then $a \mid c$ and $b \mid c$.
    e. If $a \mid b$ and $b \mid a$, then $a = b$.

20. Prove the test for divisibility by 9 for any five-digit whole number.

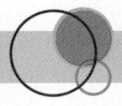

Assessment 4-1B

1. Classify each of the following as true or false. If false, tell why.
   a. 5 is a multiple of 20.
   b. 10 is a divisor of 30.
   c. $8 \mid 32$.
   d. 10 is divisible by 1.
   e. 30 is a factor of 6.
   f. 6 is a multiple of 20.

2. Using divisibility tests, answer each of the following:
   a. Six friends win with a lottery ticket. The payoff is $242,800. Can the money be divided evenly?
   b. Jack owes $7812 on a new car. Can this amount be paid in 12 equal monthly installments?

3. Without using a calculator, test each of the following numbers for divisibility by 2, 3, 4, 5, 6, 8, 9, 10, and 11:
   a. 4,201,012
   b. 1001
   c. 10,001

4. Answer each of the following without actually performing the division. Explain how you did it in each case.
   a. Is 24,013 divisible by 12?
   b. Is 24,036 divisible by 12?

c. Is 17,034 divisible by 17?

d. Is $2 \cdot 3 \cdot 5 \cdot 7$ divisible by 3?

e. Is $(2 \cdot 3 \cdot 5 \cdot 7) + 1$ divisible by 6?

5. Justify each of the following:
   a. $26 | (13^4 \cdot 100)$
   b. $13 \nmid (2^4 \cdot 5^3 \cdot 26 + 1)$
   c. $2^4 \nmid (2 \cdot 4 \cdot 6 \cdot 8 \cdot 17^{10} + 1)$
   d. $2^4 | (10^4 + 6^4)$

6. Justify each of the following.
   a. $a^3 | a^4$, if $a \neq 0$.
   b. $a^4 | a^{10}$, if $a \neq 0$.
   c. $a^n | a^m$, if $0 \leq n \leq m$, if $a \neq 0$.
   d. If $b | a$ and $c \neq 0$, then $bc | ac$.

7. Justify each of the following:
   a. $7 | 280$
   b. $19 | (3800 + 19)$
   c. $15 | 2^4 \cdot 3^5 \cdot 5$
   d. $19 \nmid (3800 + 37)$

8. Classify each of the following as true or false:
   a. If a number is divisible by 6, then it is divisible by 2 and by 3.
   b. If a number is divisible by 2 and 3, then it is divisible by 6.
   c. If a number is divisible by 2 and 4, then it is divisible by 8.
   d. If a number is divisible by 8, then it is divisible by 2 and 4.
   e. A number is divisible by 8 if, and only if, it is divisible by 2 and by 4.

9. Devise a test for divisibility by each of the following numbers:
   a. 16
   b. 25

10. Find all possible single digits that can be placed in the square so that the number $2821\square6$ is divisible by
    a. 2   b. 3   c. 4   d. 9

11. Without using a calculator, classify each of the following as true or false. Justify your answers.
    a. $7 | 280,021$
    b. $19 \nmid 3,800,018$
    c. $23 | 46^{10}$
    d. $23 \nmid 460,045$
    e. $23 | 460,046$

12. In a football game, a touchdown with an extra point is worth 7 points and a field goal is worth 3 points. Suppose that in a game the only scoring done by teams are touchdowns with extra points and field goals.
    a. Which of the scores 1 to 25 are impossible for a team to score?
    b. List all possible ways for a team to score 40 points.

c. A team scored 57 points with 6 touchdowns and 6 extra points. How many field goals did the team score?

13. When the two missing digits in 85_ _1 are replaced, the number is divisible by 99. What is the number?

14. Complete the following table where n is the given whole number.

n	Remainder when n Is Divided by 9	Sum of the Digits of n	Remainder when the Sum of the Digits of n Is Divided by 9
a. 31			
b. 143			
c. 345			
d. 2987			
e. 7652			

   f. Make a conjecture about the remainders when a whole number and the sum of its digits are divided by 9.

15. Use the approach outlined in exercise 15 of Assessment 4-1A to show that the following computations are wrong:
    a. $99 + 28 = 227$ b. $11,199 - 21 = 11,168$
    c. $99 \cdot 26 = 2575$

16. A palindrome is a number that reads the same forward as backward.
    a. Check the following four-digit palindromes for divisibility by 11:
       i. 4554   ii. 9339   iii. 2002   iv. 2222
    b. Are all four-digit palindromes divisible by 11? Why or why not?

17. If $16 | n$, what other whole numbers less than 16 divide n? Why?

18. If $7 | x$ and $7 | y$, classify the following as true or false.
    a. $7 | xy$ b. $7 | (5x + 2y)$
    c. $7 | (x^2 + y)$ d. $49 | xy$

19. Classify each of the following as true or false, assuming that $a, b, c,$ and d are whole numbers. If a statement is false, give a counterexample.
    a. If $d | a$ and $d | b$, then $d | (ax + by)$ for any whole number x and y.
    b. If $d \nmid a$ and $d \nmid b$, then $d \nmid (a + b)$.
    c. If $d | a^2$, then $d | a$.
    d. If $d \nmid a$, then $d \nmid a^2$.

20. Prove the test for divisibility by 9 for any four-digit number.

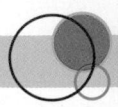

Mathematical Connections 4-1

Communication

1. A customer wants to mail a package. The postal clerk determines the cost of the package to be $18.95, but only 6¢ and 9¢ stamps are available. Can the available stamps be used for the exact amount of postage for the package? Why or why not?

2. a. Jim uses his calculator to see if a number n having eight or fewer digits is divisible by a number d. He finds that $n \div d$ has a display of 32. Does $d | n$? Why?
   b. If $n \div d$ gives a display of 16.8, does $d | n$? Why?

3. Is the area (in cm²) of each of the following rectangles divisible by 4? Explain why or why not.

   a.

   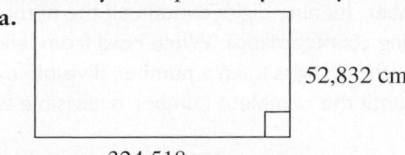

   52,832 cm

   324,518 cm

   b.

   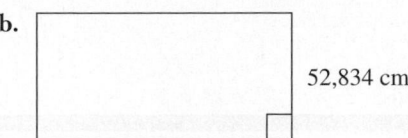

   52,834 cm

   324,514 cm

4. Can you find three consecutive natural numbers none of which is divisible by 3? Explain your answer.

5. Answer each of the following and justify your answers.
   a. If a number is not divisible by 5, can it be divisible by 10?
   b. If a number is not divisible by 10, can it be divisible by 5?

6. A number in which each digit except 0 appears exactly 3 times is divisible by 3. For example, 777,555,222 and 414,143,313 are divisible by 3. Explain why this statement is true.

7. Enter any three-digit number on the calculator; for example, enter 243. Repeat it: 243,243. Divide by 7. Divide by 11. Divide by 13. What is the answer? Try it again with any other three-digit number. Will this always work? Why?

8. Alexa claims that she can justify the divisibility test by 11. She says: *I noticed that each even power of 10 can be written as a multiple of 11 plus 1 and every odd power of 10 can be written as a multiple of 11 minus 1. In fact:*

   $$10 = 11 - 1$$
   $$10^2 = 99 + 1 = 9 \cdot 11 + 1$$
   $$10^3 = 10 \cdot 10^2 = 10(9 \cdot 11 + 1) = 90 \cdot 11 + 10$$
   $$= 90 \cdot 11 + 11 - 1 = 91 \cdot 11 - 1$$
   $$10^4 = 10^2 \cdot 10^2 = 100(9 \cdot 11 + 1)$$
   $$= 900 \cdot 11 + 9 \cdot 11 + 1 = 909 \cdot 11 + 1$$

   and so on.

   Now I look at a four-digit number abcd and proceed as in the divisibility by 3. I collect the parts that are divisible by 11 regardless of what the digits are and put together the rest, which is

   $$d - c + b - a$$

   If subtraction cannot be performed in whole numbers then add 11 to d or to b or to both.

   Complete the details of Alexa's argument and justify the test for divisibility by 11.

9. Take a four digit number written in base ten and subtract each of the following from your number. By what specific numbers can you be sure that the difference is divisible? Justify your answers.
   a. The units digit
   b. The number formed by the last two digits (that is, the tens digit followed by the units digit).
   c. The sum of the digits
   d. Answer the preceding questions for a four digit number written in base five.

10. **a.** In what bases will divisibility by 2 depend only on the units digit? Justify your answer.
    b. In what bases will divisibility by 2 depend only on the sum of the digits being even or odd? Justify your answer.

Open-Ended

11. A breakfast-food company had a contest in which numbers were placed in breakfast-food boxes. A prize of $1000 was awarded to anyone who could collect numbers whose sum was 100. The company had thousands of cards made with the following numbers on them:

    3   12   15   18   27   33   45   51   66   75   84   90

    a. If the company did not make any more cards, is there a winning combination?
    b. If the company is going to add one more number to the list and it wants to make sure the contest has at most 1000 winners, suggest a strategy for it to use.

12. How would you use concrete materials to explain to young children the following:
    a. A number being even or odd
    b. A number being divisible by 3 or not being divisible by 3
    c. If $4 \mid a$, then $2 \mid a$

Cooperative Learning

13. In your group, discuss the value of teaching various divisibility tests in middle school. If a teacher decides to discuss the various tests, how should they be introduced?

Questions from the Classroom

14. Jane claimed that a number is divisible by 4 if each of the last two digits is divisible by 4. Is this claim accurate? If not, how would you suggest that Jane change it to make it accurate?

15. A student claims that $a \mid a$ and $a \mid a$ implies $a \mid (a - a)$, and hence, $a \mid 0$. Is the student correct?

16. A student writes, "If $d \nmid a$ and $d \nmid b$, then $d \nmid (a + b)$." How do you respond?

17. Your seventh-grade class has just completed a unit on divisibility rules. One student asks why divisibility by numbers other than 3 and 9 cannot be tested by dividing the sum of the digits by the tested number. How should you respond?

18. A student says that a number with an even number of digits is divisible by 7 if, and only if, each of the numbers formed by pairing the digits into groups of two is divisible by 7. For example, 49,562,107 is divisible by 7, since each of the numbers 49, 56, 21, and 07 is divisible by 7. Is this true?

19. A student claims that a number is divisible by 24 if, and only if, it is divisible by 6 and by 4, and, in general, a number is divisible by $a \cdot b$ if, and only if, it is divisible by a and by b. What is your response?

20. A student found that all three-digit numbers of the form aba, where $a + b$ is a multiple of 7, are divisible by 7. She would like to know why this is so. How do you respond?

National Assessment of Educational Progress (NAEP) Question

Sam did the following problems.

$$2 + 1 = 3$$
$$6 + 1 = 7$$

Sam concluded that when he adds 1 to any whole number, his answer will always be odd.

Is Sam correct? _____
Exaplain your answer.

NAEP, Grade 4, 2009

Dee finds that she has an extraordinary Social Security number. Its nine digits contain all the numbers from 1 through 9. They also form a number with the following characteristics: When read from left to right, its first two digits form a number divisible by 2, its first three digits form a number divisible by 3, its first four digits form a number divisible by 4, and so on, until the complete number is divisible by 9. What is Dee's Social Security number?

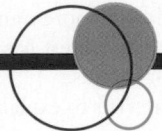

4-2 Prime and Composite Numbers

The grade 4 *Common Core Standards* states that students should be able to:

Find all factor pairs for a whole number in the range 1–100. Recognize that a whole number is a multiple of each of its factors. Determine whether a given whole number in the range 1–100 is a multiple of a given one-digit number. Determine whether a given whole number in the range 1–100 is prime or composite. (p. 29)

One method used in elementary schools to determine the positive factors of a non-zero whole number is to use squares of paper or cubes and to represent the number as a rectangle. Such a rectangle resembles a candy bar formed with small squares. The dimensions of the rectangle are divisors or factors of the number. For example, Figure 4-5 shows rectangles to represent 12.

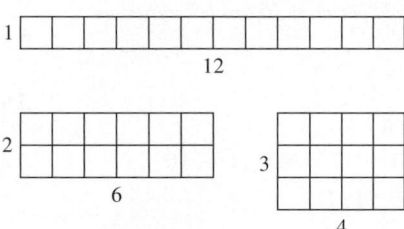

Figure 4-5

As seen in Figure 4-5 above, the number 12 has six positive divisors: 1, 2, 3, 4, 6, and 12. If rectangles were used to find the divisors of 7, then we would find only a 1×7 rectangle, as shown in Figure 4-6. Thus, 7 has exactly two divisors: 1 and 7.

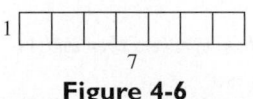

Figure 4-6

Table 4-1 illustrates the number of positive divisors of a non-zero whole number. Below each number listed across the top, we identify the numbers less than or equal to 37 that have that number of whole number divisors. For example, 12 is in the 6 column because it has six divisors, and 7 is in the 2 column because it has only two divisors.

NOW TRY THIS 4-4

  a. What patterns do you see forming in Table 4-1?
  b. Will there be other entries in the 1 column? Why?
  c. What are the next three numbers in the 3 column?
  d. Find an entry for the 7 column.
  e. What kinds of numbers have an odd number of factors? Why?

Table 4-1 Number of Positive Divisors

1	2	3	4	5	6	7	8	9
1	2	4	6	16	12		24	36
	3	9	8		18		30	
	5	25	10		20			
	7		14		28			
	11		15		32			
	13		21					
	17		22					
	19		26					
	23		27					
	29		33					
	31		34					
	37		35					

The numbers in the 2 column in Table 4-1 are of particular importance: they have exactly two positive divisors, namely, 1 and themselves. Any positive integer with exactly two distinct, positive divisors is a *prime number*, or a **prime**. Any whole number greater than 1 that has a positive factor other than 1 and itself is a *composite number*, or a **composite**. For example, 4, 6, and 16 are composites because they have positive factors other than 1 and themselves. The number 1 has only one positive factor, so it is neither prime nor composite. From the 2 column in Table 4-1, we see that the first 12 primes are 2, 3, 5, 7, 11, 13, 17, 19, 23, 29, 31, and 37. The number 2 is sometimes called the "oddest" prime because it is the only even prime.

EXAMPLE 4-9

Show that the following numbers are composite:

a. 1564 b. 2781
c. 1001 d. $3 \cdot 5 \cdot 7 \cdot 11 \cdot 13 + 1$

Solution
a. Since $2|4$, 1564 is divisible by 2 and is composite.
b. Since $3|(2 + 7 + 8 + 1)$, 2781 is divisible by 3 and is composite.
c. Since $11|[(1 + 0) - (0 + 1)]$, 1001 is divisible by 11 and is composite.
d. Because a product of odd numbers is odd (why?), $3 \cdot 5 \cdot 7 \cdot 11 \cdot 13$ is odd. If we add 1 to an odd number, the sum is even. An even number (other than 2) has a factor of 2 and is therefore composite.

NOW TRY THIS 4-5

The *FoxTrot* cartoon on p.192 concerns divisibility and prime numbers. Answer the following questions based on the cartoon.

a. Select a number at random from the cartoon and divide it by 13; then divide it by 17; then divide it by 19. Keep doing this until a number is found that when divided by 13, 17, or 19 gives you a whole-number answer. What does it mean when a whole-number answer is found?
b. The cartoonist made one mistake. The number 2261 appears in the left center of the cartoon. Explain why this number cannot be included in the cartoon.

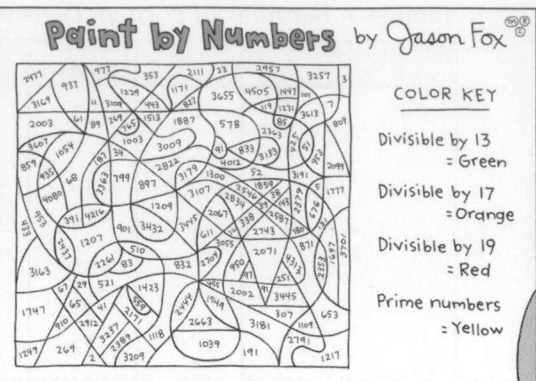

Prime Factorization

In the grade 7 *Focal Points*, we find this statement:

> Students continue to develop their understanding of multiplication and division and the structure of numbers by determining if a counting number greater than 1 is a prime, and if it is not, by factoring it into a product of primes. (p. 19)

Composite numbers can be expressed as products of two or more whole numbers greater than 1. For example, $18 = 2 \cdot 9$, $18 = 3 \cdot 6$, or $18 = 2 \cdot 3 \cdot 3$. Each expression of 18 as a product of factors is a **factorization**.

A factorization containing only prime numbers is a **prime factorization**. To find a prime factorization of a given composite number, we first rewrite the number as a product of two smaller numbers greater than 1. We continue the process, factoring the lesser numbers until all factors are primes. For example, consider 260:

$$260 = 26 \cdot 10 = (2 \cdot 13)(2 \cdot 5) = 2 \cdot 2 \cdot 5 \cdot 13 = 2^2 \cdot 5 \cdot 13$$

The procedure for finding a prime factorization of a number can be organized using a **factor tree**, as Figure 4-7(a) demonstrates. The last branches of the tree display the prime factors of 260. A second way to factor 260 is shown in Figure 4-7(b). The two trees produce the same prime factorization, except for the order in which the primes appear in the products.

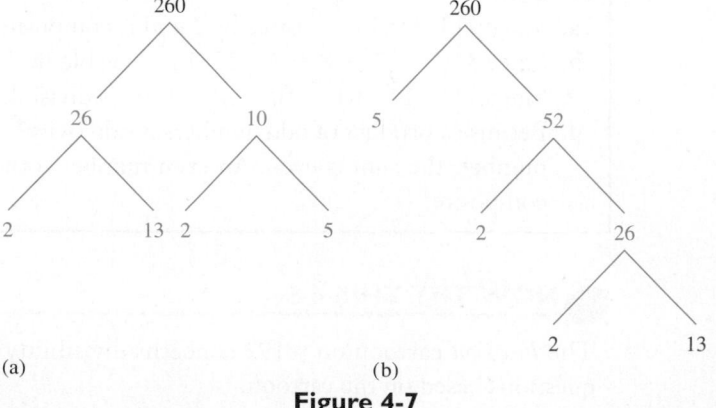

Figure 4-7

The *Fundamental Theorem of Arithmetic*, or the *Unique Factorization Theorem*, states that in general, if order is disregarded, the prime factorization of a number is unique.

> **Theorem 4-12**
>
> **Fundamental Theorem of Arithmetic**   Each composite number can be written as a product of primes in one, and only one, way except for the order of the prime factors in the product.

The Fundamental Theorem of Arithmetic assures that once we find a prime factorization of a number, a different prime factorization of the same number cannot be found. For example, consider 260. We start with the least prime, 2, and see whether it divides 260. If not, we try the next greater prime and check for divisibility by this prime. Once we find a prime that divides the number in question, we must find the quotient when the number is divided by the prime. This step in the prime factorization of 260 is shown in Figure 4-8(a). Next we check whether the prime divides the quotient. If so, we repeat the process; if not, we try the next greater prime, 3, and check to see if it divides the quotient. We see that 130 divided by 2 yields 65, as shown in Figure 4-8(b). We continue the procedure, using greater primes, until a quotient of 1 is reached. The original number is the product of all the prime divisors used. The complete procedure for 260 is shown in Figure 4-8(c). An alternative form of this procedure is shown in Figure 4-8(d).

$$
\begin{array}{ll}
2\,\lfloor\underline{260} & \\
130 & \\
\text{(a)} &
\end{array}
\qquad
\begin{array}{ll}
2\,\lfloor\underline{260} & \\
2\,\lfloor\underline{130} & \\
65 & \\
\text{(b)}
\end{array}
\qquad
\begin{array}{ll}
2\,\lfloor\underline{260} & \\
2\,\lfloor\underline{130} & \\
5\,\lfloor\underline{65} & \\
13\,\lfloor\underline{13} & \\
1 & \\
\text{(c)}
\end{array}
\qquad
\begin{array}{ll}
\lfloor\underline{260} & \\
2\,\lfloor\underline{130} & \\
2\,\lfloor\underline{65} & \\
5\,\lfloor\underline{13} & \\
13\,\rfloor\,1 & \\
\text{(d) Alternative form}
\end{array}
$$

Figure 4-8

The primes in the prime factorization of a number are typically listed in increasing order from left to right, and if a prime appears in a product more than once, exponential notation is used. Thus, the factorization of 260 is written as $2^2 \cdot 5 \cdot 13$. Prime factorization is demonstrated in the student page on page 194. Notice that the factor tree is developed in two different ways leading to the same result.

 NOW TRY THIS 4-6

Colored rods are used in the elementary-school classroom to teach many concepts. The rods vary in length from 1 cm to 10 cm. Various lengths have different colors; for example, the 5 rod is yellow. The rods and their colors are shown in Figure 4-9. A row with all the same color rods is called a *one-color train*.

a. What rods can be used to form a one-color train for 18?
b. What one-color trains are possible for 24?
c. How many one-color trains of two or more rods are possible for each prime number?
d. If a number can be represented by an all-red train, an all-green train, and an all-yellow train, what is the least number of factors it must have? What are they?

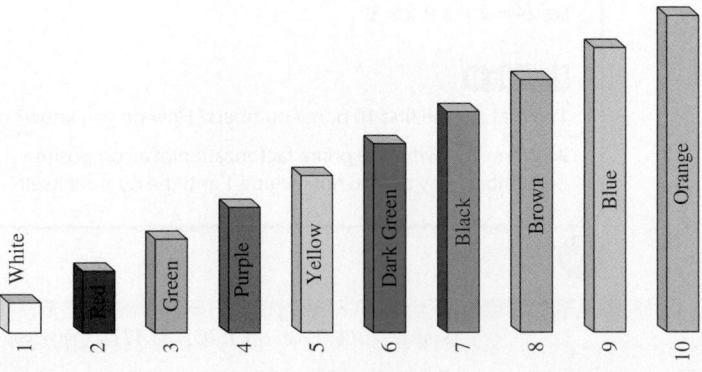

Figure 4-9

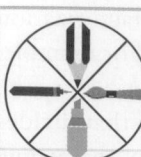

School Book Page   PRIME AND COMPOSITE NUMBERS

Lesson
4-8

Understand It!
The factors of a whole number determine if the number is prime or composite.

Prime and Composite Numbers

What are prime and composite numbers?

Every whole number greater than 1 is either a prime number or a composite number. A prime number has exactly two factors, 1 and itself. A composite number has more than two factors.

● ● ●
$1 \times 3 = 3$

● ● ● ● ● ● ● ●
$1 \times 8 = 8$

● ● ● ●
● ● ● ●
$2 \times 4 = 8$

Another Example   **How can you write a composite number as a product of prime factors?**

Write 24 as a product of prime factors.

A product of prime factors is called the prime factorization of a number. A factor tree is a diagram that shows the prime factorization of a composite number.

One Way

Step 1 Find a factor pair for 24.

24
4×6

Step 2 Write each factor that is not prime as a product of prime numbers.

24
4   6
$2 \times 2 \times 2 \times 3$

Last "branch" of the tree contains all prime numbers.

So, $24 = 2 \times 2 \times 2 \times 3$.

Another Way

You can use a different factor pair for 24.

24
3×8
$3 \times 2 \times 4$
$3 \times 2 \times 2 \times 2$

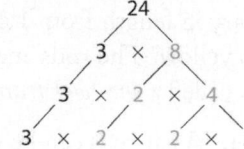 *Continue until all "branches" end in prime numbers.*

Explain It

1. What are the first 10 prime numbers? How do you know?

2. When you write the prime factorization of a composite number, why do you not include 1 and the number itself?

106

Number of Whole Number Divisors

How many whole number divisors does 24 have? The question asks for the number of divisors, not just prime divisors. To aid in the listing, we group divisors as in Figure 4-10:

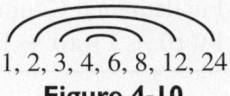

1, 2, 3, 4, 6, 8, 12, 24
Figure 4-10

The whole number divisors of 24 occur in pairs, where the product of the divisors in each pair is 24. If 3 is a divisor of 24, then 24/3, or 8, is also a divisor of 24. In general, if a whole number k is a divisor of 24, then $24/k$ is also a divisor of 24.

Another way to think of the number of whole number divisors of 24 is to consider the prime factorization $24 = 2^3 \cdot 3$. The whole number divisors of 2^3 are 2^0, 2^1, 2^2, and 2^3. The whole number divisors of 3 are 3^0 and 3^1. We know that 2^3 has $(3 + 1)$, or 4, divisors and 3^1 has $(1 + 1)$, or 2, divisors. Because each divisor of 24 is the product of a divisor of 2^3 and a divisor of 3^1, we use the Fundamental Counting Principle to conclude that 24 has $4 \cdot 2$, or 8, positive divisors. This is summarized in Table 4-2.

Table 4-2

Divisors of 2^3	$2^0 = 1$	$2^1 = 2$	$2^2 = 4$	$2^3 = 8$
Divisors of 3^1	$3^0 = 1$	$3^1 = 3$		
Divisors of $3^1 \cdot 2^3$ (Divisors of 24)	$3^0 \cdot 2^0 = 1$ $3^1 \cdot 2^0 = 3$	$3^0 \cdot 2^1 = 2$ $3^1 \cdot 2^1 = 6$	$3^0 \cdot 2^2 = 4$ $3^1 \cdot 2^2 = 12$	$3^0 \cdot 2^3 = 8$ $3^1 \cdot 2^3 = 24$

This discussion can be generalized as follows: If p is any prime and n is any whole number, then the whole number divisors of p^n are p^0, p^1, p^2, p^3, ..., p^n. Therefore, there are $(n + 1)$ whole number divisors of p^n. Now, using the Fundamental Counting Principle, we can find the number of divisors of any number whose prime factorization is known.

> **Theorem 4-13**
>
> If p and q are different primes, and n, m are whole numbers, then $p^n q^m$ has $(n + 1)(m + 1)$ whole number divisors. In general, if $p_1, p_2, \ldots, p_k$ are primes and $n_1, n_2, \ldots, n_k$ are whole numbers, then $p_1^{n_1} \cdot p_2^{n_2} \cdot \ldots \cdot p_k^{n_k}$ has $(n_1 + 1)(n_2 + 1) \cdot \ldots \cdot (n_k + 1)$ whole number divisors.

EXAMPLE 4-10

Find the number of whole number divisors of each of the following:

a. 1,000,000   b. 210^{10}

Solution

a. We first find the prime factorization of 1,000,000.

$$1,000,000 = 10^6 = (2 \cdot 5)^6 = (2 \cdot 5)(2 \cdot 5)(2 \cdot 5)(2 \cdot 5)(2 \cdot 5)(2 \cdot 5)$$
$$= (2 \cdot 2 \cdot 2 \cdot 2 \cdot 2 \cdot 2)(5 \cdot 5 \cdot 5 \cdot 5 \cdot 5 \cdot 5)$$
$$= 2^6 \cdot 5^6$$

Because 2^6 has $6 + 1$ divisors and 5^6 has $6 + 1$ divisors, then by the Fundamental Counting Principle $2^6 \cdot 5^6$ has $(6 + 1)(6 + 1)$, or 49, divisors.

b. We first find the prime factorization of 210^{10}:

$$210 = 21 \cdot 10 = 3 \cdot 7 \cdot 2 \cdot 5 = 2 \cdot 3 \cdot 5 \cdot 7,$$
$$210^{10} = (2 \cdot 3 \cdot 5 \cdot 7)^{10} = 2^{10} \cdot 3^{10} \cdot 5^{10} \cdot 7^{10}$$

By the Fundamental Counting Principle, the number of divisors of 210^{10} is $(10 + 1)(10 + 1)(10 + 1)(10 + 1) = 11^4 = 14{,}641.$

⬤ NOW TRY THIS 4-7

To determine whether it is necessary to divide 97 by 2, 3, 4, 5, 6, . . ., 96 to check whether it is prime, answer the following (justify your answers):

a. If 2 is not a divisor of 97, could any multiple of 2 be a divisor of 97?
b. If 3 is not a divisor of 97, what other numbers could not be divisors of 97?
c. If 5 is not a divisor of 97, what other numbers could not be divisors of 97?
d. If 7 is not a divisor of 97, what other numbers could not be divisors of 97?
e. Conjecture what numbers we have to check for divisibility in order to determine if 97 is prime.

Determining Whether a Number Is Prime

As depicted in the following cartoon by Sidney Harris, prime numbers have fascinated people of various backgrounds. In Now Try This 4-7, we found that to determine if a number is prime, we must check only divisibility by prime numbers less than the given number. (Why?) However, do we need to check all the primes less than the number? Suppose we want to check whether 97 is prime and we find that 2, 3, 5, and 7 do not divide 97. Could a greater prime divide 97? If p is a prime greater than 7, then $p \geq 11$. If $p \mid 97$, then $97/p$ also divides 97. However, because $p \geq 11$ then $97/p$ must be less than 9 and hence cannot divide 97. Why? We see that there is no need to check for divisibility by numbers other than 2, 3, 5, and 7. These ideas are generalized in the following theorems.

ScienceCartoonsPlus.com

> **Theorem 4-14**
>
> If d is a divisor of n, then $\dfrac{n}{d}$ is also a divisor of n.

Suppose that p is the *least* divisor of a composite number n. Such a divisor must be prime. (Why?) Then $n = pk$, $k \neq 1$. Since $k \mid n$ and p was the least divisor of n, $k \geq p$. Therefore, $n = pk \geq pp = p^2$. Since $n \geq p^2$, we have $p^2 \leq n$. This idea is summarized in the following theorem.

> **Theorem 4-15**
>
> If n is composite, then n has a prime factor p such that $p^2 \leq n$.

Theorem 4-14 can be used to help determine whether a given number is prime or composite. For example, consider the number 109. If 109 is composite, it must have a prime divisor p such that $p^2 \leq 109$. The primes whose squares do not exceed 109 are 2, 3, 5, and 7. Mentally, we can see that $2 \nmid 109$, $3 \nmid 109$, $5 \nmid 109$, and $7 \nmid 109$. Hence, 109 is prime. The argument used leads to the following theorem.

> **Theorem 4-16**
>
> If n is a whole number greater than 1 and not divisible by any prime p such that $p^2 \leq n$, then n is prime.

 **EXAMPLE 4-11**

　a. Is 397 composite or prime?
　b. Is 91 composite or prime?

Solution
　a. The possible primes p such that $p^2 \leq 397$ are 2, 3, 5, 7, 11, 13, 17, and 19. Because $2 \nmid 397$, $3 \nmid 397$, $5 \nmid 397$, $7 \nmid 397$, $11 \nmid 397$, $13 \nmid 397$, $17 \nmid 397$, and $19 \nmid 397$, the number 397 is prime.
　b. The possible primes p such that $p^2 \leq 91$ are 2, 3, 5, and 7. Because 91 is divisible by 7, it is composite.

　One way to find all the primes less than a given number is to use the *Sieve of Eratosthenes*, named after the Greek mathematician Eratosthenes (ca. 276–194 BCE). If all the whole numbers greater than 1 are considered (or placed in the sieve), the numbers that are not prime are methodically crossed out (or drop through the holes of the sieve). The remaining numbers are prime. The partial student page on page 198 illustrates this process.

　The Sieve of Eratosthenes is another way to motivate Theorem 4-16. Notice the observations from the sieve in Table 4-3 as we crossed out numbers.

Table 4-3

Prime	Observation
2	Once 2 is circled first number not crossed out that 2 divides is $4 = 2^2$.
3	Once 3 is circled first number not crossed out that 3 divides is $9 = 3^2$.
5	Once 5 is circled first number not crossed out that 5 divides is $25 = 5^2$.
7	Once 7 is circled first number not crossed out that 7 divides is $49 = 7^2$.

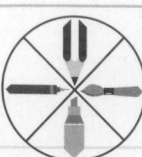

School Book Page   SIEVE OF ERATOSTHENES

Enrichment

Sieve of Eratosthenes

About 230 B.C., the Greek mathematician Eratosthenes developed a method for identifying prime numbers. The method is called the **Sieve of Eratosthenes.**

Follow the steps to identify the prime numbers from 1 to 100.

Step 1 Copy the table at the right.

Step 2 Cross out 1 because it is neither prime nor composite.

Step 3 Circle 2. Then, cross out all other multiples of 2.

Step 4 Go to the first number that is not crossed out. Circle it and cross out its other multiples.

Step 5 Repeat Step 4 until all numbers in the table are either crossed out or circled. The circled numbers are prime.

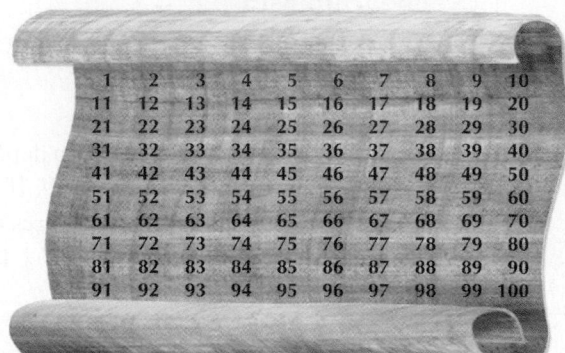

1. List the prime numbers from 1 to 100.

2. Explain why some numbers could be crossed out more than once.

All text pages available online and on CD-ROM.   **Section A Lesson 3-2**   **149**

We need not continue the procedure for 11 because the first number not crossed out that 11 divides is 11^2, or 121, and the table on the student page only goes to 100. Therefore, to test whether a number such as 137 is a prime, we first test for divisibility by all primes up to but not including the first prime whose square is greater than 137. Because $13^2 = 169$, any prime greater than or equal to 13 would give a quotient less than or equal to 13, and we have already checked these primes. This shows that when testing to see whether a number is prime, we need try as divisors only primes whose squares are less than or equal to the number being tested.

Problem Solving   Perfect Squares

A perfect square is a whole number that is a square of a whole number. For example, 81 is a perfect square because $81 = 9^2$. Show that a non-zero whole number is a perfect square if, and only if, it has an odd number of divisors.

Understanding the Problem We need to show that all perfect squares—such as 36—have an odd number of divisors, and if a number is not a perfect square—such as 24—then it has an even number of divisors. Indeed, $36 = 6^2 = 2^2 \cdot 3^2$ and it has $(2 + 1)(2 + 1)$ or 9 divisors, which is odd. In contrast, $24 = 4 \cdot 6 = 2^3 \cdot 3^1$ has $(3 + 1)(1 + 1)$ or 8 divisors, an even number. We need to prove the statement in the problem for all non-zero whole numbers.

Devising a Plan We write the prime factorization of a perfect square and use Theorem 4-12 to find an expression for the number of divisors. Similarly, if a number is not a perfect square, we characterize its prime factorization and again use Theorem 4-13.

Carrying Out the Plan Every perfect square can be written as m^2, where m is a whole number. If the prime factorization of m is $m = p_1^{n_1} p_2^{n_2} \ldots p_k^{n_k}$ then $m^2 = p_1^{2n_1} p_2^{2n_2} \ldots p_k^{2n_k}$. Theorem 4-12 tells us that the number of divisors of m^2 is $(2n_1 + 1)(2n_2 + 1) \ldots (2n_k + 1)$. Each of these factors is odd. Since a product of odd numbers is odd, the above product is odd.

If a number is not a perfect square, then the exponents to which the primes in its prime factorization are raised cannot be all even (if they were all even, the number would be a perfect square). Thus, at least one of the exponents is odd. Suppose that exponent is n_j, then by Theorem 4-13, the number of divisors would be a product in which one of the factors in that product is $n_j + 1$. Since n_j is odd, $n_j + 1$ is even. When an even whole number is multiplied by any whole number, the result is even and hence the product is even.

Looking Back We could approach the problem differently. Theorem 4-14 states that if $d \mid n$, then n/d is also a divisor of n. Consequently, if for all divisors d of n, $d \neq n/d$, then each divisor can be paired with a different divisor of n and therefore must have an even number of divisors. If for some divisor d, $d = n/d$, then $n = d^2$, and all the divisors of n except d are paired with a different divisor. Hence the number of divisors is odd. It follows that n has an even number of divisors if, and only if, n is a perfect square.

More About Primes

There are infinitely many whole numbers, infinitely many odd whole numbers, and infinitely many even whole numbers. Are there infinitely many primes? The answer to this question is not obvious. Euclid was the first to prove that there are infinitely many primes.

Mathematicians have long looked for a formula that produces only primes, but no one has found one. One result was the expression $n^2 - n + 41$, where n is a whole number. Substituting $0, 1, 2, 3, \ldots, 40$ for n in the expression always results in a prime number. However, substituting 41 for n gives $41^2 - 41 + 41$, or 41^2, a composite number. In 1998, Roland Clarkson, a 19-year-old student at California State University, showed that $2^{3021377} - 1$ is prime. The number has 909,526 digits. The full decimal expansion of the number would fill several hundred pages. Since then, more large primes have been discovered: in 2006, $2^{32,582,657} - 1$ (9,808,358 digits) and in 2008, $2^{43,112,609} - 1$ (12,978,189 digits). These are examples of *Mersenne primes*. A Mersenne prime, named after the French monk Marin Mersenne (1588–1648), is a prime of the form $2^n - 1$, where n is prime.

Another type of interesting prime is a *Sophie Germain prime*, which is an odd prime p for which $2p + 1$ is also a prime. Notice that $p = 3$ is a Sophie Germain prime, since $2 \cdot 3 + 1$, or 7, is also a prime. Check that 5, 11, and 23 are also such primes. The primes were named after the French mathematician Sophie Germain. The greatest Sophie Germain prime discovered as of 2010 is $(183,027) \cdot 2^{265,440} - 1$; it has 79,911 digits.

Historical Note

Eratosthenes (276–194 BCE) spent most of his life in Alexandria as a chief librarian. In mathematics Eratosthenes is best known for his "sieve"—a systematic procedure for isolating the prime numbers—and for a simple method for calculating the circumference of the Earth. ●

Problem Solving   How Many Bears?

A large toy store carries one kind of stuffed bear. On Monday the store sold a certain number of the stuffed bears for a total of $1843 and on Tuesday, without changing the price, the store sold a certain number of the stuffed bears for a total of $1957. How many toy bears were sold each day if the price of each bear is a whole number and greater than $1?

Understanding the Problem One day a store sold a number of stuffed bears for $1843 and on the next day a number of them for a total of $1957. We need to find the number of stuffed bears sold on each day.

Devising a Plan If x bears were sold the first day and y bears the second day, and if the price of each bear was c dollars, we would have $cx = 1843$ and $cy = 1957$. Thus, 1843 and 1957 should have a common factor—the price c. We could factor each number and find the possible factors. If the problem is to have a unique solution, the two numbers should have only one common factor other than 1. Any common factor of 1957 and 1843 will also be a factor of $1957 - 1843 = 114$ and the factors of 114 are easier to find.

Carrying Out the Plan We have $114 = 2 \cdot 57 = 2 \cdot 3 \cdot 19$. Thus, if 1957 and 1843 have a common prime factor, it must be 2, 3, or 19. But neither 2 nor 3 divides the numbers, hence the only possible common factor is 19. We divide each number by 19 and find

$$1843 = 19 \cdot 97$$
$$1957 = 19 \cdot 103$$

Notice that neither 97 nor 103 is divisible by 2, 3, 5, or 7. Hence 97 and 103 are primes (why?) and therefore the only common factor (greater than 1) of 1843 and 1957 is 19. Consequently, the price of each bear was $19. The first day 97 bears were sold and the next day 103 bears were sold.

Looking Back The problem had a unique solution because the only common factor (greater than 1) of the two numbers was 19. We could create similar problems by having the price of the item be a prime number and the number of items sold each day also be prime numbers. For example, the total sale on the first day could have been $23 \cdot 101$, or $2323 and on the second day $23 \cdot 107$, or $2461 (23, 101, and 107 are prime numbers).

To find a common factor of 1957 and 1843, we found all the common factors of $1957 - 1843 = 114 = 2 \cdot 3 \cdot 19$ and checked which of the factors of the difference was a common factor of the original numbers. We used Theorem 4-2(c): If $d \mid a$ and $d \mid b$, then $d \mid (a - b)$. This theorem assures that every common factor of a and b is also a factor of $a - b$ if $a \geq b$. Thus the set of all common factors of a and b is a subset of all the factors of $a - b$.

Historical Note

Sophie Germain (1776–1831) was born in Paris and grew up during the French Revolution. She wanted to study at the prestigious École Polytechnique but women students were not allowed. Consequently, she studied from lecture notes and from Gauss's monograph on number theory. She made major contributions to the mathematical theory of elasticity, for which she was awarded the prize of the French Academy of Sciences. Germain's work was highly regarded by Gauss, who recommended her for an honorary degree from the University of Göttingen. She died before the degree could be awarded.

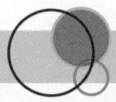

Historical Note

In the 1970s, determining large prime numbers became extremely useful in coding and decoding secret messages. In all coding and decoding, the letters of an alphabet correspond in some way to non-zero whole numbers. A "safe" coding system, in which messages are unintelligible to everyone except the intended receiver, was devised by three Massachusetts Institute of Technology scientists (Ronald Rivest, Adi Shamir, and Leonard Adleman) and is referred to as the RSA (their initials) system. The secret deciphering key consists of two large prime numbers chosen by the user. The enciphering key is the product of these two primes. Because it is extremely difficult and time consuming to factor large numbers, it is practically impossible to recover the deciphering key from a known enciphering key. In 1982, new methods for factoring large numbers were invented, which resulted in the use of even greater primes to prevent the breaking of decoding keys. ●

Assessment 4-2A

1. Find the least non-zero whole number that is divisible by three different primes.
2. Determine which of the following numbers are primes:
   a. 109   **b.** 119
   c. 33   **d.** 101
   e. 463   **f.** 97
   g. $2 \cdot 3 \cdot 5 \cdot 7 + 1$ **h.** $2 \cdot 3 \cdot 5 \cdot 7 - 1$
3. Use a factor tree to find the prime factorization for each of the following:
   a. 504   **b.** 2475   **c.** 11,250
4. **a.** Fill in the missing numbers in the following factor tree:

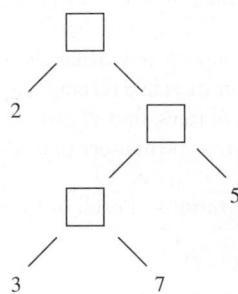

   b. How could you find the top number without finding the other two numbers?
5. What is the greatest prime that must be considered to test whether 5669 is prime?
6. Find the prime factorizations of the following:
   a. $1 \cdot 2 \cdot 3 \cdot 4 \cdot 5 \cdot 6 \cdot 7 \cdot 8 \cdot 9 \cdot 10$
   b. $10^2 \cdot 26 \cdot 49^{10}$
   c. 251
   d. 1001
7. **a.** When the U.S. flag had 48 stars, the stars formed a 6 by 8 rectangular array. In what other rectangular arrays could they have been arranged?
   b. How many rectangular arrays of stars could there be if there were only 47 states?
8. Find the least non-zero whole number divisible by all whole numbers 1 through 10.

9. **a.** Use the Fundamental Theorem of Arithmetic to justify that if $2 \mid n$ and $3 \mid n$, then $6 \mid n$.
   b. Is it always true that if $a \mid n$ and $b \mid n$, then $ab \mid n^2$? Either prove the statement or give a counterexample.
10. Find the least three-digit number that has exactly five factors.
11. If $n = 2 \cdot 3 \cdot 5 \cdot 7 \cdot 11 \cdot 13 + 1 = 30{,}031$, is $2 \cdot 3 \cdot 5 \cdot 7 \cdot 11 \cdot 13 + 1$ the prime factorization of n? Why or why not?
12. Is it possible to find non-zero whole numbers x, y, and z such that $3^x \cdot 5^y = 8^z$? Justify your answer.
13. **a.** Show that there are infinitely many composite numbers in the arithmetic sequence 4, 7, 10, 13, 16, 19, 22, ...
    b. Find infinitely many composite numbers in the sequence 1, 11, 111, 1111, ...
14. If $32n = 2^6 \cdot 3^5 \cdot 5^4 \cdot 7^3 \cdot 11^7$, explain why $2 \cdot 3 \cdot 5 \cdot 7 \cdot 11^6$ is a factor of n.
15. Is $7^4 \cdot 11^3$ a factor of $7^5 \cdot 11^3$? Explain why or why not.
16. Explain why each of the following numbers is composite:
    a. $3 \cdot 5 \cdot 7 \cdot 11 \cdot 13$
    b. $(3 \cdot 4 \cdot 5 \cdot 6 \cdot 7 \cdot 8) + 2$
    c. $(3 \cdot 5 \cdot 7 \cdot 11 \cdot 13) + 5$
    d. $10! + 7$ (*Note:* $10! = 10 \cdot 9 \cdot 8 \cdot 7 \cdot 6 \cdot 5 \cdot 4 \cdot 3 \cdot 2 \cdot 1$.)
17. Explain why $2^3 \cdot 3^2 \cdot 25^3$ is not a prime factorization and find the prime factorization of the number.
18. The prime numbers 11 and 13 are called *twin primes* because they differ by 2. (The existence of infinitely many twin primes has not been proved.) Find all the twin primes less than 200.
19. Mr. Arboreta wants to plant fruit trees in a rectangular array. For each of the following numbers of trees, find all possible numbers of rows if each row is to have the same number of trees:
    a. 36
    b. 28
    c. 17
    d. 144
20. Find the prime factorizations of each of the following:
    a. $36^{10} \cdot 49^{20} \cdot 6^{15}$
    b. $100^{60} \cdot 300^{40}$
    c. $2 \cdot 3^4 \cdot 5^{110} \cdot 7 + 4 \cdot 3^4 \cdot 5^{110}$
    d. $2 \cdot 3 \cdot 5 \cdot 7 \cdot 11 + 1$

Assessment 4-2B

1. Find the least whole number greater than 0 that is divisible by four different primes.
2. Determine which of the following numbers are primes:
   a. 89   b. 147
   c. 159   d. 187
   e. $2 \cdot 3 \cdot 5 \cdot 7 + 5$ f. $2 \cdot 3 \cdot 5 \cdot 7 - 5$
3. Use a factor tree to find the prime factorization for each of the following:
   a. 304   b. 1570   c. 9550
4. a. Fill in the missing numbers in the following factor tree:

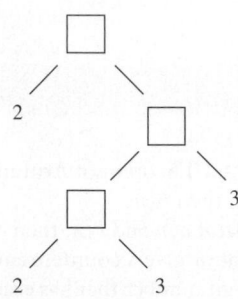

   b. How could you find the top number without finding the other two numbers?
5. What is the greatest prime you must consider to test whether 503 is prime?
6. Find the prime factorizations of the following:
   a. 1001   b. 1001^2
   c. 999^{10} d. $111^{10} - 111^9$
7. Suppose the 435 members of the House of Representatives are placed on committees consisting of more than 2 members but fewer than 30 members. Each committee is to have an equal number of members and each member is to be on only one committee.
   a. What size committees are possible?
   b. How many committees are there of each size?
8. Find the least natural number divisible by each natural number less than or equal to 12.

9. a. Use the Fundamental theorem of Arithmetic to prove that if $4|n$ and $9|n$, then $36|n$.
   b. When is it true that if $a|n$ and $b|n$, then $ab|n$? Justify your answer.
10. Find the greatest four-digit whole number that has exactly three positive factors.
11. Show that if 1 were considered a prime, every number would have more than one prime factorization.
12. Is it possible to find non-zero whole numbers x, y, and z such that $2^x \cdot 3^y = 5^z$? Why or why not?
13. Show that there are infinitely many composite numbers in the arithmetic sequence 1, 5, 9, 13, 17, ...
14. If $2N = 2^6 \cdot 3^5 \cdot 5^4 \cdot 7^3 \cdot 11^7$, explain why $2 \cdot 3 \cdot 5 \cdot 7 \cdot 11$ is a factor of N.
15. Is $3^2 \cdot 2^4$ a factor of $3^3 \cdot 2^2$? Explain why or why not.
16. Explain why each of the following numbers is composite:
    a. $7 \cdot 11 \cdot 13 \cdot 17 + 17$
    b. $10! + k$, where $k = 2, 3, 4, 5, 6, 7, 8, 9,$ or 10 (10! is the product of the whole numbers 1 through 10.)
17. Explain why $2^2 \cdot 5^3 \cdot 9^2$ is not a prime factorization and find the prime factorization of the number.
18. A prime such as 7331 is a *superprime* because any integers obtained by deleting digits from the right of 7331 are prime; namely, 733, 73, and 7.
    a. For a prime to be a superprime, what digits cannot appear in the number?
    b. Of the digits that can appear in a superprime, what digit cannot be the leftmost digit of a superprime?
    c. Find all of the two-digit superprimes.
    d. Find a three-digit superprime other than 733.
19. Gina wants to plant fruit trees in a rectangular array. For each of the following numbers of trees, find all possible numbers of rows if each row is to have the same number of trees:
    a. 15   b. 20   c. 19   d. 100
20. Find the prime factorizations of each of the following:
    a. $16^4 \cdot 81^4 \cdot 6^6$ b. $8^4 \cdot 32^5$
    c. $2^2 \cdot 3^5 \cdot 7^{55} + 2^4 \cdot 3^4 \cdot 7^{55}$

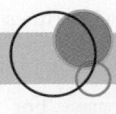

Mathematical Connections 4-2

Communication

1. Explain why the product of any three consecutive whole numbers is divisible by 6.
2. Explain why the product of any four consecutive whole numbers is divisible by 24.
3. In order to test for divisibility by 12, one student checked to determine divisibility by 3 and 4; another checked for divisibility by 2 and 6. Are both students using a correct approach to divisibility by 12? Why or why not?
4. In the Sieve of Eratosthenes for numbers less than 100, explain why, after crossing out all the multiples of 2, 3, 5, and 7, the remaining numbers are primes.
5. Let $M = 2 \cdot 3 \cdot 5 \cdot 7 + 11 \cdot 13 \cdot 17 \cdot 19$. Without multiplying, show that none of the primes less than or equal to 19 divides M.

6. A woman with a basket of eggs finds that if she removes the eggs from the basket 3 or 5 at a time, there is always 1 egg left. However, if she removes the eggs 7 at a time, there are no eggs left. If the basket holds up to 100 eggs, how many eggs does she have? Explain your reasoning.
7. Explain why, when a number is composite, its least whole number divisor, other than 1, must be prime.
8. Euclid proved that given any finite list of primes, there exists a prime not in the list. Read the following argument and answer the questions that follow.

   Let 2, 3, 5, 7, . . ., p be a list of all the primes less than or equal to a certain prime p. We will show that there exists a prime not on the list. Consider the product

$$2 \cdot 3 \cdot 5 \cdot 7 \cdot \ldots \cdot p.$$

Notice that every prime in the list divides that product. However, if we add 1 to the product; that is, form the number $N = (2 \cdot 3 \cdot 5 \cdot 7 \cdot \ldots \cdot p) + 1$, then none of the primes in the list will divide N. Notice that whether N is prime or composite, some prime q must divide N. Because no prime in the list divides N, q is not one of the primes in the list. Consequently $q > p$. We have shown that there exists a prime greater than p.

a. Explain why no prime in the list will divide N.

b. Explain why some prime must divide N.

c. Someone discovered a prime that has 65,050 digits. How does the preceding argument assure us that there exists an even larger prime?

d. Does the argument show that there are infinitely many primes? Why or why not?

e. Let $M = 2 \cdot 3 \cdot 5 \cdot 7 \cdot 11 \cdot 13 \cdot 17 \cdot 19 + 1$. Without multiplying, explain why some prime greater than 19 will divide M.

Open-Ended

9. a. In which of the following intervals are there more primes? Why?

    i. 0–99   **ii.** 100–199

b. What is the longest string of consecutive composite numbers in the intervals?

c. How many twin primes are there in each interval?

d. What patterns, if any, do you see for any of the preceding questions? Predict what might happen in other intervals.

10. A whole number is a *perfect* number if the sum of its factors (other than the number itself) is equal to the number. For example, 6 is a perfect number because its factors sum to 6; that is, $1 + 2 + 3 = 6$. An *abundant* number has factors whose sum is greater than the number itself. A *deficient* number is a number with factors whose sum is less than the number itself.

a. Classify each of the following numbers as perfect, abundant, or deficient:

    i. 12   **ii.** 28   **iii.** 35

b. Find at least one more number that is deficient and one that is abundant.

Cooperative Learning

11. A class of 23 students was using square tiles to build rectangular shapes. Each student had more than 1 tile and each had a different number of tiles. Each student was able to build only one shape of rectangle. All tiles had to be used to build a rectangle and the rectangle could not have holes. For example, a 2 by 6 rectangle uses 12 tiles and is considered the same as a 6 by 2 rectangle but is different from a 3 by 4 rectangle. The class did the activity using the least number of tiles. How many tiles did the class use? Explore the various rectangles that could be made.

Questions from the Classroom

12. Mary says that her factor tree for 72 begins with 3 and 24 so her prime factors will be different from Larry's because he is going to start with 8 and 9. What do you tell Mary?

13. Bob says that to check whether a number is prime he just uses the divisibility rules he knows for 2, 3, 4, 5, 6, 8, and 10. He says if the number is not divisible by these numbers, then it is prime. How do you respond?

14. Joe says that every odd number greater than 3 can be written as the sum of two primes. To convince the class, he wrote $7 = 2 + 5$, $5 = 2 + 3$, and $9 = 7 + 2$. How do you respond?

15. An eighth grader at the Roosevelt Middle School claims that because there are as many even numbers as odd numbers between 1 and 1000, there must be as many numbers that have an even number of whole number divisors as numbers that have an odd number of whole number divisors between 1 and 1000. Is the student correct? Why or why not?

16. A sixth-grade student argues that there are infinitely many primes because "there is no end to numbers." How do you respond?

17. A student claims that every prime greater than 3 is a term in the arithmetic sequence whose nth term is $6n + 1$ or in the arithmetic sequence whose nth term is $6n - 1$. Is this true? If so, why?

Review Problems

18. Classify the following as true or false:

a. 11 is a factor of 189.

b. 1001 is a multiple of 13.

c. $7 | 1001$ and $7 \nmid 12$ imply $7 \nmid (1001 - 12)$.

19. Check each of the following for divisibility by 2, 3, 4, 5, 6, 7, 8, 9, 10, and 11:

a. 438,162

b. 2,345,678,910

20. Prove that if a number is divisible by 12, then it is divisible by 3.

21. Could $3376 be divided exactly among either seven or eight people?

BRAIN TEASER

One Saturday Jody cut short her visit with her friend Natasha to take three other friends to a movie. "How old are they?" asked Natasha. "The product of their ages is 2450 and the sum is exactly twice your age," replied Jody. Natasha said: "I need more information." To that Jody replied, "I should have mentioned that I am at least one year younger than the oldest of my three friends." With this information Natasha found the ages of the friends. How did Natasha figure the ages of the friends and what were their ages?

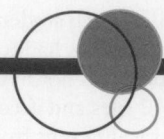

4-3 Greatest Common Divisor and Least Common Multiple

Consider the following Problem:

> Two bands are to be combined to march in a parade. A 24-member band will march behind a 30-member band. The combined bands must have the same number of columns. Each column must be the same size. What is the greatest number of columns in which they can march?

The bands could each march in 2 columns, and we would have the same number of columns, but this does not satisfy the condition of having the greatest number of columns. The number of columns must divide both 24 and 30. (Why?) Numbers that divide both 24 and 30 are 1, 2, 3, and 6. The greatest of these numbers is 6, so the bands should each march in 6 columns. The first band would have 6 columns with 4 members in each column, and the second band would have 6 columns with 5 members in each column. In this problem, we have found the greatest number that divides both 24 and 30, that is, the **greatest common divisor (GCD)** of 24 and 30. Another name for the greatest common divisor is the **greatest common factor (GCF)**.

Definition

The **greatest common divisor (GCD)** or the **greatest common factor (GCF)** of two whole numbers a and b not both 0 is the greatest whole number that divides both a and b.

In what follows we provide several methods for finding GCDs.

Colored Rods Method

We can build a model of two or more non-zero whole numbers with colored rods to determine their GCD. For example, consider finding the GCD of 6 and 8 using the 6 rod and the 8 rod, as in Figure 4-11.

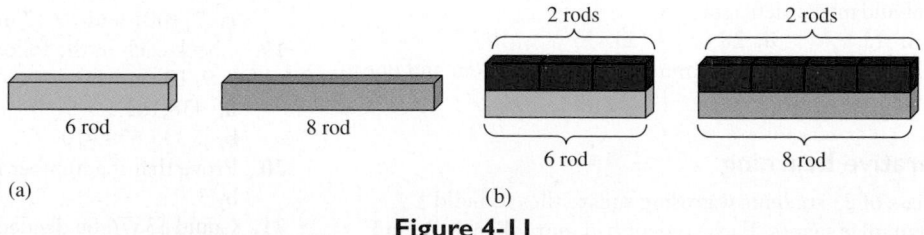

(a)

(b)

Figure 4-11

To find the GCD of 6 and 8, we must find the longest rod such that we can use multiples of that rod to build both the 6 rod and the 8 rod. The 2 rods can be used to build both the 6 and 8 rods, as shown in Figure 4-11(b); the 3 rods can be used to build the 6 rod but not the 8 rod; the 4 rods can be used to build the 8 rod but not the 6 rod; the 5 rods can be used to build neither; and the 6 rods cannot be used to build the 8 rod. Therefore, $\text{GCD}(6, 8) = 2$.

NOW TRY THIS 4-8

Explain how you could use colored rods to solve the marching bands' problem, stated at the beginning of this section.

The Intersection of Sets Method

In the *intersection of sets* method, we list all members of the set of whole number divisors of the two numbers, then find the set of all *common divisors*, and, finally, pick the *greatest* element in that set. For example, to find the GCD of 20 and 32, denote the sets of divisors of 20 and 32 by D_{20} and D_{32}, respectively.

$$D_{20} = \{1, 2, 4, 5, 10, 20\}$$

$$D_{32} = \{1, 2, 4, 8, 16, 32\}$$

The set of all common whole number divisors of 20 and 32 is

$$D_{20} \cap D_{32} = \{1, 2, 4\}$$

Because the greatest number in the set of common whole number divisors is 4, the GCD of 20 and 32 is 4, written $GCD(20, 32) = 4$.

 NOW TRY THIS 4-9

The Venn diagram in Figure 4-12 shows the factors of 24 and 40. Answer the following:

  a. What is the meaning of each of the shaded regions?
  b. Which factor is the GCD?
  c. Draw a similar Venn diagram to find the GCD of 36 and 44.

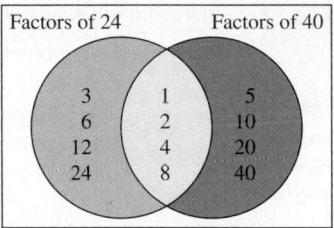

Figure 4-12

The Prime Factorization Method

The intersection of sets method is rather time consuming and tedious if the numbers have many divisors. Another, more efficient, method is the prime factorization method. To find $GCD(180, 168)$, we first consider the prime factorization of the numbers:

$$180 = 2 \cdot 2 \cdot 3 \cdot 3 \cdot 5 = (2^2 \cdot 3)3 \cdot 5$$

and

$$168 = 2 \cdot 2 \cdot 2 \cdot 3 \cdot 7 = (2^2 \cdot 3)2 \cdot 7$$

We see that 180 and 168 have two factors of 2 and one of 3 in common. These common primes divide both 180 and 168. In fact, the only numbers other than 1 that divide both 180 and 168 must have no more than two 2s and one 3 and no other prime factors in their prime factorizations. The possible common divisors are $1, 2, 2^2, 3, 2 \cdot 3$, and $2^2 \cdot 3$. Hence, the greatest common divisor of 180 and 168 is $2^2 \cdot 3$ or 12. The procedure for finding the GCD of two or more numbers by using the prime factorization method is summarized as follows:

> To find the GCD of two or more non-zero whole numbers, first find the prime factorizations of the given numbers and then identify each common prime factor of the given numbers. The GCD is the product of the common prime factors, each raised to the lowest power of that prime that occurs in any of the prime factorizations.

If we apply the prime factorization technique to finding GCD(4, 9), we see that 4 and 9 have no common prime factors. But that does not mean there is no GCD. We still have 1 as a common divisor, so GCD(4, 9) = 1. Numbers, such as 4 and 9, whose GCD is 1 are said to be **relatively prime**. Both the intersection of sets method and the prime factorization method are found on the student page. Study the page and work through the Talk About It questions at the bottom of the student page.

School Book Page   GREATEST COMMON FACTOR

Lesson 3-3

Key Idea
There are different ways to find the factors that are common to two or more numbers.

Vocabulary
• common factor
• greatest common factor (GCF)
• prime factorization (p. 147)

Think It Through
• I can **use factors** to identify equal groups for sharing.
• I can **make an organized list** to find the common factors and GCF.

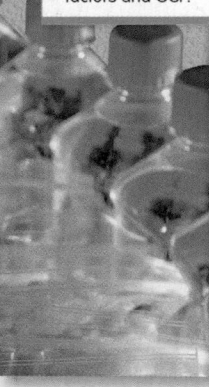

Greatest Common Factor

▶ **LEARN**

How can you use factors?

Janelle is making snack packs for a group hike. Each pack should have the same number of bags of trail mix and the same number of bottles of water. What is the greatest number of snack packs that she can make with no refreshments left over?

To solve this problem, you need to find the numbers that are factors of both 60 and 90. These are the **common factors** of 60 and 90. The **greatest common factor (GCF)** is the *greatest* number that is a factor of both 60 and 90.

Hike Refresments
60 bags of trail mix
90 bottles of water

Example

Find the GCF of 60 and 90.

One Way

List the factors of each number.

60: 1, 2, 3, 4, 5, 6, 10, 12, 15, 20, 30, 60
90: 1, 2, 3, 5, 6, 9, 10, 15, 18, 30, 45, 90

Circle pairs of common factors. Select the greatest one.

60: 1, 2, 3, 4, 5, 6, 10, 12, 15, 20, 30, 60
90: 1, 2, 3, 5, 6, 9, 10, 15, 18, 30, 45, 90

The GCF is 30.

Another Way

Use prime factorization.

$60 = 2 \times 2 \times 3 \times 5$
$90 = 2 \times 3 \times 3 \times 5$

Find the product of the common prime factors. If there are no common prime factors, the GCF is 1.

$60 = 2 \times 2 \times 3 \times 5$
$90 = 2 \times 3 \times 3 \times 5$ $2 \times 3 \times 5 = 30$

The GCF is 30.

The greatest number of snack packs she can make is 30.

EXAMPLE 4-12

Find each of the following:

a. $GCD(108, 72)$

b. $GCD(0, 13)$

c. $GCD(x, y)$ if $x = 2^3 \cdot 7^2 \cdot 11 \cdot 13$ and $y = 2 \cdot 7^3 \cdot 13 \cdot 17$

d. $GCD(x, y, z)$ if $z = 2^2 \cdot 7$, using x and y from (c)

e. $GCD(x, y)$, where $x = 5^4 \cdot 13^{10}$ and $y = 3^{10} \cdot 11^{20}$

Solution

a. Since $108 = 2^2 \cdot 3^3$ and $72 = 2^3 \cdot 3^2$, it follows that $GCD(108, 72) = 2^2 \cdot 3^2 = 36$.

b. Because $13 | 0$ and $13 | 13$, it follows that $GCD(0, 13) = 13$.

c. $GCD(x, y) = 2 \cdot 7^2 \cdot 13 = 1274$.

d. Because $x = 2^3 \cdot 7^2 \cdot 11 \cdot 13$, $y = 2 \cdot 7^3 \cdot 13 \cdot 17$, and $z = 2^2 \cdot 7$, then $GCD(x, y, z) = 2 \cdot 7 = 14$. Notice that $GCD(x, y, z)$ can also be obtained by finding the GCD of z and 1274, the answer from (c).

e. Because x and y have no common prime factors, $GCD(x, y) = 1$.

Calculator Method

Calculators with a $\boxed{\text{Simp}}$ key can be used to find the GCD of two numbers. For example, to find $GCD(120, 180)$, use the following sequence of buttons to start: First, press $\boxed{1}\boxed{2}\boxed{0}\boxed{/}\boxed{1}\boxed{8}\boxed{0}$ $\boxed{\text{Simp}}$ $\boxed{=}$ to obtain the display $\boxed{\text{N/D} \rightarrow \text{n/d } 60/90}$. By pressing the $\boxed{x \cdot y}$ button, we see $\boxed{2}$ on the display as a common divisor of 120 and 180. By pressing the $\boxed{x \cdot y}$ button again and pressing $\boxed{\text{Simp}}$ $\boxed{=}\boxed{x \cdot y}$, we see 2 again as a factor. The process is repeated to reveal 3 and 5 as other common factors. The GCD of 120 and 180 is the product of the common prime factors $2 \cdot 2 \cdot 3 \cdot 5$, or 60.

Some calculators have a built-in GCD feature, probably found in the *MATH* menu. With this feature, select GCD and enter the numbers separated by a comma and enclosed within parentheses; for example, $GCD(120, 180)$. When the $\boxed{=}$ is pressed, the GCD of 60 will be displayed.

Euclidean Algorithm Method

For large numbers a more efficient method than factorization for finding the GCD is available. For example, suppose we want to find $GCD(676, 221)$. If we could find two smaller numbers whose GCD is the same as $GCD(676, 221)$, the task would be easier. From Theorem 4-2(c), every divisor of 676 and 221 is also a divisor of $676 - 221$ and 221. Conversely, every divisor of $676 - 221$ and 221 is also a divisor of 676 and 221. Thus, the set of all the common divisors of 676 and 221 is the same as the set of all common divisors of $676 - 221$ and 221. Consequently, $GCD(676, 221) = GCD(676 - 221, 221)$. This process can be continued to subtract three 221s from 676 so that $GCD(676, 221) = GCD(676 - 3 \cdot 221, 221) = GCD(13, 221)$. Because 13 is a prime, we can conclude now that the GCD is either 1 or 13. Since $13 | 221$, $GCD(13, 221) = 13$. However it is often advantageous to continue the process until the remainder 0 is obtained: $GCD(13, 221) = GCD(221, 13) = GCD(221 - 17 \cdot 13, 13) = GCD(0, 13) = 13$.

Consequently GCD(676, 221) = 13. To determine how many 221s can be subtracted from 676, and how many 13s from 221 we could have divided as follows:

$$\begin{array}{cc} 3 \text{ R } 13 & 17 \text{ R } 0 \\ 221\overline{)676} & 13\overline{)221} \end{array}$$

When 0 is reached as a remainder, the divisions are complete. Based on this illustration, the following generalization is outlined in Theorem 4-16.

Theorem 4-17

If a and b are any whole numbers with $b \neq 0$ and $a \geq b$, then $\text{GCD}(a, b) = \text{GCD}(r, b)$, where r is the remainder when a is divided by b.

REMARK Because $\text{GCD}(x, y) = \text{GCD}(y, x)$ for all whole numbers x and y not both 0, Theorem 4-17 can be written

$$\text{GCD}(a, b) = \text{GCD}(b, r)$$

We have seen that GCD(676, 221) = GCD(676 − 221, 221)

$$= \text{GCD}(676 - 3 \cdot 221, 221).$$

This is true, in general, as stated in Theorem 4-18:

Theorem 4-18

If a and b are any non-zero whole numbers and $a \geq b$, then:

a. $\text{GCD}(a, b) = \text{GCD}(a - b, b)$

b. $\text{GCD}(a, b) = \text{GCD}(a - kb, b)$ for any whole number k for which $a - kb$ is a whole number.

Finding the GCD of two numbers by repeatedly using Theorem 4-17 until the remainder 0 is reached is referred to as the **Euclidean algorithm**. This method is found in Book VII of Euclid's *Elements* (300 BCE). Figure 4-13 is a flowchart for using the Euclidean algorithm.

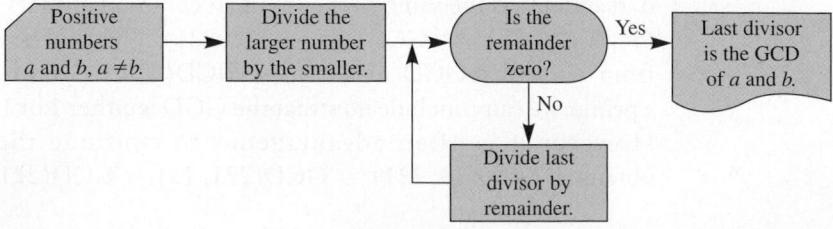

Figure 4-13

EXAMPLE 4-13

Use the Euclidean algorithm to find GCD(10764, 2300).

Solution

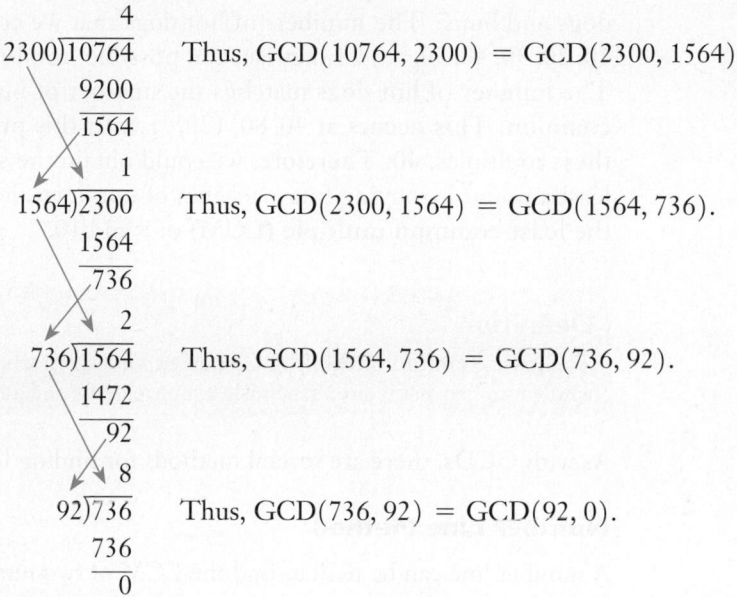

$$4$$
$$2300 \overline{)10764} \qquad \text{Thus, } \mathrm{GCD}(10764, 2300) = \mathrm{GCD}(2300, 1564).$$
$$\underline{9200}$$
$$1564$$

$$1$$
$$1564 \overline{)2300} \qquad \text{Thus, } \mathrm{GCD}(2300, 1564) = \mathrm{GCD}(1564, 736).$$
$$\underline{1564}$$
$$736$$

$$2$$
$$736 \overline{)1564} \qquad \text{Thus, } \mathrm{GCD}(1564, 736) = \mathrm{GCD}(736, 92).$$
$$\underline{1472}$$
$$92$$

$$8$$
$$92 \overline{)736} \qquad \text{Thus, } \mathrm{GCD}(736, 92) = \mathrm{GCD}(92, 0).$$
$$\underline{736}$$
$$0$$

Because $\mathrm{GCD}(92, 0) = 92$, it follows that $\mathrm{GCD}(10764, 2300) = 92$.

The procedure for finding the GCD by using the Euclidean algorithm can be stopped at any step at which the GCD is obvious.

Sometimes shortcuts can be used to find the GCD of two or more numbers, as in the following example.

EXAMPLE 4-14

Find each of the following:

a. GCD(134791, 6341, 6339)
b. The GCD of any two consecutive whole numbers.

Solution

a. Any common divisor of three numbers is also a common divisor of any two of them (why?). Consequently, the GCD of three numbers cannot be greater than the GCD of any two of the numbers. The numbers 6341 and 6339 are close to each other and therefore it is easy to find their GCD:

$$\mathrm{GCD}(6341, 6339) = \mathrm{GCD}(6341 - 6339, 6339)$$
$$= \mathrm{GCD}(2, 6339)$$
$$= 1$$

Because GCD(134791, 6341, 6339) cannot be greater than 1, it follows that it must equal 1.

b. $\mathrm{GCD}(4, 5) = 1$, $\mathrm{GCD}(5, 6) = 1$, $\mathrm{GCD}(6, 7) = 1$, and $\mathrm{GCD}(99, 100) = 1$. It seems that the GCD of any two consecutive whole numbers is 1. To justify this conjecture, we show that for all whole numbers n, $\mathrm{GCD}(n, n + 1) = 1$. Using Theorem 4-17 we have:

$$\mathrm{GCD}(n, n + 1) = \mathrm{GCD}(n + 1, n) = \mathrm{GCD}(n + 1 - n, n)$$
$$= \mathrm{GCD}(1, n)$$
$$= 1$$

Least Common Multiple

Hot dogs are usually sold 10 to a package, while hot dog buns are usually sold 8 to a package. This mismatch causes troubles when one is trying to match hot dogs and buns. What is the least number of packages of each we could order so that there is an equal number of hot dogs and buns? The numbers of hot dogs that we could have are the multiples of 10, that is, 10, 20, 30, 40, 50, ... Likewise, the possible numbers of buns are 8, 16, 24, 32, 40, 48, ... The number of hot dogs matches the number of buns whenever 10 and 8 have multiples in common. This occurs at 40, 80, 120, ... In this problem, we are interested in the least of these multiples, 40. Therefore, we could obtain the same number of hot dogs and buns in the least amount by buying four packages of hot dogs and five packages of buns. The answer 40 is the **least common multiple (LCM)** of 8 and 10.

> ### Definition
> The **least common multiple (LCM)** of two non-zero whole numbers a and b is the least non-zero whole number that is simultaneously a multiple of a and a multiple of b.

As with GCDs, there are several methods for finding least common multiples.

Number Line Method

A number line can be used to find the LCM of two numbers. For example, to find LCM(3, 4), we show the multiples of 3 and 4 on the number line using intervals of 3 and 4, as shown in Figure 4-14.

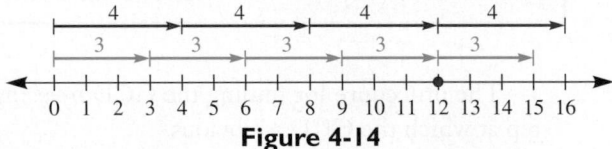

Figure 4-14

Beginning at 0, the arrows do not coincide until the point 12 on the number line. If the line were continued, the arrows would coincide again at 24, 36, 48, and so on. There are an infinite number of common multiples of 3 and 4, but the least common multiple is 12. This number line approach is instructive and promotes understanding but is not practical for large numbers.

Colored Rods Method

Colored rods are used to determine the LCM of two numbers. For example, consider the 3 rod and the 4 rod in Figure 4-15(a). Build trains of 3 rods and 4 rods until they are the same length, as shown in Figure 4-15(b). The LCM is the common length of the train.

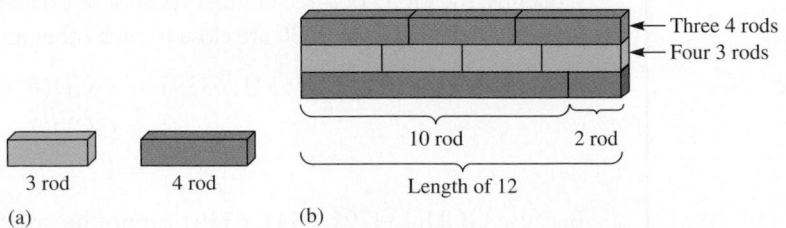

Figure 4-15

The Intersection of Sets Method

In the *intersection of sets* method, we first find the set of all positive *multiples* of both the first and second numbers, then find the set of all *common multiples* of both numbers, and finally pick the *least* element in that set. For example, to find the LCM of 8 and 12, we denote the sets of non-zero whole number multiples of 8 and 12 by M_8 and M_{12}, respectively.

$$M_8 = \{8, 16, 24, 32, 40, 48, 56, 64, 72, \ldots\}$$

$$M_{12} = \{12, 24, 36, 48, 60, 72, 84, 96, 108, \ldots\}$$

The set of common multiples is

$$M_8 \cap M_{12} = \{24, 48, 72, \ldots\}.$$

Because the least number in $M_8 \cap M_{12}$ is 24, the LCM of 8 and 12 is 24, written LCM(8, 12) $= 24$.

NOW TRY THIS 4-10

Draw a Venn diagram showing M_8 and M_{12} and show how to find LCM(8, 12) using the diagram.

The Prime Factorization Method

The intersection of sets method for finding the LCM is often lengthy, especially when it is used to find the LCM of three or more non-zero whole numbers. Another, more efficient, method for finding the LCM of several numbers is the *prime factorization method.* For example, to find LCM(40, 12), we first find the prime factorizations of 40 and 12, namely, $2^3 \cdot 5$ and $2^2 \cdot 3$, respectively.

If $m = $ LCM(40, 12), then m is a multiple of 40 and must contain both 2^3 and 5 as factors. Also, m is a multiple of 12 and must contain 2^2 and 3 as factors. Since 2^3 is a multiple of 2^2, then $m = 2^3 \cdot 5 \cdot 3 = 120$. In general, we have the following:

> To find the LCM of two non-zero whole numbers, we first find the prime factorization of each number. Then we take each of the primes that are factors of either of the given numbers. The LCM is the product of these primes, each raised to the greatest power of the prime that occurs in either of the prime factorizations.

EXAMPLE 4-15

Find the LCM of 2520 and 10,530.

Solution

$$2520 = 2^3 \cdot 3^2 \cdot 5 \cdot 7$$

$$10,530 = 2 \cdot 3^4 \cdot 5 \cdot 13$$

$$\text{LCM}(2520, 10,530) = 2^3 \cdot 3^4 \cdot 5 \cdot 7 \cdot 13 = 294,840$$

The prime factorization method can also be used to find the LCM of more than two numbers. For example, to find LCM(12, 108, 120), we proceed as follows:

$$12 = 2^2 \cdot 3$$

$$108 = 2^2 \cdot 3^3$$

$$120 = 2^3 \cdot 3 \cdot 5$$

Then, LCM(12, 108, 120) $= 2^3 \cdot 3^3 \cdot 5 = 1080$.

The GCD-LCM Product Method

To see the connection between the GCD and LCM, we consider the GCD and LCM of 24 and 30. The prime factorizations of these numbers are

$$24 = 2^3 \cdot 3$$

$$30 = 2 \cdot 3 \cdot 5$$

Figure 4-16 shows a diagram with the prime factors.

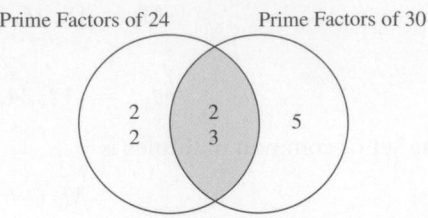

Prime Factors of 24   Prime Factors of 30

Figure 4-16

$\text{GCD}(24, 30) = 2 \cdot 3$ and is the product of the factors in the shaded region, and $\text{LCM}(24, 30) = 2^3 \cdot 3 \cdot 5$ is the product of the prime factors in the combined regions. Also:

$$\text{GCD}(24, 30) \cdot \text{LCM}(24, 30) = (2 \cdot 3)(2^3 \cdot 3 \cdot 5) = (2^3 \cdot 3)(2 \cdot 3 \cdot 5) = 24 \cdot 30$$

This shows that the product of the GCD and LCM of 24 and 30 is equal to $24 \cdot 30$. In general, the connection between the GCD and LCM of any pair of non-zero whole numbers is given by Theorem 4-19.

> **Theorem 4-19**
>
> For any two non-zero whole numbers a and b,
>
> $$\text{GCD}(a, b) \cdot \text{LCM}(a, b) = ab.$$

Theorem 4-19 can be justified in several ways. A specific example suggesting how the theorem might be proved is given below:

$$\text{Let} \quad a = 5^{13} \cdot 7^{20} \cdot 11^4 \quad \text{and} \quad b = 5^{10} \cdot 7^{25} \cdot 11^6 \cdot 13.$$

Then,

$$\text{LCM}(a, b) = 5^{13} \cdot 7^{25} \cdot 11^6 \cdot 13 \quad \text{and} \quad \text{GCD}(a, b) = 5^{10} \cdot 7^{20} \cdot 11^4.$$

We have $\text{LCM}(a, b) \cdot \text{GCD}(a, b) = 5^{13+10} \cdot 7^{25+20} \cdot 11^{6+4} \cdot 13$ and $ab = 5^{13+10} \cdot 7^{20+25} \cdot 11^{4+6} \cdot 13$.

For the preceding values of a and b, Theorem 4-19 is true. However, in general we reason as follows: in the product $\text{LCM}(a, b) \cdot \text{GCD}(a, b)$, we have all the powers of the primes appearing in a or in b, because for the LCM we take the greater of the powers of the common primes and for the GCD the lesser. Also in ab we have all the powers. Hence, Theorem 4-18 is true in general.

The Euclidean Algorithm Method

Theorem 4-19 is useful for finding the LCM of two numbers a and b when their prime factorizations are not easy to find. $\text{GCD}(a, b)$ can be found by the Euclidean algorithm, the product ab can be found by multiplication, and $\text{LCM}(a, b)$ can be found by division.

EXAMPLE 4-16

a. Find $\text{LCM}(731, 952)$.
b. If $b \mid a$ find $\text{LCM}(a, b)$ in terms of a or b.

Solution

a. By the Euclidean algorithm, $\text{GCD}(731, 952) = 17$. By Theorem 4-19,

$$17 \cdot \text{LCM}(731, 952) = 731 \cdot 952.$$

Consequently,

$$\text{LCM}(731, 952) = \frac{731 \cdot 952}{17} = 40{,}936.$$

b. Because $b \mid a$, a is a multiple of b, thus $\text{LCM}(a, b) = a$.

The Division-by-Primes Method

Another procedure for finding the LCM of several non-zero whole numbers involves *division by primes*. For example, to find LCM(12, 75, 120), start with the least prime that divides at least one of the given numbers and divide as follows:

$$2 \underline{|12, 75, 120}$$
$$6, 75, \ 60$$

Because 2 does not divide 75, we simply bring down the 75. To obtain the LCM using this procedure, we continue the division process until the row of answers consists of relatively prime numbers.

$$
\begin{array}{r|rrr}
2 & 12, & 75, & 120 \\
\hline
2 & 6, & 75, & 60 \\
\hline
2 & 3, & 75, & 30 \\
\hline
3 & 3, & 75, & 15 \\
\hline
5 & 1, & 25, & 5 \\
\hline
 & 1, & 5, & 1 \\
\end{array}
$$

Thus, LCM(12, 75, 120) $= 2 \cdot 2 \cdot 2 \cdot 3 \cdot 5 \cdot 1 \cdot 5 \cdot 1 = 2^3 \cdot 3 \cdot 5^2 = 600$.

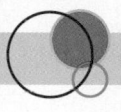

Assessment 4-3A

1. Find the GCD and the LCM for each of the following using the intersection-of-sets method:
   a. 18 and 10   b. 24 and 36
   c. 8, 24, and 52   d. 7 and 9
2. Find the GCD and the LCM for each of the following using the prime factorization method:
   a. 132 and 504   b. 65 and 1690
   c. 96, 900, and 630   d. 108 and 360
3. Find the GCD for each of the following using the Euclidean algorithm:
   a. 220 and 2924   b. 14,595 and 10,856
4. Find the LCM for each of the following using any method:
   a. 24 and 36
   b. 72, 90 and 96
   c. 90, 105 and 315
   d. 9^{100} and 25^{100}
5. Find the LCM for each of the following pairs of numbers using Theorem 4-19 and the answers from exercise 3:
   a. 220 and 2924
   b. 14,595 and 10,856
6. In Quinn's dormitory room, there are three snooze-alarm clocks, each of which is set at a different time. Clock A goes off every 15 min, clock B goes off every 40 min, and clock C goes off every 60 min. If all three clocks go off at 6:00 A.M., answer the following:
   a. How long will it be before the clocks go off simultaneously again after 6:00 A.M.?
   b. Would the answer to (a) be different if clock B went off every 15 min and clock A went off every 40 min?
7. Use colored rods to find the GCD and the LCM of 6 and 10.
8. Midas has 120 gold coins and 144 silver coins. He wants to place his gold coins and his silver coins in stacks so that there are the same number of coins in each stack. What is the greatest number of coins that he can place in each stack?

9. By selling cookies at 24¢ each, José made enough money to buy several cans of pop costing 45¢ per can. If he had no money left over after buying the pop, what is the least number of cookies he could have sold?
10. Two bike riders ride around in a circular path. The first rider completes one round in 12 min and the second rider completes it in 18 min. If they both start at the same place and the same time and go in the same direction, after how many minutes will they meet again at the starting place?
11. Three motorcyclists ride around a circular race course starting at the same place and the same time. The first passes the starting point every 12 min, the second every 18 min, and the third every 16 min. After how many minutes will all three pass the starting point again at the same time? Explain your reasoning.
12. Assume a and b are natural numbers and answer the following:
    a. If GCD$(a, b) = 1$, find LCM(a, b).
    b. Find GCD(a, a) and LCM(a, a).
    c. Find GCD(a^2, a) and LCM(a^2, a).
    d. If $a | b$, find GCD(a, b) and LCM(a, b).
13. Classify each of the following as true or false where a and b are whole numbers:
    a. If GCD$(a, b) = 1$, then a and b cannot both be even.
    b. If GCD$(a, b) = 2$, then both a and b are even.
    c. If a and b are even, then GCD$(a, b) = 2$.
14. To find GCD(24, 20, 12), it is possible to find GCD(24, 20), which is 4, and then find GCD(4, 12), which is 4. Use this approach and the Euclidean algorithm to find
    a. GCD(120, 75, 105)
    b. GCD(34578, 4618, 4619)
15. Show that 97,219,988,751 and 4 are relatively prime.
16. The radio station gave away a discount coupon for every twelfth and thirteenth caller. Every twentieth caller received free concert tickets. Which caller was first to get both a coupon and a concert ticket?

17. Determine how many complete revolutions gear 2 in the following must make before the arrows are lined up again.

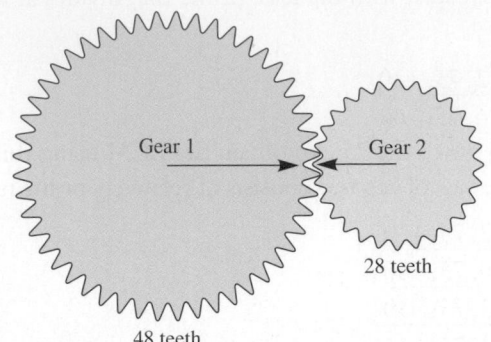

Gear 1

Gear 2

28 teeth

48 teeth

18. Diagrams can be used to show factors of two or more numbers. Draw diagrams to show the prime factors for each of the following sets of three numbers:
    a. 10, 15, 60   b. 8, 16, 24
19. Find all non-zero whole numbers x such that $GCD(49, x) = 1$ and $1 \leq x \leq 49$.
20. What are the factors of 4^{10}?
21. In algebra it is often necessary to factor an expression as much as possible. For example, $a^3b^2 + a^2b^3 = a^2b^2(a + b)$ and no further factoring is possible without knowing the values of a and b. Notice that a^2b^2 is the GCD of a^3b^2 and a^2b^3. Factor each of the following as much as possible:
    a. $12x^4y^3 + 18x^3y^4$
    b. $12x^3y^2z^2 + 18x^2y^4z^3 + 24x^4y^3z^4$
22. Factor 1 billion into a product of two numbers, neither of which contains any zeros.

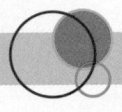

Assessment 4-3B

1. Find the GCD and the LCM for each of the following using the intersection-of-sets method:
   a. 12 and 18   b. 18 and 36
   c. 12, 18, and 24   d. 6 and 11
2. Find the GCD and the LCM for each of the following using the prime factorization method:
   a. 11 and 19   b. 140 and 320
   c. 800, 75, and 450   d. 104 and 320
3. Find the GCD for each of the following using the Euclidean algorithm:
   a. 14,560 and 8250   b. 8424 and 2520
4. Find the LCM for each of the following using any method:
   a. 25 and 36
   b. 82 and 90 and 50
   c. 80 and 105 and 315
   d. 8^{100} and 50^{100}
5. Find the LCM for each of the following pairs of numbers using Theorem 4-19 and the answers from problem 3:
   a. 14,560 and 8250
   b. 8424 and 2520
6. A movie rental store gives a free popcorn to every fourth customer and a free movie rental to every sixth customer. Which customer was the first to win both prizes?
7. Use colored rods to find the GCD and the LCM of 4 and 10.
8. A nursery has 240 cedar trees and 288 pine trees. The manager wants to arrange each type of tree in rows so that there are the same number of trees in each row. What is the greatest number of trees that can be placed in each row?
9. Bill and Sue both work at night. Bill has every sixth night off and Sue has every eighth night off. If they are both off tonight, how many nights will it be before they are both off again?
10. Bijous I and II start their movies at 7:00 P.M. The movie at Bijou I takes 75 min, while the movie at Bijou II takes 90 min. If the shows run continuously, when will they start at the same time again?
11. The principal of Valley Elementary School wants to divide each of the three fourth-grade classes into small same-size groups with at least 2 students in each. If the classes have 18, 24, and 36 students, respectively, what size groups are possible?
12. Assume a and b are natural numbers and answer the following:
    a. If a and b are two different primes, find $GCD(a, b)$ and $LCM(a, b)$.
    b. What is the relationship between a and b if $GCD(a, b) = a$?
    c. What is the relationship between a and b if $LCM(a, b) = a$?
13. Classify each of the following as true or false for all natural numbers a and b.
    a. $LCM(a, b) | GCD(a, b)$.
    b. $LCM(a, b) | ab$.
    c. $GCD(a, b) \leq a$
    d. $LCM(a, b) \geq a$
14. To find $GCD(24, 20, 12)$, it is possible to find first $GCD(24, 20)$, which is 4, and then find $GCD(4, 12)$, which is 4. Use this approach and the Euclidean algorithm to find
    a. $GCD(180, 240, 306)$
    b. $GCD(5284, 1250, 1280)$
15. Show that 181,345,913 and 11 are relatively prime.
16. Larry and Mary bought a special 360-day joint membership to a tennis club. Larry will use the club every other day, and Mary will use the club every third day. They both use the club on the first day. How many days will neither person use the club in the 360 days?
17. Determine how many complete revolutions each gear in the following must make before the arrows are lined up again:

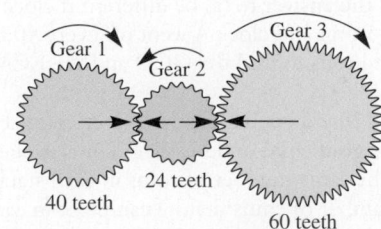

Gear 1   Gear 2   Gear 3

24 teeth

40 teeth   60 teeth

18. Diagrams can be used to show factors of two or more numbers. Draw diagrams to show the prime factors for each of the following sets of three numbers:
    a. 12, 14, 70
    b. 6, 8, 18
19. Find all natural numbers x such that $GCD(25, x) = 1$ and $1 \le x \le 25$.
20. What are the factors of 9^{10}?
21. In algebra it is often necessary to factor an expression as much as possible. For example, $a^3b^2 + a^2b^3 = a^2b^2(a + b)$ and no further factoring is possible without knowing the

values of a and b. Notice that a^2b^2 is the GCD of a^3b^2 and a^2b^3. Factor each of the following as much as possible:
    a. $18x^4y^3 + 12x^2y^4$
    b. $18x^3y^2z^3 + 48x^2y^3z^2 + 12x^4y^3z^4$
22. If you find the sum of any two-digit number and the number formed by reversing its digits, the resulting number is always divisible by which three positive non-zero whole numbers?

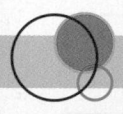

Mathematical Connections 4-3

Communication

1. Can two non-zero whole numbers have a greatest common multiple? Explain your answer.
2. Describe to a sixth-grade student the difference between a divisor and a multiple.
3. Is it true that $GCD(a, b, c) \cdot LCM(a, b, c) = abc$? Explain your answer.
4. A rectangular plot of land is 558 m by 1212 m. A surveyor needs to divide the plot into the largest possible square plots of the same size, being a whole number of meters long. What is the size of each square and how many square plots can be created? Explain your reasoning.
5. Suppose that $GCD(a, b, c) = 1$. Is it necessarily true that $GCD(a, b) = GCD(b, c) = 1$? Explain your reasoning.
6. Suppose $GCD(a, b) = GCD(b, c) = 2$. Does that always imply that $GCD(a, b, c) = 2$? Justify your answer.
7. How can you tell from the prime factorization of two numbers if their LCM equals the product of the numbers? Explain your reasoning.
8. Can the LCM of two non-zero whole numbers ever be greater than the product of the numbers? Explain your reasoning.
9. Let $GCD(m, n) = g$ and $LCM(m, n) = l$. Jackie conjectures that $GCD(m + n, l) = g$ for all whole numbers m and n. Check Jackie's conjecture for three different pairs of integers.

Open-Ended

10. Make up a word problem that can be solved by finding the GCD and another that can be solved by finding the LCM. Solve your problems and explain why you are sure that your approach is correct.
11. Find three pairs of numbers for which the LCM of the numbers in a pair is less than the product of the two numbers.
12. Describe infinitely many pairs of numbers whose GCD is
    a. 2
    b. 6
    c. 91

Cooperative Learning

13. a. In your group, discuss whether the Euclidean algorithm for finding the GCD of two numbers should be introduced in middle school (to all students? to some?). Why or why not?

    b. If you decide that it should be introduced in middle school, discuss how it should be introduced. Report your group's decision to the class.

Questions from the Classroom

14. Alba asked why we don't talk about the LCD (least common divisor) and GCM (greatest common multiple). How do you respond?
15. A student says that for any two non-zero whole numbers a and b, $GCD(a, b)$ divides $LCM(a, b)$ and, hence, $GCD(a, b) < LCM(a, b)$. Is the student correct? Why or why not?
16. A student wants to know how many whole numbers between 1 and 10,000 inclusive are either multiples of 3 or multiples of 5. She wonders if it is correct to find the number of those whole numbers that are multiples of 3 and add the number of those that are multiples of 5. How do you respond?

Review Problems

17. Find two whole numbers x and y such that

$$xy = 1,000,000$$

and neither x nor y contains any zeros as digits.
18. Fill each blank space with a single digit that makes the corresponding statement true. Find all possible answers.
    a. $3 | 83\_51$
    b. $11 | 8\_691$
    c. $23 | 103\_6$
19. Is 3111 a prime? Justify your answer.
20. Find a number that has exactly six prime factors.
21. Produce the least positive number that is divisible by 2, 3, 4, 5, 6, 7, 8, 9, 10, and 11.
22. What is the greatest prime that must be used to determine if 2089 is prime?

National Assessment of Educational Progress (NAEP) Question

The least common multiple of 8, 12, and a third number is 120. Which of the following could be the third number?
    a. 15
    b. 16
    c. 24
    d. 32
    e. 48

NAEP, Grade 8, 1990

BRAIN TEASER

For any $n \times m$ rectangle such that $\mathrm{GCD}(n, m) = 1$, find a rule for determining the number of unit squares (1×1) that a diagonal passes through. For example, in the drawings in Figure 4-17, the diagonal passes through 8 and 6 unit squares, respectively.

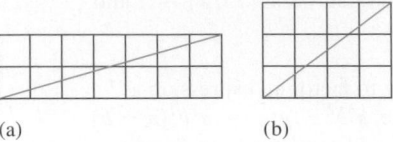

(a)   (b)

Figure 4-17

Hint for Solving the Preliminary Problem

If Jacob had n marbles, then $n + 1$ is a common multiple of 4, 5, and 7.

Chapter 4 Summary

4-1 Divisibility	Pages

Definition of "Divides"

- If a and b are any whole numbers and $b \neq 0$, then b divides a, written $b \mid a$, if, and only if, there is a whole number q such that $a = bq$. — 177
- If a and b are any whole numbers and $b \neq 0$, then b does not divide a, written $b \nmid a$, if, and only if, the remainder when a is divided by b is not 0. — 177
- **Theorem:** For any whole numbers a and d, if $d \mid a$ and n is any whole number, then $d \mid na$. — 178
- **Theorem:** For any whole numbers a, b, and d. — 178
  - If $d \mid a$ and $d \mid b$, then $d \mid (a + b)$
  - If $d \mid a$ and $d \nmid b$, then $d \nmid (a + b)$
  - If $d \mid a$, $a \mid b$, and $a \geq b$, then $d \mid (a - b)$
  - If $d \mid a$, $d \nmid b$, and $a \geq b$, then $d \nmid (a - b)$
  - If $d \nmid a$, $d \mid b$, and $a \geq b$, then $d \nmid (a - b)$

Divisibility Rules

- **Divisibility Tests** for 2, 5, and 10
  - A whole number is divisible by 2 if, and only if, its unit digit is divisible by 2; that is if and only if, the unit digit is even. — 180
  - A whole number is divisible by 5 if, and only if, its unit digit is divisible by 5; that is, if, and only if, the units digit is 0 or 5. — 180
  - A whole number is divisible by 10 if, and only if, its unit digit is divisible by 10; that is, if, and only if, the units digit is 0. — 180
- **Divisibility Tests** for 4, 8, 3, 9, 11, and 6
  - A whole number is divisible by 4 if, and only if, the last two digits of the number represent a number divisible by 4. — 181
  - A whole number is divisible by 8 if, and only if, the last three digits of the number represent a number divisible by 8. — 182

    o A whole number is divisible by 3 if, and only if, the sum of its digits is divisible by 3.   183

    o A whole number is divisible by 9 if, and only if, the sum of its digits is divisible by 9.   183

    o A whole number is divisible by 11 if, and only if, the difference between the sum of the digits in the places that are even powers of 10 and the sum of the digits in the places that are odd powers of 10 is divisible by 11 (subtract the smaller sum from the greater one).   184

    o A whole number is divisible by 6 if, and only if, the number is divisible by 2 and 3.   184

4-2 Prime and Composite Numbers

Prime Factorization
- Any whole number with exactly two divisors is **prime**. A whole number greater than 1 that is not prime is **composite**.   192
- A factorization containing only prime numbers is a **prime factorization**.   192
- **Fundamental Theorem of Arithmetic:** Each composite number can be written as a product of primes in one, and only one way, except for the order of the prime factors in the product.   192

Number of Divisors   195
- **Theorem:** If p and q are different primes and n, m are whole numbers, then $p^n q^m$ has $(n + 1)(m + 1)$ number divisors. In general, if $p_1, p_2, \ldots, p_k$ are primes and $n_1, n_2, \ldots, n_k$ are whole numbers, then $p_1^{n_1} \cdot p_2^{n_2} \cdot \ldots \cdot p_k^{n_k}$ has $(n_1 + 1)$ $(n_2 + 1) \cdot \ldots \cdot (n_k + 1)$ whole number divisors.

Determining whether a number is prime

- **Theorem:** If d is a divisor of n, then $\dfrac{n}{d}$ is also a divisor of n.   197
- **Theorem:** If n is composite, then n has a prime factor p such that $p^2 \le n$.   197
- **Theorem:** If n is a whole number greater than 1 and not divisible by any prime p such that $p^2 \le n$, then n is prime.   197
- **Sieve of Eratosthenes** is a method of finding all the primes less than a given number.   197

4-3 Greatest Common Divisor and Least Common Multiple

The Greatest Common Divisor (GCD) of two whole numbers a and b not both 0, is the greatest whole number, that divides both a and b.   204
- Colored rods method   204
- Intersection of sets method   205
- Prime factorization method   205
- Calculator method   206
- Euclidean algorithm method   206
  - o If a and b are any whole numbers not both 0, and $a \ge b$, then
    $\mathrm{GCD}(a, b) = \mathrm{GCD}(a - b, b)$
    $\mathrm{GCD}(a, b) = \mathrm{GCD}(a - kb, b)$ for any whole number k for which $a - kb$ is a whole number.   208
  - o If a and b are any whole numbers not both 0, and $a \ge b$, then $\mathrm{GCD}(a, b) = \mathrm{GCD}(b, r)$ where r is the remainder when a is divided by b.   208

The Least Common Multiple (LCM) of two non-zero whole numbers a and b is the least natural number that is a multiple of a and also of b.   210
- Number line method   210
- Colored rods method   210

- Intersection of sets method
  In this method, first find the set of all positive multiples of both the first and second number, then find the set of all common multiples of both numbers, and finally, pick the least element in that set.   210
- Prime factorization method
  First, find the prime factorization of each number. Then, take each of the primes that are factors of either of the given numbers. The LCM is the product of these primes, raised to the greatest power of the prime that occurs in either of the prime factorizations.   211
- The GCD-LCM product method is based on the fact that
  $GCD(a, b) \cdot LCM(a, b) = ab$ 212
- The Euclidean algorithm method is based on finding $GCD(a, b)$ using the Euclidean algorithm and then using the above property.   212
- Division by primes method.   213

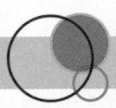

Chapter 4 Review

1. Classify each of the following as true or false:
   a. $8 \mid 4$
   b. $0 \mid 4$
   c. $4 \mid 0$
   d. If a number is divisible by 4 and by 6, then it is divisible by 24.
   e. If a number is not divisible by 12, then it is not divisible by 3.
2. Classify each of the following as true or false for whole numbers x and y. If false, show a counterexample.
   a. If $7 \mid x$ and $7 \nmid y$, then $7 \nmid xy$.
   b. If $d \nmid (a + b)$, then $d \nmid a$ and $d \nmid b$.
   c. If $d \mid (a + b)$ and $d \nmid a$, then $d \nmid b$.
   d. If $d \mid (x + y)$ and $d \mid x$, then $d \mid y$.
   e. If $4 \nmid x$ and $4 \nmid y$, then $4 \nmid xy$.
3. Test each of the following numbers for divisibility by 2, 3, 4, 5, 6, 8, 9, and 11:
   a. 83,160
   b. 83,193
4. Assume that 10,007 is prime. Without actually dividing 10,024 by 17, prove that 10,024 is not divisible by 17.
5. Fill each blank with one digit to make each of the following true (find all the possible answers):
   a. $6 \mid 87\_4$
   b. $24 \mid 4\_856$
   c. $29 \mid 87\_\_4$
6. A student claims that the sum of five consecutive whole numbers is always divisible by 5.
   a. Check the student's claim for a few cases.
   b. Prove or disprove the student's claim.

7. Determine whether each of the following numbers is prime or composite:
   a. 143
   b. 223
8. How can you tell if a number is divisible by 24? Check 4152 for divisibility by 24.
9. Is the LCM of two numbers always greater than the GCD of the numbers? Justify your answer.
10. Explain how to find the LCM of three numbers with the help of the Euclidean algorithm.
11. To find if the number $2 \cdot 3 \cdot 5 \cdot 7 + 11 \cdot 13$ is prime, a student finds that the number equals 353. She checks that $17 \nmid 353$ and $19^2 > 353$ and without further checking, claims that 353 is prime. Explain why the student is correct.
12. Find the GCD for each of the following:
    a. 24 and 52
    b. 5767 and 4453
13. Find the LCM for each of the following:
    a. $2^3 \cdot 5^2 \cdot 7^3$, $2 \cdot 5^3 \cdot 7^2 \cdot 13$ and $2^4 \cdot 5 \cdot 7^4 \cdot 29$
    b. 278 and 279
14. Construct a number that has exactly five positive divisors. Explain your construction.
15. Find all the positive divisors of 144.
16. Find the prime factorization of each of the following:
    a. 172
    b. 288
    c. 260
    d. 111

17. Find the least non-zero whole number that is divisible by every non-zero whole number less than or equal to 12.

18. Candy bars priced at 50¢ each were not selling, so the price was reduced. Then they all sold in one day for a total of $31.93. What was the reduced price of each candy bar?

19. Two bells ring at 8:00 A.M. For the remainder of the day, one bell rings every half hour and the other bell rings every 45 min. What time will it be when the bells ring together again?

20. If the GCD of two positive whole numbers is 1, what can you say about the LCM of the two numbers? Explain your reasoning.

21. If there were to be 9 boys and 6 girls at a party and the host wanted each to be given exactly the same number of candies that could be bought in packages containing 12 candies, what is the fewest number of packages that could be bought?

22. Jane and Ramon are running laps on a track. If they start at the same time and place and go in the same direction, with Jane running a lap in 5 min and Ramon running a lap in 3 min, how long will it take for them to be at the starting place at the same time if they continue to run at these speeds?

23. June, an owner of a coffee stand, marked down the price of a latte between 7:00 A.M. and 8:00 A.M. from $2.00 a cup. If she grossed $98.69 from the latte sale and we know that she never sells a latte for less than a dollar, how many lattes did she sell between 7:00 A.M. and 8:00 A.M.? Explain your reasoning. (*Note:* $71 \mid 9869$.)

24. Find the prime factorizations of each of the following:
    a. 6^{10}
    b. 34^n
    c. 97^4
    d. $8^4 \cdot 6^3 \cdot 26^2$
    e. $2^3 \cdot 3^2 + 2^4 \cdot 3^3 \cdot 7$
    f. $2^4 \cdot 3 \cdot 5^7 + 2^4 \cdot 5^6$

25. What are the possible remainders when a prime number greater than 3 is divided by 12? Justify your answer.

26. Prove the test for divisibility by 9 using a three-digit number n such that $n = a \cdot 10^2 + b \cdot 10 + c$.

27. The triplet 3, 5, 7 consists of consecutive odd whole numbers that are all prime. Give a convincing argument that this is the only triplet of consecutive odd integers that are all prime. (*Hint:* use the division algorithm.)

28. a. Explain why if a number is composite its least divisor greater than 1 must be prime.
    b. Prove that if d and n are whole numbers and if $d \mid n$, then $\left(\dfrac{n}{d}\right) \Big| n$.

Integers

Preliminary Problem

In the 1400s, European merchants used integers to label barrels of flour. For example, a barrel labeled $^+3$ meant the barrel was 3 lb overweight, while a barrel labeled $^-5$ meant the barrel was 5 lb underweight. A worker who had labeled 100-lb barrels turned in a weight sheet knowing that the total of the weight was off 53 lbs. He had used only 5s and 6s to label the barrels, but he forgot how many of the labels were positive and how many were negative. He was certain that there were fewer than 20 barrels. How might the barrels have been labeled?

If needed, see Hint on page 255.

The *PSSM* student expectations for grades 3–5 and 6–8, respectively, include the following:

- explore numbers less than 0 by extending the number line and through familiar applications. (p. 148)
- develop meaning for integers and represent and compare quantities with them. (p. 214)

In addition, the *PSSM* points out that in grades 6–8:

Middle-grades students should also work with integers. In lower grades, students may have connected negative integers in appropriate ways to informal knowledge derived from everyday experiences, such as below-zero winter temperatures or lost yards on football plays. In the middle grades, students should extend these initial understandings of integers. Positive and negative integers should be seen as useful for noting relative changes or values. Students can also appreciate the utility of negative integers when they work with equations whose solution requires them, such as $2x + 7 = 1$. (pp. 217–18)

Additionally, grade 6 *Common Core Standards* say that students

should understand that positive and negative numbers are used together to describe quantities having opposite directions or values. (p. 43)

For example, Mount Everest is 29,028 ft above sea level, and the Dead Sea is 1293 ft below sea level. We may symbolize these elevations as 29,028 and ⁻1293.

In mathematics, the need for negative integers arises because subtractions cannot always be performed in the set of whole numbers as Linus has tried to do in the following cartoon.

To compute $4 - 6$ using the definition of subtraction for whole numbers, we must find a whole number n such that $6 + n = 4$. There is no such whole number n. To perform the computation, we must invent a new number, a *negative integer*. If we attempt to calculate $4 - 6$ on a number line, we must draw intervals to the left of 0. In Figure 5-1, $4 - 6$ is pictured as an arrow that starts at 0 and ends 2 units to the left of 0. The new number that corresponds to the point 2 units to the left of 0 is *negative two*, symbolized by ⁻2.

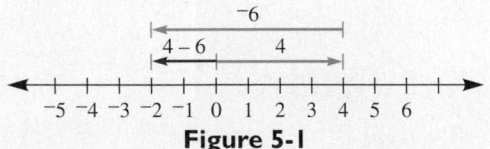

Figure 5-1

Other numbers to the left of 0 are created similarly. The new set of numbers $\{⁻1, ⁻2, ⁻3, ⁻4, \ldots\}$ is the set of **negative integers**. The set $\{1, 2, 3, 4, \ldots\}$ is the set of **positive integers**. The integer 0 is neither positive nor negative.

Historical Note

The Hindu mathematician Brahmagupta (ca. 598–665 CE) provided the first systematic treatment of negative numbers and of zero. Only about 1000 years later did the Italian mathematician Gerolamo Cardano (1501–1576) consider negative solutions of equations. Still uncomfortable with the concept of negative numbers, he called them "fictitious" numbers.   ●

> **Definition**
> The union of the set of negative integers, the set of positive integers, and $\{0\}$ is the set of **integers**, denoted by I.
>
> $$I = \{\ldots, ^-4, ^-3, ^-2, ^-1, 0, 1, 2, 3, 4, \ldots\}$$

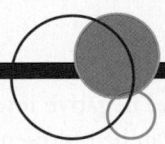

5-1 Integers and the Operations of Addition and Subtraction

Representations of Integers

It is unfortunate that we use the symbol "−" to indicate both a subtraction and a negative sign. To reduce confusion between the uses of this symbol in this text, a raised "−" sign is used for negative numbers, as in $^-2$, and for the opposite of a number (described later), as in ^-x, in contrast to the lower sign for subtraction. To emphasize that an integer is positive, sometimes a raised plus sign is used, as in $^+3$. In this text, we use the plus sign for addition only and write $^+3$ simply as 3.

The negative integers are **opposites** of the positive integers. For example, the opposite of 5 is $^-5$. Similarly, the positive integers are the opposites of the negative integers. Because the opposite of 4 is denoted $^-4$, the opposite of $^-4$ can be denoted $^-(^-4)$, which equals 4. The opposite of 0 is 0. In the set of integers I, every element has an opposite that is also in I. Using addition of integers, we shall soon see that when an opposite of an integer is added to the integer the sum is 0. In fact, ^-a can be defined as the solution of $x + a = 0$.

EXAMPLE 5-1

For each of the following, find the opposite of x:

  a. $x = 3$ **b.** $x = ^-5$ **c.** $x = 0$

Solution
  a. $^-x = ^-3$
  b. $^-x = ^-(^-5) = 5$
  c. $^-x = ^-0 = 0$

The value of ^-x in Example 5-1(b) is 5. Note that ^-x is the opposite of x and might *not represent a negative number*. In other words, x is a variable that can be replaced by some number either positive, zero, or negative. *Note: ^-x is read "the opposite of x" not "minus x" or "negative x."*

In grade 6 *Common Core Standards* as in Example 5-1(b), students should "recognize that the opposite of the opposite of a number is the number itself." (p. 43)

Historical Note

The dash has not always been used for both the subtraction operation and the negative sign. Other notations were developed but never adopted universally. One such notation was used by Abu al-Khwarizmi (ca. 825), who indicated a negative number by placing a small circle over it. For example, $^-4$ was recorded as $\overset{\circ}{4}$. The Hindus denoted a negative number by enclosing it in a circle; for example, $^-4$ was recorded as ④. The symbols + and − first appeared in print in European mathematics in the late fifteenth century, at which time the symbols referred not to addition or subtraction nor positive or negative numbers, but to surpluses and deficits in business problems.

We next investigate informal ways to introduce operations on integers beginning with the addition of integers.

Integer Addition

E-Manipulative
Activity

Use the *Token Addition* activity for additional practice using the chip or charged-field models.

Hands-on materials can be an aid when working with integers. Several models are presented to motivate integer addition.

Chip Model for Addition

In the chip model, positive integers are represented by black chips and negative integers by red chips. One red chip neutralizes one black chip. Hence, the integer $^-1$ can be represented by 1 red chip, or 2 red and 1 black, or 3 red and 2 black, and so on. Similarly, every integer can be represented in many ways using chips, as shown in Figure 5-2.

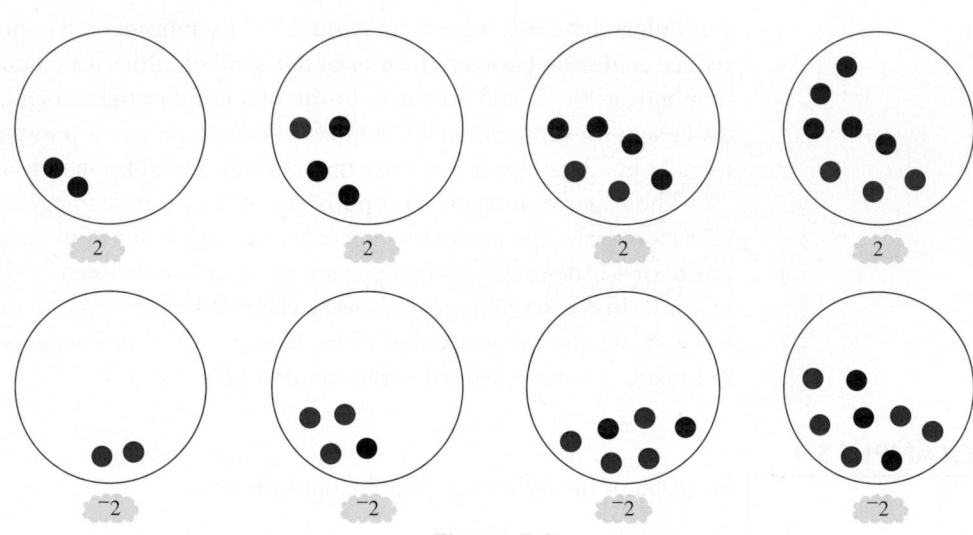

Figure 5-2

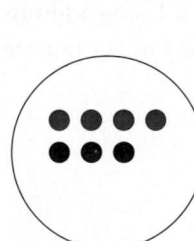

$^-4 + 3 = ^-1$

Figure 5-3

Figure 5-3 shows a chip model for the addition $^-4 + 3$. Place 4 red chips together with 3 black chips. Because 3 red chips neutralize 3 black ones, Figure 5-3 represents the equivalent of 1 red chip or $^-1$.

Charged-Field Model for Addition

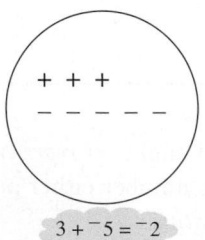

$3 + ^-5 = ^-2$

Figure 5-4

A model similar to the chip model uses positive and negative charges. A field has 0 charge if it has the same number of positive $(+)$ and negative $(-)$ charges. As in the chip model, a given integer can be represented in many ways using the charged-field model. Figure 5-4 uses the model for $3 + ^-5$. Because 3 positive charges "neutralize" 3 negative charges, the net result is 2 negative ones. Hence, $3 + ^-5 = ^-2$.

Number-Line Model for Addition

Another model for addition of integers involves a number line, and it can be introduced with the idea of a hiker walking the number line, as seen on the student page on page 225. Without the hiker, $^-3 + ^-5$ can be pictured as in Figure 5-5.

E-Manipulative
Activity

Use the *Number Line* activity to illustrate the number-line model.

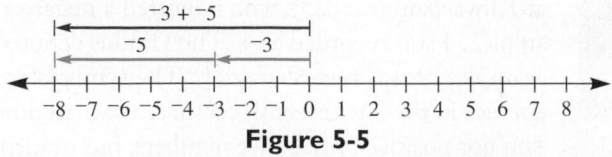

Figure 5-5

School Book Page ADDING INTEGERS

 **Adding Integers**

✓ Check Skills You'll Need

1. Vocabulary Review
Find two integers with an *absolute value* of 6.

Find each absolute value.

2. |15| **3.** |−12|

4. |−8| **5.** |8|

GO for Help
Lesson 11-1

What You'll Learn

To add integers and to solve problems by adding integers

Why Learn This?

Sometimes you add positive and negative integers, such as yards gained and yards lost in a football game.

You can use a number line to model the addition of integers. You start at 0, facing the positive direction. You move forward for a positive integer and backward for a negative integer. Here is how to model 3 + 2.

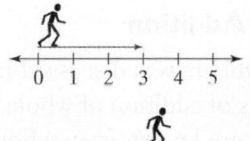

Start at 0. Face the positive direction.
Move forward 3 units for 3.

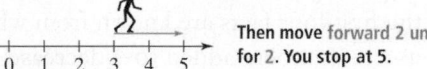

Then move forward 2 units for 2. You stop at 5.

EXAMPLE Using a Number Line to Add Integers

1 Use a number line to find −3 + (−2).

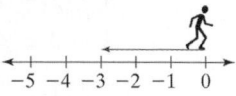

Start at 0, and face the positive direction.
Move backward 3 units for −3.

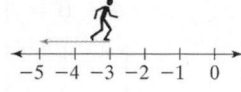

Then move backward 2 units for −2. You stop at −5.

So −3 + (−2) = −5.

524 Chapter 11 Integers

Figure 5-6 similarly depicts integer addition of 3 + ⁻5.

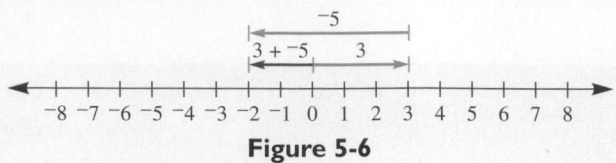

Figure 5-6

NOW TRY THIS 5-1

a. Explain whether the sum of two negative integers is always negative.
b. Explain whether the sum of a positive and a negative integer is positive or negative and why.
c. Use a number line to add 6 + (⁻8) + (⁻2).

Example 5-2 involves a thermometer with a scale in the form of a vertical number line.

EXAMPLE 5-2

The temperature was ⁻4°C. In an hour, it rose 10°C. What is the new temperature?

Solution   Figure 5-7 shows that the new temperature is 6°C and that ⁻4 + 10 = 6.

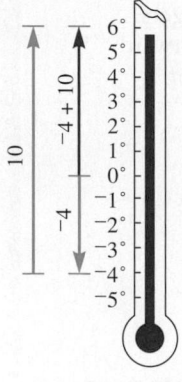

Figure 5-7

Pattern Model for Addition

Addition of whole numbers was discussed in Chapter 3. Addition of integers can also be motivated by using patterns of addition of whole numbers. Notice in the left column of the following list, the first four facts are known from whole number addition. Also notice that the 4 stays fixed and as the numbers added to 4 decrease by 1, the sum decreases by 1. Following this pattern, 4 + ⁻1 = 3 and we can continue the pattern as shown. Similar reasoning can be used to complete the computations in the right column, where ⁻2 stays fixed and the other numbers decrease by 1 each time.

$$4 + 3 = 7 \qquad ⁻2 + 4 = 2$$
$$4 + 2 = 6 \qquad ⁻2 + 3 = 1$$
$$4 + 1 = 5 \qquad ⁻2 + 2 = 0$$
$$4 + 0 = 4 \qquad ⁻2 + 1 = ⁻1$$
$$4 + ⁻1 = 3 \qquad ⁻2 + 0 = ⁻2$$
$$4 + ⁻2 = 2 \qquad ⁻2 + ⁻1 = ⁻3$$
$$4 + ⁻3 = 1 \qquad ⁻2 + ⁻2 = ⁻4$$
$$4 + ⁻4 = 0 \qquad ⁻2 + ⁻3 = ⁻5$$
$$4 + ⁻5 = ⁻1 \qquad ⁻2 + ⁻4 = ⁻6$$
$$4 + ⁻6 = ⁻2 \qquad ⁻2 + ⁻5 = ⁻7$$

However, the inductive reasoning with patterns does not constitute a proof that the sums are correct.

Absolute Value

Because 4 and ⁻4 are opposites, they are on opposite sides of 0 on the number line and are the same distance (4 units) from 0, as shown in Figure 5-8.

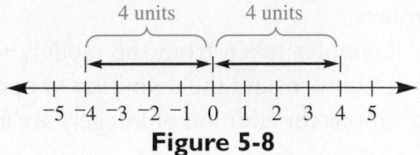

Figure 5-8

Distance is always a positive number or zero. The distance between the point corresponding to an integer and 0 is the **absolute value** of the integer. Thus, the absolute value of both 4 and ⁻4 is 4, written $|4| = 4$ and $|^-4| = 4$, respectively. Notice that if $x \geq 0$, then $|x| = x$, and if $x < 0$, then ^-x is positive. Therefore, we have the following:

Definition of Absolute Value

For any integer x,
$$|x| = x \quad \text{if } x \geq 0$$
$$|x| = {}^-x \quad \text{if } x < 0$$

In grade 6 *Common Core Standards* we find that students should

> understand the absolute value of a rational number [integer] as its distance from 0 on the number line; interpret absolute value as magnitude for a positive or negative quantity in a real-world situation. (p. 43)

EXAMPLE 5-3

Evaluate each of the following:

  a. $|20|$
  b. $|^-5|$
  c. $|0|$
  d. $^-|^-3|$
  e. $|2 + {}^-5|$

Solution

  a. $|20| = 20$
  b. $|^-5| = 5$
  c. $|0| = 0$
  d. $^-|^-3| = {}^-3$
  e. $|2 + {}^-5| = |^-3| = 3$

NOW TRY THIS 5-2

Write each of the following in simplest form without the absolute value notation in the final answer. Show your work.

  a. $|x| + x \quad$ if $x \leq 0$
  b. $^-|x| + x \quad$ if $x \leq 0$
  c. $^-|x| + x \quad$ if $x \geq 0$

Integer Addition—Definitions

Addition of two integers can be defined in cases using what we have learned with the hands-on models in this section. And at the same time, because the set of integers includes the set of whole numbers, the definition of addition of integers should not conflict with the addition of whole numbers.

Examples taken from the models are seen in Table 5-1. With a knowledge of addition of whole numbers and the definition of opposites for integers, along with the model examples, generalizations for addition of integers are also given in Table 5-1.

Table 5-1

Examples from Models	Generalizations
$^-2 + 0 = \,^-2 = 0 + \,^-2$	$m + 0 = m = 0 + m$ for all integers m
$4 + 3 = 7$	If m and n are non-negative integers, they are added as whole numbers.
$^-3 + \,^-5 = \,^-(3 + 5)$	If $m \geq 0$ and $n \geq 0$ $^-m + \,^-n = \,^-(m + n)$.
$4 + \,^-3 = 4 - 3$	If $n > 0$ and $m \geq n$ $m + \,^-n = m - n$.
$3 + \,^-5 = \,^-(5 - 3)$	If $n > m$ and $m > 0$ $m + \,^-n = \,^-(n - m)$.

The generalizations for addition of integers in Table 5.1 do not conflict with the addition of whole numbers and are used in the following definition.

Definition of Addition of Integers

Let m and n be integers.

a. $0 + m = m = m + 0$.
b. If $m \geq 0$ and $n \geq 0$, then $m + n$ is defined using addition of whole numbers.
c. If $m \geq 0$ and $n \geq 0$, then $^-m + \,^-n = \,^-(m + n)$.
d. If $n > 0$ and $m \geq n$, then $m + \,^-n = m - n$.
e. If $n > m$ and $m > 0$, then $m + \,^-n = \,^-(n - m)$.

The above definition of integers is not the only one that could be used. On the following partial student page, we see in the lower right that absolute values can be used in the addition of integers.

Using absolute values, integer addition can be accomplished in the following ways:

To add two integers with the same sign, add the absolute values of the integers. The sum has the same sign as the integers being added.

To add two integers with different signs, subtract the lesser absolute value from the greater one. The sum has the same sign as the integer with the greater absolute value.

School Book Page   ADDING INTEGERS

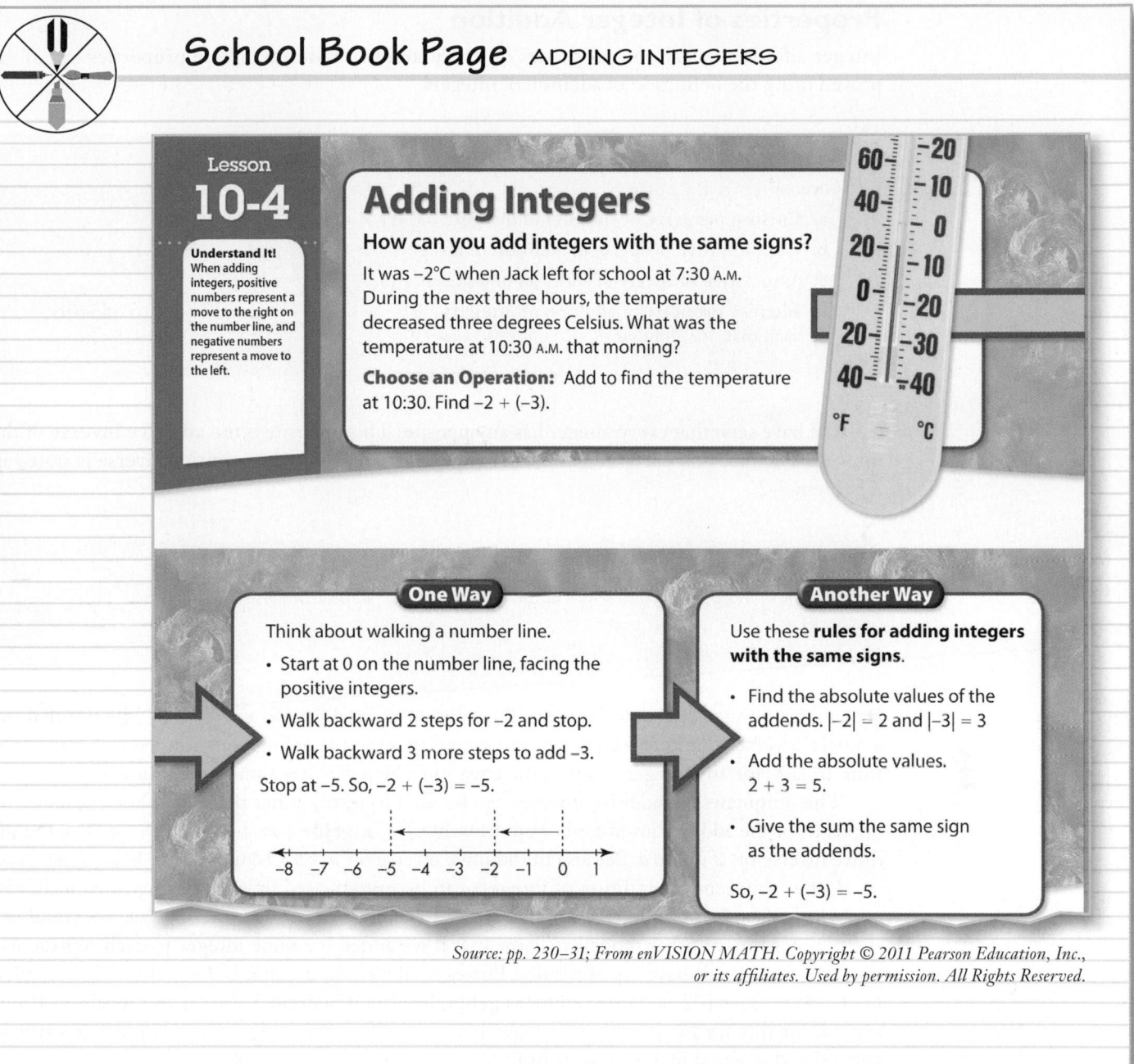

Lesson
10-4

Understand It!
When adding integers, positive numbers represent a move to the right on the number line, and negative numbers represent a move to the left.

Adding Integers

How can you add integers with the same signs?

It was –2°C when Jack left for school at 7:30 A.M. During the next three hours, the temperature decreased three degrees Celsius. What was the temperature at 10:30 A.M. that morning?

Choose an Operation: Add to find the temperature at 10:30. Find –2 + (–3).

One Way

Think about walking a number line.

- Start at 0 on the number line, facing the positive integers.
- Walk backward 2 steps for –2 and stop.
- Walk backward 3 more steps to add –3.

Stop at –5. So, –2 + (–3) = –5.

Another Way

Use these **rules for adding integers with the same signs**.

- Find the absolute values of the addends. |–2| = 2 and |–3| = 3
- Add the absolute values. 2 + 3 = 5.
- Give the sum the same sign as the addends.

So, –2 + (–3) = –5.

Source: pp. 230–31; From enVISION MATH. Copyright © 2011 Pearson Education, Inc., or its affiliates. Used by permission. All Rights Reserved.

An alternative definition of integer addition using absolute values is given below.

Alternative Definition of Integer Addition

If m is any integer, then $m + 0 = m = 0 + m$.

If m and n are positive integers, then $m + n$ is the sum of whole numbers m and n.

If m and n are negative integers, then $m + n = {}^{-}(|m| + |n|)$.

If m is a positive integer and n is a negative integer, then $m + n$ is defined as follows:

If $|m| > |n|$, then $m + n = |m| - |n|$.

If $|m| < |n|$, then $m + n = {}^{-}(|m| - |n|)$.

Properties of Integer Addition

Integer addition has all the properties of whole number addition. These properties can all be proved using the definition of addition of integers.

Theorem 5-1:   Properties of Integer Addition

Given integers a, b, and c:

    a. Closure property of addition of integers $a + b$ is a unique integer.

    b. Commutative property of addition of integers $a + b = b + a$.

    c. Associative property of addition of integers $(a + b) + c = a + (b + c)$.

    d. Identity property of addition of integers   0 is the unique integer, the **additive identity**, such that, for all integers a, $0 + a = a = a + 0$.

We have seen that every integer has an opposite. This opposite is the **additive inverse** of the integer. The fact that each integer has a unique (one and only one) additive inverse is stated in Theorem 5-2.

Theorem 5-2:   Additive Inverse Property

For every integer a, there exists a unique integer ^-a, the additive inverse of a, such that $a + \ ^-a = 0 = \ ^-a + a$.

By definition the additive inverse, ^-a, is the solution of the equation $x + a = 0$. The fact that the additive inverse is unique is equivalent to saying that the preceding equation has only one solution. In fact, for any integers a and b, the equation $x + a = b$ has a unique solution, $b + \ ^-a$.

The uniqueness of additive inverses can be used to justify other theorems. For example, the opposite, or the additive inverse, of ^-a can be written $^-(^-a)$. However, because $a + \ ^-a = 0$, the additive inverse of ^-a is also a. Because the additive inverse of ^-a must be unique, we have $^-(^-a) = a$.

Other theorems of addition of integers can be investigated by considering previously discussed notions. For example, if we had two representations of the same integer, we could use the number line or other models to show that if we added the same integer to each representation, we would still have equal results. For example, we know that if $3 + \ ^-4 = \ ^-2 + 1$, then $(3 + \ ^-4) + 3 = (^-2 + 1) + 3$. This is generalized in Theorem 5-3(b) as the Addition Property of Equality for Integers. Also, we saw that $^-2 + \ ^-4 = \ ^-(2 + 4)$. This relationship is true in general and is stated in Theorem 5-3(c).

Theorem 5-3

For any integers, a, b, and c:

    a. $^-(^-a) = a$

    b. Addition Property of Equality for Integers: If $a = b$, then $a + c = b + c$.

    c. $^-a + \ ^-b = \ ^-(a + b)$

We prove Theorem 5-3(c) as follows: By definition $^-(a + b)$ is the additive inverse of $(a + b)$; that is, $(a + b) + \ ^-(a + b) = 0$. If we could show that $^-a + \ ^-b$ is also the additive inverse of $a + b$, the uniqueness of the additive inverse implies that $^-(a + b)$ and $^-a + \ ^-b$ are equal. To show that $^-a + \ ^-b$ is also the additive inverse of $a + b$, we need only to show that their sum is

0. Using the associative and commutative properties of integer addition and the definition of the additive inverse we have the following:

$$(a + b) + (\ ^-a + \ ^-b) = (a + b) + (\ ^-b + \ ^-a)$$
$$= [(a + b) + \ ^-b] + \ ^-a$$
$$= [a + (b + \ ^-b)] + \ ^-a$$
$$= (a + 0) + \ ^-a$$
$$= a + \ ^-a$$
$$= 0$$

Now we have $(a + b) + (\ ^-a + \ ^-b) = 0$.
Also, $(a + b) + \ ^-(a + b) = 0$.
Hence, $^-(a + b) = \ ^-a + \ ^-b$.

EXAMPLE 5-4

Find the additive inverse of each of the following:

   a. $^-(3 + x)$ **b.** $a + \ ^-4$ **c.** $^-3 + \ ^-x$

Solution
  a. $3 + x$
  b. $^-(a + \ ^-4)$, which can be written as $^-a + \ ^-(\ ^-4)$, or $^-a + 4$
  c. $^-(\ ^-3 + \ ^-x)$, which can be written $^-(\ ^-3) + \ ^-(\ ^-x)$, or $3 + x$

Integer Subtraction

As with integer addition, we explore several models for integer subtraction.

Chip Model for Subtraction

To find $3 - \ ^-2$, we want to subtract $^-2$ (take away 2 red chips) from 3 black chips. As seen in Figure 5-9(a), if we just have 3 black chips, we cannot take 2 red ones away. Therefore, we need to represent 3 so that at least 2 red chips are present. Recall that 1 red chip neutralizes 1 black chip and so adding a black chip and a red chip is the same as adding 0, and the problem does not change. Because we need 2 red chips, we can add 2 black chips and 2 red chips without changing the problem. In Figure 5-9(b), we now see 3 represented using 5 black chips and 2 red chips. Now when the 2 red chips are "taken away," in Figure 5-9(c), 5 black chips are left and hence, $3 - \ ^-2 = 5$.

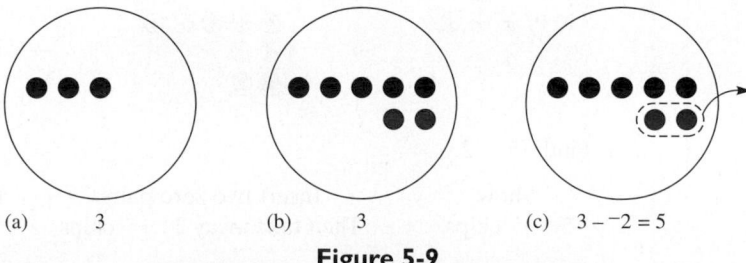

    (a)   3   (b)   3   (c) $3 - \ ^-2 = 5$

Figure 5-9

Charged-Field Model for Subtraction

Integer subtraction can be modeled with a charged field. For example, consider $^-3 - \ ^-5$. To subtract $^-5$ from $^-3$, we first represent $^-3$ in Figure 5-10(a) so that at least 5 negative charges are present as in Figure 5-10(b). To subtract $^-5$, remove the 5 negative charges as in Figure 5-10(c), leaving 2 positive charges, as in Figure 5-10(d). Hence, $^-3 - \ ^-5 = 2$.

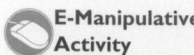
E-Manipulative
Activity

Use *Token Subtraction* for additional practice using the chip and charged field models.

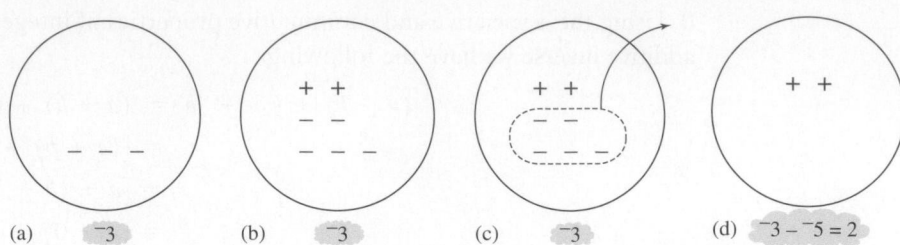

(a) ⁻3   (b) ⁻3   (c) ⁻3   (d) ⁻3 − ⁻5 = 2

Figure 5-10

    The chip model and the charged-field model are combined on the student page.

School Book Page ACTIVITY LAB

11-4a Activity Lab   Hands On

Modeling Subtraction of Integers

ACTIVITY

1. Find −5 − (−2).

Show   Take away   There are 3 "−" chips left.
5 "−" chips.   2 "−" chips.   So −5 − (−2) = −3.

● ● ● ● ●   →   ● ● ● ⦅● ●⦆   →   ● ● ●

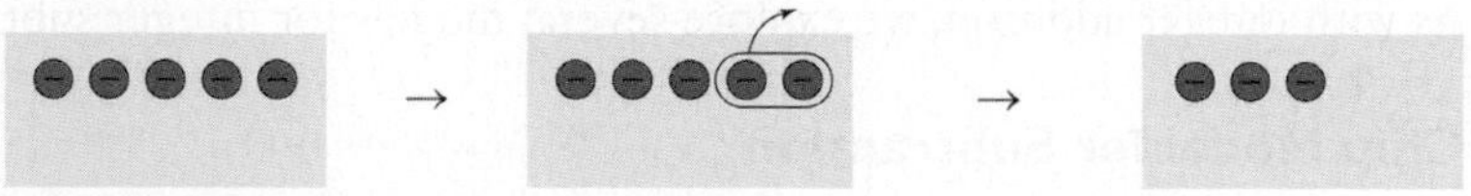

Remember that ⊕ and ● are a zero pair. Sometimes you need to insert zero pairs in order to subtract.

ACTIVITY

2. Find 5 − (−2).

Show   Insert two zero pairs.   There are 7 "+" chips left.
5 "+" chips.   Then take away 2 "−" chips.   So 5 − (−2) = 7.

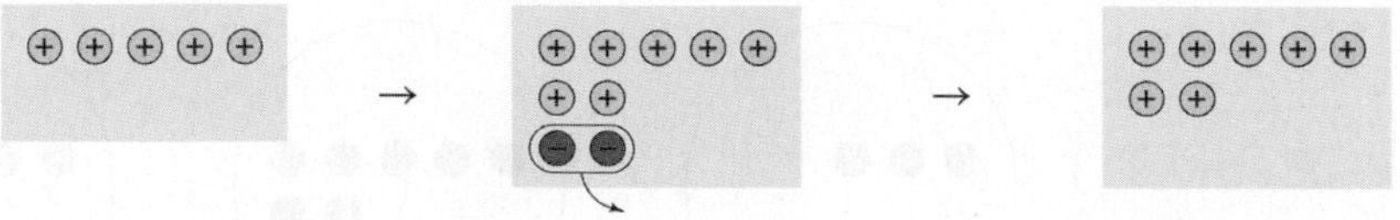

⊕ ⊕ ⊕ ⊕ ⊕   →   ⊕ ⊕ ⊕ ⊕ ⊕   →   ⊕ ⊕ ⊕ ⊕ ⊕
                 ⊕ ⊕   ⊕ ⊕
                 ● ●

3. Find −5 − 2.

Show   Insert two zero pairs.   There are 7 "−" chips left.
5 "−" chips.   Then take away 2 "+" chips.   So −5 − 2 = −7.

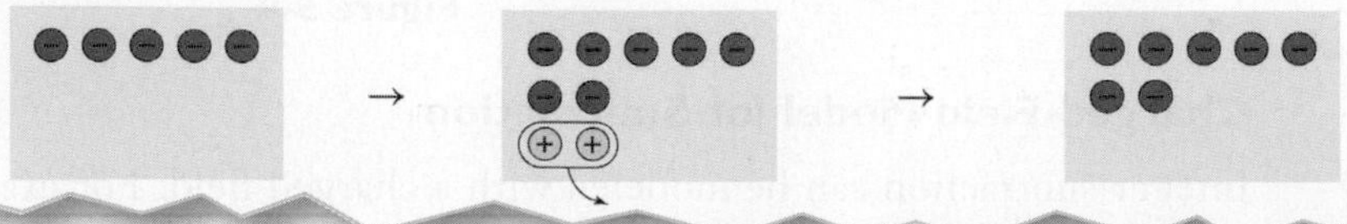

● ● ● ● ●   →   ● ● ● ● ●   →   ● ● ● ● ●
                ● ●   ● ●
                ⊕ ⊕

Number-Line Model for Subtraction

The number-line model used for integer addition can also be used to model integer subtraction. While integer addition is modeled by maintaining the same direction and moving forward or backward depending on whether a positive or negative integer is added, subtraction is modeled by turning around. To see how this works, examine the following partial student page.

School Book Page SUBTRACTING INTEGERS

11-4 Subtracting Integers

EXAMPLE **Using a Number Line to Subtract**

1 Use a number line to find $3 - (-2)$.

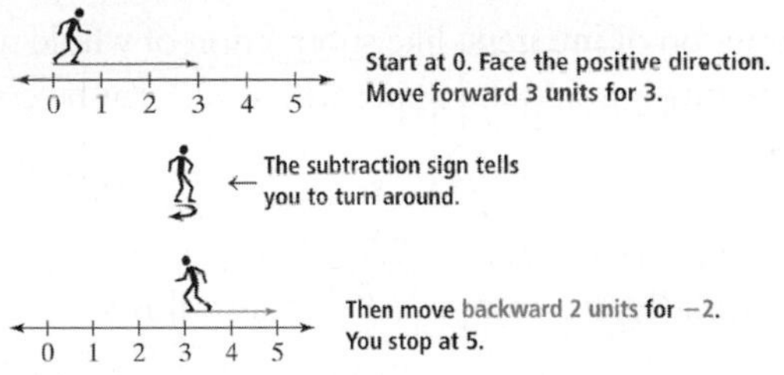

Start at 0. Face the positive direction.
Move forward 3 units for 3.

The subtraction sign tells you to turn around.

Then move backward 2 units for -2.
You stop at 5.

So $3 - (-2) = 5$.

✓ **Quick Check**

1. **a.** Find $5 - (-1)$. **b.** Find $-3 - 3$.

Source: p. 530; From MIDDLE GRADES MATH (Course 1) Copyright © 2010 Pearson Education, Inc., or its affiliates. Used by permission. All Rights Reserved.

NOW TRY THIS 5-3

Suppose a mail carrier brings us three letters, one with a check for $25 and the other two with bills for $15 and $20, respectively. We record this as $25 + {}^-15 + {}^-20$, or ${}^-10$; that is, we are $10 poorer. Suppose that the next day we find out that the bill for $20 was actually intended for someone else and therefore we give it back to the delivery person. We record the new balance as

$${}^-10 - {}^-20$$

or as

$$25 + {}^-15 + {}^-20 - {}^-20$$

which equals $25 + {}^-15$, or 10.

For each of the following, make up a mail delivery story and explain how the story helps to find the answer.

a. $23 + {}^-13 + {}^-12$
b. $18 - {}^-37$

Pattern Model for Subtraction

By using inductive reasoning, we can find the difference of two integers by considering the following patterns, where we start with subtractions we already know how to do. Both the following pattern on the left and the pattern on the right start with $3 - 2 = 1$.

$$
\begin{array}{ll}
3 - 2 = 1 \qquad & 3 - 2 = 1 \\
3 - 3 = 0 & 3 - 1 = 2 \\
3 - 4 = ? & 3 - 0 = 3 \\
3 - 5 = ? & 3 - {}^-1 = ?
\end{array}
$$

In the pattern on the left, the difference decreases by 1. If we continue the pattern, we have $3 - 4 = {}^-1$ and $3 - 5 = {}^-2$. In the pattern on the right, the difference increases by 1. If we continue the pattern, we have $3 - {}^-1 = 4$ and $3 - {}^-2 = 5$.

Defining Integer Subtraction

Subtraction of integers, like subtraction of whole numbers, can be defined in terms of addition. Using the missing-addend approach, $5 - 3$ can be computed by finding a whole number n as follows:

$$5 - 3 = n \quad \text{if, and only if,} \quad 5 = 3 + n$$

Because $3 + 2 = 5, n = 2$.

Similarly, we compute $3 - 5$ as follows:

$$3 - 5 = n \quad \text{if, and only if,} \quad 3 = 5 + n$$

Because $5 + {}^-2 = 3, n = {}^-2$. In general, for integers a and b, we have the following definition of *subtraction*.

Definition of Integer Subtraction

For integers a and b, $a - b$ is the unique integer n such that $a = b + n$.

Addition "undoes" subtraction; that is, $(a - b) + b = a$. Also, subtraction "undoes" addition; that is, $(a + b) - b = a$.

EXAMPLE 5-5

Use the definition of subtraction to compute the following:

a. $3 - 10$
b. ${}^-2 - 10$

Solution

a. Let $3 - 10 = n$. Then $10 + n = 3$, so $n = {}^-7$. Therefore, $3 - 10 = {}^-7$.
b. Let ${}^-2 - 10 = n$. Then $10 + n = {}^-2$, so $n = {}^-12$. Therefore, ${}^-2 - 10 = {}^-12$.

Subtraction Using Adding the Opposite Approach

In grade 7 *Common Core Standards*, we find that students should "understand subtraction of rational numbers [integers] as adding the additive inverse, $p - q = p + {}^-q$." (p. 48)

This approach is discussed by the student on the partial student page.

School Book Page   SUBTRACTING INTEGERS

What is a rule for subtracting integers?

A class made the table at the right to show some sums and differences they found on number lines.

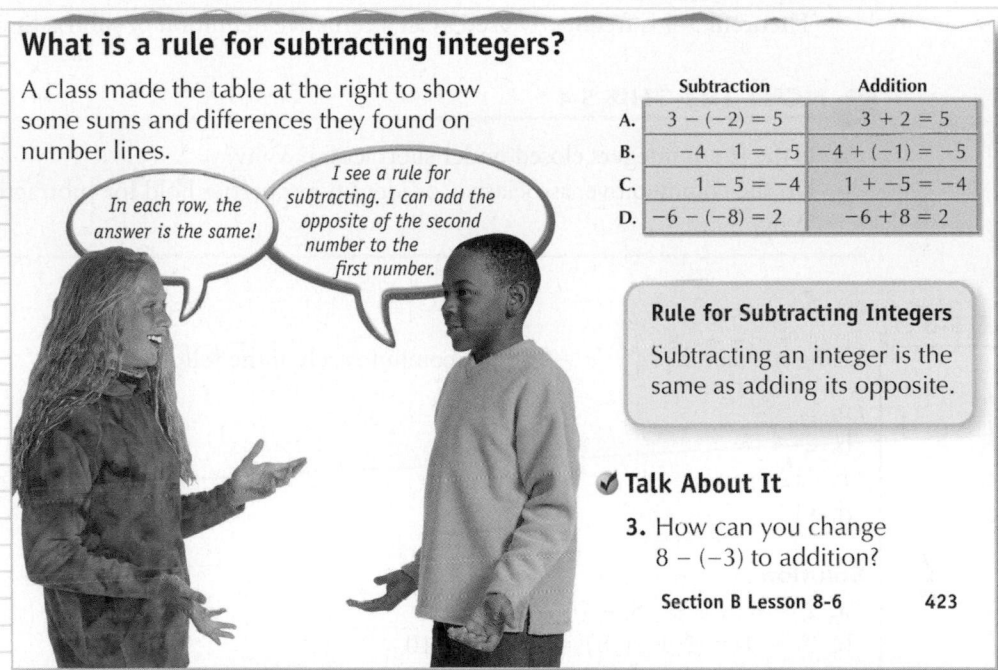

> *In each row, the answer is the same!*

> *I see a rule for subtracting. I can add the opposite of the second number to the first number.*

	Subtraction	Addition
A.	$3 - (-2) = 5$	$3 + 2 = 5$
B.	$-4 - 1 = -5$	$-4 + (-1) = -5$
C.	$1 - 5 = -4$	$1 + -5 = -4$
D.	$-6 - (-8) = 2$	$-6 + 8 = 2$

Rule for Subtracting Integers

Subtracting an integer is the same as adding its opposite.

✓ **Talk About It**

3. How can you change $8 - (-3)$ to addition?

Section B Lesson 8-6   423

Source: p. 423; From SCOTT FORESMAN-ADDISON WESLEY MATHEMATICS.
Copyright © 2008 Pearson Education, Inc., or its affiliates. Used by permission. All Rights Reserved.

From previous work we know that $3 - 5 = {}^-2$ and $3 + {}^-5 = {}^-2$. Hence, $3 - 5 = 3 + {}^-5$. In general, the following is true.

Theorem 5-4

For all integers a and b, $a - b = a + {}^-b$.

The preceding theorem can be justified using the fact that the equation $b + x = a$ has a unique solution for x. From the definition of subtraction, the solution of the equation is $a - b$. To show that $a - b = a + {}^-b$, we show that $a + {}^-b$ is also a solution by checking whether $b + (a + {}^-b) = a$:

$$b + (a + {}^-b) = b + ({}^-b + a)$$
$$= (b + {}^-b) + a$$
$$= 0 + a$$
$$= a$$

Consequently, $a - b = a + {}^-b$.

Theorem 5-4 is frequently used as an alternative definition of subtraction.

NOW TRY THIS 5-4

a. Is the set of integers closed under subtraction? Why?

b. Do the commutative, associative, or identity properties hold for subtraction of integers? Why?

EXAMPLE 5-6

Using the fact that $a - b = a + {}^-b$, compute each of the following:

a. $2 - 8$
b. $2 - {}^-8$
c. ${}^-12 - {}^-5$
d. ${}^-12 - 5$

Solution

a. $2 - 8 = 2 + {}^-8 = {}^-6$
b. $2 - {}^-8 = 2 + {}^-({}^-8) = 2 + 8 = 10$
c. ${}^-12 - {}^-5 = {}^-12 + {}^-({}^-5) = {}^-12 + 5 = {}^-7$
d. ${}^-12 - 5 = {}^-12 + {}^-5 = {}^-17$

EXAMPLE 5-7

Write expressions equal to each of the following without parentheses.

a. ${}^-(b - c)$
b. $a - (b + c)$

Solution

a. ${}^-(b - c) = {}^-(b + {}^-c) = {}^-b + {}^-({}^-c) = {}^-b + c$
b. $a - (b + c) = a + {}^-(b + c) = a + ({}^-b + {}^-c) = (a + {}^-b) + {}^-c = a + {}^-b + {}^-c$

 **EXAMPLE 5-8**

Simplify each of the following:

a. $2 - (5 - x)$ b. $5 - (x - 3)$ c. $^-(x - y) - y$

Solution

a. $2 - (5 - x) = 2 + {}^-(5 + {}^-x)$
              $= 2 + {}^-5 + {}^-({}^-x)$
              $= 2 + {}^-5 + x$
              $= {}^-3 + x$ or $x + {}^-3$ or $x - 3$

b. $5 - (x - 3) = 5 + {}^-(x + {}^-3)$
              $= 5 + {}^-x + {}^-({}^-3)$
              $= 5 + {}^-x + 3$
              $= 8 + {}^-x$
              $= 8 - x$

c. $^-(x - y) - y = {}^-(x + {}^-y) + {}^-y$
              $= [{}^-x + {}^-({}^-y)] + {}^-y$
              $= ({}^-x + y) + {}^-y$
              $= {}^-x + (y + {}^-y)$
              $= {}^-x + 0$
              $= {}^-x$

Subtraction of Integers on a Calculator

 Many calculators have a change of sign key, either $\boxed{\text{CHS}}$ or $\boxed{+/-}$. Other calculators use $\boxed{(-)}$, a key that allows computation with integers. For example, to compute $8 - ({}^-3)$, we would press $\boxed{8}\boxed{-}\boxed{3}\boxed{+/-}\boxed{=}$. Investigate what happens if you press $\boxed{8}\boxed{-}\boxed{-}\boxed{3}\boxed{=}$.

Order of Operations

Subtraction in the set of integers is neither commutative nor associative, as illustrated in these counterexamples:

$$5 - 3 \neq 3 - 5 \quad \text{because} \quad 2 \neq {}^-2$$
$$(3 - 15) - 8 \neq 3 - (15 - 8) \quad \text{because} \quad {}^-20 \neq {}^-4$$

An expression such as $3 - 15 - 8$ is ambiguous unless we know in which order to perform the subtractions. Mathematicians agree that $3 - 15 - 8$ means $(3 - 15) - 8$; that is, the subtractions in $3 - 15 - 8$ are performed in order from left to right. Similarly, $3 - 4 + 5$ means $(3 - 4) + 5$ and not $3 - (4 + 5)$. Thus, $(a - b) - c$ may be written without parentheses as $a - b - c$. Order of operations for integers will be revisited after multiplication and division are discussed.

 **EXAMPLE 5-9**

Compute each of the following:

a. $2 - 5 - 5$ b. $3 - 7 + 3$ c. $3 - (7 - 3)$

Solution

a. $2 - 5 - 5 = {}^-3 - 5 = {}^-8$
b. $3 - 7 + 3 = {}^-4 + 3 = {}^-1$
c. $3 - (7 - 3) = 3 - 4 = {}^-1$

Assessment 5-1A

1. Find the additive inverse of each of the following integers. Write the answer in the simplest possible form.
   a. 2
   b. $^-5$
   c. m
   d. 0
   e. ^-m
   f. $a + b$

2. Simplify each of the following:
   a. $^-(^-2)$
   b. $^-(^-m)$
   c. $^-0$

3. Evaluate each of the following:
   a. $|^-5|$
   b. $|10|$
   c. $^-|^-5|$
   d. $^-|5|$

4. Demonstrate each of the following additions using the charged-field or chip model:
   a. $5 + ^-3$
   b. $^-2 + 3$
   c. $^-3 + 2$
   d. $^-3 + ^-2$

5. Demonstrate each of the additions in exercise 4 using a number-line model.

6. Compute each of the following using $a - b = a + ^-b$:
   a. $3 - ^-2$
   b. $^-3 - 2$
   c. $^-3 - ^-2$

7. Answer each part of exercise 6 using subtraction with the missing-addend approach.

8. Write an addition that corresponds to each of the following sentences and then answer the question:
   a. A certain stock dropped 17 points and the following day gained 10 points. What was the net change in the stock's worth?
   b. The temperature was $^-10°C$ and then it rose by 8°C. What is the new temperature?
   c. The plane was at 5000 ft and dropped 100 ft. What is the new altitude of the plane?

9. On January 1, Jane's bank balance was $300. During the month, she wrote checks for $45, $55, $165, $35, and $100 and made deposits of $75, $25, and $400.
   a. If a check is represented by a negative integer and a deposit by a positive integer, express Jane's transactions as a sum of positive and negative integers.
   b. What was the balance in Jane's account at the end of the month?

10. Use a number-line model to find the following:
    a. $^-4 - ^-1$
    b. $^-4 - ^-3$

11. Use patterns to show the following:
    a. $^-4 - ^-1 = ^-3$
    b. $^-2 - 1 = ^-3$

12. Do exercise 11 using the charged field model.

13. Compute each of the following:
    a. $^-2 + (3 - 10)$
    b. $[8 - ^-5] - 10$
    c. $^-2 - 7 + 10$

14. In each of the following, write both a subtraction and an addition problem that correspond to the question and then answer the questions:
    a. The temperature is 55°F and is supposed to drop 60°F by midnight. What is the expected midnight temperature?
    b. Moses has overdraft privileges at his bank. If he had $200 in his checking account and he wrote a $220 check, what is his balance?

15. Simplify each of the following as much as possible. Show all work.
    a. $3 - (2 - 4x)$
    b. $x - (^-x - y)$

16. For which integers a, b, and c does $a - b - c = a - (b - c)$? Justify your answer.

17. Let W stand for the set of whole numbers, I the set of integers, I^+ the set of positive integers, and I^- the set of negative integers. Find each of the following:
    a. $W \cup I$
    b. $W \cap I$
    c. $I^+ \cup I^-$
    d. $I^+ \cap I^-$
    e. $W - I$

18. Place the integers $^-4, ^-3, ^-2, ^-1, 0, 1, 2, 3, 4$ in the grid to make a magic square.

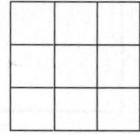

19. Let $y = ^-x - 1$. Find the value of y (in parts (a)–(d)) when x has the following values:
    a. $^-1$
    b. 100
    c. $^-2$
    d. ^-a in terms of a
    e. For which values of x will y be 3?

20. Find all integers x, if there are any, such that the following are true:
    a. ^-x is positive.
    b. ^-x is negative.
    c. $^-x - 1$ is positive.
    d. $|x| = 2$

21. Let $y = |1 - x|$. Find the value of y (in parts (a)–(b)) for each given value of x.
    a. 10
    b. $^-1$
    c. Find all values of x for which y is 1.

22. In each of the following, find all integers x satisfying the given equation:
    a. $|x - 6| = 6$
    b. $|x| + 2 = 10$
    c. $|^-x| = |x|$

23. Determine how many integers there are between, not including, the following given integers:
    a. 10 and 100
    b. $^-30$ and $^-10$

24. An arithmetic sequence may have a positive or negative difference. In each of the following arithmetic sequences, find the difference and write the next two terms:
    a. $0, ^-3, ^-6, ^-9$
    b. $x + y, x, x - y$

25. Classify each of the following as true or false. If false, show a counterexample that makes it false.
    a. $|^-x| = |x|$
    b. $|x - y| = |y - x|$
    c. $|^-x + ^-y| = |x + y|$

26. Solve the following equations:
    a. $x + 7 = 3$
    b. $^-10 + x = ^-7$
    c. $^-x = 5$

27. Complete each of the following integer arithmetic problems on your calculator, making use of the change-of-sign key.
    a. $^-12 + ^-6$ b. $^-12 + 6$
    c. $27 + ^-5$ d. $^-12 - 6$
    e. $16 - ^-7$

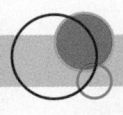

Assessment 5-1B

1. Find the additive inverse of each of the following integers. Write the answer in the simplest possible form.
    a. 3   b. $^-4$
    c. q d. 6
    e. ^-n f. $3 + x$

2. Simplify each of the following:
    a. $^-(^-5)$ b. $^-(^-x)$

3. Evaluate each of the following:
    a. $|^-3|$ b. $|15|$ c. $^-|^-3|$ d. $^-|6|$

4. Demonstrate each of the following additions using the charged-field or chip model:
    a. $^-2 + 5$ b. $^-5 + 2$
    c. $^-3 + ^-3$ d. $6 + ^-4$

5. Demonstrate each of the additions in exercise 4 using a number-line model.

6. Compute each of the following using $a - b = a + ^-b$.
    a. $^-3 - 5$ b. $5 - (^-3)$ c. $^-2 - ^-3$

7. Answer each part of exercise 6 using the missing-addend approach.

8. Write an addition that corresponds to each of the following sentences and then answer the question:
    a. A visitor in a Las Vegas casino lost $200, won $100, and then lost $50. What is the change in the gambler's net worth?
    b. In four downs, the football team lost 2 yd, gained 7 yd, gained 0 yd, and lost 8 yd. What is the total gain or loss?
    c. In a game of Triominoes, Jack's scores in five successive turns are 17, $^-8$, $^-9$, 14, and 45. What is his total at the end of five turns?

9. a. The largest bubble chamber in the world is 15 ft in diameter and contains 7259 gal of liquid hydrogen at a temperature of $^-247°C$. If the temperature is dropped by $11°C$ per hour for 2 consecutive hours, what is the new temperature?
    b. The greatest recorded temperature ranges in the world are around the "cold pole" in Siberia. Temperatures in Verkhoyansk have varied from $^-94°F$ to $98°F$. What is the difference between the high and low temperatures in Verkhoyansk?

10. Use a number-line model to find the following:
    a. $^-3 - ^-2$ b. $^-4 - 3$

11. Use patterns to show the following:
    a. $^-2 - ^-3 = 1$ b. $^-3 - 2 = ^-5$

12. Do exercise 11 using the charged-field model.

13. Compute each of the following:
    a. $^-2 - (7 + 10)$
    b. $8 - 11 - 10$
    c. $^-2 - 7 + 3$

14. Motor oils protect car engines over a range of temperatures. These oils have names like 10W–40 or 5W–30. The following graph shows the temperatures, in degrees Fahrenheit, at which the engine is protected by a particular oil. Using the graph, find which oils can be used for the following temperatures:
    a. Between $^-5°$ and $90°$
    b. Below $^-20°$
    c. Between $^-10°$ and $50°$
    d. From $^-20°$ to over $100°$
    e. From $^-8°$ to $90°$

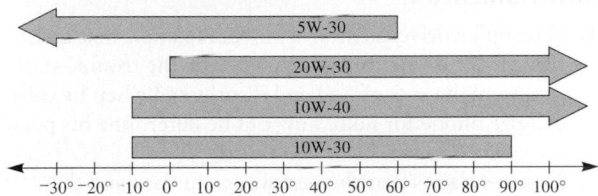

15. Simplify each of the following as much as possible. Show all work.
    a. $4x - 2 - 3x$ b. $4x - (2 - 3x)$

16. a. Prove that $^-x - y = ^-y - x$; for all integers x and y.
    b. Does part (a) imply that subtraction is commutative? Explain.

17. Let W stand for the set of whole numbers, I the set of integers, I^+ the set of positive integers, and I^- the set of negative integers. Find each of the following:
    a. $W - I^+$ b. $W - I^-$
    c. $I \cap I$ d. $I - W$

18. Complete the magic square using the following integers: $^-13$, $^-10$, $^-7$, $^-4$, 2, 5, 8, 11.

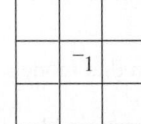

19. Let $y = ^-3x - 2$. Find the value of y (in parts (a)–(d)) for each of the following values of x:
    a. $^-1$ b. 100
    c. $^-2$ d. ^-a in terms of a
    e. For which values of x will y be $^-11$?

20. Find all integers x, if there are any, such that the following are true:
    a. $^-|x| = 2$.
    b. $^-|x|$ is negative.
    c. $^-|x|$ is positive.
    d. $^-x + 1$ is positive.
    e. $^-x - 1$ is negative.

21. Let $y = |x - 5|$. Find the value of y (in parts (a)–(b) for each of the following values of x:
    a. 10   b. $^-1$
    c. All the values of x for which y is 7
22. By the definition of absolute value, the equation $y = |x|$ can be written as follows:

$$y = \begin{cases} x, & \text{if } x \geq 0 \\ ^-x, & \text{if } x < 0 \end{cases}$$

    Write $y = |x - 6|$ in a similar way without absolute value.
23. Determine how many integers there are between, not including, the following given integers:
    a. $^-10$ and 10   b. x and y (if $x < y$)

24. An arithmetic sequence may have a positive or negative difference. In each of the following arithmetic sequences, find the difference and write the next two terms:
    a. 7, 3, $^-1$, $^-5$ b. $1 - 3x, 1 - x, 1 + x$
25. Classify each of the following as true or false. If false, show a counterexample that makes it false.
    a. $|x^2| = x^2$ b. $|x^3| = x^3$
    c. $|x^3| = x^2|x|$
26. Solve the following equations:
    a. $^-x + 5 = 7$ b. $1 - x = ^-13$ c. $^-x - 8 = ^-9$
27. Estimate each of the following and then use a calculator to find the actual answer:
    a. $343 + ^-42 - 402$ b. $^-1992 + 3005 - 497$
    c. $992 - ^-10{,}003 - 101$ d. $^-301 - ^-1303 + 4993$

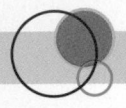

Mathematical Connections 5-1

Communication

1. A turnpike driver had car trouble. He knew that he had driven 12 mi from milepost 68 before the trouble started. Assuming he is confused and disoriented when he calls on his cellular phone for help, how can he determine his possible location? Explain.
2. Dolores claims that the best way to understand that $a - b = a + ^-b$, for all integers a and b, is to show that when you add b to each expression you get equal answers. Explain why you think Dolores is making this claim.
3. Explain why $b - a$ and $a - b$ are additive inverses of each other.
4. The absolute value of an integer is never negative. Does this contradict the fact that the absolute value of x could be equal to ^-x? Explain why or why not.
5. If an integer a is pictured on the number line, then the distance from the point on the number line that represents the integer to the origin is $|a|$. Using this idea, explain why $|a - b|$ is the distance between the points that represent the integers a and b.
6. Recall the definition of less than for whole numbers using addition and define $a < b$ when a and b are any integers. Use your definition to show that $^-8 < ^-7$.

Open-Ended

7. Describe a realistic word problem that models $^-50 + (^-85) - (^-30)$.
8. In a library some floors are below ground level and others are above ground level. If the ground-level floor is designated the zero floor, design a system to number the floors.
9. a. I am choosing an integer. I then subtract 10 from the integer, take the opposite of the result, add $^-3$, and find the opposite of the new result. My result is $^-3$. What is the original number?
   b. Judy wants to do the activity in part (a) with her classmates. Each classmate probably chooses a different number and Judy wants to tell each classmate quickly what number was chosen. Judy figures out that the only thing she needs to do is to add 7 to each answer she gets. Does this always work? Explain why or why not.
   c. Come up with your own "trick" similar to the one in part (b) that works for each answer you get from your classmates.

Cooperative Learning

10. Examine several elementary mathematics textbooks. Report on how addition and subtraction of integers is treated, and on how various properties are justified. Discuss in your group how the treatment of addition and subtraction of integers presented in this section compares to the treatment in elementary textbooks.
11. Look at several history of mathematics books and the Internet and report in your group on when and how negative integers were introduced first.
12. Number each card in a set of 21 3-by-5 cards with an integer from $^-10$ to 10. Lay the cards on the floor to form a number line. Choose someone from your group to act like the hiker on the student pages for number line addition and subtraction. Give directions to walk the number line to solve exercises 6 and 10 in Problem Set 5-lA. Try other addition and subtraction problems to make sure that the number line model is understood by everyone in your group and could be used in an elementary classroom.

Questions from the Classroom

13. A fourth-grade student devised the following algorithm for subtracting $84 - 27$:

    4 minus 7 equals negative 3.

$$\begin{array}{r} 84 \\ -\ 27 \\ \hline ^-3 \end{array}$$

    80 minus 20 equals 60.

$$\begin{array}{r} 84 \\ -\ 27 \\ \hline ^-3 \\ 60 \end{array}$$

    60 plus negative 3 equals 57.

$$\begin{array}{r} 84 \\ -\ 27 \\ \hline ^-3 \\ +\ 60 \\ \hline 57 \end{array}$$

Thus the answer is 57. How should you respond as a teacher? Will this technique always work?

14. An eighth-grade student claims she can prove that subtraction of integers is commutative. She points out that if a and b are integers, then $a - b = a + {}^-b$. Since addition is commutative, so is subtraction. What is your response?

15. A student had the following picture of an integer and its opposite. Other students in the class objected, saying that ^-a should be to the left of 0. How do you respond?

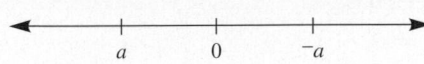

Trends in Mathematics and Science Study (TIMSS) Question

When Tracy left for school, the temperature was minus [negative] 3 degrees.

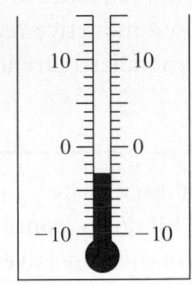

At recess, the temperature was 5 degrees.

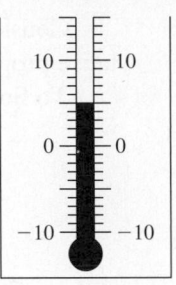

How many degrees did the temperature rise?
a. 2 degrees **b.** 3 degrees **c.** 5 degrees **d.** 8 degrees

TIMSS, Grade 4, 2003

Place either $+$ or $-$ into each box so that this expression has the largest possible total.

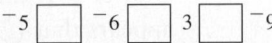

TIMSS, Grade 8, 2007

National Assessment of Educational Progress (NAEP) Question

Paco had 32 trading cards. He gave N trading cards to his friend. Which expression tells how many trading cards Paco has now?
a. $32 + N$ **b.** $32 - N$ **c.** $N - 32$ **d.** $32 \div N$

NAEP, Grade 4, 2007

BRAIN TEASER 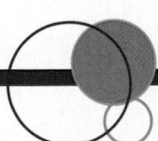 If the digits 1 through 9 are written in order, it is possible to place plus and minus signs between the numbers or to use no operation symbol at all to obtain a total of 100. For example, $1 + 2 + 3 - 4 + 5 + 6 + 78 + 9 = 100$. Can you obtain a total of 100 using fewer plus or minus signs than in the given example? Note that digits, such as 7 and 8 in the example, may be combined.

5-2 Multiplication and Division of Integers

Multiplication of integers is approached through a variety of models: *patterns, charged-field, chip,* and *number-line.* Reasoning with these models is inductive and is used to motivate the operations.

Patterns Model for Multiplication of Integers

We may approach multiplication of integers using repeated addition. For example, if a running-back lost 2 yd on each of three carries in a football game, then he had a net loss of $^-2 + {}^-2 + {}^-2$, or $^-6$, yards. Since $^-2 + {}^-2 + {}^-2$ can be written as $3(^-2)$, using repeated addition, we have $3(^-2) = {}^-6$.

Consider ($^-$2)3. It is meaningless to say that there are $^-$2 threes in a sum. But if the commutative property of multiplication is to hold for all integers, then ($^-$2)3 = 3($^-$2) = $^-$6.

To find ($^-$3)($^-$2), we develop the following pattern:

$$3(^-2) = ^-6$$
$$2(^-2) = ^-4$$
$$1(^-2) = ^-2$$
$$0(^-2) = 0$$
$$^-1(^-2) = ?$$
$$^-2(^-2) = ?$$
$$^-3(^-2) = ?$$

The first four products, $^-$6, $^-$4, $^-$2, and 0, are terms in an arithmetic sequence with a fixed difference of 2. If the pattern continues, the next three terms in the sequence are 2, 4, and 6. Thus it appears that ($^-$3)($^-$2) = 6. Likewise, ($^-$2)($^-$3) = 6. We used inductive reasoning to reach the results but will use deductive reasoning later to justify the generalized product.

NOW TRY THIS 5-5

On a spreadsheet, in column A enter 5 as the first entry and then write a formula to add $^-$1 to 5 for the second entry. Then add $^-$1 to the second entry and fill down, continuing the pattern. In column B, repeat the process. In column C, find the product of the respective entries in columns A and B. What patterns do you observe?

Next multiplication of integers is approached using the chip model, the charged-field model, and the number-line model. In all of these models we start with 0, represented in various ways.

Chip Model and the Charged-Field Model for Multiplication

The chip model and the charged-field *model* can both be used to illustrate multiplication of integers. Consider Figure 5-11(a), where 3($^-$2) is pictured using a chip model. The product 3($^-$2) is interpreted as putting in 3 groups of 2 red chips each. In Figure 5-11(b), 3($^-$2) is pictured as 3 groups of 2 negative charges.

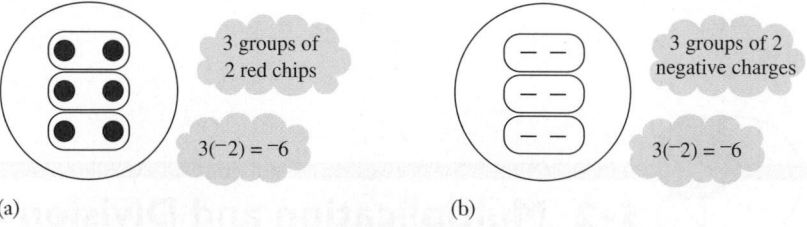

(a)   (b)

Figure 5-11

To find ($^-$3)($^-$2) using the chip model, we interpret the signs as follows: $^-$3 is taken to mean "*remove 3 groups of*"; $^-$2 is taken to mean "*2 red chips.*" To do this, we start with a value of 0 that includes at least 6 red chips, as shown in Figure 5-12(a). When we remove 6 red chips, we are left with 6 black chips. The result is a positive 6, so ($^-$3)($^-$2) = 6. Similar reasoning can be used in Figure 5-12(b) with the charged-field model.

Number-Line Model

As with addition and subtraction, we demonstrate multiplication by using a hiker moving along a number line, according to the following rules:

1. Traveling to the left (west) means moving in the negative direction, and traveling to the right (east) means moving in the positive direction.

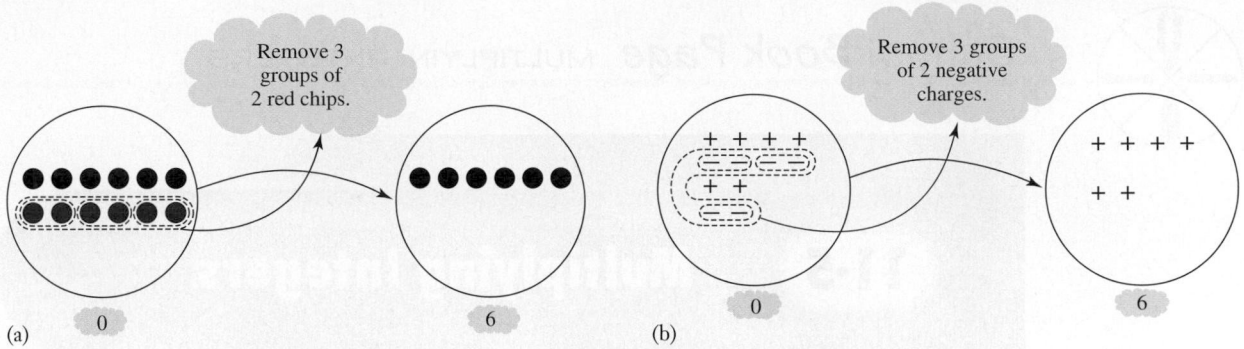

Figure 5-12

2. Time in the future is denoted by a positive value, and time in the past is denoted by a negative value.

Consider the number line shown in Figure 5-13. Various cases using this number line are given next.

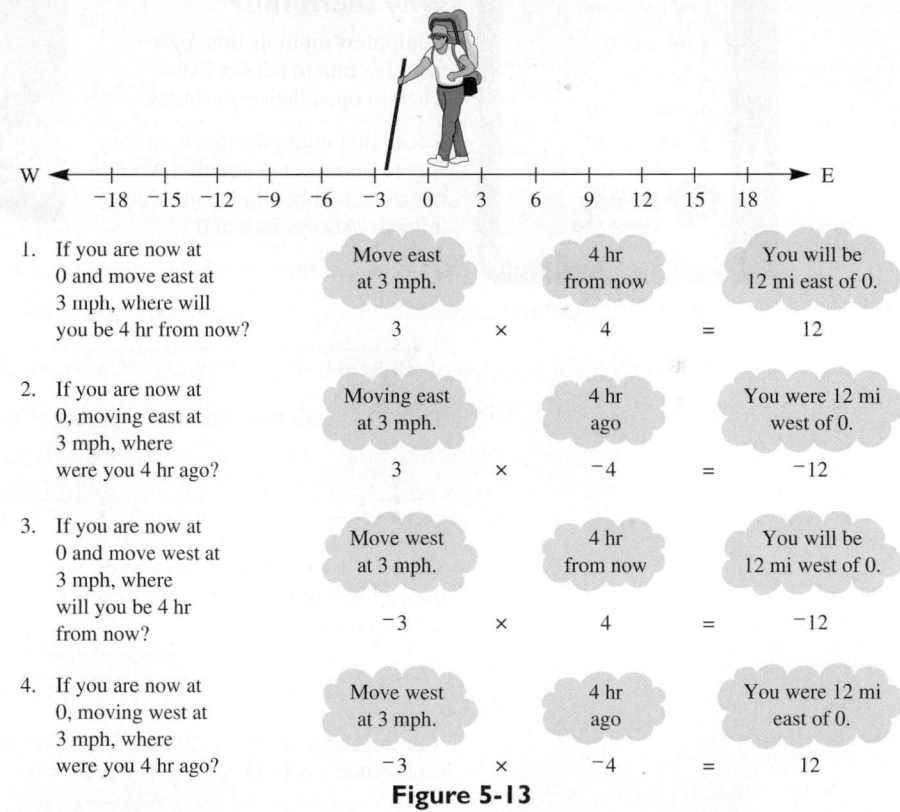

1. If you are now at 0 and move east at 3 mph, where will you be 4 hr from now?

   Move east at 3 mph.   4 hr from now   You will be 12 mi east of 0.

   3   ×   4   =   12

2. If you are now at 0, moving east at 3 mph, where were you 4 hr ago?

   Moving east at 3 mph.   4 hr ago   You were 12 mi west of 0.

   3   ×   ⁻4   =   ⁻12

3. If you are now at 0 and move west at 3 mph, where will you be 4 hr from now?

   Move west at 3 mph.   4 hr from now   You will be 12 mi west of 0.

   ⁻3   ×   4   =   ⁻12

4. If you are now at 0, moving west at 3 mph, where were you 4 hr ago?

   Move west at 3 mph.   4 hr ago   You were 12 mi east of 0.

   ⁻3   ×   ⁻4   =   12

Figure 5-13

An alternative use of the number line to show multiplication of integers is seen on the following student page.

The models presented earlier in this section suggest ways to define the multiplication of integers. We give a formal definition using absolute values in the following:

Definition of Integer Multiplication

Let a and b be any integers.

1. If either $a \geq 0$ and $b \geq 0$, or $a \leq 0$ and $b \leq 0$, then a and b are whole numbers with the product ab.
2. If $a \leq 0$ and $b \leq 0$, then $ab = |a||b|$.
3. If one of a or b is less than 0 while the other is greater than or equal to 0, then $ab = {}^{-}|a||b|$.

School Book Page　MULTIPLYING INTEGERS

Multiplying Integers

Check Skills You'll Need

1. Vocabulary Review
The sum of two negative integers is always ___?___ .

Find each sum.

2. −4 + (−4)

3. 32 + 32

4. −14 + (−14)

5. −45 + (−45)

GO for Help
Lesson 11-3

What You'll Learn

To multiply integers and to solve problems by multiplying integers

Why Learn This?

Computers multiply time by a negative rate to tell skydivers when to open their parachutes.

Recall that multiplication is an easy way to do repeated addition. You can use a number line to multiply integers. Always start at 0.

3×2 means three groups of 2 each: $3 \times 2 = 6$.

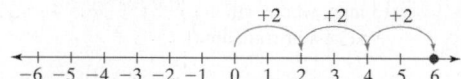

$3 \times (-2)$ means three groups of −2 each: $3 \times (-2) = -6$.

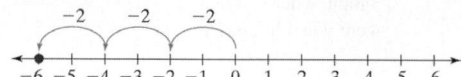

The integers 3 and −3 are opposites. You can think of -3×2 as the opposite of three groups of 2 each. So $-3 \times 2 = -6$.

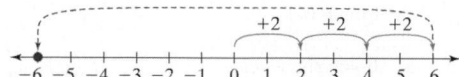

You can think of $-3 \times (-2)$ as the opposite of three groups of −2 each. Since $3 \times (-2) = -6$, $-3 \times (-2) = 6$.

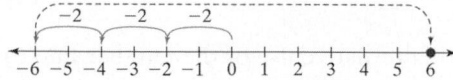

534　Chapter 11　Integers

Properties of Integer Multiplication

In grade 7 *Common Core Standards*, students should

> understand that multiplication is extended . . . by requiring that operations continue to satisfy the properties of operations [of addition and multiplication on whole numbers], particularly the distributive property leading to products such as $(^-1)(^-1) = 1$ and the rules for multiplying signed numbers. (p. 48)

The set of integers has properties under multiplication analogous to those of the set of whole numbers under multiplication and are summarized in Theorem 5-5.

Theorem 5-5: Properties of Integer Multiplication

The set of integers I satisfies the following properties of multiplication for all integers $a, b, c \in I$:

Closure property of multiplication of integers ab is a unique integer.

Commutative property of multiplication of integers $ab = ba$.

Associative property of multiplication of integers $(ab)c = a(bc)$.

Identity property of multiplication of integers   1 is the unique integer such that for all integers a, $1 \cdot a = a = a \cdot 1$.

Distributive properties of multiplication over addition for integers $a(b + c) = ab + ac$ and $(b + c)a = ba + ca$.

Zero multiplication property of integers   0 is the unique integer such that for all integers a, $a \cdot 0 = 0 = 0 \cdot a$.

Theorem 5-6 is another convenient theorem that follows from the definition of multiplication of integers and the properties in Theorem 5-5.

Theorem 5-6

For every integer a, $(^-1)a = {}^-a$.

It is important to keep in mind that $(^-1)a = {}^-a$ is true for all integers a. Thus if we substitute $^-1$ for a, we get $(^-1)(^-1) = {}^-(^-1)$. Because $^-(^-1) = 1$, we have justification for the fact that $(^-1)(^-1) = 1$. Using this result, the preceding property, and the properties of integers listed earlier, we show that for all integers a and b, $(^-a)b = {}^-(ab)$ and $(^-a)(^-b) = ab$ as follows:

$$(^-a)b = [(^-1)a]b$$
$$= (^-1)(ab)$$
$$= {}^-(ab)$$

Also:

$$(^-a)(^-b) = [(^-1)a][(^-1)b]$$
$$= [(^-1)(^-1)](ab)$$
$$= 1(ab)$$
$$= ab$$

We have established the following theorem:

> **Theorem 5-7**
>
> For all integers a and b,
>
>   **a.** $(^-a)b = {}^-(ab)$
>
>   **b.** $(^-a)(^-b) = ab$

The distributive property of multiplication over subtraction follows from the distributive property of multiplication over addition:

$$a(b - c) = a(b + {}^-c)$$
$$= ab + a(^-c)$$
$$= ab + {}^-(ac)$$
$$= ab - ac$$

Consequently, $a(b - c) = ab - ac$. From this and the commutative property of multiplication, it follows that $(b - c)a = ba - ca$.

> **Theorem 5-8:   Distributive Property of Multiplication over Subtraction for Integers**
>
> For all integers a, b, and c,
>
>   **a.** $a(b - c) = ab - ac$
>
>   **b.** $(b - c)a = ba - ca$

EXAMPLE 5-10

Simplify each of the following so that there are no parentheses in the final answer:

  a. $(^-3)(x - 2)$ **b.** $(a + b)(a - b)$

Solution

  a. $(^-3)(x - 2) = (^-3)x - (^-3)(2) = {}^-3x - (^-6) = {}^-3x + {}^-(^-6) = {}^-3x + 6$

  b. $(a + b)(a - b) = (a + b)a - (a + b)b$
$$= (a^2 + ba) - (ab + b^2)$$
$$= a^2 + ba + {}^-(ab + b^2)$$
$$= a^2 + ab + {}^-(ab) + {}^-b^2 \quad \text{(Note: } {}^-b^2 \text{ means } {}^-(b^2).)$$
$$= a^2 + 0 + {}^-b^2$$
$$= a^2 - b^2$$

  Thus, $(a + b)(a - b) = a^2 - b^2$.

The result $(a + b)(a - b) = a^2 - b^2$ in Example 5-10(b) is the **difference-of-squares** formula.

Historical Note

The development of the set of integers was over time and no single person is given credit for that development. However, George Cantor (1845–1918) built on known sets of numbers including the integers when he invented the cardinal numbers of infinite sets. Many opposed this creation. One German critic, Leopold Kronecker (1823–1891) wrote that "God made the integers and all the rest is the work of man." Ironically after Cantor's death, a leading mathematician, David Hilbert (1862–1943) wrote, "No one shall expel us from the paradise which Cantor has created for us."

EXAMPLE 5-11

Use the difference-of-squares formula to simplify the following:

a. $(4 + b)(4 - b)$ **b.** $(^-4 + b)(^-4 - b)$ **c.** $(x + 3)^2 - (x - 3)^2$

Solution

a. $(4 + b)(4 - b) = 4^2 - b^2 = 16 - b^2$
b. $(^-4 + b)(^-4 - b) = (^-4)^2 - b^2 = 16 - b^2$
c. $(x + 3)^2 - (x - 3)^2 = [(x + 3) + (x - 3)][(x + 3) - (x - 3)]$
$$= 2x(x + 3 - x + 3)$$
$$= 2x(6)$$
$$= 12x$$

NOW TRY THIS 5-6

Determine how to use the difference-of-squares formula to compute the following mentally:

a. $101 \cdot 99$ **b.** $22 \cdot 18$ **c.** $24 \cdot 36$ **d.** $998 \cdot 1002$

When the distributive property of multiplication over subtraction is written in reverse order as

$$ab - ac = a(b - c) \quad \text{and} \quad ba - ca = (b - c)a$$

and similarly for addition, the expressions on the right of each equation are in *factored* form. We say that the common factor a has been *factored out*. Both the difference-of-squares formula and the distributive properties of multiplication over addition and subtraction can be used for factoring.

EXAMPLE 5-12

Factor each of the following completely:

a. $x^2 - 9$ **b.** $(x + y)^2 - z^2$ **c.** $^-3x + 5xy$ **d.** $3x - 6$ **e.** $5x^2 - 2x^2$

Solution

a. $x^2 - 9 = x^2 - 3^2 = (x + 3)(x - 3)$
b. $(x + y)^2 - z^2 = (x + y + z)(x + y - z)$
c. $^-3x + 5xy = ^-3x + 5yx = (^-3 + 5y)x$
d. $3x - 6 = 3x - 3 \cdot 2 = 3(x - 2)$
e. $5x^2 - 2x^2 = (5 - 2)x^2 = 3x^2$

Integer Division

In the set of whole numbers, where $b \neq 0$, $a \div b$ is the unique whole number c such that $a = bc$. If such a whole number c does not exist, then $a \div b$ is undefined. Division on the set of integers is defined analogously with division by 0 undefined.

Definition of Integer Division

For all integers a and b, $a \div b$ is the unique integer c, if it exists, such that $a = bc$.

Notice that $a \div b$, if it exists, is the solution of $a = bx$. If $a \div b$ is an integer, then we can extend the concept of divisibility introduced in Chapter 4 as follows:

Definition of Divisibility for Integers

If $a \div b$ is a unique integer, then a is **divisible** by b, or equivalently b **divides** a written as $b \,|\, a$.

EXAMPLE 5-13

Use the definition of integer division, if possible, to evaluate each of the following:

a. $12 \div (^-4)$ **b.** $^-12 \div 4$ **c.** $^-12 \div (^-4)$
d. $^-12 \div 5$ **e.** $(ab) \div b, b \neq 0$ **f.** $(ab) \div a, a \neq 0$

Solution

a. Let $12 \div (^-4) = c$. Then $12 = ^-4c$ and consequently $c = ^-3$. Thus, $12 \div (^-4) = ^-3$.
b. Let $^-12 \div 4 = c$. Then $^-12 = 4c$ and therefore $c = ^-3$. Thus, $^-12 \div 4 = ^-3$.
c. Let $^-12 \div (^-4) = c$. Then $^-12 = ^-4c$ and consequently $c = 3$. Thus, $^-12 \div (^-4) = 3$.
d. Let $^-12 \div 5 = c$. Then $^-12 = 5c$. Because no integer c exists to satisfy this equation (why?), we say that $^-12 \div 5$ is undefined over the set of integers.
e. Let $(ab) \div b = x$. Then $ab = bx$ and consequently $x = a$.
f. Let $(ab) \div a = x$. Then $ab = ax$ and hence $x = b$.

Example 5-13 suggests that *the quotient of two negative integers, if it exists, is a positive integer and the quotient of a positive and a negative integer, if it exists, or of a negative and a positive integer, if it exists, is negative.*

 NOW TRY THIS 5-7

Division by 0 in the set of whole numbers is undefined. Use the definition of division for integers to show that dividing by 0 is not possible.

Algorithm for Division Extended

The division process for whole numbers can be extended to work with integers as seen in the following:

$$
\begin{array}{r}
752 \\
\hline
2 \\
50 \\
^-200 \\
900 \\
17\overline{)12985} \\
-15300 \\
\hline
^-2315 \\
-(^-3400) \\
\hline
1085 \\
-1050 \\
\hline
35 \\
-34 \\
\hline
1
\end{array}
$$

Though not efficient as a "normal" division, the end result that $12{,}985 = 17 \cdot 752 + 1$ still holds with 752 as the quotient and 1 as the remainder.

Furthermore, the Euclidean Algorithm approach from Chapter 4 for finding the greatest common divisor of two integers can be extended as seen in the Now Try This 5-8.

 NOW TRY THIS 5-8

Show that $\text{GCD}(206, 23) = \text{GCD}(206 - 9 \cdot 23, 23) = \text{GCD}(^-1, 23) = 1$.

Order of Operations on Integers

When addition, subtraction, multiplication, division, and exponentiation appear without parentheses, exponentiation is done first in order from right to left, then multiplications and divisions in the order of their appearance from left to right, and then additions and subtractions in the order of their appearance from left to right. Arithmetic operations that appear inside parentheses must be done first.

EXAMPLE 5-14

Evaluate each of the following:

a. $2 - 5 \cdot 4 + 1$ b. $(2 - 5)4 + 1$ c. $2 - 3 \cdot 4 + 5 \cdot 2 - 1 + 5$

d. $2 + 16 \div 4 \cdot 2 + 8$ e. $(^-3)^4$ f. $^-3^4$

Solution

a. $2 - 5 \cdot 4 + 1 = 2 - 20 + 1 = {}^-18 + 1 = {}^-17$

b. $(2 - 5)4 + 1 = (^-3)4 + 1 = {}^-12 + 1 = {}^-11$

c. $2 - 3 \cdot 4 + 5 \cdot 2 - 1 + 5 = 2 - 12 + 10 - 1 + 5 = 4$

d. $2 + 16 \div 4 \cdot 2 + 8 = 2 + 4 \cdot 2 + 8 = 2 + 8 + 8 = 10 + 8 = 18$

e. $(^-3)^4 = (^-3)(^-3)(^-3)(^-3) = 81$

f. $^-3^4 = {}^-(3^4) = {}^-(81) = {}^-81$

Notice that from Example 5-14(e) and (f), we have $(^-3)^4 \neq {}^-3^4$. By convention, $(^-x)^4$ means $(^-x)(^-x)(^-x)(^-x)$ and $^-x^4$ means $^-(x^4)$ and equals $^-(x \cdot x \cdot x \cdot x)$.

With the meaning of exponentiation given, the properties of exponents given in Chapter 3 hold for all integers when their exponents are whole numbers. Negative integer exponents will be explored in Chapter 6.

Ordering Integers

As with whole numbers, a number line as shown in Figure 5-14 can be used to describe greater than and less than relations for the set of integers. Because $^-5$ is to the left of $^-3$ on the number line, we say that "$^-5$ is less than $^-3$," and we write $^-5 < {}^-3$. We can also say that "$^-3$ is greater than $^-5$," and we can write $^-3 > {}^-5$.

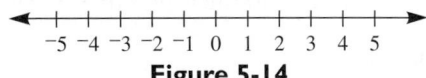

$$^-5 \quad ^-4 \quad ^-3 \quad ^-2 \quad ^-1 \quad 0 \quad 1 \quad 2 \quad 3 \quad 4 \quad 5$$

Figure 5-14

Notice that since $^-5$ is to the left of $^-3$, there is a positive integer that can be added to $^-5$ to get $^-3$, namely, 2. Thus, $^-5 < {}^-3$ because $^-5 + 2 = {}^-3$. The definition of *less than* for integers is similar to that for whole numbers.

Definition of Less Than for Integers

For all integers a and b, a is less than b, written $a < b$, if, and only if, there exists a positive integer k such that $a + k = b$.

The last equation implies that $k = b - a$. Thus we have proved the following theorem:

Theorem 5-9

For integers a and b, $a < b$ (or equivalently, $b > a$) if, and only if, $b - a$ is equal to a positive integer; that is, $b - a$ is greater than 0.

Using this theorem, $^-5 < ^-3$ because $^-3 - (^-5) = ^-3 + ^-(^-5) = ^-3 + 5 = 2 > 0$. The preceding theorem can be used to justify each of the following.

Theorem 5-10

let x, y, n be any integers.

   a. If $x < y$ and n is any integer, then $x + n < y + n$.
   b. If $x < y$, then $^-x > ^-y$.
   c. If $x < y$ and $n > 0$, then $nx < ny$.
   d. If $x < y$ and $n < 0$, then $nx > ny$.

The justifications are given next.

   a. To show that $x + n < y + n$, we need to show that $(y + n) - (x + n) > 0$. We have $y + n - (x + n) = y + n - x - n = y - x$. Because $x < y$, $y - x > 0$, and we have $y + n - (x + n) > 0$; hence $x + n < y + n$.

   b. To show that $^-x > ^-y$, we need to show that $^-x - (^-y) > 0$. We have $^-x - (^-y) = ^-x + ^-(^-y) = ^-x + y = y + ^-x = y - x$. Because $x < y$, $y - x > 0$, we have $^-x - (^-y) > 0$, and hence $^-x > ^-y$.

   c. To show that $nx < ny$, we need to show that $ny - nx > 0$. We have $ny - nx = n(y - x)$. Because n is a positive integer, $x < y$, and $y - x > 0$, $n(y - x)$ must also be positive. Because $ny - nx > 0$, we have $nx < ny$.

   d. To show that $nx > ny$, we show that $nx - ny > 0$. We have $nx - ny = n(x - y)$. Since $y - x > 0$, $x - y < 0$ (why?). Because $n < 0$ and $x - y < 0$, $n(x - y)$ is positive. Thus, $nx - ny > 0$ and hence, $nx > ny$.

EXAMPLE 5-15

Use the theorems developed above to find all integers x that satisfy parts (a) and (b).

   a. $x + 3 < ^-2$
   b. $^-x - 3 < 5$
   c. If $x \leq ^-2$, find the values of $5 - 3x$.

Solution

   a. If $x + 3 < ^-2$, then by Theorem 5–10(a),

$$x + 3 + ^-3 < ^-2 + ^-3$$
$$x < ^-5, \quad x \text{ is an integer.}$$

We can also write the solution set (the set of all solutions) as $\{^-6, ^-7, ^-8, ^-9, \ldots\}$.

    Strictly speaking, we have only shown that every x that satisfies the first inequality also satisfies $x < ^-5$. To be sure that $x < ^-5$ represents all the solutions, we need to show the converse; that is, if $x < ^-5$ then $x + 3 < ^-2$. This can be easily done by adding 3 to both sides of $x < ^-5$.

   b. If $^-x - 3 < 5$, then

$$^-x - 3 + 3 < 5 + 3$$
$$^-x < 8$$
$$^-(^-x) > ^-8 \quad \text{by Theorem 5–10(b)}$$
$$x > ^-8, \quad x \text{ is an integer; that is, all the integers in the set } \{^-7, ^-6, ^-5, \ldots\}.$$

c. If $x \leq {}^-2$, then to find an inequality for $5 - 3x$, we proceed as follows:

$$^-3x \geq {}^-3({}^-2)$$
$$^-3x \geq 6$$
$$5 + {}^-3x \geq 5 + 6$$
$$5 - 3x \geq 5 + 6$$
$$5 - 3x \geq 11; \text{ that is, all integers in the set } \{11, 12, 13, 14, \ldots\}.$$

BRAIN TEASER

Express each of the numbers from 1 through 10 using four 4s and any operations. For example,

$$1 = 44 \div 44, \text{ or}$$
$$1 = (4 \div 4)^{44}, \text{ or}$$
$$1 = {}^-4 + 4 + (4 \div 4)$$

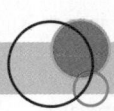

Assessment 5-2A

1. Use patterns to show that $({}^-1)({}^-1) = 1$.
2. Use the charged-field model to show that $({}^-4)({}^-2) = 8$.
3. Use the number-line model to show that $2({}^-4) = {}^-8$.
4. In each of the following charged field models, the encircled charges are removed. Write the corresponding integer multiplication problem with its solution based on the model.

   a. 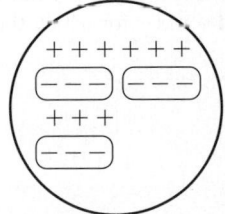   **b.**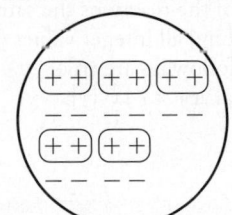

5. The number of students eating in the school cafeteria has been decreasing at the rate of 20 per year for many years. Assuming this trend continues, write a multiplication problem that describes the change in the number of students eating in the school cafeteria for each of the following:
   a. The change over the next 4 years
   b. The situation 4 years ago
   c. The change over the next n years
   d. The situation n years ago
6. Use the definition of division to find each quotient, if possible. If a quotient is not defined, explain why.
   a. $^-40 \div {}^-8$ **b.** $^-143 \div 13$ **c.** $^-5 \div 0$
7. Evaluate each of the following, if possible:
   a. $({}^-10 \div {}^-2)({}^-2)$
   b. $({}^-10 \cdot 5) \div 5$
   c. $^-8 \div ({}^-8 + 8)$
   d. $({}^-6 + 6) \div ({}^-2 + 2)$
   e. $|{}^-24| \div [4(9 - 15)]$
8. Evaluate each of the following products and then, if possible, write two division statements that are equivalent to the given multiplication statement. If two division statements are not possible, explain why.
   a. $({}^-6)5$

   b. $({}^-5)({}^-4)$
   c. $({}^-3)0$
9. In each of the following, x and y are integers; $y \neq 0$. Use the definition of division in terms of multiplication to perform the indicated operations. Write your answers in simplest form.
   a. $(4x) \div 4$
   b. $(xy) \div y$
10. In a lab, the temperature of various chemical reactions is changing by a fixed number of degrees per minute. Write a numeric expression that describes each of the following if all times are on the same day.
    a. The temperature at 8:00 P.M. is 32°C. If it drops 3°C per minute, what will the temperature be at 8:30 P.M.?
    b. The temperature at 8:20 P.M. is 0°C. If it has dropped 4°C per minute, what was the temperature at 7:55 P.M.?
    c. The temperature at 8:00 P.M. is $^-$20°C. If it has dropped 4°C per minute, what was the temperature at 7:30 P.M.?
    d. The temperature at 8:00 P.M. is 25°C. If it has been increasing every minute by 3°C, what was the temperature at 7:40 P.M.?
11. If it was predicted that the farmland acreage lost to family dwellings over the next 9 years would be 12,000 acres per year, how much acreage would be lost to homes during this time period?
12. Show that the distributive property of multiplication over addition, $a(b + c) = ab + ac$, is true for each of the following values of a, b, and c:
    a. $a = {}^-1, b = {}^-5, c = {}^-2$ **b.** $a = {}^-3, b = {}^-3, c = 2$
13. Compute each of the following:
    a. $({}^-2)^3$ **b.** $({}^-2)^4$
    c. $({}^-10)^5 \div ({}^-10)^2$ **d.** $({}^-3)^5 \div ({}^-3)$
    e. $({}^-1)^{50}$ **f.** $({}^-1)^{151}$
    g. $^-2 + 3 \cdot 5 - 1$ **h.** $10 - 3 \cdot 7 - 4({}^-2) + 3$
14. Compute the following without using a calculator:
    a. $({}^-2)^{64} - 2^{64}$ **b.** $^-2^8 + 2^8$
    c. $^-({}^-2)^5 + 0 \cdot 9 - |7 - 15| - 15$

15. If x is an integer and $x \neq 0$, which of the following are always positive and which are always negative?
    a. $^-x^2$ b. x^2 c. $(^-x)^2$
    d. $^-x^3$ e. $(^-x)^3$

16. Which of the expressions in exercise 15 are equal to each other for all values of x?

17. Identify the property of integers being illustrated in each of the following:
    a. $(^-3)(4 + 5) = (4 + 5)(^-3)$
    b. $(^-4)(^-7) \in I$
    c. $5[4(^-3)] = (5 \cdot 4)(^-3)$
    d. $(^-9)[5 + (^-8)] = (^-9)5 + (^-9)(^-8)$

18. Simplify each of the following:
    a. $(^-x)(^-y)$ b. $^-2x(^-y)$
    c. $^-2(^-x + y) + x + y$ d. ^-1x

19. Multiply each of the following and combine terms where possible:
    a. $^-2(x - y)$ b. $x(x - y)$
    c. $^-x(x - y)$ d. $^-2(x + y - z)$

20. Find all integers x (if any) that make each of the following true:
    a. $^-3x = 6$ b. $^-3x = ^-6$
    c. $^-2x = 0$ d. $5x = ^-30$
    e. $x \div 3 = ^-12$ f. $x \div (^-3) = ^-2$
    g. $x \div (^-x) = ^-1$ h. $0 \div x = 0$

21. Solve each of the following for x:
    a. $^-3x - 8 = 7$ b. $^-2(5x - 3) = 26$
    c. $3x - x - 2x = 3$ d. $^-2(5x - 6) - 30 = ^-x$
    e. $x^2 = 4$ f. $(x - 1)^2 = 9$
    g. $(x - 1)^2 = (x + 3)^2$ h. $(x - 1)(x + 3) = 0$

22. Use the difference-of-squares formula to simplify each of the following, if possible:
    a. $52 \cdot 48$
    b. $(5 - 100)(5 + 100)$
    c. $(^-x - y)(^-x + y)$

23. Factor each of the following expressions completely.
    a. $3x + 5x$ b. $xy + x$
    c. $x^2 + xy$ d. $3xy + 2x - xz$
    e. $abc + ab - a$ f. $16 - a^2$
    g. $4x^2 - 25y^2$

24. a. Use the distributive property of multiplication over addition or over subtraction and other properties to show that
    $$(a - b)^2 = a^2 - 2ab + b^2$$
    b. Use your results from (a) to compute each of the following mentally:
       i. 98^2 (*Hint:* Write $98 = 100 - 2$.)
       ii. 99^2
       iii. 997^2

25. In each of the following, find the next two terms. Assume the sequence is arithmetic or geometric, and find its common difference or ratio and the nth term.
    a. $^-10, ^-7, ^-4, ^-1, 2, 5, \_, \_$
    b. $^-2, ^-4, ^-8, ^-16, ^-32, ^-64, \_, \_$
    c. $2, ^-2^2, 2^3, ^-2^4, 2^5, ^-2^6, \_, \_$

26. Find the first five terms of the sequence whose nth term is given.
    a. $n^2 - 10$ b. $^-5n + 3$ c. $(^-2)^n - 1$

27. Tira noticed that every 30 sec, the temperature of a chemical reaction in her lab was decreasing by the same number of degrees. Initially, the temperature was $28°C$ and 5 min later, $^-12°C$. In a second experiment, Tira noticed that the temperature of the chemical reaction was initially $^-57°C$ and was decreasing by $3°C$ every minute. If she started the two experiments at the same time, when were the temperatures of the reactions the same? What was that temperature?

28. Find all integer values (if any) of x and y for which the following are true:
    a. $xy = ^-|x||y|$ b. $^-x^2 = x^2$

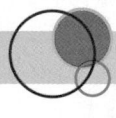

Assessment 5-2B

1. Use patterns to show that $(^-2)(^-2) = 4$.
2. Use the charged-field model to show that $(^-2)(^-3) = 6$.
3. Use the number-line model to show that $2(^-3) = ^-6$.
4. In each of the following charged-field models, the encircled charges are removed. Write the corresponding integer multiplication problem with its solution based on the model.

a.
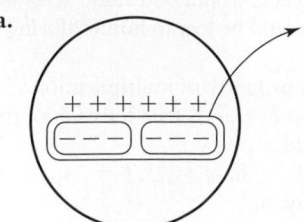

b.
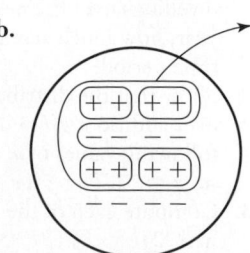

5. The number of students taking sophomore mathematics at Union College has been decreasing at the rate of 20 per year for many years. Assuming this trend continues, write a multiplication problem that describes the change in the number of sophomores taking mathematics at Union for each of the following:
    a. The change over the next 5 years
    b. The situation 5 years ago
    c. The change over the next n years
    d. The situation n years ago

6. Use the definition of division to find each quotient, if possible. If a quotient is not defined, explain why.
    a. $143 \div (^-11)$ b. $0 \div (^-5)$ c. $0 \div 0$

7. Evaluate each of the following, if possible:
    a. $(a \div b)b$
    b. $(ab) \div b$
    c. $(^-8 + 8) \div 8$
    d. $(^-23 - ^-7) \div 4$
    e. $|^-28| \div (2|^-7|)$

8. Evaluate each of the following products and then, if possible, write two division statements that are equivalent to the given multiplication statement. If two division statements are not possible, explain why.
    a. $(^-5)4$ b. $(^-4)(^-3)$ c. $0 \cdot 0$

9. In each of the following, x and y are integers. Use the definition of division in terms of multiplication to perform the indicated operations. Write your answers in simplest form.
   a. $(^-4x) \div x$
   b. $(^-10x + 5) \div 5$

10. In a lab, the temperature of various chemical reactions is changing by a fixed number of degrees per minute. Write a numeric expression that describes each of the following:
   a. The temperature at 8:00 A.M. is $^-5°$C. If it increases by d degrees per minute, what will the temperature be m minutes later?
   b. The temperature at 8:00 P.M. is $0°$C. If it has been dropping by d degrees per minute, what was the temperature m minutes before?
   c. The temperature at 8:00 P.M. is $20°$C. If it has been increasing every minute by d degrees, what was the temperature m minutes before?

11. a. On each of four consecutive plays in a football game, a team lost 11 yd. If lost yardage is interpreted as a negative integer, write the information as a product of integers and determine the total number of yards lost.
   b. If Jack Jones lost a total of 66 yd in 11 plays, how many yards, on the average, did he lose on each play?

12. Show that the distributive property of multiplication over addition, $a(b + c) = ab + ac$, is true for each of the following values of a, b, and c:
   a. $a = ^-5, b = 2, c = ^-6$
   b. $a = ^-2, b = ^-3, c = 4$

13. Compute each of the following:
   a. $10 - 3 - 12$ b. $10 - (3 - 12)$
   c. $(^-3)^2$ d. $^-3^2$
   e. $^-5^2 + 3(^-2)^2$ f. $^-2^3$
   g. $(^-2)^5$ h. $^-2^4$

14. Compute the following without using a calculator:
   a. $^-2^{63} + 2^{64}$
   b. $^-|^-6| - 8^2 + (^-1)^{49} \cdot 48 \div (^-4) \cdot 3 + (^-5)^3$

15. If x is an integer and $x \neq 0$, which of the following are always positive and which are always negative?
   a. $^-x^4$ b. $(^-x)^4$
   c. x^4 d. x
   e. ^-x

16. Which of the expressions in exercise 15 are equal to each other for all values of x?

17. Identify the property of integers being illustrated in each of the following:
   a. $(^-2)(3) \in I$
   b. $(^-4)0 = 0$
   c. $^-2(3 + 4) = ^-2(3) + (^-2)4$
   d. $(^-2)3 = 3(^-2)$

18. Simplify each of the following:
   a. $x - 2(^-y)$ b. $a - (a - b)(^-1)$
   c. $y - (y - x)(^-2)$ d. $^-1(x - y) + x$

19. Multiply each of the following and combine terms where possible:
   a. $^-x(x - y - 3)$
   b. $(^-5 - x)(5 + x)$
   c. $(x - y - 1)(x + y + 1)$

20. Find all integers x (if any) that make each of the following true:
   a. $x \div 0 = 1$
   b. $x^2 = 9$
   c. $x^2 = ^-9$
   d. $^-x \div ^-x = 1$
   e. $^-x^2$ is negative.
   f. $^-(1 - x) = x - 1$
   g. $x - 3x = ^-2x$
   h. $^-3(x + 2) = ^-3x + 6$

21. Solve the following for x or find the values of the indicated expression:
   a. $(2x - 1)^2 = (1 - 2x)^2$
   b. $x^3 = ^-2^9$
   c. $^-6x > ^-x + 20$
   d. $^-5(x - 3) > ^-5$
   e. If $x > ^-2$, find the values of $3 - 5x$.
   f. If $x < 0$, find the values of $2 - 7x$.

22. Use the difference-of-squares formula to simplify or compute each of the following, if possible:
   a. $(2 + 3x)(2 - 3x)$
   b. $(x - 1)(1 + x)$
   c. $213^2 - 13^2$

23. Factor each of the following expressions completely and then simplify, if possible:
   a. $ax + 2x$
   b. $ax - 2x$
   c. $3x - 4x + 7x$
   d. $3x^2 + xy - x$
   e. $(a + b)(c + 1) - (a + b)$
   f. $x^2 - 9y^2$
   g. $(x^2 - y^2) + x + y$

24. a. Show that $(a - 1)^2 = a^2 - 2a + 1$.
   b. Use the result of part (a) to find the following:
      i. 99^2
      ii. 199^2

25. In each of the following, find the next two terms. Assume the sequence is arithmetic or geometric, and find its common difference or ratio and the nth term.
   a. $10, 7, 4, 1, ^-2, ^-5, \_, \_$
   b. $^-2, 4, ^-8, 16, ^-32, 64, \_, \_$

26. Find the first five terms of the sequence whose nth term is given.
   a. $(^-2)^n + 2^n$
   b. $n^2(^-1)^n$
   c. $|10 - n^2|$

27. Jon has two checking accounts. In the first one, he is $120 overdrawn, and in the second, his balance is $300. If he deposits $40 every day in the first account but withdraws $20 daily from the second account, after how many days will the balance in each account be the same? Explain your solution. [Bank fees are ignored.]

28. If x and y are integers, classify each of the following as true or false. If true, explain why. If false, give a counterexample.
   a. $|x + y| = |x| + |y|$
   b. $|xy| = |x||y|$
   c. $|x^2| = x^2$
   d. $|x|^2 = x^2$

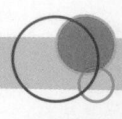

Mathematical Connections 5-2

Communication

1. Can $(^-x - y)(x + y)$ be multiplied by using the difference-of-squares formula? Explain why or why not.
2. Kahlil said that using the equation $(a + b)^2 = a^2 + 2ab + b^2$, he can find a similar equation for $(a - b)^2$. Examine his argument. If it is correct, supply any missing steps or justifications; if it is incorrect, point out why.

$$(a - b)^2 = [a + (^-b)]^2$$
$$= a^2 + 2a(^-b) + (^-b)^2$$
$$= a^2 - 2ab + b^2$$

3. Nancy gave the following argument to show that $(^-a)b = ^-(ab)$, for all integers a and b: *I know that* $(^-1)a = ^-a$ and a could be a variable to be replaced by ab so that $(^-1)(ab) = ^-(ab)$ and $(^-1)(ab) = (^-1 \cdot a)b$ by the associative property. Also $(^-1 \cdot a)b = ^-(a)b$. Therefore, $(^-a)b = ^-(ab)$. If the argument is valid, give reasons for each step; if it is not valid, explain why.
4. Hosni gave the following argument that $^-(a + b) = ^-a + ^-b$, for all integers a and b. If the argument is correct, supply the missing reasons. If it is incorrect, explain why.

$$^-(a + b) = (^-1)(a + b)$$
$$= (^-1)a + (^-1)b$$
$$= ^-a + ^-b$$

5. The Swiss mathematician Leonhard Euler (1707–1783) argued that $(^-1)(^-1) = 1$ as follows: "The result must be either $^-1$ or 1. If it is $^-1$, then $(^-1)(^-1) = ^-1$. Because $^-1 = (^-1)1$, we have $(^-1)(^-1) = (^-1)1$. Now dividing both sides of the last equation by $^-1$ we get $^-1 = 1$, which of course cannot be true. Hence $(^-1)(^-1)$ must be equal to 1."
   a. What is your reaction to this argument? Is it logical? Why or why not?
   b. Can Euler's approach be used to justify other properties of integers? Explain.
6. Jill asks each of her classmates to choose a number, then multiply the number by $^-3$, add 2 to the product, multiply the result by $^-2$, and then subtract 14. Finally, each student is asked to divide the result by 6 and record the answer. When Jill gets an answer from a classmate, she just adds 3 to it in her head and announces the number that classmate originally chose. How did Jill know to add 3 to each answer?

Open-Ended

7. On a national mathematics competition, scoring is accomplished using the formula 4 times the number answered correctly minus the number answered incorrectly. In this scheme, problems left blank are considered neither correct nor incorrect. Devise a scenario that would allow a student to have a negative score.

8. Select a current middle-school text that introduces multiplication and division of integers and discuss any models that were used and how effective you think they would be with a group of students.
9. Choose two integers and illustrate the algorithm for division on page 248. Be careful to keep up with positive and negative signs in the computation.

Cooperative Learning

10. Devise a scheme for determining a grade-point average for a college student that allows negative quality points for a failing grade.
    a. Use your scheme to determine possible grades for students with positive, zero, and negative grade-point averages.
    b. Compare your scheme with that of another class group and write a rationale for the best scheme.
11. a. How would you introduce multiplication of integers in a middle-school class and how would you explain that a product of two negative numbers is positive? Write a rationale for your approach.
    b. Present your answers and compare them to those of another class group and together decide the most appropriate way to introduce the concepts.
12. In your group, discuss each person's favorite approach to justify $(^-1)(^-1) = 1$.

Questions from the Classroom

13. A seventh-grade student does not believe that $^-5 < ^-2$. The student argues that a debt of $5 is greater than a debt of $2. How do you respond?
14. A student computes $^-8 - 2(^-3)$ by writing $^-10(^-3) = 30$. How would you help this student?
15. A student says that his father showed him a very simple method for dealing with expressions like $^-(a - b + 1)$ and $x - (2x - 3)$. The rule is, if there is a negative sign before the parentheses, change the signs of the expressions inside the parentheses. Thus, $^-(a - b + 1) = ^-a + b - 1$ and $x - (2x - 3) = x - 2x + 3$. What is your response?
16. Betty used the charged field model to show that $^-2(^-3) = 6$. She said that this proves that any negative integer times a negative integer is a positive integer. How do you respond?

Review Problems

17. Illustrate $^-8 + ^-5$ on a number line.
18. Find the additive inverse of each of the following:
    a. $^-5$ b. 7   c. 0
19. Compute each of the following:
    a. $|^-14|$ b. $|^-14| + 7$
    c. $8 - |^-12|$ d. $|11| + |^-11|$

Hint for Preliminary Problem

There are unknown numbers of barrels that are 5 or 6 lbs over- or underweight. There are no more than 20 barrels. And the weight is known to be incorrect by 53 lb. We do not know whether the 53 lb is overweight or underweight. Thus, we are looking for all combinations of 6 and 5, ⁻6 and 5, 6 and ⁻5, or ⁻6 and ⁻5 weights that add to 53 or ⁻53. And the total number of combinations in each case has to be less than 20. One way to approach this problem is to make a table and consider the possibilities.

Chapter 5 Summary

5-1 Integers and the Operations of Addition and Subtraction	Pages
Basic Concepts of Integers	
• The set of **integers** is $I = \{\ldots, {}^-3, {}^-2, {}^-1, 0, 1, 2, 3, \ldots\}$.	223
• The set of **negative integers** is $\{{}^-1, {}^-2, {}^-3, \ldots\}$.	222
• The set of **positive integers** is $\{1, 2, 3, 4, \ldots\}$.	222
• Negative integers arc **opposites** of positive integers.	223
Models for Integer Addition	
• Chip model	224
• Charged-field model	224
• Number-line model	224–226
• Pattern model	226
Absolute Value	
• Dcfinition: For any integer x, $|x| = x$ if $x \geq 0$, and $|x| = {}^-x$ if $x \leq 0$.	227
• The **absolute value** of a number is its distance from 0.	227
Definition of **Integer Addition**: Let m and n be integers.	228
• $0 + m = m = m + 0$.	
• If $m \geq 0$ and $n \geq 0$, then $m + n$ is defined using addition of whole numbers.	
• If $m \geq 0$ and $n \geq 0$, then ${}^-m + {}^-n = {}^-(m + n)$.	
• If $n > 0$ and $m \geq n$, then $m + {}^-n = m - n$.	
• If $n > m$ and $m > 0$, then $m + {}^-n = {}^-(n - m)$.	
Alternative Definition of **Integer Addition**	229
• If m is any integer, then $m + 0 = m = 0 + m$.	
• If m and n are positive integers, then $m + n$ is the sum of whole numbers m and n.	
• If m and n are negative integers, then $m + n = {}^-(|m| + |n|)$.	
• If m is a positive integer and n is a negative integer, then $m + n$ is defined as follows:	
o If $|m| > |n|$, then $m + n = |m| - |n|$.	
o If $|m| < |n|$, then $m + n = {}^-(|m| - |n|)$.	
Properties of Integer Addition	
• **Theorem: (Closure property of addition of integers):** For all integers, a and b, $a + b$ is a unique integer.	230
• **Theorem: (Commutative property of addition of integers):** For all integers a and b, $a + b = b + a$.	230
• **Theorem: (Associative property of addition of integers):** For all integers, $a, b,$ and c, $(a + b) + c = a + (b + c)$.	230
• **Theorem: (Identity property of addition of integers):** 0 is the unique integer, the additive identity, such that for all integers a, $0 + a = a = a + 0$.	230

5-2 Multiplication and Division of Integers

Theorems about Integers

• **Theorem:** For every integer a, $(^-1)(a) = {}^-a$.	245
• **Theorem:** For all integers a and b, $(^-a)b = {}^-(ab)$.	246
• **Theorem:** For all integers a and b, $(^-a)(^-b) = ab$.	246
• **Theorem:** Difference-of-squares: $a^2 - b^2 = (a + b)(a - b)$	247

Definition of **Integer Division:** If a and b are any integers, then $a \div b$ is the unique integer c, if it exists, such that $a = bc$.   247

Definition of **Divisibility for Integers**: If $a \div b$ is unique integer, then a is **divisible by** b, or equivalently b **divides** a written as $b \mid a$.   247

Order of Operations   237, 249
  • When addition, subtraction, multiplication, division, and exponentiation appear without parentheses, exponentiation is done first in order from right to left, and then multiplications and divisions in the order of their appearance from left to right, and then additions and subtractions in order of their appearance from left to right.
  • Arithmetic operations inside parentheses must be done first.   249

Definition of **Less Than** for Integers: For any integers a and b, a is less than b, written $a < b$, if, and only if, there exists a positive integer k such that $a + k = b$.   249

Theorems about Integers and Inequalities

• **Theorem:** For integers, a and b, $a < b$ (or equivalently, $b > a$) if, and only if, $b - a$ is equal to a positive integer; that is, $b - a > 0$.	249
• **Theorem:** For integers, x and y, if $x < y$ and n is any integer, then $x + n < y + n$.	250
• **Theorem:** For integers, x and y, if $x < y$, then $^-x > {}^-y$.	250
• **Theorem:** For integers, x, y, and n, if $x < y$ and $n > 0$, then $nx < ny$.	250
• **Theorem:** For integers, x, y and n, if $x < y$ and $n < 0$, then $nx > ny$.	250

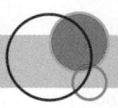

Chapter 5 Review

1. Find the additive inverse of each of the following:
   a. 3
   b. ^-a
   c. $^-2 + 3$
   d. $x + y$
   e. $^-x + y$
   f. $^-x - y$
   g. $(^-2)^5$
   h. $^-2^5$

2. Perform each of the following operations:
   a. $(^-2 + {}^-8) + 3$
   b. $^-2 - (^-5) + 5$
   c. $^-3(^-2) + 2$
   d. $^-3(^-5 + 5)$
   e. $^-40 \div (^-5)$
   f. $(^-25 \div 5)(^-3)$

3. For each of the following, find all integer values of x (if there are any) that make the given equation true:
   a. $^-x + 3 = 0$
   b. $^-2x = 10$
   c. $0 \div (^-x) = 0$
   d. $^-x \div 0 = {}^-1$
   e. $3x - 1 = {}^-124$
   f. $^-2x + 3x = x$

4. Use a pattern approach to explain why $(^-2)(^-3) = 6$.

5. In each of the following chip models, the encircled chips are removed. Write the corresponding integer problem with its solution.

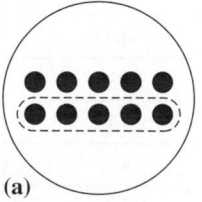
(a)

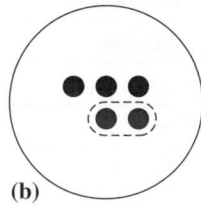
(b)

6. Simplify each of the following expressions:
   a. ^-1x
   b. $(^-1)(x - y)$
   c. $2x - (1 - x)$
   d. $(^-x)^2 + x^2$
   e. $(^-x)^3 + x^3$
   f. $(^-3 - x)(3 + x)$
   g. $(^-2 - x)(^-2 + x)$

7. Factor each of the following expressions and then simplify, if possible:
   a. $x - 3x$
   b. $x^2 + x$
   c. $x^2 - 36$
   d. $81y^4 - 16x^4$
   e. $5 + 5x$
   f. $(x - y)(x + 1) - (x - y)$

8. Classify each of the following as true or false (all letters represent integers). Justify your answers.
   a. $|x|$ always is positive.
   b. For all x and y, $|x + y| = |x| + |y|$.
   c. If $a < {}^-b$, then $a < 0$.
   d. For all x and y, $(x - y)^2 = (y - x)^2$.

9. Find a counterexample to disprove each of the following properties on the set of integers:
   a. Commutative property for division
   b. Associative property for subtraction
   c. Closure for division
   d. Distributive property of division over subtraction

10. Solve each of the following for x, where x is an integer:
   a. $x + 3 = {}^-x - 17$
   b. $2x = {}^-2^{100}$
   c. $2^{10}x = 2^{99}$
   d. ${}^-x = x$
   e. $|{}^-x| = 3$
   f. $|x| = {}^-x$
   g. $|x| > 3$
   h. $(x - 1)^2 = 100$

11. Write the first six terms of each of the sequences whose nth term is
   a. $({}^-1)^n$
   b. $({}^-2)^n$
   c. ${}^-2 - 3n$

12. In each part of exercise 11, if a sequence is arithmetic, find its difference, and if it is geometric, find its ratio.

13. On a college test, students receive 4 points for every question answered correctly and a student receives a penalty of 7 points when a question is answered incorrectly. On this particular test, Terry answered 87 questions correctly and 46 questions incorrectly. What is Terry's score for this test?

14. The following states' lowest temperatures on record are given below. Order the states according to their lowest temperatures from lowest to highest of those given.

State	Temperature in degree Celsius
Alabama	${}^-33$
Arizona	${}^-40$
California	${}^-43$
Iowa	${}^-44$
Nevada	${}^-46$
Ohio	${}^-39$
Oregon	${}^-48$
West Virginia	${}^-38$

15. In 1971, the lowest recorded temperature, ${}^-62°C$, in the United States was in Prospect Creek Camp, Alaska. That same year, Arizona recorded a low temperature of ${}^-40°C$ at Hawley Lake. Alaska's temperature was recorded at an elevation of 1100 ft above sea level while Arizona's was recorded at 8180 ft above sea level.
   a. What is the temperature difference from Alaska to Arizona?

   b. What is the elevation difference from Arizona to Alaska where the low temperatures were measured?
   c. Hawaii's lowest temperature on record is ${}^-11°C$. What would you expect to be true about the elevation of the site where this temperature was measured?

16. Two states record their lowest points below sea level, one at ${}^-282$ ft and one at ${}^-8$ ft.
   a. What is the difference in elevation between the lowest points of the two states?
   b. Which two states do you think these are and why?

17. The average depth of the Pacific Ocean is 3963 m. Interpret this number in relation to sea level.

18. In the metric system there are two official temperature scales: degree Celsius and kelvin. The kelvin temperature scale is obtained by shifting the Celsius scale so that 0 kelvin corresponds to absolute zero, the absence of any heat whatsoever. An equation approximating the relationship between the two scales follows:

$$\text{kelvin} \approx \text{degree Celsius} + 273°$$

   a. Write an equation to find degree Celsius in terms of degrees kelvin.
   b. If the boiling point of water in Celsius is 100°, what is the comparable temperature in kelvin?
   c. ${}^-40°$ is the point on a thermometer at which both degree Celsius and degree Fahrenheit are the same. What is this temperature in kelvin?

19. In the Battle of Gettysburg, it was estimated that there were 75,000 Confederate forces engaged and 82,289 Union forces engaged. The number of casualties were estimated at 23,049 for the Union and 28,063 for the Confederates.
   a. How might positive and negative integers be used to describe these numbers?
   b. Using positive and negative numbers, describe the total casualties.

20. The drawing below depicts an elevator. Explain what ${}^-2$ might mean as a floor number.

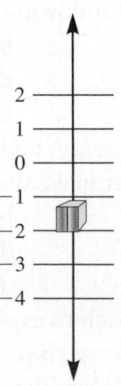

Rational Numbers and Proportional Reasoning

Preliminary Problem

The Doll Store has three times as many female dolls as male dolls. If the store sells 30 of the female dolls and no male dolls, then the ratio of remaining female dolls to male dolls is 1:2. How many male dolls does the store have?

If needed, see Hint on page 324.

Integers such as $^-5$ were invented to solve equations like $x + 5 = 0$. Similarly, a new type of number is needed to solve an equation like $2x = 1$. We need notation for this new number. If multiplication is to work with this new number as with whole numbers, then $2x = x + x$, so $x + x = 1$. In other words, the number created must have the property that when added to itself the result is 1.

The number invented to solve the equation is *one-half*, denoted $\frac{1}{2}$. It is an element of the set of numbers of the form $\frac{a}{b}$, where $b \neq 0$ and a and b are integers. More generally, numbers of the form $\frac{a}{b}$ are solutions to equations of the form $bx = a$. This set, Q, is the set of **rational numbers** and is denoted as follows:

$$Q = \left\{ \frac{a}{b} \mid a \text{ and } b \text{ are integers and } b \neq 0 \right\}$$

Q is a subset of another set of numbers called *fractions*. Fractions are of the form $\frac{a}{b}$, where $b \neq 0$ but a and b are not necessarily integers. For example, $\frac{1}{\sqrt{2}}$ is a fraction but not a rational number.

The fact that $b \neq 0$ is always necessary because division by 0 is undefined.

As indicated in the *PSSM* excerpt below, children may come to understand fractions through concrete activities.

> Beyond understanding whole numbers, young children can be encouraged to understand and represent commonly used fractions in context, such as 1/2 of a cookie or 1/8 of a pizza, and to see fractions as part of a unit whole or of a collection. Teachers should help students develop an understanding of fractions as division of numbers. And in the middle grades, in part as a basis for their work with proportionality, students need to solidify their understanding of fractions as numbers. (p. 33)

Because children frequently think of a fraction as two numbers separated by a line, rather than as a single number, it is important to emphasize that a fraction such as $\frac{3}{4}$ represents a single number.

In elementary school, rational numbers are restricted to numbers in the form $\frac{a}{b}$, where a and b are whole numbers and $b \neq 0$.

6-1 The Set of Rational Numbers

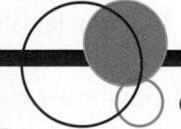

(a) Area model

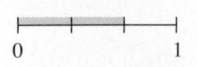

(b) Number-line model

In the rational number $\frac{a}{b}$, a is the **numerator** and b is the **denominator**. The rational number $\frac{a}{b}$ may also be represented as a/b or as $a \div b$. The word *fraction* is derived from the Latin word *fractus*, meaning "broken." The word *numerator* comes from a Latin word meaning "numberer," and *denominator* comes from a Latin word meaning "namer." Table 6-1 shows several uses of rational numbers.

In the grade 3 *Focal Points* we find:

> Students develop an understanding of the meanings and uses of fractions to represent parts of a whole, parts of a set, or points or distances on a number line. They understand that the size of a fractional part is relative to the size of the whole, and they use fractions to represent numbers that are equal to, or less than, or greater than 1. (p. 15)

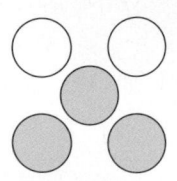

(c) Set model
Figure 6-1

Figure 6-1 illustrates the use of rational numbers as part of a whole, a distance on a number line, and as part of a given set. For example, in the *area model* in Figure 6-1(a), one part out of

Table 6-1

Use	Example
Division problem or solution to a multiplication problem	The solution to $2x = 3$ is $\frac{3}{2}$.
Partition, or part, of a whole	Joe received $\frac{1}{2}$ of Mary's salary each month for alimony.
Ratio	The ratio of Republicans to Democrats on a Senate committee is three to five.
Probability	When you toss a fair coin, the probability of getting heads is $\frac{1}{2}$.

three congruent parts, or $\frac{1}{3}$ of the largest rectangle, is shaded. In Figure 6-1(b), two parts out of three congruent parts, or $\frac{2}{3}$ of the unit segment of a number line, are shaded. In Figure 6-1(c), three circles out of a set of five congruent circles, or $\frac{3}{5}$ of the circles, are shaded.

Early exposure to rational numbers as fractions usually takes the form of description rather than mathematical notation. We hear phrases such as "one-half of a pizza," "one-third of a cake," or "three-fourths of a pie." We encounter such questions as "If three identical fruit bars are distributed equally among four friends, how much does each get?" The answer is that each receives $\frac{3}{4}$ of a bar.

The English words used for denominators of rational numbers are the same words we use to tell "order", for example, the *fourth* person in a line and the glass is three-fourths full. This causes confusion for students learning about fractions. In contrast, in Chinese $\frac{3}{4}$ is read "out of four parts, (take) three." The Chinese model enforces the idea of partitioning quantities into equal parts and choosing some number of these parts. The concept of sharing quantities and comparing sizes of shares provides entry points to introduce students to rational numbers.

When rational numbers are introduced as fractions that represent a part of a whole, we must pay attention to the whole from which a rational number is derived. For example, if we talk about $\frac{3}{4}$ of a pizza, then the amount of pizza is determined both by the fractional part, $\frac{3}{4}$, and the size

---

◯– Historical Note

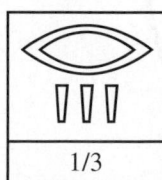

1/3

The early Egyptian numeration system had symbols for fractions with numerators of 1. Most fractions with numerators other than 1 were expressed as a sum of different fractions with numerators of 1, for example, $\frac{7}{12} = \frac{1}{3} + \frac{1}{4}$.

Fractions with denominator 60 or powers of 60 were common in ancient Babylon about 2000 BCE, where 12,35 meant $12 + \frac{35}{60}$. The method was later adopted by the Greek astronomer Ptolemy (approximately 125 CE). The same method was also used in Islamic and European countries and is presently used in the measurements of angles, where $13° 19' 47''$ means $13 + \frac{19}{60} + \frac{47}{60^2}$ degrees.

The modern notation for fractions—a bar between numerator and denominator—is of Hindu origin. It came into general use in Europe in sixteenth-century books. ●

of the pizza. Three-fourths of a large pizza is certainly more than three-fourths of a small pizza. Attention must be paid to the context and the size of the *whole* being considered.

To understand the meaning of any fraction, $\frac{a}{b}$, where $a, b \in W, b \neq 0$, using the parts-to-whole model, we must consider each of the following:

1. The *whole* being considered.
2. The number b of equal-size parts into which the whole has been divided.
3. The number a of parts of the whole that are selected.

A fraction $\frac{a}{b}$, where $0 \leq a < b$, is a **proper fraction**. For example, $\frac{4}{7}$ is a proper fraction, but $\frac{7}{4}, \frac{4}{4}$, and $\frac{9}{7}$ are not; $\frac{7}{4}$ is an **improper fraction**. In general $\frac{a}{b}$ is an improper fraction if $a \geq b > 0$.

Every integer n can be represented as a rational number because $n = \frac{nk}{k}$, where k is any non-zero integer. In particular, $0 = \frac{0 \cdot k}{k} = \frac{0}{k}$.

 NOW TRY THIS 6-1

Draw a Venn diagram to show the relationship among natural numbers, whole numbers, integers, and rational numbers.

The cartoon shows that the concept of a fraction can be confusing for some children because of the unit being considered.

THE FAMILY CIRCUS   **By Bil Keane**

6-2
©2006 Bil Keane, Inc.
Dist. by King Features Synd.
www.familycircus.com

JEFF and Bil KEANE

"How come a quarter is worth
25 cents, and a quarter of an
hour is only 15 minutes?"

Family Circus copyright © 2006 Bil Keane, Inc. Distributed by King Features Syndicate

NOW TRY THIS 6-2

a. How would you answer Billy's question in the cartoon?

b. Jim claims that $\frac{1}{3} > \frac{1}{2}$ because in Figure 6-2 the shaded

portion for $\frac{1}{3}$ is larger than the shaded portion

for $\frac{1}{2}$. How would you help him?

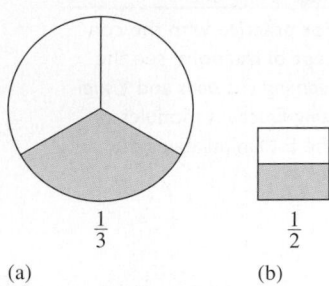

(a) $\frac{1}{3}$ (b) $\frac{1}{2}$

Figure 6-2

Rational Numbers on a Number Line

In the grade 3 *Common Core Standards* we find the following standard:

> Represent a fraction $\frac{a}{b}$ on a number line by marking off a lengths of $\frac{1}{b}$ from 0. Recognize that the resulting
> interval has size $\frac{a}{b}$ and that its endpoint locates the number $\frac{a}{b}$ on the number line. (p. 24)

Once the integers 0 and 1 are assigned to points on a line, the unit segment is defined and every other rational number is assigned to a specific point. For example, to represent $\frac{3}{4}$ on the number line, we divide the segment from 0 to 1 into 4 segments of equal length and mark the line accordingly. Then, starting from 0, we count 3 of these segments and stop at the mark corresponding to the right endpoint of the third segment to obtain the point assigned to the rational number $\frac{3}{4}$. Figure 6-3 shows the points that correspond to $^{-}2$, $\frac{^{-}5}{4}$, $^{-}1$, $\frac{^{-}3}{4}$, 0, $\frac{3}{4}$, 1, $\frac{5}{4}$, and 2.

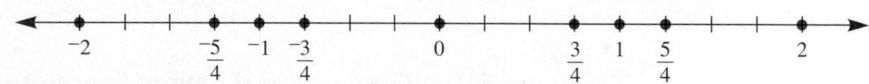

$^{-}2$ $\frac{^{-}5}{4}$ $^{-}1$ $\frac{^{-}3}{4}$ 0 $\frac{3}{4}$ 1 $\frac{5}{4}$ 2

Figure 6-3

NOW TRY THIS 6-3

In Figure 6-3, locate points that correspond to $\frac{1}{2}, \frac{^{-}1}{2}, \frac{3}{2}$, and $\frac{^{-}7}{4}$.

Equivalent or Equal Fractions

The grade 4 *Common Core Standards*, state that students should be able to:

> Explain why a fraction $\frac{a}{b}$ is equivalent to a fraction $\frac{na}{nb}$ by using visual fraction models with attention to how
> the number and the size of the parts differ even though the two fractions themselves are the same size. Use
> the principle to recognize and generate equivalent fractions. (p. 30)

Fractions can be introduced in the classroom through a concrete activity such as paperfolding. In Figure 6-4(a), 1 of 3 congruent parts, or $\frac{1}{3}$, is shaded. In this case, the whole is the rectangle. In Figure 6-4(b), each of the thirds has been folded in half so that now we have 6 sections, and

**E-Manipulatives
Activity**

For practice with the concept of fractions, see the *Naming Fractions* and *Visualizing Fractions* modules on the E-Manipulatives disc.

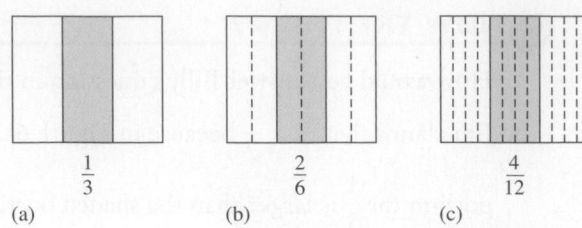

$$\frac{1}{3}$$ $$\frac{2}{6}$$ $$\frac{4}{12}$$

(a)   (b)   (c)

Figure 6-4

2 of 6 congruent parts, or $\frac{2}{6}$, are shaded. Thus, both $\frac{1}{3}$ and $\frac{2}{6}$ represent exactly the same shaded portion. Although the symbols $\frac{1}{3}$ and $\frac{2}{6}$ do not look alike, they represent the same rational number and are **equivalent fractions** or **equal fractions**. Equivalent fractions are numbers that represent the same point on a number line. Because they represent equal amounts, we write $\frac{1}{3} = \frac{2}{6}$ and say that "$\frac{1}{3}$ equals $\frac{2}{6}$."

Figure 6-4(c) shows the rectangle with each of the original thirds folded into 4 equal parts with 4 of the 12 parts now shaded. Thus, $\frac{1}{3}$ is equal to $\frac{4}{12}$ because the same portion of the model is shaded. Similarly, we could illustrate that $\frac{1}{3}, \frac{2}{6}, \frac{3}{9}, \frac{4}{12}, \frac{5}{15}, \ldots$ are equal.

Fraction strips can be used for generating equivalent fractions, as seen on the student page. Part (a) on the student page shows that $\frac{1}{2} = \frac{3}{6} = \frac{6}{12}$. Also on the student page, in Example A we see another way to find equivalent fractions. This technique makes use of the Fundamental Law of Fractions, which can be stated as follows: *The value of a fraction does not change if its numerator and denominator are multiplied by the same nonzero integer.* Under certain assumptions this Law of Fractions can be proved and is stated as a theorem.

Theorem 6-1:   Fundamental Law of Fractions

If $\frac{a}{b}$ is a fraction and n a nonzero rational number, then, $\frac{a}{b} = \frac{an}{bn}$.

In Theorem 6–1 it can be shown that n can be any nonzero number. This version of the Fundamental Law of Fractions is used later in this book. Theorem 6-1 implies that if d is a common factor of a and b then $\frac{a}{b} = \frac{a \div d}{b \div d}$. At this point, some students justify the Fundamental Law of Fractions as follows:

$$\frac{an}{bn} = \frac{a \cdot n}{b \cdot n} = \frac{a}{b} \cdot \frac{n}{n} = \frac{a}{b} \cdot 1 = \frac{a}{b}$$

This approach as seen on the student page on p. 265 is correct, however, we have not yet discussed multiplication of fractions. This justification will only be valid after multiplication of fractions is introduced.

 NOW TRY THIS 6-4

Explain why, in the Fundamental Law of Fractions, n must be nonzero.

School Book Page EQUIVALENT FRACTIONS

Lesson 3-7

Key Idea
A part of a whole or of a set can be named by equivalent fractions.

Vocabulary
• equivalent fractions
• common factor (p. 150)
• least common denominator (LCD)
• greatest common factor (GCF) (p. 150)
• simplest form

Materials
• fraction strips or ⚙ **tools**

Equivalent Fractions

LEARN

 **Activity**

What are equivalent fractions?

Fractions that name the same amount are called **equivalent fractions**.

a. Use fraction strips to identify one or more fractions equivalent to each fraction below.

$\frac{3}{4}$ $\frac{8}{12}$ $\frac{6}{10}$ $\frac{5}{6}$

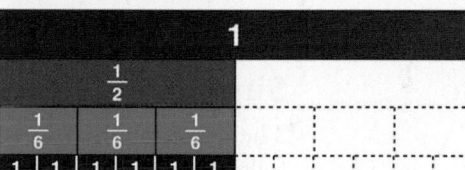

$\frac{1}{2} = \frac{3}{6} = \frac{6}{12}$

b. What two equivalent fractions are shown by this model? Explain.

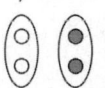

TEST TALK

Think It Through
I can **use objects** to help me identify equivalent fractions.

How can you find equivalent fractions?

Example A

Find two fractions that are equivalent to $\frac{18}{24}$.

One Way
Use multiplication.

Multiply both the numerator and denominator by the same nonzero number.

The number 2 is easy to use, so multiply the numerator and denominator by 2.

$\frac{18}{24} = \frac{18 \times 2}{24 \times 2} = \frac{36}{48}$

Another Way
Use division.

Divide both the numerator and denominator by the same nonzero number.

The number 3 is a common factor, so divide the numerator and denominator by 3.

$\frac{18}{24} = \frac{18 \div 3}{24 \div 3} = \frac{6}{8}$

So, $\frac{18}{24}$, $\frac{36}{48}$, and $\frac{6}{8}$ are all equivalent fractions.

164

From the Fundamental Law of Fractions, $\dfrac{7}{^-15} = \dfrac{^-7}{15}$ because $\dfrac{7}{^-15} = \dfrac{7(^-1)}{^-15(^-1)} = \dfrac{^-7}{15}$.

Similarly, $\dfrac{a}{^-b} = \dfrac{^-a}{b}$. **The form $\dfrac{^-a}{b}$, where b is a positive number, is usually preferred.**

EXAMPLE 6-1

Find a value for x such that $\dfrac{12}{42} = \dfrac{x}{210}$.

Solution   Because $210 \div 42 = 5$, we use the Fundamental Law of Fractions to obtain $\dfrac{12}{42} = \dfrac{12 \cdot 5}{42 \cdot 5} = \dfrac{60}{210}$. Hence, $\dfrac{x}{210} = \dfrac{60}{210}$, and $x = 60$.

Alternative approach: $\dfrac{12}{42} = \dfrac{2 \cdot 6}{7 \cdot 6} = \dfrac{2}{7} = \dfrac{2 \cdot 30}{7 \cdot 30} = \dfrac{60}{210}$. Therefore $x = 60$.

Simplifying Fractions

The Fundamental Law of Fractions justifies the process of **simplifying fractions**. Consider the following:

$$\frac{60}{210} = \frac{6 \cdot 10}{21 \cdot 10} = \frac{6}{21}$$

Also,

$$\frac{6}{21} = \frac{2 \cdot 3}{7 \cdot 3} = \frac{2}{7}$$

We can simplify $\dfrac{60}{210}$ because the numerator and denominator have a common factor of 10.

We can simplify $\dfrac{6}{21}$ because 6 and 21 have a common factor of 3. However, we cannot simplify $\dfrac{2}{7}$ because 2 and 7 have no positive common factor other than 1. We could also simplify $\dfrac{60}{210}$ in one step: $\dfrac{60}{210} = \dfrac{2 \cdot 30}{7 \cdot 30} = \dfrac{2}{7}$. Notice that $\dfrac{2}{7}$ is the **simplest form** of $\dfrac{60}{210}$ because both 60 and 210 have been divided by their greatest common divisor, 30. To write a fraction $\dfrac{a}{b}$ in simplest form, that is, in **lowest terms**, we divide both a and b by GCD(a, b).

> **Definition of Simplest Form**
>
> A rational number $\dfrac{a}{b}$ is in simplest form if, and only if, $b > 0$ and GCD$(a, b) = 1$, that is, if a and b have no common factor greater than 1.

Scientific/fraction calculators can simplify fractions. For example, to simplify $\dfrac{6}{12}$, we enter

$\boxed{6}\,\boxed{/}\,\boxed{1}\,\boxed{2}$ and press $\boxed{\text{SIMP}}\boxed{=}$, and $\dfrac{3}{6}$ appears on the screen. At this point, an indicator tells us that

this is not in simplest form, so we press $\boxed{\text{SIMP}}\boxed{=}$ again to obtain $\dfrac{1}{2}$. At any time, we can view the

factor that was removed by pressing the $\boxed{\text{x}\bigcirc\text{y}}$ key.

The Fundamental Law of Fractions is used to simplify algebraic expressions, as seen in the following example.

EXAMPLE 6-2

Write each of the following fractions in simplest form if they are not already so:

a. $\dfrac{28ab^2}{42a^2b^2}$ b. $\dfrac{(a + b)^2}{3a + 3b}$ c. $\dfrac{x^2 + x}{x + 1}$ d. $\dfrac{3 + x^2}{3x^2}$

e. $\dfrac{3 + 3x^2}{3x^2}$ f. $\dfrac{a^2 - b^2}{a - b}$ g. $\dfrac{a^2 + b^2}{a + b}$

Solution

a. $\dfrac{28\,ab^2}{42a^2b^2} = \dfrac{2(14ab^2)}{3a(14ab^2)} = \dfrac{2}{3a}$

b. $\dfrac{(a + b)^2}{3a + 3b} = \dfrac{(a + b)(a + b)}{3(a + b)} = \dfrac{a + b}{3}$

c. $\dfrac{x^2 + x}{x + 1} = \dfrac{x(x + 1)}{x + 1} = \dfrac{x(x + 1)}{1(x + 1)} = \dfrac{x}{1} = x$

d. $\dfrac{3 + x^2}{3x^2}$ cannot be simplified because $3 + x^2$ and $3x^2$ have no factors in common except 1.

e. $\dfrac{3 + 3x^2}{3x^2} = \dfrac{3(1 + x^2)}{3x^2} = \dfrac{1 + x^2}{x^2}$

f. Recall the difference of squares formula from Chapter 5: $a^2 - b^2 = (a - b)(a + b)$. Thus,

$$\dfrac{a^2 - b^2}{a - b} = \dfrac{(a - b)(a + b)}{(a - b)1} = \dfrac{a + b}{1} = a + b.$$

g. The fraction is already in simplest form because $a^2 + b^2$ does not have $(a + b)$ as a factor. Notice that $a^2 + b^2 \neq (a + b)^2$.

REMARK

When an algebraic expression is written as a fraction, the denominator is not 0. Thus, even after the fraction is simplified, this restriction has to be maintained. For example, in part (c) of Example 6-2, $\dfrac{x^2 + x}{x + 1} = x$ if $x \neq {}^-1$, and in part (f) the result holds if $a - b \neq 0$; that is, if $a \neq b$.

Some students think of the Fundamental Law of Fractions as a *cancellation property* and "simplify" an expression like $\dfrac{6 + a^2}{3a}$ by thinking of it as $\dfrac{2 \cdot 3 + a \cdot a}{3a}$ and "canceling" equal numbers in the products to obtain $2 + a$ as the answer. Emphasizing the factor approach that neither 3 nor a is a factor of $6 + a^2$ may help to avoid such mistakes.

Equality of Fractions

We can use three methods to show that two fractions, such as $\dfrac{12}{42}$ and $\dfrac{10}{35}$, are equal.

1. Simplify both fractions to simplest form:

$$\dfrac{12}{42} = \dfrac{2^2 \cdot 3}{2 \cdot 3 \cdot 7} = \dfrac{2}{7} \quad \text{and} \quad \dfrac{10}{35} = \dfrac{5 \cdot 2}{5 \cdot 7} = \dfrac{2}{7}$$

Thus,

$$\frac{12}{42} = \frac{10}{35}$$

2. Rewrite both fractions with the same least common denominator. Since LCM(42, 35) = 210, then

$$\frac{12}{42} = \frac{60}{210} \quad \text{and} \quad \frac{10}{35} = \frac{60}{210}.$$

Thus,

$$\frac{12}{42} = \frac{10}{35}.$$

3. Rewrite both fractions with a common denominator (not necessarily the least). A common multiple of 42 and 35 may be found by finding the product 42 · 35, or 1470. Now,

$$\frac{12}{42} = \frac{420}{1470} \quad \text{and} \quad \frac{10}{35} = \frac{420}{1470}.$$

Hence,

$$\frac{12}{42} = \frac{10}{35}.$$

The third method suggests a general algorithm for determining whether two fractions $\frac{a}{b}$ and $\frac{c}{d}$ are equal. Rewrite both fractions with common denominator bd, that is,

$$\frac{a}{b} = \frac{ad}{bd} \quad \text{and} \quad \frac{c}{d} = \frac{bc}{bd}.$$

Because the denominators are the same, $\frac{ad}{bd} = \frac{bc}{bd}$ if, and only if, $ad = bc$. For example, $\frac{24}{36} = \frac{6}{9}$ because 24 · 9 = 216 = 36 · 6. In general, the following theorem results.

E-Manipulative Activity

Additional practice with ordering fractions is available in the *Ranking Fractions* module on the E-Manipulative disc.

> **Theorem 6-2:   Equality of Fractions**
>
> Two fractions $\frac{a}{b}$ and $\frac{c}{d}$, $b, d \neq 0$, are equal if, and only if, $ad = bc$.

Using a calculator, determine whether two fractions are equal by using Theorem 6-2. Since both $\boxed{2}\,\boxed{\times}\,\boxed{2}\,\boxed{1}\,\boxed{9}\,\boxed{6}\,\boxed{=}$ and $\boxed{4}\,\boxed{\times}\,\boxed{1}\,\boxed{0}\,\boxed{9}\,\boxed{8}\,\boxed{=}$ yield a display of 4392, we see that $\frac{2}{4} = \frac{1098}{2196}$.

Ordering Rational Numbers

As discussed in the grade 3 *Focal Points*:

> Students solve problems that involve comparing and ordering fractions by using models, benchmark fractions, or common numerators or denominators. They understand and use models, including the number line, to identify equivalent fractions. (p. 15)

First consider the comparison of fractions with like denominators. Children know that $\frac{7}{8} > \frac{5}{8}$

because if a pizza is divided into 8 parts of equal size, then 7 parts of the pizza is more than 5 parts.

Similarly, $\frac{3}{7} < \frac{4}{7}$. Thus, given two fractions with common positive denominators, the one with the greater numerator is the greater fraction. To make ordering of rational numbers consistent with the ordering of whole numbers and integers we have the following.

> ### Definition of Greater Than for Rational Numbers with Like Denominators
>
> If a, b, and c are integers and $b > 0$, then $\frac{a}{b} > \frac{c}{b}$ if, and only if, $a > c$.

● NOW TRY THIS 6-5

Determine whether the following is true: If a, b, c are integers and $b < 0$, then $\frac{a}{b} > \frac{c}{b}$ if, and only if, $a > c$.

To compare fractions with unlike denominators, some students may incorrectly reason that $\frac{1}{8} > \frac{1}{7}$ because 8 is greater than 7. In another case, they might falsely believe that $\frac{6}{7}$ is equal to $\frac{7}{8}$ because in both fractions the difference between the numerator and the denominator is 1. Comparing positive fractions with unlike denominators may be aided by using fraction strips to compare the fractions visually. For example, consider the fractions $\frac{4}{5}$ and $\frac{11}{12}$ shown in Figure 6-5.

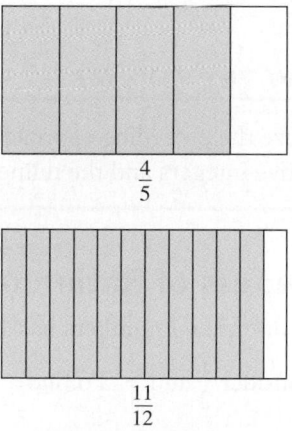

Figure 6-5

From Figure 6-5, students see that each fraction is one piece less than the same-size whole unit. However, they see that the missing piece for $\frac{11}{12}$ is smaller than the missing piece for $\frac{4}{5}$, so $\frac{11}{12}$ must be greater than $\frac{4}{5}$.

Comparing any fractions with unlike denominators can be accomplished by rewriting the fractions with the same positive common denominator. Using the common denominator bd, we can write the fractions $\frac{a}{b}$ and $\frac{c}{d}$ as $\frac{ad}{bd}$ and $\frac{bc}{bd}$. Because $b > 0$ and $d > 0$, then $bd > 0$; and we have the following:

$$\frac{a}{b} > \frac{c}{d} \text{ if, and only if, } \quad \frac{ad}{bd} > \frac{bc}{bd} \quad \text{and} \quad \frac{ad}{bd} > \frac{bc}{bd} \text{ if, and only if, } ad > bc$$

Therefore, we have the following theorem.

> **Theorem 6-3**
>
> If a, b, c, and d are integers with $b > 0$ and $d > 0$, then $\frac{a}{b} > \frac{c}{d}$ if, and only if, $ad > bc$.

 NOW TRY THIS 6-6

Order the fractions $\frac{3}{4}, \frac{9}{16}, \frac{5}{8}, \frac{2}{3}, \frac{-3}{8}, \frac{-6}{11}$, and $\frac{-4}{9}$ from least to greatest.

Next consider two positive fractions with numerators that are the same. For example, consider $\frac{3}{4}$ and $\frac{3}{10}$. If the whole is the same for both fractions, this means that we have three $\frac{1}{4}$s and three $\frac{1}{10}$s. Because $\frac{1}{4}$ is greater than $\frac{1}{10}$, then three of the larger parts is greater than three of the smaller parts, so $\frac{3}{4} > \frac{3}{10}$.

 NOW TRY THIS 6-7

Generalize the preceding approach for comparing fractions whose numerators and denominators are positive integers and the numerators are equal.

Denseness of Rational Numbers

The set of rational numbers has a property unlike the set of whole numbers and the set of integers. Consider $\frac{1}{2}$ and $\frac{2}{3}$. To find a rational number between $\frac{1}{2}$ and $\frac{2}{3}$, we first rewrite the fractions with a common denominator, as $\frac{3}{6}$ and $\frac{4}{6}$. Because there is no whole number between the numerators 3 and 4, we next find two fractions equal, respectively, to $\frac{1}{2}$ and $\frac{2}{3}$ with greater denominators. For example, $\frac{1}{2} = \frac{6}{12}$ and $\frac{2}{3} = \frac{8}{12}$, and $\frac{7}{12}$ is between the two fractions $\frac{6}{12}$ and $\frac{8}{12}$. So $\frac{7}{12}$ is between $\frac{1}{2}$ and $\frac{2}{3}$. This property is generalized as follows and stated as a theorem.

> **Theorem 6-4:   Denseness Property for Rational Numbers**
>
> Given two different rational numbers $\frac{a}{b}$ and $\frac{c}{d}$, there is another rational number between these two numbers.

NOW TRY THIS 6-8

Explain why there are infinitely many rational numbers between any two rational numbers.

EXAMPLE 6-3

a. Find two fractions between $\frac{7}{18}$ and $\frac{1}{2}$.

b. Show that the sequence $\frac{1}{2}, \frac{2}{3}, \frac{3}{4}, \frac{4}{5}, \ldots, \frac{n}{n+1} \ldots$, where $n \in W$, is an *increasing sequence;* that is, that each term starting from the second term is greater than the preceding term.

Solution

a. Because $\frac{1}{2} = \frac{1 \cdot 9}{2 \cdot 9} = \frac{9}{18}$, we see that $\frac{8}{18}$, or $\frac{4}{9}$, is between $\frac{7}{18}$ and $\frac{9}{18}$. To find another fraction between the given fractions, we find two fractions equal to $\frac{7}{18}$ and $\frac{9}{18}$, respectively, but with greater denominators; for example, $\frac{7}{18} = \frac{14}{36}$ and $\frac{9}{18} = \frac{18}{36}$.

We now see that $\frac{15}{36}, \frac{16}{36}$, and $\frac{17}{36}$ are all between $\frac{14}{36}$ and $\frac{18}{36}$ and thus between $\frac{7}{18}$ and $\frac{1}{2}$.

b. Because the nth term of the sequence is $\frac{n}{n+1}$, the next term is $\frac{n+1}{(n+1)+1}$, or $\frac{n+1}{n+2}$. We need to show that for all whole numbers n, $\frac{n+1}{n+2} > \frac{n}{n+1}$.

The inequality will be true if, and only if, $(n+1)(n+1) > n(n+2)$. This inequality is equivalent to

$$n^2 + 2n + 1 > n^2 + 2n$$
$$2n + 1 > 2n$$
$$1 > 0, \text{ which is true.}$$

Therefore we have an increasing sequence.

Another way to find a number between any two rational numbers involves adding numerators and adding denominators. In Example 6-3(a), to find a number between $\frac{7}{18}$ and $\frac{1}{2}$ we could add the numerators and add the denominators to produce $\frac{7+1}{18+2} = \frac{8}{20}$. We see that $\frac{7}{18} < \frac{8}{20}$ because $140 < 144$. Also, $\frac{8}{20} < \frac{1}{2}$ because $16 < 20$. The general property is stated in the following theorem.

Theorem 6-5

If $\frac{a}{b}$ and $\frac{c}{d}$ are any rational numbers with positive denominators, where $\frac{a}{b} < \frac{c}{d}$. Then,

$$\frac{a}{b} < \frac{a+c}{b+d} < \frac{c}{d}.$$

◉ NOW TRY THIS 6-9

Prove Theorem 6–5; that is, if $\frac{a}{b}$ and $\frac{c}{d}$ are any rational numbers with positive denominators,

where $\frac{a}{b} < \frac{c}{d}$, then $\frac{a}{b} < \frac{a+c}{b+d} < \frac{c}{d}$. $\left(\textit{Hint: Prove that } \frac{a}{b} < \frac{a+c}{b+d} \text{ and } \frac{a+c}{b+d} < \frac{c}{d}.\right)$

The proof of Theorem 6-5 suggested in Now Try This 6-9 also proves Theorem 6-4.

Assessment 6-1A

1. Write a sentence that illustrates the use of $\frac{7}{8}$ in each of the following ways:
   a. As a division problem
   b. As part of a whole
   c. As a ratio

2. For each of the following, write a fraction to approximate the shaded portion as part of the whole:

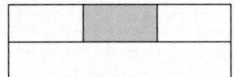
a.

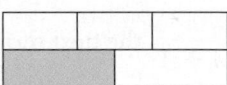
b.

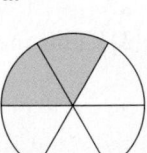
c.

d.

3. For each of the following four squares, write a fraction to approximate the shaded portion. What property of fractions does the diagram illustrate?

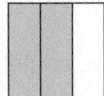 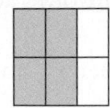 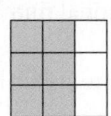 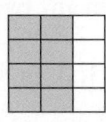
   a.   b.   c.   d.

4. Based on your observations could the shaded portions in the following figures represent the indicated fractions? Tell why.

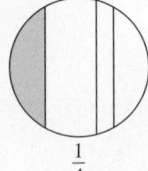
   a. $\frac{1}{4}$

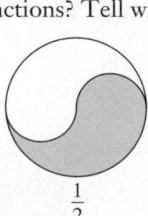
   b. $\frac{3}{4}$ c. $\frac{1}{2}$

5. In each case, subdivide the *whole* shown on the right to show the equivalent fraction.

a.
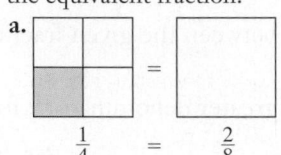
$\frac{1}{4} = \frac{2}{8}$

b.
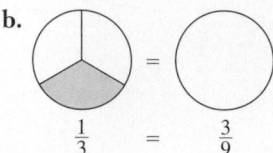
$\frac{1}{3} = \frac{3}{9}$

c.
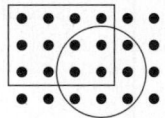
$\frac{1}{2} = \frac{3}{6}$

6. Referring to the figure, represent each of the following as a fraction:

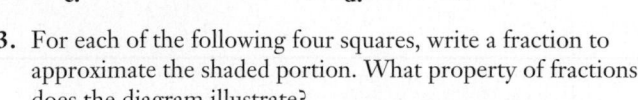

   a. The dots in the interior of the circle as a part of all the dots
   b. The dots in the interior of the rectangle as a part of all the dots
   c. The dots in the intersection of the interiors of the rectangle and the circle as a part of all the dots
   d. The dots outside the circular region but inside the rectangular region as part of all the dots

7. For each of the following, write three fractions equal to the given fraction:
   a. $\frac{2}{9}$ b. $\frac{^-2}{5}$
   c. $\frac{0}{3}$ d. $\frac{a}{2}$

8. Find the simplest form for each of the following fractions:
   a. $\frac{156}{93}$ b. $\frac{27}{45}$ c. $\frac{^-65}{91}$

9. For each of the following, choose the expression in parentheses that equals or describes best the given fraction:
   a. $\frac{0}{0}$ (1, undefined, 0)
   b. $\frac{5}{0}$ (undefined, 5, 0)

c. $\frac{0}{5}$ (undefined, 5, 0)

d. $\frac{2 + a}{a}$ (2, 3, cannot be simplified)

e. $\frac{15 + x}{3x}$ $\left(\frac{5 + x}{x}, 5, \text{cannot be simplified}\right)$

10. Find the simplest form for each of the following:

   a. $\frac{a^2 - b^2}{3a + 3b}$ **b.** $\frac{14x^2y}{63xy^2}$

11. Determine whether the following pairs are equal:

   a. $\frac{3}{8}$ and $\frac{375}{1000}$ **b.** $\frac{18}{54}$ and $\frac{23}{69}$

12. Determine whether the following pairs are equal by changing both to the same denominator:

   a. $\frac{10}{16}$ and $\frac{12}{18}$ **b.** $\frac{^-21}{86}$ and $\frac{^-51}{215}$

13. Draw an area model to show that $\frac{3}{4} = \frac{6}{8}$.

14. If a fraction is equal to $\frac{3}{4}$ and the sum of the numerator and denominator is 84, what is the fraction?

15. Mr. Gomez filled his car's 16 gal gas tank. He took a trip and used 6 gal of gas. Draw an arrow in the following figure to show what his gas gauge looked like after the trip:

16. Solve for x in each of the following:

   a. $\frac{2}{3} = \frac{x}{16}$ **b.** $\frac{3}{4} = \frac{^-27}{x}$

17. For each of the following pairs of fractions, replace the comma with the correct symbol ($<$, $=$, $>$) to make a true statement:

   a. $\frac{7}{8}, \frac{5}{6}$ **b.** $2\frac{4}{5}, 2\frac{3}{6}$ **c.** $\frac{^-7}{8}, \frac{^-4}{5}$

18. Arrange each of the following in decreasing order:

   a. $\frac{11}{22}, \frac{11}{16}, \frac{11}{13}$

   b. $\frac{^-1}{5}, \frac{^-19}{36}, \frac{^-17}{30}$

19. Show that the sequence $\frac{1}{3}, \frac{2}{4}, \frac{3}{5}, \frac{4}{6}, \frac{5}{7}, \frac{6}{8}, \ldots$ (in which each successive term is obtained from the previous term by adding 1 to the numerator and the denominator), is an increasing sequence; that is, show that each term in the sequence is greater than the preceding one.

20. For each of the following, find two rational numbers between the given fractions:

   a. $\frac{3}{7}$ and $\frac{4}{7}$

   b. $\frac{^-7}{9}$ and $\frac{^-8}{9}$

21. A scale on a map is 12 mi to the inch. What is the airline mileage between two cities that are 38 in. apart on the map?

22. **a.** 6 oz is what part of a pound? A ton?
   b. A dime is what fraction of a dollar?
   c. 15 min is what fraction of an hour?
   d. 8 hr is what fraction of a day?

23. Based on your visual observation write a fraction to represent the shaded portion.

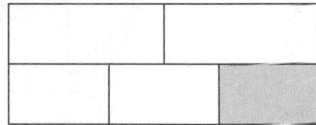

24. Mr. Gonzales and Ms. Price gave the same test to their fifth-grade classes. In Mr. Gonzales's class, 20 out of 25 students passed the test, and in Ms. Price's class, 24 out of 30 students passed the test. One of Ms. Price's students heard about the results of the tests and claimed that the classes did equally well. Is the student right? Explain.

25. In Amy's algebra class, 6 of the 31 students received As on a test. The same test was given to Bren's class and 5 of the 23 students received As. Which class had the higher rate of As?

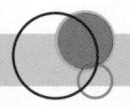

Assessment 6-1B

1. Write a sentence that illustrates the use of $\frac{7}{10}$ in each of the following ways:
   a. As a division problem
   b. As part of a whole
   c. As a ratio

2. For each of the following, write a fraction to approximate the shaded portion of the whole:

   a. [number line from 0 to 1]   **b.** ○ ○ ○ ● ● ● ● ● ● ● ● ●

   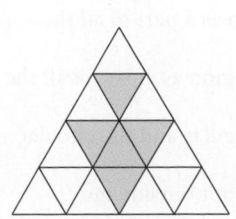

   c.   **d.**

   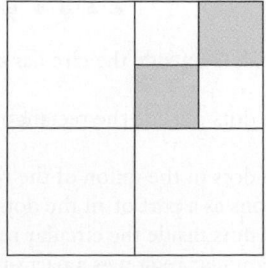

3. Complete each of the following figures so that it illustrates $\frac{3}{5}$:

   a.   b.

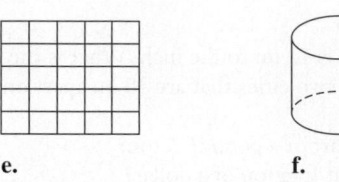

   c.

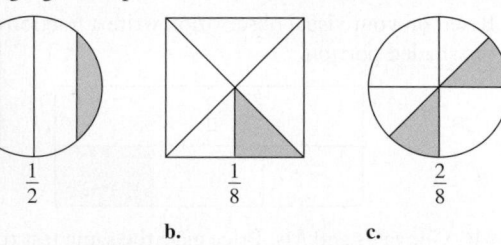

   d.

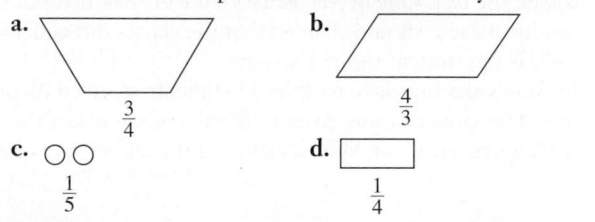

   e.   f.

4. Based on your observations, could the shaded portions in the following figures represent the indicated fractions? Tell why.

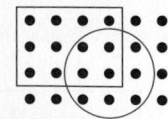

   $\frac{1}{2}$ $\frac{1}{8}$ $\frac{2}{8}$

   a.   b.   c.

5. If each of the following models represents the given fraction, draw a model that represents the *whole*. Shade your answer.

   a.   b.

   $\frac{3}{4}$ $\frac{4}{3}$

   c. ⚬⚬   d. ▭

   $\frac{1}{5}$ $\frac{1}{4}$

6. Referring to the figure, represent each of the following as a fraction:

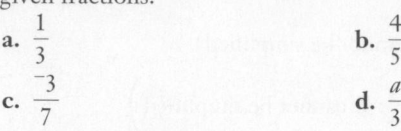

   a. The dots outside the circular region as a part of all the dots
   b. The dots outside the rectangular region as a part of all the dots
   c. The dots in the union of the rectangular and the circular regions as a part of all the dots
   d. The dots inside the circular region but outside the rectangular region as a part of all the dots.

7. For each of the following, write three fractions equal to the given fractions.

   a. $\frac{1}{3}$ b. $\frac{4}{5}$

   c. $\frac{-3}{7}$ d. $\frac{a}{3}$

8. Find the simplest form for each of the following fractions:

   a. $\frac{0}{68}$ b. $\frac{84^2}{91^2}$ c. $\frac{662}{703}$

9. For each of the following, choose the expression in parentheses that equals or describes best the given fraction:

   a. $\frac{6+x}{3x}$ $\left(\frac{2+x}{x}, 3, \text{cannot be simplified}\right)$

   b. $\frac{2^6+2^5}{2^4+2^7}$ $\left(1, \frac{2}{3}, \text{cannot be simplified}\right)$

   c. $\frac{2^{100}+2^{98}}{2^{100}-2^{98}}$ $\left(2^{196}, \frac{5}{3}, \text{too large to simplify}\right)$

10. Find the simplest form for each of the following:

   a. $\frac{a^2+ab}{a+b}$ b. $\frac{a}{3a+ab}$

11. Determine whether the following pairs are equal:

   a. $\frac{6}{16}$ and $\frac{3,750}{10,000}$ b. $\frac{17}{27}$ and $\frac{25}{45}$

12. Determine whether the following pairs are equal by changing both to the same denominator:

   a. $\frac{3}{-12}$ and $\frac{-36}{144}$ b. $\frac{-21}{430}$ and $\frac{-51}{215}$

13. Draw an area model to show that $\frac{2}{3}=\frac{6}{9}$.

14. A board is needed that is exactly $\frac{11}{32}$ in. wide to fit a hole. Can a board that is $\frac{3}{8}$ in. be shaved down to fit the hole? If so, how much must be shaved from the board?

15. The following two parking meters are next to each other with the times left as shown. Which meter has more time left on it? How much more?

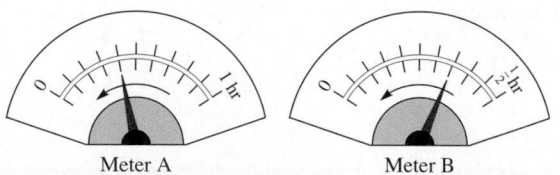

   Meter A   Meter B

16. Solve for x in each of the following:

   a. $\frac{2}{3}=\frac{x}{18}$

   b. $\frac{3}{x}=\frac{3x}{x^2}$

17. For each of the following pairs of fractions, replace the comma with the correct symbol ($<, =, >$) to make a true statement:

   a. $\frac{1}{-7}, \frac{1}{-8}$ b. $\frac{2}{5}, \frac{4}{10}$ c. $\frac{0}{7}, \frac{0}{17}$

18. Show that the sequence $\frac{2}{1}, \frac{3}{2}, \frac{4}{3}, \frac{5}{4}, \frac{6}{5}, \frac{7}{6}, \ldots$ is a decreasing sequence; that is, show that each term in the sequence is less than the preceding one.

19. If $0 < a < b$, $c > 0$, and $d > 0$, compare the size of $\dfrac{c}{d}$ with $\dfrac{ac}{bd}$.

20. For each of the following, find two rational numbers between the given fractions:
    a. $\dfrac{5}{6}$ and $\dfrac{83}{100}$
    b. $\dfrac{^-1}{3}$ and $\dfrac{3}{4}$

21. A scale on a map is 120 mi to the inch. What is the airline mileage between two cities that are $\dfrac{3}{4}$ in. apart on the map?

22. a. 12 oz is what part of a pound?
    b. A nickel is what fraction of a dollar?
    c. 25 min is what fraction of an hour?
    d. 16 hr is what fraction of a 24 hr day?

23. Read each measurement as shown on the following ruler:

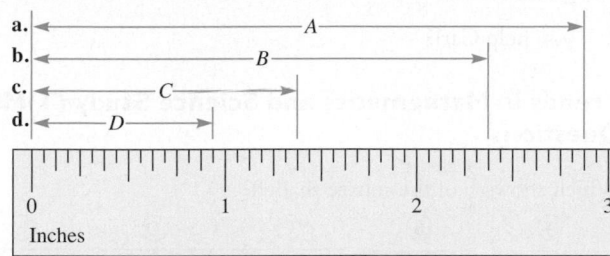

24. Determine whether each of the following is always correct. If not, find when it is true. Explain.
    a. $\dfrac{ab + c}{b} = a + c$
    b. $\dfrac{a + b}{a + c} = \dfrac{b}{c}$
    c. $\dfrac{ab + ac}{ac} = \dfrac{b + c}{c}$
    d. $\dfrac{a^2 + a}{a} = a + 1$

25. Consider the following number grid. The circled numbers form a rhombus (that is, all sides are the same length).

1	2	3	4	5	6	7	8	9	10
11	12	13	14	15	16	17	18	19	20
21	22	23	24	25	26	27	28	29	30
31	32	33	34	35	36	37	38	39	40
41	42	43	44	45	46	47	48	49	50

   a. If A is the sum of the four circled numbers and B is the sum of the four interior numbers, find $\dfrac{A}{B}$.

   b. Form a rhombus by circling the numbers 6, 18, 37, and 25. If A and B are defined as in (a), find $\dfrac{A}{B}$.

   c. How do the answers in (a) and (b) compare? Why does this happen?

26. a. If $\dfrac{a}{c} = \dfrac{b}{c}$, what must be true?
    b. If $\dfrac{a}{b} = \dfrac{a}{c}$, what must be true?

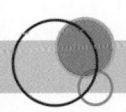

Mathematical Connections 6-1

Communication

1. Jane has a recipe that calls for 4 c of flour. She wants to make $\dfrac{3}{4}$ of the recipe. Instead of determining directly how many cups are needed for the new recipe, she fills $\dfrac{3}{4}$ of a cup 4 times. Explain why Jane's method works.

2. In each of two different fourth-grade classes, $\dfrac{1}{3}$ of the members are girls. Does each class have the same number of girls? Explain.

3. Consider the set of all fractions equal to $\dfrac{1}{2}$. If you take any 10 of those fractions, add their numerators to obtain the numerator of a new fraction and add their denominators to obtain the denominator of a new fraction, how does the new fraction relate to $\dfrac{1}{2}$? Generalize what you found and explain.

4. Should fractions always be reduced to their simplest form? Why or why not?

Open-Ended

5. Make three statements about yourself or your environment and use fractions in each. Explain why your statements are true. For example, your parents have three children, two of whom live at home; hence $\dfrac{2}{3}$ of their children live at home.

Cooperative Learning

6. Assume the tallest person in your group is 1 unit tall and do the following:
   a. Find rational numbers to approximately represent the heights of other members of the group.
   b. Make a number line and plot the rational number for each person ordered according to height.

7. Obtain tangram pieces or build them and cut them out as shown. Answer each of the following.
   a. If the area of the entire square is 1 square unit, find the area of each tangram piece.
   b. If the area of piece *a* is 1 square unit, find the area of each tangram piece.

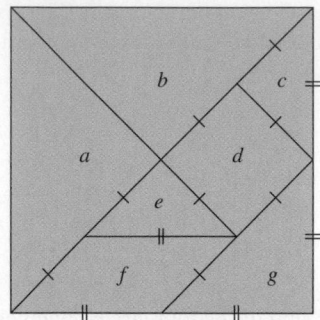

Questions from the Classroom

8. A student asks if $\dfrac{0}{6}$ is in its simplest form. How do you respond?

9. A student writes $\dfrac{15}{53} < \dfrac{1}{3}$ because $3 \cdot 15 < 53 \cdot 1$. Another student writes $\dfrac{1\cancel{5}}{\cancel{5}3} = \dfrac{1}{3}$. Where is the fallacy?

10. A student claims that there are no numbers between $\dfrac{999}{1000}$ and 1 because they are so close together. What is your response?

11. A student simplified the fraction $\dfrac{m+n}{p+n}$ to $\dfrac{m}{p}$. How would you help this student?

12. A student argued that a pizza cut into 12 pieces was more than a pizza cut into 6 pieces. How would you respond?

13. Ann claims that she cannot show $\dfrac{3}{4}$ of the following faces because some are big and some are small. What do you tell her?

14. A student claims that $\dfrac{2}{3} = \dfrac{6}{7}$ because if you add 4 to both the top and the bottom of a fraction, the fraction does not change. How do you respond?

15. How would you respond to each of the following students?
    a. Iris claims that if we have two positive rational numbers, the one with the greater numerator is the greater.
    b. Shirley claims that if we have two positive rational numbers, the one with the greater denominator is the lesser.

16. Steve claims that the shaded portions cannot represent $\dfrac{2}{3}$ since there are 10 circles shaded and $\dfrac{2}{3}$ is less than 1. How do you respond?

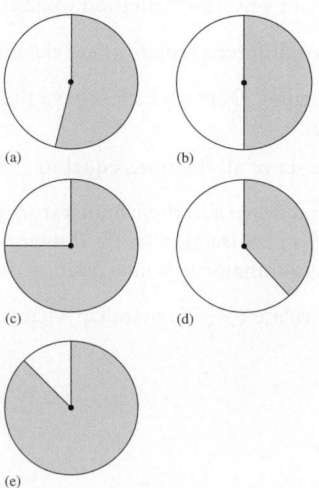

17. Daryl says that each piece of the pie shown represents $\dfrac{1}{3}$ of the pie. How do you respond?

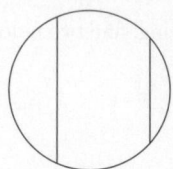

18. a. Cassidy noticed that $5 > 4$, but $\dfrac{1}{5} < \dfrac{1}{4}$. In general she thinks that if *a* and *b* are positive integers, and $a > b$, then $\dfrac{1}{a} < \dfrac{1}{b}$. She would like to know if this is always true. How do you respond?
    b. Cassidy would like to know if her discovery in part (a) is true when *a* and *b* are negative integers, and if so, why or why not. How do you respond?

19. Carl says that $\dfrac{3}{8} > \dfrac{2}{3}$ because $3 > 2$ and $8 > 3$. How would you help Carl?

Trends in Mathematics and Science Study (TIMSS) Questions

Which shows $\dfrac{2}{3}$ of the square shaded?

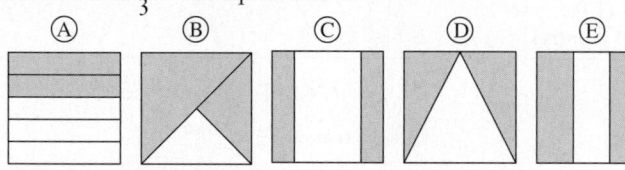

TIMSS, Grade 4, 2003

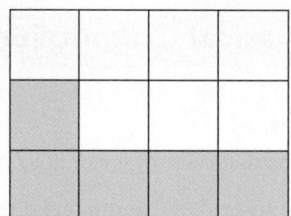

Which circle has approximately the same fraction of its area shaded as the rectangle above?

(a) (b) (c) (d)

(e)

TIMSS, Grade 8, 2007

National Assessment of Educational Progress (NAEP) Question

What fraction of the figure is shaded?

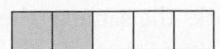

NAEP, Grade 4, 2007

In which of the following are the three fractions arranged from least to greatest?

a. $\dfrac{2}{7}, \dfrac{1}{2}, \dfrac{5}{9}$ **b.** $\dfrac{1}{2}, \dfrac{2}{7}, \dfrac{5}{9}$ **c.** $\dfrac{1}{2}, \dfrac{5}{9}, \dfrac{2}{7}$

d. $\dfrac{5}{9}, \dfrac{1}{2}, \dfrac{2}{7}$ **e.** $\dfrac{5}{9}, \dfrac{2}{7}, \dfrac{1}{2}$

NAEP, Grade 8, 2007

BRAIN TEASER

In an old Sam Loyd puzzle, a watch is described as having stopped when the minute and hour hands formed a straight line and the second hand was not on 12. At what times can this happen?

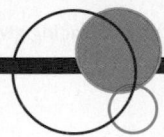

6-2 Addition, Subtraction, and Estimation with Rational Numbers

Addition and subtraction of rational numbers is very much like addition and subtraction of whole numbers and integers. We first demonstrate the addition of two rational numbers with like denominators, $\dfrac{2}{5} + \dfrac{1}{5}$, using an area model in Figure 6-6(a) and a number line model in Figure 6-6(b).

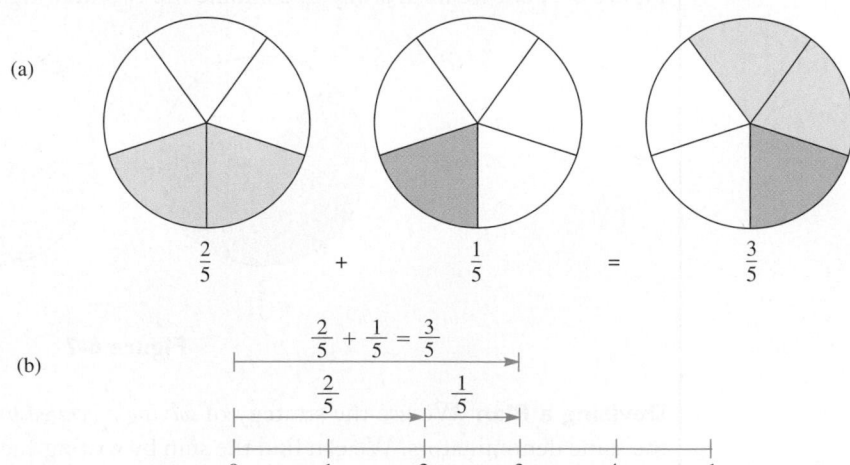

Figure 6-6

Why does the area model in Figure 6-6(a) make sense? Suppose that someone gives us $\dfrac{2}{5}$ of a pie to start with and then gives us another $\dfrac{1}{5}$ of the pie. In Figure 6-6(a), $\dfrac{2}{5}$ is represented by

2 pieces when the pie is cut into 5 equal-size pieces and $\frac{1}{5}$ is represented by 1 piece of the 5 equal-size pieces. So all together you have $2 + 1 = 3$ pieces of the 5 equal-size pieces, or $\frac{3}{5}$ of the total (whole) pie. The number line model in Figure 6-6(b) works the same as the number line model for whole numbers.

The ideas illustrated in Figure 6-6 can be applied to the sum of two rational numbers with like denominators and are summarized in the following definition.

Definition of Addition of Rational Numbers with Like Denominators

If $\frac{a}{b}$ and $\frac{c}{b}$ are rational numbers, then $\frac{a}{b} + \frac{c}{b} = \frac{a + c}{b}$.

Next we consider the addition of two rational numbers with unlike denominators. In the grade 5 *Common Core Standards* we find that students should be able to:

Add and subtract fractions with unlike denominators (including mixed numbers) by replacing given fractions with equivalent fractions in such a way as to produce an equivalent sum or difference of fractions with like denominators. (p. 36)

We use Pólya's four-step process to develop this concept.

Problem Solving   **Adding Rational Numbers Problem**

Determine how to add the rational numbers $\frac{2}{3}$ and $\frac{1}{4}$.

Understanding the Problem We model $\frac{2}{3}$ and $\frac{1}{4}$ as parts of the same-sized whole, as seen in Figure 6-7, but we need a way to combine the two drawings to find the sum.

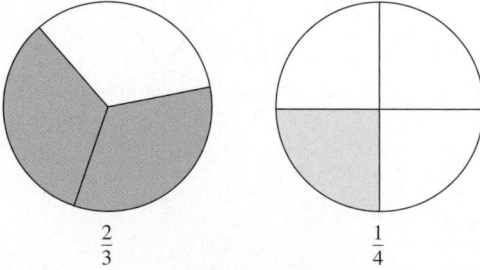

$$\frac{2}{3} \qquad \frac{1}{4}$$

Figure 6-7

Devising a Plan We use the strategy of *solving a related problem*: adding rational numbers with the same denominators. We can find the sum by writing each fraction with a common denominator and then completing the computation.

Carrying Out the Plan From earlier work in this chapter, we know that $\frac{2}{3}$ has infinitely many representations, including $\frac{4}{6}, \frac{6}{9}, \frac{8}{12}$, and so on. Also $\frac{1}{4}$ has infinitely many representations, including $\frac{2}{8}, \frac{3}{12}, \frac{4}{16}$, and so on. We see that $\frac{8}{12}$ and $\frac{3}{12}$ have the same denominator. One is

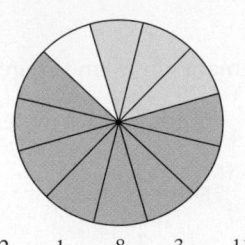

$$\frac{2}{3} + \frac{1}{4} = \frac{8}{12} + \frac{3}{12} = \frac{11}{12}$$

Figure 6-8

8 parts of 12 equal parts, while the other is 3 parts of 12 equal parts. Consequently, the sum is $\frac{2}{3} + \frac{1}{4} = \frac{8}{12} + \frac{3}{12} = \frac{11}{12}$. Figure 6-8 illustrates the addition.

Looking Back To add two rational numbers with unlike denominators, we considered equal rational numbers with like denominators. A common denominator for $\frac{2}{3}$ and $\frac{1}{4}$ is 12. This is also the least common denominator, or the LCM of 3 and 4. To add two fractions with unequal denominators such as $\frac{5}{12}$ and $\frac{7}{18}$, we could find equal fractions with the denominator LCM(12, 18), or 36. However, any common denominator will work as well, for example, 72 or even $12 \cdot 18$.

By considering the sum $\frac{2}{3} + \frac{1}{4} = \frac{2 \cdot 4}{3 \cdot 4} + \frac{1 \cdot 3}{4 \cdot 3} = \frac{8}{12} + \frac{3}{12} = \frac{11}{12}$, we can generalize to finding the sum of two rational numbers with unlike denominators, as in the following.

Definition of Addition of Rational Numbers with Unlike Denominators

If $\frac{a}{b}$ and $\frac{c}{d}$ are rational numbers, then $\frac{a}{b} + \frac{c}{d} = \frac{ad + bc}{bd}$.

There is no need to write the preceding as a definition because it can be proven and hence is a theorem. However, it is listed as a definition in many books.

REMARK The definition of addition of rational numbers with like denominators can be applied to unlike denominators as seen below:

$$\frac{a}{b} + \frac{c}{b} = \frac{ab + cb}{b \cdot b} = \frac{(a + c)b}{b \cdot b} = \frac{a + c}{b}$$

EXAMPLE 6-4

Find each of the following sums:

a. $\dfrac{2}{15} + \dfrac{4}{21}$ **b.** $\dfrac{2}{^{-}3} + \dfrac{1}{5}$ **c.** $\left(\dfrac{3}{4} + \dfrac{1}{5}\right) + \dfrac{1}{6}$ **d.** $\dfrac{3}{x} + \dfrac{4}{y}$ **e.** $\dfrac{2}{a^2 b} + \dfrac{3}{ab^2}$

Solution

a. Because $LCM(15, 21) = 3 \cdot 5 \cdot 7$, then $\dfrac{2}{15} + \dfrac{4}{21} = \dfrac{2 \cdot 7}{15 \cdot 7} + \dfrac{4 \cdot 5}{21 \cdot 5} = \dfrac{14}{105} + \dfrac{20}{105} = \dfrac{34}{105}$.

b. $\dfrac{2}{^{-}3} + \dfrac{1}{5} = \dfrac{(2)(5) + (^{-}3)(1)}{(^{-}3)(5)} = \dfrac{10 + ^{-}3}{^{-}15} = \dfrac{7}{^{-}15} = \dfrac{7(^{-}1)}{^{-}15(^{-}1)} = \dfrac{^{-}7}{15}$.

c. $\dfrac{3}{4} + \dfrac{1}{5} = \dfrac{3 \cdot 5 + 4 \cdot 1}{4 \cdot 5} = \dfrac{19}{20}$; hence, $\left(\dfrac{3}{4} + \dfrac{1}{5}\right) + \dfrac{1}{6} = \dfrac{19}{20} + \dfrac{1}{6} = \dfrac{19 \cdot 6 + 20 \cdot 1}{20 \cdot 6} = \dfrac{134}{120}$, or $\dfrac{67}{60}$.

d. $\dfrac{3}{x} + \dfrac{4}{y} = \dfrac{3y}{xy} + \dfrac{4x}{xy} = \dfrac{3y + 4x}{xy}$.

e. $LCM(a^2 b, ab^2) = a^2 b^2$; $\dfrac{2}{a^2 b} + \dfrac{3}{ab^2} = \dfrac{2b}{(a^2 b)b} + \dfrac{3a}{a(ab^2)} = \dfrac{2b + 3a}{a^2 b^2}$.

Mixed Numbers

In everyday life, we often use **mixed numbers**, that is, numbers that are made up of an integer and a proper fraction. For example, Figure 6-9 shows that the nail is $2\frac{3}{4}$ in. long. The mixed number $2\frac{3}{4}$ means $2 + \frac{3}{4}$. It is sometimes inferred that $2\frac{3}{4}$ means 2 times $\frac{3}{4}$, since xy means $x \cdot y$, but this is not correct. Also, the number $^-4\frac{3}{4}$ means $^-\left(4\frac{3}{4}\right)$, or $^-4 - \frac{3}{4}$, not $^-4 + \frac{3}{4}$.

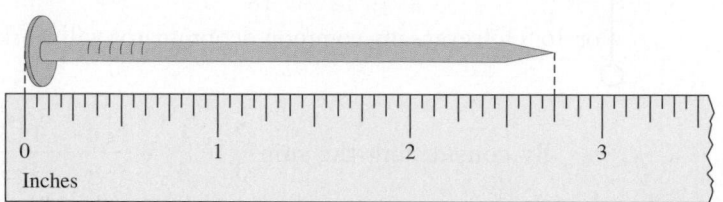

Figure 6-9

In a National Assessment of Educational Progress (NAEP) test, students were given the following question:

$$5\frac{1}{4} \text{ is the same as:}$$

(a) $5 + \frac{1}{4}$ (b) $5 - \frac{1}{4}$ (c) $5 \times \frac{1}{4}$ (d) $5 \div \frac{1}{4}$

Only 47% of the seventh graders chose the correct response, (a), and only 44% of the eleventh graders chose the correct response.

A mixed number is a rational number, and therefore it can always be written in the form $\frac{a}{b}$. For example,

$$2\frac{3}{4} = 2 + \frac{3}{4} = \frac{2}{1} + \frac{3}{4} = \frac{2 \cdot 4 + 1 \cdot 3}{1 \cdot 4} = \frac{8 + 3}{4} = \frac{11}{4}.$$

EXAMPLE 6-5

Change each of the following mixed numbers to the form $\frac{a}{b}$, where a and b are integers:

a. $4\frac{1}{3}$ **b.** $^-3\frac{2}{5}$

Solution

a. $4\frac{1}{3} = 4 + \frac{1}{3} = \frac{4}{1} + \frac{1}{3} = \frac{4 \cdot 3 + 1 \cdot 1}{1 \cdot 3} = \frac{12 + 1}{3} = \frac{13}{3}.$

b. $^-3\frac{2}{5} = ^-\left(3 + \frac{2}{5}\right) = ^-\left(\frac{3}{1} + \frac{2}{5}\right) = ^-\left(\frac{3 \cdot 5 + 1 \cdot 2}{1 \cdot 5}\right) = ^-\left(\frac{17}{5}\right) = \frac{^-17}{5}.$

NOW TRY THIS 6-10

Use the process in Example 6-5 to write $2\frac{3}{4} + 5\frac{3}{8}$ as a mixed number.

EXAMPLE 6-6

Change $\dfrac{39}{5}$ to a mixed number.

Solution   We divide 39 by 5 and use the division algorithm as follows:

$$\frac{39}{5} = \frac{7 \cdot 5 + 4}{5} = \frac{7 \cdot 5}{5} + \frac{4}{5} = 7 + \frac{4}{5} = 7\frac{4}{5}$$

In elementary schools, problems like Example 6-6 are usually computed using division, as follows:

$$\begin{array}{r} 5 \\ 5\overline{)29} \\ \underline{25} \\ 4 \end{array}$$

Hence, $\dfrac{29}{5} = 5 + \dfrac{4}{5} = 5\dfrac{4}{5}$.

Scientific/fraction calculators can change improper fractions to mixed numbers. For example, if we enter $\boxed{2}\boxed{9}\boxed{/}\boxed{5}$ and press $\boxed{Ab/c}$, then $5 \sqcup 4/5$ appears, which means $5\dfrac{4}{5}$.

We can also use scientific/fraction calculators to add mixed numbers. For example, to add $2\dfrac{4}{5} + 3\dfrac{5}{6}$, we enter $\boxed{2}\boxed{\text{Unit}}\boxed{4}\boxed{/}\boxed{5}\boxed{+}\boxed{3}\boxed{\text{Unit}}\boxed{5}\boxed{/}\boxed{6}\boxed{=}$, and the display reads $5 \sqcup 49/30$. We then press $\boxed{Ab/c}$ to obtain $6 \sqcup 19/30$, which means $6\dfrac{19}{30}$.

Adding Mixed Numbers

Because mixed numbers are rational numbers, the methods of adding rationals can be used to include mixed numbers. The student page shows a method for computing sums of mixed numbers.

NOW TRY THIS 6-11

a. Compute $8\dfrac{1}{4} + 6\dfrac{1}{2}$ from the student page on p. 282 by converting each number to an improper fraction, performing the addition, and then converting the improper fraction to a mixed number.

b. Compute the Quick Check problem on the student page.

Properties of Addition for Rational Numbers

Rational numbers have the following properties for addition: *closure*, *commutative*, *associative*, *additive identity*, and *additive inverse*. To emphasize the additive inverse property of rational numbers, we state it explicitly, as follows.

Theorem 6-6:   Additive Inverse Property of Rational Numbers

For any rational number $\dfrac{a}{b}$, there exists a unique rational number $-\dfrac{a}{b}$, the additive inverse of $\dfrac{a}{b}$, such that

$$\frac{a}{b} + \left(-\frac{a}{b}\right) = 0 = \left(-\frac{a}{b}\right) + \frac{a}{b}$$

School Book Page   ADDING MIXED NUMBERS

5-4   Adding Mixed Numbers

✓ Check Skills You'll Need

1. **Vocabulary Review** Explain how you know that $\frac{36}{15}$ is an *improper fraction*.

Write each fraction as a mixed number in simplest form.

2. $\frac{8}{6}$ 3. $\frac{15}{6}$

4. $\frac{7}{4}$ 5. $\frac{25}{10}$

 **for Help**
Lesson 4-6

What You'll Learn
To add mixed numbers with and without renaming

Why Learn This?
When you say, "I'll be there in an hour and a half," or, "I had band practice until quarter after five," you are using mixed numbers to talk about time.

You can find the sum of mixed numbers by adding the whole number and fraction parts separately. Then you combine the two parts to find the total.

EXAMPLE   Adding Mixed Numbers

❶ You spent $8\frac{1}{4}$ hours on Saturday and $6\frac{1}{2}$ hours on Sunday working on a science project. How long did you work on the project?

Estimate $8\frac{1}{4} + 6\frac{1}{2} \approx 8 + 7 = 15$

$$
\begin{array}{rcl}
8\frac{1}{4} & \rightarrow & 8\frac{1}{4} \\
+6\frac{1}{2} & \rightarrow & +6\frac{2}{4} \\
\hline
& & 14\frac{3}{4}
\end{array}
$$

The LCD is 4. Write the fractions with the same denominator.

Add the whole numbers. Then add the fractions.

You work a total of $14\frac{3}{4}$ hours on your science project.

Check for Reasonableness $14\frac{3}{4}$ is close to the estimate of 15. The answer is reasonable.

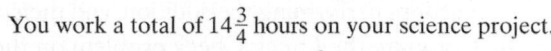

✓ Quick Check

1. A giant tortoise traveled $2\frac{1}{3}$ yards and stopped. Then it traveled $3\frac{1}{2}$ yards. Find the total distance the giant tortoise traveled.

228   Chapter 5   Adding and Subtracting Fractions

Another form of $-\dfrac{a}{b}$ can be found by considering the sum $\dfrac{a}{b} + \dfrac{^-a}{b}$. Because

$$\frac{a}{b} + \frac{^-a}{b} = \frac{a + ^-a}{b} = \frac{0}{b} = 0,$$

it follows that $-\dfrac{a}{b}$ and $\dfrac{^-a}{b}$ are both additive inverses of $\dfrac{a}{b}$, so $-\dfrac{a}{b} = \dfrac{^-a}{b}$.

EXAMPLE 6-7

Find the additive inverses for each of the following:

a. $\dfrac{3}{5}$ b. $\dfrac{^-5}{11}$ c. $4\dfrac{1}{2}$

Solution

a. $-\dfrac{3}{5}$ or $\dfrac{^-3}{5}$ b. $-\left(\dfrac{^-5}{11}\right) = \dfrac{^-(^-5)}{11} = \dfrac{5}{11}$ c. $^-4\dfrac{1}{2}$, or $\dfrac{^-9}{2}$

Properties of the additive inverse for rational numbers are analogous to those of the additive inverse for integers, as shown in Table 6-2.

Table 6-2

Integers	Rational Numbers
1. $^-(^-a) = a$	1. $-\left(\dfrac{^-a}{b}\right) = \dfrac{a}{b}$
2. $^-(a + b) = ^-a + ^-b$	2. $^-\left(\dfrac{a}{b} + \dfrac{c}{d}\right) = \dfrac{^-a}{b} + \dfrac{^-c}{d}$

The set of rational numbers has the addition property of equality, which says that the same number can be added to both sides of an equation.

> **Theorem 6-7: Addition Property of Equality**
>
> If $\dfrac{a}{b}$ and $\dfrac{c}{d}$ are any rational numbers such that $\dfrac{a}{b} = \dfrac{c}{d}$, and if $\dfrac{e}{f}$ is any rational number, then
>
> $$\frac{a}{b} + \frac{e}{f} = \frac{c}{d} + \frac{e}{f}.$$

Subtraction of Rational Numbers

In elementary school, subtraction of rational numbers is usually introduced by using a take-away model. If we have $\dfrac{6}{7}$ of a pizza and $\dfrac{2}{7}$ of the original pizza is taken away, $\dfrac{4}{7}$ of the pizza remains; that is, $\dfrac{6}{7} - \dfrac{2}{7} = \dfrac{6 - 2}{7} = \dfrac{4}{7}$. In general, subtraction of rational numbers with like denominators is determined as follows:

$$\frac{a}{b} - \frac{c}{b} = \frac{a - c}{b}$$

As with integers, a number line can be used to model subtraction of non-negative rational numbers. If a line is marked off in units of length $\frac{1}{b}$ and $a \geq c$, then $\frac{a}{b} - \frac{c}{b}$ is equal to $(a - c)$ units of length $\frac{1}{b}$, which implies that $\frac{a}{b} - \frac{c}{b} = \frac{a - c}{b}$. When the denominators are not the same we can perform the subtraction by finding a common denominator. For example,

$$\frac{3}{4} - \frac{2}{3} = \frac{3 \cdot 3}{4 \cdot 3} - \frac{2 \cdot 4}{3 \cdot 4} = \frac{9}{12} - \frac{8}{12} = \frac{9 - 8}{12} = \frac{1}{12}.$$

Subtraction of rational numbers, like subtraction of integers, can be defined in terms of addition as follows.

Definition of Subtraction of Rational Numbers in Terms of Addition

If $\frac{a}{b}$ and $\frac{c}{d}$ are any rational numbers, then $\frac{a}{b} - \frac{c}{d}$ is the unique rational number $\frac{e}{f}$ such that $\frac{a}{b} = \frac{c}{d} + \frac{e}{f}$.

As with integers, we can see that subtraction of rational numbers can be performed by adding the additive inverses as stated on the following theorem.

Theorem 6-8

If $\frac{a}{b}$ and $\frac{c}{d}$ are any rational numbers, then $\frac{a}{b} - \frac{c}{d} = \frac{a}{b} + \frac{^-c}{d}$.

Now, using Theorem 6-8, we obtain the following:

$$\frac{a}{b} - \frac{c}{d} = \frac{a}{b} + \frac{^-c}{d}$$

$$= \frac{ad + b(^-c)}{bd}$$

$$= \frac{ad + {^-}(bc)}{bd}$$

$$= \frac{ad - bc}{bd}$$

We proved the following theorem which is sometimes given as a definition of subtraction.

Theorem 6-9

If $\frac{a}{b}$ and $\frac{c}{d}$ are any rational numbers, then $\frac{a}{b} - \frac{c}{d} = \frac{ad - bc}{bd}$.

EXAMPLE 6-8

Find each difference in the following:

a. $\frac{5}{8} - \frac{1}{4}$ **b.** $5\frac{1}{3} - 2\frac{3}{4}$

Solution

a. One approach is to find the LCM for the fractions. Because LCM(8, 4) = 8, we have

$$\frac{5}{8} - \frac{1}{4} = \frac{5}{8} - \frac{2}{8} = \frac{3}{8}.$$

An alternative approach follows:

$$\frac{5}{8} - \frac{1}{4} = \frac{5 \cdot 4 - 8 \cdot 1}{8 \cdot 4} = \frac{20 - 8}{32} = \frac{12}{32}, \text{ or } \frac{3}{8}.$$

b. Two methods of solution are given:

$$
\begin{array}{l}
5\dfrac{1}{3} = \ 5\dfrac{4}{12} = \ 4 + 1\dfrac{4}{12} = \ \ 4\dfrac{16}{12} \\[2mm]
\underline{-2\dfrac{3}{4}} = \underline{-2\dfrac{9}{12}} = \underline{-2\dfrac{9}{12}} = \ \underline{-2\dfrac{9}{12}} \\[2mm]
\hphantom{5\dfrac{1}{3} = \ 5\dfrac{4}{12} = \ 4 + 1\dfrac{4}{12} = \ \ } 2\dfrac{7}{12}
\end{array}
\qquad
\begin{array}{l}
5\dfrac{1}{3} - 2\dfrac{3}{4} = \dfrac{16}{3} - \dfrac{11}{4} \\[3mm]
\hphantom{5\dfrac{1}{3} - 2\dfrac{3}{4} } = \dfrac{16 \cdot 4 - 3 \cdot 11}{3 \cdot 4} \\[3mm]
\hphantom{5\dfrac{1}{3} - 2\dfrac{3}{4} } = \dfrac{64 - 33}{12} \\[3mm]
\hphantom{5\dfrac{1}{3} - 2\dfrac{3}{4} } = \dfrac{31}{12}, \text{ or } 2\dfrac{7}{12}
\end{array}
$$

The following examples show the use of fractions in algebra.

EXAMPLE 6-9

Add or subtract each of the following. Write your answer in simplest form.

a. $\dfrac{x}{2} + \dfrac{x}{3}$

b. $\dfrac{2 - x}{6 - 3x} + \dfrac{4 - 2x}{3x - 6}$

c. $\dfrac{2}{a + b} - \dfrac{2}{a - b}$

d. $\dfrac{1}{x} - \dfrac{1}{2x^2}$

Solution

a. $\dfrac{x}{2} + \dfrac{x}{3} = \dfrac{3x}{3 \cdot 2} + \dfrac{2x}{2 \cdot 3}$

$\hphantom{\dfrac{x}{2} + \dfrac{x}{3} } = \dfrac{3x + 2x}{6} = \dfrac{5x}{6}$

b. We first write each fraction in simplest form:

$$\frac{2 - x}{6 - 3x} = \frac{2 - x}{3(2 - x)} = \frac{1(2 - x)}{3(2 - x)} = \frac{1}{3}$$

$$\frac{4 - 2x}{3x - 6} = \frac{{}^-2(x - 2)}{3(x - 2)} = \frac{{}^-2}{3}$$

Thus, if $x \neq 2$, the sum is $\dfrac{1}{3} + \dfrac{{}^-2}{3} = \dfrac{{}^-1}{3}$.

c. Using Theorem 6-9:

$$\frac{2}{a + b} - \frac{2}{a - b} = \frac{2(a - b) - 2(a + b)}{(a + b)(a - b)}$$

$$\hphantom{\frac{2}{a + b} - \frac{2}{a - b} } = \frac{2a - 2b - 2a - 2b}{(a + b)(a - b)}$$

$$\hphantom{\frac{2}{a + b} - \frac{2}{a - b} } = \frac{{}^-4b}{(a + b)(a - b)} \text{ or } \frac{{}^-4b}{a^2 - b^2}$$

d. $\dfrac{1}{x} - \dfrac{1}{2x^2} = \dfrac{2x \cdot 1}{2x \cdot x} - \dfrac{1}{2x^2}$

$\hphantom{\dfrac{1}{x} - \dfrac{1}{2x^2} } = \dfrac{2x}{2x^2} - \dfrac{1}{2x^2}$

$\hphantom{\dfrac{1}{x} - \dfrac{1}{2x^2} } = \dfrac{2x - 1}{2x^2}$

Estimation with Rational Numbers

Estimation helps us make practical decisions in our everyday lives. For example, suppose we need to double a recipe that calls for $\frac{3}{4}$ of a cup of flour. Will we need more or less than a cup of flour?

Many of the estimation and mental math techniques that we learned to use with whole numbers also work with rational numbers.

The grade 5 *Common Core Standards* calls for students to "use benchmark fraction and number sense of fractions to estimate mentally and assess the reasonableness of answers." Estimation plays an important role in judging the reasonableness of computations. Students do not necessarily have this skill. For example on the second NAEP examination, when asked to estimate $\frac{12}{13} + \frac{7}{8}$ on a multiple-choice exam, only 24% of 13-year-old students said the answer was close to 2. Most said it was close to 1, 19, or 21. These incorrect estimates suggest common computational errors in adding fractions and a lack of understanding of the operation being carried out. These incorrect estimates also suggest a lack of number sense.

NOW TRY THIS 6-12

A student added $\frac{3}{4} + \frac{1}{2}$ and obtained $\frac{4}{6}$. How would you use estimation to show that this answer could not be correct?

Sometimes to obtain an estimate it is desirable to round fractions to a ***convenient*** or ***benchmark*** fraction, such as $\frac{1}{2}, \frac{1}{3}, \frac{1}{4}, \frac{1}{5}, \frac{2}{3}, \frac{3}{4}$, or 1. For example, if a student had 59 correct answers out of 80 questions, the student answered $\frac{59}{80}$ of the questions correctly, which is approximately $\frac{60}{80}$, or $\frac{3}{4}$. We know $\frac{60}{80}$ is greater than $\frac{59}{80}$. The estimate $\frac{3}{4}$ for $\frac{59}{80}$ is a high estimate. In a similar way, we can estimate $\frac{31}{90}$ by $\frac{30}{90}$, or $\frac{1}{3}$. In this case, the estimate of $\frac{1}{3}$ is a low estimate of $\frac{31}{90}$.

EXAMPLE 6-10

A sixth-grade class is collecting cans to take to the recycling center. Becky's group brought the following amounts (in pounds). About how many pounds does her group have all together?

$$1\frac{1}{8}, 3\frac{4}{10}, 5\frac{7}{8}, \frac{6}{10}$$

Solution   We can estimate the amount by using front-end estimation with the adjustment made by using $0, \frac{1}{2}$, and 1 as benchmark fractions. The front-end estimate is $(1 + 3 + 5)$, or 9. The adjustment is $\left(0 + \frac{1}{2} + 1 + \frac{1}{2} \right)$, or 2. An adjusted estimate would be $9 + 2$ or 11 lb.

EXAMPLE 6-11

Estimate each of the following:

a. $\frac{27}{13} + \frac{10}{9}$ **b.** $3\frac{9}{10} + 2\frac{7}{8} + \frac{11}{12}$

Solution

a. Because $\frac{27}{13}$ is slightly more than 2 and $\frac{10}{9}$ is slightly more than 1, an estimate might be 3. We know the estimate is low.

b. We first add the whole-number parts to obtain $3 + 2$, or 5. Because each of the fractions, $\frac{9}{10}, \frac{7}{8}$, and $\frac{11}{12}$, is close to but less than 1, their sum is close to but less than 3. The approximate answer is $5 + 3$ or 8.

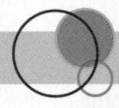

Assessment 6-2A

1. Perform the following additions or subtractions:

   a. $\frac{1}{2} + \frac{2}{3}$

   b. $\frac{4}{12} - \frac{2}{3}$

   c. $\frac{5}{x} + \frac{-3}{y}$

   d. $\frac{-3}{2x^2y} + \frac{5}{2xy^2} + \frac{7}{x^2}$

   e. $\frac{5}{6} + 2\frac{1}{8}$

   f. $-4\frac{1}{2} - 3\frac{1}{6}$

2. Change each of the following fractions to a mixed number:

   a. $\frac{56}{3}$ b. $-\frac{293}{100}$

3. Change each of the following mixed numbers to a fraction in the form $\frac{a}{b}$, where a and b are integers and $b \neq 0$:

   a. $6\frac{3}{4}$ b. $-3\frac{5}{8}$

4. Approximate each of the following situations with a benchmark fraction. Explain your reasoning. Tell whether your estimate is high or low.
   a. Giorgio had 15 base hits out of 46 times at bat.
   b. Ruth made 7 goals out of 41 shots.
   c. Laura answered 62 problems correctly out of 80.
   d. Jonathan made 9 baskets out of 19.

5. Compute $7\frac{1}{4} + 3\frac{5}{12} - 2\frac{1}{3}$.

6. Use the information in the table to answer each of the following questions:

Team	Games Played	Games Won
Ducks	22	10
Beavers	19	10
Tigers	28	9
Bears	23	8
Lions	27	7
Wildcats	25	6
Badgers	21	5

a. Which team won more than $\frac{1}{2}$ of its games and was closest to winning $\frac{1}{2}$ of its games?

b. Which team won less than $\frac{1}{2}$ of its games and was closest to winning $\frac{1}{2}$ of its games?

c. Which team won more than $\frac{1}{3}$ of its games and was closest to winning $\frac{1}{3}$ of its games?

7. Sort the following fraction cards into the ovals by estimating in which oval the fraction belongs:

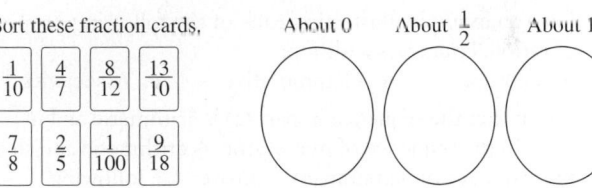

8. Approximate each of the following fractions by $0, \frac{1}{4}, \frac{1}{2}, \frac{3}{4}$, or 1. Tell whether your estimate is high or low.

   a. $\frac{19}{39}$ b. $\frac{3}{197}$

   c. $\frac{150}{201}$ d. $\frac{8}{9}$

9. Without actually finding the exact answer, state which of the numbers given in parentheses in the following is the best approximation for the given sum or difference:

   a. $\frac{6}{13} + \frac{7}{15} + \frac{11}{23} + \frac{17}{35} \left(1, 2, 3, 3\frac{1}{2}\right)$

   b. $\frac{30}{41} + \frac{1}{1000} + \frac{3}{2000} \left(\frac{3}{8}, \frac{3}{4}, 1, 2\right)$

10. Compute each of the following mentally:

    a. $1 - \frac{3}{4}$ b. $3\frac{3}{8} + 2\frac{1}{4} - 5\frac{5}{8}$

11. The following ruler has regions marked M, A, T, H:

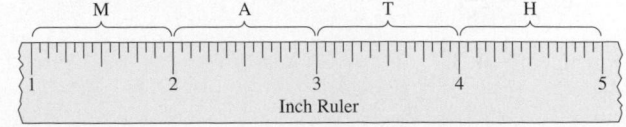

Use mental mathematics and estimation to determine into which region on the ruler each of the following falls. For example, $\frac{12}{5}$ in. falls into region A:

a. $\frac{20}{8}$ in.

b. $\frac{36}{8}$ in.

c. $\frac{60}{16}$ in.

d. $\frac{18}{4}$ in.

12. Use *clustering* to estimate the following sum:

$$3\frac{1}{3} + 3\frac{1}{5} + 2\frac{7}{8} + 2\frac{2}{9}$$

13. A class consists of $\frac{2}{5}$ freshmen, $\frac{1}{4}$ sophomores, and $\frac{1}{10}$ juniors; the rest are seniors. What fraction of the class is seniors?

14. A clerk sold three pieces of one type of ribbon to different customers. One piece was $\frac{1}{3}$ yd long, another was $2\frac{3}{4}$ yd long, and the third was $3\frac{1}{2}$ yd long. What was the total length of that type of ribbon sold?

15. Martine bought $8\frac{3}{4}$ yd of fabric. She wants to make a skirt using $1\frac{7}{8}$ yd, pants using $2\frac{3}{8}$ yd, and a vest using $1\frac{2}{3}$ yd.

How much fabric will be left over?

16. Give an example illustrating each of the following properties of rational numbers addition:
    a. Closure   **b.** Commutative   **c.** Associative

17. Given that the sequence in part (a) is arithmetic and in part (b) the sequence of numerators is arithmetic and so is the sequence of denominators, answer the following:
    (i) Write three more terms of each sequence.
    (ii) Is the sequence in part (b) arithmetic? Justify your answer.

a. $\frac{1}{4}, \frac{1}{2}, \frac{3}{4}, 1, \frac{5}{4}, \cdots$

b. $\frac{1}{2}, \frac{2}{3}, \frac{3}{4}, \frac{4}{5}, \frac{5}{6}, \cdots$

18. Insert five fractions between the numbers 1 and 2 so that the seven numbers (including 1 and 2) constitute part of an arithmetic sequence.

19. **a.** Check that each of the following is true:

   (i) $\frac{1}{3} = \frac{1}{4} + \frac{1}{3 \cdot 4}$

   (ii) $\frac{1}{4} = \frac{1}{5} + \frac{1}{4 \cdot 5}$

   (iii) $\frac{1}{5} = \frac{1}{6} + \frac{1}{5 \cdot 6}$

   b. Based on the examples in (a), write $\frac{1}{n}$ as a sum of two unit fractions; that is, as a sum of fractions with numerator 1.

20. Solve for x in each of the following:
    a. $x + 2\frac{1}{2} = 3\frac{1}{3}$
    b. $x - 2\frac{2}{3} = \frac{5}{6}$

21. Find each sum or difference; simplify if possible.
    a. $\frac{3x}{xy^2} + \frac{y}{x^2}$
    b. $\frac{a}{xy^2} - \frac{b}{xyz}$
    c. $\frac{a^2}{a^2 - b^2} - \frac{a - b}{a + b}$

22. Al runs $\frac{5}{8}$ mi in 10 minutes. Bill runs $\frac{7}{8}$ mi in 10 minutes. If both runners continue to run at the same rate, how much farther can Bill run than Al in 20 minutes?

23. One recipe calls for $1\frac{3}{4}$ cups of milk and a second recipe calls for $1\frac{1}{2}$ cups of milk. If you only have 3 cups of milk, can you make both recipes? Why?

Assessment 6-2B

1. Perform the following additions or subtractions:
   a. $\frac{^-1}{2} + \frac{2}{3}$
   b. $\frac{5}{12} - \frac{2}{3}$
   c. $\frac{5}{4x} + \frac{^-3}{2y}$
   d. $\frac{^-3}{2x^2y^2} + \frac{5}{2xy^2} + \frac{7}{x^2y}$
   e. $\frac{5}{6} - 2\frac{1}{8}$
   f. $^-4\frac{1}{2} + 3\frac{1}{6}$

2. Change each of the following fractions to a mixed number:
   a. $\frac{14}{5}$ **b.** $-\frac{47}{8}$

3. Change each of the following mixed numbers to a fraction in the form $\frac{a}{b}$, where a and b are integers and $b \neq 0$.
   a. $7\frac{1}{2}$ **b.** $^-4\frac{2}{3}$

4. Place the numbers 2, 5, 6, and 8 in the following boxes to make the equation true:

$$\frac{\square}{\square} + \frac{\square}{\square} = \frac{23}{24}$$

5. Compute $5\frac{1}{3} + 5\frac{5}{6} - 3\frac{1}{9}$.

6. Use the information in the table to answer each of the following questions:

Team	Games Played	Games Won
Ducks	22	10
Beavers	19	10
Tigers	28	9
Bears	23	8
Lions	27	7
Wildcats	25	6
Badgers	21	5

a. Which team won less than $\frac{1}{3}$ of its games and was closest to winning $\frac{1}{3}$ of its games?

b. Which team won more than $\frac{1}{4}$ of its games and was closest to winning of $\frac{1}{4}$ of its games?

c. Which teams won less than $\frac{1}{4}$ of their games?

7. Sort the following fraction cards into the ovals by estimating in which oval the fraction belongs:

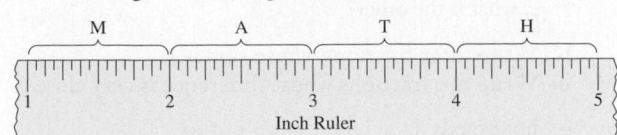

Sort these fraction cards.
$\frac{14}{16}$ $\frac{1}{100}$ $\frac{36}{70}$ $\frac{19}{36}$ $\frac{1}{30}$ $\frac{7}{800}$

About 0   About $\frac{1}{2}$ About 1

8. Approximate each of the following fractions by $0, \frac{1}{4}, \frac{1}{2}, \frac{3}{4},$ or 1. Tell whether your estimate is high or low.

a. $\frac{113}{100}$ b. $\frac{3}{1978}$

c. $\frac{150}{198}$ d. $\frac{8}{9}$

9. Without actually finding the exact answer, state which of the numbers given in parentheses in the following is the best approximation for the given sum or difference:

a. $\frac{2}{13} + \frac{7}{15} + \frac{12}{23} + \frac{33}{35} \left(1, 2, 3, 3\frac{1}{2}\right)$

b. $\frac{30}{41} + \frac{220}{1000} + \frac{5}{2000} \left(\frac{3}{8}, \frac{3}{4}, 1, 2\right)$

10. Compute each of the following mentally:

a. $6 - \frac{7}{8}$ b. $2\frac{3}{5} + 4\frac{1}{10} + 3\frac{3}{10}$

11. The following ruler has regions marked M, A, T, H:

M   A   T   H

```
|1        2        3        4        5|
            Inch Ruler
```

Use mental mathematics and estimation to determine into which region on the ruler each of the following falls. For example, $\frac{12}{5}$ in. falls into region A.

a. $\frac{9}{8}$ in.   b. $\frac{18}{8}$ in.

c. $\frac{50}{16}$ in.   d. $\frac{17}{4}$ in.

12. A class consists of $\frac{1}{4}$ freshmen, $\frac{1}{5}$ sophomores, and $\frac{1}{10}$ juniors; the rest are seniors. What fraction of the class is seniors?

13. The Naturals Company sells its products in many countries. The following two circle graphs show the fractions of the company's earnings for 2009 and 2012. Based on this information, answer the following questions:

a. In 2009, how much greater was the fraction of sales for Japan than for Canada?

b. In 2012, how much less was the fraction of sales for England than for the United States?

c. How much greater was the fraction of total sales for the United States in 2012 than in 2009?

d. Is it true that the amount of sales in dollars in Australia was less in 2009 than in 2012? Why?

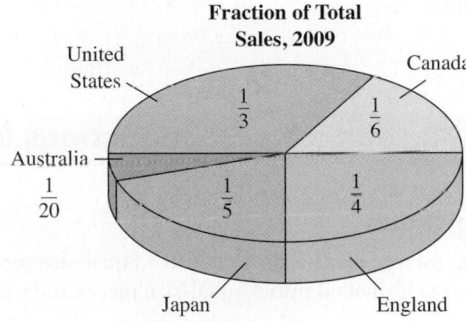

Fraction of Total Sales, 2009

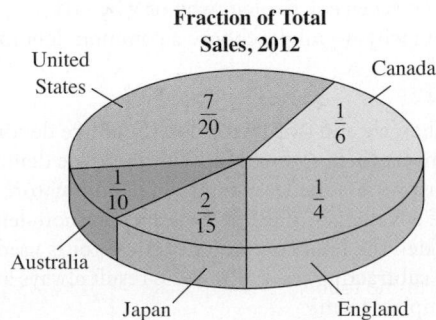

Fraction of Total Sales, 2012

14. A recipe requires $3\frac{1}{2}$ cups of milk. Ran put in $1\frac{3}{4}$ cups and emptied the container. How much more milk does he need to put in?

15. A $15\frac{3}{4}$ in. board is cut in a single cut from a $38\frac{1}{4}$ in. board. The saw cut takes $\frac{3}{8}$ in. How much of the $38\frac{1}{4}$ in. board is left after cutting?

16. Students from Rattlesnake School formed four teams to collect cans for recycling during the months of April and May. The students received 31¢ for each pound of cans. A record of their efforts follows:

Number of Pounds Collected

	Team 1	Team 2	Team 3	Team 4
April	$28\frac{3}{4}$	$32\frac{7}{8}$	$28\frac{1}{2}$	$35\frac{3}{16}$
May	$33\frac{1}{3}$	$28\frac{5}{12}$	$25\frac{3}{4}$	$41\frac{1}{2}$

   a. Which team collected the most for the 2-month period? How much did they collect?

   b. What was the difference in the total pounds collected in April and the total pounds collected in May?

17. In each of the following sequences, the numerators form an arithmetic sequence. Write three more terms of the sequence and determine which of the sequences are arithmetic, and which are not. Justify your answers.

   a. $\frac{2}{3}, \frac{5}{3}, \frac{8}{3}, \frac{11}{3}, \frac{14}{3}, \ldots$ b. $\frac{5}{4}, \frac{3}{4}, \frac{1}{4}, \frac{^-1}{4}, \frac{^-3}{4}, \ldots$

18. Insert four fractions between the numbers 1 and 3 so that the six numbers (including 1 and 3) constitute part of an arithmetic sequence.

19. Solve for x in each of the following:

   a. $x - \frac{5}{6} = \frac{2}{3}$

   b. $x - \frac{7}{2^3 \cdot 3^2} = \frac{5}{2^2 \cdot 3^2}$

20. Find each sum or difference.

   a. $\frac{3x}{xy^2} + \frac{a}{x^2}$

   b. $\frac{a}{xy^2} - \frac{b}{xyz} + \frac{1}{x^2y}$

21. Joe has $\frac{3}{4}$ cup of paint in a container. He uses $\frac{1}{3}$ cup on a project and then adds another $\frac{1}{2}$ cup. How much paint does he have now?

22. a. Find $\frac{1}{2} + \frac{1}{4}$.

   b. Find $\frac{1}{2} + \frac{1}{4} + \frac{1}{8}$.

   c. Find $\frac{1}{2} + \frac{1}{4} + \frac{1}{8} + \frac{1}{16}$.

   d. If you continue in this pattern with powers of 2 in the denominator, will the sum ever become greater than 1? Why?

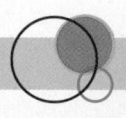

Mathematical Connections 6-2

Communication

1. Suppose a large pizza is divided into 3 equal-size pieces and a small pizza is divided into 4 equal-size pieces and you get 1 piece from each pizza. Does $\frac{1}{3} + \frac{1}{4}$ represent the amount that you received? Explain why or why not.

2. Explain why we might choose a common denominator to add $\frac{1}{3} + \frac{3}{4}$.

3. a. When we add two fractions with unlike denominators and convert them to fractions with the same denominator, must we use the least common denominator? What are the advantages of using the least common denominator?

   b. When the least common denominator is used in adding or subtracting fractions, is the result always a fraction in simplest form?

4. Explain why we can do the following to convert $5\frac{3}{4}$ to a mixed number:

$$\frac{5 \cdot 4 + 3}{4} = \frac{23}{4}$$

5. Kara spent $\frac{1}{2}$ of her allowance on Saturday and $\frac{1}{3}$ of what she had left on Sunday. Can this situation be modeled as $\frac{1}{2} - \frac{1}{3}$? Explain why or why not.

6. Compute $3\frac{3}{4} + 5\frac{1}{3}$ in two different ways and leave your answer as a mixed number. Tell which way you prefer and why.

7. Explain whether each of the following properties hold for subtraction of rational numbers.

   a. Closure   b. Commutative

   c. Associative   d. Identity

   e. Inverse

8. Explain an error pattern in each of the following:

   a. $\frac{13}{35} = \frac{1}{5}, \quad \frac{27}{73} = \frac{2}{3}, \quad \frac{16}{64} = \frac{1}{4}$

   b. $\frac{4}{5} + \frac{2}{3} = \frac{6}{8}, \quad \frac{2}{5} + \frac{3}{4} = \frac{5}{9}, \quad \frac{7}{8} + \frac{1}{3} = \frac{8}{11}$

   c. $8\frac{3}{8} - 6\frac{1}{4} = 2\frac{2}{4}, \quad 5\frac{3}{8} - 2\frac{2}{3} = 3\frac{1}{5}, \quad 2\frac{2}{7} - 1\frac{1}{3} = 1\frac{1}{4}$

Open-Ended

9. Write a story problem for $\frac{2}{3} - \frac{1}{4}$.

10. a. Write two fractions whose sum is 1. If one of the fractions is $\frac{a}{b}$, what is the other?

   b. Write three fractions whose sum is 1.

   c. Write two fractions whose difference is very close to 1 but not exactly 1.

11. a. With the exception of $\frac{2}{3}$, the Egyptians used only unit fractions (fractions that have numerators of 1). Every unit fraction can be expressed as the sum of two unit fractions in more than one way, for example, $\frac{1}{2} = \frac{1}{4} + \frac{1}{4}$ and $\frac{1}{2} = \frac{1}{3} + \frac{1}{6}$. Find at least two different unit fraction representations for each of the following:

  (i) $\frac{1}{3}$

  (ii) $\frac{1}{7}$

  b. Show that $\frac{1}{n} - \frac{1}{n + 1} = \frac{1}{n(n + 1)}$

  c. Rewrite the equation in part (b) as a sum and then use the sum to answer the question in part (a).

  d. Write $\frac{1}{17}$ as a sum of two different unit fractions.

Cooperative Learning

12. Interview 10 people and ask them if and when they add and subtract fractions in their lives. Combine their responses with those of the rest of the class to get a view of how "ordinary" people must use computation of rational numbers in their daily lives.

Questions from the Classroom

13. Kendra showed that $\frac{1}{3} + \frac{3}{4} = \frac{4}{7}$ by using the following figure. How would you help her?

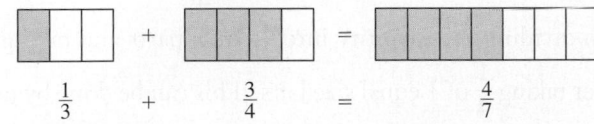

$$\frac{1}{3} \quad + \quad \frac{3}{4} \quad = \quad \frac{4}{7}$$

14. To show $2\frac{3}{4} = \frac{11}{4}$, the teacher drew the following picture. Ken said this shows a picture of $\frac{11}{12}$, not $\frac{11}{4}$. What is Ken thinking and how should the teacher respond?

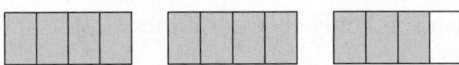

15. Sally claims that it is easier to add two fractions if she adds the numerators and then adds the denominators. How can you help her?

16. Jill claims that for positive fractions, $\frac{a}{b} + \frac{a}{c} = \frac{a}{b + c}$ because the fractions have a common numerator. How do you respond?

Review Problems

17. Simplify each rational number if possible.

  a. $\frac{14}{21}$ **b.** $\frac{117}{153}$

c. $\frac{5^2}{7^2}$

d. $\frac{a^2 + a}{1 + a}$

e. $\frac{a^2 + 1}{a + 1}$

f. $\frac{a^2 - b^2}{a - b}$

18. Determine whether the fractions in each of the following pairs are equal:

  a. $\frac{a^2}{b}$ and $\frac{a^2 b}{b^2}$

  b. $\frac{377}{400}$ and $\frac{378}{401}$

  c. $\frac{0}{10}$ and $\frac{0}{^-10}$

  d. $\frac{a}{b}$ and $\frac{a + 1}{b + 1}$, where $a \neq b$

Trends in Mathematics and Science Study (TIMSS) Question

Janis, Maija, and their mother were eating a cake. Janis ate $\frac{1}{2}$ of the cake. Maija ate $\frac{1}{4}$ of the cake. Their mother ate $\frac{1}{4}$ of the cake. How much of the cake is left?

  a. $\frac{3}{4}$

  b. $\frac{1}{2}$

  c. $\frac{1}{4}$

  d. None

TIMSS, Grade 4, 2007

Which shows a correct procedure for finding $\frac{1}{5} - \frac{1}{3}$?

  a. $\frac{1}{5} - \frac{1}{3} = \frac{1 - 1}{5 - 3}$

  b. $\frac{1}{5} - \frac{1}{3} = \frac{1}{5 - 3}$

  c. $\frac{1}{5} - \frac{1}{3} = \frac{5 - 3}{5 \times 3}$

  d. $\frac{1}{5} - \frac{1}{3} = \frac{3 - 5}{5 \times 3}$

TIMSS, Grade 8, 2007

Tickets for a concert cost either 10 zeds, 15 zeds, or 30 zeds. Of the 900 tickets sold, $\frac{1}{5}$ cost 30 zeds each and $\frac{2}{3}$ cost 15 zeds each.

What FRACTION of the tickets sold for 10 zeds each?

Answer: _____

TIMSS, Grade 8, 2007

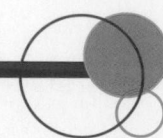

6-3 Multiplication and Division of Rational Numbers

Multiplication of Rational Numbers

To motivate the definition of multiplication of rational numbers, we use the interpretation of multiplication as repeated addition. Using repeated addition, we can interpret $3\left(\dfrac{3}{4}\right)$ as follows:

$$3\left(\frac{3}{4}\right) = \frac{3}{4} + \frac{3}{4} + \frac{3}{4} = \frac{9}{4} = 2\frac{1}{4}$$

The area model in Figure 6-10 is another way to calculate this product.

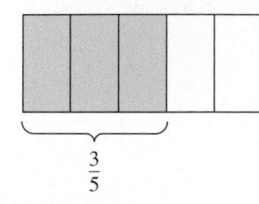

$$3\left(\frac{3}{4}\right) = \qquad \frac{3}{4} \quad + \quad \frac{3}{4} \quad + \quad \frac{3}{4} \quad = \quad \frac{9}{4}, \quad \text{or} \quad 2\frac{1}{4}$$

Figure 6-10

We next consider $\left(\dfrac{3}{4}\right)3$. How should this product be interpreted? If the commutative property of multiplication of rational numbers is to hold, then $\left(\dfrac{3}{4}\right)3 = 3\left(\dfrac{3}{4}\right) = \dfrac{9}{4}$.

Next, we consider another interpretation of multiplication. What is $\dfrac{3}{4}$ of 3? Recall that $\dfrac{3}{4}$ of a quantity is the amount resulting from dividing the quantity into 4 equal parts and taking 3 of these parts. To see what $\dfrac{1}{4}$ of 3 is consider taking $\dfrac{1}{4}$ of 3 equal size bars. This can be done by taking $\dfrac{1}{4}$ of each of the 3 bars, that is, $\dfrac{1}{4} + \dfrac{1}{4} + \dfrac{1}{4} = \dfrac{3}{4}$. Thus $\dfrac{3}{4}$ of 3 is $3 \cdot \dfrac{3}{4}$. Thus we can interpret $\dfrac{3}{4} \cdot 3$ as $\dfrac{3}{4}$ of 3.

If forests once covered about $\dfrac{3}{5}$ of Earth's land and only about $\dfrac{1}{2}$ of these forests remain, what fraction of Earth is covered with forests today? We need to find $\dfrac{1}{2}$ of $\dfrac{3}{5}$, and can use an area model to find the answer.

Figure 6-11(a) shows a rectangle representing the *whole* separated into fifths, with $\dfrac{3}{5}$ shaded. To find $\dfrac{1}{2}$ of $\dfrac{3}{5}$, we divide the shaded portion of the rectangle in Figure 6-11(a) into two congruent parts and take one of those parts. The result would be the green portion of Figure 6-11(b). However, the green portion represents 3 parts out of 10, or $\dfrac{3}{10}$, of the whole. Thus,

$$\frac{1}{2} \cdot \frac{3}{5} = \frac{3}{10} = \frac{1 \cdot 3}{2 \cdot 5}.$$

(a)

(b)

Figure 6-11

An area model like the one in Figure 6-11 is used on the student page on page 294. Note that in part (g) two fractions with numerators other than 1 are used. First, $\frac{5}{6}$ of the whole is shaded vertically in red, then $\frac{3}{4}$ of the whole is shaded horizontally in yellow. This results in $\frac{3}{4}$ of the $\frac{5}{6}$ or $\frac{3}{4} \cdot \frac{5}{6}$ being shaded in orange. We see that 15 of the 24 equal-sized parts are shaded orange, showing that $\frac{3}{4} \cdot \frac{5}{6} = \frac{3 \cdot 5}{4 \cdot 6} = \frac{15}{24}$.

This discussion leads to the following definition of multiplication for rational numbers.

Definition of Multiplication of Rational Numbers

If $\frac{a}{b}$ and $\frac{c}{d}$ are any rational numbers, then $\frac{a}{b} \cdot \frac{c}{d} = \frac{ac}{bd}$.

 **EXAMPLE 6-12**

If $\frac{5}{6}$ of the population of a certain city are college graduates and $\frac{7}{11}$ of the city's college graduates are female, what fraction of the population of that city is female college graduates?

Solution The fraction should be $\frac{7}{11}$ of $\frac{5}{6}$, or $\frac{7}{11} \cdot \frac{5}{6} = \frac{7 \cdot 5}{11 \cdot 6} = \frac{35}{66}$.

The fraction of the population that is female college graduates is $\frac{35}{66}$.

Properties of Multiplication of Rational Numbers

Multiplication of rational numbers has properties analogous to the properties of multiplication of integers. These include the following properties for multiplication: closure, commutative, associative, and multiplicative identity. When we expand from the set of integers to the set of rationals, we pick up an additional property; that is, the multiplicative inverse property. For emphasis, we list the last two properties.

Theorem 6-10: Multiplicative Identity and Multiplicative Inverse Properties of Rational Numbers

a. The number 1 is the unique number such that for every rational number $\frac{a}{b}$,

$$1 \cdot \left(\frac{a}{b}\right) = \frac{a}{b} = \left(\frac{a}{b}\right) \cdot 1.$$

b. For any nonzero rational number $\frac{a}{b}$, $\frac{b}{a}$, $a \neq 0$, is the unique rational number such that

$$\frac{a}{b} \cdot \frac{b}{a} = 1 = \frac{b}{a} \cdot \frac{a}{b}.$$

$\frac{b}{a}$ is the multiplicative inverse of $\frac{a}{b}$ and $\frac{b}{a}$ is also called the **reciprocal** of $\frac{a}{b}$.

School Book Page   MULTIPLYING FRACTIONS

Lesson 8-12

Key Idea
You can multiply two fractions to find a fraction of a fraction.

Materials
• square sheets of paper
• red and yellow colored pencils
or 🔵 **tools**

Think It Through
• I can **make a model** to show the problem.
• I can **look for patterns** in the products to find a multiplication rule.

Multiplying Fractions

LEARN

Activity

How can you use a model to find the product of two fractions?

You can use paper folding to find $\frac{1}{2}$ of $\frac{3}{4}$.

a. Fold a square sheet of paper vertically in half.

b. Fold the paper vertically in half again. Each section is what fraction of the sheet of paper?

c. Now fold the paper in half horizontally. Each section is what fraction of the sheet of paper?

d. Shade $\frac{3}{4}$ of the vertical sections red and $\frac{1}{2}$ of the horizontal sections yellow.

e. The part that is shaded both red and yellow shows $\frac{1}{2} \times \frac{3}{4}$. What is $\frac{1}{2} \times \frac{3}{4}$?

f. Fold paper to find each product.

$\frac{1}{2} \times \frac{1}{2} = \blacksquare$ $\frac{1}{2} \times \frac{1}{4} = \blacksquare$ $\frac{1}{4} \times \frac{1}{4} = \blacksquare$ $\frac{3}{4} \times \frac{3}{4} = \blacksquare$ $\frac{1}{2} \times \frac{5}{8} = \frac{\blacksquare}{\blacksquare}$

g. Use the pictures to find the products.

$\frac{3}{4} \times \frac{1}{3} = \blacksquare$ $\frac{1}{3} \times \frac{4}{5} = \blacksquare$ $\frac{3}{4} \times \frac{5}{6} = \frac{\blacksquare}{\blacksquare}$

h. Study the numerators and denominators in each problem in parts e and f. What pattern do you see between the fractions multiplied and their products?

496

EXAMPLE 6-13

Find the multiplicative inverse, if possible, for each of the following rational numbers:

a. $\dfrac{2}{3}$ b. $\dfrac{^-2}{5}$ c. 4   d. 0   e. $6\dfrac{1}{2}$

Solution

a. $\dfrac{3}{2}$

b. $\dfrac{5}{^-2}$, or $\dfrac{^-5}{2}$

c. Because $4 = \dfrac{4}{1}$, the multiplicative inverse of 4 is $\dfrac{1}{4}$.

d. Even though $0 = \dfrac{0}{1}, \dfrac{1}{0}$ is undefined; there is no multiplicative inverse of 0.

e. Because $6\dfrac{1}{2} = \dfrac{13}{2}$, the multiplicative inverse of $6\dfrac{1}{2}$ is $\dfrac{2}{13}$.

Multiplication and addition are connected through the distributive property of multiplication over addition. Also there are multiplication properties of equality and inequality for rational numbers and a multiplication property of zero similar to those for whole numbers and integers. These properties can be proved using the definition of operations on rational numbers and properties of integers. They are stated in the following theorem.

Theorem 6-11:   Properties of Rational Number Operations

a. **Distributive Properties of Multiplication Over Addition and Subtraction for Rational Numbers**

If $\dfrac{a}{b}, \dfrac{c}{d}$, and $\dfrac{e}{f}$ are rational numbers, then

$$\dfrac{a}{b}\left(\dfrac{c}{d} + \dfrac{e}{f}\right) = \dfrac{a}{b} \cdot \dfrac{c}{d} + \dfrac{a}{b} \cdot \dfrac{e}{f}; \dfrac{a}{b}\left(\dfrac{c}{d} - \dfrac{e}{f}\right) = \dfrac{a}{b} \cdot \dfrac{c}{d} - \dfrac{a}{b} \cdot \dfrac{e}{f}.$$

b. **Multiplication Property of Equality for Rational Numbers**

If $\dfrac{a}{b}, \dfrac{c}{d}$, and $\dfrac{e}{f}$ are rational numbers such that $\dfrac{a}{b} = \dfrac{c}{d}$, then $\dfrac{a}{b} \cdot \dfrac{e}{f} = \dfrac{c}{d} \cdot \dfrac{e}{f}$.

c. **Multiplication Properties of Inequality for Rational Numbers**

(i) If $\dfrac{a}{b} > \dfrac{c}{d}$ and $\dfrac{e}{f} > 0$, then $\dfrac{a}{b} \cdot \dfrac{e}{f} > \dfrac{c}{d} \cdot \dfrac{e}{f}$.

(ii) If $\dfrac{a}{b} > \dfrac{c}{d}$ and $\dfrac{e}{f} < 0$, then $\dfrac{a}{b} \cdot \dfrac{e}{f} < \dfrac{c}{d} \cdot \dfrac{e}{f}$.

d. **Multiplication Property of Zero for Rational Numbers**

If $\dfrac{a}{b}$ is a rational number, then $\dfrac{a}{b} \cdot 0 = 0 = 0 \cdot \dfrac{a}{b}$.

EXAMPLE 6-14

A bicycle is on sale at $\frac{3}{4}$ of its original price. If the sale price is \$330, what was the original price?

Solution   Let x be the original price. Then $\frac{3}{4}$ of the original price is $\frac{3}{4}x$. Because the sale price is \$330, we have $\frac{3}{4}x = 330$. Solving for x gives

$$\frac{4}{3} \cdot \frac{3}{4}x = \frac{4}{3} \cdot 330$$
$$1 \cdot x = 440$$
$$x = 440.$$

Thus, the original price was \$440.

An alternative approach, which does not use algebra, follows. Because $\frac{3}{4} = 3\left(\frac{1}{4}\right)$, then 3 of the $\frac{1}{4}$ parts is \$330 and 1 of the $\frac{1}{4}$ parts is \$110. If 1 of the $\frac{1}{4}$ parts is \$110, then 4 of these is $4 \cdot \$110$, or \$440.

Multiplication with Mixed Numbers

In the *Peanuts* cartoon, Sally is having trouble multiplying two mixed numbers. If she used an estimate to check whether her answer was reasonable, she would notice that if she multiplies two numbers that are both less than 3, the answer must be less than 9.

Peanuts copyright © 1988 and 2011 Peanuts Worldwide LLC. Distributed by Universal Uclick. Reprinted with permission. All rights reserved.

One way to multiply $2\frac{1}{2} \cdot 2\frac{1}{2}$ is to change the mixed numbers to improper fractions and use the definition of multiplication, as shown here.

$$2\frac{1}{2} \cdot 2\frac{1}{2} = \frac{5}{2} \cdot \frac{5}{2} = \frac{25}{4}$$

We could then change $\frac{25}{4}$ to the mixed number $6\frac{1}{4}$.

Another way to multiply mixed numbers uses the distributive property of multiplication over addition, as seen below.

$$2\frac{1}{2} \cdot 2\frac{1}{2} = \left(2 + \frac{1}{2}\right)\left(2 + \frac{1}{2}\right)$$
$$= \left(2 + \frac{1}{2}\right)2 + \left(2 + \frac{1}{2}\right)\frac{1}{2}$$
$$= 2 \cdot 2 + \frac{1}{2} \cdot 2 + 2 \cdot \frac{1}{2} + \frac{1}{2} \cdot \frac{1}{2}$$

$$= 4 + 1 + 1 + \frac{1}{4}$$

$$= 6 + \frac{1}{4}$$

$$= 6\frac{1}{4}$$

Multiplication of fractions enables us to obtain equivalent fractions, to perform addition and subtraction of fractions, as well as to solve equations in a different way, as shown in the following example.

EXAMPLE 6-15

Use the definition of multiplication of fractions and its properties to justify the following:

a. The Fundamental Law of Fractions, $\frac{a}{b} = \frac{an}{bn}$ if $b \neq 0, n \neq 0$.

b. Addition of fractions using a common denominator.

Solution

a. $\dfrac{a}{b} = \dfrac{a}{b} \cdot 1 = \dfrac{a}{b} \cdot \dfrac{n}{n} = \dfrac{an}{bn}$

b. $\dfrac{a}{b} + \dfrac{c}{d} = \dfrac{a}{b} \cdot \dfrac{d}{d} + \dfrac{c}{d} \cdot \dfrac{b}{b}$

$$= \dfrac{ad}{bd} + \dfrac{bc}{bd}$$

$$= \dfrac{ad + bc}{bd}$$

Division of Rational Numbers

In the grade 6 *Common Core Standards*, we find the following concerning division of rational numbers:

> Apply and extend previous understandings of multiplication and division to divide fractions by fractions.
>
> 1. Interpret and compute quotients of fractions and solve word problems involving division of fractions by fractions, e.g., by using visual fraction models and equations to represent the problem. (p. 42)

We apply and extend division of whole numbers by recalling that $6 \div 3$ means "How many 3s are there in 6?" We found that $6 \div 3 = 2$ because $3 \cdot 2 = 6$ and, in general, when $a, b, c \in W$, then $a \div b = c$, if, and only if, c is the unique whole number such that $bc = a$. Consider $3 \div \left(\dfrac{1}{2}\right)$, which is equivalent to finding how many halves there are in 3. We see from the area model in Figure 6-12 that there are 6 half pieces in the 3 whole pieces. We record this as $3 \div \left(\dfrac{1}{2}\right) = 6$. This is true because $\left(\dfrac{1}{2}\right)6 = 3$.

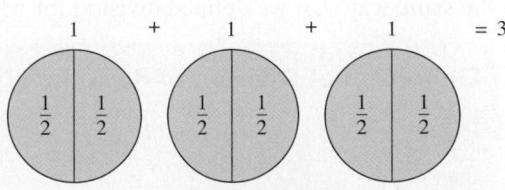

Figure 6-12

Another way to show that $3 \div \left(\frac{1}{2}\right) = 6$ is on a ruler. In Figure 6-13 we see that there are six $\frac{1}{2}$s in 3.

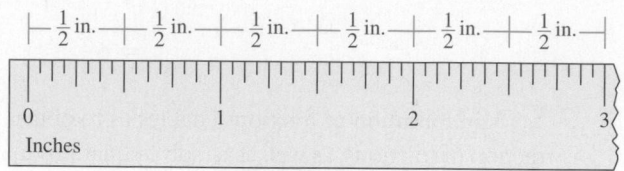

Figure 6-13

With whole numbers one way to think about division was in terms of *repeated subtraction*. We found that $6 \div 2 = 3$ because 2 could be subtracted from 6 three times; that is, $6 - 3 \cdot 2 = 0$. Similarly, with $3 \div \frac{1}{2}$, we want to know how many halves can be subtracted from 3. Because $3 - 6\left(\frac{1}{2}\right) = 0$, we know that $3 \div \frac{1}{2} = 6$.

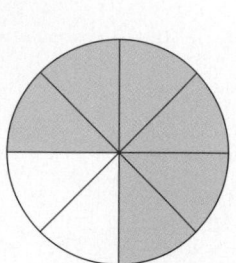

Figure 6-14

Next, consider $\frac{3}{4} \div \frac{1}{8}$. This means "How many $\frac{1}{8}$s are in $\frac{3}{4}$?" Figure 6-14 shows that there are six $\frac{1}{8}$s in the shaded portion, which represents $\frac{3}{4}$ of the whole. Therefore, $\frac{3}{4} \div \frac{1}{8} = 6$. This is true because $\left(\frac{1}{8}\right)6 = \frac{3}{4}$. Using repeated subtraction, we see that $\frac{3}{4} \div \frac{1}{8} = \frac{6}{8} \div \frac{1}{8}$ and that $\frac{6}{8} - 6\left(\frac{1}{8}\right) = 0$, so $\frac{3}{4} \div \frac{1}{8} = 6$.

The measurement or number line model may be used to understand division of fractions. For example, consider $\frac{7}{8} \div \frac{3}{4}$. First we draw a measurement or number line divided into eighths, as shown in Figure 6-15. Next we want to know how many $\frac{3}{4}$s there are in $\frac{7}{8}$. The bar of length $\frac{3}{4}$ is made up of 6 equal-size pieces of length $\frac{1}{8}$. We see that there is at least one length of $\frac{3}{4}$ in $\frac{7}{8}$. If we put another bar of length $\frac{3}{4}$ on the number line, we see there is 1 more of the 6 equal-length segments needed to make $\frac{7}{8}$. Therefore, the answer is $1\frac{1}{6}$, or $\frac{7}{6}$.

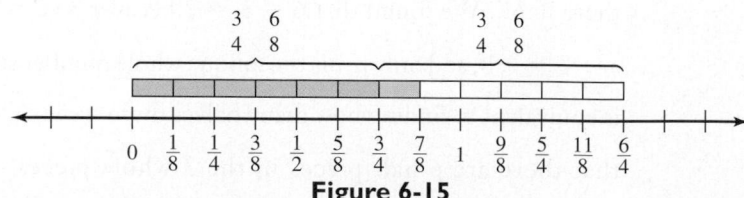

Figure 6-15

In the previous examples, we saw a relationship between division and multiplication of rational numbers. We can define division for rational numbers formally in terms of multiplication in the same way that we defined division for whole numbers.

Definition of Division of Rational Numbers

If $\frac{a}{b}$ and $\frac{c}{d}$ are any rational numbers, then $\frac{a}{b} \div \frac{c}{d} = \frac{e}{f}$ if, and only if, $\frac{e}{f}$ is the unique rational number such that $\frac{c}{d} \cdot \frac{e}{f} = \frac{a}{b}$.

In the above definition of division, $\frac{c}{d} \neq 0$ because in that case there is no unique $\frac{e}{f}$ satisfying the definition.

NOW TRY THIS 6-13

Students often confuse division by 2 and division by $\frac{1}{2}$. Notice that

$$a \div 2 = \frac{a}{2} = \frac{1}{2}a, \text{ but } a \div \frac{1}{2} = x \text{ if, and only if, } \frac{1}{2}x = a, \ 2\left(\frac{1}{2}x\right) = 2a, \ x = 2a.$$

Write a real-life story that will help students see the difference between division by 2 and division by $\frac{1}{2}$.

Algorithm for Division of Rational Numbers

As we see in the *Peanuts* cartoon, Peppermint Patty doesn't understand why the algorithm for division of fractions works. She is not alone in her confusion, and we will explain why "to divide fractions we use the reciprocal and multiply."

Does the method in the cartoon, often called the *invert-and-multiply* method, make sense based on what we know about rational numbers? Consider what such a division might mean. For example, using the definition of division of rational numbers,

$$\frac{2}{3} \div \frac{5}{7} = x \quad \text{implies} \quad \frac{2}{3} = \frac{5}{7}x.$$

To solve for x, we multiply both sides of the equation by $\frac{7}{5}$, the reciprocal of $\frac{5}{7}$. Thus,

$$\frac{7}{5} \cdot \frac{2}{3} = \frac{7}{5}\left(\frac{5}{7}x\right) = \left(\frac{7}{5} \cdot \frac{5}{7}\right)x = 1 \cdot x = x.$$

Therefore, $\frac{2}{3} \div \frac{5}{7} = \frac{2}{3} \cdot \frac{7}{5}$. This illustrates Peppermint Patty's rule of "use the reciprocal and multiply."

A traditional justification of this rule follows. The algorithm for division of fractions is usually justified in the middle grades by using the Fundamental Law of Fractions, $\frac{a}{b} = \frac{ac}{bc}$, where a, b, and c are all fractions, or equivalently, the identity property of multiplication. For example,

$$\frac{2}{3} \div \frac{5}{7} = \frac{\frac{2}{3}}{\frac{5}{7}} = \frac{\frac{2}{3}}{\frac{5}{7}} \cdot 1 = \frac{\frac{2}{3} \cdot \frac{7}{5}}{\frac{5}{7} \cdot \frac{7}{5}} = \frac{\frac{2}{3} \cdot \frac{7}{5}}{1} = \frac{2}{3} \cdot \frac{7}{5}.$$

Thus,

$$\frac{2}{3} \div \frac{5}{7} = \frac{2}{3} \cdot \frac{7}{5}.$$

 NOW TRY THIS 6-14

Use an argument similar to the preceding one to show that, in general, if $\frac{a}{b}$ and $\frac{c}{d}$ are rational numbers and $\frac{c}{d} \neq 0$, then $\frac{a}{b} \div \frac{c}{d} = \frac{a}{b} \cdot \frac{d}{c}$.

We summarize the algorithm as follows:

> **Theorem 6-12:   Algorithm for Division of Fractions**
>
> If $\frac{a}{b}$ and $\frac{c}{d}$ are any rational numbers and $\frac{c}{d} \neq 0$, then
>
> $$\frac{a}{b} \div \frac{c}{d} = \frac{a}{b} \cdot \frac{d}{c}.$$

Alternative Algorithm for Division of Rational Numbers

An alternative algorithm for division of fractions can be found by first dividing fractions that have equal denominators. For example, $\frac{9}{10} \div \frac{3}{10} = \frac{9}{10} \cdot \frac{10}{3} = \frac{9}{3}$ and $\frac{15}{23} \div \frac{5}{23} = \frac{15}{23} \cdot \frac{23}{5} = \frac{15}{5}$.

These examples suggest that when two fractions with the same denominator are divided, the result can be obtained by dividing the numerator of the first fraction by the numerator of the second; that is, $\frac{a}{b} \div \frac{c}{b} = \frac{a}{c}$. To divide fractions with different denominators, we rename the fractions so that the denominators are equal. Thus,

$$\frac{a}{b} \div \frac{c}{d} = \frac{ad}{bd} \div \frac{bc}{bd} = \frac{ad}{bd} \cdot \frac{bd}{bc} = \frac{ad}{bc}.$$

 NOW TRY THIS 6-15

Show that $\frac{a}{b} \div \frac{c}{d}$ and $\frac{a \div c}{b \div d}$ are equivalent.

The next three examples illustrate the use of division of rational numbers.

EXAMPLE 6-16

A radio station provides 36 min for public service announcements for every 24 hr of broadcasting.

   a. What part of the 24-hr broadcasting day is allotted to public service announcements?

   b. How many $\frac{3}{4}$-min public service announcements can be allowed in the 36 min?

Solution

a. There are 60 min in an hour and $60 \cdot 24$ min in the broadcasting day. Thus, $36/(60 \cdot 24)$, or $\frac{1}{40}$, of the day is allotted for the announcements.

b. $36/\left(\frac{3}{4}\right) = 36\left(\frac{4}{3}\right)$, or 48, announcements are allowed.

EXAMPLE 6-17

We have $35\frac{1}{2}$ yd of material available to make towels. Each towel requires $\frac{3}{8}$ yd of material.

a. How many towels can be made?

b. How much material will be left over?

Solution

a. We need to find the integer part of the answer to $35\frac{1}{2} \div \frac{3}{8}$. The division follows:

$$35\frac{1}{2} \div \frac{3}{8} = \frac{71}{2} \cdot \frac{8}{3} = \frac{284}{3} = 94\frac{2}{3}$$

Thus we can make 94 towels.

b. Because the division in (a) was by $\frac{3}{8}$, the amount of material left over is $\frac{2}{3}$ of $\frac{3}{8}$, or $\frac{2}{3} \cdot \frac{3}{8}$, or $\frac{1}{4}$ yd. This can also be answered by noting that the $\frac{2}{3}$ in part (a) is two-thirds of a towel, which requires $\frac{2}{3}$ of $\frac{3}{8}$ yd of material.

EXAMPLE 6-18

A bookstore has a shelf that is $37\frac{1}{2}$ in. long. Each book that is to be placed on the shelf is $1\frac{1}{4}$ in. thick. How many books can be placed on the shelf?

Solution   We need to find how many $1\frac{1}{4}$s there are in $37\frac{1}{2}$. Thus,

$$\frac{37\frac{1}{2}}{1\frac{1}{4}} = \frac{75}{2} \div \frac{5}{4}$$
$$= \frac{75}{2} \cdot \frac{4}{5}$$
$$= \frac{300}{10}$$
$$= 30$$

Note that we could compute $\frac{75}{2} \cdot \frac{4}{5}$ by first eliminating common factors; that is, $\frac{75}{2} \cdot \frac{4}{5} = \frac{15 \cdot 2}{1 \cdot 1} = \frac{30}{1} = 30$.

Therefore 30 books can be placed on the shelf.

Estimation and Mental Math with Rational Numbers

Estimation and mental math strategies that were developed with whole numbers can also be used with rational numbers.

EXAMPLE 6-19

Use mental math to find

a. $(12 \cdot 25)\dfrac{1}{4}$ **b.** $\left(5\dfrac{1}{6}\right)12$ **c.** $\dfrac{4}{5}(20)$

Solution   Each of the following is a possible approach:

a. $(12 \cdot 25)\dfrac{1}{4} = 25\left(12 \cdot \dfrac{1}{4}\right) = 25 \cdot 3 = 75$

b. $\left(5\dfrac{1}{6}\right)12 = \left(5 + \dfrac{1}{6}\right)12 = 5 \cdot 12 + \dfrac{1}{6} \cdot 12 = 60 + 2 = 62$

c. $\dfrac{4}{5}(20) = 4\left(\dfrac{1}{5} \cdot 20\right) = 4 \cdot 4 = 16$

EXAMPLE 6-20

Estimate each of the following:

a. $3\dfrac{1}{4} \cdot 7\dfrac{8}{9}$ **b.** $24\dfrac{5}{7} \div 4\dfrac{1}{8}$

Solution
   a. Using rounding, the product will be close to $3 \cdot 8 = 24$. If we use the range strategy, we can say the product must be between $3 \cdot 7 = 21$ and $4 \cdot 8 = 32$.
   b. We can use compatible numbers a and think of the estimate as $24 \div 4 = 6$ or $25 \div 5 = 5$.

Extending the Notion of Exponents

Recall that a^m was defined for any whole number a and any natural number m as the product of m a's. We define a^m for any rational number a in a similar way as follows.

> **Definition of a to the mth Power**
>
> $a^m = \underbrace{a \cdot a \cdot a \cdot \ \ldots \ \cdot a}_{m \text{ factors}}$, where a is any rational number and m is any natural number.

From the definition, $a^3 \cdot a^2 = (a \cdot a \cdot a)(a \cdot a) = a^{3+2}$. In a similar way, it follows that

1. If a is any rational number and m and n are natural numbers, $a^m \cdot a^n = a^{m+n}$.

   If this property is to be true for all whole numbers m and n, then because $a^1 \cdot a^0 = a^{1+0} = a^1$, we must have $a^0 = 1$. Hence, it is useful to give meaning to a^0 when $a \neq 0$ as follows.

2. For any nonzero rational number a, $a^0 = 1$.

   If $a^m \cdot a^n = a^{m+n}$ is extended to all integer powers of a, then how should a^{-3} be defined? If (1) is to be true for all integers m and n, then $a^{-3} \cdot a^3 = a^{-3+3} = a^0 = 1$. Therefore, $a^{-3} = 1/a^3$. This is true in general and we have the following.

3. For any nonzero rational number a and any natural number n, $a^{-n} = \dfrac{1}{a^n}$.

In elementary grades the definition of a^{-n} is typically motivated by looking at patterns. Notice that as the following exponents decrease by 1, the numbers on the right are divided by 10. Thus the pattern might be continued, as shown.

$$10^3 = 10 \cdot 10 \cdot 10$$
$$10^2 = 10 \cdot 10$$
$$10^1 = 10$$
$$10^0 = 1$$
$$10^{-1} = \frac{1}{10} = \frac{1}{10^1}$$
$$10^{-2} = \frac{1}{10} \cdot \frac{1}{10} = \frac{1}{10^2}$$
$$10^{-3} = \frac{1}{10^2} \cdot \frac{1}{10} = \frac{1}{10^3}$$

If the pattern is extended in this way, then we would predict that $10^{-n} = \dfrac{1}{10^n}$. Notice that this is inductive reasoning and hence is not a mathematical justification.

Consider whether the property $a^m \cdot a^n = a^{m+n}$ can be extended to include all powers of a, where the exponents are integers. For example, is it true that $2^4 \cdot 2^{-3} = 2^{4+\,-3} = 2^1$? The definitions of 2^{-3} and the properties of nonnegative exponents ensure this is true, as shown next.

$$2^4 \cdot 2^{-3} = 2^4 \cdot \frac{1}{2^3} = \frac{2^4}{2^3} = \frac{2^1 \cdot 2^3}{2^3} = 2^1$$

Similarly, $2^{-4} \cdot 2^{-3} = 2^{-4+\,-3} = 2^{-7}$ is true because

$$2^{-4} \cdot 2^{-3} = \frac{1}{2^4} \cdot \frac{1}{2^3} = \frac{1 \cdot 1}{2^4 \cdot 2^3} = \frac{1}{2^{4+3}} = \frac{1}{2^7} = 2^{-7}.$$

In general, with integer exponents, the following theorem holds.

Theorem 6-13

For any nonzero rational number a and any integers m and n, $a^m \cdot a^n = a^{m+n}$.

If $a = 0$ then Theorem 6-13 is invalid if m is negative because then we would be dividing by zero.

Other properties of exponents can be developed by using the properties of rational numbers. For example,

$$\frac{2^5}{2^3} = \frac{2^3 \cdot 2^2}{2^3} = 2^2 = 2^{5-3} \qquad \frac{2^5}{2^8} = \frac{2^5}{2^5 \cdot 2^3} = \frac{1}{2^3} = 2^{-3} = 2^{5-8}.$$

With integer exponents, the following theorem holds.

Theorem 6-14

For any nonzero rational number a and any integers m and n, $\dfrac{a^m}{a^n} = a^{m-n}$.

At times, we may encounter an expression like $(2^4)^3$. This expression can be written as a single power of 2 as follows:

$$(2^4)^3 = 2^4 \cdot 2^4 \cdot 2^4 = 2^{4+4+4} = 2^{3 \cdot 4} = 2^{12}$$

In general, if a is any rational number and m and n are positive integers, then

$$(a^m)^n = \underbrace{a^m \cdot a^m \cdot a^m \cdot \ldots \cdot a^m}_{n \text{ factors}} = a^{\overbrace{m+m+\ldots+m}^{n \text{ terms}}} = a^{nm} = a^{mn}.$$

Does this theorem hold for negative-integer exponents? For example, does $(2^3)^{-4} = 2^{(3)(-4)} = 2^{-12}$? The answer is yes because $(2^3)^{-4} = \dfrac{1}{(2^3)^4} = \dfrac{1}{2^{12}} = 2^{-12}$. Also, $(2^{-3})^4 = \left(\dfrac{1}{2^3}\right)^4 = \dfrac{1}{2^3} \cdot \dfrac{1}{2^3} \cdot \dfrac{1}{2^3} \cdot \dfrac{1}{2^3} = \dfrac{1^4}{(2^3)^4} = \dfrac{1}{2^{12}} = 2^{-12}$.

> **Theorem 6-15**
>
> For any nonzero rational number a and any integers m and n,
> $$(a^m)^n = a^{mn}.$$

If $a = 0$, Theorem 6-15 is valid as long as neither of m and n is negative.

Using the definitions and theorems developed, we derive additional properties, for example:

$$\left(\frac{2}{3}\right)^4 = \frac{2}{3} \cdot \frac{2}{3} \cdot \frac{2}{3} \cdot \frac{2}{3} = \frac{2 \cdot 2 \cdot 2 \cdot 2}{3 \cdot 3 \cdot 3 \cdot 3} = \frac{2^4}{3^4}.$$

This property is generalized as follows.

> **Theorem 6-16**
>
> For any nonzero rational number $\dfrac{a}{b}$ and any integer m,
> $$\left(\frac{a}{b}\right)^m = \frac{a^m}{b^m}.$$

From the definition of negative exponents, the preceding theorem, and division of fractions, we have

$$\left(\frac{a}{b}\right)^{-m} = \frac{1}{\left(\dfrac{a}{b}\right)^m} = \frac{1}{\dfrac{a^m}{b^m}} = \frac{b^m}{a^m} = \left(\frac{b}{a}\right)^m.$$

> **Theorem 6-17**
>
> For any nonzero rational number $\dfrac{a}{b}$ and any integer m, $\left(\dfrac{a}{b}\right)^{-m} = \left(\dfrac{b}{a}\right)^m$.

A property similar to the one in Theorem 6-16 holds for multiplication. For example,

$$(2 \cdot 3)^{-3} = \frac{1}{(2 \cdot 3)^3} = \frac{1}{2^3 \cdot 3^3} = \left(\frac{1}{2^3}\right) \cdot \left(\frac{1}{3^3}\right) = 2^{-3} \cdot 3^{-3}.$$

This property is generalized as follows.

> **Theorem 6-18**
>
> For any nonzero rational numbers a and b and any integer m,
> $$(a \cdot b)^m = a^m \cdot b^m.$$

If a or $b = 0$ and $m > 0$, the theorem is still true.

The properties of exponents are summarized below.

> **Theorem 6-19:   Properties of Exponents**
>
> For any nonzero rational numbers a and b and integers m and n, the following are true.
>
> **a.** $a^0 = 1$
>
> **b.** $a^{-m} = \dfrac{1}{a^m}$
>
> **c.** $a^m \cdot a^n = a^{m+n}$
>
> **d.** $\dfrac{a^m}{a^n} = a^{m-n}$
>
> **e.** $(a^m)^n = a^{nm}$
>
> **f.** $\left(\dfrac{a}{b}\right)^m = \dfrac{a^m}{b^m}$
>
> **g.** $\left(\dfrac{a}{b}\right)^{-m} = \left(\dfrac{b}{a}\right)^m$
>
> **h.** $(ab)^m = a^m b^m$

Notice that property (h) is for multiplication. Analogous properties do not hold for addition and subtraction. For example, in general, $(a + b)^{-1} \neq a^{-1} + b^{-1}$. To see why, a numerical example is sufficient, but it is instructive to write each side with positive exponents:

$$(a + b)^{-1} = \frac{1}{a + b}$$

$$a^{-1} + b^{-1} = \frac{1}{a} + \frac{1}{b}$$

We know from addition of fractions, $\dfrac{1}{a + b} \neq \dfrac{1}{a} + \dfrac{1}{b}$. Therefore, $(a + b)^{-1} \neq a^{-1} + b^{-1}$.

EXAMPLE 6-21

In each of the following, show each equality or inequality is true in general for nonzero rational numbers x, a, and b.

a. $(^{-}x)^{-2} \neq {}^{-}x^{-2}$
b. $(^{-}x)^{-3} = {}^{-}x^{-3}$
c. $ab^{-1} \neq (ab)^{-1}$ if $a \neq 1$ or $a \neq -1$
d. $(a^{-2}b^{-2})^{-1} = a^2 b^2$
e. $(a^{-2} + b^{-2})^{-1} \neq a^2 + b^2$

Solution

a. $(^{-}x)^{-2} = \dfrac{1}{(^{-}x)^2} = \dfrac{1}{x^2}$

$${}^{-}x^{-2} = -(x^{-2}) = {}^{-}\left(\frac{1}{x^2}\right) = \frac{^{-}1}{x^2}$$

Hence, $(^{-}x)^{-2} \neq {}^{-}x^{-2}$.

b. $(^-x)^{-3} = \dfrac{1}{(^-x)^3} = \dfrac{1}{^-x^3} = {}^-\left(\dfrac{1}{x^3}\right)$

$^-x^{-3} = {}^-(x^{-3}) = {}^-\left(\dfrac{1}{x^3}\right)$

Hence, $(^-x)^{-3} = {}^-x^{-3}$.

c. $ab^{-1} = a(b^{-1}) = a \cdot \dfrac{1}{b} = \dfrac{a}{b}$, but $(ab)^{-1} = \dfrac{1}{ab}$. Hence, $ab^{-1} \neq (ab)^{-1}$.

d. $(a^{-2}b^{-2})^{-1} = (a^{-2})^{-1}(b^{-2})^{-1}$
$= a^{(^-2)(^-1)}b^{(^-2)(^-1)}$
$= a^2 b^2$

e. $(a^{-2} + b^{-2})^{-1} = \left(\dfrac{1}{a^2} + \dfrac{1}{b^2}\right)^{-1} = \left(\dfrac{a^2 + b^2}{a^2 b^2}\right)^{-1} = \dfrac{a^2 b^2}{a^2 + b^2} \neq a^2 + b^2$

Observe that all the properties of exponents refer to powers with either the same base or the same exponent. To evaluate expressions using exponents where different bases and powers are used, perform all the computations or rewrite the expressions in either the same base or the same exponent, if possible. For example, $\dfrac{27^4}{81^3}$ can be rewritten $\dfrac{27^4}{81^3} = \dfrac{(3^3)^4}{(3^4)^3} = \dfrac{3^{12}}{3^{12}} = 1$.

EXAMPLE 6-22

Perform the following computations and leave your answers without negative exponents.

a. $16^2 \cdot 8^{-3}$
b. $20^2 \div 2^4$
c. $(10^{-1} + 5 \cdot 10^{-2} + 3 \cdot 10^{-3})10^3$
d. $(x^3 y^{-2})^{-4}$

Solution

a. $16^2 \cdot 8^{-3} = (2^4)^2 \cdot (2^3)^{-3} = 2^8 \cdot 2^{-9} = 2^{8 + ^-9} = 2^{-1} = \dfrac{1}{2}$

b. $\dfrac{20^2}{2^4} = \dfrac{(2^2 \cdot 5)^2}{2^4} = \dfrac{2^4 \cdot 5^2}{2^4} = 5^2 \text{ or } 25$

c. $(10^{-1} + 5 \cdot 10^{-2} + 3 \cdot 10^{-3})10^3 = 10^{-1} \cdot 10^3 + 5 \cdot 10^{-2} \cdot 10^3 + 3 \cdot 10^{-3} \cdot 10^3$
$= 10^{-1+3} + 5 \cdot 10^{-2+3} + 3 \cdot 10^{-3+3}$
$= 10^2 + 5 \cdot 10^1 + 3 \cdot 10^0$
$= 153$

d. $(x^3 y^{-2})^{-4} = (x^3)^{-4} \cdot (y^{-2})^{-4} = x^{3 \cdot (^-4)} \cdot y^{(^-2)(^-4)} = x^{-12}y^8 = \left(\dfrac{1}{x^{12}}\right)y^8 = \dfrac{y^8}{x^{12}}$, if $x, y \neq 0$

BRAIN TEASER

A castle in the faraway land of Aluossim was surrounded by four moats. One day, the castle was attacked and captured by a fierce tribe from the north. Guards were stationed at each bridge. Prince Juan was allowed to take a number of bags of gold from the castle as he went into exile. However, the guard at the first bridge demanded half the bags of gold plus one more bag. Juan met this demand and proceeded to the next bridge. The guards at the second, third, and fourth bridges made identical demands, all of which the prince met. When Juan finally crossed all the bridges, a single bag of gold was left. With how many bags did Juan start?

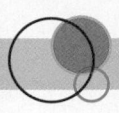

Assessment 6-3A

1. In the following figures, a unit rectangle is used to illustrate the product of two fractions. Name the fractions and their products.

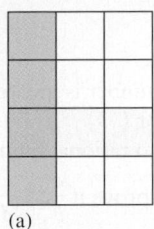 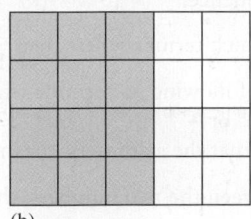

       (a)   (b)

2. Use a rectangular region to illustrate each of the following products:
   a. $\dfrac{1}{3} \cdot \dfrac{3}{4}$
   b. $\dfrac{2}{3} \cdot \dfrac{1}{5}$

3. Find each of the following products. Write your answers in simplest form.
   a. $\dfrac{49}{65} \cdot \dfrac{26}{98}$ b. $\dfrac{a}{b} \cdot \dfrac{b^2}{a^2}$ c. $\dfrac{xy}{z} \cdot \dfrac{z^2 a}{x^3 y^2}$

4. Use the distributive property of multiplication over addition to find each product.
   a. $4\dfrac{1}{2} \cdot 2\dfrac{1}{3}$ $\left[\text{Hint:} \left(4 + \dfrac{1}{2} \right)\left(2 + \dfrac{1}{3} \right). \right]$
   b. $3\dfrac{1}{3} \cdot 2\dfrac{1}{2}$

5. Find the multiplicative inverse of each of the following:
   a. $\dfrac{^-1}{3}$ b. $3\dfrac{1}{3}$
   c. $\dfrac{x}{y}$, if $\dfrac{x}{y} \neq 0$ d. $^-7$

6. Solve for x in each of the following:
   a. $\dfrac{2}{3}x = \dfrac{7}{6}$ b. $\dfrac{3}{4} \div x = \dfrac{1}{2}$
   c. $\dfrac{5}{6} + \dfrac{2}{3}x = \dfrac{3}{4}$ d. $\dfrac{2x}{3} - \dfrac{1}{4} = \dfrac{x}{6} + \dfrac{1}{2}$

7. Show that the following properties do not hold for the division of rational numbers:
   a. Commutative
   b. Associative

8. Compute the following mentally. Find the exact answers.
   a. $3\dfrac{1}{4} \cdot 8$ b. $7\dfrac{1}{4} \cdot 4$
   c. $9\dfrac{1}{5} \cdot 10$ d. $8 \cdot 2\dfrac{1}{4}$

9. Choose the number that best approximates each of the following from among the numbers in parentheses:
   a. $3\dfrac{11}{12} \cdot 5\dfrac{3}{100}$ $(8, 20, 15, 16)$
   b. $2\dfrac{1}{10} \cdot 7\dfrac{7}{8}$ $(16, 14, 4, 3)$
   c. $\dfrac{1}{101} \div \dfrac{1}{103}$ $\left(0, 1, \dfrac{1}{2}, \dfrac{1}{4} \right)$

10. Without actually doing the computations, choose the phrase in parentheses that correctly describes each:
    a. $\dfrac{13}{14} \cdot \dfrac{17}{19}$ (greater than 1, less than 1)
    b. $3\dfrac{2}{7} \div 5\dfrac{1}{9}$ (greater than 1, less than 1)
    c. $4\dfrac{1}{3} \div 2\dfrac{3}{100}$ (greater than 2, less than 2)

11. A sewing project requires $6\dfrac{1}{8}$ yd of material that sells for 62¢ per yard and $3\dfrac{1}{4}$ yd of material that sells for 81¢ per yard. Choose the best estimate for the cost of the project:
    a. Between $2.00 and $4.00   b. Between $4.00 and $6.00
    c. Between $6.00 and $8.00   d. Between $8.00 and $10.00

12. Five-eighths of the students at Salem State College live in dormitories. If 6000 students at the college live in dormitories, how many students are there in the college?

13. Alberto owns $\dfrac{5}{9}$ of the stock in the N.W. Tofu Company. His sister Renatta owns half as much stock as Alberto. What part of the stock is owned by neither Alberto nor Renatta?

14. A suit is on sale for $180.00. What was the original price of the suit if the discount was $\dfrac{1}{4}$ of the original price?

15. John took all his money out of his savings account. He spent $50.00 on a radio and $\dfrac{3}{5}$ of what remained on presents. Half of what was left he put back in his checking account, and the remaining $35.00 he donated to charity. How much money did John originally have in his savings account?

16. Al gives $\dfrac{1}{2}$ of his marbles to Bev. Bev gives $\dfrac{1}{2}$ of these to Carl. Carl gives $\dfrac{1}{2}$ of these to Dani. If Dani was given four marbles, how many did Al have originally?

17. Write each of the following in simplest form using positive exponents in the final answer:
    a. $3^{-7} \cdot 3^{-6}$ b. $3^7 \cdot 3^6$
    c. $5^{15} \div 5^4$ d. $5^{15} \div 5^{-4}$
    e. $(^-5)^{-2}$ f. $\dfrac{a^2}{a^{-3}}$

18. Write each of the following in simplest form using positive exponents in the final answer:
    a. $\left(\dfrac{1}{2} \right)^3 \cdot \left(\dfrac{1}{2} \right)^7$ b. $\left(\dfrac{1}{2} \right)^9 \div \left(\dfrac{1}{2} \right)^6$
    c. $\left(\dfrac{2}{3} \right)^5 \cdot \left(\dfrac{4}{9} \right)^2$ d. $\left(\dfrac{3}{5} \right)^7 \div \left(\dfrac{3}{5} \right)^7$

19. If a and b are rational numbers, with $a \neq 0$ and $b \neq 0$, and if m and n are integers, which of the following are true and which are false? Justify your answers.
    a. $a^m \cdot b^n = (ab)^{m+n}$ b. $a^m \cdot b^n = (ab)^{mn}$
    c. $a^m \cdot b^m = (ab)^{2m}$ d. $(ab)^0 = 1$
    e. $(a + b)^m = a^m + b^m$ f. $(a + b)^{-m} = \dfrac{1}{a^m} + \dfrac{1}{b^m}$

20. Solve for the integer n in each of the following:

   a. $2^n = 32$ **b.** $n^2 = 36$

   c. $2^n \cdot 2^7 = 2^5$ **d.** $2^n \cdot 2^7 = 8$

21. Solve each of the following inequalities for x, where x is an integer:

   a. $3^x \le 9$ **b.** $25^x < 125$

   c. $3^{2x} > 27$ **d.** $4^x > 1$

22. Determine which fraction in each of the following pairs is greater:

   a. $\left(\dfrac{1}{2}\right)^3$ or $\left(\dfrac{1}{2}\right)^4$ **b.** $\left(\dfrac{3}{4}\right)^{10}$ or $\left(\dfrac{3}{4}\right)^8$

   c. $\left(\dfrac{4}{3}\right)^{10}$ or $\left(\dfrac{4}{3}\right)^8$ **d.** $\left(\dfrac{3}{4}\right)^{10}$ or $\left(\dfrac{4}{5}\right)^{10}$

23. Let $S = \dfrac{1}{2} + \dfrac{1}{2^2} + \dfrac{1}{2^3} + \ldots + \dfrac{1}{2^{64}}$.

   a. Use the distributive property of multiplication over addition to find an expression for $2S$.

   b. Show that $2S - S = S = 1 - \left(\dfrac{1}{2}\right)^{64}$.

c. Find a simple expression for the sum

$$\frac{1}{2} + \frac{1}{2^2} + \frac{1}{2^3} + \ldots + \frac{1}{2^n}$$

24. If the nth term of a sequence is $3 \cdot 2^{-n}$, answer the following:

   a. Find the first five terms.

   b. Show that the first five terms are in a geometric sequence.

   c. Which terms are less than $\dfrac{3}{1000}$?

25. In the following, determine which number is greater:

   a. 32^{50} or 4^{100} **b.** $(^-27)^{-15}$ or $(^-3)^{-75}$

26. Show that the arithmetic mean of two rational numbers is between the two numbers; that is, prove if $\dfrac{a}{b} < \dfrac{c}{d}$ then

$$\frac{a}{b} < \frac{1}{2}\left(\frac{a}{b} + \frac{c}{d}\right) < \frac{c}{d}.$$

Assessment 6-3B

1. In the following figure a unit rectangle is used to illustrate the product of two fractions. Name the fractions and their products.

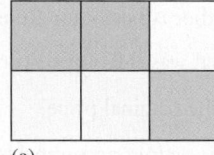   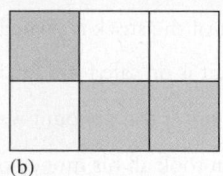

    (a)   (b)

2. Use a rectangular region to illustrate each of the following products:

   a. $\dfrac{2}{5} \cdot \dfrac{1}{3}$ **b.** $\dfrac{2}{3} \cdot \dfrac{2}{3}$

3. Find each of the following products of rational numbers. Write your answers in simplest form.

   a. $2\dfrac{1}{3} \cdot 3\dfrac{3}{4}$ **b.** $\dfrac{22}{7} \cdot 4\dfrac{2}{3}$

   c. $\dfrac{^-5}{2} \cdot 2\dfrac{1}{2}$ **d.** $2\dfrac{3}{4} \cdot 2\dfrac{1}{3}$

   e. $\dfrac{a^2}{b^3} \cdot \dfrac{b^2}{a^3}$ **f.** $\dfrac{x^3 y^2}{z} \cdot \dfrac{z}{x^2 y}$

4. Use the distributive property to find each product of rational numbers.

   a. $2\dfrac{1}{3} \cdot 4\dfrac{3}{5}$ **b.** $\left(\dfrac{x}{y} + 1\right)\left(\dfrac{y}{x} - 1\right)$ **c.** $248\dfrac{2}{5} \cdot 100\dfrac{1}{8}$

5. Find the multiplicative inverse of each of the following:

   a. $\dfrac{6}{7}$ **b.** 8 **c.** $4\dfrac{1}{5}$ **d.** $^-1\dfrac{1}{2}$

6. Solve for x in each of the following:

   a. $\dfrac{2}{3}x = \dfrac{11}{6}$ **b.** $\dfrac{3}{4} \div x = \dfrac{1}{3}$

   c. $\dfrac{5}{6} - \dfrac{2}{3}x = \dfrac{3}{4}$ **d.** $\dfrac{2x}{3} + \dfrac{1}{4} = \dfrac{x}{6} - \dfrac{1}{2}$

7. Find a fraction such that if you add the denominator to the numerator and place the sum over the original denominator, the new fraction has triple the value of the original fraction.

8. Compute the following mentally. Find the exact answers.

   a. $3\dfrac{1}{2} \cdot 8$ **b.** $7\dfrac{3}{4} \cdot 4$

   c. $9\dfrac{1}{5} \cdot 6$ **d.** $8 \cdot 2\dfrac{1}{3}$

   e. $3 \div \dfrac{1}{2}$ **f.** $3\dfrac{1}{2} \div \dfrac{1}{2}$

   g. $3 \div \dfrac{1}{3}$ **h.** $4\dfrac{1}{2} \div 2$

9. Choose the number that best approximates each of the following from among the numbers in parentheses:

   a. $20\dfrac{2}{3} \div 9\dfrac{7}{8} \left(2, 180, \dfrac{1}{2}, 10\right)$

   b. $3\dfrac{1}{20} \cdot 7\dfrac{77}{100} \left(21, 24, \dfrac{1}{20}, 32\right)$

   c. $\dfrac{1}{10^3} \div \dfrac{1}{1001} \left(\dfrac{1}{10^3}, 1, 1001, 0\right)$

10. Without actually doing the computations, choose the phrase in parentheses that correctly describes each:

   a. $4\dfrac{1}{3} \div 2\dfrac{13}{100}$ (greater than 2, less than 2)

   b. $16 \div 4\dfrac{3}{18}$ (greater than 4, less than 4)

   c. $16 \div 3\dfrac{8}{9}$ (greater than 4, less than 4)

11. When you multiply a certain number by 3 and then subtract $\dfrac{7}{18}$, you get the same result as when you multiply the number by 2 and add $\dfrac{5}{12}$. What is the number?

12. Di Paloma University had a faculty reduction and lost $\dfrac{1}{5}$ of its faculty. If 320 faculty members were left after the reduction, how many members were there originally?

13. A person has $29\frac{1}{2}$ yd of material available to make doll outfits. Each outfit requires $\frac{3}{4}$ yd of material.
    a. How many outfits can be made?
    b. How much material will be left over?

14. Every employee's salary at the Sunrise Software Company increases each year by $\frac{1}{10}$ of that person's salary the previous year.
    a. If Martha's present annual salary is $100,000.00, what will her salary be in 2 yr?
    b. If Aaron's present salary is $99,000.00, what was his salary 1 yr ago?
    c. If Juanita's present salary is $363,000.00, what was her salary 2 yr ago?

15. Jasmine is reading a book. She has finished $\frac{3}{4}$ of the book and has 82 pages left to read. How many pages has she read?

16. Peter, Paul, and Mary start at the same time walking around a circular track in the same direction. Peter takes $\frac{1}{2}$ hr to walk around the track. Paul takes $\frac{5}{12}$ hr, and Mary takes $\frac{1}{3}$ hr.
    a. How many minutes does it take each person to walk around the track?
    b. How many times will each person go around the track before all three meet again at the starting line?

17. Write each of the following rational numbers in simplest form using positive exponents in the final answer:
    a. $\left(\frac{1}{3}\right)^{-1}$
    b. $\frac{a^{-3}}{a}$
    c. $\frac{(a^{-4})^3}{a^{-4}}$
    d. $\frac{a}{a^{-1}}$
    e. $\frac{a^{-3}}{a^{-2}}$

18. Write each of the following in simplest form using positive exponents in the final answer:
    a. $\left(\frac{1}{2}\right)^{10} \div \left(\frac{1}{2}\right)^2$
    b. $\left(\frac{2}{3}\right)^5 \left(\frac{4}{9}\right)^{-2}$
    c. $\left(\frac{3}{5}\right)^7 \div \left(\frac{5}{3}\right)^4$
    d. $\left[\left(\frac{5}{6}\right)^7\right]^3$

19. If a and b are rational numbers, with $a \neq 0$ and $b \neq 0$, and if m and n are integers, which of the following are true and which are false? Justify your answers.
    a. $\frac{a^m}{b^n} = \left(\frac{a}{b}\right)^{m-n}$
    b. $(ab)^{-m} = \frac{1}{a^m} \cdot \frac{1}{b^m}$
    c. $\left(\frac{2}{a^{-1}+b^{-1}}\right)^{-1} = \frac{1}{2} \cdot \frac{1}{a+b}$
    d. $2(a^{-1}+b^{-1})^{-1} = \frac{2ab}{a+b}$
    e. $a^{mn} = a^m \cdot a^n$
    f. $\left(\frac{a}{b}\right)^{-1} = \frac{b}{a}$

20. Solve, if possible, for n where n is an integer in each of the following:
    a. $2^n = {}^-32$
    b. $n^3 = \frac{-1}{27}$
    c. $2^n \cdot 2^7 = 1024$
    d. $2^n \cdot 2^7 = 64$
    e. $(2+n)^2 = 2^2 + n^2$
    f. $3^n = 27^5$

21. Solve each of the following inequalities for x, where x is an integer:
    a. $3^x \geq 81$
    b. $4^x \geq 8$
    c. $3^{2x} \leq 27$
    d. $2^x < 1$

22. Determine which fraction in each of the following pairs is greater:
    a. $\left(\frac{4}{3}\right)^{10}$ or $\left(\frac{4}{3}\right)^8$
    b. $\left(\frac{3}{4}\right)^{10}$ or $\left(\frac{4}{5}\right)^{10}$
    c. $\left(\frac{4}{3}\right)^{10}$ or $\left(\frac{5}{4}\right)^{10}$
    d. $\left(\frac{3}{4}\right)^{100}$ or $\left(\frac{3}{4} \cdot \frac{9}{10}\right)^{100}$

23. Let $S = \frac{1}{3} + \frac{1}{3^2} + \frac{1}{3^3} + \ldots + \frac{1}{3^{64}}$.
    a. Use the distributive property of multiplication over addition to find an expression for $3S$.
    b. Show that $3S - S = 2S = 1 - \left(\frac{1}{3}\right)^{64}$.
    c. Find a simple expression for the sum
    $$\frac{1}{3} + \frac{1}{3^2} + \frac{1}{3^3} + \ldots + \frac{1}{3^n}.$$

24. In an arithmetic sequence, the first term is 1 and the hundredth term is 2. Find the following:
    a. The 50th term
    b. The sum of the first 50 terms

25. In the following, determine which number is greater:
    a. 32^{100} or 4^{200}
    b. $({}^-27)^{15}$ or $({}^-3)^{-50}$

26. There is a simple method for squaring any number that consists of a whole number and $\frac{1}{2}$. For example:
    $$\left(3\frac{1}{2}\right)^2 = 3 \cdot 4 + \left(\frac{1}{2}\right)^2 = 12\frac{1}{4}; \quad \left(4\frac{1}{2}\right)^2 = 4 \cdot 5 +$$
    $$\left(\frac{1}{2}\right)^2 = 20\frac{1}{4}; \quad \left(5\frac{1}{2}\right)^2 = 5 \cdot 6 + \left(\frac{1}{2}\right)^2 = 30\frac{1}{4}.$$
    a. Write a statement for $\left(n + \frac{1}{2}\right)^2$ that generalizes these examples, where n is a whole number.
    b. Justify this procedure.

27. Brandy bought a horse for $270.00 and immediately started paying for his keep. She sold the horse for $540.00. Considering the cost of his keep she found that she had lost an amount equal to half of what she paid for the horse plus one-fourth of the cost of his keep. How much did Brandy lose on the horse?

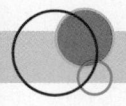

Mathematical Connections 6-3

Communication

1. Amy says that dividing a number by $\frac{1}{2}$ is the same as taking half of a number. How do you respond?

2. Noah says that dividing a number by 2 is the same as multiplying it by $\frac{1}{2}$. He wants to know if he is right, and if so, why. How do you respond?

3. Suppose you divide a natural number, n, by a positive rational number less than 1. Will the answer always be less than n, sometimes less than n, or never less than n? Why?

4. If the fractions represented by points C and D on the following number line are multiplied, what point best represents the product? Explain why.

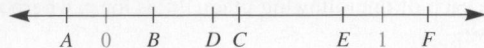
$$A \quad 0 \quad B \quad D\ C \quad E \quad 1 \quad F$$

5. If the product of two numbers is 1 and one of the numbers is greater than 1, what do you know about the other number? Explain your answer.

6. a. Amal can finish a job in $\frac{1}{2}$ of a day working by herself. Her son Sharif can finish the same job in $\frac{1}{4}$ of a day working alone. How long will it take to finish the job if they work together?

   b. If Amal can finish a job in a hours and Sharif in b hours, then how long will it take to finish the job if they work together?

Open-Ended

7. In the book *Knowing and Teaching Elementary Mathematics*, Liping Ma presents the following scenario:

   Imagine that you are teaching division with fractions. To make this situation meaningful for kids, something that many teachers try to do is relate mathematics to other things. Sometimes they try to come up with real-world situations or story problems to show the application of some particular piece of content. What would you say would be a good story or model for $1\frac{3}{4} \div \frac{1}{2}$? (p. 55)*

   a. How would you respond to her question?

   b. If possible, obtain a copy of Ma's book and read how U.S. teachers responded to this task compared to Chinese teachers. Report your findings.

8. a. Solve the problem in the cartoon below.

   b. Would you use the problem in the cartoon in your class? Why or why not?

Cooperative Learning

9. Choose a brick building on your campus. Measure the height of one brick and the thickness of mortar between bricks. Estimate the height of the building and then calculate the height of the building. Were rational numbers used in your computations?

Questions from the Classroom

10. Bente says to do the problem $12\frac{1}{4} \div 3\frac{3}{4}$ you just find
    $12 \div 3 = 4$ and $\frac{1}{4} \div \frac{3}{4} = \frac{1}{3}$ to get $4\frac{1}{3}$. How do you respond?

11. Carl says that every rational number has a multiplicative inverse. How do you respond?

12. Dani says that if we have $\frac{3}{4} \cdot \frac{2}{5}$, we could just do $\frac{3}{5} \cdot \frac{2}{4} = \frac{3}{5} \cdot \frac{1}{2} = \frac{3}{10}$. Is she correct? Explain why.

13. Joel says that $2\frac{2}{5} \cdot 3\frac{4}{5} = 2\frac{4}{5} \cdot 3\frac{2}{5}$ because multiplication is commutative. Is he right? Explain why or why not.

14. Jim is not sure when to use multiplication by a fraction and when to use division. He has the following list of problems. How would you help him solve these problems in a way that would enable him to solve similar problems on his own?

    a. $\frac{3}{4}$ of a package of sugar fills $\frac{1}{2}$ c. How many cups of sugar are in a full package of sugar?

    b. How many packages of sugar will fill 2 c?

    c. If $\frac{1}{3}$ c sugar is required to make two loaves of challah, how many cups of sugar are needed for three loaves?

    d. If $\frac{3}{4}$ c sugar is required for 1 gal of punch, how many gallons can be made with 2 c of sugar?

    e. If you have $22\frac{3}{8}$ in. of ribbon, and need $1\frac{1}{4}$ in. to decorate one doll, how many dolls can be decorated, and how much ribbon will be left over?

15. When working on the problem of simplifying
    $$\frac{3}{4} \cdot \frac{1}{2} \cdot \frac{2}{3}$$
    a student did the following:
    $$\frac{3}{4} \cdot \frac{1}{2} \cdot \frac{2}{3} = \left(\frac{3 \cdot 1}{4 \cdot 2}\right)\left(\frac{3 \cdot 2}{4 \cdot 3}\right) = \frac{3}{8} \cdot \frac{6}{12} = \frac{18}{96}$$
    What was the error, if any?

16. A student claims that division always makes things smaller so $5 \div \left(\frac{1}{2}\right)$ cannot 10 because 10 is greater than the number 5 she started with. How do you respond?

17. A student simplified the fraction $\frac{m + n}{p + n}$ to $\frac{m}{p}$. How would you help this student?

18. Jillian says she learned that 17 divided by 5 can be written as $17 \div 5 = 3$ R2, but she thinks that writing $17 \div 5 = \dfrac{17}{5} = 3\dfrac{2}{5}$ is much better. How do you respond?

19. Tyto would like to know how a story problem can help him to figure out the answer to $4\dfrac{2}{3} \div \dfrac{1}{3}$ without using the invert-and-multiply algorithm. How do you respond?

20. Fran claimed that $2\dfrac{1}{2} \cdot \dfrac{3}{5} = 2\dfrac{3}{10}$. What did Fran do and how would you help her?

Review Problems

21. Perform each of the following computations. Leave your answers in simplest form or as mixed numbers.

   a. $\dfrac{^-3}{16} + \dfrac{7}{4}$

   b. $\dfrac{1}{6} + \dfrac{^-4}{9} + \dfrac{5}{3}$

   c. $\dfrac{^-5}{2^3 \cdot 3^2} - \dfrac{^-5}{2 \cdot 3^3}$

   d. $3\dfrac{4}{5} + 4\dfrac{5}{6}$

   e. $5\dfrac{1}{6} - 3\dfrac{5}{8}$

   f. $^-4\dfrac{1}{3} - 5\dfrac{5}{12}$

22. Each student at Sussex Elementary School takes one foreign language. Two-thirds of the students take Spanish, $\dfrac{1}{9}$ take French, $\dfrac{1}{18}$ take German, and the rest take some other foreign language. If there are 720 students in the school, how many do not take Spanish, French, or German?

Trends in Mathematics and Science Study (TIMSS) Questions

There are 600 balls in a box, and $\dfrac{1}{3}$ of the balls are red.

How many red balls are in the box?

TIMSS, Grade 4, 2003

Dana makes a large batch of cranberry bread that is one and a half times the original recipe. If the original recipe requires $\dfrac{3}{4}$ cup of sugar, how many cups of sugar are required for the bread Dana is making?

  a. $\dfrac{3}{8}$ **b.** $1\dfrac{1}{8}$ **c.** $1\dfrac{1}{4}$ **d.** $1\dfrac{3}{8}$

TIMSS, Grade 8, 2007

National Assessment of Educational Progress (NAEP) Questions

Both figures below show the same scale. The marks on the scale have no labels except the zero point.

The weight of the cheese is $\dfrac{1}{2}$ pound. What is the total weight of the two apples?

NAEP, Grade 8, 2007

Jim has $\dfrac{3}{4}$ of a yard of string which he wishes to divide into pieces each $\dfrac{1}{8}$ of a yard long. How many pieces will he have?

  a. 3   **b.** 4   **c.** 6   **d.** 8

NAEP, Grade 8, 2003

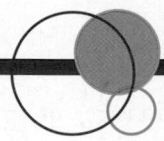

6-4 Ratios, Proportions, and Proportional Reasoning

Proportional reasoning is an extremely important concept taught in grades K–8. Proportionality has connections to most, if not all, of the other foundational middle-school topics and can provide a context to study these topics. In *PSSM* we find:

> Working with proportions is a major focus proposed in these Standards for the middle grades. Students should become proficient in creating ratios to make comparisons in situations that involve pairs of numbers, as in the following problem:
>
> If three packages of cocoa make fifteen cups of hot chocolate, how many packages are needed to make sixty cups? (p. 34)

Further, in the grade 7 *Focal Points* we find:

> Students extend their work with ratios to develop an understanding of proportionality that they apply to solve single and multistep problems in numerous contexts. (p. 19)

Ratios are encountered in everyday life. For example, there may be a 2-to-3 ratio of Democrats to Republicans on a certain legislative committee, a friend may be given a speeding ticket for driving 69 miles per hour, or eggs may cost $1.40 a dozen. Each of these illustrates a **ratio**.

Definition of Ratio

A **ratio**, $\frac{a}{b}$ or $a:b$, where a and b are rational numbers, is a comparison of two quantities a and b expressed as the number of times one is greater than or less than the other.

A ratio of $1:3$ for boys to girls in a class means that the number of boys is $\frac{1}{3}$ that of girls; that is, there is 1 boy for every 3 girls. We could also say that the ratio of girls to boys is $3:1$, or that there are 3 times as many girls as boys. Ratios can represent **part-to-whole** or **whole-to-part** comparisons. For example, if the ratio of boys to girls in a class is $1:3$, then the ratio of boys (part) to children (whole) is $1:4$. If there are b boys and g girls, then $g = 3b$; that is, $\frac{1}{3} = \frac{b}{3b}$. Also, the ratio of boys to the entire class is $\frac{b}{b + g} = \frac{b}{b + 3b} = \frac{b}{4b} = \frac{1}{4}$. We could also say that the ratio of all children (whole) to boys (part) is $4:1$. Some ratios give **part-to-part** comparisons, such as the ratio of the number of boys to girls or the number of students to one teacher. For example, a school might say that the average ratio of students to teachers cannot exceed $24:1$.

The ratio of $1:3$ for boys to girls in a class does not tell us how many boys and how many girls there are in the class. It only tells us the relative size of the groups. There could be 2 boys and 6 girls or 3 boys and 9 girls or 4 boys and 12 girls or some other numbers that give a fraction equivalent to $\frac{1}{3}$.

EXAMPLE 6-23

There were 7 males and 12 females in the Dew Drop Inn on Monday evening. In the game room next door were 14 males and 24 females.

  a. Express the number of males to females at the inn as a ratio (part-to-part).
  b. Express the number of males to females at the game room as a ratio (part-to-part).
  c. Express the number of males in the game room to the number of people in the game room as a ratio (part-to-whole).

Solution

  a. The ratio is $\frac{7}{12}$.   **b.** The ratio is $\frac{14}{24}$, or $\frac{7}{12}$.   **c.** The ratio is $\frac{14}{38}$, or $\frac{7}{19}$.

Proportions

In a study of sixth graders conducted by Harel and colleagues (1994)*, children were shown a picture of a carton of orange juice and were told that the orange juice was made from orange concentrate and water. Then they were shown two glasses—a large glass and a small glass—and they were

---

*Harel, G., M. Bahr, R. Lesh, and T. Post. "Invariance of Ratio: The Case of Children's Anticipatory Scheme for Constency of Taste," *Journal of Research in Mathematics Education* 25(July 1994): 324–345.

told that both glasses were filled with orange juice from the carton. They were then asked if the orange juice from each of the two glasses would taste equally "orangey" or if one would taste more "orangey." About half of the students said they were not equally orangey. Of those about half the students said the larger glass would be more "orangey" and about half said the smaller glass would be more "orangey." These students were thinking of only one quantity—the water alone or the orange concentrate alone. For example, one student said that the larger glass would be more "orangey" because the glass is bigger and it would hold more orange concentrate, while another said the small glass would be more "orangey" because it has less water but more orange per ounce.

Suppose Recipe A for an orange drink calls for 2 cans of orange concentrate for every 3 cans of water. We could say that the ratio of cans of orange concentrate to cans of water is 2:3. We represent this pictorially as in Figure 6-16(a), where O represents a can of orange concentrate and W represents a can of water. In Figure 6-16(b) and (c), we continue the process of adding 2 cans of orange concentrate for every 3 cans of water.

Recipe A

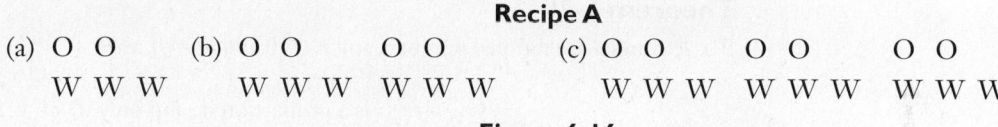

Figure 6-16

From Figure 6-16 we could develop and continue the **ratio table**, as shown in Table 6-3.

Table 6-3

Cans of Orange Concentrate	2	4	6	8	10	12
Cans of Water	3	6	9	?	?	?

In Table 6-3, the ratios 2/3 and 4/6 are equal. The equation $2/3 = 4/6$ is a proportion. In general, we have the following definition.

Definition of a Proportion

A **proportion** is a statement that two given ratios are equal.

If Recipe B calls for 4 cans of orange concentrate for every 8 cans of water, then the ratio of cans of orange concentrate to cans of water for this recipe is 4:8. We picture this in Figure 6-17(a).

Recipe B

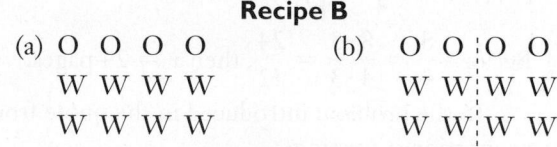

Figure 6-17

Which of the two recipes produces a drink that tastes more "orangey"? In Figure 6-16(a), we see that in Recipe A there are 2 cans of orange concentrate for every 3 cans of water. In Figure 6-17(a), we see that in Recipe B there are 4 cans of orange concentrate for every 8 cans of water.

To compare the two recipes, we need either the same number of cans of orange concentrate or the same number of cans of water. Either is possible. Figure 6-16(b) shows that for Recipe A there are 4 cans of orange concentrate for every 6 cans of water. In Recipe B, for 4 cans of orange concentrate there are 8 cans of water. Recipe B calls for more water per 4 cans of orange concentrate, so it is less "orangey." An alternative is to observe that in Figure 6-17(b), Recipe B shows that there are 2 cans of orange concentrate for every 4 cans of water. We compare this with Figure 6-16(a), showing 2 cans of orange concentrate for every 3 cans of water, and reach the same conclusion.

From our work in Section 6-1, we know that $2/3 = 4/6$ because $2 \cdot 6 = 3 \cdot 4$. Hence $2/3 = 4/6$ is a proportion. Also $2/3 \neq 4/8$ because $2 \cdot 8 \neq 3 \cdot 4$ and this is not a proportion. In general, we have the following theorem that follows from Theorem 6-2 developed in Section 6-1.

Theorem 6-20

If a, b, c, and d are rational numbers and $b \neq 0$ and $d \neq 0$, then

$$\frac{a}{b} = \frac{c}{d} \quad \text{is a proportion if, and only if, } ad = bc.$$

Students in the lower grades typically experience problems that are *additive*. Consider the problem below.

Allie and Bente type at the same speed. Allie started typing first. When Allie had typed 8 pages, Bente had typed 4 pages. When Bente has typed 10 pages, how many has Allie typed?

This is an example of an *additive* relationship. Students should reason that since the two people type at the same speed, when Bente has typed an additional 6 pages, Allie should have also typed an additional 6 pages, so she should have typed $8 + 6$, or 14, pages.

Next consider the following problem:

Carl can type 8 pages for every 4 pages that Dan can type. If Dan has typed 12 pages, how many pages has Carl typed?

If students try an *additive* approach, they will conclude that since Dan has typed 8 more pages than in the original relationship, then Carl should have typed an additional 8 pages for a total of 16 pages. However, the correct reasoning is that since Carl types twice as fast as Dan he will type twice as many pages as Dan. Therefore, when Dan has typed 12 pages, Carl has typed 24 pages. The relationship between the ratios is *multiplicative*. Another way to solve this problem is to set up the proportion $\frac{8}{4} = \frac{x}{12}$, where x is the number of pages that Carl will type, and solve for x. Because $\frac{8}{4} = \frac{8 \cdot 3}{4 \cdot 3} = \frac{24}{12}$, then $x = 24$ pages.

In the problem introduced in the quote from *PSSM* on page 311 of this text, one term in the proportion is missing:

$$\begin{array}{c} \text{packages} \rightarrow \\ \text{cups} \rightarrow \end{array} \frac{3}{15} = \frac{x}{60}$$

One way to solve the equation is to multiply both sides by 60, as follows:

$$\frac{3}{15} \cdot 60 = \frac{x}{60} \cdot 60$$

$$3 \cdot 4 = x$$

$$12 = x$$

Therefore, 12 packages of cocoa are needed to make 60 cups of hot chocolate.

We could also use the Fundamental Law of Fractions to solve the above problem: $\frac{3}{15} = \frac{3 \cdot 4}{15 \cdot 4} = \frac{12}{60}$, so $x = 12$. Another method of solution uses Theorem 6-20. This is often called the *cross-multiplication method*. The equation $\frac{3}{15} = \frac{x}{60}$ is a proportion if, and only if,

$$3 \cdot 60 = 15x$$
$$180 = 15x$$
$$12 = x.$$

EXAMPLE 6-24

If there are 3 cars for every 8 students at a high school, how many cars are there for 1200 students?

Solution   We use the strategy of *setting up a table*, as shown in Table 6-4

Table 6-4

Number of cars	3	x
Number of students	8	1200

The ratio of cars to students is always the same:

$$\begin{array}{l} \text{Cars} \;\;\;\;\to \\ \text{Students} \to \end{array} \frac{3}{8} = \frac{x}{1200}$$
$$3 \cdot 1200 = 8x$$
$$3600 = 8x$$
$$450 = x$$

Thus, there are 450 cars.

Next consider two car rental companies where the rates for 1–4 days are given in Table 6-5.

Table 6-5

(a) Ace Car Rental		(b) Better Car Rental	
Days	Cost	Days	Cost
1	$20	1	$20
2	$40	2	$35
3	$60	3	$48
4	$80	4	$52

The first two days for Ace Car Rental rates can be used to write a proportion because $\frac{1 \, day}{\$20} = \frac{2 \, days}{\$40}$. In a proportion, the units of measure must be in the same relative positions. In this case, the numbers of days are in the numerators and the costs are in the denominators.

For the Better Car Rental we see that $\frac{1 \, day}{\$20} \neq \frac{2 \, days}{\$35}$ so a proportion is not formed.

Consider Table 6-6 which is a ratio table built from the values for Ace Car Rental.

Table 6-6

Days (d)	1	2	3	4
Cost (c)	20	40	60	80

The ratios $\frac{d}{c}$ are all equal, that is, $\frac{1}{20} = \frac{2}{40} = \frac{3}{60} = \frac{4}{80}$. Thus, each pair of ratios forms a proportion. In this case, $\frac{d}{c} = \frac{1}{20}$ for all values of c and d. This is expressed by saying that *d is proportional to c* or *d varies proportionally to c* or *d varies directly with c.* In this case, $d = \frac{1}{20}c$ for every c and d. The number $\frac{1}{20}$ is the **constant of proportionality.** We can say that *gas used by a car is proportional to the miles traveled* or *lottery profits vary directly with the number of tickets sold.*

Definition of Constant of Proportionality

If the variables x and y are related by the equality $y = kx$ $\left(\text{or } k = \frac{y}{x}\right)$, then **y is proportional to x** and **k** is the **constant of proportionality** between y and x.

A central idea in proportional reasoning is that a relationship between two quantities is such that the ratio of one quantity to the other remains unchanged as the numerical values of both quantities change.

It is important to remember that in the ratio $a:b$ or $\frac{a}{b}$, a and b do not have to be whole numbers. For example, if in Eugene, Oregon, $\frac{7}{10}$ of the population exercise regularly, then $\frac{3}{10}$ of the population do not exercise regularly, and the ratio of those who do to those who do not is $\frac{7}{10} : \frac{3}{10}$. This ratio can be written $7:3$.

It is important to pay special attention to units of measure when working with proportions. For example, if a turtle travels 5 in. every 10 sec, how many feet does it travel in 50 sec? If units of measure are ignored, we might set up the following proportion:

$$\frac{5}{10} = \frac{x}{50}$$

In this proportion the units of measure are not listed. A more informative proportion that often prevents errors is the following:

$$\frac{5 \text{ in.}}{10 \text{ sec}} = \frac{x \text{ in.}}{50 \text{ sec}}$$

This implies that $x = 25$. Consequently, since 12 in. = 1 ft, the turtle travels $\frac{25}{12}$ ft, or $2\frac{1}{12}$ ft, or 2 ft 1 in.

PSSM points out the following concerning the cross-multiplication method of solving proportions:

Instruction in solving proportions should include methods that have a strong intuitive basis. The so-called cross-multiplication method can be developed meaningfully if it arises naturally in students' work, but it can also have unfortunate side effects when students do not adequately understand when the method is appropriate to use. Other approaches to solving proportions are often more intuitive and also quite powerful. For example, when trying to decide which is the better buy—12 tickets for $15.00 or 20 tickets for $23.00—students might choose to use a scaling strategy (finding the cost for a common number of tickets) or a unit-rate strategy (finding the cost for one ticket). (p. 221)

The **scaling strategy** for solving the problem would involve finding the cost for a common number of tickets. Because LCM(12, 20) = 60, we could choose to find the cost of 60 tickets under each plan.

In the first plan, since 12 tickets cost $15.00, then 60 tickets cost $75.00.

In the second plan, since 20 tickets cost $23.00, then 60 tickets cost $69.00.

Therefore, the second plan is a better buy.

The **unit-rate strategy** for solving this problem involves finding the cost of one ticket under each plan and then comparing unit costs.

In the first plan, since 12 tickets cost $15.00, then 1 ticket costs $1.25.

In the second plan, since 20 tickets cost $23.00, then 1 ticket costs $1.15.

The grade 6 *Common Core Standards* states the following:

Solve unit rate problems including those involving unit pricing and constant speed. For example, if it took 7 hours to mow 4 lawns, then at that rate how many lawns could be mowed in 35 hours? (p. 42).

 The partial student page shows how a proportion can be solved using *equivalent fractions* and how the same problem could be solved using *unit rates*.

**NOW TRY THIS 6-16**

a. Work the problem posed in the grade 6 *Common Core Standards*.
b. Work the problem on the bottom of the partial student page on page 318.

EXAMPLE 6-25

Kai, Paulus, and Judy made $2520.00 for painting a house. Kai worked 30 hr, Paulus worked 50 hr, and Judy worked 60 hr. They divided the money in proportion to the number of hours worked. If they all earn the same rate of pay, how much did each earn?

Solution   Let x be the unit rate or the rate of pay per hour. Then, $30x$ denotes the amount of money that Kai received. Paulus received $50x$ because then, and only then, will the ratios of the amounts be the same as 30:50, as required. Similarly, Judy received $60x$. Because the total amount of money received is $30x + 50x + 60x$, we have

$$30x + 50x + 60x = 2520$$
$$140x = 2520$$
$$x = 18. \text{ (dollars per hour)}$$

Hence,

$$\text{Kai received } 30x = 30 \cdot 18, \text{ or } \$540.00$$
$$\text{Paulus received } 50x = 50 \cdot 18, \text{ or } \$900.00$$
$$\text{Judy received } 60x = 60 \cdot 18, \text{ or } \$1080.00$$

Dividing each of the amounts by 18 shows that the proportion is as required.

School Book Page   MORE THAN ONE WAY

● More Than One Way

A package of 50 blank CDs costs $25. However, the store has run out of 50-packs. The manager agrees to sell you packages of 12 at the same unit price. How much should a 12-pack of CDs cost?

Jessica's Method

I'll set up a proportion and use equivalent fractions to solve it.

$$\frac{50 \text{ CDs}}{\$25} = \frac{12 \text{ CDs}}{x \text{ dollars}}$$

$$\frac{2}{1} = \frac{12}{x}$$

$$\frac{2}{1} = \frac{12}{x}$$

$$x = 6$$

A pack of 12 blank CDs should cost $6.

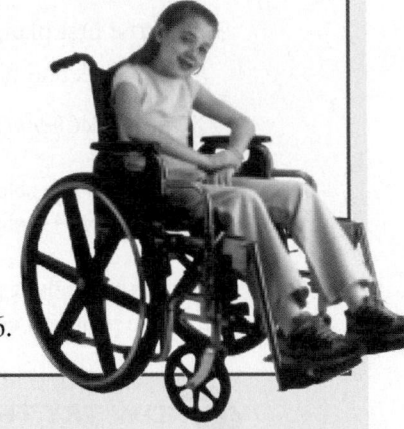

Michael's Method

I'll find the unit rate for the cost of one CD. Then I'll multiply.

$$\frac{\text{Cost}}{\text{Quantity}} \rightarrow \frac{25 \div 50}{50 \div 50} = \frac{0.5}{1}$$

Each CD costs $.50. Twelve CDs cost $12 \times \$.50 = \$6.00$.

A pack of 12 blank CDs should cost $6.

Choose a Method

An ad says "3 movies for $18." At that rate, what is the cost of 5 movies? Describe your method and explain why you chose it.

Consider the proportion $\frac{15}{30} = \frac{3}{6}$. Because the ratios in the proportion are equal and because equal nonzero fractions have equal reciprocals, it follows that $\frac{30}{15} = \frac{6}{3}$. Also notice that the proportions are true because each results in $15 \cdot 6 = 30 \cdot 3$. In general, we have the following theorem.

> **Theorem 6-21**
>
> For any rational numbers $\frac{a}{b}$ and $\frac{c}{d}$, with $a \neq 0$ and $c \neq 0$, $\frac{a}{b} = \frac{c}{d}$ if, and only if, $\frac{b}{a} = \frac{d}{c}$.

Consider $\frac{15}{30} = \frac{3}{6}$ again. Notice that $\frac{15}{3} = \frac{30}{6}$; that is, the ratio of the numerators is equal to the ratio of the corresponding denominators. In general, we have the following theorem.

> **Theorem 6-22**
>
> For any rational numbers $\frac{a}{b}$ and $\frac{c}{d}$, $c \neq 0$, $\frac{a}{b} = \frac{c}{d}$ if, and only if, $\frac{a}{c} = \frac{b}{d}$.

Scale Drawings

Ratio and proportions are used in scale drawings. For example, if the scale is 1:300, then the length of 1 cm in such a drawing represents 300 cm, or 3 m in true size. The **scale** is the ratio of the size of the drawing to the size of the object. The following example shows the use of scale drawings.

EXAMPLE 6-26

The floor plan of the main floor of a house in Figure 6-18 is drawn in the scale of 1:300. Find the dimensions in meters of the living room.

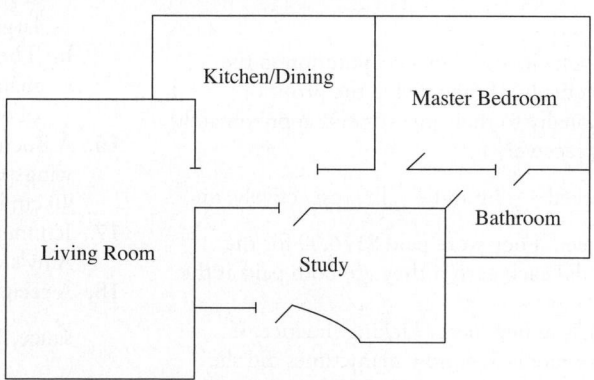

Figure 6-18

Solution In Figure 6-18, the dimensions of the living room measured with a centimeter ruler are approximately 3.7 cm by 2.5 cm. Because the scale is 1:300, 1 cm in the drawing represents 300 cm, or 3 m in true size. Hence, 3.7 cm represents $3.7 \cdot 3$, or 11.1 m, and 2.5 cm represents $2.5 \cdot 3$, or 7.5 m. Hence, the dimensions of the living room are approximately 11.1 m by 7.5 m.

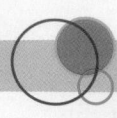

Assessment 6-4A

1. Answer the following regarding the English alphabet:
   a. Determine the ratio of vowels to consonants.
   b. What is the ratio of consonants to vowels?
   c. What is the ratio of consonants to letters in the English alphabet?
   d. Write a word that has a ratio of 2:3 of vowels to consonants.

2. Solve for x in each of the following proportions:
   a. $\dfrac{12}{x} = \dfrac{18}{45}$ b. $\dfrac{x}{7} = \dfrac{^-10}{21}$
   c. $\dfrac{5}{7} = \dfrac{3x}{98}$ d. $3\dfrac{1}{2}$ is to 5 as x is to 15.

3. a. If the ratio of boys to girls in a class is $2:3$, what is the ratio of boys to all the students in the class? Why?
   b. If the ratio of boys to girls in a class is $m:n$, what is the ratio of boys to all the students in the class?
   c. If $\dfrac{3}{5}$ of the class are girls, what is the ratio of girls to boys?

4. There are approximately 2 lb of muscle for every 5 lb of body weight. For a 90-lb person, approximately how much of the weight is muscle?

5. Which is a better buy—4 grapefruits for 80¢ or 12 grapefruits for $1.80?

6. On a map, $\dfrac{1}{3}$ in. represents 5 mi. If New York and Aluossim are 18 in. apart on the map, what is the actual distance between them?

7. David reads 40 pages of a book in 50 min. How many pages should he be able to read in 80 min if he reads at a constant rate?

8. Two numbers are in the ratio $3:4$. Find the numbers if
   a. their sum is 98.
   b. their product is 768.

9. Gary, Bill, and Carmella invested in a corporation in the ratio of $2:3:5$, respectively. If they divide the profit of $82,000.00 proportionally to their investment, approximately how much will each receive?

10. Sheila and Dora worked $3\dfrac{1}{2}$ hr and $4\dfrac{1}{2}$ hr, respectively, on a programming project. They were paid $176.00 for the project. How much did each earn if they are both paid at the same rate?

11. Vonna scored 75 goals in her soccer kicking practice. If her success-to-failure rate is $5:4$, how many times did she attempt a goal?

12. Express each of the following as a ratio $\dfrac{a}{b}$, where a and b are whole numbers:
   a. $\dfrac{1}{6} : 1$
   b. $\dfrac{1}{3} : \dfrac{1}{3}$
   c. $\dfrac{1}{6} : \dfrac{2}{7}$

13. Use Theorems 6–21 and 6–22 to write three other proportions that follow from the following proportion:
$$\frac{12¢}{36\text{ oz}} = \frac{16¢}{48\text{ oz}}$$

14. The *rise* and *span* for a house roof are identified as shown on the drawing. The *pitch* of a roof is the ratio of the rise to the half-span.
   a. If the rise is 10 ft and the span is 28 ft, what is the pitch?

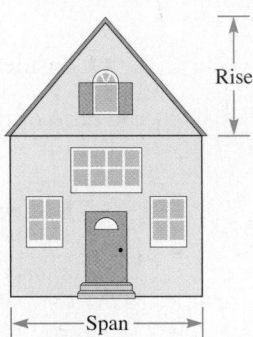

Rise

Span

   b. If the span is 16 ft and the pitch is $\dfrac{3}{4}$, what is the rise?

15. Gear ratios are used in industry. A gear ratio is the comparison of the number of teeth on two gears. When two gears are meshed, the revolutions per minute (rpm) are inversely proportional to the number of teeth; that is,
$$\frac{\text{rpm of large gear}}{\text{rpm of small gear}} = \frac{\text{Number of teeth on small gear}}{\text{Number of teeth on large gear}}$$
   a. The rpm ratio of the large gear to the small gear is $4:6$. If the small gear has 18 teeth, how many teeth does the large gear have?
   b. The large gear revolves at 200 rpm and has 60 teeth. How many teeth are there on the small gear, which has an rpm of 600?

16. A Boeing 747 jet is approximately 230 ft long and has a wingspan of 195 ft. If a scale model of the plane is about 40 cm long, what is the model's wingspan?

17. Jennifer weighs 160 lb on Earth and 416 lb on Jupiter. Find Amy's weight on Jupiter if she weighs 120 lb on Earth.

18. A recipe calls for 1 tsp of mustard seeds, 3 c of tomato sauce, $1\dfrac{1}{2}$ c of chopped scallions, and $3\dfrac{1}{4}$ c of beans. If one ingredient is altered as specified, how must the other ingredients be changed to keep the proportions the same?
   a. 2 c of tomato sauce
   b. 1 c of chopped scallions
   c. $1\dfrac{3}{4}$ c of beans

19. The electrical resistance of a wire, measured in ohms (Ω), is proportional to the length of the wire. If the electrical resistance of a 5-ft wire is 4.2 Ω, what is the resistance of 18 ft of the same wire?

20. In a photograph of a father and his daughter, the daughter's height is 2.3 cm and the father's height is 5.8 cm. If the father is actually 188 cm tall, how tall is the daughter?

21. The amount of gold in jewelry and other products is measured in karats (K), where 24K represents pure gold. The mark 14K on a chain indicates that the ratio between the mass of the gold in the chain and the mass of the chain is 14:24. If a gold ring is marked 18K and it weighs 0.4 oz, what is the value of the gold in the ring if pure gold is valued at $1800 per oz?

22. If Amber is paid $8.00 per hour for typing, the table shows how much she earns.

Hours (*h*)	1	2	3	4	5
Wages (*w*)	$8	$16	$24	$32	$40

  a. How much would Amber make for a 40-hr work week?
  b. What is the constant of proportionality?

23. a. In Room A there are 1 man and 2 women; in Room B there are 2 men and 4 women; and in Room C there are 5 men and 10 women. If all the people in Rooms B and C go to Room A, what will be the ratio of men to women in Room A?
  b. Prove the following generalization of the proportions used in (a):

$$\text{If } \frac{a}{b} = \frac{c}{d} = \frac{e}{f}, \text{ then } \frac{a}{b} = \frac{c}{d} = \frac{e}{f} = \frac{a+c+e}{b+d+f}.$$

Assessment 6-4B

1. Answer the following regarding the letters in the word *Mississippi*.
  a. Determine the ratio of vowels to consonants.
  b. What is the ratio of consonants to vowels?
  c. What is the ratio of consonants to letters in the word?

2. Solve for *x* in each of the following proportions:
  a. $\dfrac{5}{x} = \dfrac{30}{42}$
  b. $\dfrac{x}{8} = \dfrac{^-12}{32}$
  c. $\dfrac{7}{8} = \dfrac{3x}{48}$
  d. $3\dfrac{1}{2}$ is to 8 as *x* is to 24

3. There are 5 adult drivers to each teenage driver in Aluossim. If there are 12,345 adult drivers in Aluossim, how many teenage drivers are there?

4. A candle is 30 in. long. After burning for 12 min, the candle is 25 in. long. If it continues to burn at the same rate, how long will it take for the whole candle to burn?

5. A rectangular yard has a width-to-length ratio of 5:9. If the distance around the yard is 2800 ft, what are the dimensions of the yard?

6. A grasshopper can jump 20 times its length. If jumping ability in humans (height) were proportional to a grasshopper's (length), how far could a 6-ft-tall person jump?

7. Jim found out that after working for 9 months he had earned 6 days of vacation time. How many days per year does he earn at this rate?

8. At Rattlesnake School the teacher–student ratio is 1:30. If the school has 1200 students, how many additional teachers must be hired to reduce the ratio to 1:20?

9. At a particular time, the ratio of the height of an object that is perpendicular to the ground to the length of its shadow is the same for all objects. If a 30-ft tree casts a shadow of 12 ft, how tall is a tree that casts a shadow of 14 ft?

10. The following table shows several possible widths *W* and corresponding lengths *L* of a rectangle whose area is 10 ft².

Width (*W*) (Feet)	Length (*L*) (Feet)	Area (Square Feet)
0.5	20	$0.5 \cdot 20 = 10$
1	10	$1 \cdot 10 = 10$
2	5	$2 \cdot 5 = 10$
2.5	4	$2.5 \cdot 4 = 10$
4	2.5	$4 \cdot 2.5 = 10$
5	2	$5 \cdot 2 = 10$
10	1	$10 \cdot 1 = 10$
20	0.5	$20 \cdot 0.5 = 10$

Area = 10 ft² (rectangle with length *L* and width *W*)

  a. Use the values in the table and some additional values to graph the length *L* on the vertical axis versus the width *W* on the horizontal axis.
  b. What is the algebraic relationship between *L* and *W*?
  c. Write an equation to show that *W* varies directly with *L*.

11. Find three sets of *x*- and *y*-values for the following:

$$\frac{4 \text{ tickets}}{\$20} = \frac{x \text{ tickets}}{\$y}$$

12. If rent is $850.00 for each 2 weeks, how much is the rent for 7 weeks?

13. Leonardo da Vinci in his drawing *Vitruvian Man* showed that the man's armspan was equal to the man's height. Some other ratios are listed below.

$$\frac{\text{Length of hand}}{\text{Length of foot}} = \frac{7}{9}$$

$$\frac{\text{Distance from elbow to end of hand}}{\text{Distance from shoulder to elbow}} = \frac{8}{5}$$

$$\frac{\text{Length of hand}}{\text{Length of big toe}} = \frac{14}{3}$$

Using the ratios above, answer the following:

a. If the length of a big toe is 6 cm, how long should the hand be?

b. If a hand is 21 cm, how long is the foot?

c. If the distance from the elbow to the end of the hand is 20 in., what is the distance from the shoulder to the elbow?

14. On a city map, a rectangular park has a length of 4 in. If the actual length and width of the park are 500 ft and 300 ft, respectively, how wide is the park on the map?

15. Jim's car will travel 240 mi on 15 gal of gas. How far can he expect to go on 3 gal of gas?

16. Some model railroads use an *O* scale in replicas of actual trains. The *O* scale uses the ratio 1 in./48 in. How many feet long is the actual locomotive if an *O* scale replica is 18 in. long?

17. a. On an American flag, what is the ratio of stars to stripes?

b. What is the ratio of stripes to stars?

18. On an American flag, the ratio of the length of the flag to its width must be 19:10.

a. If a flag is to be $9\frac{1}{2}$ ft long, how wide should it be?

b. The flag that was placed on the Moon measured 5 ft by 3 ft. Does this ratio form a proportion with the official length-to-width ratio? Why?

19. If $\frac{x}{y} = \frac{a}{b}$, $a \neq 0$, $x \neq 0$, is true, what other proportions do you know are true?

20. If a certain recipe takes $1\frac{1}{2}$ c of flour and 4 c of milk, how much milk should be used if the cook only has 1 c of flour?

21. Prove that if $\frac{a}{b} = \frac{c}{d}$, $a \neq {}^{-}b$, and $a \neq b$, then the following are true:

a. $\frac{a + b}{b} = \frac{c + d}{d}$ $\left(Hint: \frac{a}{b} + 1 = \frac{c}{d} + 1 \right)$

b. $\frac{a}{a + b} = \frac{c}{c + d}$ **c.** $\frac{a - b}{a + b} = \frac{c - d}{c + d}$

22. To estimate the number of fish in a lake, scientists use a tagging and recapturing technique. A number of fish are captured, tagged, and then released back into the lake. After a while, some fish are captured and the number of tagged fish is counted.

Let *T* be the total number of fish captured, tagged, and released into the lake, *n* the number of fish in a recaptured sample, and *t* the number of fish found tagged in that sample. Finally, let *x* be the number of fish in the lake. The assumption is that the ratio between tagged fish and the total number of fish in any sample is approximately the same and hence scientists assume $\frac{t}{n} = \frac{T}{x}$. Suppose 173 fish were captured, tagged, and released. Then 68 fish were recaptured and among them 21 were found to be tagged. Estimate the number of fish in the lake.

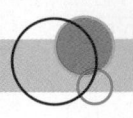

Mathematical Connections 6-4

Communication

1. Iris has found some dinosaur bones and a fossil footprint. The length of the footprint is 40 cm, the length of the thigh bone is 100 cm, and the length of the body is 700 cm.

a. What is the ratio of the footprint's length to the dinosaur's length?

b. Iris found a new track that she believes was made by the same species of dinosaur. If the footprint was 30 cm long and if the same ratio of foot length to body length holds, how long is the dinosaur?

c. In the same area, Iris also found a 50-cm thigh bone. Do you think this thigh bone belonged to the same dinosaur that made the 30-cm footprint that Iris found? Why or why not?

2. Suppose a 10-in. circular pizza costs $4.00. To find the price, *x*, of a 14-in. circular pizza, is it correct to set up the proportion $\frac{x}{4} = \frac{14}{10}$? Why or why not?

3. If $\frac{a}{b}$ is a proper fraction and $\frac{a}{b} = \frac{c}{d}$, then show that $\frac{d}{c - d} = \frac{b}{a - b}$.

4. Nell said she can tell just by looking at the ratios 15:7 and 15:8 that these do not form a proportion. Is she correct? Why?

5. Sol had photographs that were 4 in. by 6 in., 5 in. by 7 in., and 8 in. by 10 in. Do the dimensions vary proportionately? Explain why.

6. Can $\frac{a}{b}$ and $\frac{a + b}{b}$ ever form a proportion? Why?

7. In a condo complex, $\frac{2}{3}$ of the men were married to $\frac{3}{4}$ of the women. What is the ratio of married people to the total adult population of the condo complex? Explain how you can obtain this ratio without knowing the actual number of men or women.

Open-Ended

8. Write a paragraph in which you use the terms *ratio* and *proportion* correctly.

9. List three real-world situations that involve ratio and proportion.

10. Find examples of ratios in a newspaper.

11. Research the golden ratio that the Greeks used in the design of the Parthenon. Write a report on this ratio and include a drawing of a golden rectangle.

Cooperative Learning

12. In *Gulliver's Travels* by Jonathan Swift we find the following:

    The seamstresses took my measure as I lay on the ground, one standing at my neck and another at mid-leg, with a strong cord extended, that each held by the end, while the third measured the length of the cord with a rule of an inch long. Then they measured my right thumb and desired no more; for by a mathematical computation, that twice around the thumb is once around the wrist, and so on to the neck and the waist; and with the help of my old shirt, which I displayed on the ground before them for a pattern, they fitted me exactly.

    a. Explore the measurements of those in your group to see if you believe the ratios mentioned for Gulliver.
    b. Suppose the distance around a person's thumb is 9 cm. What is the distance around the person's neck?
    c. What ratio could be used to compare a person's height to armspan?
    d. Do you think there is a ratio between foot length and height? If so, what might it be?
    e. Estimate other body ratios and then see how close you are to actual measurements.

Questions from the Classroom

13. Mary is working with measurements and writes the following proportion:

    $$12 \text{ in.}/1 \text{ ft} = 5 \text{ ft}/60 \text{ in.}$$

    How would you help her?
14. Nora said she can use division to decide whether two ratios form a proportion; for example, $32:8$ and $40:10$ form a proportion because $32 \div 8 = 4$ and $40 \div 10 = 4$. Is she correct? Why?
15. Al is 5 ft tall and has a shadow that is 18 in. long. At the same time, a tree has a shadow that is 15 ft long. Al sets up and solves the proportion as follows:

    $$\frac{5 \text{ ft}}{15 \text{ ft}} = \frac{18 \text{ in.}}{x \text{ in.}}, \quad \text{so } x = 54 \text{ in.}$$

    How would you help him?
16. Amy's friend told her the ratio of girls to boys in her new class is $5:6$. Amy was very surprised to think her friend's class had only 11 students. What do you tell her?
17. Manday read that the arm of the Statue of Liberty is 42 ft long. She would like to know how long the Statue of Liberty's nose is. How would you advise her to proceed?

Trends in Mathematics and Science Study (TIMSS) Questions

For every soft drink bottle that Fred collected, Maria collected 3. Fred collected a total of 9 soft drink bottles. How many did Maria collect?

a. 3
b. 12
c. 13
d. 27

TIMSS, Grade 4, 2003

There are 30 students in a class. The ratio of boys to girls in the class is $2:3$. How many boys are there in the class?

a. 6
b. 12
c. 18
d. 20

TIMSS, Grade 8, 2007

Three brothers, Bob, Dan, and Mark, receive a gift of 45,000 zeds from their father. The money is shared between the brothers in proportion to the number of children each one has. Bob has 2 children, Dan has 3 children, and Mark has 4 children. How many zeds does Mark get?

a. 5,000
b. 10,000
c. 15,000
d. 20,000

TIMSS, Grade 8, 2003

National Assessment of Educational Progress (NAEP) Question

Sarah has a part-time job at Better Burgers restaurant and is paid $5.50 for each hour she works. She has made the chart below to reflect her earnings but needs your help to complete it.

a. Fill in the missing entries in the chart.

Hours Worked	Money Earned (in dollars)
1	$5.50
4	
	$38.50
$7\frac{3}{4}$	$42.63

b. If Sarah works h hours, then in terms of h, how much will she earn?

NAEP, Grade 8, 2007

BRAIN TEASER

A woman's will decreed that her cats be shared among her three daughters as follows: $\frac{1}{2}$ of the cats to the eldest daughter, $\frac{1}{3}$ of the cats to the middle daughter, and $\frac{1}{9}$ of the cats to the youngest daughter.

Since the woman had 17 cats, the daughters decided that they could not carry out their mother's wishes. The judge who held the will agreed to lend the daughters a cat so that they could share the cats as their mother wished. Now, $\frac{1}{2}$ of 18 is 9; $\frac{1}{3}$ of 18 is 6; and $\frac{1}{9}$ of 18 is 2. Since $9 + 6 + 2 = 17$, the daughters were able to divide the 17 cats and return the borrowed cat. They obviously did not need the extra cat to carry out their mother's bequest, but they could not divide 17 into halves, thirds, and ninths. Has the woman's will really been followed?

Hint for Solving the Preliminary Problem

Originally the ratio of female dolls (F) to male dolls (M) is 3:1. This implies $F = 3M$. Use this information to set up a proportion with the new number of dolls and the new ratio.

Chapter 6 Summary

6-1 The Set of Rational Numbers	Pages
Numbers of the form $\frac{a}{b}$, where a and b are integers and $b \neq 0$, are **rational numbers**. a is the *numerator* and b is the *denominator*.	260
A rational number can be used as follows: • A division problem or the solution to a multiplication problem • A partition, or part, of a whole • A ratio • A probability	261
A fraction $\frac{a}{b}$, where $0 \leq a < b$, is a **proper fraction**. If $a \geq b > 0$, $\frac{a}{b}$ is an **improper fraction**.	262
Theorem: Fundamental Law of Fractions: For any fraction $\frac{a}{b}$ and any nonzero rational number $n \neq 0$, $\frac{a}{b} = \frac{an}{bn}$.	264
A rational number $\frac{a}{b}$ is in **simplest form** if, and only if, $b > 0$ and $\text{GCD}(a, b) = 1$, that is, if a and b have no common factor greater than 1.	266
Theorem: Equality of Fractions: Two fractions $\frac{a}{b}$ and $\frac{c}{d}$, $b, d \neq 0$, are **equal** (**equivalent**) if, and only if, $ad = bc$.	268
If a, b, and c are integers and $b > 0$, then $\frac{a}{b} > \frac{c}{b}$ if, and only if, $a > c$.	269

Theorem: If a, b, c, and d are integers with $b > 0$ and $d > 0$, then $\frac{a}{b} > \frac{c}{d}$ if, and only if, $ad > bc$.　　　270

Theorem: Denseness Property: Given two different rational numbers $\frac{a}{b}$ and $\frac{c}{d}$, there is another rational number between these two numbers.　　　270

Theorem: If $\frac{a}{b}$ and $\frac{c}{d}$ are any rational numbers with positive denominators, where $\frac{a}{b} < \frac{c}{d}$, then $\frac{a}{b} < \frac{a + c}{b + d} < \frac{c}{d}$.　　　271

6-2 Addition, Subtraction, and Estimation with Rational Numbers

If $\frac{a}{b}$ and $\frac{c}{b}$ are rational numbers, then $\frac{a}{b} + \frac{c}{b} = \frac{a + c}{b}$.　　　278

If $\frac{a}{b}$ and $\frac{c}{d}$ are rational numbers, then $\frac{a}{b} + \frac{c}{d} = \frac{ad + bc}{bd}$.　　　279

A **mixed number** is made up of an integer and a proper fraction.　　　280

Theorem: For any rational number $\frac{a}{b}$, there exists a unique rational number $-\frac{a}{b}$, the **additive inverse** of $\frac{a}{b}$, such that $\frac{a}{b} + \left(-\frac{a}{b}\right) = 0 = \left(-\frac{a}{b}\right) + \frac{a}{b}$.　　　281

Theorem: Addition Property of Equality: If $\frac{a}{b}$ and $\frac{c}{d}$ are any rational numbers such that $\frac{a}{b} = \frac{c}{d}$, and if $\frac{e}{f}$ is any rational number, then $\frac{a}{b} + \frac{e}{f} = \frac{c}{d} + \frac{e}{f}$.　　　283

If $\frac{a}{b}$ and $\frac{c}{d}$ are any rational numbers, then $\frac{a}{b} - \frac{c}{d}$ is the unique rational number $\frac{e}{f}$ such that $\frac{a}{b} = \frac{c}{d} + \frac{e}{f}$.　　　284

Theorem: If $\frac{a}{b}$ and $\frac{c}{d}$ are any rational numbers, then $\frac{a}{b} - \frac{c}{d} = \frac{a}{b} + \frac{^-c}{d}$.　　　284

Theorem: If $\frac{a}{b}$ and $\frac{c}{d}$ are any rational numbers, then $\frac{a}{b} - \frac{c}{d} = \frac{ad - bc}{bd}$.　　　284

Sometimes to obtain an estimate it is desirable to round fractions to a **convenient** or **benchmark** fraction such as $\frac{1}{2}, \frac{1}{3}, \frac{1}{4}, \frac{1}{5}, \frac{2}{3}, \frac{3}{4}$, or 1.　　　286

6-3 Multiplication and Division of Rational Numbers

If $\frac{a}{b}$ and $\frac{c}{d}$ are any rational numbers, then $\frac{a}{b} \cdot \frac{c}{d} = \frac{ac}{bd}$.　　　293

Theorem: Multiplicative Identity: The number 1 is the unique number such that for every rational number $\frac{a}{b}$, $1 \cdot \left(\frac{a}{b}\right) = \frac{a}{b} = \left(\frac{a}{b}\right) \cdot 1$.　　　293

Theorem: Multiplicative Inverse: For any nonzero rational number $\frac{a}{b}$, $\frac{b}{a}$, $a \neq 0$ is the unique rational number such that $\frac{a}{b} \cdot \frac{b}{a} = 1 = \frac{b}{a} \cdot \frac{a}{b}$.　　　293

Theorem: Distributive Properties of Multiplication Over Addition and Subtraction:

If $\dfrac{a}{b}, \dfrac{c}{d}$, and $\dfrac{e}{f}$ are any rational numbers, then

a. $\dfrac{a}{b}\left(\dfrac{c}{d} + \dfrac{e}{f}\right) = \dfrac{a}{b} \cdot \dfrac{c}{d} + \dfrac{a}{b} \cdot \dfrac{e}{f};$

b. $\dfrac{a}{b}\left(\dfrac{c}{d} - \dfrac{e}{f}\right) = \dfrac{a}{b} \cdot \dfrac{c}{d} - \dfrac{a}{b} \cdot \dfrac{e}{f}.$ 295

Theorem: Multiplication Property of Equality: If $\dfrac{a}{b}, \dfrac{c}{d}$, and $\dfrac{e}{f}$ are any rational

numbers such that $\dfrac{a}{b} = \dfrac{c}{d}$, then $\dfrac{a}{b} \cdot \dfrac{e}{f} = \dfrac{c}{d} \cdot \dfrac{e}{f}.$ 295

Theorem: Multiplication Properties of Inequality:

- If $\dfrac{a}{b} > \dfrac{c}{d}$ and $\dfrac{e}{f} > 0$, then $\dfrac{a}{b} \cdot \dfrac{e}{f} > \dfrac{c}{d} \cdot \dfrac{e}{f}.$ 295

- If $\dfrac{a}{b} > \dfrac{c}{d}$ and $\dfrac{e}{f} < 0$, then $\dfrac{a}{b} \cdot \dfrac{e}{f} < \dfrac{c}{d} \cdot \dfrac{e}{f}.$ 295

Theorem: Multiplication Property of Zero: If $\dfrac{a}{b}$ is a rational number, then $\dfrac{a}{b} \cdot 0 = 0 = 0 \cdot \dfrac{a}{b}.$ 295

If $\dfrac{a}{b}$ and $\dfrac{c}{d}$ are rational numbers, then $\dfrac{a}{b} \div \dfrac{c}{d} = \dfrac{e}{f}$ if, and only if, $\dfrac{e}{f}$ is the unique

rational number such that $\dfrac{c}{d} \cdot \dfrac{e}{f} = \dfrac{a}{b}.$ 298

Theorem: Algorithm for Division of Fractions: If $\dfrac{a}{b}$ and $\dfrac{c}{d}$ are any rational numbers and

$\dfrac{c}{d} \neq 0$, then $\dfrac{a}{b} \div \dfrac{c}{d} = \dfrac{a}{b} \cdot \dfrac{d}{c}.$ 300

$a^m = \underbrace{a \cdot a \cdot a \cdot \ldots \cdot a}_{m \text{ factors}}$, where m is a positive integer and a is a rational number   302

Theorem: Properties of Exponents: For any nonzero rational numbers a and b and integers m and n,   305

- $a^0 = 1$

- $a^{-m} = \dfrac{1}{a^m}$

- $a^m \cdot a^n = a^{m+n}$

- $\dfrac{a^m}{a^n} = a^{m-n}$

- $(a^m)^n = a^{mn}$

- $\left(\dfrac{a}{b}\right)^m = \dfrac{a^m}{b^m}$

- $(ab)^m = a^m \cdot b^m$

- $\left(\dfrac{a}{b}\right)^{-m} = \left(\dfrac{b}{a}\right)^m$

6-4 Ratios, Proportions, and Proportional Reasoning

A **ratio**, written $\dfrac{a}{b}$ or $a:b$, where a and b are rational numbers, is a comparison of two

quantities a and b expressed as the number of times one is greater than or less than the other.   312

A **proportion** is a statement that two given ratios are equal. 313

Theorem: If a, b, c, and d are rational numbers and $b \neq 0$ and $d \neq 0$, then $\dfrac{a}{b} = \dfrac{c}{d}$ if, and only if, $ad = bc$. 314

If the variables x and y are related by the equality $y = kx$ $\left(\text{or } k = \dfrac{y}{x}\right)$, then y is **proportional to** x and k is the **constant of proportionality** between y and x. 316

Theorem: For any rational numbers $\dfrac{a}{b}$ and $\dfrac{c}{d}$, with $a \neq 0$ and $c \neq 0$, $\dfrac{a}{b} = \dfrac{c}{d}$ if, and only if, $\dfrac{b}{a} = \dfrac{d}{c}$. 319

Theorem: For any rational numbers $\dfrac{a}{b}$ and $\dfrac{c}{d}$, $c \neq 0$, $\dfrac{a}{b} = \dfrac{c}{d}$ if, and only if, $\dfrac{a}{c} = \dfrac{b}{d}$. 319

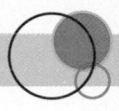

Chapter 6 Review

1. For each of the following, draw a diagram illustrating the fraction:

   a. $\dfrac{3}{4}$

   b. $\dfrac{2}{3}$

   c. $\dfrac{3}{4} \cdot \dfrac{2}{3}$

2. Write three rational numbers equal to $\dfrac{5}{6}$.

3. Write each of the following rational numbers in simplest form:

   a. $\dfrac{24}{28}$

   b. $\dfrac{ax^2}{bx}$

   c. $\dfrac{0}{17}$

   d. $\dfrac{45}{81}$

   e. $\dfrac{b^2 + bx}{b + x}$

   f. $\dfrac{16}{216}$

   g. $\dfrac{x + a}{x - a}$

   h. $\dfrac{xa}{x + a}$

4. In each of the following pairs, replace the comma with $>$, $<$, or $=$ to make a true statement:

   a. $\dfrac{6}{10}, \dfrac{120}{200}$

   b. $\dfrac{{}^-3}{4}, \dfrac{{}^-5}{6}$

   c. $\left(\dfrac{4}{5}\right)^{10}, \left(\dfrac{4}{5}\right)^{20}$

   d. $\left(1 + \dfrac{1}{3}\right)^2, \left(1 + \dfrac{1}{3}\right)^3$

5. Find the additive and multiplicative inverses for each of the following:

   a. 3

   b. $3\dfrac{1}{7}$

   c. $\dfrac{5}{6}$

   d. $-\dfrac{3}{4}$

6. Order the following numbers from least to greatest:

   $^-1\dfrac{7}{8}, 0, \, ^-2\dfrac{1}{3}, \dfrac{69}{140}, \dfrac{71}{140}, \left(\dfrac{71}{140}\right)^{300}, \dfrac{1}{2}, \left(\dfrac{74}{73}\right)^{300}$

7. Can $\dfrac{4}{5} \cdot \dfrac{7}{8} \cdot \dfrac{5}{14}$ be written as $\dfrac{4}{8} \cdot \dfrac{7}{14} \cdot \dfrac{5}{5}$ to obtain the same answer? Why or why not?

8. Use mental math to compute the following. Explain your method.

   a. $\dfrac{1}{3} \cdot (8 \cdot 9)$

   b. $36 \cdot 1\dfrac{5}{6}$

9. John has $54\dfrac{1}{4}$ yd of material.

   a. If he needs to cut the cloth into pieces that are $3\dfrac{1}{12}$ yd long, how many pieces can he cut?

   b. How much material will be left over?

10. Without actually performing the given operations, choose the most appropriate estimate (among the numbers in parentheses) for the following expressions:

    a. $\dfrac{30\dfrac{3}{8}}{4\dfrac{1}{9}} \cdot \dfrac{8\dfrac{1}{3}}{3\dfrac{8}{9}}$ $(15, 20, 8)$

    b. $\left[\dfrac{3}{800} + \dfrac{4}{5000} + \dfrac{15}{6}\right] \cdot 6$ $(15, 0, 132)$

    c. $\dfrac{1}{407} \div \dfrac{1}{1609}$ $\left(\dfrac{1}{4}, 4, 0\right)$

11. Justify the invert-and-multiply algorithm for division of rational numbers in two different ways.

12. Write a story problem that models $4\frac{5}{8} \div \frac{1}{2}$. Answer the problem by drawing appropriate diagrams.

13. Find two rational numbers between $\frac{3}{4}$ and $\frac{4}{5}$.

14. Suppose the $\boxed{\div}$ button on your calculator is broken, but the $\boxed{1/x}$ button works. Explain how you could compute $504792/23$.

15. Jim is starting a diet. When he arrived home, he ate $\frac{1}{3}$ of the half of a pizza that was left from the previous night. The whole pizza contains approximately 2000 calories. How many calories did Jim consume?

16. If a person got heads on a flip of a fair coin one-half the time and obtained 376 heads, how many times was the coin flipped?

17. If a person obtained 240 heads when flipping a coin 1000 times, what fraction of the time did the person obtain heads? Put the answer in simplest form.

18. If the University of New Mexico won $\frac{3}{4}$ of its women's basketball games and $\frac{5}{8}$ of its men's basketball games, explain whether it is reasonable to say that the university won $\frac{3}{4} + \frac{5}{8}$ of its basketball games.

19. The carvings of the faces at Mount Rushmore in South Dakota measure 60 ft from chin to forehead. If the distance from chin to forehead is typically 9 in., and the distance between the pupils of the eyes is typically $2\frac{1}{2}$ in., what is the approximate distance between the pupils on the carving of George Washington's head?

20. A student argues that the following fraction is not a rational number because it is not the quotient of two integers:

$$\frac{\frac{2}{3}}{\frac{3}{4}}$$

How would you respond?

21. Molly wants to fertilize 12 acres of park land. If it takes $9\frac{1}{3}$ bags for each acre, how many bags does she need?

22. If $\frac{2}{3}$ of all students in the academy are female and $\frac{2}{5}$ of those are blondes, what fraction describes the number of blond females in the academy?

23. Explain which is greater: $\frac{^-11}{9}$ or $\frac{^-12}{10}$.

24. Solve for x in each of the following:
    a. $7^x = 343$
    b. $2^{-3x} = \frac{1}{512}$

25. Solve for x in each of the following:
    a. $2x - \frac{5}{3} = \frac{5}{6}$
    b. $x + 2\frac{1}{2} = 5\frac{2}{3}$
    c. $\frac{20 + x}{x} = \frac{4}{5}$
    d. $2x + 4 = 3x - \frac{1}{3}$

26. Write each of the following in simplest form. Leave all answers with positive exponents.
    a. $\dfrac{(x^3 a^{-1})^{-2}}{xa^{-1}}$
    b. $\left(\dfrac{x^2 y^{-2}}{x^{-3} y^2}\right)^{-2}$

27. Find each sum or difference.
    a. $\dfrac{3a}{xy^2} + \dfrac{b}{x^2 y^2}$
    b. $\dfrac{5}{xy^2} - \dfrac{2}{3x}$
    c. $\dfrac{a}{x^3 y^2 z} - \dfrac{b}{xyz}$
    d. $\dfrac{7}{2^3 3^2} + \dfrac{5}{2^2 3^3}$

28. Mike drew the following picture to find out how many pieces of ribbon $\frac{1}{2}$ yd long could be cut from a strip of ribbon $1\frac{3}{4}$ yd long.

(1)	(2)	(3)	left over	

$0 \quad \frac{1}{4} \quad \frac{1}{2} \quad \frac{3}{4} \quad 1 \quad 1\frac{1}{4} \quad 1\frac{1}{2} \quad 1\frac{3}{4} \quad 2$

From the picture he concluded that $1\frac{3}{4} \div \frac{1}{2}$ is 3 pieces with $\frac{1}{4}$ yd left over, so the answer is $3\frac{1}{4}$ pieces. He checked this using the algorithm $\frac{7}{4} \cdot \frac{2}{1} = \frac{14}{4} = 3\frac{1}{2}$ and is confused why he has two different answers. How would you help him?

29. If a, b, and c are nonzero integers, express the following as a rational number in simplest form.

$$\left(\frac{a^{-1} + b^{-1} + c^{-1}}{2}\right)^{-1}$$

30. Tom tossed a coin 30 times and got 17 heads.
    a. What is the ratio of heads to coin tosses?
    b. What is the ratio of heads to tails?
    c. What is the ratio of tails to heads?

31. Which bottle of juice is a better buy (cost per ounce): 48 fl oz for $3.05 or 64 fl oz for $3.60?

32. Eighteen-karat gold contains 18 parts (grams) gold and 6 parts (grams) other metals. Amy's new ring contains 12 parts gold and 3 parts other metals. Is the ring 18-karat gold? Why?

33. Solve for x in each of the following:

a. $\dfrac{15}{12} = \dfrac{21}{x}$ b. $\dfrac{20}{35} = \dfrac{110}{x}$

c. $\dfrac{\frac{1}{2}}{\frac{1}{3}} = \dfrac{\frac{3}{2}}{x}$

34. A recipe for fruit salad serves 4 people. It calls for 3 oranges and 16 grapes. How many oranges and grapes do you need to serve 11 people?

35. If the scale on a drawing of a house is 1 cm $= 2\frac{1}{2}$ m, what is the length of the house if it measures 3 cm on the scale drawing?

36. In water (H_2O), the ratio of the weight of oxygen to the weight of hydrogen is approximately 8:1. How many ounces of hydrogen are in 1 lb of water?

37. A manufacturer produces the same kind of computer chip in two plants. In the first plant, the ratio of defective chips to good chips is 15:100 and in the second plant, that ratio is 12:100. A buyer of a large number of chips is aware that some come from the first plant and some from the second. However, she is not aware of how many come from each. The buyer would like to know the ratio of defective chips to good chips in any given order. Can she determine that ratio? If so, explain how. If not, explain why not.

38. Suppose the ratio of the lengths of the sides in two squares is 1:r. What is the ratio of their areas? ($A = s^2$)

39. The Grizzlies won 18 games and lost 7.
a. What is the ratio of games won to games lost?
b. What is the ratio of games won to games played?

40. Express each of the following as a ratio $\dfrac{a}{b}$ where a and b are whole numbers.
a. $\dfrac{1}{5} : 1$
b. $\dfrac{2}{5} : \dfrac{3}{4}$

41. The ratio of boys to girls in Mr. Good's class is 3 to 5, the ratio of boys to girls in Ms. Garcia's is the same, and you know that there are 15 girls in Ms. Garcia's class. How many boys are in Ms. Garcia's class?

Decimals: Rational Numbers and Percent

Preliminary Problem

A store marks up an item by adding 30% to the wholesale price. Later the store has a sale and marks the item down 30% from the retail price. Is the store breaking even on the item, making a profit, or losing money on the item?

If needed, see Hint on page 385.

In grades 6–8 *PSSM*, we find that students should

- work flexibly with fractions, decimals, and percents to solve problems;
- compare and order fractions, decimals, and percents efficiently and find their approximate locations on a number line; . . .
- develop an understanding of large numbers and recognize and appropriately use exponential, scientific, and calculator notation; . . .
- understand the meaning and effects of arithmetic operations with fractions, decimals, and integers. (p. 214)

In grade 4 *Focal Points*, students are expected to relate "their understanding of fractions to reading and writing decimals, comparing and ordering decimals, and estimating decimal or fractional amounts in problem solving. They connect equivalent fractions and decimals by comparing models to symbols and locating equivalent symbols on the number line." (p. 31)

In grade 5 *Common Core Standards* we find that students should:

add, subtract, multiply, and divide decimals to hundredths, using concrete models or drawings and strategies based on place value, properties of operations, and/or the relationship between addition and subtraction; relate the strategy to a written method and explain the reasoning used. (p. 35)

In this chapter, we develop the understanding that teachers need to help students reach these expectations. To begin, we look at a brief history of decimals.

The word *decimal* comes from Latin *decem*, meaning "ten." Although the Hindu-Arabic numeration system discussed in Chapter 2 was used in many places around the sixth century, the extension of the system to decimals by Dutchman Simon Stevin did not take place until about a thousand years later. The only significant improvement in the system since Stevin's time has been in notation. Even today there is no universally accepted form of writing a decimal. For example, in the United States, we write 6.75; in England, this number is written 6·75; and in Germany and France, it is written 6,75.

In Section 7-1, we explore relationships between fractions and decimals and see how decimals are an extension of the base ten system. Later in the chapter, we consider operations on decimals, properties of decimals, and percents.

⊙ Historical Note

In 1584, Simon Stevin (1548–1620), a Dutch quartermaster general, wrote *La Thiende (The Tenth)*, giving rules for computing with decimals. He not only stated these rules but also suggested practical applications for decimals and recommended that his government adopt them. To show place value, Stevin used circled numerals between digits. For example, he wrote 0.4789 as 4 ① 7 ② 8③ 9 ④ .

THIENDE.   13

HET ANDER DEEL

DER THIENDE VANDE
WERCKINCHE.

I. VOORSTEL VANDE
VERGADERINGHE.

Wefende ghegeven Thiendetalen te ver-
gaderen: hare Somme te vinden.

T'GHEGHEVEN. Het fijn drie oirdens van
Thiendetalen, welcker eerfte 27 ⓪ 8 ① 4 ②
7 ③ , de tweede, 37 ⓪ 6 ① 7 ② 5 ③ , de derde,
875 ⓪ 7 ① 8 ② 2 ③ , TBEGHEERDE. Wy
moeten haer Somme vinden . WERCKING.
Men fal de ghegheven ghe-
talen in oirden ftellen als
hier neven, die vergaderen-
de naer de ghemeene manie
re der vergaderinghe van
heelegetalen aldus:

	⓪	①	②	③
	2 7	8	4	7
	3 7	6	7	5
	8 7 5	7	8	2
	9 4 1	3	0	4

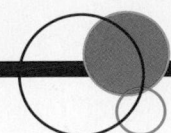

7-1 Introduction to Decimals

Any number that is represented using place value and powers of 10 is a **decimal.** The word *decimal* comes from the Latin *decem,* meaning "ten." Thus integers written in base ten are technically decimals although most people do not think of them as such. We encounter decimals when dealing with money. For example, if a bike costs $128.95, the dot in $128.95 is the **decimal point.** Because $0.95 is $\frac{95}{100}$ of a dollar, $128.95 = 128 + \frac{95}{100}$ dollars. Also because $0.95 is 9 dimes and 5 cents, 1 dime is $\frac{1}{10}$ of a dollar and 1 cent is $\frac{1}{100}$ of a dollar, we see that $0.95 is $9 \cdot \frac{1}{10} + 5 \cdot \frac{1}{100}$ dollars. Consequently,

$$128.95 = 1 \cdot 10^2 + 2 \cdot 10 + 8 \cdot 1 + 9 \cdot \frac{1}{10} + 5 \cdot \frac{1}{10^2}$$

The digits in 128.95 correspond, respectively, to the following place values: 10^2, 10, 1, $\frac{1}{10}$, and $\frac{1}{10^2}$. Similarly, 12.61843 represents

$$1 \cdot 10 + 2 \cdot 1 + \frac{6}{10^1} + \frac{1}{10^2} + \frac{8}{10^3} + \frac{4}{10^4} + \frac{3}{10^5}, \quad \text{or} \quad 12\frac{61,843}{100,000}$$

Note that $\frac{61,843}{100,000}$ represents a division of two whole numbers when the divisor is a power of 10, in this case, 10^5. The decimal representation for $\frac{61,843}{100,000}$ is 0.61843.

Technically the decimal 12.61843 is read "twelve and sixty-one thousand eight hundred forty-three hundred-thousandths" although most people will say "twelve point six, one, eight, four, three." Each place of a decimal may be named by its power of 10. For example, the places of 12.61843 can be named as shown in Table 7-1.

Table 7-1

1	2	•	6	1	8	4	3
Tens	Units	*and*	Tenths	Hundredths	Thousandths	Ten-thousandths	Hundred-thousandths

Decimals can be introduced with concrete materials. For example, suppose that a long in the base-ten block set represents 1 unit instead of letting the cube represent 1 unit. Then the cube represents $\frac{1}{10}$ and 5.4 could be represented as in Figure 7-1(a).

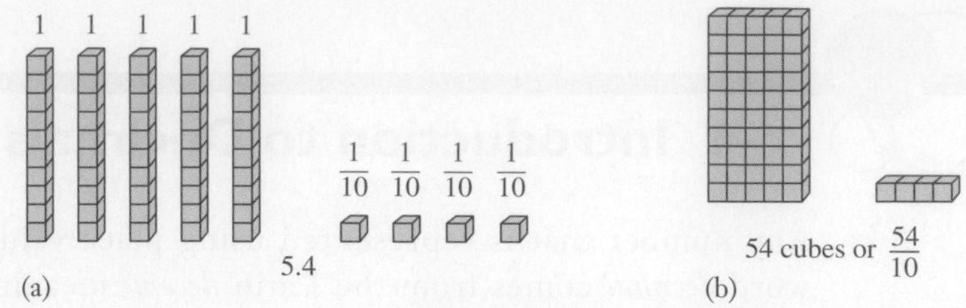

54 cubes or $\frac{54}{10}$

(a)   5.4   (b)

Figure 7-1

We could use Figure 7-1(b) to obtain a different interpretation of 5.4 and see that 5.4 is also equivalent to 54 tenths, or $\frac{54}{10}$. This equivalence can be stated symbolically as

$$5.4 = 5 + 0.4 = 5 + \frac{4}{10} = \frac{50}{10} + \frac{4}{10} = \frac{54}{10}$$

This approach gives students a concrete connection between fractions and decimals.

NOW TRY THIS 7-1

In a set of base-ten blocks, let 1 flat represent 1 unit.

a. What does 1 long represent?
b. What does 1 cube represent?
c. Represent 1.23 using the blocks with these equivalences.

To represent a decimal such as 2.235, think of a block shown in Figure 7-2(a) as a unit. Then a flat represents $\frac{1}{10}$, a long represents $\frac{1}{100}$, and a cube represents $\frac{1}{1000}$. Using these objects, we show a representation of 2.235 in Figure 7-2(b).

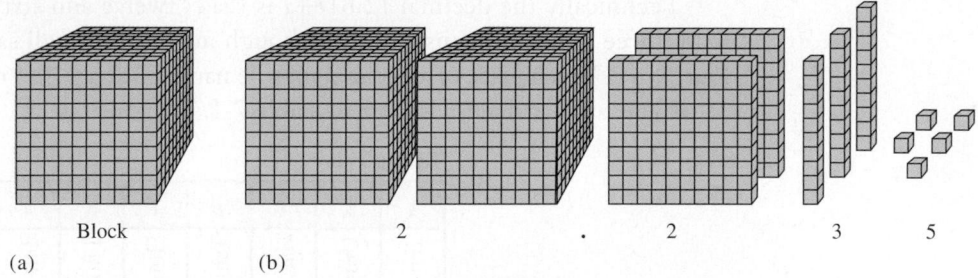

Block

(a)   (b)   2   .   2   3   5

Figure 7-2

Table 7-2 shows other examples of decimals, their fractional meanings, and their common fraction notations.

Table 7-2

Decimal	Expanded Fractional Meaning	Common Fraction Notation
5.3	$5 + \dfrac{3}{10}$	$5\dfrac{3}{10}$, or $\dfrac{53}{10}$
0.02	$0 + \dfrac{0}{10} + \dfrac{2}{100}$	$\dfrac{2}{100}$
2.0103	$2 + \dfrac{0}{10} + \dfrac{1}{100} + \dfrac{0}{1000} + \dfrac{3}{10,000}$	$2\dfrac{103}{10,000}$, or $\dfrac{20,103}{10,000}$
$^-3.6$	$-\left(3 + \dfrac{6}{10}\right)$	$^-3\dfrac{6}{10}$, or $-\dfrac{36}{10}$

Decimals can also be written in expanded (or standard) form using place value and negative exponents. For example,

$$12.61843 = 1 \cdot 10^1 + 2 \cdot 10^0 + 6 \cdot 10^{-1} + 1 \cdot 10^{-2} + 8 \cdot 10^{-3} + 4 \cdot 10^{-4} + 3 \cdot 10^{-5}$$

From the student page on Decimal Place Value, we see that in grade 4, students learn different representations of decimals. Work number 2 in the Talk About It section.

School Book Page DECIMAL PLACE VALUE

Lesson 11-2

Key Idea
There are many ways to represent decimal numbers.

Think It Through
I can **use objects**, **draw pictures**, or **make a chart** to represent 1.48.

Decimal Place Value

LEARN

What are some ways to represent decimals?

Here are different ways to represent 1.48.

Number line:

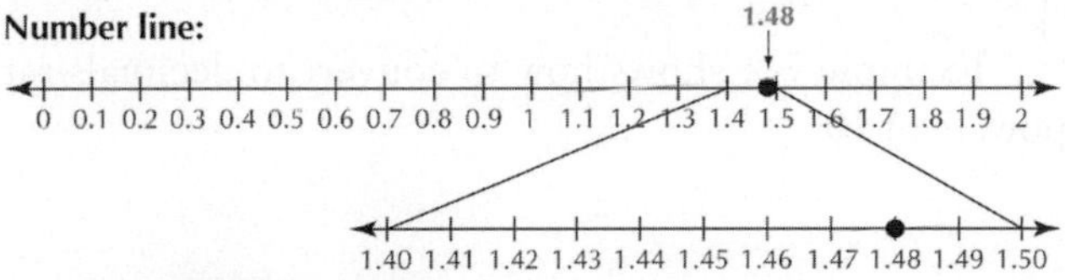

1.48

0 0.1 0.2 0.3 0.4 0.5 0.6 0.7 0.8 0.9 1 1.1 1.2 1.3 1.4 1.5 1.6 1.7 1.8 1.9 2

1.40 1.41 1.42 1.43 1.44 1.45 1.46 1.47 1.48 1.49 1.50

Grids:

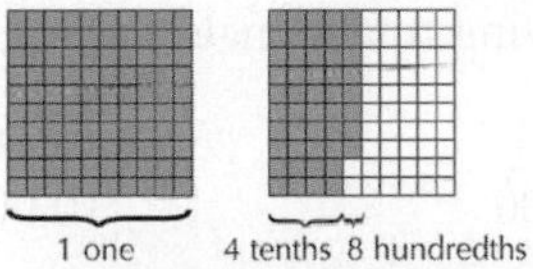

1 one   4 tenths 8 hundredths

Place-value chart:

tens	ones		tenths	hundredths
	1	.	4	8

Expanded form: 1 + 0.4 + 0.08

Standard form: 1.48

Word form: One and forty-eight hundredths

> **Example**
>
> Write the word form and the expanded form for 5.02. Then, tell the value of the red digit.
>
> *Word form:* five and two hundredths
>
> *Expanded form:* 5 + 0.02
>
> The red digit is in the hundredths place, so its value is 2 hundredths, 0.02.

*I write the word **and** for the decimal point.*

✔ **Talk About It**

1. Which digit is in the tenths place in 1.48?

2. Explain how to locate 1.48 on a number line.

Take It to the NET
More Examples
www.scottforesman.com

628

> **✔ WARM UP**
>
> Tell the value of the red digit for each number.
>
> 1. 8,264   2. 17,932
>
> 3. 2,925   4. 5,924

NOW TRY THIS 7-2

Decimals can sometimes cause confusion. In the *Blondie* cartoon, the cartoonist intended the price of tomatoes to be increased. Is that what the cartoon says? Give other examples of this type of mistake involving dollars and cents.

Blondie copyright © 1989 and 2011 King Features Syndicate

Example 7-1 shows how to convert to decimals rational numbers whose denominators are powers of 10.

 EXAMPLE 7-1

Convert each of the following to decimals:

a. $\dfrac{25}{10}$ **b.** $\dfrac{56}{100}$ **c.** $\dfrac{205}{10,000}$

Solution

a. $\dfrac{25}{10} = \dfrac{2 \cdot 10 + 5}{10} = \dfrac{2 \cdot 10}{10} + \dfrac{5}{10} = 2 + \dfrac{5}{10} = 2.5$

b. $\dfrac{56}{100} = \dfrac{5 \cdot 10 + 6}{10^2} = \dfrac{5 \cdot 10}{10^2} + \dfrac{6}{10^2} = 0 + \dfrac{5}{10} + \dfrac{6}{10^2} = 0.56$

c. $\dfrac{205}{10,000} = \dfrac{2 \cdot 10^2 + 0 \cdot 10 + 5}{10^4} = \dfrac{2 \cdot 10^2}{10^4} + \dfrac{0 \cdot 10}{10^4} + \dfrac{5}{10^4}$

$\quad = \dfrac{2}{10^2} + \dfrac{0}{10^3} + \dfrac{5}{10^4} = 0 + \dfrac{0}{10^1} + \dfrac{2}{10^2} + \dfrac{0}{10^3} + \dfrac{5}{10^4} = 0.0205$

The conversions in Example 7-1 can also be performed using negative exponents. For example,

$$\frac{25}{10} = (2 \cdot 10 + 5) \cdot 10^{-1} = (2 \cdot 10) \cdot 10^{-1} + 5 \cdot 10^{-1} = 2 \cdot (10 \cdot 10^{-1}) + 5 \cdot 10^{-1}$$

$$= 2 + 5 \cdot 10^{-1} = 2.5$$

 The ideas in Example 7-1 are reinforced with a calculator. In Example 7-1(b), press ⑤ ⑥ ÷ ① ⓪ ⓪ ＝ and watch the display. Next divide by 10 again and look at the new placement of the decimal point. Once more, divide by 10 (which amounts to dividing the original number, 56, by

10,000) and note the placement of the decimal point. This leads to the general rule for dividing or multiplying a decimal by a positive integer power of 10:

To divide a decimal by 10^n where n is a positive integer, start at the decimal point, count n place values to the left, appending zeros if necessary, and insert the decimal point to the left of the nth place value counted.

To multiply a decimal by 10^n where n is a positive integer, move the decimal point n places to the right attaching zeros if needed.

The fractions in Example 7-1 are easy to convert to decimals because the denominators are powers of 10. If the denominator of a fraction is not a power of 10, as in $\frac{3}{5}$, we use the problem-solving strategy of *converting the problem to one that we already know how to do*. First, we rewrite $\frac{3}{5}$ as a fraction in which the denominator is a power of 10, and then we convert the fraction to a decimal.

$$\frac{3}{5} = \frac{2 \cdot 3}{2 \cdot 5} = \frac{6}{10} = 0.6$$

The reason for multiplying the numerator and the denominator by 2 is apparent when we observe that $10 = 5 \cdot 2$ or $2 \cdot 5$.

NOW TRY THIS 7-3

Convert the following fractions to decimals without using a calculator.

a. $\frac{5}{2}$

b. $\frac{3}{8}$ (*Hint:* $\frac{3}{8} = \frac{3}{2^3}$. By what power of 5 must one multiply to make the denominator a power of 10?)

c. $\frac{3}{20}$ (*Hint:* $\frac{3}{20} = \frac{3}{2^2 \cdot 5}$. By what power of 5 must one multiply to make the denominator a power of 10?)

When the denominator of a fraction is a power of 10, then converting the fraction to a decimal is a simple use of division of a whole number by a power of 10. Any power of 10 is of the form 10^n, and in general, because $10^n = (2 \cdot 5)^n = 2^n \cdot 5^n$, the prime factorization of the denominator that is a power of 10^n must be $2^n \cdot 5^n$. We use this idea in Example 7-2.

EXAMPLE 7-2

Express each of the following as decimals:

a. $\frac{7}{2^6}$ b. $\frac{1}{2^3 \cdot 5^4}$ c. $\frac{1}{125}$ d. $\frac{7}{250}$

Solution

a. $\frac{7}{2^6} = \frac{7 \cdot 5^6}{2^6 \cdot 5^6} = \frac{7 \cdot 15{,}625}{(2 \cdot 5)^6} = \frac{109{,}375}{10^6} = 0.109375$

b. $\frac{1}{2^3 \cdot 5^4} = \frac{1 \cdot 2^1}{2^3 \cdot 5^4 \cdot 2^1} = \frac{2}{2^4 \cdot 5^4} = \frac{2}{(2 \cdot 5)^4} = \frac{2}{10^4} = 0.0002$

c. $\frac{1}{125} = \frac{1}{5^3} = \frac{1 \cdot 2^3}{5^3 \cdot 2^3} = \frac{8}{(5 \cdot 2)^3} = \frac{8}{10^3} = 0.008$

d. $\frac{7}{250} = \frac{7}{2 \cdot 5^3} = \frac{7 \cdot 2^2}{2 \cdot 5^3 \cdot 2^2} = \frac{28}{(2 \cdot 5)^3} = \frac{28}{10^3} = 0.028$

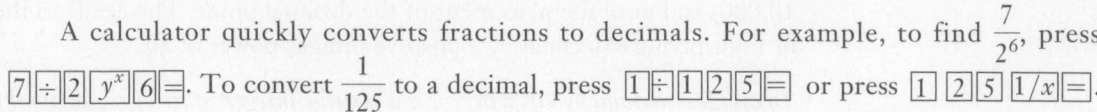

A calculator quickly converts fractions to decimals. For example, to find $\frac{7}{2^6}$, press $\boxed{7}\boxed{\div}\boxed{2}\boxed{y^x}\boxed{6}\boxed{=}$. To convert $\frac{1}{125}$ to a decimal, press $\boxed{1}\boxed{\div}\boxed{1}\boxed{2}\boxed{5}\boxed{=}$ or press $\boxed{1}\boxed{2}\boxed{5}\boxed{1/x}\boxed{=}$.

The above calculator answers as well as the answers in Example 7-2 are illustrations of **terminating decimals**—*decimals that can be written with only a finite number of places to the right of the decimal point.* Not every rational number can be written as a terminating decimal. For example, if we attempt to rewrite $\frac{2}{11}$ as a terminating decimal using the method just developed, we try to find a natural number b such that the following holds:

$$\frac{2}{11} = \frac{2b}{11b}, \quad \text{where } 11b \text{ is a power of } 10$$

By the Fundamental Theorem of Arithmetic, the only prime factors of a power of 10 are 2 and 5. Because $11b$ has 11 as a factor, we cannot write $11b$ as a power of 10. Therefore it seems that $\frac{2}{11}$ cannot be written as a terminating decimal. A similar argument using the Fundamental Theorem of Arithmetic holds in general, so we have the following theorem.

Theorem 7-1

A rational number $\frac{a}{b}$ in simplest form can be written as a terminating decimal if, and only if, the prime factorization of the denominator contains no primes other than 2 or 5.

EXAMPLE 7-3

Which of the following fractions can be written as terminating decimals?

   a. $\frac{7}{8}$ **b.** $\frac{11}{250}$ **c.** $\frac{21}{28}$ **d.** $\frac{37}{768}$

Solution

  a. Because the denominator, 8, is 2^3 and the fraction is in simplest form, $\frac{7}{8}$ can be written as a terminating decimal.

  b. Because the denominator, 250, is $2 \cdot 5^3$ and the fraction is in simplest form, $\frac{11}{250}$ can be written as a terminating decimal.

  c. $\frac{21}{28}$ can be written in simplest form as $\frac{21}{28} = \frac{3 \cdot 7}{2^2 \cdot 7} = \frac{3}{2^2}$. The denominator is now 2^2, so $\frac{21}{28}$ can be written as a terminating decimal.

  d. In simplest form, the fraction $\frac{37}{768} = \frac{37}{2^8 \cdot 3}$ has a denominator with a factor of 3, so $\frac{37}{768}$ cannot be written as a terminating decimal.

REMARK As Example 7-3(c) shows, to determine whether a rational number $\frac{a}{b}$ can be represented as a terminating decimal, we consider the prime factorization of the denominator *only* if the fraction is in simplest form.

Ordering Terminating Decimals

Given two numbers the one located farther to the right on a number line is the greater. A terminating decimal is easily located on a number line because it can be represented as a rational number number $\frac{a}{b}$, where $b \neq 0$, and b is a power of 10. For example, consider 0.56, or $\frac{56}{100}$. One way to think about $\frac{56}{100}$ is as the rightmost endpoint of the 56th part of a unit segment divided into 100 equal parts, as in Figure 7-3.

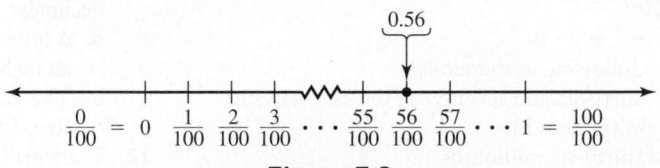

Figure 7-3

Two terminating decimals can be ordered by converting each to rational numbers in the form $\frac{a}{b}$ and determining which is greater. For example, because $0.36 = \frac{36}{100}$, $0.9 = 0.90 = \frac{90}{100}$, and $\frac{90}{100} > \frac{36}{100}$, it follows that $0.9 > 0.36$. One could also tell that $0.9 > 0.36$ because $0.90 is 90¢ and $0.36 is 36¢ and 90¢ > 36¢.

Comparing 0.36 and 0.9 by converting them to "money" suggests a method for comparing any two decimals. For example, consider the decimals 0.345 and 0.1474. Note that 0.345 is a decimal with thousandths as the least place value, and 0.1474 has ten-thousandths as its least place value. To compare the decimals, we could write both with ten-thousandths as the least place value. Note that 0.345 could be written as 0.3450. Now we have 0.3450 and 0.1474 both with ten-thousandths as the least place value. These decimals could be written as follows:

$$\frac{3450}{10,000} \quad \text{and} \quad \frac{1474}{10,000}$$

These two fractions can be compared by looking only at the numerators. Because 3450 is greater than 1474, $\frac{3450}{10,000} > \frac{1474}{10,000}$, and hence $0.345 > 0.1474$.

This process is frequently shortened by looking only at the decimals lined up as follows:

$$0.3450 = 0.345$$
$$0.1474 = 0.1474$$

And with the decimals lined up in this fashion, 3 tenths is greater than 1 tenth. Thus, $0.345 > 0.1474$. This suggests a way to order terminating decimals without conversion to fractions. The steps used to compare terminating decimals are like those for comparing whole numbers:

1. Line up the numbers by place value and append zeros if necessary.
2. Start at the left and find the first place where the face values are different.
3. Compare these digits. The digit with the greater face value in this place represents the greater of the two numbers.

Assessment 7-1A

1. Write each of the following as a sum in expanded place value form:
   a. 0.023 b. 206.06
   c. 312.0103 d. 0.000132
2. Rewrite each of the following as decimals:
   a. $4 \cdot 10^3 + 3 \cdot 10^2 + 5 \cdot 10 + 6 + 7 \cdot 10^{-1} + 8 \cdot 10^{-2}$
   b. $4 \cdot 10^3 + 6 \cdot 10^{-1} + 8 \cdot 10^{-3}$
   c. $4 \cdot 10^4 + 3 \cdot 10^{-2}$
   d. $2 \cdot 10^{-1} + 4 \cdot 10^{-4} + 7 \cdot 10^{-7}$
3. Write each of the following as numerals:
   a. Five hundred thirty-six and seventy-six ten-thousandths
   b. Three and eight thousandths
   c. Four hundred thirty-six millionths
   d. Five million and two tenths
4. Write each of the following in words:
   a. 0.34
   b. 20.34
   c. 2.034
   d. 0.000034
5. Write each of the following terminating decimals as $\frac{a}{b}$ in simplest form where $a, b \in I$ and $b \neq 0$.
   a. 0.436 b. 25.16
   c. ⁻316.027 d. 28.1902
   e. ⁻4.3 f. ⁻62.01
6. Mentally determine which of the following represent terminating decimals:
   a. $\frac{4}{5}$ b. $\frac{61}{2^2 \cdot 5}$
   c. $\frac{3}{6}$ d. $\frac{1}{2^5}$
   e. $\frac{4}{34}$ f. $\frac{133}{625}$
7. Where possible, write each of the numbers in exercise 6 as a terminating decimal.
8. Seven minutes is part of an hour. If 7 min were to be expressed as a decimal part of an hour, explain whether it would be a terminating decimal.

9. Given the U.S. monetary system, what reason can you think of for having coins only for a penny, nickel, dime, quarter, and half-dollar as coins less than $1.00?
10. In each of the following, order the decimals from greatest to least:
    a. 13.4919, 13.492, 13.49183, 13.49199
    b. ⁻1.453, ⁻1.45, ⁻1.4053, ⁻1.493
11. Write the numbers in each of the following sentences as decimals:
    a. A mite has body length of about fourteen thousandths of an inch.
    b. The Earth goes around the Sun once every three hundred sixty-five and twenty-four hundredths days.
12. Use a grid with 100 squares and represent 0.32. Explain your representation.
13. If the decimals 0.804, 0.84, and 0.8399 are arranged on a typical horizontal number line, which is furthest to the right?
14. Write a decimal number that has a ten-thousandths place and is between 8.34 and 8.341.
15. a. Show that between any two terminating decimals, there is another terminating decimal.
    b. Argue that part (a) can be used to show that there are infinitely many terminating decimals between any two terminating decimals.
16. If "decimals" in other number bases work the same as in base ten, explain the meaning of the following: 3.145_{six}.
17. The five top swimmers in an event had the following times:

Emily	64.54 sec	Kathy	64.02 sec
Molly	64.46 sec	Rhonda	63.54 sec
Martha	63.59 sec		

    List them in the order they placed with the "best" time first (least time is best here).
18. Suppose a carpenter's ruler was marked in thirty-seconds. What would be the fraction and its decimal equivalent for the mark on the ruler between $\frac{1}{16}$ and $\frac{2}{16}$?

Assessment 7-1B

1. Write each of the following as a sum in expanded place value form:
   a. 0.045 b. 103.03
   c. 245.6701 d. 0.00034
2. Rewrite each of the following as decimals:
   a. $5 \cdot 10^3 + 2 \cdot 10^2 + 4 \cdot 10^{-1}$
   b. $4 \cdot 10^{-3} + 2 \cdot 10^4$
   c. $2 \cdot 10^2 + 3 \cdot 10^4$
   d. $10^{-3} + 10^{-5}$
3. Write each of the following as numerals:
   a. Two thousand twenty-seven thousandths
   b. Two thousand and twenty-seven thousandths
   c. Two thousand twenty and seven thousandths
   d. Four hundred-thousandths

4. Write each of the following in words:
   a. 0.45
   b. 2.035
   c. 45.0006
   d. 0.0000445
5. Write each of the following terminating decimals as $\frac{a}{b}$ in simplest form where $a, b \in I$ and $b \neq 0$.
   a. 28.32 b. 34.1736 c. ⁻27.32
6. Mentally determine which of the following represent terminating decimals:
   a. $\frac{4}{8}$ b. $\frac{1}{2^6}$ c. $\frac{137}{625}$
   d. $\frac{1}{17}$ e. $\frac{3}{25}$ f. $\frac{14}{35}$

7. Where possible, write each of the numbers in exercise 6 as a terminating decimal.

8. What whole number of minutes (less than 60) could be expressed as terminating decimal parts of an hour?

9. If in a set of base ten-blocks, one block represented $\frac{1}{10}$, what is the value of each of the following?
   a. 1 cube
   b. 1 flat
   c. 1 long
   d. 3 blocks, 1 long, and 4 cubes

10. In each of the following, order the decimals from least to greatest:
    a. 24.9419, 24.942, 24.94189, 24.94199
    b. ⁻34.25, ⁻34.251, ⁻34.205, ⁻34.2519

11. Write the numbers of each of the following sentences as decimals:
    a. A flea has a body length about one sixteenth of an inch.
    b. Venus goes around the sun about once every 224 and $\frac{7006}{10,000}$ days.

12. Use a grid with 100 squares and represent 0.23.

13. If the decimals 0.8114, 0.8119, 0.82 are arranged on a typical number line, which is furthest to the right?

14. Write a decimal that is between 8.345 and 8.3456.

15. Argue that there are infinitely many terminating decimals between any two specific terminating decimals such as 0.0625 and 0.125.

16. If "decimals" in other number bases work the same as in base ten, explain the meaning of the following: 0.00334_{seven}.

17. The five top swimmers in an event had the following times:

Eddie	63.51 sec	Karl	62.99 sec
Marius	63.43 sec	Ricky	62.51 sec
Michael	62.56 sec		

List them in the order they placed with the "best" time first (least time is best here).

18. A normal carpenter's rule is marked off in sixteenths. What would be the values at each of the marks in decimal form?

Mathematical Connections 7-1

Communication

1. Using Simon Stevin's notation (see Historical Note), how would you write the following:
   a. 0.3256
   b. 0.0032

2. If 1 mL is 0.001 L, how should 18 mL be expressed as a terminating decimal number of liters (L stands for liter)?

3. Explain whether 1 day can be expressed as a terminating decimal part of a 365-day year.

4. Using the number line model to depict terminating decimals, explain whether you think that there is a greatest terminating decimal less than 1.

5. Explain why you think a sign on a copy machine reading ".05¢ a copy" is put up by mistake.

6. Explain how you could use base-ten blocks to represent two and three hundred forty-five thousandths.

7. Explain why in Theorem 7-1 the rational number must be in simplest form before examining the denominator.

8. Explain why appending any number of zeros to the right of a finite decimal does not change its value, for example, 0.34 = 0.340 = 0.3400.

Open-Ended

9. Determine how decimal notation is symbolized in different countries.

10. Examine three elementary school textbooks and report how the introductions of the topics of exponents and decimals differ, if they do. Report the grade level that decimals are introduced.

Cooperative Learning

11. Using base-five blocks similar to base-ten blocks, determine how you might introduce a "decimal" notation in base five.

12. Simon Stevin is credited with propagating the decimal system. In small groups, research the history of decimals to find contributions of the Arabs, the Chinese, and the people of Renaissance Europe. Explain whether you believe Stevin "invented" the decimal system.

Questions from the Classroom

13. A student claims that 0.86 is greater than 0.9 because 86 is greater than 9. How do you respond?

14. A student argues that fractions should no longer be taught once students learn how to work with decimal numbers. How do you respond?

15. A student claims that 0.304 = 0.34. How do you respond?

16. A student claims that because 0.1, 0.01, 0.001, . . . is a geometric sequence so is 1.1, 1.01, 1.001, ... How do you respond?

National Assessment of Educational Progress (NAEP) Question

What is 4 hundredths written in decimal notation?
   a. 0.004
   b. 0.04
   c. 0.400
   d. 4.00

NAEP, Grade 8, 2007

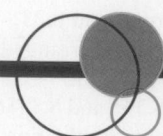

7-2 Operations on Decimals

Grade 5 *Common Core Standards* expects students to:

add, subtract, multiply, and divide decimals to hundredths, using concrete models or drawings and strategies based on place value, properties of operations, and/or the relationship between addition and subtraction; relate the strategy to a written method and explain the reasoning used. (p. 35)

To develop an algorithm for addition of terminating decimals, consider the sum 2.16 + 1.73. In elementary school, base-ten blocks are recommended to demonstrate such an addition problem. Figure 7-4 shows how the addition can be performed.

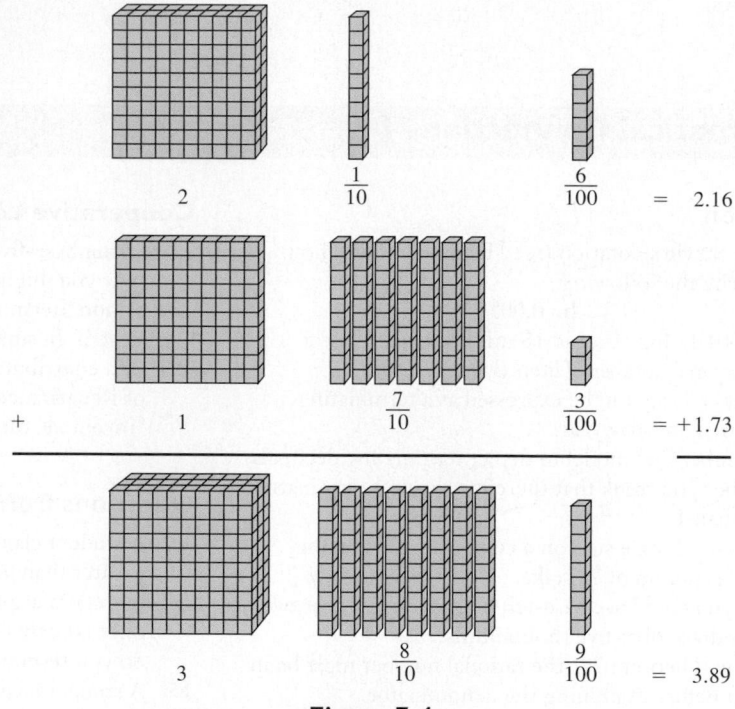

Figure 7-4

The computation in Figure 7-4 can be explained by *changing it to a problem we already know how to solve*; that is, to a sum involving fractions. We then use the commutative and associative properties of addition to aid in the computation as follows:

$$2.16 + 1.73 = \left(2 + \frac{1}{10} + \frac{6}{100} \right) + \left(1 + \frac{7}{10} + \frac{3}{100} \right)$$

$$= (2 + 1) + \left(\frac{1}{10} + \frac{7}{10} \right) + \left(\frac{6}{100} + \frac{3}{100} \right)$$

$$= 3 + \frac{8}{10} + \frac{9}{100}$$

$$= 3.89$$

A different way of looking at the addition $2.16 + 1.73$ links to a different representation of the decimals as fractions with denominators that are powers of 10 in the following:

$$2.16 = \frac{216}{100} \text{ and } 1.73 = \frac{173}{100} \text{ so } 2.16 + 1.73 = \frac{216}{100} + \frac{173}{100}$$

$$= \frac{216 + 173}{100} = \frac{389}{100} = 3.89.$$

Similarly, $2.1 + 1.73 = \frac{21}{10} + \frac{173}{100} = \frac{210}{100} + \frac{173}{100} = \frac{383}{100} = 3.83.$ Alternatively,

$$2.1 + 1.73 = 2.10 + 1.73 = \frac{210}{100} + \frac{173}{100} = \frac{383}{100} = 3.83.$$

In each of these cases, the decimals are converted to fractions with the same denominator that is a power of 10. Once this is done, the numerators can be added as whole numbers to form the numerator of the answer fraction, which can then be converted back to a decimal. If these decimals are written as shown below, we "line up the decimal points" and add as we do with whole numbers, placing the decimal point in the appropriate place in the answer.

$$\begin{array}{ccc}
2.16 & 2.1 & 2.10 \\
+1.73 \quad \text{and} & +1.73 \quad \text{or} & +1.73 \\
\hline
3.89 & 3.83 & 3.83
\end{array}$$

 In the preceding computations, we add units to units, tenths to tenths, and hundredths to hundredths. The lining-up-the-decimal-points technique works for both addition and subtraction, as demonstrated on the student page (page 344).

Multiplying Decimals

Just as in the presented algorithms for adding terminating decimals by representing them as fractions, we develop and explain an algorithm for multiplication of decimals. Consider the product $4.62 \cdot 2.4$:

$$(4.62)(2.4) = \frac{462}{100} \cdot \frac{24}{10} = \frac{462}{10^2} \cdot \frac{24}{10^1} = \frac{462 \cdot 24}{10^2 \cdot 10^1} = \frac{11{,}088}{10^3} = 11.088$$

The answer to this computation was obtained by multiplying the whole numbers 462 and 24 and then dividing the result by 10^3.

 On the student page (page 345), observe how multiplication of a whole number times a decimal can be accomplished by first shading 0.6 three times, once each in salmon, green, and purple. And then 0.04 is shaded three times in similar colors, as seen on the rightmost part of the Activity. The product is the sum of all the shaded portions: $0.6 + 0.6 + 0.6 + 0.04 + 0.04 + 0.04$ or 1.92. In this manner, multiplication of a whole number times a decimal is interpreted almost exactly as the multiplication of a whole number by a whole number. Consider the shading method in the Activity at the bottom of the student page to show why 0.5×0.7 is 0.35.

An algorithm for multiplying decimals can be stated as follows:

If there are n digits to the right of the decimal point in one number and m digits to the right of the decimal point in a second number, multiply the two numbers, ignoring the decimals, and then place the decimal point so that there are $n + m$ digits to the right of the decimal point in the product.

School Book Page

ADDING AND SUBTRACTING
WHOLE NUMBERS AND DECIMALS

Lesson 2-5

Key Idea
Adding and subtracting decimals is similar to adding and subtracting whole numbers.

Adding and Subtracting Whole Numbers and Decimals

LEARN

How can you add decimals?

When the starting gun is fired, relay runners can lose 0.2 second in reaction time, and another 1.8 seconds to go from a standing start to running speed. About how many seconds pass before running speed is reached?

Think It Through
I **remember** that annexing zeros does not change the decimal's value.

Example A

Find 0.2 + 1.8.
Estimate: 0 + 2 = 2

		What You **Think**	What You **Write**
STEP 1	Write the numbers, lining up decimal points.		0.2 + 1.8
STEP 2	Add the tenths. Regroup if necessary. Write the decimal point in your answer.	+ ▊ =	0.2 + 1.8 2.0

The answer and estimate match, so the answer is reasonable.

Example B

Find 5.6 + 2.973.
Estimate: 6 + 3 = 9

```
  5.600
+ 2.973
-------
  8.573
```
Write the numbers, lining up the decimal points. Add thousandths, hundredths, tenths. Regroup if necessary. Write the decimal point in your answer.

Since 8.573 is close to 9, the answer is reasonable.

Example C

Use a calculator to find 338.09 + 517.3.

Estimate: 300 + 500 = 800

Press: 338.09 ⊕ 517.3 [ENTER/=]

Display: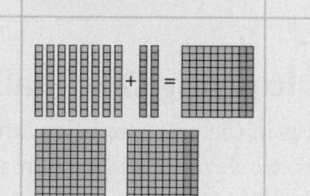
```
855.39
```

Since 855.39 is close to 800, the answer in the display is reasonable.

86

School Book Page

MULTIPLYING WHOLE NUMBERS AND DECIMALS

Lesson 2-6

Key Idea
Multiplying decimals is similar to multiplying whole numbers. You just need to know where to place the decimal point in the product.

Materials
• 10-by-10 decimal grids or

 **tools**

• colored pencils or markers

Think It Through
I can **use models** to show multiplication of decimals.

Multiplying Whole Numbers and Decimals

▶ **LEARN**

How can you multiply a whole number by a decimal?

You can use the same methods for multiplying decimals as you use for multiplying whole numbers.

Activity

a. To find 3 × 0.64, shade the tenths for the decimal number on a grid. Do this 3 times, using a different color each time.

b. Using the same 3 colors, shade three groups of the hundredths for the decimal number.

3 × 0.64 = 192 hundredths = 1.92

c. Count all of the shaded hundredths.

d. Use grids to find 1.2 × 2 and 2 × 0.18.

How can you multiply a decimal by a decimal?

Activity

a. To find 0.5 × 0.7, shade 5 columns using the same color to show 5 tenths.

b. Shade 7 rows in another color to show 7 tenths.

c. Count the hundredths in the overlapping shaded area.

0.5 × 0.7 = 35 hundredths = 0.35

d. Use grids to find 0.4 × 0.7 and 1.6 × 0.3.

90

EXAMPLE 7-4

Compute each of the following:

a. $(6.2)(1.43)$ **b.** $(0.02)(0.013)$ **c.** $(1000)(3.6)$

Solution

a.

$$
\begin{array}{r}
1.4\,3 \quad \text{(2 digits after the decimal point)} \\
\times\,6.2 \quad \text{(1 digit after the decimal point)} \\
\hline
2\,8\,6 \\
8\,5\,8 \\
\hline
8.8\,6\,6 \quad \text{(2 + 1, or 3 digits after the decimal point)}
\end{array}
$$

b.

$$
\begin{array}{r}
0.0\,1\,3 \\
\times\,\ \ 0.0\,2 \\
\hline
0.0\,0\,0\,2\,6
\end{array}
$$

c.

$$
\begin{array}{r}
3.6 \\
\times\,1\,0\,0\,0 \\
\hline
3\,6\,0\,0.0
\end{array}
$$

NOW TRY THIS 7-4

Example 7-4(c) suggests that multiplication by 1000, or 10^3, results in moving the decimal point in the product three places to the right. **(a)** Explain why this is true using expanded notation and the distributive property of multiplication over addition for rational numbers. **(b)** In general, how does multiplication by 10^n, where n is a positive integer, affect the product? Why?

Scientific Notation

 Many calculators display the decimals for the fractions $\dfrac{3}{45{,}689}$ and $\dfrac{5}{76{,}146}$ as $\boxed{6.5661319 \quad -05}$ and $\boxed{6.566333 \quad -05}$, respectively. The displays are in **scientific notation**. The first display is a notation for $6.5661319 \cdot 10^{-5}$ and the second for $6.566333 \cdot 10^{-5}$.

Scientists use scientific notation to handle either very small or very large numbers. For example, "the Sun is 93,000,000 mi from Earth" is expressed as "the Sun is $9.3 \cdot 10^7$ mi from Earth." A micron, a metric unit of measure that is 0.000001 m, is written $1 \cdot 10^{-6}$ m.

> ### Definition of Scientific Notation
>
> In **scientific notation**, a positive number is written as the product of a number greater than or equal to 1 and less than 10 and an integer power of 10. To write a negative number in scientific notation, treat the number as a positive number and adjoin the negative sign in front of the result.

The following numbers are in scientific notation:

$$8.3 \cdot 10^8, \quad 1.2 \cdot 10^{10}, \quad {}^{-}7.32 \cdot 10^8, \quad \text{and} \quad 7.84 \cdot 10^{-6}$$

The numbers $0.43 \cdot 10^9$ and $12.3 \cdot 10^{-6}$ are not in scientific notation because 0.43 and 12.3 are not greater than or equal to 1 and less than 10. To write a number like 934.5 in scientific notation, we divide by 10^2 to get 9.345 and then multiply by 10^2 to retain the value of the original number:

$$934.5 = \left(\frac{934.5}{10^2}\right)10^2 = 9.345 \cdot 10^2$$

This amounts to moving the decimal point two places to the left (dividing by 10^2) and then multiplying by 10^2. Similarly, to write 0.000078 in scientific notation, we first multiply by 10^5 to obtain 7.8 and then divide by 10^5 (or equivalently multiply by 10^{-5}) to keep the original value:

$$0.000078 = (0.000078 \cdot 10^5)10^{-5} = 7.8 \cdot 10^{-5}$$

This amounts to moving the decimal point five places to the right and multiplying by 10^{-5}.

EXAMPLE 7-5

Write each of the following in scientific notation:

a. 413,682,000   b. 0.0000231   c. 83.7   d. $^-$10,000,000   e. $0.34 \cdot 10^{-6}$

Solution

a. $413,682,000 = \left(\dfrac{413,682,000}{10^8}\right)10^8 = 4.13682 \cdot 10^8$

b. $0.0000231 = (0.0000231 \cdot 10^5)10^{-5} = 2.31 \cdot 10^{-5}$

c. $83.7 = \left(\dfrac{83.7}{10^1}\right)10^1 = 8.37 \cdot 10^1$

d. $^-10,000,000 = \dfrac{^-10,000,000}{10^7} \cdot 10^7 = {^-1} \cdot 10^7$

e. $0.34 \cdot 10^{-6} = (0.34 \cdot (10 \cdot 10^{-1})) \cdot 10^{-6} = (0.34 \cdot 10) \cdot 10^{-1} \cdot 10^{-6} = 3.4 \cdot 10^{-7}$

EXAMPLE 7-6

Convert the following to standard numerals:

a. $6.84 \cdot 10^{-5}$ b. $3.12 \cdot 10^7$ c. $^-4.08 \cdot 10^4$

Solution

a. $6.84 \cdot 10^{-5} = 6.84\left(\dfrac{1}{10^5}\right) = 0.0000684$

b. $3.12 \cdot 10^7 = 31,200,000$

c. $^-4.08 \cdot 10^4 = {^-}40,800$

Computations involving scientific notation make use of the laws of exponents. For example, $(5.6 \cdot 10^5)(6 \cdot 10^4)$ can be rewritten $(5.6 \cdot 6)(10^5 \cdot 10^4) = 33.6 \cdot 10^9$, which is $3.36 \cdot 10^{10}$ in scientific notation. Also,

$$(2.35 \cdot 10^{-15})(2 \cdot 10^8) = (2.35 \cdot 2)(10^{-15} \cdot 10^8) = 4.7 \cdot 10^{-7}$$

Calculators with an $\boxed{EE}$ key can be used to represent numbers in scientific notation. For example, to find $(5.2 \cdot 10^{16})(9.37 \cdot 10^4)$, press

$\boxed{5}\ \boxed{.}\ \boxed{2}\ \boxed{EE}\ \boxed{1}\ \boxed{6}\ \boxed{\times}\ \boxed{9}\ \boxed{.}\ \boxed{3}\ \boxed{7}\ \boxed{EE}\ \boxed{4}\ \boxed{=}$

Dividing Decimals

The set of non-zero whole numbers and the set of non-zero integers are not closed under division. However, the set of non-zero rational numbers is closed under division (why?).

The following examples deal with terminating decimals and in the next section we discuss repeating decimals. Consider $75.45 \div 3$:

$$75.45 \div 3 = \frac{7545}{100} \div \frac{3}{1} = \frac{7545}{100} \cdot \frac{1}{3} = \frac{7545}{3} \cdot \frac{1}{100} = 2515 \cdot \frac{1}{100} = 25.15$$

By writing the dividend as a fraction whose denominator is a power of 10 and using the commutative property of multiplication, the division is changed into a division of whole numbers $(7545 \div 3)$ times a power of 10, namely, $\dfrac{1}{100}$, or 10^{-2}. A different method of accomplishing the change of the division into whole number division is to multiply the dividend and the divisor by the same power of 10, as follows:

$$75.45 \div 3 = \frac{75.45}{3} = \frac{75.45 \cdot 10^2}{3 \cdot 10^2} = \frac{7545}{300} = 25\frac{45}{300} = 25\frac{15}{100} = 25 \cdot 15$$

When the divisor is a whole number as shown above, the division can be handled as with whole number division and the decimal point placed directly over the decimal point in the dividend. When the divisor is not a whole number, as in $1.2032 \div 0.32$, we can obtain a whole-number divisor by expressing the quotient as a fraction and then multiplying the numerator and denominator of the fraction by a power of 10 as seen in the activity at the bottom of the student page (page 349). This corresponds to rewriting the division problem in form (a) as an equivalent problem in form (b), as follows:

(a) $0.32\overline{)1.2032}$ (b) $32\overline{)120.32}$

In elementary school texts, this process is usually described as "moving" the decimal point two places to the right in both the dividend and the divisor. It is usually indicated with arrows, as shown in the following.

$$
\begin{array}{r}
3.76 \\
0.3\,2\overline{)1.2\,0\,3\,2} \\
9\,6 \\
\hline
2\,4\,3 \\
2\,2\,4 \\
\hline
1\,9\,2 \\
1\,9\,2 \\
\hline
0
\end{array}
$$

Multiply divisor and dividend by 100.

As seen above, in some cases of division of decimals, just as in division of whole numbers, the remainder for the division is 0.

EXAMPLE 7-7

Compute each of the following:

a. $13.169 \div 0.13$ **b.** $9 \div 0.75$

Solution

a.
$$
\begin{array}{r}
1\,0\,1.3 \\
0.1\,3\overline{)1\,3.1\,6\,9} \\
1\,3 \\
\hline
1\,6 \\
1\,3 \\
\hline
3\,9 \\
3\,9 \\
\hline
0
\end{array}
$$

b.
$$
\begin{array}{r}
1\,2 \\
0.7\,5\overline{)9.0\,0} \\
7\,5 \\
\hline
1\,5\,0 \\
1\,5\,0 \\
\hline
0
\end{array}
$$

In Example 7-7(b), we appended two zeros in the dividend because $\dfrac{9}{0.75} = \dfrac{9 \cdot 100}{0.75 \cdot 100} = \dfrac{900}{75}$.

School Book Page   DIVIDING BY A DECIMAL

Lesson 2-9

Key Idea
To divide by a decimal, you need to change the divisor to a whole number.

Materials
• decimal models
  or  tools

Think It Through
I can **use models** to show division by decimals.

Dividing by a Decimal

▶ LEARN

How can you divide by a decimal?

Nihad and his friends divide the cost of a $2.70 bag of nuts. Each person contributes $0.90. How many people will share the nuts?

Activity

You can use the same methods for dividing decimals as you use for dividing whole numbers.

a. Find 2.7 ÷ 0.9.

b. Circle groups of 0.9.

c. Count the circled groups.

d. Use decimal models to find 3.0 ÷ 0.3 and 2.5 ÷ 0.5.

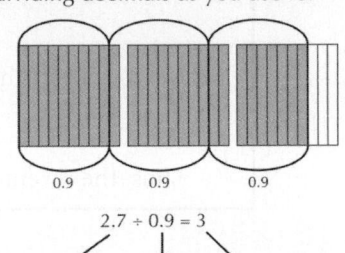

$$2.7 \div 0.9 = 3$$

dividend   divisor   quotient

How can you change a decimal divisor into a whole number divisor without changing the quotient?

Activity

a. Study the patterns below. Compare the first and second columns.

Pattern 1	
80 ÷ 0.4 = 200	800 ÷ 4 = 200
8 ÷ 0.4 = 20	80 ÷ 4 = 20
0.8 ÷ 0.4 = 2	8 ÷ 4 = 2
0.08 ÷ 0.4 = 0.2	0.8 ÷ 4 = 0.2

Pattern 2	
80 ÷ 0.04 = 2,000	8,000 ÷ 4 = 2,000
8 ÷ 0.04 = 200	800 ÷ 4 = 200
0.8 ÷ 0.04 = 20	80 ÷ 4 = 20
0.08 ÷ 0.04 = 2	8 ÷ 4 = 2

b. For Pattern 1, what happens to the quotient when you multiply the dividend and divisor by 10?

c. What number were the dividend and divisor multiplied by in Pattern 2?

d. Complete the table at the right.

Dividend	Divisor	Multiply Dividend and Divisor by:	Quotient
5.2	0.013	1,000	
5.2	0.13		40
5.2	1.3		

100

EXAMPLE 7-8

An owner of a gasoline station must collect a gasoline tax of $0.11 on each gallon of gasoline sold. One week, the owner paid $1595.00 in gasoline taxes. The pump price of a gallon of gas that week was $3.35.

a. How many gallons of gas were sold during the week?
b. What was the revenue after taxes for the week?

Solution

a. To find the number of gallons of gas sold during the week, we must divide the total gas tax bill by the amount of the tax per gallon:

$$\frac{1595}{0.11} = \frac{159500}{11} = 14,500$$

Thus, 14,500 gallons were sold.

b. To obtain the revenue after taxes, first determine the revenue before taxes. To do so, multiply the number of gallons sold by the cost per gallon:

$$(14,500)(\$3.35) = \$48,575$$

Next, subtract the gasoline taxes from the total revenue:

$$\$48,575 - \$1,595 = \$46,980$$

Thus, the revenue after gasoline taxes is $46,980.

NOW TRY THIS 7-5

Use decimal division to help Hi, in the *Hi and Lois* cartoon, determine the best buy.

Hi & Lois copyright © 1989 and 2011 King Features Syndicate

Next consider a division resulting in a non-zero remainder as shown below:

a.
```
         3.6 4  Quotient
0.3 3)1.2 0 3 2
        9 9
        2 1 3
        1 9 8
          1 5 2
          1 3 2
            2 0  Remainder
```

b.
```
        3 6 4  Quotient
3 3)1 2 0 3 2
      9 9
      2 1 3
      1 9 8
        1 5 2
        1 3 2
          2 0  Remainder
```

Note that in part (a), all digits in the dividend are used and 20 is shown at the bottom of the division. This division could be treated as in the division of part (b), where the quotient is 364 and the remainder is 20. In part (a), the quotient is 3.64 and the remainder shown as 20 actually represents 0.0020, as can be seen by the original decimal alignment. This could be verified by checking: $3.64 \cdot 0.33 + 0.0020 = 1.2032$. In the next section we will show how to write $\dfrac{12032}{33}$ as an infinite repeating decimal.

Mental Computation

Some of the tools used for mental computations with whole numbers can be used to perform mental computations with decimals, as seen in the following:

1. *Breaking and bridging*

$$1.5 + 3.7 + 4.48 \qquad\qquad 1.5 + 3$$

$$= 4.5 + 0.7 + 4.48 \qquad\qquad 4.5 + 0.7$$

$$= 5.2 + 4.48 \qquad\qquad 5.2 + 4$$

$$= 9.2 + 0.48 = 9.68 \qquad\qquad 9.2 + 0.48$$

2. *Using compatible numbers*
   Decimal numbers are compatible when they add up to a whole number.

   $$
   \begin{array}{ll}
   7.91 & \quad 12 \qquad\qquad 7.91 + 4.09 \\
   3.85 & \\
   4.09 & \quad + 4 \qquad\qquad 3.85 + 0.15 \\
   + 0.15 & \\
   & \quad 16 \qquad\qquad 12 + 4
   \end{array}
   $$

3. *Making compatible numbers*

   $$
   \begin{array}{rcl}
   9.27 & = & 9.25 + 0.02 \\
   + 3.79 & = & 3.75 + 0.04 \\
   \hline
   & & 13.00 + 0.06 = 13.06
   \end{array}
   $$

4. *Balancing with decimals in subtraction*

   $$
   \begin{array}{c}
   4.63 \\
   - 1.97
   \end{array}
   \text{ or }
   \begin{array}{rcr}
   4.63 + 0.03 & = & 4.66 \\
   -(1.97 + 0.03) & = & -2.00 \\
   \hline
   & & 2.66
   \end{array}
   $$

5. *Balancing with decimals in division*

   $$0.25\overline{)8}$$
   $$\times 4 \qquad \times 4$$
   $$1\overline{)32}$$

REMARK Balancing with decimals in division uses the property $\dfrac{a}{b} = \dfrac{ac}{bc}$, if $c \neq 0$.

Rounding Decimals

Frequently, it is not necessary to know the exact numerical answer to a question. For example, if we want to know the distance to the Moon or the population of metropolitan New York City, the approximate answers of 239,000 mi and 21,200,000 people, respectively, may be adequate.

Often a situation determines how you should round. For example, suppose a purchase came to $38.65 and the cashier used a calculator to figure out the 6% sales tax by multiplying $0.06 \cdot 38.65$. The display showed 2.319. Because the display is between 2.31 and 2.32 and it is closer to 2.32, the cashier rounds up the sales tax to $2.32.

Suppose a display of 8.7345649 needs to be reported to the nearest hundredth. The display is between 8.73 and 8.74 but is closer to 8.73, so we round it down to 8.73. Next suppose the number 6.8675 needs to be rounded to the nearest thousandth. Notice that 6.8675 is exactly halfway between 6.867 and 6.868. *In such cases, it is common practice to round up* and therefore the answer to the nearest thousandth is 6.868. We write this as $6.8675 \approx 6.868$ and say 6.8675 is approximately equal to 6.868.

 **EXAMPLE 7-9**

Round each of the following numbers:

  a. 7.456 to the nearest hundredth
  b. 7.456 to the nearest tenth
  c. 7.456 to the nearest unit
  d. 7456 to the nearest thousand
  e. 745 to the nearest ten
  f. 74.56 to the nearest ten

Solution
  a. $7.456 \approx 7.46$
  b. $7.456 \approx 7.5$
  c. $7.456 \approx 7$
  d. $7456 \approx 7000$
  e. $745 \approx 750$
  f. $74.56 \approx 70$

     Rounding can also be done on some calculators using the $\boxed{\text{FIX}}$ key. If we want the number 2.3669 to be rounded to thousandths, we enter $\boxed{\text{FIX}}\,\boxed{3}$. The display will show 0.000. If we then enter 2.3669 and press the $\boxed{=}$ key, the display will show 2.367.

Estimating Decimal Computations Using Rounding

Rounded numbers can be useful for estimating answers to computations. For example, consider each of the following:

1. Karly goes to the grocery store to buy items that cost the following amounts. She estimates the total cost by rounding each amount to the nearest dollar and adding the rounded numbers.

$$
\begin{array}{rcr}
\$2.39 & \rightarrow & \$2 \\
0.89 & \rightarrow & 1 \\
6.13 & \rightarrow & 6 \\
4.75 & \rightarrow & 5 \\
+\,5.05 & \rightarrow & \underline{5} \\
& & \$19
\end{array}
$$

Thus, Karly's estimate for her grocery bill is $19.

2. Karly's bill for car repairs was $72.80, and she has a coupon for $17.50 off. She can estimate her total cost by rounding each amount to the nearest 10 dollars and subtracting.

$$\begin{array}{rcr} \$72.80 & \to & \$70 \\ -\ 17.50 & \to & -\ 20 \\ \hline & & \$50 \end{array}$$

Thus, an estimate for the repair bill is $50. Notice that this is not a very good estimate. A better estimate is obtained by finding $73 - 17$, which is $56.

3. Karly sees a flash of lightning and hears the thunder 3.2 sec later. She knows that sound travels at 0.33 km/sec. She may estimate the distance she is from the lightning by rounding the time to the nearest unit and the speed to the nearest tenth and multiplying.

$$\begin{array}{rcr} 0.33 & \to & 0.3 \\ \times\ 3.2 & \to & \times\ 3 \\ \hline & & 0.9 \end{array}$$

Thus, Karly estimates that she is approximately 0.9 km from the lightning.

An alternative approach is to recognize that $0.33 \approx \dfrac{1}{3}$ and 3.2 is close to 3.3, so an approximation using compatible numbers is $\left(\dfrac{1}{3}\right)3.3$, or 1.1 km.

4. Karly wants to estimate the number of miles she gets per gallon of gas. If she had driven 298 mi and it took 12.4 gal to fill the gas tank, she rounds and divides as follows:

$$12.4\overline{)298} \quad \to \quad 12\overline{)300}^{\,25}$$

Thus, she got about 25 mi/gal.

5. Johanna wanted to place the decimal point in a product resulting when she bought 21.45 lb of sesame seeds to grind for tahini at $3.40 per pound. The multiplication (without a decimal) resulted in 7293. She knew that very rough estimates could be obtained by rounding 21.45 to 20 and $3.40 to $3. The product had to be in the neighborhood of 20×3, or 60. So her placement of the decimal point was $72.93.

> ● **NOW TRY THIS 7-6**

Other estimation strategies, such as front-end, clustering, and grouping to convenient numbers, that we investigated with whole numbers also work with decimals. Take the grocery store bill in part 1 on the preceding page and use a front-end-with-adjustment strategy to estimate the bill.

Round-off Errors

Round-off errors are typically compounded when computations are involved. For example, if two distances are 42.6 mi and 22.4 mi rounded to the nearest tenth, then the sum of the distances appears to be $42.6 + 22.4$, or 65.0 mi. To the nearest hundredth, the distances might have been more accurately reported as 42.55 and 22.35 mi, respectively. The sum of these distances is 64.9 mi. Alternatively, the original distances may have been rounded from 42.64 and 22.44 mi. The sum now is 65.08, or 65.1 rounded to the nearest tenth. The original sum of 65.0 mi is between 64.9 and 65.1 mi, but the exact answer need not be 65.0 mi to the nearest tenth.

The *greatest possible error* in measuring is defined as one half of that measuring unit. For example, if we measure a length to be 4.7 cm then because the measurement was made to the nearest tenth, the greatest possible error is $\frac{1}{2}$ of 0.1 cm or 0.05 cm.

Similar errors arise in other arithmetic operations. *When computations are done with approximate numbers, the final result should not be reported using more significant digits than the number used with the fewest significant digits.* Non-zero digits are always significant. Zeros before other digits are non-significant. Zeros between other non-zero digits are significant. Zeros to the right of a decimal point are significant.

● **NOW TRY THIS 7-7**

In the *Hi and Lois* cartoon that follows,

    a. Estimate whether Trixie's calculation is accurate.
    b. To the nearest hundred-millionth, how many times is Trixie older than her twin brother?

Hi & Lois copyright © 2011 King Features Syndicate

 **Assessment 7-2A**

1. If Maura went to the store and bought a chair for $17.95, a lawn rake for $13.59, a spade for $14.86, a lawn mower for $179.98, and two six-packs of mineral water for $2.43 each, what was the bill?

2. **a.** Complete the following magic square; that is, make the sum of every row, column, and diagonal the same:

8.2		
3.7	5.5	
	9.1	2.8

    b. If each cell of the magic square has 0.85 added to it,
      i. Is the square still magic?
      ii. If the answer to part (i) is "yes," what is the sum of each row?

3. Karin bought 25 lb of peaches at $4.00/lb and 15 lb of apples at $2.00/lb. If she wanted to buy 10 more pounds of another kind of fruit to make the average price per pound equal to $3.50, what price should she pay for the additional 10 lb?

4. Automobile engines were once measured in cubic inches (in.^{3}) but are now usually measured in cubic centimeters (cm^3). Susan's 1963 Thunderbird has a 390 in.^{3} engine. If 2.54 cm is equivalent to 1 in., approximately how many cubic centimeters is this?

5. A stock rose $0.24 in the market on Thursday. If the resulting price was $73.245, what was the price of the stock before the rise?

6. A U.S. $1 bill was valued at 0.9826 Canadian dollars on March 18, 2011. What was the value of 27.32 American dollars in Canadian dollars that day?

7. A kilowatt hour means 1000 watts of electricity are being used continuously for 1 hr. The electric utility company in Laura's town charges $0.06715 for each kilowatt hour used. Laura heats her house with three electric wall heaters that use 1200 watts each.
    a. How much does it cost to heat her house for 1 day?
    b. How many hours would a 75-watt lightbulb have to stay on to result in $1 for electricity charges?

8. If one liter is 4.224 cups. How many liters does it take to hold 36.5 cups?

9. If each of the following sequences is arithmetic, continue the decimal patterns:
   a. 0.9, 1.8, 2.7, 3.6, 4.5, ___, ___, ___
   b. 0.3, 0.5, 0.7, 0.9, 1.1, ___, ___, ___

10. If the first term of a geometric sequence is 0.9 and its ratio is 0.2, what is the sum of the first five terms?

11. Interpret 0.2222 as a sum of a finite geometric sequence whose first term is 0.2. (*Hint:* Write 0.2222 as the sum of fractions whose denominators are powers of 10.)

12. Estimate the placement of each of the following on the given number line by placing the letter for each computation in the appropriate box:
    a. $0.3 \div 0.31$
    b. $0.3 \cdot 0.31$

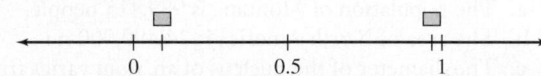

13. A bank statement from a local bank shows that a checking account has a balance of $83.62. The balance recorded in the checkbook shows only $21.69. After checking the canceled checks against the record of these checks, the customer finds that the bank has not yet recorded six checks in the amounts of $3.21, $14.56, $12.44, $6.98, $9.51, and $7.49. Is the bank record correct? (Assume the person's checkbook records *are* correct.)

14. Convert each of the following to standard numerals:
    a. $3.2 \cdot 10^{-9}$
    b. $3.2 \cdot 10^{9}$
    c. $4.2 \cdot 10^{-1}$
    d. $6.2 \cdot 10^{5}$

15. Write the numerals in each of the following sentences in scientific notation:
    a. The diameter of Earth is about 12,700,000 m.
    b. The distance from Pluto to the Sun is about 4,486,000,000 km.
    c. Each year, about 50,000,000 cans are discarded in the United States.

16. Solve the following for x where x is a decimal:
    a. $8.56 = 3 - 2x$
    b. $2.3x - 2 = x + 2.55$

17. Write the numerals in each of the following sentences in standard form:
    a. The mass of a dust particle is $7.53 \cdot 10^{-10}$ g.
    b. The speed of light is approximately $2.98 \cdot 10^{5}$ km/sec
    c. Jupiter is approximately $7.7857 \cdot 10^{8}$ km from the Sun.

18. Write the results of each of the following in scientific notation:
    a. $(8 \cdot 10^{12})(6 \cdot 10^{15})$
    b. $(16 \cdot 10^{12}) \div (4 \cdot 10^{5})$
    c. $(5 \cdot 10^{8})(6 \cdot 10^{9}) \div (15 \cdot 10^{15})$

19. Round each of the following numbers as specified:
    a. 203.651 to the nearest hundred
    b. 203.651 to the nearest ten
    c. 203.651 to the nearest unit
    d. 203.651 to the nearest tenth
    e. 203.651 to the nearest hundredth

20. Jane's car travels 243 mi on 12 gal of gas. How many miles to the gallon does her car get?

21. Audrey wants to buy some camera equipment to take pictures on her daughter's birthday. To estimate the total cost, she rounds each price to the nearest dollar and adds the rounded prices. What is her estimate for the items listed?

Camera	$54.56
Film	$4.50
Case	$17.85

22. Estimate the sum or difference in each of the following by using (i) rounding and (ii) front-end estimation. Then perform the computations to see how close your estimates are to the actual answers.

    a.   65.84
         24.29
         12.18
       + 19.75

    b.   89.47
       − 32.16

    c.   5.85
         6.13
         9.10
       + 4.32

    d.   223.75
       − 87.60

23. Find the least and the greatest possible products for the expression using the digits 1 through 9. Each digit may be used only once in each case.

    $$\square.\square \times \square$$

24. Some digits in the following number are covered by squares:

    $$4\square\square3\square.\square\square8\square$$

    If each of the digits 1 through 9 is used exactly once in the number, determine the greatest possible number.

25. Iris worked a 40-hour week at $8.25/hr. Mentally compute her salary for the week and explain how you did it.

26. Mentally compute the number to fill in the blank in each of the following:
    a. $8.4 \cdot 6 = 4.2 \cdot$ ___
    b. $10.2 \div 0.3 = 20.4 \div$ ___
    c. $ab = (a/2) \cdot$ ___
    d. $a \div b = 2a \div$ ___

27. Which of the following result in equal quotients?
    i. $7 \div 0.25$
    ii. $70 \div 2.5$
    iii. $0.7 \div 0.25$
    iv. $700 \div 25$

28. a. Fill in the parentheses in each of the following to write a true equation:

    $$2 \cdot 1 + 0.25 = (\ )^2$$
    $$3 \cdot 2 + 0.25 = (\ )^2$$

    Conjecture what the next two equations in this pattern will be.

    b. Generalize your answer in part (a) by filling in an appropriate expression in the equation

    $$n(n - 1) + 0.25 = (\ )^2$$

    where $n = 2, 3, 4, \ldots$.

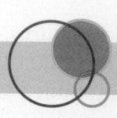

Assessment 7-2B

1. At a store, Samuel bought a bouquet for $4.99, a candy bar for ¢79, a memory stick for $49.99, and a bottle of water for $1.49. What was his bill?
2. **a.** Complete the following magic square; that is, make the sum of every row, column, and diagonal the same:

7.5		
3.0	4.8	
	8.4	2.1

    b. If each cell of the original magic square is multiplied by 0.5, is the square still magic? If "yes," what is the sum of each row?
3. Keith bought 30 lb of nuts at $3.00/lb and 20 lb of nuts at $5.00/lb. If he wanted to buy 10 more pounds of a different kind of nut to make the average price per pound equal to $4.50, what price should he pay for the additional 10 lb?
4. Automobile engines were once measured in cubic inches (in.^{3}) but are now usually measured in cubic centimeters (cm^3). Dan's 1991 Taurus has a 3000 cm^3 engine. If 2.54 cm is equivalent to 1 in., approximately how many cubic inches is this?
5. A stock's price dropped from $63.28 per share to $27.45. What was the loss on a single share of the stock?
6. A U.S. $1 bill was valued at 1.0079 Canadian dollars on February 10, 2011. What was the value of 28.43 American dollars in Canadian dollars that day?
7. Eugene Water & Electric Board charges $0.07104 for each kilowatt-hour used. Terry used her computer for 3hr 24 min while the computer itself used 45 watts and the computer monitor used 35 watts. What was the charge for the computer usage?
8. If one quart is 4 cups. How many quarts in 18.5 cups?
9. If each of the following sequences is either arithmetic or geometric, continue the decimal patterns:
    a. 1, 0.5, 0.25, 0.125, _____, _____, _____
    b. 0.2, 1.5, 2.8, 4.1, 5.4, ____, ____, ____
10. If the first term of a finite geometric sequence is 0.4 and its ratio is 0.3, what is the sum of the first five terms?
11. Interpret the decimal 0.3333333 as a sum of a finite geometric sequence whose first term is 0.3. (*Hint:* Write 0.3333333 as the sum of fractions whose denominators are powers of 10.)
12. Estimate the placement of each of the following on the given number line by placing the letter for each computation in the appropriate box:
    a. 0.3 + 0.31
    b. 0.3 − 0.31

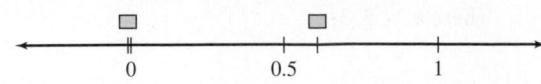

13. Mary Kim invested $964 in 18 shares of stock. A month later, she sold the 18 shares at $61.48 per share. She also invested

in 350 shares of another stock for a total of $27,422.50. She sold this stock for $85.35 a share and paid $495 in total commissions. What was Mary Kim's profit or loss on the transactions to the nearest dollar?
14. Convert each of the following to standard numerals:
    a. $3.5 \cdot 10^7$
    b. $3.5 \cdot 10^{-7}$
    c. $^-(2.4 \cdot 10^{-3})$
15. Write the numerals in each of the following sentences in scientific notation:
    a. The population of Montana is 989,415 people.
    b. The area of North America is 24,490,000 mi^2.
    c. The diameter of the nucleus of an atom varies from about 1.6 to 15 times 10^{-15} m.
16. Solve the following for x where x is a decimal:
    a. $2x + 1.3 = 4.1$
    b. $4.2 - 3x = 10.2$
17. Write the numerals in each of the following sentences in standard form:
    a. A computer requires $4.4 \cdot 10^{-6}$ sec to do an addition problem.
    b. There are about $1.99 \cdot 10^4$ km of coastline in the United States.
    c. Earth has existed for approximately $3 \cdot 10^9$ yr.
18. Write the results of each of the following in scientific notation:
    a. $(5 \cdot 10^7)(7 \cdot 10^{12})$
    b. $(^-13 \cdot 10^4) \div 65$
    c. $(3 \cdot 10^7)(4 \cdot 10^5) \div (6 \cdot 10^{-7})$
19. Round each of the following numbers as specified:
    a. 715.04 to the nearest hundred
    b. 715.04 to the nearest tenth
    c. 715.04 to the nearest unit
    d. 715.04 to the nearest ten
    e. 715.04 to the nearest thousand
20. Jane drives at a constant speed of 55.5 mph. How far should she expect to drive in $\frac{3}{4}$ hr?
21. Use estimation to choose a decimal to multiply by 9 in order to get within 1 of 93. Explain how you made your choice and check your estimate.
22. Estimate the sum or difference in each of the following by using (i) rounding and (ii) front-end estimation. Then perform the computation to see how close your estimates are to the actual answers.

    a.   47.62
           27.99
           13.14
        + 7.61

    b.   79.86
         − 27.37

    c.   5.85
         6.17
         9.1
      + 4.23

    d.   232.65
       − 78.92

23. Find the least and the greatest possible products for the expression using the digits 1 through 9. Each digit may be used only once in each case.

$$\square.\square \times \square.\square$$

24. Some digits in the following number are covered by squares:

$$4\square\square3\square.\square\square8\square$$

If each of the digits 1 through 9 is used exactly once in the number, determine the least possible number.

25. Iris worked a 40-hour week at \$6.25/hr. Mentally compute her salary for the week and explain how you did it.

26. Mentally compute the number to fill in the blank in each of the following:
    a. $12.4 \cdot 7 = 6.2 \cdot$ _____
    b. $12.4 \div 0.2 =$ _____ $\div 0.1$
    c. $ab = (a \cdot 10^{-1}) \cdot$ _____
    d. $12.3 = 10^{-2} \cdot$ _____

27. Which of the following result in equal quotients?
    a. $9 \div 0.35$
    b. $90 \div 3.5$

c. $900 \div 35$
d. $0.9 \div 0.035$

28. a. Fill in the parentheses in each of the following to write a true equation:

$$1 \cdot 2 + 0.25 = (\ )^2$$
$$2 \cdot 3 + 0.25 = (\ )^2$$

Conjecture what the next two equations in this pattern will be.

b. Do the computations to determine if your next two equations are correct.

c. Generalize your answer in part (a) by filling in an appropriate expression in the equation

$$n(n + 1) + 0.25 = (\ )^2$$

where $n = 1, 2, 3, \ldots$.

Mathematical Connections 7-2

Communication

1. Give an example of a balanced checkbook where entries are incorrect.
2. How is multiplication of decimals like multiplication of whole numbers? How is it different?
3. Why are estimation skills important in dividing decimals?
4. In the text, multiplication and division were done using both fractional and decimal forms. Discuss the advantages and disadvantages of each.
5. Explain why subtraction of terminating decimals can be accomplished by lining up the decimal points, subtracting as if the numbers were whole numbers, and then placing the decimal point in the difference.
6. Is a product of two positive decimals each less than 1 always less than each of the decimals? Justify your answer.

Open-Ended

7. Find several examples of the use of decimals in the newspaper. Tell whether you think the numbers are exact or estimates. Also tell why you think decimals were used instead of fractions.
8. How could a calculator be used to develop or reinforce the understanding of multiplication of decimals?
9. Do you think that decimals should always be converted to fractions to explain computational procedures?

Cooperative Learning

10. In your group, decide on all the prerequisite skills that students need before learning to perform arithmetic operations on decimals.
11. In your group, each person will write a test of 5 questions that examines students' ability to estimate products of decimals. Next, compare the tests to choose the best questions, and cooperate to devise a single test.

Questions from the Classroom

12. A student multiplies $(6.5)(8.5)$ to obtain the following:

$$
\begin{array}{r}
8.5 \\
\times\ 6.5 \\
\hline
425 \\
510 \\
\hline
55.25
\end{array}
$$

However, when the student multiplies $8\frac{1}{2} \cdot 6\frac{1}{2}$, she obtains the following:

$$
\begin{array}{r}
8\frac{1}{2} \\
\times\ 6\frac{1}{2} \\
\hline
4\frac{1}{4} \quad \left(\frac{1}{2} \cdot 8\frac{1}{2}\right) \\
48 \quad (6 \cdot 8) \\
\hline
52\frac{1}{4}
\end{array}
$$

How is this possible?

13. A student tries to calculate $0.999^{10,000}$ on a calculator and finds the answer to be $4.5173346 \cdot 10^{-5}$. The student wonders how it could be that a number like 0.999, so close to 1, when raised to 10,000 power could result in a number close to 0. How do you respond?

14. How would you respond to the following:
    a. A student claims that $\frac{9443}{9444}$ and $\frac{9444}{9445}$ are equal because each equals aproximately 0.9998941.
    b. Another student claims that the fractions are not equal and wants to know if there is any way the same calculator can determine which is greater.

Review Problems

15. Write 14.0479 in expanded form.

16. Without dividing, determine which of the following represent terminating decimals:

    a. $\dfrac{24}{36}$

    b. $\dfrac{49}{56}$

17. If the denominator of a fraction is 26, is it possible that the fraction could be written as a terminating decimal? Why or why not?

18. $\dfrac{35}{56}$ can be written as a decimal that terminates. Explain why.

Trends in Mathematics and Science Study (TIMSS) Questions

What is the sum of 2.5 and 3.8?

    a. 5.3
    b. 6.3
    c. 6.4
    d. 9.5

TIMSS, Grade 4, 2007

National Assessment of Educational Progress (NAEP) Questions

It costs $0.25 to operate a clothes dryer for 10 minutes at a laundromat. What is the total cost to operate one clothes dryer for 30 minutes, a second for 40 minutes, and a third for 50 minutes?

  a. $3.25   **b.** $3.00   **c.** $2.75   **d.** $2.00   **e.** $1.20

Add the numbers $\dfrac{7}{10}$, $\dfrac{7}{100}$, and $\dfrac{7}{1,000}$. Write this sum as a decimal.

NAEP, Grade 8, 2007

**$2.45

Sales Tax Table

Amount of Sales	Amount of Tax
$6.00	$0.36
6.20	0.37
6.40	0.38
6.60	0.40
6.80	0.41
7.00	0.42
7.20	0.43
7.40	0.44
7.60	0.46
7.80	0.47
8.00	0.48

Carlos bought the cereal and milk shown. Use the table to find out the total amount Carlos spent, including tax.

Total amount spent: _____

Show how you found your answer.

NAEP, Grade 4, 2007

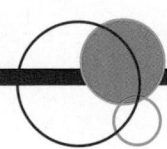

7-3 Nonterminating Decimals

In grade 7 *Focal Points*, we find:

> Students now use division to express any fraction as a decimal, including fractions that they must represent with infinite decimals. (p. 38)

Also in grade 7 *Common Core Standards* we find that student should be able: "to convert a rational number to a decimal using long division; know that the decimal form of a rational number terminates or eventually repeats." (p. 49)

Earlier in the chapter, we developed procedures for converting some rational numbers to decimals. For example, $\dfrac{7}{8}$ can be written as a terminating decimal as follows:

$$\frac{7}{8} = \frac{7}{2^3} = \frac{7 \cdot 5^3}{2^3 \cdot 5^3} = \frac{875}{1000} = 0.875$$

The decimal for $\frac{7}{8}$ can also be found by division:

$$
\begin{array}{r}
0.875 \\
8\overline{)7.000} \\
\underline{6\ 4}\ \ \ \ \\
60\ \ \\
\underline{56}\ \ \\
40 \\
\underline{40} \\
0
\end{array}
$$

However, we showed that $\frac{2}{11}$ was not a terminating decimal. In the following, we investigate nonterminating decimals.

Repeating Decimals

If we use a calculator to find a decimal representation for $\frac{2}{11}$, the calculator may display 0.1818181. It seems that the block of numbers 18 repeats. To examine what digits, if any, the calculator did not display, consider the following division:

$$
\begin{array}{r}
0.18 \\
11\overline{)2.00} \\
\underline{1\ 1}\ \ \\
90 \\
\underline{88} \\
2
\end{array}
$$

At this point, if the division is continued, the division pattern repeats, since the remainder 2 divided by 11 repeats the division. Thus the quotient is 0.181818.... A decimal of this type is a **repeating decimal**, and the repeating block of digits is the **repetend**. The repeating decimal is written $0.\overline{18}$, where the bar indicates that the block of digits underneath is repeated continuously. A repeating decimal such as $0.\overline{18}$ is the infinite sum $0.18 + 0.0018 + 0.000018 + \ldots$. Beacause we can't add infinitely many numbers, we need to define the meaning of such a sum. We interpret it by looking at partial finite sums: $0.18, 0.18 + 0.0018, 0.18 + 0.0018 + 0.000018, \ldots$. The more terms we take the closer and closer do the partial sums get to a specific number; in the above example $\frac{2}{11}$. The value of the infinite sum is defined as that number.

EXAMPLE 7-10

Use a calculator to convert the following to decimals:

a. $\frac{1}{7}$

b. $\frac{2}{13}$

Solution   When we use a calculator to divide, it seems that the division pattern repeats.

a. $\frac{1}{7} = 0.\overline{142857}$

b. $\frac{2}{13} = 0.\overline{153846}$

To see why in Example 7-10 the division pattern repeats as predicted, consider the following divisions:

a.
$$
\begin{array}{r}
0.142857 \\
7)\overline{1.000000} \\
\underline{7} \\
30 \\
\underline{28} \\
20 \\
\underline{14} \\
60 \\
\underline{56} \\
40 \\
\underline{35} \\
50 \\
\underline{49} \\
1
\end{array}
$$

b.
$$
\begin{array}{r}
0.153846 \\
13)\overline{2.000000} \\
\underline{1\,3} \\
70 \\
\underline{65} \\
50 \\
\underline{39} \\
110 \\
\underline{104} \\
60 \\
\underline{52} \\
80 \\
\underline{78} \\
2
\end{array}
$$

In $\frac{1}{7}$, the remainders obtained in the division are 3, 2, 6, 4, 5, and 1. These are all the possible non-zero remainders that can be obtained when dividing by 7. If we had obtained a remainder of 0, the decimal would terminate. Consequently, the seventh division cannot produce a new remainder. Whenever a remainder recurs, the process repeats itself. Using similar reasoning, we could predict that the repetend for $\frac{2}{13}$ could not have more than 12 digits, because there are only 12 possible non-zero remainders. However, one of the remainders could repeat sooner than that, which was actually the case in part (b). In general, if $\frac{a}{b}$ is any rational number in simplest form with $b \neq 0$ and $b > a$, and it does not represent a terminating decimal, the repetend has at most $b - 1$ digits. Therefore, a *rational number may always be represented either as a terminating decimal or as a repeating decimal.*

Students frequently forget to simplify a given rational number before determining the possible length of the repetend. For example, consider $\frac{6}{21}$. Even though $21 = 3 \cdot 7$ the maximum possible length of the repetend is not 20. Because $\frac{6}{21} = \frac{2}{7}$, the maximum possible length of the repetend is 6.

● NOW TRY THIS 7-8

a. Write $\frac{1}{9}$ as a decimal.

b. Based on your answer in part (a), mentally compute the decimal representation for each of the following.

(i) $\frac{2}{9}$ (ii) $\frac{3}{9}$ (iii) $\frac{5}{9}$ (iv) $\frac{8}{9}$

EXAMPLE 7-11

Use a calculator to convert $\dfrac{1}{17}$ to a repeating decimal.

Solution   In using a calculator, if we press $\boxed{1}\,\boxed{\div}\,\boxed{1}\,\boxed{7}\,\boxed{=}$, we obtain the following, shown as part of a division problem:

$$\begin{array}{r} 0.0588235 \\ 17\overline{)1.} \end{array}$$

Without knowing whether the calculator has an internal round-off feature and with the calculator's having an eight-digit display, we find the greatest number of digits to be trusted in the quotient is six following the decimal point. (Why?) If we use those six places and multiply 0.058823 times 17, we may continue the operation as follows:

$$\boxed{\cdot}\,\boxed{0}\,\boxed{5}\,\boxed{8}\,\boxed{8}\,\boxed{2}\,\boxed{3}\,\boxed{\times}\,\boxed{1}\,\boxed{7}\,\boxed{=}$$

We then obtain 0.999991, which we may place in the preceding division:

$$\begin{array}{r} 0.0\,5\,8\,8\,2\,3 \\ 17\overline{)1.0\,0\,0\,0\,0\,0} \\ 9\,9\,9\,9\,9\,1 \\ \hline 9 \end{array}$$

Next, we divide 9 by 17 to obtain 0.5294118. Again ignoring the rightmost digit, we continue as before, completing the division as follows, where the repeating pattern is apparent:

$$\begin{array}{r} 0.0588235294117647058\,8235 \\ 17\overline{)1.000000000000000000000000} \\ 999991 \\ \hline 9000000 \\ 8999987 \\ \hline 13000000 \\ 12999985 \\ \hline 15 \end{array}$$

Thus, $\dfrac{1}{17} = 0.\overline{0588235294117647}$, and the repetend is 16 digits long.

Example 7-11 suggests a method to show that every rational number can be written as a decimal. Additionally, Example 7-11 illustrates how a calculator with only a finite display of digits can be used to do division beyond what the calculator was designed to do. A common suggestion for elementary students is that they only use the division algorithm selectively to learn the process. The calculator example given and similar ones are used to see whether we truly understand the division process and the place values involved along the way.

Writing a Repeating Decimal in the Form $\dfrac{a}{b}$, where $a, b \in I, b \neq 0$

We have already considered how to write terminating decimals in the form $\dfrac{a}{b}$, where a, b are integers and $b \neq 0$. For example,

$$0.55 = \frac{55}{10^2} = \frac{55}{100}.$$

To write $0.\overline{5}$ in a similar way, we see that because the repeating decimal has infinitely many digits, the denominator cannot be written as a single power of 10. To overcome this difficulty, we must somehow eliminate the infinitely repeating part of the decimal. If we let $n = 0.\overline{5}$, then the *subgoal* is to write an equation for n without a repeating decimal. It can be shown that $10(0.555\ldots) = 5.555\ldots = 5.\overline{5}$. Hence, $10n = 5.\overline{5}$. Using this information, we subtract the corresponding sides of the equations to obtain an equation whose solution can be written without a repeating decimal.

$$
\begin{aligned}
10n &= 5.\overline{5} \\
- \quad n &= 0.\overline{5} \\
\hline
9n &= 5 \\
n &= \frac{5}{9}
\end{aligned}
$$

Performing the subtraction gives an equation that contains only integers. The repeating blocks "cancel" each other. Thus, $0.\overline{5} = \frac{5}{9}$. This result can be checked by performing the division $5 \div 9$.

 NOW TRY THIS 7-9

Is $0.\overline{9}$ equal to 1 or is it less than 1? Justify your answer.

Suppose a decimal has a repetend of more than one digit such as $0.\overline{235}$. One approach is to multiply the decimal $0.\overline{235}$ by 10^3, since there is a three-digit repetend. Let $n = 0.\overline{235}$. Our *subgoal* is to write an equation for n without the repeating decimal:

$$
\begin{aligned}
1000n &= 235.\overline{235} \\
- \quad n &= 0.\overline{235} \\
\hline
999n &= 235 \\
n &= \frac{235}{999}
\end{aligned}
$$

Thus, $0.\overline{235} = \frac{235}{999}$.

We generalize the above method by first noticing that because $0.\overline{5}$ repeats in one-digit blocks to write it in the form $\frac{a}{b}$, we first multiply by 10^1. Because $0.\overline{235}$ repeats in three-digit blocks, we first multiply by 10^3. In general, *if the repetend is immediately to the right of the decimal point, first multiply by 10^m, where m is the number of digits in the repetend, and then continue as in the preceding cases.*

The above approach to finding the fraction equivalent of a repeating decimal is seen on the partial student page (page 363).

Now, suppose the repeating block does *not* occur immediately after the decimal point. For example, let $n = 2.3\overline{45}$. A strategy for solving this problem is to *change it to a related problem* we know how to solve; that is, change it to a problem where the repeating block immediately follows the decimal point. This becomes a *subgoal*. To accomplish this, we multiply both sides by 10:

$$
\begin{aligned}
n &= 2.3\overline{45} \\
10n &= 23.\overline{45}
\end{aligned}
$$

We now proceed as with previous problems. Because $10n = 23.\overline{45}$ and the number of digits in the repetend is 2, we multiply by 10^2 as follows:

$$
100(10n) = 2345.\overline{45}
$$

Thus,

$$1000n = 2345.\overline{45}$$
$$- \quad 10n = \quad 23.\overline{45}$$
$$990n = 2322$$
$$n = \frac{2322}{990}, \text{ or } \frac{129}{55}$$

Hence, $2.3\overline{45} = \frac{2322}{990}$, or $\frac{129}{55}$. We can check that the answer is correct by using a calculator to convert $\frac{129}{59}$ back to decimal form.

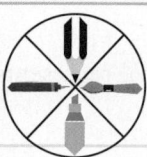

Activity Lab REPEATING DECIMALS

2-2b Activity Lab

Repeating Decimals

In Lesson 2-2, you learned how to write a terminating decimal as a fraction. You use algebra to write a repeating decimal as a fraction.

EXAMPLE Writing a Repeating Decimal as a Fraction

In a recent survey, $0.\overline{45}$ of those asked chose blue as their favorite color. Write $0.\overline{45}$ as a fraction in simplest form.

Step 1 Represent the given decimal with a variable.

$n = 0.\overline{45}$

Step 2 Multiply by 10^n, where n = the number of digits that repeat. In this case, multiply by 10^2, or 100, because the repeating part of the decimal is 45.

$100n = 45.\overline{45}$

Step 3 Subtract to eliminate the repeating part.

$$100n = \quad 45.454545\ldots$$
$$- \quad n = - \quad 0.454545\ldots \quad \leftarrow \text{Use the Subtraction Property of Equality.}$$
$$99n = \quad 45.000000\ldots \quad \leftarrow \text{Simplify.}$$
$$99n = \quad 45$$

Step 4 Solve the new equation.

$$\frac{99n}{99} = \frac{45}{99} \quad \leftarrow \text{Divide each side by 99.}$$
$$n = \frac{45}{99} = \frac{5}{11} \quad \leftarrow \text{Simplify using the GCF, 9.}$$

● The repeating decimal $0.\overline{45}$ equals $\frac{5}{11}$.

A Surprising Result

To find the $\frac{a}{b}$ form of $0.\overline{9}$, we proceed as follows. Let $n = 0.\overline{9}$, then $10n = 9.\overline{9}$. Next we subtract and continue to solve for n:

$$10n = 9.\overline{9}$$
$$- n = 0.\overline{9}$$
$$9n = 9$$
$$n = 1$$

Hence, $0.\overline{9} = 1$. This approach to the problem may not be convincing. Another approach to show that $0.\overline{9}$ is really another name for 1 is shown next:

(a) $\dfrac{1}{3} = 0.33333333\ldots$ **(b)** $\dfrac{2}{3} = 0.66666666\ldots$

Adding equations (a) and (b), we have $1 = 0.99999999\ldots$ or $0.\overline{9}$. This last decimal represents the infinite sum $\dfrac{9}{10} + \dfrac{9}{10^2} + \dfrac{9}{10^3} + \ldots$.

Still another approach is as follows:

$$\frac{1}{9} = 0.11111111\ldots$$
$$1 = 9 \cdot \frac{1}{9} = 9 \cdot 0.11111111\ldots$$
$$= 0.999999\ldots$$

Some may prefer a visual approach to show that $0.\overline{9} = 1$. Consider the number line in Figure 7-5(a). Most would agree that $0.\overline{9}$ would be between 0.9 and 1.0, so we start there. Then $0.\overline{9}$ is between 0.99 and 1.0, as in Figure 7-5(b).

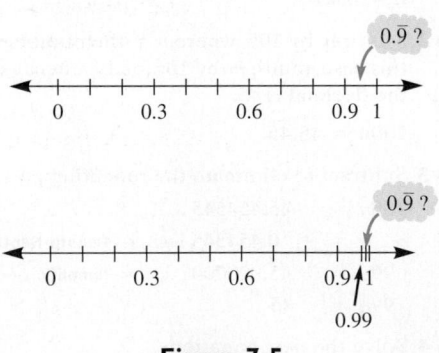

Figure 7-5

Then we can proceed similarly to argue that $0.\overline{9}$ would be between 0.999 and 1, and so forth. Next, it is reasonable to ask if there is any tiny amount a such that $0.\overline{9} + a = 1$? The answer has to be no. (Why?) If there is no such number a, then $0.\overline{9}$ cannot be less than 1. Beacuse $0.\overline{9}$ cannot be greater than 1, we know that $0.\overline{9} = 1$. In more advanced mathematics courses, sums like $0.\overline{9}$, or $0.9 + 0.09 + 0.009 + \ldots$, are defined as the *limits of finite sums*.

Ordering Repeating Decimals

We now know that any repeating decimal can be written as a rational number in the form $\frac{a}{b}$, where $b \neq 0$. Thus, any repeating decimal can be represented on a number line in a manner similar to the way that terminating decimals can be placed on the number line. Also, because we have seen that between any two rational numbers in $\frac{a}{b}$ form there are infinitely many more rational numbers of that form (denseness property), it is reasonable that there should be infinitely many repeating decimals between any two other decimals.

To order repeating decimals, we consider where a repeating decimal might lie on a number line or we compare them using place value in a manner similar to how we ordered terminating decimals. For example, to order repeating decimals such as $1.\overline{3478}$ and $1.34\overline{7821}$ we write the decimals one under the other, in their equivalent forms without the bars, and line up the decimal points (or place values) as follows:

$$1.34783478\ldots$$
$$1.34782178\ldots$$

The digit to the left of the decimal points and the first four digits after the decimal points are the same in each of the numbers. However, since the digit in the hundred-thousandths place of the top number, which is 3, is greater than the digit 2 in the hundred-thousandths place of the bottom number, $1.\overline{3478}$ is greater than $1.34\overline{7821}$.

It is easy to compare two fractions, such as $\frac{21}{43}$ and $\frac{37}{75}$, using a calculator. We convert each to a decimal and then compare the decimals.

$$\boxed{2}\,\boxed{1}\,\boxed{\div}\,\boxed{4}\,\boxed{3}\,\boxed{=} \rightarrow 0.4883721$$
$$\boxed{3}\,\boxed{7}\,\boxed{\div}\,\boxed{7}\,\boxed{5}\,\boxed{=} \rightarrow 0.4933333$$

Examining the digits in the hundredths place, we see that

$$\frac{37}{75} > \frac{21}{43}$$

EXAMPLE 7-12

Find a rational number in decimal form between $0.\overline{35}$ and $0.\overline{351}$.

Solution First, line up the decimals.

$$0.353535\ldots$$
$$0.351351\ldots$$

Then, to find a decimal between these two, observe that starting from the left, the first place at which the two numbers differ is the thousandths place. Clearly, one decimal between these two is 0.352. Others include 0.3514, $0.35\overline{15}$, and 0.35136. In fact, there are infinitely many others.

NOW TRY THIS 7-10

Given the terminating decimal 0.36, find two repeating decimals, one less than 0.36 but no more than 0.01 less, and one greater than 0.36, but no more than 0.01 greater.

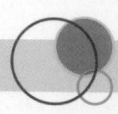

Assessment 7-3A

1. Find the decimal representation for each of the following:
   - a. $\dfrac{4}{9}$
   - b. $\dfrac{2}{7}$
   - c. $\dfrac{3}{11}$
   - d. $\dfrac{1}{15}$
   - e. $\dfrac{2}{75}$
   - f. $\dfrac{1}{99}$
   - g. $\dfrac{5}{6}$
   - h. $\dfrac{1}{13}$
   - i. $\dfrac{1}{21}$
   - j. $\dfrac{3}{19}$

2. Convert each of the following repeating decimals to $\dfrac{a}{b}$ form, where a, b are integers and $b \neq 0$.
   - a. $0.\overline{4}$
   - b. $0.\overline{61}$
   - c. $1.3\overline{96}$
   - d. $0.\overline{55}$
   - e. $^-2.3\overline{4}$
   - f. $^-0.0\overline{2}$

3. Express 1 min as a repeating decimal part of an hour.

4. Order the following decimals from greatest to least:

   $$^-1.4\overline{54},\ ^-1.\overline{454},\ ^-1.\overline{45},\ ^-1.45\overline{4},\ ^-1.454$$

5. Find three more terms for the following sequence:

   $$0,\ 0.5,\ 0.\overline{6},\ 0.75,\ 0.8,\ 0.8\overline{3},\ \underline{\hspace{1cm}},\ \underline{\hspace{1cm}},\ \underline{\hspace{1cm}}$$

6. Write $(0.\overline{5})^2$ as a repeating decimal.

7. Give an argument why $3\dfrac{1}{7}$ must be a repeating decimal.

8. Suppose $a = 0.\overline{32}$ and $b = 0.\overline{123}$. Find $a + b$ by adding from left to right. How many digits are in the repetend of the sum?

9. Explain whether a terminating decimal could ever be written as a repeating decimal.

10. Find three decimals between each of the following pairs of decimals:
    - a. $3.\overline{2}$ and 3.22
    - b. $462.\overline{24}$ and 462.243

11. Find the decimal halfway between the decimals: $0.\overline{4}$ and 0.5.

12. a. Find three rational numbers between $\dfrac{3}{4}$ and $0.7\overline{5}$.

    b. Find three rational numbers between $\dfrac{1}{3}$ and $0.\overline{34}$.

13. a. What is the 21st digit in the decimal expansion of $\dfrac{3}{7}$?

    b. What is the 5280th digit in the decimal expansion of $\dfrac{1}{17}$?

14. a. Write each of the following as a fraction in the form $\dfrac{a}{b}$, where a and b are integers and $b \neq 0$:
    - i. $0.\overline{1}$
    - ii. $0.\overline{01}$
    - iii. $0.\overline{001}$

    b. What fraction would you expect for $0.\overline{0001}$?

    c. Mentally compute the decimal equivalent for $\dfrac{1}{90}$.

15. Use the fact that $0.\overline{9} = 1$ to find the terminating decimal equal to the following:
    - a. $0.0\overline{9}$
    - b. $0.3\overline{9}$
    - c. $9.\overline{99}$

16. Use the fact that $0.\overline{1} = \dfrac{1}{9}$ to mentally convert each of the following into fractions:
    - a. $0.\overline{2}$
    - b. $0.\overline{3}$
    - c. $9.\overline{9}$

17. Use the fact that $0.\overline{01} = \dfrac{1}{99}$ and $0.\overline{001} = \dfrac{1}{999}$ to mentally convert each of the following into fractions:
    - a. $0.\overline{05}$
    - b. $0.\overline{003}$

18. Find the sum of the finite geometric sequence whose first term is 0.4, whose ratio is 0.5, and which has five terms.

19. Find a rational number in the form $\dfrac{a}{b}$, where a and b are integers and $b \neq 0$, for the following repeating decimals:
    - a. $0.2\overline{9}$
    - b. $2.0\overline{29}$

20. Consider the repeating decimals $0.\overline{235}$ and $0.\overline{2356}$. How many places do you expect in the repetend of the sum of the two decimals? Why?

21. Find values of x such that each of the following is true. Write answers both in $\dfrac{a}{b}$ form, where a and b are integers and $b \neq 0$, and as decimals.
    - a. $1 - 3x = 8$
    - b. $1 = 3x + 8$
    - c. $1 = 8 - 3x$

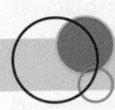

Assessment 7-3B

1. Find the decimal representation for each of the following:
   - a. $\dfrac{2}{3}$
   - b. $\dfrac{7}{9}$
   - c. $\dfrac{1}{24}$
   - d. $\dfrac{3}{60}$
   - e. $\dfrac{2}{99}$
   - f. $\dfrac{7}{6}$
   - g. $\dfrac{2}{21}$
   - h. $\dfrac{4}{19}$

2. Convert each of the following repeating decimals to $\dfrac{a}{b}$ form, where a and b are integers and $b \neq 0$:
   - a. $0.\overline{7}$
   - b. $0.\overline{46}$
   - c. $2.\overline{37}$
   - d. $2.3\overline{4}$
   - e. $^-4.3\overline{4}$
   - f. $^-0.0\overline{3}$

3. Express 1 sec as a repeating decimal part of an hour.

4. Order the following decimals from least to greatest:
$$^-4.34, \, ^-4.\overline{34}, \, ^-4.3\overline{4}, \, ^-4.3\overline{43}, \, ^-4.4\overline{34}$$

5. List the next three terms in the following arithmetic sequence:
$$0, 0.\overline{3}, 0.\overline{6}, 1, 1.\overline{3}, \underline{\quad} \, \underline{\quad} \, \underline{\quad}$$

6. Write $0.4\overline{9} \cdot 0.\overline{62}$ as a repeating decimal.

7. Give an argument why $\dfrac{2}{26}$ must be a repeating decimal.
What can you say about length of the repetend without performing the division.?

8. Find $a + b$ if $a = 1.2\overline{34}$ and $b = 0.\overline{1234}$. Is the answer a rational number? How many digits are in the repetend?

9. Explain whether a repeating decimal could ever be written as a terminating decimal.

10. Find three decimals between each of the following pairs of decimals.
    a. $4.\overline{3}$ and 4.3
    b. $203.\overline{76}$ and 203.7

11. Find the decimal halfway between: $0.\overline{9}$ and 1.1.

12. Find three rational numbers between the following:
    a. $\dfrac{2}{3}$ and 0.67
    b. $\dfrac{2}{3}$ and $0.6\overline{7}$

13. What is the 23rd decimal in the expansion of $\dfrac{1}{17}$?

14. a. Write each of the following as a fraction in the form $\dfrac{a}{b}$, where a and b are integers and $b \neq 0$:
       i. $0.\overline{2}$ ii. $0.0\overline{2}$ iii. $0.00\overline{2}$

b. What fraction would you expect for $0.000\overline{2}$?

c. Mentally compute the decimal equivalent for $\dfrac{4}{90}$.

15. Use the fact that $0.\overline{9} = 1$, to find the terminating decimal equal to the following:
    a. $1.\overline{9}$ b. $0.00\overline{9}$ c. $0.3\overline{9}$

16. Use the fact that $0.\overline{1} = \dfrac{1}{9}$, $0.\overline{01} = \dfrac{1}{99}$, and $0.\overline{001} = \dfrac{1}{999}$ to convert each of the following mentally to the form $\dfrac{a}{b}$, where a and b are integers and $b \neq 0$:
    a. $0.\overline{4}$ b. $0.\overline{12}$ c. $0.\overline{111}$

17. Use the fact that $0.\overline{01} = \dfrac{1}{99}$ and $0.\overline{001} = \dfrac{1}{999}$ to convert each of the following into a rational number $\dfrac{a}{b}$ where a and b are integers and $b \neq 0$.
    a. $3.\overline{25}$ b. $3.\overline{125}$

18. Find the sum of the finite geometric sequence whose first term is 0.1, whose ratio is 0.3, and which has four terms.

19. Find the $\dfrac{a}{b}$ form, where a and b are integers and $b \neq 0$, for each of the following:
    a. $0.\overline{29}$ b. $0.000\overline{29}$

20. Consider the repeating decimals, $0.\overline{23}$ and $0.\overline{235}$. How many places do you expect in the repetend of the sum of the two decimals? Why?

21. Find values of x such that each of the following is true. Write answers both in $\dfrac{a}{b}$ form, where a and b are integers and $b \neq 0$, and as decimals.
    a. $3x = 8$ b. $3x + 1 = 8$ c. $3x - 1 = 8$

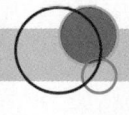

Mathematical Connections 7-3

Communication

1. a. If a grocery store advertised three lemons for $2.00, what is the cost of one lemon?
   b. If you choose to buy exactly one lemon at the cost given in part (a), what will the grocery store charge?
   c. How is the store treating the repeating decimal cost of one lemon?
   d. Explain whether or not a grocery store would ever use a repeating decimal as a cost for an item.
   e. Explain whether you think cash registers ever work with repeating decimals.

2. A friend claims that every finite decimal is equal to some infinite decimal. Is the claim true? Explain why or why not.

3. Some addition problems are easier to compute with fractions and some are easier to do with decimals. For example,
$\dfrac{1}{7} + \dfrac{5}{7}$ is easier to compute than $0.\overline{142857} + 0.\overline{714285}$ and
$0.4 + 0.25$ is easier to compute than $\dfrac{2}{5} + \dfrac{1}{4}$. Describe
situations in which you think it would be easier to compute the additions with fractions than with decimals, and vice versa.

Open-Ended

4. Notice that $\dfrac{1}{7} = 0.\overline{142857}$, $\dfrac{2}{7} = 0.\overline{285714}$, $\dfrac{3}{7} = 0.\overline{428571}$,
$\dfrac{4}{7} = 0.\overline{571428}$, $\dfrac{5}{7} = 0.\overline{714285}$, and $\dfrac{6}{7} = 0.\overline{857142}$.
   a. Describe a common property that all of these repeating decimals share.
   b. Suppose you memorized the decimal form for $\dfrac{1}{7}$. How could you quickly find the answers for the decimal expansion of the rest of the preceding fractions? Describe as many ways as you can.
   c. Find some other fractions that behave like $\dfrac{1}{7}$. In what way is the behavior similar?

5. a. Show that every integer can be written as a decimal.
   b. Multiply each of the following, giving your answer as a decimal in simplest form.
      i. $2 \cdot 0.\overline{3}$ ii. $3 \cdot 0.\overline{3}$ iii. $3 \cdot 0.\overline{35}$

c. Explain whether the traditional algorithm for multiplying decimals can be used for multiplying repeating decimals.

d. Explain whether repeating decimals can be multiplied.

6. a. Does your calculator allow you to enter repeating decimals? If so, in what form?

b. Explain whether repeating decimal arithmetic can be performed on your calculator.

7. Explain whether we would rather have the solution to $3x = 7$ expressed as a fraction or as a repeating decimal.

Cooperative Learning

8. Choose a partner and play the following game. Write a repeating decimal of the form $0.\overline{abcdef}$. Tell your partner that the decimal is of that form but do not reveal the specific values for the digits. Your partner's objective is to find your repeating decimal. Your partner is allowed to ask you for the values of six digits that are at the 100th or greater places after the decimal point but not the digits in consecutive places. For example, your opponent may ask for the 100th, 200th, 300th, ... digits but may not ask for the 100th and 101st digit. Switch roles at least once. After playing the game, discuss in your group a strategy for asking your partner the least number of questions that will allow you to find your partner's repetend.

Questions from the Classroom

9. A student argues that repeating decimals are of little value because no calculator will handle computations with repeating decimals and in real life decimals do not have an infinite number of digits. How do you respond?

10. Samantha says that base two numerals such as 0.1_{two} and 0.101_{two} do not exist. How would you convince her that is untrue?

11. A student asks if a sum of two repeating decimals is always a repeating decimal. How do you respond?

12. A student needs to find $0.3333 \cdot 48$ without using a calculator. She finds $\frac{1}{3}$ of 48 and gets 16 as the answer. How do you respond?

Review Problems

13. Jill received a bonus totaled $27,849.50 and deposited it in a new checking account. She wrote checks for $1520.63, $723.30, and $2843.62. What was the balance in her new checking account?

14. The speed of light is approximately 186,000 mi/sec. It takes light from the nearest star, Alpha Centauri, approximately 4 yrs to reach Earth. How many miles away is Alpha Centauri from Earth? Express the answer in scientific notation.

15. Find the product of 0.22 and 0.35 on a calculator. How does the placement of the decimal point in the answer on the calculator compare with the placement of the decimal point using the rule in this chapter? Explain.

16. a. Find a number to add to $^{-}0.023$ to obtain a sum greater than 3 but less than 4.

b. Find a number to subtract from $^{-}0.023$ to obtain a difference greater than 3 but less than 4.

c. Find a number to multiply times 0.023 to obtain a product greater than 3 but less than 4.

d. Find a number to divide into 0.023 to obtain a quotient greater than 3 but less than 4.

Trends in Mathematics and Science Study (TIMSS) Questions

In which list are the numbers ordered from greatest to least?

  a. 0.233, 0.3, 0.32, 0.332

  b. 0.3, 0.32, 0.332, 0.233

  c. 0.32, 0.233, 0.332, 0.3

  d. 0.332, 0.32, 0.3, 0.233

TIMSS, Grade 8, 2003

In which of these pairs of numbers is 2.25 larger than the first number but smaller than the second number?

  a. 1 and 2

  b. 2 and $\frac{5}{2}$

  c. $\frac{5}{2}$ and $\frac{11}{4}$

  d. $\frac{11}{4}$ and 3

TIMSS, Grade 8, 2003

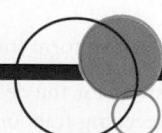

7-4 Percents and Interest

In grade 6 *Common Core Standards* students are expected to be able to accomplish the following:

> Find a percent of a quantity as a rate per 100 (e.g., 30% of a quantity means 30/100 times the quantity); solve problems involving finding the whole, given a part and the percent. (p. 42)

Percents are very useful in conveying information. People hear that there is a 60% chance of rain or that their savings account is drawing 2% annual interest. Percents are special kinds of fractions—namely, fractions with a denominator of 100. The word **percent** comes from the

Latin phrase *per centum*, which means *per hundred*. A bank that pays 2% annual simple interest on a savings account pays $2 for each $100 in the account for one year; that is, it pays 2/100 of whatever amount is in the account for one year. The symbol % indicates percent; 2% means 2 for each 100. Hence, to find 2% of $400, we determine how many hundreds are in 400. There are 4 hundreds in 400, so 2% of 400 is $2 \cdot 4 = 8$. Therefore, 2% of $400 = $8.

E-Manipulative Activity

For more help with percents using a hundreds grid, see the *Percents* module.

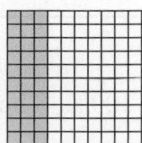

Figure 7-6

> **Definition of Percent**
>
> $$n\% = \frac{n}{100}, \text{ where } n \text{ is any non-negative number}$$

Thus, $n\%$ of a quantity is $\frac{n}{100}$ of the quantity that is $\frac{n}{100}$ times the quantity. Therefore, 1% is one hundredth of a whole and 100% represents the entire quantity, whereas 200% represents $\frac{200}{100}$, or 2 times, the given quantity. Percents can be illustrated by using a hundreds grid. For example, what percent of the grid is shaded in Figure 7-6? Because 30 out of the 100, or $\frac{30}{100}$, of the squares are shaded, we say that 30% of the grid is shaded (or similarly, 70% of the grid is not shaded).

 NOW TRY THIS 7-11

In each part of Figure 7-7 write the fraction in lowest terms and the equivalent percent that represents the shaded portion of the large square.

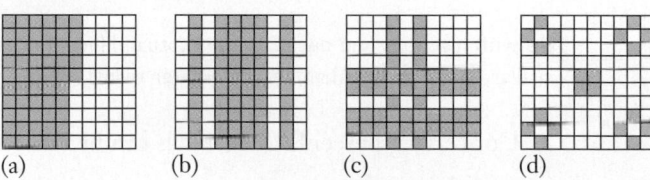

    (a)   (b)   (c)   (d)

Figure 7-7

Because $n\% = \frac{n}{100}$, one way to convert a number to a percent is to write it as a fraction with denominator 100; the numerator gives the amount of the percent. For example, $\frac{3}{4} = \frac{3 \cdot 25}{4 \cdot 25} = \frac{75}{100}$. Hence, $\frac{3}{4} = 75\%$.

EXAMPLE 7-13

Write each of the following as a percent:

  a. 0.03   **b.** $0.\overline{3}$ **c.** 1.2   **d.** 0.00042

  e. 1   **f.** $\frac{3}{5}$ **g.** $\frac{2}{3}$ **h.** $2\frac{1}{7}$

Solution

  a. $0.03 = \frac{3}{100} = 3\%$

  b. $0.\overline{3} = \frac{33.\overline{3}}{100} = 33.\overline{3}\%$

  c. $1.2 = \frac{120}{100} = 120\%$

d. $0.00042 = \dfrac{0.042}{100} = 0.042\%$

e. $1 = \dfrac{100}{100} = 100\%$

f. $\dfrac{3}{5} = 100\left[\dfrac{\left(\dfrac{3}{5}\right)}{100}\right] = \dfrac{60}{100} = 60\%$

g. $\dfrac{2}{3} = 100\left[\dfrac{\left(\dfrac{2}{3}\right)}{100}\right] = \dfrac{\left(\dfrac{200}{3}\right)}{100} = \dfrac{66.\overline{6}}{100} = 66.\overline{6}\%$

h. $2\dfrac{1}{7} = 100\left[\dfrac{\left(2\dfrac{1}{7}\right)}{100}\right] = \dfrac{\left(\dfrac{1500}{7}\right)}{100} = \dfrac{1500}{7}\%, \text{ or } 214\dfrac{2}{7}\%.$

Still another way to convert a number to a percent is to recall that $1 = 100\%$. Thus, for example, $\dfrac{3}{4} = \dfrac{3}{4}$ of $1 = \dfrac{3}{4} \cdot 1 = \dfrac{3}{4} \cdot 100\% = 75\%$.

The % symbol is crucial in identifying the meaning of a number. For example, $\dfrac{1}{2}$ and $\dfrac{1}{2}\%$ are different numbers: $\dfrac{1}{2} = 50\%$, which is not equal to $\dfrac{1}{2}\%$. Similary, 0.01 is different from 0.01%, which is 0.0001.

In grades 6–8 of *PSSM*, we find the following:

> As with fractions and decimals, conceptual difficulties need to be carefully addressed in instruction. In particular, percents less than 1% and greater than 100% are often challenging. (p. 217)

Converting percents to decimals can be done by writing the percent as a fraction in the form $\dfrac{n}{100}$ and then converting the fraction to a decimal. Percents less than 1% and percents greater than 100% are examined in Example 7-14.

EXAMPLE 7-14

Write each of the following percents as a decimal:

a. 5% **b.** 6.3% **c.** 100% **d.** 250% **e.** $\dfrac{2}{3}\%$ **f.** $33\dfrac{1}{3}\%$

Solution

a. $5\% = \dfrac{5}{100} = 0.05$

b. $6.3\% = \dfrac{6.3}{100} = 0.063$

c. $100\% = \dfrac{100}{100} = 1$

d. $250\% = \dfrac{250}{100} = 2.5$

e. $\dfrac{2}{3}\% = \dfrac{\dfrac{2}{3}}{100} = \dfrac{0.\overline{6}}{100} = 0.00\overline{6}$

f. $33\dfrac{1}{3}\% = \dfrac{33\dfrac{1}{3}}{100} = \dfrac{33.\overline{3}}{100} = 0.\overline{3}$

Another approach to converting percent to a decimal is first to convert 1% to a decimal. Because $1\% = \dfrac{1}{100} = 0.01$, we conclude that $5\% = 5 \cdot 1\% = 5 \cdot 0.01 = 0.05$ and that $6.3\% = 6.3 \cdot 0.01 = 0.063$.

 NOW TRY THIS 7-12

a. Investigate how your calculator handles percents and tell what the calculator does when the $\boxed{\%}$ key is pushed.

b. Use your calculator to change $\frac{1}{3}$ to a percent.

Applications Involving Percent

In the grade 7 *Focal Points* we find the following:

> They [students] use ratio and proportionality to solve a wide variety of percent problems, including problems involving discounts, interest, taxes, tips, and percent increase or decrease. (p. 19)

Application problems that involve percents usually take one of the following forms:

1. Finding a percent of a number
2. Finding what percent one number is of another
3. Finding a number when a percent of that number is known

Before we consider examples illustrating these forms, recall what it means to find a fraction "of" a number. For example, $\frac{2}{3}$ of 70 means $\frac{2}{3} \cdot 70$. Similarly, to find 40% of 70, we have $\frac{40}{100}$ of 70, which means $\frac{40}{100} \cdot 70$, or $0.40 \cdot 70 = 28$.

A different way to think about 40% of 70 is to consider that 70 represents 100 parts (or the whole) and 40% requires only 40 of those 100 parts. For example, if

$$100 \text{ parts} = 70$$

$$1 \text{ part} = \left(\frac{70}{100}\right), \text{ or } 0.7$$

$$40 \text{ parts} = 40(0.7), \text{ or } 28$$

Thus, 40% of 70 = 28.

A percent bar can be used as a model for understanding what 100% of a number means as well as understanding other percents. In Figure 7-8, consider the percent bar that represents 100% of the whole with 40% of the whole shaded. Note that 100% of the bar represents 70.

Table 7-3

Percentage	Bar Length
0%	0
10%	
20%	
30%	
40%	?
50%	35
60%	
70%	
80%	
90%	
100%	70

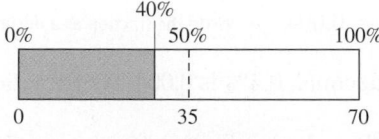

Figure 7-8

Also, half of the percent bar (50% denoted by the dotted segment) represents half of 70, or 35. Thus, we know that 40% of the bar (or 40% of 70) is less than 35. In fact, if the top of the bar is thought of as being marked off in 1% intervals, there are 100 intervals marking whole numbers of percentages. If at the same time the bottom of the bar is considered to be marked in intervals of 1, there would be only 70 intervals marked at the bottom. Where would you expect the two sets of intervals to align?

Suppose that we know that as in Table 7-3, 0% of 70 is 0; 50% of 70 is 35; and 100% of 70 is 70. What percentages of 70 are 10%, 20%, 30%, and so on? If there are 100 intervals marking percentages compared to only 70 intervals marking the corresponding length, there must be a ratio of $\frac{100}{70}$, or $\frac{10}{7}$. Thus, 10% of 70 is 7; 20% of 70 is $2 \cdot 7$, or 14; and so on. Hence, 40% of 70 is $4 \cdot 7$, or 28.

Percents can be greater than 100%. For example, if our resting heart rate is considered the base unit, then this would be 100% of your resting heart rate. Increasing your heart rate would lead to a rate that is greater than 100%. The partial student page below investigates percents less than 1% and greater than 100%.

EXAMPLE 7-15

A house that sells for $92,000 requires a 20% down payment. What is the amount of the down payment?

Solution   The amount of the down payment is 20% of $92,000, or 0.20 · $92,000 = $18,400.

School Book Page   EXTENSION

Extension   **For Use With Lesson 7-9**

Percents Under 1% or Over 100%

Percents can be less than 1% or greater than 100%.

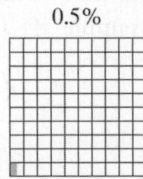

0.5%   100%   +   5%   =   105%

Less than 1%   Greater than 100%

EXAMPLES

❶ Write 0.4% as a decimal.

$0.4\% = \dfrac{0.4}{100}$ ← Write the percent as a fraction.

$= 0.004$ ← Write the fraction as a decimal.

As a decimal, 0.4% is 0.004. As a fraction, 0.4% is $\dfrac{4}{1,000}$.

❷ **Nutrition** A vitamin supplement provides 150% of the Recommended Daily Allowance (RDA) of vitamin C. The RDA is 60 milligrams. How many milligrams of vitamin C are in the vitamin supplement?

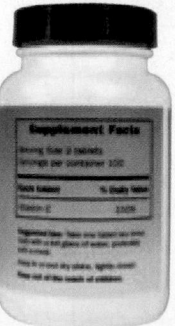

150% of $60 = 1.50 \times 60$ ← Write the percent as a decimal.

$= 90$ ← Multiply.

The vitamin supplement contains 90 milligrams of vitamin C.

EXAMPLE 7-16

If Alberto has 45 correct answers on an 80-question test, what percent of his answers are correct?

Solution   Alberto has $\dfrac{45}{80}$ of the answers correct. To find the percent of correct answers, we need to convert $\dfrac{45}{80}$ to a percent. We can do this by multiplying the fraction by 100 and attaching the % symbol, as follows:

$$\frac{45}{80} = 100 \cdot \frac{45}{80}\%$$
$$= 56.25\%$$

Thus, 56.25% of the answers are correct.

EXAMPLE 7-17

Forty-two percent of the parents of the schoolchildren in the Paxson School District are employed at Di Paloma University. If the number of parents employed by the university is 168, how many parents are in the school district?

Solution   Let n be the number of parents in the school district. Then 42% of n is 168. We translate this information into an equation and solve for n.

$$42\% \text{ of } n = 168$$
$$\frac{42}{100}n = 168$$
$$0.42n = 168$$
$$n = \frac{168}{0.42} = 400$$

There are 400 parents in the school district.

Example 7-17 can be solved using a proportion. Forty-two percent, or $\dfrac{42}{100}$, of the parents are employed at the university. If n is the total number of parents, then $\dfrac{168}{n}$ also represents the fraction of parents employed there. Thus,

$$\frac{42}{100} = \frac{168}{n}$$
$$42n = 100 \cdot 168$$
$$n = \frac{16,800}{42} = 400$$

We can also solve the problem as follows:

$$42\% \text{ of } n \text{ is } 168$$
$$1\% \text{ of } n \text{ is } \frac{168}{42}$$
$$100\% \text{ of } n \text{ is } 100\left(\frac{168}{42}\right)$$

Therefore,

$$n = 100\left(\frac{168}{42}\right), \text{ or } 400$$

EXAMPLE 7-18

Kelly bought a bicycle and a year later sold it for 20% less than what she paid for it. If she sold the bike for $144, what did she pay for it?

Solution   We are looking for the original price, P, that Kelly paid for the bike. We know that she sold the bike for $144 and that this included a 20% loss. Thus, we can write the following equation:

$$144 = P - \text{Kelly's loss}$$

Because Kelly's loss is 20% of P, we proceed as follows:

$$144 = P - 20\% \cdot P$$
$$144 = P - 0.20P$$
$$144 = (1 - 0.20)P$$
$$144 = 0.80P$$
$$\frac{144}{0.80} = P$$
$$180 = P$$

Thus, she paid $180 for the bike.

Alternatively, we can argue that since Kelly lost 20% of the original price, she sold the bike for $100\% - 20\%$ or 80% of the original price. Thus,

$$144 = 80\% \text{ of } P$$
$$144 = 0.8P$$
$$P = 180.$$

We can also approach the problem as follows:

$$80\% \text{ of } P \text{ is } 144$$
$$1\% \text{ of } P \text{ is } \frac{144}{80}$$
$$100\% \text{ of } P \text{ is } 100 \cdot \frac{144}{80} \text{ or } 180$$

EXAMPLE 7-19

Westerner's Clothing Store advertised a suit for 10% off, for a savings of $15. Later, the manager marked the suit at 30% off the original price. What is the amount of the current discount?

Solution   A 10% discount amounts to a $15 savings. We could find the amount of the current discount if we knew the original price P. Thus, finding the original price becomes our *subgoal*. Because 10% of P is $15, we have the following:

$$10\% \cdot P = 15$$
$$0.10P = 15$$
$$P = 150$$

To find the current discount, we calculate 30% of $150. Because $0.30 \cdot \$150 = \$45$, the amount of the 30% discount is $45.

In the *Looking Back* stage of problem solving, we check the answer and look for other ways to solve the problem. A different approach leads to a more efficient solution and confirms the answer. If 10% of the price is $15, then 30% of the price is 3 times $15, or $45.

NOW TRY THIS 7-13

In the following cartoon, compute the percentage and number of slices for the portions with olives, plain, and with onions and green peppers.

Rhymes with Orange copyright © 2001 and 2011 Hilary B Price. King Features Syndicate

Mental Math with Percents

Mental math may be helpful when working with percents. Two techniques follow:

1. *Using fraction equivalents*
    Knowing fraction equivalents for some percents can make some computations easier. Table 7-4 gives several fraction equivalents.

Table 7-4

Percent	25%	50%	75%	$33\frac{1}{3}\%$	$66\frac{2}{3}\%$	10%	1%
Fraction Equivalent	$\frac{1}{4}$	$\frac{1}{2}$	$\frac{3}{4}$	$\frac{1}{3}$	$\frac{2}{3}$	$\frac{1}{10}$	$\frac{1}{100}$

These equivalents can be used in such computations as the following:

$$50\% \text{ of } 80 = \left(\frac{1}{2}\right)80 = 40$$

$$66\frac{2}{3}\% \text{ of } 90 = \left(\frac{2}{3}\right)90 = 60$$

2. *Using a known percent*
    Frequently, we may not know a percent of something, but we know a close percent of it. For example, to find 55% of 62, we might do the following:

$$50\% \text{ of } 62 = \left(\frac{1}{2}\right)(62) = 31$$

$$5\% \text{ of } 62 = \left(\frac{1}{2}\right)(10\%)(62) = \left(\frac{1}{2}\right)(6.2) = 3.1$$

Adding, we see that 55% of 62 is 31 + 3.1 = 34.1.

Estimations with Percents

Estimations with percents can be used to determine whether answers are reasonable. Following are two examples:

1. To estimate 27% of 598, note that 27% of 598 is a little more than 25% of 598, but 25% of 598 is approximately the same as 25% of 600, or $\frac{1}{4}$ of 600, or 150. Here, we have adjusted 27% downward and 598 upward, so 150 should be a reasonable estimate. A better estimate might be obtained by estimating 30% of 600 and then subtracting 3% of 600 to obtain 27% of 600, giving 180 − 18, or 162.

2. To estimate 148% of 500, note that 148% of 500 should be slightly less than 150% of 500. 150% of 500 is 1.5(500) = 750. Thus, 148% of 500 should be a little less than 750.

EXAMPLE 7-20

Laura wants to buy a blouse originally priced at $26.50 but now on sale at 40% off. She has $17 in her wallet and wonders if she has enough cash. How can she mentally find out? (Ignore the sales tax.)

Solution It is easier to find 40% of $25 (versus $26.50) mentally. One way is to find 10% of $25, which is $2.50. Now, 40% is 4 times that much, that is, 4 · $2.50, or $10. Thus, Laura estimates that the blouse will cost $26.50 − $10, or $16.50. Since the actual discount is greater than $10 (40% of 26.50 is greater than 40% of 25), Laura will have to pay less than $16.50 for the blouse and, hence, she has enough cash.

Sometimes it may not be clear which operations to perform with percent. The following example investigates this.

EXAMPLE 7-21

Which of the following statements could be true and which are false? Explain your answers.

a. Leonardo got a 10% raise at the end of his first year on the job and a 10% raise after another year. His total raise was 20% of his original salary.
b. Jung and Dina paid 45% of their first department store bill of $620 and 48% of the second department store bill of $380. They paid 45% + 48% = 93% of the total bill of $1000.
c. Bill spent 25% of his salary on food and 40% on housing. Bill spent 25% + 40% = 65% of his salary on food and housing.
d. In Bordertown, 65% of the adult population works in town, 25% works across the border, 15% is unemployed, and everyone in town is in exactly one of these categories.
e. In Clean City, the fine for various polluting activities is a certain percentage of one's monthly income. The fine for smoking in public places is 40%, for driving a polluting car is 50%, and for littering is 30%. Mr. Schmutz committed all three polluting crimes in one day and paid a fine of 120% of his monthly income.

Solution
a. In applications, percent has meaning only when it represents part of a quantity. For example, 10% of a quantity plus another 10% of the same quantity is 20% of that quantity. In Leonardo's case, the first 10% raise was calculated based on his original salary and the second 10% raise was calculated on his new salary. Consequently, the percentages cannot be added, and the statement is false. He received a 21% raise.
b. The last statement does not make sense; 45% of one bill plus 48% of the other bill is not 93% of the total bill because the bills are different.

c. Because the percentages are of the same quantity, the statement is true.

d. Because the percentages are of the same quantity, that is, the number of adults, we can add them: 65% + 25% + 15% = 105%. But 105% of the population accounts for more (5% more) than the town's population, which is impossible. Hence, the statement is false.

e. Again, the percentages are of the same quantity; that is, the individual's monthly income. Hence, we can add them: 120% of one's monthly income is a stiff fine, but possible.

Computing Interest

When a bank advertises a $5\frac{1}{2}\%$ interest rate on a savings account, the **interest** is the amount of money the bank will pay for using your money. The original amount deposited or borrowed is the **principal**. The percent used to determine the interest is the **interest rate**. Interest rates are given for specific periods of time, such as years, months, or days. Interest computed on the original principal is **simple interest**. For example, suppose we borrow $5000 from a company at an annual simple interest rate of 9% for 1 yr. The interest we owe on the loan for 1 yr is 9% of $5000, or 0.09 · $5000. In general, *if a principal, P, is invested at an annual interest rate of r, then the simple interest after 1 yr is Pr · 1; after t years, it is Prt*. Thus, if I represents simple interest, we have

$$I = Prt$$

The amount needed to pay off a $5000 loan at 9% annual simple interest for 1 yr is the $5000 borrowed plus the interest on the $5000; that is, 5000 + 5000 · 0.09, or $5450. In general, *an* **amount** (*or* **balance**) *A is equal to the principal P plus the interest I*; that is,

$$A = P + I = P + Prt = P(1 + rt).$$

EXAMPLE 7-22

Vera opened a savings account that pays simple interest at the rate of $5\frac{1}{4}\%$ per year. If she deposits $2000 and makes no other deposits, find the interest and the final amount for the following time periods:

a. 1 yr   **b.** 90 days

Solution

a. To find the interest for 1 yr, we proceed as follows:

$$I = \$2000 \cdot 5\frac{1}{4}\% \cdot 1 = \$2000 \cdot 0.0525 \cdot 1 = \$105$$

Her final amount at the end of 1 yr is

$$\$2000 + \$105 = \$2105$$

b. When the interest rate is annual and the interest period is given in days, we represent the time as a fractional part of a year by dividing the number of days by 365. Thus,

$$I = \$2000 \cdot 5\frac{1}{4}\% \cdot \frac{90}{365}$$

$$= \$2000 \cdot 0.0525 \cdot \frac{90}{365} \approx \$25.89$$

Hence,

$$A \approx \$2000 + \$25.89$$

$$A \approx \$2025.89$$

Thus, Vera's amount after 90 days is approximately $2026.

EXAMPLE 7-23

Find the annual interest rate if a principal of $10,000 increased to $10,900 at the end of 1 yr.

Solution   Let the annual interest rate be x%. We know that x% of $10,000 is the increase. Because the increase is $10,900 - $10,000 = $900, we use the strategy of *writing an equation* for x as follows:

$$x\% \text{ of } 10{,}000 = 900$$

$$\frac{x}{100} \cdot 10{,}000 = 900$$

$$x = 9$$

Thus, the annual interest rate is 9%. We can also solve this problem mentally by asking, "What percent of 10,000 is 900?" Because 1% of 10,000 is 100, to obtain 900, we take 9% of 10,000.

Compound Interest

In business transactions, interest is sometimes calculated daily (365 times a year). In the case of savings, the earned interest is added daily to the principal, and each day the interest is earned on a different amount; that is, it is earned on the previous interest as well as the principal. Interest earned in this way is **compound interest**. Compounding is usually done annually (once a year), semiannually (twice a year), quarterly (4 times a year), or monthly (12 times a year). Though common in the past for banks to compound interest quarterly, today banks may compound interest monthly, daily, or even continuously. However, even when the interest is compounded, it is given as an annual rate. For example, if the annual rate is 6% compounded monthly, the interest per month is $\frac{6}{12}$%, or 0.5%. If it is compounded daily, the interest per day is $\frac{6}{365}$%. In general, *the interest rate per period is the annual interest rate divided by the number of periods in a year.*

If we invest $100 at 8% annual interest compounded quarterly, how much will we have in the account after 1 yr? The quarterly interest rate is $\frac{1}{4} \cdot 8$%, or 2%. It seems that we would have to calculate the interest 4 times. But we can also reason as follows. If at the beginning of any of the four periods there are x dollars in the account, at the end of that period there will be

$$x + 2\% \text{ of } x = x + 0.02x$$
$$= x(1 + 0.02)$$
$$= x(1.02) \text{ dollars}$$

Hence, to find the amount at the end of any period, we need only multiply the amount at the beginning of the period by 1.02. From Table 7-5, we see that the amount at the end of the fourth period is $100 \cdot 1.02^4$. On a scientific calculator, we can find the amount using $\boxed{1}\,\boxed{0}\,\boxed{0}\,\boxed{\times}\,\boxed{1}\,\boxed{.}\,\boxed{0}\,\boxed{2}\,\boxed{y^x}\,\boxed{4}\,\boxed{=}$. The calculator displays 108.24322. Thus, the amount at the end of 1 yr is approximately $108.24.

Table 7-5

Period	Initial Amount	Final Amount
1	100	$100 \cdot 1.02$
2	$100 \cdot 1.02$	$(100 \cdot 1.02)1.02$, or $100 \cdot 1.02^2$
3	$100 \cdot 1.02^2$	$(100 \cdot 1.02^2)1.02$, or $100 \cdot 1.02^3$
4	$100 \cdot 1.02^3$	$(100 \cdot 1.02^3)1.02$, or $100 \cdot 1.02^4$

Finding the final amount at the end of the nth period amounts to finding the nth term of a geometric sequence whose first term is $100 \cdot 1.02$ (amount at the end of the first period) and whose ratio is 1.02. Thus, the amount at the end of the nth period is given by $(100 \cdot 1.02)(1.02)^{n-1} = 100 \cdot 1.02^n$. We can generalize this discussion. If the principal is P and the interest rate per period is r, then the amount A after n periods is $P(1 + r)(1 + r)^{n-1}$, or $P(1 + r)^n$. Therefore, the *formula for computing the amount at the end of the nth period is*

$$A = P(1 + r)^n$$

For convenient comparison, banks are required to report the **effective annual** yield, also called **annual** percentage yield or APY. The effective annual yield on an investment is the simple interest rate that after 1 yr would pay the same amount as the given compound rate compounded at given intervals.

EXAMPLE 7-24

Suppose you deposit $1000 in a savings account that pays 6% annual interest compounded quarterly.

a. What is the balance at the end of 1 yr?
b. What is the *effective annual yield* on this investment?

Solution

a. An annual interest rate of 6% earns $\dfrac{1}{4}$ of 6%, or an interest rate of $\dfrac{0.06}{4}$, in 1 quarter. Because there are 4 periods, we have the following:

$$A = 1000\left(1 + \frac{0.06}{4}\right)^4 \approx \$1061.36$$

The balance at the end of 1 yr is approximately $1061.36.

b. Because the interest earned is $1061.36 - \$1000.00 = \$61.36$, the effective annual yield can be computed by using the simple interest formula, $I = Prt$.

$$61.36 = 1000 \cdot r \cdot 1$$

$$\frac{61.36}{1000} = r$$

$$0.06136 = r$$

$$6.136\% = r$$

The effective annual yield is 6.136%.

EXAMPLE 7-25

To save for their child's college education, a couple deposits $3000 into an account that pays 7% annual interest compounded daily. Find the amount in this account after 8 yr.

Solution The principal in the problem is $3000, the daily rate i is $0.07/365$, and the number of compounding periods is $8 \cdot 365$, or 2920. Thus we have

$$A = \$3000\left(1 + \frac{0.07}{365}\right)^{2920} \approx \$5251.74$$

Thus, the amount in the account is approximately $5251.74.

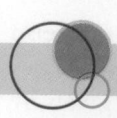

Assessment 7-4A

1. Express each of the following as a percent:
   a. 7.89
   b. 193.1
   c. $\frac{5}{6}$
   d. $\frac{1}{8}$
   e. $\frac{5}{8}$
   f. $\frac{4}{5}$

2. Convert each of the following percents to a decimal:
   a. 16%
   b. $\frac{1}{5}\%$
   c. $13\frac{2}{3}\%$
   d. $\frac{1}{3}\%$

3. Fill in the following blanks to find other expressions for 4%:
   a. _____ for every 100
   b. _____ for every 50
   c. 1 for every _____
   d. 8 for every _____
   e. 0.5 for every _____

4. Answer each of the following:
   a. What is 6% of 34?
   b. 17 is what percent of 34?
   c. 18 is 30% of what number?
   d. What is 7% of 49?

5. a. Write a fraction representing 5% of x.
   b. If 10% of an amount is a, what is the amount in terms of a?

6. Marc had 84 boxes of candy to sell. He sold 75% of the boxes. How many did he sell?

7. Gail received a 7% raise last year. If her salary is now $27,285, what was her salary last year?

8. Joe sold 180 newspapers out of 200. Bill sold 85% of his 260 newspapers. Ron sold 212 newspapers, 80% of those he had.
   a. Who sold the most newspapers? How many?
   b. Who sold the greatest percentage of his newspapers? What percent?
   c. Who started with the greatest number of newspapers? How many?

9. If a dress that normally sells for $35 is on sale for $28, what is the "percent off"? (This could be called a *percent of decrease*, or a *discount*.)

10. Mort bought his house in 2006 for $359,000. It was recently appraised at $195,000. What is the approximate *percent of decrease* in value to the nearest percent?

11. Sally bought a dress marked 20% off. If the regular price was $28.00, what was the sale price?

12. An airline ticket costs $320 without the tax. If the tax rate is 5%, what is the total bill for the airline ticket?

13. Bill got 52 correct answers on an 80-question test. What percent of the questions did he answer incorrectly?

14. If $66\frac{2}{3}\%$ of 1800 employees favored a new insurance program. How many employees favored the new program?

15. Which represents the greater percent: $\frac{325}{500}$ or $\frac{600}{1000}$? How can you tell?

16. An advertisement reads that if you buy 10 items, you get 20% off your total purchase price. You need 8 items that cost $9.50 each.
   a. How much would 8 items cost? 10 items?
   b. Is it more economical to buy 8 items or 10 items?

17. John paid $330 for a new mountain bicycle to sell in his shop. He wants to price it so that he can offer a 10% discount and still make 20% of the price he paid for it. At what price should the bike be marked?

18. Solve each of the following using mental mathematics:
   a. 15% of $22
   b. 20% of $120
   c. 5% of $38
   d. 25% of $98

19. A crew consists of one apprentice, one journeyman, and one master carpenter. The crew receives a check for $4200 for a job they just finished. A journeyman makes 200% of what an apprentice makes, and a master makes 150% of what a journeyman makes. How much does each person in the crew earn?

20. a. In an incoming freshman class of 500 students, only 20 claimed to be math majors. What percent of the freshman class is this?
   b. When the survey was repeated the next year, 5% of nonmath majors had decided to switch and become math majors.
      i. How many math majors are there now?
      ii. What percent of the former freshman class do they represent?

21. Ms. Price has received a 10% raise in salary in each of the last 2 yr. If her annual salary this year is $100,000, what was her salary 2 yr ago, rounded to the nearest penny?

22. When the U.S. Congress was sent a $2.57 trillion budget, it was reported that one would have to purchase a $100 item every second for 815 years to spend that much money.
   a. Decide whether or not you believe this report.
   b. Assuming that exactly 815 years are required to spend the entire $2.57 trillion, what percentage of the money is spent each year?

23. If you wanted to spend 25% of your monthly salary on entertainment and 56% of the salary on rent, could those amounts be $500 and $950? Why or why not?

24. An organization has 100,000 members. A bylaw change can be made at the annual business meeting held once each year, and a bylaw change must be approved by a majority of those

attending the meeting. The chair of the meeting cannot vote unless there is a tie vote but does count as an attendee at the meeting.

a. With these rules, what is the minimum number required at the meeting to make a bylaw change?

b. Based on your answer to part (a), what percentage of the membership can change the bylaws of the organization?

25. A tip in a restaurant has been typically figured at 15% of the total bill.

a. If the bill is $30, what would be the typical tip?

b. If the patron receiving the bill gave a tip that was half the bill, what is the percentage of the tip?

c. If the patron receiving the bill gave a tip that was equal to the bill, what is the percentage of the tip?

26. Suppose the percent bar below shows the number of students in a school who do not favor dress codes. How many students are in the school?

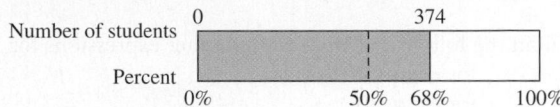

27. Complete the following compound interest chart.

Compounding Period	Principal	Annual Rate	Length of Time (Years)	Interest Rate per Period	Number of Periods	Amount of Interest Paid	Total Amount in Account
a. Semiannual	$1000	6%	2		4		
b. Quarterly	$1000	8%	3		12		
c. Monthly	$1000	10%	5		60		
d. Daily	$1000	12%	4		1460		

28. Ms. Jackson borrowed $42,000 at 8.75% annual simple interest. If exactly 1 yr later she was able to repay the loan without penalty, how much interest would she owe?

29. Falafel King will need $50,000 in 5 yr for a new addition. To meet this goal, the company deposits money in an account today that pays 3% annual interest compounded quarterly. Find the amount that should be invested to total $50,000 in 5 yr.

30. To save for their retirement, a couple deposits $4000 in an account that pays 5.9% annual interest compounded quarterly. What will be the value of their investment after 20 yr?

31. A chocolate bar costs $1.35 and the price continues to rise at a rate of 11% a year for the next 6 yr. What will be the price of the chocolate bar be at the end of 6 yr?

32. Adrien and Jarrell deposit $300 on January 1 in a holiday savings account that pays 1.1% per month interest. What is the effective annual yield?

33. The price of a house depreciates each month at the rate 0.2% for a period of n months. The value of the house each months represents what type of sequence?

34. An amount of $3000 was deposited in a bank at a rate of 2% annual interest compounded quarterly for 3 yr. The rate then increased to 3% annual interest and was compounded quarterly for the next 3 yr. If no money was withdrawn, what was the balance at the end of this time?

35. The New Age Savings Bank advertises 4% annual interest rates compounded daily, while the Pay More Bank pays 5.2% annual interest compounded annually. Which bank offers a better rate for a customer who plans to leave her money in for exactly 1 yr?

36. Amy is charged 12% annual interest compounded monthly on the unpaid balance of a $2000 loan. She did not make any payments for 2 yr. Her friend said the amount she owed had more than doubled. Is this correct? How much does she now owe?

Assessment 7-4B

1. Express each of the following as a percent:

a. 0.032

b. 0.2

c. $\dfrac{3}{20}$

d. $\dfrac{13}{8}$

e. $\dfrac{1}{6}$

f. $\dfrac{1}{40}$

2. Convert each of the following percents to a decimal:
   a. $4\frac{1}{2}\%$
   b. $\frac{2}{7}\%$
   c. 125%
   d. $\frac{1}{4}\%$

3. Fill in the following blanks to find other expressions for 5%:
   a. _____ for every 100
   b. _____ for every 50
   c. 1 for every _____
   d. 8 for every _____
   e. 0.5 for every _____

4. Answer each of the following:
   a. 63 is 30% of what number?
   b. What is 7% of 150?
   c. 61.5 is what percent of 20.5?
   d. 16 is 40% of what number?

5. a. Write a fraction representing 0.5% of x.
   b. If 0.1% of an amount is a, what is the amount in terms of a?

6. What is the sale price of a softball if the regular price is $6.80 and there is a 25% discount?

7. Brandy received a 10% raise last year. If her salary is now $60,000, what was her salary last year?

8. A line segment X is 3 in. long. This segment represents 50% of another segment Y. Find the length of each of the following.
   a. A segment that represents 100% of segment Y.
   b. A segment that represents 25% of the segment in part (a).
   c. A segment that represents 150% of segment in part (a).

9. A used car originally cost $1700. One year later, it was worth $1400. What is the percentage of depreciation?

10. On a certain day in Glacier Park, 728 eagles were counted. Five years later, 594 were counted. What was the percentage of decrease in the number of eagles counted?

11. A salesperson earns a weekly salary of $900 plus a commission rate of 4% on all sales. What did the person make for a week with total sales of $1800?

12. If you buy a new bicycle for $380 and the sales tax is 9%, what is your total bill?

13. The number of known living species is about 1.7 million. About 4500 species are mammals. What percent of known living species are mammals?

14. Jim bought two shirts that were originally marked at $40 each. One shirt was discounted 20% and the other was discounted 25%. The sales tax was 4.5%. How much did he spend in all?

15. Without multiplying out find which number represents the greater percent: 0.625 or $(0.625)^2$?

16. a. It is recommended that no more than 30% of your calorie intake should be from fat. If you consumed about 2400 calories daily, what is the maximum amount of fat calories you should consume?
    b. If one cookie contains 140 calories, and 70 calories in the cookie are fat calories, could you eat 3 cookies and not exceed the recommended amount of fat calories for the day?

17. The price of a suit that sells for $200 is reduced by 25%. By what percent must the price of the suit be increased to bring the price back to $200?

18. Solve each of the following using mental mathematics:
    a. 15% of $42
    b. 20% of $280
    c. 5% of $28
    d. 25% of $84

19. If a $\frac{1}{4}$-cup serving of Crunchies breakfast food has 0.5% of the minimum daily requirement of vitamin C, how many cups would you have to eat to obtain the minimum daily requirement of vitamin C?

20. The car Elsie bought 1 yr ago has depreciated by $1116.88, which is 12.13% of the price she paid for it. How much did she pay for the car, to the nearest dollar?

21. If you add 20% of a number to the number itself, what percent of the result would you have to subtract to get the original number back?

22. If we build a 10 × 10 model with blocks, as shown in the following figure, and paint the entire model, what percent of the cubes will have each of the following?
    a. Four faces painted
    b. Three faces painted
    c. Two faces painted

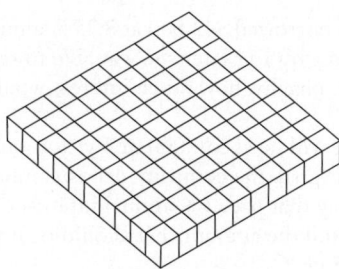

23. If 70% of the girls in a class wanted to have a prom and 60% of the boys wanted a prom, is it possible that only 50% of the students in the class wanted a prom? Explain your answer.

24. There are 80 coins in a piggy bank of which 20% are quarters. What is the least possible amount of money that could be in the piggy bank?

25. If 70% of the girls in a class wanted to have a prom and 40% of the boys wanted a prom, is it possible that only 50% of the students in the class wanted a prom? Explain your answer.

26. According to a *TV Guide* survey, 46% of people in the United States said they would not stop watching television for anything less than a million dollars. Use the percent bar for the U.S. population to estimate the number of people who would not stop watching television for anything less than a million dollars.

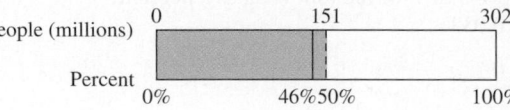

27. Complete the following compound interest chart.

Compounding Period	Principal	Annual Rate	Length of Time (Years)	Interest Rate per Period	Number of Periods	Amount of Interest Paid	Total Amount in Account
a. Semiannual	$1000	4%	2				
b. Quarterly	$1000	6%	3				
c. Monthly	$1000	18%	5				
d. Daily	$1000	18%	4				

28. A man collected $28,500 on a loan of $25,000 he made 4 yr ago. If he charged simple interest, what was the rate he charged?

29. Green Energy store will need $100,000 in 5 yr for a new addition. To meet this goal, the company deposits money in an account that pays 3% annual interest compounded quarterly. Find the amount that should be invested to total $100,000 in 5 yr.

30. Sara invested money at a bank that paid 3.5% annual interest compounded quarterly. If she had $4650 at the end of 4 yr, what was her initial investment?

31. A bank pays 2% annual interest compounded daily. What is the value of $10,000 after 15 yr?

32. If a saving account pays 0.5% per month interest, what is the effective annual yield?

33. If interest is compounded annually at 4% on a savings account for a period of n years and the interest remains in the account, the amount every year would represent what type of sequence?

34. If a publishing company signed an agreement to allow a textbook (1st edition with 500 pages) to expand over several editions to 1000 pages and the book was growing at approximately 10% in the number of pages over each edition, how many editions could be published before it reached the contractual limit?

35. Al invests $1000 at 6% annual interest compounded daily and Betty invests $1000 at 7% simple interest. After how many whole years will Al's investments be worth more than Betty's investment?

36. The number of trees in a rain forest decreases each month by 0.5%. If the forest has approximately $2.34 \cdot 10^9$ trees, how many trees will be left after 20 yr?

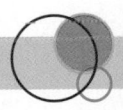

Mathematical Connections 7-4

Communication

1. Use mental math to find 11% of 850. Explain your method.
2. Does 0.4 = 0.4%? Explain.
3. What does it mean to reach 125% of your savings goal?
4. Is 4% of 98 the same as 98% of 4? Explain.
5. a. If 25% of a number is 55, is the number greater than or less than 55? Explain.
   b. If 150% of a number is 55, is the number greater than or less than 55? Explain.
6. Can 35% of one number be greater than 55% of another number? Explain.
7. Why does one picture have so much more shaded area when they both show 50%?

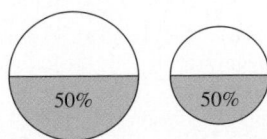

8. Why is it possible to have an increase of 150% in price but not a 150% decrease in price?

9. Two equal amounts of money were invested in two different stocks. The value of the first stock increased by 15% the first year and then decreased by 15% the second year. The second stock decreased by 15% the first year and increased by 15% the second year. Was one investment better than the other? Explain your reasoning.

10. Because of a recession, the value of a new house depreciated 10% each year for 3 yr in a row. Then, for the next 3 yr, the value of the house increased 10% each year. Had the value of the house increased or decreased after 6 yr? Explain.

11. Determine the number of years (to the nearest tenth) it would take for any amount of money to double if it were deposited at a 2% annual interest rate compounded annually. Explain your reasoning.

12. Each year a car's value depreciated 20% from the previous year. Mike claims that after 5 yr the car would depreciate 100% and would not be worth anything. Is Mike correct? Explain why or why not. If not, find the actual percent the car would depreciate after 5 yr.

Open-Ended

13. Write and solve a word problem whose solution involves the following. If any of these tasks is impossible, explain why.
    a. Addition of percent
    b. Subtraction of percent
    c. Multiplication of percent
    d. Division of percent
    e. A percent whose decimal representation is raised to the second power
    f. A percent greater than 100

14. Look at newspapers and magazines for information given in percents.
    a. Based on your findings, write a problem that involves social science as well as mathematics.
    b. Write a clear solution to your problem in (a).

15. Write a percentage problem whose answer is the solution of each of the following equations:
    a. $\dfrac{37}{100} = \dfrac{115}{x}$
    b. $\dfrac{p}{100} = \dfrac{a}{x}$

16. The effect of depreciation can be computed using a formula similar to the formula for compound interest.
    a. Assume depreciation is the same each month. Write a problem involving depreciation and solve it.
    b. Develop a general formula for depreciation defining what each variable in the formula stands for.

17. Find four large cities around the world and an approximate percentage rate of population growth for the cities. Estimate the population in each of the four cities in 25 yr.

18. State different situations that do not involve money in which a formula like the one for compound interest is used. In each case, state a related problem and write its solution.

Cooperative Learning

19. Find the percentage of students in your class who engage in each of the following activities:

    a. **Studying and Doing Homework**

Number of Hours per Week (h)	Percent
$h < 1$	
$1 \leq h < 3$	
$3 \leq h < 5$	
$5 \leq h < 10$	
$h \geq 10$	
Total	

    b. **Watching TV**

Number of Hours per Week (h)	Percent
$h < 1$	
$1 \leq h < 5$	
$5 \leq h < 10$	
$h \geq 10$	
Total	

    c. Did your totals add up to 100% in each table? Why or why not?

20. The federal Truth in Lending Act, passed in 1969, requires lending institutions to quote an annual percentage rate (APR) that helps consumers compare the true cost of loans regardless of how each lending institution computes the interest and adds on costs.
    a. Call different banks and ask for their APR on some loans and the meaning of APR.
    b. Based on your findings in (a), write a definition of APR.
    c. Use the information given by your credit card (you may need to call the bank) and compute the APR on cash advances. Is your answer the same as that given by the bank? Compare the APR for different credit cards.

Questions from the Classroom

21. A student says that $3\frac{1}{4}\% = 0.03 + 0.25 = 0.28$. Is this correct? Why?

22. A student argues that a $p\%$ increase in salary followed by a $q\%$ decrease is equivalent to a $q\%$ decrease followed by a $p\%$ increase because of the commutative property of multiplication. How do you respond?

23. A student argues that $0.01\% = 0.01$ because in 0.01%, the percent is already written as a decimal. How do you respond?

24. Noel read that women used to make 75¢ for every dollar that men made. She says that this means that men were paid 25% more than women. Is she correct? Why?

25. A student claims that if the value of an item increases by 100% each year from its value the previous year and if the original price is d dollars, then the value after n years will be $d \cdot 2^n$ dollars. Is the student correct? Why or why not?

Review Problems

26. a. Human bones make up 0.18 of a person's body weight. How much do the bones of a 120 lb person weigh?

b. Muscles make up about 0.4 of a person's body weight. How much do the muscles of a 120 lb person weigh?

27. Write each of the following decimals in the form $\frac{a}{b}$, where $a, b \in I, b \neq 0$:

a. 16.72
b. 0.003
c. ⁻5.07
d. 0.123

28. Write a repeating decimal equal to each of the following without using more than one zero:

a. 5
b. 5.1
c. $\frac{1}{2}$

29. Write 0.00024 as a fraction in simplest form.

30. Write $0.\overline{24}$ as a fraction in simplest form.

31. Write each of the following as a standard numeral:
a. $2.08 \cdot 10^5$ **b.** $3.8 \cdot 10^{-4}$

National Assessment of Educational Progress (NAEP) Question

Term	1	2	3	4
Fraction	1/2	2/3	3/4	4/5

If the list of fractions above continues in the same pattern, which term will be equal to 0.95?

a. The 100th
b. The 95th
c. The 20th
d. The 19th
e. The 15th

NAEP, Grade 8, 2003

There were 90 employees in a company last year. This year the number of employees increased by 10 percent. How many employees are in the company this year?

a. 9
b. 81
c. 91
d. 99
e. 100

NAEP, Grade 8, 2005

Trends in Mathematics and Science Study (TIMSS) Questions

Experts say that 25% of all serious bicycle accidents involve head injuries and that, of all head injuries, 80% are fatal. What percentage of all serious bicycle accidents involve fatal head injuries?

a. 16%
b. 20%
c. 55%
d. 105%

TIMSS, Grade 8, 1995

Last year there were 1172 students at Beaton High School. This year there are 15 percent more students than last year. Approximately how many students are at Beaton High School this year?

a. 1800
b. 1600
c. 1500
d. 1400
e. 1300

TIMSS, Grade 8, 1995

 **BRAIN TEASER** The crust of a certain pumpkin pie is 25% of the pie. By what percent should the amount of crust be reduced in order to make it constitute 20% of the pie?

 **Hint for Solving the Preliminary Problem**

If the wholesale price of the item is p, then the retail price is 130% of p. Compare the wholesale price to the discounted price of the item.

Chapter 7 Summary

7-1 Introduction to Decimals	Pages

Any number represented using place value and powers of 10 is a **decimal**.

- Each place of a decimal may be named by its power of 10. | 333
- Decimals can be written in expanded form. | 333

  For example: | 333

$$12.618 = 1 \cdot 10 + 2 \cdot 10^0 + 6 \cdot 10^{-1} + 1 \cdot 10^{-2} + 8 \cdot 10^{-3}$$

The dot in 12.618 is the **decimal point**. | 333

- Dividing or multiplying a **terminating** decimal by 10^n where n is a positive integer | 337
  - To divide a decimal by 10^n, start at the decimal point, count n place values left—annexing zeros if necessary—and insert the decimal point to the left of the nth place value counted.
  - To multiply a decimal by 10^n, follow the above direction, replacing "left" with "right."

- **Theorem:** A rational number $\frac{a}{b}$ in simplest form can be written as a terminating decimal if, and only if, the prime factorization of the denominator contains no primes other than 2 or 5. | 338

- Ordering terminating decimals | 339
  - To compare two decimals, we write them with a common denominator and compare the numerators.
  - To compare two decimals without converting them to rational numbers in the form $\frac{a}{b}$ where a and b are integers and $b \neq 0$, we first line up the numbers by place value. Next we start at the left and find the first place where the face values are different. Compare these digits. The digit with the greater face value in this place represents the greater of the two numbers.

7-2 Operations on Decimals

Algorithm for addition and subtraction of terminating decimals can be developed using base ten blocks. | 342

Addition and subtraction of terminating decimals can be performed by writing the decimals in expanded notation and performing addition or subtraction of fractions. | 343

Addition and subtraction of terminating decimals can be performed by writing each decimal as a rational number with common denominators that are powers of 10. | 343

Addition and subtraction of terminating decimals can be performed by lining up the columns and using the usual addition or subtraction algorithm for whole numbers. | 343

Multiplication can be performed by writing each decimal as a fraction. | 343

An algorithm for multiplying terminating decimals | 343

- If there are n digits to the right of the decimal point in one number and m digits to the right of the decimal point in a second number, multiply the two numbers, ignoring the decimals, and then place the decimal point so that there are $n + m$ digits to the right of the decimal point in the product.

Estimations with percents can be used to check if an answer is reasonable.	376
Computing Interest	377
• The original amount deposited is the **principal**.	377
• **Interest** is the amount of money the bank will pay or charge.	377
• Interest computed on the original principal is **simple interest**.	377
o **Simple interest** is computed using the formula $I = Prt$, where I is the interest, P is the principal, r is the annual interest rate, and t is the time in years.	377
• **Compound interest** is interest calculated on both the principal and the accrued interest.	378
o Compound interest is computed using the formula $A = P(1 + r)^n$, where A is the balance, P is the principal, r is the interest rate per period, and n is the number of periods.	379

Chapter 7 Review

1. **a.** On the number line, find the decimals that correspond to points A, B, and C.
   b. Indicate by D the point that corresponds to 0.09 and by E the point that corresponds to 0.15.

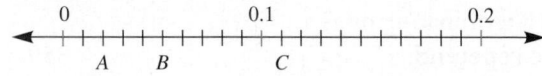

2. Write each of the following as a rational number in the form $\frac{a}{b}$, where a and b are integers and $b \neq 0$:
   a. 32.012   **b.** 0.00103

3. Describe a test to determine whether a fraction can be written as a terminating decimal without actually performing the division.

4. A board is 442.4 cm long. How many shelves can be cut from it if each shelf is 55.3 cm long? (Disregard the width of the cuts.)

5. Write each of the following as a decimal:
   a. $\frac{4}{7}$ **b.** $\frac{1}{8}$ **c.** $\frac{2}{3}$ **d.** $\frac{5}{8}$

6. Write each of the following as a fraction in simplest form:
   a. 0.28   **b.** ⁻6.07   **c.** $0.\overline{3}$ **d.** $2.0\overline{8}$

7. Round each of the following numbers as specified:
   a. 307.625 to the nearest hundredth
   b. 307.625 to the nearest tenth
   c. 307.625 to the nearest unit
   d. 307.625 to the nearest hundred

8. Rewrite each of the following in scientific notation:
   a. 426,000
   b. $324 \cdot 10^{-6}$

   c. 0.00000237
   d. ⁻0.325

9. Order the following decimals from greatest to least:
   $1.4\overline{519}$, $1.45\overline{19}$, 1.4519, $1.45\overline{19}$,   ⁻0.134,   ⁻0.13$\overline{401}$,   0.13$\overline{401}$

10. Write each of the following in scientific notation without using a calculator:
   a. 1783411.56
   b. $\dfrac{347}{10^8}$
   c. $49.3 \cdot 10^8$
   d. $29.4 \cdot \dfrac{10^{12}}{10^{-4}}$
   e. $0.47 \cdot 1000^{12}$
   f. $\dfrac{3}{5^9}$

11. **a.** Find five decimals between 0.1 and 0.11 and order them from greatest to least.
    b. Find four decimals between 0 and 0.1 listed from least to greatest so that each decimal starting from the second is twice as large as the preceding one.
    c. Find four decimals between 0.1 and 0.2 and list them in increasing order so that the first one is halfway between 0.1 and 0.2, the second halfway between the first and 0.2, the third halfway between the second and 0.2, and similarly for the fourth one.

12. Answer each of the following:
    a. 6 is what percent of 24?
    b. What is 320% of 60?
    c. 17 is 30% of what number?
    d. 0.2 is what percent of 1?

13. Change each of the following to a percent:

    a. $\dfrac{1}{8}$ b. $\dfrac{3}{40}$

    c. 6.27   d. 0.0123

    e. $\dfrac{3}{2}$

14. Change each of the following percents to a decimal:

    a. 60%   b. $\dfrac{2}{3}\%$ c. 100%

15. Sandy received a dividend that equals 11% of the value of her investment. If her dividend was $1020.80, how much was her investment?

16. Five computers in a shipment of 150 were found to be defective. What percent of the computers were defective?

17. On a mathematics examination, a student missed 8 of 70 questions. What percent of the questions, rounded to the nearest tenth of a percent, did the student answer correctly?

18. A used car costs $3450 at present. This is 60% of the cost 4 yr ago. What was the cost of the car 4 yr ago?

19. If, on a purchase of one new suit, you are offered successive discounts of 5%, 10%, or 20% in any order you wish, what order should you choose?

20. Jane bought a bicycle and sold it for 30% more than she paid for it. She sold it for $104. How much did she pay for it?

21. The student bookstore had a textbook for sale at $89.95. A student found the book on eBay for $62.00. If the student bought the book on eBay, what percentage of the cost of the bookstore book did she save?

22. When a store had a 60% off sale, Dori had a coupon for an additional 40% off any item and thought she should be able to obtain the dress that she wanted for free. If you were the store manager, how would you explain the mathematics of the situation to her?

23. A company was offered a $30,000 loan at a 12.5% annual simple interest rate for 4 yr. Find the simple interest due on the loan at the end of 4 yr.

24. A fund pays 14% annual interest compounded quarterly. What is the value of a $10,000 investment after 3 yr?

Real Numbers and Algebraic Thinking

Preliminary Problem

As countries develop high-speed trains, rail issues that have plagued engineers for years are arising. To avoid sections of rail connected with large expansion joints of the past, rails are welded together in long sections. If not restrained, rails develop sun kink; that is, they lengthen in hot weather and shrink in cold weather. Clips anchor the welded rails to concrete or steel sleepers (the newer version of cross ties). Without these clips, a one-mile length of track might expand 2 ft and buckle in an arc. Estimate the height of the buckle.

If needed, see Hint on page 465.

The development of number systems was inspired by such commercial uses as negative numbers to indicate debts, ways to expand knowledge such as the use of rational numbers to think about parts of a whole, and base ten decimals to think about very small numbers related to fractions. The development of real numbers was comparable. The Pythagoreans knew a right triangle with two sides of length 1 had to have a new type of number to describe the length of the third side. Ancient Greeks struggled with finding a number to describe the ratio of the circumference of a circle to its diameter. These examples involve irrational numbers, but the formal definition of a real number came long after such numbers began to be used.

The grade 8 *Common Core Standards* recognize that students need to know that "there are numbers that are not rational, and [how to] approximate them by rational numbers." (p. 54) An example of the use of real (and, in this case, rational) numbers is seen in the *FoxTrot* cartoon below.

FoxTrot copyright © 2002 and 2011 Bill Amend. Reprinted with permission of Universal Uclick. All rights reserved.

In this chapter we explore real numbers and examine aspects of algebra including the analysis and solution of linear equations, pairs of simultaneous linear equations, and other algebraic notions including functions, and the Cartesion Coordinate system. This study is reflected in the grade 8 *Common Core Standards* in the following:

- Work with radicals and integer exponents
- Analyze and solve linear equations and pairs of simultaneous linear equations. (p. 53)

8-1 Real Numbers

In Chapter 7, we saw that every rational number can be expressed either as a repeating decimal or as a terminating decimal. But the ancient Greeks discovered numbers that are not rational; that is, numbers with a decimal representation that neither terminates nor repeats. To find such decimals, consider the characteristics they must have:

1. There must be an infinite number of nonzero digits to the right of the decimal point.
2. There cannot be a repeating block of digits (a *repetend*).

◯- Historical Note

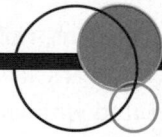

Although the Pythagoreans seem to have known about some irrational numbers, it was not until 1872 that German mathematician Richard Dedekind introduced a method of constructing irrational numbers from sets of rational numbers by showing that an irrational number could be thought of as a cut in a number line that separates the set of rational numbers into two disjoint non-empty subsets, one of which has no least element and the other of which has no greatest element, with all elements of one set less than the elements of the other set.

One way to construct a nonterminating, nonrepeating decimal is to devise a pattern of infinite digits so that there is no repetend. For example, consider the number 0.1010010001.... If the pattern continues, the next groups of digits are four zeros followed by 1, five zeros followed by 1, and so on. Because this decimal is nonterminating and nonrepeating, it cannot represent a rational number. Numbers that are not rational numbers are **irrational numbers**, and there are infinitely many of them. In the mid-eighteenth century, it was proved that the ratio of the circumference of a circle to its diameter, symbolized by π **(pi)**, is an irrational number. The numbers $\frac{22}{7}$, 3.14, or 3.14159 are rational number approximations of π. The value of π has been computed to over 5 trillion decimal places and the digits appear to be essentially random.

The set of real numbers includes both rational and irrational numbers.

Definition of Real Numbers

A **real number** is any number, either rational or irrational, that can be written as a decimal.

Square Roots

Irrational numbers occur in the study of area. For example, to find the area of a square as in Figure 8-1(a), we could use the formula $A = s^2$, where A is the area and s is the length of a side of the square. If a side of a square is 3 cm long as in Figure 8-1(b), then the area of the square is 9 cm^2 (square centimeters). Conversely, we can use the formula to find the length s of a side of a square, given its area. If the area of a square is 25 cm^2 as in Figure 8-1(c), then

$$s^2 = 25$$
$$s = 5 \text{ or } s = {}^-5$$

Each of these solutions is a **square root** of 25. However, because lengths are always non-negative, 5 is the only possible solution. The positive solution of $s^2 = 25$ (namely, 5) is the **principal square root**

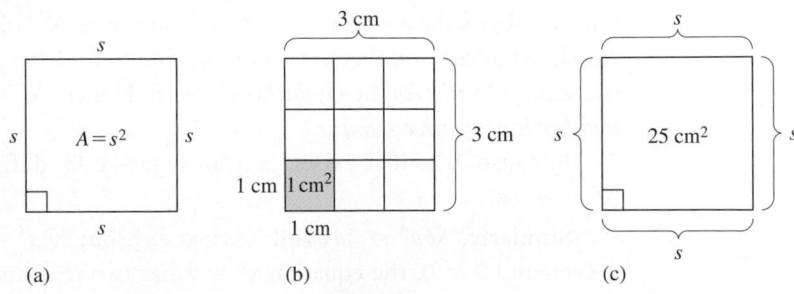

(a)   (b)   (c)

Figure 8-1

Historical Note

In 2010, Shigeru Kondo used Alexander J. Yee's y-cruncher to calculate the value of π to over 5 trillion decimal places. The primary computation took 90 days on Kondo's desktop computer.

of 25 and is denoted $\sqrt{25}$. Similarly, the principal square root of 2 is denoted $\sqrt{2}$. Note that $\sqrt{16} \neq {}^-4$ because $^-4$ is not the principal square root of 16. Can you find $\sqrt{0}$?

> **Definition of the Principal Square Root**
>
> If a is any non-negative number, the **principal square root** of a (denoted $\sqrt{a}$) is the non-negative number b such that $b^2 = a$.

 **EXAMPLE 8-1**

Find the following:

a. The square roots of 144
b. The principal square root of 144
c. $\sqrt{\dfrac{4}{9}}$

Solution

a. The square roots of 144 are 12 and $^-12$.
b. The principal square root of 144 is 12.
c. $\sqrt{\dfrac{4}{9}} = \dfrac{2}{3}$

Other Roots

We have seen that the positive solution to $s^2 = 25$ is denoted $\sqrt{25}$. Similarly, the positive solution to $s^4 = 25$ is denoted $\sqrt[4]{25}$. In general, if n is even, the positive solution to $x^n = 25$ is $\sqrt[n]{25}$ and is the principal ***n*th root** of 25. The number n is the **index**. Note that in the expression $\sqrt{25}$, the index 2 is understood and not expressed. In general, *the positive solution to* $x^n = b$, *where b is non-negative, is* $\sqrt[n]{b}$.

Substituting $\sqrt[n]{b}$ for x in the equation $x^n = b$ gives the following:

$$(\sqrt[n]{b})^n = b$$

If b is negative, $\sqrt[n]{b}$ may not be a real number. For example, consider $\sqrt[4]{^-16}$. If $\sqrt[4]{^-16} = x$, then $x^4 = {}^-16$. Because any nonzero real number raised to the fourth power is positive, there is no real-number solution to $x^4 = {}^-16$ and therefore $\sqrt[4]{^-16}$ is not a real number. Similarly, if we are restricted to real numbers, it is not possible to find *any* even root of a negative number. However, the value $^-2$ satisfies the equation $x^3 = {}^-8$. Hence, $\sqrt[3]{^-8} = {}^-2$. *In general, the odd root of a negative number is a negative number.*

Because $\sqrt{a}$, if it exists, is non-negative by definition, $\sqrt{(^-3)^2} = \sqrt{9} = 3$. In general, $\sqrt{a^2} = |a|$.

Similarly, $\sqrt[4]{a^4} = |a|$ and $\sqrt[6]{a^6} = |a|$, but $\sqrt[3]{a^3} = a$ for all a. (Why?) Notice that when n is even and $b > 0$, the equation $x^n = b$ has two real-number solutions, $\sqrt[n]{b}$ and $^-\sqrt[n]{b}$. If n is odd, the equation has only one real-number solution, $\sqrt[n]{b}$, for any real number b.

Historical Note

Evaluating square roots may have been known by Vedic Hindu scholars before 600 BCE. The *Sulbasutra* (rule of chords), Sanskrit texts, contain approximations for some square roots that are incredibly accurate. The discovery of irrational numbers by members of the Pythagorean Society was so disturbing that they decided to keep the matter secret. Legend has it that a society member was drowned because he relayed the secret to persons outside the society. In 1525, Christoff Rudolff, a German mathematician, became the first to use the symbol $\sqrt{}$ for a radical or a root.

Irrationality of Square Roots and Other Roots

Some square roots are rational numbers. Others, like $\sqrt{2}$, are irrational numbers. To see this, we note that $1^2 = 1$ and $2^2 = 4$ and that there is no whole number s such that $s^2 = 2$. Is there a rational number $\dfrac{a}{b}$ such that $\left(\dfrac{a}{b}\right)^2 = 2$? To decide, we use the strategy of *indirect reasoning*. If we assume there is such a rational number $\dfrac{a}{b}$, then the following must be true:

$$\left(\frac{a}{b}\right)^2 = 2$$

$$\frac{a^2}{b^2} = 2$$

$$a^2 = 2b^2$$

If $a^2 = 2b^2$, then by the Fundamental Theorem of Arithmetic, the prime factorizations of a^2 and $2b^2$ are the same. In particular, the prime 2 appears the same number of times in the prime factorization of a^2 as it does in the factorization of $2b^2$. Because $b^2 = bb$, no matter how many times 2 appears in the prime factorization of b, it appears twice as many times in bb.

Also, a^2 has an even number of 2s for the same reason b^2 does. In $2b^2$, another factor of 2 is introduced, resulting in an odd number of 2s in the prime factorization of $2b^2$. Because $a^2 = 2b^2$, we have 2 appearing both an odd number of times and an even number of times on different sides of the equality and thus a contradiction.

This contradiction could have been caused only by the assumption that $\sqrt{2}$ is a rational number. Consequently, $\sqrt{2}$ must be an irrational number. We can use a similar argument to show that $\sqrt{3}$ is irrational or $\sqrt{n}$ is irrational, where n is a whole number but not the square of another whole number.

Many irrational numbers can be interpreted geometrically. For example, we can find a point on a number line to represent $\sqrt{2}$ by using the **Pythagorean Theorem**. That is, if a and b are the lengths of the shorter sides (legs) of a right triangle and c is the length of the longer side (hypotenuse), then $a^2 + b^2 = c^2$, as shown in Figure 8-2.

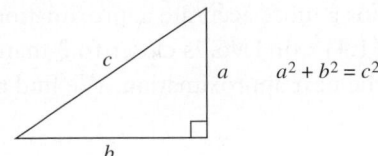

$$a^2 + b^2 = c^2$$

Figure 8-2

Figure 8-3 shows a segment 1 unit long constructed perpendicular to a number line at point P. Thus two sides of the triangle shown are each 1 unit long. If $a = b = 1$, then $c^2 = 2$ and $c = \sqrt{2}$. To find a point on the number line that corresponds to $\sqrt{2}$, we need to find a point Q on the number line such that the distance from 0 to Q is $\sqrt{2}$. Because $\sqrt{2}$ is the length of the hypotenuse, the point Q can be found by marking an arc with center 0 and radius c. The intersection of the positive number line with the arc is point Q.

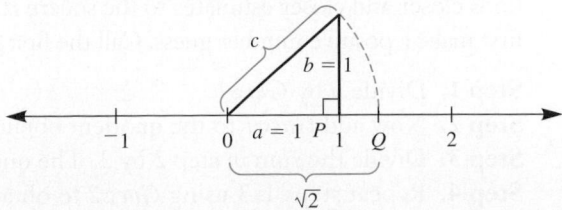

Figure 8-3

Similarly, other principal square roots can be constructed, as shown in Figure 8-4 and placed on a number line using the method of Figure 8-3. Also their opposites can then be constructed as in Figure 8-4.

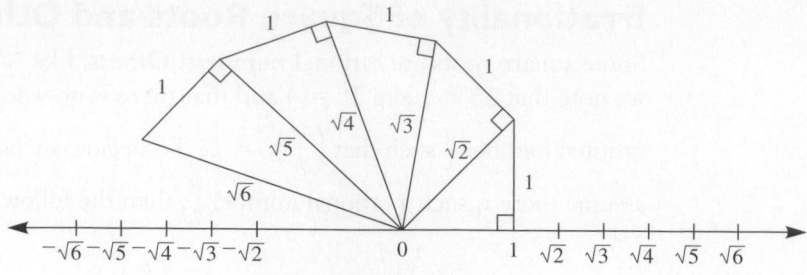

Figure 8-4

Estimating a Square Root

From Figure 8-4, we see that $\sqrt{2}$ must have a value between 1 and 2; that is, $1 < \sqrt{2} < 2$. To obtain a closer approximation of $\sqrt{2}$, we attempt to "squeeze" $\sqrt{2}$ between two numbers that are between 1 and 2. Because $(1.4)^2 = 1.96$ and $(1.5)^2 = 2.25$, it follows that $1.4 < \sqrt{2} < 1.5$. Because a^2 can be interpreted as the area of a square with side of length a, this discussion can be pictured geometrically, as in Figure 8-5.

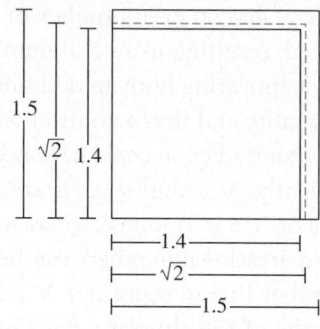

Figure 8-5

For a more accurate approximation for $\sqrt{2}$, we can continue this squeezing process. We see that $(1.4)^2$, or 1.96, is closer to 2 than is $(1.5)^2$, or 2.25, so we choose numbers closer to 1.4 to find the next approximation. We find the following:

$$(1.42)^2 = 2.0164$$
$$(1.41)^2 = 1.9981.$$

Thus, $1.41 < \sqrt{2} < 1.42$. We can continue this process until we obtain the desired approximation. If a calculator has a square-root key, we can obtain an approximation directly.

NOW TRY THIS 8-1

One algorithm for calculating square roots is sometimes attributed to Archimedes though the Babylonians had a similar method. The method is also referred to as Newton's method. The algorithm finds closer and closer estimates to the square root. To find the square root of a positive number, n, first make a positive number guess. Call the first guess *Guess1*. Now compute as follows:

Step 1. Divide n by *Guess1*.
Step 2. Now add *Guess1* to the quotient obtained in step 1.
Step 3. Divide the sum in step 2 by 2. The quotient becomes *Guess2*.
Step 4. Repeat steps 1–3 using *Guess2* to obtain successive guesses or until the desired accuracy is achieved.

   a. Use the method described to find the square root of 13 to four decimal places.
   b. Write the steps for the algorithm in a recursive formula.

The System of Real Numbers, Operations, and Their Properties

The **set of real numbers**, *R, is the union of the set of rational numbers and the set of irrational numbers.* Real numbers represented as decimals can be terminating, repeating, or nonterminating and nonrepeating.

Every integer is a rational number as well as a real number. Every rational number is a real number, but not every real number is rational, as has been shown with $\sqrt{2}$. The relationships among sets of numbers are summarized in the diagram in Figure 8-6.

$$\text{Reals} = \{x\,|\,x \text{ is a decimal}\}$$

Irrationals $= \{x\,|\,x$ is a decimal that neither repeats nor terminates$\}$

$$\text{Rationals} = \left\{x\,\middle|\,x = \frac{a}{b}, \text{ where } a, b \in I \text{ and } b \neq 0\right\}$$
$$= \{x\,|\,x \text{ is a repeating or terminating decimal}\}$$

Integers $= \{\ldots, ^-3, ^-2, ^-1, 0, 1, 2, 3, \ldots\}$

Whole Numbers $= \{0, 1, 2, 3, 4, \ldots\}$

Natural Numbers $= \{1, 2, 3, 4, 5, \ldots\}$

Figure 8-6

The set of *fractions can now be extended to include all real numbers of the form $\frac{a}{b}$, where a and b are real numbers with $b \neq 0$, such as $\frac{\sqrt{3}}{5}$.* The following partial student page depicts fractions differently than we do in this text. From this page, would $\sqrt{\frac{1}{5}}$ be considered a fraction? Consider the question in Now Try This 8-2.

School Book Page

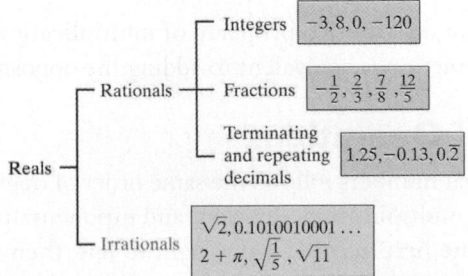

Irrational numbers are numbers that cannot be written in the form $\frac{a}{b}$, where a is any integer and b is any nonzero integer. Rational and irrational numbers form the set of **real numbers.**

The diagram below shows the relationships among sets of numbers.

> Reals
> — Rationals
>   — Integers $-3, 8, 0, -120$
>   — Fractions $-\frac{1}{2}, \frac{2}{3}, \frac{7}{8}, \frac{12}{5}$
>   — Terminating and repeating decimals $1.25, -0.13, 0.\overline{2}$
> — Irrationals $\sqrt{2}, 0.1010010001\ldots$ $2 + \pi, \sqrt{\frac{1}{5}}, \sqrt{11}$

The decimal digits of irrational numbers do not terminate or repeat. The decimal digits of $\pi = 3.14159265359 \ldots$ do not terminate or repeat, because π is an irrational number. Irrational numbers can also include decimals that have a pattern in their digits, like $0.02022022202222 \ldots$

For any integer n that is not a perfect square, $\sqrt{n}$ is irrational.

NOW TRY THIS 8-2

On the student page (p. 397), fractions are given as a subset of the set of rational numbers. $\sqrt{\dfrac{1}{5}}$ appears under the set of irrational numbers. This is equivalent to $\dfrac{\sqrt{1}}{\sqrt{5}}$, or $\dfrac{1}{\sqrt{5}}$. Explain whether $\dfrac{1}{\sqrt{5}}$ represents a fraction:

   a. Under the definition given on the student book page.
   b. Under the definition given in this text.

Addition, subtraction, multiplication, and division are defined on the set of real numbers in such a way that all the properties of these operations on rationals still hold. The properties are summarized next.

Theorem 8-1:   **Properties of Real Numbers under Addition and Multiplication and the Denseness Property**

Closure properties   For real numbers a and b, $a + b$ and ab are unique real numbers.

Commutative properties   For real numbers a and b, $a + b = b + a$ and $ab = ba$.

Associative properties   For real numbers a, b, and c, $a + (b + c) = (a + b) + c$ and $a(bc) = (ab)c$.

Identity properties   The number 0 is the unique additive identity and 1 is the unique multiplicative identity such that, for any real number a, $0 + a = a = a + 0$ and $1(a) = a = a(1)$.

Inverse properties
(a) For every real number a, ^{-}a is its unique additive inverse; that is, $a + {}^{-}a = 0 = {}^{-}a + a$.
(b) For every nonzero real number a, $\dfrac{1}{a}$ (or a^{-1}) is its unique multiplicative inverse; that is, $a\left(\dfrac{1}{a}\right) = 1 = \left(\dfrac{1}{a}\right)a$.

Distributive property of multiplication over addition   For real numbers a, b, and c, $a(b + c) = ab + ac$.

Multiplication property of zero   For all real numbers a, $a \cdot 0 = 0 = 0 \cdot a$.

Denseness property for real numbers   For real numbers a and b with $a < b$, there exists a real number c such that $a < c < b$.

Note that the distributive property of multiplication over subtraction could be included, but because subtraction is equivalent to adding the opposite, it is not listed separately.

Order of Operations

The set of real numbers follows the same order of operations given earlier; that is, when addition, subtraction, multiplication, division, and exponentiation appear without parentheses, exponentiation is done first in order from right to left, then multiplications and divisions in the order of their appearance from left to right, and then additions and subtractions in the order of their

---

Historical Note

By the denseness property, we can always find a real number between any two other real numbers. This leads to infinitely many real numbers. In 1874 Georg Cantor showed that the set of all real numbers cannot be put in a one-to-one correspondence with the set of natural numbers and in the process continued to compare infinities.

appearance from left to right. Arithmetic operations that appear inside parentheses must be done first. Examine the implied and real order of operations in Now Try This 8-3.

Radicals and Rational Exponents

Scientific calculators have a $\boxed{y^x}$ key with which we can find the values of expressions like $3.41^{2/3}$ and $4^{1/2}$. What does $4^{1/2}$ mean? By extending the properties of exponents previously developed for integer exponents,

$$4^{1/2} \cdot 4^{1/2} = 4^{(1/2 + 1/2)}.$$

Thus,

$$(4^{1/2})^2 = 4^1 = 4$$

and

$$4^{1/2} = \sqrt{4}.$$

The number $4^{1/2}$ is assumed to be the principal square root of 4, that is, $4^{1/2} = \sqrt{4}$.
In general, if x is a non-negative real number, then $x^{1/2} = \sqrt{x}$. Similarly, $(x^{1/3})^3 = x^{(1/3)3} = x^1$ and $x^{1/3} = \sqrt[3]{x}$. This discussion leads to the following:

1. $x^{1/n} = \sqrt[n]{x}$, when $\sqrt[n]{x}$ is meaningful.
2. $(x^m)^{1/n} = \sqrt[n]{x^m}$ if $\text{GCD}(m, n) = 1$.
3. $x^{m/n} = \sqrt[n]{x^m}$, if $\text{GCD}(m, n) = 1$.

 NOW TRY THIS 8-3

In the *FoxTrot* cartoon in the chapter introduction, Jason has engineered the answering machine to give a message to "press the square root of 1296 minus the cube root of 13,824 times 17.5 minus the 4th root of 1,908,029,761." Answer the following:

a. Typically, does an answering machine say press some number to leave a message?
b. Is there a corresponding number on the telephone to the one Jason's engineered answering machine asks the caller to press to leave a message? Use estimation skills to decide the answer here.
c. Use a correct order of operations to determine the answer to the machine's computation.
d. If we perform the computations exactly in the order given on the machine, what number should be pressed?
e. Use the answers in parts (c) and (d) to explain whether there was a correct order of operations followed for the arithmetic tasks Jason has engineered.

More about Properties of Exponents

It can be shown that the properties of integer exponents also hold for rational exponents. These properties are equivalent to the corresponding properties of radicals *if the expressions involving radicals are meaningful.*

Let r and s be any rational numbers, x and y be any real numbers, and n be any nonzero integer. Restrictions must be placed on each of the following to make some expressions meaningful.

a. $x^{-r} = \dfrac{1}{x^r}$.

b. $(xy)^r = x^r y^r$ implies $(xy)^{1/n} = x^{1/n} y^{1/n}$ and $\sqrt[n]{xy} = \sqrt[n]{x}\sqrt[n]{y}$.

c. $\left(\dfrac{x}{y}\right)^r = \dfrac{x^r}{y^r}$ implies $\left(\dfrac{x}{y}\right)^{1/n} = \dfrac{x^{1/n}}{y^{1/n}}$ and $\sqrt[n]{\dfrac{x}{y}} = \dfrac{\sqrt[n]{x}}{\sqrt[n]{y}}$.

d. $(x^r)^s = x^{rs}$ implies $(x^{1/n})^p = x^{p/n}$ and hence $(\sqrt[n]{x})^p = \sqrt[n]{x^p}$.

The preceding properties can be used to write equivalent expressions for the roots of many numbers. For example, $\sqrt{96} = \sqrt{16 \cdot 6} = \sqrt{16}\sqrt{6} = 4\sqrt{6}$. Similarly, $\sqrt[3]{54} = \sqrt[3]{27 \cdot 2} = \sqrt[3]{27}\sqrt[3]{2} = 3\sqrt[3]{2}$.

EXAMPLE 8-2

Simplify each of the following if possible:

a. $16^{1/4}$ **b.** $16^{5/4}$ **c.** $(^-8)^{1/3}$ **d.** $125^{-4/3}$ **e.** $(^-16)^{1/4}$

Solution

a. $16^{1/4} = (2^4)^{1/4} = 2^1 = 2$, or $16^{1/4} = \sqrt[4]{16} = 2$

b. $16^{5/4} = 16^{(1/4)5} = (16^{1/4})^5 = 2^5 = 32$

c. $(^-8)^{1/3} = ((^-2)^3)^{1/3} = (^-2)^1 = ^-2$ or $(^-8)^{1/3} = \sqrt[3]{^-8} = ^-2$

d. $125^{-4/3} = (5^3)^{-4/3} = 5^{-4} = \dfrac{1}{5^4} = \dfrac{1}{625}$

e. Because every real number raised to the fourth power is positive, $\sqrt[4]{^-16}$ is not a real number. Consequently, $(^-16)^{1/4}$ is not a real number. A simplification is not possible.

NOW TRY THIS 8-4

Compute $\sqrt[8]{10}$ on a calculator using the following sequence of keys:

$$\boxed{10}\ \boxed{\sqrt{\ }}\ \boxed{\sqrt{\ }}\ \boxed{\sqrt{\ }}$$

a. Explain why this approach works.

b. For what values of n can $\sqrt[n]{10}$ be computed using only the $\boxed{\sqrt{\ }}$ key? Why?

NOW TRY THIS 8-5

The properties of integer exponents were extended to real number exponents. Consider a base with a decimal exponent such as $8^{0.\overline{3}}$ or $8^{0.101001\cdots}$. Write an explanation for a possible meaning of these decimal exponents.

Assessment 8-1A

1. Write an irrational number whose digits are 2s and 3s.
2. Use the Pythagorean Theorem to find x.

   a.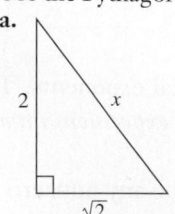

   b.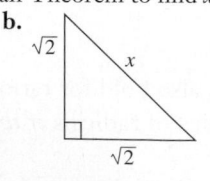

   c.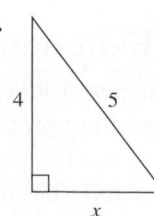

3. Arrange the following real numbers in order from greatest to least:

   $0.9, \quad 0.\overline{9}, \quad 0.\overline{98}, \quad 0.9\overline{88}, \quad 0.9\overline{98}, \quad 0.\overline{898}, \quad \sqrt{0.98}$

4. Determine which of the following represent irrational numbers:

   a. $\sqrt{51}$ **b.** $\sqrt{64}$ **c.** $\sqrt{324}$
   d. $\sqrt{325}$ **e.** $2 + 3\sqrt{2}$ **f.** $\sqrt{2} \div 5$

5. If possible, find the principal square roots for each of the following without using a calculator:

   a. 225   **b.** 169
   c. $^-81$ **d.** 625

6. Find the approximate square roots for each of the following, rounded to hundredths, by using the squeezing method:

   a. 7   **b.** 0.0120

7. Classify each of the following as true or false. If false, give a counterexample.

   a. The sum of any rational number and any irrational number is a rational number.

   b. The sum of any two irrational numbers is an irrational number.

   c. The product of any two irrational numbers is an irrational number.

   d. The difference of two irrational numbers could be a rational number.

8. Find three irrational numbers between each of the following pairs:
   a. 1 and 3
   b. $0.\overline{53}$ and $0.\overline{54}$
   c. 0.5 and 0.6

9. Based on your answer in exercise 8, argue that there are infinitely many irrational numbers.

10. If R is the set of real numbers, Q is the set of rational numbers, I is the set of integers, W is the set of whole numbers, and S is the set of irrational numbers, find each of the following:
    a. $Q \cup S$
    b. $Q \cap S$
    c. $Q \cap R$
    d. $S \cap W$
    e. $W \cup R$
    f. $Q \cup R$

11. If the following letters correspond to the sets listed in exercise 10, complete the table by placing checkmarks in the appropriate columns. (N is the set of natural numbers.)

	N	I	Q	R	S
a. 6.7			✔	✔	
b. 5					
c. $\sqrt{2}$					
d. $^-5$					
e. $3\frac{1}{7}$					

12. If the following letters correspond to the sets listed in exercise 10, put a checkmark under each set of numbers for which a solution to the problem exists. (N is the set of natural numbers.)

	N	I	Q	R	S
a. $x^2 + 1 = 5$					
b. $2x - 1 = 32$					
c. $x^2 = 3$					
d. $\sqrt{x} = ^-1$					
e. $\frac{3}{4}x = 0.\overline{4}$					

13. Determine for what real values of x, if any, each of the following statements is true:
    a. $\sqrt{x} = 8$
    b. $\sqrt{x} = ^-8$
    c. $\sqrt{^-x} = 8$
    d. $\sqrt{^-x} = ^-8$

e. $\sqrt{x} > 0$
f. $\sqrt{x} < 0$

14. Write each of the following roots in the form $a\sqrt{b}$, where a and b are integers and b has the least value possible:
    a. $\sqrt{180}$
    b. $\sqrt{363}$
    c. $\sqrt{252}$

15. Write each of the following in the simplest form or as $a\sqrt[n]{b}$, where a and b are integers, $b > 0$, and b has the least value possible:
    a. $\sqrt[3]{^-54}$
    b. $\sqrt[3]{96}$
    c. $\sqrt[3]{250}$
    d. $\sqrt[5]{^-243}$

16. If each of the following is a part of a geometric sequence, find the missing terms:
    a. 5, _____, _____, 10
    b. 2, _____, _____, _____, 1

17. A diagonal brace is placed in a 4 ft × 5 ft rectangular gate. What is the length of the brace to the nearest tenth of a foot? (*Hint:* Use the Pythagorean Theorem.)

18. The expression $2^{10} \cdot 16^t$ approximates the number of bacteria after t hr.
    a. What is the initial number of bacteria, that is, the number when $t = 0$?
    b. After $\frac{1}{4}$ hr, how many bacteria are there?
    c. After $\frac{1}{2}$ hr, how many bacteria are there?

19. Solve for x in the following, where x is a real number:
    a. $3^x = 81$
    b. $4^x = 8$
    c. $128^{^-x} = 16$
    d. $\left(\frac{4}{9}\right)^{3x} = \frac{32}{243}$
    e. $2\sqrt{3} = \sqrt{x}$

20. Classify each of the following numbers as rational or irrational:
    a. $\sqrt{2} - \frac{2}{\sqrt{2}}$
    b. $(\sqrt{2})^{^-4}$

21. Figure 8-4 showed a way to construct the square root of any positive integer and place it on a number line. Describe where $\sqrt{342}$ would be on a number line.

22. Describe how you could decide whether the following is true without using a calculator.
    $$\frac{-2}{\sqrt{3}} > \frac{-3}{\sqrt{5}}.$$

Assessment 8-1B

1. Write an irrational number whose digits are 4s and 5s.
2. Use the Pythagorean Theorem to find x.

a.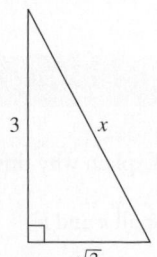

b.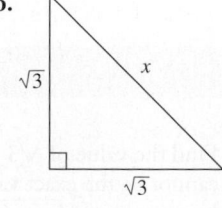

c.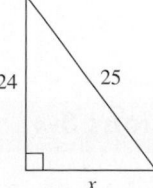

3. Arrange the following real numbers in order from greatest to least:

$$0.8, 0.\overline{8}, 0.8\overline{9}, 0.889, \sqrt{0.7744}$$

4. Determine which of the following represent irrational numbers:
   a. $\sqrt{78}$ b. $\sqrt{81}$ c. $\sqrt[3]{343}$
   d. $3 + \sqrt{81}$ e. $2 \div \sqrt{2}$
5. If possible, find whole numbers that are the square roots for each of the following without using a calculator:
   a. 256   b. 324
   c. $^-25$ d. 1024
6. Find the approximate square root for each of the following, rounded to hundredths, by using the squeezing method:
   a. 20.3   b. 1.64
7. Classify each of the following as true or false. If false, give a counterexample.
   a. The sum of any two rational numbers is a rational number.
   b. The difference of any two irrational numbers is an irrational number.
   c. The product of any rational number and any irrational number is an irrational number.
8. Find three irrational numbers between each of the following pairs:
   a. 3 and 4
   b. $7.0\overline{5}$ and $7.0\overline{6}$
   c. 0.7 and 0.8
9. Knowing that $\sqrt{2}$ is an irrational number, argue that $\dfrac{\sqrt{2}}{2}$ is also an irrational number.
10. If R is the set of real numbers, Q is the set of rational numbers, I is the set of integers, W is the set of whole numbers, N is the set of natural numbers, and S is the set of irrational numbers, answer each of the following:
    a. $Q \cap I$
    b. $S - Q$
    c. $R \cup S$
    d. Which of the sets could be a universal set for the rest of the sets?
    e. If R is the set of real numbers, how would you describe $\overline{S}$, where S is the set of irrational numbers?
11. If the following letters correspond to the sets listed in exercise 10, complete the following table by placing checkmarks in the appropriate columns:

	N	I	Q	R	S
a. $\sqrt{3}$					
b. $4\dfrac{1}{2}$					
c. $^-3\dfrac{1}{7}$					

12. If the following letters correspond to the sets listed in exercise 10, put a checkmark under each set of numbers for which a solution to the problem exists:

	N	I	Q	R	S
a. $x^2 + 2 = 4$					
b. $1 - 2x = 32$					
c. $x^3 = 4$					
d. $\sqrt{x} = ^-2$					
e. $0.\overline{7}x = 5$					

13. Determine for what real values of x, if any, each of the following statements is true:
    a. $\sqrt{x} = 7$ b. $\sqrt{x} = ^-7$
    c. $\sqrt{^-x} = 7$ d. $^-\sqrt{x} = 7$
    e. $^-\sqrt{x} = ^-7$
14. Write each of the following square roots in the form $a\sqrt{b}$, where a and b are integers and b has the least value possible:
    a. $\sqrt{360}$ b. $\sqrt{40}$ c. $\sqrt{240}$
15. Write each of the following in the simplest form or as $a\sqrt[n]{b}$, where a and b are integers and $b > 0$ and b is as small as possible:
    a. $\sqrt[3]{^-102}$ b. $\sqrt[6]{64}$ c. $\sqrt[3]{64}$
16. If each of the following is a part of a geometric sequence, find the missing terms:
    a. 4, _____, _____, 8
    b. 1, _____, _____, 2
17. If the two shorter sides of a right triangle have lengths 5 and 12, how long is the hypotenuse (the longest side)?
18. In the expression 8^t, let t represents time in hours.
    a. What is the value of the expression when $t = 0$?
    b. After $\dfrac{1}{3}$ hour, how many bacteria are there?
19. Solve for x in the following, where x is a real number:
    a. $2^x = 64$ b. $4^x = 64$
    c. $2^{-x} = 64$ d. $\sqrt{\dfrac{3x}{2}} = 36$
    e. $\dfrac{3}{\sqrt{3}} = \sqrt{x}$
20. Classify each of the following numbers as rational or irrational:
    a. $\dfrac{1}{1 + \sqrt{2}}$ b. $\dfrac{4}{\sqrt{2}} - \sqrt{2}$
21. If $a > 2$, describe where $\dfrac{1}{\sqrt{a}}$ would be located on a number line.
22. Without a calculator, describe how you would decide whether the following is true: $\dfrac{2}{\sqrt{3}} < \dfrac{3}{\sqrt{5}}$.

Mathematical Connections 8-1

Communication

1. A mathematician once described the set of rational numbers as the stars and the set of irrational numbers as the black in the night sky. What do you think the mathematician meant?

2. Find the value of $\sqrt{3}$ on a calculator. Explain why this cannot be the exact value of $\sqrt{3}$.
3. Is it true that $\sqrt{a + b} = \sqrt{a} + \sqrt{b}$ for all a and b? Explain.

4. Pi (π) is an irrational number. Could $\pi = \dfrac{22}{7}$? Why or why not?

5. Without using a calculator or doing any computation, determine if $\sqrt{13} = 3.60\overline{5}$. Explain why or why not.

6. Without using a calculator, order the following. Explain your reasoning.

$$(4/25)^{-1/3}, \ (25/4)^{1/3}, \ (4/25)^{-1/4}$$

Open-Ended

7. The sequence 1, 1.01, 1.001, 1.0001, . . . is an infinite sequence of rational numbers.
   a. Write several other infinite sequences of rational numbers.
   b. Write an infinite sequence of irrational numbers.

8. a. Place five irrational numbers between $\dfrac{1}{2}$ and $\dfrac{3}{4}$.
   b. Write an infinite sequence of irrational numbers all of whose terms are between $\dfrac{1}{2}$ and $\dfrac{3}{4}$.

Cooperative Learning

9. Let each member of a group choose a number between 0 and 1 on a calculator and check what happens when the $\boxed{x^2}$ key is pressed in succession until it is clear that there is no reason to go on.
   a. Compare your answers and write a conjecture based on what you observe.
   b. Use other keys on the calculator in a similar way. Describe the process and state a corresponding conjecture.
   c. Why do you get the result you do in parts (a) and (b)?

10. A calculator displays the following: $(3.7)^{2.4} = 23.103838$. In your group, discuss the meaning of the expression $(3.7)^{2.4}$ in view of what you know about exponents. Compare your findings with those of other groups.

Questions from the Classroom

11. Jim asked, if $\sqrt{2}$ can be written as $\dfrac{\sqrt{2}}{1}$, why is it not rational? How would you answer him?

12. Maria says that $1 + \sqrt{2}$ is not a number because the sum cannot be completed. How do you respond?

13. A student claims that $5 = {}^{-}5$ because both are square roots of 25. How do you help her?

14. Jose says that the equation $\sqrt{{}^{-}x} = 3$ has no solution, since the square root of a negative number does not exist. How would you help him?

15. A student argues that in the real world, irrational numbers are never used. How do you respond?

National Assessment of Educational Progress (NAEP) Question

Term	1	2	3	4
Fraction	1/2	2/3	3/4	4/5

If the list of fractions above continues in the same pattern, which term will be equal to 0.95?
a. The 100th
b. The 95th
c. The 20th
d. The 19th
e. The 15th

NAEP, Grade 8, 2003

BRAIN TEASER

Many schools celebrate Pi Day on March 14. Use a decimal approximation to suggest why the day is used and at what time the celebration might start. Explain your reasoning.

BRAIN TEASER

What is wrong with the following "proof" that ${}^{-}1 = 1$?

$$\sqrt{a}\sqrt{b} = \sqrt{ab}$$

Let $a = b = {}^{-}1$.

$$\sqrt{{}^{-}1}\sqrt{{}^{-}1} = \sqrt{({}^{-}1)({}^{-}1)} = \sqrt{1} = 1$$

But also $\sqrt{{}^{-}1}\sqrt{{}^{-}1} = {}^{-}1$.

Thus, ${}^{-}1 = 1$.

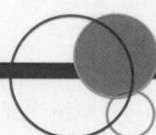

8-2 Variables

Algebraic thinking is important in mathematics at all levels—from the early grades on. In this section, we focus not only on patterns (introduced in Chapter 1) but on other features of algebraic thinking as well, including solving equations, word problems, functions, and graphing.

In the past, schoolchildren were not introduced to algebra until at least late middle school. Today, however, we realize the importance of integrating algebraic thinking and problem solving at all levels, beginning with kindergarten. In fact, as Carpenter* and others (2003) point out, algebraic thinking must be taught to *all* students.

PSSM recommends that pre-K–2 students be able to:

- use concrete, pictorial, and verbal representations to develop an understanding of invented and conventional symbolic notations;
- model situations that involve the addition and subtraction of whole numbers, using objects, pictures, and symbols. (p. 90)

In grades 3–5, students should be able to:

- represent the idea of a variable as an unknown quantity using a letter or a symbol;
- express mathematical relationships using equations. (p. 158)

And students in grades 6–8 should be able to:

- develop an initial conceptual understanding of different uses of variables;
- recognize and generate equivalent forms for simple algebraic expressions and solve linear equations. (p. 222)

Grade 6 *Focal Points* states that students should be taught to:

- write mathematical expressions and equations that correspond to given situations;
- evaluate expressions;
- use expressions and formulas to solve problems;
- understand that variables represent numbers whose exact values are not yet specified;
- use variables appropriately;
- understand that expressions in different forms can be equivalent;
- rewrite an expression to represent a quantity in a different way;
- know that the solutions of an equation are the values of the variables that make the equation true. (p. 18)

Historical Note

al-Khowarizmi

Fibonacci

The word *algebra* comes from the book *Hidab al-jabr wa'l muqabalah*, written by Mohammed ibn Musa al-Khowarizmi (ca. 825 CE). In his book he synthesized Hindu work on the notions of algebra and used the words *jabr* and *muqubalah* to designate two basic operations in solving equations: *jabr* meant to transpose subtracted terms to the other side of the equation; *muqubalah* meant to cancel like terms on opposite sides of the equation.

Another major contributor to the development of algebra was Diophantus (ca. 200–284 CE). His *Arithmetica* is the most prominent work on algebra in Greek mathematics. About 900 years later, Leonardo di Pisa (Fibonacci) (ca. 1170–1250) introduced algebra to Europe.

A third major contributor to algebra was François Viète (1540–1603), known as "the father of modern algebra," who introduced the first systematic algebraic notation in his book *In Artem Analyticam Isagoge.*   ●

---

*Carpenter, T. P., M. L. Franke, and L. Levi. *Thinking Mathematically: Integrating Arithmetic and Algebra in Elementary School.* Portsmouth, NH: Heinemann (2003).

Grade 6, *Common Core Standards* suggests that students:

> use variables to represent numbers and write expressions when solving a real-world or mathematical problem.... (p. 44)

Algebra is a branch of mathematics in which symbols—usually letters—represent numbers or members of a given set. Elementary algebra is used to generalize arithmetic. For example, the fact that $7 + (3 + 5) = (7 + 3) + 5$, or that $9 + (3 + 8) = (9 + 3) + 8$, are special cases of $a + (b + c) = (a + b) + c$, where a, b, and c are numbers from a given set—for example, whole numbers, integers, rational numbers, or real numbers. Similarly, $2 + 3 = 3 + 2$ and $2 \cdot 3 = 3 \cdot 2$ are special cases of $a + b = b + a$ and $a \cdot b = b \cdot a$ for all whole numbers a and b.

A major concept of algebraic thinking is that of **variable**. In basic arithmetic we have fixed numbers, or **constants**, as in $4 + 3 = 7$, but in algebra we have values that vary—hence the term *variable*.

- A variable may stand for a missing element or for an unknown, as in $x + 2 = 5$.
- A variable can represent more than one thing. For example, in a group of children, we could say that their heights vary with their ages. If h represents height and a represents age, then both h and a can have different values for different children in the group. Here a variable represents a changing quantity.
- A variable can be used in generalizations of patterns. If we were to use actual values instead of variables, the instructions would apply only in a limited set of situations.
- A variable can be an element of a set, or a set itself. For example, in the definition of the intersection of two sets $A \cap B = \{x \mid x \in A \text{ and } x \in B\}$, x is any element that belongs to both sets.

To apply algebra in solving problems, we frequently need to translate given information into a mathematical expression involving variables designated by letters or words. In all such examples, we may name the variables as we choose.

In the student page (p. 406), simple word statements are translated into **algebraic expressions**.

EXAMPLE 8-3

In each of the following, translate the given information into a symbolic expression involving quantities designated by letters:

 a. 2 more than a number
 b. $\sqrt{2}$ greater than a number
 c. 2/3 less than a number
 d. 2 times a number
 e. A number times itself
 f. The cost of renting a car for any number of days if the charge per day is $40
 g. The distance a car traveled at a constant speed of 65 mph for any number of hours

Historical Note

Mary Everest Boole

Mary Everest Boole (1832–1916), born in England and raised in France, was a self-taught mathematician and is most well-known for her works on mathematics and science education. In *Philosophy and Fun of Algebra* (London: C. W. Daniel, LTD, 1909), a book for children, she writes:

> But when we come to the end of our arithmetic we do not content ourselves with guesses; we proceed to algebra—that is to say, to dealing logically with the fact of our own ignorance. . . .
> Instead of guessing whether we are to call it nine, or seven, or a hundred and twenty, or a thousand and fifty, let us agree to call it *x*, and let us always remember that *x* stands for the Unknown. . . . This method of solving problems by honest confession of one's ignorance is called Algebra.

School Book Page VARIABLE AND EXPRESSIONS

Lesson 1-13

Algebra

Key Idea
Relationships among quantities can be written using algebra.

Vocabulary
• variable
• evaluate

Variables and Expressions

LEARN

How can you write an algebraic expression?

✔ **WARM UP**

Use symbols to write the expression.

1. sum of 16 and 29

2. difference of 216 and 89

Example A

Nita bought some candles costing $4 each. How can you represent their total cost?

Make a table to show the cost for different quantities of candles. Use a letter such as n to represent the number of candles. Because n represents a quantity whose value can vary, it is called a **variable.**

The total cost of the candles is represented by $4 \times n$ or $4n$.

Number of Candles	Total Cost ($)
1	4×1
2	4×2
3	4×3
4	4×4
⋮	⋮
n	$4 \times n$

An **algebraic expression** is a mathematical expression containing variables, numbers, and operation symbols. Before you write an algebraic expression, identify the operation. The table below shows how two or more word phrases can refer to an operation.

Word Phrase	Operation	Algebraic Expression
the **sum** of 9 and a number n a number m **increased** by 8 six **more than** a number t **add** eighteen to a number h	Addition	$9 + n$ $m + 8$ $t + 6$ $h + 18$
the **difference** of 12 and a number n seven **less than** a number y ten **decreased** by a number p	Subtraction	$12 - n$ $y - 7$ $10 - p$
the **product** of 4 and a number k fifteen **times** a number t two **multiplied** by a number m	Multiplication	$4k$ $15t$ $2m$
the **quotient** of a number divided by five twenty-five **divided** by a number m	Division	$\dfrac{a}{5}$ $\dfrac{25}{m}$

h. One weekend, a store sold twice as many CDs as full-size DVDs and 25 fewer mini DVDs than CDs. If the store sold d full-size DVDs, how many mini DVDs and CDs did it sell?

i. French fries have about 12 calories apiece. A hamburger has about 600 calories. Akiva is on a diet of 2000 calories per day. If he ate f french fries and one hamburger, how many more calories can he consume that day?

Solution In parts (a)–(e), let n represent a real number.

a. $n + 2$ b. $n + \sqrt{2}$ c. $n - 2/3$ d. $2n$ e. n^2

f. If d is the number of days, the cost of renting the car for d days at $40 per day is $40d$ dollars.

g. If b is the number of hours traveled at 65 mph, the total distance traveled in b hours is $65b$ miles.

h. Because d full-size DVDs were sold, twice as many CDs as full-size DVDs implies $2d$ CDs. Thus, 25 fewer mini DVDs than CDs implies $2d - 25$ mini DVDs.

i. First, find how many calories Akiva consumed eating f french fries and one hamburger. Then, to find how many more calories he can consume, subtract this expression from 2000.

1 french fry $\rightarrow$ 12 calories
f french fries $\rightarrow$ $12f$ calories

Therefore, the number of calories in f french fries and one hamburger is $600 + 12f$. The number of calories left for the day is $2000 - (600 + 12f)$, or $2000 - 600 - 12f$, or $1400 - 12f$.

EXAMPLE 8-4

A teacher asked her class to do the following:

Take any real number and add 15 to it. Now multiply that sum by 4. Next subtract 8 and divide the difference by 4. Now subtract 12 from the quotient and tell me the answer. I will tell you the original number.

Analyze the instructions to see whether the teacher can determine the original number.

Solution Translate the information into an algebraic expression as seen in Table 8-1.

Table 8-1

Instructions	Discussion	Symbols/Algebraic Expressions
Take any real number.	Since any number is used, we need a variable to represent the number. Let n be that variable.	n
Add 15 to it.	We are told to add 15 to "it." "It" refers to the variable n.	$n + 15$
Multiply that sum by 4.	We are told to multiply "that sum" by 4. "That sum" is $n + 15$.	$4(n + 15)$
Subtract 8.	We are told to subtract 8 from the product.	$4(n + 15) - 8$
Divide the difference by 4.	The difference is $4(n + 15) - 8$. Divide it by 4.	$\dfrac{4(n + 15) - 8}{4}$
Subtract 12 from the quotient and tell me the answer.	We are told to subtract 12 from the quotient.	$\dfrac{4(n + 15) - 8}{4} - 12$

Translating what the teacher asked the class to do results in the algebraic expression $\dfrac{4(n + 15) - 8}{4} - 12$. We are asked to determine whether the teacher can take a given student answer and produce the original number. We use the strategy of *working backward* to help with this determination. Suppose a student gives the teacher an answer of r. Think about how r was obtained and reverse the steps in Table 8-1. Just before obtaining "r," the student had subtracted 12. To reverse that operation, add 12 to obtain $r + 12$. Prior to that, the student had divided by 4. To reverse that, multiply by 4 to obtain $4r + 48$. To get that result, the student had subtracted 8, so now add 8 to obtain $4r + 56$. Just prior to that, the student had multiplied by 4, so now divide $4r + 56$ by 4 to obtain $r + 14$. The first operation had been to add 15, so now subtract 15 from $r + 14$ to get $r - 1$. Thus, the teacher knows when the student tells the final result of r, it is 1 more than the number with which the student started, or the number with which the student started, n, is the result minus 1.

This can be shown as follows:

$$\frac{4(n + 15) - 8}{4} - 12 = \frac{4(n + 15 - 2)}{4} - 12$$
$$= (n + 13) - 12$$
$$= n + 1$$

EXAMPLE 8-5

Figure 8-7 shows the first three terms of a sequence of figures containing small square tiles. Some of the tiles are shaded. Notice that the first figure has one shaded tile. The second figure has $2 \cdot 2$, or 2^2, shaded tiles. The third figure has $3 \cdot 3$, or 3^2, shaded tiles. If this pattern of having shaded squares surrounded by white borders continues, answer the following:

  a. How many shaded tiles are there in the nth figure?
  b. How many white tiles are there in the nth figure?

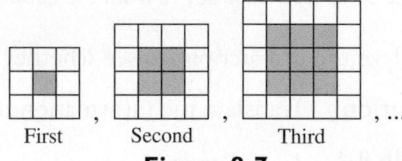

First   ,   Second   ,   Third   , ...

Figure 8-7

Solution
  a. The squares of shaded tiles have sides with increasing lengths 1, 2, 3, and so on. In the nth figure, the length of a side of the shaded region would be n. Hence, the nth figure has n^2 shaded tiles.
  b. One way to think about the number of white tiles is to recognize that the number of white tiles on a side is 2 more than n, or $n + 2$. The number of white tiles could be 4 times $(n + 2)$, less any overlapping counting. In this case, each corner tile would be duplicated so 4 white tiles are overcounted giving us $4(n + 2) - 4$, or $4n + 4$, white tiles.

  Another way to count the white tiles in the nth figure is to count the total number of tiles and then subtract the number of shaded tiles from this total. The number of white tiles on the bottom side of the nth square is $n + 2$, and the number of the shaded tiles on a side is n. Thus, the number of white tiles is $(n + 2)^2 - n^2$. This answer is equal to $4n + 4$ obtained earlier. (Why?)

NOW TRY THIS 8-6

  a. Another way to count the white tiles in Example 8-5 is to first remove the four white corner tiles and then count the number of remaining white tiles. Complete this approach.
  b. Noah has some white square tiles and some blue square tiles. They are all the same size. He first makes a row of white tiles and then surrounds the white tiles with a single layer of blue tiles, as shown in Figure 8-8.

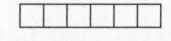

Figure 8-8

How many blue tiles does he need:

   i. to surround a row of 100 white tiles?

  ii. to surround a row of n white tiles?

Generalizations for Arithmetic Sequences

Example 8-5(b) is an illustration of a arithmetic sequence. Substituting $n = 1, 2, 3, \ldots$, we see that the first term is 8, the second term is 12, the third term is 16, and so on, generating the arithmetic sequence: $8, 12, 16, 20, \ldots, 4n + 4$. Terms can also be found by adding 4 to successive multiples of 4. Work with arithmetic sequences can be generalized using algebraic thinking and functional notation.

    As in chapter 1, an arithmetic sequence is determined by its first term and the difference. Suppose the first term of an arithmetic sequence is a_1 and the difference is d. The strategy of *making a table* can be used to find the general term for the sequence: $a_1, a_1 + d, a_1 + 2d, a_1 + 3d, \ldots$ as shown in Table 8-2.

Table 8-2

Number of Term	Term
1	a_1
2	$a_2 = a_1 + d = a_1 + (2 - 1)d$
3	$a_3 = a_1 + 2d = a_1 + (3 - 1)d$
4	$a_4 = a_1 + 3d = a_1 + (4 - 1)d$
5	$a_5 = a_1 + 4d = a_1 + (5 - 1)d$
$\vdots$	$\vdots$
n	$a_n = a_1 + (n - 1)d$

Observe that the number of ds in the given terms is 1 less than the number of the term. This pattern continues since we add d to get the next term. Thus, the *nth term of any arithmetic sequence with first term a_1 and difference d is given by $a_n = a_1 + (n - 1)d$, where n is a natural number*.

    For example, in the arithmetic sequence $5, 9, 13, 17, 21, 25, \ldots$, the first term is 5 and the difference is 4. Thus, the nth term is given by $a_n = 5 + (n - 1)4$. Simplifying, we obtain $a_n = 5 + (n - 1)4 = 5 + 4n - 4 = 4n + 1$.

EXAMPLE 8-6

In an arithmetic sequence with the second term 11 and the fifth term 23, find the 100th term.

Solution   In a sequence with first term a_1, second term 11, and fifth term 23, we are to find the 100th term. We know that

$$a_2 = 11 = a_1 + d$$
$$a_5 = a_1 + (5 - 1)d = 23, \text{ or}$$
$$a_1 + 4d = 23.$$

Because we know that $11 = a_1 + d$, we write

$$a_1 + 4d = (a_1 + d) + 3d = 23$$
$$11 + 3d = 23$$
$$3d = 12$$
$$d = 4.$$

    To find the first term, we have $11 = a_1 + d$, so

$$11 = a_1 + 4, \text{ or } a_1 = 7.$$

To find the 100th term, we know that $a_{100} = a_1 + (100 - 1)d$, or that $a_{100} = 7 + (100 - 1)4 = 403$.

The Fibonacci Sequence and Spreadsheets

Variables are commonly used in spreadsheets. To compute the 50th term of the Fibonacci sequence, 1, 1, 2, 3, 5, 8, 13, . . ., in which the first two terms are 1, 1 and each subsequent term is the sum of the two preceding terms, could take a very long time by hand or even by calculator. However, using a spreadsheet, any desired term of the Fibonacci sequence and the previous term appear instantly. The partial student page shows how to create the Fibonacci sequence on a spreadsheet using two variables, *A*1 and *A*2. Each term of a Fibonacci sequence is a **Fibonacci number**.

School Book Page   LEARNING WITH TECHNOLOGY

Learning with Technology

Spreadsheet/Data/Grapher eTool: Generating a Sequence

Almost 800 years ago, an Italian mathematician named Leonardo Fibonacci discovered this sequence of numbers: 1, 1, 2, 3, 5, 8, 13, 21, 34, 55, 89, 144, 233, 377, 610,…

Beginning with the number 1, each number in the sequence is the sum of the previous two numbers.
1 + 1 = **2**, 1 + 2 = **3**, 2 + 3 = **5**, 3 + 5 = **8**, 5 + 8 = **13**,
8 + 13 = **21**, 13 + 21 = **34**, and so on.

Create a spreadsheet that will generate the first 32 terms of the Fibonacci sequence. Copy the formula in A5 to cells A6–A35, and B6 to cells B7–B35.

Leonardo Fibonacci

1. Do you think there is an infinite number of terms in the sequence? Explain.

2. Change the number in cell C2 to 3. What happens to the numbers in column B? How does the sequence depend on the numbers in cells B2 and C2?

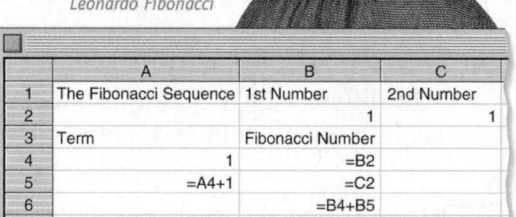

	A	B	C
1	The Fibonacci Sequence	1st Number	2nd Number
2		1	1
3	Term	Fibonacci Number	
4	1	=B2	
5	=A4+1	=C2	
6		=B4+B5	

All text pages available online and on CD-ROM.   **Section B Lesson 3-6**   **163**

The standard mathematical way to represent a Fibonacci sequence has F_1 as the first term, F_2 as the second term, and in general, F_n as the nth term. The two terms before the nth, F_n, are F_{n-1} and F_{n-2}. With this notation, the rule for generating the Fibonacci sequence can be written as follows:

$$F_n = F_{n-1} + F_{n-2}, \text{ for } n = 3, 4, \ldots$$

This rule cannot be applied to the first two Fibonacci numbers. Because $F_1 = 1$ and $F_2 = 1$, then $F_3 = 1 + 1$, or 2. The beginning terms, or seeds, $F_1 = 1$ and $F_2 = 1$ and the rule $F_n = F_{n-1} + F_{n-2}$ provide an example of a recursive definition because the rule for the sequence defines each term after the first two using previous term in the same sequence. Using a spreadsheet, we generate as many terms of the sequence as we wish.

Generalizations for Geometric Sequences

It is possible to find the nth term, a_n, of any geometric sequence when given the first term and the ratio as mentioned in Chapter 1. For example, in the geometric sequence $3, 12, 48, 192, \ldots$, the first term is 3, and the ratio is 4. To generalize geometric sequences, let the first term be a_1 and the ratio be r, so that the third term is $a_1 r^2$, and the fourth term is $a_1 r^3$ as seen in Table 8-3.

Table 8-3

Number of Term	Term
1	a_1
2	$a_1 r$
3	$a_1 rr = a_1 r^2$
4	$a_1 r^2 r = a_1 r^3$
5	$a_1 r^3 r = a_1 r^4$
$\vdots$	$\vdots$
n	$a_1 r^{(n-1)}$

The power of r in each term is 1 less than the number of the term. This pattern continues since we multiply by r to get the next term. Thus, *the nth term of a geometric sequence, a_n, is $a_1 r^{n-1}$.* In the geometric sequence $3, 12, 48, 192, \ldots$, the first term is 3, and the ratio is 4, so the nth term $a_n = 3 \cdot 4^{n-1}$.

EXAMPLE 8-7

a. Find the nth term of the following geometric sequence: $2, 3, 9/2, 27/4, \ldots$.
b. If a geometric sequence has first term 3 and ratio $\sqrt{2}$, find its 10th term.

Solution

a. The ratio $3/2$ of the sequence can be found by dividing two consecutive terms. Thus, the formula for the nth term is

$$a_n = 2(3/2)^{n-1}.$$

b. The 10th term is computed as follows:

$$a_{10} = 3(\sqrt{2})^{10-1} = 3(\sqrt{2})^9 = 3(2^{9/2}) = 3(2^{8/2})(2^{1/2}) = 48\sqrt{2}.$$

NOW TRY THIS 8-7

a. Two bacteria are in a dish. The number of bacteria triples every hour. Following this pattern, find the number of bacteria in the dish after 10 hours and after n hours.
b. Suppose that instead of increasing geometrically as in part(a), the number of bacteria increases arithmetically by 3 bacteria each hour. Compare the geometric and arithmetic growth after 10 hours and after n hours. Comment on the difference in growth of a geometric sequence versus an arithmetic sequence.

More Algebraic Thinking

Algebraic thinking can occur in different ways. One example that uses pictures is seen in Example 8-8.

EXAMPLE 8-8

At a local farmer's market, three purchases were made for the prices shown in Figure 8-9. What is the cost of each object?

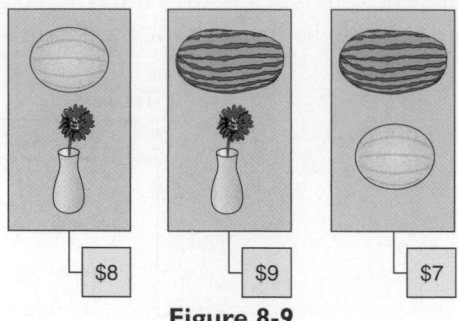

Figure 8-9

Solution   Approaches to this problem may vary. For example, if the objects in the first two purchases are put together, the total cost would be $8 + $9, or $17. That cost would be for two vases and one each of the cantaloupe and watermelon, as in Figure 8-10.

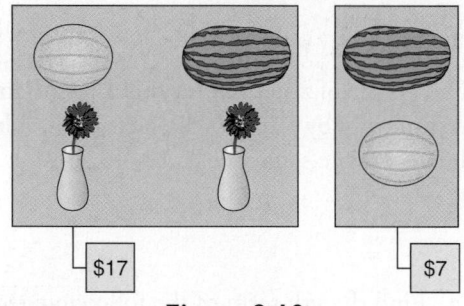

Figure 8-10

Now if the cantaloupe and watermelon are taken away from that total, then according to the cost of those two objects from the tag on the right, the cost should be reduced to $10 for two vases. That means each of the two vases costs $5. This in turn tells us that the cantaloupe costs $8 − $5, or $3, and the watermelon costs $9 − $5, or $4.

The solution in Example 8-8 could involve the strategy of *writing an equation*. But we need a basic knowledge of solving equations, which is discussed in Section 8-3.

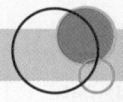

Assessment 8-2A

1. Translate the following statements into algebraic expressions:
   a. The third term of an arithmetic sequence whose first term is 10 and whose difference is d
   b. 10 less than twice a number
   c. 10 times the square of a number
   d. The difference between the square of a number and twice the number

2. a. Translate the following instructions into an algebraic expression: Take any number, add $\sqrt{3}$ to it, multiply the sum by 7, subtract 14, and divide the difference by 7. Finally, subtract the original number.
   b. Simplify your answer in part (a).

3. In the tile pattern in the sequence of figures shown, each figure starting from the second has two more blue squares than the preceding one and represents an arithmetic sequence. Answer the following:

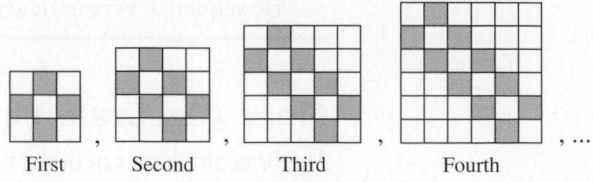

First   Second   Third   Fourth

a. How many blue tiles are there in the nth figure if this arithmetic sequence continues?

b. If the pattern of large square tiles with shaded blue ones continue, how many white tiles are there in the nth figure?

4. In the following, write an expression in terms of the given variable that represents the indicated quantity.

   a. The cost of having a plumber spend h hr at your house if the plumber charges $20.00 for coming to the house and $25.00 per hour for labor

   b. The amount of money in cents in a jar containing d dimes and some nickels and quarters if there are 3 times as many nickels as dimes and twice as many quarters as nickels

   c. The sum of three consecutive integers if the least integer is x

   d. The amount of bacteria after n min if the initial amount of bacteria is q and the amount of bacteria doubles every minute (*Hint:* The answer should contain q as well as n.)

   e. The temperature after t hr if the initial temperature is 40°F and each hour it drops by 3°F

   f. Pawel's total earning after 3 yr if the first year his salary was s dollars, the second year it was $5000.00 higher, and the third year it was twice as much as the second year

   g. The sum of three consecutive odd natural numbers if the least is x

   h. The sum of three consecutive natural numbers if the middle is m

5. If the number of professors in a college is P and the number of students S, and there are 20 times as many students as professors, write an algebraic equation that shows this relationship.

6. If g is the number of girls in a class and b the number of boys and there are five more girls than boys in the class, write an algebraic equation that shows this relationship.

7. Ryan is building matchstick sequences as shown. How many matchsticks will he use for the nth figure if the backwards L shapes continue?

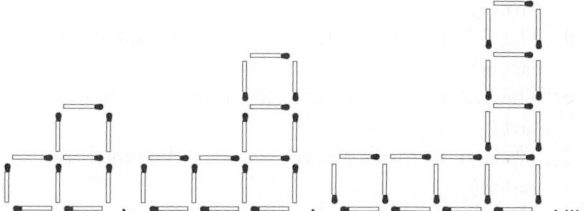

8. Write an algebraic equation relating the variables described in each of the following situations:

   a. The pay, P, for t hr if you are paid $8.00 an hour

   b. The pay, P, for t hr if you are paid $15.00 for the first hour and $10 for each additional hour

9. For a particular event, a student pays $5.00 per ticket and a nonstudent pays $13 per ticket. If x students and 100 nonstudents buy tickets, find the total revenue from the sale of the tickets in terms of x.

10. Suppose a will decreed that three siblings will each receive a cash inheritance according to the following: The eldest receives 3 times as much as the youngest, and twice as much as the middle sibling. Answer the following.

    a. If the youngest sibling receives x, how much do the other two receive in terms of x?

    b. If the middle sibling receives y, how much do the other two receive in terms of y?

    c. If the oldest sibling receives z, how much do the other two receive in terms of z?

11. In each of the following arithmetic sequences find (i) the 100th term, and (ii) the nth term.

    a. $^-3, ^-7, ^-11, \ldots$

    b. $1, 1 + \sqrt{2}, 1 + 2\sqrt{2}, 1 + 3\sqrt{2}, \ldots$

    c. $\pi + 0.5, \pi + 2.5, \pi + 4.5, \ldots$

12. In each of the following geometric sequences, find (i) the 15th term, and (ii) the nth term.

    a. $3, 3\sqrt{2}, 6, 6\sqrt{2}, \ldots$

    b. $\pi, 1, 1/\pi, \ldots$

    c. $1 + \sqrt{2}, 2 + \sqrt{2}, 2 + 2\sqrt{2}, \ldots$

13. In an arithmetic sequence, find the first term if the 10th term is 14 and the difference is $\sqrt{2}$.

14. In a geometric sequence, find the first term if the 9th term is 2048 and the ratio is $\sqrt{2}$.

15. Find the following:

    a. The first seven terms of the Fibonacci sequence with seeds: 1, 1

    b. The sum of the first three terms of the sequence in part (a)

    c. The sum of the first four terms of the sequence in part (a)

    d. The sum of the first five terms of the sequence in part (a)

    e. The sum of the first six terms of the sequence in part (a)

    f. The sum of the first seven terms of the sequence in part (a)

    g. A pattern for the sums in parts (b)–(f).

    h. A rule for your pattern in part (g) using the notation for Fibonacci numbers.

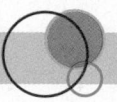

Assessment 8-2B

1. Translate the following statements into algebraic expressions:

   a. 10 more than a number

   b. 10 less than a number

   c. 10 times a number

   d. The sum of a number and 10

   e. The difference between the square of a number and the number

2. Translate the following into an algebraic expression:

   a. Take any number, add 25 to it, multiply the sum by 3, subtract 60, and divide the difference by 3. Finally, add 5.

   b. Simplify your answer in part (a).

3. Discover a possible tile pattern in the following sequence and answer the following:

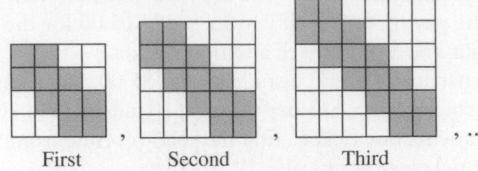

First   Second   Third   , ...

   a. How many shaded tiles are there in the nth figure of your pattern if the arithmetic sequence of shaded tiles continues?
   b. How many white tiles are there in the nth figure of your pattern?

4. In the following, write an expression in terms of the given variables that represents the indicated quantity:
   a. The cost of having a plumber spend h hr at your house if the plumber charges $30 for coming to the house and $$x$ per hour for labor
   b. The amount of money in cents in a jar containing some nickels and d dimes and some quarters if there are 4 times as many nickels as dimes and twice as many quarters as nickels
   c. The sum of three consecutive integers if the greatest integer is x
   d. The amount of bacteria after n min if the initial amount of bacteria is q and the amount of bacteria triples every 30 sec. (*Hint:* The answer should contain q as well as n.)
   e. The temperature t hr ago if the present temperature is 40°F and each hour it drops by 3°F
   f. Pawel's total earnings after 3 yr if the first year his salary was s dollars, the second year it was $5000 higher, and the third year it was twice as much as the first year
   g. The sum of three consecutive even whole numbers if the greatest is x

5. If a school has w women and m men and you know that there are 100 more men than women, write an algebraic equation relating w and m.

6. Suppose there are c chairs and t tables in a classroom and there are 15 more chairs than tables. Write an algebraic equation relating c and t.

7. Ryan is building matchstick square sequences so that one square is added to the right each time, as shown. How many matchsticks will he use for the nth figure and for the figure one before the nth?

8. Write an algebraic equation relating the variables described in each of the following situations:
   a. The pay, P, for t hr if you are paid $$d$ an hour
   b. The pay, P, for t hr if you are paid $15 for the first hour and $$k$ for each additional hour

   c. The total pay, P, for a visit and t hr of gardening if you are paid $20 for the visit and $10 for each hour of gardening
   d. The total cost, C, of membership in a health club that charges a $300 initiation fee and $4 for each of n days attended
   e. The cost, C, of renting a midsized car for 1 day of driving m mi if the rent is $30 per day plus 35¢ per mile

9. A teacher instructed her class as follows:

Take any odd number, multiply it by 4, add 16, and divide the result by 2. Subtract 7 from the quotient and tell me your answer. I will tell you the original number.

Explain how the teacher was able to tell each student's original number.

10. Matt has twice as many stickers as David. If David has d stickers and Matt m stickers, and Matt gives David 10 stickers, how many stickers does each have in terms of d?

11. In each of the following arithmetic sequences find (i) the 100th term, and (ii) the nth term.
    a. $3, {}^-1, {}^-5, \ldots$
    b. $1, 1 - \sqrt{2}, 1 - 2\sqrt{2}, 1 - 3\sqrt{2}, \ldots$
    c. $0.5 + \pi, 0.5 + 2\pi, 0.5 + 3\pi, \ldots$

12. In each of the following geometric sequences, find (i) the 15th term, and (ii) the nth term.
    a. $2, 2\sqrt{2}, 4, 4\sqrt{2}, \ldots$
    b. $\pi, {}^-\pi, \pi, \ldots$
    c. $1 - \sqrt{2}, 2 - \sqrt{2}, 2 - 2\sqrt{2}, \ldots$

13. In an arithmetic sequence, find the first term if the 10th term is 16 and the difference is $\sqrt{2}$.

14. In a geometric sequence, find the first term if the 9th term is 4096 and the ratio is $1/\sqrt{2}$.

15. Find the following:
    a. The first seven terms of the Fibonacci-type sequence with seeds: 1, 2
    b. The sum of the first three terms of the sequence in part (a)
    c. The sum of the first four terms of the sequence in part (a)
    d. The sum of the first five terms of the sequence in part (a)
    e. The sum of the first six terms of the sequence in part (a)
    f. The sum of the first seven terms of the sequence in part (a)
    g. A pattern for the sums in parts (b)–(f).
    h. A rule for your pattern in part (g) using the notation for Fibonacci numbers.

Mathematical Connections 8-2

Communication

1. Students were asked to write an algebraic expression for the sum of three consecutive natural numbers. One student wrote $x + (x + 1) + (x + 2) = 3x + 3$. Another wrote $(x - 1) + x + (x + 1) = 3x$. Explain who is correct and why.
2. Explain why a variable is used to generalize an arithmetic sequence.

Open-Ended

3. A teacher instructed her class to take any number and perform a series of computations using that number. The teacher was able to tell each student's original number by subtracting 1 from the student's answer. Create similar instructions for students so that the teacher needs to do only the following to obtain the student's original number:
   a. Add 1 to the answer.
   b. Multiply the answer by 2.
   c. Multiply the answer by 1.
4. Give an example of an arithmetic sequence that might be considered as "growing faster" than a specific geometric sequence.

Cooperative Learning

5. Examine several elementary school textbooks for grades 1 through 5 and report on which algebraic concepts involving variables are introduced at each grade level.

Questions from the Classroom

6. A student claims that the sum of five consecutive integers is equal to 5 times the middle integer and would like to know if this is always true, and if so, why. He would like to know if the statement generalizes to the sum of five consecutive terms in any arithmetic sequence. How do you respond?
7. A student writes $a(bc) = (ab)(ac)$. How do you respond?
8. A student wonders if sets can ever be considered as variables. What do you tell her?
9. A student thinks that if A and B are sets, then the statements $A \cup B = B \cup A$ and $A \cap B = B \cap A$ are algebraic generalizations of set properties in a way similar to the statements $a + b = b + a$ and $ab = ba$ are generalizations of arithmetic properties of numbers. How do you respond?

Review Problems

10. Find two rational numbers and two irrational numbers between 1.41 and $\sqrt{2}$.
11. Show whether $\sqrt[3]{729}$ is a representation for an integer.

12. Write a paragraph explaining whether it is possible for the sum of two rational numbers to be an irrational number.
13. What is the simplest representation for $\sqrt{0.\overline{9}}$?
14. Write an arithmetic sequence that has an irrational number as a difference.

Trends in Mathematics and Science Study (TIMSS) Questions

$\square$ represents the number of magazines that Lina reads each week. Which of these represents the total number of magazines that Lina reads in 6 weeks?
   a. $6 + \square$
   b. $6 \times \square$
   c. $\square + 6$
   d. $(\square + \square) \times 6$

TIMSS, Grade 4, 2003

Which is equivalent to $4x - x + 7y - 2y$?
   a. 9
   b. $9xy$
   c. $4 + 5y$
   d. $3x + 5y$

TIMSS, Grade 8, 2007

The first pipe is x meters long. The second pipe is y times as long as the first one. How long is the second pipe?
   a. xy meters
   b. $x + y$ meters
   c. $\dfrac{x}{y}$ meters
   d. $\dfrac{y}{x}$ meters

TIMSS, Grade 8, 2008

National Assessment of Educational Progress (NAEP) Question

N stands for the number of hours of sleep Ken gets each night. Which of the following represents the number of hours of sleep Ken gets in 1 week?
   a. $N + 7$
   b. $N - 7$
   c. $N \times 7$
   d. $N \div 7$

NAEP, Grade 4, 2005

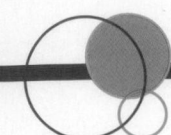

8-3 Equations

With variables w and c, we consider equations such as $w + c = \sqrt{7}$. The equal sign indicates that the values on both sides of the equation are the same even though they do not look the same.

To solve equations, we need several properties of equality. Children discover many of these by using a balance scale. For example, consider two objects a and b of the same weight on the balances, as in Figure 8-11(a). If the balance is level, then $a = b$. When we add an equal amount of weight, c, to both sides, the balance is still level, as in Figure 8-11(b).

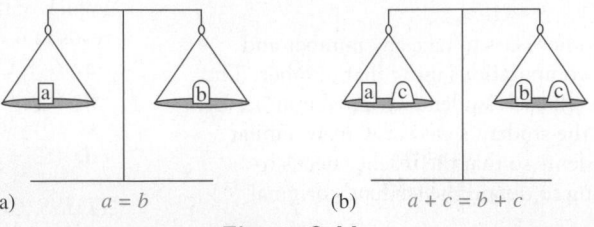

(a) $a = b$ (b) $a + c = b + c$

Figure 8-11

This demonstrates that *if $a = b$, then $a + c = b + c$*, which is *the addition property of equality*.

Similarly, if the scale is balanced with amounts a and b, as in Figure 8-12(a), and we put additional a's on one side and an equal number of b's on the other side, the scale remains level, as in Figure 8-12(b).

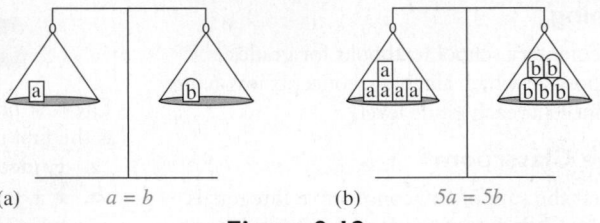

(a) $a = b$ (b) $5a = 5b$

Figure 8-12

Figure 8-12 suggests that *if c is any natural number and $a = b$, then $ac = bc$*. While this is true, it can be extended to any real number c. The extension is the *multiplication property of equality*. These properties are summarized in the next theorem.

> **Theorem 8-2:   The Addition and Multiplication Properties of Equality**
> **a.** For any real numbers a, b, and c, if $a = b$, then $a + c = b + c$.
> **b.** For any real numbers a, b, and c, if $a = b$, then $ac = bc$.

Equality is not affected if we substitute a number for its equal. This property is referred to as the **substitution property**. Examples of substitution follow:

a. If $a + b = c + d$ and $d = 5$, then $a + b = c + 5$.
b. If $a + b = c + d$, $b = e$, and $d = f$, then $a + e = c + f$.

Using the substitution property we can see that equations can be added or subtracted "side by side"; that is, we have the following.

> **Theorem 8-3:   Addition and Subtraction Property of Equations**
> If $a = b$ and $c = d$, then $a + c = b + d$ and $a - c = b - d$.

Theorem 8-2 implies that we may add the same real number to both sides of an equation or multiply both sides of the equation by the same real number without affecting the equality. If $a + c = b + c$ and $ac = bc$, the cancellation properties of equality can be developed by adding ^-c to both sides of the first equation and multiplying both sides by $\frac{1}{c}$, where $c \neq 0$, in the second.

Theorem 8-4:   Cancellation Properties of Equality

   a. For any real numbers a, b, and c, if $a + c = b + c$, then $a = b$.

   b. For any real numbers a, b, and c, with $c \neq 0$, if $ac = bc$, then $a = b$.

Suppose $ab = 0$. If $ab = 0$, then at least one factor, a or b, must be 0. Thus, we have the following: *For any real numbers a and b, if ab = 0, then a = 0 or b = 0.*

An algebraic use is seen when we find a solution to an equation like $(x - 3)(x - 5) = 0$. We know that $x - 3 = 0$ or $x - 5 = 0$, Hence $x = 3$ or $x = 5$ the above theorems and properties for real numbers also hold for algebraic expressions.

As mentioned in an earlier chapter, when using the commutative property of multiplication, each of the distributive properties can be written in the equivalent forms:

$$(b + c)a = ba + ca, \text{ and}$$

$$(b - c)a = ba - ca$$

When the distributive properties are written from right to left, we refer to them as *factoring*. Thus, $ab + ac = a(b + c)$ and $ab - ac = a(b - c)$. We say that a has been "factored out."

Solving Equations with One Variable

Finding solutions to equations is a major part of algebra. Use of tangible objects can increase students' engagement and comprehension when they work with equations. A balance-scale model fosters understanding of the basic concepts used in solving equations and inequalities.

For example, consider Figure 8-13. If we release the pan on the left, what happens? Upon release, the scale tilts down on the right side and we have an *inequality*, $2 \cdot 3 < 3 + (2 \cdot 2)$.

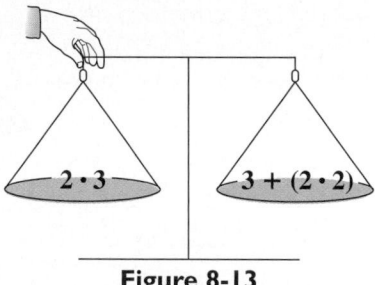

Figure 8-13

Next consider Figure 8-14. If we release the pan, then the sides will balance and we have the *equality* $2 \cdot 3 = (1 + 1) + 4$.

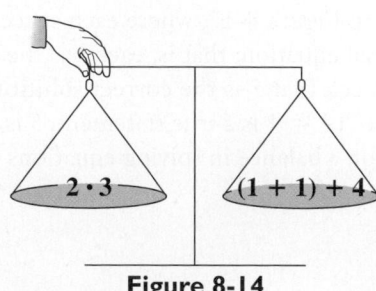

Figure 8-14

A balance scale can also be used to reinforce the idea of a replacement set for a variable. Name some solutions in Figure 8-15 that keep the scale balanced. For example, $3(5/2)$ balances $2(5/2)$, $3\sqrt{16}$ balances $2 \cdot 6$, or 12, and so on.

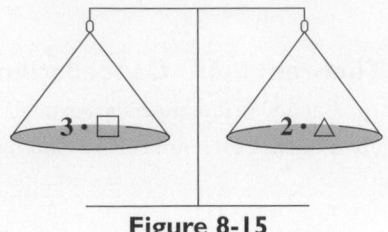

Figure 8-15

Other types of balance scale problems may help students with algebra. Work through Now Try This 8-8 before proceeding.

● NOW TRY THIS 8-8

What are the values of $\square$ and $\triangle$ in parts (a) and (b) below?

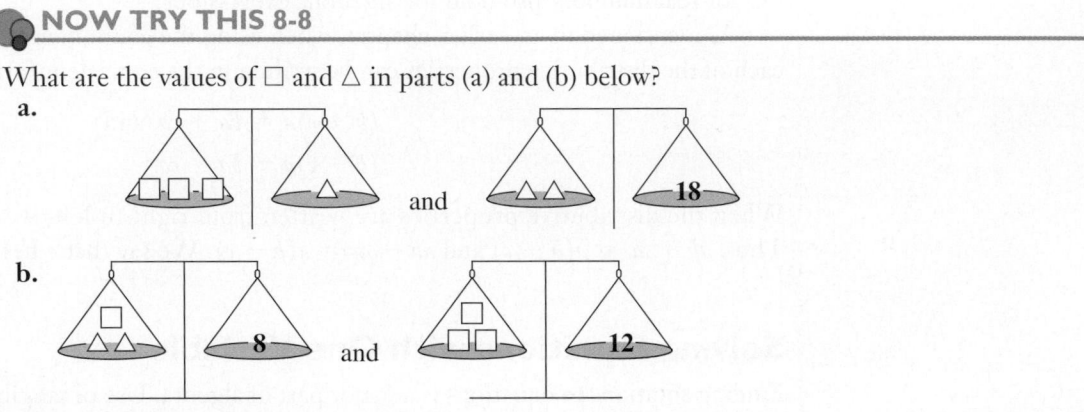

To solve equations, we may think of the properties of equality used on a balance pan. Consider $3x - 14 = 1$. Put the equal expressions on the opposite pans of the balance scale. Because the expressions are equal, the pans should be level, as in Figure 8-16.

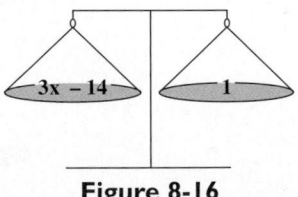

Figure 8-16

To solve for x, we use the properties of equality to manipulate the expressions on the scale so that after each step, the scale remains level and, at the final step, only an x remains on one side of the scale. The number on the other side of the scale represents the solution to the original equation. To find x in the equation of Figure 8-16, consider the scales pictured in successive steps in Figure 8-17, where each successive scale represents an equation that is equivalent to the original equation; that is, each has the same solution as the original. The last scale shows $x = 5$. To check that 5 is the correct solution, we substitute 5 for x in the original equation. Because $3 \cdot 5 - 14 = 1$ is a true statement, 5 is the solution to the original equation. Concrete objects are used on a balance in solving equations on the student page (420).

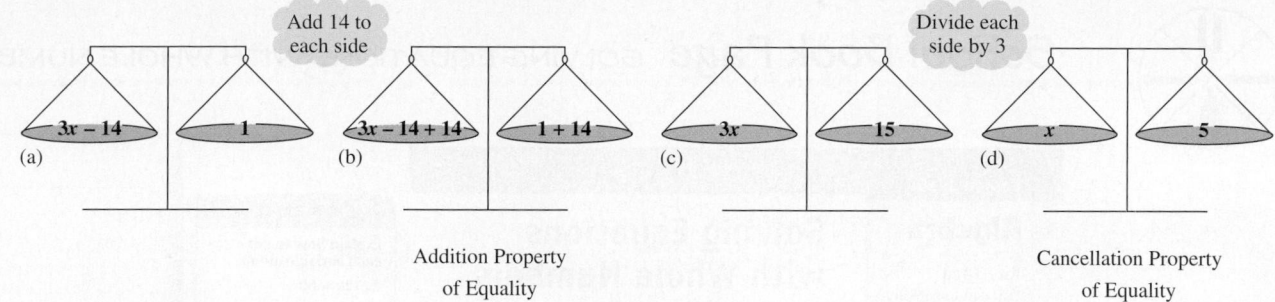

Add 14 to each side			Divide each side by 3

(a) $3x - 14$ | 1 (b) $3x - 14 + 14$ | $1 + 14$ (c) $3x$ | 15 (d) x | 5

	Addition Property of Equality		Cancellation Property of Equality

Figure 8-17

EXAMPLE 8-9

Solve each of the following for x:

a. $x + \sqrt[3]{4} = 20$
b. $3x = x + \sqrt{10}$
c. $4x + 5x = 99$
d. $4(x + 3) + 5(x + 3) = 99$

Solution

a.
$$x + \sqrt[3]{4} = 20$$
$$(x + \sqrt[3]{4}) - \sqrt[3]{4} = 20 - \sqrt[3]{4}$$
$$x = 20 - \sqrt[3]{4}$$

b.
$$3x = x + \sqrt{10}$$
$$3x - x = (x + \sqrt{10}) - x$$
$$x(3 - 1) = (\sqrt{10} + x) - x$$
$$2x = \sqrt{10}$$
$$x = \sqrt{10}/2$$

c.
$$4x + 5x = 99$$
$$(4 + 5)x = 99$$
$$9x = 99$$
$$x = 11$$

d.
$$4(x + 3) + 5(x + 3) = 99$$
$$(4 + 5)(x + 3) = 99 \ (\text{Why?})$$
$$9(x + 3) = 99$$
$$x + 3 = 11$$
$$x = 8$$

Historical Note

Mary Fairfax Somerville

Born in Scotland, Mary Fairfax Somerville (1780–1872) first studied simple arithmetic at the age of 13. At about this time she saw mysterious symbols in a women's fashion magazine and, after persuading her brother's tutor to purchase some elementary literature for her on the subject, began her study of algebra. As a young mother and widow, she obtained a library providing her with a background in mathematics. Throughout her life, Somerville distinguished herself as a skilled scientific writer, publishing a number of works, including *Molecular and Microscopic Science* when she was 89. In her autobiography Somerville wrote of how she "was sometimes annoyed when in the midst of a difficult problem" a visitor would enter. Shortly before her death she wrote

I am now in my ninety-second year, . . . , I am extremely deaf, and my memory of ordinary events, and especially of the names of people, is failing, but not for mathematical and scientific subjects. I am still able to read books on the higher algebra for four or five hours in the morning and even to solve the problems. Sometimes I find them difficult, but my old obstinacy remains, for if I do not succeed today, I attack them again tomorrow.

School Book Page   SOLVING EQUATIONS WITH WHOLE NUMBERS

Lesson 1-15

Algebra

Key Idea
You can use inverse operations and the properties of equality to solve equations.

Vocabulary
• equation (p. 44)
• inverse operations (p. 45)
• properties of equality (p. 44)

TEST TALK

Think It Through
I can **think of a pan balance** to help solve the problem.

Solving Equations with Whole Numbers

LEARN

✓ WARM UP
Explain how to get each variable alone.
1. $12x = 60$
2. $d - 10 = 10$
3. $32 = 8 + a$

How can you solve an equation?

When you **solve** an equation, you find the value of the variable that makes the equation true.

Example A

Wynn sold 6 sketches, each for the same amount, and made $180 in sales. How much did he charge for each sketch?

Let s equal the amount for each sketch.

Then the equation is $6s = 180$.

What You Write	Balancing the Pans	
$6s = 180$		The pans are balanced.
$6s \div 6 = 180 \div 6$		180 has been separated into 6 equal parts.
$s = 30$		Each s equals 30.

Wynn charged $30 for each sketch.

✓ Talk About It

1. Why was each side of the equation in Example A divided by 6?

How can you check your answer?

To check your answer, substitute it for the variable in the original equation. In Example A, substitute 30 for s in $6s = 180$.

Check: $6s = 180$
            $6(30) = 180$
            $180 = 180$ When both sides of the equation can be simplified to the same number, the value of the variable is correct.

48

Application Problems

Figure 8-18 demonstrates a method for solving application problems with a third grade example: formulate the problem with a mathematical model, solve that mathematical model, and interpret the solution in terms of the original problem.

Application Problem Amy earned \$12 baby-sitting and \$5 washing the car. How much did she earn all together?	→	*Mathematical Model* $12 + 5 = ?$
		↓
Original Problem Interpretation Amy's earnings totaled \$17.	←	*Mathematical Solution* $12 + 5 = 17$

Figure 8-18

We apply Pólya's four-step problem-solving process to solving word problems with algebraic thinking. In Understanding the Problem, we identify what is given and what is to be found. In Devising a Plan, we assign letters to the unknown quantities and try to translate the information in the problem into a model involving equations. In Carrying Out the Plan, we solve the equations or inequalities. In Looking Back, we interpret and check the solution in terms of the original problem.

Problem Solving Overdue Books

Bruno has five books overdue at the library. The fine for overdue books is 10¢ a day per book. He remembers that he checked out an astronomy book a week before he checked out four novels. If his total fine was \$8.70, how long was each book overdue?

Understanding the Problem Bruno has five books overdue. The astronomy book was checked out seven days before the four novels. The fine per day for each book is 10¢, and the total fine was \$8.70. We need to find out how many days each book is overdue.

Devising a Plan Let d be the number of days that each of the four novels is overdue. The astronomy book was due seven days $(d + 7)$ before the novels. To *write an equation* for d, we express the total fine in two ways. The total fine is \$8.70. This fine in cents equals the fine for the astronomy book plus the fine for the four novels.

$$\text{Fine for each of the novels} = \underbrace{\text{Fine per day}}_{10}\ \underbrace{\text{times}}_{\cdot}\ \underbrace{\text{number of overdue days}}_{d}$$

$$\begin{aligned}\text{Fine for the four novels} &= \underbrace{\text{1 day's fine for novels}}_{4\cdot 10}\ \underbrace{\text{times}}_{\cdot}\ \underbrace{\text{number of overdue days}}_{d}\\ &= (4\cdot 10)d\\ &= 40d\end{aligned}$$

$$\begin{aligned}\text{Fine for the astronomy book} &= \underbrace{\text{Fine per day}}_{10}\ \underbrace{\text{times}}_{\cdot}\ \underbrace{\text{number of overdue days}}_{(d+7)}\\ &= 10(d + 7)\end{aligned}$$

Because each of the expressions is in cents, we need to write the total fine of \$8.70 as 870¢ to produce the following:

$$\begin{matrix}\text{Fine for the four novels} & + & \text{Fine for the astronomy book} & = & \text{Total fine}\\ 40d & + & 10(d+7) & = & 870\end{matrix}$$

Carrying Out the Plan Solve the equation for d.

$$\begin{aligned}40d + 10(d + 7) &= 870\\ 40d + 10d + 70 &= 870\\ 50d + 70 &= 870\\ 50d &= 800\\ d &= 16\end{aligned}$$

Thus, each of the four novels was 16 days overdue, and the astronomy book was overdue $d + 7$, or 23, days.

Looking Back To check the answer, follow the original information. Each of the four novels was 16 days overdue, and the astronomy book was 23 days overdue. Because the fine was 10¢ per day per book, the fine for each of the novels was $16 \cdot 10$¢, or 160¢. Hence, the fine for all four novels was $4 \cdot 160$¢, or 640¢. The fine for the astronomy book was $23 \cdot 10$¢, or 230¢. Consequently, the total fine was 640¢ $+ 230$¢, or 870¢, which agrees with the given information of $8.70 as the total fine.

Problem Solving Newspaper Delivery

In a small town, 3 children deliver all the newspapers. Abby delivers 3 times as many papers as Bob, and Connie delivers 13 more than Abby. If the 3 children deliver a total of 496 papers, how many papers does each deliver?

Understanding the Problem The problem asks for the number of papers each child delivers. It compares the number of papers that each child delivers as well as the total number of papers delivered in the town.

Devising a Plan Let a, b, and c be the number of papers delivered by Abby, Bob, and Connie, respectively. We translate the given information into *algebraic equations* as follows:

Abby delivers 3 times as many papers as Bob: $a = 3b$.

Connie delivers 13 more papers than Abby: $c = a + 13$.

Total delivery is 496: $a + b + c = 496$.

To reduce the number of variables, substitute $3b$ for a in the second and third equations:

$$c = a + 13 \qquad \text{becomes} \qquad c = 3b + 13.$$
$$a + b + c = 496 \quad \text{becomes} \quad 3b + b + c = 496.$$

Next, make an equation in one variable, b, by substituting $3b + 13$ for c in the equation $3b + b + c = 496$; solve for b; and then find a and c.

Carrying Out the Plan

$$3b + b + 3b + 13 = 496$$
$$7b + 13 = 496$$
$$7b = 483$$
$$b = 69$$

Thus, $a = 3b = 3 \cdot 69 = 207$. Also, $c = a + 13 = 207 + 13 = 220$. So, Abby delivers 207 papers, Bob delivers 69 papers, and Connie delivers 220 papers.

Looking Back To check the answers, follow the original information, using $a = 207$, $b = 69$, and $c = 220$. The information in the first sentence, "Abby delivers 3 times as many papers as Bob," checks, since $207 = 3 \cdot 69$. The second sentence, "Connie delivers 13 more papers than Abby," is true because $220 = 207 + 13$. The information on the total delivery checks, since $207 + 69 + 220 = 496$.

🔴 NOW TRY THIS 8-9

Solve the *Newspaper Delivery* problem above by introducing only one unknown for the number of newspapers Bob delivered.

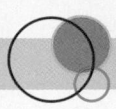

Assessment 8-3A

1. Consider the balances

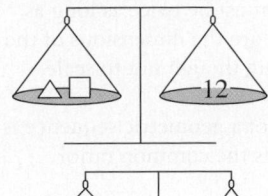

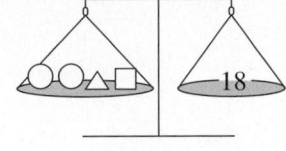

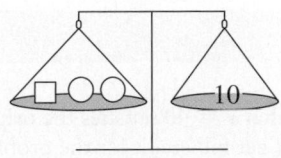

What is the value of each shape? Explain why.

2. Solve each of the following, if possible:
   a. $x - \sqrt{3} = 21$
   b. $2x + 5/2 = x + 25$
   c. $2x + {}^-5 = 3x - 4$
   d. $5(2x + 1) + 7(2x + 1) = 84$
   e. $3(2x - 6) = 4(2x - 6)$

Solve exercises 3 through 10 by setting up and solving an equation.

3. Ryan is building matchstick square sequences so that one square is added to the right each time a new figure is formed, as shown. He used 67 matchsticks to form the last figure in his sequence. How many squares are in this last figure?

4. For a particular event, 812 tickets were sold for a total of $1912. If students paid $2 per ticket and nonstudents paid $3 per ticket, how many student tickets were sold?

5. An estate of $486,000 is left to three siblings. The eldest receives 3 times as much as the youngest. The middle sibling receives $14,000 more than the youngest. How much did each receive?

6. A 10 ft board is to be cut into three pieces, two equal-length ones and the third 3 in. shorter than each of the other two. If the cutting does not result in any length being lost, how long are the pieces?

7. A box contains 67 coins, only dimes and nickels. The amount of money in the box is $4.20. How many dimes and how many nickels are in the box?

8. Miriam is 10 years older than Ricardo. Two years ago, Miriam was 3 times as old as Ricardo is now. How old are they now?

9. In a college, 15 times as many undergraduate students as graduate students are enrolled. If the total student enrollment at the college is 10,000, how many graduate students are there?

10. A farmer has 700 yd of fencing to enclose a rectangular pasture for her goats. Since one side of the pasture borders a river, that side does not need to be fenced. Side b must be twice as long as side a. Find the dimensions of the rectangular pasture.

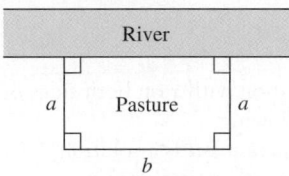

11. The sum of two consecutive terms in the arithmetic sequence 1, 4, 7, 10, . . . is 299; find these two terms.

Assessment 8-3B

1. Consider the balances

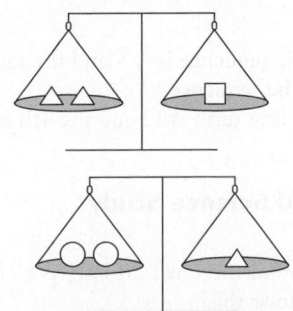

   a. Which shape weighs the most? Tell why.
   b. Which shape weighs the least? Tell why.

2. Solve each of the following, if possible:
   a. $3x + 13 = 2x + \sqrt{100}$
   b. $2x + 5 = 2(x + 5)$
   c. $7(3x + 6) + 5(3x + 6) = \sqrt{144}$
   d. $22 - x = 3x + \sqrt{6}$
   e. $22 - (2x - 6) = 3(2x - 6) + 6$
   f. $5(2x - 10) = 4(2x - 10)$

Solve exercises 3 through 11 by setting up and solving an equation.

3. Ryan is building matchstick square sequences, as shown. He used 599 matchsticks to form the last two figures in his sequence. How many matchsticks did he use in each of the last two figures?

4. At the Out-Rage Benefit Concert, 723 tickets were sold for $3/student and $5/non-student. The benefit raised $2815. How many non-student tickets were sold?

5. An estate of $1,000,000 is left to four siblings. The eldest is to receive twice as much as the youngest. The other two siblings are each to receive $16,000 more than the youngest. How much will each receive?

6. Ten years from now Alex's age will be 3 times her present age. Find Alex's age now.

7. Matt has twice as many stickers as David. How many stickers must Matt give David so that they will each have 120 stickers? Check that your answer is correct.

8. Miriam is four years older than Ricardo. Ten years ago Miriam was 3 times as old as Ricardo was then.
   a. How old are they now?
   b. Determine whether your answer is correct by checking that it satisfies the conditions of the problem.

9. In a college there are 13 times as many students as professors. If together the students and professors number 28,000, how many students are there in the college?

10. A farmer has 800 yd of fencing to enclose a rectangular pasture. One side of the pasture borders a river and does not need to be fenced. If side a must be twice as long as side b parallel to the river, what are the dimensions of the rectangular pasture? (see drawing though not to scale, exercise 10, p. 423)

11. The sum of the first two terms of a geometric sequence is 100 times the first term. What is the common ratio?

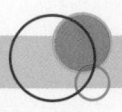

Mathematical Connections 8-3

Communication

1. Students were asked to find three consecutive whole numbers whose sum is 393. One student wrote the equation $x + (x + 1) + (x + 2) = 393$. Another wrote $(x - 1) + x + (x + 1) = 393$. Explain whether either approach works to find the answer to the question.

2. Explain how to solve the equation $3x + 5 = 5x - 3$ using a balance scale.

Open-Ended

3. Create an equation with x on both sides of the equation for each of the following.
   a. Every whole number is a solution.
   b. No whole number is a solution.
   c. 0 is a solution.

Cooperative Learning

4. Examine several elementary school textbooks for grades 1 through 5 and report on how algebraic concepts involving equations are introduced in each grade level.

5. Examine several elementry textbooks for grades 5–8 to see how spreadsheets are used to solve equations.

Questions from the Classroom

6. A student claims that the equation $3x = 5x$ has no solution because $3 \neq 5$. How do you respond?

7. A student claims that because in the following problem we need to find three unknown quantities, he must set up equations with three unknowns. How do you respond?
   Abby delivers twice as many papers as Jillian, and Brandy delivers 50 more papers than Abby. How many papers does each deliver if the total number of papers delivered is 550?

8. A student was told that in order to check a solution to a word problem like the one in exercise 7, it is not enough to check that the solution found satisfies the equation set up, but rather that it is necessary to check the answer against the original problem. She would like to know why. How do you respond?

9. On a test, a student was asked to solve the equation $4x + 5 = 3(x + 15)$. He proceeded as follows:
   $$4x + 5 = 3x + 45 = x + 5 = 45 = x = 40$$

Hence, $x = 40$. He checked that $x = 40$ satisfies the original equation; however, he did not get full credit for the problem and wants to know why. How do you respond?

Review Problems

10. If the number of sophomores, juniors, and seniors combined is denoted by x and it is 3 times the number of freshmen, denoted by y, write an algebraic equation that shows the relationship.

11. Write the sum of five consecutive even numbers if the middle one is n. Simplify your answer.

12. If Julie has twice as many CDs as Jack and Tira has 3 times as many as Julie, write an algebraic expression for the number of CDs each has in terms of one variable.

13. Write an algebraic equation relating the variables described in each of the following:
   a. The pay, P, for t hr if you are paid $30 for the first hour and $5 more than the preceding hour for each hour thereafter.
   b. Jimmy's total pay P after 4 years if the first year his salary was d dollars and then each year thereafter his salary is twice as much as in the preceding year.

14. Factor each of the following expressions and simplify as much as possible.
   a. $x^2 - 9$
   b. $x^2 - 5$
   c. $3(x + 5) - 4(x + 5)$
   d. $\sqrt{7x} - \sqrt{2x} + x$

15. If the first term of a geometric sequence is $\sqrt{5}$ and the ratio is 0.5, find the sixth term of the sequence.

16. In a geometric sequence, the first term is 12 and the 4th term is $\sqrt{5}$. Find the ratio.

Trends in Mathematics and Science Study (TIMSS) Questions

Ali had 50 apples. He sold some and then had 20 left. Which of these is a number sentence that shows this?
   a. $\Box - 20 = 50$ b. $20 - \Box = 50$
   c. $\Box - 50 = 20$ d. $50 - \Box = 20$

TIMSS, Grade 4, 2003

The objects on the scale make it balance exactly. On the left pan there is a 1 kg weight (mass) and half a brick. On the right pan there is one brick.

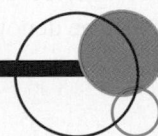

What is the weight (mass) of one brick?

a. 0.5 kg   **b.** 1 kg   **c.** 2 kg   **d.** 3 kg

TIMSS, Grade 8, 2003

$$3(2x - 1) + 2x = 21$$

What is the value of x?

a. -3

b. $-\dfrac{11}{4}$

c. $\dfrac{11}{4}$

d. 3

TIMSS, Grade 8, 2007

8-4 Functions

The concept of a function is central to all of mathematics and particularly to algebra, as elaborated in the following excerpt from grade 8 *Common Core Standards*. We find that students should:

- Define, evaluate, and compare functions.
- Use functions to model relationships between quantities. (p. 53)

Functions can model many real-world phenomena. In this section, we explore different ways to represent functions—as *rules, machines, equations, arrow diagrams, tables, ordered pairs, and graphs.* It is important that students see a variety of ways of representing functions.

Functions as Rules Between Two Sets

The following is an example of a game called "guess my rule," often used to introduce the concept of a function.

> When Tom said 2, Noah said 5. When Dick said 4, Noah said 7. When Mary said 10, Noah said 13. When Liz said 6, what did Noah say? What is Noah's rule?

The answer to the first question may be 9, and the rule could be "Take the original number and add 3"; that is, for any number n, Noah's answer is $n + 3$.

EXAMPLE 8-10

Guess the teacher's rule for the following responses:

a.

Student	Teacher
1	3
0	0
4	12
10	30

b.

Student	Teacher
2	5
3	7
5	11
10	21

c.

Student	Teacher
2	0
4	0
7	1
21	1

Solution

a. The teacher's rule could be "Multiply the given number n by 3;" that is, $3n$.

b. The teacher's rule could be "Double the original number n and add 1;" that is, $2n + 1$.

c. The teacher's rule could be "If the number n is even, answer 0; if the number is odd, answer 1." Another possible rule is "If the number is less than 5, answer 0; if greater than or equal to 5, answer 1."

Note that in Example 8-10, the rule connects the set describing what the student says to the set describing the teacher responses.

Functions as Machines

Another way to prepare students for the concept of a function is by using a "function machine." The following partial student page shows an example of a function machine. What goes into the machine is referred to as *input* and what comes out as *output*. On the student page, if the input to the function, f, is 3, the output is 12. For any input element x, the output could be denoted as $f(x)$, read "f of x." For the function in the example on the student page, one possibility is to write it as $f(x) = 4x$, where x is a real number.

School Book Page   WHY LEARN THIS?

Why Learn This?

Pretend you have a machine. You can put any number, or input, into the machine. The machine performs an operation on the number and provides a result, or output. A **function** is a rule that assigns exactly one output value to each input value.

Suppose you tell the machine to multiply by 4. A function table, such as the one at the right, shows the input and output values.

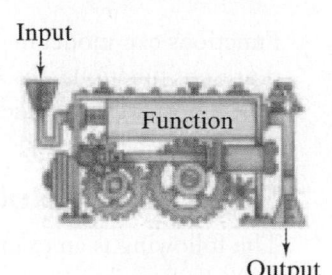

Input

Function

Output

Input	Output
3	12
−7	−28

Historical Note

Leonhard Euler

The Babylonians of Mesopotamia (ca. 2000 BCE) developed a precursor to a function. To them, a function was a table or a correspondence.

In his book *Geometry* (1637), René Descartes (1596–1650) used functions to describe many mathematical relationships. Almost 50 years after the publication of Descartes's book, Gottfried Wilhelm Leibniz (1646–1716) introduced the term *function*. Function was further formalized by Leonhard Euler (pronounced "oiler," 1707–1783), who introduced the notation $y = f(x)$. In the early twenty-first century, on most graphing calculators, $Y_1, Y_2, Y_3, \ldots$ serve as function notations where Y_1 acts like $f(x)$ if the function rule is written in terms of x.

EXAMPLE 8-11

Consider the function machine in Figure 8-19. For the function named *f*, what happens when the numbers 0, 1, 3, 4, and 6 are input?

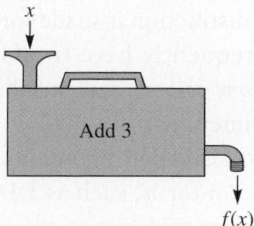

Figure 8-19

Solution For the given values of *x*, the corresponding values $f(x)$ are described in Table 8-4.

Table 8-4

x	f(x)
0	3
1	4
3	6
4	7
6	9

Functions as Equations

We write the equation $f(x) = x + 3$ to depict the function in Example 8-11. The output values can be obtained by substituting the values 0, 1, 3, 4, and 6 for *x* in $f(x) = x + 3$, as shown:

$$f(0) = 0 + 3 = 3$$
$$f(1) = 1 + 3 = 4$$
$$f(3) = 3 + 3 = 6$$
$$f(4) = 4 + 3 = 7$$
$$f(6) = 6 + 3 = 9$$

In many applications, both the inputs and the outputs of a function machine are numbers. However, inputs and outputs can be any objects. For example, consider a particular candy machine that accepts only 25¢, 50¢, and 75¢ and outputs one of three types of candy with costs of 25¢, 50¢, and 75¢, respectively. A function machine associates *exactly one output with each input*. If we enter some element *x* as input and obtain $f(x)$ as output, then every time we enter the same *x* as input, we obtain the same $f(x)$ as output. The idea of a function machine associating exactly one output with each input according to some rule leads to a definition of a function as a relation between two sets as seen below.

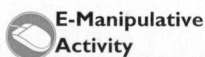
E-Manipulative Activity

Additional practice with function machines can be found in *Function Machine* on the E-Manipulative Disc.

Definition of Function

A **function** from set *A* to set *B* is a correspondence from *A* to *B* in which each element of *A* is paired with one, and only one, element of *B*.

The set, *A*, of all allowable inputs in the definition is the **domain** of the function. Normally, *if no domain is given to describe a function, then the domain is assumed to contain all elements for which the rule is meaningful.* Set *B*, the **codomain**, is any set that includes all the possible outputs. The set of all outputs is the **range** of the function. Set *B* includes the range and could be the range itself. The distinction is made for convenience sake, since sometimes the range is not easy to find. Students frequently have trouble with the language of functions (for example, *image, domain, range,* and *one-to-one*), which subsequently impacts their ability to work with graphical representations of functions.

A typical calculator contains many functions. For example, the $\boxed{\pi}$ button always displays an approximation for π, such as 3.1415927; the $\boxed{+/-}$ button either displays a negative sign in front of a number or removes an existing negative sign; and the $\boxed{x^2}$ and $\boxed{\sqrt{}}$ buttons square numbers and take the principal square root of numbers, respectively.

Not all input-output machines are function machines. Consider the machine in Figure 8-20. For any *natural-number* input *x*, the machine outputs a number that is less than *x*. If, for example, the number 10 is input, the machine may output 9, since 9 is less than 10. If 10 is input again, the machine may output 3, since 3 is less than 10. Such a machine is not a function machine because the same input may have different outputs.

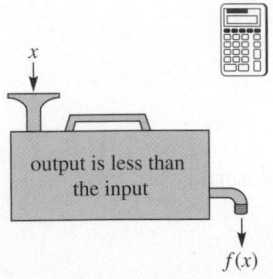

Figure 8-20

EXAMPLE 8-12

A bicycle manufacturer incurs a daily fixed cost of $1400 for overhead expenses and a cost of $500 per bike manufactured.

a. Find the total cost $C(x)$ of manufacturing *x* bikes in a day.
b. If the manufacturer sells each bike for $700, and the profit (or loss) in producing and selling *x* bikes in a day is $P(x)$, find $P(x)$ in terms of *x*.
c. Find the break-even point, that is, the number of bikes, *x*, produced and sold at which break even occurs (to break even means neither to make a profit nor a loss).

Solution
a. Since the cost of producing a single bike is $500, the cost of producing *x* bikes is $500x$ dollars. Because of the fixed cost of $1400 per day, the total cost, $C(x)$ in dollars, of producing *x* bikes in a given day is $C(x) = 500x + 1400$.
b. $P(x) =$ Income from selling *x* bikes $-$ total cost of manufacturing *x* bikes

$$P(x) = 700x - (500x + 1400)$$
$$= 200x - 1400$$

c. We need the number of bikes *x* to be produced so that $P(x) = 0$; that is, we need to solve $200x - 1400 = 0$.

$$200x - 1400 = 0$$
$$200x = 1400$$
$$x = \frac{1400}{200} \text{ or } 7$$

Thus, the manufacturer needs to produce and sell 7 bikes to break even.

Functions as Arrow Diagrams

Arrow diagrams can be used to examine whether a correspondence represents a function. This representation is normally used when sets *A* and *B* are finite sets with few elements. Example 8-13 shows how arrow diagrams can be used to examine both functions and nonfunctions.

EXAMPLE 8-13

Which, if any, of the parts of Figure 8-21 exhibit a function from A to B? If a correspondence is a function from A to B, find the range of the function.

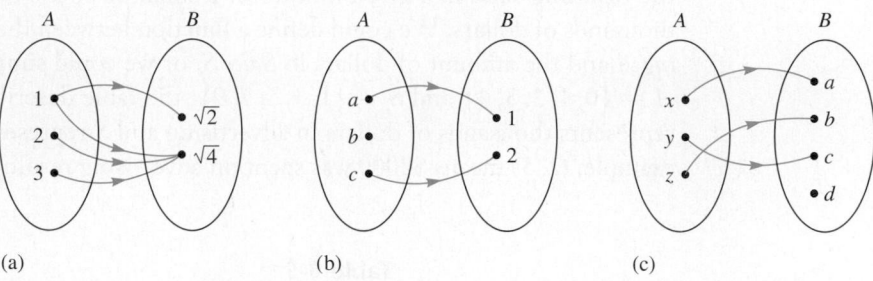

(a)   (b)   (c)

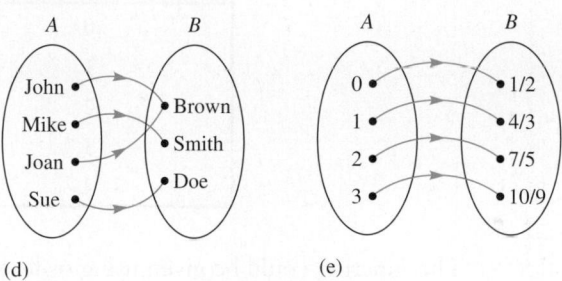

(d)   (e)

Figure 8-21

Solution

a. Figure 8-21(a) does not define a function from A to B, since the element 1 is paired with both $\sqrt{2}$ and $\sqrt{4}$.

b. Figure 8-21(b) does not define a function from A to B, since the element b is not paired with any element of B. (It is a function from a subset of A to B.)

c. Figure 8-21(c) does define a function from A to B, since there is one, and only one, arrow leaving each element of A. The fact that d, an element of B, is not paired with any element in the domain does not violate the definition. The range is $\{a, b, c\}$ and does not include d because d is not an output of this function, as no element of A is paired with d.

d. Figure 8-21(d) illustrates a function from A to B, since there is one, and only one, arrow leaving each element in A. It does not matter that an element of set B, Brown, has two arrows pointing to it. The range is {Brown, Smith, Doe}.

e. Figure 8-21(e) illustrates a function from A to B whose range is $\left\{\dfrac{1}{2}, \dfrac{4}{3}, \dfrac{7}{5}, \dfrac{10}{9}\right\}$.

Figure 8-21(e) also illustrates a one-to-one correspondence between A and B. In fact, any one-to-one correspondence between A and B defines a function from A to B as well as a function from B to A.

NOW TRY THIS 8-10

Determine which of the following are functions from the set of natural numbers to {0, 1}.

a. For every natural-number input, the output is 0.

b. For every natural-number input, the output is 0 if the input is an even number, and the output is 1 if the input is an odd number.

Functions as Tables and Ordered Pairs

Another way to describe a function is with a table as was seen earlier on the student page (p. 426). Consider the information in Table 8-5 relating the amount spent on advertising and the resulting sales in a given month for a small business. Note that the information is given in thousands of dollars. We could define a function between the amount of dollars spent in *Advertising A* and the amount of dollars in *Sales S*, or we could simply define the function as follows: If $A = \{0, 1, 2, 3, 4\}$ and $S = \{1, 3, 5, 7, 9\}$, the table describes a function from A to S, where A represents thousands of dollars in advertising and S represents thousands of dollars in sales. For example, $(2, 5)$ means \$2000 was spent on advertising resulting in \$5000 in sales.

Table 8-5

Amount of Advertising (in $1000s)	Amount of Sales (in $1000s)
0	1
1	3
2	5
3	7
4	9

The function could be given using ordered pairs as $(0, 1), (1, 3), (2, 5), (3, 7)$, and $(4, 9)$.

EXAMPLE 8-14

Which of the following sets of ordered pairs represent functions? If a set represents a function, give its domain and range. If it does not, explain why.

a. $\{(1, 2), (1, 3), (2, 3), (3, 4)\}$

b. $\left\{ \left(1, \frac{1}{2}\right), \left(2, \frac{1}{3}\right), \left(3, \frac{1}{4}\right), \left(4, \frac{1}{5}\right) \right\}$

c. $\{(1, 0), (2, 0), (3, 0), (4, 4)\}$

d. $\{(a, b) \mid a \in N \text{ and } b = 2a\}$

Solution

a. This is not a function because the input 1 has two different outputs.

b. This is a function with domain $\{1, 2, 3, 4\}$. Because the range is the set of outputs corresponding to these inputs set of outputs, the range is $\left\{ \frac{1}{2}, \frac{1}{3}, \frac{1}{4}, \frac{1}{5} \right\}$.

c. This is a function with domain $\{1, 2, 3, 4\}$ and range $\{0, 4\}$. The output 0 appears more than once, but this does not contradict the definition of a function in that each input corresponds to only one output.

d. This is a function with domain N and range E, the set of all even natural numbers.

Functions as Graphs

One widely recognized representation of a function is as a graph. Graphs appear in many media. In earlier chapters and in this chapter, we examined arithmetic and geometric sequences whose domains were the set of natural numbers. For example in Figure 8-22(a), we see a partial graph of the arithmetic sequence: $3, 5, 7, 9, 11, \ldots, 2n + 1, \ldots$. The graph of this sequence is depicted on the grid with points given as $(1, 3), (2, 5), (3, 7), (4, 9)$, and $(5, 11)$. Each ordered pair (a, b) is paired with a point on the grid. The horizontal axis in this case is used for the inputs (the numbers of the terms) and the vertical scale depicts the outputs (the terms of the sequence).

In general, to plot an ordered pair (a, b), we move to the point a (the **abscissa**) on the horizontal scale, and then move to the point b (the **ordinate**) along the vertical line that goes through a.

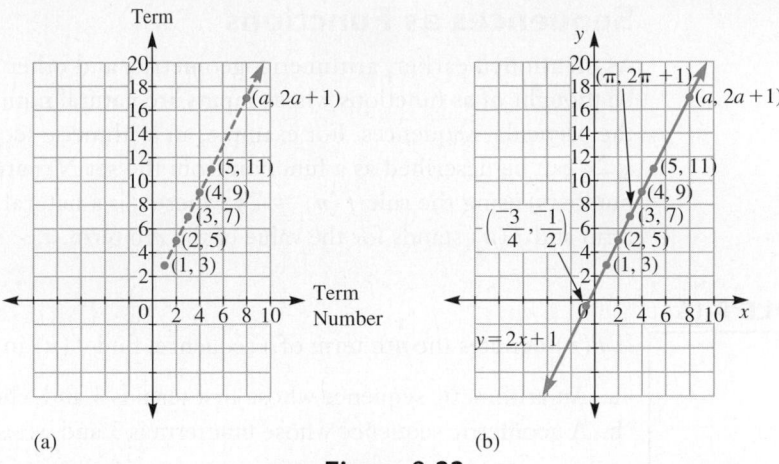

Figure 8-22

For example, to mark the point corresponding to $(1, 3)$, we start at 1 on the horizontal scale and move up 3 units on the vertical grid line through 1. Marking the point that corresponds to an ordered pair is referred to as **graphing** the ordered pair. The set of all points corresponding to all ordered pairs is the **graph** of the function or relation. In the graph of 8-22(a) the points are connected by a dashed ray to emphasize that they lie on a straight line, but not every point on the ray belongs to the graph.

Using all real numbers in the domain of the function $y = 2x + 1$ is depicted in Figure 8-22(b) and the graph is drawn as a solid line because for every real number a, there is a corresponding real number $2a + 1$ resulting in the ordered pair $(a, 2a + 1)$ which lies on that line. For example, $(\pi, 2\pi + 1)$ and $(^-3/4, ^-1/2)$ are points on the line. Each of the domain and range in this case is the set of real numbers. Additionally, the horizontal scale in this case is the **x-axis**; the vertical scale is the **y-axis**.

EXAMPLE 8-15

Explain why a telephone company would not set rates for telephone calls as depicted on the graph in Figure 8-23.

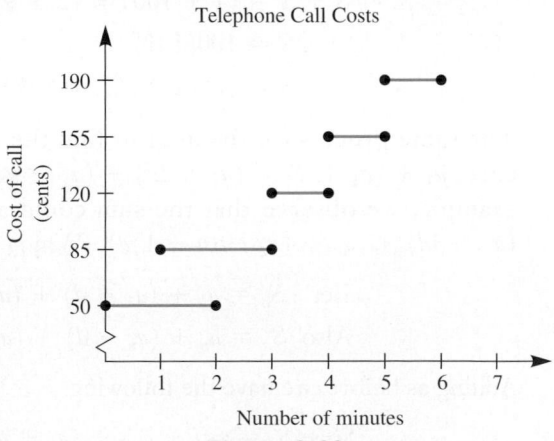

Telephone Call Costs

Figure 8-23

Solution The graph does not depict a function. For example, a customer could be charged either $0.50 or $0.85 for a 2-min call; hence, not every input has a unique output. This graph represents a relation (discussed later in the section) whose domain is the set of real numbers greater than or equal to 0 and whose codomain is the set of nonnegative real numbers. The range here is the set $\{50, 85, 120, 155, 190\}$.

Sequences as Functions

As mentioned earlier, arithmetic, geometric, and other sequences introduced in Chapter 1 can be thought of as functions whose inputs are natural numbers and whose outputs are the terms of the particular sequences. For example, an arithmetic sequence 2, 4, 6, 8, . . . , whose nth term a_n is $2n$ can be described as a function from the set N (natural numbers) to the set E (even natural numbers) using the rule $f(n) = 2n$, where n is a natural number representing the number of the term and $f(n)$ stands for the value of the nth term, a_n.

EXAMPLE 8-16

If $f(n)$ denotes the nth term of a sequence, find $f(n)$ in terms of n for each of the following:

   a. An arithmetic sequence whose first term is 3 and whose difference is 3
   b. A geometric sequence whose first term is 3 and whose ratio is 3.

Solution

   a. The first term is 3, the second term is $3 + 3$, or $2 \cdot 3$, the third is $2 \cdot 3 + 3$, or $3 \cdot 3$, the fourth term is $3 \cdot 3 + 3$, or $4 \cdot 3$, the nth term is $3n$, and hence $f(n) = 3n$, where n is a natural number.
   b. The first term is 3, the second $3 \cdot 3$, or 3^2, the third $3 \cdot 3^2$, or 3^3, and so on. Hence, the nth term is 3^n and therefore $f(n) = 3^n$, where n is a natural number.

Sums of Sequences as Functions

Gauss's problem in Chapter 1 to find the sum of $1 + 2 + 3 + \ldots + 100$ is an example of finding the sum of an arithmetic sequence with 100 terms. Recall that one way to find this sum was to treat it as follows:

$$\text{Let} \quad S = \quad 1 + \ 2 + \ 3 + \ldots + 100.$$
$$\text{Also} \quad S = 100 + 99 + 98 + \ldots + \quad 1.$$

Adding equals to equals we have the following:

$$S + S = (1 + 100) + (2 + 99) + (3 + 98) + \ldots + (100 + 1)$$
$$2S = 100(101)$$
$$S = 100(101)/2, \text{ or } 5050.$$

The same process can be used to find the sum of the first n terms of an arithmetic sequence: $a_1 + (a_1 + d) + (a_1 + 2d) + (a_1 + 3d) + \ldots + (a_1 + (n-1)d)$. To follow the above example, we observe that the sum could also be written as $a_n + (a_n - d) + (a_n - 2d) + (a_n - 3d) + \ldots + (a_n - (n-1)d)$. (Why?)

$$\text{Let} \quad S_n = a_1 + (a_1 + d) + (a_1 + 2d) + \cdots + (a_1 + (n-1)d).$$
$$\text{Also} \quad S_n = a_n + (a_n - d) + (a_n - 2d) + \cdots + (a_n - (n-1)d).$$

Adding as before, we have the following:

$$S_n + S_n = (a_1 + a_n) + (a_1 + a_n) + (a_1 + a_n) + \cdots + (a_1 + a_n).$$

Thus $2S_n = n(a_1 + a_n)$.

$$S_n = \left(\frac{n}{2}\right)(a_1 + a_n).$$

We summarize the sum of an arithmetic sequence in the following:

Theorem 8-5

The sum S_n of the first n terms of an arithmetic sequence with first term a_1 and nth term a_n is given by $S_n = \left(\dfrac{a_1 + a_n}{2}\right)n$.

Because $a_n = a_1 + (n - 1)d$, we may substitute for a_n to obtain an equivalent expression for the sum as follows:

$$S_n = \left(\frac{a_1 + a_1 + (n - 1)d}{2}\right)n, \text{ or } \left(\frac{2a_1 + (n - 1)d}{2}\right)n.$$

This last expression is sometimes written as $S_n = na_1 + n(n - 1)(d/2)$.

The sum can be thought of as a function of n where n is the number of terms of the sequence. In this case the domain is the set of natural numbers and the output for input n is the sum of the first n terms of the sequence.

EXAMPLE 8-17

Find the sum of the first 100 terms of the following arithmetic sequence: $3, 7, 11, 15, 19, \ldots$.

Solution   The arithmetic sequence given has first term 3 and difference $7 - 3$, or 4. The sum of the first 100 terms is $100(3) + 100(100 - 1)(4/2)$, or 20,100.

Similarly we can find the sum of the first n terms of a geometric sequence. Consider the geometric sequence: $a_1, a_1r, a_1r^2, a_1r^3, \ldots, a_1r^{(n-1)}$.

$$\text{Let} \quad S_n = a_1 + a_1r + a_1r^2 + \ldots + a_1r^{(n-1)}$$

Because the terms are very similar, suppose we multiply both sides of the equation by r obtaining $rS_n = a_1r + a_1rr + a_1r^2r + \ldots + a_1rr^{(n-1)}$, or $rS = a_1r + a_1r^2 + a_1r^3 + \ldots + a_1r^n$.

$$\text{Then } S_n = a_1 + a_1r + a_1r^2 + \ldots + a_1r^{(n-1)}$$
$$rS_n = \qquad a_1r + a_1r^2 + a_1r^3 + \ldots + a_1r^n$$

Subtracting, we obtain:

$$S_n - rS_n = a_1 - a_1r^n$$
$$S_n(1 - r) = a_1(1 - r^n), \text{ or}$$
$$S_n = a_1(1 - r^n)/(1 - r).$$

Note that in the above formula, $r \neq 1$ because division by 0 is not defined.

To summarize, we have the following:

Theorem 8-6

The sum of the first n terms of a geometric sequence whose first term is a_1, and whose ratio is $r \neq 1$ is $S_n = a_1(1 - r^n)/(1 - r)$.

If the sum of n terms of a geometric sequence is multiplied by $(^-1)/(^-1)$, it can be written as $S_n = a_1\left(\dfrac{r^n - 1}{r - 1}\right)$. As with the arithmetic sequence sum, the sum of the first n terms of a geometric sequence is a function of n, the number of terms.

> **REMARK** Students should compare the sum of a finite number of terms in a geometric sequence with the sum of an infinite number of terms.

 **EXAMPLE 8-18**

Find the sum of the first 10 terms of the geometric sequence: $3, -\dfrac{3}{2}, \dfrac{3}{4}, -\dfrac{3}{8}, \ldots$.

Solution The geometric sequence has first term 3, and ratio $\dfrac{-\frac{3}{2}}{3} = -\dfrac{1}{2}$. Thus the sum of

the first 10 terms is $S_{10} = 3\left(\dfrac{1 - \left(\frac{-1}{2}\right)^{10}}{1 - \frac{-1}{2}}\right)$, or $\dfrac{1023}{512}$.

Composition of Functions

Consider the function machines in Figure 8-24. If 2 is entered in the top machine, then $f(2) = 2 + 4 = 6$. The number 6 is then entered in the second machine and $g(6) = 2 \cdot 6 = 12$. The functions in Figure 8-24 illustrate the **composition of two functions**. In the composition of two functions, the range of the first function must be a subset of the domain of the second function.

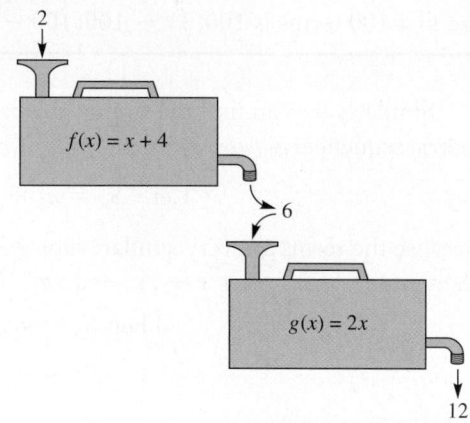

Figure 8-24

If the first function f is followed by a second function g, as in Figure 8-24, we symbolize the composition of the functions as $g \circ f$. If we input 3 in the function machines of Figure 8-24, then the output is symbolized by $(g \circ f)(3)$. Because f acts first on 3, to compute $(g \circ f)(3)$ we find $f(3) = 3 + 4 = 7$ and then $g(7) = 2 \cdot 7 = 14$. Hence, $(g \circ f)(3) = 14$. Observe that $(g \circ f)(3) = g(f(3))$. Also $(g \circ f)(x) = g(f(x)) = 2f(x) = 2(x + 4)$ and hence $g(f(3)) = 2(3 + 4) = 14$.

EXAMPLE 8-19

If $f(x) = 2x + 3$ and $g(x) = x - 3$, find the following:

  a. $(f \circ g)(3)$ **b.** $(g \circ f)(3)$ **c.** $(f \circ g)(x)$ **d.** $(g \circ f)(x)$

Solution
  a. $(f \circ g)(3) = f(g(3)) = f(3 - 3) = f(0) = 2 \cdot 0 + 3 = 3$
  b. $(g \circ f)(3) = g(f(3)) = g(2 \cdot 3 + 3) = g(9) = 9 - 3 = 6$
  c. $(f \circ g)(x) = f(g(x)) = 2g(x) + 3 = 2(x - 3) + 3 = 2x - 6 + 3 = 2x - 3$
  d. $(g \circ f)(x) = g(f(x)) = f(x) - 3 = (2x + 3) - 3 = 2x$

Example 8-19 shows that composition of functions is not commutative, since $(f \circ g)(3) \neq (g \circ f)(3)$.

Calculator Representation of a Function

A function can be represented in a variety of ways: pictures of sets with arrows, function machines, tables, equations, or graphs. Depending on the situation, one representation may be more useful than another. For example, if the domain of a function has many elements, a table is not a convenient representation. Graphing calculators can be used to display a graph of most functions given by equations with specified domains.

A sketch of the function $f(x) = 2x + 1$ is shown in Figure 8-25. A graphing calculator is available for free download on a computer. Note that on this graphing calculator $f(x)$ is depicted as y_1, and the domain was automatically chosen.

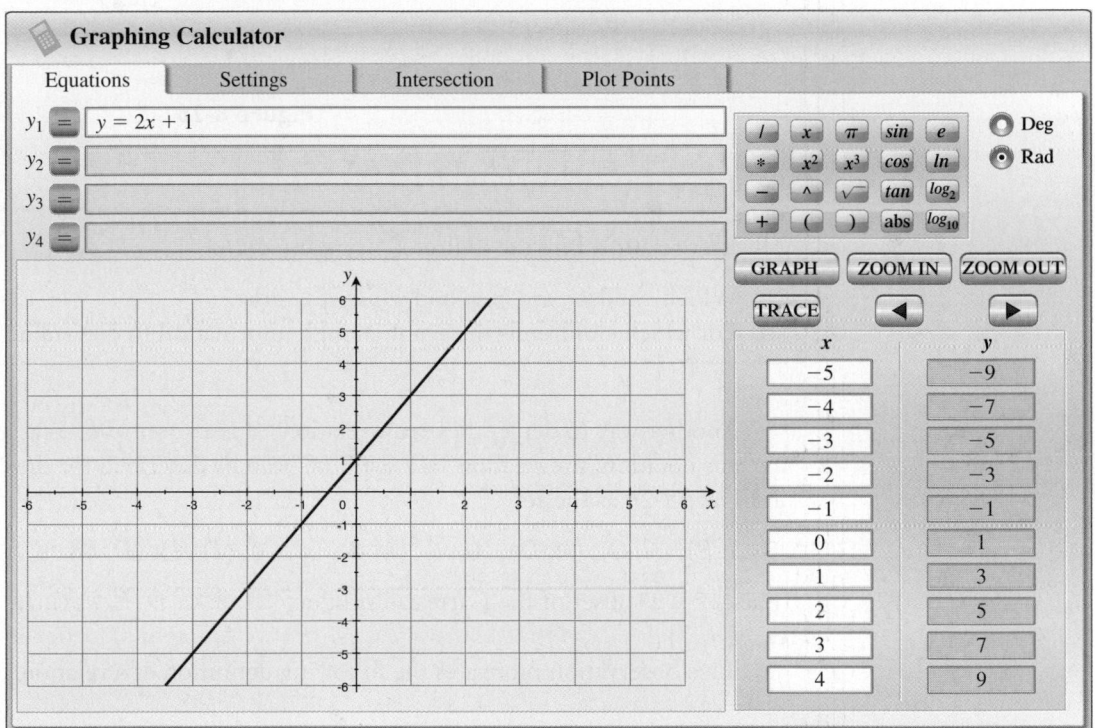

Figure 8-25

 NOW TRY THIS 8-11

Use a graphing calculator to sketch the graphs of $y = 2x + b$ for three choices of b. What do the graphs seem to have in common? Why?

Relations

A function from set A to set B is a form of *relation from set A to set B*. In a relation from A to B, there is a correspondence between elements of A and elements of B, but we do not require that each element of A be paired with one, and only one, element of B. Consequently, any set of ordered pairs is a relation. With appropriate set definitions, examples of relations include the following:

"is a daughter of"   "is the same color as"
"is less than"   "is greater than or equal to"

Consider the relation "is a sister of." Figure 8-26 illustrates this relation among children on a playground, with letters A through J representing the childrens. An arrow from I to J indicates that I "is a sister of" J. Notice the arrows from F to G and from G to F, which indicate that F is a sister of G and G is a sister of F. This implies that F and G are girls. On the other hand, the absence of an arrow from J to I implies that J is not a sister of I. Thus, I is a girl and J is a boy.

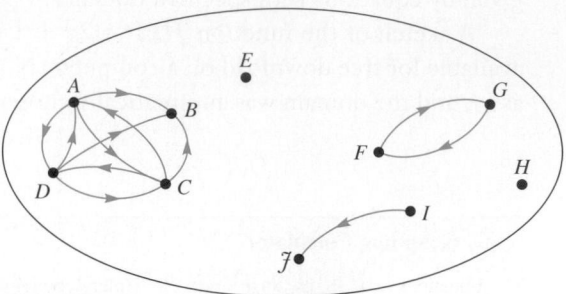

Figure 8-26

 NOW TRY THIS 8-12

All sister relationships are indicated in Figure 8-26.

   a. Which children are boys and which are girls?
   b. For which children is there not enough information to determine gender?

Another way to depict the arrow relation "A is a sister of B" is as an ordered pair (A, B). Using this notation, the relation "is a sister of" can be described for the children on the playground in Figure 8-26 as the set

$$\{(A, B), (A, C), (A, D), (C, A), (C, B), (C, D), (D, A), (D, B), (D, C), (F, G), (G, F), (I, J)\}$$

This set is a subset of the Cartesian product $\{A, B, C, D, E, F, G, H, I, J\} \times \{A, B, C, D, E, F, G, H, I, J\}$.

This observation motivates the following definition of a relation.

Defintion of a Relation from Set A to Set B

Given any two sets A and B, a **relation** from A to B is a subset of $A \times B$; that is, R is a relation from set A to set B if, and only if, $R \subseteq A \times B$.

In the definition, if $A = B$, we say that the **relation is on A**.

Properties of Relations

Figure 8-27 represents a set of children in a small group. The children have drawn all possible arrows representing the relation "has the same first letter in his or her name as." Three properties of relations are illustrated in Figure 8-27.

Definition of the Reflexive Property

A relation R on a set X is **reflexive** if, and only if, for every element $a \in X$, a is related to a; that is, for every $a \in X$, $(a, a) \in R$.

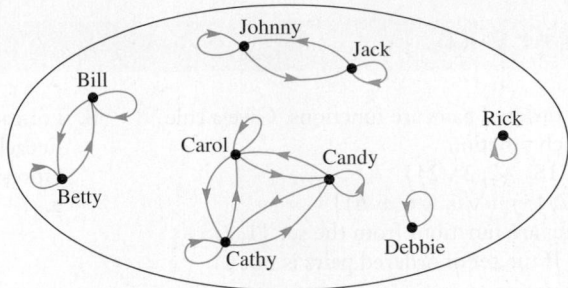

Figure 8-27

In the diagram, there is a loop at every point. For example, Rick has the same first initial as himself, namely R. A relation such as "is taller than" is not reflexive because people cannot be taller than themselves.

Definition of the Symmetric Property

A relation R on a set X is **symmetric** if, and only if, for all elements a and b in X, whenever a is related to b, then b also is related to a; that is, if $(a, b) \in R$, then $(b, a) \in R$.

In terms of the diagram, every pair of points that has an arrow headed in one direction also has a return arrow. For example, if Bill has the same first initial as Betty, then Betty has the same first initial as Bill. A relation such as "is a brother of" is not symmetric since Dick can be a brother of Jane, but Jane cannot be a brother of Dick.

Definition of the Transitive Property

A relation R on a set X is **transitive** if, and only if, for all elements a, b, and c of X, whenever a is related to b and b is related to c, then a is related to c. That is, if $(a, b) \in R$ and $(b, c) \in R$, then $(a, c) \in R$.

In the definitions for the symmetric and transitive properties, a, b, and c are variables and do not have to be different.

The relation in Figure 8-27 is transitive. For example, if Carol has the same first initial as Candy, and Candy has the same first initial as Cathy, then Carol has the same first initial as Cathy. A relation such as "is the father of" is not transitive since, if Tom Jones is the father of Tom Jones, Jr. and Tom Jones, Jr. is the father of Joe Jones, then Tom Jones is not the father of Joe Jones. He is, instead, the grandfather of Joe Jones.

The relation "is the same color as" is reflexive, symmetric, and transitive. In general, relations that satisfy all three properties are **equivalence relations**.

Definition of the Equivalence Relation

An **equivalence relation** is any relation R that satisfies the reflexive, symmetric, and transitive properties.

The most natural equivalence relation encountered in elementary school is "is equal to" on the set of all numbers. In subsequent chapters, we see other examples of equivalence relations, such as congruence and similarity in geometry.

Assessment 8-4A

1. The following sets of ordered pairs are functions. Give a rule that could describe each function.
   a. $\{(2, 4), (3, 6), (9, 18), (2, 2\sqrt{2})\}$
   b. $\{(2, 8), (5, 11), (7, 13), (\sqrt{6}, \sqrt{6} + 6)\}$

2. Which of the following are functions from the set $\{1, 2, 3\}$ to the set $\{a, b, c, d\}$? If the set of ordered pairs is not a function, explain why not.
   a. $\{(1, a), (2, b), (3, c), (1, d)\}$
   b. $\{(1, a), (2, b), (3, a)\}$

3. a. Draw an arrow diagram of a function with domain $\{1, 2, 3, 4, 5\}$ and range $\{a, b\}$.
   b. How many possible functions are there for part (a)?

4. Suppose $f(x) = 2x + 1$ and the domain is $\{0, 1, \sqrt{2}, \sqrt{3}, 4\}$. Describe the function in the following ways:
   a. Draw an arrow diagram involving two sets.
   b. Use ordered pairs.
   c. Make a table.
   d. Draw a graph to depict the function.

5. Determine which of the following are functions from the set of real numbers, R, or a subset of R, to R. If your answer is that it is not a function, explain why not.
   a. $f(x) = 2$ for all $x \in R$
   b. $f(x) = \sqrt{x}$

6. a. Make an arrow diagram for each of the following:
   (i) Rule: "when doubled is"

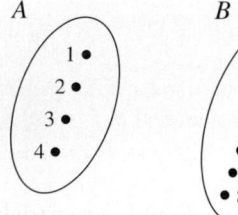

   A   B

   (ii) Rule: "is greater than"

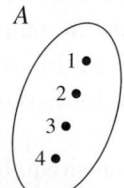 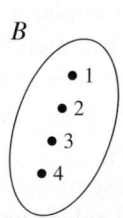

   A   B

   b. Which, if any, of the parts in (a) exhibits a function from A to B? If it is a function, tell why and find the range of the function.

7. The dosage of a certain drug is related to the weight of a child as follows: 50 mg of the drug and an additional 15 mg for each 2 lb or fraction of 2 lb of body weight above 30 lb. Sketch the graph of the dosage as a function of the weight of a child for children who weigh between 20 and 40 lb.

8. For each of the following, guess what might be Latifah's rule. In each case, if n is your input and $L(n)$ is Latifah's answer, express $L(n)$ in terms of n.

   a.
You	Latifah
3	8
4	11
5	14
10	29

   b.
You	Latifah
0	1
3	10
5	26
8	65

9. In *PSSM* for grades 6–8, the "Algebra" section (p. 229) poses the following problem. Quick-Talk advertises monthly cellular phone service for $0.50 a minute for the first 60 minutes but only $0.10 a minute for each minute thereafter. Quick-Talk charges for the exact amount of time used. Answer the following:
   a. Make one graph showing the cost per minute as a function of number of minutes and the other showing the total cost for calls as a function of the number of minutes up to 100 min.
   b. If you connect the points in the second graph in part (a), what kind of assumption needs to be made about the way the telephone company charges phone calls?
   c. Why does the total cost for calls consist of two line segments? Why is one part steeper than the other?
   d. The function representing the total cost for calls as a function of number of minutes talked can be represented by two equations. Write these equations.

10. For each of the following sequences (either arithmetic or geometric), find a possible function $f(n)$ whose domain is the set of natural numbers and whose outputs are the terms of the sequence.
    a. 3, 8, 13, 18, 23, . . .
    b. 3, 9, 27, 81, 243, . . .

11. Consider the following two function machines. Find the final output for each of the following inputs:
    a. 5   b. 10
    c. $\sqrt{7}$ d. 0

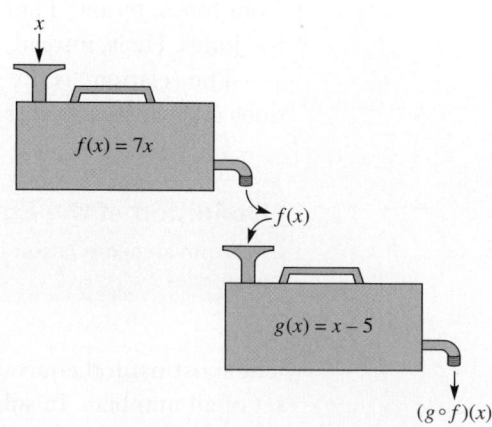

12. Let $t(n)$ represent the nth term of a sequence for $n \in N$.
Answer the following:
  a. If $t(n) = 4n - 3$, determine which of the following are
  output values of the function:
   (i) 1   **(ii)** 385   **(iii)** 389   **(iv)** 392
  b. If $t(n) = n^2$, determine which of the following are output
  values of the function:
   (i) 0   **(ii)** 25   **(iii)** 625   **(iv)** 1000   **(v)** 90
  c. If $t(n) = n(n - 1)$, determine which of the following are
  in the range of the function:
   (i) 0   **(ii)** 2   **(iii)** 20   **(iv)** 999

13. Consider a function machine that accepts inputs as ordered
pairs. Suppose the components of the ordered pairs are real
numbers and the first component is the length of a rectangle
and the second is its width. The following machine computes
the perimeter (the distance around a figure) of the rectangle.
Thus, for a rectangle whose length, l, is 3 and whose width,
w, is 2, the input is $(3, 2)$ and the output is $2 \cdot 3 + 2 \cdot 2$, or 10.

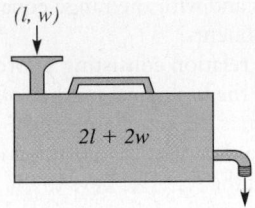

  a. For each of the following inputs, find the corresponding
  output: $(1, 7), (2, 6), (6, 2), (\sqrt{5}, \sqrt{5})$.
  b. Find the set of all the inputs for which the output is 20.
  c. What is the domain and the range of the function?

14. The following graph shows the relationship between the
number of cars on a certain road and the time of day for
times between 5:00 A.M. and 9:00 A.M.:

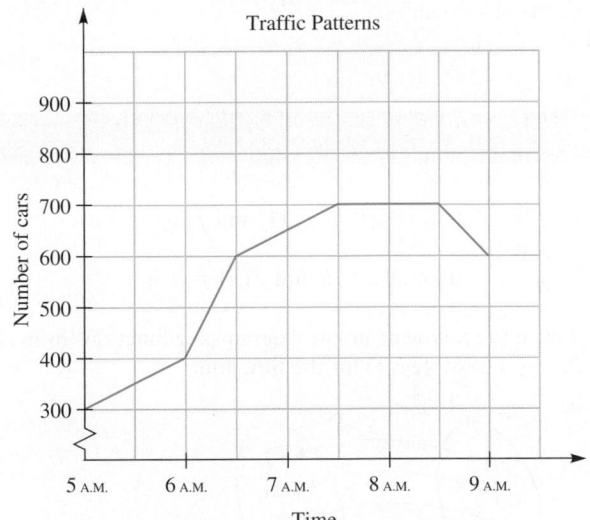

  a. What was the increase in the number of cars on the road
  between 6:30 A.M. and 7:00 A.M.?
  b. During which half hour was the increase in the number of
  cars the greatest?
  c. What was the increase in the number of cars between
  8:00 A.M. and 8:30 A.M.?

  d. During which half hour(s) did the number of cars
  decrease? By how much?
  e. The graph for this problem is composed of segments
  rather than just points. Why do you think segments are
  used here instead of just points?

15. A ball is shot out of a cannon at ground level. Its height H in
feet after t sec is given by the function $H(t) = 128t - 16t^2$.
  a. Find $H(2), H(6), H(3)$, and $H(5)$. Why are some of the
  outputs equal?
  b. Graph the function and from the graph find at what
  instant the ball is at its highest point. What is its height at
  that instant?
  c. How long does it take for the ball to hit the ground?
  d. What is the domain of H?
  e. What is the range of H?

16. For each of the following sequences of matchstick figures, let
$S(n)$ be the function giving the total number of matchsticks
in the nth figure.
  a. For each of the following, find the total number of
  matchsticks in the fourth figure.
  b. For each of the following, find as simple a formula as
  possible for $S(n)$ in terms of n.
   (i)

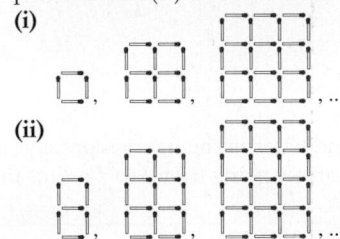

   (ii)

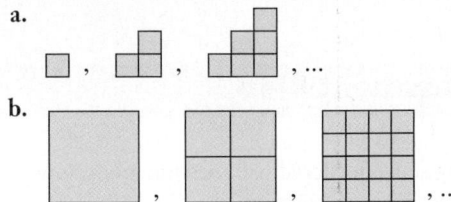

17. Assume the pattern continues for each of the following
sequences of square tile figures. Let $S(n)$ be the function giving
the total number of tiles in the nth figure. For each of the
following, find a formula for $S(n)$ in terms of n. In part (b), each
square is divided into four squares in the subsequent figure.
  a.

  ☐ , ⊟ , ⊞⊟ , ...

  b.

  ▢ , ▢ , ▢ , ...

18. A function can be represented as a set of ordered pairs where
the set of all the first components is the domain and where
the set of all the second components is the range. Is the
converse also true? That is, is every set of ordered pairs a
function whose domain is the set of first components and
whose range is the set of second components? Justify your
answer.

19. Which of the following equations or inequalities represent
functions and which do not? In each case x and y are real
numbers. Justify your answers.
  a. $x + y = 2$
  b. $x - y < 2$
  c. $y = x^3 + x$
  d. $xy = 2$

20. Which of the following are graphs of functions and which are not? Justify your answers.

a.
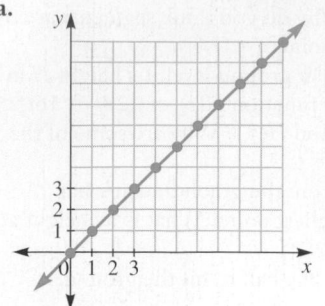

b.
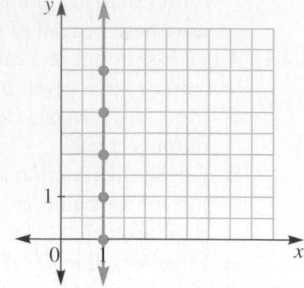

c.
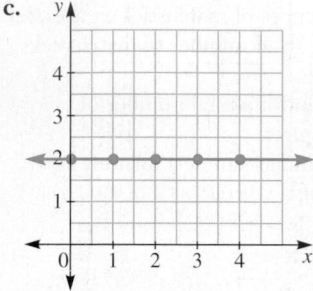

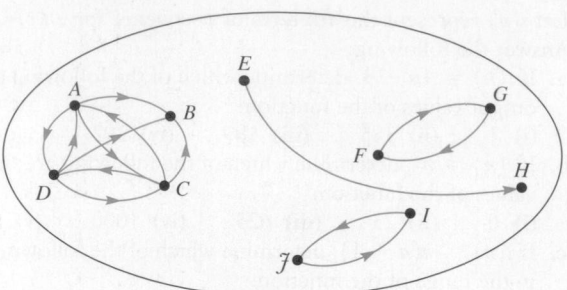

a. Based on the information in the figure, who are definitely girls and who are definitely boys?

b. Suppose we write "*A* is the sister of *B*" as an ordered pair (A, B). Based on the information in the diagram, write the set of all such ordered pairs.

c. Is the set of all ordered pairs in (b) a function with the domain equal to the set of all first components of the ordered pairs and with the range equal to the set of all second components?

22. a. Consider the relation consisting of ordered pairs (x, y) such that *y* is the biological mother of *x*. Is this a function whose domain is the set of all people?

b. Like in part (a) but now *y* is a brother of *x*. Is the relation a function from the set of all boys to the set of all boys?

23. Tell whether each of the following is reflexive, symmetric, or transitive on the set of all people. Which are equivalence relations?

a. "Is a parent of"

b. "Is the same age as"

c. "Has the same last name as"

d. "Is the same height as"

e. "Is married to"

f. "Lives within 10 mi of "

g. "Is older than"

21. Suppose each point and letter in the figure represents a child on a playground, and an arrow going from *I* to *J* means that *I* "is the sister of" *J*.

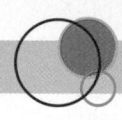 **Assessment 8-4B**

1. The following sets of ordered pairs are functions. Give a rule that could describe each function.
  a. $\{(5, 3), (7, 5), (11, 9), (\sqrt{3}, \sqrt{3} - 2)\}$
  b. $\{(2, 5), (3, 10), (4, 17), (\sqrt{3}, 4)\}$

2. Which of the following are functions from the set $\{1, 2, 3\}$ to the set $\{a, b, c, d\}$? If the set of ordered pairs is not a function, explain why not.
  a. $\{(1, c), (3, d)\}$ **b.** $\{(1, a), (1, b), (1, c)\}$

3. a. Draw an arrow diagram of a function with domain $\{1, 2, 3\}$ and range $\{a, b\}$.
  b. How many possible functions are there for part (a)?

4. Suppose $f(x) = 2(x + 1)$ and the domain is $\{0, 1, \sqrt{2}, \sqrt{3}, 4\}$. Describe the function in the following ways:
  a. Draw an arrow diagram involving two sets.
  b. Use ordered pairs.
  c. Make a table.
  d. Draw a graph to depict the function.

5. Determine which of the following are functions from the set of real numbers, *R*, or a subset of *R*, to *R*. If your answer is that it is not a function, explain why not.

a. $f(x) = 0$ if $x \in \{0, 1, 2, 3\}$, and $f(x) = 3$ if $x \notin \{0, 1, 2, 3\}$

b. $f(x) = 0$ for all $x \in R$ and $f(x) = 1$ if $x \in \{3, 4, 5, 6, \ldots\}$

6. Given the following arrow diagrams for functions from *A* to *B*, give a possible rule for the function:

a.
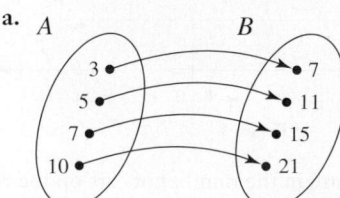

b.
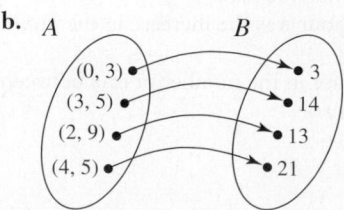

7. According to wildlife experts, the rate at which crickets chirp is a function of the temperature; specifically, $C = T - 40$, where C is the number of chirps every 15 sec and T is the temperature in degrees Fahrenheit.
   a. How many chirps does the cricket make per second if the temperature is 70°F?
   b. What is the temperature if the cricket chirps 40 times in 1 min?

8. For each of the following, guess what might be Latifah's rule. In each case, if n is your input and $L(n)$ is Latifah's answer, express $L(n)$ in terms of n.

   a.

You	Latifah
6	42
0	0
8	72
2	6

   b.

You	Latifah
0	1
1	2
5	32
6	64
10	1024

9. *PSSM* for grades 6–8 points out that "in their study of algebra, middle-grades students should encounter questions that focus on quantities that change" (p. 229). It poses the following problem:
   ChitChat charges $0.45 a minute for cellular phone calls. The cost per minute does not change, but the total cost changes as the telephone is used.

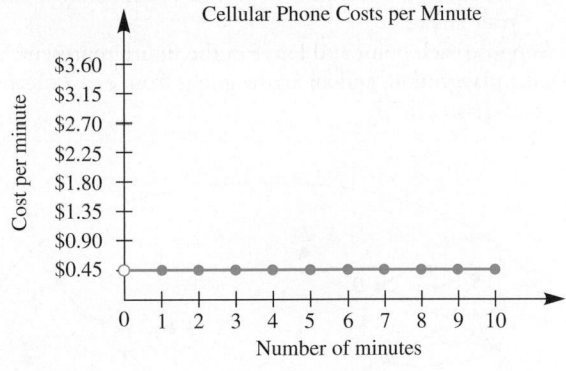

Cellular Phone Costs per Minute

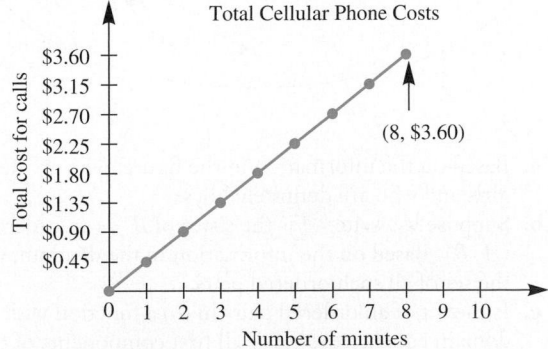

Total Cellular Phone Costs

(8, $3.60)

   a. When the number of minutes is 6, what do the values of the corresponding point on each graph represent?
   b. What kind of assumption about the charges needs to be made to allow the connection of the points on each graph? Explain.
   c. If time in minutes is t and the cost in dollars for calls is c, write c as a function of t for each graph.

10. For each of the following (arithmetic or geometric) sequences, discover the pattern and find a function whose

domain is the set of natural numbers and whose outputs are the terms of the sequence:
   a. 2, 4, 6, 8, 10,...   b. 1, 3, 9, 27, 81, ...

11. Consider two function machines that are placed as shown. Find the final output for each of the following inputs:
   a. 5   b. $\sqrt{3}$ c. 10   d. a

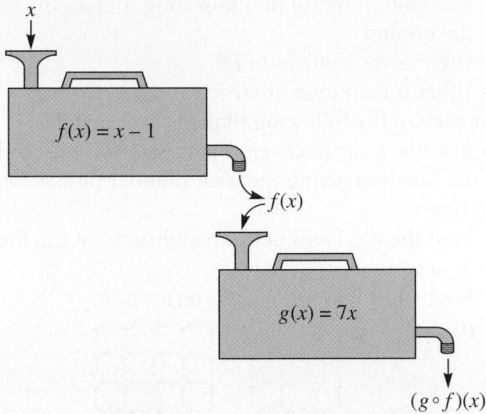

12. Let $t(n)$ represent the nth term of a sequence for $n \in N$.
   a. If $t(n) = n^2$, determine which of the following are output values of the function:
      (i) 1   (ii) 4   (iii) 9   (iv) 10   (v) 900
   b. If $t(n) = n(n + 1)$, determine which of the following are in the range of the function:
      (i) 2   (ii) 12   (iii) 2550   (iv) 2600

13. Consider a function machine that accepts inputs as ordered pairs. Suppose the components of the ordered pairs are natural numbers and the first component is the length of a rectangle and the second is its width. The following machine computes the perimeter (the distance around a figure) of the rectangle. Thus, for a rectangle whose length, l, is 3 and whose width, w, is 1, the input is (3, 1) and the output is $2 \cdot 3 + 2 \cdot 1$, or 8.

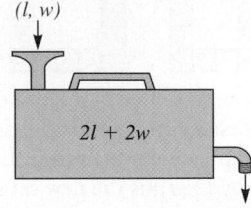

   a. For each of the following inputs, find the corresponding output: $(1, 4), (2, 1), (1, 2), (\sqrt{3}, \sqrt{3}), (x, y)$.
   b. Find the set of all the inputs for which the output is 20.
   c. Is (2, 2) a possible output? Explain.

14. A health club charges a one-time initiation fee of $100.00 plus a membership fee of $40.00 per month.
   a. Write an expression for the cost function $C(x)$ that gives the total cost for membership at the health club for x months.
   b. Draw the graph of the function in (a).
   c. The health club decided to give its members an option of a higher initiation fee but a lower monthly membership charge. If the initiation fee is $300.00 and the monthly membership fee is $30.00, use a different color and draw on the same set of axes the cost graph under this plan.
   d. Determine after how many months the second plan is less expensive for the member.

15. A ball is shot straight up at ground level. Its height H in feet after t sec is given by the function $H(t) = 128t - 16t^2$.

  a. Graph the function and from the graph find at what instant the ball is at its highest point. What is its height at that instant?

  b. Find from the graph all t such that $H(t) = H(1)$.

  c. Use your graph to find how long it takes the ball to hit the ground.

  d. What is the domain of H?

  e. What is the range of H?

16. For each of the following sequences of matchstick figures, assume that your discovered pattern continues and let $S(n)$ be the function giving the total number of matchsticks in the nth figure.

  a. Find the total number of matchsticks in the fourth figure.

  b. Find a formula for $S(n)$ in terms of n.

  (i)

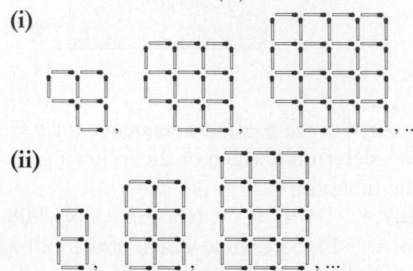

  (ii)

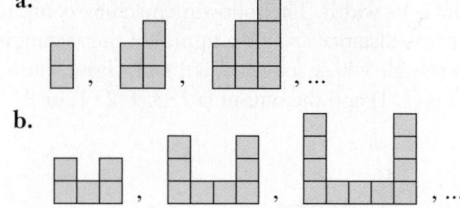

17. Assume the pattern continues for each of the following sequences of square tile figures. Let $S(n)$ be the function giving the total number of tiles in the nth figure. For each of the following, find a formula for $S(n)$ in terms of n.

  a.

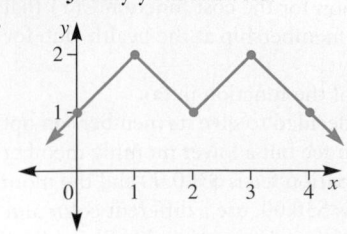

  b.

18. A function can be represented as a set of ordered pairs where the set of all the first components is the domain and the set of all the second components is the range. If each ordered pair (a, b) is replaced by (b, a), is the new set still a function?

19. Which of the following equations or inequalities represent functions and which do not? In each case x and y are real numbers. Justify your answers.

  a. $x - y = 2$ **b.** $x + y < 20$

  c. $y = 2x^2$ **d.** $y = x^3 - 1$

20. Which of the following are graphs of functions and which are not? Justify your answers.

  a.

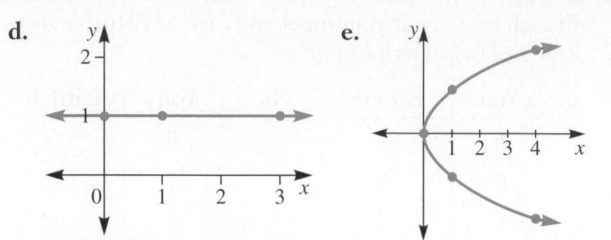

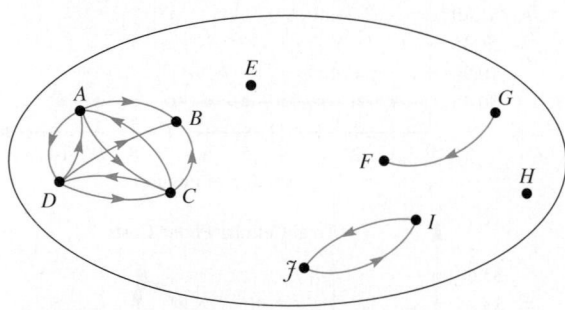

21. **a.** Which of the following relations from the set W of whole numbers to W have the symmetric property? Justify your answers.

    (i) $x + y = 10$

    (ii) $x - y = 100$

    (iii) $xy = 100$

    (iv) $y = x$

    (v) $y = x^2$

  b. Which of the relations in part (a) are functions? Justify your answer.

22. Suppose each point and letter in the figure represents a child on a playground, and an arrow going from I to J means that I "is the sister of" J.

  a. Based on the information in the figure, who are definitely girls and who are definitely boys?

  b. Suppose we write "A is the sister of B" as an ordered pair (A, B). Based on the information in the diagram, write the set of all such ordered pairs.

  c. Is the set of all ordered pairs in (b) a function with the domain equal to the set of all first components of the ordered pairs and with the range equal to the set of all second components?

23. Which of the following are functions and which are relations but not functions from the set of first components of the ordered pairs to the set of second components?

  a. {(Montana, Helena), (Oregon, Salem), (Illinois, Springfield), (Arkansas, Little Rock)}

b. {(Pennsylvania, Philadelphia), (New York, Albany), (New York, Niagara Falls), (Florida, Ft. Lauderdale)}

c. {(x, y) | x resides in Birmingham, Alabama, and x is the mother of y, where y is a U.S. resident}

d. {(1, 1), (2, 4), (3, 9), (4, 16)}

e. {(x, y) | x and y are natural numbers and $x + y$ is an even number}

24. Tell whether each of the following is reflexive, symmetric, or transitive on the set of subsets of a nonempty set. Which are equivalence relations?

a. "Is equal to"

b. "Is a proper subset of"

c. "Is not equal to"

d. "Has the same cardinal number as"

Mathematical Connections 8-4

Communication

1. Does the diagram define a function from A to B? Why or why not?

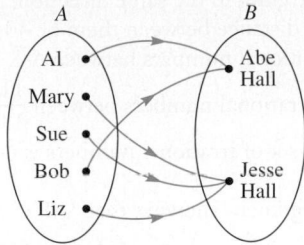

2. Is a one-to-one correspondence a function? Explain your answer and give an example.

3. Which of the following are functions from A to B? If your answer is "not a function," explain why not.

a. A is the set of mathematics faculty at the university. B is the set of all mathematics classes. To each mathematics faculty member, we associate the class that person is teaching during a given term.

b. A is the set of mathematics classes at the university. B is the set of mathematics faculty. To each mathematics class, we associate the teacher who is teaching the class.

c. A is the set of all U.S. senators and B is the set of all senate committees. We associate each senator to the committee of which the senator is chairperson.

4. If S is the set of students in Ms. Carmel's class, and A is any subset of S, we define: $f(A) = \overline{A}$ (where $\overline{A}$ is the complement of A). Notice that the input in this function is a subset of S and the output is a subset of S. Answer the following:

a. Explain why f is a function and describe the domain and the range of f. *

b. If there are 20 children in the class, what are the number of elements in the domain and the number in the range? Explain.

c. Is the function in this question a one-to-one correspondence? Justify your answer.

Open-Ended

5. Examine several newspapers and magazines and describe at least three examples of functions that you find. What is the domain and range of each function?

6. Give at least three examples of functions from A to B where neither A nor B is a set of numbers.

7. Draw a sequence of matchstick figures and describe the pattern in words. Find as simple an expression as possible for $S(n)$, the total number of matchsticks in the nth figure.

8. A function whose output is always the same regardless of the input is a *constant function*. Give several examples of constant functions from real life.

9. A function whose output is the same as its input is an *identity function*. Give several concrete examples of identity functions.

Cooperative Learning

10. Each person in a group picks a natural number and uses it as an input in the following function machine:

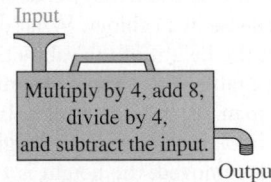

a. Compare your answers. Based on the answers, make a conjecture about the range of the function.

b. Based on your answer in (a), graph the function.

c. Write the function in the simplest possible way using $f(x)$ notation.

d. Justify your conjecture in (a).

e. Make up similar function machines and try different inputs in your group.

f. Devise a function machine in which the machine performs several operations but the output is always the same as the input. Exchange your answer with someone in the group and check that the other person's function machine performs as required.

11. In a group of four, work through the following. You will need a metric tape or meterstick.

a. Place your mathematics book on a desk and measure the distance (to the nearest centimeter) from the floor to the top of the book. Record the distance.

b. Place a second mathematics book on top of the first and measure the distance (to the nearest centimeter) from the floor to the top of the second book. Record the distance.

c. Continue this procedure for all four of your mathematics books and complete the following table and graph:

Number of books	Distance from floor
1	
2	
3	
4	

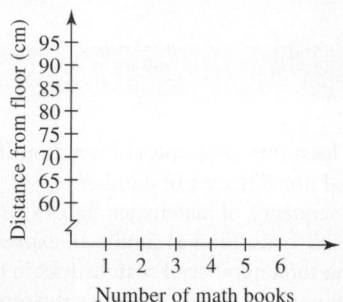

d. Without measuring, what is the distance from the floor with 0 books? 5 books?
e. Write a rule or function for $d(x)$, where $d(x)$ is the distance above the floor to the top of the stack of books and x is the number of books.
f. Suppose the distance from the floor to the ceiling is 2.5 m. If you stack the books as described above, how many books would be needed to reach the ceiling?
g. The function $h(x) = 34x + 70$ represents the height of another stack of x mathematics books (in centimeters) of the same thickness in a cabinet. What does the function tell you about the height of the cabinet?
h. Suppose that a table with a stack of similar mathematics books (more than 10) is 200 cm high. If the top mathematics book is removed, the height is 197 cm. If a second book is removed, the height is 194 cm. What is the height if 5 books are removed?
i. Write a function $h(x)$ for the height of the stack in part (h) after x books are removed.

Questions from the Classroom

12. A student claims that the following machine does not represent a function machine because it accepts two inputs at once rather than a single input. How do you respond?

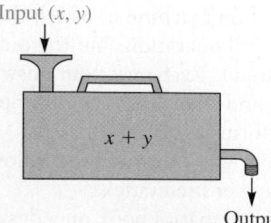

13. A student claims that the following does not represent a function, since all the values of x correspond to the same number.

x	0	1	2	3	4	5
y	1	1	1	1	1	1

How do you respond?

14. A student thinks that the function $f(x) = 3x + 5$ with domain the set of real numbers is a one-to-one correspondence and he would like to know why. How do you respond?
15. A student wants to know why sometimes it is incorrect to connect points on the graph of a function. How do you respond?

Review Problems

16. Solve the following equations for x, if possible:
    a. $3x - 1 = x + 99$
    b. $2(5x + 1) - 11 = x + 9$
    c. $3(x - 1) = 2(x - 1) + 99$
    d. $5(2x - 6) = 3(2x - 6)$
17. Solve the following problem by setting up an appropriate equation:

    Two cars, each traveling at a constant speed—one 60 mph and the other 70 mph—start at the same time from the same point traveling in the same direction. After how many hours will the distance between them be 40 mi?
18. a. Find two rational numbers between $\sqrt{3}$ and 2.

    b. Find two irrational numbers between $\frac{11}{13}$ and $\frac{12}{13}$.
19. Show that the set of irrational numbers is not closed under addition.
20. Develop an argument showing that $\sqrt{2} + 2$ is an irrational number.

National Assessment of Educational Progress (NAEP) Questions

In	Out
2	5
3	7
4	9
5	11
15	31
38	

The table shows how the "In" numbers are related to the "Out" numbers. When 38 goes in, what number comes out?
a. 41
b. 51
c. 54
d. 77

NAEP, Grade 4, 2007

Each figure in the pattern below is made of hexagons that measure 1 centimeter on each side.

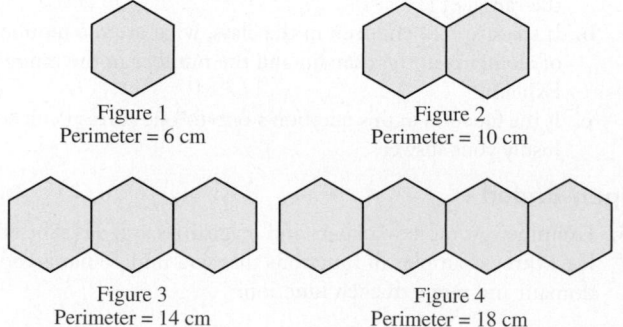

Figure 1
Perimeter = 6 cm

Figure 2
Perimeter = 10 cm

Figure 3
Perimeter = 14 cm

Figure 4
Perimeter = 18 cm

If the pattern of adding one hexagon to each figure is continued, what will be the perimeter of the 25th figure in the pattern? Show how you got your answer.

NAEP, Grade 8, 2007

In the equation $y = 4x$, if the value of x is increased by 2, what is the effect on the value of y?
  a. It is 8 more than the original amount.
  b. It is 6 more than the original amount.
  c. It is 2 more than the original amount.
  d. It is 16 times the original amount.
  e. It is 8 times the original amount.

NAEP, Grade 8, 2007

Trends in Mathematics and Science Study (TIMSS) Questions

A number machine takes a number and operates on it.
When the Input Number is 5, the Output Number is 9, as shown below.

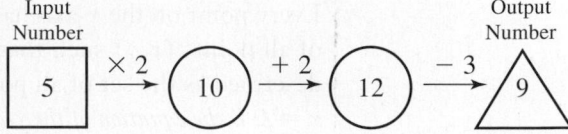

When the Input Number is 7, which of these is the Output Number?
  a. 11   **b.** 13   **c.** 14   **d.** 25

TIMSS, Grade 4, 2003

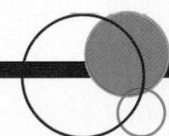

8-5 Equations in a Cartesian Coordinate System

The Cartesian coordinate system (named for René Descartes) enables us to combine geometry and algebra. A **Cartesian coordinate system**, as used in Section 8-4, is constructed by placing two number lines perpendicular to each other, as shown in Figure 8-28. The intersection point of the two lines is the **origin**, the horizontal line is the **x-axis**, and the vertical line is the **y-axis**. As noted earlier, the location of any point P can be described by an ordered pair of numbers (a, b), where a perpendicular from P to the x-axis intersects at a point with coordinate a and a perpendicular from P to the y-axis intersects at a point with coordinate b; the point is identified as $P(a, b)$. A line is **perpendicular** to another line if they form a 90° angle (right angle). There is a one-to-one correspondence between all the points in the plane and all the ordered pairs of real numbers. For example, in Figure 8-28, R has coordinates $(^-4, ^-3)$, written $R(^-4, ^-3)$.

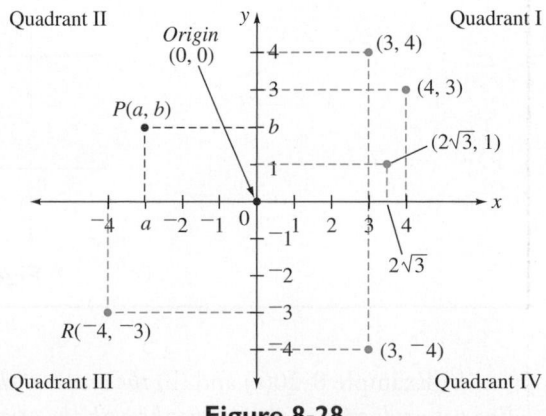

Figure 8-28

In Figure 8-28, the axes (x- and y-axis) separate the plane into four **quadrants** numbered counterclockwise.

Equations of Vertical and Horizontal Lines

Every point on the x-axis has a y-coordinate of zero. Thus, the x-axis can be described as the set of all points (x, y) such that $y = 0$. *The x-axis has equation $y = 0$.* Similarly, the y-axis can be described as the set of all points (x, y) such that $x = 0$ and y is an arbitrary real number. Thus, $x = 0$ *is the equation of the y-axis.*

EXAMPLE 8-20

Sketch the graph for each of the following on a Cartesian coordinate system.

a. $x = 2$
b. $y = 3$
c. $x < 2$ and $y = 3$

Solution

a. The equation $x = 2$ represents the set of all points (x, y) for which $x = 2$ and y is any real number, as in Figure 8-29(a).

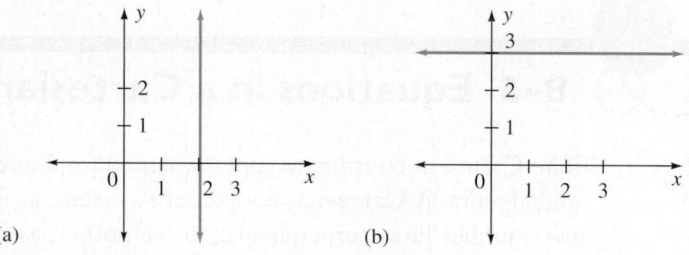

(a)   (b)

Figure 8-29

b. The equation $y = 3$ represents the set of all points (x, y) for which $y = 3$ and x is any real number, as in Figure 8-29(b).

c. The statements in part (c) represent the set of all points (x, y) for which $x < 2$, but $y = 3$. The set is part of a line, as shown in Figure 8-30. Note that the hollow dot at $(2, 3)$ indicates that this point is not included in the solution set.

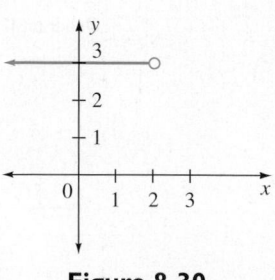

Figure 8-30

In Example 8-20(a) and (b) *the graph of the equation $x = a$, where a is some real number, is the line perpendicular to the x-axis through the point with coordinates $(a, 0)$. Similarly, the graph of the equation $y = b$ is the line perpendicular to the y-axis through the point with coordinates $(0, b)$.*

Equations of Lines

The arithmetic sequence 4, 7, 10, 13,... in Table 8-6 has xth term $3x + 1$. If the number of the term is the x-coordinate and the corresponding term the y-coordinate, the set of points appear to lie on a line that is parallel to neither the x-axis nor the y-axis, as in Figure 8-31.

Table 8-6

Number of Term	Term
1	4
2	7
3	10
4	13
.	.
.	.
.	.
x	$3x + 1$

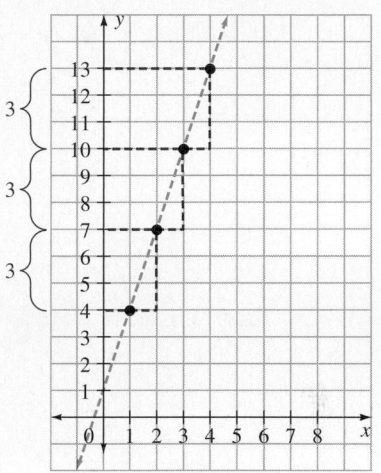

Figure 8-31

In Table 8-6, we see that there is a difference of 3 in the y-coordinates of the marked points. Correspondingly, there is a difference of 1 in the number of the term. Hence, we might say that the rate of change of the term with respect to the number of the term is 3, or the ratio is 3 to 1. Does something comparable happen with other arithmetic sequences?

To help answer this question, we consider the following sequences in Table 8-7.

Table 8-7

Number of Term x	$1x$	$2x$	$\frac{1}{2}x$	$(^-1)x$	$(^-2)x$
1	1	2	$\frac{1}{2}$	$^-1$	$^-2$
2	2	4	1	$^-2$	$^-4$
3	3	6	$\frac{3}{2}$	$^-3$	$^-6$
4	4	8	2	$^-4$	$^-8$
5	5	10	$\frac{5}{2}$	$^-5$	$^-10$
6	6	12	3	$^-6$	$^-12$
.	.	.	.	.	.
.	.	.	.	.	.
.	.	.	.	.	.
x	x	$2x$	$\frac{1}{2}x$	^-x	^-2x

Figure 8-32 shows the sets of ordered pairs (x, y) plotted on a graph so that the number of the term is the x-coordinate and the corresponding term appearing in a particular column in Table 8-7, is the y-coordinate. Again the separate sets of points appear to lie on straight lines.

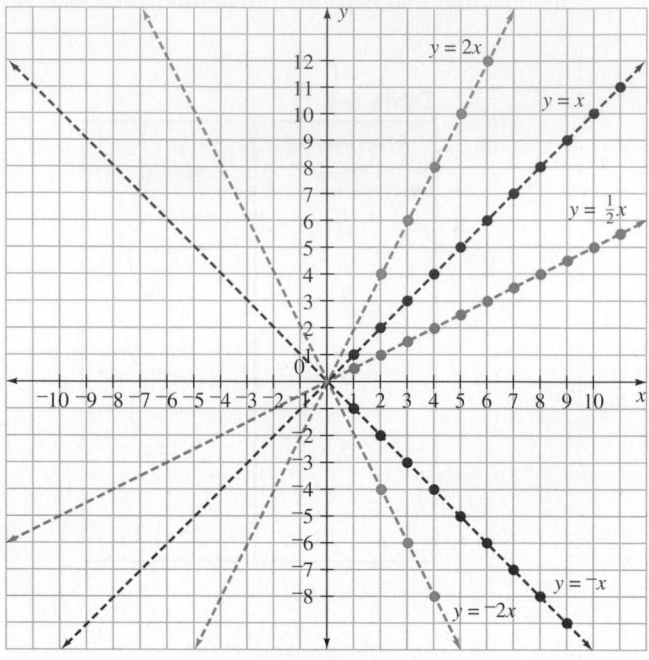

Figure 8-32

All dashed lines in Figure 8-32 have equations of the form $y = mx$, where m takes the values $2, 1, \frac{1}{2}, {}^{-}1,$ and ${}^{-}2$. If we consider the equations but allow x to be any real number, then all points along a given dashed line are connected and all the points on that line satisfy the corresponding equation. The number m is a measure of steepness and is the **slope** of the line whose equation is $y = mx$. The graph goes up from left to right (increases) if m is positive, and it goes down from left to right (decreases) if m is negative.

> **NOW TRY THIS 8-13**
>
> **a.** In the equation $y = mx$, if m is 0, what happens to the line?
> **b.** What happens to the line as m increases from 0?
> **c.** What happens to the line when m decreases from 0?

All dashed lines in Figure 8-32 pass through the origin. This is true for any line whose equation is $y = mx$. If $x = 0$, then $y = m(0) = 0$ and $(0, 0)$ is a point on the graph of $y = mx$. Conversely, it is possible to show that any nonvertical line passing through the origin has an equation of the form $y = mx$ for some value of m.

EXAMPLE 8-21

Find the equation of the line that contains $(0, 0)$ and $(2, 3)$.

Solution   The line goes through the origin so its equation has the form $y = mx$. To find the equation of the line, we find the value of m. The line contains $(2, 3)$, so we substitute 2 for x and 3 for y in the equation $y = mx$ to obtain $3 = m(x)$, and thus $m = \frac{3}{2}$. The required equation is $y = \frac{3}{2}x$.

Next, we consider equations of the form $y = mx + b$, where b is a real number. We do this by first examining the graphs of $y = x + 2$ and $y = x$. Given the graph of $y = x$, we obtain the graph of $y = x + 2$ by "raising" each point on the graph of $y = x$ by 2 units. This is shown in Figure 8-33(a). Similarly, to sketch the graph of $y = x - 2$, we first draw the graph of $y = x$ and then lower each point vertically by 2 units, as shown in Figure 8-33(a).

The graphs of $y = x$, $y = x + 2$, and $y = x - 2$ are parallel lines. In general, *for a given value of m, the graph of $y = mx + b$ is a straight line through $(0, b)$ and parallel to the line whose equation is $y = mx$.*

Further, the graph of the line $y = mx + b$, can be obtained from the graph of $y = mx$ by sliding $y = mx$ up (or down) b units depending on the value of b, as shown in Figure 8-33(b). In general, *any two parallel lines have the same slope or are vertical lines with no slope.*

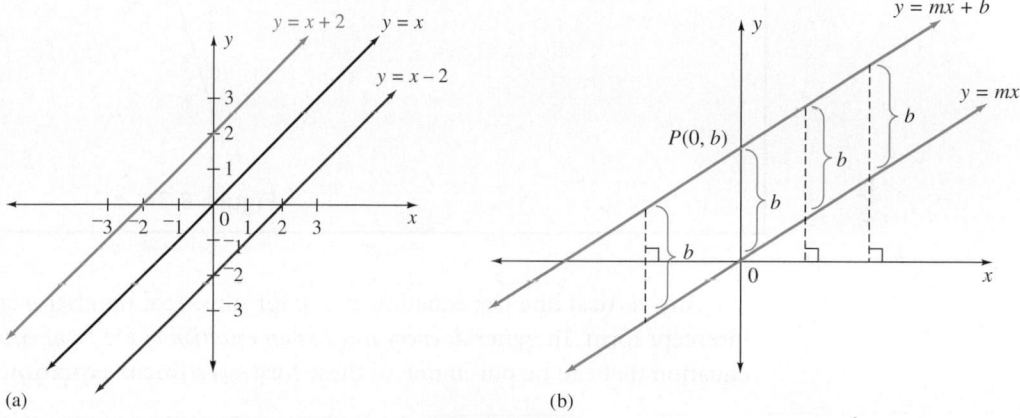

Figure 8-33

The graph of $y = mx + b$ in Figure 8-33(b) crosses the y-axis at point $P(0, b)$. The value of y at the point of intersection of any line with the y-axis is the **y-intercept**. Thus, b is the y-intercept of $y = mx + b$; this form of a **linear equation** is the **slope-intercept form**. Similarly, the value of x at the point of intersection of a line with the x-axis is the **x-intercept**.

EXAMPLE 8-22

Given the equation $y - 3x = {}^-6$, answer the following:

  a. Find the slope of the line.
  b. Find the y-intercept.
  c. Find the x-intercept.
  d. Sketch the graph of the equation.

Solution

  a. We write the equation in the form $y = mx + b$ by adding $3x$ to both sides of the given equation: $y = 3x + ({}^-6) = 3x - 6$. Hence, the slope is 3.
  b. The form $y = 3x + ({}^-6)$ shows that $b = {}^-6$, which is the y-intercept. The y-intercept can also be found directly by substituting $x = 0$ in the equation and finding the corresponding value of y.
  c. The x-intercept is the x-coordinate of the point where the graph intersects the x-axis. At that point, $y = 0$. Substituting 0 for y in $y = 3x - 6$ gives 2 as the x-intercept.
  d. Knowing the y-intercept and the x-intercept gives us two points, $(0, {}^-6)$ and $(2, 0)$, on the line. We may plot these points and draw the line through them to obtain the desired graph in Figure 8-34.

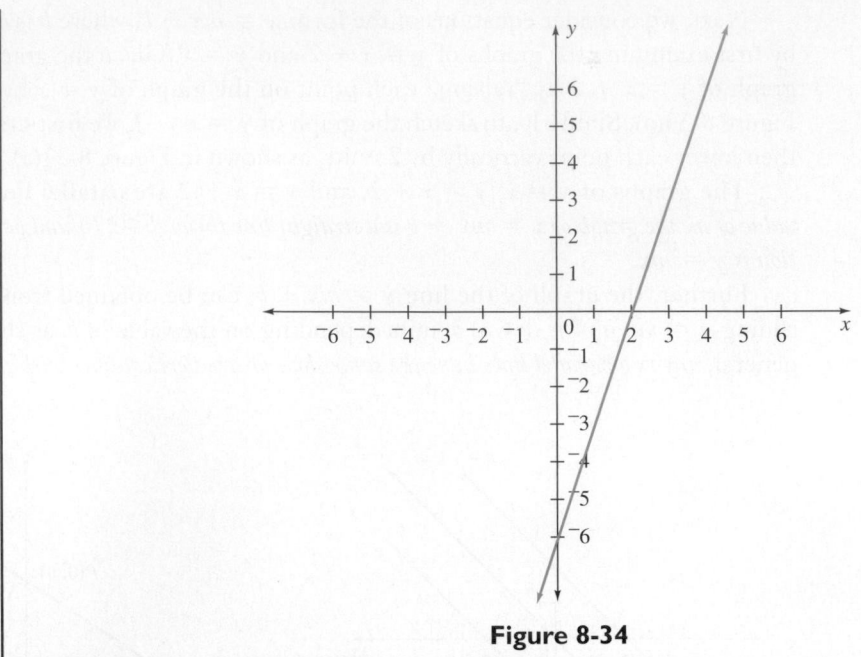

Figure 8-34

Any vertical line has equation $x = a$ for some real number a and cannot be written in slope-intercept form. In general, *every line has an equation of the form either $y = mx + b$ or $x = a$.* Any equation that can be put in one of these forms is a **linear equation**.

Theorem 8-7: Equation of a Line

Every line has an equation of the form either $y = mx + b$ or $x = a$, where m is the slope and b is the y-intercept. In the equation $x = a$, a is the **x-intercept**.

NOW TRY THIS 8-14

The nth term of an arithmetic sequence is given as $a_n = a_1 + (n - 1)d$. Explain why this is considered a *linear relationship*.

Determining Slope

We defined the slope of a line with equation $y = mx + b$ to be m. The slope is a measure of steepness of a line. A different way to discuss the steepness of a line is to consider the rate of change in y-values in relation to their corresponding x-values. In Figure 8-35, line k has a greater rate of change than line ℓ. In other words, line k rises higher than line ℓ for the same horizontal run.

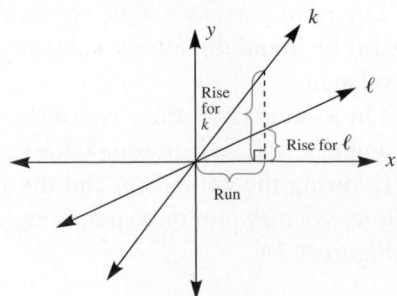

Figure 8-35

Thus, we could express the rate of change, the steepness, or the slope, as the ratio $\dfrac{\text{Change in } y\text{-values}}{\text{Corresponding change in } x\text{-values}}$. The rate is frequently expressed as $\dfrac{\text{Rise}}{\text{Run}}$.

We generalize to find the slope of a line $\overleftrightarrow{AB}$ as the change in y-coordinates divided by the corresponding change in x-coordinates of any two points on $\overleftrightarrow{AB}$. The difference $x_2 - x_1$ is the *run*, and the difference $y_2 - y_1$ is the *rise*. The slope formula can be interpreted with coordinates, as shown in Figure 8-36. The ratio $\dfrac{y_2 - y_1}{x_2 - x_1}$ is always the same, regardless of which two points on a given nonvertical line are chosen.

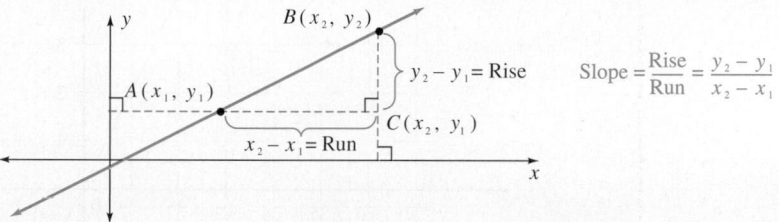

Figure 8-36

The discussion of slope is summarized in the following definition for the slope of a line.

Definition of Slope

Given two points $A(x_1, y_1)$ and $B(x_2, y_2)$ with $x_1 \neq x_2$, the slope m of the line $\overleftrightarrow{AB}$ is

$$m = \frac{y_2 - y_1}{x_2 - x_1}$$

In the above definition, if we multiply both the numerator and the denominator on the right side of the slope formula by $^-1$, we obtain

$$m = \frac{y_2 - y_1}{x_2 - x_1} = \frac{(y_2 - y_1)(^-1)}{(x_2 - x_1)(^-1)} = \frac{y_1 - y_2}{x_1 - x_2}.$$

Thus, it does not matter which point is named (x_1, y_1) and which is named (x_2, y_2); *the order of the coordinates in the subtraction must be the same.*

NOW TRY THIS 8-15

a. Use the slope formula to find the slope of any horizontal line.
b. What happens when we attempt to use the slope formula for a vertical line? What is your conclusion about the slope of a vertical line?

When a line is inclined downward from the left to right, the slope is negative. This is illustrated in Figure 8-37, where the graph of the line $y = {}^-2x$ is shown.

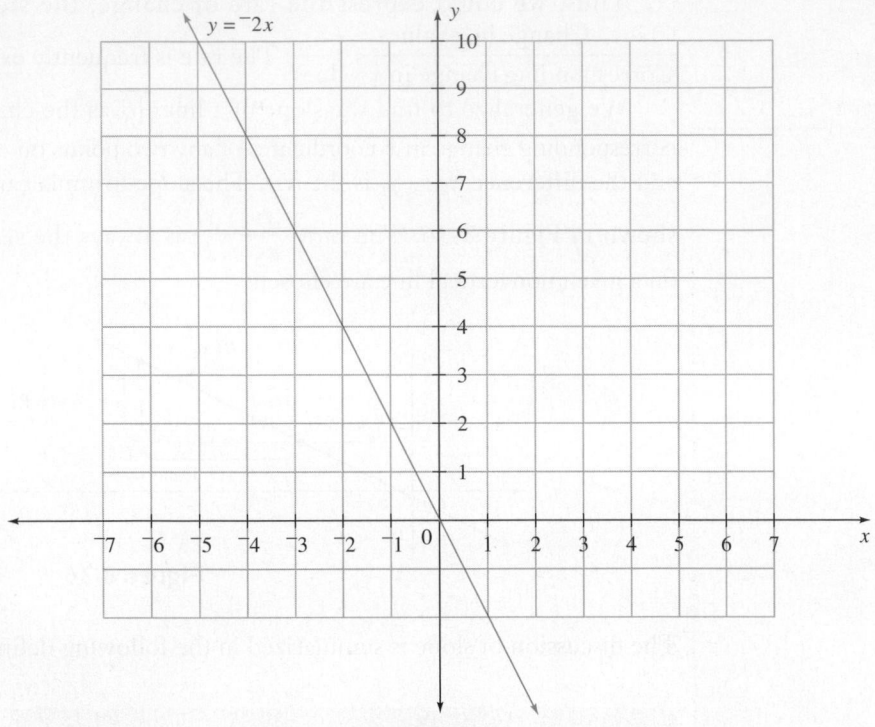

Figure 8-37

EXAMPLE 8-23

a. Given $A(3, 1)$ and $B(5, 4)$, find the slope of $\overleftrightarrow{AB}$.
b. Find the slope of the line passing through the points $A(^-3, 4)$ and $B(^-1, 0)$.
c. In Figure 8-38, find the slope of $\overleftrightarrow{OA}$.

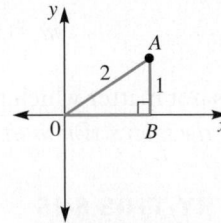

Figure 8-38

Solution

a. $m = \dfrac{4 - 1}{5 - 3} = \dfrac{3}{2}$,   or $\dfrac{1 - 4}{3 - 5} = \dfrac{^-3}{^-2} = \dfrac{3}{2}$

b. $m = \dfrac{4 - 0}{^-3 - (^-1)} = \dfrac{4}{^-2} = {}^-2$,   or $\dfrac{0 - 4}{^-1 - (^-3)} = \dfrac{^-4}{2} = {}^-2$

c. OB is the run while AB, or 1, is the rise.
$(OB)^2 + 1^2 = 2^2$ in right triangle OAB. Using the Pythagorean theorem.
$(OB)^2 + 1 = 4$
$(OB)^2 = 4 - 1$, or 3
$OB = \sqrt{3}$ (Why?)

The slope of $\overleftrightarrow{OA}$ is $\dfrac{1}{\sqrt{3}}$, which is sometimes written as $\dfrac{1}{\sqrt{3}} \cdot \dfrac{\sqrt{3}}{\sqrt{3}}$, or $\dfrac{\sqrt{3}}{3}$.

Given two points on a nonvertical line, we use the slope formula to find the slope of the line and its equation as demonstrated in the following example.

EXAMPLE 8-24

In Figure 8-39, the points $(^-4, 0)$ and $(1, 4)$ are on the line ℓ. Find:

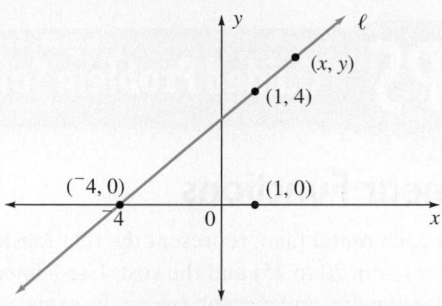

Figure 8-39

   a. The slope of the line
   b. The equation of the line

Solution

   a. $m = \dfrac{4 - 0}{1 - (^-4)} = \dfrac{4}{5}$

   b. Any point (x, y) on ℓ different from $(^-4, 0)$ can be found using the slope $4/5$ determined by points $(^-4, 0)$ and (x, y). Thus,

$$\frac{y - 0}{x - (^-4)} = \frac{4}{5}$$

$$y = \frac{4}{5}(x + 4).$$

The equation of the line is $y = \dfrac{4}{5}x + \dfrac{16}{5}$.

Systems of Linear Equations

The mathematical descriptions of many problems involve more than one equation, each having more than one unknown. To solve such problems, we must find a common solution to the equations, if it exists. An example is given on the following student page. (p. 454)

EXAMPLE 8-25

May Chin paid $18.00 for three soyburgers and twelve orders of fries. Another time she had paid $12 for four soyburgers and four orders of fries. Assume the prices have not changed. Set up a system of equations with two unknowns representing the prices of a soyburger and an order of fries, respectively.

Solution   Let x be the price in dollars of a soyburger and y be the price in dollars of an order of fries. Three soyburgers cost $3x$ dollars, and twelve orders of fries cost $12y$ dollars. Because May paid $18.00 for her entire order, we have $3x + 12y = 18$, or, after dividing each side by 3, $x + 4y = 6$. Similarly, $4x + 4y = 12$, or $x + y = 3$.

School Book Page GUIDED PROBLEM SOLVING

GPS Guided Problem Solving

Linear Functions

For each rental plan, represent the relationship between the number of miles (from 20 to 45) and the cost. Use a linear function, a table of ordered pairs, and a graph (using the same coordinate grid). What conclusions can you draw about these plans?

What You Might Think

What do I know? What do I want to find out?

How can I write a function rule for each plan?

How can I make a table and graph?

What conclusions can be stated?

What You Might Write

Plan 1 costs $15 plus $.25 per mile. Plan 2 costs $8 plus $.45 per mile. I want to compare the two plans using function rules, tables, and graphs.

Let m = the number of miles driven. Let C = the cost of the rental in dollars.

Plan 1: $C_1 = \$.25m + 15$

Plan 2: $C_2 = \$.45m + 8$

I can use the function rule to get data points for the table. Then I can graph those points.

m	C_1	C_2
20	20	17
30	22.5	21.5
35	23.75	23.75
45	26.25	28.25

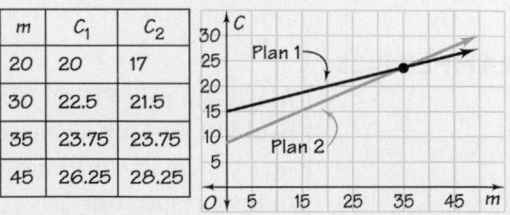

The lines intersect at (35, 23.75). Plan 2 is better if you drive less than 35 miles; otherwise, Plan 1 is better.

An ordered pair satisfying the two linear equations in Example 8-25 is a point that belongs to each of the lines. Figure 8-40 shows the graphs of $x + 4y = 6$ and $x + y = 3$. The costs are the common values x and y that satisfy each equation, that is, a point on each of the lines. The two lines appear to intersect at $(2, 1)$. Thus, $(2, 1)$ seems to be the solution of the given system of equations. This solution can be checked by substituting 2 for x and 1 for y in each equation. Because two distinct lines intersect in only one point, $(2, 1)$ is the only solution to the system. Therefore, in Example 8-25 a soyburger costs \$2 and fries cost \$1.

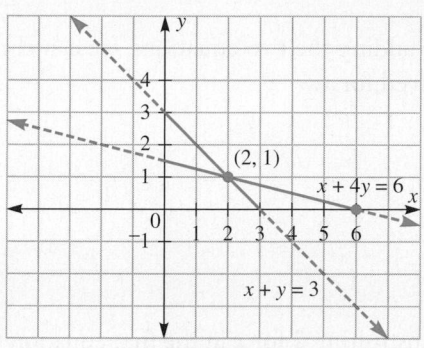

Figure 8-40

Because costs of the soyburger and an order of fries cannot be negative, parts of the lines are dashed.

Substitution Method

There are algebraic methods for solving systems of linear equations as in Example 8-25. Example 8-26 demonstrates one such algebraic method: the **substitution method**.

EXAMPLE 8-26

Solve the following system:

$$3x - 4y = 5$$
$$2x + 5y = 1$$

Solution   First, rewrite one equation, expressing y in terms of x.

$$y = \frac{3x - 5}{4}$$

Because we are looking for x and y that satisfy each equation, we substitute $\dfrac{3x - 5}{4}$ for y in the other original equation and solve the resulting equation for x.

$$2x + 5\left(\frac{3x - 5}{4}\right) = 1$$
$$4\left[2x + 5\left(\frac{3x - 5}{4}\right)\right] = 4 \cdot 1$$
$$8x + 5(3x - 5) = 4$$
$$8x + 15x - 25 = 4$$
$$23x = 4 + 25$$
$$x = \frac{29}{23}$$

Substituting $\dfrac{29}{23}$ for x in $y = \dfrac{3x - 5}{4}$ gives $y = \dfrac{^-7}{23}$. Hence, $x = \dfrac{29}{23}$ and $y = \dfrac{^-7}{23}$. This can be checked by substituting the values for x and y in the original equations.

Elimination Method

The **elimination method** for solving two equations with two unknowns is based on eliminating one of the variables by adding or subtracting the original or equivalent equations. For example, consider the following system:

$$x - y = {}^-3$$
$$x + y = 7$$

By adding the two equations, we can eliminate the variable y. The resulting equation can then be solved for x.

$$
\begin{array}{rcl}
x - y &=& {}^-3 \\
x + y &=& 7 \\
\hline
2x &=& 4 \\
x &=& 2
\end{array}
$$

Substituting 2 for x in the first equation (either equation may be used) gives $y = 5$. Checking this result shows that $x = 2$ and $y = 5$, or $(2, 5)$, is the solution to the system.

Often, another operation is required before equations are added so that an unknown can be eliminated. For example, consider the following system:

$$3x + 2y = 5$$
$$5x - 4y = 3$$

Neither adding nor subtracting the equations eliminates an unknown. However, if the first equation contained $4y$ rather than $2y$, the variable y could be eliminated by adding. To obtain $4y$ in the first equation, we multiply both sides of the equation by 2 to obtain the equivalent equation $6x + 4y = 10$. Adding the equations in the equivalent system gives the following:

$$
\begin{array}{rcl}
6x + 4y &=& 10 \\
5x - 4y &=& 3 \\
\hline
11x &=& 13 \\
x &=& \dfrac{13}{11}
\end{array}
$$

To find the corresponding value of y, we substitute $\dfrac{13}{11}$ for x in either of the original equations and solve for y.

Thus $\quad y = \dfrac{8}{11}$.

Consequently, $\left(\dfrac{13}{11}, \dfrac{8}{11} \right)$ is the solution of the original system. This solution, as always, should be checked by substitution in the *original* equations.

Solve Paige's problem in the *FoxTrot* cartoon algebraically.

Solutions to Various Systems of Linear Equations

All examples thus far have had unique solutions. However, other situations may arise. Geometrically, a system of two linear equations can be characterized as follows:

1. *The system has a unique solution if, and only if, the graphs of the equations intersect in a single point.*
2. *The system has no solution if, and only if, the equations represent distinct parallel lines. Two distinct lines in a plane are* **parallel** *if they do not intersect.*
3. *The system has infinitely many solutions if, and only if, the equations represent the same line.* [*A line is considered parallel to itself.*

Consider the following system and assume it has a solution.

$$2x - 3y = 1$$
$$^-4x + 6y = 5$$

In an attempt to solve for x, we multiply the first equation by 2 and then add as follows:

$$4x - 6y = 2$$
$$\underline{^-4x + 6y = 5}$$
$$0 = 7$$

A false statement results from the assumption that the system had a solution. That assumption caused a false statement; therefore, the assumption itself must be false. Hence, the system has no solution. In other words, the solution set is $\varnothing$. This situation arises if, and only if, the corresponding lines are parallel and different.

Next, consider the following system:

$$2x - 3y = 1$$
$$^-4x + 6y = ^-2$$

To solve this system, we multiply the first equation by 2 and add as follows:

$$
\begin{array}{r}
4x - 6y = 2 \\
^-4x + 6y = ^-2 \\
\hline
0 = 0
\end{array}
$$

The resulting statement, $0 = 0$, is always true. In the original system, if $^-2$ had been multiplied times both sides of the first equation, the result would be the second equation. Thus the graphs are exactly the same or the graph is a single line. All pairs of x and y that correspond to points on the line $2x - 3y = 1$ fit $^-4x + 6y = ^-2$.

Thus, one way to determine that a system has infinitely many solutions is by checking whether each of the original equations represents the same line.

EXAMPLE 8-27

Identify each of the following systems as having a unique solution, no solution, or infinitely many solutions:

a. $2x - 3y = 5$
$\dfrac{1}{2}x - y = 1$

b. $\dfrac{x}{3} - \dfrac{y}{4} = 1$
$3y - 4x + 12 = 0$

c. $6x - 9y = 5$
$^-8x + 12y = 7$

Different methods are illustrated in the solutions.

Solution   One method is to attempt to solve each system. Another method is to write each equation in the slope-intercept form and interpret the system geometrically.

a. *First method.*   To eliminate x, multiply the second equation by $^-4$ and add the equations.

$$
\begin{array}{r}
2x - 3y = 5 \\
^-2x + 4y = ^-4 \\
\hline
y = 1
\end{array}
$$

Substituting 1 for y in either equation gives $x = 4$. Thus, $(4, 1)$ is the unique solution of the system.

Second method.   In slope-intercept form, the first equation is $y = \dfrac{2}{3}x - \dfrac{5}{3}$. The second equation is $y = \dfrac{1}{2}x - 1$. The slopes of the corresponding lines are $\dfrac{2}{3}$ and $\dfrac{1}{2}$, respectively. Consequently, the lines are distinct and are not parallel and, therefore, intersect in a single point whose coordinates are the unique solution to the original system.

b. *First method.*   Multiply the first equation by 12 and rewrite the second equation as $^-4x + 3y = ^-12$. Then, adding the resulting equations gives the following:

$$
\begin{array}{r}
4x - 3y = 12 \\
^-4x + 3y = ^-12 \\
\hline
0 = 0
\end{array}
$$

This implies that the two equations represent the same line, and the original system has infinitely many solutions (all the points on the line).

Second method.   In slope-intercept form, both equations are

$$y = \frac{4}{3}x - 4.$$

Thus, the two lines are identical, so the system has infinitely many solutions.

c. *First method.*   To eliminate *y*, multiply the first equation by 4 and the second by 3; then, add the resulting equations.

$$24x - 36y = 20$$
$$\overline{24x + 36y = 21}$$
$$0 = 41$$

No pair of numbers satisfies $0 \cdot x + 0 \cdot y = 41$, so this equation has no solutions, and, consequently, the original system has no solutions.

Second method.   In slope-intercept form, the first equation is $y = \frac{2}{3}x - \frac{5}{9}$. The second equation is $y = \frac{2}{3}x + \frac{7}{12}$. The corresponding lines have the same slope, $\frac{2}{3}$, but different *y*-intercepts. Consequently, the lines are distinct and parallel, and the original system has no solution.

 NOW TRY THIS 8-17

Find all solutions (if any) of each of the following systems

a. $x - y = 1$
     $2x - y = 5$

b. $2x - y = 1$
     $2y - 4x = 3$

c. $2x - 3y = 1$
     $6y - 4x = 2$

Fitting a Line to Data

In many practical situations, a relationship between two variables comes from collected data such as from population or business surveys. When the data are graphed, there may not be a single line that goes through all of the points, but the points may appear to approximate, or "follow," a straight line. In such cases, it is useful to find the equation of what seems to be the **trend line**. Knowing the equation of such a line enables us to predict an outcome.

There are several approaches to define, and hence find, the trend line. We take a graphical approach as follows:

1. Choose a line that seems to follow the given points so that there are about an equal number of points below the line as above the line.
2. Determine two convenient points on the line and approximate the *x*- and *y*-coordinates of these points.
3. Use the points in (2) to determine the equation of the line.

Example 8-28 illustrates the idea.

EXAMPLE 8-28

A shirt manufacturer noticed that the number of units sold depends on the price charged. The data in Table 8-8 show the number of units sold for a given price per unit.

a. Find the equation of a line that seems to fit the data best.

b. Use the equation in (a) to predict the number of units that will be sold if the price per unit is $45.

Table 8-8

Price per Unit (Dollars)	Number of Units Sold (Thousands)
50	200
44	250
41	300
33	380
31	400
24.5	450
20	500
14.5	550

Solution

a. Figure 8-41(a) shows the graph of the data displayed in Table 8-8. Figure 8-41(b) shows a line that seems to fit the data so that approximately the same number of points are below the line as above the line. We choose the points $(50, 200)$ and $(20, 500)$, which are on the line in Figure 8-41(b).

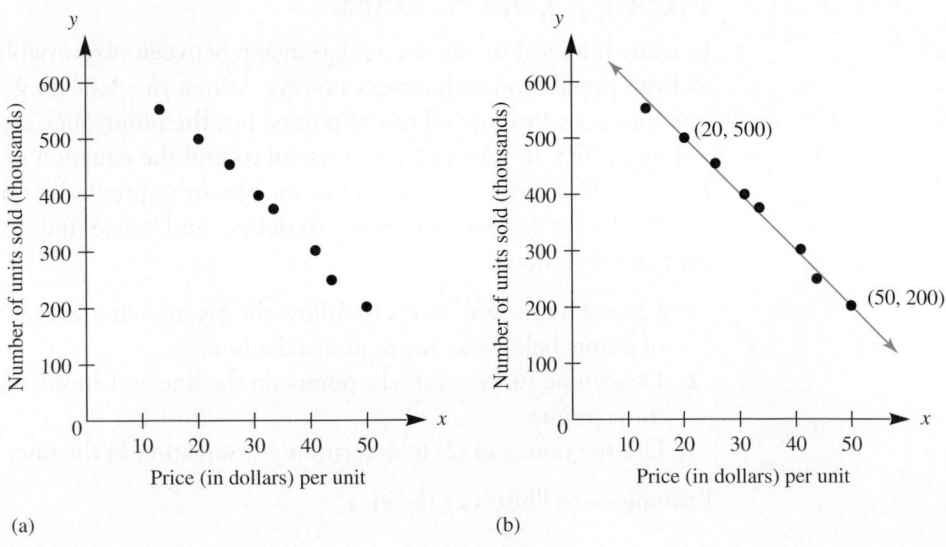

(a)   (b)

Figure 8-41

To find the equation of the line, we may find m and b in the equation $y = mx + b$. Substituting the points $(50, 200)$ and $(20, 500)$ into this equation, we obtain the following:

$$200 = 50m + b$$

$$500 = 20m + b$$

We express b in terms of m in the first equation as follows:

$$b = 200 - 50m$$

Substituting $200 - 50m$ for b in the second equation, we solve for m:

$$500 = 20m + 200 - 50m$$
$$300 = {}^-30m$$
$${}^-10 = m$$

Substituting this value for m, we obtain

$$b = 200 - 50({}^-10) = 700$$

Consequently, the equation of the trend line is $y = {}^-10x + 700$.

b. Using the equation in (a), substitute $x = 45$ to obtain $y = {}^-10(45) + 700$, or $y = 250$. Thus, we predict that 250,000 units will be sold if the price per unit is $45.

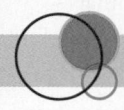

Assessment 8-5A

1. The graph of $y = mx$ is given in the following figure. Sketch the graphs for each of the following on the same figure. Explain your answers.
   a. $y = mx + 3$
   b. $y = mx - 3$

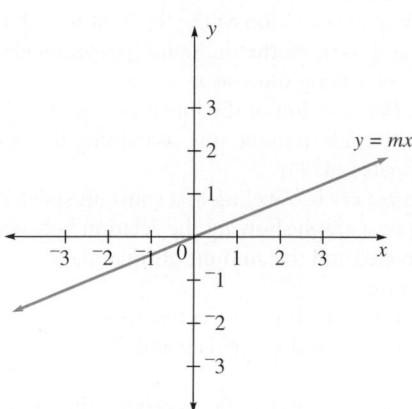

2. Sketch the graphs for each of the following equations:
   a. $y = \dfrac{{}^-3}{4}x + 3$
   b. $y = {}^-3$
   c. $y = 15x - 30$
3. Find the x-intercept and y-intercept for the equations in exercise 2, if they exist.

4. In the following figure, part (i) shows a dual-scale thermometer and part (ii) shows the corresponding points plotted on a graph:

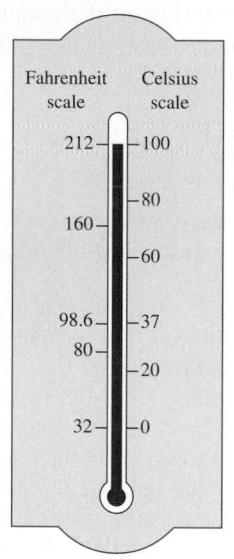

(i)

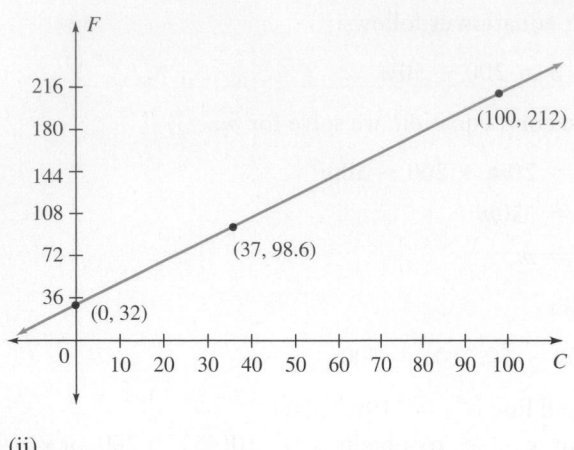

(ii)

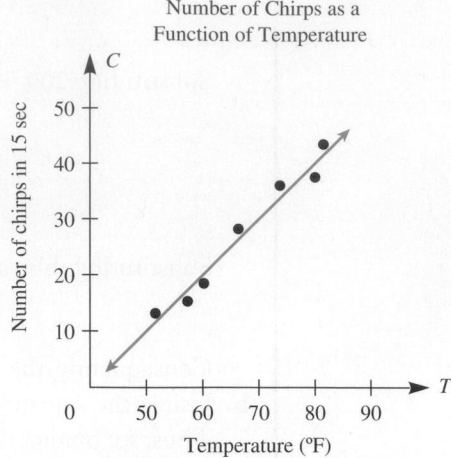

Number of Chirps as a Function of Temperature

a. Use two of the points on the graph to develop a formula for conversion from degrees Celsius (C) to degrees Fahrenheit (F).

b. Use your answer in (a) to find a formula for converting from degrees F to degrees C.

c. Is there a point where degree Celsius and degree Fahrenheit are the same? If so, find it.

5. Write each of the following equations in slope-intercept form and identify the slope and y-intercept:
   a. $3y - x = 0$ b. $x + y = 3$ c. $x = 3y$

6. For each of the following, write the equation of the line determined by the given pair of points in slope-intercept form or in the form $x = a$:
   a. $(^-4, 3)$ and $(1, ^-2)$
   b. $(0, 0)$ and $(2, 1)$
   c. $\left(0, \dfrac{-1}{2}\right)$ and $\left(\dfrac{1}{2}, 0\right)$

7. Find the coordinates of two other points **collinear** (on the same line) with each of the following pairs of given points:
   a. $P(2, 2)$, $Q(4, 2)$ b. $P(0, 0)$, $Q(0, 1)$

8. For each of the following, give as much information as possible about x and y:
   a. The ordered pairs $(^-2, 0)$, $(^-2, 1)$, and (x, y) represent collinear points.
   b. The ordered pair (x, y) is in the fourth quadrant.

9. Consider the lines through $P(2, 4)$ that each form 90° angles with the x- and y-axes, respectively. Find the area and the perimeter of the rectangle formed by these lines and the axes.

10. Find the equations for each of the following:
    a. The line containing $P(3, 0)$ and perpendicular to the x-axis
    b. The line containing $P(^-4, 5)$ and parallel to the x-axis

11. For each of the following, find the slope, if it exists, of the line determined by the given pair of points:
    a. $(4, 3)$ and $(^-5, 0)$ b. $(\sqrt{5}, 2)$ and $(1, 2)$
    c. (a, a) and (b, b)

12. Write the equation of each line in exercise 11.

13. Wildlife experts found that the number of chirps a cricket makes in a 15-sec interval is related to the temperature T in degrees Fahrenheit, as shown in the following graph:

a. If C is the number of chirps in 15 sec, write a formula for C in terms of T (temperature in degrees Fahrenheit) that seems to fit the data best.

b. Use the equation in (a) to predict the number of chirps in 15 sec when the temperature is 90°.

c. If N is the number of chirps per minute, write a formula for N in terms of T.

14. An equilateral triangle, each of whose side lengths is 6, is placed so that one side lies along the x-axis of a coordinate system and one vertex is at the origin.
    a. Write the coordinates of the midpoint of the side along the x-axis.
    b. Write the equation of the line that is both perpendicular to and contains the midpoint (perpendicular bisector) of the side along the x-axis.

15. Write the equation of the line meeting the following condition: A horizontal line containing the point with coordinates $(3, 4)$.

16. Suppose a car is traveling at a constant speed of 60 mph.
    a. Draw a graph showing the relation between the distance traveled and the amount of time that it takes to travel the distance.
    b. What is the slope of the line that you drew in part (a)?

17. a. Graph the following data and find the equation of a trend line.
    b. Use your answer in (a) to predict the value of y when $x = 10$.

x	y
1	8.1
2	9.9
3	12
4	14.1
5	15.9
6	18
7	19.9

18. Solve each of the following systems, if possible. Indicate whether the system has a unique solution, infinitely many solutions, or no solution.
    a. $y = 3x - 1$
       $y = x + 3$

b. $3x + 4y = {}^-17$
  $2x + 3y = {}^-13$

19. The owner of a 5000-gal oil truck loads the truck with gasoline and kerosene. The profit on each gallon of gasoline is 13¢ and on each gallon of kerosene it is 12¢. How many gallons of each fuel did the owner load if the profit was $640.00?

20. Josephine's bank contains 27 coins. If all the coins are either dimes or quarters and the value of the coins is $5.25, how many of each kind of coin are there?

Assessment 8-5B

1. The graph of $y = mx + 5$ is given in the following figure. Sketch the graphs for each of the following equations on the same figure. Explain your answers.
   a. $y = mx$ **b.** $y = mx - 5$

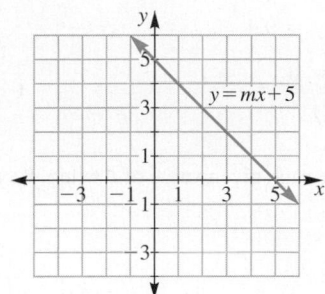

2. Sketch the graphs for each of the following equations:
   a. $x = {}^-2$
   b. $y = 3x - 1$
   c. $y = \dfrac{1}{20}x$

3. Find the x-intercept and y-intercept for the equations in exercise 2, if they exist.

4. On a ruler that measures both metric and English units, we see that 1 m = 39.37 in. This implies that 2 m = 78.74 in. Use this information to write an equation to convert x in. to y meters.

5. Write each of the following equations in slope-intercept form and identify the slope and y-intercept:
   a. $\dfrac{x}{3} + \dfrac{y}{4} = 1$
   b. $3x - 4y + 7 = 0$
   c. $x - y = 4(x - y)$

6. For each of the following, write the equation of the line determined by the given pair of points in slope-intercept form or in the form $x = a$:
   a. $(0, 1)$ and $(2, 1)$
   b. $(2, 1)$ and $(2, {}^-1)$
   c. $({}^-a, 0)$ and $(a, 0), a \neq 0$

7. Find the coordinates of two other points **collinear** (on the same line) with each of the following pairs of given points:
   a. $P({}^-1, 0), Q({}^-1, 2)$
   b. $P(0, 0), Q(1, 1)$

8. Give as much information as possible about x and y in the following:
   a. The ordered pairs $({}^-2, 1), (0, 1)$, and (x, y) represent collinear points.
   b. $(x, y), (0, 0)$ and $({}^-1, {}^-1)$ are collinear.

9. A rectangle has two vertices on the x-axis, two vertices on the y-axis, and one vertex at the point with coordinates $(4, 6)$.

a. Make a sketch showing that there is a rectangle with these characteristics.
   b. Write the coordinates of each of the other three vertices.
   c. Write the equations of the diagonals of the rectangle.
   d. Find the coordinates of the point of intersection of the diagonals.

10. Find the equations for each of the following:
   a. The line containing $P(0, {}^-2)$ and parallel to the x-axis
   b. The line containing $P({}^-4, 5)$ and parallel to the y-axis

11. For each of the following, find the slope, if it exists, of the line determined by the given pair of points:
   a. $({}^-4, 1)$ and $(5, 2)$
   b. $({}^-3, 81)$ and $({}^-3, 198)$
   c. $(1.0001, 12)$ and $(1, 10)$

12. Write the equation of each line in exercise 11.

13. At the end of 10 mo, the balance of an account earning annual simple interest is $2100.00.
   a. If, at the end of 18 mo, the balance is $2180.00, how much money was originally in the account?
   b. What is the annual rate of interest?

14. A triangle has vertices with coordinates $({}^-4, 6), (2, {}^-6)$, and $(8, 10)$.
   a. Write the equations of the lines containing each of the sides of the triangle.
   b. Prove that your equations are correct by finding the coordinates of the points of intersection of each pair of sides of the triangle.

15. Write the equation of the line meeting the following condition: A vertical line containing the point with coordinates $({}^-7, {}^-8)$.

16. Suppose a car is traveling at a constant speed of 45 mph.
   a. Draw a graph showing the relation between the distance traveled and the amount of time that it takes to travel the distance.
   b. What is the slope of the line you drew in part (a)?

17. a. Graph the following data and find the equation of a trend line.
   b. Use your answer in part (a) to predict the value of y when $x = 7$.

x	y
${}^-1$	${}^-7$
0	${}^-4.8$
1	${}^-3.5$
2	${}^-1$
3	1
4	4.9

18. Solve each of the following systems, if possible. Indicate whether the system has a unique solution, infinitely many solutions, or no solution.
    a. $2x - 6y = 7$
       $3x - 9y = 10$
    b. $4x - 6y = 1$
       $6x - 9y = 1.5$

19. The vertices of a triangle are given by $(0, 0)$, $(10, 0)$, and $(6, 8)$. Show that the segments connecting $(5, 0)$ and $(6, 8)$, $(10, 0)$ and $(3, 4)$, and $(0, 0)$ and $(8, 4)$ intersect at a common point.

Mathematical Connections 8-5

Communication

1. In this chapter, an arithmetic sequence was associated with a linear graph. Explain whether a geometric sequence should be associated with a linear graph.
2. Explain why two lines with the same slope and different y intercepts are parallel.

Open-Ended

3. Look for data in newspapers, magazines, or books whose graphs appear to be close to linear and find the equations of the lines that you think best fit the data.
4. a. Write equations of two lines that intersect but when graphed look parallel.
   b. At what point do those two lines intersect?

Cooperative Learning

5. Play the following game between your group and another group. Each group makes up four linear equations that have a common property and presents the equations to the other group. For example, one group could present the equations $2x - y = 0$, $4x - 2y = 3$, $y - 2x = 3$, and $3y - 6x = 5$. If the second group discovers a common property that the equations share, such as the graphs of the equations are four parallel lines, they get 1 point. Each group takes a specified number of turns.

Questions from the Classroom

6. A student would like to know why it is impossible to find the slope of a vertical line. How do you respond?
7. A student argues that it is possible that there is no trend line for a set of data. Could the student be correct? Give an example to make your point to the student.
8. Jonah tried to solve the equation $^-5x + y = 20$ by adding 5 to both sides. He wrote $5 - 5x + y = 5 + 20$ or $0 \cdot x + y = 25$ and finally $y = 25$. How would you help Jonah?
9. Jill would like to know why two lines with an undefined slope are parallel. How would you respond?

Review Problems

10. Write the following with algebraic expressions or equations:
    a. The cube root of a number is 3 less than its square.
    b. The sum of the squares of two numbers is 36.
    c. $0.\overline{9}$ has 1 as its square root.

11. Give an approximation of $\sqrt{6}$ correct to hundredths.
12. Solve the following for x:
    a. $x\sqrt{2} - 3y = 0.\overline{4}$
    b. $x^2 - 81 = 0$
    c. $3x < ^-\sqrt{7}$

13. Find $f(x)$ for each given value of x when $f(x) = x\sqrt{7} - \sqrt{7}$.
    a. 3
    b. $\sqrt{7}$
    c. $^-4$

14. If $f(x) = 12$ when $f(x) = 3x - \sqrt{2}$, find x.
15. What are the domains for each of the following functions?
    a. $f(x) = \sqrt{x + 1}$
    b. $f(x) = \sqrt{^-x}$

Trends in Mathematics and Science Study (TIMSS) Question

A straight line passes through the points $(2, 3)$ and $(4, 7)$. Which of these points is also on the line?
a. $(0, 2)$
b. $(1, 2)$
c. $(2, 4)$
d. $(3, 5)$
e. $(4, 5)$

TIMSS, Grade 8, 2003

National Assessment of Educational Progress (NAEP) Question

Which of the following is the graph of the line with equation $y = ^-2x + 1$?

a.

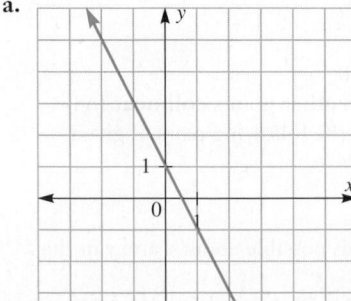

b.

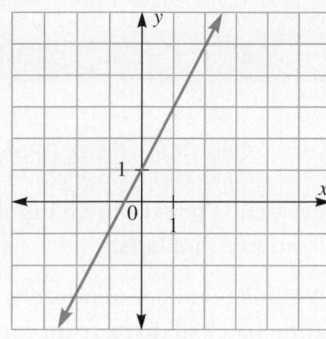

c.

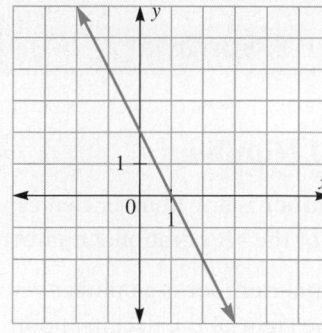

d.

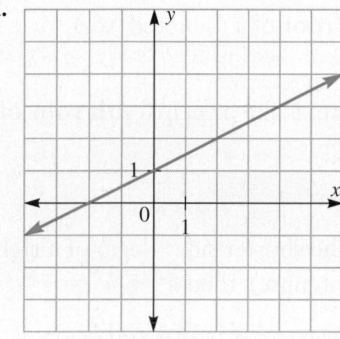

e.

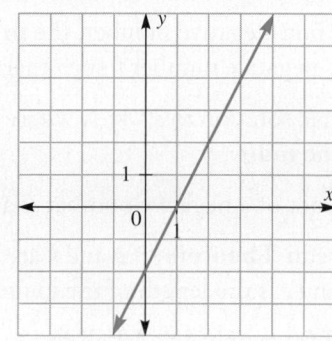

NAEP, Grade 8, 2007

BRAIN TEASER

When teaching how to solve a system of linear equations, Ms. Whippledorf divided the class into four groups and asked each to solve one of the following systems.

$$5x + 6y = 7 \qquad x + 2y = 3 \qquad 11x + 12y = 13 \qquad {}^-4x + {}^-3y = {}^-2$$
$$8x + 9y = 10 \qquad 4x + 5y = 6 \qquad 14x + 15y = 16 \qquad {}^-1x + 0y = 1$$

When the groups finished, they were surprised at the answers. Next, Ms. Whippledorf asked each student for a system similar to the above, and the class was even more surprised at the answers. What surprised the students each time?

Hint for the Preliminary Problem

The arc of the rail can be approximated by two sides of a triangle as seen below.

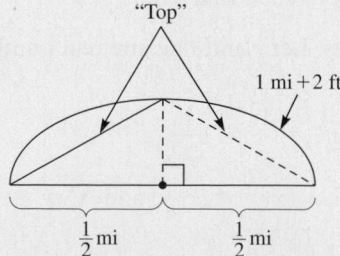

"Top"

1 mi + 2 ft

$\frac{1}{2}$ mi $\frac{1}{2}$ mi

Assume that the length of the expanded rail (1 mi + 2 ft) is evenly divided between the two sides forming the "top" of the triangle. And assume that the original length of the rail is separated into two equal parts along the base of the triangle. Use this information and the Pythagorean Theorem to find the height of the buckle.

Chapter 8 Summary

8-1 Real Numbers	**Pages**
A **real number** is any number that can be written as a decimal. The set of real numbers is the union of the set of rational numbers and the set of irrational numbers.	393
An **irrational number** is a number whose decimal has an infinite number of nonzero digits to the right of the decimal point and has no repeating block of digits from some point.	393
If a is any non-negative number, the **principal square root** of a (denoted $\sqrt{a}$) is the non-negative number b such that $b^2 = a$.	393–394
The positive solution to $x^n = b$, where b is non-negative, is the principal **nth root of b**, or $\sqrt[n]{b}$. n is the **index**.	394
The odd root of a negative number is a negative number.	394
Pythagorean Theorem: If a and b are the lengths of the shorter sides (legs) of a right triangle, and c is the length of the longer side (the hypotenuse), then $a^2 + b^2 = c^2$.	395
The set of **fractions** includes all real numbers of the form a/b, where a and b are real numbers with $b \neq 0$.	397
Properties of Real Numbers and Properties of Real Numbers under Addition and Multiplication	398
• **Theorem: (Closure properties)**: For real numbers a and b, $a + b$ and ab are unique real numbers.	
• **Theorem: (Commutative properties)**: For real numbers a and b, $a + b = b + a$ and $ab = ba$.	398
• **Theorem: (Associative properties)**: For real numbers a, b, and c, $(a + b) + c = a + (b + c)$ and $(ab)c = a(bc)$.	398
• **Theorem: (Identity properties)**: The number 0 is the unique real number, the **additive identity**, such that for all real numbers a, $0 + a = a = a + 0$, and 1 is the unique real number, the **multiplicative identity**, such that for all real numbers a, $1a = a = a(1)$.	398
• **Theorem: Inverse properties**: (1) For every real number a, ^-a is its unique **additive inverse**; that is, $a + {^-a} = 0 = {^-a} + a$. (2) For every non-zero real number a, $1/a$ is its unique **multiplicative inverse**; that is, $a\left(\dfrac{1}{a}\right) = 1 = \left(\dfrac{1}{a}\right)a$.	398
• **Theorem: (Distributive property** of **multiplication over addition)**: For real numbers a, b, and c, $a(b + c) = ab + ac$.	398
• **Theorem: (Denseness property)**: For real numbers a and b such that $a < b$, there exists a real number c such that $a < c < b$.	398
Properties of Exponents: Let r and s be any real numbers with x and y real numbers (where meaningful).	
• $x^{-r} = \dfrac{1}{x^r}$ implies $x^{-\frac{1}{n}} = \dfrac{1}{x^{\frac{1}{n}}} = \dfrac{1}{\sqrt[n]{x}}$	399
• $(xy)^r = x^r y^r$ implies $(xy)^{\frac{1}{n}} = x^{\frac{1}{n}} y^{\frac{1}{n}}$ and $\sqrt[n]{xy} = \sqrt[n]{x}\sqrt[n]{y}$	399
• $\left(\dfrac{x}{y}\right)^r = \dfrac{x^r}{y^r}$ implies $\left(\dfrac{x}{y}\right)^{\frac{1}{n}} = \dfrac{x^{\frac{1}{n}}}{y^{\frac{1}{n}}}$ and $\sqrt[n]{\dfrac{x}{y}} = \dfrac{\sqrt[n]{x}}{\sqrt[n]{y}}$	399
• $(x^r)^s = x^{rs}$ implies $\left(x^{\frac{1}{p}}\right)^n = x^{\frac{n}{p}}$ and hence, $\left(\sqrt[p]{x}\right)^p = \sqrt[p]{x^p}$	399

8-2 Variables

A fixed number is a **constant**.	405
A **variable** may (1) represent a missing element or an unknown; (2) stand for more than one thing; (3) be used in generalizations of patterns; and (4) be an element of a set, or a set itself.	405
The *n*th **term of the Fibonacci sequence** with seeds 1, 1 is $F_n = F_{n-1} + F_{n-2}$. where $n = 3, 4, 5, \ldots$	409
The *n*th **term of an arithmetic sequence** with first term a_1 and difference d is given by $a_n = a_1 + (n-1)d$.	410
The sum of the first n terms of an arithmetic sequence is $S_n = n\left(\dfrac{a_1 + a_n}{2}\right)$ or $na_1 + n(n-1)d/2$.	412–413
The *n*th **term of a geometric sequence** with first term a_1 and ratio r is given by $a_n = a_1 r^{n-1}$.	413
The sum of the first n terms of a geometric sequence is $S_n = a_1\left(\dfrac{1 - r^n}{1 - r}\right)$ where $r \neq 1$, or $a_1\left(\dfrac{r^n - 1}{r - 1}\right)$.	413–414

8-3 Equations

Properties of Equality	416
• **Theorem: (Addition Property of Equality):** For any real numbers a, b, and c, if $a = b$, then $a + c = b + c$.	
• **Theorem: (Multiplication Property of Equality):** For any real numbers a, b, and c, if $a = b$, then $ac = bc$.	416
• **Theorem: (Cancellation Properties of Equality):**	
o For any real numbers a, b, and c, if $a + c = b + c$, then $a = b$.	416
o For any real numbers a, b, and c with $c \neq 0$, if $ac = bc$, then $a = b$.	416
Property of Equations	
• **Theorem: (Addition and Subtraction Property of equations)** If $a = b$ and $c = d$, then $a + c = b + d$ and $a - c = b - d$.	417

8-4 Functions

A **function** from set A to set B is a correspondence from A to B in which each element of A is paired with one, and only one, element of B. Set A is the **domain**; set B is the **codomain**; the set of images of set A is the **range**.	425–428
Functions may be modeled as rules between two sets, as machines relating inputs and outputs, as equations relating two variables, as tables relating two variables, as arrow diagrams as ordered pairs, or as graphs in two dimensions.	425–431
One operation on functions is the **composition of the functions**; for example, $g \circ f(x) = g(f(x))$.	434
A **relation from set A to set B** is a subset of $A \times B$; that is, R is a relation from set A to set B if, and only if, $R \subseteq A \times B$.	435–436

Properties of Relations

- A relation R on a set X is **reflexive** if, and only if, for every element $a \in X$, a is related to a; that is, for every $a \in X$, $(a, a) \in R$.   436
- A relation R on a set X is **symmetric** if, and only if, for all elements $a, b \in X$, whenever a is related to b, b is related to a; that is, for every $(a, b) \in R$, $(b, a) \in R$.   437
- A relation R on a set X is **transitive**, if and only if, for all elements a, b, and c, whenever a is related to b and b is related to c, then a is related to c; that is, if $(a, b) \in R$ and $(b, c) \in R$, then $(a, c) \in R$.   437
- An **equivalence relation** is any relation R that satisfies the reflexive, symmetric, and transitive properties.   437

8-5 Equation in a Cartesian Coordinate System

A **Cartesian coordinate system** is constructed by placing two number lines perpendicular to each other. The intersection point of the two lines is the **origin**. The lines are the ***x***- and ***y*-axes**.   445

The x-coordinate in an ordered pair is the **abscissa**, and the y-coordinate is the **ordinate**.

Every line has an equation of the form either $y = mx + b$ (the **slope-intercept** form where m is the **slope** and b is the **y-intercept**) or $x = a$. In the equation $x = a$, a is the **x-intercept**.   445

Given two points $A(x_1, y_1)$ and $B(x_2, y_2)$ with $x_1 \neq x_2$, the **slope** m of the line

through points A and B is $m = \dfrac{y_2 - y_1}{x_2 - x_1}$.   449

A systems of equations in two unknowns may be solved by elimination or substitution.   455–456

A system of two equations in two unknowns may be characterized by the following:

- The system has a unique solution if, and only if, the graphs of the equation intersect in a single point.
- The system has no solution if, and only if, the equations represent distinct parallel lines.   457
- The system has infinitely many solutions if, and only if, the equations represent the same line.

A **trend line** is a line that approximates a set of data.   459

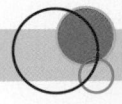

Chapter 8 Review

1. Classify each of the following as rational or irrational (assume the patterns shown continue):
   a. $2.19119911999119999119\ldots$
   b. $\dfrac{1}{\sqrt{2}}$
   c. $\dfrac{4}{9}$
   d. $0.0011001100110011\ldots$
   e. $0.001100011000011\ldots$

2. Write each of the following in the form $a\sqrt{b}$ or $a\sqrt[n]{b}$, where a and b are positive integers and b has the least value possible:
   a. $\sqrt{242}$
   b. $\sqrt{288}$
   c. $\sqrt{180}$
   d. $\sqrt[3]{162}$

3. Answer each of the following and explain your answers:
   a. Is the set of irrational numbers closed under addition?
   b. Is the set of irrational numbers closed under subtraction?

c. Is the set of irrational numbers closed under multiplication?

d. Is the set of irrational numbers closed under division?

4. Find an approximation for $\sqrt{23}$ correct to three decimal places without using the $\boxed{y^x}$ or the $\boxed{\sqrt{}}$ keys.

5. Approximate $\sqrt[3]{2}$ to two decimal places by using the squeezing method.

6. Each of the following is a geometric sequence. Find the missing terms.

a. 5, _____, 10

b. 1, _____, _____, _____, 1/4

7. In a geometric sequence, if the nth term is $\sqrt{7}(^-1)^n$, what is the 10th term?

8. If the first two terms of an arithmetic sequence are 1 and $\sqrt{2}$, what is an algebraic expression for the nth term?

9. There are 13 times as many students as professors at a college. Use S for the number of students and P for the number of professors to represent the given information.

10. Write a sentence that gives the same information as the following equation: $A = 103B$, where A is the number of girls in a neighborhood and B is the number of boys.

11. Write an equation to find the number of feet given the number of yards (let f be the number of feet and y be the number of yards).

12. The sum of a set of n whole numbers is S. If each number is multiplied by 10 and then decreased by 10, what is the sum of the new set in terms of n and S?

13. I am thinking of a whole number. If I divide it by 13, then multiply the answer by 12, then subtract 20, and then add 89, I end up with 93. What was my original number?

14. a. Think of a number.

Add 17.

Double the result.

Subtract 4.

Double the result.

Add 20.

Divide by 4.

Subtract 20.

Your answer will be your original number. Explain how this trick works.

b. Fill in two more steps that will take you back to your original number.

Think of a number.

Add 18.

Multiply by 4.

Subtract 7.

.

.

.

15. Find all the values of x that satisfy the following equations:

a. $4x - 2 = 3x + \sqrt{10}$

b. $4(x - 12) = 2x + 10$

c. $4(7x - 21) = 14(7x - 21)$

d. $2(3x + 5) = 6x + 11$

e. $3(x + 1) + 1 = 3x + 4$

16. Mike has 3 times as many baseball cards as Jordan, who has twice as many cards as Paige. Together, the three children have 999 cards. Find how many cards each child has.

17. Jeannie has 10 books overdue at the library. She remembers she checked out 2 science books two weeks before she checked out 8 children's books. The daily fine per book is $0.20. If her total fine was $11.60, how long was each book overdue?

18. Three children deliver all the newspapers in a small town. Jacobo delivers twice as many papers as Dahlia, who delivers 100 more papers than Rashid. If altogether 500 papers are delivered, how many papers does each child deliver?

19. Which of the following sets of ordered pairs are functions from the set of first components to the set of second components?

a. $\{(a, b), (c, d), (e, a), (f, g)\}$

b. $\{(a, b), (a, c), (b, b), (b, c)\}$

c. $\{(a, b), (b, a)\}$

20. Given the following function rules and the domains, find the associated ranges:

a. $f(x) = x + 3$; domain $= \{0, 1, 2, 3\}$

b. $f(x) = 3x - 1$; domain $= \{5, 10, 15, 20\}$

c. $f(x) = x^2$; domain $= \{0, 1, 2, 3, 4\}$

d. $f(x) = x^2 + 3x + 5$; domain $= \{0, 1, 2\}$

21. Which of the following correspondences from A to B describe a function? If a correspondence is a function, find its range. Justify your answers.

a. A is the set of college students, and B is the set of majors. To each college student corresponds his or her major.

b. A is the set of books in the library, and B is the set N of natural numbers. To each book corresponds the number of pages in the book.

c. $A = \{(a, b) \mid a \in N \text{ and } b \in N\}$, and $B = N$. To each element of A corresponds the number $4a + 2b$.

d. $A = N$ and $B = N$. If x is even, then $f(x) = 0$, and if x is odd, then $f(x) = 1$.

e. $A = N$ and $B = N$. To each natural number corresponds the sum of its digits.

22. A health club charges an initiation fee of $200, which gives 1 month of free membership, and then charges $55 per month.

a. If $C(x)$ is the total cost of membership in the club for x months, express $C(x)$ in terms of x.

b. Graph $C(x)$ for the first 12 months.

c. Use the graph in (b) to find when the total cost of membership in the club will exceed $600.

d. When will the total cost of membership exceed $6000?

23. If the rule for the function is $f(x) = 4x - 5$ and $f(x) = 15$ is the output, what is the input?

24. Which of the following graphs represent functions? Tell why.

a.

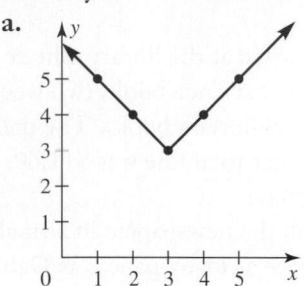

b.

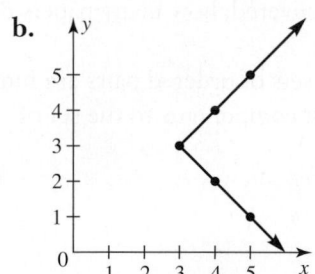

c.

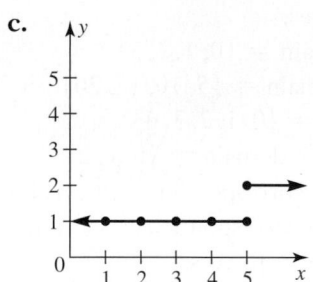

25. a. Jilly is building towers with cubes, placing one cube on top of another and painting the tower (including the top and the bottom, but not the faces touching each other). Find the number of square faces that Jilly needs to paint for towers made of 1, 2, 3, 4, 5, and 6 cubes by filling in the following table:

# of Cubes	# of Squares to Paint
1	6
2	10
3	
4	
5	
6	

b. Graph the information you found in part (a) where the number of cubes in a tower is on the horizontal x-axis and the number of squares to be painted is on the vertical axis.

c. If x is the number of cubes in a tower and y is the corresponding number of squares to be painted, write an equation that gives y as a function of x.

d. Is the graph describing the number of squares as a function of the number of cubes used a straight line?

26. Graph each of the following lines:

a. $y = {}^-2x + 5$

b. $y = {}^-\left(\dfrac{1}{5}\right)x + 7$

c. $y = {}^-x\sqrt{2}$

27. Use graphs to estimate the solution of the following system of linear equations:

$$y = {}^-1x + 6$$
$$y = 2x - 7$$

28. Solve the following equations for real number solution(s):

a. ${}^-3x + 12 = 23$

b. $\dfrac{2}{3}x + 18 = 42$

c. $x^3 = {}^-1$

d. $4x^2 - 33 = 3$

e. $(2x - 7)(3x + 2) = 0$

f. $(x - 3)(x + 2) = 0$

g. $(2x - 1)(x - 1) = 0$

29. What would be the third term of a geometric sequence whose first term is π and whose ratio is $\dfrac{1}{\pi}$?

30. Explain whether or not you would ever expect a price in a store to be an irrational number.

Probability

Preliminary Problem

A bag contains 3 blue marbles, 4 red marbles, and 3 yellow marbles. The probability of drawing a blue marble out of the bag is 3/10 or 30%. How many of what color of marbles must be added to the bag so that the probability of a blue marble being drawn at random from the bag is 75%?

If needed, see Hint on page 535.

Probability is used in predicting sales, planning political campaigns, determining insurance premiums, making investment decisions, and testing experimental drugs. Some uses of probability in conversations include the following:

What is the probability that the Chicago Cubs will win the World Series?

There is no chance you will get a raise.

There is a 50% chance of rain today.

In the grades 6–8 *PSSM* there is a call for an increased emphasis on probability and we find the following:

> Although the computation of probabilities can appear to be simple work with fractions, students must grapple with many conceptual challenges in order to understand probability … it is useful for students to make predictions and then compare the predictions with actual outcomes. (p. 254)

In grade 8 *Focal Points*, we find that students are expected to "use proportionality and a basic understanding of probability to make and test conjectures about the results of experiments and simulations." (p. 39) Additionally in pre-K–8 *Focal Points*, students are expected to "compute probabilities for simple compound events, using such methods as organized lists, tree diagrams, and models." (p. 39)

In this chapter, we introduce all topics in the preceding expectations along with several others. We use tree diagrams and geometric probabilities (area models) to solve problems and to analyze games that involve spinners, cards, and dice. We introduce counting techniques and discuss the role of simulations in probability.

9-1 How Probabilities Are Determined

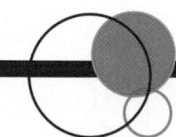

Outcomes
H

T

Figure 9-1

Probabilities are ratios, expressed as fractions, decimals, or percents, determined by considering results or outcomes of experiments. An **experiment** is an activity whose results can be observed and recorded. Each of the possible results of an experiment is an **outcome**. If we toss a typical coin that cannot land on its edge, there are two distinct possible outcomes: heads (H) and tails (T).

A set of all possible outcomes for an experiment is a **sample space**. The outcomes in the sample space cannot overlap. In a single coin toss of a typical coin, the sample space S is given by $S = \{H, T\}$. The sample space can be modeled by a **tree diagram**, as shown in Figure 9-1. Each outcome of the experiment is designated by a separate branch in the tree diagram. The sample

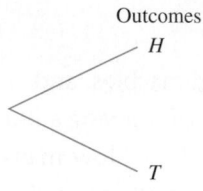

Historical Note

Probability is mentioned in a 1477 commentary on Dante's *Divine Comedy*, but most historians think that probability originated in an unfinished dice game. The French mathematician Blaise Pascal (1623–1669) was asked by his friend Chevalier de Méré, a professional gambler, how to divide the stakes if two players start, but fail to complete, a game consisting of five matches in which the winner is the one who wins three out of five matches and players were to divide the stakes according to their chances of winning the game. Pascal shared the problem with Pierre de Fermat (1601–1665) and together they solved the problem, which led to the development of probability. Since the work of the French mathematician Pierre Simon de Laplace (1749–1827), probability theory has become a major part of mathematics.   ●

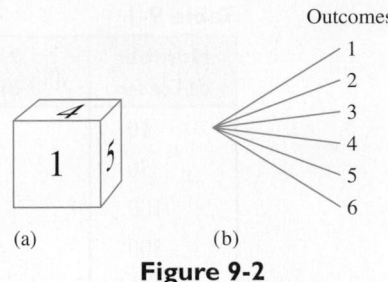

Outcomes
1
2
3
4
5
6

(a)   (b)

Figure 9-2

space S for rolling a standard die as in Figure 9-2(a), is $S = \{1, 2, 3, 4, 5, 6\}$. Figure 9-2(b) gives a tree diagram for that sample space.

Any subset of a sample space is an **event**. For example, the set of all even-numbered rolls of a die $\{2, 4, 6\}$ is an event and is a subset of all possible rolls $\{1, 2, 3, 4, 5, 6\}$. Because events are subsets of a sample space and a sample space is a set, we commonly use letters such as A, B, C to represent events. We see this in Example 9-1.

EXAMPLE 9-1

Suppose an experiment consists of drawing 1 slip of paper from a jar containing 12 slips of paper, each with a different month of the year written on it. Find each of the following:

  a. Sample space S for the experiment
  b. Event A consisting of outcomes having a month beginning with J
  c. Event B consisting of outcomes having the name of a month that has exactly four letters
  d. Event C consisting of outcomes having a month that begins with M or N.

Solution
  a. $S = \{$January, February, March, April, May, June, July, August, September, October, November, December$\}$
  b. $A = \{$January, June, July$\}$
  c. $B = \{$June, July$\}$
  d. $C = \{$March, May, November$\}$

Determining Probabilities

According to the grade 7 *Common Core Standards* students should be able to;

> Approximate the probability of a chance event by collecting data on the chance process that produces it and observing its long-run relative frequency, and predict the approximate relative frequency given the probability. *For example, when rolling a number cube 600 times, predict that a 3 or 6 would be rolled roughly 200 times, but probably not exactly 200 times.* (p. 51)

Around 1900, the English statistician Karl Pearson collected data on the chance process of coin tossing. He tossed a coin 24,000 times and recorded 12,012 heads. During World War II, John Kerrich, a Dane and a prisoner of war, tossed a coin 10,000 times. A subset of his results is in Table 9-1. The *relative frequency* column on the right is obtained by dividing the number of heads by the number of tosses of the coin.

As the number of Kerrich's tosses increased, he obtained heads close to half the time. The relative frequency for Pearson's 24,000 tosses gave a similar result of 12,012/24,000, or approximately $\frac{1}{2}$. Kerrich used the relative frequency interpretation of probability. In this interpretation, *the probability of an event is the long-run fraction of times that an event will occur given many repetitions under identical circumstances.*

Table 9-1

Number of Tosses	Number of Heads	Relative Frequency (rounded)
10	4	0.400
50	25	0.500
100	44	0.440
500	255	0.510
1,000	502	0.502
5,000	2,533	0.507
8,000	4,034	0.504
10,000	5,067	0.507

When a probability is determined by observing outcomes of experiments, it is **experimental**, or **empirical**, probability. The exact number of heads that occurs when a fair coin is tossed a few times cannot be predicted accurately. A *fair coin* is a coin that on each toss is just as likely to land "heads" as it is to land "tails." When a coin is tossed many times and the fraction (or ratio) of heads is near $\frac{1}{2}$, then we assume the coin is a fair coin and say; "the probability of heads occurring is one half."

Probabilities only suggest what will happen in the "long run." This concept is called *The Law of Large Numbers* or sometimes *Bernoulli's Theorem*, and is given here as Theorem 9-1.

Theorem 9-1:   Law of Large Numbers (Bernoulli's Theorem)

If an experiment is repeated a large number of times, the *experimental* or *empirical* probability of a particular outcome approaches a fixed number as the number of repetitions increases.

In this text, by "probability" we mean *theoretical probability*. We assign **theoretical probabilities** to events, not outcomes, unless the event is a simple event; that is, an event with one outcome or the outcome is completed after one step. For example, we could argue that since an ideal coin marked heads (H) and tails (T) is symmetric and has two sides, then each side should appear about the same number of times if the coin is tossed many times. Again we would conclude that if $A = \{H\}$ and $B = \{T\}$ from the sample space $S = \{H, T\}$, then

$$P(A) = P(B) = \frac{1}{2}.$$

When one outcome is just as likely as another, as in coin tossing, the outcomes are **equally likely**. If an experiment is repeated many times, the experimental probability of the event's occurring should be approximately equal to the theoretical probability of the event's occurring.

⊙ Historical Note

Jakob Bernoulli (1654–1705) was the first of the Bernoulli family who pursued the study of mathematics. After taking his degree in theology, Jakob traveled for 7 years studying mathematics. He returned to Basel, Switzerland, in 1683 and in 1687 became the chair of mathematics at the University of Basel. His greatest and most original work, *Ars Conjectandi* (*The Art of Conjecturing*), laid the foundation for the modern theory of probability.   ●

A *fair* die is a die that is just as likely to land showing any of the numerals 1 through 6 on any toss. Its sample space S is given by $S = \{1, 2, 3, 4, 5, 6\}$, and $P(1) = P(2) = P(3) = P(4) = P(5) = P(6) = \frac{1}{6}$. The probability of rolling an even number, that is, the probability of the event $E = \{2, 4, 6\}$, is $\frac{3}{6}$, or $\frac{1}{2}$. For a sample space with equally likely outcomes, the probability of an event A can be defined as follows:

> **Definition of Probability of an Event with Equally Likely Outcomes**
> For an experiment with non-empty sample space S with equally likely outcomes, the **probability of an event A** is given by
> $$P(A) = \frac{\text{Number of elements of } A}{\text{Number of elements of } S} = \frac{n(A)}{n(S)}$$

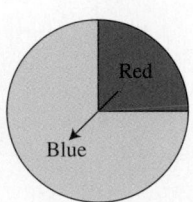

Figure 9-3

The above definition of the probability of an event applies only to a sample space that has equally likely outcomes. If each possible outcome of the sample space is equally likely, the sample space is a **uniform sample space**. Thus, in a uniform sample space with n outcomes, the probability of each outcome is $\frac{1}{n}$. Trying to apply the definition of probability given above to a sample space with outcomes that are not equally likely (nonuniform) leads to incorrect conclusions. For example, the sample space for spinning the spinner in Figure 9-3 is given by $S = \{$Red, Blue$\}$, but Blue is more likely to occur than Red. If the spinner were spun 100 times, we could reasonably expect that about $\frac{1}{4}$ of 100, or 25, of the outcomes would be Red, whereas about $\frac{3}{4}$ of 100, or 75, of the outcomes would be Blue.

NOW TRY THIS 9-1

a. In an experiment of tossing a fair coin once, what is the sum of the probabilities of all the distinct outcomes in the sample space?
b. In an experiment of tossing a fair die once, what is the sum of the probabilities of all the distinct outcomes in the sample space?
c. Does the sum of the probabilities of all the distinct outcomes of any sample space always result in the same number?
d. What is the probability of the spinner in Figure 9-4 landing on Red? Why?

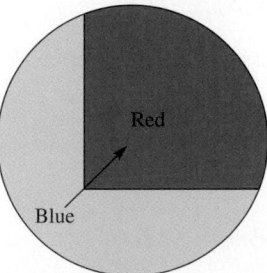

Figure 9-4

In the grade 7 *Common Core Standards* we find that students should be able to:

understand that the probability of a chance event is a number between 0 and 1 that expresses the likelihood of the event occurring. Larger numbers indicate greater likelihood. A probability near 0 indicates an unlikely event, a probability around 1/2 indicates an event that is neither unlikely nor likely, and a probability near 1 indicates a likely event. (p. 50)

EXAMPLE 9-2

Let $S = \{1, 2, 3, 4, 5, \ldots, 25\}$. If a number is chosen **at random**, that is, with the same chance of being drawn as all other numbers in the set, calculate each of the following probabilities:

   a. the event A that an even number is drawn
   b. the event B that a number less than 10 and greater than 20 is drawn
   c. the event C that a number less than 26 is drawn
   d. the event D that a prime number is drawn
   e. the event E that a number both even and prime is drawn

Solution   Each of the 25 numbers in set S has an equal chance of being drawn.

   a. $A = \{2, 4, 6, 8, 10, 12, 14, 16, 18, 20, 22, 24\}$, so $n(A) = 12$. Thus, $P(A) = \dfrac{n(A)}{n(S)} = \dfrac{12}{25}$.

   b. $B = \varnothing$, so $n(B) = 0$. Thus, $P(B) = \dfrac{0}{25} = 0$.

   c. $C = S$ and $n(C) = 25$. Thus, $P(C) = \dfrac{25}{25} = 1$.

   d. $D = \{2, 3, 5, 7, 11, 13, 17, 19, 23\}$, so $n(D) = 9$. Thus, $P(D) = \dfrac{n(D)}{n(S)} = \dfrac{9}{25}$.

   e. $E = \{2\}$, so $n(E) = 1$. Thus, $P(E) = \dfrac{1}{25}$.

In Example 9-2(b), event B is the empty set. An event such as B that has no outcomes is an **impossible event** *and has probability* 0. If the word *and* were replaced by *or* in Example 9-2(b), then event B would no longer be the empty set. If event B is the empty set, $\varnothing$, then it has no elements and its probability is the ratio $P(B) = \dfrac{n(\varnothing)}{n(S)} = \dfrac{0}{n(S)} = 0$. (Note that an assumption is made that the sample space does not have 0 elements.) In Example 9-2(c), event C consists of drawing a number less than 26 on a single draw. Because every number in S is less than 26, $P(C) = \dfrac{25}{25} = 1$. An event that has probability 1 is a **certain event**.

If A is a subset of the sample space S, the greatest number of elements that A can have is the number of elements in the sample space, S. So the probability of A in this case would be $P(A) = \dfrac{n(A)}{n(S)} = \dfrac{n(S)}{n(S)} = 1$. For any set A that is a proper subset of S, $n(A)$ is greater than or equal to 0 but less than $n(S)$. Hence $P(A) = \dfrac{n(A)}{n(S)} < 1$. We summarize this discussion in the following theorem.

Theorem 9-2

If A is any event and S is the sample space, then $0 \leq P(A) \leq 1$.

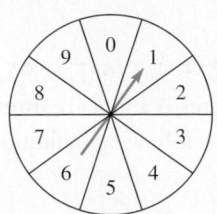

Figure 9-5

Consider one spin of the number wheel shown in Figure 9-5 where each sector of the circle is determined by angles that are the same size. For this experiment, $S = \{0, 1, 2, 3, 4, 5, 6, 7, 8, 9\}$. If A is the event of spinning a number in the set $\{0, 1, 2, 3, 4\}$ and B is the event of spinning a number in the set $\{5, 7\}$, then using the definition of probability of an event with equally likely outcomes, $P(A) = \dfrac{n(A)}{n(S)} = \dfrac{5}{10}$ and $P(B) = \dfrac{n(B)}{n(S)} = \dfrac{2}{10}$. The probability of an event can be found by adding the probabilities of disjoint events representing the various outcomes in the set. For example, event $B = \{5, 7\}$ can be represented as the union of two disjoint events; that is, spinning a 5 or spinning a 7. Then $P(B)$ can be found by adding the probabilities of each event.

$$P(B) = \frac{1}{10} + \frac{1}{10} = \frac{2}{10}$$

Likewise,

$$P(A) = \frac{1}{10} + \frac{1}{10} + \frac{1}{10} + \frac{1}{10} + \frac{1}{10} = \frac{5}{10}.$$

These are special cases of the following theorem, which holds for all finite probability spaces.

> **Theorem 9-3**
>
> The probability of an event is equal to the sum of the probabilities of disjoint events whose union is the event.

EXAMPLE 9-3

If we draw a card at random from an ordinary deck of playing cards, find the probability that the

a. card is an ace.
b. card is an ace or a queen.

Solution

a. There are 52 cards in a deck, 4 of which are aces. If event A is drawing an ace, then $A = \{\spadesuit, \clubsuit, \diamondsuit, \heartsuit\}$. We use the definition of probability of events with equally likely outcomes to compute the following:

$$P(A) = \frac{n(A)}{n(S)} = \frac{4}{52}$$

An alternative approach is to find the sum of the probabilities of the disjoint events containing each of the outcomes in the event, where the probability of the event of drawing any single ace from the deck is $\dfrac{1}{52}$:

$$P(A) = \frac{1}{52} + \frac{1}{52} + \frac{1}{52} + \frac{1}{52} = \frac{4}{52}$$

b. From part (a), $P(A) = \dfrac{4}{52}$. By similar reasoning if event B is drawing a queen then $P(B) = \dfrac{4}{52}$. Therefore the probability of the union of these disjoint sets is $\dfrac{4}{52} + \dfrac{4}{52} = \dfrac{8}{52}$.

Mutually Exclusive Events

Consider one spin of the wheel in Figure 9-6(a). For this experiment, $S = \{0, 1, 2, 3, 4, 5, 6, 7, 8, 9\}$. If $A = \{0, 1, 2, 3, 4\}$ and $B = \{5, 7\}$, then $A \cap B = \varnothing$. (See Figure 9-6(b).) Events A and B are **mutually exclusive** events. If event A occurs, then event B cannot occur, and we have the following definition.

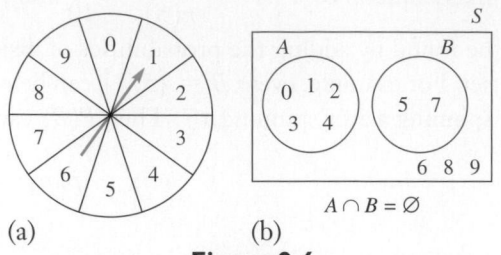

(a)　　　　　(b)

$A \cap B = \varnothing$

Figure 9-6

Definition of Mutually Exclusive Events

Events A and B are **mutually exclusive** if they have no elements in common; that is, $A \cap B = \varnothing$.

Each outcome in the above sample space $S = \{0, 1, 2, 3, 4, 5, 6, 7, 8, 9\}$ is equally likely. The probability of the event with a single outcome in S is $\frac{1}{10}$. With mutually exclusive events A and B given, if we write the probability of A or B as $P(A \cup B)$, we have the following:

$$P(A \cup B) = \frac{n(A \cup B)}{n(S)} = \frac{7}{10} = \frac{5 + 2}{10} = \frac{5}{10} + \frac{2}{10}$$

$$= \frac{n(A)}{n(S)} + \frac{n(B)}{n(S)} = P(A) + P(B)$$

The result developed in this example is true for all mutually exclusive events. In general, we have the following theorem.

Theorem 9-4

If events A and B are mutually exclusive, then $P(A \text{ or } B) = P(A \cup B) = P(A) + P(B)$.

For a sample space with equally likely outcomes, this theorem follows immediately from the fact that if $A \cap B = \varnothing$, then $n(A \cup B) = n(A) + n(B)$.

Complementary Events

If the weather forecaster tells us that the probability of rain is 25%, what is the probability that it will not rain? These two events—rain and not rain—are **complements** of each other. Therefore, if the probability of rain is 25%, or $\frac{1}{4}$, the probability it will not rain is $100\% - 25\% = 75\%$, or $1 - \frac{1}{4} = \frac{3}{4}$. Notice that $P(\text{no rain}) = 1 - P(\text{rain})$. The two events "rain" and "no rain" are mutually exclusive because if one happens, the other cannot. Two mutually exclusive events whose union is the sample space are **complementary events**. If A is an event, the complement of A, written $\overline{A}$, is also an event. For example, consider the event $A = \{2, 4\}$ of tossing a 2 or a 4 using a standard die. The complement of A is the set $\overline{A} = \{1, 3, 5, 6\}$. Because the sample space is $S = \{1, 2, 3, 4, 5, 6\}$, we have $P(A) = \dfrac{2}{6}$ and $P(\overline{A}) = \dfrac{4}{6}$. Notice that

$P(A) + P(\overline{A}) = \dfrac{2}{6} + \dfrac{4}{6} = 1$. This is true in general for any set A and its complement, $\overline{A}$.

Theorem 9-5

If A is an event and $\overline{A}$ is its complement, then

$$P(\overline{A}) = 1 - P(A)$$

Non-Mutually Exclusive Events

Consider the spinner in Figure 9-7.

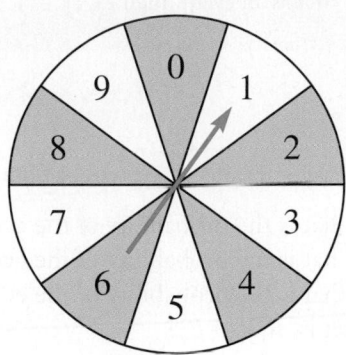

Figure 9-7

Let E be the event of spinning an even number and T the event of spinning a number divisible by 3; that is,

$$E = \{0, 2, 4, 6, 8\}$$
$$T = \{0, 3, 6, 9\}$$

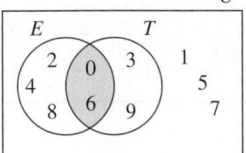

Figure 9-8

The event of spinning an even number and a number divisible by 3 is denoted $E \cap T$. Because $E \cap T = \{0, 6\}$, as seen in Figure 9-8, E and T are not mutually exclusive and $P(E \cup T) \neq P(E) + P(T)$. However, because each outcome is equally likely, we compute the probability of E or T as follows:

$$P(E \cup T) = \frac{n(E \cup T)}{n(S)}$$

Because $E \cup T = \{0, 2, 4, 6, 8, 3, 9\}$, $n(E \cup T) = 7$. Because, $n(S) = 10$, we have $P(E \cup T) = \dfrac{7}{10}$.

In general, we can also compute the probability of E or T by using the following result from Chapter 2:

$$n(E \cup T) = n(E) + n(T) - n(E \cap T)$$

Therefore, for a sample space with equally likely outcomes,

$$P(E \cup T) = \frac{n(E \cup T)}{n(S)}$$
$$= \frac{n(E) + n(T) - n(E \cap T)}{n(S)}$$

$$= \frac{n(E)}{n(S)} + \frac{n(T)}{n(S)} - \frac{n(E \cap T)}{n(S)}$$

$$= P(E) + P(T) - P(E \cap T)$$

This result, although given for events in a sample space with equally likely outcomes, is true in general. This and other results of probability are summarized next.

Summary of Probability Properties

1. $P(\varnothing) = 0$ (impossible event).
2. $P(S) = 1$, where S is the sample space (certain event).
3. For any event $A, 0 \leq P(A) \leq 1$.
4. If A and B are mutually exclusive events, then $P(A \cup B) = P(A) + P(B)$.
5. If A and B are any events, then $P(A \cup B) = P(A) + P(B) - P(A \cap B)$.
6. If A is an event, then $P(\overline{A}) = 1 - P(A)$.

EXAMPLE 9-4

A golf bag contains 2 red tees, 4 blue tees, and 5 white tees.

a. What is the probability of the event R that a tee drawn at random is red?
b. What is the probability of the event "not R," that is, that a tee drawn at random is not red?
c. What is the probability of the event that a tee drawn at random is either red (R) or blue (B); that is, $P(R \cup B)$?

Solution

a. Because the bag contains a total of $2 + 4 + 5$, or 11, tees and 2 tees are red, $P(R) = \frac{2}{11}$.

b. The bag contains 11 tees and 9 are not red, so the probability of "not R" is $\frac{9}{11}$. Also, notice that $P(\overline{R}) = 1 - P(R) = 1 - \frac{2}{11} = \frac{9}{11}$.

c. The bag contains 2 red tees and 4 blue tees and $R \cap B = \varnothing$, so $P(R \cup B) = \frac{2}{11} + \frac{4}{11}$, or $\frac{6}{11}$.

On the student p. 481, events are categorized as "less likely" if the probability of the event is near 0 and "more likely" if the event has probability close to 1. Read through the Examples.

Notice that on the student page the probability of an event is the ratio of the "number of favorable outcomes" compared to the "total number of possible outcomes." This suggests the following alternate definition.

Alternate Definition of Probability of an Event with Equally Likely Outcomes

$$P(\text{event}) = \frac{\text{number of favorable outcomes}}{\text{total number of possible outcomes}}$$

The examples so far in Section 9-1 could likely have been done without listing the sample space. However, listing the sample space helped make sure we accounted for all the possible outcomes and prepared us for more complicated probability problems in the problem set and Section 9-2.

School Book Page

All probabilities range from 0 to 1. The probability of rolling a 7 on a number cube is 0, so that is an *impossible* event. The probability of rolling a positive integer less than 7 is 1, so that is a *certain* event.

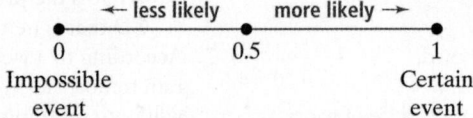

← less likely   more likely →

0   0.5   1
Impossible   Certain
event   event

The **complement** of an event is the collection of outcomes not contained in the event. The sum of the probabilities of an event and its complement is 1. So $P(\text{event}) + P(\text{not event}) = 1$.

EXAMPLES **Finding Probabilities From 0 to 1**

2 **Clothes** The picture shows the jeans in Juanita's closet. She selects a pair of jeans with her eyes shut. Find $P(\text{dark color})$.

There are 8 possible outcomes. Since there are 3 black pairs and 2 blue pairs, the event *dark color* has 5 favorable outcomes.

$P(\text{dark color}) = \dfrac{5}{8}$ ← number of favorable outcomes
← total number of possible outcomes

3 Refer to Juanita's closet. Find $P(\text{red})$.

The event *red* has no favorable outcome.

$P(\text{red}) = \dfrac{0}{8}$, or 0 ← number of favorable outcomes
← total number of possible outcomes

4 Refer to Juanita's closet. Find $P(\text{not dark color})$.

$P(\text{dark color}) + P(\text{not dark color}) = 1$ ← **The sum of probabilities of an event and its complement is 1.**

$\dfrac{5}{8} + P(\text{not dark color}) = 1$ ← **Substitute $\frac{5}{8}$ for $P(\text{dark color})$.**

$\dfrac{5}{8} - \dfrac{5}{8} + P(\text{not dark color}) = 1 - \dfrac{5}{8}$ ← **Subtract $\frac{5}{8}$ from each side.**

$P(\text{not dark color}) = \dfrac{3}{8}$ ← **Simplify.**

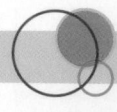

Assessment 9-1A

1. Consider the experiment of drawing a single card from a standard deck of cards and determine which of the following are uniform sample spaces; that is, sample spaces with equally likely outcomes:
    a. {face card, not face card}
    b. {club, diamond, heart, spade}
    c. {black, red}
    d. {king, queen, jack, ace, even-numbered card, odd-numbered card}

2. Each letter of the alphabet is written on a separate piece of paper and placed in a box and then one piece is drawn at random.
    a. What is the probability that the selected piece of paper has a vowel written on it? (Assume y is not a vowel.)
    b. What is the probability that it has a consonant written on it?
    c. What is the probability that the paper has a vowel on it or a letter from the word *probability*?

3. The following spinner is spun:

    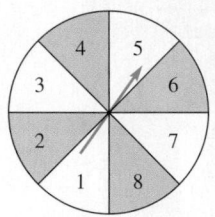

    Find the probabilities of obtaining each of the following events:
    a. A: the number is a factor of 35
    b. B: the number is a multiple of 3
    c. C: the number is even
    d. D: the number is 6 or 2
    e. E: the number is 11
    f. F: the number is composite
    g. G: the number is neither a prime nor a composite

4. A card is selected from an ordinary deck of 52 cards. Find the probabilities for the events consisting of the outcomes below.
    a. Red card
    b. Face card
    c. Red card or a 10
    d. Queen
    e. Not a queen
    f. Face card or a club
    g. Face card and a club
    h. Not a face card and not a club

5. Suppose a drawer contains six black socks, four brown socks, and two green socks. We draw one sock from the drawer and it is equally likely that any one of the socks is drawn. Find the probabilities of the outcomes below.
    a. Sock is brown
    b. Sock is either black or green
    c. Sock is red
    d. Sock is not black
    e. We reach into the drawer without looking to pull out four socks. What is the probability that we get two socks of the same color?

6. Riena has six unmarked CDs in a box, one disk dedicated to each of English, mathematics, French, American history, chemistry, and computer science.
    a. If she chooses a CD at random, what is the probability of the event that she chooses the English CD?
    b. What is the probability of the event that she chooses a CD that is neither mathematics nor chemistry?

7. According to a weather report, there is a 30% chance it will rain tomorrow. What is the probability of the event that it will not rain tomorrow? Explain your answer.

8. A roulette wheel has 38 slots around the rim, 36 slots are numbered from 1 to 36. Half of these 36 slots are red, and the other half are black. The remaining 2 slots are numbered 0 and 00 and are green. As the roulette wheel is spun in one direction, a small ivory ball is rolled along the rim in the opposite direction. The ball has an equally likely chance of falling into any one of the 38 slots. Find the probabilities of the events listed below.
    a. Ball lands in a black slot
    b. Ball lands on 0 or 00
    c. Ball does not land on a number from 1 through 12
    d. Ball lands on an odd number or on a green slot

9. If the roulette wheel in exercise 8 is spun 190 times, predict about how many times the ball can be expected to land on 0 or 00.

10. In Sentinel High School, there are 350 freshmen, 320 sophomores, 310 juniors, and 400 seniors. If a student is chosen at random from the student body to represent the school, what is the probability of the event that the chosen student is a freshman?

11. If A and B are mutually exclusive and if $P(A) = 0.3$ and $P(B) = 0.4$, what is $P(A \cup B)$?

12. A calculus class is composed of 35 men and 45 women. There are 20 business majors, 30 biology majors, 10 computer science majors, and 20 mathematics majors. No person has a double major. If a single student is chosen from the class, what is the probability of the event that the student is the following:
    a. female
    b. a computer science major
    c. not a mathematics major
    d. a computer science major or a mathematics major

13. A box contains five white balls, three black balls, and two red balls.
    a. If only red balls are added to the box, how many must be added so that the probability of the event of drawing a red ball is $\frac{3}{4}$?
    b. If only black balls are added to the original box, how many balls must be added so that the probability of the event of drawing a white ball is $\frac{1}{4}$?

14. Zoe is playing a game in which she draws one ball from one of the boxes shown. She wins if she draws a white ball from either box #1 or box #2. She says that in order to maximize

her chances of winning she will always pick box #2 because it has more white balls. Is she correct? Why?

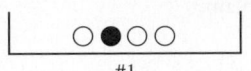
#1

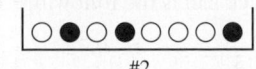
#2

15. If you flipped a fair coin 15 times and got 15 heads, what is the probability of the event of getting a head on the 16th toss? Explain your answer.

16. Consider the letters in the word *numbers*. If one letter is drawn at random, find the probabilities of the events listed below.
   a. Letter is a vowel.
   b. Letter is a consonant.

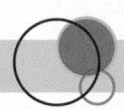

Assessment 9-1B

1. When a thumbtack is dropped, it will land point up (⊥) or point down (⋏). This experiment was repeated 80 times with the following results: point up: 56 times; point down: 24 times.
   a. What is the experimental probability that the thumbtack will land point up?
   b. What is the experimental probability that the thumbtack will land point down?
   c. If you were to try this experiment another 80 times, would you get the same results? Why?
   d. Would you expect to get nearly the same results on a second trial? Why?

2. An experiment consists of selecting the last digit of a telephone number from a telephone book. Assume that each of the 10 digits is equally likely to appear as a last digit. List each of the following:
   a. The sample space
   b. The event consisting of outcomes that the digit is less than 5
   c. The event consisting of outcomes that the digit is odd
   d. The event consisting of outcomes that the digit is not 2
   e. Find the probability of each of the events in (b) through (d).

3. In a refrigerator there are 16 bottles of diet soda, 8 bottles of regular soda, and 4 bottles of water. Find the probabilities for the events listed below.
   a. Choosing a bottle of diet soda when a bottle is chosen at random
   b. Choosing a bottle of regular soda when a bottle is chosen at random
   c. Choosing a bottle of water when a bottle is chosen at random

4. A bag contains n cards, each having one of the consecutive numbers $(1, 2, 3, 4, \ldots, n)$ written on it, with each number being used once. The probability of drawing a card with a number less than or equal to 10 is $\frac{4}{10}$. How many cards are in the bag?

5. Determine whether each player has an equal chance of winning each of the following games:
   a. Toss a fair coin. If heads appears, I win; if tails appears, you lose.
   b. Toss a fair coin. If heads appears, I win; otherwise, you win.
   c. Toss a fair die numbered 1 through 6. If 1 appears, I win; if 6 appears, you win.
   d. Toss a fair die numbered 1 through 6. If an even number appears, I win; if an odd number appears, you win.
   e. Toss a fair die numbered 1 through 6. If a number greater than or equal to 3 appears, I win; otherwise, you win.
   f. Toss two fair dice numbered 1 through 6. If a 1 appears on each die, I win; if a 6 appears on each die, you win.
   g. Toss two fair dice numbered 1 through 6. If the sum is 3, I win; if the sum is 2, you win.

6. In each of the following, sketch a single spinner with the following characteristics:
   a. The outcomes M, A, T, and H are each equally likely.
   b. The outcomes are R, A, and T with $P(R) = \frac{3}{4}$, $P(A) = \frac{1}{8}$, and $P(T) = \frac{1}{8}$.

7. In the game of "Between," two cards are dealt and not replaced in the deck. You then pick a third card from the deck. To win, you must pick a card that has a value between the other two cards. The order of values is 2, 3, 4, 5, 6, 7, 8, 9, 10, J, Q, K, A, where the letters represent a jack, queen, king, and ace, respectively. Determine the probability of your winning if the first two cards dealt are the following:
   a. a 5 and a jack
   b. a 2 and a king
   c. a 5 and a 6

8. Calculators, watches, scoreboards, and many other devices display numbers using arrays like the following. The device lights up different parts of the array (any segment lettered a through g) to display any single digit 0 through 9. Suppose a digit is chosen at random. Determine the probabilities for the events below. (*Hint:* Use a digital watch to see which segments are lit up for the different numbers.)

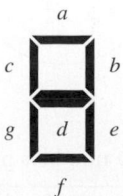

   a. Segment a will be lit.
   b. Segment b will be lit.
   c. Segments e and b will be lit.
   d. Segment e or b will be lit.

9. Let the universal set, U, be the set of students at Central High, A be the set of students taking algebra, and C be the set of students taking chemistry. If a student is selected at random, describe in words what is meant by each of the following probabilities:
   a. $P(A \cup C)$ **b.** $P(A \cap C)$ **c.** $1 - P(C)$

10. A box contains 25% black balls and 75% white balls. The same number of black balls as were in the box are added (so the new number of black balls is twice the original number). A ball is now drawn from the box at random. What is the probability of the event that the ball chosen is black?

11. A teacher gave students three questions and wrote that she would randomly choose one of the three questions for an exam. If Harry prepared for only one of the questions, what is the probability that he did not study for the question that the teacher put on the exam?

12. Nanci has 15 files in a filing cabinet labeled from A to O, one file per label. If she chooses one file at random, find the probabilities for the following events.
   a. A file with the label B.
   b. A file not labeled B.

13. Suppose we have a box containing five white balls, three black balls, and two red balls. Is it possible to add the same number of balls of each color to the box so that when a ball is drawn at random, the probability of the event that it is a black ball is the following? Explain.
   a. $\dfrac{1}{3}$ **b.** 0.32

14. Sylvia decided to simulate the probability of the birth of a boy or a girl by spinning a spinner with the numbers 1–7 on the seven equally divided sectors of the spinner. Suppose the birth of a boy is represented by spinning an odd number and a girl is represented by spinning an even number. If the chance of the birth of a boy is approximately equal to the chance for the birth of a girl, explain whether the spinner described is a good tool to simulate the birth.

15. Describe each of the following as likely or unlikely:
   a. The probability that a current U.S. Supreme Court Justice chosen at random is a man.
   b. The probability that a current member of the U.S. Senate chosen at random is African-American.

16. Explain whether you think that when dialing a seven-digit phone number without its area code, each digit 0–9 has an equal chance of being chosen as the lead number.

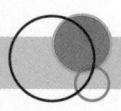

Mathematical Connections 9-1

Communication

1. If events A and B are from the same sample space, and if $P(A) = 0.8$ and $P(B) = 0.9$, can events A and B be mutually exclusive? Explain.

2. Bobbie says that when she shoots a free throw in basketball, she will either make it or miss it. Because there are only two outcomes and one of them is making a basket, Bobbie claims the probability of her making a free throw is $\dfrac{1}{2}$. Explain whether Bobbie's reasoning is correct.

3. If the following spinner is spun 100 times, Joe claims that the number 4 will occur most often because the greatest area of the spinner is covered by the number 4. What would you tell Joe about his conjecture? What is the probability that a 4 will occur on any spin?

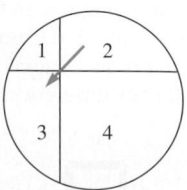

Open-Ended

4. Prepare a card with the numbers printed on it as shown.

   | 1 | 2 | 3 | 4 |

   Show the card to 10 people and ask each of them to pick a single digit listed on the card and tell you what number they picked. Record the results. Pool the results with 4 other people and answer the following questions:
   a. What is the probability of selecting a 3 from the card if the choice was made at random?
   b. Based on the data you gathered, what is the experimental probability of selecting a 3 based on the sample of 10? Based on the sample of 50?
   c. Why might the theoretical probability not agree with the experimental probability?
   d. Conjecture what will happen if the numbers 1, 2, 3, 4, 5 are printed on the card and a similar experiment is repeated. Test the conjecture.

5. Select any book, go to the first complete paragraph in it, and count the number of words the paragraph has. Now count the number of words that start with a vowel. If the paragraph has fewer than 100 words, continue to count words until there are more than 100 words from the start. What is the experimental probability that a word chosen at random from the book starts with a vowel? Open the book to any page and choose a paragraph with more than 100 words. Count the words. Predict how many words on the page start with a vowel, and then count to see how close the prediction is.

6. List three real-world situations that do not involve weather or gambling where probability might be used.

7. If possible, for each of the following letters, describe an event, that has the approximate probability marked by the letter on the probability line.

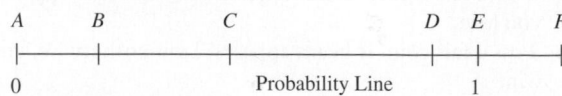

Cooperative Learning

8. Form groups of three or four students. Each group has a pair of dice. Each player needs 18 markers and a sheet of paper with a gameboard drawn on it similar to the following:

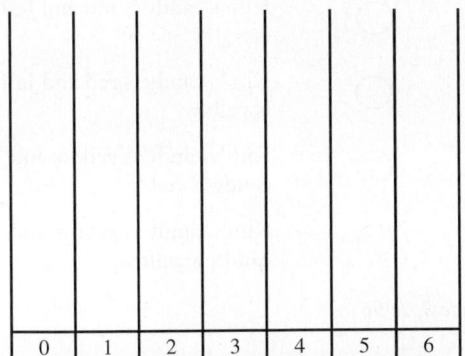

At the beginning of each game, each player places his or her markers on the boards in any arrangement above the numbers. Two players then take turns rolling the dice. The result of each roll is the difference between the greater and the smaller of the two numbers. For example, if a 6 and a 4 were rolled, the difference would be $6 - 4$, or 2. All players who have a marker on the number that represents the difference can remove one marker. Only one marker can be removed each roll. The first player to remove all of the markers is the winner. The game may be stopped if it is clear that there can be no winner.

 a. Play the game twice. What differences seem to occur most often? Least often?
 b. Roll the dice 20 times and record how often the various differences occur. Using this information, explain how to distribute the 18 markers to win.
 c. Compute the theoretical probabilities for each possible difference.
 d. Use the answers to (c) to explain how to arrange the markers to obtain the best chance of winning.

9. The following game is played with two players and a single fair die numbered 1 through 6. The die is rolled and one person receives a score that is the square of the number appearing on the die. The other person will receive a score of 4 times the value showing on the die. The person with the greater score wins.

 a. Play the game several times to see if it appears to be a *fair game*, that is, a game in which each player has an equal chance of winning.
 b. Determine if this is a fair game. If it is not fair, who has the advantage? Explain the answer.

10. Complete the following activities.

 a. Suppose a paper cup is tossed in the air. The different ways it can land are shown below. Toss a cup 100 times and record each result. From this information, calculate the experimental probability of each outcome. Do the outcomes appear to be equally likely? Using experimental probabilities, predict how many times the cup will land on its side if tossed 100 times.

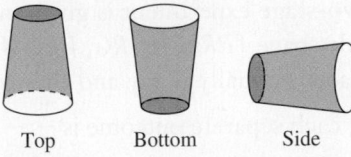

Top　　　Bottom　　　Side

 b. Toss a fair coin 200 times and record the results. From this information, calculate the experimental probability of getting heads on a particular toss. Does the experimental result agree with the expected theoretical probability of $\frac{1}{2}$?

 c. Hold a coin upright on its edge under your forefinger on a hard surface and then spin it with your other finger so that it spins before landing. Repeat this experiment 100 times and calculate the experimental probability of the coin's landing on heads on a particular spin. Compare your experimental probabilities with those in part (b).

Questions from the Classroom

11. A student claims that if a fair coin is tossed and comes up heads 5 times in a row, then, according to the law of averages, the probability of tails on the next toss is greater than the probability of heads. What is your reply?

12. A student observes the following spinner and claims that the color red has the highest probability of appearing, since there are two red areas on the spinner. What is your reply?

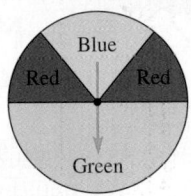

13. A student tosses a coin 3 times, and a head appears each time. The student concludes that the coin is not fair. What is your response?

14. A student wonders why probabilities cannot be negative. What is your response?

15. A student claims that "if the probability of an event is $\frac{3}{5}$, then there are three ways the event can occur and only five elements in the sample space." How do you respond?

Trends in Mathematics and Science Study (TIMSS) Questions

In an eighth-grade class of 30 students, the probability that a student chosen at random will be less than 13 years old is $\frac{1}{5}$. How many students in the class are less than 13 years old?

 a. Two　　　　　b. Three
 c. Four　　　　d. Five
 e. Six

TIMSS, Grade 8, 2003

In a school there were 1200 students (boys and girls). A sample of 100 students was selected at random, and 45 boys were found in the sample. Which of these is most likely to be the number of boys in the school?

 a. 450　　　　　b. 500
 c. 540　　　　d. 600

TIMSS, Grade 8, 2003

National Assessment of Educational Progress (NAEP) Questions

Each of the 6 faces of a certain cube is labeled either R or S. When the cube is tossed, the probability of the cube landing with an R face up is $\frac{1}{3}$.

How many faces are labeled R?
  a. Five
  b. Four
  c. Three
  d. Two
  e. One

NAEP, Grade 8, 2005

A bag contains two red candies and one yellow candy. Kim takes out one candy and eats it, and then Jeff takes out one candy. For

each sentence below, fill in the circle to indicate whether it is possible or not possible.

Possible	Not Possible	
◯	◯	Kim's candy is red and Jeff's candy is red.
◯	◯	Kim's candy is red and Jeff's candy is yellow.
◯	◯	Kim's candy is yellow and Jeff's candy is red.
◯	◯	Kim's candy is yellow and Jeff's candy is yellow.

NAEP, Grade 8, 1996

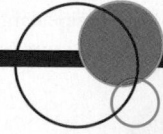

9-2 Multistage Experiments with Tree Diagrams and Geometric Probabilities

In Section 9-1, we considered **one-stage experiments**, that is, experiments that were over after one step. For example, drawing one ball at random from the box containing a red, white, and green ball in Figure 9-9(a) is a one-stage experiment. A tree diagram for this experiment is shown in Figure 9-9(b).

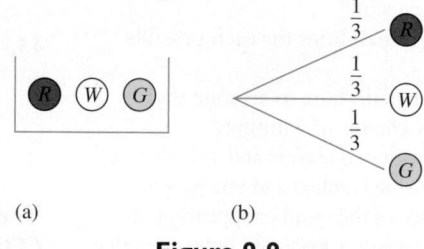

(a)   (b)

Figure 9-9

According to the grade 7 *Common Core Standards*, students should be able to:

> Find probabilities of compound events using organized lists, tables, tree diagrams, and simulation.
>
> **a.** Understand that, just as with simple events, the probability of a compound event is the fraction of outcomes in the sample space for which the compound event occurs.
>
> **b.** Represent sample spaces for compound events using methods such as organized lists, tables, and tree diagrams. For an event described in everyday language (e.g., "rolling double sixes"), identify the outcomes in the sample space which compose the event. (p. 51)

Next we consider **multistage experiments**. For example, a ball is drawn from the box in Figure 9-9(a) and its color is recorded. Then the ball is *replaced*, and a second ball is drawn and its color is recorded. A tree diagram for this two-stage experiment is given in Figure 9-10. The tree diagram can be used to generate the sample space {*RR, RW, RG, WR, WW, WG, GR, GW, GG*}. Each of the outcomes in the sample space is equally likely and there are nine total outcomes, so the probability of the event containing each separate outcome is $\frac{1}{9}$.

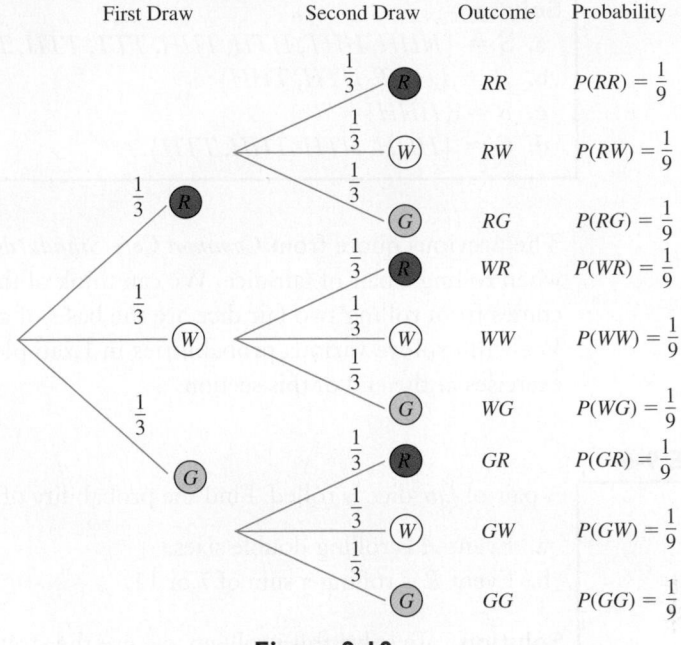

Figure 9-10

An alternative way to generate the sample space is to use a table, as shown in Table 9-2.

Table 9-2

		Second Draw	
	R	**W**	**G**
R	RR	RW	RG
W	WR	WW	WG
G	GR	GW	GG

First Draw

NOW TRY THIS 9-2

1. Use the tree diagram in Figure 9-10 to determine the probability of each of the following events when two balls are drawn from the box.
   a. Both balls are red
   b. No ball is red
   c. At least one ball is red
   d. At most one ball is red
   e. Both balls are the same color
2. Answer each part of question 1 using Table 9-2.

EXAMPLE 9-5

Suppose we toss a fair coin 3 times and record the results. Find each of the following:

   a. the sample space for this experiment
   b. the event A of tossing 2 heads and 1 tail
   c. the event B of tossing no tails
   d. the event C of tossing a head on the last toss

Solution
  a. $S = \{HHH, HHT, HTH, THH, TTT, TTH, THT, HTT\}$
  b. $A = \{HHT, HTH, THH\}$
  c. $B = \{HHH\}$
  d. $C = \{HHH, HTH, THH, TTH\}$

The previous quote from *Common Core Standards* mentions the event of "rolling double sixes" when rolling a pair of fair dice. We can think of this as a two-stage experiment. The various outcomes from rolling two fair dice are the basis of a gambling game commonly known as "craps." We will explore various probabilities in Example 9-6 and explain the game in the assessment exercises at the end of this section.

EXAMPLE 9-6

A pair of *fair* dice is rolled. Find the probability of each of the following events.

  a. Event A is rolling double sixes.
  b. Event B is rolling a sum of 7 or 11.

Solution   To solve this problem, we use the strategy of *making a table*. Figure 9-11(a) shows all possible outcomes of tossing the dice. We know that there are 6 possible results from tossing the first die and 6 from tossing the second die, so by the Fundamental Counting Principle, there are $6 \cdot 6$, or 36, entries.

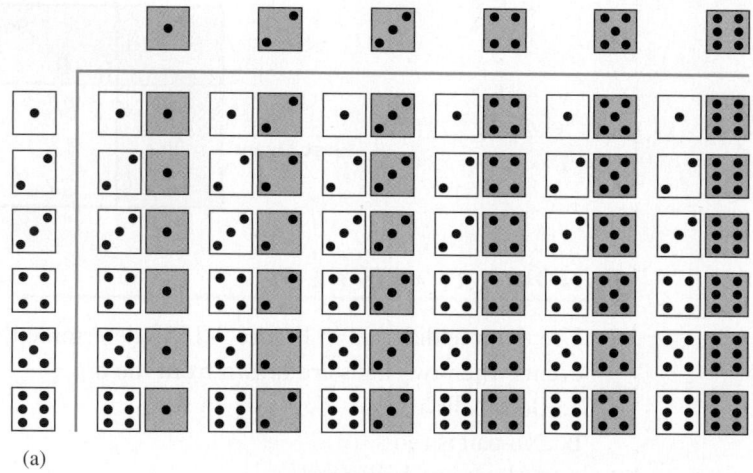

(a)

Number on Second Die

	1	2	3	4	5	6
1	(1, 1)	(1, 2)	(1, 3)	(1, 4)	(1, 5)	(1, 6)
2	(2, 1)	(2, 2)	(2, 3)	(2, 4)	(2, 5)	(2, 6)
3	(3, 1)	(3, 2)	(3, 3)	(3, 4)	(3, 5)	(3, 6)
4	(4, 1)	(4, 2)	(4, 3)	(4, 4)	(4, 5)	(4, 6)
5	(5, 1)	(5, 2)	(5, 3)	(5, 4)	(5, 5)	(5, 6)
6	(6, 1)	(6, 2)	(6, 3)	(6, 4)	(6, 5)	(6, 6)

Number on First Die

(b)

Number on Second Die

	1	2	3	4	5	6
1	2	3	4	5	6	7
2	3	4	5	6	7	8
3	4	5	6	7	8	9
4	5	6	7	8	9	10
5	6	7	8	9	10	11
6	7	8	9	10	11	12

Number on First Die

Possible Sums

(c)

Figure 9-11

a. In Figure 9-11(a) we see that event A, "double sixes," appears only one time out of 36 possible outcomes. Therefore $P(A) = \dfrac{1}{36}$.

b. It may be easier to read the results when they are recorded as ordered pairs, as in Figure 9-11(b), where the first component represents the number on the first die and the second component represents the number on the second die. We show the possible sums from rolling the pair of dice in Figure 9-11(c). In Figure 9-11(c) we see that a sum of 7 appears 6 times. Hence, the event "a sum of 7" arises from the following subset of the set of ordered pairs in Figure 9-11(b):

$$\{(6, 1), (5, 2), (4, 3), (3, 4), (2, 5), (1, 6)\}$$

Each outcome in this set is equally likely and hence $P(\text{a sum of 7}) = \dfrac{6}{36}$. Similarly, $P(\text{a sum of 11}) = \dfrac{2}{36}$. The probabilities of each of the possible sums can be calculated in the same way.

From Figure 9-11(c), the sample space for the experiment is $\{2, 3, 4, 5, 6, 7, 8, 9, 10, 11, 12\}$ but the sample space is not uniform; that is, the probabilities of the given sums are not equal. A **probability distribution** as in Table 9-3 summarizes these probabilities.

Table 9-3

Outcome	2	3	4	5	6	7	8	9	10	11	12
Probability	$\dfrac{1}{36}$	$\dfrac{2}{36}$	$\dfrac{3}{36}$	$\dfrac{4}{36}$	$\dfrac{5}{36}$	$\dfrac{6}{36}$	$\dfrac{5}{36}$	$\dfrac{4}{36}$	$\dfrac{3}{36}$	$\dfrac{2}{36}$	$\dfrac{1}{36}$

The probability of event B "rolling a sum of 7 or 11" is given by $P(B) = P(\text{sum of 7 or sum of 11}) = P(\text{sum of 7}) + P(\text{sum of 11}) = \dfrac{6}{36} + \dfrac{2}{36} = \dfrac{8}{36}$.

EXAMPLE 9-7

A fair pair of dice is rolled. Let E be the event of rolling a sum that is an even number and F the event of rolling a sum that is a prime number. Find the probability of rolling a sum that is even *or* prime, that is, $P(E \cup F)$.

Solution To solve this problem, we use Table 9-3. Note that $E \cup F = \{2, 4, 6, 8, 10, 12, 3, 5, 7, 11\}$. Therefore,

$$\begin{aligned}
P(E \cup F) &= P(2) + P(4) + P(6) + P(8) + P(10) + P(12) + P(3) \\
&\quad + P(5) + P(7) + P(11) \\
&= \frac{1}{36} + \frac{3}{36} + \frac{5}{36} + \frac{5}{36} + \frac{3}{36} + \frac{1}{36} + \frac{2}{36} + \frac{4}{36} + \frac{6}{36} + \frac{2}{36} \\
&= \frac{32}{36} \text{ or } \frac{8}{9}.
\end{aligned}$$

Another approach is to use a property of probabilities. We know that events E and F are not mutually exclusive because $E = \{2, 4, 6, 8, 10, 12\}$, $F = \{2, 3, 5, 7, 11\}$, and $E \cap F = \{2\}$. Therefore,

$$P(E \cup F) = P(E) + P(F) - P(E \cap F)$$
$$= \frac{18}{36} + \frac{15}{36} - \frac{1}{36}$$
$$= \frac{32}{36} \text{ or } \frac{8}{9}.$$

E-Manipulative Activity

Additional practice with probability problems can be found in *Spinner Statistics*.

A third approach to finding $P(E \cup F)$ is to find $P(\overline{E \cup F})$ and subtract this probability from 1. Because $E \cup F = \{2, 3, 4, 5, 6, 7, 8, 10, 11, 12\}$, then $\overline{E \cup F} = \{9\}$ and $P(\overline{E \cup F}) = \frac{4}{36}$. Hence, $P(E \cup F) = 1 - P(\overline{E \cup F}) = 1 - \frac{4}{36} = \frac{32}{36} \text{ or } \frac{8}{9}.$

More Multistage Experiments

The box in Figure 9-12(a) contains one black ball and two white balls. If a ball is drawn at random from the box and the color recorded, a tree diagram for the experiment might look like the one in Figure 9-12(b). Because each ball has the same chance of being drawn, we may combine the branches and obtain the tree diagram shown in Figure 9-12(c). Combining branches in this way is a common practice because it simplifies tree diagrams.

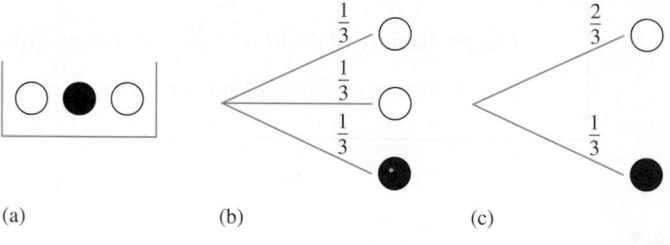

(a)   (b)   (c)

Figure 9-12

Now suppose a ball is drawn at random from the box in Figure 9-12(a) and its color recorded. The ball is then *replaced*, and a second ball is drawn and its color recorded. The sample space for this two-stage experiment may be recorded using ordered pairs as $\{(\bullet, \bullet), (\bullet, \bigcirc), (\bigcirc, \bullet), (\bigcirc, \bigcirc)\}$ or, more commonly, as $\{\bullet\ \bullet, \bullet\ \bigcirc, \bigcirc\ \bullet, \bigcirc\ \bigcirc\}$, as shown in the tree diagram in Figure 9-13.

To assign the probability of the outcomes in this experiment, consider the path for the outcome $\bullet\ \bigcirc$. In the first draw, the probability of the event of obtaining a black ball is $\frac{1}{3}$. Then, the probability of obtaining a white ball in the second draw is $\frac{2}{3}$. Thus we expect to obtain a black ball on the first draw $\frac{1}{3}$ of the time, and then on the second draw to obtain a white ball $\frac{2}{3}$ of those times that we obtained a black ball on the first draw; that is, $\frac{2}{3}$ of $\frac{1}{3}$, or $\frac{2}{3} \cdot \frac{1}{3}$. This product can be obtained by multiplying the probabilities along the branches used for the path leading to $\bullet\ \bigcirc$; that is, $\frac{1}{3} \cdot \frac{2}{3}$, or $\frac{2}{9}$. The probabilities shown in Figure 9-13 can be

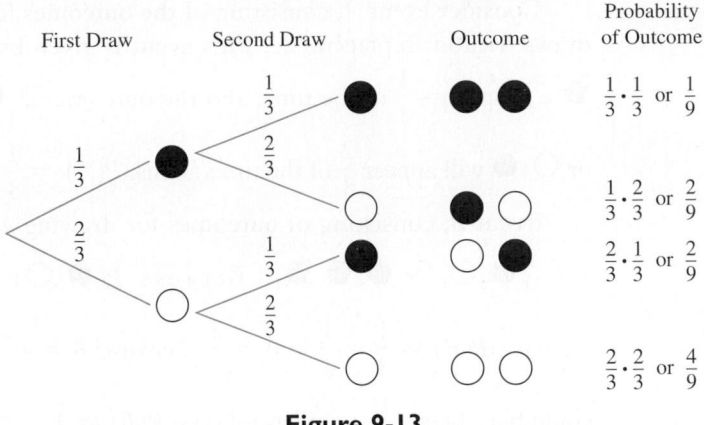

Figure 9-13

obtained by following the paths leading to each of the four outcomes and multiplying the probabilities along the paths. The sum of the probabilities on all the branches from any point always equals 1, and the sum of the probabilities for all the possible outcomes must also be 1. This procedure is an instance of the following general theorem.

Theorem 9-6:   Multiplication Rule for Probabilities for Tree Diagrams

For all multistage experiments, the probability of the outcome along any path of a tree diagram is equal to the product of all the probabilities along the path.

Look again at the box pictured in Figure 9-12(a). This time, suppose two balls are drawn one-by-one *without replacement*. A tree diagram for this experiment, along with the set of possible outcomes, is shown in Figure 9-14. The branch showing the second draw, with probability $\frac{0}{2}$, is added here for completeness. As seen later, the branch could be deleted, as could any unneeded branches of a tree diagram. Compare this figure to the one in Figure 9-13. How are they different?

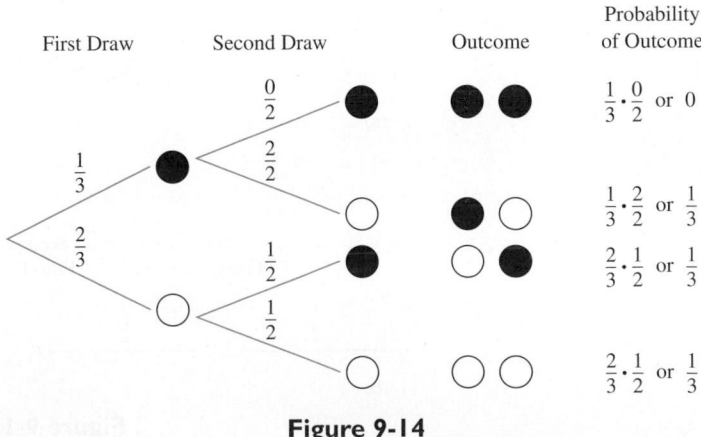

Figure 9-14

The denominators of the fractions in the second draw are all 2. Because the draws are made without replacement, only two balls remain for the second draw.

Consider event A, consisting of the outcomes for drawing exactly one black ball in the two draws without replacement. This event is given by $A = \{\bullet \bigcirc, \bigcirc \bullet\}$. Since the outcome $\bullet \bigcirc$ appears $\frac{1}{3}$ of the time, and the outcome $\bigcirc \bullet$ appears $\frac{1}{3}$ of the time, then either $\bullet \bigcirc$ or $\bigcirc \bullet$ will appear $\frac{2}{3}$ of the time. Thus, $P(A) = \frac{1}{3} + \frac{1}{3} = \frac{2}{3}$.

Event B, consisting of outcomes for drawing *at least* one black ball, could be recorded as $B = \{\bullet \bigcirc, \bigcirc \bullet, \bullet \bullet\}$. Because $P(\bullet \bigcirc) = \frac{1}{3}$, $P(\bigcirc \bullet) = \frac{1}{3}$, and $P(\bullet \bullet) = 0$, we have $P(B) = \frac{1}{3} + \frac{1}{3} + 0 = \frac{2}{3}$. Because $\overline{B} = \{\bigcirc \bigcirc\}$ and $P(\overline{B}) = \frac{1}{3}$, the probability of B could have been computed as follows: $P(B) = 1 - P(\overline{B}) = 1 - \frac{1}{3} = \frac{2}{3}$.

Independent Events

In Figure 9-14, the fact that the first ball was not replaced affects the probability of the color of the second ball drawn. When the occurrence or nonoccurrence of event A has no influence on the outcome of event B, the events A and B are **independent**. For example, if two coins are flipped and if event E_1 is obtaining a head on the first coin and E_2 is obtaining a tail on the second coin, then E_1 and E_2 are independent events because one event has no influence on the second. Figure 9-15(a) depicts flipping the coins. In Figure 9-15(b), the tree diagram is abbreviated to show the branch of interest. Notice that $P(E_1) = \frac{1}{2}$, $P(E_2) = \frac{1}{2}$, and $P(E_1 \cap E_2) = \frac{1}{4}$. So in this case, $P(E_1 \cap E_2) = P(E_1) \cdot P(E_2)$.

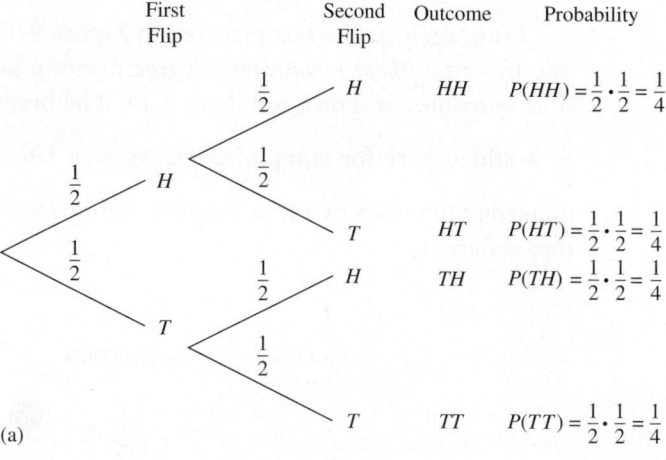

Figure 9-15

Next, consider two boxes as in Figure 9-16. Box 1 contains two white and two black balls, and box 2 contains two white balls and three black balls. We draw one ball from each box and record their respective colors. Let B_1 be the event of drawing a black ball from

Box 1   Box 2

$P(B_1) = \frac{2}{4}, \text{ or } \frac{1}{2}$ $P(B_2) = \frac{3}{5}$

Figure 9-16

box 1 and B_2 the event of drawing a black ball from box 2. B_1 and B_2 are independent. Suppose a ball is drawn from each box and we are interested in the probability of the event that each ball is black, that is, $P(B_1 \cap B_2)$. We know that $P(B_1) = \frac{2}{4}$, or $\frac{1}{2}$, and $P(B_2) = \frac{3}{5}$.

Is $P(B_1 \cap B_2) = P(B_1) \cdot P(B_2)$ in this case as well? We can answer this question by computing $P(B_1 \cap B_2)$ using a familiar approach. If we consider all the black balls to be different and all the white balls to be different, there are $4 \cdot 5$ different pairs of balls. (Why?) Among these pairs, we are interested in pairs consisting only of black balls. There are $2 \cdot 3$ such pairs. (Why?) Hence,

$$P(B_1 \cap B_2) = \frac{2 \cdot 3}{4 \cdot 5} = \frac{2}{4} \cdot \frac{3}{5} = \frac{1}{2} \cdot \frac{3}{5}$$

Thus we see that for the independent events B_1 and B_2, we have $P(B_1 \cap B_2) = P(B_1) \cdot P(B_2)$. This discussion can be generalized as follows.

Theorem 9-7

For any independent events E_1 and E_2,

$$P(E_1 \cap E_2) = P(E_1) \cdot P(E_2).$$

Students often assign a higher probability to the conjunction of two events than to either of the two events individually. For example, students rate the probability of "being 55 and having a heart attack" as more likely than the probability of either "being 55" or "having a heart attack."

 The student page on page 494 gives examples of problems that involve independent and dependent events where **dependent events** are defined to be not independent events. Answer question 1 at the bottom of the student page.

NOW TRY THIS 9-3

In the following cartoon, assume that the events are independent and that there is a 30% chance of rain tonight and a 60% chance of rain tomorrow.

  a. What is the probability that it will rain both times?
  b. What is the probability that it will not rain either of the times?
  c. What is the probability that it will rain exactly one of the times?
  d. What is the probability that it will rain at least one of the times?
  e. In real life, do you think that the events of rain tonight and rain tomorrow are independent events? Why or why not?

Wizard of Id copyright © 2011 by John L. Hart FLP/Distributed by Creators Syndicate

School Book Page INDEPENDENT EVENTS

Think It Through
I need to **remember vocabulary terms.** Mutually exclusive events cannot happen at the same time. Independent and dependent events can happen at the same time.

Multiplying Probabilities

LEARN

How do you find the probability that two events will occur?

Roxanne draws one marble from this bag without looking. Then she draws a second marble.

Example A

What is the probability that Roxanne will get a red marble each time if she replaces the first marble before she draws the second one?

Since she replaces the first marble before the second draw, the outcome of the first draw has no effect on the outcome of the second draw. The two draws are **independent events.**

To find P(red, red), find the probability of each event and multiply.

P(red, red) $= P$(red on 1st draw) $\times P$(red on 2nd draw)

$$= \frac{3}{5} \times \frac{3}{5} = \frac{9}{25} = 36\%$$

1st draw

P(red) $= \frac{3}{5}$

2nd draw

P(red) $= \frac{3}{5}$

Example B

What is the probability that Roxanne will get a red marble each time if she does not replace the first marble before she draws second one?

Since the outcome of the first draw affects the outcome of the second draw, the two draws are **dependent events.**

To find P(red, red), find the probability of each event and multiply.

P(red, red) $= P$(red on 1st draw) $\times P$(red on 2nd draw)

$$= \frac{3}{5} \times \frac{1}{2} = \frac{3}{10} = 30\%$$

1st draw

P(red) $= \frac{3}{5}$

2nd draw

P(red) $= \frac{1}{2}$

✓ **Talk About It**

1. In Example B, why is the probability of the first draw different from the second draw?

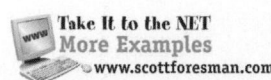
Take It to the NET
More Examples
www.scottforesman.com

EXAMPLE 9-8

Figure 9-17 shows a box with 11 letters. Some letters are repeated. Suppose 4 letters are drawn at random from the box one-by-one without replacement. What is the probability of the event consisting of the outcome *BABY*, with the letters chosen in exactly the order given?

$$\boxed{\text{P R O B A B I L I T Y}}$$

Figure 9-17

Solution The portion of the tree branch leading to the outcome *BABY* is shown in Figure 9-18.

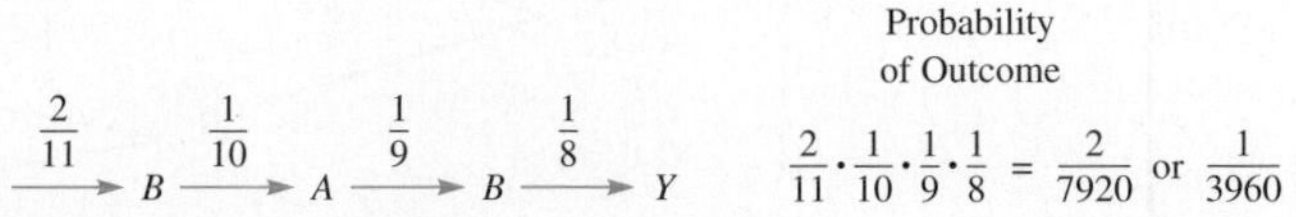

Probability
of Outcome

$$\xrightarrow{\frac{2}{11}} B \xrightarrow{\frac{1}{10}} A \xrightarrow{\frac{1}{9}} B \xrightarrow{\frac{1}{8}} Y \qquad \frac{2}{11}\cdot\frac{1}{10}\cdot\frac{1}{9}\cdot\frac{1}{8} = \frac{2}{7920} \text{ or } \frac{1}{3960}$$

Figure 9-18

The probability of the event that the first letter is B is $\frac{2}{11}$ because there are 2 B's out of 11 letters. The probability of the event of the second B is $\frac{1}{9}$ because there are 9 letters left after 1 B and 1 A have been chosen. Then, $P(BABY)$ is $\frac{2}{7920}$ or $\frac{1}{3960}$, as shown.

In Example 9-8, suppose 4 letters are drawn one-by-one from the box and the letters are replaced after each drawing. In this case, the branch needed to find $P(BABY)$ in the order drawn is pictured in Figure 9-19. Then, $P(BABY) = \frac{2}{11}\cdot\frac{1}{11}\cdot\frac{2}{11}\cdot\frac{1}{11}$, or $\frac{4}{14,641}$.

Probability
of Outcome

$$\xrightarrow{\frac{2}{11}} B \xrightarrow{\frac{1}{11}} A \xrightarrow{\frac{2}{11}} B \xrightarrow{\frac{1}{11}} Y \qquad \frac{2}{11}\cdot\frac{1}{11}\cdot\frac{2}{11}\cdot\frac{1}{11} \text{ or } \frac{4}{14,641}$$

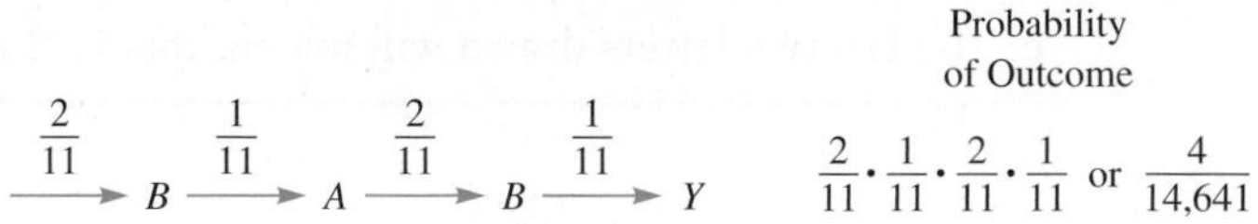

Figure 9-19

EXAMPLE 9-9

Consider the three boxes in Figure 9-20. A letter is drawn from box 1 and placed in box 2. Then, a letter is drawn from box 2 and placed in box 3. Finally, a letter is drawn from box 3. What is the probability of the event that the letter drawn from box 3 is *B*?

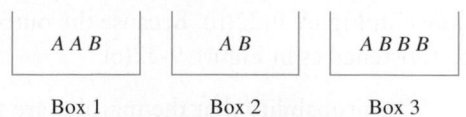

Box 1   Box 2   Box 3

Figure 9-20

Solution A tree diagram for this experiment is given in Figure 9-21. The denominators in the second stage are 3 rather than 2 because in this stage, there are now three letters in box 2. The denominators in the third stage are 5 because in this stage, there are five letters in box 3. To find the probability that a B is drawn from box 3, add the probabilities for the outcomes *AAB*, *ABB*, *BAB*, and *BBB* that make up this event.

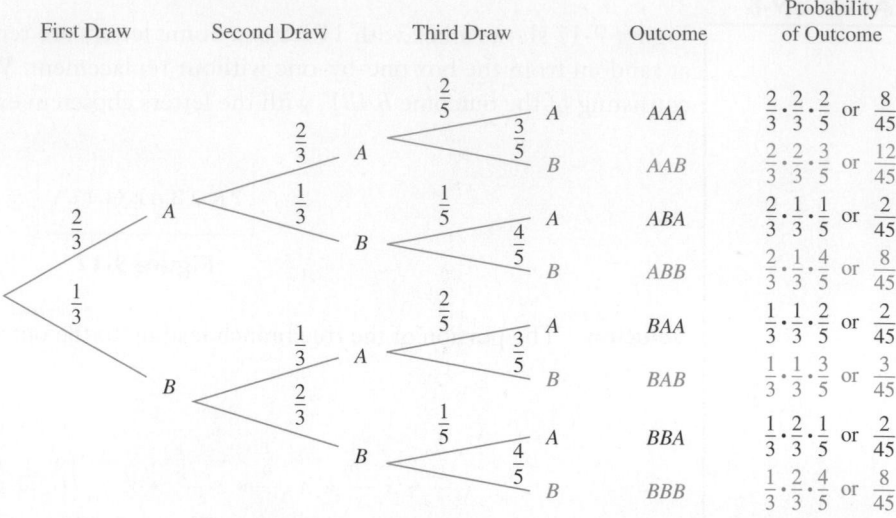

Figure 9-21

Thus, the probability of the event of obtaining a *B* on the draw from box 3 in this experiment is

$$\frac{12}{45} + \frac{8}{45} + \frac{3}{45} + \frac{8}{45} = \frac{31}{45}$$

NOW TRY THIS 9-4

Suppose that in Example 9-9 it is known that the letter *A* was drawn on the first draw. What is the probability of the events that

a. the last letter drawn is a *B*?
b. the last letter drawn is an *A*?
c. the last two letters drawn will match; that is, 2 *A*'s or 2 *B*'s?

Modeling Games

We use models to analyze games that involve probability. Consider the following game, which Arthur and Gwen play: There are two black marbles and one white marble in a box, as in Figure 9-22(a). Gwen mixes the marbles, and Arthur draws two marbles at random without replacement. If the two marbles match, Arthur wins; otherwise, Gwen wins. Does each player have an equal chance of winning? We *develop a model* to analyze the game. One possible model is a tree diagram, as shown in Figure 9-22(b). Because the outcome ○ ○ cannot happen, the tree diagram could also be shortened as in Figure 9-22(c).

The probability that the marbles are the same color is $\frac{1}{3}$ + 0, or $\frac{1}{3}$, and the probability that they are not the same color is $\frac{1}{3}$ + $\frac{1}{3}$, or $\frac{2}{3}$. Because $\frac{1}{3}$ ≠ $\frac{2}{3}$, the players do not have the same chance of winning.

An alternative model for analyzing this game is given in Figure 9-23(a), where the black and white marbles are shown along with the possible ways of drawing two marbles. Each line segment in the diagram represents one pair of marbles that could be drawn. *S* indicates that the marbles in the pair are the same color, and *D* indicates that the marbles are different colors. Because there are two *D*'s in Figure 9-23(a), we see that the probability of drawing two different-colored

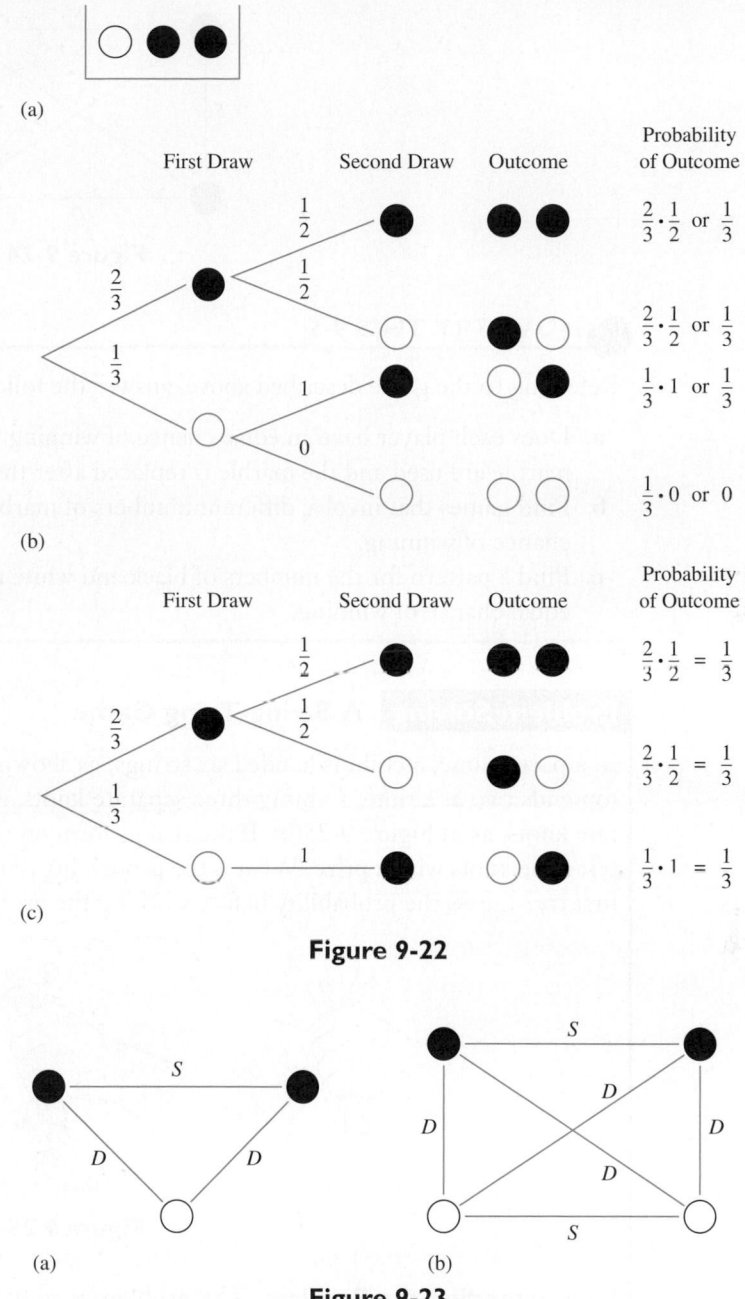

Figure 9-22

Figure 9-23

marbles is $\frac{2}{3}$. Likewise, the probability of drawing two marbles of the same color is $\frac{1}{3}$. Because $\frac{2}{3} \neq \frac{1}{3}$, the players do not have an equal chance of winning.

Will adding another white marble give each player an equal chance of winning? With two white and two black marbles, we have the model in Figure 9-23(b). Therefore, $P(D) = \frac{4}{6}$, or $\frac{2}{3}$, and $P(S) = \frac{2}{6}$, or $\frac{1}{3}$. We see that adding a white marble does not change the probabilities.

Next, consider a game with the same rules but using three black marbles and one white marble. Figure 9-24 shows a model for this situation. Thus, the probability of drawing two marbles of the same color is $\frac{3}{6}$, and the probability of drawing two marbles of different colors is $\frac{3}{6}$. Finally, we have a game in which each player has an equal chance of winning.

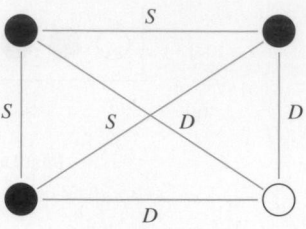

Figure 9-24

NOW TRY THIS 9-5

Referring to the game described above, answer the following:

a. Does each player have an equal chance of winning if only one white marble and one black marble are used and the marble is replaced after the first draw?

b. Find games that involve different numbers of marbles in which each player has an equal chance of winning.

c. Find a pattern for the numbers of black and white marbles that allow each player to have an equal chance of winning.

Problem Solving A String-Tying Game

In a party game, a child is handed six strings, as shown in Figure 9-25(a). Another child ties the top ends two at a time, forming three separate knots, and the bottom ends, forming three separate knots, as in Figure 9-25(b). If the strings form one closed ring, as in Figure 9-25(c), the child tying the knots wins a prize. What is the probability of the event that the child wins a prize on the first try? Guess the probability before working the problem.

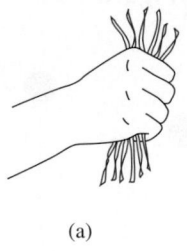

(a)   (b)   (c)

Figure 9.25

Understanding the Problem The problem is to determine the probability of the event that one closed ring will be formed. One closed ring means that all six pieces are joined end-to-end to form one, and only one, ring, as shown in Figure 9-25(c).

Devising a Plan Figure 9-26(a) shows what happens when the ends of the strings of one set are tied in pairs at the top. Notice that no matter in what order those ends are tied, the result appears as in Figure 9-26(a).

(a)   (b)   (c)

Figure 9-26

Then, the other ends are tied in a three-stage experiment. If we pick any string in the first stage, then there are five choices for its mate. Four of these choices are favorable choices for forming a ring. (Why?) Thus, the probability of the event of forming a favorable first tie is $\frac{4}{5}$. Figure 9-26(b) shows a favorable tie at the first stage.

For any one of the remaining four strings, there are three choices for its mate. Two of these choices are favorable ones. (Why?) Thus, the probability of the event of forming a favorable second tie after forming a favorable first tie is $\frac{2}{3}$. Figure 9-26(c) shows a favorable tie at the second stage.

Now, two ends remain. Since nothing can go wrong at the third stage, the probability of the event of making a favorable tie after making favorable ties at the first two stages is 1. If we use the probabilities completed at each stage and a single branch of a tree diagram, we can calculate the probability of performing three successful ties in a row and hence the probability of the event of forming one closed ring.

Carrying Out the Plan If we let S represent a successful tie at each stage, then the branch of the tree with which we are concerned is the one shown in Figure 9-27.

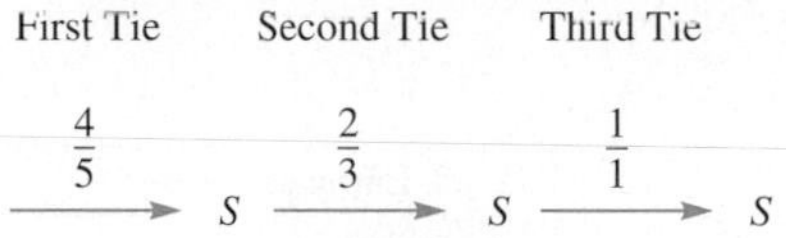

First Tie   Second Tie   Third Tie

$$\xrightarrow{\frac{4}{5}} S \xrightarrow{\frac{2}{3}} S \xrightarrow{\frac{1}{1}} S$$

Figure 9-27

Thus, the probability of the event of forming one ring is $P(\text{ring}) = \frac{4}{5} \cdot \frac{2}{3} \cdot \frac{1}{1} = \frac{8}{15}$.

Looking Back The probability of the event that a child will form a ring on the first try is $\frac{8}{15}$. A class might simulate this problem several times with strings to see how the fraction of successes compares with the theoretical probability of $\frac{8}{15}$.

Related problems that could be posed for solution include the following:

1. If a child fails to get a ring 10 times in a row, the child may not play again. What is the probability of such a streak of bad luck?
2. If the number of strings is reduced to three and the rule is that an upper end must be tied to a lower end, what is the probability of the event of forming a single ring?
3. If the number of strings is three, but an upper end can be tied to either an upper or a lower end, what is the probability of the event of forming a single ring?
4. What is the probability of the event of forming three rings in the original problem?
5. What is the probability of the event of forming two rings in the original problem?

Geometric Probability (Area Models)

A probability model that uses geometric shapes is an **area model**. When area models are used to determine probabilities geometrically, outcomes are associated with points chosen at random in a geometric region that represents a sample space. This process is often called finding **geometric probabilities**.

Suppose we throw darts at a square target 2 units long on a side and divided into 4 congruent triangles, as shown in Figure 9-28. If the dart must hit the target somewhere and if all spots can be hit with equal probability, what is the probability of the event that the dart will land in the shaded region? The entire target, which has an area of $2 \cdot 2$ or 4 square units, represents the sample

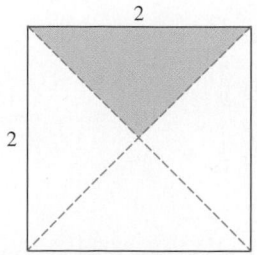

Figure 9-28

space. The shaded area is the event of a successful toss. The area of the shaded part is $\frac{1}{4}$ of the sample space. The probability of the dart's landing in the shaded region is the ratio of the area of the event to the area of the sample space, or $\frac{1}{4}$. In general when computing the probability of an event geometrically, we find the ratio of the area of the region representing an event to the area of the region representing the sample space.

Problem Solving   A Quiz-Show Game

On a quiz show, a contestant stands at the entrance to a maze that opens into two rooms, labeled *A* and *B* in Figure 9-29. The master of ceremonies' assistant is to place a new car in one room and a donkey in the other. The contestant must walk through the maze into one of the rooms and will win whatever is in that room. If the contestant makes each decision in the maze at random, in which room should the assistant place the car to give the contestant the best chance to win?

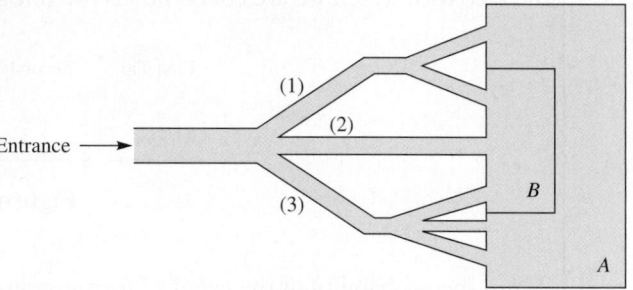

Figure 9-29

Understanding the Problem The contestant must first choose one of the paths marked 1, 2, or 3 and then choose another path as she proceeds through the maze. To determine the room the contestant is most likely to choose, the assistant must be able to determine the probability of the contestant's reaching each room.

Devising a Plan One way to determine where the car should be placed is to *model the choices with a tree diagram* and to compute the probabilities along the branches of the tree.

Carrying Out the Plan A tree diagram for the maze is shown in Figure 9-30, along with the possible outcomes and the probabilities of each branch.

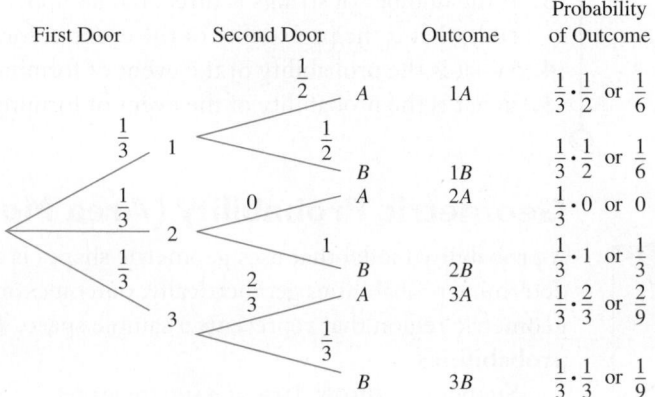

Figure 9-30

The probability of the event that room A is chosen is $\frac{1}{6} + 0 + \frac{2}{9} = \frac{7}{18}$. Hence the probability of the event that room B is chosen is $1 - \frac{7}{18} = \frac{11}{18}$. Thus room B has the greater probability of being chosen. This is where the car should be placed for the contestant to have the best chance of winning it.

Looking Back An alternative model for this problem and for many probability problems is an area model. The rectangle in Figure 9-31(a) represents the first three choices that the contestant can make. Because each choice is equally likely, each is represented by an equal area. If the contestant chooses the upper path, then rooms A and B have an equal chance of being chosen. If she chooses the middle path, then only room B can be entered. If she chooses the lower path, then room A is entered $\frac{2}{3}$ of the time. This can be expressed in terms of the area model shown in Figure 9-31(b). Dividing the rectangle into pieces of equal area, we obtain the model in Figure 9-31(c), in which the area representing room B is shaded. Because the area representing room B is greater than the area representing room A, room B has the greater probability of being chosen. Figure 9-31(c) can enable us to find the probability of choosing room B. Because the shaded area consists of 11 rectangles out of a total of 18 rectangles, the probability of the event of choosing room B is $\frac{11}{18}$. We can vary the problem by changing the maze or by changing the locations of the rooms.

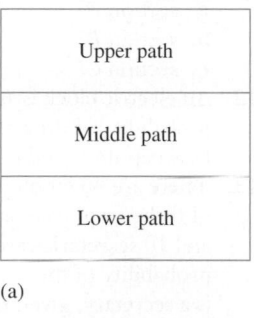

(a)

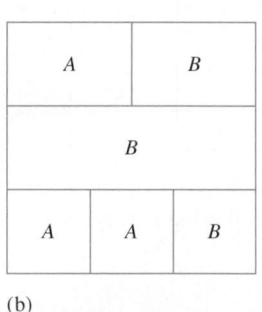

(b)

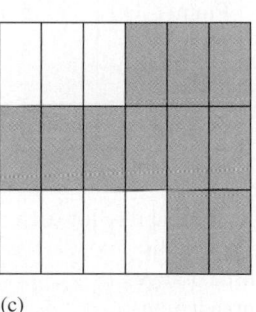

(c)

Figure 9-31

Assessment 9-2A

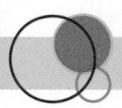

1. **a.** A box contains three white balls and two black balls. A ball is drawn at random from the box and not replaced. Then a second ball is drawn from the box. Draw a tree diagram for this experiment and find the probability of the event that the two balls are of different colors.

   b. Suppose that a ball is drawn at random from the box in part (a), its color is recorded, and then the ball is put back in the box. Draw a tree diagram for this experiment and find the probability of the event that the two balls are of different colors.
2. A box contains six letters, shown as follows. What is the probability of the event *DAN* is obtained in that order if three letters are drawn one-by-one:
   a. With replacement?

   b. Without replacement?

   $\boxed{\text{R A N D O M}}$

3. An executive committee consisted of 10 members: 4 women and 6 men. Three members were selected at random to be sent to a meeting in Hawaii. A blindfolded woman drew 3 of the 10 names from a hat to determine who would go. All 3 names drawn were women's. What was the probability of such luck?
4. Following are three boxes containing balls. Draw a ball from box 1 and place it in box 2. Then draw a ball from box 2 and place it in box 3. Finally, draw a ball from box 3.

   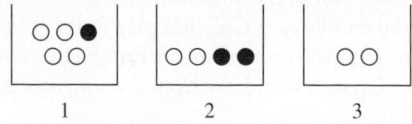

   1   2   3

a. What is the probability of the event that the last ball, drawn from box 3, is white?

b. What is the probability of the event that the last ball drawn is black?

5. Assume the probability is $\frac{1}{2}$ that a child born at any time is a boy and births are independent. What is the probability of the event that if a family has four children, they are all boys?

6. A box contains five slips of paper. Each slip has one of the numbers 4, 6, 7, 8, or 9 written on it and all numbers are used. Al reaches into the box and draws two slips and adds the two numbers. If the sum is even, Al wins a prize.

a. What is the probability of the event that Al wins a prize?

b. Does the probability change if the two numbers are multiplied? Explain.

7. The following shows the numbers of symbols on each of the three dials of a standard slot machine:

Symbol	Dial 1	Dial 2	Dial 3
Bar	1	3	1
Bell	1	3	3
Plum	5	1	5
Orange	3	6	7
Cherry	7	7	0
Lemon	3	0	4
Total	20	20	20

Find the probability for each of the following events:
a. Three plums
b. Three oranges
c. Three lemons
d. No plums

8. You play a game in which you first choose one of the two spinners shown. You then spin your spinner and a second person spins the other spinner. The one with the greater number wins.

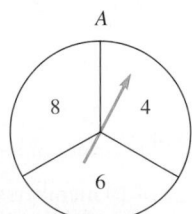 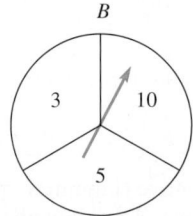

a. Which spinner should you choose? Why?

b. Notice that the sum of the numbers on each spinner is 18. Design two spinners with unequal sums so that choosing the spinner with the lesser sum will give the player a greater probability of winning.

9. If a person takes a five-question true-false test, what is the probability of the event that the score is 100% correct if the person guesses on every question?

10. Rattlesnake and Paxson Colleges play four games against each other in a chess tournament. Rob Fisher, the chess whiz from Paxson, withdrew from the tournament, so the probabilities that Rattlesnake and Paxson will win each game are $\frac{2}{3}$ and $\frac{1}{3}$, respectively. Determine the probabilities of the following events:

a. Paxson loses all four games.

b. The match is a draw with each school winning two games.

11. Consider the following dartboard: (Assume that all quadrilaterals are squares and that the x's represent equal lengths.)

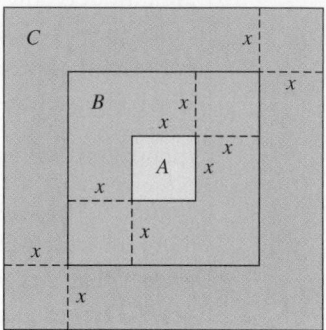

If a dart may land in any region on the board with probability determined by the ratio of the area of the chosen region to the whole, what is the probability that it will land in
a. section A?
b. section B?
c. section C?

12. An electric clock is stopped by a power failure. What is the probability of the event that the second hand is stopped between the 3 and the 4?

13. There are 40 employees in a certain firm. We know that 28 of these employees are males, 2 of these males are secretaries, and 10 secretaries are employed by the firm. What is the probability of the event that an employee chosen at random is a secretary, given that the person is a male?

14. Four blue socks, four white socks, and four gray socks are mixed in a drawer. You pull out two socks, one at a time, without looking.

a. Draw a tree diagram along with the possible outcomes and the probabilities of each branch.

b. What is the probability of the event of getting a pair of socks of the same color?

c. What is the probability of the event of getting two gray socks?

d. Suppose that, instead of pulling out two socks, you pull out four socks. What is the probability of the event of getting two socks of the same color?

15. When you toss a quarter 4 times, what is the probability of the event that you get
a. at least as many heads as tails?
b. at least as many tails as heads?

16. A manufacturer found that among 500 randomly selected smoke detectors only 450 worked properly. Based on this information, how many smoke detectors would you have to install to be sure that the probability that at least one of them will work will be at least 99.99%?

17. Bob leaves the top of Snow Mountain for his last ski run of the day. There are 6 trails to the base of the mountain. He would like to end up at the lodge but it is snowing and he cannot tell one trail from another. If he chooses a path at

random at each fork, what is the probability of the event that he will end up at the lodge?

18. Carolyn will win a large prize if she wins two tennis games in a row out of three games. She is to play alternately against Billie and Bobby. She may choose to play Billie-Bobby-Billie or Bobby-Billie-Bobby. Assume she wins against Billie 50% of the time and against Bobby 80% of the time. Which alternative should she choose, and why?

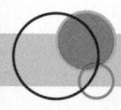

Assessment 9-2B

1. Suppose an experiment consists of spinning X and then spinning Y, as follows:

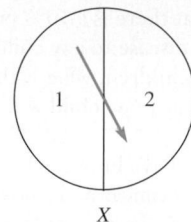 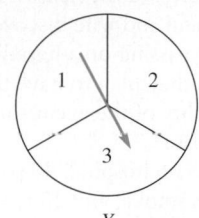

Find the following:
   a. The sample space S for the experiment
   b. The event A consisting of outcomes from spinning an even number followed by an even number
   c. The event B consisting of outcomes from spinning at least one 2
   d. The event C consisting of outcomes from spinning exactly one 2

2. Following are three boxes containing letters:

|MATH| |AND| |HISTORY|
   1   2   3

   a. From box 1, three letters are drawn one-by-one without replacement and recorded in order. What is the probability that the outcome is *HAT*?
   b. From box 1, three letters are drawn one-by-one with replacement and recorded in order. What is the probability that the outcome is *HAT*?
   c. One letter is drawn at random from box 1, then another from box 2, and then another from box 3, with the results recorded in order. What is the probability that the outcome is *HAT*?
   d. If a box is chosen at random and then a letter is drawn at random from the box, what is the probability of the event that the outcome is *A*?

3. A penny, a nickel, a dime, and a quarter are tossed. What is the probability of the event of obtaining at least three heads on the tosses?

4. Two boxes with letters follow. Choose a box and draw three letters at random, one-by-one, without replacement. If the outcome is SOS, you win a prize.

|SOS| |SOSSOS|
   1   2

   a. Which box should you choose?
   b. Which box would you choose if the letters are to be drawn with replacement?

5. An assembly line has two inspectors. The probability that the first inspector will miss a defective item is 0.05. If the defective item passes the first inspector, the probability that the second inspector will miss it is 0.01. What is the probability of the event that a defective item will pass by both inspectors?

6. Following are two boxes containing black and white balls. A ball is drawn at random from box 1. Then a ball is drawn at random from box 2, and the colors of balls from both boxes are recorded in order.

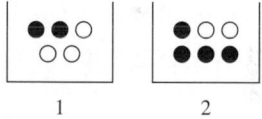

   Find the probability of each of the following events:
   a. two white balls
   b. at least one black ball
   c. at most one black ball
   d. ● ○ or ○ ●

7. The following questions refer to a very popular dice game, craps, in which a player rolls two dice:
   a. Rolling a sum of 7 or 11 on the first roll of the dice is a win. What is the probability of the event of winning on the first roll?
   b. Rolling a sum of 2, 3, or 12 on the first roll of the dice is a loss. What is the probability of the event of losing on the first roll?
   c. Rolling a sum of 4, 5, 6, 8, 9, or 10 on the first roll is neither a win nor a loss. What is the probability of the event of neither winning nor losing on the first roll?
   d. After rolling a sum of 4, 5, 6, 8, 9, or 10, a player must roll the same sum again before rolling a sum of 7. Which sum—4, 5, 6, 8, 9, or 10—has the highest probability of occurring again?
   e. What is the probability of the event of rolling a sum of 1 on any roll of the dice?
   f. What is the probability of the event of rolling a sum less than 13 on any roll of the dice?
   g. If the two dice are rolled 60 times, predict about how many times a sum of 7 will be rolled.

8. Suppose we spin the following spinner with the first spin giving the numerator and the second spin giving the

denominator of a fraction. What is the probability of the event that the fraction will be greater than $1\frac{1}{2}$?

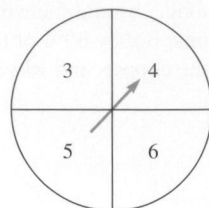

9. Brittany is going to ascend a four-step staircase. At any time, she is just as likely to stride up one step or two steps. Find the probability of the event that she will ascend the four steps in
   a. two strides   **b.** three strides   **c.** four strides
10. An experiment consists of spinning the spinner shown and then flipping a coin with sides numbered 1 and 2.

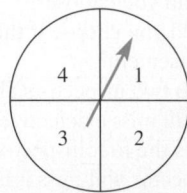

   What is the probability of the following events:
   a. the number on the spinner will be greater than the number on the coin?
   b. the outcome will consist of two consecutive integers in any order?
11. In the following square dartboard, suppose a dart is equally likely to land in any small square of the board.

E									
	D								
		C							
			B						
				A					

   Points are given as follows:

Region	Points
A	10
B	8
C	6
D	4
E	2

   a. What is the total area of the board?
   b. What is the probability of the event of a dart's landing in each lettered region of the board?

   c. If two darts are tossed, what is the probability of the event of scoring 20 points?
   d. What is the probability of the event that the dart will land in neither *D* nor *E*?
12. The combinations on the lockers at the high school consist of three numbers, each ranging from 0 to 39. If a combination is chosen at random, what is the probability of the event that the first two numbers are each multiples of 9 and the third number is a multiple of 4?
13. The following box contains the 11 letters shown. The letters are drawn one-by-one without replacement, and the results are recorded in order. Find the probability of the event of the outcome MISSISSIPPI.

   $$\boxed{\text{M I I I I P P S S S S}}$$

14. The land area of Earth is approximately 57,500,000 mi². The water area of Earth is approximately 139,600,000 mi². If a meteor lands at random on the planet, what is the probability, to the nearest tenth, that it will hit water?
15. A husband and wife discover that there is a 10% probability of their passing on a hereditary disease to any child they have. If they plan to have three children, what is the probability of the event that at least one child will inherit the disease?
16. At a certain hospital, 40 patients have lung cancer, 30 patients smoke, and 25 have lung cancer and smoke. Suppose the hospital contains 200 patients. If a patient chosen at random is known to smoke, what is the probability of the event that the patient has lung cancer?
17. Solve the Quiz-Show Game in this section by replacing Figure 9-29 with the following maze:

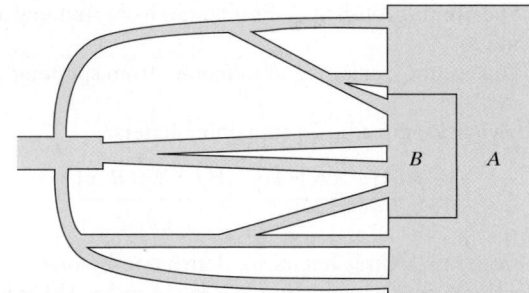

18. Jane has two tennis serves, a hard serve and a soft serve. Her hard serve is in (a good serve) 50% of the time, and her soft serve is in (good) 75% of the time. If her hard serve is in, she wins 75% of her points. If her soft serve is in, she wins 50% of her points. Since she is allowed to re-serve one time if her first serve is out, what should her serving strategy be? That is, should she serve hard followed by soft; both hard; soft followed by hard; or both soft? Note that two bad serves is a lost point.
19. In a certain population of caribou, the probability of an animal's being sickly is $\frac{1}{20}$. If a caribou is sickly, the probability of its being eaten by wolves is $\frac{1}{3}$. If a caribou is not sickly, the probability of its being eaten by wolves is $\frac{1}{150}$.

   If a caribou is chosen at random from the herd, what is the probability of the event that it will be eaten by wolves?

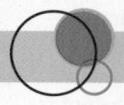

Mathematical Connections 9-2

Communication

1. Jim rolled a fair die 5 times and obtained a 3 every time. He concluded that on the next roll, a 3 is more likely to occur than the other numbers. Explain whether this is true.

2. A witness to a crime observed that the criminal had blond hair and blue eyes and drove a red car. When the police look for a suspect, is the probability greater that they will arrest someone with blond hair and blue eyes or that they will arrest someone with blond hair and blue eyes who drives a red car? Explain your answer.

3. You are given three white balls, one red ball, and two identical boxes. You are asked to distribute the balls in the boxes in any way you like. You then are asked to select a box (after the boxes have been shuffled) and without looking pick a ball at random from that box. If the ball is red, you win a prize. How should you distribute the balls in the boxes to maximize your chances of winning? Justify your reasoning.

Open-Ended

4. Make up a game in which the players have an equal chance of winning and that involves rolling two regular dice.

5. How can the faces of two cubes be numbered so that when they are rolled, the resulting sum is a number 1 to 12 inclusive and each sum has the same probability?

6. Use graph paper to design a dartboard such that the probability of hitting a certain part of the board is $\frac{3}{5}$. Explain your reasoning.

Cooperative Learning

7. Use two spinners divided into four equal areas numbered 1–4. A player spins both spinners and computes the product of the two numbers. If the product is 1, 2, 3, or 4, player A wins. In the same way, player B wins if the product is 6, 8, 9, 12, or 16. The game ends when 20 plays have been made. Each player receives 1 point for each win, and the game winner is the person with the most points when the game ends.
   a. Do you think the game looks fair; that is, both players have the same chance of winning? Why or why not? If not, who do you think has the best chance of winning?
   b. Play the game to see if it seems fair. Do you think it is fair based on playing it? Why?
   c. Complete the following table to determine possible products.

2nd spin

×	1	2	3	4
1				
2				
3			12	
4				

1st spin

   d. Based on the table in part (c), is this a fair game? Explain why.
   e. Replace the spinners in the preceding game with ones divided into six equal areas with the numbers 1, 2, 3, 4, 5,

and 6 in the regions. Design a similar game and decide what products each player should use so that this is a fair game. Explain how your game is fair.

8. Play the following game in pairs. One player chooses one of four equally likely outcomes from the set {HH, HT, TH, TT}, which is the sample space for the experiment of tossing a fair coin twice. The other player then chooses one of the other outcomes. A coin is flipped until either player's choice appears. For example, the first player chooses TT and the second player chooses HT. If the first two flips yield TH, then no one wins and the game continues. If, after five flips, the string THHHT appears, the second player is the winner because the sequence HT finally appeared. Play the game 10 times. Does each player appear to have the same chance of winning? Analyze the game for the case in which the first player chooses TT and the second HT, and explain whether the game is fair for all choices made by the two players. (*Hint:* Find the probability that the first player wins the game by showing that the first player will win if, and only if, "tails" appears on the first and on the second flip.)

9. Consider the three spinners A, B, and C shown in the following figure:

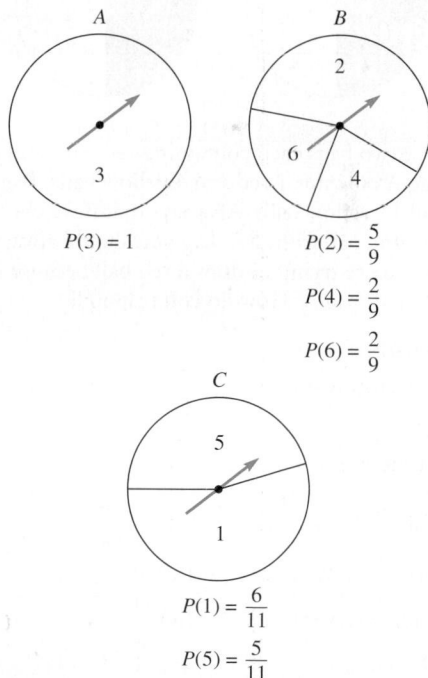

$P(3) = 1$

$P(2) = \frac{5}{9}$

$P(4) = \frac{2}{9}$

$P(6) = \frac{2}{9}$

$P(1) = \frac{6}{11}$

$P(5) = \frac{5}{11}$

   a. Suppose there are only two players and that the first player chooses a spinner, then the second player chooses a different spinner, and each person spins his or her spinner, with the greater number winning. Play the game several times to get a feeling for it. Determine if each player appears to have the same chances of winning the game. If not, which spinner should you choose in order to win?
   b. This time play the same game with three players. If each player must choose a different spinner and spin it, is the winning strategy the same as it was in (a)? Why or why not?

Questions from the Classroom

10. An experiment consists of tossing a fair coin twice. The student reasons that there are three possible outcomes: two heads, one head and one tail, or two tails. Thus, $P(HH) = \frac{1}{3}$. What is your reply?

11. A student would like to know the difference between two events being independent and two events being mutually exclusive. How would you answer her?

12. In response to the question "If a fair die is rolled twice, what is the probability of the event of rolling a pair of 5s?" a student replies, "One-third, because $\frac{1}{6} + \frac{1}{6} = \frac{1}{3}$." How do you respond?

13. A student is not sure when to add and when to multiply probabilities. How do you respond?

14. Alberto is to spin the spinners shown and compute the probability of two blacks. He looks at the spinners and says the answer is $\frac{1}{2}$ because $\frac{1}{2}$ of the areas of the circles are black. How do you respond?

 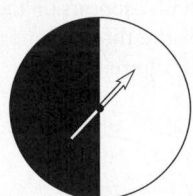

15. There are two bags each containing red balls and yellow balls. Bag A contains 1 red and 4 yellow balls. Bag B contains 3 red and 13 yellow balls. Alva says that if we choose one bag and draw one ball from that bag you should always choose bag B if you are trying to draw a red ball because it has more red balls than bag A. How do you respond?

Review Problems

16. Match the following phrase to the probability that describes it:

a. A certain event   (i) $\dfrac{1}{1000}$

b. An impossible event   (ii) $\dfrac{999}{1000}$

c. A very likely event   (iii) 0

d. An unlikely event   (iv) $\dfrac{1}{2}$

e. A 50% chance   (v) 1

17. A date in the month of April is chosen at random. Find the probability of the event of the date's being each of the following:
a. April 7
b. April 31
c. Before April 20

18. Three men were walking down a street talking when they met a fourth man. If the fourth man knew that two of the men always lied and the third always told the truth and he asked the three a question, what is the probability that he got a truthful answer when one man answered?

Trends in Mathematics and Science Study (TIMSS) Questions

The figure below shows a spinner with 24 sectors. When someone spins the arrow, it is equally likely to stop on any sector.

$\frac{1}{8}$ of the sectors are blue, $\frac{1}{24}$ are purple, $\frac{1}{2}$ are orange, and $\frac{1}{3}$ are red. If a person spins the arrow, on which color sector is the spinner LEAST likely to stop?

a. blue
b. purple
c. orange
d. red

TIMSS, Grade 8, 2003

Sophie has a bag in which there are 16 marbles: 8 are red and 8 are black marbles. She draws 2 marbles from the bag and does not put them back. Both marbles are black. She then draws a third marble out of the bag. What can you say about the likely color of this third marble?

a. It is more likely to be red than black.

b. It is more likely to be black than red.

c. It is equally likely to be red or black.

d. You cannot tell if red or black is more likely.

TIMSS, Grade 8, 2007

National Assessment for Educational Progress (NAEP) Question

A package of candies contained only 10 red candies, 10 blue candies, and 10 green candies. Bill shook up the package, opened it, and started taking out one candy at a time and eating it. The first 2 candies he took out and ate were blue. Bill thinks the probability of getting a blue candy on his third try is $\frac{10}{30}$ or $\frac{1}{3}$. Is Bill correct or incorrect?

NAEP, Grade 8, 2005

Bradley Efron, a Stanford University statistician, designed a set of nonstandard dice whose faces are numbered as shown in Figure 9-32. The dice are to be used in a game in which each player chooses a die and then rolls it. Whoever rolls the greater number is the winner. What strategy should you use so that you have the best chance of winning this game?

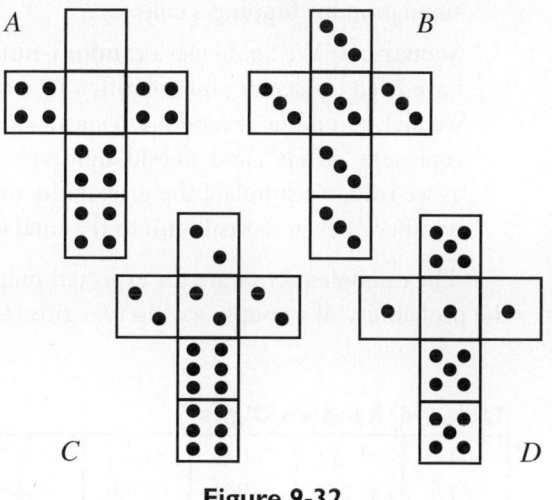

Figure 9-32

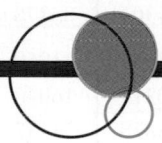

9-3 Using Simulations in Probability

Students may use simulations to study phenomena too complex to analyze by other means. A **simulation** is a technique used to act out a situation by conducting experiments whose outcomes are analogous to the original situation. Using simulations, students typically estimate a probability using many trials rather than determine probabilities theoretically. Simulations can take various forms.

In grades 6–8 *PSSM*, we find the following:

If simulations are used, teachers need to help students understand what the simulation data represent and how they relate to the problem situation, such as flipping coins. (p. 254)

In Section 9-1, Table 9-1, we saw partial results of John Kerrich's coin tosses. It would take considerable time to toss a coin 10,000 times as he did. But suppose we could simulate those tosses without having to flip a coin that many times. If that could be done, then we could make predictions based on the simulation, which, if the simulation was correct, would be very much like the experimental results Kerrich found. So what makes sense to simulate a coin toss?

Because the probabilities of tossing a fair coin and obtaining a head or a tail are each $\frac{1}{2}$, it is sensible to devise a system that has two outcomes, each of which is expected to have a probability of $\frac{1}{2}$. Consider the following possibilities (and there are many more):

Scenario 1: Take an ordinary deck of playing cards. There are 52 cards (without jokers) in the deck, half are red and half are black. One simulation for the coin toss is simply to draw a card at random from the deck. Assign red to mean "a head appears," and assign black to mean "a tail appears." The probability of the event of drawing a red (or obtaining a head) is $\frac{1}{2}$, and the probability of obtaining a black (or a tail) is $\frac{1}{2}$. No coin has to be used, but the flipping of a coin could be simulated by drawing a card.

Scenario 2: Take an ordinary die with six faces, half are even and half are odd. We assign an even number as a "head," and an odd number as a "tail." Rolling the die then becomes a simulation for flipping a coin.

Scenario 3: We could use a **random-number table** as in Table 9-4. The digits in the table have been chosen at random, often by a computer or a calculator. To simulate the coin toss, we pick a number at random to start, and then read across the table, letting an even digit represent "heads", and an odd digit represent "tails." By keeping a record of what we obtain as we read, we simulate the probability of the event of tossing heads as the ratio of the number of even digits found to the total number read.

The examples above are for expected single outcomes of an experiment. Suppose we simulate the probability of a couple having two girls (*GG*) in an expected family. If we assume the birth of

Table 9-4 Random Digits

36422	93239	76046	81114	77412	86557	19549	98473	15221	87856
78496	47197	37961	67568	14861	61077	85210	51264	49975	71785
95384	59596	05081	39968	80495	00192	94679	18307	16265	48888
37957	89199	10816	24260	52302	69592	55019	94127	71721	70673
31422	27529	95051	83157	96377	33723	52902	51302	86370	50452
07443	15346	40653	84238	24430	88834	77318	07486	33950	61598
41348	86255	92715	96656	49693	99286	83447	20215	16040	41085
12398	95111	45663	55020	57159	58010	43162	98878	73337	35571
77229	92095	44305	09285	73256	02968	31129	66588	48126	52700
61175	53014	60304	13976	96312	42442	96713	43940	92516	81421
16825	27482	97858	05642	88047	68960	52991	67703	29805	42701
84656	03089	05166	67571	25545	26603	40243	55482	38341	97782
03872	31767	23729	89523	73654	24626	78393	77172	41328	95633
40488	70426	04034	46618	55102	93408	10965	69744	80766	14889
98322	25528	43808	05935	78338	77881	90139	72375	50624	91385
13366	52764	02407	14202	74172	58770	65348	24115	44277	96735
86711	27764	86789	43800	87582	09298	17880	75507	35217	08352
53886	50358	62738	91783	71944	90221	79403	75139	09102	77826
99348	21186	42266	01531	44325	61042	13453	61917	90426	12437
49985	08787	59448	82680	52929	19077	98518	06251	58451	91140
49807	32863	69984	20102	09523	47827	08374	79849	19352	62726
46569	00365	23591	44317	55054	99835	20633	66215	46668	53587
09988	44203	43532	54538	16619	45444	11957	69184	98398	96508
32916	00567	82881	59753	54761	39404	90756	91760	18698	42852
93285	32297	27254	27198	99093	97821	46277	10439	30389	45372
03222	39951	12738	50303	25017	84207	52123	88637	19369	58289
87002	61789	96250	99337	14144	00027	43542	87030	14773	73087
68840	94259	01961	42552	91843	33855	00824	48733	81297	80411
88323	28828	64765	08244	53077	50897	91937	08871	91517	19668
55170	71062	64159	79364	53088	21536	39451	95649	65256	23950

a boy and the birth of a girl are equally likely, and assume successive births are independent, we could use the random-digit table with an even digit representing a girl and an odd digit representing a boy. Because there are two children, we need to consider pairs of digits. If we examine 100 pairs, then the simulated probability of *GG* will be the number of pairs of even digits divided by 100, the total number of pairs considered.

 EXAMPLE 9-10

A baseball player, Reggie, has a batting average of 0.400. This gives his theoretical probability of getting a hit on any particular time at bat. Estimate the probability of the event that he will get at least one hit in his next three times at bat.

Solution   We use a random-digit table to simulate this situation. We choose a starting point and place the random digits in groups of three. Because Reggie's probability of getting a hit on any particular time at bat is 0.400, we could use the occurrence of four numbers from 0 through 9 to represent a hit. Suppose a hit is represented by 0, 1, 2, and 3. At least one hit is obtained in three times at bat if, in any sequence of three digits, a 0, 1, 2, or 3 appears. Data for 50 trials are given next:

780	862	760	580	783	720	590	506	021	366
848	118	073	077	042	254	063	667	374	153
377	883	573	683	780	115	662	591	685	274
279	652	754	909	754	892	310	673	964	351
803	034	799	915	059	006	774	640	298	961

We see that a 0, 1, 2, or 3 appears in 42 out of the 50 trials. Thus, an estimate for the probability of the event of at least one hit on Reggie's next three times at bat is $\dfrac{42}{50}$.

 NOW TRY THIS 9-6

    a. Use the random-digit table to estimate the probability of the event that a couple that plans to have three children will have two girls and one boy.

    b. Determine the theoretical probability of the event discribed in part (a).

    c. Should the answers in (a) and (b) always be exactly the same? Why?

According to the grade 7 *Common Core Standards*, students should be able to:

> Design and use a simulation to generate frequencies for compound events. *For example, use random digits as a simulation tool to approximate the answer to the question: If 40% of donors have type A blood, what is the probability that it will take at least 4 donors to find one with type A blood?* (p. 51)

The student page on page 510 explores this exact type of problem. The example on the student page shows how random numbers can be used to simulate blood type problems.

School Book Page ACTIVITY LAB

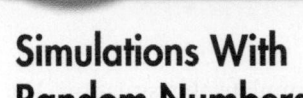 **Activity Lab**   Technology

Simulations With Random Numbers

You can generate random numbers with a graphing calculator or a computer to simulate some situations.

To generate a group of ten digits on a graphing calculator, select the **rand** option from the PRB menu. Each time you press ENTER you will get a different random number. If a random number contains nine digits, use 0 for the tenth digit. When reordering random numbers, ignore the decimal point.

rand	.606334928
rand	.9518983326
rand	.2209784733
rand	.5972865589

EXAMPLE

Blood Types About 10% of people in the United States have type-B blood. Find the experimental probability that exactly one of the next two donors at a hospital will have type-B blood.

Since 10% of the people have type-B blood, let 10%, or one out of ten digits, represent this group of people. Let 0 represent type-B blood and the remaining nine digits represent the other blood types.

Group the digits of a random number into pairs to represent two donors.

60	63	34	92	80
95	18	98	33	26
22	09	78	47	33
59	72	86	55	89

Any pair with exactly one 0 represents one of two people with type-B blood. There are three such pairs in this list.

● Based on 20 trials, the experimental probability is $\frac{3}{20}$, or 15%.

Exercises

Generate random numbers to simulate each problem.

1. **Blood Types** About 40% of people in the United States have type-A blood. Find the experimental probability that exactly one of the next two donors at a blood drive will have type-A blood.

2. Choose *coin, number cube, spinner,* or *calculator* to simulate each probability. Justify your choice.
   a. 30% chance of rain   b. random date is Saturday
   c. 1 in 3 chance of winning

484   Activity Lab   Simulations With Random Numbers

NOW TRY THIS 9-7

Work through the exercises on the student page.

The following simulation problem concerns how many chips there are in chocolate chip cookies.

EXAMPLE 9-11

Suppose Lucy makes enough batter for exactly 100 chocolate chip cookies and mixes 100 chocolate chips into the batter. If the chips are distributed at random and Charlie chooses a cookie at random from the 100 cookies, estimate the probability that it will contain exactly one chocolate chip.

Solution We use a simulation to estimate the probability of choosing a cookie with exactly one chocolate chip. We construct a 10×10 grid, as in Figure 9-33(a), to represent the 100 cookies Lucy made. Each square (cookie) can be associated with some ordered pair, where the first component is for the horizontal scale and the second is for the vertical scale. For example, the squares $(0, 2)$ and $(5, 3)$ are pictured in Figure 9-33(a). Using the random-digit Table 9-4 we close our eyes and then take a pencil and point to one number to start. Look at the number and the number immediately following it. Consider these numbers an ordered pair and continue until we obtain 100 ordered pairs to represent the locations of the 100 chips. For example, suppose we start at a 3 and the numbers following 3 are as follows:

39968   80495   00192...

Then the ordered pairs would be given as $(3, 9)$, $(9, 6)$, $(8, 8)$, $(0, 4)$, and so on. Use each pair of numbers as the coordinates for the square (cookie) and place a tally on the grid to represent each chip, as shown in Figure 9-33(b). We estimate the probability of the event that a cookie has exactly one chip by counting the number of squares with exactly one tally and dividing by 100.

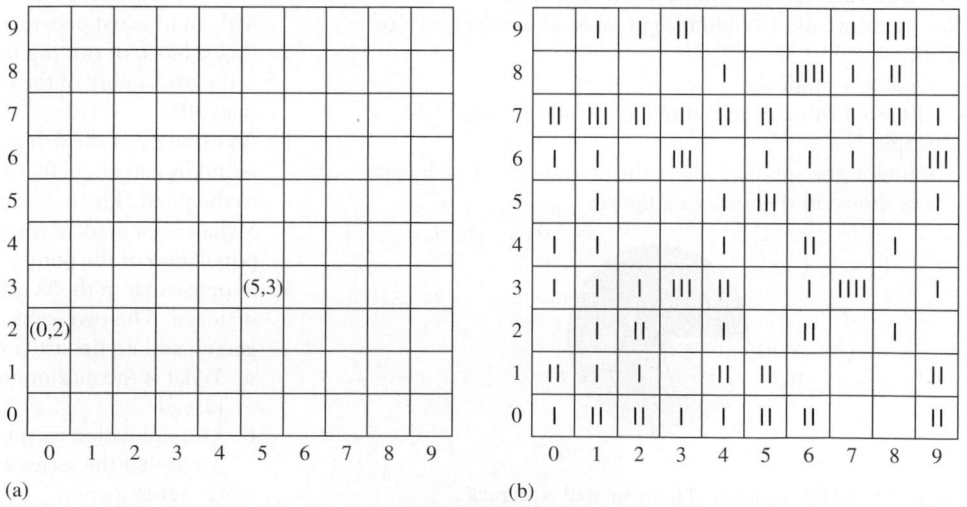

Figure 9-33

Table 9-5 shows the results of one simulation. Thus an estimate for the probability of Charlie receiving a cookie with exactly one chip is $\dfrac{34}{100}$.

Table 9-5

Number of Chips	Number of Cookies
0	38
1	34
2	20
3	6
≥ 4	2

The results given in Table 9-6 were obtained by theoretical methods.

Table 9-6

Number of Chips	Number of Cookies
0	36.8
1	36.8
2	18.4
3	6.1
≥ 4	1.9

Assessment 9-3A

1. Could we use a thumb tack to simulate the birth of boys and girls? Why or why not?
2. How could we use a random-number table to estimate the probability that two cards drawn from a standard deck of cards with replacement will be of the same suit?
3. How might we use a random-digit table to simulate each of the following?
   a. Tossing a single die
   b. Choosing three people at random from a group of 20 people
   c. Spinning the spinner, where the probability of each color is as shown in the following figure:

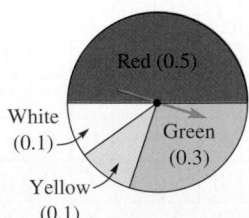

4. A school has 500 students. The principal is to pick 30 students at random from the school to go to the Rose Bowl. How can this be done using a random-digit table?
5. In a certain city, the probability of the event that it will rain on a certain day is 0.8 if it rained the day before. The probability of the event that it will be dry on a certain day is 0.3 if it was dry the day before. It is now Sunday, and it is raining. Use the random-digit table to simulate the weather for the rest of the week.

6. It is reported that 15% of people who come into contact with a person infected with strep throat contract the disease. How might we use the random-digit table to simulate the probability that at least one child in a three-child family will catch the disease, given that each child has come into contact with an infected person?
7. Pick a block of two digits from the random-digit table. What is the probability of the event that the block picked is less than 30?
8. An estimate of the fish population of a certain pond was found by catching 200 fish and marking and returning them to the pond. The next day, 300 fish were caught, of which 50 had been marked the previous day. Estimate the fish population of the pond.
9. Suppose that in the World Series, the two teams are evenly matched. The two teams play until one team wins four games, and no ties are possible.
   a. What is the maximum number of games that could be played?
   b. Use simulation to approximate the probabilities of the event that the series will end in (i) four games and in (ii) seven games.
10. Assume Carmen Smith, a basketball player, makes free throws with 80% probability of success and is placed in a one-and-one situation where she is given a second foul shot if, and only if, the first shot goes through the basket. If you assume successive shots are independent, simulate the 25 attempts from the foul line in one-and-one situations to determine how many times we would expect Carmen to score 0 points, 1 point, and 2 points.

Assessment 9-3B

1. How could we use a spinner as shown below to simulate the birth of boys and girls?

2. How could we use a random-digit generator or random-number table to simulate the probability of the event of rain if you knew that 60% of the time with conditions as you have today, it will rain?
3. How could you use a random-digit generator or a random-number table to simulate choosing 4 people at random from a group of 25 people?
4. In a school with 200 students, the band director chooses four people at random to carry a banner in front of the marching band. If she uses a random-digit table or a random-number generator, how could this be done?
5. Simulate tossing two fair coins 25 times and obtaining the outcome *TT*, where *T* represents tails. What is your simulated probability and what method was used?
6. An estimate of the frog population in a certain pond was found by catching 30 frogs, marking them, and returning them to the pond. The next day, 50 frogs were caught, of which 14 had been marked the previous day. Estimate the frog population of the pond.
7. Suppose a dot is placed at random in a 10 × 10 graph grid in which squares have been numbered from 1 to 100 with no number repeated. Now simulate choosing 100 numbers at random between 1 and 100 inclusive. How many numbers might you expect to match the square with the dot?
8. In a class of seven students, a teacher spins a seven-sectored spinner (with equal-sized sectors) to determine which students to ask questions. Determine about how many times a student can expect to be called on when 100 questions are asked.
9. Use a random digit table to estimate the probability of the event that at least 2 people in a group of 5 people have the same zodiac sign. There are 12 zodiac signs. Assume each sign is equally likely for each person.
10. In the United States, about 45% of the people have blood type O. Assuming that donors arrive independently and randomly at a local blood bank, use a simulation to answer each of the following:
    a. If 10 donors come to the blood bank one day, what is the probability of the event that at least four have blood type O?
    b. How many donors on average should the bank expect to see in order to obtain exactly four people with blood type O?

Mathematical Connections 9-3

Communication

1. In an attempt to reduce the growth of its population, China instituted a policy limiting a family to one child. Rural Chinese suggested revising the policy to limit families to one son. Assuming the suggested policy is adopted and that any birth is as likely to produce a boy as a girl, and that successive births are independent explain how to use simulation to answer the following:
   a. What would be the average family size?
   b. What would be the ratio of newborn boys to newborn girls?
2. Consider a "walk" on the following grid starting out at the origin O and "walking" 1 unit (block) north, and at each intersection turning left with probability $\frac{1}{2}$, turning right with probability $\frac{1}{6}$, and moving straight with probability $\frac{1}{3}$. Explain how to simulate the "walk" using a regular six-sided die.

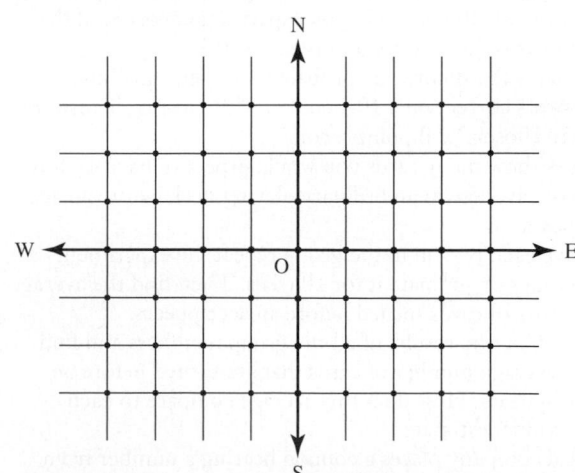

Open-Ended

3. What is the probability of the event that in a group of five people chosen at random, at least two will have birthdays in the same month? Assume all months are equally likely for

birth days. Design a simulation for this situation and try it 10 times.

4. The probability of the event of the home team's winning a basketball game is 80%. Describe a simulation of the probability that the home team will win three home games in a row.

5. Montana duck hunters are all perfect shots. Ten Montana hunters are in a duck blind when 10 ducks fly over. All 10 hunters pick a duck at random to shoot at, and all 10 hunters fire at the same time. How many ducks could be expected to escape, on the average, if this experiment were repeated a large number of times? How could this situation be simulated?

Cooperative Learning

6. The sixth-grade class decided that an ideal family of four children has two boys and two girls.
   a. As a group, design a simulation to determine the probability of the event of two boys and two girls in a family with four kids.
   b. Have each person in the group try the simulation 25 times and compare the probabilities.
   c. Combine the results of all the group members and use this information to find a simulated probability.
   d. Compute the theoretical probability of the event of having two boys and two girls in a family of four and compare your answer to the simulated probability.

7. Have you ever wondered how you would score on a 10-item true-false test if you guessed at every answer?
   a. Simulate a score by tossing a coin 10 times, with heads representing true and tails representing false. Check the answers by using the following key to score the test.

1	2	3	4	5	6	7	8	9	10
F	T	F	F	T	F	T	T	T	F

   b. Combine all results in the group to find the average (mean) number of correct answers for the group.
   c. How many items would you expect to get correct if the number of items were 30 instead of 10?
   d. What is the theoretical probability of getting all the answers correct on a 10-item true-false test if all answers were chosen by flipping a coin?

8. a. Guess how many cards you would expect to have to draw on the average in an ordinary playing deck before an ace appears.
   b. Have each person in the group repeat the experiment 10 times or simulate it for 10 trials. Then find the average number of cards turned before an ace appears.
   c. Combine the results of all the group members and find the average number of cards that are turned before an ace appears. How does this average compare to each individual estimate?

9. A cereal company places a coupon bearing a number from 1 to 9 in each box of cereal. If the numbers are distributed at random in the boxes, estimate the number of boxes you would have to purchase in order to obtain all nine numbers. Explain how the random-digit table could be used to estimate the number of coupons. Each person in the group should simulate 10 trials, and the results in the group should be combined to find the estimate.

10. As a group, explore the random-digit features on graphing calculators.
    a. On some graphing calculators, if we choose the MATH menu and then select PRB, which stands for PROBABILITY, we find **RANDINT**, the random integer feature. RANDINT(generates a random integer within a specified range. It requires two inputs, which are the upper and lower boundaries for the integers. For example, RANDINT(1, 10) generates a random integer from 1 through 10.
       i. How could we use RANDINT(to simulate tossing a single die?
       ii. How could we use RANDINT(to simulate the sum of the numbers when tossing two dice?
    b. Some graphing calculators have a **RAND** function, a random-digit generator. RAND generates a random number greater than 0 and less than 1. For example, RAND might produce the numbers .5956605, .049599836, or .876572691. To have RAND produce random numbers from 1 to 10 as in (1), we enter int (10 *RAND) + 1. The **int** (greatest integer) feature is found in the MATH menu under NUM. The feature int returns the greatest integer less than or equal to a number.
       i. How could we use RAND to simulate tossing a single die?
       ii. How could we use RAND to simulate tossing two dice and taking the sum of the numbers?
    c. Use one of these random features to simulate tossing two dice 30 times. Based on our simulation, what is the probability of the event that a sum of 7 will occur?

Questions from the Classroom

11. Maximilian said that he could make up an experimental probability answer to a homework question about simulations and no one would ever know. How do you react?

Review Problems

12. In a two-person game, four coins are tossed. If exactly two heads come up, one player wins. If anything else comes up, the other player wins. Does each player have an equal chance of winning the game? Explain why or why not.

13. A single card is drawn from an ordinary deck. What is the probability of each of the following events?
    a. Club
    b. Queen and a spade
    c. Not a queen
    d. Not a heart
    e. Spade or a heart
    f. 6 of diamonds
    g. Queen or a spade
    h. Either red or black

14. From a sack containing seven red marbles, eight blue marbles, and four white marbles, marbles are drawn at random for several experiments. Determine the probability of each of the following events:
    a. One marble drawn at random is either red or blue.
    b. The first draw is red and the second is blue, where one marble is drawn at random, its color is recorded, the marble is replaced, and another marble is drawn.
    c. The event in (b) where the first marble is not replaced.

15. A basketball player shoots free throws and makes them with probability $\frac{1}{3}$. If you assume successive shots are independent, what is the probability the player will miss three in a row?

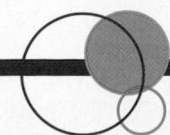

9-4 Odds, Conditional Probability, and Expected Value

Computing Odds

People talk about the *odds in favor of* and the *odds against* a particular event's happening. In the *Grass Roots* cartoon, a teacher was reported to have said that the odds against winning the lottery are greater than 55 million to 1. What does this mean?

Grass Roots copyright © Stan Lynde 2011

To see what this means, we think about a simpler set of odds. When the **odds in favor** of the president being reelected are 4 to 1, this refers to how likely the president is to win the election relative to how likely the president is to lose. The probability of the event of the president's winning is 4 times the probability of the event of the president's losing. If W represents the event the president wins the election and L represents the event the president loses, then

$$P(W) = 4P(L)$$

$$\frac{P(W)}{P(L)} = \frac{4}{1}, \text{ or } 4:1$$

Because W and L are complements of each other, $L = \overline{W}$, and $P(\overline{W}) = 1 - P(W)$ so

$$\frac{P(W)}{P(\overline{W})} = \frac{P(W)}{1 - P(W)} = \frac{4}{1}, \text{ or } 4:1$$

The **odds against** the president's winning are how likely the president is to lose relative to how likely the president is to win. Using the preceding information, we have

$$\frac{P(L)}{P(W)} = \frac{P(\overline{W})}{P(W)} = \frac{1 - P(W)}{P(W)} = \frac{1}{4}, \text{ or } 1:4$$

Formally, odds are defined as follows.

> ### Definition of Odds
>
> Let $P(A)$ be the probability that event A occurs and $P(\overline{A})$ be the probability that event A does not occur. Then the **odds in favor** of event A occuring are
>
> $$\frac{P(A)}{P(\overline{A})} \quad \text{or} \quad \frac{P(A)}{1 - P(A)}$$
>
> and the **odds against** event A occuring are
>
> $$\frac{P(\overline{A})}{P(A)} \quad \text{or} \quad \frac{1 - P(A)}{P(A)}$$

If the odds in favor of an event are 4 : 1, this means that if circumstances surrounding an event are to occur many times, we expect that for every 5 times the event occurs, we expect 4 of those times for it to happen and 1 time that it would not happen. This interpretation gives the following intuitive results:

$$P(\text{event happens}) = \frac{4}{5}$$

$$P(\text{event does not happen}) = \frac{1}{5}$$

We could have started with probabilities and worked to get odds as follows:

$$\frac{P(\text{event happens})}{P(\text{event does not happen})} = \frac{\frac{4}{5}}{\frac{1}{5}} = \frac{4}{5} \div \frac{1}{5} = \frac{4}{5} \cdot \frac{5}{1} = \frac{4}{1}$$

This process can be generalized. Because the probabilities are ratios, we can determine the odds in favor of or against an event without knowing the number of outcomes in the event. When odds are calculated for equally likely outcomes, we have the following:

$$\text{Odds in favor of an event } A: \frac{P(A)}{P(\overline{A})} = \frac{n(A)}{n(S)} \div \frac{n(\overline{A})}{n(S)} = \frac{n(A)}{n(S)} \cdot \frac{n(S)}{n(\overline{A})} = \frac{n(A)}{n(\overline{A})}$$

$$\text{Odds against an event } A: \frac{n(\overline{A})}{n(A)}$$

Thus, in the case of equally likely outcomes, we have

$$\text{Odds in favor} = \frac{\text{Number of favorable outcomes}}{\text{Number of unfavorable outcomes}}$$

$$\text{Odds against} = \frac{\text{Number of unfavorable outcomes}}{\text{Number of favorable outcomes}}$$

When we roll a die, the number of ways of rolling a 4 in one throw of a die is 1, and the number of ways of not rolling a 4 is 5. Thus the odds in favor of rolling a 4 are 1 to 5.

 **EXAMPLE 9-12**

For each of the following, find the odds in favor of the event occurring:

   a. rolling a number less than 5 on a die
   b. tossing heads on a fair coin
   c. drawing an ace from an ordinary 52-card deck
   d. drawing a heart from an ordinary 52-card deck

Solution

   a. The probability of the event of rolling a number less than 5 is $\frac{4}{6}$; the probability of the event of rolling a number not less than 5 is $\frac{2}{6}$. The odds in favor of rolling a number less than 5 are $\left(\frac{4}{6}\right) \div \left(\frac{2}{6}\right)$, or $4:2$, or $2:1$.

   b. $P(H) = \frac{1}{2}$ and $P(\overline{H}) = \frac{1}{2}$. The odds in favor of getting heads are $\left(\frac{1}{2}\right) \div \left(\frac{1}{2}\right)$, or $1:1$.

   c. The probability of the event of drawing an ace is $\frac{4}{52}$, and the probability of the event of not drawing an ace is $\frac{48}{52}$. The odds in favor of drawing an ace are $\left(\frac{4}{52}\right) \div \left(\frac{48}{52}\right)$, or $4:48$, or $1:12$.

   d. The probability of the event of drawing a heart is $\frac{13}{52}$, or $\frac{1}{4}$, and the probability of the event of not drawing a heart is $\frac{39}{52}$, or $\frac{3}{4}$. The odds in favor of drawing a heart are $\left(\frac{13}{52}\right) \div \left(\frac{39}{52}\right) = \frac{13}{39}$, or $13:39$, or $1:3$.

 **NOW TRY THIS 9-8**

In Example 9-12(a), there are 4 ways to roll a number less than 5 on a die (favorable outcomes) and 2 ways of not rolling a number less than 5 (unfavorable outcomes), so the odds in favor of rolling a number less than 5 are $4:2$, or $2:1$. Work the other three parts of Example 9-12 using this approach.

 EXAMPLE 9-13

In the following cartoon, find the probability of the event of making totally black copies if the odds are 3 to 1 against making totally black copies:

www.CartoonStock.com

Solution   Let B represent the event of making a totally black copy and $\overline{B}$ represent not making a totally black copy. Because the odds against making totally black copies are 3:1, we have

$$\frac{P(\overline{B})}{1 - P(\overline{B})} = \frac{3}{1}$$

$$P(\overline{B}) = 3(1 - P(\overline{B}))$$

$$P(\overline{B}) = 3 - 3P(\overline{B})$$

$$4P(\overline{B}) = 3$$

$$P(\overline{B}) = \frac{3}{4}$$

$$P(B) = 1 - P(\overline{B}) = 1 - \frac{3}{4}$$

$$P(B) = \frac{1}{4}$$

In Example 9-13, the odds in favor of making totally black copies are $1:3$. Therefore, the ratio of favorable outcomes to unfavorable outcomes is $\frac{1}{3}$. Thus, in a sample space modeling this situation with four elements, there would be 1 favorable outcome and 3 unfavorable outcomes and $P(\text{black copies}) = \frac{1}{1 + 3} = \frac{1}{4}$. In general, we have the following.

Theorem 9-8

If the *odds in favor* of event E are $m:n$, then $P(E) = \dfrac{m}{m + n}$.

If the *odds against* event E are $m:n$, then $P(E) = \dfrac{n}{m + n}$.

Conditional Probabilities

When the sample space of an experiment is affected by additional information, the new sample space often is reduced in size. For example, suppose we toss a fair coin 3 times and consider the following events:

A: getting a tail on the first toss

B: getting a tail on all 3 tosses

Let S be the sample space in Figure 9-34. We see that $P(A) = \dfrac{4}{8} = \dfrac{1}{2}$ and $P(B) = \dfrac{1}{8}$.

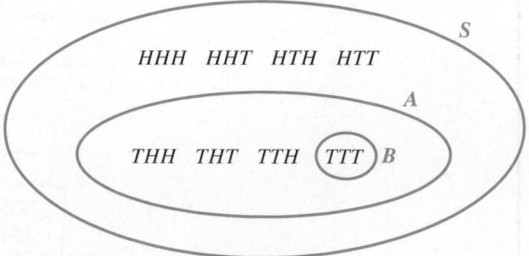

Figure 9-34

If event A has occurred (that is, a tail occurred on the first toss), how does that change $P(B)$? The sample space is now reduced to $\{THH, THT, TTH, TTT\}$. Now the probability of the event that all three tosses are tails given that the first toss is a tail is $\frac{1}{4}$. (Note that $A \cap B = \{TTT\}$.) The notation that we use for this situation is $P(B \mid A)$, read "the probability of the event B given A," and we write $P(B \mid A) = \frac{1}{4}$. Note that

$$P(B \mid A) = \frac{1}{4} = \frac{\frac{1}{8}}{\frac{4}{8}} = \frac{P(A \cap B)}{P(A)}$$

This is true in general and we have the following.

Definition of Conditional Probability

If A and B are events in sample space S and $P(A) \neq 0$, then the **conditional probability** that event B occurs given that event A has occurred is given by

$$P(B \mid A) = \frac{P(A \cap B)}{P(A)}$$

EXAMPLE 9-14

What is the probability of the event of rolling a 6 on a fair die if we know that the roll is an even number?

Solution   If event B is rolling a 6 and event A is rolling an even number, then

$$P(B \mid A) = \frac{P(A \cap B)}{P(A)} = \frac{\frac{1}{6}}{\frac{1}{2}} = \frac{1}{3}.$$

It is not really necessary to use the formula for conditional probability to answer the question because the new sample space is $\{2, 4, 6\}$, and so $P(6) = \frac{1}{3}$.

Expected Value

Racetracks use odds for betting purposes. If the odds against Fast Jack are $3 : 1$, this means the track will pay \$3 for every \$1 you bet. If Fast Jack wins, then for a \$5 bet, the track will return your \$5 plus $3 \cdot \$5$, or \$15, for a total of \$20. The $3 : 1$ odds means in theory that bettors expect Fast Jack to lose 3 out of 4 times in this situation. If the odds at racetracks were accurate, bettors would receive an even return for their money; that is, bettors would not expect to win or lose money in the long run. For example, if the stated odds of $3 : 1$ against Fast Jack were accurate, then the probability of Fast Jack's losing the race would be $\frac{3}{4}$ and of Fast Jack's winning, $\frac{1}{4}$. If we compute the expected average winnings (expected value) over the long run, the gain is \$3 for every \$1 bet for a win, and a loss of \$1 otherwise. The expected value, E, is computed as follows:

$$E = 3\left(\frac{1}{4}\right) + {}^{-}1\left(\frac{3}{4}\right) = 0$$

Therefore, the expected value is \$0. If the racetrack gave accurate odds, it could not stay in business because it could not cover its expenses and make a profit. This is why the track overestimates the horses' chances of winning by about 20% and uses "house odds."

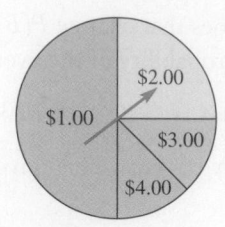

Figure 9-35

As another example of using expected value, consider the spinner in Figure 9-35; the payoff is in each sector of the circle. Using area models, we can assign the following probabilities to each region:

$$P(\$1.00) = \frac{1}{2} \qquad P(\$2.00) = \frac{1}{4} \qquad P(\$3.00) = \frac{1}{8} \qquad P(\$4.00) = \frac{1}{8}$$

Should the owner of this spinner expect to make money over an extended period of time if the charge is $2.00 per spin?

To determine the average payoff over the long run, we find the product of the probability of the event of landing on the payoff and the payoff itself and then find the sum of the products. This computation is given by

$$E = \left(\frac{1}{2}\right)1 + \left(\frac{1}{4}\right)2 + \left(\frac{1}{8}\right)3 + \left(\frac{1}{8}\right)4 = 1.875$$

The owner can expect to pay out about $1.88 per spin. This is less than the $2.00 charge, so the owner should make a profit if the spinner is used many times. The sum of the products in this example, $1.875, is the **expected value**, or **mathematical expectation**, of the experiment of spinning the wheel in Figure 9-35. The owner's expected average earnings are $2.00 − $1.875 = $0.125 per spin, and the player's expected average earnings (loss) are ⁻$0.125. Note that the expected value does not need to be one of the individual possible outcomes so might never be the actual amount won on any trial.

The expected value is an average of winnings over the long run. Expected value can be used to predict the average result of an experiment when it is repeated many times. But *an expected value cannot be used to determine the outcome of any single experiment.*

Definition of Expected Value

If, in an experiment, the possible outcomes are numbers $a_1, a_2, \ldots, a_n$, occurring with probabilities $p_1, p_2, \ldots, p_n$, respectively, then the **expected value (mathematical expectation)** E is given by the equation

$$E = a_1 \cdot p_1 + a_2 \cdot p_2 + a_3 \cdot p_3 + \ldots + a_n \cdot p_n$$

When payoffs are involved and the expected value minus cost to play a game of chance is $0, the game is a **fair game**. Gambling casinos and lotteries make sure that the games are not fair or they could not stay in business.

Previously, a *fair game* was defined as a game in which each player has an equal chance of winning. If there are two players and each has the same probability of winning a given number of dollars from the other, it follows that the expected value minus the cost to play (net winning) for each player is $0.

EXAMPLE 9-15

Suppose it costs $5.00 to play the following game. Two coins are tossed. We receive $10 if two heads occur, $5 if exactly one head occurs, and nothing if no heads appear. Is this a fair game? That is, are the net winnings $0?

Solution   Before we determine the net winning, recall that $P(HH) = \frac{1}{4}$, $P(HT \text{ or } TH) = \frac{1}{2}$, and $P(TT) = \frac{1}{4}$. To find the expected value, we perform the following computation:

$$E = \left(\frac{1}{4}\right)(\$10) + \left(\frac{1}{2}\right)(\$5) + \left(\frac{1}{4}\right)(\$0) = \$5$$

Since the cost of the game is $5 then the net winnings are $0 so this is a fair game.

Problem Solving **A Coin-Tossing Game**

Al and Betsy played a coin-tossing game in which a fair coin was tossed until a total of either three heads or three tails occurred. Al was to win when a total of three heads were tossed, and Betsy was to win when a total of three tails were tossed. Each bet $50 on the game. If the coin was lost when two heads and one tail had occurred, how should the stakes be fairly split if the game is not continued?

Understanding the Problem Al and Betsy each bet $50 on a coin-tossing game in which a fair coin was to be tossed. Al was to win when a total of three heads was obtained; Betsy was to win when a total of three tails was obtained. When two heads and one tail had occured, the coin was lost. The problem is how to split the stakes fairly.

Different people could have different interpretations of what "splitting fairly" means. Possibly, though, the best interpretation is to split the pot in proportion to the probabilities of each player's winning the game when play was halted. We must calculate the expected value for each player and split the pot accordingly.

Devising a Plan A third head would make Al the winner, whereas Betsy needs two more tails to win. A tree diagram that simulates the completion of the game allows us to find the probability of the events of each player winning the game. Once we find the probabilities, all we need to do is multiply the probabilities by the amount of the pot, $100, to determine each player's fair share.

Carrying Out the Plan The tree diagram in Figure 9-36 shows the possibilities for game winners if the game is completed. We can find the probabilities of the event of each player winning as follows:

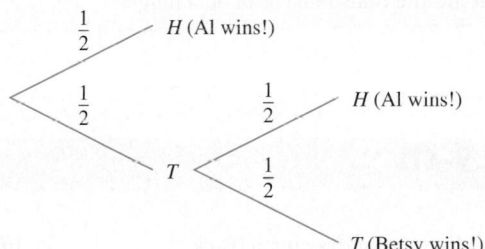

Figure 9-36

$$P(\text{Betsy wins}) = \frac{1}{2} \cdot \frac{1}{2} = \frac{1}{4}$$

$$P(\text{Al wins}) = 1 - \frac{1}{4} = \frac{3}{4}$$

Hence, the fair way to split the stakes is for Al to receive $\frac{3}{4}$ of $100, or $75, and Betsy should receive $\frac{1}{4}$ of $100, or $25.

Looking Back The problem could be made even more interesting by assuming that the coin is not fair so that the probability is not $\frac{1}{2}$ for each branch in the tree diagram. Other possibilities arise if the players have unequal amounts of money in the pot or if more tosses are required in order to win.

BRAIN TEASER

Al tosses one quarter and at the same time Betty tosses two quarters. What is the probability that Betty gets the same number of heads as Al?

Assessment 9-4A

1. **a.** What are the odds in favor of drawing a face card from an ordinary deck of playing cards?
   b. What are the odds against drawing a face card?

2. On a single roll of a pair of dice, what are the odds against rolling a sum of 7?

3. If the probability of a boy's being born is $\frac{1}{2}$, and a family plans to have four children, what are the odds against having all boys?

4. If the odds against Deborah's winning first prize in a chess tournament are 3 to 5, what is the probability of the event that she will win first prize?

5. If the probability of the event that a randomly chosen household has a cat is 0.27, what are the odds against a chosen household having a cat?

6. From a set of eight marbles, five red and three white, we choose one at random. What are the odds in favor of choosing a red marble?

7. In exercise 6, what are the odds against choosing a red marble?

8. If the odds in favor of achieving a Grand Slam golf sweep are 2 : 9, what is the probability of the event of achieving the sweep?

9. A regular die has the numbers 1–6, respectively, on its faces. When a die is tossed, what are the odds in favor of obtaining a prime number?

10. If the probability of an event happening is $\frac{88}{93}$, what are the odds against its happening?

11. If the probability of the event of rain for the day is 90%, what are the odds in favor of rain?

12. What are the odds in favor of rolling double sixes on a single roll of a fair pair of dice?

13. A container has three white balls and two red balls. A first ball is drawn at random and not replaced. Then a second ball is drawn. Given the following conditions, what is the probability of the event that the second ball was red?
    a. The first ball was white.
    b. The first ball was red.

14. Suppose five quarters, five dimes, five nickels, and ten pennies are in a box. One coin is selected at random. What is the expected value of the amount of money drawn from the bag?

15. If the odds in favor of Fast Leg winning a horse race are 5 to 2 and the first prize is $14,000.00, what is the expected value of Fast Leg's monetary winnings in this race?

16. Suppose it costs $8.00 to roll a pair of dice. We get paid the sum of the numbers in dollars that appear on the dice. Is it a fair game?

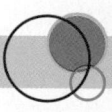

Assessment 9-4B

1. **a.** Susan said that the odds in favor of drawing a black card from a normal deck of 52 cards was 2 : 1. Do you agree?
   b. Explain what is meant by: "the odds in favor of drawing a black card from the normal deck of 52 cards."

2. Diane tossed a coin 9 times and got 9 tails. Assume that Diane's coin is fair and answer each of the following questions:
   a. What is the probability of the event of tossing a tail on the 10th toss?
   b. What is the probability of the event of tossing 10 more tails in a row?
   c. What are the odds against tossing 10 more tails in a row?

3. If a family has one girl and plans to have another child, answer the following if the probability of the event of a girl being born is 1/2:
   a. What is the probability of the event of the second child's being a boy?
   b. What are the odds in favor of having one girl and one boy given their current family?

4. Similar to the cartoon at the beginning of Section 9-4, suppose the odds against winning the lottery are exactly 55 million to 1. What is the probability of the event of winning the lottery given these odds?

5. What are the odds in favor of tossing at least two heads if a fair coin is tossed 3 times?

6. What are the odds in favor of randomly drawing the letter *S* from the letters in the word *MISSISSIPPI*?

7. In exercise 6, what are the odds against choosing the letter *S*?

8. If a whole number less than 10 is chosen at random, what are the odds in favor of the number being greater than 5?

9. If the probability of an event is $0.\overline{3}$, what are the odds against the event?

10. The following spinner is spun. Given the conditions listed below, what is the probability of the event that the spinner

    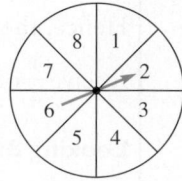

    a. lands on an odd number?
    b. lands on a number divisible by 3?
    c. does not land on 5, 6, or 7?
    d. lands on a number less than 4?

11. If the probability of the event of rain for the day is 60%, what are the odds against rain?

12. On an American roulette wheel, half of the slots numbered 1 through 36 are red and half are black. Two slots, numbered 0 and 00, are green. What are the odds against a red slot's coming up on any spin of the wheel?

13. On a roulette wheel are 36 slots numbered 1 through 36 and 2 slots numbered 0 and 00. You can bet on a single number. If the ball lands on your number, you receive 35 chips plus the chip you played.
    a. What is the probability of the event that the ball will land on 17?
    b. What are the odds against the ball landing on 17?
    c. If each chip is worth $1, what is the expected value for a player who plays the number 17 for a long time?
14. We play a game in which two dice are rolled. If a sum of 7 appears, we win $10; otherwise, we lose $2.00. If we intend to play this game for a long time, should we expect to make money, lose money, or come out about even? Explain.
15. Suppose a standard six-sided die is rolled and you receive $1 for every dot showing on the top of the die. What should the cost of playing the game be in order to make it a fair game?
16. A family with 3 children can have 0, 1, 2, or 3 girls. What is the expected value for the number of girls in a family of these children assuming the probability of a girl being born is $\frac{1}{2}$?

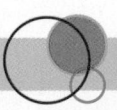

Mathematical Connections 9-4

Communication

1. Explain the difference between odds and probability.
2. A game involves tossing two coins. A player wins $1.00 if both tosses result in heads. What should you pay to play this game in order to make it a fair game? Explain your reasoning.

Open-Ended

3. Suppose we toss two fair coins. Design a fair game with a payoff based on the results of the toss.
4. An insurance company sells a policy that pays $50,000.00 in case of accidental death. According to company figures, the rate of accidental death is 47 per 100,000 population each year. What annual premium should the company charge for this coverage?
5. Write a game-type problem about odds and payoffs so that the odds in favor of an event are 2 : 3 and the game is a fair game.

Cooperative Learning

6. As a group, design a game that involves cards, dice, or spinners.
    a. Write the rules so that any person who wants to play can understand the game.
    b. Write a description explaining whether the game is fair.
    c. Calculate the odds in favor of each player's winning.
    d. If betting is involved, discuss expected values.
    e. Exchange games and alalyze them in groups.

Questions from the Classroom

7. A student claims that if the odds in favor of winning a game are $a:b$, then out of every $a + b$ games she would win a games. Hence, the probability of the event of winning the game is $\frac{a}{a + b}$. Is the student's reasoning correct? Why or why not?
8. A student wants to know why if the odds in favor of an event are 3:4, the probability of the event occurring is not $\frac{3}{4}$. How do you respond?

9. Maria wants 51% of her class to vote for her. To achieve this, she decides that there is a probability of $\frac{1}{3}$ that an individual student will not vote. Also, she promises 24 of the 48 students that she will do what they individually want if she is elected so that she is reasonably sure they will vote for her. Would you advise her that she is on safe ground for winning with this strategy?

Review Problems

10. Refer to the following spinners and write the sample space for each of the following experiments:

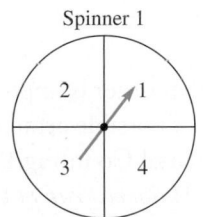
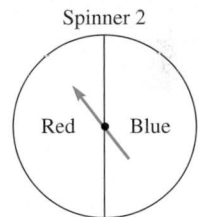

    a. Spin spinner 1 once.
    b. Spin spinner 2 once.
    c. Spin spinner 1 once and then spin spinner 2 once.
    d. Spin spinner 2 once and then roll a normal die.
    e. Spin spinner 1 twice.
    f. Spin spinner 2 twice.
11. Draw a spinner with two sections, red and blue, such that the probability of the event of getting (Blue, Blue) on two spins is $\frac{25}{36}$.
12. Find the probability of the event of getting two vowels when someone draws two letters from the English alphabet with replacement.

BRAIN TEASER

It is your first day of class; your class has 40 students. A friend who does not know any students in your class bets you that at least 2 of them share a birthday (month and day). What are your friend's chances of winning the bet?

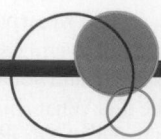

9-5 Using Permutations and Combinations in Probability

Permutations of Unlike Objects

An arrangement of things in a definite order with no repetitions is a **permutation**. For example, different arrangements of the three letters R, A, and T are seen in Figure 9-37.

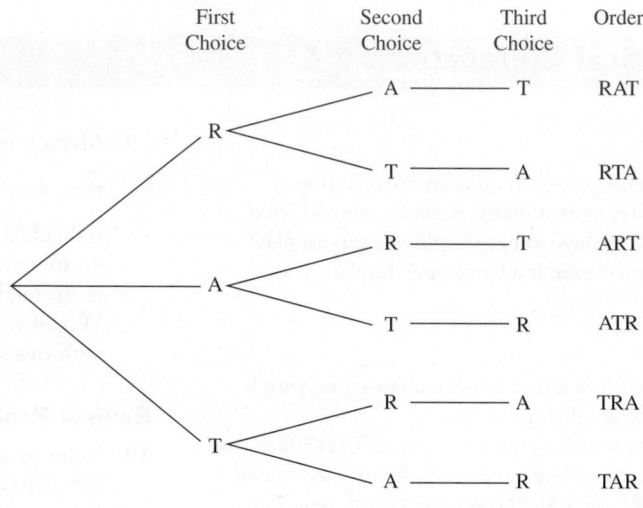

Figure 9-37

Notice that order is important and there are no repetitions in the arrangements. Determining the number of possible arrangements of the three letters without making a list can be done using the Fundamental Counting Principle. Recall that the Fundamental Counting Principle states that *if an event M can occur in m ways and, after M has occurred, event N can occur in n ways, then event M followed by event N can occur in mn ways.* There are three ways to choose the first letter, and for each of these choices there are two ways to choose the second letter. Thus far there are $3 \cdot 2$ ways to choose the first two letters. Next, for each of the $3 \cdot 2$ choices there is one way to choose the third letter, so there are $3 \cdot 2 \cdot 1$, or 6, ways to arrange the letters. It is common to record the number of permutations of three objects taken three at a time as $_3P_3$. Therefore, $_3P_3 = 6$.

Consider how many ways the owner of an ice cream parlor can display 10 flavors in a row along the front of the display case. The first position can be filled in 10 ways, the second position in 9 ways, the third position in 8 ways, and so on. By the Fundamental Counting Principle, there are $10 \cdot 9 \cdot 8 \cdot 7 \cdot 6 \cdot 5 \cdot 4 \cdot 3 \cdot 2 \cdot 1$, or 3,628,800, ways to display the flavors. If there were 16 flavors, there would be $16 \cdot 15 \cdot 14 \cdot 13 \cdot \ldots \cdot 3 \cdot 2 \cdot 1$ ways to arrange them.

In general, *if there are n objects, then the number of possible ways to arrange the objects in a row is the product of all the natural numbers from n to 1, inclusive.* This expression, **n factorial**, is denoted **n!** as shown next.

$$n! = n(n - 1)(n - 2) \cdot \ldots \cdot 3 \cdot 2 \cdot 1$$

For example, $5! = 5 \cdot 4 \cdot 3 \cdot 2 \cdot 1$; $3! = 3 \cdot 2 \cdot 1$; and $1! = 1$.

Many calculators have a factorial key such as $\boxed{x!}$. To use this key, enter a whole number and then press the factorial key. For example, to compute 5! press $\boxed{5}$ $\boxed{x!}$ and 120 will appear on the display.

Consider the set of people in a small club, {Al, Betty, Carl, Dan}. How many ways are there to elect a president and a secretary for the club? One way to answer the question is to agree that the choice "Al, Betty" denotes Al as president and Betty as secretary, while the choice "Betty, Al" indicates that Betty is president and Al is secretary. Order is important and no repetitions are possible. Consequently, counting the number of possibilities is a permutation problem. Since there are four ways of choosing a president and then for each of these choices there are three ways of choosing a secretary, by the Fundamental Counting Principle, there are $4 \cdot 3$, or 12, ways of choosing a president and a secretary. Choosing two officers from a club of four is a permutation of four people chosen two at a time. The number of possible permutations of four objects taken two at a time, denoted $_4P_2$, may be counted using the Fundamental Counting Principle. Therefore, we have $_4P_2 = 4 \cdot 3$, or 12.

NOW TRY THIS 9-9

a. Write $_nP_2$, $_nP_3$, and $_nP_4$ in terms of n.
b. Based on your answers in part (a), write $_nP_r$ in terms of n and r.
c. In the club mentioned above, how many ways are there to elect a president, vice president, and secretary?

The number of permutations can be written in terms of factorials. Consider the number of permutations of 20 objects chosen 3 at a time:

$$
\begin{aligned}
_{20}P_3 &= 20 \cdot 19 \cdot 18 \\
&= \frac{20 \cdot 19 \cdot 18 \cdot (17 \cdot \ldots \cdot 3 \cdot 2 \cdot 1)}{17 \cdot \ldots \cdot 3 \cdot 2 \cdot 1} \\
&= \frac{20!}{17!} \\
&= \frac{20!}{(20 - 3)!}
\end{aligned}
$$

This can be generalized as follows:

Permutation of Objects in a Set

The number of permutations of r objects chosen from a set of n objects, where $0 \leq r \leq n$, is denoted by $_nP_r$ and is given by

$$
_nP_r = \frac{n!}{(n - r)!}
$$

Recall that $_nP_n$ is the number of permutations of n objects chosen n at a time, that is, the number of ways of rearranging all n objects in a row. We have seen that this number is $n!$. If we use the formula for $_nP_r$ to compute $_nP_n$, we obtain

$$
_nP_n = \frac{n!}{(n - n)!} = \frac{n!}{0!}
$$

Consequently, $n! = n!/0!$. To make this equation true, *we define 0! to be 1*.

 Many calculators, especially graphing calculators, can calculate the number of permutations of n objects taken r at a time. This feature is usually denoted $\boxed{_nP_r}$. To use this key, enter the value of n, then press $\boxed{_nP_r}$, followed by the value of r. If you then press $\boxed{=}$ or $\boxed{\text{ENTER}}$, the number of permutations is displayed.

NOW TRY THIS 9-10

a. Use a factorial key $\boxed{x!}$ on a calculator to compute $\dfrac{100!}{98!}$. What happens? Why?

b. Without using a calculator, use the definition of factorials to compute the expression in part (a).

EXAMPLE 9-16

a. A baseball team has nine players. Find the number of ways the manager can arrange the batting order.
b. Find the number of ways of choosing three initials from the alphabet if none of the letters can be repeated.

Solution

a. Because there are nine ways to choose the first batter, eight ways to choose the second batter, and so on, there are $9 \cdot 8 \cdot 7 \cdot \ldots \cdot 2 \cdot 1 = 9!$, or 362,880 ways of arranging the batting order.

   Using the formula for permutations, we have $_9P_9 = \dfrac{9!}{0!} = 362,880$.

b. There are 26 ways of choosing the first letter, 25 ways of choosing the second letter, and 24 ways of choosing the third letter. Hence, there are $26 \cdot 25 \cdot 24$, or 15,600, ways of choosing the three letters. Alternatively, if we use the formula for permutations, we have

$$_{26}P_3 = \frac{26!}{23!} = \frac{26 \cdot 25 \cdot 24 \cdot (23 \cdot 22 \cdot 21 \cdot \ldots \cdot 1)}{23 \cdot 22 \cdot 21 \cdot \ldots \cdot 1}$$

$$= 26 \cdot 25 \cdot 24$$

$$= 15,600$$

Permutations Involving Like Objects

In the previous counting examples, each object to be counted was distinct. Suppose we wanted to arrange the letters in the word *ZOO*. How many choices would we have? If the *O*'s were distinguishable, a tree diagram, as in Figure 9-38, suggests that there might be $3 \cdot 2 \cdot 1 = 3!$, or 6, possibilities. However, looking at the list of possibilities shows that *ZOO*, *OZO*, and *OOZ* each appear twice because the *O*'s are not different. We need to determine how to remove the duplication in arrangements such as this where some objects are the same. To eliminate the duplication, we divide the number of arrangements shown by the number of ways the two *O*'s can be rearranged, which is 2!. Consequently, there are $\dfrac{3!}{2!}$, or 3, ways of arranging the letters in *ZOO*.

The arrangements are *ZOO*, *OZO*, and *OOZ*.

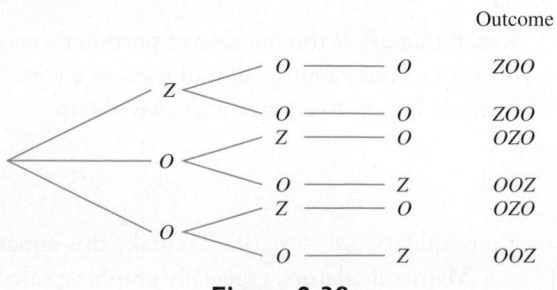

Figure 9-38

This discussion is generalized in the following:

> ### Permutations of Like Objects
>
> If there are n objects, of which r_1 are alike, r_2 are alike, and so on through r_k, then the number of different arrangements of all n objects, where alike objects are indistinguishable, is equal to
>
> $$\frac{n!}{r_1! \cdot r_2! \cdot r_3! \cdot \ldots \cdot r_k!}$$

EXAMPLE 9-17

Find the number of rearrangements of the letters in each of the following words:

    a. *bubble*   **b.** *statistics*

Solution

  a. There are 6 letters with *b* repeated 3 times. Hence, the number of arrangements is

$$\frac{6!}{3!} = 6 \cdot 5 \cdot 4 = 120$$

  b. There are 10 letters in the word *statistics*, with three *s*'s, three *t*'s, and two *i*'s in the word. Hence, the number of arrangements is

$$\frac{10!}{3! \cdot 3! \cdot 2!} = \frac{10 \cdot 9 \cdot 8 \cdot 7 \cdot 6 \cdot 5 \cdot 4 \cdot 3 \cdot 2 \cdot 1}{3 \cdot 2 \cdot 1 \cdot 3 \cdot 2 \cdot 1 \cdot 2 \cdot 1} = 50{,}400$$

Combinations

Reconsider the club {Al, Betty, Carl, Dan}. Suppose a two-person committee is selected with no chair. In this case, order is not important, and an Al-Betty choice is the same as a Betty-Al choice. An arrangement of objects in which the order makes no difference is a **combination**. A comparison of the results of electing a president and a secretary for the club and the results of simply selecting a two-person committee are shown in Figure 9-39. Because each two permutations "shrink" into one combination, we see that the number of combinations is the number of permutations divided by 2, or

$$\frac{4 \cdot 3}{2} = 6$$

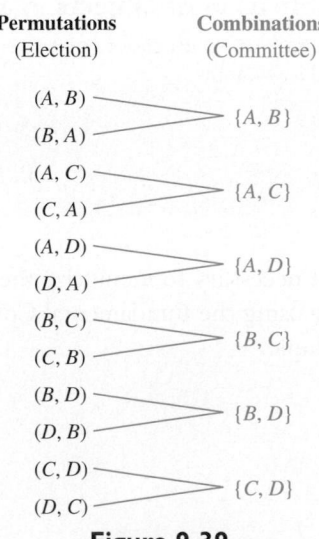

Figure 9-39

In how many ways can a committee of three people be selected from the club {Al, Betty, Carl, Dan}? To solve this problem, we proceed as we did earlier and find the number of ways to select three people from a group of four for three offices, say president, vice president, and secretary (a permutation problem) and then use this result to see how many combinations of people are possible for the committee. Figure 9-40 shows two examples from the whole list.

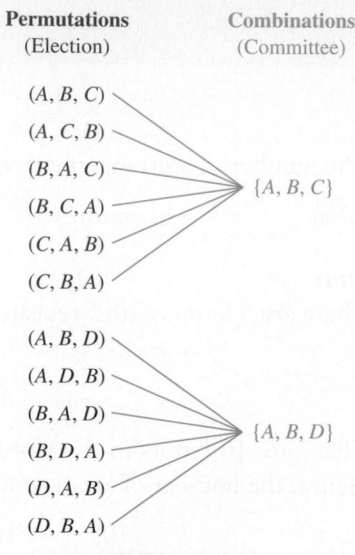

Figure 9-40

By the Fundamental Counting Principle, if order is important, the number of ways to choose three people from the list of four is $4 \cdot 3 \cdot 2$, or 24. However, with each triple chosen, there are 3!, or 6, ways to rearrange the triple, as seen in Figure 9-40. Therefore, there are 3! times as many permutations as combinations, or equivalently, each 3!, or 6, permutations "shrink" into one combination. Therefore, to find the number of combinations, we divide the number of permutations, 24, by 3!, or 6, to obtain 4. The four committees are $\{A, B, C\}$, $\{A, B, D\}$, $\{A, C, D\}$, and $\{B, C, D\}$.

To find the number of combinations possible in a counting problem, first use the Fundamental Counting Principle to find the number of permutations and then divide by the number of ways in which each choice can be arranged. A general formula follows:

> ### Combination of Objects in a Set
> The number of combinations of r objects chosen from a set of n objects, where $0 \leq r \leq n$ is denoted by $_nC_r$ and is given by:
>
> $$_nC_r = \frac{_nP_r}{_rP_r} = \frac{\frac{n!}{(n-r)!}}{r!} = \frac{n!}{r!(n-r)!}$$

It is not necessary to memorize the formula above; we can always find the number of combinations by using the Fundamental Counting Principle and the reasoning developed in the committee example.

EXAMPLE 9-18

The Library of Science Book Club offers 3 free books from a list of 42 books. How many possible combinations are there?

Solution Order is not important, so this is a combination problem. By the Fundamental Counting Principle, there are $42 \cdot 41 \cdot 40$ ways to choose the 3 free books. Because each set of 3 books could be rearranged in $3 \cdot 2 \cdot 1$ ways, there is an extra factor of 3! in the original $42 \cdot 41 \cdot 40$ ways. Therefore, the number of combinations possible for 3 books is

$$\frac{42 \cdot 41 \cdot 40}{3!} = 11,480$$

If we use the formula for $_nC_r$, we have $_{42}C_3 = \frac{42!}{3! \cdot 39!} = \frac{42 \cdot 41 \cdot 40 \cdot 39!}{3 \cdot 2 \cdot 1 \cdot 39!} = \frac{42 \cdot 41 \cdot 40}{6} = 11,480.$

EXAMPLE 9-19

At the beginning of the first semester of a mathematics class for elementary school teachers, each of the class's 25 students shook hands with each other student exactly once. How many handshakes took place?

Solution Since the handshake between persons A and B is the same as that between persons B and A, this is a problem of choosing combinations of 25 people 2 at a time. There are

$$_{25}C_2 = \frac{25 \cdot 24}{2!} = 300$$

different handshakes.

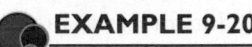

EXAMPLE 9-20

Given a class of 12 girls and 10 boys, answer each of the following:

a. In how many ways can a committee of 5 consisting of 3 girls and 2 boys be chosen?
b. What is the probability of the event that a committee of 5, chosen at random from the class, consists of 3 girls and 2 boys?
c. How many of the possible committees of 5 have no boys?
d. What is the probability of the event that a committee of 5, chosen at random from the class, consists only of girls?

Solution

a. Based on the information given, we do not assign special functions to members on a committee and, hence, the order of the children on a committee does not matter. From 12 girls we can choose 3 girls in $_{12}C_3$ ways. Each of these choices can be paired with $_{10}C_2$ combinations of boys. By the Fundamental Counting Principle, the total number of committees is

$$_{12}C_3 \cdot {}_{10}C_2 = \frac{12 \cdot 11 \cdot 10}{3!} \cdot \frac{10 \cdot 9}{2} = 9900$$

b. The total number of committees of 5 is $_{22}C_5$, or 26,334. Using part (a), we find the probability of the event that a committee of 5 will consist of 3 girls and 2 boys to be

$$\frac{_{12}C_3 \cdot {}_{10}C_2}{_{22}C_5} = \frac{9900}{26,334} \approx 0.376$$

c. The number of ways to choose 0 boys from the 10 boys and 5 girls from the 12 girls in the class is

$$_{10}C_0 \cdot {}_{12}C_5 = 1 \cdot {}_{12}C_5 = \frac{12 \cdot 11 \cdot 10 \cdot 9 \cdot 8}{5 \cdot 4 \cdot 3 \cdot 2 \cdot 1} = 792$$

d. $\frac{_{12}C_5}{_{22}C_5} = \frac{792}{26,334} \approx 0.030$

Problem Solving   **A True-False Test Problem**

In the following *Peanuts* cartoon, suppose Peppermint Patty took a six-question true-false test. If she answered each question true or false at random, what is the probability of the event that she answered exactly 50% of the questions correctly?

Peanuts copyright © 1980 and 2011 Peanuts Worldwide LLC. Distributed by Universal Uclick. Reprinted with permission. All rights reserved.

Understanding the Problem A score of 50% indicates that Peppermint Patty answered $\frac{1}{2}$ of the six questions, or three questions, correctly. She answered the questions true or false at random, so the probability that she answered a given question correctly is $\frac{1}{2}$. We are asked to determine the probability of the event that Patty answered exactly three of the questions correctly.

Devising a Plan We do not know which three questions Patty missed. She could have missed any three out of six on the test. Suppose she answered questions 2, 4, and 5 incorrectly. In this case, she would have answered questions 1, 3, and 6 correctly. We can compute the probability of answering questions 1, 3, and 6 correctly by using a branch of a tree diagram, as in Figure 9-41, where C represents a correct answer and I represents an incorrect answer.

$$
\begin{array}{ccccccc}
\text{Question:} & 1 & 2 & 3 & 4 & 5 & 6 \\
& \xrightarrow{\frac{1}{2}} C \xrightarrow{\frac{1}{2}} & I \xrightarrow{\frac{1}{2}} & C \xrightarrow{\frac{1}{2}} & I \xrightarrow{\frac{1}{2}} & I \xrightarrow{\frac{1}{2}} & C
\end{array}
$$

Probability of Outcome: $\left(\frac{1}{2}\right)^6$

Figure 9-41

Multiplying the probabilities along the branches, we obtain $\left(\frac{1}{2}\right)^6$ as the probability of answering questions 1 through 6 in the following way: $C I C I I C$. There are other ways to answer exactly three questions correctly: for example, $C C C I I I$. The probability of answering questions 1 through 6 in this way is also $\left(\frac{1}{2}\right)^6$. The number of ways to answer exactly three of the questions correctly is simply the number of ways of arranging three C's and three I's in a row, which is also the number of ways of choosing three correct questions out of six, that is, $_6C_3$. Because all these arrangements give Patty a score of 50%, the desired probability is the sum of the probabilities for each arrangement.

Carrying Out the Plan There are $_6C_3$, or 20, sets of answers similar to the one in Figure 9-41, with three correct and three incorrect answers. The product of the probabilities for each of these sets of answers is $\left(\frac{1}{2}\right)^6$, so the sum of the probabilities for all 20 sets is $20\left(\frac{1}{2}\right)^6$, or 0.3125. Thus, Peppermint Patty has a probability of 0.3125 of obtaining a score of exactly 50% on the test.

Looking Back It may seem strange to learn that the probability of the event of obtaining a score of 50% on a six-question true-false test is not close to $\frac{1}{2}$. As an extension of the problem, suppose a passing score is a score of at least 70%. Now what is the probability of the event that Peppermint Patty will pass? What is the probability of the event of her obtaining a score of at least 50% on the test? If the test is a six-question multiple-choice test with five alternative answers for each question, what is the probability of the event of obtaining a score of at least 50% by random guessing?

Problem Solving | **Matching Letters to Envelopes**

Stephen placed three letters in envelopes while he was having a telephone conversation. He addressed the envelopes and sealed them without checking whether each letter was in the correct envelope. What is the probability of the event that each of the letters was inserted correctly?

Understanding the Problem Stephen sealed three letters in addressed envelopes without checking to see if each was in the correct envelope. We are to determine the probability of the event that each of the three letters was placed correctly. This probability could be found if we knew the sample space, or at least how many elements are in the sample space.

Devising a Plan To aid in solving the problem, we represent the respective letters as *a*, *b*, and *c* and the respective addressed envelopes as *A*, *B*, and *C*. For example, a correctly placed letter *a* would be in envelope *A*. To construct the sample space, we use the strategy of *making a table*. The table should show all the possible permutations of letters in envelopes. Once the table is completed, we can determine the probability of the event that each letter is placed correctly.

Carrying Out the Plan Table 9-7 is constructed by using the envelope labels *A*, *B*, and *C* as headings and listing all ways that letters *a*, *b*, and *c* could be placed in the envelopes. The first arrangement in the list is the only one out of six in which each of the envelopes is labeled correctly, so the probability of the event that each envelope is labeled correctly is $\frac{1}{6}$.

Table 9-7

A	B	C	
a	*b*	*c*	} Envelopes
a	*c*	*b*	
b	*a*	*c*	
b	*c*	*a*	} Envelope contents
c	*a*	*b*	
c	*b*	*a*	

Looking Back We also could have used a counting argument to solve the problem. Given an envelope, there is only one correct letter to place in the envelope. Thus, there is $1 \cdot 1 \cdot 1$ or one correct way to place the letters in the envelopes. By the Fundamental Counting Principle, there are $3 \cdot 2 \cdot 1$ ways of choosing the letters to place in the envelopes, so the probability of the event of having the letters correctly placed is $\frac{1}{6}$. Is the probability of having each letter placed incorrectly the same as the probability of having each letter placed correctly? A first guess might be that the probabilities are the same, but that is not true. Why?

Assessment 9-5A

1. The eighth-grade class at a grade school has 16 girls and 14 boys. How many different boy-girl partners can be arranged?

2. The telephone prefix for a university is 243. The prefix is followed by four digits. How many different telephone numbers are possible before a new prefix is needed?

3. Carlin's Pizza House offers 3 kinds of salad, 15 kinds of pizza, and 4 kinds of dessert. How many different three-course meals consisting of a salad, pizza, and dessert can be ordered?

4. Decide whether each of the following is true:
   a. $6! = 6 \cdot 5!$
   b. $3! + 3! = 6!$
   c. $\dfrac{6!}{3!} = 2!$

5. Find the number of ways to arrange the letters in the following words:
   a. *SCRAMBLE*
   b. *PERMUTATION*

6. How many two-person committees can be formed from a group of six people?

7. Assume a class has 30 members.
   a. In how many ways can a president, a vice president, and a secretary be selected?
   b. How many committees of three people can be chosen?

8. A five-volume numbered set of books is placed randomly on a shelf. What is the probability of the event that the books will be numbered in the correct order from left to right?

9. There are 10 points in a plane, no 3 of them on a line. How many straight lines can be drawn if each line is drawn through a pair of points?

10. Sally has four red flags, three green flags, and two white flags. Each arrangement of flags is a different signal. How many nine-flag signals can she run up a flagpole?

11. At a party, 28 handshakes took place. Each person shook hands exactly once with each of the others present. How many people were at the party?

12. How many different 5-card hands can be dealt from a standard deck of 52 playing cards?

13. In a certain lottery game, 54 numbers are randomly mixed and 6 are selected. A person must pick all 6 numbers to win. Order is not important. What is the probability of winning?

14. From a group of 10 boys and 12 girls, a committee of 4 students is chosen at random. What is the probability of the following events:
    a. all 4 members on the committee will be girls?
    b. all 4 members of the committee will be boys?
    c. there will be at least 1 girl on the committee?

15. From a group of 20 Britons, 21 Italians, and 4 Danes, a committee of 8 people is chosen at random. Expressing your answers using the notation for combinations, find the probability that
    a. the committee will consist of 2 Britons, 4 Italians, and 2 Danes.
    b. the committee will have no Britons.
    c. there will be at least one Briton on the committee.
    d. all members of the committee will be Britons.

16. Stephen placed five letters in envelopes while he was watching television and addressed the envelopes without checking whether each letter was in the correct envelope. What is the probability of the event that all five letters were in the correct envelopes?

17. A company is setting up four-digit ID numbers for employees.
    a. How many four-digit numbers are there if numbers can start with 0 and digits can be repeated?
    b. How many four-digit numbers are there if numbers can start with 0 and all the digits must be different?
    c. If you randomly assign a four-digit ID number that can start with 0 and digits can be repeated, what is the probability that all the digits are even?

18. A company president presents identical awards to four out of six finalists for the awards. In how many ways can she present the awards?

19. Your English teacher asks that you read any three of the eight books on his reading list. How many choices do you have for the set of three books you read?

Assessment 9-5B

1. If a coin is tossed 5 times, in how many different ways can the sequence of heads and tails appear?

2. Radio stations in the United States have call letters that begin with either *K* or *W*. Some have three letters; others have four letters.
   a. How many three-letter call letters are possible?
   b. How many four-letter call letters are possible?

3. Three men and four women line up at a checkout counter at a store. In how many ways can they line up if we consider only their gender?

4. Classify each of the following as true or false:
   a. $\dfrac{6!}{3} = 2!$
   b. $\dfrac{6!}{5!} = 6!$
   c. $\dfrac{6!}{4!2!} = 15$
   d. $n!(n + 1) = (n + 1)!$

5. Find the number of ways to arrange the letters in the following words:
   a. *OHIO*
   b. *ALABAMA*
   c. *ILLINOIS*
   d. *MISSISSIPPI*
   e. *TENNESSEE*

6. In a car race, there are 6 Chevrolets, 4 Fords, and 2 Hondas. In how many ways can the 12 cars finish if we consider only the makes of the cars?

7. A basketball coach was criticized in the newspaper for not trying out every combination of players. If the team roster has 12 players, how many 5-player combinations are possible?

8. Seven performers *A, B, C, D, E, F,* and *G* are to appear in a talent contest. The order of appearance is by random selection. Find the probability of the following events:
   a. *C* will perform first
   b. *F* or *G* will perform first
   c. They will appear in the order *C, D, E, A, B, F, G.*

9. Find the number of shortest paths from point *A* to point *B* along the edges of the cubes in each of the following. (For example, in (a) one shortest path is *A-C-D-B*.)

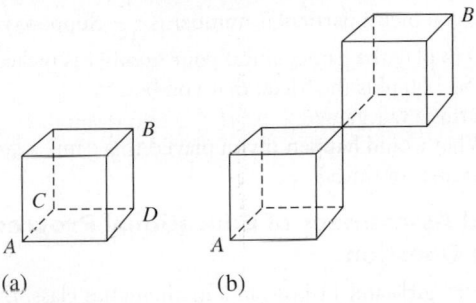

   (a)   (b)

10. License plates in a certain state have three letters followed by three digits. How many different plates are possible if no repetitions of letters or digits are allowed?

11. How many different 12-person juries can be selected from a pool of 24 people?

12. Social Security numbers are in the form ###-##-####, where each symbol represents a number 0 through 9. How many Social Security numbers are possible using this format?

13. A fair die is rolled 8 times. What is the probability of the event of getting:
   a. 1 on each of the 8 rolls?
   b. 6 exactly twice in the 8 rolls?
   c. 6 at least once in the 8 rolls?

14. A committee of three people is selected at random from a set consisting of seven Americans, five French people, and three English people.
   a. What is the probability of the event that the committee consists of all Americans?
   b. What is the probability of the event that the committee has no Americans?

15. A club selects an executive committee of 5 and from the 5, one of the group becomes the president. From a membership of 32, how many ways are there to choose a president?

16. Assume the probability of the event of a basketball player making a free throw successfully at any time in a game is $\frac{2}{3}$. If the player attempts 10 free throws in a game, what is the probability that exactly 6 are made?

17. What are the odds against a royal flush in poker, that is, a 10, jack, queen, king, and ace all of the same suit?

18. At the American Kennel Club, there are 36 dogs entered in a show where there will be a first-, second-, and third-place award. How many possibilities are there for these awards?

19. Two fair dice are rolled 5 times and the sum of the numbers that come up on each roll is recorded. Find the probability of the event of getting:
   a. a sum of 7 on each of the five rolls.
   b. a sum of 7 exactly twice in the five rolls.

Mathematical Connections 9-5

Communication

1. The terms *Fundamental Counting Principle*, *permutations*, and *combinations* are all used to work with counting problems. In your own words, explain how all these terms are related and how they are used.

2. a. A bicycle lock has three reels, each of which contains the numbers 0 through 9. To open the lock, you must enter the numbers in the correct order, such as 369 or 455, where one number is chosen from each reel. How many different possibilities are there for the numbers to open the lock? Explain how you arrived at your answer.
   b. The lock in the cartoon is a *combination* lock. Explain why this is probably not a good name for this lock for someone who has studied counting problems.

REAL LIFE ADVENTURES by Gary Wise and Lance Aldrich

The hacksaw industry would be in serious trouble without the combination-lock industry.

3. **a.** Ten people are to be seated on 10 chairs in a line. Among them is a family of 3 that does not want to be separated. How many different seating arrangements are possible? Explain how you arrived at your answer.

   b. How many possible seating arrangements are there in part (a) in which the family members do not all sit together? Explain how you arrived at your answer.

4. In how many ways can five couples be seated in a row of 10 chairs if no couple is separated? Explain how you arrived at your answer.

Open-Ended

5. Suppose the Department of Motor Vehicles uses six digits and the numbers 0 through 9 to create its license plates. Numbers can be repeated.

   a. How many license plates are possible?

   b. If you were in charge of making license plates for the state of California, describe the method you would use to ensure you would have enough license plates.

Cooperative Learning

6. The following triangular array of numbers is a part of **Pascal's triangle:**

													Row
						1							(0)
					1		1						(1)
				1		2		1					(2)
			1		3		3		1				(3)
		1		4		6		4		1			(4)
	1		5		10		10		5		1		(5)
1		6		15		20		15		6		1	(6)

   a. In your group, decide how the triangle was constructed and complete the next two rows.

   b. Describe at least three number patterns in Pascal's triangle.

   c. Find the sum of the numbers in each row. Predict the sum of the numbers in row 10.

   d. The entries in row 2 are just $_2C_0$, $_2C_1$, and $_2C_2$. Have different members of your group investigate whether a similar pattern holds for other rows in Pascal's triangle.

   e. Describe how you could use combinations to find any entry in Pascal's triangle.

Questions from the Classroom

7. A student does not understand the meaning of $_4P_0$. He wants to know how we can consider permutations of four objects chosen zero at a time. How do you respond?

8. A student wants to know why, if we can define 0! as 1, we cannot define $\frac{1}{0}$ as 1. How do you respond?

9. A student claims that he does not need to define permutations because he knows the Fundamental Counting Principle. What do you say?

Review Problems

10. Two cards are drawn at random without replacement from a deck of 52 cards. What is the probability of the event that
    a. at least 1 card is an ace?
    b. exactly 1 card is red?

11. If two regular dice are tossed, what is the probability of the event of tossing a sum greater than 10?

12. Two coins are tossed. You win $5.00 if both coins are heads and $3.00 if both coins are tails and lose $4.00 if the coins do not match. What is the expected value of this game? Is this a fair game?

13. On a roulette wheel, the probability of the event of winning when you pick a particular number is $\frac{1}{38}$. Suppose you bet $1.00 to play the game, and if your number is picked, you get back $35.00 plus the $1.00 that you bet.
    a. Is this a fair game?
    b. What would happen if you played this game a large number of times?

National Assessment of Educational Progress (NAEP) Question

There are 15 girls and 11 boys in a mathematics class. If a student is selected at random to run an errand, what is the probability that a boy will be selected?

   a. $\frac{4}{26}$

   b. $\frac{11}{26}$

   c. $\frac{15}{26}$

   d. $\frac{11}{15}$

   e. $\frac{15}{11}$

NAEP, Grade 8, 1990

BRAIN TEASER

An airplane can complete its flight if at least $\frac{1}{2}$ of its engines are working. If the probability that an engine fails is 0.01 and each engine failures does not depend on another engine, what is the probability of the event of a successful flight if the plane has
a. two engines?   **b.** four engines?

Hint for Solving the Preliminary Problem

In the bag there are only 3 out of 10 or 30% of the marbles that are blue. To obtain the 75% probability, we need to add blue marbles to the bag. There are several ways to solve this problem. One is to use *guess and check* and add a number of blue marbles and then compute the percentage of blue marbles, thus obtaining the probability of drawing a blue marble at each check. Based on this computation, the number of blue marbles to be added can be adjusted until the 75% probability is reached.

Algebra can be used in the proportion $\dfrac{n + 3}{n + 10}$ where n is the number of blue marbles to be added. The

ratio $\dfrac{n + 3}{n + 10}$ is the probability that must equal 75%.

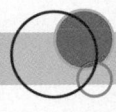

Chapter 9 Summary

9-1 How Probabilities Are Determined	**Pages**
An **experiment** is an activity whose results can be observed and recorded.	472
A **sample space** is the set of all possible outcomes for an experiment.	472
A **tree diagram** can be used to model a sample space.	472–473
Probabilities can be determined **experimentally (empirically)** or **theoretically**.	474
Theorem: Law of Large Numbers If an experiment is repeated a large number of times, the *experimental* probability of a particular outcome approaches a fixed number as the number of repetitions increases.	474
Outcomes are **equally likely** if one outcome is as likely to occur as another.	474
For an experiment with non-empty sample space S with equally likely outcomes, the **probability of an event** A from sample space S is given by $$P(A) = \frac{n(A)}{n(S)}$$	475
If each possible outcome of the sample space is equally likely, the sample space is a **uniform sample space**.	475
An **impossible event** is an event with a probability of 0. An impossible event can never occur.	476
A **certain event** is an event with a probability of 1. A certain event is sure to happen.	476
Theorem: If A is an event, then $0 \le P(A) \le 1$.	476
Theorem: The probability of an event is equal to the sum of the probabilities of disjoint events whose union is the event.	477
Two events are **mutually exclusive** if, and only if, only one of the events can occur at any given time—that is, if, and only if, the events are disjoint.	478

Theorem: If events A and B are mutually exclusive, then $P(A \text{ or } B) = P(A \cup B) = P(A) + P(B)$.	478
Two mutually exclusive events whose union is the sample space are **complementary events**.	478
Theorem: If A is an event and $\overline{A}$ is its complement, then $P(\overline{A}) = 1 - P(A)$.	479
If an event has equally likely outcomes, then $P(\text{event}) = \dfrac{\text{number of favorable outcomes}}{\text{total number of outcomes}}$	480

9-2 Multistage Experiments with Tree Diagrams and Geometric Probabilities

One-stage experiments are experiments that are over after one step. **Multistage experiments** take two or more steps.	486
Theorem: **Multiplication Rule for Probabilities** For all multistage experiments, the probability of the outcome along any path of a tree diagram is equal to the product of all the probabilities along the path.	491
Theorem: If events E_1 and E_2 are **independent**—that is, the occurrence of one has no influence on the occurrence of the other—then $P(E_1 \cap E_2) = P(E_1) \cdot P(E_2)$.	493
Dependent events are events that are not independent.	493
When **area models** are used to determine probabilities geometrically, outcomes are associated with points chosen at random in a geometric region that represents a sample space. This process is called finding **geometric probabilities**.	499

9-3 Using Simulations in Probability

A **simulation** is a technique used to act out a situation by conducting experiments whose outcomes are analogous to the original situation.	507
Simulations can play an important part in probability. Fair coins, dice, spinners, and random-digit tables are useful in performing simulations.	508
A **random-number table** is a table of digits that have been chosen at random.	508

9-4 Odds, Conditional Probability, and Expected Value

Let $P(A)$ be the probability that event A occurs and $P(\overline{A})$ be the probability that event A does not occur. If the case of equally likely outcomes,	516
the **odds in favor** of an event A are	
$$\frac{P(A)}{P(\overline{A})} = \frac{P(A)}{1 - P(A)} = \frac{\text{Number of ways } A \text{ can occur}}{\text{Number of ways } A \text{ cannot occur}}.$$	516
the **odds against** an event A are	
$$\frac{P(\overline{A})}{P(A)} = \frac{P(\overline{A})}{1 - P(\overline{A})} = \frac{\text{Number of ways } \overline{A} \text{ can occur}}{\text{Number of ways } \overline{A} \text{ cannot occur}}.$$	516
Theorem: If the odds in favor of event E are $m:n$, then $P(E) = \dfrac{m}{m + n}$.	518
If the odds against event E are $m:n$, then $P(E) = \dfrac{n}{m + n}$.	

If A and B are events in sample space S and $P(A) \neq 0$, then the **conditional probability** that event B occurs given that event A has occurred is

519

$$P(B|A) = \frac{P(A \cap B)}{P(A)}$$

If, in an experiment, the possible outcomes are numbers $a_1, a_2, \ldots, a_n$, occurring with probabilities $p_1, p_2, \ldots, p_n$, respectively, then the **expected value (mathematical expectation)** E is given by the equation

520

$$E = a_1 \cdot p_1 + a_2 \cdot p_2 + a_3 \cdot p_3 + \ldots + a_n \cdot p_n$$

A **fair game** is a game in which the expected net winnings or expected value is 0. A game is fair if each player has equal probability of winning.

520

9-5 Using Permutations and Combinations in Probability

Fundamental Counting Principle If an event M can occur in m ways and, after it has occurred, event N can occur in n ways, then event M followed by event N can occur in mn ways.

524

The expression $n!$, called n **factorial**, represents the product of all the natural numbers less than or equal to n. $0!$ is defined as 1.

524

Permutations are arrangements in which order is important. The number of permutations of r elements chosen from a set of n elements, where $0 \leq r \leq n$, is denoted by $_nP_r$ and is given by

525

$$_nP_r = \frac{n!}{(n-r)!}$$

Permutations of Like Objects If a set contains n elements, of which r_1 are of one kind, r_2 are of another kind, and so on through r_k, then the number of different arrangements of all n elements is equal to

527

$$\frac{n!}{r_1! \cdot r_2! \cdot r_3! \cdot \ldots \cdot r_k!}$$

Combinations are arrangements in which order is *not* important. To find the number of combinations possible, first use the Fundamental Counting Principle to find the number of permutations and then divide by the number of ways in which each choice can be arranged.

528

The number of combinations of r objects chosen from a set of n objects, where $0 \leq r \leq n$ is denoted by $_nC_r$ and is given by:

$$_nC_r = \frac{_nP_r}{_rP_r} = \frac{n!}{r!(n-r)!}$$

Chapter 9 Review

1. A coin is flipped 3 times and heads (H) or tails (T) are recorded.
   a. List all the elements in the sample space.
   b. List the elements in the event "at least two heads appear."
   c. Find the probability that the event in part (b) occurs.

2. Suppose the names of the days of the week are placed in a box and one name is drawn at random.
   a. List the sample space for this experiment.
   b. List the event consisting of outcomes that the day drawn starts with the letter T.
   c. What is the probability of the event of drawing a day that starts with T?

3. If you have a jar of 1000 red and blue jelly beans and you know that $P(\text{Blue}) = \dfrac{4}{5}$ and $P(\text{Red}) = \dfrac{1}{8}$, list several mathematical observations you can make about the number of beans in the jar.

4. In the 2000 presidential election, George W. Bush received 50,460,110 votes and Albert A. Gore received 51,003,926. In this election, if a voter for either Bush or Gore is chosen at random, answer the following:
   a. What is the probability of the event that the voter opted for Bush?
   b. What is the probability that the voter opted for Gore?
   c. What are the odds in favour of this voter not voting for Bush?

5. A box contains three red balls, five black balls, and four white balls. Suppose one ball is drawn at random. Find the probability of each of the following events:
   a. Black ball is drawn.
   b. Black or a white ball is drawn.
   c. Neither a red nor a white ball is drawn.
   d. Red ball is not drawn.
   e. Black ball and a white ball are drawn.
   f. Black or white or red ball is drawn.

6. One card is selected at random from an ordinary set of 52 cards. Find the probability of each of the following events:
   a. Club is drawn.
   b. Spade and a 5 are drawn.
   c. Heart or a face card is drawn.
   d. Jack is not drawn.

7. A box contains five colored balls and four white balls. If three balls are drawn one by one, find the probability of the event that they are all white if the draws are made as follows:
   a. With replacement
   b. Without replacement

8. Consider the following two boxes. If a letter is drawn from box 1 and placed into box 2 and then a letter is drawn from box 2, what is the probability of the event that the letter is an *L*?

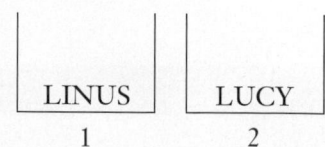

9. Use the following boxes for a two-stage experiment. First select a box at random and then select a letter at random from the box. What is the probability of the event of drawing an *A*?

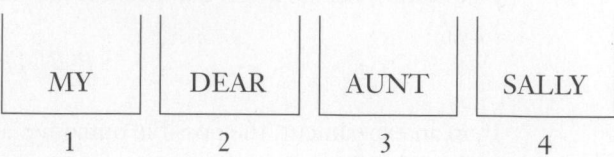

10. Consider the following boxes. Draw a ball from box 1 and put it into box 2. Then draw a ball from box 2 and put it into box 3. Finally, draw a ball from box 3. Construct a tree diagram for this experiment and calculate the probability of the event that the last ball chosen is black.

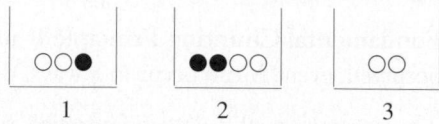

11. What are the odds in favor of drawing a jack when one card is drawn from an ordinary deck of playing cards?

12. A fair normal die is rolled once. What are the odds against rolling a prime number?

13. If the odds in favor of a particular event are 3 to 5, what is the probability that the event will occur?

14. A game consists of rolling two dice. Rolling double 1s pays $7.20. Rolling double 6s pays $3.60. Any other roll pays nothing. What is the expected value for this game?

15. A total of 3000 tickets have been sold for a drawing. If one ticket is drawn for a single prize of $1000.00, what is a fair price for a ticket?

16. In a special raffle, a ticket costs $2.00. You mark any four digits on a card (repetition and 0 are allowed). If you select the winning number, you win $15,000.00. What is the expected value?

17. How many four-digit numbers can be formed if the first digit cannot be 0 and the last digit must be 2?

18. A club consists of 10 members. In how many different ways can a group of 3 people be selected to go on a European trip?

19. Find the number of ways that 4 flags can be displayed on a flagpole, one above the other, if 10 different flags are available.

20. Five women live together in an apartment. Two have blue eyes. If two of the women are chosen at random, what is the probability of the event that they both have blue eyes?

21. Five evenly matched horses (Applefarm, Bandy, Cash, Deadbeat, and Egglegs) run in a race.
    a. In how many ways can the first-, second-, and third-place horses be determined?
    b. Find the probability that Deadbeat finishes first and Bandy finishes second in the race.
    c. Find the probability of the event that the first-, second-, and third-place horses are Deadbeat, Egglegs, and Cash, in that order.

22. Al and Ruby each roll an ordinary die once. What is the probability of the event that the number of Ruby's roll is greater than the number of Al's roll?

23. Amy has a quiz on which she is to answer any three of the five questions. If she is equally well versed on all questions and chooses three questions at random, what is the probability of the event that question 1 is not chosen?

24. How many batting lineups are there for the nine players of a baseball team if the center fielder must bat fourth and the pitcher last?

25. On a certain street are three traffic lights. At any given time, the probability of the event that a light is green is 0.3. What is the probability of the event that a person will hit all three lights when they are green if they represent independent events?

26. A three-stage rocket has the following probabilities for failure: The probability for failure at stage one is $\frac{1}{6}$; at stage two, $\frac{1}{8}$; and at stage three, $\frac{1}{10}$. What is the probability of the event of a successful flight, given that the first stage was successful and all stages are independent?

27. How could each of the following be simulated by using a random-digit table?
    a. Tossing a fair die
    b. Picking 3 different months at random from the 12 months of the year
    c. Spinning the spinner shown

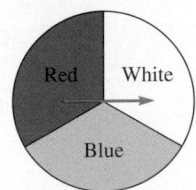

28. Otto says that if you toss three coins you get $3H$, $2H$, $1H$, or $0H$, so the probability of the event of getting three heads is $\frac{1}{4}$. How do you respond?

29. If a dart is thrown at the following tangram dartboard and we assume the dart lands at random on the board, what is the probability of the event of its landing in each of the following areas?
    a. Area A
    b. Area B
    c. Area C

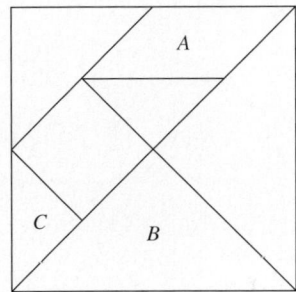

30. The points M, N, O, P, and Q in the following figure represent exits on a highway (the numbers represent miles). An accident occurs at random between points M and Q. What is the probability of the event that it has occurred between N and O?

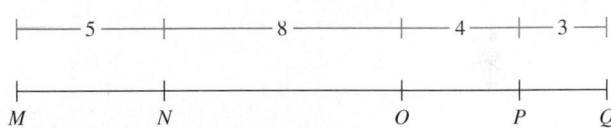

31. If three cards are drawn at random from a standard deck of 52 playing cards, what is the probability that at least one of the cards is a face card: that is, one of the cards is a king, queen, or jack?

Data Analysis/Statistics: An Introduction

Preliminary Problem

At a class reunion five friends were talking and wanted to know their mean salary. However, not one was willing to tell any of the others any specific salary information. They had a calculator and needed to find a way to compute the mean salary without anyone revealing any person's salary to the other four. How could it be done?

If needed, see Hint on page 622.

Data analysis usually refers to an informal approach to statistics. It is a relatively new term in mathematics. *Statistics* once referred to numerical information about state or political territories; it comes from the Latin *statisticus*, meaning "of the state." Today, much of statistics involves making sense of data.

In PSSM, we find the following:

Instructional programs from prekindergarten through grade 12 should enable all students to

- formulate questions that can be addressed with data and collect, organize, and display relevant data to answer them;
- select and use appropriate statistical methods to analyze data;
- develop and evaluate inferences and predictions that are based on data . . . . (p. 48)

Focal Points recommends that from the earliest grades students "pose questions and gather data about themselves and their surroundings" (kindergarten, p. 26), "represent data using concrete objects, pictures, and graphs" (grade 1, p. 26), and "describe parts of the data and the set of data as a whole to determine what the data show" (grades kindergarten and 1, p. 26). Students in grades 3–5 are expected to "design investigations to address a question and consider how data-collection methods affect the nature of the data set," "collect data using observations, surveys, and experiments," and "represent data using tables and graphs such as line plots, bar graphs, and line graphs." (p. 33) Additionally students in grade 5 are expected to "recognize the differences in representing categorical and numerical data." (p. 33) In grade 7, students are expected to "formulate questions, design studies, and collect data about a characteristic shared by two populations or different characteristics within one population." (p. 39)

In *Guidelines for Assessment and Instruction in Statistics Education (GAISE) Report: A PreK–12 Curriculum Framework* (March 2005) (hereafter referred to as the *Statistics Framework*), presented to the American Statistical Association, recommendations are made for statistics education that complement *PSSM*. In the *Statistics Framework*, specific data analysis recommendations are divided into three parts—A, B, and C—with the more intuitive parts for early grades being in part A and the more advanced ideas being in part C. According to the *Statistics Framework*, "Sound statistical reasoning skills take a long time to develop. . . . The surest way to reach the necessary skill level is to begin the educational process in the elementary grades and keep strengthening and expanding these skills throughout the middle and high school years." (p. 3)

In both *PSSM* and *Focal Points*, we see that students in the early grades should explore the basic ideas of statistics by collecting data, organizing the data pictorially, and then interpreting information. The ideas of gathering, representing, and analyzing data are expanded in the later grades. In this chapter, we deal with categorical and numerical data, representations of data, and key statistical concepts including measures of central tendency and of variation. Additionally, some uses and misuses of statistics are discussed.

◯⊢ **Historical Note**

The seventeenth-century work of John Graunt (1620–1674) and the nineteenth-century work of Adolph Quetelet (1796–1874) involved making predictions from data. Graunt dealt with birth and death records; Quetelet dealt with crime and mortality rates. Florence Nightingale (1820–1910) worked with mortality tables during the Crimean War to improve hospital care. Other notables in data collection and analysis include Sir Francis Galton (1822–1911), Gregor Mendel (1822–1884), Ronald Fisher (1890–1962) and Andrei Nikolaevich Kolmogorov (1903–1987). ●

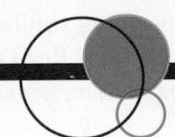

10-1 Designing Experiments/Collecting Data

Variability

The grade 6 *Common Core Standards* point out that students should be able to:

> Recognize a statistical question as one that anticipates variability in the data related to the question and accounts for it in the answers. *For example, "How old am I?" is not a statistical question, but "How old are the students in my school?" is a statistical question because one anticipates variability in students' ages.* (p. 45)

Much of the study of statistics deals with **variability** of data, or the amount by which the pieces of data in a data set differ. It is important to understand different types of variability when designing experiments and examining data. Among the types identified are the following:

Measurement variability: For example, suppose 20 students measure the length of a board. In a perfect world, all the measurements would be the same. However, simple human fallibility will produce different results. Questions for the statistician might be "How much difference could be expected in such a measurement? Are the different measures drastically different or close enough to be considered acceptable?"

Natural variability: It has been said that nature abhors a vacuum, but it might also be said that nature loves differences. For example, any two individuals are different; even genetically identical twins differ in personality, aptitude, and so on.

Induced variability: Induced variability is frequently studied to see, for example, how individuals react to certain stimuli or how bean plants grow based on the amount of food, water, and sunlight they receive.

Sampling variability: If we want to choose, for example, a set of college students to see how they react to a question, we could do so in many ways. Some selections might elicit a reaction that could be judged as representative of the entire student population; other selection methods might produce a very biased result. Suppose only one class is interviewed; would you expect the same reaction from a graduate class in mathematics as a freshman class in English composition?

In each type of variability, the role of context is important when considering how much variability is acceptable. To make generalizations from a sample, one might want to choose a group that could be deemed highly representative of the population from which it was taken. If generalizations are not to be "widespread," then a very different sample might be acceptable. For example, if we wanted to know the favorite pet of Mr. Carter's fourth-grade class at Russell Elementary School, the entire class (the **population**) might be polled. The population in this case is the **sample**. The only generalization that can be made is about the class itself. That generalization is a fact of the poll and cannot be extended to the remainder of the school population.

The grade 7 *Common Core Standards* point out that students should be able to:

1. Understand that statistics can be used to gain information about a population by examining a sample of the population; generalizations about a population from a sample are valid only if the sample is representative of that population. Understand that random sampling tends to produce representative samples and support valid inferences.

2. Use data from a random sample to draw inferences about a population with an unknown characteristic of interest. Generate multiple samples (or simulated samples) of the same size to gauge the variation in estimates or predictions. *For example, estimate the mean word length in a book by randomly sampling words from the book; predict the winner of a school election based on randomly sampled survey data. Gauge how far off the estimate or prediction might be.* (p. 50)

A **random sample** as mentioned in part (2) above is a subset of the population where each member of the population has the same chance of being selected.

In this section we consider issues of formulating questions and designing studies to collect and analyze data that can be used to find possible resolutions to various issues. Read the student page on page 545 and consider the *Quick Check* to better understand *population*, *sample*, and *random* sample.

Underlying Issues in Designing Studies

In the following cartoon, Billy wants to check the advertising claim on a can of nuts that less than half of the nuts are peanuts.

Family Circus copyright © 2007 Bil Keane, Inc. Distributed by King Features Syndicate

Billy can count the number of nuts and the types of nuts in a specific can to check the claim. However, the company might still be allowed to make that claim without violating regulations of the Federal Trade Commission (FTC). For students in the lower grades, Billy's approach is very reasonable. In the middle and higher grades, it would not be sufficient. How, then, do we approach advertising and other issues that may be considered statistically at different grade levels or as adults?

From the FTC standpoint, the claim on the can label can legitimately be made if the company has evidence that, in general, the claim is true, the claim does not "injure" the consumer, and the consumer would likely buy the product even if the claim were not true in every instance, such as in Billy's can. Evaluating a claim such as this from a company's standpoint might rely on the selection of a random sample of cans from the production line and having a very high percentage of the cans from the sample meet the advertised standard. Few can visit assembly lines to check advertising claims. However, students can become very aware of the process necessary to evaluate statistical claims. The following framework for statistical problem solving was suggested for the classroom by Franklin and Garfield (2006, p. 350):*

a. Formulate questions	**b.** Collect data
c. Analyze the data	**d.** Interpret the results

We consider each in turn.

Formulating Questions

For any given problem where data are needed to determine either an answer or an approach to an answer, it is important that the problem is clarified to the extent that meaningful data can be collected. For example, if we wanted to collect data on the appearance of an average fifth grader, we must clarify what "the appearance of an average" fifth grader means. By appearance, do we mean clothing; do we

---

*Franklin, C., and J. Garfield. "The GAISE Project: Developing Statistics Education Guidelines for Grades Pre-K-12 and College Courses." In *Thinking and Reasoning with Data and Chance, 68th Yearbook*. Reston, VA: National Council of Teachers of Mathematics, 2006.

School Book Page RANDOM SAMPLES AND SURVEYS

 **11-4** ## Random Samples and Surveys

✓ Check Skills You'll Need

1. Vocabulary Review
   A ? is a ratio that compares a number to 100.

Write each ratio as a percent.

2. 4 out of 5

3. 10 out of 40

4. 14 out of 200

GO for Help
Lesson 6-1

What You'll Learn

To identify a random sample and to write a survey question

🔊 **New Vocabulary** population, sample, random sample, biased question

Why Learn This?

You can use a survey to gather information from a group of people. Pollsters use surveys to understand group preferences.

A **population** is a group of objects or people. The population of an election is all the people who vote in that election. It is not practical to ask all the voters how they expect to vote. Pollsters select a **sample,** or a part of the population. A sample is called a **random sample** when each member of a population has the same chance of being selected.

EXAMPLE Identifying a Random Sample

1 You survey customers at a mall. You want to know which stores they shop at the most. Which sample is more likely to be random? Explain.

a. You survey shoppers in a computer store.

   Customers that shop in a particular store may not represent all the shoppers in the entire mall. This sample is not random.

b. You walk around the mall and survey shoppers.

   By walking around, you give everyone in the mall the same chance to be surveyed. This sample is more likely to be random.

✓ Quick Check

1. You survey a store's customers. You ask why they chose the store. Which sample is more likely to be random? Explain.
   a. You survey 20 people at the entrance from 5:00 P.M. to 8:00 P.M.
   b. You survey 20 people at the entrance throughout the day.

550 Chapter 11 Displaying and Analyzing Data

mean size; or do we mean something entirely different? If we mean size, then what exactly is meant by size? Is it height? Is it weight? Is it a combination of height and weight? Or is it something else? The point is that we must have a clear understanding of exactly what we want to know when we mention the "appearance of an average fifth grader." And we may need several questions to get at what we want. All of that has to be decided before data is collected to try to determine an answer.

In the early grades, teachers, not students, may pose questions to be examined and restrict data collection to the classroom or possibly to families. In middle grades, students pose their own questions and collect data beyond the classroom, and in the higher grades, students may pose questions from which generalizations can be made. As pointed out on the student page on page 547 we must be careful when formulating questions so that they are not **biased**. Read through the *EXAMPLE* on the student page to see *biased* and *fair* questions.

NOW TRY THIS 10-1

On the student page, on page 547 work through the *Quick Check* and *Check Your Understanding*.

A formulation of unbiased questions on questionnaires for surveys requires careful consideration. Example 10-1 poses a question about survey methods.

EXAMPLE 10-1

The IM Reliable Polling Company plans to conduct a survey of college students to determine their favorite movies in 2011–2012. What are some survey methods that could be used to obtain information from the students?

Solution   Answers vary. One possibility is to have randomly selected students list their favorite movies in 2011–2012. A second possibility is to provide a list of movies from the time period and ask randomly selected students to rate them from most favored to least favored using a numerical scale.

In Example 10-1, the type of questions asked will have some influence on the responses. For example, if a list of movies is provided, then someone who knows nothing about the movies may rate them anyway and return the survey. All such possibilities have to be considered when adopting a survey method.

In early grades, interpreting the results of a study may lead to no generalizations, but students could begin to consider predictions as long as teachers help them recognize that those predictions may not be accurate. In middle grades, students should begin to recognize that observations of a few cases may not be indicative of an entire population and begin to understand the difference between association and cause and effect; that is, association does not imply cause and effect. At the higher grade levels, students can make generalizations, recognize the effect of randomization on results, and begin to know how strong associations are between variables (Franklin and Garfield 2006, pp. 353–354).*

We consider the questions below to see differences in interpretation at each of the levels.

Early Grades: How many pets does each student in the class have?

Middle Grades: Does the number of each student's pets change from Mr. Smith's grade 1 class to Mrs. Oneida's grade 2 class?

Higher Grades: Do students have more pets per student as they progress from grade 1 to grade 2?

Collect Data

Once questions are formulated, the next step is to develop a plan for collecting the data needed to answer the questions. In the process we identify types of variability to determine a minimal

---

*Franklin, C., and J. Garfield. "The GAISE Project: Developing Statistics Education Guidelines for Grades Pre-K-12 and College Courses." In *Thinking and Reasoning with Data and Chance, 68th Yearbook*. Reston, VA: National Council of Teachers of Mathematics, 2006.

School Book Page   BIASED QUESTIONS

Vocabulary Tip

Bias means "slant." A biased question slants the answers in one direction.

When you conduct a survey, ask questions that do not influence the answer. A **biased question** is a question that makes an unjustified assumption or makes some answers appear better than others.

EXAMPLE   **Identifying Biased Questions**

2 Music Is each question *biased* or *fair?* Explain.

a. "Do you think that soothing classical music is more pleasing than the loud, obnoxious pop music that teenagers listen to?"

This question is biased against pop music. It implies that all pop music is loud and that only teenagers listen to it. The adjectives "soothing" and "obnoxious" may also influence responses.

b. "Which do you think is the most common age group of people who like pop music?"

This question is fair. It does not assume that listeners of pop music fall into only one age group.

c. "Do you prefer classical music or pop music?"

This question is fair. It does not make any assumptions about classical music, pop music, or people.

✓ Quick Check

2. Is each question *biased* or *fair?* Explain.
   a. Do you prefer greasy meat or healthy vegetables on your pizza?
   b. Which pizza topping do you like best?

✓ Check Your Understanding

Vocabulary Match each statement with the appropriate term.

1. a group of objects or people

2. makes some answers appear better

3. gives members of a group the same chance to be selected

A. biased question
B. random sample
C. population

You want to determine the favorite spectator sport of seventh-graders at your school. You ask the first 20 seventh-graders who arrive at a soccer game, "Is soccer your favorite sport to watch?"

4. What was the population of your survey? What was the sample?

5. The survey (was, was not) random.

6. You used a (biased, fair) question.

11-4 Random Samples and Surveys   **551**

expectation from eventual results. When we think about collecting data, an immediate question is from whom or where the data have to be collected. Are we talking about data from humans, from animals, even from data based on other data? The list could go on. The context is most important. To continue the example about the appearance of an average fifth grader, do we want to examine medical records to make this determination? Do we want to choose a sample group of students and make generalizations from the sample? If we want a sample, what geographical factors, if any, might need to be considered? Do all fifth graders "have the same appearance" in Montana? in the United States? in the world? In any event, a plan has to be devised and then implemented to collect the desired data.

In the early grades, there will likely be few plans to address variability in data collected because questions are fairly simple. In the middle grades, students may begin to use random selection for samples and choose students from the entire school as subjects. In later grades, experimental designs for study will include randomization to a wider audience. Using the same pet data questions as before, we illustrate collection methods for the different levels.

Early Grades: How many pets does each student in the class have?

The number of pets for each student is collected and recorded.

Middle Grades: Does the number of each student's pets change from Mr. Smith's grade 1 class to Mrs. Oneida's grade 2 class?

A simple random sample of the two classes might be used to compare the two classes. An important consideration is how to choose the samples.

Higher Grades: Do students have more pets per student as they progress from grade 1 to grade 2?

A random sample from first- and second-grade classes across a school system or a state might be used to help answer this question. Just how this is to be done is one of the major issues.

Analyze the Data

In analyzing collected data, we must make a decision about how to display the data or whether to report numbers to summarize them. Sections 10-2 and 10-3 deal with graphics for displaying data, and Section 10-4 discusses summaries of data. Often, decisions about how to analyze data are made with the data in hand. However, a good plan for collecting data should include some consideration of how the data will be analyzed once it is collected.

In the early grades, individual data may be compared to other individual data or to all the collected data for the entire group. In the middle grades, students may consider distributions and their associated properties. For example, groups may be compared to other groups using box plots (discussed in the next section). It is possible that students at this level may recognize sampling errors when analyzing data or begin to associate one variable with another. For example, students may consider the relationship between age and weight of children. In higher grades, students will analyze variability by considering numerical summaries of data.

Using the pet data questions, types of data analysis and display are considered at different levels.

Early Grades: How many pets does each student in the class have?

Simple bar graphs or graphical displays may be used to illustrate the data.

Middle Grades: Does the number of each student's pets change from Mr. Smith's grade 1 class to Mrs. Oneida's grade 2 class?

Double bar graphs might be used on the same scale so that differences can be seen. Also double stem and leaf plots might be used at this stage.

Higher Grades: Do students have more pets per student as they progress from grade 1 to grade 2?

Here, the bar graphs of the middle grades might be used or the stem and leaf plots (discussed in the next section) might be used, but students would also consider margins of error as a result of sampling techniques.

Interpret the Results

The results of data analysis must be interpreted. Some interpretations are clear-cut; others sometimes misuse data, a topic discussed in the final section of the chapter. In any event, the interpretation must be related to the original questions.

Early Grades: How many pets does each student in the class have?

An interpretation at this level might be as simple as reading and understanding a graph of the data.

Middle Grades: Does the number of each student's pets change from Mr. Smith's grade 1 class to Mrs. Oneida's grade 2 class?

From double bar graphs or the stem and leaf plots, or possibly even box plots, students can compare the numbers and might acknowledge that this is only for the given classes and not generalizable.

Higher Grades: Do students have more pets per student as they progress from grade 1 to grade 2?

The interpretation at this level might include all the interpretation at the middle-grades level but would consider generalizing across schools or the state.

Assessments 10-1A and 10-1B are basically communication exercises designed to interact about designing experiments and collecting data.

Assessment 10-1A

1. A second-grade class has a project to determine how many houses are on the blocks where they live. Formulate questions that might be asked, how the data might be collected, how it might be analyzed, and how it might be interpreted.
2. A middle-school class decides to determine if adults were more active than sixth graders. What are possible questions for investigation?
3. An eighth-grade class wants to determine if the water quality in a school is better on one floor of the classroom building than another. What are some considerations?
4. To determine the most popular book among students at an elementary school, which sample is more likely to be random? Why?
   a. Ask students from different grades at the school.
   b. Ask a group of friends sitting in the library.
5. Tell whether each of the following questions is biased or fair. Why?
   a. What is your favorite subject in school?
   b. Do you like the calm, soothing, beautiful ocean?
6. A group of adults remarked that they could "tell the difference between Coke and Pepsi." How might students design a study to test this claim if the students were
   a. elementary-grade students?
   b. middle-grade students?
7. If a class were to investigate how many countries students have visited, what are potential issues to be resolved before the investigation begins?

8. In 1936, the *Literary Digest* predicted from questionnaires sent to owners of telephones and automobiles that Alfred Landon would win the U.S. presidency over Franklin Roosevelt. Clearly that prediction was incorrect. What are some likely errors made by the *Literary Digest*?
9. In a set of student evaluations of their professors, one university lists the "three most positive comments and the three most negative comments" on its web site. Discuss this listing and its usefulness in evaluating a professor's class.
10. If you wanted to choose a sample of 50 from a school of 400, do you think it would be better to have the students line up in a row and choose the first 50 in line or to put all the names in a hat and draw out the first 50?
11. If a bar graph like the one shown here were used to depict data for second graders, what observations might be realistic for students to make in analyzing the graph?

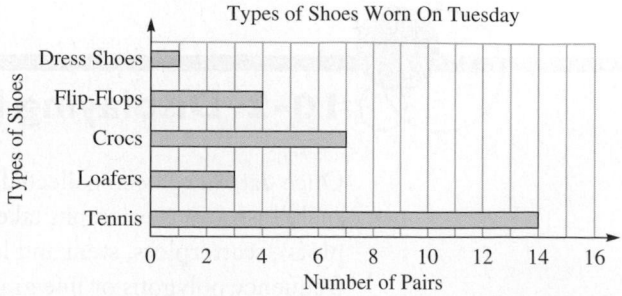

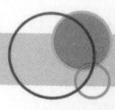

Assessment 10-1B

1. Suppose a fourth-grade class had a project to determine how many people were in their families. Describe questions that might be asked, how the data might be collected, how it might be analyzed, and how it might be interpreted.

2. An elementary school class decides to determine whether adult family members were lazier than students. What are possible questions for investigation?

3. A seventh-grade class wants to determine if the students on one floor of the school are more affluent than those on another floor. What are some types of questions to consider?

4. To determine student computer use, which sample is more likely to be random?
   (i) Ask students who are eating lunch.
   (ii) Ask students entering the library on Friday night.

5. Tell whether each of the following questions is *biased* or *fair*.
   a. Given the great tradition of space exploration in the United States, do you favor continued funding for space flights?
   b. Do you think middle school students should be required to wear uniforms?

6. Some adults said different types of chocolate had distinctive flavors. How might students design a study to test whether the adults could tell the difference between types of chocolate if the students were
   a. elementary-grade students?
   b. middle-grade students?

7. A class wants to investigate how many different states students have visited. What are some of the potential issues to be resolved before the investigation begins?

8. Many predictions for the winner of the 2008 Super Bowl were incorrect. According to *The Arizona Republic* of February 3, 2008, "Giants pull off Super upset: With bruising defense and 1 brilliant drive, Giants stun Patriots." The predictions were based on the fact that the New England Patriots had won 18 games in a row. What data might be studied to help predict the winner before the game?

9. A university professor used a rating scale to evaluate her classes: 1. Excellent class; 2. Good class; 3. Above-average class. Discuss the rating scale and its usefulness in evaluating a class.

10. Suppose a poll was to be conducted in an elementary school by interviewing a representative sample of 50 students. If the interveiwer sat outside the cafeteria when it opened and talked to the first 50 students, explain whether or not a representative sample would be determined.

11. A teacher used a large grid printed on plastic, labeled the axes as kinds of shoes and number of wearers, respectively, had his students put one of their shoes on the grid, and proceeded to analyze the data with the class. What types of observation might it be reasonable for first graders to make?

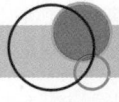

Mathematical Connections 10-1

Communication

1. Describe a problem that could be solved by using methods discussed in this section.

2. Read the data analysis section of *PSSM* for grades 3–5 and discuss why students at these grade levels should be studying data analysis.

Open-Ended

3. Design a grade-appropriate question that students could investigate using statistics.

4. Suggest how students at different grade levels might investigate the length of words in a textbook.

Cooperative Learning

5. Examine grade-school books to see what types of data analysis are introduced and when they are introduced.

Questions from the Classroom

6. A student asks if the precision with which manufacturers must calibrate their tools is at all related to statistics. How do you respond?

7. Mariah read that deaths due to stampedes occurred primarily in non-U.S. countries at rock concerts and argued with her parents that she would be perfectly safe at the running of the bulls in Pamplona, Spain. If you were her parents, how would you reason with her?

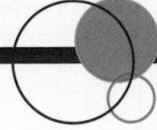

10-2 Displaying Data: Part I

Once data have been collected, their organization and visual illustrations are important. Such visual illustrations or graphs take many forms: pictographs, circle graphs (pie charts), dot plots (line plots), scatterplots, stem and leaf plots, box plots, frequency tables, histograms, bar graphs, and frequency polygons or line graphs. A *graph* is a picture that displays data. Graphs are used to tell a story. In the *Herman* cartoon, we see a graph being used to display some particular data to make a point to an audience. What message is the presenter trying to get across? What labels might appear on the vertical and horizontal axes?

"That's the last time I go on vacation."

What message is the *Bureau of Labor Statistics* trying to get across in the following graph?

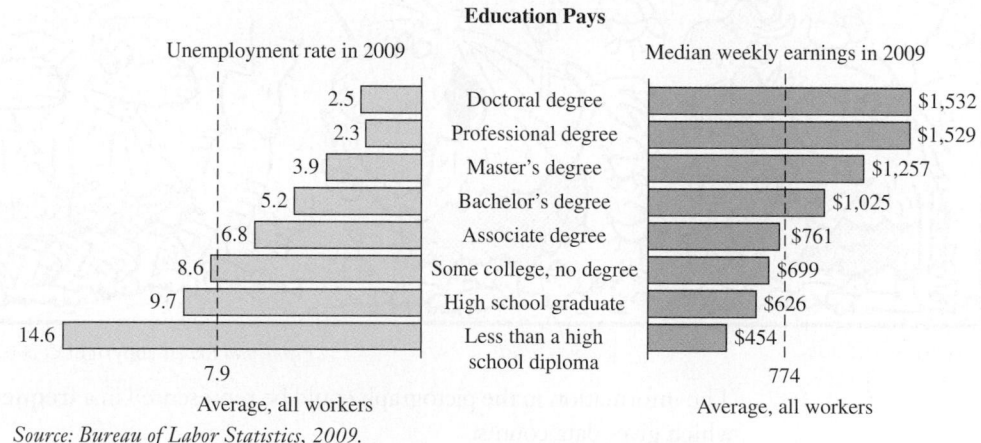

Education Pays

Unemployment rate in 2009		Median weekly earnings in 2009
2.5	Doctoral degree	$1,532
2.3	Professional degree	$1,529
3.9	Master's degree	$1,257
5.2	Bachelor's degree	$1,025
6.8	Associate degree	$761
8.6	Some college, no degree	$699
9.7	High school graduate	$626
14.6	Less than a high school diploma	$454
7.9 Average, all workers		774 Average, all workers

Source: Bureau of Labor Statistics, 2009.

Describing data frequently involves reading information from graphical displays, tables, lists, and so on. In *PSSM*, we find the following:

> A fundamental idea in prekindergarten through grade 2 is that data can be organized or ordered and that this "picture" of the data provides information about the phenomenon or question. In grades 3–5, students should develop skill in representing their data, often using bar graphs, tables, or line plots. . . . Students in grades 6–8 should begin to compare the effectiveness of various types of displays in organizing the data for further analysis or in presenting the data clearly to an audience. (p. 49)

Data: Categorical and Numerical

Statistical thinking begins in early grades with a need to know such things as "the most popular" pet, the favorite type of shoe, the most used color to paint, and so on. **Data** may be collected to find answers to such questions. Data collected may be either *categorical* or *numerical* depending on the questions being answered. For example, according to the *Statistics Framework*, in the elementary grades, students may

> be interested in the favorite type of music among students at a certain grade level. . . . The class might investigate the question: *What type of music is most popular among students?* . . . The characteristic, favorite music type is a categorical variable—each child in that grade would be placed in a particular nonnumerical category based on his or her favorite music type. The resulting data are often called Categorical Data. (p. 22)

Categorical data are data that represent characteristics of objects or individuals in groups (or categories), such as black or white, inside or outside, male or female. **Numerical data** are data collected on numerical variables. For example, in grade school, students may ask whether there is a difference in the distance that girls and boys can jump. The distance jumped is a numerical variable and the collected data are *numerical*.

Some representations are used for both categorical and numerical data. We see different representations for both types of data in the following subsections.

Pictographs

An elementary student might use a picture graph, or **pictograph**, to represent tallies of categories such as their favorite pets. Pictographs are often seen in newspapers and magazines. In the *Frank and Ernest* cartoon, the projected results of the third quarter's hunts are seen in a pictograph. With only what is pictured, what is the most common animal that the hunters bag?

Frank and Ernest

Frank and Ernest copyright © 2007 by Thaves. Reprinted by permission.

The information in the pictograph could be represented in a **frequency table**, as seen in Table 10-1, which gives data counts.

Table 10-1

Type	Frequency
	2
	5
	3
	2

In Figure 10-1, categorical data might be seen in the determination of the month in which the most newspapers were recycled. The month is the *category*. In a pictograph, a symbol or an icon is used to represent a quantity of items. A *legend* tells what the symbol represents. Pictographs are used frequently to show comparisons of outputs (Figure 10-1). A major disadvantage of pictographs is evident in Figure 10-1. The month of September contains a partial bundle of newspapers. It is impossible to tell from the graph the weight of that partial bundle with any accuracy.

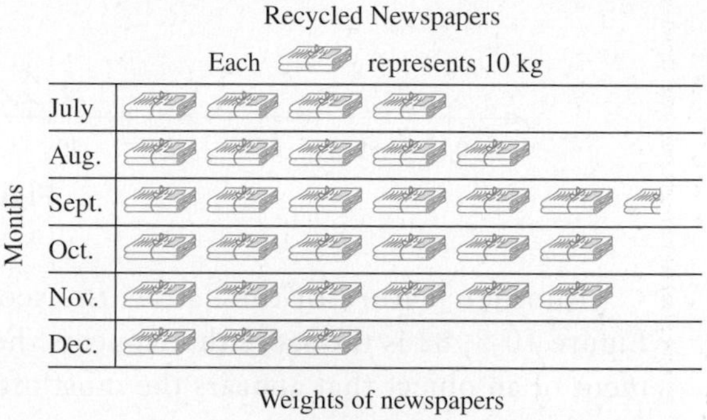

Recycled Newspapers

Each ▱ represents 10 kg

Months	
July	▱ ▱ ▱ ▱
Aug.	▱ ▱ ▱ ▱ ▱
Sept.	▱ ▱ ▱ ▱ ▱ ▱ ▱
Oct.	▱ ▱ ▱ ▱ ▱ ▱
Nov.	▱ ▱ ▱ ▱ ▱ ▱
Dec.	▱ ▱ ▱

Weights of newspapers

Figure 10-1

The number of students in each teacher's fifth-grade class at Hillview School is depicted in the tabular representation in Table 10-2. Each teacher is a category and the frequency/count table provides a method to summarize the categorical data. Figure 10-2 depicts the information in a pictograph. All graphs need titles; and if applicable, legends should be shown.

Table 10-2

Teacher	Frequency or Count
Ames	20
Ball	20
Cox	15
Day	25
Eves	15
Fagin	10

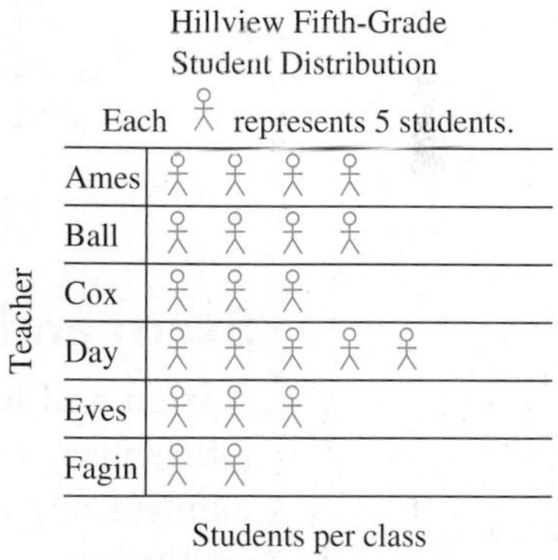

Hillview Fifth-Grade
Student Distribution

Each ⚲ represents 5 students.

Teacher	
Ames	⚲ ⚲ ⚲ ⚲
Ball	⚲ ⚲ ⚲ ⚲
Cox	⚲ ⚲ ⚲
Day	⚲ ⚲ ⚲ ⚲ ⚲
Eves	⚲ ⚲ ⚲
Fagin	⚲ ⚲

Students per class

Figure 10-2

Dot Plots (Line Plots)

Next we examine a **dot plot**, sometimes called a **line plot**. (The names *dot plots* and *line plots* have been used synonymously in the past, but current usage favors dot plots.) Dot plots provide a quick, simple way of organizing data. Typically, we use them when there is only one group of data with fewer than 50 values.

Suppose the 30 students in Abel's class received the following test scores:

82 97 70 72 83 75 76 84 76 88 80 81 81 52 82

82 73 98 83 72 84 84 76 85 86 78 97 97 82 77

A dot plot for the class scores consists of a horizontal number line on which each score is denoted by a dot, or an ×, above the corresponding number-line value, as shown in Figure 10-3. The number of ×'s above each score indicates how many times each score occurred.

Figure 10-3 yields information about Abel's class exam scores. For example, three students scored 76 and four scored greater than 90. We also see that the low score was 52 and the high score was 98. Several features of the data become more obvious when dot plots are used. For example, outliers, clusters, and gaps are apparent. An **outlier** is a data point whose value is significantly greater or less than other values, such as the score of 52 in Figure 10-3. (Outliers are discussed in more detail in a later section.) A **cluster** is an isolated group of points, such as the one located at the scores 97 and 98. A **gap** is a large space between points, such as the one between 88 and 97.

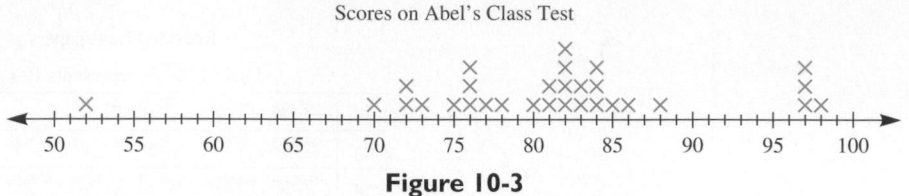

Figure 10-3

Another feature of the data is the score that appears most often in the data set. In Figure 10-3, 82 is the score that appears the most number of times. The count or measurement of an object that appears the most often is the **mode**. The mode is discussed in more detail later.

If a dot plot is constructed on grid paper, then shading in the squares with ✕'s and adding a vertical axis depicting the scale allows the formation of a *histogram*, as in Figure 10-4. (Histograms are discussed in more detail later in this section.) The break in the horizontal axis is denoted by a squiggle and indicates that a part of the number line is missing.

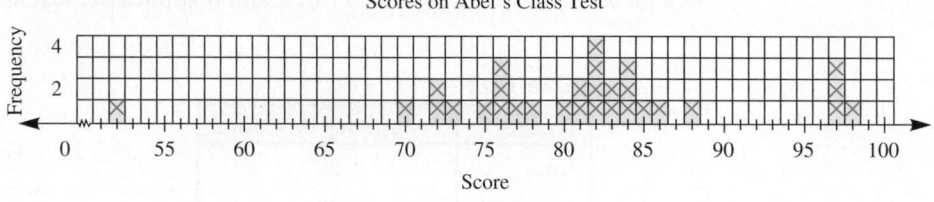

Figure 10-4

Stem and Leaf Plots

A **stem and leaf plot** is akin to a dot plot, but the scale is usually vertical, and digits are used rather than ✕'s. A stem and leaf plot of test scores for Abel's class is shown in Figure 10-5. The numbers on the left side of the vertical segment are the **stems**, and the numbers on the right side are the **leaves**. In Figure 10-5, the stems are the tens digits of the scores on the test and the leaves are the units digits. In this case, the legend, "9|7 represents 97," shows how to read the plot.

The data in a stem and leaf plot, as in Figure 10-5, are in order from least to greatest on a given row. This is an **ordered stem and leaf plot**. Stem and leaf plots are sometimes first constructed with data in the order given. To make them useful, most users then order the data.

⊙ **Historical Note** _____

Stem and leaf plots evolved from Arthur Bowley's work in the early 1900s and are a useful tool in exploratory data analysis. John Tukey (1915–2000), an American statistician, published *Exploratory Data Analysis* in 1977. This publication led to the common usage of stem and leaf plots in data analysis.   ●

John Tukey

Scores for Abel's Class

5	2
6	
7	0223566678
8	011222233444568
9	7778

9|7 represents 97

Figure 10-5

There is no unique way to construct stem and leaf plots. Smaller numbers are usually placed at the top so that when the plot is turned counterclockwise 90°, it resembles a bar graph or a histogram (discussed later in this section). Important advantages of stem and leaf plots are that they can be created by hand rather easily and they do not become unmanageable when the number of values becomes large. Moreover, no original values are lost in a stem and leaf plot. For example, in the stem and leaf plot of Figure 10-5, we know that one score was 75 and that exactly three students scored 97. A disadvantage of stem and leaf plots is that we do not see some significant information; for example, we know from such a plot that a student scored 88, but we do not see which one.

Following is a summary of how to construct a stem and leaf plot:

1. Find the high and low values of the data.
2. Decide on the stems.
3. List the stems in a column from least to greatest.
4. Use each piece of data to create leaves to the right of the stems on the appropriate rows.
5. If the plot is to be ordered, list the leaves in order from least to greatest.
6. Add a legend identifying the values represented by the stems and leaves.
7. Add a title explaining what the graph is about.

NOW TRY THIS 10-2

Construct a stem and leaf plot using the data in Table 10-3, which lists the presidents of the United States and their ages at death.

Table 10-3

President	Age at Death	President	Age at Death	President	Age at Death
George Washington	67	Franklin Pierce	64	Woodrow Wilson	67
John Adams	90	James Buchanan	77	Warren Harding	57
Thomas Jefferson	83	Abraham Lincoln	56	Calvin Coolidge	60
James Madison	85	Andrew Johnson	66	Herbert Hoover	90
James Monroe	73	Ulysses Grant	63	Franklin Roosevelt	63
John Q. Adams	80	Rutherford Hayes	70	Harry Truman	88
Andrew Jackson	78	James Garfield	49	Dwight Eisenhower	78
Martin Van Buren	79	Chester Arthur	57	John Kennedy	46
William H. Harrison	68	Grover Cleveland	71	Lyndon Johnson	64
John Tyler	71	Benjamin Harrison	67	Richard Nixon	81
James K. Polk	53	William McKinley	58	Gerald Ford	93
Zachary Taylor	65	Theodore Roosevelt	60	Ronald Reagan	93
Millard Fillmore	74	William Taft	72		

Back-to-Back Stem and Leaf Plots

If two sets of related data with a similar number of data values are to be compared, a *back-to-back stem and leaf plot* can be used. In this case, two plots are made: one with leaves to the right, and one with leaves to the left. For example, if Abel gave the same test to two classes, he might prepare a back-to-back stem and leaf plot, as shown in Figure 10-6, where the data for the first class are on the left and for the second class are on the right.

Abel's Class Test Scores

Second-period Class			Fifth-period Class
20	5	2	
531	6	24	
99987542	7	1257	
875420	8	4456999	
0|5| represents	1	9	2457	|5|2 represents a
a score of 50		10	0	score of 52

Figure 10-6

 NOW TRY THIS 10-3

In the stem and leaf plots in Figure 10-6, which class do you think did better on the test? Why?

 **EXAMPLE 10-2**

Group the presidents in Table 10-3 into two groups, the first consisting of George Washington to Rutherford. Hayes and the second consisting of James Garfield to Ronald Reagan.

a. Create a back-to-back stem and leaf plot of the two groups and see if there appears to be a difference in ages at death between the two groups.
b. Which group of presidents seems to have lived longer?

Solution
a. Because the ages at death vary from 46 to 93, the stems vary from 4 to 9. In Figure 10-7, the first 19 presidents are listed on the left and the remaining 19 on the right.

Ages of Presidents at Death

Early Presidents			Later Presidents
	4	96	
63	5	787	
364587	6	707034	
0741983	7	128	
3|8| represents	053	8	81	|6|7 represents
83 years old	0	9	033	67 years old

Figure 10-7

b. The early presidents seem, on average, to have lived longer because the ages at the high end, especially in the 70s through 90s, come more often from the early presidents. The ages at the lower end come more often from the later presidents. For the stems in the 50s and 60s, the numbers of leaves are about equal.

A stem and leaf plot shows how wide a range of values the data cover, where the values are concentrated, whether the data have any symmetry, where gaps in the data are, and whether any data points are decidedly different from the rest of the data.

Grouped Frequency Tables

The stem and leaf plot in Figure 10-5 naturally groups scores into intervals or **classes**. For the data in Figure 10-5, the following classes are used: 40–49, 50–59, 60–69, 70–79, 80–89, and 90–99. Each class has interval size 10 because 10 different scores can fall within each interval. Students often incorrectly report the interval size as 9 because $59 - 50 = 9$.

A **grouped frequency table** shows how many times data occur in a range. For example, consider the data in Table 10-3 for the ages of the presidents at death. These results are summarized in Table 10-4.

Table 10-4

Ages at Death	Tally	Frequency
40–49	||	2
50–59	HHT	5
60–69	HHT HHT ||	12
70–79	HHT HHT	10
80–89	HHT	5
90–99	||||	4
	Total	38

Table 10-4 shows that 12 presidents had ages at death in the interval 60–69, but it does not show the particular ages at which the presidents died. As the interval size increases, information is lost. Choices of interval size may vary. Classes should be chosen to accommodate all the data and each item should fit into only one class; that is, the classes should not overlap. Data from frequency tables can be graphed, as will be shown next.

Histograms and Bar Graphs

The data in Table 10-4 could be pictured graphically using a **histogram**. Figure 10-8(a) shows a histogram of the frequencies in Table 10-4. A histogram is made up of adjoining rectangles,

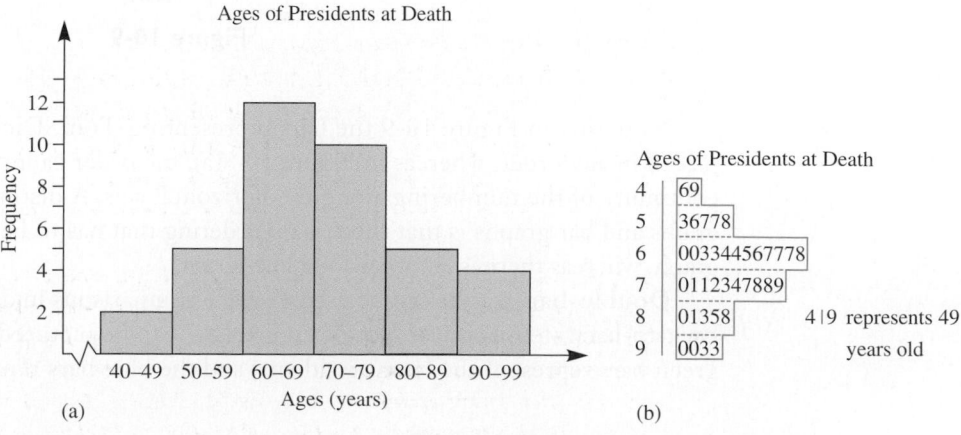

(a)

(b)

Figure 10-8

or bars. In this case, the death ages are shown on the horizontal axis and the numbers along the vertical axis give the scale for the frequency. The frequencies of the death ages are shown by the bars, which are all the same width. The scale on the vertical axis must be uniform. In some reports the midpoints of the bars in a histogram are marked instead of the intervals.

Histograms can be made easily from single-sided stem and leaf plots. For example, if we take the stem and leaf plot in Figure 10-8(b) and enclose each row (set of leaves) in a bar, we will have what looks like a histogram. Histograms show gaps and clusters just as stem and leaf plots do. However, with a histogram we cannot retrieve data as we can in a stem and leaf plot. Another disadvantage of a histogram is that it is often necessary to estimate the heights of the bars. The student page on page 559 shows a dot plot (line plot), a frequency table, and a histogram.

NOW TRY THIS 10-4

a. Work *Quick Checks* 2 and 3 on the student page.
b. In the histogram for Atlantic Ocean hurricanes on the student page do you think there should be space on the *x*-axis to represent the interval 0–2? Why?

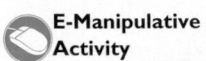
E-Manipulative Activity

Use the activity *Bar Graphs* to create and analyze data in a bar graph.

A **bar graph** is somewhat like a histogram. It typically has spaces between the bars and is used to depict categorical data. A bar graph showing the heights in centimeters of five students appears in Figure 10-9. The height of each bar represents the height in centimeters of each student named on the horizontal axis. Each space between bars is usually one-half the width of a bar.

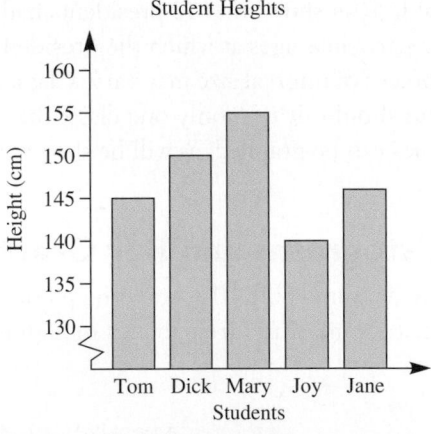

Figure 10-9

Note that in Figure 10-9 the bars representing Tom, Dick, Mary, Joy, and Jane could be placed in any order, whereas in Figure 10-8(a), the order cannot be changed without losing the continuity of the numbering along the horizontal axis. A distinguishing feature between histograms and bar graphs is that there is no ordering that has to be done among the bars of the bar graph, whereas there is an order for a histogram.

Double-bar graphs can be used to make comparisons in data. For example, the data in the back-to-back stem and leaf plot of Figure 10-7 can be pictured as shown in Figure 10-10. The green bars represent the later presidents, and the blue bars represent the earlier presidents.

School Book Page   MAKING LINE PLOTS AND HISTOGRAMS

A **line plot** is a graph that shows the shape of a data set by stacking **✗**'s above each data value on a number line.

EXAMPLE   Making a Line Plot

2 Make a line plot of the data in Example 1.

Step 1 Draw a number line from the least to the greatest value (from 2 to 11).

Step 2 Write an **✗** above each value for each time the value occurs in the data.

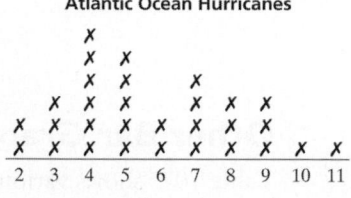

Atlantic Ocean Hurricanes

✔ **Quick Check**

2. Make a line plot of the number of students in math classes: 24 27 21 25 25 28 22 23 25 25 28 22 23 25 22 24 25 28 27 22.

Vocabulary Tip

The *histo* in histogram is short for *history*.

A **histogram** is a bar graph with no spaces between the bars. The height of each bar shows the frequency of data within that interval. The intervals of a histogram are of equal size and do not overlap.

EXAMPLE   Making a Histogram

3 Make a histogram of the data in Example 1.

Make a frequency table. Use the equal-sized intervals 2–3, 4–5, 6–7, 8–9, and 10–11. Then make a histogram.

Atlantic Ocean Hurricanes

Number	Frequency
2–3	////
4–5	//// //// /
6–7	//// /
8–9	//// /
10–11	//

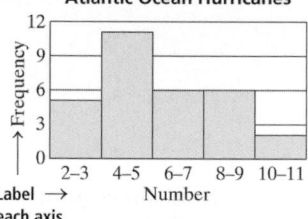

Atlantic Ocean Hurricanes

✔ **Quick Check**

3. Make a histogram of the ages of employees at a retail store: 28 20 44 72 65 40 59 29 22 36 28 61 30 27 33 55 48 24 28 32.

11-1 Reporting Frequency   **533**

Ages of Presidents at Death

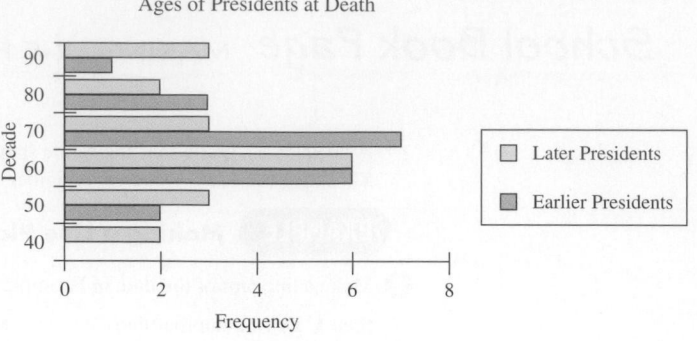

Figure 10-10

Other Bar Graphs

Table 10-5 shows various types of shoes worn by students in one class and the approximate percentages of students who wore them. Figure 10-11 depicts a **percentage bar graph** with that same data.

Table 10-5

Shoe Type	Frequency	*Percentage
Dress	1	4%
Flip-Flops	4	14%
Crocs	7	24%
Loafers	3	10%
Tennis	14	48%

*Note: percentages are approximate to sum to 100%.

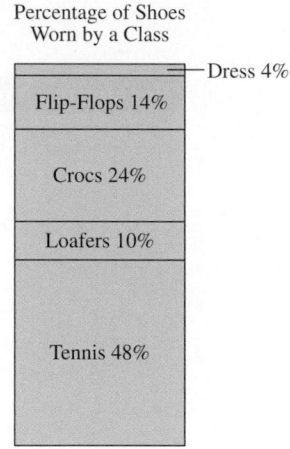

Figure 10-11

Table 10-6 shows information about the expenditures of a business over a period of years. These data are depicted in a **stacked bar graph** in Figure 10-12.

Table 10-6

Years	Materials	Labor
1970–1979	$ 795,000.00	$1,500,000.00
1980–1989	$ 950,000.00	$1,900,000.00
1990–1999	$1,230,000.00	$2,400,000.00
2000–2009	$1,500,000.00	$2,400,000.00

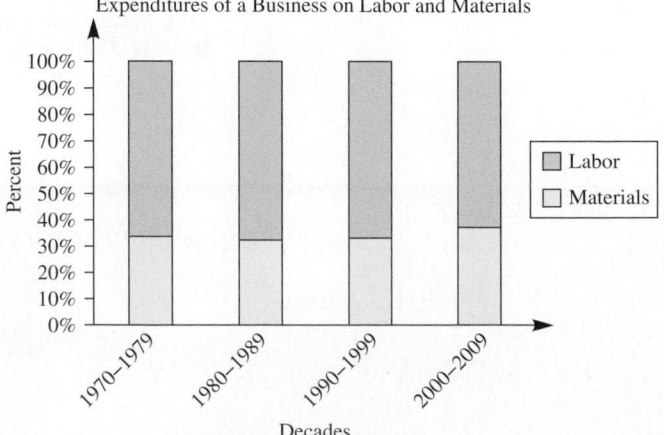

Figure 10-12

Circle Graphs (Pie Charts)

Another type of graph used to represent categorical data is the circle graph. A **circle graph**, or **pie chart**, consists of a circular region partitioned into disjoint sectors, with each sector representing a part or percentage of the whole. A circle graph shows how parts are related to the whole. An example of a circle graph is given in Figure 10-13.

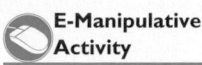

Use the activity *Pie Graphs* to create and analyze data in a pie chart.

U.S. Population by Age, 2008

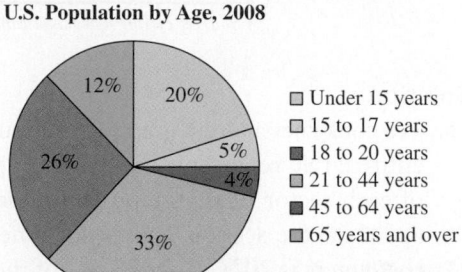

☐ Under 15 years
☐ 15 to 17 years
■ 18 to 20 years
☐ 21 to 44 years
■ 45 to 64 years
☐ 65 years and over

Source: U.S. Census Bureau. Current Population Survey, Annual Social and Economic Supplement, 2009.

Figure 10-13

In Figure 10-13 the measure of the central angle (an angle whose vertex is at the center of the circle) of each sector of the graph is proportional to the fraction or percentage of the population the section represents. For example, the measure of the angle for the sector for the under-15 group is approximately 20% of the circle. Because the entire circle is 360°, then 20% of 360°, which is approximately 72°, should be devoted to the under-15 group. Similarly, we can compute the number of degrees for each age group, as shown in Table 10-7.

Table 10-7

Age	Approximate Percent	Approximate Degrees
Under 15	20	72
15–17	5	18
18–20	4	14
21–44	33	119
45–64	26	94
65 and Over	12	43
Total	100	360

Source: U.S. Census Bureau, 2009.

EXAMPLE 10-3

The information in Table 10-8 is based on the survival and death rates of the crew and passengers in the various ticket classes on the *Titanic* when it sank. The numbers in the table reflect the numbers of people who survived or died in each class.

a. Construct circle graphs (pie charts) for the survival group and the death group to make comparisons.

b. Construct stacked bar graphs for the survival group and the death group to make comparisons.

Table 10-8

	Class				
	1st	2nd	3rd	Crew	Total
Survival	202	118	178	212	710
Death	123	167	528	673	1491

Source: HistoryontheNet.com

Solution

a. We construct a circle graph for the survival passengers based on the data in Table 10-8. The entire circle represents the total 710 people who survived. The measure of the central angle of each sector of the graph is proportional to the fraction or percentage of the passengers and crew the section represents. The measure of the angle for the sector for the first-class passengers is 202/710 or 28.5% of the circle. Because the entire circle is 360°, then 202/710 or 28.5% of 360° should be devoted to the first-class group. Similarly, we can compute the number of degrees for each group listed in Table 10-8. This is given in Table 10-9.

Table 10-9

	Survival Group		
Class	**Ratio**	**Approximate Percent**	**Approximate Degrees**
1st	$\dfrac{202}{710}$	28.5%	103°
2nd	$\dfrac{118}{710}$	16.6%	60°
3rd	$\dfrac{178}{710}$	25.1%	90°
Crew	$\dfrac{212}{710}$	29.9%	108°

The percents and degrees in Table 10-9 are only approximate. Appropriate software or a compass and protractor can be used to draw the sectors in the circle graph. Similar reasoning can be used to construct a circle graph for the death groups of the *Titanic*. The two circle graphs are given in Figure 10-14. The circle graphs show that the chance of survival depended on ticket class.

b. For the stacked bar graphs, each bar represents the whole (100%) and is divided proportionally based on the percentages given in Figure 10-14. Again, we see from the graph that the rate of survival seemed to depend on the ticket class of a passenger.

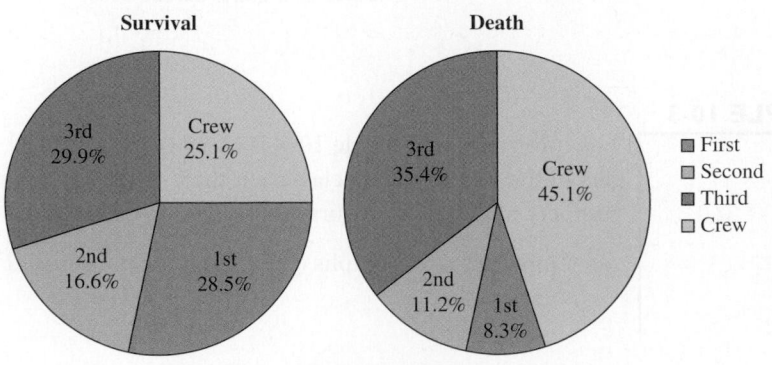

Figure 10-14

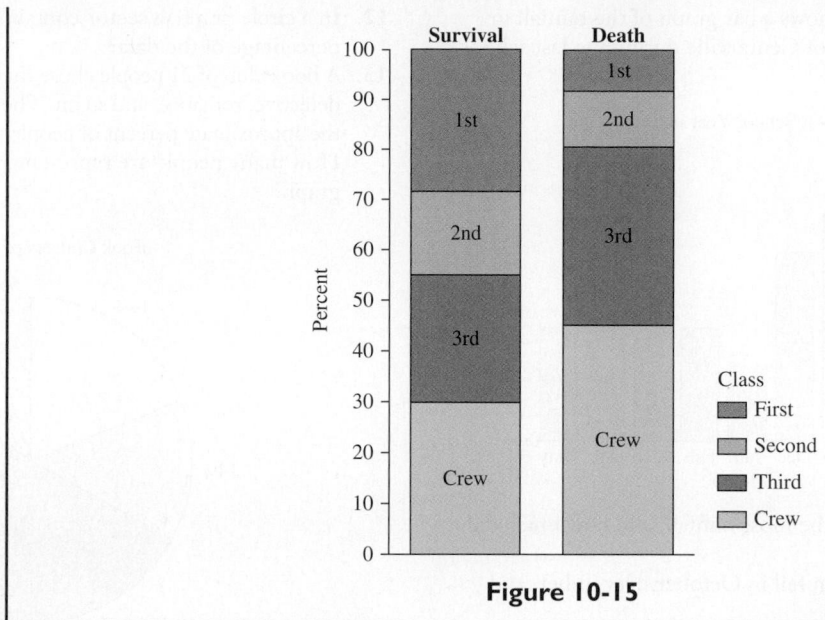

Figure 10-15

Assessment 10-2A

1. The following pictograph shows the approximate number of pieces of mail processed at the Townville post office over a five-day week.
   a. On what day were the most pieces processed?
   b. About how many more pieces of mail were processed on Tuesday than on Monday?
   c. How many symbols would be used to represent 3000 pieces of mail?

   Number of Pieces of Mail Processed by the Townville Post Office

Monday	▭ ▭ ▭
Tuesday	▭ ▭ ▭ ▭ ▭
Wednesday	▭ ▭ ▭ ▭
Thursday	▭ ▭ ▭ ▭
Friday	▭ ▭ ▭ ▭

   Each ▭ represents 1000 pieces of mail

2. Make a pictograph to represent the categorical data in the following table. Use 🥛 to represent 10 glasses of lemonade sold.

Glasses of Lemonade Sold

Day	Tally	Frequency
Monday	⊦⊦⊦⊦ ⊦⊦⊦⊦ ⊦⊦⊦⊦	15
Tuesday	⊦⊦⊦⊦ ⊦⊦⊦⊦ ⊦⊦⊦⊦ ⊦⊦⊦⊦	20
Wednesday	⊦⊦⊦⊦ ⊦⊦⊦⊦ ⊦⊦⊦⊦ ⊦⊦⊦⊦ ⊦⊦⊦⊦ ⊦⊦⊦⊦	30
Thursday	⊦⊦⊦⊦	5
Friday	⊦⊦⊦⊦ ⊦⊦⊦⊦	10

3. Students reported the number of songs they downloaded per month on their iPods. The results are given below. Draw a dot plot to display these data.

   $$3, 2, 5, 8, 2, 0, 5, 7, 2, 4, 6, 10, 6, 4, 3, 2, 4, 5, 7, 4$$

4. The following stem and leaf plot gives the weights in pounds of all 15 students in the Algebra 1 class at East Junior High:

   Weights of Students in East Junior High Algebra 1 Class

7	24
8	112578
9	2478
10	3
11	
12	35

   10|3 represents 103 lb

   a. Write the weights of the 15 students.
   b. What is the weight of the lightest student in the class?
   c. What is the weight of the heaviest student in the class?

5. Draw a histogram based on the stem and leaf plot in exercise 4.

6. Toss a coin 30 times.
   a. Construct a dot plot for the data.
   b. Draw a bar graph for the data.

7. The following figure shows a bar graph of the rainfall in centimeters in the city of Centerville during the last school year:

Rainfall Last School Year in Centerville

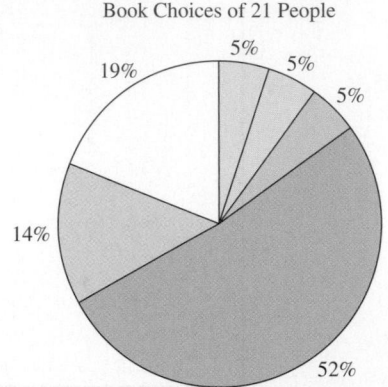

   a. Which month had the most rainfall, and how much did it have?
   b. How much total rain fell in October, December, and January?

8. HKM Company employs 40 people of the following ages:

34	58	21	63	48	52	24	52	37	23
23	34	45	46	23	26	21	18	41	27
23	45	32	63	20	19	21	23	54	62
41	32	26	41	25	18	23	34	29	26

   a. Draw a stem and leaf plot for the data.
   b. Are more employees in their 40s or in their 50s?
   c. How many employees are less than 30 years old?
   d. What percentage of the people are 50 years or older?

9. Five coins are tossed 64 times. A distribution for the number of heads obtained is shown in the following table. Draw a bar graph for the data.

Number of Heads	0	1	2	3	4	5
Frequency	2	10	20	20	10	2

10. If a 3-in. long rectangular bar represents 100% of a population, use a percentage bar graph to represent the budget of a family whose total monthly income is $4500 and who spends the following:

Rent	$1800
Food	$1500
Transportation	$200
Entertainment	$400
Utilities	$300
Savings	All that is left

11. Seventh graders at Russell School were asked to pick their favorite type of book from four possibilities. The following are the percentages of students that picked each type of book.

Favorite Types of Books

Mysteries	Biographies	Fiction	Humor
20%	15%	60%	5%

Construct a circle graph for the data.

12. In a circle graph, a sector containing 45° represents what percentage of the data?

13. A book club of 21 people chose their favorite book type, such as detective, romance, and so on. The following circle graph shows the approximate percent of people with a particular preference. How many people are represented in each sector of the circle graph?

Book Choices of 21 People

14. a. If the number of people reading mysteries in different age groups in a survey are depicted as in the bar graph given, explain whether or not you think that the graph is accurate and why.
    b. Based on the data in the graph, in what age decade do people read the most?

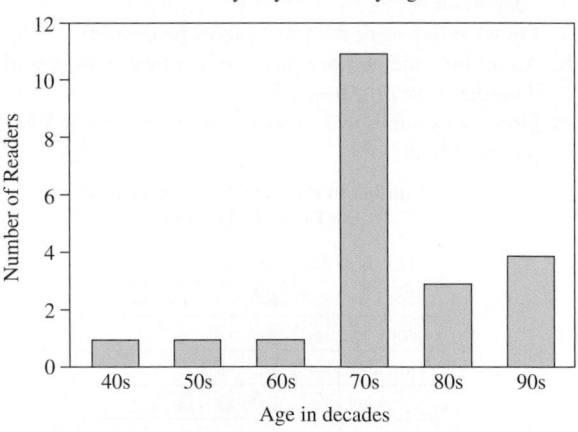

15. The following are the amounts (to the nearest dollar) paid by 25 students for textbooks during the fall term:

35	42	37	60	50
42	50	16	58	39
33	39	23	53	51
48	41	49	62	40
45	37	62	30	23

   a. Draw an ordered stem and leaf plot to illustrate the data.
   b. Construct a grouped frequency table for the data, starting the first class at $15.00 with intervals of $5.00 each.
   c. Draw a bar graph of the data.

16. The following bar graph shows the life expectancies for men and women:

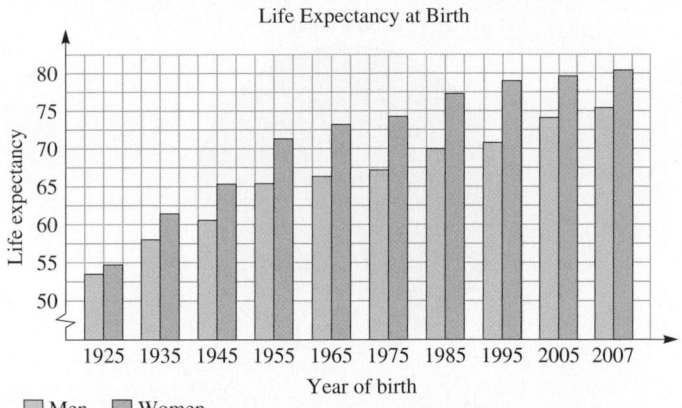

Life Expectancy at Birth

a. Whose life expectancy has changed the most since 1925?
b. In 1925, about how much longer was a woman expected to live than a man?
c. In 2007, about how much longer was a woman expected to live than a man?

17. A percentage bar graph is drawn to depict the information in the table. If the bar is 8 cm, how long is each piece?

Savings	Rent	Food	Auto Payment	Tuition
10%	30%	12%	27%	x%

18. The following bar graph shows the 2003–04 average teacher salaries reported by the National Education Association (NEA) in 2005.

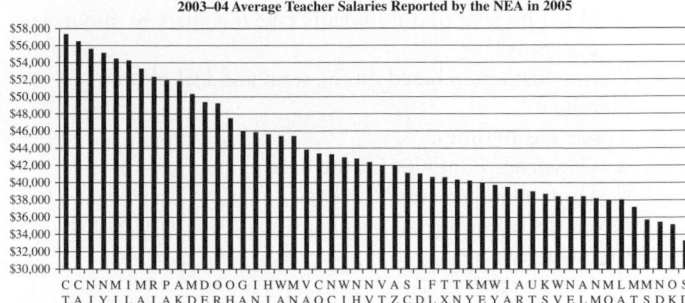

2003–04 Average Teacher Salaries Reported by the NEA in 2005

Source: NEA Teacher Salary Analysis, updated 2006. Provided courtesy of Jerry Moore, myshortpencil.com.

a. Which state pays the highest average teacher salaries?
b. Which state pays the lowest average teacher salaries?
c. Estimate the median average salary; that is, the salary where half the states pay more and half the states pay less.
d. About what percent more in average salary does Connecticut pay than South Dakota?

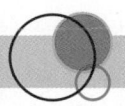

Assessment 10-2B

1. The following pictograph shows the approximate attendance at four Big Sky High School winter sporting events. Estimate how many attended each event.

Attendance at Big Sky High School Sporting Events

Boy's soccer	⬤ ⬤ ⬤ ⬤ ◗
Girl's soccer	⬤ ⬤ ◣
Boy's basketball	⬤ ⬤ ◖ ⬤ ⬤ ◖
Girl's basketball	⬤ ⬤ ⬤ ⬤

Each ⬤ represents 1000 people

2. Display the following information about the status of selected hardware sold in the United States in 2010 in a pictograph where a computer icon represents $5 million in sales.

Hardware	Sales (in millions of dollars)
Computers	26,060
Smartphones	20,607
Digital video recorders	4668
Digital cameras	2442

Hardware	Sales (in millions of dollars)
Camcorders	6345
Portable media/ MP3 players	5400

Source: Consumer Electronics Association, 2010.

3. Following are the ages of the 30 students from Washington School who participated in the city track meet. Draw a dot plot to represent these data.

10	10	11	10	13	8	10	13	14	9
14	13	10	14	11	9	13	10	11	12
11	12	14	13	12	8	13	14	9	14

4. The time it takes 15 students to get ready for school is shown in the stem and leaf plot below.

Time to Get Ready for School (in minutes)

1	5
2	0 5 8
3	0 0 0 5 9
4	0 2 5
5	0 2
6	5

6|5 represents 65 minutes

a. What is the least amount of time for an individual to get ready?

b. What is the greatest amount of time for an individual to get ready?

c. What percent of the students take less than 40 minutes to get ready?

5. Draw a histogram based on the stem and leaf plot in exercise 4.

6. Toss a die 30 times.
   a. Construct a dot plot for the data.
   b. Draw a bar graph for the data.

7. Given the following bar graph, estimate the length of the following rivers:
   a. Mississippi
   b. Columbia

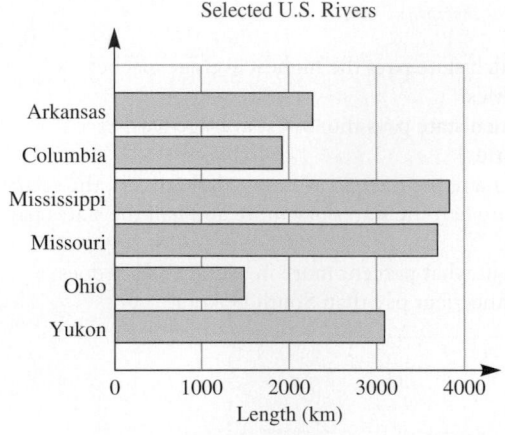

Selected U.S. Rivers

8. The data below show the number of car thefts in a city for a period of 20 days.

   50, 62, 69, 56, 52, 66, 53, 49, 72, 35
   65, 53, 78, 66, 55, 48, 82, 78, 72, 75

   a. Draw an ordered stem and leaf plot for the data.
   b. What percent of the 20 days had more than 70 cars stolen?

9. A die is tossed 50 times. A distribution for the results is given in the table below. Draw a bar graph for the data.

Number on Die	1	2	3	4	5	6
Frequency	7	9	8	7	9	10

10. The following table shows the grade distribution for the final examination in the mathematics course for elementary teachers. Draw a percentage bar graph for the data.

Grade	Frequency
A	6
B	10
C	36
D	6
F	2

11. Draw a circle graph for the data in exercise 10.

12. Use the circle graph to answer the following questions:

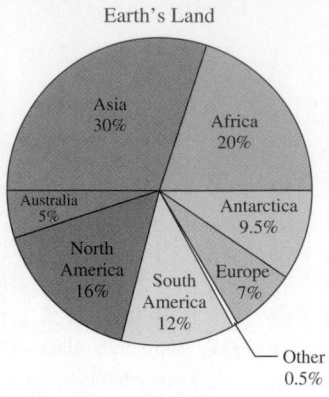

Earth's Land

a. Which is the largest continent?
b. Which continent is about twice the size of Antarctica?
c. How does Africa compare in size to Asia?
d. Which two continents make up about half of Earth's surface?
e. What is the ratio of the size of Australia to North America?
f. If Europe has approximately 4.1 million mi^2 of land, what is the total area of the land on Earth?

13. In a circle graph, a sector containing 32° represents what percentage of the data?

14. The following horizontal bar graph gives the top speeds of several animals:

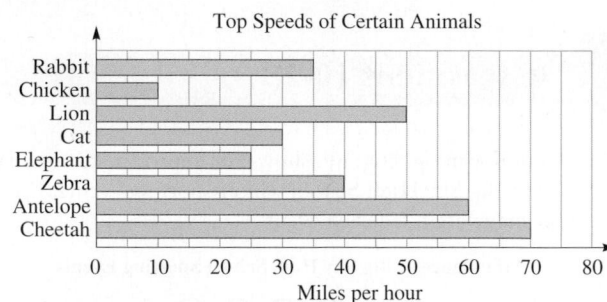

Top Speeds of Certain Animals

a. Which is the slowest animal shown?
b. How fast can a chicken run?
c. Which animal can run twice as fast as a rabbit?
d. Can a lion outrun a zebra?

15. A list of presidents, with the number of children for each, follows:

   1. Washington, 0
   2. J. Adams, 5
   3. Jefferson, 6
   4. Madison, 0
   5. Monroe, 2
   6. J. Q. Adams, 4
   7. Jackson, 0
   8. Van Buren, 4
   9. W. H. Harrison, 10
   10. Tyler, 15

   11. Polk, 0
   12. Taylor, 6
   13. Fillmore, 2
   14. Pierce, 3
   15. Buchanan, 0
   16. Lincoln, 4
   17. A. Johnson, 5
   18. Grant, 4
   19. Hayes, 8
   20. Garfield, 7

21. Arthur, 3
22. Cleveland, 5
23. B. Harrison, 3
24. McKinley, 2
25. T. Roosevelt, 6
26. Taft, 3
27. Wilson, 3
28. Harding, 0
29. Coolidge, 2
30. Hoover, 2
31. F. D. Roosevelt, 6
32. Truman, 1

33. Eisenhower, 2
34. Kennedy, 4
35. L. B. Johnson, 2
36. Nixon, 2
37. Ford, 4
38. Carter, 4
39. Reagan, 4
40. G. Bush, 5
41. Clinton, 1
42. G. W. Bush, 2
43. B. Obama, 2

a. Construct a dot plot for these data.
b. Make a frequency table for these data.
c. What is the most frequent number of children?

16. The double bar graph below shows the *2011 Quarterly Report of Income and Expenses for Acme Toy Company.*

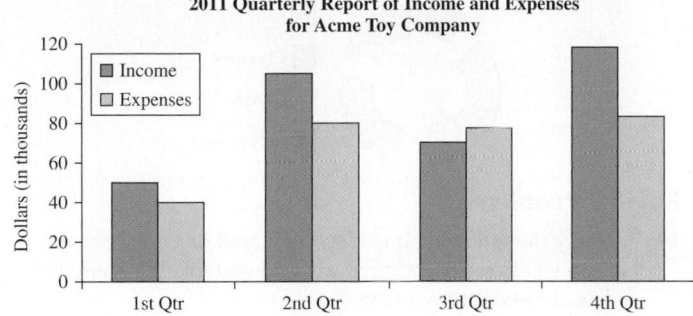

2011 Quarterly Report of Income and Expenses for Acme Toy Company

a. In what quarters did the company experience its greatest profit and the greatest loss?
b. Overall, did Acme Toy Company experience a profit or a loss for the year. why?
c. Is it true that if Acme's income is over $60,000, then it's expenses are over $60,000?

17. The table below depicts the number of deaths in the United States from Acquired Immunodeficiency Syndrome (AIDS) by age in 2003, 2005, and 2007.
    a. Choose and construct a graph to display the data.
    b. Are any patterns of difference evident in the comparison of the two groups of data?

Age (yr)	2003	2005	2007
<13	23	7	8
13–19	45	56	53
20–29	694	614	720
30–39	4217	3231	3016
40–49	7037	6632	7019
50–59	3851	4164	5102
≥60	1537	1613	2170

Source: Centers for Disease Control and Prevention, 2008.

18. The histogram below shows the number of books read by 20 people in a book club over a 3-month period.

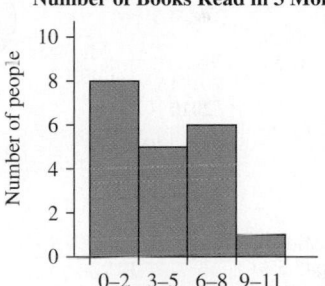

Number of Books Read in 3 Months

a. How many people read 3–5 books during this period?
b. How many people read more than 5 books?
c. What would you estimate for the average number of books read by a person in this club?

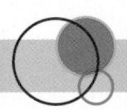

Mathematical Connections 10-2

Communication

1. a. Discuss when a pictograph might be more appropriate than a circle graph.
   b. Discuss when a circle graph might be more appropriate than a bar graph.
   c. Give an example of a set of data for which a stem and leaf plot would be more informative than a histogram.
2. Explain whether a circle graph would change if the amount of data in each category were doubled.
3. Explain why the sum of the percents in a circle graph should always be 100%. How could it happen that the sum is only close to 100%?
4. The federal budget for one year is typically depicted with one type of visual representation. Which one is used and why?
5. Tell whether it is appropriate to use a bar graph for each of the following. If so, draw the appropriate graph.
   a. U.S. population

Year	U.S. Population
1920	105,710,620
1930	122,775,046
1940	131,669,275
1950	150,697,361
1960	179,323,175
1970	203,302,031
1980	226,542,203
1990	248,765,170
2000	281,421,906
2010	308,745,538

b. Continents of the world

Continent	Area in Square Miles (mi^2)
Africa	11,694,000
Antarctica	5,100,000
Asia	16,968,000
Australia	2,966,000
Europe	4,066,000
North America	9,363,000
South America	6,886,000

6. Car sales from 1990 to 2010 for a large city are given below. Discuss the trend in car sales based on the information in the given circle graphs.

Car Sales by Vehicle Size and Type

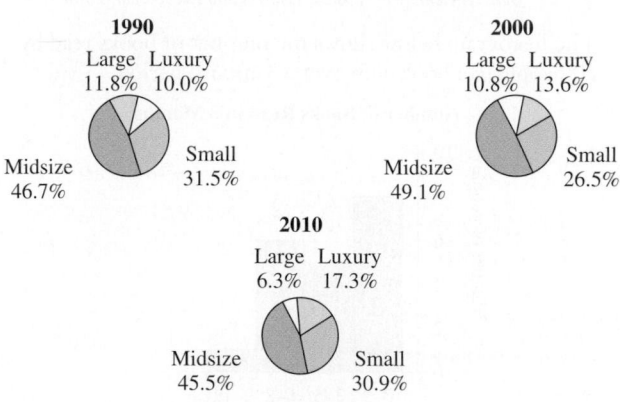

Open Ended

7. Find six recent examples of different types of visual representations of data in your newspaper. Are the representations appropriate? Explain.

8. Choose a topic, describe how you would go about collecting data on the topic, and then explain how you would display your data in a graph. Tell why you chose the particular graph.

9. A graph similar to the following one was depicted on a package of cigarettes in Canada.

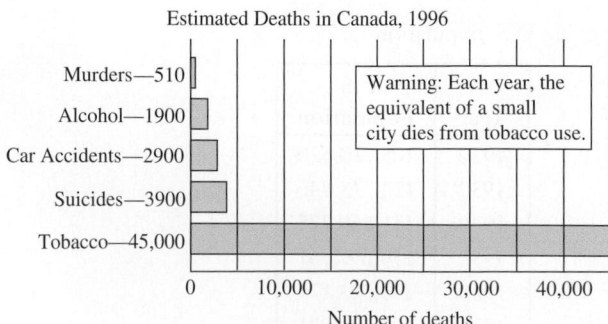

a. Make a circle graph of the data.
b. Explain which graph, bar graph or circle graph, you think is more effective.

Cooperative Learning

10. Decide on and give a rationale for the type of graph that you would use to show the percentage of their working time that professors spend on teaching, service, and research.

11. Choose one page of this text. Find the word length of every word on the page. Draw a graph depicting these data. If you chose another page of the book at random, what would you expect the most common word length to be? Why?

Questions from the Classroom

12. Jackson asks whether a stem and leaf plot should be constructed with the greatest numbers at the bottom and the least ones at the top and why this should or should not be done. How do you respond?

13. Aliene says that she constructed the following graph on a spreadsheet and wants to know if it is an acceptable type of graph to be used to depict data regarding the types of cars owned by a certain subset of the population. How do you react?

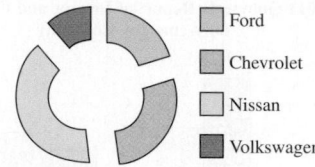

Review Problems

14. Classify the following questions as biased or fair:
   a. Do you agree or disagree with the statement "There is a need for stricter gun laws"?
   b. Many people have said there is a need for stricter laws on dangerous weapons. Do you agree?
   c. Do you like this pair of pants?
   d. You don't like this pair of pants, do you?
   e. Do you agree that this new law is a problem?
   f. Do you agree or disagree that this new law is a problem?

15. A restaurant wants to sample customers to find out their impressions of the restaurant. Which technique would provide better feedback? Why?
   i. A postcard is placed with every other bill and diners choose whether to fill it out later and mail it back.
   ii. A pollster is hired to stand outside the restaurant and ask every other bill payer some questions.

National Assessment of Educational Progress (NAEP) Questions

Final Test Scores	
Score	Number of Students
95	50
90	120
85	170
80	60
75	10

Use the information in the preceding table to complete the following bar graph.

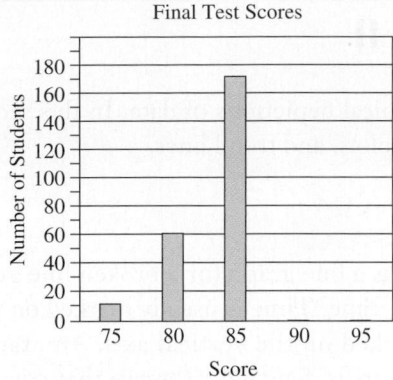

Final Test Scores

NAEP, Grade 4, 2003

- Fred planted 8 trees.
- Yolanda planted 12 trees.

Make a pictograph of the information above. Use 🌳 to represent 2 trees.

TREES PLANTED	
Fred	
Yolanda	

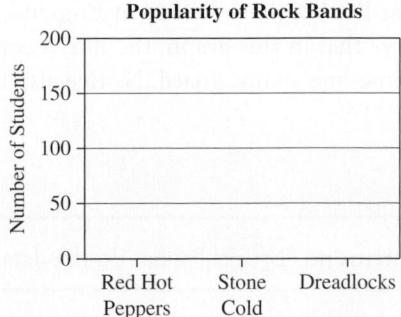

🌳 = 2 trees

NAEP, Grade 4, 2009

Trends in Mathematics and Science Study (TIMSS) Questions

The results of a survey of 200 students are shown in the pie chart.

Popularity of Rock Bands

Dreadlocks 30%
Red Hot Peppers 25%
Stone Cold 45%

Make a bar char showing the number of students in each category in the pie chart.

Popularity of Rock Bands

TIMSS, Grade 8, 2007

Four students watched the traffic passing their school for 1 hour. The table shows what they saw:

Type of Vehicle	Number
Cars	60
Bicycles	30
Buses	10
Trucks	20

Each student drew a graph to show the results. Which graph shows the results correctly?

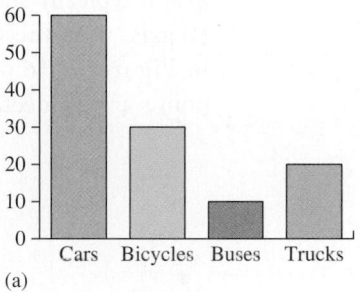

(a)

Cars 🚗🚗🚗🚗🚗
Bicycles 🚗🚗🚗🚗
Buses 🚗
Trucks 🚗🚗

1 wheel = 10 vehicles

(b)

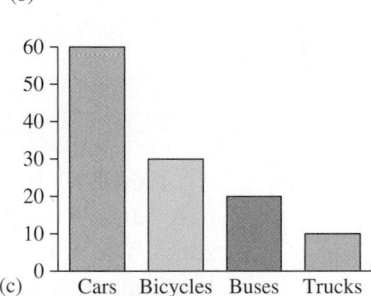

(c)

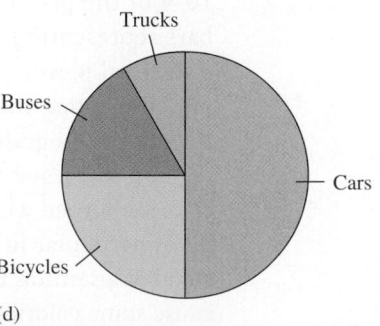

(d)

TIMSS, Grade 8, 2007

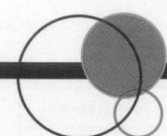

10-3 Displaying Data: Part II

In the previous section, we examined some graphical depictions of data. In this section, we explore other graphical displays: line graphs, scatterplots, and trend lines.

Line Graphs

A graphical form used to present numerical data is a line graph (or a broken line graph). A **line graph** typically shows the trend of a variable over time. Time is usually marked on the horizontal axis, with the variable being considered marked on the vertical axis. An example is seen in Figure 10-16, where Sanna's weight over 9 years is depicted. Observe that consecutive data points are connected by line segments.

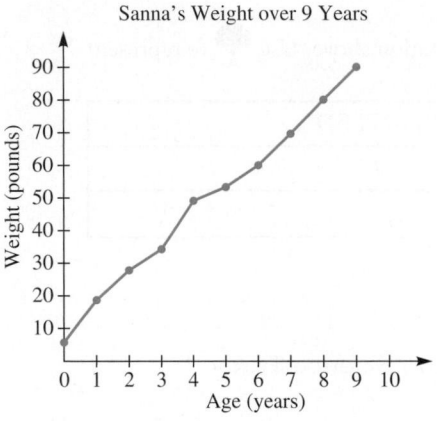

Figure 10-16

Figure 10-16 is our first example of a depiction of **continuous data**. Though the dots mark only Sanna's weight at given years, we know that she had a weight at every age along the horizontal axis. It is for this reason that it makes sense to connect the given data points with a set of line segments. Much data examined thus far have been **discrete data**. For example, in Figure 10-9 of the previous section, it would make no sense to connect the points at the tops of the bars representing the heights of five students. These are discrete data that essentially stand as individual pieces of information. They are graphed together to examine the set of information, but connecting them would have no mathematical or statistical meaning. It is sometimes done to "prettify" a graph.

On the grade 5 student page on page 571, a line graph is drawn that depicts the number of calories burned when a person is standing over a period of time. This type of information would be of particular interest to guards, for example at Buckingham Palace in England, where they stand at attention for long periods of time. Observe that in this graph, the line is continuous because some calories are being burned the entire time one stands guard. Notice also that this line is not "broken." Answer questions 1–3.

● NOW TRY THIS 10-5

Because the line depicted on the student page contains no "breaks," what do the data imply?

School Book Page LINE GRAPHS

Lesson 5-3

Key Idea
Line graphs are
used to compare
data over time.

Vocabulary
• line graph
• trend
• axes (p. 263)

Materials
• grid paper or
 **tools**

Think It Through
I can look at **the
way the graph
rises or falls** to
determine the
trend.

266

Line Graphs

LEARN

How do you read a line graph?

A **line graph** uses data points connected
with line segments to represent data
collected over time.

A line graph is often used to show
a **trend** or general direction in data.
The appearance of the graph indicates
if the data numbers are increasing
or decreasing.

The graph below shows how many
calories are burned over a period of
4 hours. To read the graph, locate
a point and read the numbers on
both **axes.**

WARM UP

1. 240 − 138

2. 260 + 400

3. 220 − 160

To Determine a Trend
• If the part of a line
between two points is
rising from left to right,
the data numbers are
increasing.
• Similarly, if the part of
the line between the two
points is falling from left
to the right, the data
numbers are decreasing.

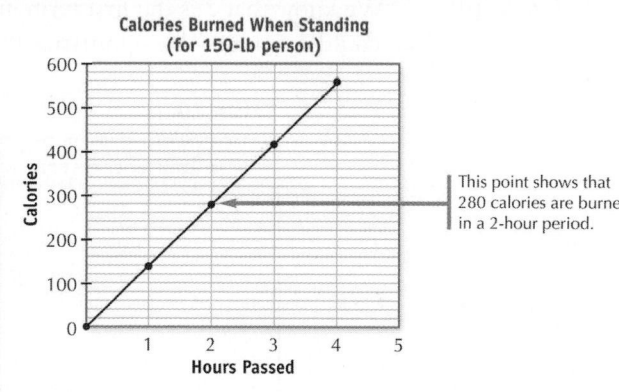

This point shows that
280 calories are burned
in a 2-hour period.

✔ Talk About It

1. About how many calories are burned by
   a 150-lb person who stands for 3 hours?

2. Is the graph rising or falling from left to
   right? Describe the trend that is shown.

3. **Representations** A 150-lb person burns
   100 calories each hour when sitting. If
   after 2 hours of standing up, a person of
   this weight were to then sit for 2 hours,
   how would the graph change?

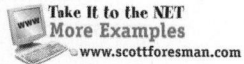
**Take It to the NET
More Examples**
www.scottforesman.com

Scatterplots

Sometimes a relationship between variables cannot be easily depicted by even a broken line. Frequently, a **scatterplot** is used. Figure 10-17 shows a scatterplot depicting the relationship between the number of hours studied and quiz scores. The highest score is a 10 and the lowest is 1.

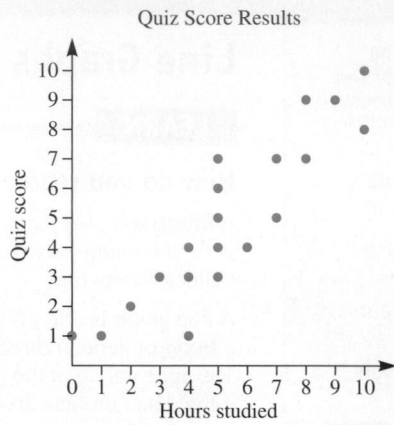

Figure 10-17

In a scatterplot like that in Figure 10-17, we frequently try to discern patterns, but before that problem is approached, we consider a simpler discrete set of data that second or third graders may recognize. Imagine an arithmetic sequence determined by skip counting by 2 to obtain 2, 4, 6, 8, 10, ... We know that 2 is the first term in the sequence, 4 is the second term, 6 is the third term, and we could continue. We summarize the data in Table 10-10.

Table 10-10

Number of the Term	Term
1	2
2	4
3	6
4	8
5	10
6	12
7	14
8	16
9	18
10	20

In earlier chapters, we described the value of the term using two different methods: a recursive method, and a closed-form method. In the recursive method, we recognized the following:

First term is 2.
Each next term is 2 more than the previous term.

In the closed-form method, we observed the constant difference of 2 in the terms and learned that this was an arithmetic sequence such that the value of the term was twice the number of the term. We wrote this as $f(n) = 2n$, where n was the number of the term and y is the value of the term.

Now consider what might happen if we plotted the same data with a scatterplot, as in Figure 10-18.

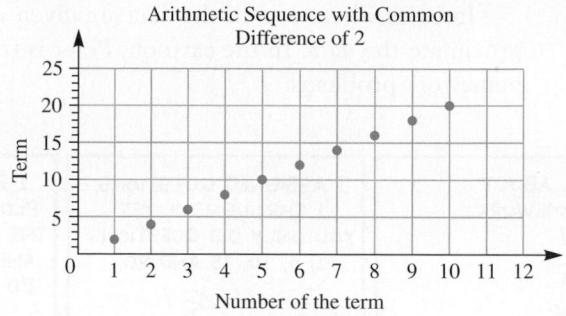

Figure 10-18

Note that the graph of Figure 10-18 is depicted as a set of discrete data in a scatterplot. But this scatterplot is unlike the one of Figure 10-17. All of the points in this scatterplot appear to lie along a line. If we draw a representation of the line, as in Figure 10-19, then we have depicted the data as a set of continuous data.

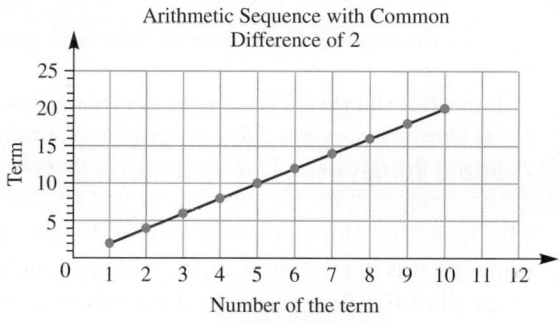

Figure 10-19

Putting together what we know from earlier chapters with what we have learned about graphs, we could say that a point would lie along this line (or be on this line) if the term is twice the term number. The second, or "y-coordinate," is twice the first, or "x-coordinate." (Note that this description makes sense only if the vertical axis is the "y-axis" and the horizontal axis is the "x-axis.") Generalizing to continuous data, we have $y = 2x$ for all x's that are real numbers. With this final representation, we moved from the discrete case, and in the process lost the sequence notation, but for any value of x given, we could predict the corresponding value of y. The line itself is a **trend line**, *a line that closely fits the data and can be used to describe it.* In this case, all the given data fall on this line and we have a complete fit of the data by the line.

NOW TRY THIS 10-6

Consider a sequence of data points formed using the following recursive formula:

The first term is $^-10$ and thereafter each term is 4 more than the previous term.

a. Construct a table of values for this recursive relation.
b. Draw a scatterplot for the relation with the trend line.
c. Identify by formula the "best" trend line for the data.

In cases where not all the data are given we try to find a trend line that can be used to approximate the data. In the cartoon, Peter is trying to use this idea to reduce the number of his homework problems.

In the grade 8 *Common Core Standards* we find that students should:

Know that straight lines are widely used to model relationships between two quantitative variables. For scatter plots that suggest a linear association, informally fit a straight line, and informally assess the model fit by judging the closeness of the data points to the line. (p. 56)

In the scatterplot of Figure 10-20(a) for the quiz scores, is it possible to identify a "good" trend line for this data? And if so, how might it be done? Clearly, there is no single line that can contain all of the data points in the graph of Figure 10-20(a). However, we could consider moving a line across the data to define a line that follows the general pattern of the data. If we used this technique with Figure 10-20(a), we might produce a trend line as seen in Figure 10-20(b). There could be a better trend line as seen in Figure 10-21. Could we find the equation of this line?

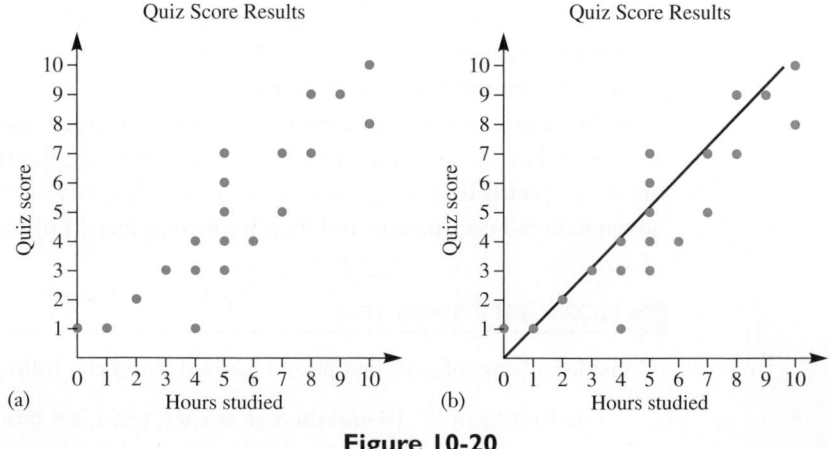

Figure 10-20

Attempting to find an equation of the line in Figure 10-20(b), we realize that the line approximately goes through points with coordinates (2, 2) and (7, 7). The slope of the line through the points is $\frac{7-2}{7-2}$ or 1. The line also goes through (0, 0) so the *y*-intercept is 0 and the line has equation $y = 1 \cdot x + 0$ or $y = x$.

Quiz Score Results

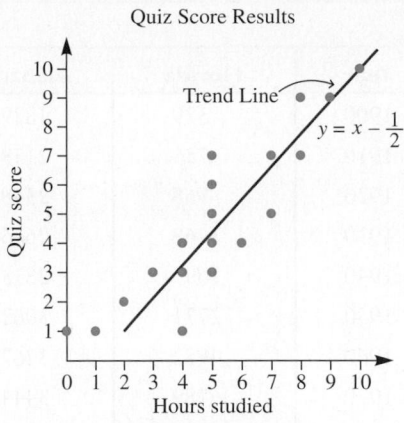

Figure 10-21

If we plot the equation $y = x$ on the graph, we see that it is a bit higher than a "better" trend line, as shown in Figure 10-21. We could "lower" the line in Figure 10-20(b) by "moving" it down $\frac{1}{2}$ unit, yielding the equation $y = x - \frac{1}{2}$, as seen in Figure 10-21.

A trend line can be used to make predictions. If a trend line slopes up from left to right, as in Figure 10-21, then there is a ***positive association*** between the number of hours studied and the quiz score. From the trend line in Figure 10-21, we would predict that students who studied 7 hours might score about $6\frac{1}{2}$.

In Figure 10-21, note that the student who studied 4 hours and received a score of 1 did no better than the student who did not study. Although there is a possible association between studying and scoring well on a quiz, we cannot deduce cause and effect based on scatterplots and trend lines.

If the trend line slopes downward to the right, we can also make predictions; there is a ***negative association***. If the points do not approximately fall about any line, there is ***no association***. Scatterplots also show clusters of points and outliers. Examples of various associations are given in Figure 10-22.

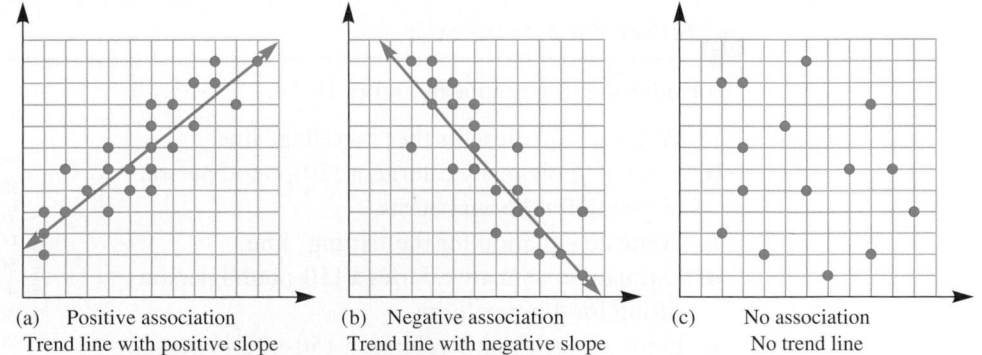

(a) Positive association
Trend line with positive slope

(b) Negative association
Trend line with negative slope

(c) No association
No trend line

Figure 10-22

Multiple-Line Graphs

Frequently, **multiple-line graphs** are used to demonstrate different sets of data where comparisons may be made. For example, Figure 10-23 shows the population growth in Florida, Georgia, Alabama, Mississippi, and Louisana between 1900 and 1990. Notice the difference in population growth for Florida over this time period in comparison to the other states.

State					
Year	Florida	Alabama	Georgia	N. Carolina	S. Carolina
1900	529	1829	2216	1894	1340
1910	753	2138	2609	2206	1515
1920	968	2348	2896	2559	1684
1930	1468	2646	2909	3170	1739
1940	1897	2833	3124	3572	1900
1950	2771	3062	3445	4062	2117
1960	4952	3267	3943	4556	2383
1970	6789	3444	4590	5082	2591
1980	9746	3894	5463	5882	3122
1990	12938	4040	6478	6632	3486

Population (thousands) (left axis label for table)

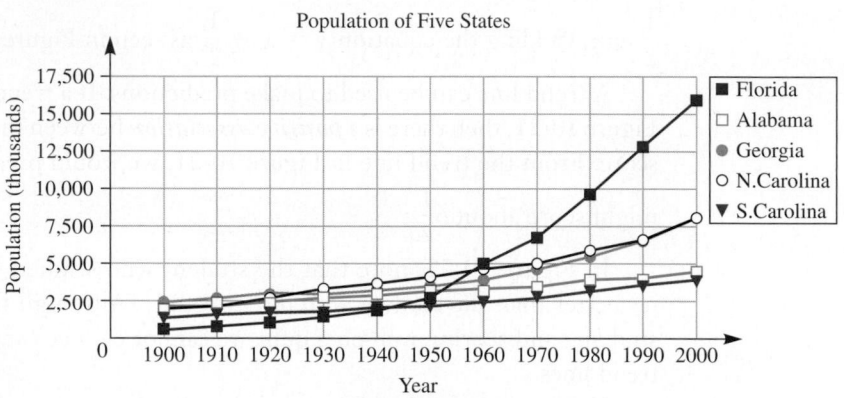

Population of Five States

Figure 10-23

 The student page on page 577 shows double bar graphs and double line graphs depicting the same data. Answer questions 1–4 on the student page.

 NOW TRY THIS 10-7

Consider the line graphs in Figure 10-24.

a. Write an equation for the "bicycling" line.
b. Estimate how many calories a 150-pound person bicycling for 2 hours burns.
c. Write an equation for the "sitting" line.
d. Estimate how many calories a 150-pound person sitting for 4 hours burns.
e. Estimate how many calories a 150-pound person bicycling for 4 hours and then sitting for 3 hours burns.
f. Describe a line graph for burning calories if a 150-pound person sits for an hour and then bikes for 3 hours.

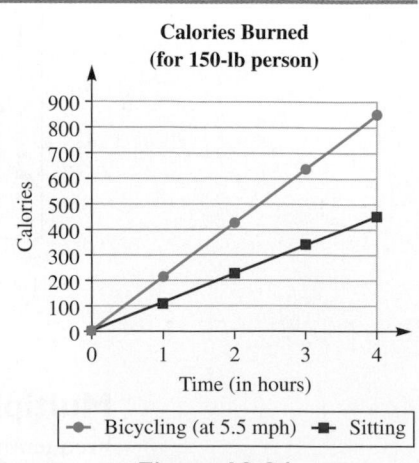

Calories Burned
(for 150-lb person)

Bicycling (at 5.5 mph)   Sitting

Figure 10-24

School Book Page DOUBLE BAR AND LINE GRAPHS

Extension

For Use With Lesson 2-4

Double Bar and Line Graphs

You can plot two data sets on the same graph to compare them easily.

EXAMPLE

A bookstore tracks sales of cooking and travel books. Make a double bar graph and a double line graph of the data at the right.

Use a different color for each data set in the graphs.

Books Sold Each Month

Month	Jan.	Feb.	Mar.	Apr.
Cooking	86	98	112	110
Travel	100	106	88	102

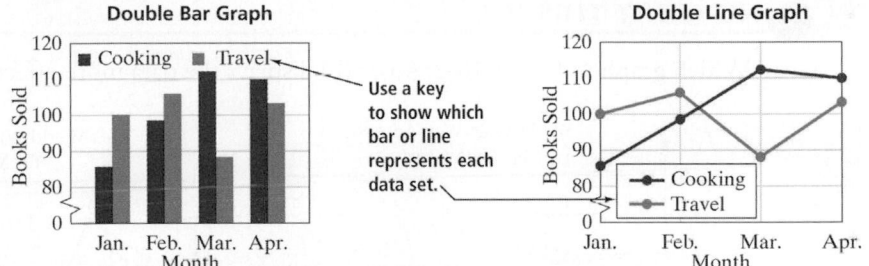

Use a key to show which bar or line represents each data set.

Exercises

Use the example above for Exercises 1 and 2.

1. Which graph shows most clearly how book sales changed from January to February?

2. Which graph shows most clearly the differences in sales between the two types of books?

Use the table at the right for Exercises 3 and 4.

3. Make a double bar graph. Show the differences between the numbers of endangered plants and animals.

4. Make a double line graph. Show how the numbers of endangered plants and animals have changed over time.

Endangered Species in the United States

Year	1985	1990	1995	2000	2005
Plants	93	179	432	592	599
Animals	207	263	324	379	389

Source: U.S. Fish and Wildlife Service.
Go to PHSchool.com for a data update.
Web Code: aqg-9041

Extension Double Bar and Line Graphs **79**

Choosing a Data Display

Choosing an appropriate data display is not always easy, as seen in the student page on page 579. Each type of graph is suitable for presenting certain kinds of data. In this chapter, we explored pictographs, dot plots, stem and leaf plots, histograms, bar graphs, line graphs, scatterplots, and circle graphs (pie charts). Some types of display and uses follow:

Bar graph—Used to compare numbers of data items in grouped categories; order of categories does not matter except for convenience.

Histogram—Used to compare numbers of data items grouped in numerical intervals; order matters in the data depicted.

Stem and leaf plot—Used to show each value in a data set and to group values into intervals.

Scatterplot—Used to show the relationship between two variables.

Line graph—Used to show how data values change over time; normally used for continuous data.

Circle graph—Used to show the division of a whole into parts.

 Study the example on the student page on page 579 and work the exercises to determine the best type of data display for each of the situations.

> ● **NOW TRY THIS 10-8**
>
> ---
>
> **a.** Which graph in Figure 10-25(a) or (b) displays the data more effectively? Why?

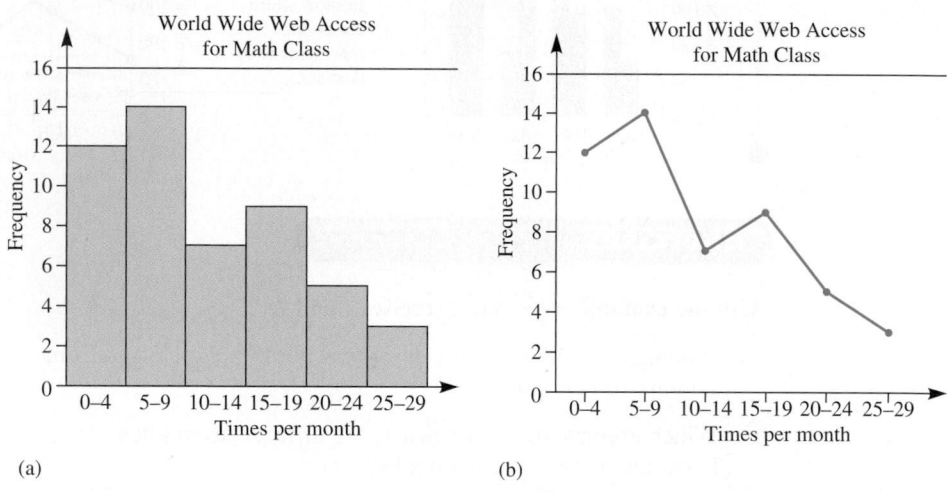

Figure 10-25

> *Note:* Figure 10-25(b) was formed by connecting the midpoints of the tops of the bars in Figure 10-25(a).
>
> **b.** Explain whether connecting the dots with line segments as in Figure 10-25(b) is meaningful.
> **c.** To show each of the following, which graph is the best choice: line graph, bar graph, or circle graph? Why?
> > **i.** The percentage of a college student's budget devoted to housing, clothing, food, tuition and books, taxes
> > **ii.** Showing the change in the cost of living over the past 12 months

Activity Lab CHOOSING THE BEST DISPLAY

11-3b **Activity Lab**   **Data Analysis**

Choosing the Best Display

When you display data, you should consider which type of display best represents the data.

EXAMPLE

A city council conducts a survey to decide how to develop a new public park. The council asks people to choose one of four park uses, as shown in the table. Choose the best display to represent the data. Explain your choice.

City Park Use Survey Results

Use	18 and Under	Over 18
Tennis	72	86
Basketball	114	95
Skate Park	173	57
Garden	48	139

Step 1 Summarize the purpose of the data display. The display must compare data for two age groups.

Step 2 Narrow down your options. There are too many responses to use a line plot. Each person chose one use for the park. Since there is no overlap, you can eliminate a Venn diagram. You are not showing the distribution of data over a range, so you can eliminate a stem-and-leaf plot. There is no change over time, so you can eliminate a line graph.

Step 3 Choose one of the remaining displays. Both a circle graph and a double bar graph can compare parts to wholes. In this situation, the table compares two sets of data across the same four categories, so the best option is a double bar graph.

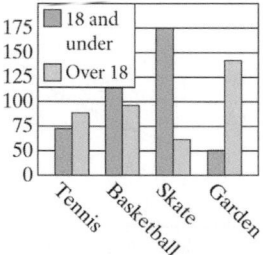

Exercises

Reasoning Match each data set at the left with the best type of data display at the right. Explain your choices.

1. the number of students who hold part-time jobs, tutor, play sports, or do a combination of all three activities

2. favorite pizza topping, by percent of students in a class

3. the popularity of different bikes among boys and among girls

4. changes in desktop and laptop computer prices over time

5. the height in centimeters of each student in a class

6. the number of letters in the first names of students in your class

A. line plot
B. double line graph
C. double bar graph
D. stem-and-leaf plot
E. circle graph
F. Venn diagram

548   **Activity Lab**   Choosing the Best Display

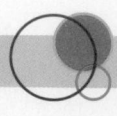

Assessment 10-3A

1. The following graph shows how the value of a car depreciates each year. This graph allows us to find the trade-in value of a car for each of 5 years. The percents given in the graph are based on the selling price of the new car.

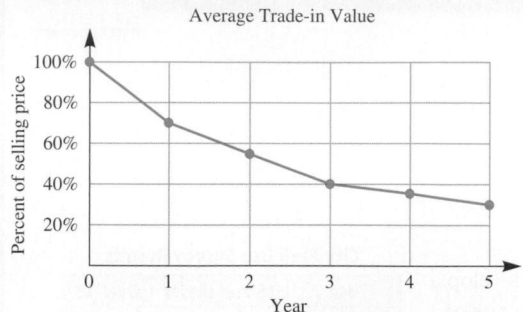

Average Trade-in Value

a. What is the approximate trade-in value of a $12,000 car after 1 yr?
b. How much has a $20,000 car depreciated after 5 yr?
c. What is the approximate trade-in value of a $20,000 car after 4 yr?
d. Dani wants to trade in her car before it loses half its value. When should she do this?

2. The graph below shows the population of the United States in the years 1790–1830.

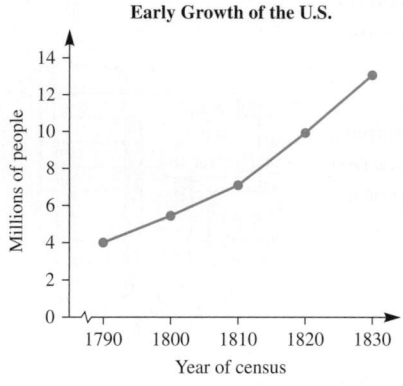

Early Growth of the U.S.

a. Estimate what year the population was about 6,000,000.
b. Estimate the population in 1830.
c. Did the U.S. population increase more between 1790 and 1800 or between 1810 and 1820? How can you tell?

3. The graph below shows the number of snow shovels sold in winter months at a store in 2010 and 2011.

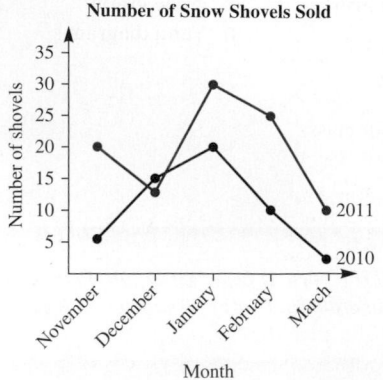

Number of Snow Shovels Sold

a. During which month did the store sell the most snow shovels in 2011?
b. During which month did the store sell the fewest snow shovels in 2010?
c. Based on snow shovel sales, which year, 2010 or 2011, seems to have had the most snow in the store's locale? Why?

4. On the partial student page shown, answer questions 8–9.

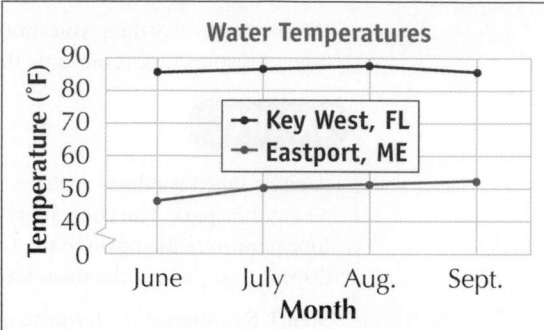

Water Temperatures

8. During which month is the water temperature in Eastport the highest?

   A. June   C. August

   B. July   D. September

9. Which statement is TRUE?

   A. The water temperature is higher in Maine than in Florida.

   B. The water temperature in both cities increases from June to September.

   C. In the summer, the water temperature in Eastport is greater than 50°F.

   D. The water temperature in Key West decreases from August to September.

Source: p. 684; From MATHEMATICS. Copyright © 2008 Pearson Education, Inc., or its affiliates. Used by permission. All Rights Reserved.

5. Coach Lewis kept track of the basketball team's jumping records for a 10-year period, as shown below. Draw a line graph for the data.

Year	2003	2004	2005	2006	2007	2008
Record (nearest in.)	65	67	67	68	70	74

Year	2009	2010	2011	2012
Record (nearest in.)	77	78	80	81

6. Refer to the following scatterplot regarding movie attendance in a certain city:

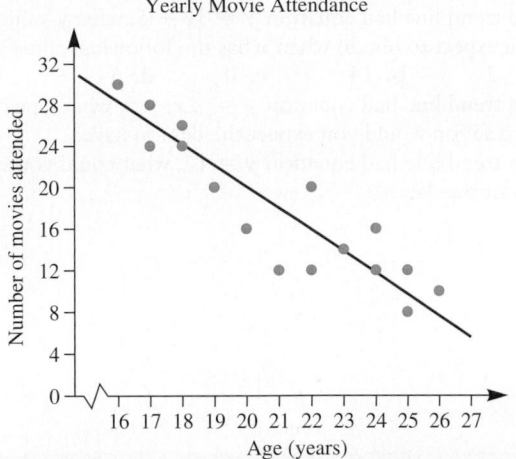

Yearly Movie Attendance

  a. What type of association exists for these data?
  b. About how many movies does an average 25-year-old attend?
  c. From the data in the scatterplot, conjecture how old you think a person is who attends 16 movies a year.

7. Given an "add 4" sequence with a first term of 2, do the following:
  a. Plot this sequence as a scatterplot.
  b. Sketch a trend line.
  c. Find an equation for the trend line you sketched.

Note: In exercises 8 and 9, numbers marking axes are moved to show dots.

8. For each of the scatterplots below, answer the following:
  i. What type of association, if any, can you identify?
  ii. Is there an identifiable trend line? If so, sketch it.
  a.

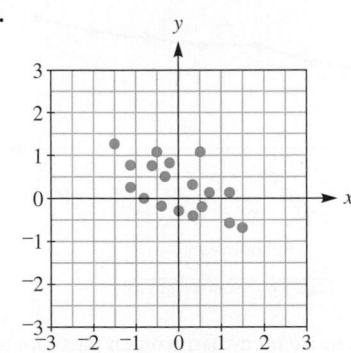

  b.

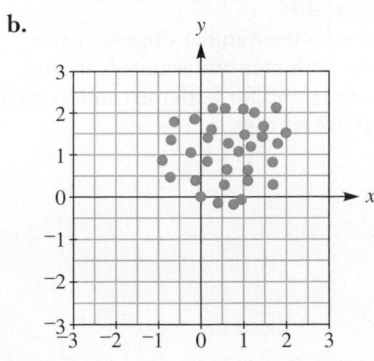

9. In the following scatterplots, estimate an equation of the trend line pictured:
  a.

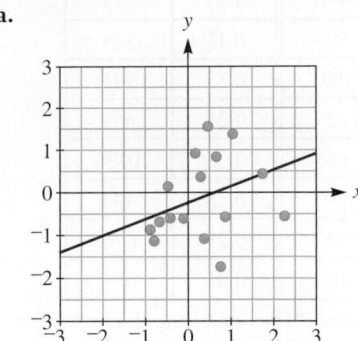

  b.

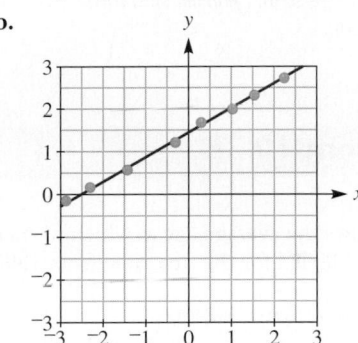

  c.

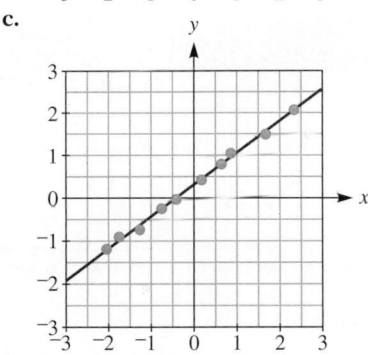

10. The following data show the cost of various diamonds per carat:

Carats	Cost($)	Carats	Cost($)
0.17	355	0.17	353
0.16	328	0.18	438
0.17	350	0.17	318
0.18	325	0.18	419
0.25	642	0.17	346
0.16	342	0.15	315
0.15	322	0.17	350
0.19	485	0.32	918
0.21	483	0.32	919
0.15	323	0.15	298
0.18	462	0.16	339
0.28	823	0.16	338
0.16	336	0.23	595

(continued)

Carats	Cost($)	Carats	Cost($)
0.2	498	0.23	553
0.23	595	0.17	345
0.29	860	0.33	945
0.12	223	0.25	655
0.26	663	0.35	1086
0.25	750	0.18	443
0.27	720	0.25	678
0.18	468	0.25	675

Source: Singfat Chu. "Diamond Ring Pricing Using Linear Regression." *Journal of Statistics Education* 4 (1996).

Draw a scatterplot of the data and determine if there is a trend line that could be used to predict the cost of 0.5 carat of a diamond.

11. If a trend line had equation $y = 3x + 5$, what y-value would you expect to obtain when x has the following values?
    a. $^-2$ **b.** 14   **c.** 0   **d.** 5
12. If a trend line had equation $y = {}^-2x - 5$, what type of association would you expect the data to have?
13. If a trend line had equation $y = 12$, what could you say about the data?

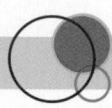

Assessment 10-3B

1. The graph below shows the average age at which women in the United States married for the first time from 1890 to 2010.

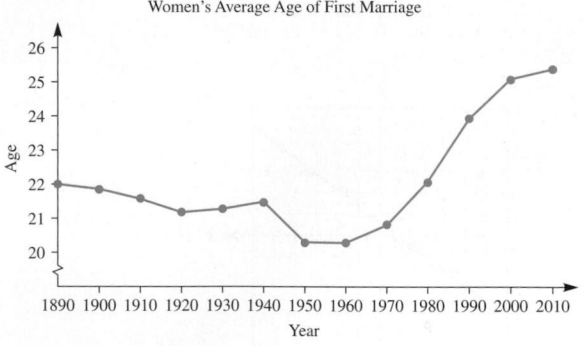

Source: U.S. Census Bureau, 2010.

   a. What was the approximate average age for a woman's first marriage in
   i. 1890   **ii.** 1950   **iii.** 2010
   b. During what 10-year period was there the greatest decrease in the average age of first marriage?
   c. During what 10-year period was there the greatest increase in the average age of first marriage?
2. Answer the following questions based on the line graph below.

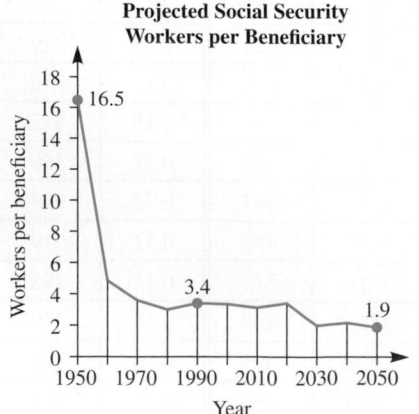

Source: Social Security Administration

   a. What does the line graph tell us about Social Security after 2020?
   b. About how many workers were there per beneficiary in 1960?
   c. About how many workers will there be per beneficiary in 2030?
3. The graph below shows line graphs for participation in National Collegiate Athletic Association (NCAA) sports for both men and women.

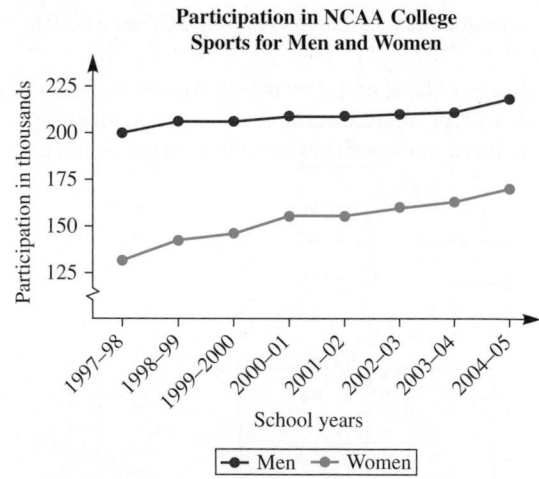

   a. About how many more men than women participated in NCAA sports in 2004–05?
   b. Between what years did the number of participants slightly decrease for both men and women's sports?
   c. What seems to be the trend for both men and women's participation in NCAA sports?

4. On the partial student page shown, answer questions 7–9.

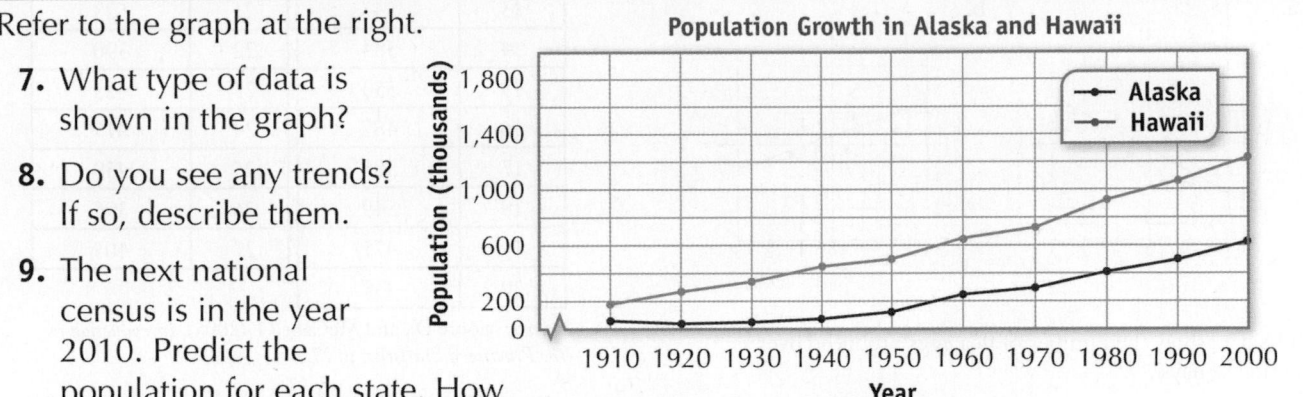

Refer to the graph at the right.

7. What type of data is shown in the graph?

8. Do you see any trends? If so, describe them.

9. The next national census is in the year 2010. Predict the population for each state. How did you make your prediction?

Population Growth in Alaska and Hawaii

Source: Scott Foresman-Addison Wesley Mathematics, Grade 6, 2008 (p. 639).

5. Use the following data to draw a line graph. Use the graph to see what might be determined about the temperature of a patient in the clinic.

Time	Temperature in Degrees Fahrenheit
6:00 A.M.	96.8
7:00 A.M.	98.8
8:00 A.M.	99.9
9:00 A.M.	99.8
10:00 A.M.	100.2
11:00 A.M.	101.2
Noon	102.0

6. Refer to the following scatterplot regarding fuel efficiency.

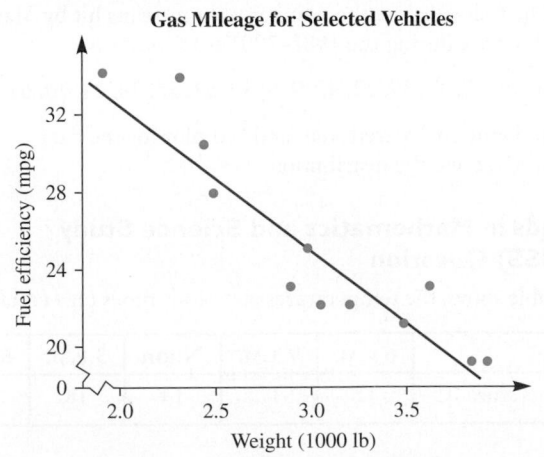

Gas Mileage for Selected Vehicles

a. What type of association exists for these data?
b. About how many miles per gallon would you expect for a 2000-pound vehicle?

7. Given an "add 5" sequence with a first term of 3, do the following:
   a. Plot this sequence as a scatterplot.
   b. Sketch a trend line.
   c. Find an equation for the trend line you sketched.

Note: In exercises 8 and 9, numbers marking the axes are moved to show dots.

8. Use the scatterplots shown to answer the following:
   i. What type of association, if any, can you identify?
   ii. Sketch a trend line for the data if one exists.

a.

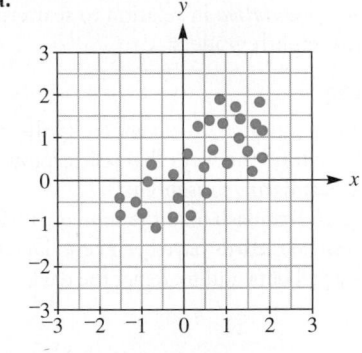

b.

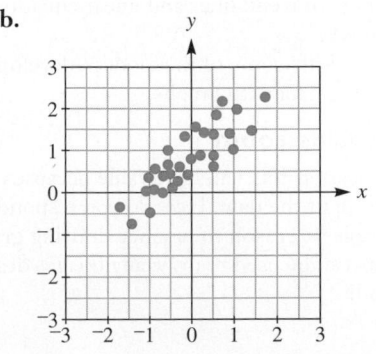

9. In the following scatterplots, find an equation to estimate trend line pictured:

a.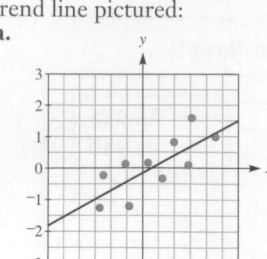

b.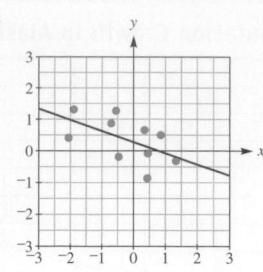

10. The data show the number of flea eggs produced over a period of days.

Day	No. of Eggs	Day	No. of Eggs
1	436	7	945
2	495	8	655
3	575	9	782
4	444	10	704
5	754	11	590
6	915	12	411

(continued)

Day	No. of Eggs	Day	No. of Eggs
13	547	21	523
14	584	22	390
15	550	23	425
16	487	24	415
17	585	25	450
18	549	26	395
19	475	27	405
20	435		

Source: Moore D., and McCabe G. (2005). *Introduction to the Practice of Statistics*, p. 27.

a. Draw a line graph of the given data.
b. Predict how many eggs there might be on day 28.

11. If a trend line has equation $y = 2x - 7$, what y-value would you expect to obtain when x has the following values?
a. $^-2$ b. 14   c. 0   d. 5

12. If a trend line has equation $y = ^-2x + 5$, what type of association would you expect the data to have?

13. If a trend line has an equation $y = 15$, what could you say about the data?

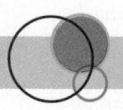

Mathematical Connections 10-3

Communications

1. Write an explanation of *association* in relation to scatterplots that could be used with eighth graders.

Cooperative Learning

2. With a group, examine a set of sixth- through eighth-grade textbooks to determine the ways that scatterplots, associations, and line graphs are presented.
3. Consider the women's 100-yd dash (and the subsequent 100-m dash) in the Olympics since records have been kept. Determine the best way to combine all data and to depict the data.

Open-Ended

4. Report on the coverage of trend lines and linear equations in an eight-grade textbook.
5. Use the Internet to identify some of the modern developers of methods of depicting data with graphs.

Questions from the Classroom

6. Jacquie argued that scatterplots had little value because rarely does a single trend line fit the data. How do you respond?
7. Merle says that there is no reason to practice drawing graphs because spreadsheets can be used to draw any images desired. How do you respond?

Review Problems

8. The Smith family drew a circle graph of their budget that contained the following:

Taxes, 20%
Rent, 32%
Food, 20%
Utilities, 5%
Gas, 13%
Miscellaneous, 12%

What would you tell the family concerning the data?

9. Adjust the miscellaneous percentage in exercise 8 to 10% and draw a stacked bar graph of the entire data set.
10. The following are the number of home runs hit by Mark McGwire during the 1982–2001 seasons.

3, 49, 32, 33, 39, 22, 42, 9, 9, 39, 52, 58, 34, 24, 70, 65, 32, 29

a. Draw an ordered stem and leaf plot for the data.
b. Describe the distribution.

Trends in Mathematics and Science Study (TIMSS) Question

The table shows the temperatures at various times on a certain day.

Time	6 A.M.	9 A.M.	Noon	3 P.M.	6 P.M.
Temperature °C	12	17	14	18	15

A graph, without a temperature scale, is drawn. Of the following, which could be the graph that shows the information given in the table?

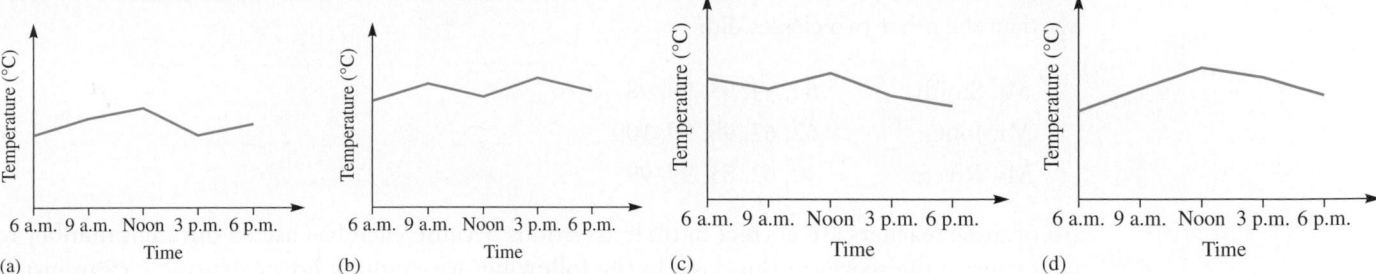

(a) (b) (c) (d)

TIMSS, Grade 8, 2007

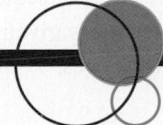

10-4 Measures of Central Tendency and Variation

In *PSSM*, we find the following by grade bands:

> In grades 3–5 all students should . . .
>
> • use measures of center, focusing on the median, and understand what each does and does not indicate about the data sets;
> • compare different representations of the same data and evaluate how well each representation shows important aspects of the data. (p. 400)
>
> In grades 6–8 all students should . . .
>
> • find, use, and interpret measures of center and spread, including mean and interquartile range. (p. 401)

The grade 6 *Common Core Standards* suggest that all students should be able to:

> Summarize numerical data sets in relation to their context, such as by: . . . Giving quantitative measures of center (median and/or mean) and variability (interquartile range and/or mean absolute deviation), as well as describing any overall pattern and any striking deviations from the overall pattern with reference to the context in which the data were gathered. (p. 45)

The media present a daily variety of data and statistics. For example, we may find that the average person's lifetime includes 6 years of eating, 4 years of cleaning, 2 years of trying to return telephone calls to people who never seem to be in, 6 months waiting at stop lights, 1 year looking for misplaced objects, and 8 months opening junk mail. In the previous section, we examined data by looking at graphs to display the overall distribution of values. In this section, we describe specific aspects of data by using a few carefully chosen numbers. These numbers will help in the analysis of data. Two important aspects of data are their *center* and their *spread*. The mean and median are **measures of central tendency** that describe where data are centered. Each of these measures is a single number that describes the data. The *range, interquartile range, variance, mean absolute deviation*, and *standard deviation* introduced later in this section describe the spread of data and should be used with measures of central tendency.

A word that is often used in statistics is *average*. To explore averages, examine the following set of data for three teachers, each of whom claims that his or her class scored better *on the average* than the other two classes did:

Mr. Smith:	62, 94, 95, 98, 98
Mr. Jones:	62, 62, 98, 99, 100
Ms. Rivera:	40, 62, 85, 99, 99

All of these teachers are correct in their assertions because each has used a different number to characterize the scores in the class. In the following, we examine how each teacher can justify the claim.

Means

A number commonly used to characterize a set of data is the **arithmetic mean**, frequently called the **average**, or the **mean**. To find the mean of scores for each of the teachers given previously, we find the sum of the scores in each case and divide by 5, the number of scores.

$$\text{Mean (Smith):} \quad \frac{62 + 94 + 95 + 98 + 98}{5} = \frac{447}{5} = 89.4$$

$$\text{Mean (Jones):} \quad \frac{62 + 62 + 98 + 99 + 100}{5} = \frac{421}{5} = 84.2$$

$$\text{Mean (Rivera):} \quad \frac{40 + 62 + 85 + 99 + 99}{5} = \frac{385}{5} = 77$$

In terms of the mean, Mr. Smith's class scored better than the others. In general, we define the *arithmetic mean* as follows.

Definition of Arithmetic Mean

The **arithmetic mean** of the numbers $x_1, x_2, \ldots, x_n$, denoted $\bar{x}$ and read "x bar," is given by

$$\bar{x} = \frac{x_1 + x_2 + x_3 + \ldots + x_n}{n}$$

Understanding the Mean as a Balance Point

Because the mean is a widely used measure of central tendency, we provide a model for thinking about it. Suppose a student at a rural school reports that the mean number of pets for the six students in a group is 5. Do we know anything about the distribution of these pets? All six students could have exactly five pets, as shown in the dot plot in Figure 10-26(a).

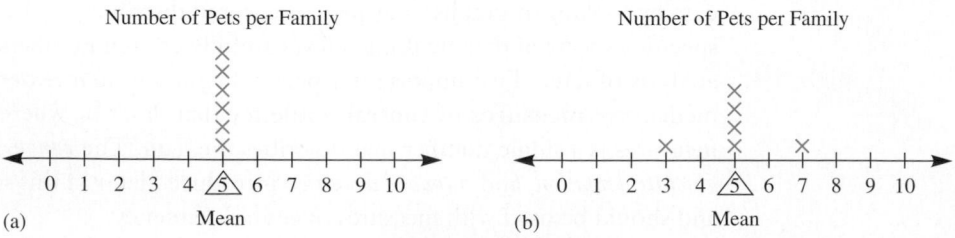

Figure 10-26

If we change the dot plot as shown in Figure 10-26(b), the mean is still 5. Notice that the new dot plot could be obtained by moving one value from Figure 10-26(a) 2 units to the right and then balancing this by moving one value 2 units to the left. We can think of the mean as a *balance point*.

Consider Figure 10-27, which shows the number of children for each family in a group. The mean of 5 is the balance point where the sum of the distances from the mean to the data points above the mean equals the sum of the distances from the mean to the data points below the mean. The sum of the distances above the mean is $3 + 5$, or 8. The sum of the distances below the mean is $1 + 2 + 2 + 3$, or 8. In this case, we see that the data are centered about the mean, but the mean does not belong to the set of data.

Number of Children per Family

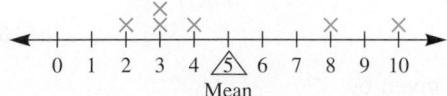

Figure 10-27

Knowing that the mean is a balance point as depicted in Figure 10-27 also shows the importance of not reporting the mean as a single number to summarize data. Using the data in Figure 10-27, it would be possible to rearrange the data to have the same mean but be spread very differently. For example, the data, independent of context, is given as Case 1 in Table 10-11. Case 2 of Table 10-11 has 5 subtracted from each of the data points to the left of the mean while 10 is added to each of the data points to the right of the mean. Note that the means of each of the sets of data in the two cases are exactly the same but the data are differently spread. Later, we consider the mean absolute deviation and the standard deviation, either of which could be used along with the mean to describe the spread of the data.

Table 10-11

	Case 1	Case 2
	2	⁻3
	3	⁻2
	3	⁻2
	4	⁻1
	8	18
	10	20
Mean	$(2 + 3 + 3 + 4 + 8 + 10)/6 = 5$	$(^-3 + {}^-2 + {}^-2 + {}^-1 + 18 + 20)/6 = 5$

NOW TRY THIS 10-9

a. A litter of six puppies was born with a mean weight of 7 lb. List two possibilities for the weights of the pups.

b. Could the mean of a set of scores ever be equal to the greatest score? The least score? Explain your answers.

c. Using the data in Figure 10-27, move the data points until a single "balance point" (the mean) can be found or estimated. For example, move the value 10 two units left to 8 and counterbalance that by moving the value 2 two units right to 4. Continue until a single balance point is determined. What is the mean?

Medians

The value exactly in the middle of an ordered set of numbers is the **median**. To find the median for the teachers' scores, we arrange each of their scores in increasing or decreasing order and pick the middle score.

Median (Smith):	62, 94, (95), 98, 98	median = 95
Median (Jones):	62, 62, (98), 99, 100	median = 98
Median (Rivera):	40, 62, (85), 99, 99	median = 85

In terms of the median, Mr. Jones's class scored better than the others.

With an odd number of scores, as in the present example, the median is the middle score. With an even number of scores, however, the median is defined as the mean of the middle two scores. Thus, to find the median, we add the middle two scores and divide by 2. For example, the median of the scores

$$64, 68, \boxed{70, 74,} \; 82, 90$$

is given by

$$\frac{70 + 74}{2}, \text{ or } 72$$

In general, to find the median for a set of n numbers, we proceed as follows:

1. Arrange the numbers in order from least to greatest.
2. a. If n is odd, the median is the middle number.
   b. If n is even, the median is the mean of the two middle numbers.

Just as the mean needs a measure of spread to help it describe a set of data, the median needs such a measure as well. For example, if we consider the data of Jones's students' scores as 62, 62, 98, 99, and 100, the median of 98. If the scores were 32, 32, 98, 99, and 100, the median is still 98 but the two sets of scores are very different. A median is often reported with the *interquartile range*, a measure of spread that shows where the middle 50% of the scores lie with the median in that range. The interquartile range is discussed later in this section. The two together form a much better pair to describe the data than the median alone.

Modes

The **mode** of a set of data is sometimes reported as a measure of central tendency (where the data is centered), but when it is reported in that form, it is frequently being misused. *The mode of a set of data is the number that appears most frequently, if there is one.* Examples of this follow. A mode is frequently reported with categorical data. Note that in some distributions, no number appears more than once. In other distributions, there may be more than one mode. Saying that there is no mode does not say that there are no measures of central tendency.

The set of scores 64, 79, 80, 82, 90 has no mode. Some would say that this set of data has five modes, but because no one score appears more than another, we prefer to say there is no mode. The set of scores 64, 75, 75, 82, 90, 90, 98 is **bimodal** (two modes) because both 75 and 90 are modes. It is possible for a set of data to have too many modes for this type of number to be useful in describing the data. For the three classes listed previously, if the mode were used as the criterion for an average, Ms. Rivera's class scored better than the others. This example shows a misuse of both *mode* and *average*.

Mode (Smith):	62, 94, 95, 98, 98	mode = 98
Mode (Jones):	62, 62, 98, 99, 100	mode = 62
Mode (Rivera):	40, 62, 85, 99, 99	mode = 99

EXAMPLE 10-4

Find (a) the mean, (b) the median, and (c) the mode for the following collection of data:

$$60 \quad 60 \quad 70 \quad 95 \quad 95 \quad 100$$

Solution

a. $\bar{x} = \dfrac{60 + 60 + 70 + 95 + 95 + 100}{6} = \dfrac{480}{6} = 80$

b. The median is $\dfrac{70 + 95}{2}$, or 82.5.

c. The set of data is bimodal and has both 60 and 95 as modes.

 NOW TRY THIS 10-10

When the data values are all the same, the mean, median, and mode are all the same. Describe a situation in which not all the data points are the same, but the mean, median, and mode are still the same.

 NOW TRY THIS 10-11

a. Suppose the average number of children per family for the employees of the university in a certain city is 2.58. Could this be a mean? Median? Mode? Explain why.

b. Answer the questions in (a) if the average number of children was reported to be 2.5.

Choosing the Most Appropriate Measure of Central Tendency

Although the *mean* is a commonly used measure of central tendency to describe a set of data, it may not always be the most appropriate choice.

EXAMPLE 10-5

Suppose a company employs 20 people. The president of the company earns $200,000, the vice president earns $75,000, and 18 part-time employees earn $10,000 each. Is the mean the best number to choose to represent the "average" salary for the company?

Solution   The mean salary for this company is

$$\frac{\$200{,}000 + \$75{,}000 + 18(\$10{,}000)}{20} = \frac{\$455{,}000}{20} = \$22{,}750$$

In this case, the mean salary of $22,750 is not representative. Either the median or mode, both of which are $10,000, would describe the typical salary better.

In Example 10-5, notice that *the mean is affected by extreme values*. In most cases, the *median* is not affected by extreme values. The median, however, can also be misleading, as shown in the following example.

EXAMPLE 10-6

Suppose nine students make the following scores on a test:

$$30, 35, 40, 40, 92, 92, 93, 98, 99$$

Is the median the best "average" to represent the set of scores?

Solution   The median score is 92. From that score, one might infer that the individuals all scored very well, yet 92 is certainly not a typical score. In this case, the mean of approximately 69 might be more appropriate than the median. However, with the spread of the scores, neither is very appropriate for this distribution.

As mentioned earlier, the *mode*, too, can be misleading in describing a set of data with very few items that occur frequently, as shown in the following example.

EXAMPLE 10-7

Is the mode an appropriate "average" for the following test scores?

$$40, 42, 50, 62, 63, 65, 98, 98$$

Solution   The mode of the set of scores is 98 because this score occurs most frequently. The score of 98 is not representative of the set of data because of the large spread of scores and the much lower mean (and median).

The choice of which number to use to represent a particular set of data is not always easy. In the example involving the three teachers, each teacher chose the number that best suited his or her claim. The measure of central tendency should always be specified along with a measure of spread (*Statistics Framework*, p. 569).

NOW TRY THIS 10-12

Mr. Ramirez and Ms. Jonsey gave tests to their classes with the results seen in Table 10-12.

Table 10-12

	Overall Mean	Mean for Females	Mean for Males	Percent Females
Ramirez	218	230	205	
Jonsey	221	224		88

What are the missing entries? For an overall mean, Ms. Jonsey's class is higher, but for females, the mean is higher in Mr. Ramirez's class. Is it possible that the mean for males is higher in Mr. Ramirez's class as well?

Problem Solving   **The Missing Grades**

Students of Dr. Van Horn were asked to keep track of their own grades. One day, Dr. Van Horn asked the students to report their grades. One student had lost the papers but claims to remember the grades on four of six assignments: 100, 82, 74, and 60. In addition, the student remembered that the mean of all six papers was 69, and the other two papers had identical grades. What were the grades on the other two homework papers?

Understanding the Problem The student had scores of 100, 82, 74, and 60 on four of six papers. The mean of all six papers was 69, and two identical scores were missing.

Devising a Plan To find the missing grades, we use the strategy of *writing an equation*. The mean is obtained by finding the sum of the scores and then dividing by the number of scores, which is 6. So if we let x stand for each of the two missing grades, we have

$$69 = \frac{100 + 82 + 74 + 60 + x + x}{6}$$

Carrying Out the Plan We now solve the equation as follows:

$$69 = \frac{100 + 82 + 74 + 60 + x + x}{6}$$

$$69 = \frac{316 + 2x}{6}$$

$$49 = x$$

Since the solution to the equation is $x = 49$, each of the two missing scores was 49.

Looking Back The answer of 49 seems reasonable. We can check this by computing the mean of the scores 100, 82, 74, 60, 49, 49 and showing that it is 69.

Measures of Spread

The grade 6 *Common Core Standards* suggests that all students:

> Recognize that a measure of center for a numerical data set summarizes all of its values with a single number, while a measure of variation describes how its values vary with a single number. (p. 45)

The mean and median are measures of central tendency that provide limited information about a whole distribution of data. For example, if you sit in a sauna for 30 minutes and then in a refrigerated room for 30 minutes, an average temperature of your surroundings for that hour might sound comfortable. To tell how much the data are scattered, we develop measures of *spread* or *dispersion*. Perhaps the easiest way to measure spread is the **range**, the difference between the greatest and the least values in a data set. For example, the range in the set of data 1, 3, 7, 8, 10 is $10 - 1 = 9$. However, just because the ranges of two sets of data are the same, the data do not have to have the same dispersion. For example, the data set 1, 10, 10, 10, 10 also has a range of 9 and is spread quite differently from the first collection of data. For this reason we need other measures of spread besides the range.

Another measure of spread is the **interquartile range (IQR)**. The IQR is the range of the middle half of the data. Consider the following set of test scores:

20   25   40   50   50   60   70   75   80   80   90   100   100

The range for this set of scores is $100 - 20 = 80$. The median score for this set of data is 70. We mark this location with a vertical bar between the 7 and the 0 and circle the data point for emphasis, as shown.

20   25   40   50   50   60   ⟨7 | 0⟩   75   80   80   90   100   100

Next, we consider only the data values to the left of the **vertical bar** and draw another vertical bar where the median of those values is located:

20   25   40 | 50   50   60

The score of $(40 + 50)/2 = 45$ is the median of the lower half of the scores and is the **lower quartile**. The lower quartile, or the **first quartile**, is denoted Q_1. Approximately one quarter, or 25%, of the scores lie at or below Q_1. Similarly, we can find the upper, or third, quartile (Q_3), which is $(80 + 90)/2$, or 85. The **upper quartile (Q_3)** is the median of the upper half of all scores. Approximately three-quarters, or 75%, of the scores lie at or below Q_3. The median is the **second quartile** or Q_2. Thus we have divided the scores into four groups of three scores each:

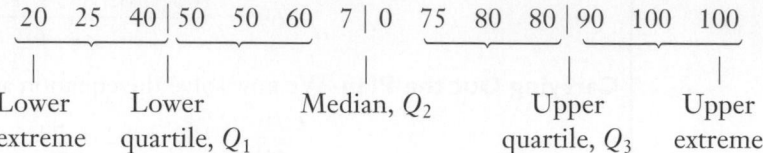

20 25 40 | 50 50 60 7 | 0 75 80 80 | 90 100 100

Lower extreme   Lower quartile, Q_1 Median, Q_2 Upper quartile, Q_3 Upper extreme

The interquartile range (IQR) represents the difference between the upper quartile and the lower quartile. In this case, $IQR = 85 - 45 = 40$. The IQR is a useful measure of spread because it is less influenced by extreme values. The IQR represents the difference of Q_1 and Q_3 and contains approximately the middle 50% of the values.

The interquartile range is the measure of spread most often reported with the median. This is done because the median marks the center of the data. However, as we saw earlier, the same median could be reported with two sets of data that have very different spreads. With the interquartile range reported along with the median, not only do we know the middle, we know how spread out the middle 50% of the data are. If they have a wide spread, then we are aware of it and can assess and use the data accordingly. For example, in the data set just presented, the median of 70 should be reported with the interquartile range of 40.

NOW TRY THIS 10-13

Create a set of data that ranges from 0 to 100 and is widely spread yet has median 70 and an interquartile range of 50.

Box Plots

A **box plot** (or a **box-and-whisker plot**) is a way to display data visually and draw informal conclusions. Box plots show only certain data; they are visual representations of the *five-number summary* of the data. The five numbers are the median, the upper and lower quartiles (and, hence, the interquartile range information), and the least and greatest values in the distribution. The center, the spread, and the overall range are immediately evident by looking at the plot.

To construct a box plot, we need the data's median, upper and lower quartiles, and extremes. To construct the box, we draw bars at the quartiles. We draw segments from each end of the box to the extreme values to form the whiskers. The box plot can be either vertical or horizontal. A vertical version of the box plot for the given data is shown in Figure 10-28.

The box plot gives a fairly clear picture of the spread of the data examined earlier. If we look at the graph in Figure 10-28, we can see that the median is 70, the maximum value is 100, the minimum value is 20, and the upper and lower quartiles are 45 and 85.

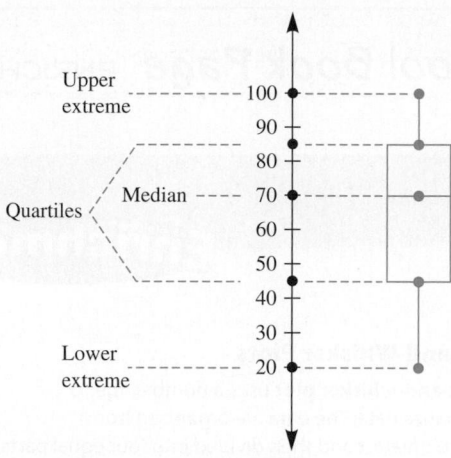

Figure 10-28

 The student page on page 594 outlines the steps for constructing a box-and-whisker plot. Read through the page and work through the Practice problems.

EXAMPLE 10-8

What are the minimum and maximum values, the median, and the lower and upper quartiles of the data set whose box plot is shown in Figure 10-29?

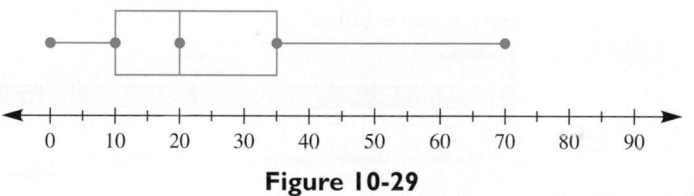

Figure 10-29

Solution   The minimum value is 0, the maximum value is 70, the median is 20, the lower quartile is 10, and the upper quartile is 35.

Outliers

An *outlier* is a value widely separated from the rest of a group of data. For example, in a set of scores such as

$$91 \quad 92 \quad 92 \quad 93 \quad 93 \quad 93 \quad 94$$

all data are grouped close together and no values are widely separated. However, in a set of scores such as

$$21 \quad 92 \quad 92 \quad 93 \quad 93 \quad 93 \quad 95 \quad 150$$

both 21 and 150 are widely separated from the rest of the data. These values are potential outliers. The upper and lower extreme values are not necessarily outliers. In data such as

$$75 \quad 90 \quad 91 \quad 92 \quad 92 \quad 93 \quad 93$$

it is not easy to decide, so we develop a convention for determining outliers. *An **outlier** is any value that is more than 1.5 times the interquartile range above the upper quartile or below the lower quartile.* Statisticians sometimes use values different from 1.5 to determine outliers.

It is common practice to indicate outliers in box plots with asterisks. Whiskers are then drawn to the extreme points that are not *outliers.* To investigate how this works, consider Example 10-9.

School Book Page ENRICHMENT

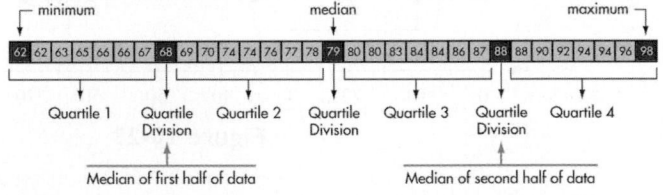

Box-and-Whisker Plots

A **box-and-whisker plot** uses a number line to summarize data. The data are organized from least to greatest and then divided into four equal parts, called **quartiles**.

Follow the steps to make a box-and-whisker plot of the test scores.

Example: Mr. Garcia's math class took a test. The scores for his 31 students were 74, 86, 94, 67, 79, 63, 83, 88, 78, 62, 70, 92, 98, 80, 68, 62, 90, 84, 94, 66, 65, 96, 76, 66, 80, 74, 87, 84, 69, 77, and 88.

Step 1 Write the data in order from least to greatest.

| 62 | 62 | 63 | 65 | 66 | 66 | 67 | 68 | 69 | 70 | 74 | 74 | 76 | 77 | 78 | 79 | 80 | 80 | 83 | 84 | 84 | 86 | 87 | 88 | 88 | 90 | 92 | 94 | 94 | 96 | 98 |

Step 2 Find the median of the data. It divides the data into halves. Then find the median for each of these halves. These are the quartile divisions. They divide the data into four quartiles.

minimum   median   maximum

| 62 | 62 | 63 | 65 | 66 | 66 | 67 | 68 | 69 | 70 | 74 | 74 | 76 | 77 | 78 | 79 | 80 | 80 | 83 | 84 | 84 | 86 | 87 | 88 | 88 | 90 | 92 | 94 | 94 | 96 | 98 |

Quartile 1   Quartile   Quartile 2   Quartile   Quartile 3   Quartile   Quartile 4
               Division   Division   Division

Median of first half of data   Median of second half of data

Step 3 Use the minimum, and maximum to draw a number line. Then draw a *box* using the middle two quartiles and label the median. The part of the number line showing the 1st and 4th quartiles are *whiskers*.

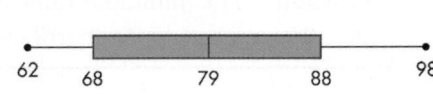

62   68   79   88   98

Practice

Make a box-and-whisker plot for each set of data.

1. Miss Hanson's students earned the following scores on their science tests:

   73, 78, 66, 61, 85, 90, 99, 76, 64, 70, 72, 72, 93, 81, 71, 79, 85, 89, 84, 75, 79, 91, 82

2. Jamell is a runner on the track team. Here are his times in minutes for the last season:

   9, 13, 12, 15, 9, 11, 10, 16, 13, 14, 10, 15, 14, 14, 13, 16, 10

Lesson 19-6   497

EXAMPLE 10-9

Draw a box plot of the data in Table 10-13 and identify possible outliers.

Table 10-13 Final Medal Standings for Top 20 Countries—2008 Summer Olympics

United States	110	
China	100	
Russia	72	
Great Britain	47	
Australia	46	
Germany	41	$Q_3 = 43.5$
France	40	
South Korea	31	
Italy	28	
Ukraine	27	
Japan	25	Median $(Q_2) = 26$
Cuba	24	
Belarus	19	
Spain	18	
Canada	18	$Q_1 = 17$
Netherlands	16	
Brazil	15	
Kenya	14	
Kazakhstan	13	
Jamaica	11	

Solution The extreme scores are 110 and 11, the median is 26, $Q_1 = 17$, and $Q_3 = 43.5$. The IQR is $43.5 - 17 = 26.5$. Outliers are scores that are greater than $43.5 + 1.5(26.5)$, or 83.25, or less than $17 - 1.5(26.5)$, or $^-22.75$. Therefore, in this data 110 and 100 are the only outliers. A box plot is given in Figure 10-30. The whisker stops at the extreme point 11 on the lower end and at 72 on the upper end. Outliers are indicated with asterisks.

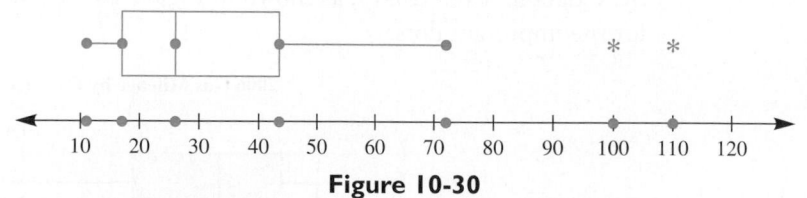

Figure 10-30

Comparing Sets of Data

Box plots are used primarily for large sets of data or for comparing several distributions. The stem and leaf plot is usually a much clearer display for a single distribution. Parallel box plots drawn using the same number line provide the easiest comparison of medians, extreme scores, and the quartiles for the sets of data. As an example, we construct parallel box plots comparing the data in Table 10-14.

Table 10-14 Gas Mileage by Car Size

Compact		Midsize		SUV	
48	23	24	17	21	18
45	22	23	19	19	28
40	21	22	22	18	21
31	20	21	21	30	20
30	19	20	20	28	20
28	45	19	20	27	20
27	18	19	17	21	19
26	19	19	18	20	19
25	20	18	18	19	19
24	21	18	18	18	19

Source: U. S. Department of Energy and U. S. Environmental Protection Agency, 2007.

Before constructing parallel box plots, we find the five important values for each group of data. These values are given in Table 10-15.

Table 10-15

Value	Compact	Midsize	SUV
Maximum	48	24	30
Q_3	30.5	21	21
Median	24.5	19	20
Q_1	20.5	18	19
Minimum	18	17	18

In this example, the IQR for compact is 10, for midsize is 3, and for SUV is 2.

Next we draw the horizontal scale and construct the box plots for the compact, midsize, and SUV data in Table 10-15, as shown in Figure 10-31. Grid lines are used to emphasize the value for the important dots.

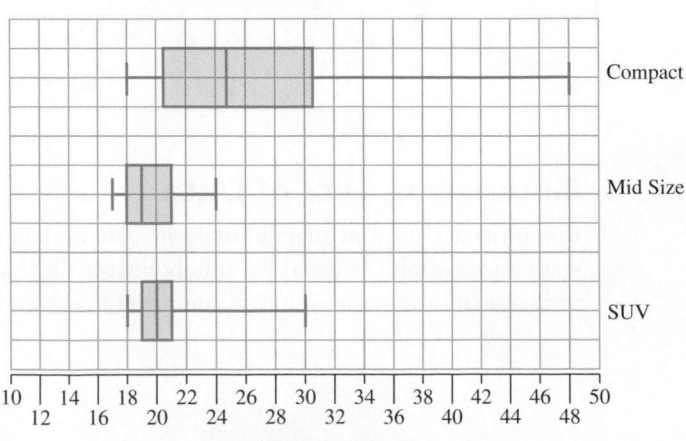

2006 Gas Mileage by Car Size

Figure 10-31

From the box plots we see that the median gas mileage for compacts is higher than that for midsize or SUV cars. However, the median for SUVs is greater than that for midsize cars. The data for compact cars is more spread out than for the other two groups and the midsize cars have the least spread. More than 50% of the compact cars have better gas mileage than 100% of the midsize cars. The length of the whiskers suggest that we might check for outliers in the compact and SUV groups.

Although we cannot spot clusters or gaps in box plots as we can with stem and leaf or line plots, we can more easily compare data from different sets. With box plots, we do not need to have sets of data that are approximately the same size, as we did for back-to-back stem and leaf plots. To compare data from two or more sets using their box plots, we first study the boxes to see whether they are located in approximately the same places. Next, we consider the lengths of the boxes to see whether the variability of the data is about the same. We also check whether the median, the quartiles, and the extreme values in one set are greater than those in another set. If they are, the data in the first set are greater than those of the other set, no matter how we compare them. If they are not, we can continue to study the data for other similarities and differences.

 NOW TRY THIS 10-14

a. Using only the data from Table 10-15, decide whether there are outliers.
b. Redraw Figure 10-31 showing outliers.

Variation: Mean Absolute Deviation, Variance, and Standard Deviation

Earlier in the section, we noted that a measure of spread is needed when data are summarized with a single number, such as the mean or median. There are many measures of spread that are generally discussed and each has a different use. To examine these measures of spread, we start again with a simple test-score example.

Suppose Professors Abel and Babel each taught a section of a graduate statistics course and each had six students. Both professors gave the same final exam. The results, along with the means for each group of scores, are given in Table 10-16, with stem and leaf plots in Figure 10-32(a) and (b), respectively. As the stem and leaf plots show, the sets of data are very different. The first is more spread out, or varies more, than the second. However, each set has

Table 10-16

Abel's Class Scores	Babel's Class Scores
100	70
80	70
70	60
50	60
50	60
10	40
$\bar{x} = \dfrac{360}{6} = 60$	$\bar{x} = \dfrac{360}{6} = 60$

Professor Abel's Class Scores

```
 1 | 0
 2
 3
 4
 5 | 00
 6 | 00
 7 | 0      7 | 0 represents
 8 | 0      a score of 70
 9
10 | 0
```

(a)

Professor Babel's Class Scores

```
 4 | 0
 5 |
 6 | 000
 7 | 00     7 | 0 represents
           a score of 70
```

(b)

Figure 10-32

60 as the mean. Each median also equals 60. Although the mean and the median for these two groups are the same, the two distributions of scores are very different.

As we have seen, there are many ways to measure the spread of data. The simplest way is to find the range. The range for Professor Abel's class is $100 - 10$, or 90. The range for Professor Babel's class is $70 - 40$, or 30. If we use the range as the measure of dispersion, we see that Abel's class is much more spread out than Babel's class. If we use the interquartile range, the IQR for Abel's class is 30, and for Babel's class, 10. Again, these measures of spread show more of a spread for Abel's class than for Babel's. The disadvantage of the range is that it uses only extreme values.

Mean Absolute Deviation

One of the most basic ways to measure the spread of data is to measure the distance that each data point is away from the mean. The absolute value is one method used for finding distance. The **mean absolute deviation (MAD)** makes use of the absolute value to find the distance each data point is away from the mean; then the mean of those distances is found to give an "average distance from the mean" for each of the points. For example, we find the mean absolute deviation of the scores by using the following steps:

- Measure the distance from the mean by calculating the score minus the mean.
- Find the absolute value of each difference.
- Sum those absolute values (the absolute deviation).
- Find the mean absolute deviation (MAD) by dividing the sum by the number of scores.

Table 10-17 contains a sample set of data. Compute the mean absolute deviation.

Table 10-17

Test Scores	|Test Score − Mean|
99	$\|99 - 83.2\|$, or 15.8
67	$\|67 - 83.2\|$, or 16.2
84	$\|84 - 83.2\|$, or 0.8
99	$\|99 - 83.2\|$, or 15.8
67	$\|67 - 83.2\|$, or 16.2
$\bar{x} = \dfrac{(99 + 67 + 84 + 99 + 67)}{5} = 83.2$	Sum of absolute deviations = 64.8
	$\text{MAD} = \dfrac{64.8}{5}$, or 12.96

Definition of Mean Absolute Deviation (MAD)

The **mean absolute deviation (MAD)** for the numbers $x_1, x_2, x_3, \ldots, x_n$, where $\bar{x}$ is the mean of the numbers, is:

$$\text{MAD} = \frac{|x_1 - \bar{x}| + |x_2 - \bar{x}| + \ldots + |x_n - \bar{x}|}{n}$$

Visual pictures of the MAD for the given set of test scores are given in Figure 10-33(a) and (b).

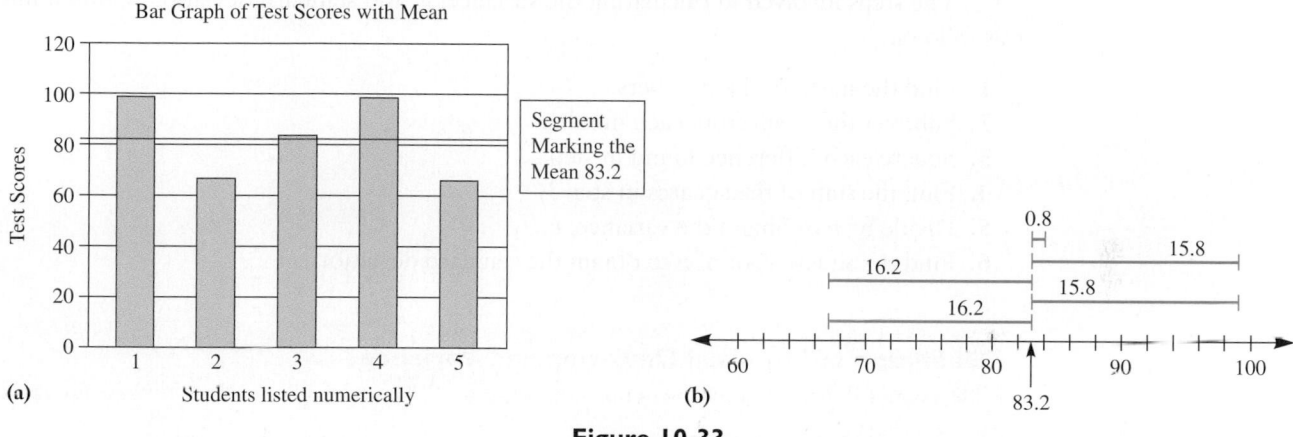

Figure 10-33

The MAD is a rough measure of the "average" distance that a score is from the mean and hence gives an idea of how far away from the mean the test scores are. The MAD is recommended for use in *Statistics Framework* (p. 43), and serves as a precursor to the standard deviation. The MAD does give a measure of spread that works well for some data sets. It handles fairly large deviations in some sets. Consider the data for Abel from Table 10-16 and displayed in Table 10-18.

Table 10-18

Abel's Class Scores	|Abel's Scores $-$ 60|	Babel's Class Scores	|Babel's Scores $-$ 60|
100	40	70	10
80	20	70	10
70	10	60	0
50	10	60	0
50	10	60	0
10	50	40	20
Mean = 60	MAD = $\frac{140}{6}$, or approximately 23.3	Mean = 60	MAD = $\frac{40}{6}$, or approximately 6.7

Using the MAD on the two sets of scores, it is easy to see that even though the means are 60, the spreads are very different. For Abel's class, the scores are an average of 23.3 points from the mean, whereas in Babel's class, the scores are an average of 6.7 points from the mean. Babel's test scores are "less spread out."

Statisticians commonly use two other measures of spread: the *variance* and the *standard deviation*. These measures are also based on how far the scores are from the mean. To find out how far each value differs from the mean, we subtract each value in the data from the mean to obtain the deviation. Some deviations may be positive, and others may be negative. Because the mean is the balance point, the total of the deviations above the mean equals the total of the deviations below the mean. (The mean of the deviations is 0 because the sum of the deviations is 0.) Squaring the deviations makes them all positive. The mean of the squared deviations is the **variance**. Because the variance involves squaring the deviations, it does not have the same units of measurement as the original observations. For example, lengths measured in feet have a variance measured in square feet. To obtain the same units as the original observations, we take the square root of the variance and obtain the **standard deviation**.

The steps involved in calculating the variance, v, and standard deviation, s, of n numbers are as follows:

1. Find the mean of the numbers.
2. Subtract the mean from each number.
3. Square each difference found in step 2.
4. Find the sum of the squares in step 3.
5. Divide by n to obtain the variance, v.
6. Find the square root of v to obtain the standard deviation, s.

Definition of Standard Deviation and Variance

The **standard deviation** for the numbers $x_1, x_2, x_3, \ldots, x_n$, where $\bar{x}$ is the mean of these numbers and v is the **variance**, is:

$$s = \sqrt{v} = \sqrt{\frac{(x_1 - \bar{x})^2 + (x_2 - \bar{x})^2 + (x_3 - \bar{x})^2 + \ldots + (x_n - \bar{x})^2}{n}}$$

In some textbooks, this formula for standard deviation and variance involves division by $n - 1$ instead of by n. Division by $n - 1$ is more useful for advanced work in statistics.

The variances and standard deviations for the final exam data from the classes of Professors Abel and Babel are calculated by using Table 10-19(a) and (b), respectively.

Table 10-19 (a) Abel's Class Scores

x	$x - \bar{x}$	$(x - \bar{x})^2$
100	40	1600
80	20	400
70	10	100
50	⁻10	100
50	⁻10	100
10	⁻50	2500
Totals: 360	0	4800

(b) Babel's Class Scores

x	$x - \bar{x}$	$(x - \bar{x})^2$
70	10	100
70	10	100
60	0	0
60	0	0
60	0	0
40	⁻20	400
Totals: 360	0	600

$$\bar{x} = \frac{360}{6} = 60$$

$$v = \frac{4800}{6} = 800$$

$$s = \sqrt{800} \approx 28.3$$

$$\bar{x} = \frac{360}{6} = 60$$

$$v = \frac{600}{6} = 100$$

$$s = \sqrt{100} = 10$$

Values far from the mean on either side will have greater positive squared deviations, whereas values close to the mean will have lesser positive squared deviations. Therefore, the standard deviation is a greater number when the values from a set of data are widely spread and a lesser number (close to 0) when the data values are close together.

EXAMPLE 10-10

Professor Abel gave two group exams. Exam A had grades of 0, 0, 0, 100, 100, 100, and exam B had grades of 50, 50, 50, 50, 50, 50. Find the following for each exam:

a. Mean   **b.** Range   **c.** Mean absolute deviation
d. Standard deviation   **e.** Median   **f.** Interquartile range

Solution

a. The means for exams A and B are each 50.
b. The range for exams A and B are 100 and 0, respectively.
c. The mean absolute deviations for the two exams are as follows:

$$\text{MAD}_A = \frac{|0 - 50| + |0 - 50| + |0 - 50| + |100 - 50| + |100 - 50| + |100 - 50|}{6} = 50$$

$$\text{MAD}_B = \frac{|50 - 50| + |50 - 50| + |50 - 50| + |50 - 50| + |50 - 50| + |50 - 50|}{6} = 0$$

d. The standard deviations for exams A and B are as follows:

$$s_A = \sqrt{\frac{3(0 - 50)^2 + 3(100 - 50)^2}{6}} = 50$$

$$s_B = \sqrt{\frac{6(50 - 50)^2}{6}} = 0$$

e. The medians for the exams are each 50.
f. The interquartile range for exams A and B, respectively, are 100 and 0.

In Example 10-10, exam A has mean 50, mean absolute deviation of 50, and standard deviation of 50. Exam A also had median 50 and interquartile range 100. Reporting the three measures of spread together in each case demonstrates that the scores are widely spread away from the mean and median. Exam B has mean 50, mean absolute deviation 0, and standard deviation of 0. Its median is 50 with interquartile range 0. All of the descriptors for Exam B show that there is very little spread of data from the mean and median. (In this case there is none.) Example 10-10 was chosen with extreme scores to illustrate what can happen with measures of central tendency and measures of spread. Normally, one can tell more about the distribution of data points when the mean, mean absolute deviation, or standard deviation are reported together, as well as when the median and interquartile range are reported together.

Normal Distributions

To better understand how standard deviations are used as measures of spread, we next consider normal distributions. The graphs of normal distributions are the bell-shaped curves or *normal curves*. Normal distribution arises naturally in many real-world situations. Human heights, gestation, test scores, IQ scores, experimental measurements all have approximate normal distributions if a large population is considered.

A **normal curve** is a smooth, bell-shaped curve that depicts frequency values distributed symmetrically about the mean. Also, the mean, median, and mode all have the same value. The normal curve is a theoretical distribution that extends infinitely in both directions. It gets closer and closer to the x-axis but never reaches it. On a normal curve, about 68% of the values lie within 1 standard deviation of the mean, about 95% lie within 2 standard deviations, and about 99.8% are within 3 standard deviations. The percentages represent approximations of the total percent of area under the curve. The curve and the percentages are illustrated in Figure 10-34.

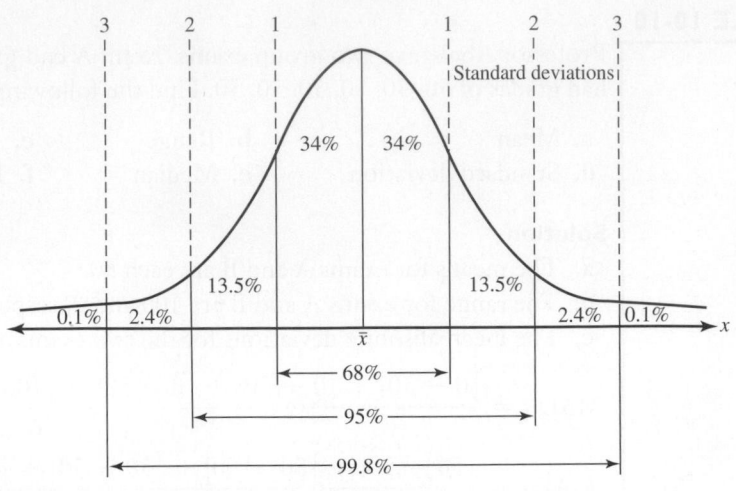

Figure 10-34

Suppose the area under the curve represents the population of the United States. Psychologists claim that the mean IQ is 100 and the standard deviation is 15. They also claim that an IQ score of over 130 represents a superior score. Because 130 is equal to the mean plus 2 standard deviations, we see from Figure 10-34 that only 2.5% of the population fall into this category.

EXAMPLE 10-11

When a standardized test was scored, there was a mean of 500 and a standard deviation of 100. Suppose that 10,000 students took the test and their scores had a bell-shaped distribution, making it possible to use a normal curve to approximate the distribution.

  a. How many scored between 400 and 600?
  b. How many scored between 300 and 700?
  c. How many scored between 200 and 800?
  d. How many scored above 800?

Solution
  a. Since 1 standard deviation on either side of the mean is from 400 to 600, about 68% of the scores fall in this interval. Thus, 0.68(10,000), or 6800, students scored between 400 and 600.
  b. About 95% of 10,000, or 9500, students scored between 300 and 700.
  c. About 99.8% of 10,000, or 9980, students scored between 200 and 800.
  d. About 0.1% of 10,000, or 10, students scored above 800.

About 0.2%, or 20, of the students' scores in Example 10-11 fall outside 3 standard deviations. About 10 of these students did very well on the test, and about 10 students did very poorly.

Application of the Normal Curve

Suppose that a group of students asked their teacher to grade "on a curve." If the teacher gave a test to 200 students and the mean on the test was 71, with a standard deviation of 7, the graph in Figure 10-35 shows how the grades could be assigned. In Figure 10-35, the teacher has used the normal curve in grading. (The use of the normal curve presupposes that the teacher had a bell-shaped distribution of scores and also that the teacher arbitrarily decided to use the lines marking standard deviations to determine the boundaries of the A's, B's, C's, D's, and F's.) Thus, based on the normal curve in Figure 10-35, Table 10-20 shows the range of grades that the teacher might assign if the grades are rounded. Students who ask their teachers to grade on the curve may wish to reconsider if the normal curve is to be used.

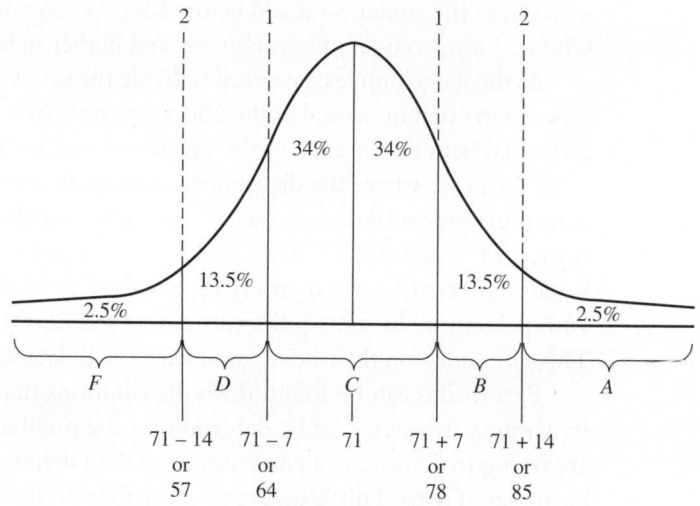

Figure 10-35

Table 10-20

Test Score	Grade	Number of People per Grade	Percentage Receiving Grade
85 and above	A	5	2.5%
78–84	B	27	13.5%
64–77	C	136	68%
57–63	D	27	13.5%
Below 57	F	5	2.5%

Percentiles

When students take a standardized test such as the ACT or SAT, their scores are often reported in **percentiles**. A percentile shows a person's score relative to other scores. For example, if a student's score is at the 82nd percentile, this means that approximately 82% of those taking the test scored lower than the student and approximately 18% had higher scores.

One application of percentiles is to make comparisons. For example, consider Kristy and Kim as they applied for a job. On the application, Kristy wrote that she finished 15th in her class. Kim reported that she finished 40th in her class. Does Kristy's class rank imply that she is a better student than Kim? Suppose that there were 50 students in Kristy's class and 400 students in Kim's class. In such a case, we need a common method of comparing their rankings. In order to

 Historical Note ————————————————————————————————

Abraham De Moivre (1667–1754), a French Huguenot, was the first to develop and study the normal curve. He was one of the first to study actuarial information, in his book *Annuities upon Lives*. De Moivre's work with the normal curve went essentially unnoticed and was developed independently by Pierre de Laplace (1749–1827) and Carl Friedrich Gauss (1777–1855), who found so many applications for the normal curve that it is referred to as the *Gaussian curve.* ●

make a balanced comparison, we can find Kristy's and Kim's percentile ranks, or percentiles, in their respective classes. In Kristy's class, there were $50 - 15$, or 35, students ranked below her, so we say that Kristy ranked at the 70th percentile $\left(\dfrac{35}{50} = 70\%\right)$. Similarly, $(400 - 40)/400$, or 90%, of the students ranked below Kim. We say that Kim ranked at the 90th percentile for her school. Thus, comparatively, Kim ranked higher in her class than Kristy did in hers.

As the name implies, percentiles divide the set of data into 100 equal parts. For example, if Tomas reports that he scored at the 50th percentile (the median, or Q_2) on the SAT, he is saying that he scored better than 50% of the people taking the test. In general, the rth percentile is denoted by P_r. In a case where the distribution is normal, we could use the percentages in Figure 10-34 to determine percentiles. For example, someone scoring 2 standard deviations above the median is at the $(0.1 + 2.4 + 13.5 + 34 + 34 + 13.5)$ percentile, or better than 97.5% of the population. Because percentiles are typically reported using whole percentages, that places the person at the 98th percentile. In some finite sets of test scores, a particular percentile may not be represented. This does not stop their being used with small data sets, but it does inhibit their usefulness.

Percentiles can be found from distributions that are not normal. This can be done by constructing a frequency table, determining the number of scores less than the score for which we are trying to determine a percentile, and then dividing that number by the total number of scores in the set of data. This is investigated further in the assessment set.

Deciles are points that divide a distribution into 10 equally spaced sections. There are nine deciles, denoted $D_1, D_2, \ldots, D_9$, and $D_1 = P_{10}, D_2 = P_{20}, \ldots, D_9 = P_{90}$. Quartiles were defined similarly earlier.

EXAMPLE 10-12

A standardized test that was distributed along a normal curve had a median (and mean) of 500 and a standard deviation of 100. The 16th percentile, P_{16}, is 400 because 400 is 1 standard deviation below the median. Find each of the following:

a. P_{50} b. P_{84}

Solution
a. Since 500 is the median, 50% of the distribution is less than 500. Thus, $P_{50} = 500$.
b. Since 600 is 1 standard deviation above the median, 84% of the distribution is less than 600. Thus, $P_{84} = 600$.

EXAMPLE 10-13

a. Ossie was ranked 25th in a class of 250. What was his percentile rank?
b. In a class of 50, Cathy has a percentile rank of 60. What is her class standing?

Solution
a. There were $250 - 25$, or 225, students ranked below Ossie. Hence, $225/250$, or 90%, of the class ranked below him. Therefore, Ossie ranked at the 90th percentile.
b. Cathy's percentile rank is 60. Thus 60% of the class ranks below her. Because 60% of 50 is 30, 30 students ranked below Cathy. Therefore, Cathy is 20th in her class.

BRAIN TEASER

At a birthday party, the honoree would not tell his age but agreed to give hints. He computed and announced that the mean age of his seven party guests was 21. When 29-year-old Jill arrived at the party, the honoree announced that the mean age of the eight people was now 22. Jack, another 29-year-old, arrived next. The honoree then added Jack's age to the set of ages of the other nine people and announced that the mean was now 27. How old was the honoree?

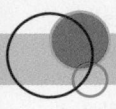

Assessment 10-4A

1. Calculate the mean, the median, and the mode for each of the following data sets:
   a. 2, 8, 7, 8, 5, 8, 10, 5
   b. 10, 12, 12, 14, 20, 16, 12, 14, 11
   c. 18, 22, 22, 17, 30, 18, 12
2. Write an example of a data set with seven data points for which the mode is not a good descriptor of the center of the data.
3. a. If three students scored 60 on a test and three students scored 80 on the same test, find each of the following related to the test scores:
   i. Mean   ii. Median   iii. Mode
   b. The mean score on a set of 20 tests is 75. What is the sum of the 20 test scores?
4. The mean for a set of 28 scores is 80. Suppose two more students take the test and score 60 and 50. What is the new mean?
5. Suppose in Selina's class there were three students who scored 100 and nine who scored 50. What is the mean of the scores in the class?
6. Suppose there were m students in a class who scored 100 and n students who scored 50. Write an algebraic expression for the mean in terms of m and n.
7. A table showing Jon's fall-quarter grades follows. Find his grade point average for the term (A = 4, B = 3, C = 2, D = 1, F = 0).

Course	Credits	Grade
Math	5	B
English	3	A
Physics	5	C
German	3	D
Handball	1	A

8. If the mean weight of seven linesmen on a team is 230 lb and the mean weight of the four backfield members is 190 lb, what is the mean weight of the 11-person team?
9. The following table gives the annual salaries of the 40 dancers in a certain troupe.
   a. Find the mean annual salary for the troupe.
   b. Find the median annual salary.
   c. Find the mode.

Salary ($)	Number of Dancers
18,000	2
22,000	4
26,000	4
35,000	3
38,000	12
44,000	8
50,000	4
80,000	2
150,000	1

10. Use the data in exercise 9 to find the following:
    a. Range
    b. Mean absolute deviation
    c. Standard deviation
    d. Interquartile range
11. Use the data in exercise 9 and the values in exercise 10 to answer the following:
    a. Which values from exercise 10 should be reported to best describe the data in exercise 9?
    b. Explain why you made the choice you did in part (a).
12. Maria filled her car's gas tank. The mileage odometer read 42,800 mi. When the odometer read 43,030, Maria filled the tank with 12 gal. At the end of the trip, she filled the tank with 18 gal and the odometer read 43,390 mi. How many miles per gallon did she get for the entire trip?
13. The youngest person in a company is 24 years old. The range of ages is 34 years. How old is the oldest person in the company?
14. To receive an A in a class, Willie needs at least a mean of 90 on five exams. Willie's grades on the first four exams were 84, 95, 86, and 94. What minimum score does he need on the fifth exam to receive an A in the class?
15. Following are box plots comparing the ticket prices of two performing arts theaters:

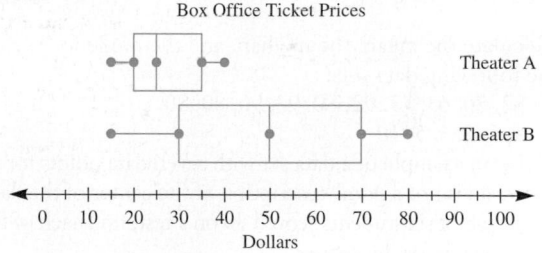

   a. What is the median ticket price for each theater?
   b. Which theater has the greatest range of prices?
   c. What is the highest ticket price at either theater?
   d. Make some statements comparing the ticket prices at the two theaters.
16. Construct a box plot for the following set of test scores. Indicate outliers, if any, with asterisks.

   20 95 40 70 90 70 80 80 90 95

17. The following table shows the heights in feet of the 8 tallest buildings in St. Louis, Missouri, and in Los Angeles, California.

Los Angeles	St. Louis
858	593
750	588
735	540
699	434
625	420
620	398
578	392
571	375

a. Draw horizontal box plots to compare the data.
b. Based on your box plots from (a), make some comparisons of the heights of the buildings in the two cities.
18. What is the standard deviation of the heights of seven trapeze artists if their heights are 175 cm, 182 cm, 190 cm, 180 cm, 192 cm, 172 cm, and 190 cm?
19. In a Math 131 class at DiPaloma University, the grades on the first exam were as follows:

$$96 \quad 71 \quad 43 \quad 77 \quad 75 \quad 76 \quad 61$$
$$83 \quad 71 \quad 58 \quad 97 \quad 76 \quad 74 \quad 91$$
$$74 \quad 71 \quad 77 \quad 83 \quad 87 \quad 93 \quad 79$$

a. Find the mean.
b. Find the median.
c. Find the mode.
d. Find the IQR.
e. Find the variance of the scores.
f. Find the standard deviation of the scores.
g. Find the mean absolute deviation of the scores.
20. Assume that the heights of American women are approximately normally distributed, with a mean of 65.5 in. and a standard deviation of 2.5 in. Within what range are the heights of 95% of American women?

21. Assume a normal distribution and that the average phone call in a certain town lasted 4 min, with a standard deviation of 1 min. What percentage of the calls lasted less than 1 min?
22. If a standardized test with scores distributed normally has a mean of 65 and a standard deviation of 12, find the following:
a. Q_2 b. P_{16} c. P_{84} d. D_5
23. For certain workers, the mean wage is $5.00/hr, with a standard deviation of $0.50. If a worker is chosen at random, what is the probability that the worker's wage is between $4.50 and $5.50? Assume a normal distribution of wages.
24. On a certain exam, the mean is 72 and the standard deviation is 9. If a grade of A is given to any student who scores at least 2 standard deviations above the mean, what is the lowest score that a person could receive and still get an A?
25. A standardized test given to 10,000 students had scores normally distributed. Al scored 648. The mean was 518 and the standard deviation was 130. About how many students scored below him?
26. In a class of 30 students, Al has a rank of 12. At what percentile is he?
27. Jack was 70th in a class of 200, whereas Jill who is in the same class has a percentile rank of 70. Which student has the higher standing in the class? Why?

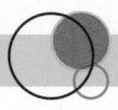

Assessment 10-4B

1. Calculate the mean, the median, and the mode for each of the following data sets:
a. 82, 80, 63, 75, 92, 80, 92, 90, 80, 80
b. 5, 5, 5, 5, 5, 10
2. Write an example of a data set with seven data points for which the mean is not a good description of the center of the data.
3. a. If each of six students scored 80 on a test, find each of the following for the scores:
   i. Mean   ii. Median   iii. Mode
b. Make up another set of six scores that are not all the same but in which the mean, median, and mode are all 80.
4. The tram at a ski area has a capacity of 50 people with a load limit of 7500 lb. What is the mean weight of the passengers if the tram is loaded to its weight limit and capacity?
5. The names and ages for each person in a family of five follow:

Name	Dick	Jane	Kirk	Jean	Scott
Age	40	36	8	6	2

a. What is the mean age?
b. Find the mean of the ages 5 yr from now.
c. Find the mean 10 yr from now.
d. Describe the relationships among the means found in (a), (b), and (c).
6. Suppose there were n students in a class; h of them scored 100 and the rest scored 50. Write an algebraic expression for the mean for the class in terms of n and h.
7. a. Mr. Alberto wanted to count the score on a term paper as 60% of a final grade, homework as 25% of the grade, and the final exam as the remainder of the grade. A student in the class made 85 on the term paper, had a 78 average on homework, and scored 90 on the final exam. What

number could Mr. Alberto use to determine the student's grade if he used his grading scheme?
b. Write an algebraic expression to generalize the scoring procedure for Mr. Alberto's class to allow him to use any percentages he likes for the scoring scheme. Clearly describe what the variables represent.
8. If 99 people had a mean income of $12,000, how much is the mean income increased by the addition of a single income of $200,000?
9. Refer to the following chart. In a gymnastics competition, each competitor receives six scores. The highest and lowest scores are eliminated, and the official score is the mean of the four remaining scores.

Gymnast	Scores					
Balance Beam						
Meta	9.2	9.2	9.1	9.3	9.8	9.6
Lisa	9.3	9.1	9.4	9.6	9.9	9.4
Olga	9.4	9.5	9.6	9.6	9.9	9.6
Uneven Bars						
Meta	9.2	9.1	9.3	9.2	9.4	9.5
Lisa	10.0	9.8	9.9	9.7	9.9	9.8
Olga	9.4	9.6	9.5	9.4	9.4	9.4
Floor Exercises						
Meta	9.7	9.8	9.4	9.8	9.8	9.7
Lisa	10.0	9.9	9.8	10.0	9.7	10.0
Olga	9.4	9.3	9.6	9.4	9.5	9.4

a. If the only events in the competition are the balance beam, the uneven bars, and the floor exercise, find the winner of each event.

b. Find the overall winner of the competition if the overall winner is the person with the highest combined official scores.

10. Choose the data set(s) that fit the descriptions given in each of the following:

a. The mean is 6.
   The range is 6.
   Set *A*: 3, 5, 7, 9
   Set *B*: 2, 4, 6, 8
   Set *C*: 2, 3, 4, 15

b. The mean is 11.
   The median is 11.
   The mode is 11.
   Set *A*: 9, 10, 10, 11, 12, 12, 13
   Set *B*: 11, 11, 11, 11, 11, 11, 11
   Set *C*: 9, 11, 11, 11, 11, 12, 12

c. The mean is 3.
   The median is 3.
   It has no mode.
   Set *A*: 0, $2\frac{1}{2}$, $6\frac{1}{2}$
   Set *B*: 3, 3, 3, 3
   Set *C*: 1, 2, 4, 5

d. Match the data set with the box plot shown below.

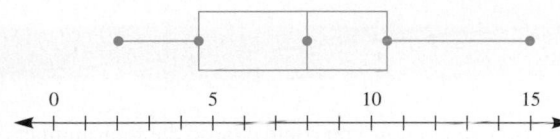

   Set *A*: 2, 3, 4, 4, 6, 6, 7, 15
   Set *B*: 2, 3, 6, 6, 8, 9, 12, 14, 15
   Set *C*: 2, 4.5, 8, 13, 15
   Set *D*: 2, 3, 6, 6, 8, 9, 10, 11, 15

11. The mean of five numbers is 6. If one of the five numbers is removed, the mean becomes 7. What is the value of the number that was removed?

12. a. Find the mean and the median of the following arithmetic sequences:
   i. 1, 3, 5, 7, 9
   ii. 1, 3, 5, 7, 9, . . . , 199
   iii. 7, 10, 13, 16, . . . , 607

b. Based on your answers in (a), make a conjecture about the mean and the median of any arithmetic sequence.

13. The youngest person in a company is 21. The oldest is 66.
a. What is the range?
b. Is the range a good measure of spread for the ages in the company? Why?

14. Ginny's median score on three tests was 90. Her mean score was 92 and her range was 6. What were her three test scores?

15. Following are box plots comparing ages at first marriage for a sample of U.S. citizens.

Age at First Marriage

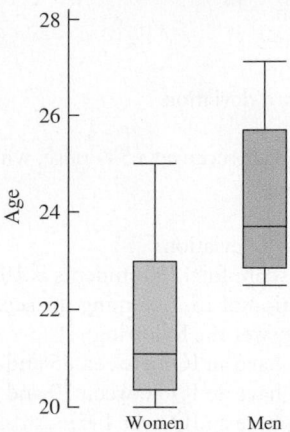

a. What is the median age of first marriage for men?
b. What is the median age of first marriage for women?
c. Which group has the greater range?
d. Compare the first marriage age for men and women.

16. Construct a box plot for the following gas mileages per gallon of various company cars:

   22 18 14 28 30 12 38 22
   30 39 20 18 14 16 10

17. The following table shows the heights in feet of the tallest 10 buildings in Los Angeles and in Minneapolis:

Los Angeles	Minneapolis
858	950
750	775
735	668
699	579
625	561
620	447
578	440
571	416
534	403
516	366

a. Draw horizontal box plots to compare the data.
b. Are there any outliers in this data? If so, which values are they?
c. Based on your box plots from (a), make some comparisons of the heights of the buildings in the two cities.

18. a. If all the numbers in a set are equal, what is the standard deviation?
b. If the standard deviation of a set of numbers is zero, must all the numbers in the set be equal?

19. In a school system, teachers start at a salary of $25,200 and have a top salary of $51,800. The teachers' union is bargaining with the school district for next year's salary increment.
a. If every teacher is given a $1000 raise, what happens to each of the following?

      i. Mean
     ii. Median
    iii. Extremes
    iv. Quartiles
     v. Standard deviation
    vi. IQR

   b. If every teacher received a 5% raise, what does this do to the following?
      i. Mean
     ii. Standard deviation

20. The mean IQ score for 1500 students is 100, with a standard deviation of 15. Assuming the scores have a normal distribution, answer the following:
   a. How many have an IQ between 85 and 115?
   b. How many have an IQ between 70 and 130?
   c. How many have an IQ over 145?

21. Sugar Plops boxes say they hold 16 oz. To make sure they do, the manufacturer fills the box to a mean weight of 16.1 oz, with a standard deviation of 0.05 oz. If the weights have a normal distribution, what percentage of the boxes actually contain 16 oz or more?

22. According to psychologists, IQs are normally distributed, with a mean of 100 and a standard deviation of 15.

   a. What percentage of the population has IQs between 100 and 130?
   b. What percentage of the population has IQs lower than 85?

23. The weights of newborn babies in a certain country are distributed normally, with a mean of approximately 105 oz and a standard deviation of 20 oz. If a newborn is selected at random, what is the probability that the baby weighs less than 125 oz?

24. A tire company tests a particular model of tire and finds the tires to be normally distributed with respect to wear. The mean is 28,000 mi and the standard deviation is 2500 mi. If 2000 tires are tested, about how many are likely to wear out before 23,000 mi?

25. A standardized mathematics test given to 10,000 students had the scores normally distributed. The mean was 500 and the standard deviation was 60. A student scoring below 440 points was deficient in mathematics. About how many students were rated deficient?

26. In a class of 25 students, Bill has a rank of 25th. At what percentile is he?

27. Jill was 80th in a class of 200, whereas Nathan, who is in the same class, has a percentile rank of 80. Which student has the higher standing in the class?

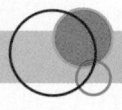

Mathematical Connections 10-4

Communication

1. Suppose you own a hat shop and decide to order hats in only *one* size for the coming season. To decide which size to order, you look at last year's sales figures, which are itemized according to size. Should you find the mean, median, or mode for the data? Why?

2. A movie chain conducts a popcorn poll in which each person entering a theater and buying a box of popcorn is asked a yes-no question. Which "average" do you think is used to report the result and why?

3. When a government agency reports the average rainfall for a state for a year, which "average" do you think it uses and why?

4. Carl had scores of 90, 95, 85, and 90 on his first four tests.
   a. Find the median, mean, and mode.
   b. Carl scored a 20 on his fifth exam. Which of the three averages would Carl want the instructor to use to compute his grade? Why?
   c. Which measure is affected most by an extreme score?

5. The mean of the five numbers given is 50:

        20 35 50 60 85

   a. Add four numbers to the list so that the mean of the nine numbers is still 50.

   b. Explain how you could choose the four numbers to add to the list so that the mean did not change.
   c. How does the mean of the four numbers you added to the list compare to the original mean of 50?

6. Selina claimed that in her class all of the scores on a test were either 100 or 50, so the mean must be 75.
   a. Explain with an example how Selina may have reasoned.
   b. Explain whether this reasoning is valid.

7. Sue drives 5 mi at 30 mph and then 5 mi at 50 mph. Is the mean speed for the trip 40 mph? Why or why not?

8. Explain why the mode could be a less-than-adequate measure of center for a data set.

9. If you were to make an argument that there were not enough women's bathrooms in a theater, what type of data would you use and why? Explain the types of measures of center and spread that you would report for your data.

10. The mean of 5, 7, 9 is 7. The mean of 67, 72, 77 is 72.
   a. Find two more examples where the mean of an ordered set of data is the "middle" data point.
   b. Suppose that a data set consists of $a_1, a_2, a_3, \ldots, a_n$, an arithmetic sequence. Explain why the mean of this data set is $\dfrac{a_1 + a_n}{2}$.

Open-Ended

11. Use the data in the following table to compare the number of people living in the United States from 1810 through 1900 and the number living in the United States from 1910 through 2000. Use any form of graphical representation to make the comparison and explain why you chose the representation that you did.

Year	Number (thousands)	Year	Number (thousands)
1810	7,239	1910	92,228
1820	9,638	1920	106,021
1830	12,866	1930	123,202
1840	17,068	1940	132,164
1850	23,191	1950	151,325
1860	31,443	1960	179,323
1870	38,558	1970	203,302
1880	50,189	1980	226,542
1890	62,979	1990	248,765
1900	76,212	2000	281,422

Data taken from *The World Almanac and Book of Facts 2002*, World Almanac Books, 2002.

Cooperative Learning

12. In small groups, determine a method of finding the number and types of graphs and statistical representations used in at least two newspapers in your campus library. Based on your findings, write a report defending which type(s) of representation should be emphasized in a journalistic statistics class.

Questions from the Classroom

13. A student asks, "If the average income of 10 people is $10,000 and one person gets a raise of $10,000, is the median, the mean, or the mode changed and, if so, by how much?"
14. Jose asks, "Why can a median number of children not be 3.8?" How do you respond?
15. Suppose the class takes a test and the following averages are obtained: mean, 80; median, 90; and mode, 70. Tom, who scored 80, would like to know if he did better than half the class. What is your response?
16. A student asks if it is possible to find the mode for data in a grouped frequency table. What is your response?
17. A student asks if she can draw any conclusions about a set of data if she knows that the mean for the data is less than the median. How do you answer?
18. A student asks if it is possible to have a standard deviation of ⁻5. How do you respond?
19. Mel's mean on 10 tests for the quarter was 89. She complained to the teacher that she should be given an A because she missed the cutoff of 90 by only a single point. Explain whether it is clear that she really missed an A by only a single point if each test was based on 100 points.

Review Problems

20. Consider the following circle graph. What is the number of degrees in each sector of the graph?

U.S. Car Sales by Vehicle Size and Type
for Ali Auto Sales

2012

Large 7%   Luxury 17%

Midsize 48%   Small 28%

21. Given the bar graph shown, answer the following:
    a. Which mountain is the highest? Approximately how high is it?
    b. Which mountains are higher than 6000 m?

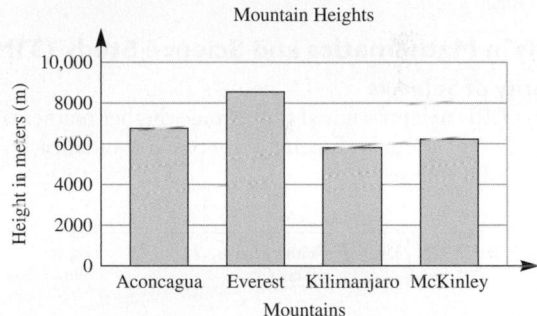

Mountain Heights

22. Following are raw test scores from a history test:

86	85	87	96	55
90	94	82	68	77
88	89	85	74	90
72	80	76	88	73
64	79	73	85	93

a. Construct an ordered stem and leaf plot for the given data.
b. Construct a grouped frequency table for these scores with intervals of 5, starting the first class at 55.
c. Draw a histogram of the data.
d. If a circle graph of the grouped data in (b) were drawn, how many degrees would be in the section representing the 85 through 89 interval?

National Assessment of Educational Progress (NAEP) Questions

Score	Number of Students
90	1
80	3
70	4
60	0
50	3

The table on the previous page shows the scores of a group of 11 students on a history test. What is the average (mean) score of the group to the nearest whole number?

NAEP, Grade 8, 2003

Tetsu rides his bicycle x miles the first day, y miles the second day, and z miles the third day. Which of the following expressions represents the average number of miles per day that Tetsu travels?

a. $x + y + z$ **b.** xyz
c. $3(x + y + z)$ **d.** $3(xyz)$
e. $(x + y + z)/3$

NAEP, Grade 8, 2003

The prices of gasoline in a certain region are \$1.41, \$1.36, \$1.57, and \$1.45 per gallon. What is the median price per gallon for gasoline in this region?

a. \$1.41   **b.** \$1.43   **c.** \$1.44   **d.** \$1.45   **e.** \$1.47

NAEP, Grade 8, 2005

Trends in Mathematics and Science Study (TIMSS)

Popularity of Subjects

A group of 10 students wanted to find out whether mathematics or history was more popular for their group. They rated each subject using the following scale.

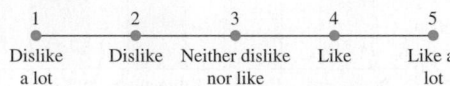

1	2	3	4	5
Dislike a lot	Dislike	Neither dislike nor like	Like	Like a lot

The table shows the results:

Students' Ratings

Student	Mathematics Rating	History Rating
Alan	1	2
Lisa	4	4
Ann	5	4
John	2	2
Connor	4	2
Georgia	3	3
Bret	2	1
Courtney	1	1
Ian	5	3
Jackson	3	2
Totals	**30**	**24**

Calculate the mean (average) rating for each subject.

Mean rating for mathematics = _____

Mean rating for history = _____

According to the ratings, which is the more popular subject for this group of students?

More popular subject: _____

TIMSS, Grade 8, 2007

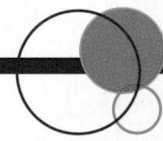

10-5 Abuses of Statistics

In *PSSM*, we find the following:

> The amount of data available to help make decisions in business, politics, research, and everyday life is staggering. Consumer surveys guide the development and marketing of products. Polls help determine political-campaign strategies, and experiments are used to evaluate the safety and efficacy of new medical treatments. Statistics are often misused to sway public opinion on issues or to misrepresent the quality and effectiveness of commercial products. (p. 48)

As noted, statistics are frequently abused. Benjamin Disraeli (1804–1881), an English prime minister, once remarked, "There are three kinds of lies: lies, damned lies, and statistics." People sometimes deliberately use statistics to mislead others. This can be seen in advertising. More often, however, the misuse of statistics is the result of misinterpreting what the data and statistics mean.

Hi & Lois copyright © 1989 King Features Syndicate

Misuses Based on Samples/Population

Consider an advertisement reporting that of the people responding to a recent survey, 98% said that Buffepain is the most effective pain reliever of headaches and arthritis. To certify that the statistics are not being misused, the following information should have been reported:

1. The questions being asked
2. The number of people surveyed
3. The number of people who responded
4. How the people who participated in the survey were chosen
5. The number and type of pain relievers tested
6. How the answers were interpreted

Without the information listed, the following situations are possible, all of which could cause the advertisement to be misleading:

1. Suppose 1,000,000 people nationwide were sent the survey, and only 50 responded. This would mean that there was only a 0.005% response, which would certainly cause mistrust in the ad.
2. Suppose a survey sentence read, "Buffepain is the best pain reliever I've tried for headaches and arthritis," and there were no questions about the kind and type of other pain relievers tried.
3. Of the 50 responding in (1), suppose 49 responses were affirmative. The 98% claim is true, but 999,950 people did not respond.
4. Suppose all the people who received the survey were chosen from a town in which the major industry was the manufacture of Buffepain. It is very doubtful that the survey would represent an unbiased sample.
5. Suppose only two "pain relievers" were tested: Buffepain, whose active ingredient is 100% aspirin, and a placebo containing only powdered sugar.

This is not to say that advertisements of this type are all misleading or dishonest, but simply that data interpretations are only as honest as their reporters.

In the Buffepain report, a primary issue deals with the survey conducted, the number of people involved in the survey, how they were chosen, and how results of the survey were interpreted. Surveys are common for gaining information from a population. However, there are classic examples in history of how surveys and survey information have been either misused or misinterpreted. A well-known example was seen in the predictions of the winner of the Harry S. Truman/Thomas E. Dewey 1948 U.S. presidential election. A leading pollster, Elmo Roper, was so confident of a loss by Truman that on September 9, 1948, approximately 2 months before the election, he announced that there would be no more polls on the election. Additionally, while Truman was still on the campaign trail, *Newsweek*, after polling 50 key political experts, stated on October 11, 1948, approximately 1 month before the election, that Dewey would win.

In the *Newsweek* survey, a very select group was surveyed. Questions that should have been asked include (1) How was the group chosen? (2) Was the group representative of the voters in the United States? (3) Could the result of this survey appropriately have been generalized to all voters? All of these types of questions are important to surveyors. What is the population to which the results are being generalized? Is the entire population to be surveyed, or is the survey given only to a sample of the population? If only a sample will be used, how is the sample chosen? Is the sample of the population randomly chosen so that each person being surveyed has an equally likely chance of being chosen? How large is the population and how large a sample must be used so that the sample is representative of the population?

NOW TRY THIS 10-15

Based on the graph in Figure 10-36, write arguments to determine whether or not the following inferences are correct:

a. 82% of the parents surveyed say that there is not too much focus on preparing for tests.
b. Only 9% of the parents say that learning is thwarted.
c. 12% of the parents say test questions are too difficult and that expectations are unreasonable.
d. At least 89% of the parents believe that schools require too many tests.
e. The vast majority of parents say that teachers are not putting too much academic pressure on their children.

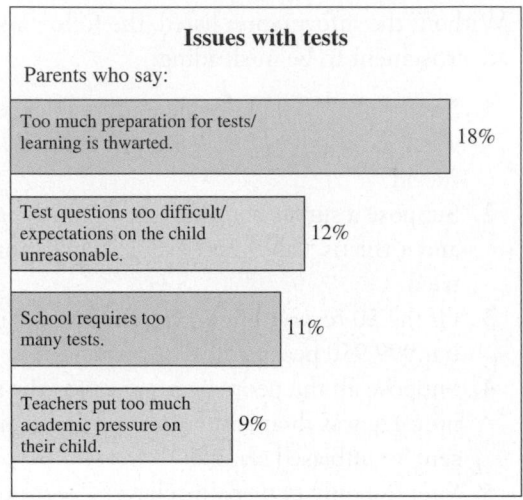

Source: Data from *USA Today*, November 21, 2000.

Figure 10-36

Misuses Based on Graphs

A different type of misuse of data and statistics involves graphs. Among the things to look for in a graph are the following. If they are not there, then the graph may be misleading.

1. Title
2. Labels on both axes of a line or bar chart and on all sections of a pie chart
3. Source of the data
4. Key to a pictograph
5. Uniform size of symbols in a pictograph
6. Scale: Does it start with zero? If not, is there a break shown?
7. Scale: Are the numbers equally spaced?

To see an example of a misleading use of graphs, consider how graphs can be used to distort data or exaggerate certain pieces of information. Graphs using a break in the vertical axis can be used to create different visual impressions, which are sometimes misleading. For example, consider the two graphs in Figure 10-37, which represent the number of girls trying out for basketball at each of three middle schools. As we can see, the graph in Figure 10-37(a) portrays a different picture from the one in Figure 10-37(b).

A line graph, histogram, or bar graph can be altered by changing the scale of the graph. For example, consider the data in Table 10-21 for the number of graduates from a community college for the years 2008 through 2012.

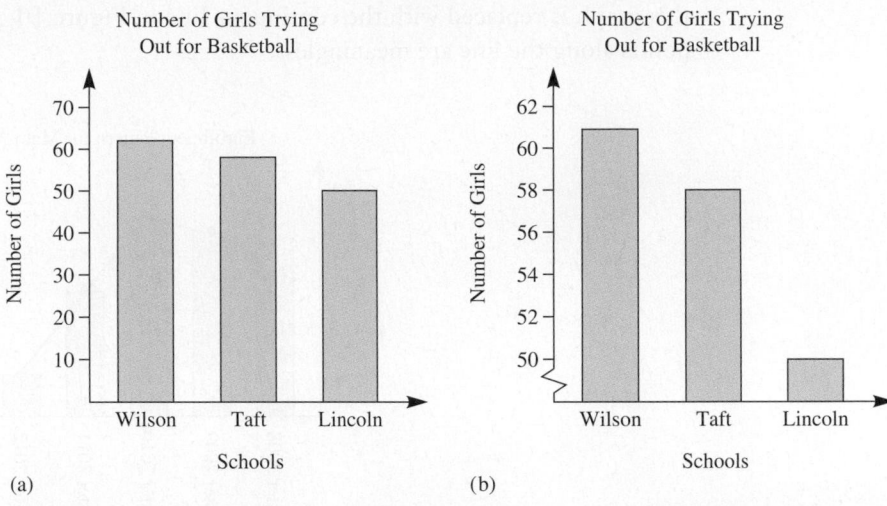

Figure 10-37

Table 10-21

Year	2008	2009	2010	2011	2012
Number of Graduates	140	180	200	210	160

The graphs in Figure 10-38(a) and (b) represent the same data, but different scales are used in each. The statistics presented are the same, but these graphs do not convey the same psychological message. In Figure 10-38(b), the spacing of the years on the horizontal axis of the graph is more spread out and the spacing of the numbers on the vertical axis is more condensed than in Figure 10-38(a). Both of these changes minimize the variability of the data. A college administrator might use a graph like the one in Figure 10-38(b) to convince people that the college was not in serious enrollment trouble.

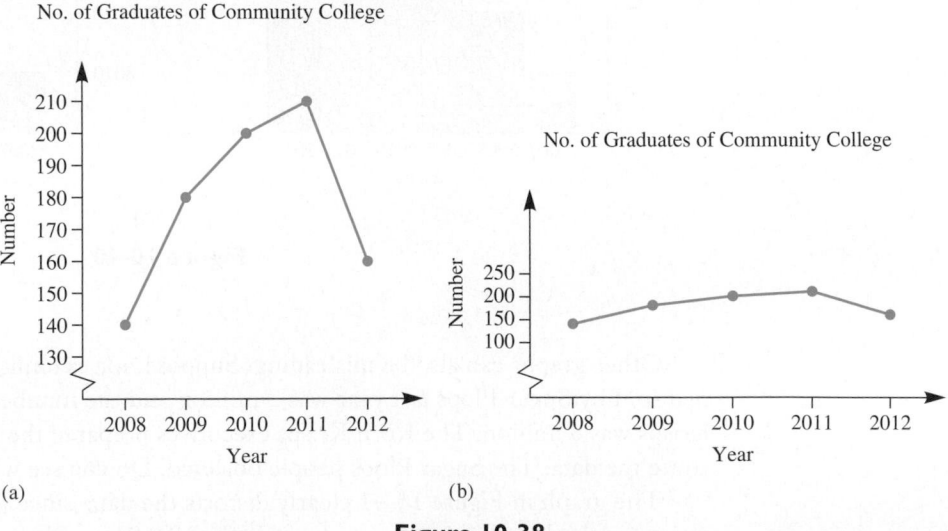

Figure 10-38

Another error that frequently occurs is the use of continuous graphs, as in Figure 10-39, to depict data that are discrete (a finite number of data values). In Figure 10-39, we see the enrollment in Math 206 for five semesters. The bar graph accurately depicts the enrollment data, but if this graph is replaced with the continuous line in Figure 10-39 but without the bars, then many points along the line are meaningless.

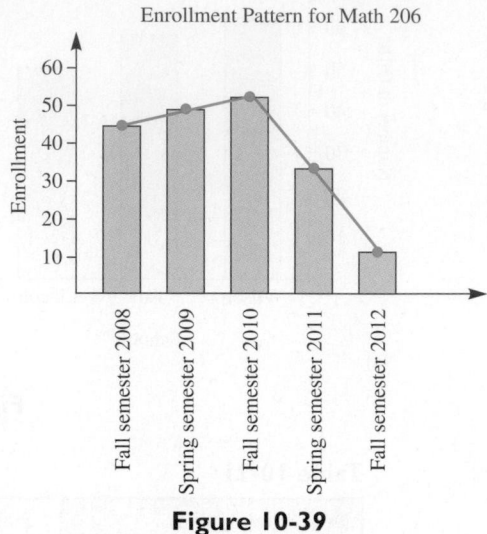

Figure 10-39

Other ways to distort graphs include omitting a scale, as in Figure 10-40(a). The scale is given in Figure 10-40(b).

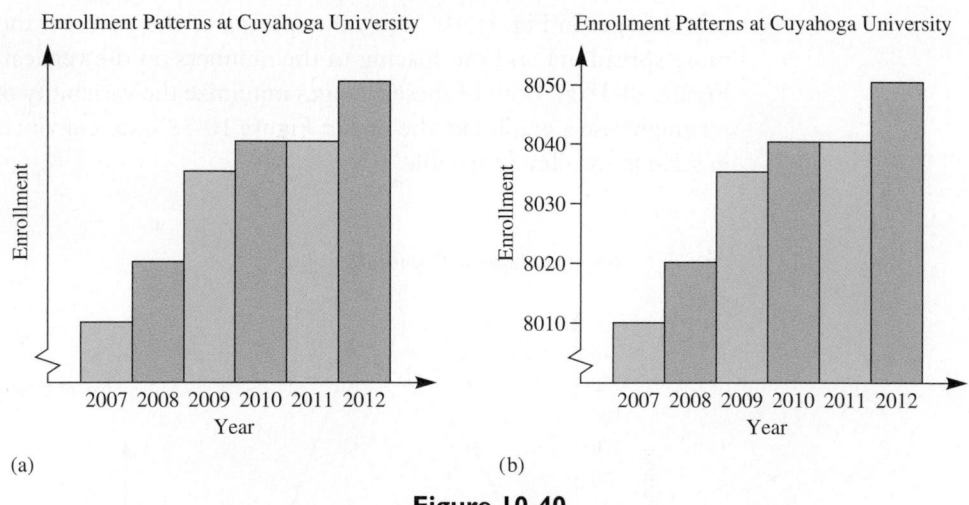

Figure 10-40

Other graphs can also be misleading. Suppose, for example, that the number of boxes of cereal sold by Sugar Plops last year was 2 million and the number of boxes of cereal sold by Korn Krisps was 8 million. The Korn Krisps executives prepared the graph in Figure 10-41 to demonstrate the data. The Sugar Plops people objected. Do you see why?

The graph in Figure 10-41 clearly distorts the data, since the figure for Korn Krisps is both 4 times as high and 4 times as wide as the bar for Sugar Plops. Thus, the area of the bar representing Korn Krisps is 16 times the comparable area representing Sugar Plops, rather than 4 times the area, as would be justified by the original data. To depict the data accurately, the length of a side of the Sugar Plops bar should be 4 units. Then four of these bars would "fit" in the Korn

Krisps bar. Figure 10-42 shows how the comparison of Sugar Plops and Korn Krisps cereals might look if the figures were made three-dimensional. The figure for Korn Krisps has a volume 64 times the volume of the Sugar Plops figure.

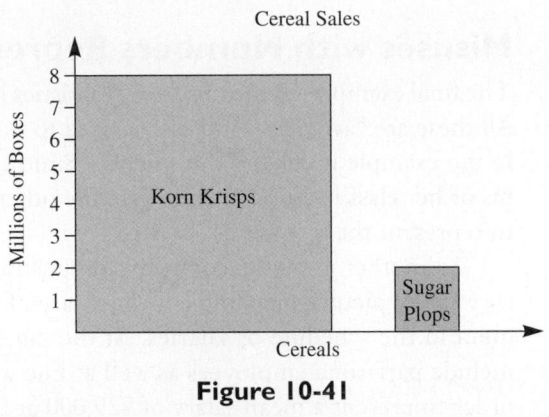

Figure 10-41

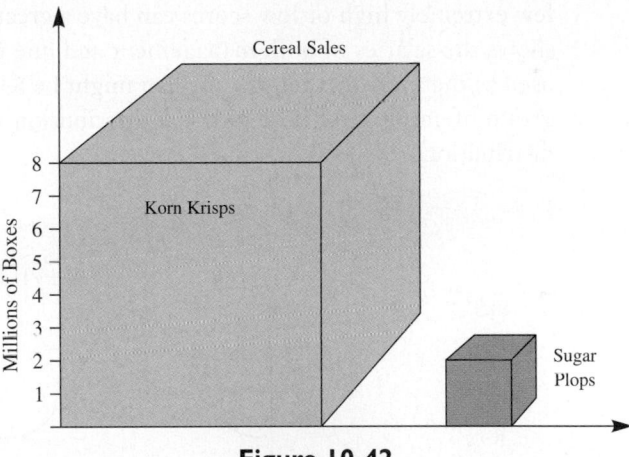

Figure 10-42

Graphs easily become misleading when attempts are made to depict them as three-dimensional. Many graphs of this type do not acknowledge either the variable thickness of the depiction or the distortion due to perspective. Observe that the 27% sector pictured in Figure 10-43(a) looks much

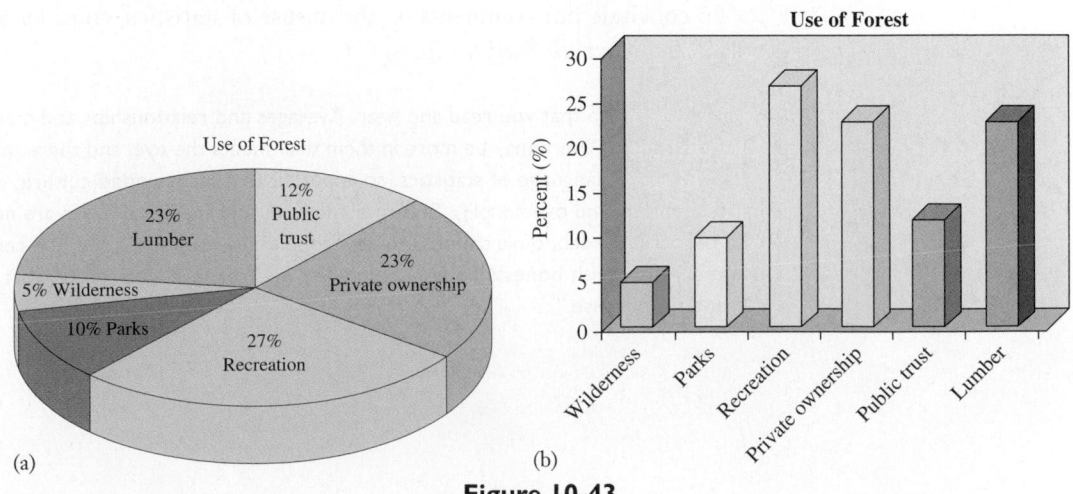

(a)   (b)

Figure 10-43

greater than the 23% sector, although they should be very nearly the same size. In Figure 10-43(b) the three-dimensional bar graph is more attractive but it is harder to read the true values of the data. These three-dimensional graphs are typically created with computer software.

Misuses with Numbers Representing Data

The final examples of the misuses of statistics involve misleading uses of mean, median, and mode. All these are "averages" and can be used to suit a person's purposes. As discussed in Section 10-4 in the example involving the teachers Smith, Jones, and Rivera, each teacher had reported that his or her class had done better than the other two. Each of the teachers used a different number to represent the test scores.

As another example, company administrators wishing to portray to prospective employees a rosy salary picture may find a mean salary of $58,000 for line workers as well as upper management in the schedule of salaries. At the same time, a union that is bargaining for salaries may include part-time employees as well as line workers and will exclude management personnel in order to present a mean salary of $29,000 at the bargaining table. The important thing to watch for when a mean is reported is disparate cases in the reference group. If the sample is small, then a few extremely high or low scores can have a great influence on the mean. Suppose Figure 10-44 shows the salaries of both management and line workers of the company. If the median is being used as the average, then the median might be $43,500, which is representative of neither major group of employees. The bimodal distribution means the median is nonrepresentative of the distribution.

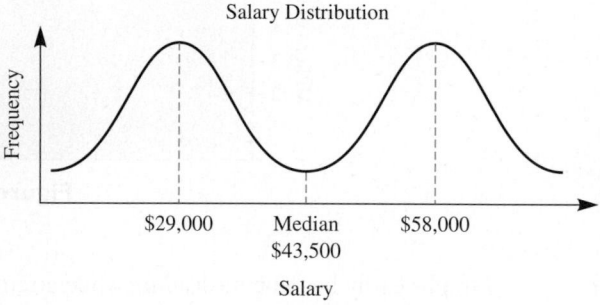

Figure 10-44

To conclude our comments on the misuse of statistics, consider a quote from Darrell Huff's book *How to Lie with Statistics* (p. 8):

> So it is with much that you read and hear. Averages and relationships and trends and graphs are not always what they seem. There may be more in them than meets the eye, and there may be a good deal less.
>
> The secret language of statistics, so appealing in a fact-minded culture, is employed to sensationalize, inflate, confuse, and oversimplify. Statistical methods and statistical terms are necessary in reporting the mass data of social and economic trends, business conditions, "opinion" polls, the census. But without writers who use the words with honesty and understanding and readers who know what they mean, the result can be semantic "nonsense."

Assessment 10-5A

This entire set of assessment items is appropriate for communication and cooperative learning. Many items are open-ended and several lend themselves to further investigation.

1. Discuss whether the following claims could be misleading. Explain why and how.
   a. A car manufacturer claims its car is quieter than a glider.
   b. A motorcycle manufacturer claims that more than 95% of its cycles sold in the United States in the last 15 yr are still on the road.
   c. A company claims its fruit juice has 10% more fruit solids than is required by U.S. government standards. (The government requires 10% fruit solids.)
   d. A brand of bread claims to be 40% fresher.

2. The city of Podunk advertised that its temperature was the ideal temperature in the country because its mean temperature was 25°C. What possible misconceptions could people draw from this advertisement?

3. A student read that 9 out of 10 pickup trucks sold in the last 10 yr are still on the road. She concluded that the average life of a pickup is around 10 yr. Is she correct?

4. What is wrong with the following line graph?

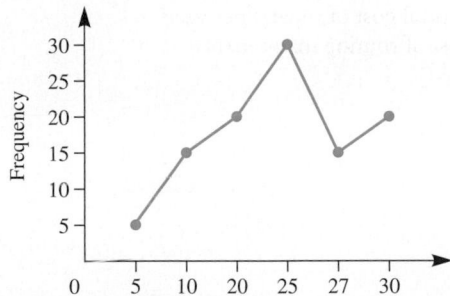

5. Doug's Dog Food Company wanted to impress the public with the magnitude of the company's growth. Sales of Doug's Dog Food had doubled from 2010 to 2011, so the company displayed the following graph, in which the radius of the base and the height of the 2011 can are double those of the 2010 can. What does the graph really show with respect to the growth of the company? (*Hint:* The volume of a cylinder is given by $V = \pi r^2 h$, where r is the radius of the base and h is the height.)

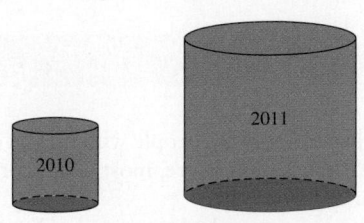

Doug's Dog Food Sales

6. Refer to the following circle graph.

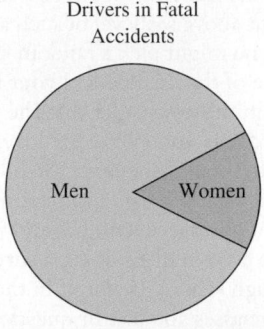

Drivers in Fatal Accidents

Ms. McNulty claims that on the basis of this information, we can conclude that men are worse drivers than women. Discuss whether you can reach that conclusion from the pictograph or you need more information. If more information is needed, what would you like to know?

7. Can you draw any valid conclusions about a set of data in which the mean is less than the median?

8. The following graph was prepared to compare prices of washing machines at three stores:

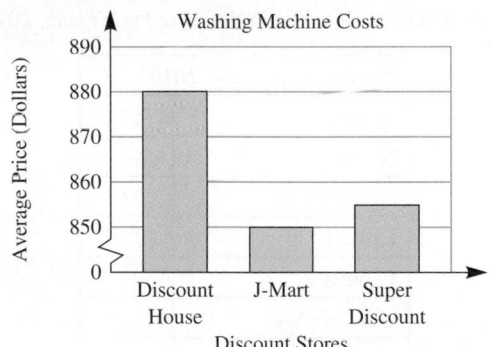

Which of the following statements is true? Explain why or why not.
   a. Prices vary widely at the three stores.
   b. The price at Discount House is 4 times as great as that at J-Mart.
   c. The prices at J-Mart and Super Discount differ by less than $10.

9. The following table gives the number of accidents per year on a certain highway for a 5-yr period:

Year	2008	2009	2010	2011	2012
Number of Accidents	24	26	30	32	38

   a. Draw a bar graph to convince people that the number of accidents is on the rise and that something should be done about it.

   b. Draw a bar graph to show that the rate of accidents is almost constant, and that nothing needs to be done.

10. Write a list of scores for which the mean and median are not representative of the list.

11. Consider a state such as Montana that has both mountains and prairies. What numbers might you report to depict the "average" height above sea level of such a state? Why?

12. Describe how you might pick a random sample of adults that is representative of the members of your town.

13. A student read in the newspaper that the pill form of a drug taken once per day is up to 92% effective in warding off the flu. She concludes that if she takes the pill 8% more times per day, she will be 100% safe from contracting the flu. Explain whether you agree with her and why.

14. **a.** Suppose you are a student doing a survey of eating habits in a high school. If you sit in the hall and ask each student who passes you in 1 hr questions about their eating habits, explain whether or not you think you have a representative sample of the school population.

   b. If you are conducting the survey mentioned in part (a) and interview students in the cafeteria at noon, explain whether or not you think that you have a representative sample of the school population.

   c. For this survey, explain how you could choose a sample to be reasonably sure that you got a representative sample of the population.

15. The following chart lists the number of complaints received about airlines as reported in the *New York Times*, 2010:

Most complaints, 2010	
American	517
Delta	403
Continental	228
United	176
Fewest	
U.S. Airways	131
Southwest Airlines	118
Alaska	62

   Discuss whether you think that American, Delta, Continental, and United are the worst airlines and U. S. Airways, Southwest, and Alaska are the best based on the above information.

16. In the graphic shown, two states are compared using their students' SAT scores and the spending ranking per pupil by those states. What are some erroneous conclusions that could be reached?

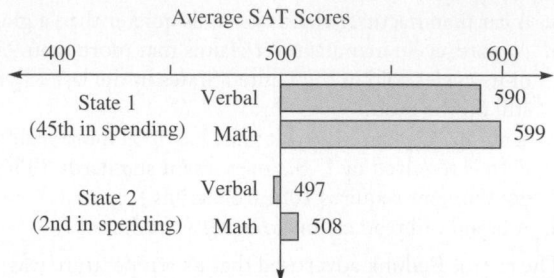

Average SAT Scores

State 1 (45th in spending): Verbal 590, Math 599
State 2 (2nd in spending): Verbal 497, Math 508

17. Which of the following pieces of information would not be helpful in deciding the type of automobile that is the most economical to drive?
   a. Range of insurance costs
   b. Modes of drivers' ages for specific vehicle types
   c. Mean miles per gallon
   d. Typical cost of repairs per year
   e. Cost of routine maintenance

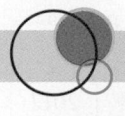

Assessment 10-5B

1. Discuss whether the following claims could be misleading. Explain why and how.
   a. A used-car dealer claims that a car she is trying to sell will get up to 30 mpg.
   b. Sudso claims that its detergent will leave your clothes brighter.
   c. A sugarless gum company claims that 8 of every 10 dentists responding to the survey recommend sugarless gum.
   d. Most accidents occur in the home. Therefore, to be safer, you should stay out of your house as much as possible.

   e. More than 95% of the people who fly to a certain city do so on Airline A. Therefore, most people prefer Airline A to other airlines.

2. General Cooster once asked a person by the side of a river if the river was too deep to ride his horse across. The person responded that the average depth was 2 ft. If General Cooster rode out across the river, what assumptions did he make on the basis of the person's information?

3. Jenny read that 80% of those responding to a survey in her school favored 2 hours of homework per night. What, if anything, might be wrong with this claim?

4. What is wrong with the following line graph?

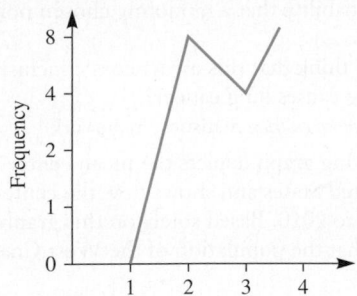

5. A histogram of the data for the number of cars sold by Acme Car Lot is given in Figure (a) below. A three-dimensional version of the histogram as shown in Figure (b) was given out by the company. What is wrong or misleading about the graph in Figure (b)?

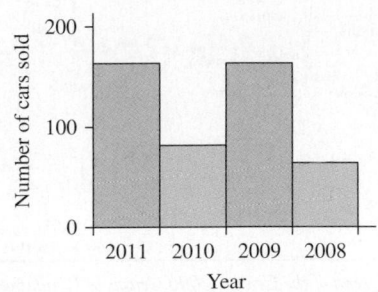

(a)

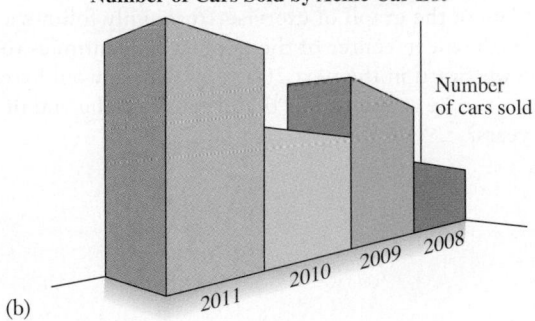

(b)

6. Suppose the following circle graphs are used to illustrate the fact that the number of elementary teaching majors at teachers' colleges has doubled between 2000 and 2009, while the percentage of male elementary teaching majors has stayed the same. What is misleading about the way the graphs are constructed?

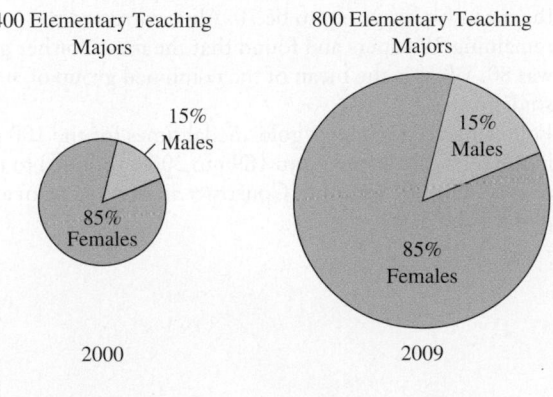

7. Jenny averaged 70 on her quizzes during the first part of the quarter and 80 on her quizzes during the second part of the quarter. When she found out that her final average for the quarter was not 75, she went to argue with her teacher. Give a possible explanation for Jenny's misunderstanding.

8. Explain what is wrong with the following graph:

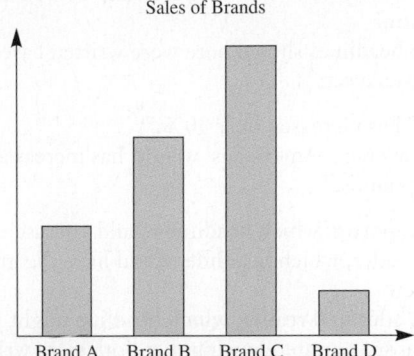

9. In a recent survey, teachers rated their mathematics textbooks as follows:

National Survey of Teachers of Mathematics

	Grades K–4 (percentage)	Grades 5–8 (percentage)
Very poor	1	2
Poor	3	5
Fair	18	16
Good	35	33
Very good	36	33
Excellent	8	10

At about the same time that the survey was being done, newspapers printed articles saying that national experts in mathematics have been very critical of mathematics textbooks at all levels.

a. What is your reaction to the survey and the reports?

b. Does the national survey agree with the newspaper articles? Explain your thinking as you react.

10. A student got 99 on a quiz and was ecstatic over the grade. What other information might you need in order to decide if the student was justified in being happy?

11. a. Use the following data to justify the amount of time that you expect to assign for weekly homework to classes in grades K–4 and grades 5–8.

National Survey of Teachers Concerning the Amount of Homework Assigned per Week

	K–4 (percentages)	5–8 (percentages)
0–30 min	48	8
31–60 min	27	21
61–90 min	13	26
91–120 min	8	24
2–3 hr	3	17
More than 3 hr	1	5

b. How might the survey data be misused to justify assigning at least 2 hr of homework per week?

12. Is it possible for a state or country to have a mean sea level that is negative? If so, what might such a region look like?

13. What are the characteristics that you think a sample might have to have to be representative of an entire population?

14. The two headlines shown here were written based on exactly the same data set:

"Obesity has increased over 30%."

"On the average, Americans' weight has increased by less than 10 pounds."

a. As a reporter, which headline would you use?
b. As a reader, which headline would have the greatest impact?
c. As an educated reader, which headline might you consider the most accurate knowing that both were written based on the same data set? Explain your answer.

15. In a British study around 1950, a group of 649 men with lung cancer were surveyed. A control group of the same size was established from a set of men who did not have lung cancer. The groups were matched according to ethnicity, age, and socioeconomic status. The statistics from the survey follow.*

	Lung Cancer Cases	Controls	Totals
Smokers	647	622	1269
Nonsmokers	2	27	29

a. What is the fraction of smokers in the group that have lung cancer?
b. What is the fraction of smokers in the control group?
c. If one person is chosen at random from each of the two groups (smokers and nonsmokers), what is the probability that a randomly chosen smoker has cancer? What is the probability that a randomly chosen non-smoker has cancer?

d. Do you think that this evidence is conclusive that smoking causes lung cancer?
(* *Problem taken from* Statistics Framework)

16. The following graph depicts the mean center of population of the United States and shows how the center has shifted from 1790 to 2010. Based solely on this graph, could you conclude that the population of the West Coast has increased since 1790?

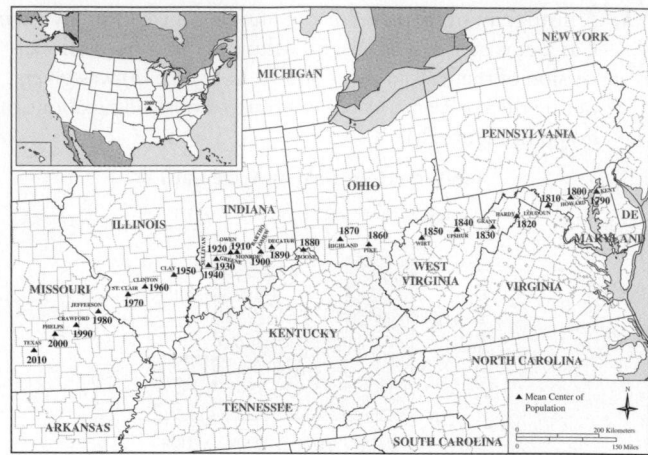

Source: U.S. Bureau of the Census, 2000 Census of Population and Housing, Population and Housing Unit Counts, United States (2000 CPH-2-1).

17. The data in the graph of exercise 16 roughly follows a line. If the mean center of the population continues to move westward in the next 200 years, where would you expect it to be at the end of that period? At the end of 400 years?

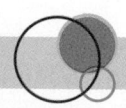

Mathematical Connections 10-5

Review Problems

1. On the English 100 exam, the scores were as follows:

    43 91 73 65
    56 77 84 91
    82 65 98 65

    a. Find the mean.
    b. Find the median.
    c. Find the mode.
    d. Find the variance.
    e. Find the standard deviation.
    f. Find the mean absolute deviation.

2. If the mean of a set of 36 scores is 27 and two more scores of 40 and 42 are added, what is the new mean?

3. On a certain exam, Tony corrected 10 papers and found the mean for his group to be 70. Alice corrected the remaining 20 papers and found that the mean for her group was 80. What is the mean of the combined group of 30 students?

4. Following are the men's gold-medal times for the 100 m run in the Olympic games from 1896 to 2008, rounded to the nearest tenth of a second. Construct an ordered stem and leaf plot for the data.

Year	Time (sec), rounded
1896	12.0
1900	11.0
1904	11.0
1908	10.8
1912	10.8
1920	10.8
1924	10.6
1928	10.8
1932	10.3
1936	10.3
1948	10.3
1952	10.4
1956	10.5
1960	10.2
1964	10.0
1968	10.0
1972	10.1
1976	10.1
1980	10.3
1984	10.0
1988	9.9
1992	9.7
1996	9.8
2000	9.9
2004	9.8
2008	9.7

5. Following are the record swimming times of the women's 100-m freestyle and 100-m butterfly in the Olympics from 1960 to 2008. Draw parallel box plots of the two sets of data to compare them.

Year	Time—100 m Freestyle (sec)	Time—100 m Butterfly (sec)
1960	61.20	69.50
1964	59.50	64.70
1968	60.00	65.50
1972	58.59	63.34
1976	55.65	60.13
1980	54.79	60.42
1984	55.92	59.26
1988	54.93	59.00
1992	54.64	58.62
1996	54.50	59.13
2000	53.83	56.61
2004	53.84	57.72
2008	53.12	56.73

National Assessment of Educational Progress (NAEP) Questions

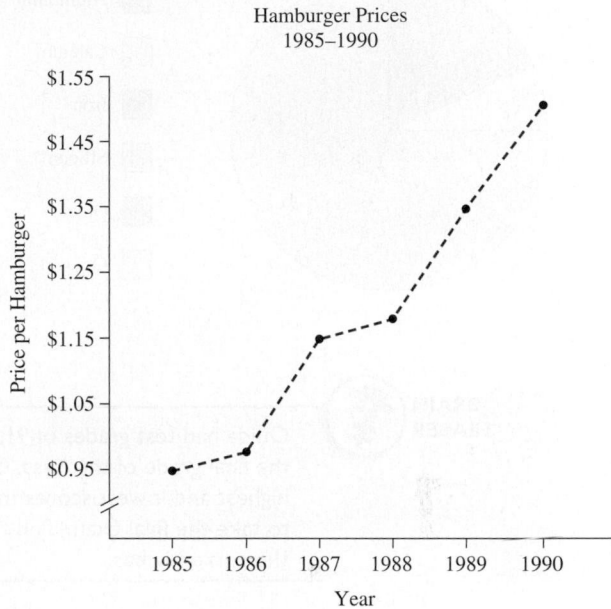

Hamburger Prices 1985–1990

According to the graph, how many times did the yearly increase of the price of a hamburger exceed 10 cents?
a. None
b. One
c. Two
d. Three
e. Four

NAEP, Grade 8, 2003

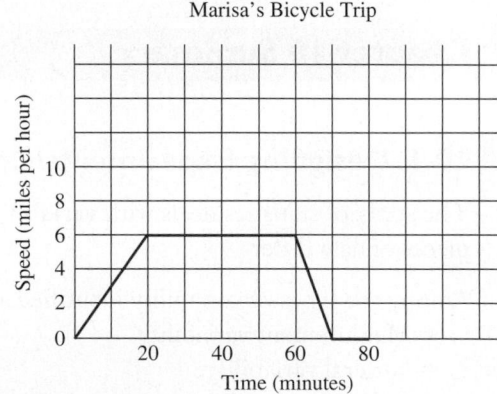

Marisa's Bicycle Trip

The graph above represents Marisa's riding speed throughout her 80-minute bicycle trip. Use the information in the graph to describe what could have happened on the trip, including her speed throughout the trip.

During the first 20 minutes, Marisa _____

_____

From 20 minutes to 60 minutes, she _____

_____

From 60 minutes to 80 minutes, she _____

_____

NAEP, Grade 8, 2003

Elements That Make Up the Earth's Crust

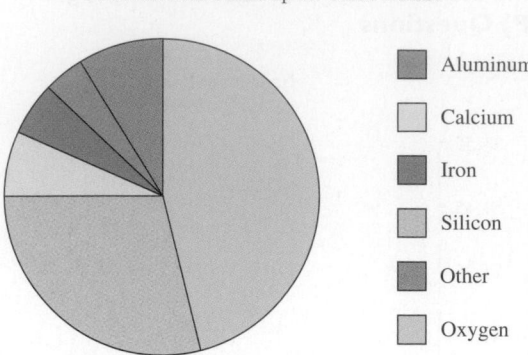

- Aluminum
- Calcium
- Iron
- Silicon
- Other
- Oxygen

According to the graph at left, which element forms the second greatest portion of the earth's crust?

a. Oxygen
b. Silicon
c. Aluminum
d. Iron
e. Calcium

NAEP, Grade 8, 2007

BRAIN TEASER Ouida had test grades of 91, 89, 83, 78, 76, and 75. Her teacher uses the mean of the test scores as the final grade of the class. Ouida has the option of keeping her current test average or replacing the highest and lowest scores that she currently has with a single score on a final exam. If Ouida chooses to take the final exam, find the lowest score she can receive and still maintain at least the mean that she currently has.

Hint for Solving the Preliminary Problem

The following process could help you find the mean salary.

The first person could have the calculator pick a large random number and then he adds his salary to the number and passes it to the next person. The next person then adds her salary to the total and passes it to the next person. Think how you could use this process to find the average salary.

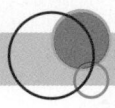

Chapter 10 Summary

10-1 Designing Experiments/Collecting Data	Pages
The study of statistics deals with **variability** of data or the amount by which pieces of data differ.	543
Among the types of variability identified are the following:	
• Measurement variability	543
• Natural variability	543
• Induced variability	543
• Sampling variability	543
A **population** is a group of objects or people. A **sample** is a part of the population or in some cases the whole population. A **random sample** is a subset of a population where each member of a population has the same chance of being selected.	543
The following framework holds for statistical problem solving:	
• Formulate the questions	544
• Collect the data	
• Analyze the data	
• Interpret the results	

10-2 Displaying Data: Part I

Data may be either **categorical** or **numerical**.
- **Categorical data** are data that represent characteristics of objects or individuals in groups (or categories).   552
- **Numerical data** are data collected on numerical variables.   552

Information can be summarized in each of the following forms
- Pictographs   552
- Frequency tables   552
- Dot plots / line plots   553
- Stem and leaf plots   554
- Grouped frequency tables   557
- Histograms   557
- Bar graphs   558
- Circle graphs/pie charts   560

10-3 Displaying Data: Part II

Information can be summarized in the following forms:
- Line graphs   570
- Scatter plots   572
  - Scatter plots can suggest a **trend line** that approximates the data.   573
  - If the trend line slopes upward from left to right, there is a **positive association**.   575
  - If the trend line slopes downward from left to right, there is a **negative association**.   575
  - If the points do not approximately fall about any line, there is **no association**.   575
- Choosing the best display for data depends upon which type of display best represents the data.   578

10-4 Measures of Central Tendency and Variation

Two important aspects of data are its **center** and its **spread**. Measures of central tendency (center) describe where data are centered.   585

The **mean** of n given numbers is the sum of the numbers divided by n.   586

The **median** of a set of numbers is the middle number if the numbers are arranged in numerical order; if there is no middle number, the median is the mean of the two middle numbers.   588

The **mode** of a set of numbers is the number or numbers that occur most frequently in the set. There may be no mode for a set of numbers.   588

Measures of **spread** or **variation** describe how much the data are scattered.   591

The **range** is the difference between the greatest and least scores.   591

Box plots, or **box-and-whisker plots**, focus attention on the median, the quartiles, and the extremes and invite comparisons among them.   592
- The **lower quartile, Q_1,** is the median of the lower half of all scores.   592
- The **upper quartile, Q_3,** is the median of the upper half of all scores.   592
- The **interquartile range (IQR)** is calculated as the difference between the upper quartile and the lower quartile.   597
- An **outlier** is any value more than 1.5 IQR above the upper quartile or more than 1.5 IQR below the lower quartile.   598

The **mean absolute deviation (MAD)** of a data set is the mean of the absolute values of the differences of the data points and the mean.

598

The **variance (ν)** of a data set is found by subtracting the mean from each value, squaring each of these differences, finding the sum of these squares, and dividing by n, where n is the number of observations.

600

The **standard deviation (s)** is equal to the square root of the variance.

600

A **normal curve** is a smooth, bell-shaped curve that depicts frequency values distributed symmetrically about the mean, as shown:

601

A **percentile** shows a person's score relative to other scores. Percentiles divide the set of data into 100 equal parts.

603

Deciles are points that divide a distribution into 10 equally spaced sections.

604

10-5 Abuses of Statistics

Statistics and graphs may be misleading and graphs can be distorted in many ways.

610

Chapter 10 Review

1. Suppose you read that "the average family in Rattlesnake Gulch has 2.41 children." What average is being used to describe the data? Explain your answer. Suppose the sentence had said 2.5? Then what are the possibilities?

2. At Bug's Bar-B-Q restaurant, the average (mean) weekly wage for full-time workers is $250. There are 10 part-time employees whose average weekly salary is $100, and the total weekly payroll is $4000. How many full-time employees are there?

3. Find the mean, the median, and the mode for each of the following groups of data:
   a. 10, 50, 30, 40, 10, 60, 10
   b. 5, 8, 6, 3, 5, 4, 3, 6, 1, 9

4. Find the range, mean absolute deviation, interquartile range, variance, and standard deviation for each set of scores in exercise 3.

5. The mass, in kilograms, of each child in Ms. Rider's class follows:

40	49	43	48	46	42	49	39	47	49
42	41	42	39	41	40	45	43	44	42

   a. Make a dot plot for the data.
   b. Make an ordered stem and leaf plot for the data.
   c. Make a frequency table for the data.
   d. Make a bar graph of the data.

6. The grades on a test for 30 students follow:

96	73	61	76	77	84
78	98	98	80	67	82
61	75	79	90	73	80
85	63	86	100	94	77
86	84	91	62	77	64

a. Make a grouped frequency table for these scores, using four classes and starting the first class at 61.

b. Draw a histogram of the grouped data.

7. The budget for the Wegetem Crime Co. is $2,000,000. Draw a circle graph to indicate how the company spends its money, where $600,000 is spent on bribes, $400,000 for legal fees, $300,000 for bail money, $300,000 for contracts, and $400,000 for public relations. Indicate percentages on your graph.

8. What, if anything, is wrong with the following bar graph?

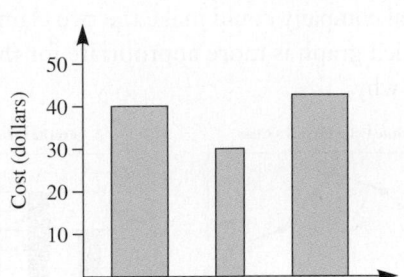

Monthly Health Club Costs

9. The mean salary of 24 people is $9000. How much will one additional salary of $80,000 increase the mean salary?

10. A cheetah can run 70 mph, a lion can run 50 mph, and a human can run 28 mph. Draw a bar graph to represent these data.

11. The life expectancies at birth for males and females in the United States are approximated in the following table:

Year	Male	Female	Year	Male	Female
1970	67.1	74.7	1989	71.7	78.5
1971	67.4	75.0	1990	71.8	78.8
1972	67.4	75.1	1991	72.0	78.9
1973	67.6	75.3	1992	72.1	78.9
1974	68.2	75.9	1993	72.1	78.9
1975	68.8	76.6	1994	72.4	79.0
1976	69.1	76.8	1995	72.5	78.9
1977	69.5	77.2	1996	73.1	79.1
1978	69.6	77.3	1997	73.6	79.2
1979	70.0	77.8	1998	73.8	79.5
1980	70.0	77.5	1999	73.9	79.4
1981	70.4	77.8	2000	74.1	79.3
1982	70.9	78.1	2001	74.2	79.4
1983	71.0	78.1	2002	74.3	79.5
1984	71.2	78.2	2003	74.5	79.6
1985	71.2	78.2	2004	74.9	79.9
1986	71.3	78.3	2005	74.9	79.9
1987	71.5	78.4	2006	75.1	80.2
1988	71.5	78.3	2007	75.4	80.4

Source: The *World Almanac*, 2011.

a. Draw back-to-back ordered stem and leaf plots to compare the data.

b. Draw box plots to compare the data.

c. Describe the spread of the data using the interquartile range.

12. Larry and Marc took the same courses last quarter. Each bet that he would receive the better grades. Their courses and grades are as follows:

Course	Larry's Grades	Marc's Grades
Math (4 credits)	A	C
Chemistry (4 credits)	A	C
English (3 credits)	B	B
Psychology (3 credits)	C	A
Tennis (1 credit)	C	A

Marc claimed that the results constituted a tie, since both received 2 A's, 1 B, and 2 C's. Larry said that he won the bet because he had the higher grade-point average for the quarter. Who is correct? (Allow 4 points for an A, 3 points for a B, 2 points for a C, 1 point for a D, and 0 points for an F.)

13. Following are the lengths in yards of the nine holes of the University Golf Course:

$$160 \quad 360 \quad 330$$
$$350 \quad 180 \quad 460$$
$$480 \quad 450 \quad 380$$

Find each of the following measures with respect to the lengths of the holes:

a. Median

b. Mode

c. Mean

d. Standard deviation

e. Range

f. Interquartile range

g. Variance

h. Mean absolute deviation

14. The speeds in miles per hour of 30 cars were checked by radar. The data are as follows:

$$62 \quad 67 \quad 69 \quad 72 \quad 75 \quad 60 \quad 58 \quad 86 \quad 74 \quad 68$$
$$56 \quad 67 \quad 82 \quad 88 \quad 90 \quad 54 \quad 67 \quad 65 \quad 64 \quad 68$$
$$74 \quad 65 \quad 58 \quad 75 \quad 67 \quad 65 \quad 66 \quad 64 \quad 45 \quad 64$$

a. Find the median.

b. Find the upper and lower quartiles.

c. Draw a box plot for the data and indicate outliers (if any) with asterisks.

d. If every person driving faster than 70 mph received a ticket, what percentage of the drivers received speeding tickets?

e. Describe the spread of the data.
f. Find P_{25}, P_{75}, and D_5.

15. The following scattergram was developed with information obtained from the girls trying out for the high school basketball team:

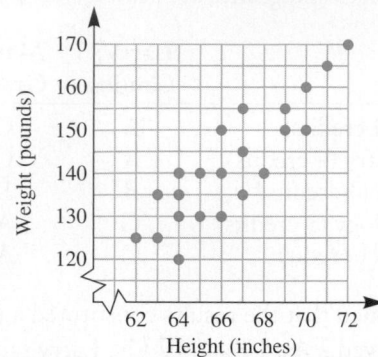

Heights and Weights
of Girls' Basketball Team Tryout Players

a. What kind of association exists between the heights and weights that are listed?
b. What is the weight of the girl who is 72 in. tall?
c. How tall is the girl who weighs 145 lb?
d. What is the mode of the heights?
e. What is the range of the weights?

16. The following table shows statistics for two grade-school basketball players. Collette is the leading scorer for the girls, while Rudy is the leading scorer for the boys.

	Collette	Rudy
Mean points per game	24	24
Mean absolute deviation (in points per game)	6	14
Points scored in the final game	36	36

a. Explain whether Collette or Rudy is the more consistent scorer.
b. For each player, identify the range of values within 1 MAD of the mean.
c. Explain which player's performance was more impressive in the last game.

17. If the test scores were distributed normally, what is the probability that a student scores more than 2 standard deviations above the mean?

18. If every person on an academic team had exactly the same score, describe the following:
a. The mean of the scores
b. The median of the scores
c. The mode of the scores
d. The standard deviation of the scores
e. The mean absolute deviation of the scores

19. If one tossed a fair die 500 times and drew a bar graph showing the frequencies of the resulting tosses, describe what you think the graph would look like.

20. A cereal company has an advertisement on one of its boxes that says, "Lose up to 6 lb in 2 weeks."
a. How much might a person who weighs 175 lb when he starts the plan expect to weigh in 1 yr if this rate could be continued?
b. Explain whether or not you believe your mathematical answer to part (a) is possible for weight loss.
c. A disclaimer on the box says that the "average weight loss is 5.0 lb." On what basis do you think that the cereal company could make the two claims?

21. Tell which graph is more appropriate for this data and explain why.

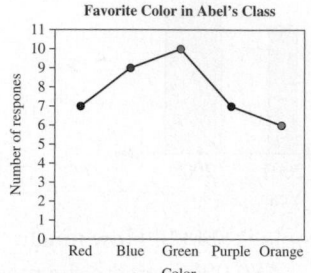
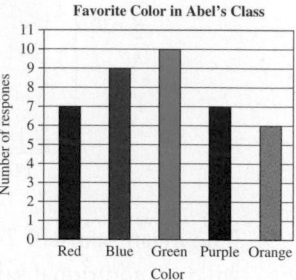

22. Explain whether or not it is reasonable to use a single number to describe the "average" depth of a swimming pool that includes a diving pool as a part of the swimming pool.

23. A box of cereal bars claims to have 138 g of contents. How might the company have determined the weight to make that claim on all comparable boxes to ensure reasonable accuracy of weight?

24. a. In box plots showing the ages of students at two different universities, the interquartile range for one was 3 (from 17 to 20) and the other was 5 (from 24 to 29). If you were a typical high school graduating senior and were trying to decide between the two universities using these data, which would you choose and why?
b. If you were a student at one of the two universities and were 23 years old, at which university might you expect to appear as a whisker in the box plot?

25. Given the figure below, what might be misleading about it?

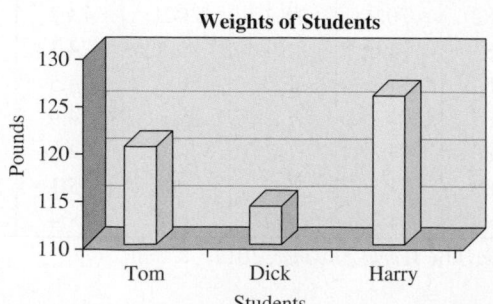

26. The Nielsen Television Index rating of 30 means that an estimated 30% of American televisions are tuned to the show with that rating. The ratings are based on the preferences of a scientifically selected sample of 1200 homes.
    a. Discuss possible ways in which viewers could bias this sample.
    b. How could networks attempt to bias the results?

27. Give examples of several ways to misuse statistics graphically.

28. Given the graph below, answer the following questions.

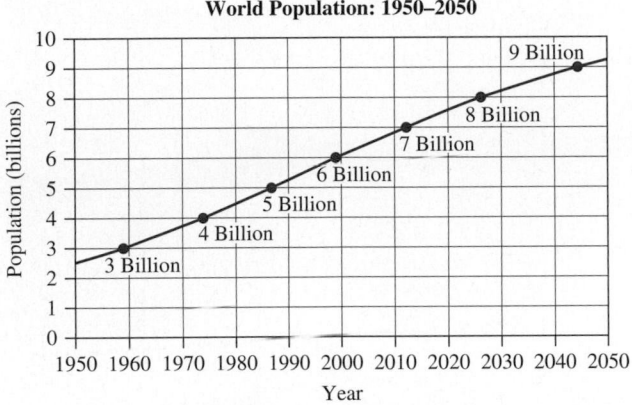

World Population: 1950–2050

Source: U.S. Census Bureau, International Data Base, December 2010 Update.

    a. What was the world population in 1959?
    b. What was the world population in 1999?
    c. What was the growth factor for the 40-year period from 1959 to 1999?
    d. Estimate the world population in 2049.
    e. What is the estimated percent growth for the 40-year period from 1999 to 2049?

29. The heights of 1000 girls at East High School were measured, and the mean was found to be 64 in., with a standard deviation of 2 in. If the heights are approximately normally distributed, about how many of the girls are
    a. over 68 in. tall?
    b. between 60 and 64 in. tall?
    c. If a girl is selected at random at East High School, what is the probability that she will be over 66 in. tall?

30. A standardized test has a mean of 600 and a standard deviation of 75. If 1000 students took the test and their scores approximated a normal curve, how many scored between 600 and 750?

31. Use the information in problem 30 to find:
    a. P_{16}
    b. D_5
    c. P_{84}

32. The circle graph below was shown on national television. What is wrong with this graph?

2012 Presidential Run GOP Candidates

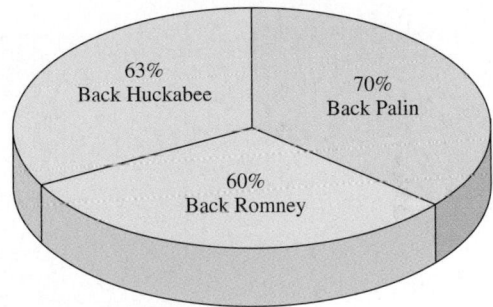

63% Back Huckabee

70% Back Palin

60% Back Romney

Introductory Geometry

Preliminary Problem

Swiss artist Max Bill (1908–1994) produced lithographs using regular polygons. The drawing starts with an equilateral triangle *ABC*, uses the length of one side of the triangle to construct a square, uses a side of the square to construct a pentagon, and so on. If *AB* = 1 m in the path $\overline{AB}, \overline{BC}, \overline{CD}, \overline{DE}, \overline{EF}, \ldots$, how long would the entire path be with 14 regular polygons complete and all sides of the last one included?

If needed, see Hint on page 695.

The origins of geometry date back to ancient Egypt, when yearly flooding of the Nile River required resurveys of surrounding land for taxation purposes. Later, the Babylonians added to geometrical knowledge with an approximation of π. Centuries later, two Greek words—*geo* ("earth") and *metron* ("measure")—were combined to give this science of land measurement a name. The consummate Greek geometer Euclid (ca. 300 BCE) pushed geometry from the realm of the practical into a theoretical one with his *Elements*.

Over the years, geometry progressed from providing the knowledge of surveying, to measurements between planets, to Mandelbrot's work with fractals.

According to *PSSM*,

> The study of geometry in grades 3–5 requires thinking and doing. As students sort, build, draw, model, trace, measure, and construct, the capacity to visualize geometric relationships will develop. At the same time they are learning to reason and to make, test, and justify conjectures about these relationships. This exploration requires access to a variety of tools, such as graph paper, rulers, pattern blocks, geoboards, and geometric solids, and is greatly enhanced by electronic tools that support exploration, such as dynamic geometry software. (p. 165)

The following are some of the different grade level expectations from *PSSM* covered in this chapter.

> In grades prekindergarten through grade 2, all students should
>
> • recognize, name, build, draw, compare, and sort two- and three-dimensional shapes;
>
> • describe attributes and parts of two- and three-dimensional shapes;
>
> • investigate and predict the results of putting together and taking apart two- and three-dimensional shapes. (p. 96)
>
> In grades 6–8, all students should
>
> • precisely describe, classify, and understand relationships among types of two- and three-dimensional objects using their defining properties;
>
> • use visual tools such as networks to represent and solve problems. (p. 232)

In the early grades, *Common Core Standards* recommend that students begin to analyze, create, compare, and compose shapes (kindergarten, p. 10), reason with shapes (grades 1–3, pp. 14, 19, 22), identify lines and angles (grade 4, p. 28), and classify two-dimensional figures (grade 5, p. 32). All of these topics and more are addressed in this chapter.

Much of the geometry instruction in U.S. schools in the past several decades is based, at least in part, on the 1950s research of two Dutch educators, Dina van Hiele-Geldof (?, 1959) and her husband, Pierre van Hiele (1909–2010). The pair identified five levels through which students may progress in developing geometric reasoning. The levels are hierarchical and sequential so that to obtain one level, students are presumed to have passed through all preceding levels. However, other research suggests that with different experiences and instruction, individual students may be at different levels with different geometrical topics. The following van Hiele levels have been adapted based on additional work by Clements and Battista[1] (1992) and Battista[2] (2007):

---

[1]Clements, D., and M. Battista. "Geometry and Special Reasoning." In *Handbook of Research on Mathematics Teaching and Learning,* edited by D. Grouws. New York: MacMillan, 1992.

[2]Battista, M. "The Development of Geometry and Spetial Thinking. "In *Second Handbook of Research on Mathematics Teaching and Learning,* edited by F. Lester, Jr. Charlotte, NC: Information Age Publishing, 2007.

Level 1: Visual-Holistic Reasoning
At this level, students identify, describe and reason about geometrical objects based on appearance as whole objects. For example, a rectangle may be described as such because it looks like a carpet runner. Some students in this level are able to identify common shapes while others are not.

Level 2: Descriptive-Analytic Reasoning
At this level, students may describe structures by analyzing shape parts; their definitions are typically not minimal; and they begin to use more formal geometric concepts to consider relationships among parts of geometric structures or figures. Still at this level, language and descriptions vary widely in sophistication.

Level 3: Relational-Inferential Reasoning
At this level, students begin to infer relationships among shapes or figures. Much of this inference is based upon empirical evidence and looking at some cases. Simple logical inferences begin to be made and increasingly students begin to develop minimal definitions, but the sophistication levels of students still vary widely.

Level 4: Formal Deductive Proof
At this level, students can produce proofs in an axiomatic system. It is at this level that most secondary high school geometry courses are pitched.

Level 5: Rigor
At this level, students understand and can use alternative axiomatic systems, typically found at the university level. (pp. 843–908)

Though the van Hieles influenced geometric study in the United States, their theory is only one of many. Prospective teachers in grades K–8 should understand that other research, beyond the van Hieles's, has been done to examine geometrical learning. For example, abstraction is the learning of objects and their properties at a very deep level that allows them to be used and operated on in both geometrical and real situations. Additionally, there is a connection between the use of natural objects and geometric objects, interwoven to allow concepts to be understood through the use of physical objects, defined as using them as models and then revisited with a geometrical structure to analyze them mathematically. The abstraction theory cuts across van Hiele levels 1–3. And finally, the interplay between physical objects and more abstract diagrams to understand and analyze in geometry presents issues in learning. Most often instruction uses physical objects to represent or to show geometrical concepts, but the formation of concepts from the objects is neglected with instruction going straight to geometrical diagrams.

Historical Note

Though legend tells us that Euclid of Alexandria (ca. 300 BCE) studied geometry for its beauty and logic, he is best known for the *Elements*, a work so systematic and encompassing that many earlier mathematical works were discarded. The *Elements*, comprising 13 books, includes not only geometry but also arithmetic and topics in algebra. In the *Elements*, Euclid set up a deductive system by starting with a set of statements, assumed to be true, and then showed that geometric discoveries followed logically from these assumptions.

11-1 Basic Notions

The fundamental building blocks of geometry are *points*, *lines*, and *planes*. Ironically, the building blocks are *undefined terms* in order to avoid circular definitions. An example of a circular definition and the frustration involved in starting without some basic, undefined notions is given in the following *B.C.* cartoon.

B.C. copyright © 1985 and 2011 by John L. Hart FLP/Distributed by Creators Syndicate

Geometric concepts are developed from undefined terms; an intuitive notion of some undefined terms appears in the illustrations of Table 11-1.

Table 11-1

Term/Symbolism	Illustration
point *A* point *B* point *C*	
line *ℓ*	
line *m*, line *AB*, $\overleftrightarrow{AB}$, or $\overleftrightarrow{BA}$	

(continued)

Table 11-1 (continued)

Term/Symbolism	Illustration
plane α	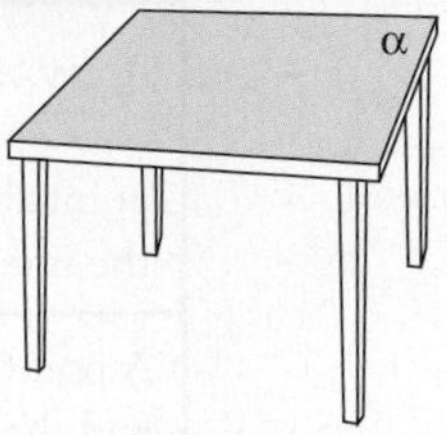
plane ABC or plane γ	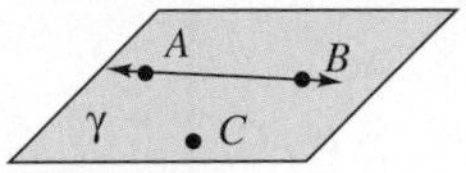

A line extends forever in two directions, but has no thickness, and *it is uniquely determined by two points; that is, given two distinct points, there is one, and only one, line that connects these points.* A basic concept in geometry is that of a *distance*. If P and Q are any two distinct points, we create a number line $\overleftrightarrow{PQ}$ such that there is one-to-one correspondence between the points on the line and real numbers. Moreover, if we correspond to P the real number 0 and to Q a positive real number, the distance between P and Q, written PQ, is the real number that corresponds to Q. Table 11-2 illustrates many commonly used terms and their meanings in geometry.

Table 11-2

Term	Illustration and Symbolism
Collinear points are points on the same line.	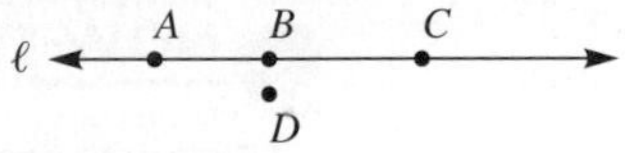 Line ℓ contains points A, B, and C. Points A, B, and C belong to line ℓ. Points A, B, and C are collinear. Points A, B, and D are not collinear (or are noncollinear).
If A, B, and C are three distinct collinear points, then B is **between** A and C if, and only if, $AB + BC = AC$.	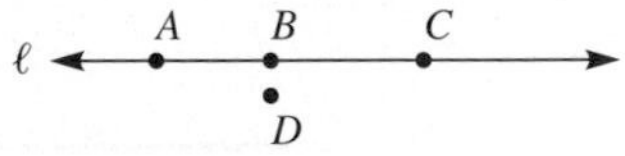 Point D is not between A and C. AB is the distance between points A and B.
A **line segment**, or **segment**, is a subset of a line that contains two points of the line and all points between those two points.	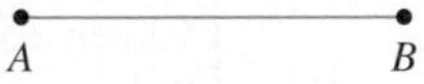 $\overline{AB}$ or $\overline{BA}$ $\overline{AB}$ or $\overline{BA}$ (closed segment)
Subsets of segments that are sometimes identified include a **closed segment** (a segment with endpoints), a **half-open** or **half-closed segment** (a segment without one endpoint) and an **open segment** (a segment without its endpoints).	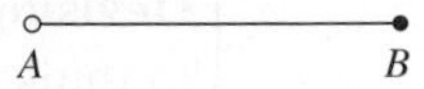 $\overline{AB}$ (half-open or half-closed segment) 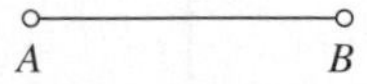 $\overline{AB}$ (open segment)

(continued)

Table 11-2 (continued)

Term	Illustration and Symbolism
A **ray**, $\overrightarrow{AB}$, is a subset of the line $\overleftrightarrow{AB}$ that contains the endpoint A, the point B, all the points between A and B, and all points C on the line such that B is between A and C.	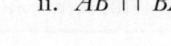 $\overrightarrow{AB}$ contains $\overline{AB}$ $\overrightarrow{BA}$ and $\overrightarrow{BC}$ are **opposite rays.**
A point separates a line into two **half-lines** and the point itself.	Half-line BA, denoted $\overset{\circ}{BA}$; half-line BC, denoted $\overset{\circ}{BC}$, and point B Point B separates line AB into half-line

NOW TRY THIS 11-1

a. Explain whether every three points must be collinear.

b. In Table 11-2, it is stated that B is between points A and C if A, B, and C are collinear and $AB + BC = AC$. If A, B, and C are three different points, is it possible to have $AB + BC = AC$ and to have the points noncollinear?

c. If we think of lines, segments, and rays as sets of points, find

  i. $\overrightarrow{AB} \cup \overrightarrow{BA}$
  ii. $\overrightarrow{AB} \cap \overrightarrow{BA}$

  iii. $\overset{\circ}{AB} \cup \overset{\circ}{BA}$
  iv. $\overleftrightarrow{AB} \cup \overline{AB}$

  v. $\overleftrightarrow{AB} \cap \overrightarrow{AB}$
  vi. $\overleftrightarrow{AB} \cap \overrightarrow{BA}$

NOW TRY THIS 11-2

As points are added on line ℓ in Figure 11-1, it is separated into parts. For example, with one point, there are two parts; with two points, there are three parts; and so on. If we do not include the points themselves, into how many parts is a line separated using n points?

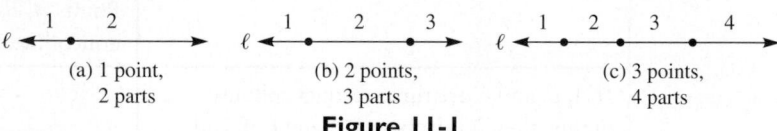

Figure 11-1

Problem Solving | **Lines Through Points**

Given 15 points, no three of which are collinear, how many lines can be drawn through the 15 points?

Understanding the Problem Because two points determine exactly one line, we must consider ways to find out how many lines are determined by the 15 points.

Devising a Plan We use the strategy of *examining related simpler cases* of the problem in order to think through the original problem. Figure 11-2 shows that 3 noncollinear points determine 3 lines, 4 determine 6 lines, 5 determine 10 lines, and 6 determine 15 lines.

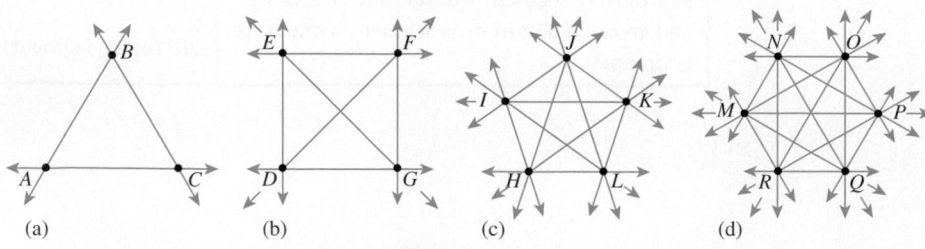

Figure 11-2

Examining Figure 11-2(d), we see that through any one of the 6 points, we can draw only 5 lines to connect to the other 5 points. If 5 lines are drawn through each point, there seem to be $6 \cdot 5$, or 30, lines. But in this approach, each line is counted twice; for example, line NO is counted as 1 of the 5 lines through N and also as 1 of the 5 lines through O. Thus, there are $\frac{6 \cdot 5}{2}$, or 15, lines in the figure. We use this information to determine the solution for 15 points.

Carrying Out the Plan Each of the 15 points can be paired with all points other than itself to determine a line; that is, each point can be paired with the other 14 points to determine 14 lines. If we do this and account for counting each line twice, we see that there should be $\frac{15 \cdot 14}{2}$, or 105, lines.

Looking Back Using this reasoning, we conclude that the number of lines determined by n points, no three of which are collinear, is $\frac{n(n-1)}{2}$.

An alternative solution to this problem uses the notion of combinations found in Chapter 9. The number of ways that n points, no 3 of which are collinear, can be chosen 2 at a time to form lines is $_nC_2$, or $\frac{n(n-1)}{2}$.

Another approach also uses the strategy of *solving a simpler problem*. Two points determine exactly 1 line. When we add a third point so that the 3 points are not collinear, as in Figure 11-2(a), we can connect that point to the 2 existing points and create 2 new lines, for a total of $1 + 2$ lines. When a fourth point is added (so that no 3 of the 4 points are collinear), it can be connected to the 3 existing points to obtain 3 additional lines, for a total of $1 + 2 + 3$ lines. Continuing in this way so that no 3 points are collinear, when the 15th point is added to the existing 14 points, 14 new lines are created for a total of $1 + 2 + 3 + \ldots + 14$ lines. Using what we know about the sum of an arithmetic sequence we get

$$1 + 2 + 3 + \ldots + 14 = \frac{14(14+1)}{2} = \frac{14 \cdot 15}{2}$$

or 105 lines.

In general, when the nth point is added, we can connect it to the $n-1$ existing points to obtain $n-1$ new lines. Using the formula for the sum of an arithmetic sequence developed in Chapter 8, the number of lines determined by n points, no 3 of which are collinear, is

$$1 + 2 + 3 + \ldots + (n-1) = \frac{(n-1)(n-1+1)}{2}, \text{ or } \frac{(n-1)n}{2}.$$

NOW TRY THIS 11-3

Use the geometric model above to find the number of handshakes that take place at a party of 20 people if each person shakes hands with everybody else at the party. (*Hint:* Think about people as points and handshakes as segments.)

Planar Notions

A plane has no thickness and it extends indefinitely. *A plane is uniquely determined by three noncollinear points.* In other words, *given three distinct noncollinear points, there is one, and only one, plane that contains these points.* Table 11-3 illustrates intuitive planar notions.

Table 11-3

Term	Illustration and Symbolism
	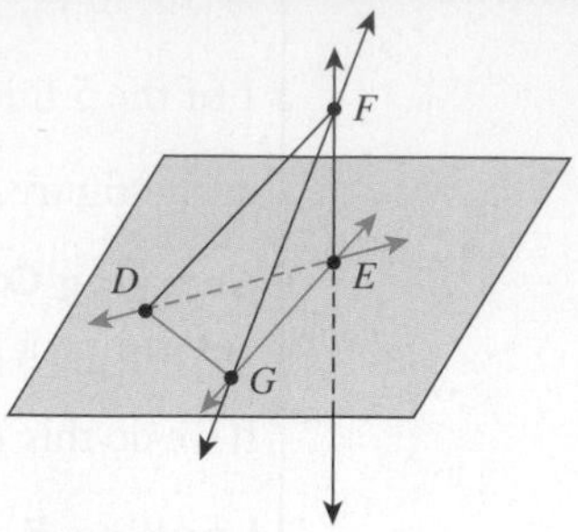
Points in the same plane are **coplanar**.	Points D, E, and G are coplanar.
Noncoplanar points cannot be placed in a single plane.	Points D, E, F, and G are noncoplanar.
Lines in the same plane are **coplanar lines**.	Lines DE, DF, and FE are coplanar. Lines DE, DG, and EG are coplanar.
Intersecting lines are lines with exactly one point in common.	Lines DE and GE are intersecting lines; they intersect at point E.
Skew lines are lines for which there is no plane that contains them; they are noncoplanar.	Lines GF and DE are skew lines.
Concurrent lines are lines that intersect in the same point.	Lines DE, EG, and EF are concurrent.
Two distinct coplanar lines are **parallel** if they have no point in common. A line is parallel to itself. Two segments or rays are parallel if they lie in parallel lines.	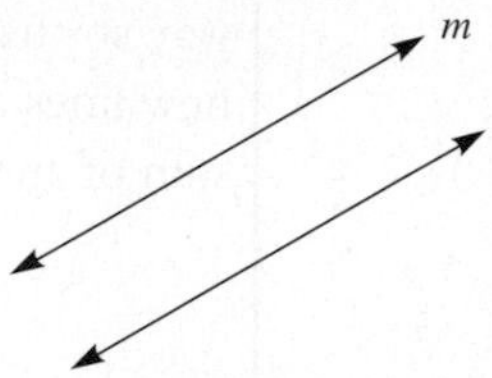 m is parallel to n, written $m\|n$; also, $m\|m$ and $n\|n$.

Table 11-4 depicts relationships among planes and lines.

Table 11-4

Term	Illustration and Symbolism
Intersecting planes have a single line in common.	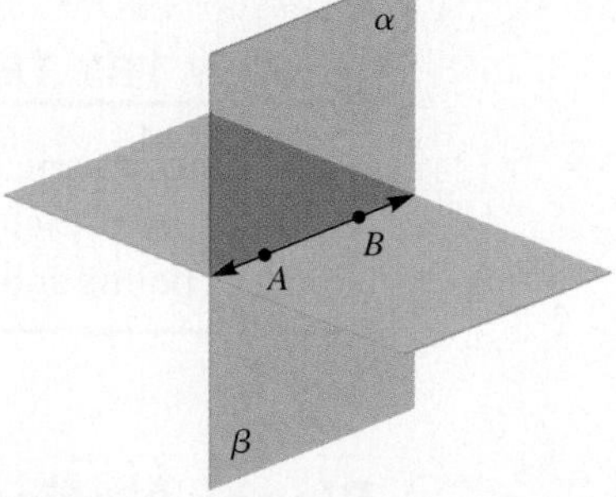 α intersects β in line AB.

(continued)

Table 11-4 *(continued)*

Term	Illustration and Symbolism
Two planes are **parallel** if they are not intersecting planes. A plane is parallel to itself. A line is either wholly contained in a plane, intersects it in one point, or has no points in common with the plane.	α is parallel to β. $\overleftrightarrow{FG} \cap \alpha = \{F\}$ $\overleftrightarrow{FG} \cap \beta = \{G\}$ $\overleftrightarrow{HG} \subseteq \beta$
A line and a plane are **parallel** if they do not intersect in one point. A line in a plane is parallel to the plane.	ℓ is parallel to α; m is parallel to α.
A line contained in a plane separates the plane into three mutually disjoint sets; two half-planes and the line itself.	ℓ separates plane α into half-planes, $A\text{-}BC$ and $D\text{-}BC$, and $\overleftrightarrow{BC}$.

NOW TRY THIS 11-4

Just as a point separates a line into parts, a line separates a plane into parts. For example, one line separates a plane into two parts (if the line is not included). Two lines can separate a plane into a maximum of 4 parts, and so on. Figure 11-3 shows the first few cases.

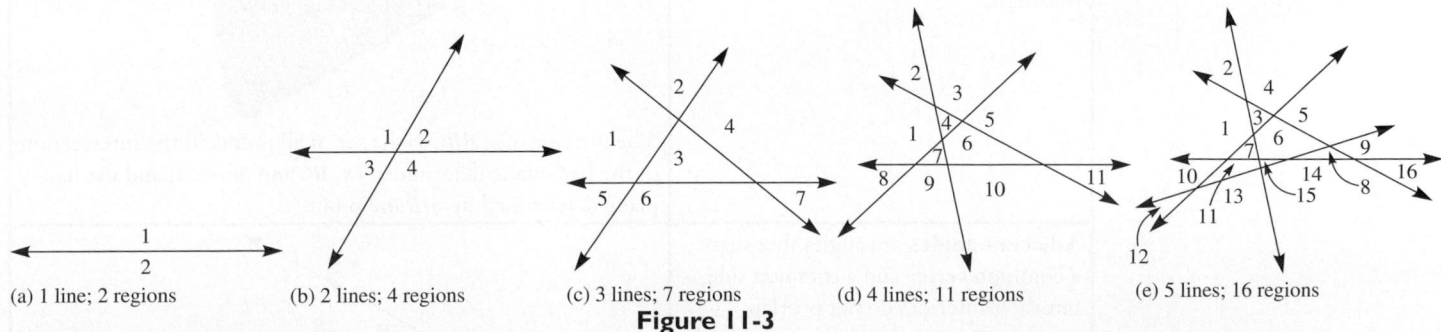

(a) 1 line; 2 regions   (b) 2 lines; 4 regions   (c) 3 lines; 7 regions   (d) 4 lines; 11 regions   (e) 5 lines; 16 regions

Figure 11-3

Conjecture a formula for the maximum number of regions into which n lines in a plane separate the plane if the lines themselves are not included.

NOW TRY THIS 11-5

a. Can skew lines have a point in common? Why?

b. Can skew lines be parallel? Why?

c. Identify skew and parallel lines marked in Figure 11-4.

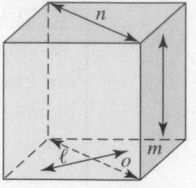

o contains a diagonal of the base

Figure 11-4

Some statements that cannot be proven are called **axioms**. Some axioms follow:

1. Axiom: There is exactly one line that contains any two distinct points.
2. Axiom: If two points lie in a plane, then the line containing the points lies in the plane.
3. Axiom: There is exactly one plane that contains any three distinct noncollinear points.

Based on undefined terms and axioms, **theorems** (statements that can be proven) are constructed. Additionally, definitions are used extensively in this work. A *definition* in mathematics is somewhat different from the everyday use of the word. In mathematics, a definition is stated as simply as possible with properties that can be proven omitted. Sometimes a geometric object can be defined in different ways, but they must be consistent in usage. In this text, there are places to examine different definitions for the same term.

Angles

The concept of an **angle** in geometry has many interpretations, including an idea of turning and a subset of a plane formed by rays. A more formal definition is given in Table 11-5.

Table 11-5

An **angle** is the union of two rays that share an endpoint. The rays of the angle are the **sides** of the angle, and the common endpoint is the **vertex** of the angle.	Vertex, A, 1 or α, Side, B, C $\angle B, \angle ABC, \angle CBA, \angle 1$, or α
The **interior of an angle** whose sides do not lie in a straight line is the region in the plane between the two sides of the angle.	A, B, Interior of $\angle ABC$, C The interior of $\angle ABC$ is the set of all points in the intersection of the half-plane determined by $\overleftrightarrow{BC}$ and point A, and the half-plane determined by $\overleftrightarrow{AB}$ and point C.
Adjacent angles are angles that share a common vertex and a common side, but their interiors do not overlap.	Q, α, R, P, β, S α and β are adjacent angles.

Angle Measurement

An angle is measured according to the amount of "opening" between its sides. The **degree** is commonly used to measure angles. A complete rotation about a point has a measure of 360 degrees, written 360°. One degree is $\frac{1}{360}$ of a complete rotation. Figure 11-5 shows that $\angle BAC$ has a measure of 30 degrees, written $m(\angle BAC) = 30°$. The measuring device pictured in the figure is a **protractor**. A degree is subdivided into 60 equal parts—**minutes**—and each minute is further divided into 60 equal parts—**seconds**. The measurement 29 degrees, 47 minutes, 13 seconds is written 29°47′13″.

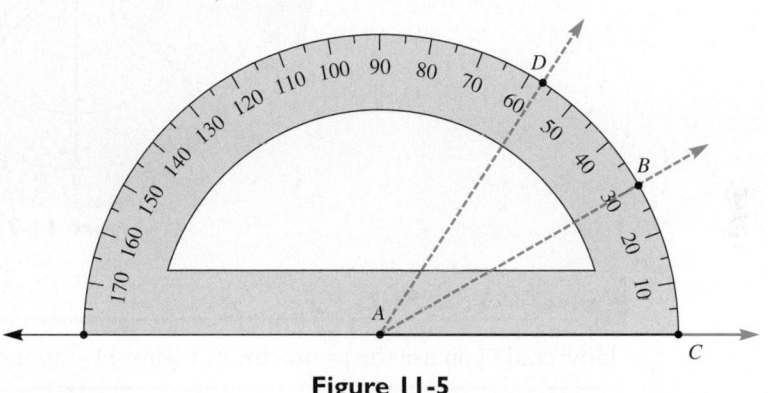

Figure 11-5

● NOW TRY THIS 11-6

Convert 8.42° to degrees, minutes, and seconds.

The protractor in Figure 11-5 shows measures from 0 degrees to 180 degrees, but by the definition of an angle from Table 11-5, there is no reason to limit the measure of an angle as the protractor shows. When we refer to an angle, as $\angle ABC$ in Figure 11-6, we typically mean the "smaller" of the two angles (pictured as $\angle 1$), but we could also refer to $\angle 2$. The "larger" opening, one whose measure is more than 180° but less than 360°, is a **reflex angle**.

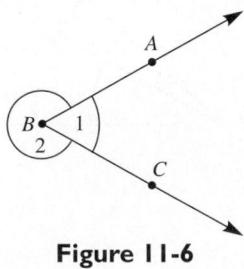

Figure 11-6

○– Historical Note

The French mathematician and astronomer Pierre Herigone (1580–1643) used a symbol for angle in 1634; the Mathematical Association of America recommended $\angle$ as the standard symbol for angle in the United States in 1923. The use of 360° to measure angles seems to date to the Babylonians (4000–3000 BCE).

To measure a reflex angle, a circular protractor as in Figure 11-7 is sometimes used. Instead of being marked from 0 to 360 degrees, this model is marked from 0 to 90 on quarters.

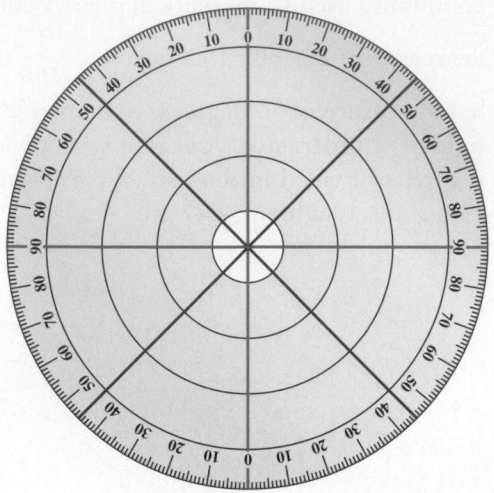

Figure 11-7

> ## NOW TRY THIS 11-7
>
> How could you use the protractor of Figure 11-7 to measure a reflex angle with 283 degrees?

EXAMPLE 11-1

a. In Figure 11-8, find the measure of $\angle BAC$ if $m(\angle 1) = 47°45'$ and $m(\angle 2) = 29°58'$.
b. Express $47°45'36''$ as a number of degrees.

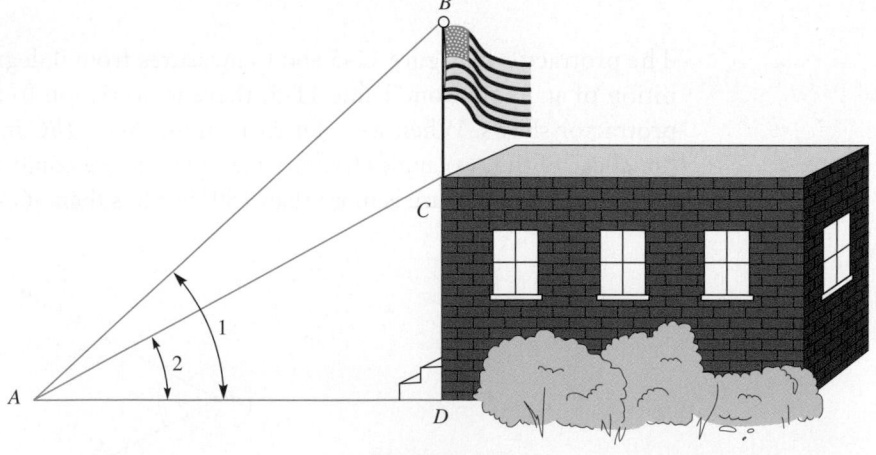

Figure 11-8

Solution

a. $m(\angle BAC) = 47°45' - 29°58'$
$= 46°(60 + 45)' - 29°58'$
$= 46°105' - 29°58'$
$= (46 - 29)° + (105 - 58)'$
$= 17°47'$

b. $45' = \left(\dfrac{45}{60}\right)^\circ = 0.75^\circ$

$36'' = \left(\dfrac{36}{60}\right)' = \left(\dfrac{1}{60} \cdot \dfrac{36}{60}\right)^\circ$

$\qquad\ = 0.01^\circ$

Thus,

$$47^\circ 45' 36'' = 47^\circ + 0.75^\circ + 0.01^\circ$$
$$= 47.76^\circ.$$

Types of Angles

We create different types of angles by paper folding. Consider the folds shown in Figure 11-9(a) and (b). A piece of paper is folded in half and then reopened. If any point on the fold line, labeled ℓ, is chosen as the vertex O, then the measure of the angle pictured is 180°. If the paper is refolded and folded once more, as shown in Figure 11-9(c), and then is reopened, as shown in Figure 11-9(d), four angles of the same size are created. Each angle has measure 90° and is a right angle. The symbol ⌐ denotes a right angle. (Why is the measure of each of these angles 90°?)

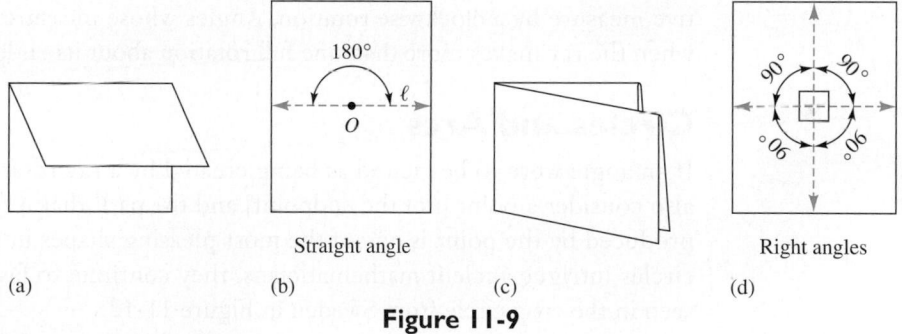

(a)	(b)	(c)	(d)

Figure 11-9

If the paper is folded as shown in Figure 11-10 and reopened, then angles α and β are formed, with measures that are less than 90° and greater than 90°, respectively. (Note that β has measure less than 180°.) Angle α is an *acute* angle, and β is an *obtuse* angle. In Figure 11-10(b), the measures of α and β add up to 180°. Any two angles the sum of whose measures is 180° are **supplementary** angles. We say that each angle is a **supplement** of the other.

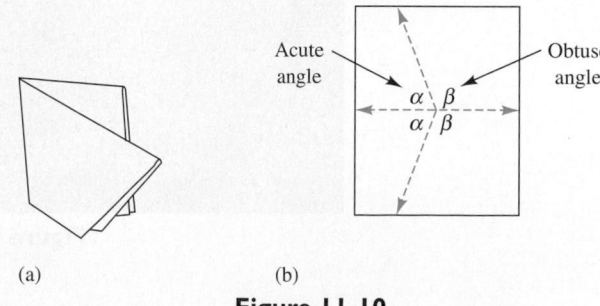

(a)	(b)

Figure 11-10

The types of planar angles are shown in Figure 11-11, along with their definitions.

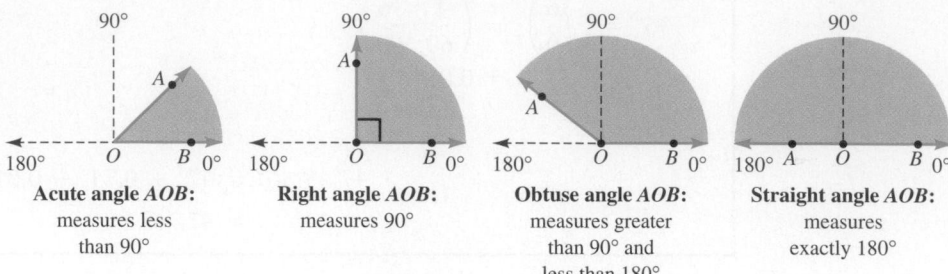

Acute angle *AOB*:
measures less
than 90°

Right angle *AOB*:
measures 90°

Obtuse angle *AOB*:
measures greater
than 90° and
less than 180°

Straight angle *AOB*:
measures
exactly 180°

Figure 11-11

A straight angle can be described without referring to degree measure: A straight angle is an angle whose sides are opposite rays forming a straight line. A model of a straight angle can be obtained by folding a page in half.

Circles and Angle Measurement

In higher mathematics and in scientific applications, an angle may be viewed as being created by a ray rotating about its endpoint. If the ray makes one full rotation, it sweeps an angle of 360°. Angles with positive measure are created by a counterclockwise rotation, and angles with negative measure by a clockwise rotation. Angles whose measures are greater than 360° are created when the ray makes more than one full rotation about its endpoint.

Circles and Arcs

If an angle were to be viewed as being created by a ray rotating around its endpoint, we might also consider a point [not the endpoint] and the path that it follows as the ray rotates. The path produced by the point is one of the most pleasing shapes in geometry, the circle. Not only did circles intrigue ancient mathematicians, they continue to fascinate people around the world as seen in the crop circle from Sweden in Figure 11-12.

Figure 11-12

Mathematically, a **circle** is the set of all points in a plane that are the same distance (the **radius**) from a given point, the **center**. An **arc** of a circle is any part of the circle that can be drawn without lifting a pencil. A circle and associated arcs are seen in Figure 11-13(a).

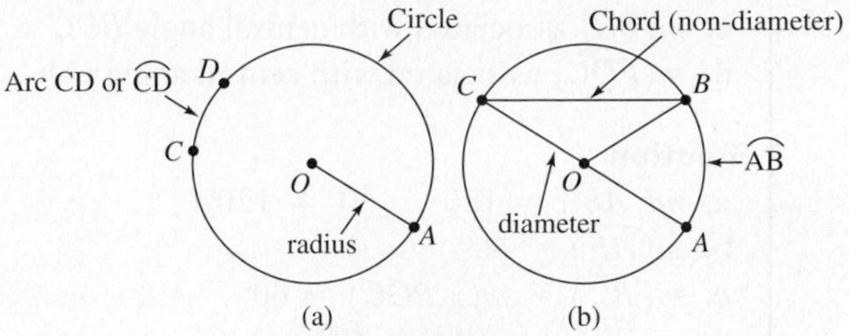

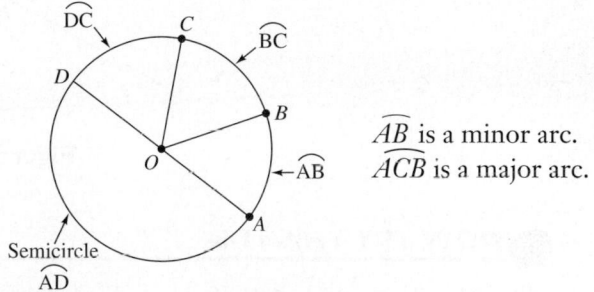

Figure 11-13

In Figure 11-13(a), we refer to circle O to indicate all points on the circle. "Radius" is a term frequently used to name any segment with one endpoint at the center of the circle and the other endpoint on the circle, as well as the length of that segment. For example $\overline{OA}$ is a radius of circle O and refers to the segment with endpoints O and A, while OA refers to the length of that segment. In a similar way, any segment with endpoints on the circle is a **chord** of the circle. If a chord contains the center of the circle, it is a **diameter** of the circle. "Diameter" can also refer to the length of the segment depending on the context in which the term is used. In Figure 11-13(b) $\overline{AC}$ is a diameter of circle O while $\overline{BC}$ is a chord but not a diameter.

Additionally, in Figure 11-13(b), arc AB is considered to have center O because it is a part of circle O. Arc AB (denoted $\overarc{AB}$) is associated with $\angle AOB$, a **central angle** of circle O. A central angle of a circle has its vertex at the center of the circle. Figure 11-14 shows some arcs and associated central angles.

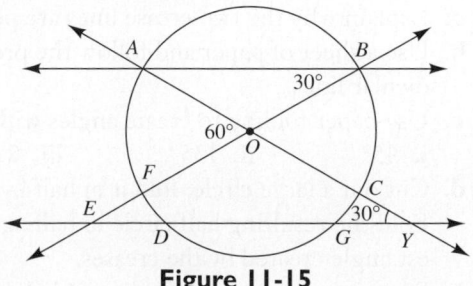

$\overarc{AB}$ is a minor arc.
$\overarc{ACB}$ is a major arc.

Figure 11-14

In Figure 11-14, central angle $\angle AOB$ determines two arcs, $\overarc{AB}$ and $\overarc{ACB}$, a minor and major arc, respectively. A **minor arc** is one determined by a central angle whose measure is less than 180° while a **major arc** is determined by a reflex angle whose measure is more than 180°. Depending on the context, *a central angle and its associated arc have the same degree measure.*

An arc determined by a diameter is a **semicircle**. A semicircle of circle O in Figure 11-14 also is considered to have center O.

EXAMPLE 11-2

Find the following in Figure 11-15.

Figure 11-15

a. $m(\overarc{AB})$ associated with central angle AOB
b. $m(\overarc{AF})$ associated with central angle AOF

    c. $m(\overset{\frown}{BC})$ associated with central angle BOC

    d. $m(\overset{\frown}{FDC})$ associated with central angle FOC

Solution

    a. $m(\overset{\frown}{AB}) = 180° - 60° = 120°$

    b. $m(\overset{\frown}{AF}) = 60°$

    c. $m(\overset{\frown}{BC}) = m(\angle BOC) = 60°$

    d. $m(\overset{\frown}{FDC}) = 180° - 60° = 120°$

Perpendicular Lines

When two lines intersect so that the angles formed are right angles, as in Figure 11-16, the lines are **perpendicular lines**. In Figure 11-16 where lines m and n are perpendicular, we write $m \perp n$. Two intersecting segments, two intersecting rays, one segment and one ray, a segment and a line, or a ray and a line that intersect are perpendicular if they lie on perpendicular lines. For example, in Figure 11-16, $\overline{AB} \perp \overline{BC}$, $\overrightarrow{BA} \perp \overrightarrow{BC}$, and $\overline{AB} \perp \overleftrightarrow{BC}$.

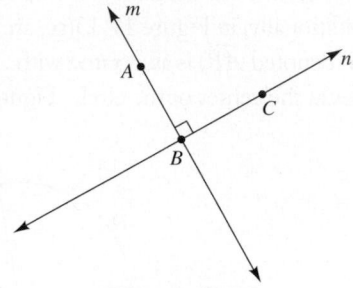

Figure 11-16

🔵 NOW TRY THIS 11-8

Consider the construction of perpendicular lines with paper folding in Figure 11-17 and answer the questions that follow:

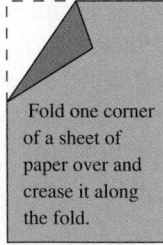
Fold one corner of a sheet of paper over and crease it along the fold.

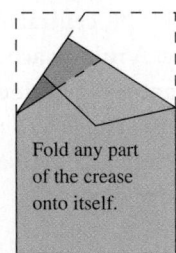
Fold any part of the crease onto itself.

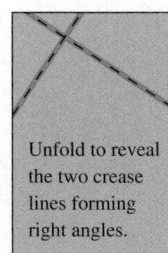
Unfold to reveal the two crease lines forming right angles.

Figure 11-17

    a. Explain why the two crease lines are perpendicular.

    b. Use a sheet of paper and follow the preceding instructions to create two pairs of perpendicular lines.

    c. Use paper folding to create angles with the following measures:

        i. 45°   **ii.** 135°   **iii.** 22°30′

    d. Cut out a large circle; fold it in half by creasing along a line through the center of the circle. Fold the resulting half circle in half again. Unfold and tell what is the measure of the smallest angle created by the creases.

    e. Suppose you continue folding in half as in part (d); what is the measure of the smallest angle created after three folds?

    f. What is the measure of the smallest angle created after n folds?

A Line Perpendicular to a Plane

If a line and a plane intersect, they can be perpendicular. For example, in Figure 11-18, planes β and γ represent two walls whose intersection is $\overleftrightarrow{AB}$. The edge $\overleftrightarrow{AB}$ is perpendicular to the floor. Also, every line in the plane of the floor (plane α) passing through point A is perpendicular to $\overleftrightarrow{AB}$. A **line perpendicular to a plane** is a line that is perpendicular to every line in the plane through its intersection with the plane.

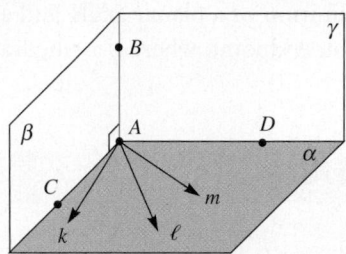

Figure 11-18

NOW TRY THIS 11-9

a. Is it possible for a line intersecting a plane to be perpendicular to exactly one line in the plane through its intersection with the plane? Explain by making an appropriate drawing.

b. Is it possible for a line intersecting a plane to be perpendicular to two distinct lines in a plane going through its point of intersection with the plane, and yet not be perpendicular to the plane?

c. Can a line be perpendicular to infinitely many lines?

d. If a line ℓ intersecting a plane α at point A is perpendicular to two distinct lines in the plane through A, what seems to be true about ℓ and α?

Perpendicular Planes and Dihedral Angles

Figure 11-19(a) shows two perpendicular planes α and β, which can be modeled by two pages of a book opened at 90°. If $\overline{AB}$ and $\overline{AC}$ represent the edges of the book, then each is perpendicular to the binding $\overline{AD}$, and $\angle BAC$ is a right angle. If P is any point on $\overline{AD}$, Q is in plane α, and S is in plane β so that $\overrightarrow{PQ} \perp \overleftrightarrow{AD}$ and $\overrightarrow{PS} \perp \overleftrightarrow{AD}$, then $\angle QPS$ is also a right angle. Since $\angle QPS$ measures 90°, we say that the **planes are perpendicular**. If we view Figure 11-19(a) as the union of two half-planes and the common line AD, we have a **dihedral angle** that measures 90°; one way to denote the angle is Q-AD-S.

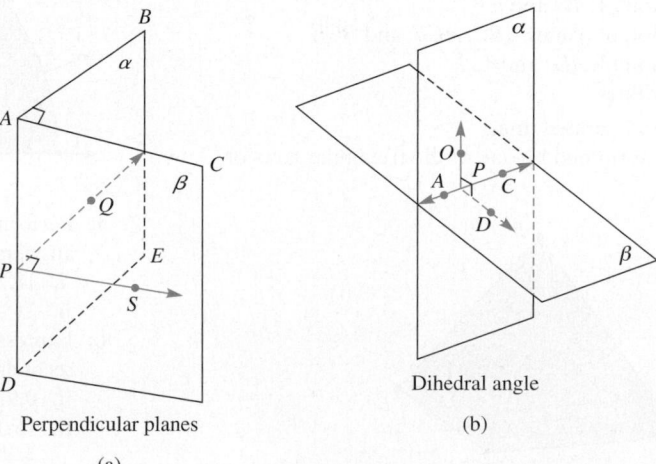

Perpendicular planes

(a)

Dihedral angle

(b)

Figure 11-19

In general, a *dihedral angle is formed by the union of two half-planes and the common line defining the half-planes.* (Remember that the common line belongs to neither half-plane.) In Figure 11-19(b), the dihedral angle *O-AC-D* is formed by the intersecting planes α and β. Note that the point *O* is in the plane α, $\overleftrightarrow{AC}$ is the edge of the dihedral angle, and point *D* is in plane β. A dihedral angle is measured by any of the associated planar angles such as ∠*OPD*, where $\overrightarrow{PO} \perp \overleftrightarrow{AC}$ and $\overrightarrow{PD} \perp \overleftrightarrow{AC}$. If any of the four dihedral angles created by the intersecting planes measures 90°, then all four dihedral angles measure 90° and the planes α and β are perpendicular. We compare the definition of a planar angle and a dihedral angle: a planar angle is a union of two rays with a common endpoint, whereas a dihedral angle is the union of two half-planes and a common line.

Assessment 11-1A

1. **a.** Points *A*, *B*, *C*, and *D* are collinear. In how many ways can the line be named using only these points? (Assume that different order means different name.)
   b. Points *A*, *B*, *C*, *D*, and *E* are coplanar and no three are collinear. In how many ways can the plane be named using only these points?

2. The following figure is a rectangular box in which *EFGH* and *ABCD* are rectangles with $\overline{BF}$ and $\overline{DH}$ perpendicular to planes *FGH* and *BCD*. Answer the following:
   a. Name two pairs of skew lines.
   b. Are $\overleftrightarrow{BD}$ and $\overleftrightarrow{FH}$ parallel, skew, or intersecting lines?
   c. Are $\overleftrightarrow{BD}$ and $\overleftrightarrow{GH}$ parallel?
   d. Find the intersection of $\overleftrightarrow{BD}$ and plane *EFG*.
   e. Explain why planes *BDH* and *FHG* are perpendicular.

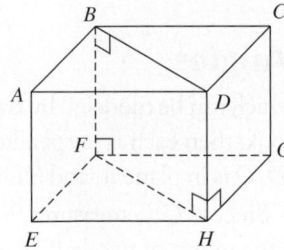

3. Use the following drawing of one of the Great Pyramids of Egypt (with square base) to find the following:
   a. The intersection of $\overline{AD}$ and $\overline{CE}$
   b. The intersection of planes *ABC*, *ACE*, and *BCE*
   c. The intersection of $\overleftrightarrow{AD}$ and $\overrightarrow{CA}$
   d. A pair of skew lines
   e. A pair of distinct parallel lines
   f. A plane not determined by one of the triangular faces or by the base

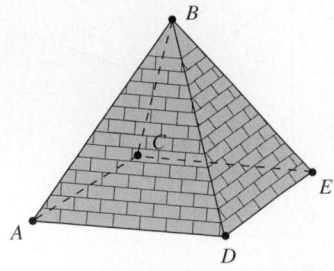

4. Determine how many acute angles are determined in the following figure:

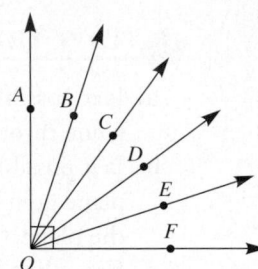

5. Identify a physical model for each of the following:
   a. Perpendicular lines
   b. An acute angle
   c. An obtuse angle

6. Find the measure of each of the following angles:
   a. ∠*EAB*   **b.** ∠*EAD*
   c. ∠*GAF*   **d.** ∠*CAF*

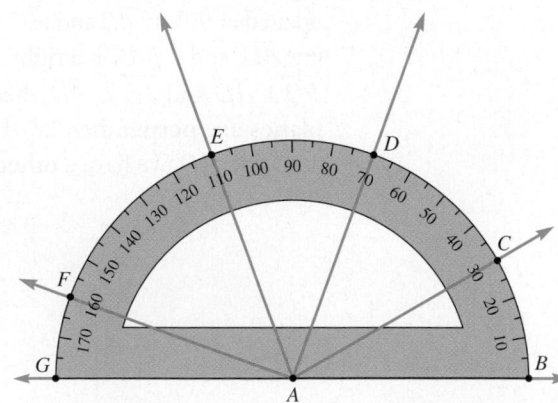

7. **a.** Perform each of the following operations. Leave your answers in simplest form.
   i. 18°35′29″ + 22°55′41″
   ii. 15°29′ − 3°45′
   b. Express the following using degrees, minutes, and seconds:
   i. 0.9°
   ii. 15.13°

8. Consider a correctly set clock that starts ticking at noon and answer the following:
   a. Find the measure of the angle swept by the hour hand by the time it reaches
      i. 3 P.M.
      ii. 12:25 P.M.
      iii. 6:50 P.M.
   b. Find the exact angle measure between the minute and the hour hands at 1:15 P.M.

9. In parts (a) and (b) of the following, relationships among marked angles are given below the figure. Find the measure of the marked angles. In part (c), find only the measure of $\angle BOC$ and tell why the exact values of x and y cannot be determined.

a.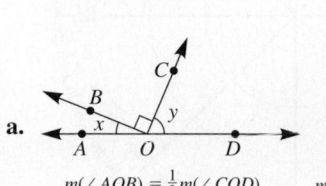

   $m(\angle AOB) = \frac{1}{3}m(\angle COD)$

b.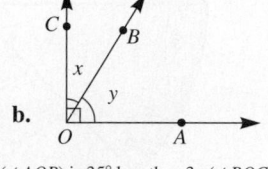

   $m(\angle AOB)$ is 35° less than $3m(\angle BOC)$

c.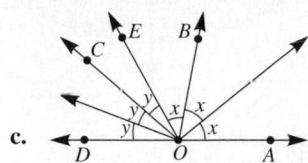

10. Given three collinear points A, B, C, with B between A and C, four different rays can be named using these points: $\overrightarrow{AB}$, $\overrightarrow{BA}$, $\overrightarrow{BC}$, and $\overrightarrow{CB}$. Determine how many different rays can be named given each of the following:
    a. Four collinear points
    b. Five collinear points
    c. n collinear points

11. Refer to the following table.

Number of Intersection Points of Coplanar Lines

	0	1	2	3	4	5
2			Not possible	Not possible	Not possible	Not possible
3					Not possible	Not possible
4						
5						
6						

(Number of Lines — vertical label)

   a. Sketch the possible intersections of the given number of lines. Three sketches are given for you.
   b. Given n lines, find a formula for determining the greatest possible number of intersection points.

12. Trace each of the following drawings. In your tracings, use dashed lines for segments that would not be seen and solid lines for segments that would be seen.

a.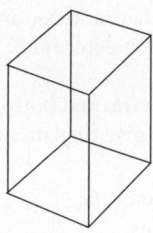

b.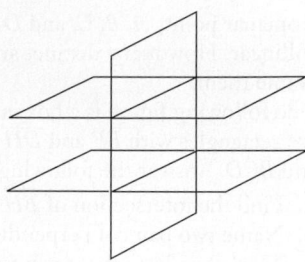

c.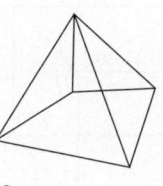

13. Draw pictures illustrating a real-world example of the following:
    a. Three planes intersecting in a common line
    b. Three planes intersecting in a common point
    c. A central angle of a circle

14. On the dot paper below, draw all possible segments parallel to $\overline{PQ}$ that have dots as endpoints.

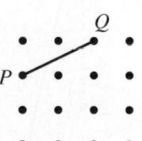

15. In the figure below, O is the center of the circle. Find the following measures:

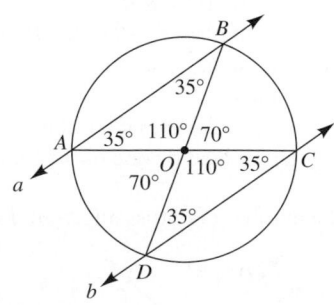

    a. $m(\overarc{BC})$
    b. $m(\overarc{CD})$
    c. $m(\overarc{AD})$
    d. $m(\overarc{AB})$
    e. $m(\overarc{ADC})$

Assessment 11-1B

1. Coplanar points *A*, *B*, *C*, and *D* are such that no three are collinear. How many distinct angles do they determine? Name them.

2. The following figure is a box in which the top and bottom are rectangles with $\overline{BF}$ and $\overline{DH}$ perpendicular to planes *FGH* and *BCD*. Answer the following:
   a. Find the intersection of $\overleftrightarrow{BH}$ and plane *DCG*.
   b. Name two pairs of perpendicular planes.
   c. Name two lines that are perpendicular to plane *EFH*.
   d. What is the measure of dihedral angle *D-HG-F*?

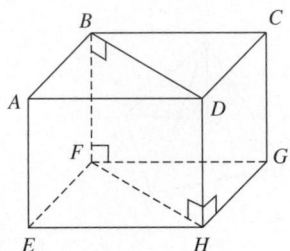

3. Use the following drawing of a triangular pyramid (Each triangle creating the pyramid determines a face.) to find the following:

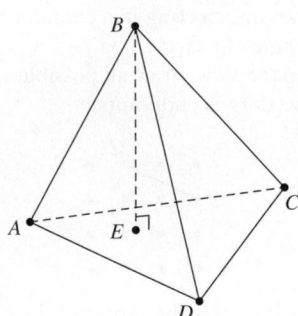

   a. The intersection of $\overline{AD}$ and $\overline{BE}$
   b. The intersection of planes *ABD* and *ACD*
   c. The intersection of $\overleftrightarrow{BC}$ and $\overleftrightarrow{AC}$
   d. A pair of skew lines
   e. A pair of distinct parallel lines if possible
   f. A plane not determined by one of the four triangular faces

4. Determine the number of obtuse angles in the following:

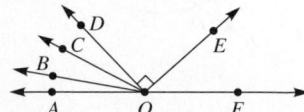

5. Identify a physical model for each of the following:
   a. Parallel lines
   b. A right angle
   c. Parallel planes

6. Find the measurement of each of the following angles:

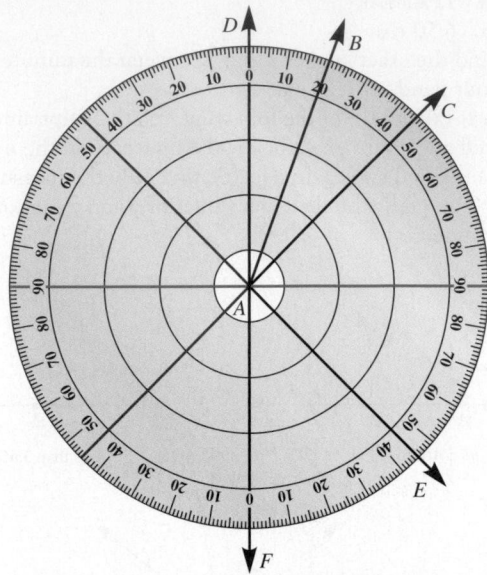

   a. ∠*BAC*   b. ∠*DAE*   c. ∠*CAF*
   d. Reflex angle, ∠*BAE*

7. a. Perform the following operations. Leave your answers in simplest form.
   i. 21°35′31″ + 49°51′32″
   ii. 93°38′14″ − 13°49′27″
   b. Express the following in degrees, minutes, and seconds, without decimals:
   i. 10.3°   ii. 15.14°

8. Consider a correctly functioning clock that starts ticking at noon and find the time between 12 noon and 1 P.M. when the angle measure between the hands is 180°.

9. In each of the following, relationships among marked angles are given below the figure. Find the measures of the marked angles.

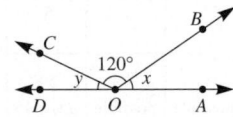

   a. $m(\angle DOC) = \frac{3}{4}m(\angle BOA)$

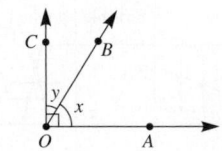

   b. $m(\angle AOB)$ is 30° less than $2m(\angle BOC)$

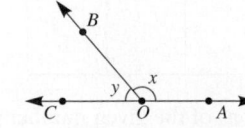

   c. $m(\angle AOB) - m(\angle BOC) = 50°$

10. a. How many planes are determined by three noncollinear points?
   b. How many planes are determined by four points, no three of which are coplanar?
   c. How many planes are determined by n points, no four of which are coplanar?

11. a. In the figure below, point A was constructed on line $\overleftrightarrow{DC}$. An arbitrary point B not on $\overleftrightarrow{DC}$ was chosen and $\overrightarrow{AB}$ drawn. Use a protractor (or geometry utility) to measure $\angle DAB$ and $\angle BAC$. Find half the measure of each angle to draw $\overrightarrow{AE}$ and $\overrightarrow{AF}$, which divide $\angle DAB$ and $\angle BAC$ in half as shown. Now measure $\angle EAF$.
   b. Repeat the construction described in part (a) for different locations of point B and conjecture what the measure of $\angle EAF$ will always be.
   c. Prove or disprove your conjecture in part (b).

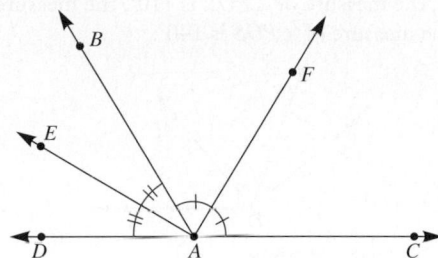

12. In each of the following pairs, determine whether the symbols name the same geometric figure:
   a. $\overrightarrow{AB}$ and $\overrightarrow{BA}$
   b. $\overline{AB}$ and $\overline{BA}$
   c. $\overleftrightarrow{AB}$ and $\overleftrightarrow{BA}$

13. Trace each of the following drawings of three-dimensional figures. In your tracings, use dashes for segments that would not be seen and solid drawings for segments that would be seen. (Different people may have different perspectives.)

a.

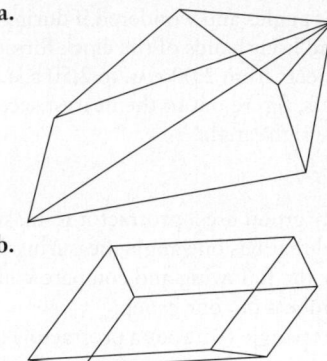

b.

14. On the dot paper below, draw all possible segments parallel to $\overline{PQ}$ that have dots as endpoints.

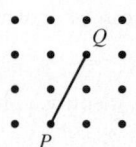

15. In the figure below, O is the center of the circle. Find the following measures.

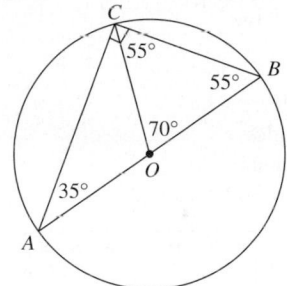

   a. $m(\angle ACO)$ **b.** $m(\overarc{AC})$ **c.** $m(\overarc{CB})$
   d. $m(\overarc{AB})$ **e.** $m(\overarc{CBA})$

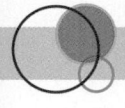

Mathematical Connections 11-1

Communication

1. Forest rangers use degree measures to identify directions and locate fires. In the drawing, a ranger at tower A observes smoke at a bearing of 149° (clockwise from the north), while another ranger at tower B observes the same source of smoke at a bearing of 250° (clockwise from the north).

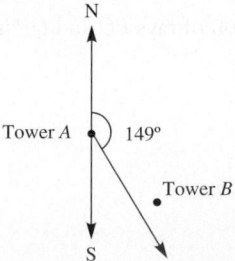

   a. On a blank sheet of paper, choose two locations for A and B and use the bearings and a protractor and a straightedge to locate the source of the smoke.
   b. Explain how the rangers find the location of the fire.

   c. Describe other situations in which location can be determined by similar methods.

2. Is it possible to locate four points in a plane such that the number of lines determined by the points is not exactly 1, 4, or 6? Explain.

3. A line n is perpendicular to plane α, and plane β contains n (n is in plane β). Must planes α and β be perpendicular? Explain why or why not.

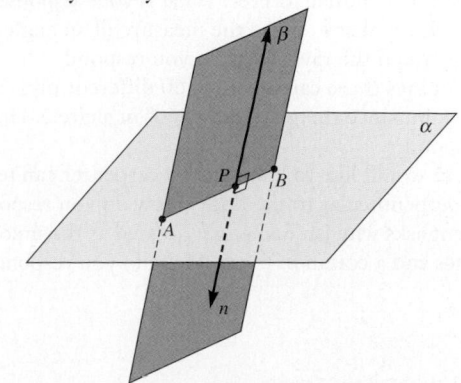

4. Mario was studying right angles and wondered if during his math class the minute and hour hands of the clock formed a right angle. If his class meets from 2:00 P.M. to 2:50 P.M., is a right angle formed? If it is, figure out to the nearest second when the hands form the right angle.

Cooperative Learning

5. Let each member of your group use a protractor to make a triangle out of cardboard that has one angle measuring 30° and another 50°. Answer the following and compare your solutions with other members of your group:
   a. Show how to use the triangle (without a protractor) to draw an angle with measure 40°.
   b. Is there more than one way to draw an angle of 40° using the triangle? Explain.
   c. What other angles can be drawn with the triangle? Why?

Open-Ended

6. Within the classroom, identify a physical object with the following:
   a. Parallel lines   b. Parallel planes
   c. Skew lines   d. Right angles
7. On a sheet of dot paper or on a geoboard like the one shown, create the following:
   a. Right angle
   b. Acute angle
   c. Obtuse angle
   d. Adjacent angles
   e. Parallel segments
   f. Intersecting segments

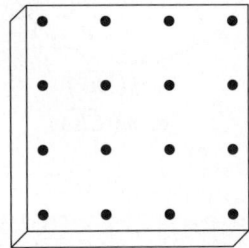

Questions from the Classroom

8. Henry claims that a line segment has a finite number of points because it has two endpoints. How do you respond?
9. A student claims that if any two planes that do not intersect are parallel, then any two lines that do not intersect should also be parallel. How do you respond?
10. A student says that it is actually impossible to measure an angle, since each angle is the union of two rays that extend infinitely and therefore continue forever. What is your response?
11. Maggie claims that to make the measure of an angle greater, you just extend the rays. How do you respond?
12. A student says there can be only 360 different rays emanating from a point since there are only 360° in a circle. How do you respond?
13. A student would like to know how a carpenter can tell if a wall is perpendicular to the floor. How do you respond?
14. A student asks why isn't an angle defined as the union of two half-lines and a common point. How do you respond?

Trends in Mathematics and Science Study (TIMSS) Questions

On the grid, draw a line parallel to line *L*.

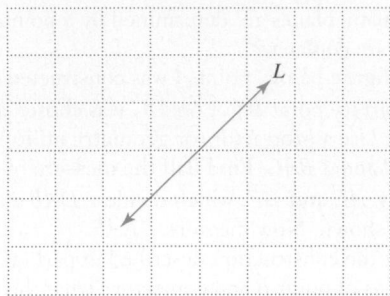

TIMSS, Grade 4, 2003

In the figure, the measure of ∠*POR* is 110°, the measure of ∠*QOS* is 90°, and the measure of ∠*POS* is 140°.

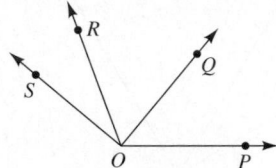

What is the measure of ∠*QOR*?

TIMSS, Grade 8, 2003

In this figure *PQ* is a straight line.

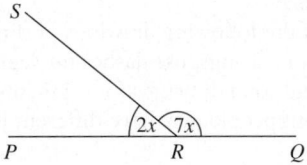

What is the degree measure of angle *PRS*?
   a. 10°   b. 20°   c. 40°   d. 70°
   e. 140°

TIMSS, Grade 8, 2007

National Assessment of Educational Progress (NAEP) Question

What is the intersection of rays *PQ* and *QP* in the figure above?
   a. Segment *PQ*
   b. Line *PQ*
   c. Point *P*
   d. Point *Q*
   e. The empty set

NAEP, Grade 8, 2007

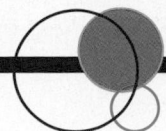

11-2 Linear Measure

In grade 4 *Common Core Standards*, students are expected to "solve problems involving measurement and conversion of measurements from a larger unit to a smaller unit" (p. 31).

In the United States, two measurement systems are used regularly: the English system and the metric system. In *PSSM*, we find the following concerning the use of the two measurement systems:

> Since the customary English system of measurement is still prevalent in the United States, students should learn both customary and metric systems and should know some rough equivalences between the metric and customary systems—for example, that a two-liter bottle of soda is a little more than half a gallon. The study of these systems begins in elementary school, and students at this level should be able to carry out simple conversions within both systems. Students should develop proficiency in these conversions in the middle grades and should learn some useful benchmarks for converting between the two systems. (pp. 45–46)

To measure a segment, we first decide on a unit of measure. In the cartoon, Jeremy's parents start measuring him in sofa lengths as the unit of measure. Early attempts at measurement lacked a standard unit and used fingers, hands, arms, and feet as units of measure. "Hands" are still used to measure the heights of horses. These early crude measurements were refined eventually and standardized by the English.

Zits copyright © 2003 Zits Partnership/Distributed by King Features Syndicate

NOW TRY THIS 11-10

a. Estimate the length of this textbook page in terms of paper clips.

b. Measure the textbook page in paper clips.

c. How close was your estimate to the measure?

d. Repeat parts (a–c) using *hands*.

e. Estimate then measure the width of a desk in hands.

f. In part (e), explain whether all students should obtain the same answer.

The English System

Originally, in the English system, a yard was the distance from the tip of the nose to the end of an outstretched arm of an adult person, and a foot was the length of a human foot. In 1893, the United States defined the yard and other units in terms of metric units. Some units of length in the English system and relationships among them are summarized in Table 11-6.

Table 11-6

Unit	Equivalent in Other Units
yard (yd)	3 ft
foot (ft)	12 in.
mile (mi)	1760 yd, or 5280 ft

Converting Units of Measure

To convert from one unit of length to another, different processes can be used. For example, to convert 5.25 mi to yards we use the facts that 1 mi = 5280 ft and 1 ft = $\frac{1}{3}$ yd. Therefore,

$$5.25 \text{ mi} = 5.25(5280) \text{ ft} = 5.25(5280)\left(\frac{1}{3}\right) \text{ yd, or } 9240 \text{ yd.}$$

Likewise, to convert 432 in. to yards we use the facts that 1 ft = $\frac{1}{3}$ yd and 1 in. = $\frac{1}{12}$ ft. Therefore,

$$432 \text{ in.} = 432\left(\frac{1}{12}\right) \text{ ft} = 432\left(\frac{1}{12}\right)\left(\frac{1}{3}\right) \text{ yd, or } 12 \text{ yd.}$$

Dimensional Analysis (Unit Analysis)

One process for converting units of measure is **dimensional analysis**. This process works with ***unit ratios*** (ratios equivalent to 1) treated as fractions. Since 1 yd = 3 ft, then $\frac{1 \text{ yd}}{3 \text{ ft}} = 1$ and $\frac{3 \text{ ft}}{1 \text{ yd}} = 1$ and these are unit ratios. Also, $\frac{5280 \text{ ft}}{1 \text{ mi}} = 1$ and is a unit ratio. Therefore, to convert 5.25 mi to yards, we have the following:

$$5.25 \text{ mi} = 5.25 \text{ mi} \cdot \frac{5280 \text{ ft}}{1 \text{ mi}} \cdot \frac{1 \text{ yd}}{3 \text{ ft}} = 9240 \text{ yd}$$

EXAMPLE 11-3

If a cheetah is clocked at 60 miles per hour (mph), what is its speed in feet per second?

Solution

$$60 \frac{\text{mi}}{\text{hr}} = 60 \frac{\text{mi}}{\text{hr}} \cdot \frac{5280 \text{ ft}}{1 \text{ mi}} \cdot \frac{1 \text{ hr}}{60 \text{ min}} \cdot \frac{1 \text{ min}}{60 \text{ sec}} = 88 \frac{\text{ft}}{\text{sec}}$$

Historical Note

The yard has been defined in many different ways over its history. For example, in 1832, it was defined as the distance between the 27th and 63rd inches on a certain brass bar made by Troughton of London. In 1856, the yard was redefined in terms of the British Bronze Yard No. 11 (which was 0.00087 in. longer than the Troughton yard). In 1893, the yard was redefined in terms of the international meter as $\frac{3600}{3937}$ of a meter through use of an 1866 U.S. law making the practice of the metric system permissible in the United States. In 1960, the meter was redefined in terms of the wavelength of light from krypton-86 and still later in terms of the distance light travels in $\frac{1}{299,792,458}$ sec. Effective July 1, 1959, the yard is defined in terms of the international yard, which in turn was based on the international definition of a meter.

Note that treating ratios as fractions in Example 11-4 allows us to use multiplication principles.

EXAMPLE 11-4

Complete each of the following:

a. 219 ft = _____ yd
b. 8432 yd = _____ mi
c. 0.2 mi = _____ ft
d. 64 in. = _____ yd

Solution

a. Because 1 ft = $\frac{1}{3}$ yd, 219 ft = $219\left(\frac{1}{3}\text{ yd}\right)$ = 73 yd. Alternatively,

$$219 \text{ ft} = 219 \text{ ft}\left(\frac{1 \text{ yd}}{3 \text{ ft}}\right) = 73 \text{ yd.}$$

b. Because 1 yd = $\frac{1}{1760}$ mi, 8432 yd = $8432\left(\frac{1}{1760}\text{ mi}\right)$ ≈ 4.79 mi. Alternatively,

$$8432 \text{ yd} = 8432 \text{ yd}\left(\frac{3 \text{ ft}}{1 \text{ yd}}\right)\left(\frac{1 \text{ mi}}{5280 \text{ ft}}\right) \approx 4.79 \text{ mi.}$$

c. 1 mi = 5280 ft. Hence, 0.2 mi = 0.2(5280 ft) = 1056 ft. Alternatively,

$$0.2 \text{ mi} = 0.2 \text{ mi} \cdot \frac{5280 \text{ ft}}{1 \text{ mi}} = 1056 \text{ ft.}$$

d. We first find a connection between yards and inches. We have 1 yd = 3 ft and
1 ft = 12 in. Hence, 1 yd = 3 ft = 3(12 in.) = 36 in. Hence, 1 in. = $\frac{1}{36}$ yd; therefore,

$$64 \text{ in.} = 64\left(\frac{1}{36}\text{ yd}\right) = \frac{16}{9}\text{ yd} \approx 1.78 \text{ yd. Alternatively, } 64 \text{ in.} = 64 \text{ in.}\left(\frac{1 \text{ ft}}{12 \text{ in.}}\right)\left(\frac{1 \text{ yd}}{3 \text{ ft}}\right) =$$

$$\frac{16}{9}\text{ yd} \approx 1.78 \text{ yd.}$$

The Metric System

At this time, the United States is the only major industrial nation in the world that continues to use the English system. However, the use of the **metric system** in the United States has been increasing, particularly in the scientific community and in industry.

Different units of length in the metric system are obtained by multiplying a base unit by a power of 10. Table 11-7 gives some of the prefixes for these units, their symbols, and the multiplication factors.

Historical Note

The metric system, a decimal system, was proposed in France in 1670 by Gabriel Mouton. However, only in 1790 did the French Academy of Sciences bring groups together to develop the system. At that time in France, there were 13 distinct measures for the *foot* ranging from 10.6 in. to 13.4 in. Recognizing the need for a standard base unit of linear measurement, the Academy chose $\frac{1}{10,000,000}$ of the distance from the equator to the North Pole on a meridian through Paris as the base unit of length and called it the *meter* (*m*), from the Greek word *metron*, meaning "to measure."

The U.S. Congress included encouragement for U.S. industrial metrication in the Omnibus Trade and Competitiveness Act of 1988 by designating the metric system as the preferred system of weights and measures for U.S. trade and commerce and by requiring each federal agency to be metric by the end of fiscal year 1992. Since 2009, all products sold in Europe (with limited exceptions) are required to have only metric units on their labels.

Table 11-7

Prefix	Symbol	Factor	
kilo	k	1000	(one thousand)
*hecto	h	100	(one hundred)
*deka	da	10	(ten)
*deci	d	0.1	(one tenth)
centi	c	0.01	(one hundredth)
milli	m	0.001	(one thousandth)

*Not commonly used

Metric prefixes, combined with the base unit meter, name different units of length. Table 11-8 gives units, the symbol for each, and their relationship to the meter.

Table 11-8

Unit	Symbol	Relationship to Base Unit
kilometer	km	1000 m
*hectometer	hm	100 m
*dekameter	dam	10 m
meter	**m**	**base unit**
*decimeter	dm	0.1 m
centimeter	cm	0.01 m
millimeter	mm	0.001 m

*Not commonly used

Other metric prefixes are used for greater and lesser quantities; for example, *mega* (1,000,000) and *micro* (0.000001). The symbols for mega and micro are M and μ, respectively.

NOW TRY THIS 11-11

If our money system used metric prefixes and the base unit was a dollar, give metric names to each of the following:

a. dime   **b.** penny   **c.** $10 bill   **d.** $100 bill   **e.** $1000 bill

Benchmarks used for estimations for a meter, a decimeter, a centimeter, and a millimeter are shown in Figure 11-20. The kilometer is commonly used for measuring longer distances: 1 km = 1000 m. Nine football fields, including end zones, laid end to end are approximately 1 km long.

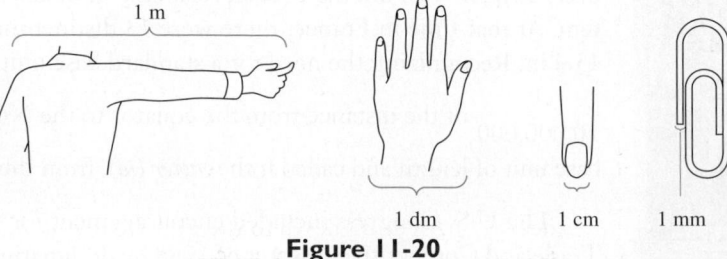

Figure 11-20

In the grade 2 *Focal Points*, we find that students:

> Understand linear measure as an iteration of units and use rulers and other measurement tools with that understanding. They understand the need for equal-length units, the use of standard units of measure (centimeter and inch), and the inverse relationship between the size of a unit and the number of units used in a particular measurement (i.e., children recognize that the smaller the unit, the more iterations they need to cover a given length). (p. 14)

As mentioned in *Focal Points*, the smaller the unit the more iterations needed to cover a given length. This motivates a need for larger units of length and the ability to convert one unit to another.

Conversions among metric lengths are accomplished by multiplying or dividing by powers of 10. As with converting dollars to cents or cents to dollars, we move the decimal point to the left or right, depending on the units. For example,

$$0.123 \text{ km} = 1.23 \text{ hm} = 12.3 \text{ dam} = 123 \text{ m} = 1230 \text{ dm} = 12,300 \text{ cm} = 123,000 \text{ mm}$$

It is possible to convert units by using the chart in Figure 11-21. We count the number of steps from one unit to the other and move the decimal point that many steps in the same direction.

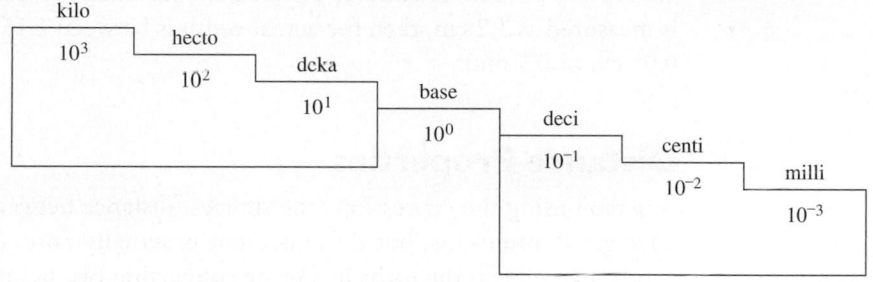

Figure 11-21

EXAMPLE 11-5

Complete each of the following:

a. 1.4 km = _____ m
b. 285 mm = _____ m
c. 0.03 km = _____ cm

Solution

a. Because 1 km = 1000 m, to change kilometers to meters, we multiply by 1000. Hence, 1.4 km = 1.4 (1000 m) = 1400 m. Alternatively,

$$1.4 \text{ km} = 1.4 \text{ km} \cdot \frac{1000 \text{ m}}{1 \text{ km}} = 1400 \text{ m}.$$

b. Because 1 mm = 0.001 m, to change from millimeters to meters, we multiply by 0.001. Thus, 285 mm = 285(0.001 m) = 0.285 m. Alternatively,

$$285 \text{ mm} = 285 \text{ mm} \cdot \frac{1 \text{ m}}{1000 \text{ mm}} = 0.285 \text{ m}.$$

c. To change kilometers to centimeters, we first multiply by 1000 to convert kilometers to meters and then multiply by 100 to convert meters to centimeters. Therefore, we move the decimal point five places to the right to obtain 0.03 km = 3000 cm. An alternative approach is to use Figure 11-20. To go from kilo to centi on the steps, we move five places to the right, so we need to move the decimal point in 0.03 five places to the right resulting in 3000.

Linear units of length or distance may be measured with rulers. Figure 11-22 shows part of a centimeter ruler.

Figure 11-22

In the grade 5 *Focal Points*, we find that, "They [students] solve problems that require attention to both approximation and precision of measurement." (p. 17)

When drawings are given, we assume that the listed measurements are accurate. When actually measuring real-world objects, such accuracy is usually impossible. Measuring distances in the real world frequently results in errors. Thus, many industries using parts from a variety of sources rely on portable calibration units to test measuring instruments used in constructing parts. Calibration helps assure that all parts fit together. To calibrate measuring instruments, industrial technicians establish the greatest possible error (GPE) allowable in order to obtain the final fit. The **greatest possible error (GPE)** of a measurement is one-half the unit used. For example, if the width of a piece of board is measured as 5 cm, the actual width must be between 4.5 cm and 5.5 cm. Therefore, the GPE for this measurement is 0.5 cm. If the width of a button is measured as 1.2 cm, then the actual width is between 1.15 cm and 1.25 cm, and so the GPE is 0.05 cm or 0.5 mm.

Distance Properties

A person using the expression "the shortest distance between two points is a straight line" may have good intentions, but the statement is actually false. (Why?) A correct statement is "the shortest among all the paths in a plane connecting two points A and B is the segment $\overline{AB}$." (The length of $\overline{AB}$ is denoted by AB.) This and other basic properties of distance are listed below:

> **Properties of Distance**
> 1. The distance between any two points A and B is greater than or equal to 0, written $AB \geq 0$.
> 2. The distance between any two points A and B is the same as the distance between B and A, written $AB = BA$.
> 3. For any three points A, B, and C, the distance between A and B plus the distance between B and C is greater than or equal to the distance between A and C, written $AB + BC \geq AC$.

In the special case where A, B, and C are collinear and B is between A and C, as in Figure 11-23(a), we have $AB + BC = AC$. Otherwise, if A, B, and C are not collinear, as in Figure 11-23(b), then they form the vertices of a triangle and $AB + BC > AC$. This inequality, $AB + BC > AC$, leads to the **Triangle Inequality**.

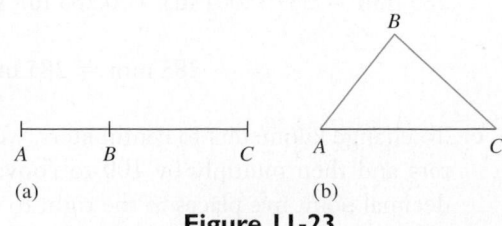

Figure 11-23

> **Theorem 11-1:   Triangle Inequality**
> The sum of the lengths of any two sides of a triangle is greater than the length of the third side.

 NOW TRY THIS 11-12

If two sides of a triangle are 31 cm and 85 cm long and the measure of the third side must be a whole number of centimeters,

   a. what is the longest the third side can be?
   b. what is the shortest the third side can be?

Perimeter of a Plane Figure

In common language, the **perimeter** of a figure is the distance around the figure and has linear measure as seen in Example 11-6.

 **EXAMPLE 11-6**

Find the perimeter of each of the shapes in Figure 11-24.

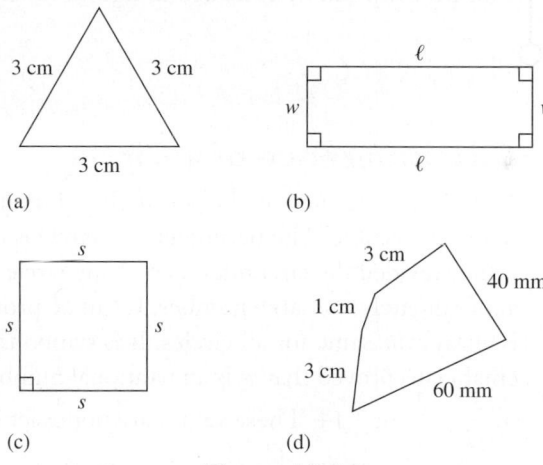

Figure 11-24

Solution
   a. The perimeter is $3(3) = 9$ cm.
   b. The perimeter is $2w + 2\ell$.
   c. The perimeter is $4s$.
   d. Because 40 mm = 4 cm and 60 mm = 6 cm, the perimeter is $1 + 3 + 4 + 6 + 3 = 17$ cm, or 170 mm.

When letters are used for distances, as in Example 11-6(b) and (c), we do not usually attach dimensions to an answer. If the perimeter is $4s$, we actually mean that it is $4s$ units of length.

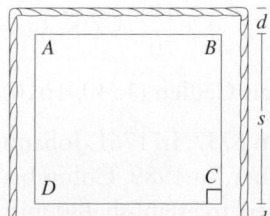

Figure 11-25

Problem Solving   **Roping a Square**

Given a square $ABCD$ with sides of length s, stretch a thin rope tightly around the square. Take the rope off, add 100 ft to it, and put the extended rope back around the square so that it is evenly spaced and the new rope makes a square around the original square, as in Figure 11-25. Find d, the distance between the squares.

Understanding the Problem We are to determine the distance between a square and a new square formed by adding 100 ft of rope to a rope that was stretched around the original square. Figure 11-25 shows the situation if the length of the side of the original square is s and the unknown distance is d.

Devising a Plan If we use variables to represent the unknowns, we can *write an equation* to model the problem. The perimeter of the new square is $4s + 100$. Another way to represent this perimeter is $4(s + 2d)$. Therefore, we have $4s + 100 = 4(s + 2d)$. We must solve this equation for d.

Carrying Out the Plan We solve the equation as shown next.

$$4s + 100 = 4(s + 2d)$$
$$4s + 100 = 4s + 8d$$
$$100 = 8d$$
$$12.5 = d$$

Therefore, the distance between the squares is 12.5 ft.

Looking Back A different way to think about the problem is to consider the individual parts that must sum to 100. The perimeter of the original square does not change in Figure 11-26, but the eight red segments sum to 100 ft so one red segment has length $\frac{100}{8}$, or 12.5 ft. Notice that the answer is the same regardless of the size of the original square. The distance is always 12.5 ft. This problem can be extended to figures other than squares.

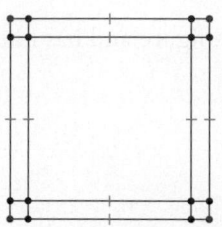

Figure 11-26

Circumference of a Circle

A circle was defined as the set of all points in a plane that are the same distance from a given point, the center. The perimeter of a circle is its **circumference**. Ancient Greeks discovered that if they divided the circumference of any circle by the length of its diameter, they always obtained approximately the same number. It can be proven that the ratio of circumference C to diameter d is always the same for all circles. It is symbolized as π (**pi**). In the late eighteenth century, mathematicians proved that π is an irrational number. For most practical purposes, π is approximated by $\frac{22}{7}, 3\frac{1}{7}$, or 3.14. These values are not exact values of π.

The relationship $\frac{C}{d} = \pi$ is used for finding the circumference of a circle and normally is written

$C = \pi d$ or $C = 2\pi r$ *because the length of the diameter d is twice the radius (r) of the circle.* The exact circumference of a circle with diameter 6 cm is 6π cm, and an approximation might be $6(3.14) = 18.84$ cm.

The following student page shows two approaches to finding circumference of a circle.

Historical Note

$\pi = 3.14159\ 26535\ 89793\ 23846\ 26433\ 83279\ 50288\ 41971\ 69399\ 37510\ 58209\ldots$

Archimedes (b. 287 BCE) approximated π by the inequality $3\frac{10}{71} < \pi < 3\frac{10}{70}$. A Chinese astronomer thought that $\pi = \frac{355}{113}$. German mathematician Ludolph van Ceulen (1540–1610) calculated π to 35 decimal places. Leonhard Euler adopted the symbol π in 1737. In 1761, Johann Lambert, an Alsatian mathematician, proved that π is an irrational number. In 1989, Columbia University mathematicians David and Gregory Chudnovsky used computers to establish 480 million digits of π and in 1995, reached the billionth digit of pi. Now the known digits of pi exceed 1 trillion and no pattern among the digits has been found.

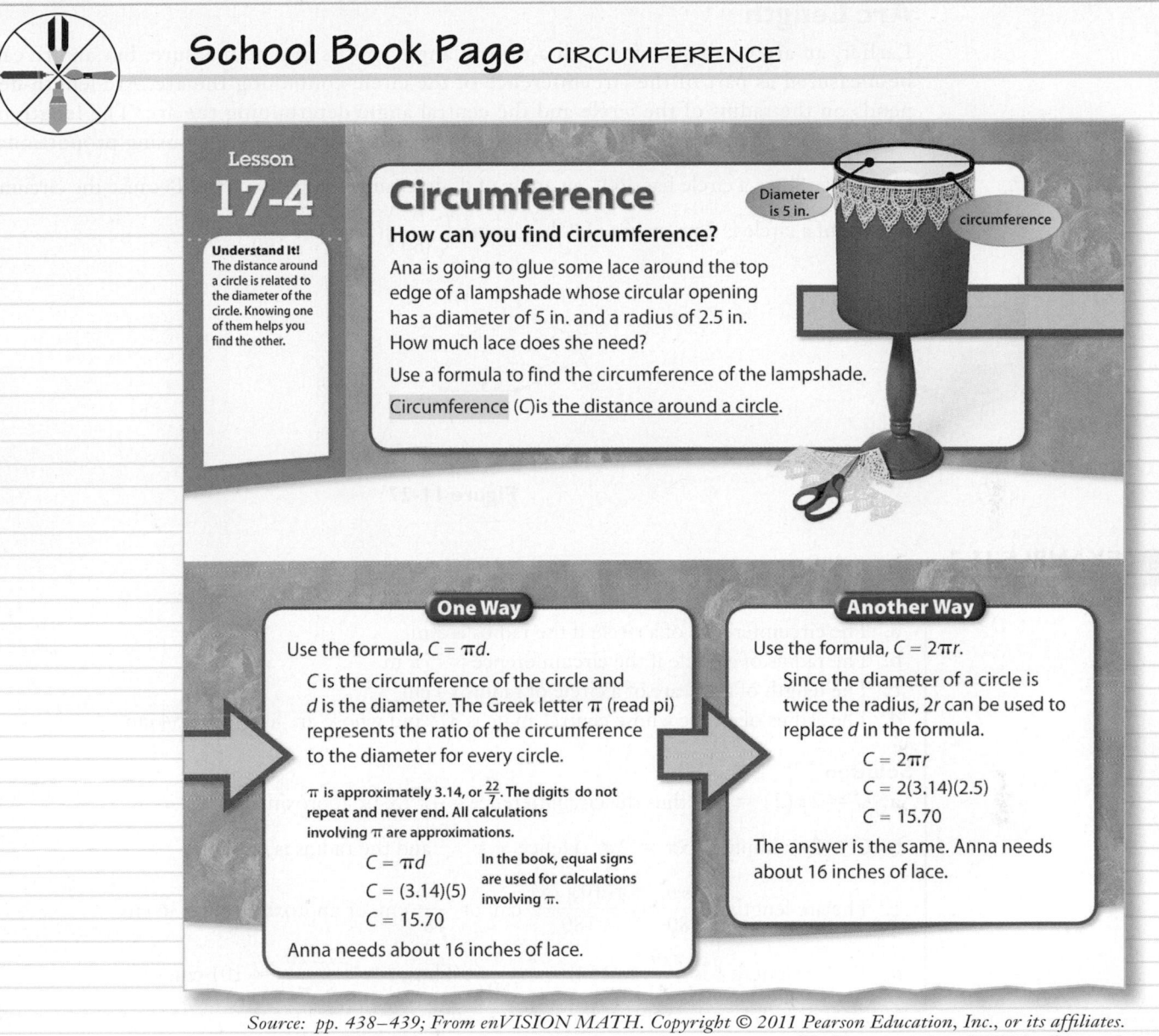

School Book Page CIRCUMFERENCE

Lesson
17-4

Understand It!
The distance around a circle is related to the diameter of the circle. Knowing one of them helps you find the other.

Circumference

How can you find circumference?

Ana is going to glue some lace around the top edge of a lampshade whose circular opening has a diameter of 5 in. and a radius of 2.5 in. How much lace does she need?

Use a formula to find the circumference of the lampshade.

Circumference (*C*)is the distance around a circle.

Diameter is 5 in.

circumference

One Way

Use the formula, $C = \pi d$.

C is the circumference of the circle and *d* is the diameter. The Greek letter π (read pi) represents the ratio of the circumference to the diameter for every circle.

π is approximately 3.14, or $\frac{22}{7}$. The digits do not repeat and never end. All calculations involving π are approximations.

$C = \pi d$
$C = (3.14)(5)$
$C = 15.70$

In the book, equal signs are used for calculations involving π.

Anna needs about 16 inches of lace.

Another Way

Use the formula, $C = 2\pi r$.

Since the diameter of a circle is twice the radius, *2r* can be used to replace *d* in the formula.

$C = 2\pi r$
$C = 2(3.14)(2.5)$
$C = 15.70$

The answer is the same. Anna needs about 16 inches of lace.

NOW TRY THIS 11-13

To approximate the value of π, we need string, a marked ruler, and several different-sized round cans or jars. Pick a can and wrap the string tightly around the can. Use a pen to mark a point on the string where the beginning of the string meets the string again. Unwrap the string and measure its length. Next, determine the diameter of the can by tracing the bottom of the can on a piece of paper. Fold the circle onto itself to find a line of symmetry. The chord determined by the line is a diameter of the circle. Measure the diameter and determine the ratio of the circumference to the diameter. Use the same units in all measurements. Repeat the experiment with at least three cans and find the average of the corresponding ratios.

In class, record all diameters and corresponding circumferences in columns A and B in a spreadsheet. Plot the graph of this data, with circumference on the vertical axis and diameter on the horizontal axis. What do we observe? Find a trend (fitted) line for the data. What is the slope of this line?

Arc Length

Earlier, an arc was associated with a central angle and its degree measure, but an arc can be measured as part of the circumference of the circle containing the arc. Arc length depends on the radius of the circle and the central angle determining the arc. The length of an arc whose central angle is $\theta°$ can be developed as in Figure 11-27 by using proportional reasoning. Since a circle has 360°, an angle of $\theta°$ determines $\dfrac{\theta}{360}$ of a circle. Because the circumference of a circle is $2\pi r$, an arc of $\theta°$ has length $\left(\dfrac{\theta}{360}\right)(2\pi r)$, or $\dfrac{\pi r \theta}{180}$.

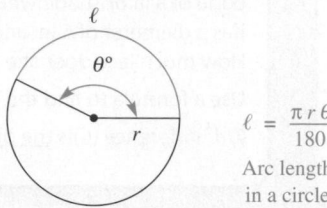

$$\ell = \frac{\pi r \theta}{180}$$

Arc length
in a circle

Figure 11-27

EXAMPLE 11-7

Find each of the following:

   a. The circumference of a circle if the radius is 2 m
   b. The radius of a circle if the circumference is 15π m
   c. The length of a 25° arc of a circle of radius 10 cm
   d. The radius of an arc whose central angle is 87° and whose arc length is 154 cm

Solution

   a. $C = 2\pi(2) = 4\pi$; thus the circumference is 4π m, or approximately 12.56 m.

   b. $C = 2\pi r$ implies $15\pi = 2\pi r$. Hence, $r = \dfrac{15}{2}$ and the radius is 7.5 m.

   c. The arc length is $\dfrac{\pi r \theta}{180} = \dfrac{\pi(10)(25)}{180}$ cm, or $\dfrac{25\pi}{18}$ cm, or approximately 4.36 cm.

   d. The arc length ℓ is $\dfrac{\pi r \theta}{180}$, so $154 = \dfrac{\pi r(87)}{180}$. Thus, $r = \dfrac{27,720}{87\pi} \approx 101$ cm.

Assessment 11-2A

1. Use the following picture of a ruler to find each of the lengths in centimeters:

   a. AB **e.** IJ
   b. DE **f.** AF
   c. CJ **g.** IC
   d. EF **h.** GB

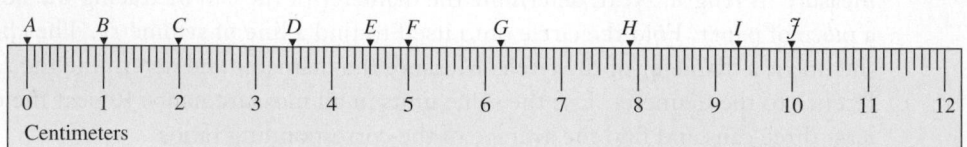

2. Estimate and then measure each of the following in terms of the units listed:
   a. The length of your desktop in cubits (elbow to outstretched fingers)
   b. The length of this page in pencil lengths
   c. The width of this book in pencil widths
3. Complete each of the following:
   a. 100 in. = _____ yd
   b. 400 yd = _____ in.
   c. 300 ft = _____ yd
   d. 372 in. = _____ ft
4. Draw segments that you estimate to be of the given lengths. Use a metric ruler to check the estimates.
   a. 10 mm
   b. 100 mm
   c. 1 cm
   d. 10 cm
5. Estimate the length of the following segment and then measure it:

   |————————————————————————————————|

   Express the measurement in each of the following units:
   a. Millimeters
   b. Centimeters
6. Choose an appropriate metric unit and estimate each of the following measures (check your estimates):
   a. The length of a pencil
   b. The diameter of a nickel
   c. The width of the top of a desk

7. Repeat exercise 6 using English measures.
8. Complete the following table:

Item	m	cm	mm
a. Length of a piece of paper		35	
b. Height of a woman	1.63		
c. Width of a filmstrip			35
d. Length of a cigarette			100
e. Length of two meter-sticks laid end to end	2		

9. For each of the following, place a decimal point in the number to make the sentence reasonable:
   a. A stack of 10 dimes is 1350 mm high.
   b. The desk is 770 m high.
   c. The distance from one side of a street to the other is 100 m.
   d. A dollar bill is 155 cm long.
10. List the following in decreasing order: 8 cm, 5218 mm, 245 cm, 91 mm, 6 m, 700 mm.
11. Draw a circle whose circumference is 4 in. as accurately as possible.
12. a. What is the length of a semicircle of a circle whose radius is 1 unit?
    b. What is the length of a semicircle of a circle whose radius is 1/2 unit?
13. Guess the perimeter in centimeters of each of the following figures and then check the estimates using a ruler:

a.   b.

14. Complete each of the following:
    a. 10 mm = _____ cm
    b. 262 m = _____ km
    c. 3 km = _____ m
    d. 30 mm = _____ m
15. Draw a triangle *ABC*. Measure the length of each of its sides in millimeters. For each of the following, tell which is greater and by how much:
    a. *AB* + *BC* or *AC*
    b. *BC* + *CA* or *AB*
    c. *AB* + *CA* or *BC*
16. Can the following be the lengths of the sides of a triangle?
    a. 23 cm, 50 cm, 60 cm
    b. 10 cm, 40 cm, 50 cm
17. Take an $8\frac{1}{2} \times 11$-in. piece of paper, fold it as shown in the following figure, and then cut the folded paper along the diagonal as shown.

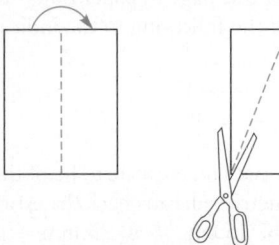

    a. Rearrange the two smaller pieces to find a triangle with the minimum perimeter.
    b. Arrange the two smaller pieces to form a triangle with the maximum perimeter.
18. For each of the following circumferences, find the radius of the circle:
    a. 12π cm
    b. 6 m
19. For each of the following, if a circle has the dimensions given, determine its circumference:
    a. 6 cm diameter
    b. $\frac{2}{\pi}$ cm radius
20. What happens to the circumference of a circle if the length of the radius is doubled?
21. Jet planes can exceed the speed of sound. A *Mach number* describes the speed of such planes. Mach 2 is twice the speed of sound. The speed of sound in air is approximately 344 m/sec.
    a. If the speed of a plane is described as Mach 2.5, what is its speed in kilometers per hour (km/hr)?
    b. If the speed of a plane is described as Mach 3, what is its speed in meters per second?
    c. Describe the speed of 5000 km/hr as a Mach number.
22. Give the greatest possible error for each of the following measurements:
    a. 23 m
    b. 3.6 cm
    c. 3.12 m

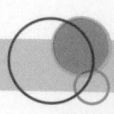

Assessment 11-2B

1. Use the following picture of a ruler to find each of the lengths in inches:

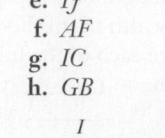

   a. *AB*   e. *IJ*
   b. *DE*   f. *AF*
   c. *CJ*   g. *IC*
   d. *EF*   h. *GB*

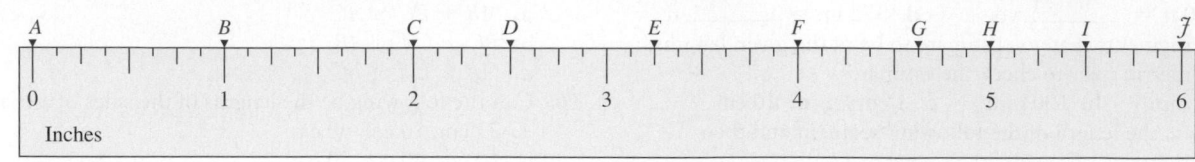

2. Estimate and then measure each of the following in terms of the units listed:
   a. The length of your desktop in mathematics book lengths
   b. The length of this page in paper dollar widths
   c. The width of this page in paper dollar lengths
3. Convert each of the following to the indicated unit.
   a. 100 in. = _____ ft
   b. 400 yd = _____ ft
   c. 300 ft = _____ in.
   d. 372 in. = _____ yd
4. Draw segments that you estimate to be of the following lengths. Use a metric ruler to check the estimates.
   a. 0.01 m   b. 15 cm   c. 35 mm   d. 150 mm
5. Estimate the length of the following segment and then measure it:

   |————————————————|

   Express the measurement in each of the following units:
   a. Millimeters
   b. Centimeters
6. Choose an appropriate metric unit and estimate each of the following measures (check your estimates):
   a. The thickness of the top of a desk
   b. The length of this page of paper
   c. The height of a door
7. Repeat exercise 6 using English measures.
8. Complete the following table:

Item	m	cm	mm
a. Width of a piece of paper		20	
b. Height of a woman	1.52		
c. Length of a pencil			90
d. Length of a baseball bat	1.1		

9. For each of the following, place a decimal point in the number to make the sentence reasonable:
   a. The basketball player is 1950 cm tall.
   b. A new piece of chalk is about 8100 cm long.
   c. The speed limit in town is 400 km/hr.
10. List the following in decreasing order: 8 m, 5218 cm, 245 cm, 91 m, 6 m, 925 mm.

11. Draw each of the following as accurately as possible:
    a. A triangle whose perimeter is 4 in.
    b. A 4-sided figure whose perimeter is 8 cm
12. The following figure is a circle whose radius is 2 cm. The diameters of the two semicircular regions inside the large circle are also 2 cm long. Compute the length of the curve that separates the shaded and white regions.

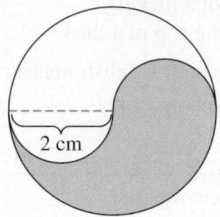

13. Guess the perimeter in centimeters of each of the following figures and then check the estimates using a ruler:

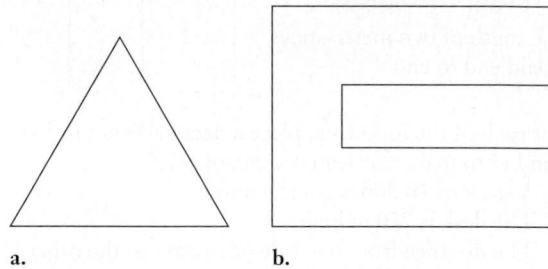

    a.   b.

14. Complete each of the following:
    a. 35 m = _____ cm
    b. 359 mm = _____ m
    c. 647 mm = _____ cm
    d. 0.1 cm = _____ mm
15. Explain in your own words why the Triangle Inequality must be true.
16. Can the following be the lengths of the sides of a triangle?
    a. 20 cm, 40 cm, 50 cm
    b. 20 cm, 40 cm, 60 cm
    c. 41 cm, 250 mm, 12 cm

17. The following figure made of 6 unit squares has a perimeter of 12 units. The figure is made in such a way that each square must share at least one complete side with another square.

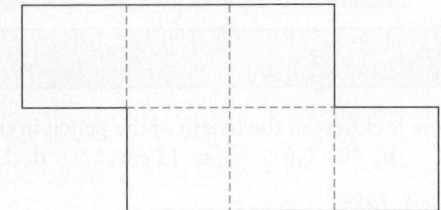

 a. Add more squares to the figure above so that the perimeter of the new figure is 18.
 b. Consider any figure made of squares where each square must share at least one complete side with another square, what is the minimum number of squares required to build a figure of perimeter 18?
 c. Under the same condition as part (b), what is the maximum number of squares possible to build a figure of perimeter 18?
18. For each of the following circumferences, find the radius of the circle:
 a. 0.67 m b. 92π cm

19. For each of the following, if a circle has the dimensions given, determine its circumference:
 a. 3-cm radius
 b. 6π-cm diameter
20. How does the radius of a circle change if the circumference is doubled?
21. Astronomers use a light-year to measure distance. A light-year is the distance light travels in 1 yr. The speed of light is approximately 300,000 km/sec.
 a. How long is 1 light-year in kilometers?
 b. The nearest star (other than the sun) is Alpha Centauri. It is 4.34 light-years from Earth. How far is that in kilometers?
 c. How long will it take a rocket traveling 60,000 km/hr to reach Alpha Centauri?
 d. How long will it take the rocket in (c) to travel to the sun if it takes approximately 8 min 19 sec for light from the sun to reach the earth?
22. Give the greatest possible error for each of the following:
 a. 136 m
 b. 3.5 ft
 c. 3.62 cm

Mathematical Connections 11-2

Communication

1. Howard Eves (1911–2004) described the creation of the metric system as one of the greatest accomplishments of the eighteenth century. Do you agree or disagree? Explain.
2. There has been considerable debate about whether the United States should change to the metric system.
 a. Based on your experiences with linear measure, what do you see as the advantages of changing?
 b. Which system do you think would be easier for children to learn? Why?
 c. What things will probably not change if the United States adopts the metric system?
3. In track, the second lane from the inside of the track is longer than the inside lane. Use this information to explain why, in running events that require a complete lap of the track, runners are lined up at the starting blocks as shown in the following figure:

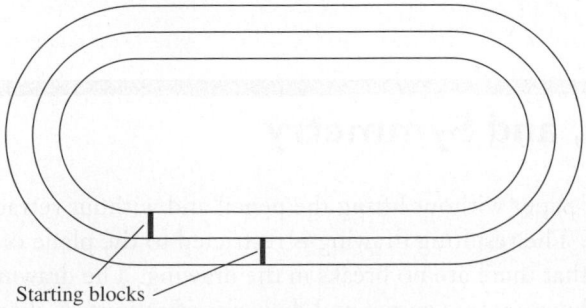

Starting blocks

Open-Ended

4. Observe that it is possible to build a triangle with toothpicks that has sides of 3, 4, and 5 toothpicks, as shown, and answer the questions that follow.

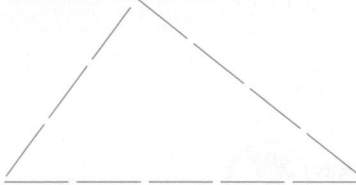

 a. Find two other triples of toothpicks that can be used as sides of a triangle and two other triples that cannot be used to create a triangle.
 b. Describe how to tell whether a given triple of numbers a, b, c can be used to construct a triangle with sides of a, b, and c toothpicks. Explain why your rule is valid.

Cooperative Learning

5. a. Help each person in the group find his or her height in centimeters.
 b. Help each person in the group find the length of his or her outstretched arms (horizontal) from fingertip to fingertip in centimeters.
 c. Compare the difference between the two measurements in parts (a) and (b). Compare the results of the group members and make a conjecture about the relationship between the two measurements.
 d. Compare your group's results with other groups to determine if they have similar findings.

6. Jerry wants to design a gold chain 60 cm long made of thin gold wire circles, each of which is the same size. He wants to use the least amount of wire and wonders what the radius of each circle should be.

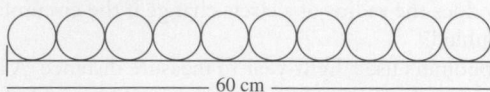

60 cm

   a. Each member of the group should choose a specific number of circles and find the length of wire needed to make a 60 cm chain with the chosen number of circles.
   b. Compare your results and make a conjecture based on the results.
   c. Justify your conjecture.

Questions from the Classroom

7. A student claims that the circumference of a semicircle is the same as that of a circle from which the semicircle was taken because one has to measure the "outer part" and the "inner part." How do you respond?
8. A student has a tennis ball can with a flat top and bottom containing three tightly fitting tennis balls. To the student's surprise, the circumference of the top of the can is longer than the height of the can. The student wants to know if this fact can be explained without performing any measurements. How do you respond?

Review Problems

9. Is it possible to have two planes perpendicular and at the same time have a planar angle with one side in each plane, a vertex on the line of intersection of the planes, and a measure of 60°? Explain your answer with a drawing.
10. If a central angle has a measure of 45°, what is the measure of its associated arc?
11. Explain why two perpendicular lines determine a plane.

Trends in Mathematics and Science Study (TIMSS) Questions

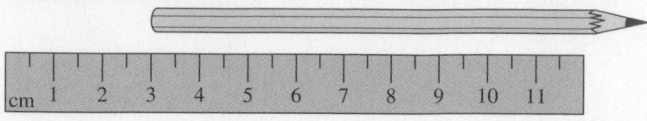

Which of these is closest to the length of the pencil in the figure?
   a. 9 cm   **b.** 10.5 cm   **c.** 12 cm   **d.** 13.5 cm

TIMSS, Grade 8, 1994

What is the ratio of the length of a side of a square to its perimeter?
   a. $\dfrac{1}{1}$ **b.** $\dfrac{1}{2}$ **c.** $\dfrac{1}{3}$ **d.** $\dfrac{1}{4}$

TIMSS, Grade 8, 1994

National Assessment of Educational Progress (NAEP) Question

Which of these units would be the best to use to measure the length of a school building?
   a. Millimeters   **b.** Centimeters
   c. Meters   **d.** Kilometers

NAEP, Grade 4, 2007

12 inches
Fred's Rope

Susan's Rope

3. If Fred's rope is 12 inches long, about how long is Susan's rope?
   1. 16 inches
   2. 20 inches
   3. 24 inches
   4. 30 inches

NAEP, Grade 4, 2007

BRAIN TEASER

Suppose a wire is stretched tightly around Earth. (The radius of Earth is approximately 6400 km.) Then suppose the wire is cut and its length is increased by 20 m. It is then placed back around the planet so that it is the same distance from Earth at every point. Could we walk under the wire?

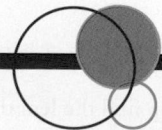

11-3 Curves, Polygons, and Symmetry

With a pencil, draw a path on a piece of paper without lifting the pencil and without retracing any part of the path except single points. The resulting drawing is restricted to the plane of the paper, and not lifting the pencil implies that there are no breaks in the drawing. The drawing is **connected** and is a **curve**. Table 11-9 shows sample curves and their classifications. A check is placed in a box if the curve has the attribute listed at the top.

Table 11-9

Curve	Simple	Closed	Polygon	Convex	Concave
	✔				
	✔				
	✔	✔			✔
	✔	✔		✔	
		✔			
	✔	✔	✔	✔	
	✔	✔	✔		✔

A **simple** curve does not intersect itself, except the starting and stopping points may be the same when the curve is traced. A **closed** curve can be drawn starting and stopping at the same point. A curve can be classified as simple, nonsimple, closed, nonclosed, and so on. **Polygons** are simple closed curves with only segments as *sides*. A point where two sides of a polygon meet is a *vertex*. **Convex** curves are simple and closed, such that *the segment connecting any two points in the interior of the curve is wholly contained in the interior of the curve.* **Concave** curves are simple, closed, and not convex; that is, it is possible for a line segment connecting two interior points to cross outside the interior of the curve.

🌑 NOW TRY THIS 11-14

Draw a curve that is neither simple nor closed.

As in Figure 11-28(a), every simple closed curve separates the plane into three disjoint subsets: the interior of the curve, the exterior of the curve, and the curve itself. Of specific interest are polygons and their interiors, together called **polygonal regions**. Figure 11-28(b) shows a polygonal region.

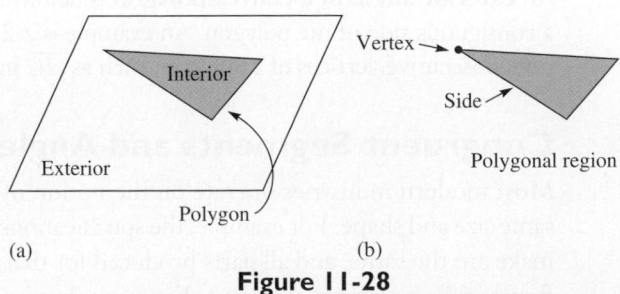

(a)   (b)

Figure 11-28

Whether a point is inside or outside a curve is not always obvious. This is explored in Now Try This 11-15.

 NOW TRY THIS 11-15

Determine whether point X is inside or outside the simple closed curve of Figure 11-29. Explain your reasoning so that it can be generalized to other simple closed curves.

Figure 11-29

More About Polygons

Polygons are classified according to the number of sides or vertices they have. For example, consider the polygons listed in Table 11-10.

Table 11-10

Polygon	Number of Sides or Vertices
Triangle	3
Quadrilateral	4
Pentagon	5
Hexagon	6
Heptagon	7
Octagon	8
Nonagon	9
Decagon	10
11-gon	11
Dodecagon	12
⋮	⋮
n-gon	n

A polygon is referred to by the capital letters that represent its consecutive vertices, such as *ABCD* or *CDAB* shown in Figure 11-30(a). Any two sides of a convex polygon having a common vertex determine an **interior angle**, or **angle of the polygon**, such as $\angle 1$ of polygon *ABCD* in Figure 11-30(a). An **exterior angle of a convex polygon** is determined by a side of the polygon and the extension of a contiguous side of the polygon. An example is $\angle 2$ in Figure 11-30(b). Any line segment connecting nonconsecutive vertices of a polygon, such as $\overline{AC}$ in Figure 11-30(a), is a **diagonal** of the polygon.

Congruent Segments and Angles

Most modern industries operate on the notion of creating **congruent parts**, parts that are of the same size and shape. For example, the specifications for the bodies of all cars of a particular model and make are the same, and all parts produced for that model are basically the same. Usually congruent figures refer to figures in a plane. For example, two line **segments** are **congruent** ($\cong$) if a tracing of

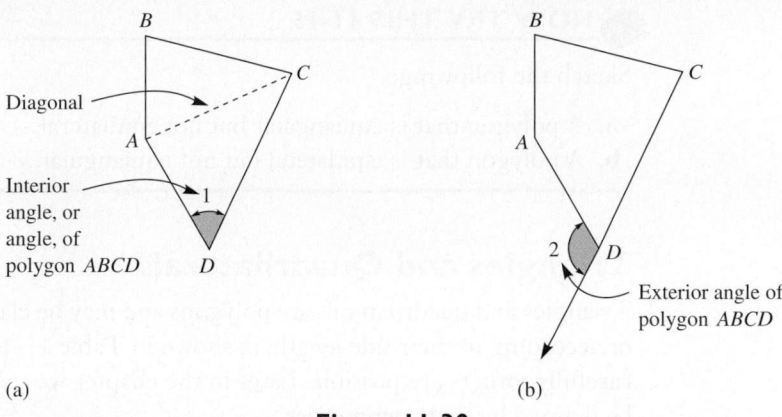

Figure 11-30

one can be fitted exactly on top of the other. If $\overline{AB}$ is congruent to $\overline{CD}$, we write $\overline{AB} \cong \overline{CD}$. Congruent segments have the same measure. Two **angles** are **congruent** if they have the same measure. Congruent segments and congruent angles are shown in Figure 11-31(a) and (b), respectively.

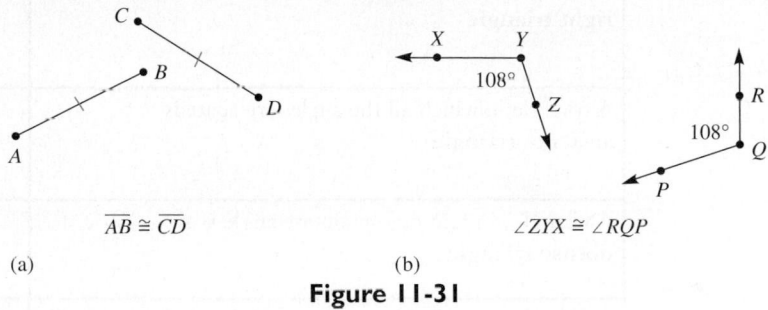

Figure 11-31

Notice in Figure 11-31(a), $\overline{AB} = \overline{BA}$, but $\overline{AB} \neq \overline{CD}$ because the segments contain different sets of points. However, $\overline{AB} \cong \overline{CD}$. In geometry, *we use the equal sign "=" for exactly the same shape, size, and location.*

Regular Polygons

Convex polygons in which all the interior angles are congruent (**equiangular**) and all the sides are congruent (**equilateral**) are **regular polygons**. Thus, a *regular polygon is both equiangular and equilateral*. A regular triangle is an equilateral triangle. A regular quadrilateral is a square. A regular pentagon and a regular hexagon are illustrated in Figure 11-32. The congruent sides and congruent angles are marked.

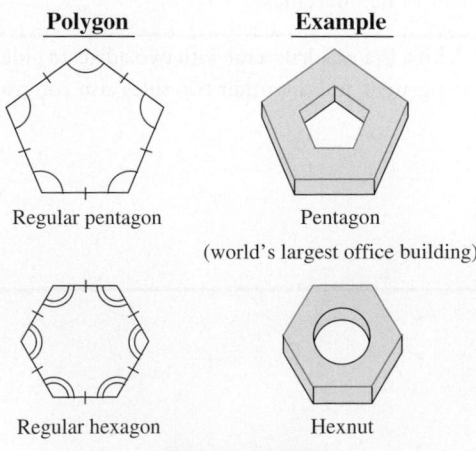

Figure 11-32

 **NOW TRY THIS 11-16**

Sketch the following:

    a. A polygon that is equiangular but not equilateral.
    b. A polygon that is equilateral but not equiangular.

Triangles and Quadrilaterals

Triangles and quadrilaterals are polygons and may be classified according to their angle measures or according to their side length, as shown in Table 11-11. Table 11-11's definitions were chosen carefully; others are possible. Later in the chapter we see that many of the definitions here could be defined by their symmetries.

Table 11-11

Definition	Illustration	Example
A triangle containing one right angle is a **right triangle**.		
A triangle in which all the angles are acute is an **acute triangle**.		YIELD
A triangle containing one obtuse angle is an **obtuse triangle**.		
A triangle with no congruent sides is a **scalene triangle.**		
A triangle with at least two congruent sides is an **isosceles triangle**.		
A triangle with three congruent sides is an **equilateral triangle**.		
A **trapezoid** is a quadrilateral with at least one pair of parallel sides.		
A **kite** is a quadrilateral with two adjacent sides congruent and the other two sides also congruent.		

(continued)

Table 11-11 (*continued*)

Definition	Illustration	Example
An **isosceles trapezoid** is a trapezoid with congruent base angles.		
A **parallelogram** is a quadrilateral in which each pair of opposite sides is parallel.		
A **rectangle** is a parallelogram with a right angle.		SPEED LIMIT **40** MPH
A **rhombus** is a parallelogram with two adjacent sides congruent.		
A **square** is a rectangle with two adjacent sides congruent.		ROAD CONSTRUCTION AHEAD

   There are different definitions for trapezoids and other figures. For example, many elementary texts define a trapezoid as a quadrilateral with *exactly* one pair of parallel sides and an isosceles triangle with exactly one pair of congruent sides. Note the definition of a trapezoid on the following partial student page. What additional arrows would have been drawn if this text's definition of trapezoid were used?

Hierarchy Among Selected Polygons

With this text's definitions there is a hierarchy as seen in Figure 11-33. Using set concepts, we note that the set of all triangles is a proper subset of the set of all polygons. Also, the set of all equilateral triangles is a proper subset of the set of all isosceles triangles. In Figure 11-33, more general terms appear above more specific ones.

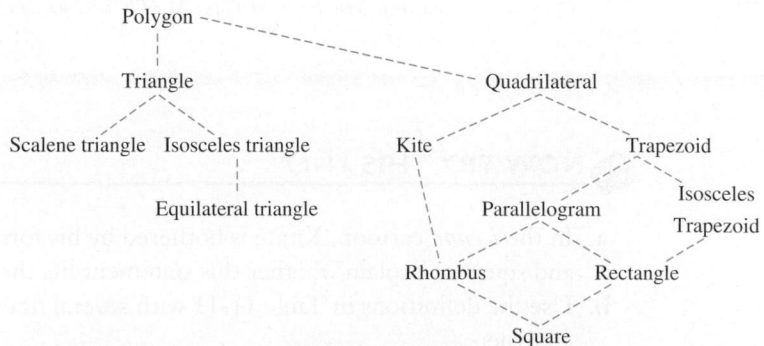

Figure 11-33

School Book Page CLASSIFYING QUADRILATERALS

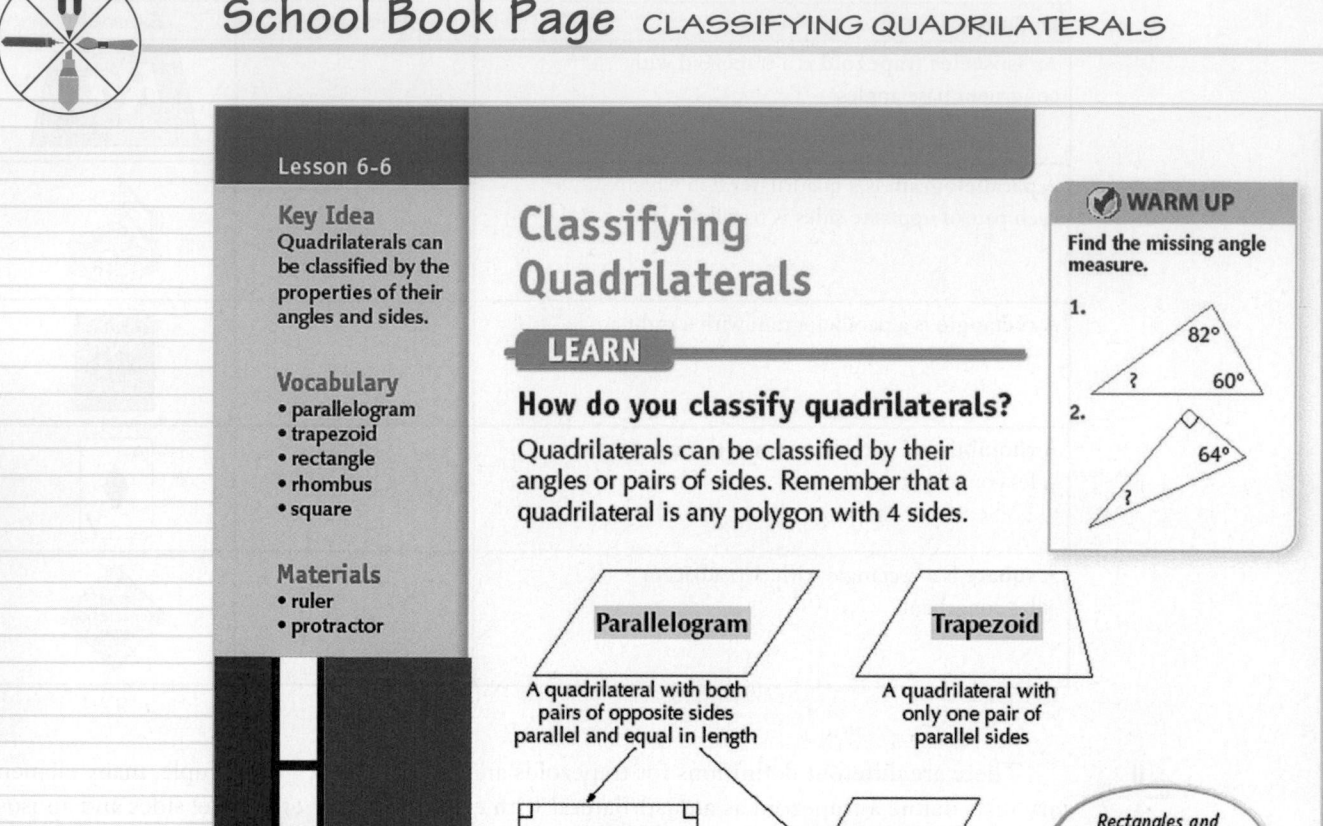

Source: p. 346; From MATHEMATICS. Copyright © 2008 Pearson Education, Inc., or its affiliates.
Used by permission. All Rights Reserved.

NOW TRY THIS 11-17

a. In the *Luann* cartoon, Knute is bothered by his former teacher's statement about rectangles and squares. Explain whether this statement fits the hierarchy of Figure 11-33.

b. Use the definitions in Table 11-11 with several drawings to decide which of the following are true:

1. An equilateral triangle is isosceles.
2. A square is a regular quadrilateral.
3. If one angle of a rhombus is a right angle, then all the angles of the rhombus are right angles.
4. A square is a rhombus with a right angle.
5. All the angles of a rectangle are right angles.

6. A rectangle is an isosceles trapezoid.
7. Some isosceles trapezoids are kites.
8. If a kite has a right angle, then it must be a square.

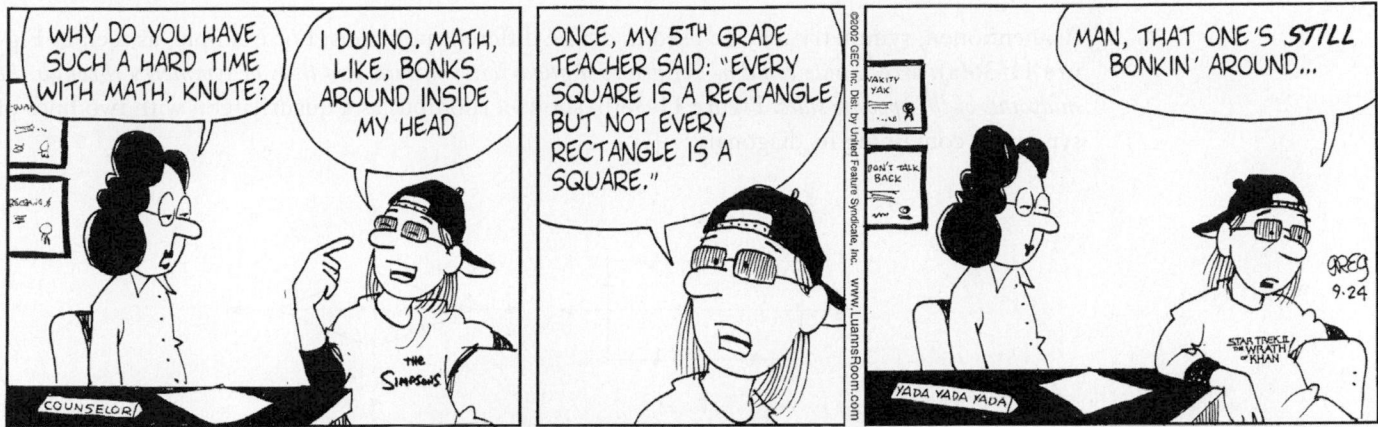

Luann copyright ©2002 and 2011 GEC Inc./Distributed by United Feature Syndicate, Inc.

Symmetry and Its Relation to Planar Figures

Paperfolding can be used to introduce the concept of symmetry. Consider Figure 11-34 where folds on the dashed lines allow the drawings to match with themselves.

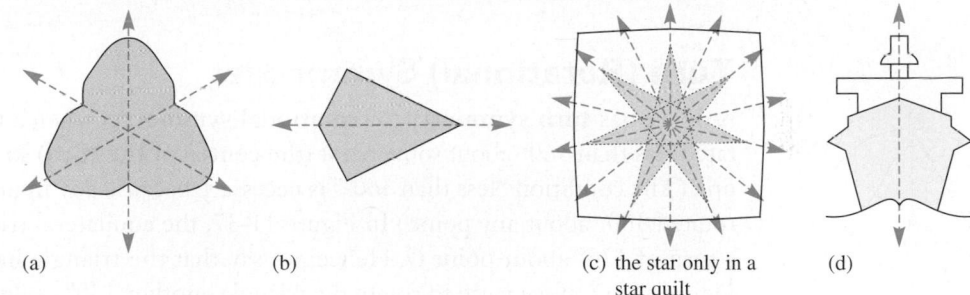

(a)   (b)   (c) the star only in a star quilt   (d)

Figure 11-34

A geometric figure has a **line of symmetry** if it is its own image when folded along the line. (This concept using paperfolding can be done much more precisely using reflections introduced in Chapter 13.)

EXAMPLE 11-8

How many lines of symmetry does each drawing in Figure 11-35 have?

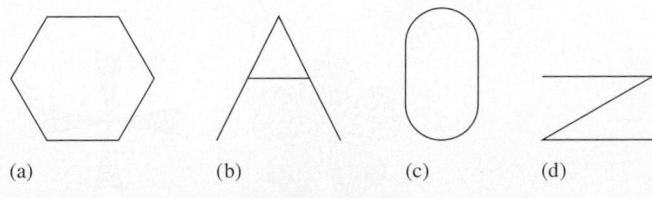

(a)   (b)   (c)   (d)

Figure 11-35

Solution

a. 6   **b.** 1   **c.** 2   **d.** 0

NOW TRY THIS 11-18

Create a figure with exactly 4 lines of symmetry.

As mentioned, symmetry may be used to define different polygons. For example, as seen in Figure 11-36(a), *a rectangle could be defined as a quadrilateral with two lines of symmetry through the midpoints of the opposite sides.* Figure 11-36(b) shows a rhombus as a quadrilateral with two lines of symmetry containing its diagonals.

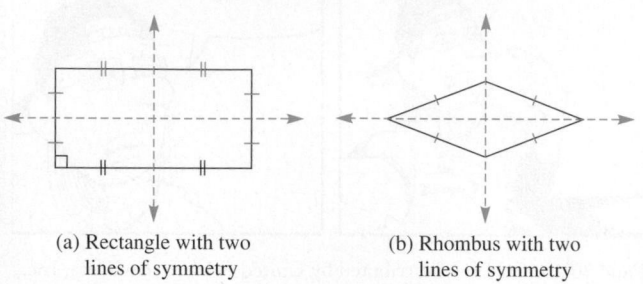

(a) Rectangle with two lines of symmetry

(b) Rhombus with two lines of symmetry

Figure 11-36

NOW TRY THIS 11-19

Use the information from Figure 11-36 and the hierarchy of Figure 11-33 to define a square in terms of line symmetry.

Turn (Rotational) Symmetries

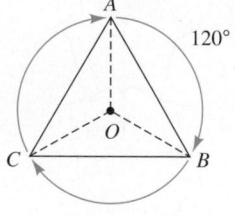

Figure 11-37

A figure has **turn symmetry** (or **rotational symmetry**) when a tracing of the figure can be rotated less than 360° about some point (the **center of the turn**) so that it matches the original figure. (The condition "less than 360°" is necessary because any figure will coincide with itself if it is rotated 360° about any point.) In Figure 11-37, the equilateral triangle coincides with itself after a turn of 120° about point O. Hence, we say that the triangle has 120° turn symmetry. Also in Figure 11-37, if we were to rotate the triangle another 120°, we would find again that it matches the original. So we can say that the triangle also has 240° turn symmetry. (*Turns counterclockwise are positive while turns clockwise are negative.*)

In general, *if a figure has α degrees of turn symmetry, it also will coincide with itself when rotated by nα degrees for any integer n*. For this reason in turn symmetry, it is sufficient to report the smallest possible positive angle measure that turns the figure onto itself. A circle has a turn symmetry by any turn around its center. Thus, *a circle has infinitely many turn symmetries*.

Other examples of figures that have turn symmetry are shown in Figure 11-38. Figures 11-38(a), (b), (c), and (d) have 72°, 90°, 180°, and 180° turn symmetries, respectively [(a) and (b) also have other turn symmetries].

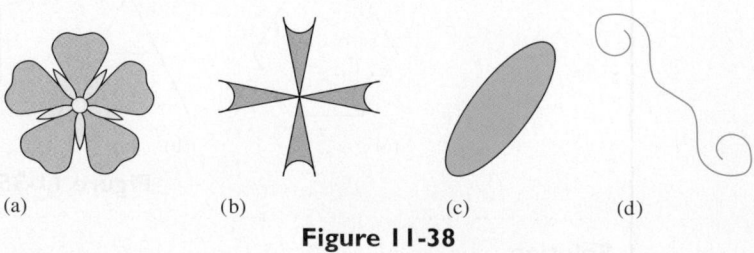

(a)   (b)   (c)   (d)

Figure 11-38

For "simple" figures, we determine whether a figure has turn symmetry by tracing it and turning the tracing about a point (the center of the figure) to see if it aligns on the figure before the tracing has turned in a complete circle, or 360°. The amount of turning is determined by measuring the *turn angle* $\angle POP'$ through which a point P is rotated around a point O to match another point P' when the figures align. Such an angle, $\angle POP'$, is labeled with points P, O, and P' in Figure 11-39 and has measure 120°. Point O, the point held fixed when the tracing is turned, is the *turn center*.

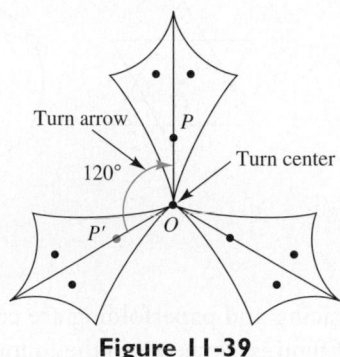

Turn arrow

P

120°

Turn center

P'

O

Figure 11-39

NOW TRY THIS 11-20

a. Describe a figure that has 2° turn symmetry.
b. Describe a square in terms of turn symmetry.

EXAMPLE 11-9

Determine the amount of the turn for the rotational symmetries of each part of Figure 11-40. Assume the turns are about the "center" of each drawing.

(a)

(c)

Figure 11-40

Solution

a. The amounts of the turns are $\dfrac{360°}{3}$ (or 120°) and 240°.

b. The amount of the turn is 180°.

c. The amounts of the turns are 60°, 120°, 180°, 240°, and 300°.

Point Symmetry

The turns in Figure 11-40(b) and (c) exemplify yet another type of symmetry, namely, point symmetry. Any figure that has 180° turn symmetry is said to have **point symmetry** about the turn center. Some figures with point symmetry are shown in Figure 11-41. The turn center of a figure with point symmetry may be found as in Figure 11-41 where X is connected to its symmetrical point X' with O being the turn center and midpoint of $\overline{XX'}$.

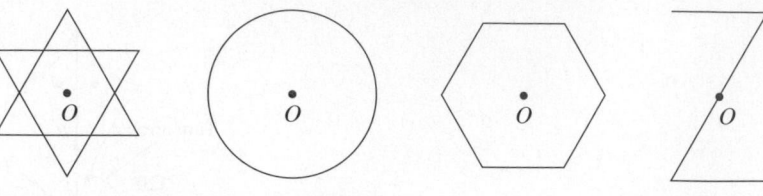

Figure 11-41

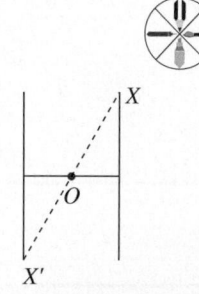

Figure 11-42

Tracing and paperfolding are commonly used in elementary school to investigate symmetries of figures, as shown in the following student page. (Observe that in part (e), symmetries are related to transformations of the plane, which are studied in Chapter 13.)

Classification of Polygons by Their Symmetries

Geometric figures in a plane can be classified according to the number of symmetries they have. Consider a triangle described as having exactly one line of symmetry and no turn symmetries. What could the triangle look like? The only possibility is a triangle in which exactly two sides are congruent, that is, an isosceles triangle that is not equilateral. The line of symmetry passes through a vertex, as shown in Figure 11-43. We can describe equilateral and scalene triangles in terms of the number of lines of symmetry they have.

A square, as in Figure 11-44, can be described as a four-sided polygon with four lines of symmetry—d_1, d_2, h, and v—and three turn symmetries about point O. In fact, we can use lines of symmetry and turn symmetries to define various types of quadrilaterals normally used in geometry.

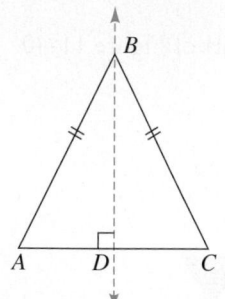

Figure 11-43

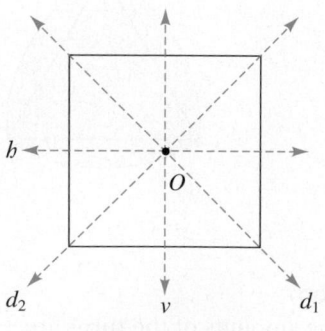

Figure 11-44

School Book Page INVESTIGATING SYMMETRY

Symmetry

LEARN

Activity

How can you describe and create symmetric figures?

An artist designed the trademark at the right for a sporting goods company. Many trademarks are **symmetric figures**. This means they can be folded into two congruent parts that fit on top of each other. The fold line is a **line of symmetry**.

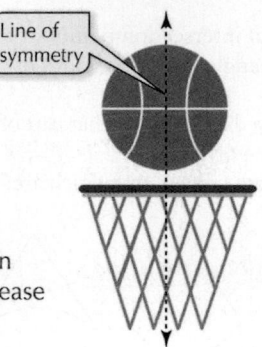

Line of symmetry

You can follow the steps below to create a design with two lines of symmetry.

a. Fold a sheet of paper in half. Then fold it in half again the other way (so the second crease is perpendicular to the first).

b. Draw a path that starts on one folded edge and ends at the other folded edge, as shown below.

c. Cut along the curve. Then open up the folded paper.

The figure you made should be symmetric with two lines of symmetry.

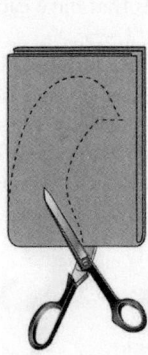

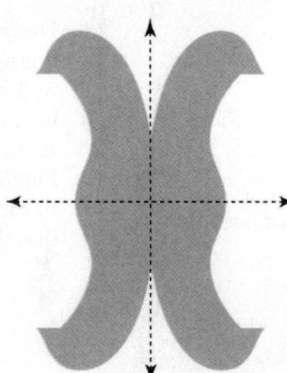

d. How many congruent parts are there in the figure you made?

e. Are the congruent parts related by slides, reflections, or turns?

✓ WARM UP

Draw an example of each figure. Then draw a flip.

1. rectangle

2. trapezoid

3. right triangle

4. obtuse triangle

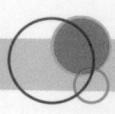

Assessment 11-3A

1. Determine for each of the following which of the figures labeled (1) through (10) can be classified under the given term:
   a. Simple closed curve   b. Polygon
   c. Convex polygon   d. Concave polygon

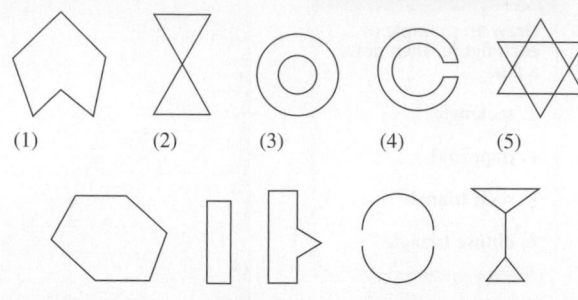

(1)   (2)   (3)   (4)   (5)

(6)   (7) (8)   (9)   (10)

2. What is the maximum number of intersection points between a quadrilateral and a triangle (where no sides of the polygons are on the same line)?
3. What type of polygon must have a diagonal such that part of the diagonal falls in the exterior of the polygon?
4. Which of the following figures are convex and which are concave? Why?

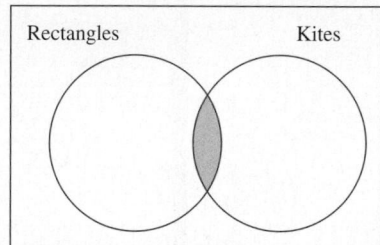

a.   b.   c.   d.

5. Describe the shaded region as simply as possible.

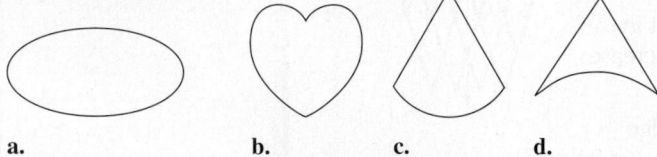

6. If possible, draw the following triangles. If it is not possible, state why.
   a. An obtuse scalene triangle
   b. A right scalene triangle
   c. An obtuse equilateral triangle
   d. A right equilateral triangle
   e. An obtuse isosceles triangle
7. Determine how many diagonals each of the following has:
   a. Pentagon   b. Decagon
   c. 20-gon   d. n-gon
8. Draw all lines of symmetry (if any exist) for each of the following:
   a.   b.

9. Identify each of the following triangles as scalene, isosceles, or equilateral (there may be more than one term that applies to these triangles):

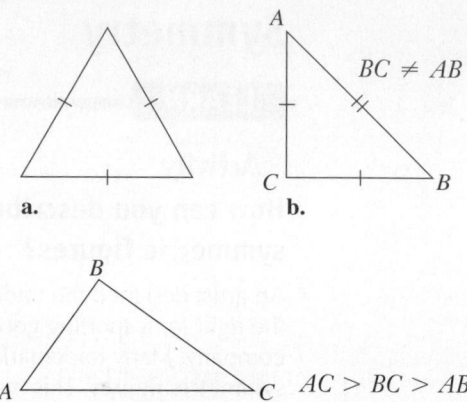

a.   b. $BC \neq AB$

c. $AC > BC > AB$

10. Various international signs have symmetries. Determine which of the following, if any, have (i) line symmetry, (ii) turn symmetry, and/or (iii) point symmetry:

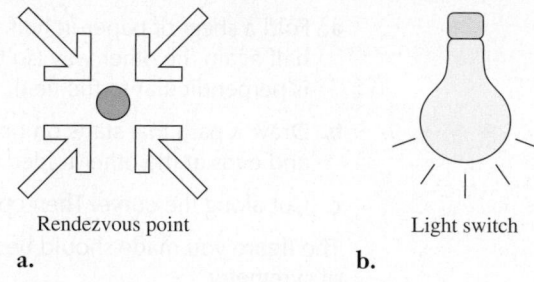

Rendezvous point   Light switch
a.   b.

11. Design symbols that have each of the following symmetries, if possible:
    a. Line symmetry but not turn symmetry
    b. Turn symmetry but not point symmetry
    c. Rotational symmetry but not line symmetry
12. Find all lines of symmetries for the figures in exercise 10.
13. In each of the following figures, complete the sketches so that they have line symmetry about ℓ:

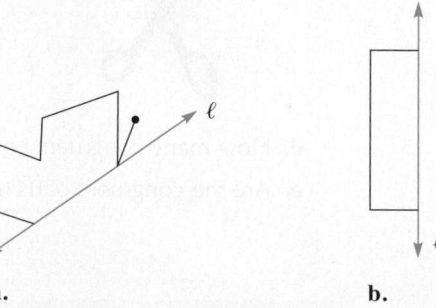

a.   b.

Assessment 11-3B

1. Determine for each of the following which of the figures (if any) labeled (1) through (10) can be classified under the given terms:
   a. Isosceles triangle
   b. Isosceles but not equilateral triangle
   c. Equilateral but not isosceles triangle
   d. Parallelogram but not a trapezoid
   e. A trapezoid but not a parallelogram
   f. A rectangle but not a square
   g. A square but not a rectangle
   h. A square but not a trapezoid
   i. A rhombus but not a kite
   j. A rhombus
   k. A kite

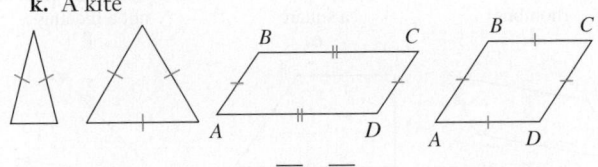

(1)   (2)   (3) $\overline{AB} \parallel \overline{CD}$, $\overline{AD} \parallel \overline{BC}$ (4) Opposite sides parallel; all sides congruent.

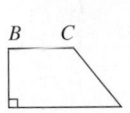 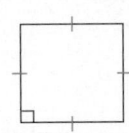 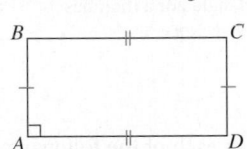

(5) $\overline{AD} \parallel \overline{BC}$; $\angle BAD$ is a right angle.   (6) Square   (7) Rectangle, $AB \neq BC$

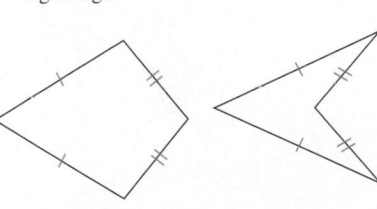 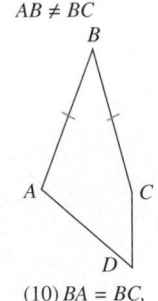

(8)   (9)   (10) $BA = BC$, $AD \neq CD$, not a trapezoid

2. What is the maximum number of intersection points between two triangles (where no sides of the triangles are on the same line)?
3. A pentagon has only two diagonals that intersect at a given vertex. Determine how many diagonals intersect at a given vertex in each of the following polygons:
   a. Hexagon   b. Decagon
   c. 20-gon   d. n-gon
4. Which of the following figures are convex and which are concave? Why?

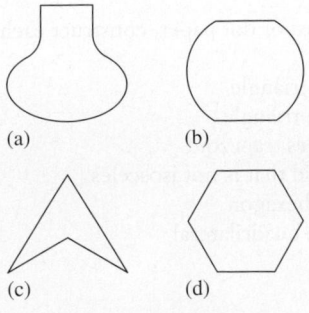

(a)   (b)
(c)   (d)

5. Describe the shaded region as simply as possible.

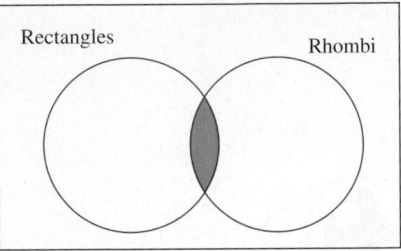

6. If possible, draw the following triangles. If it is not possible, state why.
   a. An acute scalene triangle
   b. A right isosceles triangle
   c. A scalene equiangular triangle
   d. An equilateral equiangular triangle
   e. An acute isosceles triangle
7. Determine how many diagonals each of the following has:
   a. Hexagon
   b. 11-gon
   c. 18-gon
8. Find the lines of symmetry, if any, for each of the following:
   a.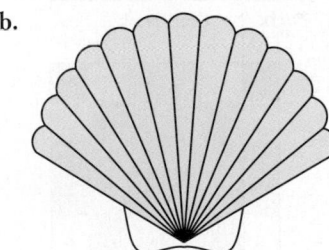
   b.

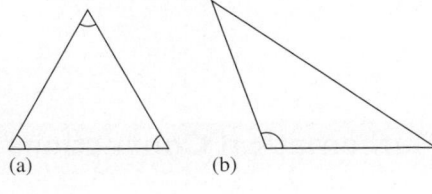

9. Identify each of the following triangles as acute, obtuse, right, or equiangular. There may be more than one term that applies to these triangles.

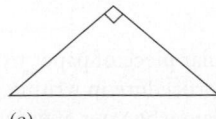

(a)   (b)

(c)

10. Various international signs have symmetries. Determine which of the following, if any, have (i) line symmetry, (ii) turn symmetry, and/or (iii) point symmetry:

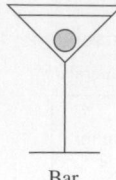

Bar

a.

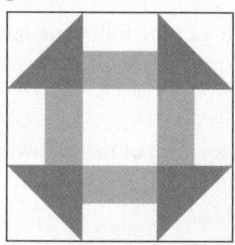

Observation deck

b.

11. Determine the types of symmetry that each separate quilt pattern below has (line, turn, point), if any.

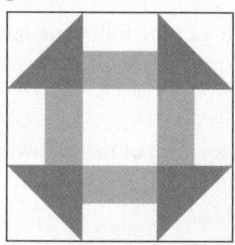

a. Churn Dash

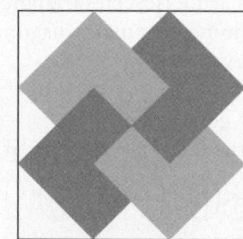

b. Card Trick

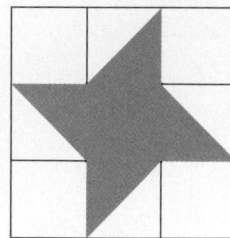

c. Friendship Star

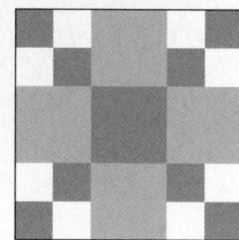

d. Linoleum

12. How many lines of symmetry exist for each of the following:

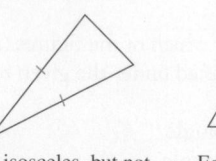

An isosceles, but not equilateral triangle
a.

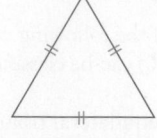

Equilateral triangle
b.

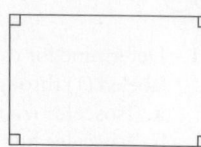

Rectangle, but not a square
c.

Kite, but not a rhombus
d.

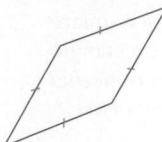

Rhombus, but not a square
e.

Isosceles trapezoid, but not a rectangle
f.

Parallelogram, but neither a rectangle nor a rhombus
g.

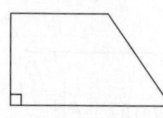

Trapezoid, but not a rectangle
h.

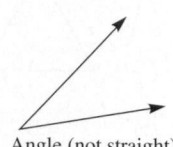

Angle (not straight)
i.

13. In each of the following figures, complete the sketches so that they have line symmetry about ℓ.

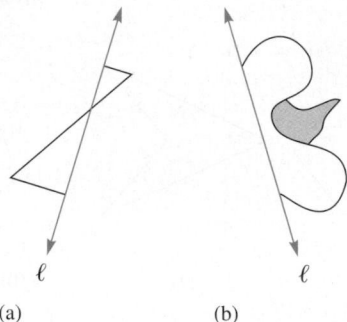

ℓ ℓ

(a)   (b)

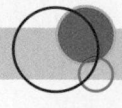 **Mathematical Connections 11-3**

Communication

1. a. Fold a rectangular piece of paper to create a square. Describe your procedure in writing and orally with a classmate. Explain why your approach creates a square.

   b. Crease the square in (a) so that the two diagonals are shown. Use paperfolding to show that the diagonals of a square are congruent and perpendicular and divide each other into congruent parts. Describe your procedure and explain why it works.

Open-Ended

2. On a geoboard or dot paper, construct each of the following:
   a. A scalene triangle
   b. An obtuse triangle
   c. An isosceles trapezoid
   d. A trapezoid that is not isosceles
   e. A convex hexagon
   f. A concave quadrilateral

g. A parallelogram

h. A rhombus that is not a square.

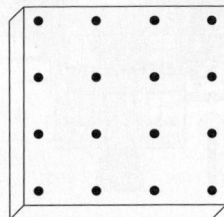

Cooperative Learning

3. Work with a partner. One constructs a figure on a geoboard or draws it on a piece of paper and identifies it. Do not show the figure to the partner but tell the partner sufficient properties of the figure to identify it. Have the partner identify the constructed figure. The partner earns 1 point if the figure is correctly identified and 2 points if a figure is found that has all the required attributes but is different from the one drawn. Each takes the same number of turns. Try this with each of the following figures:

   a. Scalene triangle

   b. Isosceles triangle

   c. Square

   d. Parallelogram

   e. Trapezoid

   f. Rectangle

   g. Regular polygon

   h. Rhombus

   i. Isosceles trapezoid

   j. A kite that is not a rhombus

4. Work with partners to create a Venn diagram with the universal set being all triangles and the subsets being isosceles, equilateral, and right triangles.

5. **a.** Use the Internet to investigate the meaning and uses of Reuleaux triangles.

   b. Explain the similarities and differences between a Reuleaux triangle and an equilateral triangle.

Questions from the Classroom

6. A student asks whether a polygon whose sides are congruent is necessarily a regular polygon and whether a polygon with all angles congruent is necessarily a regular polygon. How do you answer?

7. A student asks how to find the shortest path between two points A and B on a right circular cylinder as shown below. How do you respond?

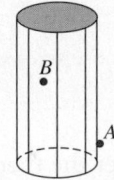

8. One student says, "My sister's high school geometry book talked about equal angles. Why don't we use the term 'equal angles' instead of 'congruent angles'?" How do you reply?

9. Jodi identifies figure (a) as a rectangle and figure (b) as a square. She claims that figure (b) is not a rectangle because it is a square. How do you respond?

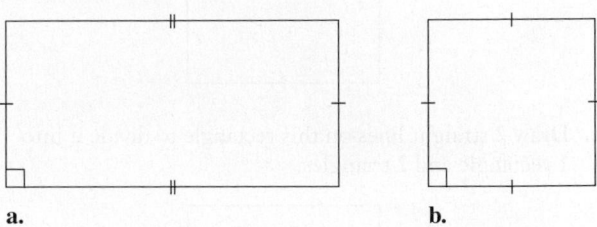

a.   **b.**

10. Millie claims that a rhombus is regular because all of its sides are congruent. How do you respond?

GSP/GeoGebra Activities

11. Use Geogebra Lab-2 to construct quadrilaterals.

Review Problems

12. If three distinct rays with the same vertex are drawn as shown in the following figure, then three angles are formed: $\angle AOB$, $\angle AOC$, and $\angle BOC$.

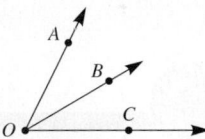

What is the maximum number of angles, measuring less than 180°, formed by using

   a. 10 distinct noncollinear rays with the same vertex?

   b. n distinct noncollinear rays with the same vertex?

13. Determine the possible intersection sets of a line and an angle.

14. Classify the following as true or false. If false, tell why.

   a. A ray has two endpoints.

   b. For any points M and N, $\overleftrightarrow{MN} = \overleftrightarrow{NM}$.

   c. Skew lines are coplanar.

   d. $MN = NM$

   e. A line segment contains an infinite number of points.

   f. If two distinct planes intersect, their intersection is a line segment.

15. Convert the following as indicated:

   a. 7 mm to _____ m

   b. 17 in. to _____ yd

   c. 4 m to _____ cm

   d. 1.7 yd to _____ in.

Trends in Mathematics and Science Study (TIMSS) Questions

   a. Draw 1 straight line on this rectangle to divide it into 2 triangles.

b. Draw 1 straight line on this rectangle to divide it into 2 rectangles.

c. Draw 2 straight lines on this rectangle to divide it into 1 rectangle and 2 triangles.

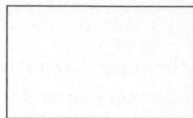

TIMSS, Grade 4, 2003

In the picture there are a number of geometric shapes, like circles, squares, rectangles, and triangles. For example, the sun looks like a circle.

Draw lines to three other different objects in the picture and write what shapes they look like.

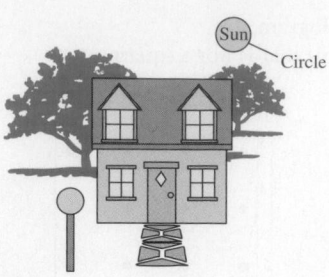

TIMSS, Grade 4, 2003

National Assessment of Educational Progress (NAEP) Question

Nick has a square piece of paper. He draws the two diagonals of the square, finds the point where they intersect, and labels that point A. Then he folds each of the four corners of the paper onto point A. What geometric shape is produced?

a. A square   **b.** A right triangle
c. An isosceles triangle   **d.** A pentagon
e. A hexagon

NAEP, Grade 8, 2007

11-4 More About Angles

In Figure 11-45 two lines intersect and form the angles marked 1, 2, 3, and 4.

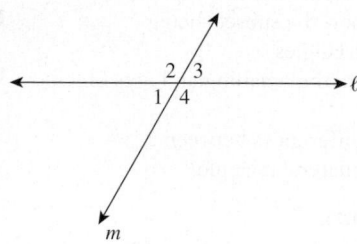

Figure 11-45

Vertical angles are pairs of angles such as $\angle 1$ and $\angle 3$ and are formed when two lines intersect. Another pair of vertical angles in Figure 11-45 is $\angle 2$ and $\angle 4$. We define vertical angles as follows:

Definition of Vertical Angles

Vertical angles, created by intersecting lines, are a pair of angles whose sides are two pairs of opposite rays.

From Figure 11-45, vertical angles appear congruent. Consider the theorem below and the following proof.

Theorem 11-2

Vertical angles are congruent.

Proof

In Figure 11-45, we must show that $\angle 1 \cong \angle 3$ and $\angle 2 \cong \angle 4$. In Figure 11-45, $m(\angle 1) + m(\angle 2) = 180°$ and $m(\angle 3) + m(\angle 2) = 180°$. Thus,

$$m(\angle 1) = 180° - m(\angle 2)$$
$$m(\angle 3) = 180° - m(\angle 2)$$

Consequently, $m\angle 1 = m\angle 3$. In a similar way we can show that $m(\angle 2) = m(\angle 4)$. ∎

Other pairs of angles appear frequently enough that it is convenient to refer to them by specific names. For example, angles are also formed when a line (a **transversal**) intersects two distinct lines. Angles formed by these lines are named according to their placement in relation to the transversal and the two given lines. Table 11-12 shows several types of angles.

Table 11-12

Supplementary angles are two angles the sum of whose measures is 180°. Each angle is a *supplement* of the other. (Supplementary angles do not need to be adjacent.)	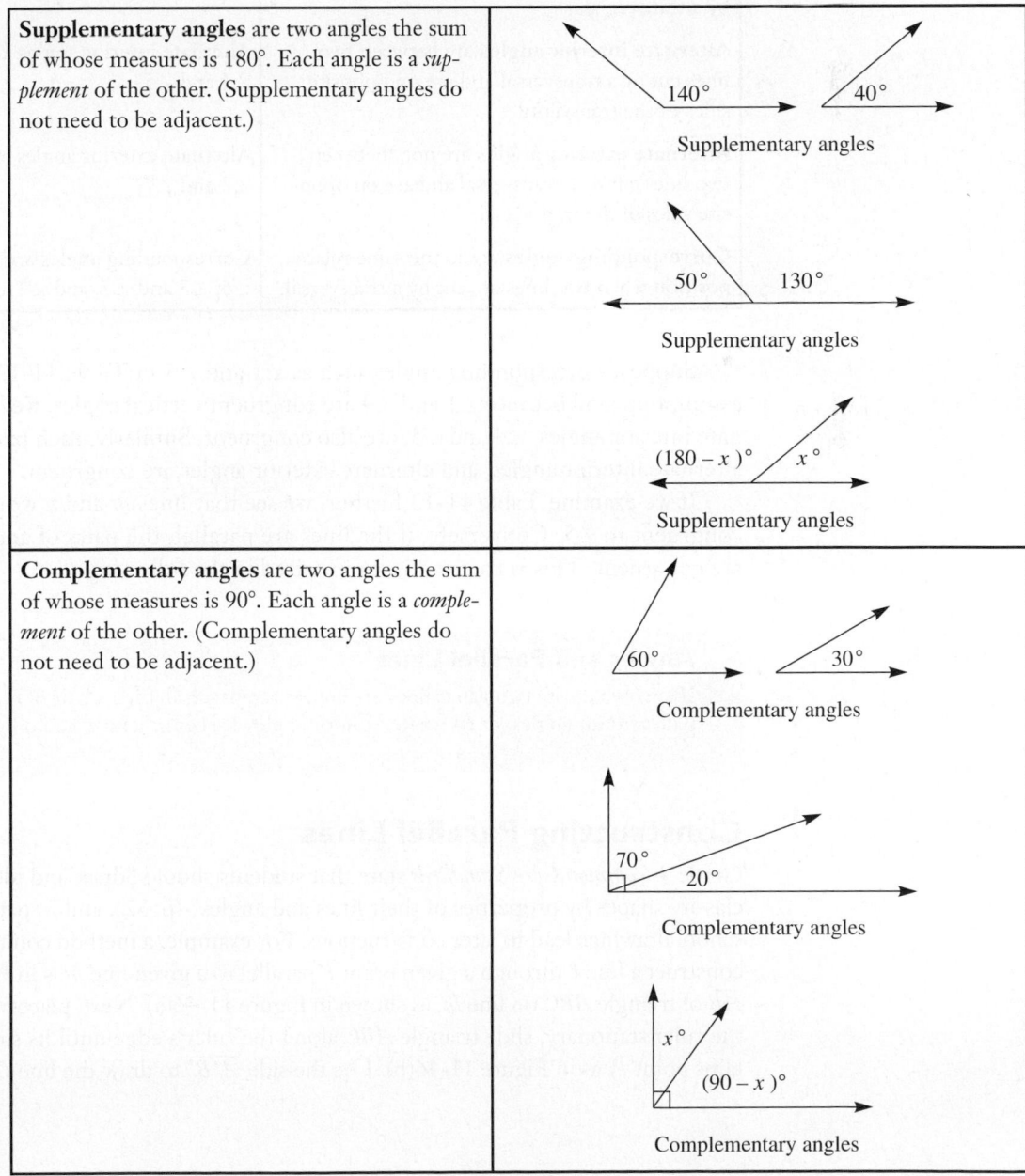
Complementary angles are two angles the sum of whose measures is 90°. Each angle is a *complement* of the other. (Complementary angles do not need to be adjacent.)	

140°   40°
Supplementary angles

50°   130°
Supplementary angles

$(180 - x)°$ $x°$
Supplementary angles

60°   30°
Complementary angles

70°   20°
Complementary angles

$x°$ $(90 - x)°$
Complementary angles

(continued)

Table 11-12 (continued)

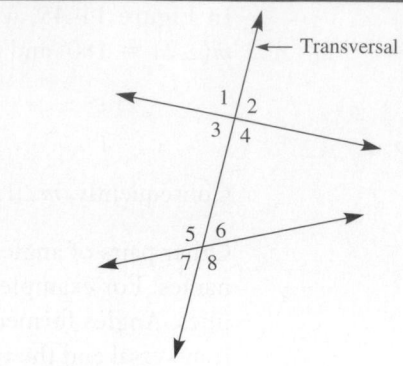

Interior angles are "between" two lines cut by a transversal.	Interior angles are $\angle 3$, $\angle 4$, $\angle 5$, and $\angle 6$.
Exterior angles are not "between" two lines cut by a transversal.	Exterior angles are $\angle 1$, $\angle 2$, $\angle 7$, and $\angle 8$.
Alternate interior angles are between two lines cut by a transversal and are on opposite sides of the transversal.	Alternate interior angles are pairs: $\angle 3$ and $\angle 6$, $\angle 4$ and $\angle 5$.
Alternate exterior angles are not "between" two lines cut by a transversal and are on opposite sides of the transversal.	Alternate exterior angles are pairs: $\angle 1$ and $\angle 8$, $\angle 2$ and $\angle 7$.
Corresponding angles are in the same relative position when two lines are cut by a transversal.	Corresponding angles are pairs: $\angle 1$ and $\angle 5$, $\angle 2$ and $\angle 6$, $\angle 3$ and $\angle 7$, and $\angle 4$ and $\angle 8$.

Suppose corresponding angles such as $\angle 1$ and $\angle 5$ in Table 11-12 are congruent. With this assumption, and because $\angle 1$ and $\angle 4$ are congruent vertical angles, we know that the pair of alternate interior angles, $\angle 4$ and $\angle 5$, are also congruent. Similarly, each pair of corresponding angles, alternate interior angles, and alternate exterior angles are congruent.

If we examine Table 11-12 further, we see that lines m and n would be parallel when $\angle 1$ is congruent to $\angle 5$. Conversely, if the lines are parallel, the pairs of angles mentioned previously are congruent. This is true and is summarized in the following:

Angles and Parallel Lines

If any two distinct coplanar lines are cut by a transversal, then a pair of corresponding angles, alternate interior angles, or alternate exterior angles are congruent if, and only if, the lines are parallel.

Constructing Parallel Lines

Grade 4 *Common Core Standards* state that students should "draw and identify lines and angles, and classify shapes by properties of their lines and angles" (p. 32), and in particular parallel lines. Early school drawings lead to later constructions. For example, a method commonly used by architects to construct a line ℓ through a given point P parallel to a given line m is in Figure 11-46. Place the side $\overline{AB}$ of triangle ABC on line m, as shown in Figure 11-46(a). Next, place a ruler on side $\overline{AC}$. Keeping the ruler stationary, slide triangle ABC along the ruler's edge until its side $\overline{AB}$ (marked $\overline{A'B'}$) contains point P, as in Figure 11-46(b). Use the side $\overline{A'B'}$ to draw the line ℓ through P parallel to m.

To show that the construction produces parallel lines, notice that when triangle *ABC* slides, the measures of its angles are unchanged. The angles of triangle *ABC* and triangle *A'B'C'* in Figure 11-46(b) are corresponding congruent angles. ∠*A* and ∠*A'* are corresponding angles formed by *m* and ℓ and the transversal $\overline{EF}$. Because corresponding angles are congruent, ℓ∥*m*. In Chapter 12, we will show how to construct parallel lines using only a compass and straightedge.

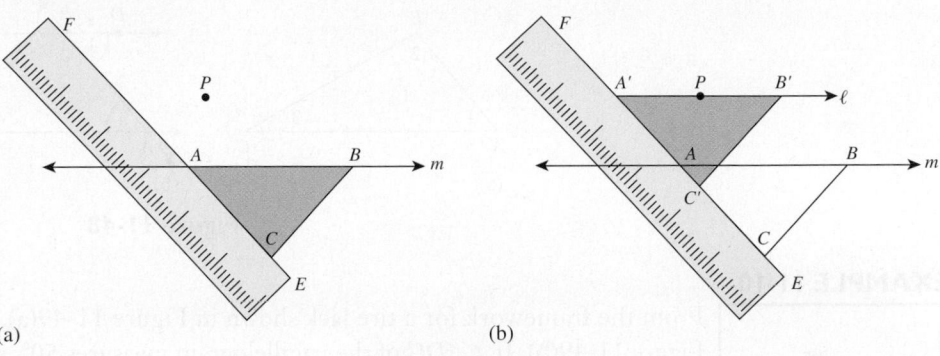

(a)   (b)

Figure 11-46

The Sum of the Measures of the Angles of a Triangle

The sum of the measures of the angles in a triangle is observed to be 180°. We see this by using a torn triangle, as shown in Figure 11-47. Angles 1, 2, and 3 of triangle *ABC* in Figure 11-47(a) are torn as pictured and then replaced as shown in Figure 11-47(b). The three angles seem to lie along a single line ℓ.

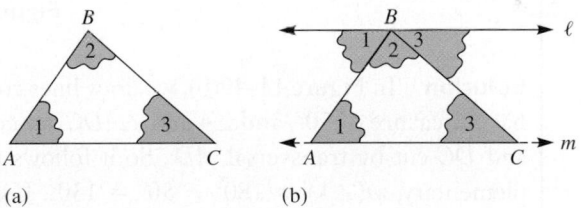

(a)   (b)

Figure 11-47

If we repeat the procedure for several different triangles, the angle measures always seem to sum to 180° or a straight angle. This conclusion is only a conjecture. We use *deductive reasoning* to prove that a statement is true using given information, previously defined and undefined terms, theorems or statements assumed to be true, and logic. A conclusion based on *deductive reasoning* must be true if the hypothesis is true. We state the triangle sum as a theorem and provide a proof.

> **Theorem 11-3**
> The sum of the measures of the interior angles of any triangle is 180°.

Proof

In Figure 11-48(a), △*ABC* has interior angles 1, 2, and 3; we prove $m(\angle 1) + m(\angle 2) + m(\angle 3) = 180°$. Motivated by the experiment in Figure 11-47, in Figure 11-48(b) we place ∠4 so that ∠4 ≅ ∠1. We next extend $\overrightarrow{BD}$ to form line ℓ. In this way, ∠5 is formed. If we could show that ∠5 ≅ ∠3, the proof would be completed. Because we constructed ∠4 so that ∠4 ≅ ∠1, we have a pair of congruent alternate interior angles created by lines ℓ and *m* and transversal $\overleftrightarrow{AB}$.

Thus, $\ell \parallel m$ and therefore $\angle 5 \cong \angle 3$, as these are alternate interior angles created by the parallel lines ℓ and m and the transversal $\overleftrightarrow{BC}$. Consequently,

$$m(\angle 1) + m(\angle 2) + m(\angle 3) = m(\angle 4) + m(\angle 2) + m(\angle 5) = 180°, \text{ or}$$

$$m(\angle 1) + m(\angle 2) + m(\angle 3) = 180°.$$

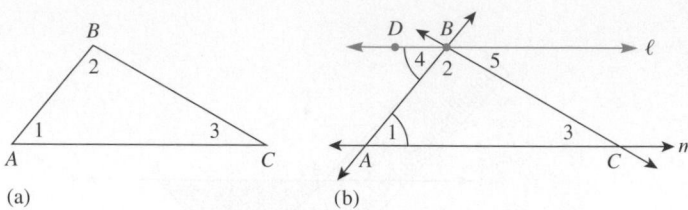

(a)   (b)

Figure 11-48   ∎

EXAMPLE 11-10

From the framework for a tire jack shown in Figure 11-49(a), parallelogram $ABCD$ is drawn in Figure 11-49(b). If $\angle ADC$ of the parallelogram measures 50°, what are the measures of the other angles of the parallelogram?

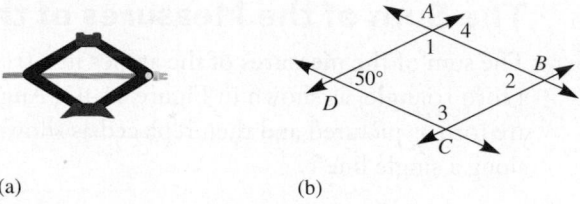

(a)   (b)

Figure 11-49

Solution   In Figure 11-49(b), we draw lines containing the sides of parallelogram $ABCD$. $\angle ADC$ has a measure of 50°, and $\angle 4$ and $\angle ADC$ are corresponding angles formed by parallel lines $\overleftrightarrow{AB}$ and $\overleftrightarrow{DC}$ cut by transversal $\overleftrightarrow{AD}$. So it follows that $m(\angle 4) = 50°$. Because $\angle 1$ and $\angle 4$ are supplementary, $m(\angle 1) = 180° - 50° = 130°$. Using similar reasoning, we find that $m(\angle 2) = 50°$ and $m(\angle 3) = 130°$.   ●

EXAMPLE 11-11

In Figure 11-50, $m \parallel n$ and k is a transversal. Explain why interior angles, $\angle 1$ and $\angle 2$ are supplementary.

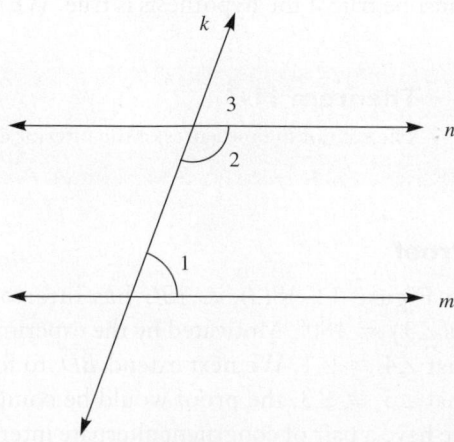

Figure 11-50

Solution Because $\angle 1$ and $\angle 3$ are corresponding angles when parallel lines m and n are cut by transversal k, $m(\angle 1) = m(\angle 3)$. Also because $\angle 2$ and $\angle 3$ are supplementary angles, $m(\angle 2) + m(\angle 3) = 180°$. Substituting $m(\angle 1)$ for $m(\angle 3)$, we have $m(\angle 2) + m(\angle 1) = 180°$. Thus, *the sum of the measures of the interior angles on the same side, formed when two parallel lines are cut by a transversal, is* 180°; *the angles are supplementary.*

The Sum of the Measures of the Interior Angles of a Convex Polygon with *n* Sides

In the following *FoxTrot* cartoon Peter "helps" Paige with her geometry problem.

FoxTrot copyright © 2005 Bill Amend. Reprinted with permission of Universal Uclick. All rights reserved.

To answer Paige's question, we study the sum of the measures of all the interior angles in any convex *n*-gon by considering several special cases. From any vertex of a polygon, diagonals can be drawn from the vertex to form adjacent, nonoverlapping triangular regions. In the quadrilateral in Figure 11-51(a), the diagonal from *B* partitions the quadrilateral into two triangles.

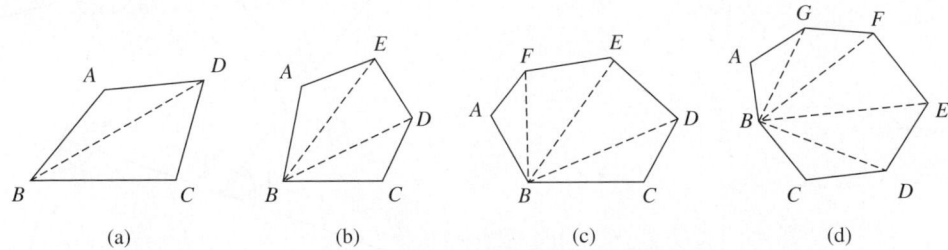

(a)   (b)   (c)   (d)

Figure 11-51

In the pentagon in Figure 11-51(b), the diagonals from *B* partition the pentagon into three triangles. In the hexagon in Figure 11-51(c), the diagonals from *B* partition the hexagon into four triangles. In each of the polygons in Figure 11-51, the number of triangles is two less than the number of sides because when we increase the number of sides by 1, we add one triangle leaving the number of triangles still 2 less than the number of sides. This is illustrated in the heptagon of Figure 11-51(d) which has one more side than the hexagon in Figure 11-51(c).

The number of triangles in a convex n-gon created by all the diagonals from a single vertex is $n - 2$. To find the sum of the measures of all the interior angles in any convex polygon, we add the measures of all the interior angles in the triangles. Since the sum of the measures of the angles in any triangle is $180°$, the sum of the measures of the interior angles in any convex n-gon is $(n - 2)180°$, proving the following theorem.

Theorem 11-4

The sum of the measures of the interior angles of any convex n-gon is $(n - 2)180°$.

The Sum of the Measures of the Exterior Angles of a Convex n-gon

Figure 11-52 shows interior and exterior angles (in blue) of a pentagon. The measures of interior angles are $\alpha_1, \alpha_2, \alpha_3, \alpha_4, \alpha_5$ and exterior angles are $\beta_1, \beta_2, \beta_3, \beta_4, \beta_5$. Since the exterior and interior angle at a vertex are supplementary and the sum of the measure of the interior angles in a pentagon is $(5 - 2)180°$, we have:

$$(\alpha_1 + \beta_1) + (\alpha_2 + \beta_2) + (\alpha_3 + \beta_3) + (\alpha_4 + \beta_4) + (\alpha_5 + \beta_5) = 5 \cdot 180°$$
$$(\alpha_1 + \alpha_2 + \alpha_3 + \alpha_4 + \alpha_5) + (\beta_1 + \beta_2 + \beta_3 + \beta_4 + \beta_5) = 5 \cdot 180°$$
$$(5 - 2)180° + (\beta_1 + \beta_2 + \beta_3 + \beta_4 + \beta_5) = 5 \cdot 180°$$
$$\beta_1 + \beta_2 + \beta_3 + \beta_4 + \beta_5 = 5 \cdot 180° - 3 \cdot 180°$$
$$= 2 \cdot 180°$$
$$= 360°$$

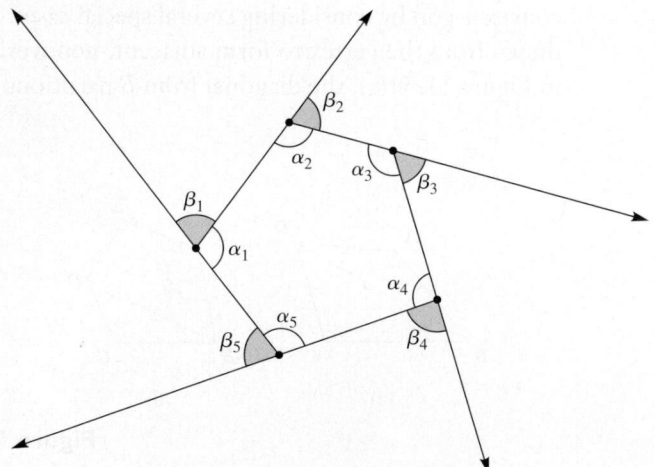

Figure 11-52

Thus the sum of the measures of the exterior angles in a pentagon is $360°$. An analogous approach works for any convex n-gon. Let S be the sum of the measures of the interior angles and E the sum of the measures of the exterior angles. We know that $S + E = n \cdot 180°$. Because $S = (n - 2)180°$, we get

$$(n - 2)180° + E = n180°$$
$$E = n180° - (n - 2)180°$$
$$= [n - (n - 2)]180°$$
$$= 2 \cdot 180° = 360°$$

Thus we have proved the following theorem.

> **Theorem 11-5**
>
> The sum of the measures of the exterior angles of any convex *n*-gon is 360°.

An intuitive justification of Theorem 11-5 is shown in Figure 11-53. Imagine walking clockwise around the convex pentagon starting at vertex *A*. At each vertex we need to turn by an exterior angle as in Figure 11-53(a). At the end of the walk we are at *A*, heading in the direction of the red arrow as in Figure 11-53(b). To return to the original starting direction we turn through one more exterior angle as in Figure 11-53(c). Thus it seems that our total turn is through 360°, whether we walk around a convex pentagon or any other convex polygon. This may be clearer if we extend the sides of the exterior angles and look at the figure from afar. Because the sides are infinite and the polygon is small, the figure will look like the one in Figure 11-53(d).

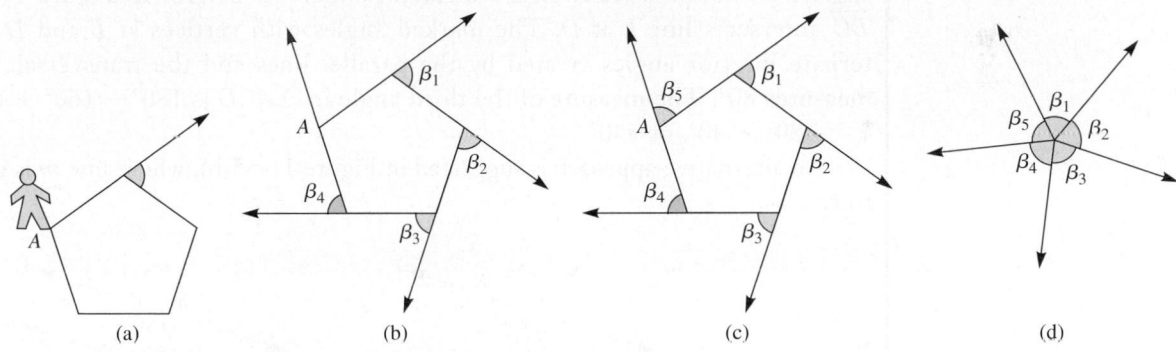

(a) (b) (c) (d)

Figure 11-53

NOW TRY THIS 11-21

a. Using the fact that the sum of the measures of the exterior angles in a convex *n*-gon is 360°, derive the formula for the sum of the interior angles.

b. Express the measure of a single interior angle of a regular *n*-gon in terms of *n*.

EXAMPLE 11-12

a. Find the measure of each interior angle of a regular decagon.

b. Find the number of sides of a regular polygon each of whose interior angles has a measure of 175°.

Solution

a. Because a decagon has 10 sides, the sum of the measures of the angles of a decagon is $10 \cdot 180° - 360°$, or 1440°. A regular decagon has 10 angles, all of which are congruent, so each one has a measure of $\dfrac{1440°}{10}$, or 144°. As an alternative solution using Now Try This 11-21, each exterior angle is $\dfrac{360°}{10}$, or 36°. Hence, each interior angle is $180° - 36°$, or 144°.

b. Each interior angle of the regular polygon is 175°. Thus, the measure of each exterior angle of the polygon is $180° - 175°$, or 5°. Because the sum of the measures of all exterior angles of a convex polygon is 360°, the number of exterior angles is $\dfrac{360}{5}$, or 72. Hence, the number of sides is 72.

EXAMPLE 11-13

In Figure 11-54, lines k and l are parallel. Measures of some of the angles are shown. Find x, the measure of $\angle BCA$.

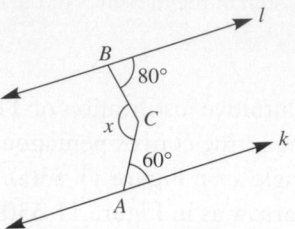

Figure 11-54

Solution Using the strategy of *examining a related problem*, if we had a transversal intersecting the parallel lines, we could consider congruent corresponding or alternate interior angles. To obtain a transversal we extend either $\overline{BC}$ or $\overline{AC}$. In Figure 11-55(a), transversal $\overleftrightarrow{BC}$ intersects line k at D. The marked angles with vertices at B and D are congruent alternate interior angles created by the parallel lines and the transversal, and hence $\angle BDA$ measures 80°. The measure of the third angle in $\triangle ACD$ is $180° - (60° + 80°)$, or 40°. Thus, $x = 180° - 40°$, or 140°.

An alternative approach is suggested in Figure 11-55(b), where line m is constructed parallel to ℓ.

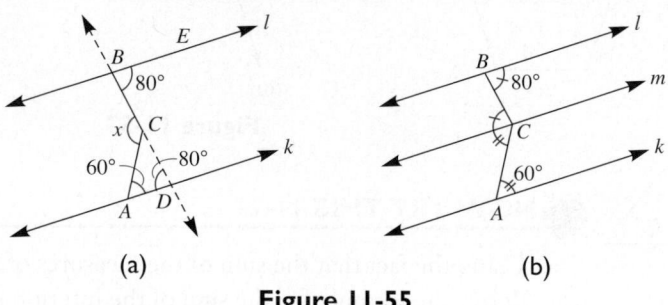

(a)   (b)

Figure 11-55

NOW TRY THIS 11-22

Walking around a polygon illustrates that the sum of the measures of the exterior angles of any convex polygon is 360°.

a. Use the same idea to find the sum of the measures of the exterior angles in the regular five-pointed star in Figure 11-56 with congruent angles at A, B, C, D, and E.

b. Find the measure of each marked angle of the star.

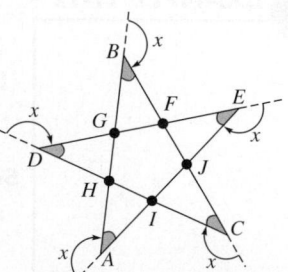

Figure 11-56

BRAIN TEASER

Sylvia, a graphic designer, needs to know the measurement of the shaded angles at the vertices of the 12-pointed star in Figure 11-57, in which all the sides and all the marked angles are congruent. How can she find the exact measurement of the angles without using a protractor?

Figure 11-57

Assessment 11-4A

1. If three lines all meet in a single point, how many pairs of vertical angles are formed?

2. Find the measure of the third angle in each of the following triangles:

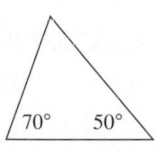
a.

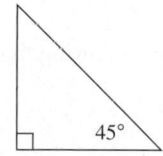
b.

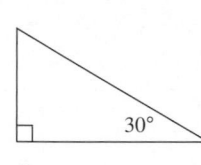
c.

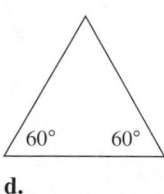
d.

3. For each of the following figures, determine whether *m* and *n* are parallel lines. Justify your answers.

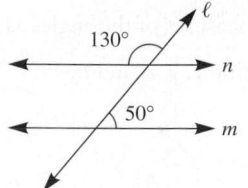
a.

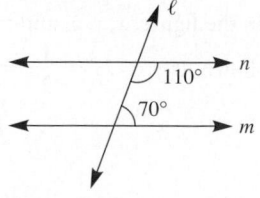
b.

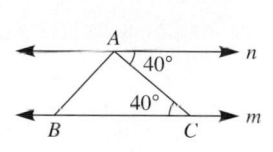
c.

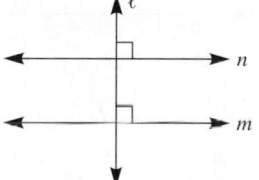
d.

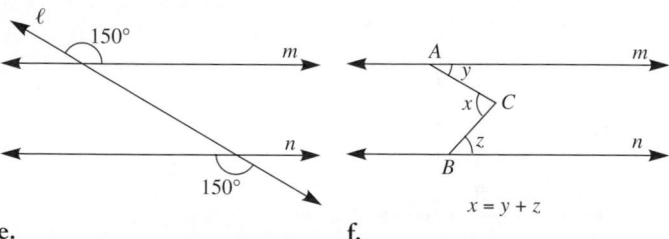
e.   **f.**

$x = y + z$

4. Two angles are complementary and the ratio of their measures is 7:2. What are the angle measures?

5. In a regular polygon, the measure of each interior angle is 162°. How many sides does the polygon have?

6. Find the sum of the measures of the marked angles in the following:

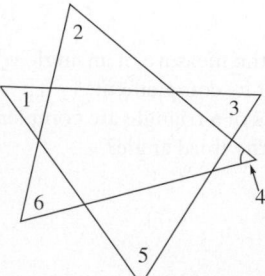

7. If the interior angles of a hexagon form an arithmetic sequence with the greatest measure being 130°, find the measure of each angle.

8. In the following figure, $\overrightarrow{DE} \parallel \overrightarrow{BC}$, $\overrightarrow{EF} \parallel \overrightarrow{AB}$, and $\overrightarrow{DF} \parallel \overrightarrow{AC}$. Also, $m(\angle 1) = 45°$ and $m(\angle 2) = 65°$. Find each of the following:
   a. $m(\angle 3)$
   b. $m(\angle D)$
   c. $m(\angle E)$
   d. $m(\angle F)$

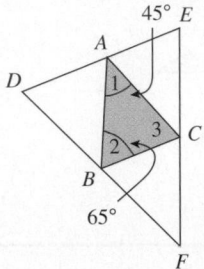

9. In the following figures, find x.

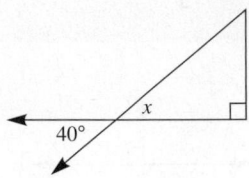

   a.

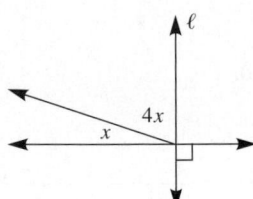

   b.

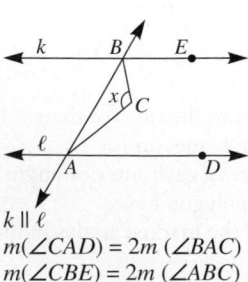

$k \parallel \ell$
$m(\angle CAD) = 2m\ (\angle BAC)$
$m(\angle CBE) = 2m\ (\angle ABC)$

   c.

10. a. Determine the measure of an angle whose measure is twice that of its complement.
    b. If two angles of a triangle are complementary, what is the measure of the third angle?

11. Find the sum of the measures of the marked angles in the following figure:

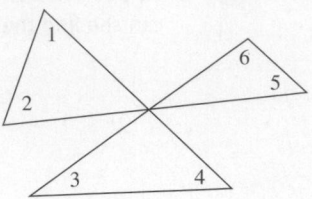

12. Find x in the following figure:

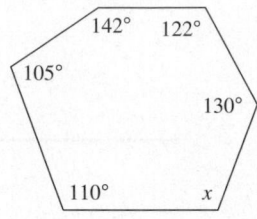

13. Find the measures of angles 1, 2, and 3 given that $TRAP$ is a trapezoid with $\overline{TR} \parallel \overline{PA}$.

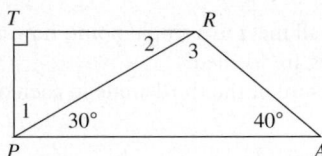

14. Refer to the following figure and answer the following:

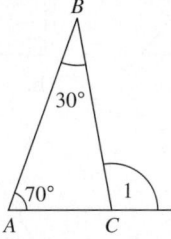

    a. Find $m(\angle 1)$.
    b. $\angle 1$ is an exterior angle of $\triangle ABC$. Use your answer in (a) to make a conjecture concerning the measure of an exterior angle of a triangle. Justify your conjecture.

15. In the figure, $x, y, z,$ and w are measures of the angles as shown. If $y = 2x = \frac{1}{2}z = \frac{1}{3}w$, find $x, y, z,$ and w.

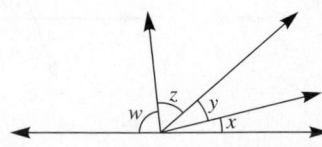

Assessment 11-4B

1. If two planes intersect in a single line forming dihedral angles, how would you define vertical dihedral angles?
2. Find the measure of the angle marked x in each of the following triangles.

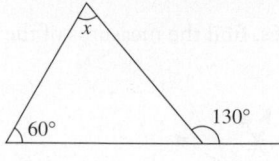

a.

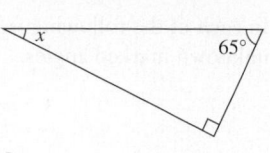

b.

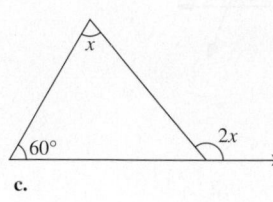

c.

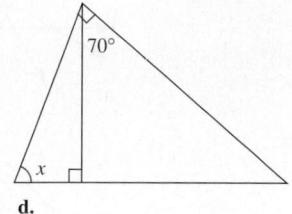

d.

3. In parts (a) and (b) prove that $k \parallel l$. In parts (c) and (d), solve for x.

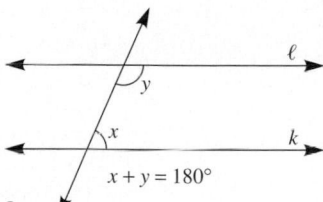

$x + y = 180°$

a.

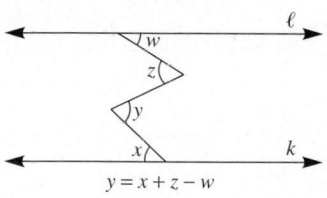

$y = x + z - w$

b.

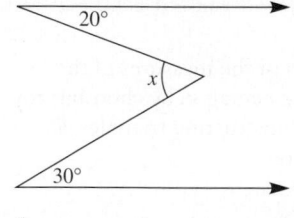

c.

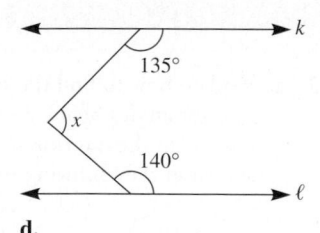

d.

4. **a.** Determine the measure of an angle whose measure is $\frac{2}{3}$ of its complement.
   b. Can two angles of a triangle be supplementary? Why or why not?
5. Find the sum of the measures of the numbered angles in the following figure:

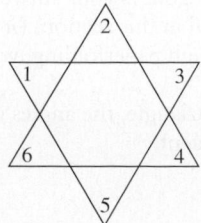

6. Find the measure of each of the interior angles of a regular dodecagon (12-sided polygon).
7. Calculate the measure of each angle of a pentagon, where the measures of the angles form an arithmetic sequence and the least measure is 60°.
8. The sides of $\triangle DEF$ are parallel to the sides of $\triangle BCA$. If the measures of two angles of $\triangle ABC$ are 60° and 70° as shown, find the measures of angles of $\triangle DEF$.

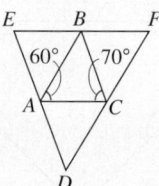

9. In each of the following figures, find the measures of the marked angles.

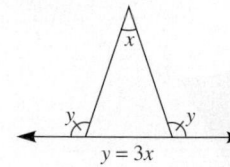

$y = 3x$

a.

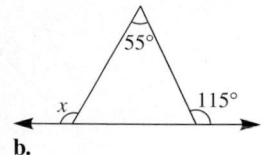

b.

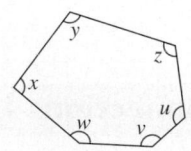

The sequence x, y, z, u, v, w is an arithmetic sequence and $v = 136°$.

c.

10. **a.** If $m(\angle ABC) = 90°$ and $\overline{BD} \perp \overline{AC}$ and $m(\angle A) = 50°$, find the measure of all the angles of $\triangle ABC$, $\triangle ADB$, and $\triangle CDB$.

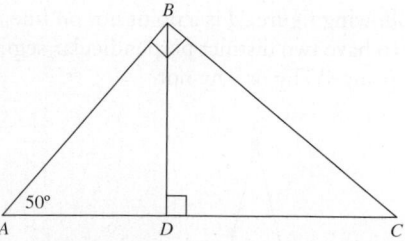

b. If $m(\angle A) = \alpha$ in (a), find the measures of all the angles of $\triangle ABC$, $\triangle ADB$, and $\triangle CDB$ in terms of α.

11. Two sides of a regular octagon are extended as shown in the following figure. Find the measure of ∠1.

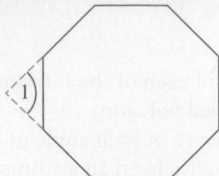

12. Home plate on a baseball field has three right angles and two other congruent angles. Refer to the following figure and find the measures of each of these two other congruent angles.

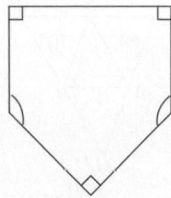

13. The measure of ∠A in △ABC is α. Congruent angles at B and C are marked. Find $m(\angle D)$ in terms of α.

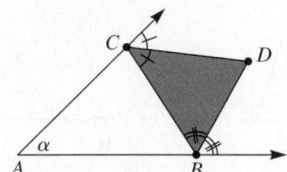

14. ABCD is a quadrilateral in which opposite angles have the same measure, as indicated in the following figure. What kind of quadrilateral is ABCD? Justify your answer.

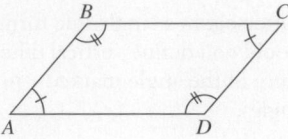

15. In each of the following figures, find the measures of the unknown marked angles.

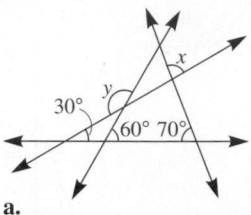

a.

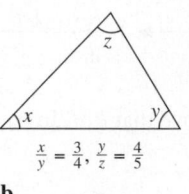

$\frac{x}{y} = \frac{3}{4}, \frac{y}{z} = \frac{4}{5}$

b.

Mathematical Connections 11-4

Communication

1. **a.** If one angle of a triangle is obtuse, can another also be obtuse? Why or why not?
   b. If one angle in a triangle is acute, can the other two angles also be acute? Why or why not?
   c. Can a triangle have two right angles? Why or why not?
   d. If a triangle has one acute angle, is the triangle necessarily acute? Why or why not?

2. In the following figure, A is a point not on line ℓ. Is it possible to have two distinct perpendicular segments from A to ℓ in a plane? Why or why not?

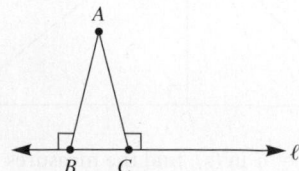

3. **a.** Explain how to find the sum of the measures of the interior angles of any convex pentagon by choosing any point P in the interior and constructing triangles, as shown in the following figure:

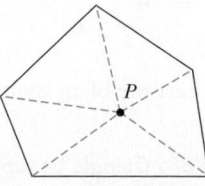

   b. Using the method suggested by the diagram in (a), explain how to find the sum of the measures of the angles of any convex n-gon. Is your answer the same as the one already obtained in this section, $(n - 2)180°$?

4. Explain how, through paperfolding, you would show each of the following:
   a. In an isosceles triangle, the angles opposite congruent sides are congruent.

b. If two angles of a triangle are congruent, the triangle is isosceles.

c. In an equilateral triangle, all the interior angles are congruent.

d. An isosceles trapezoid has two pairs of congruent angles.

5. **a.** Explain how to find the sum of the measures of the marked interior angles of a concave quadrilateral like the following:

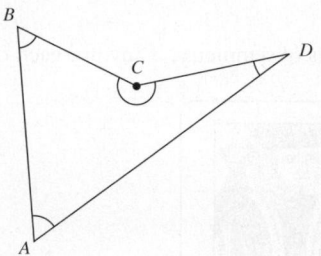

b. Conjecture whether the formula for the sum of the measures of the interior angles of a convex polygon is true for concave polygons.

c. Justify your conjecture in (b) for pentagons and hexagons and explain why your conjecture is true in general.

6. Regular hexagons have been used to tile floors. Can a floor be tiled using only regular pentagons? Why or why not?

7. Lines a and b are cut by transversals c and d. If $m(\angle 1) = m(\angle 3)$ and $m(\angle 2) = m(\angle 4)$, can you conclude that a and b are parallel? Justify your answer.

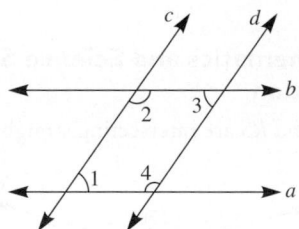

8. A beam of light from A hits the surface of a mirror at point B, is reflected, and then hits a perpendicular mirror and is reflected again. If $\angle 1 \cong \angle 2$ and $\angle 3 \cong \angle 4$, prove that the reflected beam is parallel to the incoming beam, that is, that the rays $\overrightarrow{AB}$ and $\overrightarrow{CD}$ are parallel.

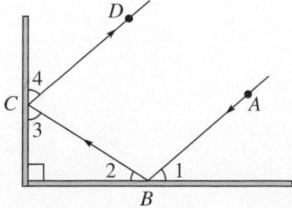

Open-Ended

9. Draw three different nonconvex polygons. When you walk around a polygon, at each vertex you need to turn either right (clockwise) or left (counterclockwise). A turn to the left is measured by a positive number of degrees and a turn to the right by a negative number of degrees. Find the sum of the measures of the turn angles of the polygons you drew. Assume you start at a vertex facing in the direction of a side, walk around the polygon, and end up at the same vertex facing in the same direction as when you started.

10. Draw three concave polygons. Measure all the interior angles, including any reflex angles. What is the sum of the interior angles of any concave polygon? Explain why.

Cooperative Learning

11. In $\triangle ABC$, $\overrightarrow{AD}$ and $\overrightarrow{BD}$ are *angle bisectors*; that is, they divide the angles at A and B respectively into congruent angles.
   a. If the measures of $\angle A$ and $\angle B$ are known, then $m(\angle D)$ can be found. Explain how.

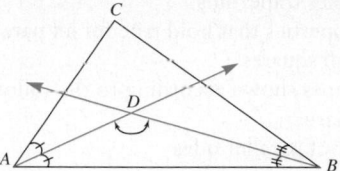

   b. Suppose the measures of $\angle A$ and $\angle B$ are not known but that of $\angle C$ is. Can $m(\angle D)$ be found? If so, find $m(\angle D)$ in terms of $m(\angle C)$. To answer this question, assign each member of your group a triangle with different angles but with the same measure for $\angle C$. Each person should compute $m(\angle D)$. Use the results to make a conjecture related to the previous question.

   c. Discuss a strategy for answering the question in (b) and write a solution to be distributed to the entire class.

12. Each person should draw a large triangle like $\triangle ABC$ in the following figure and cut it out. Fold the crease $\overline{BB'}$ so that A falls on some point A' on $\overline{AC}$. Next, unfold and fold the top B along $\overline{BB'}$ so that B falls on B'. Then fold vertices A and C along $\overline{AC}$ to match point B', as shown in the following figures:

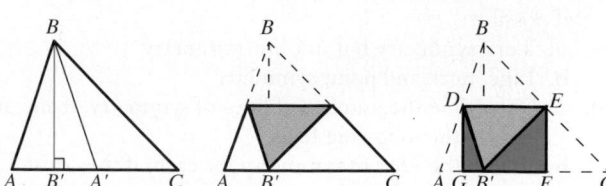

   a. Why is $\overline{BB'}$ perpendicular to $\overline{AC}$?
   b. What theorem does the folded figure illustrate? Why?
   c. The folded figure seems to be a rectangle. Explain why.
   d. What is the length of $\overline{GF}$ in terms of the length of the base $\overline{AC}$ of $\triangle ABC$? Why?

Questions from the Classroom

13. Jan, a tile designer, wants to make tiles in the shape of a convex polygon with all interior angles acute. She is wondering if there are any such polygons besides triangles. How do you respond?

14. A student wonders if there exists a convex decagon with exactly four right angles. How do you respond?

15. Rory says that because an exterior angle of a triangle has measure equal to the sum of the measures of the two nonadjacent angles (to the exterior angle), then the sum of all the exterior angles should be twice the sum of the measures of all the interior angles. Is Rory's argument correct? Explain why or why not.

Review Problems

16. In each of the following, conjecture the required properties. If this is not possible, explain why.
    a. Two properties that hold true for all rectangles but not for all rhombuses
    b. Two properties that hold true for all squares but not for all isosceles trapezoids
    c. Two properties that hold true for all parallelograms but not for all squares

17. Sort the shapes shown according to the following attributes they must have:
    a. Number of parallel sides
    b. Number of right angles
    c. Number of congruent sides

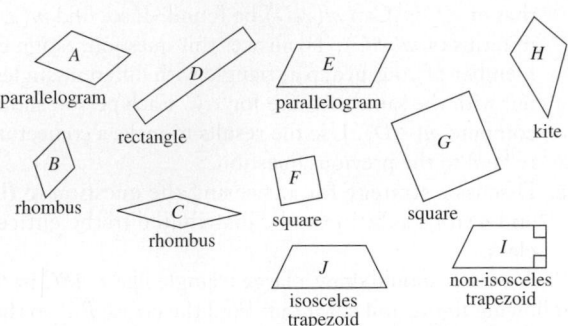

18. What geometric properties make "SOS" a good choice for the international distress symbol?

19. Design symbols that have each of the following symmetries, if possible:
    a. Turn symmetry but not line symmetry
    b. Line, turn, and point symmetry

20. a. Determine the number of lines of symmetry, if any, in each of the following flags.
    b. Sketch the lines of symmetry for each, if they exist.

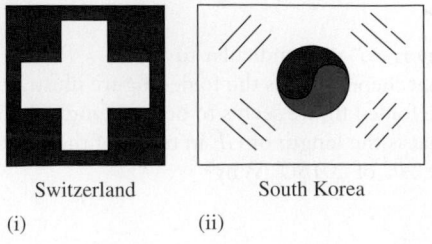

Switzerland   South Korea
(i)   (ii)

21. Explain whether the following quilt patterns have turn symmetry, and, if so, identify the turn center.

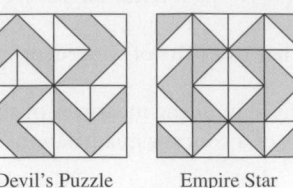

Devil's Puzzle   Empire Star
a.   b.

22. Find the lines of symmetry, if any, for each of the following:

a.

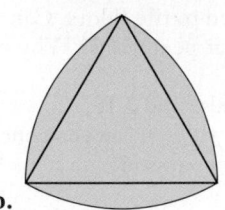

b.

Trends in Mathematics and Science Study (TIMSS) Question

In the figure, *PQ* and *RS* are intersecting straight lines.

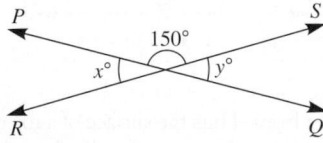

What is the value of $x + y$?
    a. 15   b. 30   c. 60   d. 180   e. 300

TIMSS, Grade 8, 2003

**National Assessment of Educational Progress
(NAEP) Question**

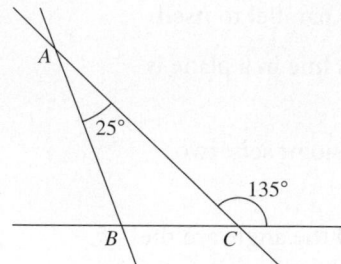

In the triangle, what is the degree measure of $\angle ABC$?
a. 45
b. 100
c. 110
d. 135
e. 160

NAEP, Grade 8, 2003

Hint for the Preliminary Problem

Consider a way to count the lengths in the path used. For example, one might think of $AB + BC$ as the sum of the lengths of two sides of the equilateral triangle beginning the path, or one might think of AB as a length of a side of the triangle, $BC + DC$ as the sum of the lengths of two sides of the square, and so on. See whether there is a pattern to make the count as easy as possible.

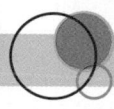

Chapter 11 Summary

11-1 Basic Notions	Pages
Fundamental building blocks of geometry are undefined terms including **points, lines,** and **planes.**	632
Collinear points are points on the same line. Point B is **between** A and C if $AB + BC = AC$.	633
A **line segment**, or **segment**, is a subset of a line that contains two points of the line and all the points between those two points. A **closed segment** is a segment with its endpoints. A **half-open** or **half-closed** segment is a segment without one endpoint An **open segment** does not contain its endpoints.	633–634
A **ray** $\overrightarrow{AB}$ is a subset of $\overleftrightarrow{AB}$ that contains the endpoint A, the point B, and all points C such that B, is between points A and C.	634
A point separates a line into two **half-lines,** and the point itself.	634
Points in the same plane are **coplanar.**	636
Noncoplanar points cannot be placed in a single plane.	636
Lines in the same plane are **coplanar lines.**	636
Intersecting lines are lines with exactly one point in common.	636
Skew lines are lines for which there is no plane that contains them.	636
Concurrent lines are lines that intersect in the same point.	636
Two distinct coplanar lines are **parallel** if they have no points in common (are not intersecting lines). A line is parallel to itself.	636
Two segments or rays are parallel if they lie in parallel lines.	636

11-2 Linear Measure

Properties of Distance:
- The distance between any two points A and B is greater than or equal to 0.
- The distance between any two points A and B is the same as the distance between B and A.
- For any three points A, B, and C, the distance between A and B plus the distance between B and C is greater than or equal to the distance between A and C.

Theorem: (Triangle Inequality): The sum of the lengths of any two sides of a triangle is greater than the length of the third side.   657

The **perimeter** of a figure is the distance around the figure; a perimeter has linear measure.   657

The perimeter of a circle is its **circumference**. If the radius of the circle is r, then its circumference C is given by $2\pi r$.   658

An arc of $\theta°$ has linear length $(\pi r\theta)/180$ units.   660

11-3 Curves, Polygons, and Symmetry

A **simple curve** does not intersect itself; except the starting and stopping points may be the same when the curve is traced.   665

A **closed curve** can be drawn starting and stopping at the same point.   665

Polygons are simple closed curves with only segments as sides.   665

Convex curves are simple and closed such that the segment connecting any two points in the interior of the curve is wholly contained in the interior of the curve.   665

Polygons along with their interiors are **polygonal regions**.   665

Concave curves are simple, closed, and not convex.   665

Two **segments** are **congruent** if a tracing of one can be fitted exactly on top of the other; they have the same measure.   666–667

Two **angles** are **congruent** if they have the same measure.   667

A **regular polygon** is both equiangular and equilateral.   667

Triangles may be classified as **right, acute,** or **obtuse** depending on angle size.   668

Triangles may be classified as **scalene, isosceles,** or **equilateral** depending on side length.   668

A **quadrilateral** is a polygon with four sides.   668–669

A **trapezoid** is a quadrilateral with at least one pair of parallel sides. An **isosceles trapezoid** is a trapezoid with congruent base angles.   668–669

A **kite** is a quadrilateral with at least one pair of adjacent sides congruent and the other two sides also congruent.   668

A **parallelogram** is a quadrilateral in which each pair of opposite sides is parallel.   669

A **rectangle** is a parallelogram with a right angle.   669

A **rhombus** is a parallelogram with two adjacent sides congruent.   669

A **square** is a rectangle with two adjacent sides congruent.   669

A geometric figure has a **line of symmetry** if it is its own image when folded along the line.   671

Chapter 11 Review

1. Refer to the following line m:

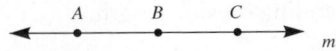

   a. Other than m list three more names for the line.
   b. Name two rays on m that have endpoint B.
   c. Find a simpler name for the intersection of rays $\overrightarrow{AB}$ and $\overrightarrow{BA}$.
   d. Find a simpler name for the intersection of rays $\overrightarrow{BA}$ and $\overleftrightarrow{AC}$.
2. In the figure, $\overleftrightarrow{PQ}$ is perpendicular to the plane α. Answer the following:
   a. Name a pair of skew lines.
   b. Using only the letters in the figure, name as many planes as possible that are perpendicular to α.

   c. What is the intersection of planes APQ and β?
   d. Is there a single plane containing A, B, P, and Q? Explain your answer.

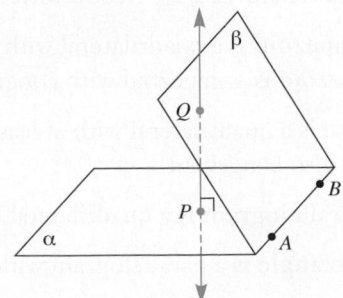

3. Find each of the following:
   a. $113°57' + 18°14'$
   b. $84°13' - 27°45'$
   c. $113°57' + 18.4°$
   d. $0.75°$ in minutes and seconds
   e. $6°48'59'' + 28°19'36''$

4. In the accompanying figure, planes α and β intersect in line AB. The dihedral angle S-AB-Q measures $90°$. The lines PS and PQ determine a new plane, γ.
   a. Why is plane γ perpendicular to α as well as to β?
   b. Why is line AB perpendicular to γ?

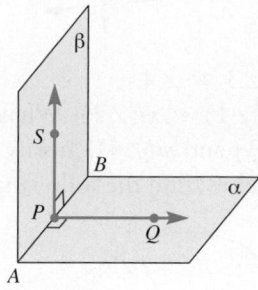

5. Draw each of the following curves:
   a. A simple closed curve
   b. A closed curve that is not simple
   c. A concave hexagon
   d. A convex decagon
   e. An equilateral pentagon that is not regular
   f. An equiangular quadrilateral that is not a square

6. a. Can a triangle have two obtuse angles? Justify your answer.
   b. Can a parallelogram have four acute angles? Justify your answer.

7. In a certain triangle, the measure of one angle is twice the measure of the smallest angle. The measure of the third angle is 7 times as great as the measure of the smallest angle. Find the measures of each of the angles in the triangle.

8. If ABC is a right triangle and $m(\angle A) = 42°$, what is the measure of the other acute angle?

9. In a certain regular polygon, the measure of each interior angle is $176°$. How many sides does the polygon have?

10. In a periscope, a pair of mirrors are parallel. If the dotted line in the following figure represents a path of light and $m(\angle 1) = m(\angle 2) = 45°$, find $m(\angle 3)$ and $m(\angle 4)$:

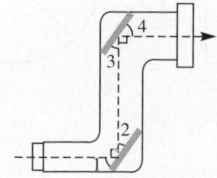

11. In the figure, ℓ is parallel to m, and $m(\angle 1) = 60°$. Find each of the following:
    a. $m(\angle 3)$ b. $m(\angle 6)$ c. $m(\angle 8)$

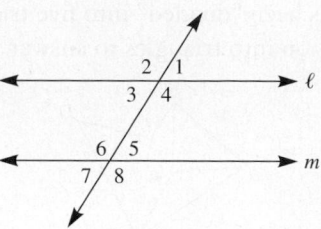

12. If $\angle ACB$ is a right angle as shown, and the other angles are as indicated in the figure, find x, the measure of $\angle DCB$.

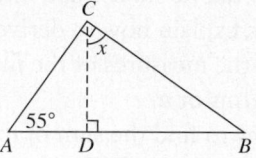

13. For each of the following, determine x if the lines a and b are parallel.

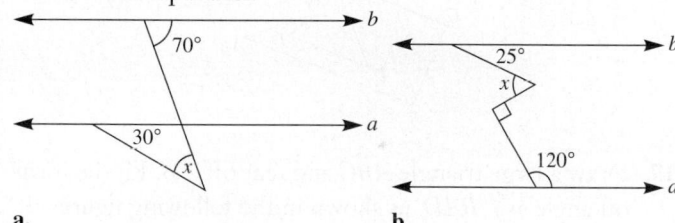

14. In the following figures, find the measures of the angles marked x and y:

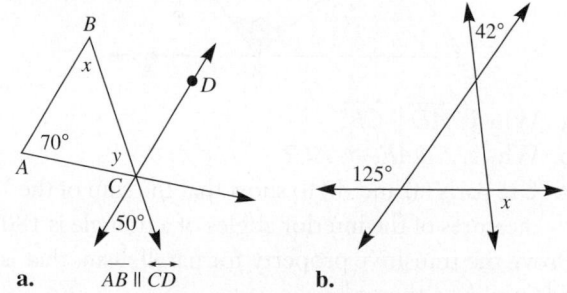

15. In $\triangle ABC$, line DE is parallel to line AB, and line DF is parallel to line AC. If $m(\angle C) = 70°$ and $m(\angle B) = 45°$, find the measures of the angles labeled 1, 2, 3, 4, and 5.

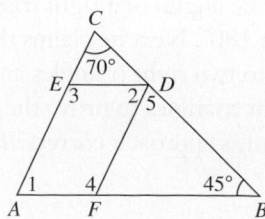

16. The polygon *ABCDE* contains a point *P* in its interior that can be joined to all the vertices by segments entirely in the interior of the polygon. In this way, the polygon has been "divided" into five triangles. Use such a division into triangles to answer the following:

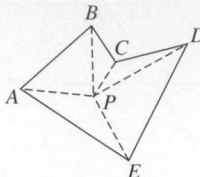

   a. Find the sum of the measures of the interior angles of the polygon *ABCDE*.
   b. If an *n*-gon can be subdivided into triangles in a similar way, explain how to derive the formula for the sum of the measures of the interior angles of the *n*-gon in terms of *n*.
   c. Explain how to find the sum of the measures of the interior angles of the following polygon:

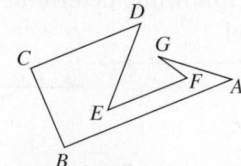

17. Draw a large triangle *ABC* and tear off ∠*B*. Fit the torn-off angle as ∠*BAD*, as shown in the following figure:

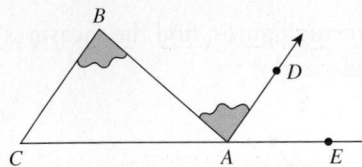

   a. Why is $\overleftrightarrow{AD} \parallel \overleftrightarrow{CB}$?
   b. Why is ∠*DAE* ≅ ∠*C*?
   c. Use parts (a) and (b) to show that the sum of the measures of the interior angles of a triangle is 180°.
18. Prove the transitive property for parallelism, that is, if $a \parallel b$ and $b \parallel c$, then $a \parallel c$.
19. Wally claims that he can easily prove that the sum of measures of the interior angles in any triangle is 180°. He says that any right triangle is half of a rectangle. Because a rectangle has four right angles, its angles add up to 360°, so the angles of a right triangle add up to half of 360°, or 180°. Next he claims that he can divide any triangle into two right triangles and use what he proved for right triangles to prove the theorem for any triangle. Is Wally's approach correct? Explain why or why not.

20. Prove that if $x = y + z$, then $a \parallel b$.

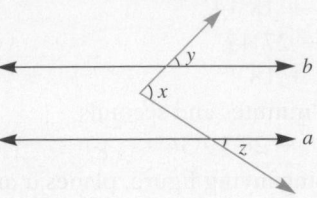

21. In the following figure, ∠1 ≅ ∠2.

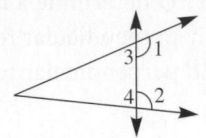

   a. Prove that ∠3 ≅ ∠4.
   b. Suppose $m(\angle 1) < m(\angle 2)$. What can you conclude about $m(\angle 3)$ and $m(\angle 4)$? Justify your answer.
22. In the figure below, find the following measures.

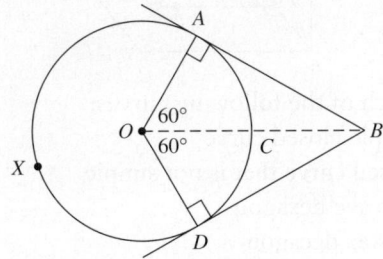

   a. $m(\angle ABO)$
   b. $m(\angle ABD)$
   c. $m(\widehat{AXD})$
23. Sketch three planes that intersect in a point
24. a. Given 10 points in the plane, no 3 of which are collinear, how many triangles can be drawn whose vertices are the given points? Explain your reasoning.
   b. Generalize and answer part (a) for *n* points.
25. In each of the following, determine the number of sides of a regular polygon with the stated property. If such a regular polygon does not exist, explain why.
   a. Each exterior angle measures 20°.
   b. Each exterior angle measures 25°.
   c. The sum of all the exterior angles is 3600°.
   d. The total number of diagonals is 4860.
26. Determine how many lines of symmetry, if any, each of the following figures has:

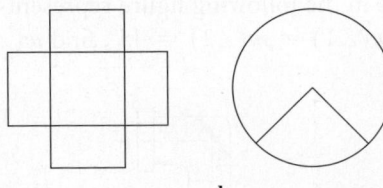

   a.   **b.**

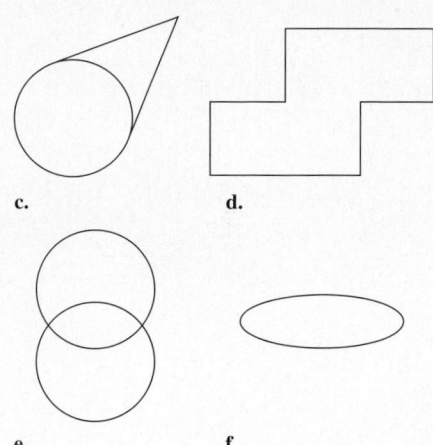

c.   d.

e.   f.

27. For each of the following figures, identify the types of symmetry (line, turn, or point) it possesses:

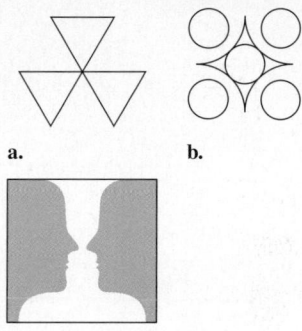

a.   b.

c.

28. In the following, draw in dotted lines for the unseen segments:

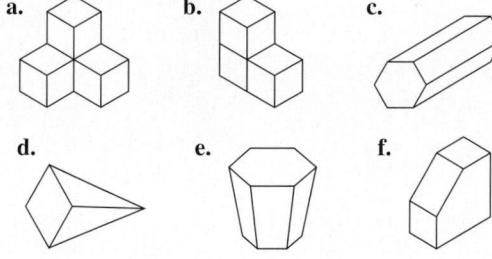

a.   b.   c.

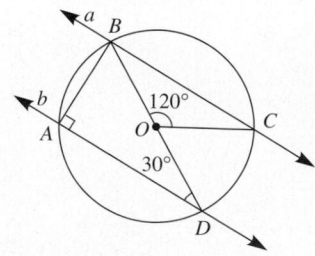

d.   e.   f.

29. In the figure below, $a \parallel b$. Find the following:

a. $m(\angle ABD)$
b. $m(\angle DBC)$
c. $m(\angle BCO)$
d. $m(\angle COD)$
e. $m(\overparen{CD})$
f. $m(\overparen{BC})$

30. Complete the following.
   a. 50 ft = _____ yd
   b. 947 yd = _____ mi
   c. 0.75 mi = _____ ft
   d. 349 in. = _____ yd
   e. 5 km = _____ m
   f. 165 cm = _____ m
   g. 52 cm = _____ mm
   h. 125 m = _____ km

31. Given three segments of length p, q, and r, where $p > q$, determine if it is possible to construct a triangle with sides of length p, q, and r in each of the following cases. Justify your answers.
   a. $p - q > r$
   b. $p - q = r$

32. The diagonal of a rectangle has measure 1.3 m, and a side of the rectangle has measure 120 cm. Find the perimeter of the rectangle.

33. The circumference of a circle is 3 m. What is its radius?

34. What is the degree measure of the central angle determining an arc whose length is 40 cm and whose radius is 18 cm?

35. Sarah reports the radius of a circle as 6 cm and its circumference as 36π cm. Explain whether she is correct.

Congruence and Similarity with Constructions

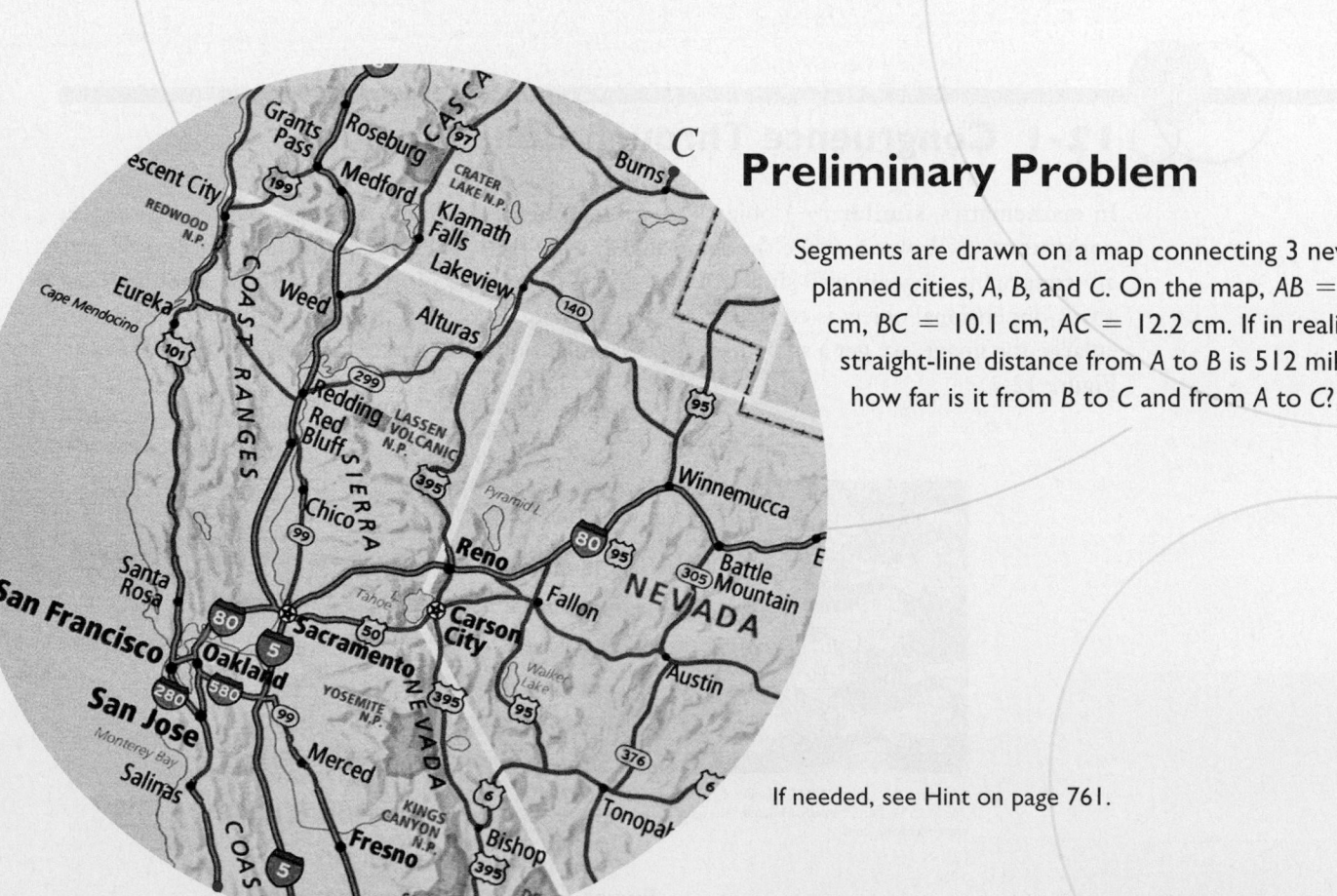

Preliminary Problem

Segments are drawn on a map connecting 3 newly planned cities, *A*, *B*, and *C*. On the map, $AB = 4.2$ cm, $BC = 10.1$ cm, $AC = 12.2$ cm. If in reality the straight-line distance from *A* to *B* is 512 miles, how far is it from *B* to *C* and from *A* to *C*?

If needed, see Hint on page 761.

Grades 3–5 *PSSM* states that students should be able to do the following:

- explore congruence and similarity (p. 396)
- make and test conjectures about geometric properties and relationships and develop logical arguments to justify conclusions (p. 396)
- recognize geometric ideas and relationships and apply them to other disciplines and to problems that arise in the classroom or in everyday life. (p. 396)

The grade 8 *Focal Points* emphasizes the use of similar triangles:

Students use fundamental facts about distance and angles to describe and analyze figures and situations in two- and three-dimensional space and to solve problems, including those with multiple steps. They prove that particular configurations of lines give rise to similar triangles because of the congruent angles created when a transversal cuts parallel lines. Students apply this reasoning about similar triangles to solve a variety of problems, including those that ask them to find heights and distances. (p. 20)

The grade 7 *Common Core Standards* states that students should:

- Draw, construct, and describe geometrical figures and describe the relationship between them. (p. 49)
- Solve problems involving scale drawings of geometric figures, including computing actual lengths and areas from a scale drawing and reproducing a scale drawing at a different scale. (p. 49)
- Draw (freehand, with ruler and protractor, and with technology) geometric shapes with given conditions. Focus on constructing triangles from three measures of angles or sides, noticing when the conditions determine a unique triangle, more than one triangle, or no triangle. (p. 50)

In this chapter, we introduce, through constructions and visualization, the concepts of congruence and similarity. We study equations of lines, investigate systems of equations both geometrically and algebraically, and consider basic trigonometry.

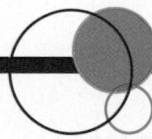

12-1 Congruence Through Constructions

In mathematics, **similar** ($\sim$) objects have the same shape but not necessarily the same size; **congruent** ($\cong$) objects have the same size as well as the same shape. Whenever two figures are congruent, they are also similar. However, the converse is not true. A 100% photocopy of a two-dimensional object is congruent to the original. When we photocopy and either reduce or enlarge the image, we get a similar object. Examples of similar and congruent objects are seen in Figure 12-1.

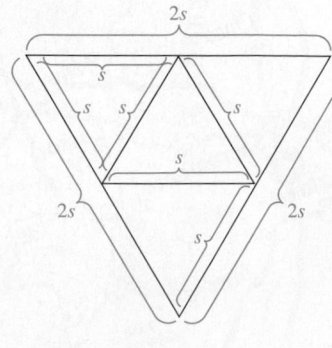

(a)   (b)

Figure 12-1

The Escher print in Figure 12-1(a) shows congruent fish and congruent birds. Figure 12-1(b) contains both congruent and similar equilateral triangles. The smaller triangles are both similar to and congruent to each other. They also are similar to the large triangle that contains them. Figure 12-1(b) is an example of a **rep-tile**, a figure that is used to construct a larger, similar figure. In Figure 12-1(b), one of the smaller equilateral triangles is a rep-tile. Figure 12-2(a) shows two circles with the same shape but different sizes. Circles with equal radii as in Figure 12-2(b) are congruent; however, all circles are similar to each other.

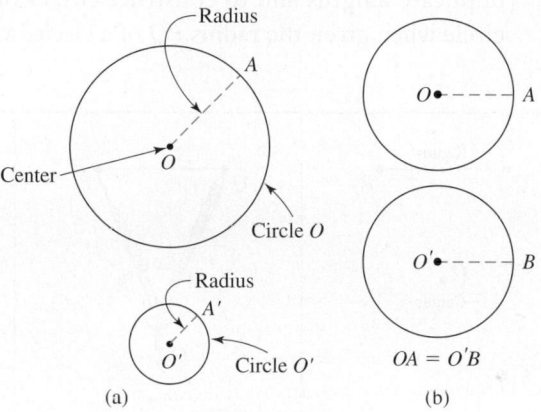

Figure 12-2

Before studying congruent and similar figures, we first consider some notation and review definitions from Chapter 11. For example, *any two line segments are congruent if and only if they have the same length, and two angles are congruent if, and only if, they have the same measure.* The length of line segment $\overline{AB}$ is denoted AB. Symbolically, we may write the following about congruent segments and angles.

Definition of Congruent Segments and Angles

$\overline{AB} \cong \overline{CD}$ if, and only if, $AB = CD$.

$\angle ABC \cong \angle DEF$ if, and only if, $m(\angle ABC) = m(\angle DEF)$.

Geometric Constructions

Geometric constructions are useful in drafting and design. Designers most frequently use computers and computer software in their work; manual tools such as the compass, ruler, triangle, and protractor are used infrequently. In Euclid's time, mathematicians required that constructions be done using only a compass and a straightedge—an unmarked ruler. Euclidean constructions follow these rules:

1. Given two points, a unique straight line can be drawn containing the points as well as a unique segment connecting the points. (This is accomplished by aligning the straightedge across the points.)
2. It is possible to extend any part of a line.
3. A circle can be drawn given its center and radius.
4. Any number of points can be chosen on a given line, segment, or circle.
5. Points of intersection of two lines, two circles, or a line and a circle can be used to construct segments, lines, or circles.
6. No other instruments (such as a marked ruler, triangle, or protractor) can be used to perform Euclidean constructions.

In reality, compass and straightedge constructions are subject to error. For example, a geometric line is an ideal line with zero width. However, a drawing of a line, no matter how sharp the pencil or how good the straightedge, has a nonzero width. Then why teach constructions using the preceding rules? There are many good reasons. We can look at this activity as a game, and games have rules. Playing games can be a source of enjoyment, and so can performing Euclidean constructions—but with the additional benefit of learning geometry in the process. Also, to use computer graphics, or computer software such as the *Geometer's Sketchpad* (*GSP*) or *GeoGebra*, one needs to be able to use strategies similar to or the same as those used in Euclidean constructions.

Ancient Greek mathematicians constructed geometric figures with a straightedge and a collapsible compass. Figure 12-3 shows a more modern compass that can be used to mark off and duplicate lengths and to construct circles or arcs with a radius of a given measure. To draw a circle when given the radius *PQ* of a circle, we follow the steps illustrated in Figure 12-3.

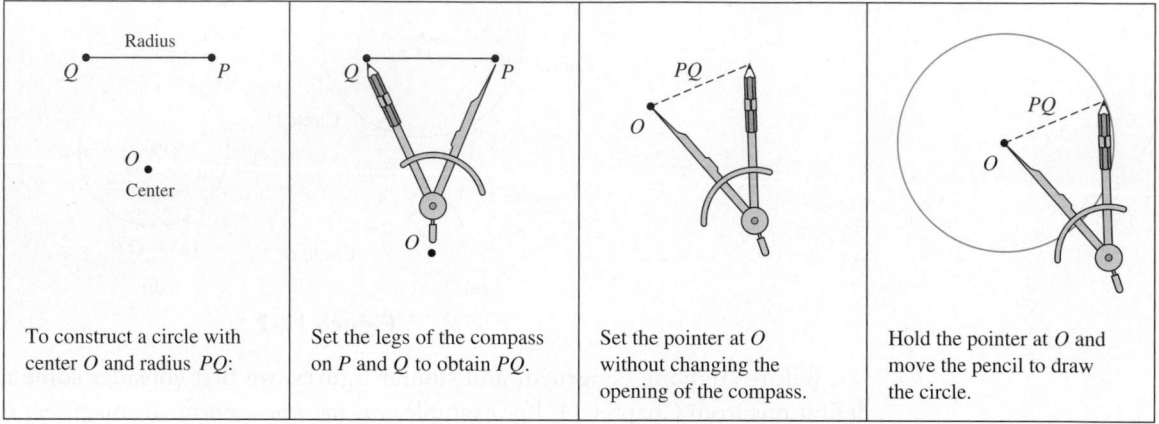

Figure 12-3 Constructing a circle given its center and radius

NOW TRY THIS 12-1

In the *Peanuts* cartoon, Sally draws circles in a plane. Describe all the possible number of intersection points two circles can have.

Constructing Segments

There are many ways to draw a segment congruent to a given segment $\overline{AB}$. A natural approach is to use a ruler, measure $\overline{AB}$, and then draw a congruent segment. A different way is to trace $\overline{AB}$ onto a piece of paper. A third method is to use a straightedge and a compass, and construct it as in Figure 12-4.

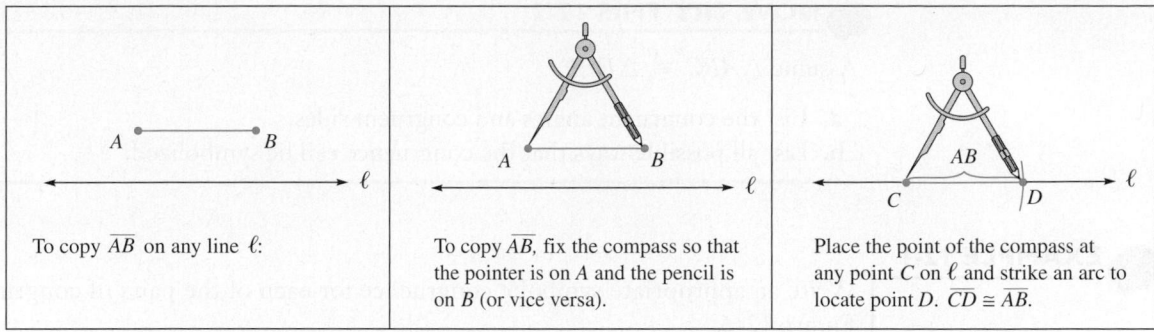

To copy $\overline{AB}$ on any line ℓ:	To copy $\overline{AB}$, fix the compass so that the pointer is on A and the pencil is on B (or vice versa).	Place the point of the compass at any point C on ℓ and strike an arc to locate point D. $\overline{CD} \cong \overline{AB}$.

Figure 12-4 Constructing a line segment congruent to a given segment

Triangle Congruence

Informally, two figures are congruent if it is possible to fit one figure onto the other so that all matching parts coincide. In Figure 12-5, $\triangle ABC$ and $\triangle A'B'C'$ have corresponding congruent parts. Tick marks are used to show congruent segments and angles in the triangles. If we were to trace $\triangle ABC$ in Figure 12-5 and put the tracing over $\triangle A'B'C'$ so that the tracing of A is over A', the tracing of B is over B', and the tracing of C is over C', then $\triangle ABC$ would coincide with $\triangle A'B'C'$. This suggests the following definition of congruent triangles.

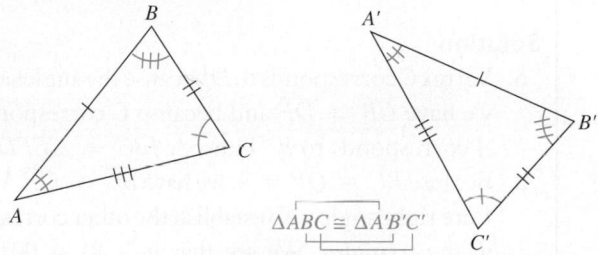

$$\triangle ABC \cong \triangle A'B'C'$$

Figure 12-5

Definition of Congruent Triangles

$\triangle ABC$ is congruent to $\triangle A'B'C'$, written $\triangle ABC \cong \triangle A'B'C'$, if, and only if, $\angle A \cong \angle A'$, $\angle B \cong \angle B'$, $\angle C \cong \angle C'$, $\overline{AB} \cong \overline{A'B'}$, $\overline{BC} \cong \overline{B'C'}$, and $\overline{AC} \cong \overline{A'C'}$.

In the definition of congruent triangles $\triangle ABC$ and $\triangle A'B'C'$, the order of the vertices is such that the listed congruent angles and congruent sides correspond, leading to the statement: "*Corresponding parts of congruent triangles are congruent*," abbreviated as *CPCTC*.

Historical Note

The straight line and circle were considered the basic geometric figures by the Greeks, and the straightedge and compass are their physical analogs. It is believed that the Greek philosopher Plato (427–347 BCE) rejected the use of mechanical devices other than the straightedge and compass for geometric constructions because use of other tools emphasized practicality rather than "ideas."

NOW TRY THIS 12-2

Assume $\triangle ABC \cong \triangle DEF$.

    a. List the congruent angles and congruent sides.
    b. List all possible ways that the congruence can be symbolized.

EXAMPLE 12-1

Write an appropriate symbolic congruence for each of the pairs of congruent triangles in Figure 12-6.

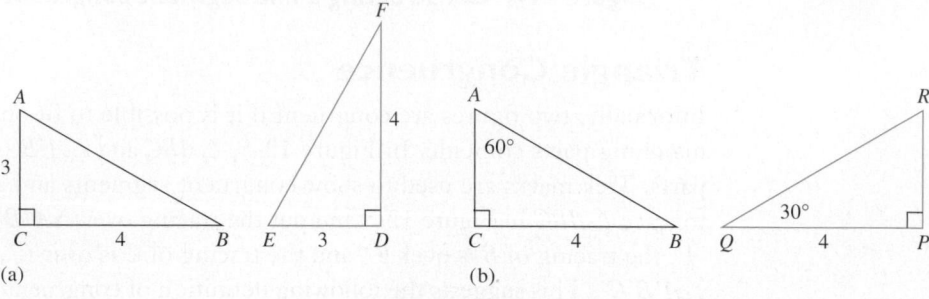

Figure 12-6

Solution

    a. Vertex C corresponds to D because the angles at C and D are right angles. Because $CB = DF = 4$, we have $\overline{CB} \cong \overline{DF}$ and because C corresponds to D, B must correspond to F. Consequently, A corresponds to E. Thus, $\triangle ABC \cong \triangle EFD$.

    b. Because $BC = QP = 4$, we have $\overline{BC} \cong \overline{QP}$. Vertex C corresponds to P because the angles at C and P are right angles. To establish the other correspondences, we first find the missing angle measures in the triangles. We see that $m(\angle B) = 90° - 60° = 30°$ and $m(\angle R) = 90° - 30° = 60°$. Because $m(\angle A) = m(\angle R) = 60°$, A corresponds to R. Because $m(\angle B) = m(\angle Q) = 30°$, B corresponds to Q. Thus, $\triangle ABC \cong \triangle RQP$.

Side, Side, Side Congruence Condition (SSS)

Calibration experts ensure that car parts on an automotive assembly line are interchangeable so that the same part fits all basic models of the same car. For cars to be congruent, the parts must be congruent. In design for automotive production, decisions have to be made about the minimal set of items to consider for eventual congruency. In considering congruence of figures in geometry, we also look for a minimal number of conditions that assure congruence.

If three sides and three angles of one triangle are congruent to the corresponding three sides and three angles of another triangle, then by definition the triangles are congruent. However, do we need to know that all six parts of one triangle are congruent to the corresponding parts of the second triangle in order to conclude that the triangles are congruent? What if we know only five parts, or four parts, or just three parts to be congruent? In this section and in the next section, we will show when three parts are sufficient to determine triangle congruency and when they are not.

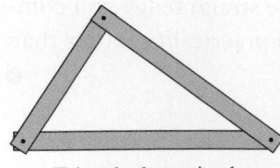

Triangle determined
by its three sides

Figure 12-7

Consider the triangle formed by attaching three segments, as in Figure 12-7. Such a triangle is *rigid*; that is, its size and shape cannot be changed. Because of this property, a manufacturer can make duplicates if the lengths of the sides are known. Once the sides are known, all the angles are automatically determined.

Many bridges and other structures that have exposed triangular frameworks demonstrate the practical use of the rigidity of triangles, as seen in Figure 12-8.

With the compass and straightedge alone, it is impossible in general to construct an angle if given only its measure. For example, an angle of measure 20° cannot be constructed with a compass and a straightedge only. Instead, a protractor or a geometry drawing utility or some other measuring tool must be used.

Side, Angle, Side Property (SAS)

We have seen that, given three segments, only one triangle can be constructed in the sense that all other triangles will be congruent to the one constructed. When given only two segments, more than one triangle can be constructed with sides congruent to these segments. Consider Figure 12-12(b), which shows three different triangles with sides congruent to the segments given in Figure 12-12(a). The length of the third side depends on the measure of the **included angle** formed by the two given sides.

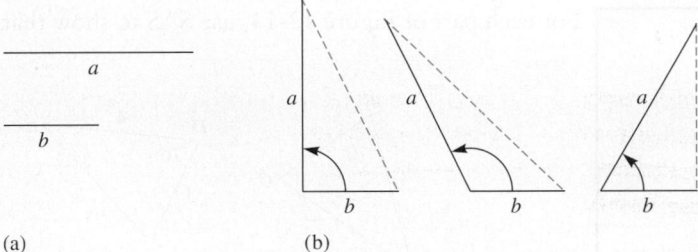

(a)   (b)

Figure 12-12

Constructions Involving Two Sides and an Included Angle of a Triangle

Figure 12-13 shows how to construct a triangle congruent to $\triangle ABC$ by using two sides $\overline{AB}$ and $\overline{AC}$ and the included angle, $\angle A$, formed by these sides. First, a ray with an arbitrary endpoint A' is drawn, and $\overline{A'C'}$ is constructed congruent to $\overline{AC}$. Then, $\angle A'$ is constructed so that $\angle A' \cong \angle A$, and B' is marked on the side of $\angle A'$ not containing C' so that $\overline{A'B'} \cong \overline{AB}$. Connecting B' and C' completes $\triangle A'B'C'$ so that $\triangle A'B'C'$ is to be congruent to $\triangle ABC$.

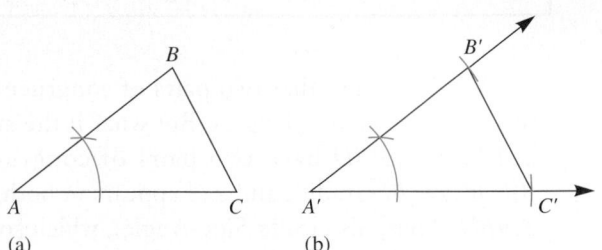

(a)   (b)

Figure 12-13

⊙ **Historical Note** ───

Three geometry compass and straightedge construction problems that concerned mathematicians for centuries were (1) trisecting a general angle, (2) duplicating a cube (constructing a cube whose volume is twice a given cube), and (3) squaring a circle (constructing a square with exactly the same area as a given circle). These constructions using only straightedge and compass were proved to be impossible in the 1800s. ●

It is the case that if we have lengths of two sides and the measure of the angle included between them, we could construct a unique triangle up to congruence (meaning that we construct infinitely many such triangles congruent to each other). We express this as **Side, Angle, Side (SAS)** congruence condition. This condition can be proved using the SSS property and is stated in the following theorem.

> ### Theorem 12-1:   Side, Angle, Side (SAS)
> If two sides and the included angle of one triangle are congruent to two sides and the included angle of another triangle, respectively, then the two triangles are congruent.

EXAMPLE 12-3

For each part of Figure 12-14, use SAS to show that the given pair of triangles are congruent.

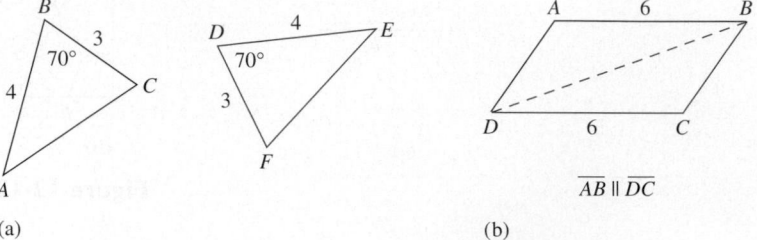

(a)   (b)

$\overline{AB} \parallel \overline{DC}$

Figure 12-14

Solution

a. $\triangle ABC \cong \triangle EDF$ by SAS because $\overline{AB} \cong \overline{ED}$, $\angle B \cong \angle D$ (the included angles), and $\overline{BC} \cong \overline{DF}$.

b. Because $\overline{AB} \cong \overline{CD}$ and $\overline{DB} \cong \overline{BD}$, we need either another side or another angle to show that the triangles are congruent. We know that $\overline{AB} \parallel \overline{DC}$. Since parallel segments $\overline{AB}$ and $\overline{DC}$ are cut by transversal $\overline{BD}$, we have congruent alternate interior angles $\angle ABD$ and $\angle CDB$. Therefore $\triangle ABD \cong \triangle CDB$ by SAS.

We have seen that two pairs of congruent sides and a pair of included angles (SAS) determine congruent triangles. But what if the angle is not an included angle? In Figure 12-15, $\triangle ABC$ and $\triangle ABD$ have two pairs of congruent sides and a common angle. They share $\overline{AB}$, $\overline{BC} \cong \overline{BD}$ (radii), and $\angle A$ appears in both triangles. However, $\triangle ABC$ is not congruent to $\triangle ABD$. Thus, SSA (Side-Side-Angle), which specifies two sides and a non-included angle, does not assure congruence.

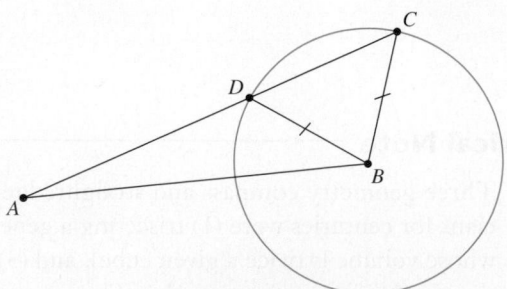

Figure 12-15

Nevertheless, if the angle in SSA is a right angle, then we have two right triangles with two pairs of congruent sides, in which the hypotenuse is opposite the right angle, as shown in Figure 12-16. In this case the triangles are congruent.

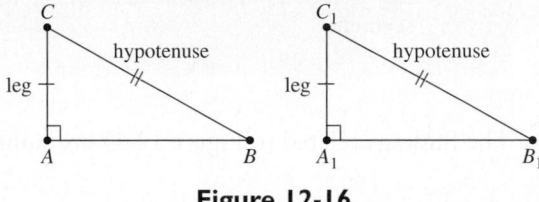

Figure 12-16

This congruence condition is referred to as the **Hypotenuse-Leg (HL)** congruence theorem.

> **Theorem 12-2:   Hypotenuse-Leg (HL)**
> If the hypotenuse and a leg of one right triangle are congruent to the hypotenuse and a leg of another right triangle, then the triangles are congruent.

It can be shown that the SSA condition is also valid when the non-included angle is obtuse.

Other congruence and noncongruence conditions for triangles are explored in Section 12-2.

Isosceles Triangle Properties

Conditions for congruence allow us to investigate a number of properties of triangles. For example, in Figure 12-17, consider isosceles triangle ABC with $\overline{AB} \cong \overline{AC}$. The symmetry of an isosceles triangle allows us to *paperfold* $\angle A$ into two congruent angles. We crease the triangle through vertex A so that vertex B "falls" on vertex C (this can be done because $\overline{AB} \cong \overline{AC}$). If point D is the intersection of the crease and $\overline{BC}$, then our folding assures that $\angle BAD \cong \angle CAD$. A ray such as $\overrightarrow{AD}$, that separates an angle into two congruent angles is the **angle bisector** of the angle. (Instead of paperfolding, we could start here by considering the angle bisector $\overrightarrow{AD}$ of $\angle A$). Thus, by SAS, $\triangle ABD \cong \triangle ACD$. Because corresponding parts must be congruent, $\angle B \cong \angle C, \overline{BD} \cong \overline{CD}$, and $\angle BDA \cong \angle CDA$. Because the last pair of angles are supplementary, it follows that each is a right angle. Consequently, $\overline{AD}$ is perpendicular to $\overline{BC}$ and bisects $\overline{BC}$. A line that is perpendicular to a segment and bisects it is the **perpendicular bisector** of the segment.

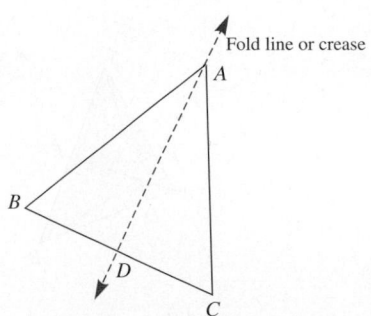

Figure 12-17

In Figure 12-17, notice that if point A is equidistant from the endpoints B and C, then A is on the perpendicular bisector of $\overline{BC}$. The converse of this statement is also true. The statement and its converse are given in Theorem 12-3.

Theorem 12-3

  a. Any point equidistant from the endpoints of a segment is on the perpendicular bisector of the segment.

  b. Any point on the perpendicular bisector of a segment is equidistant from the endpoints of the segment.

The findings related to Figure 12-17 are summarized in Theorem 12-4.

Theorem 12-4

For every isosceles triangle:

  a. the angles opposite the congruent sides are congruent (base angles of an isosceles triangle are congruent)

  b. the angle bisector of an angle formed by two congruent sides is the perpendicular bisector of the third side of the triangle.

The converse of part (a) of Theorem 12-4 is also true; that is: *If two angles of a triangle are congruent, the sides opposite these angles are congruent and the triangle is isosceles.*

Altitudes of a Triangle

An **altitude** of a triangle is the perpendicular segment from a vertex of the triangle to the line containing the opposite side of the triangle. In Figure 12-17 we see that in an isosceles triangle not only does the altitude to $\overline{BC}$ lie on the perpendicular bisector of $\overline{BC}$, but it also lies on the angle bisector of the opposit angle. Figure 12-18(a) and (b) shows the three altitudes $\overline{AE}$, $\overline{BF}$, and $\overline{CD}$ of an acute triangle and an obtuse triangle. Notice that in Figure 12-18(b), by definition the altitude $\overline{AE}$ is the perpendicular segment from the vertex A to the line containing the opposite side of the triangle. Similarly, the altitude $\overline{CD}$ is the perpendicular segment from vertex C to the line containing the opposite side, $\overline{AB}$. In Figure 12-18(a), we see that the three altitudes intersect in a single point in the interior of the triangle. In Figure 12-18(b), the three lines containing the altitudes intersect in a single point P in the exterior of the triangle.

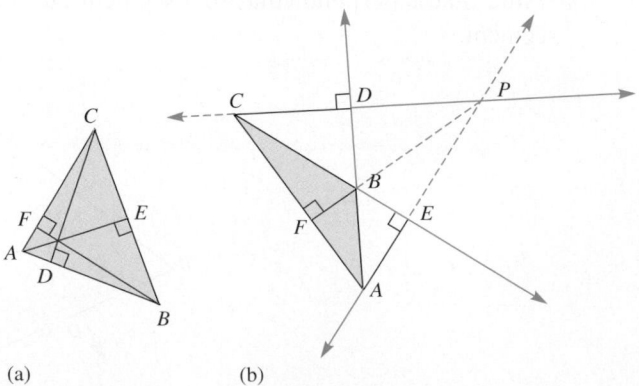

(a)   (b)

Figure 12-18

The concept of altitude and height is directly related to the concept of the **distance between a point and a line**, which is the length of the perpendicular segment from the point to the line. Thus in Figure 12-18(a), the heights *AE*, *BF*, and *CD* are respectively the distances from the vertices A, B, and C to the opposite sides. In Figure 12-18(b), the height AE of $\triangle ABC$ is the distance from A to $\overleftrightarrow{BC}$, the height CD is the distance from C to $\overleftrightarrow{AB}$, and BF is the distance from B to $\overleftrightarrow{AC}$. It can be shown that the distance from a point to a line is the shortest segment among all the segments connecting the given point to points on the line.

Construction of the Perpendicular Bisector of a Segment

Theorem 12-3 can be used to construct the perpendicular bisector of a segment by constructing any two points equidistant from the endpoints of the segment. Each point not on the segment is a vertex of an isosceles triangle, and the two points determine the perpendicular bisector of the segment. In Figure 12-19(a) we have constructed point P equidistant from A and B by drawing intersecting arcs with the same radius—one with center at A, and the other with center at B. Point Q is constructed similarly with two intersecting arcs. By Theorem 12-3(a), each point is on the perpendicular bisector of $\overline{AB}$. Because two points determine a unique line, $\overleftrightarrow{PQ}$ must be the perpendicular bisector of $\overline{AB}$. The construction can be achieved with fewer arcs as shown in Figure 12-19(b). We construct two intersecting arcs with the same radius—one with center at A and the other with center at B. The arcs intersect in points P and Q, each equidistant from A and B. Hence $\overleftrightarrow{PQ}$ is the perpendicular bisector of $\overline{AB}$. Notice that because $AP = PB = BQ = AQ$, $APBQ$ is a rhombus and the diagonals of a rhombus are perpendicular bisectors of each other.

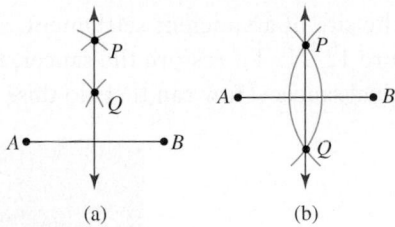

(a)   (b)

Figure 12-19

NOW TRY THIS 12-2

Theorem 12-3 and the above construction can be used to construct the perpendicular to a given line through a point on the line. Draw a line ℓ and mark a point P on the line and construct the perpendicular to the line through the point.

Construction of a Circle Circumscribed About a Triangle

In Figure 12-20(a), a circle is **circumscribed** about a given $\triangle ABC$; that is, every vertex of the triangle is on the circle ($\triangle ABC$ is **inscribed** in the circle). How can such a **circumcircle** be constructed?

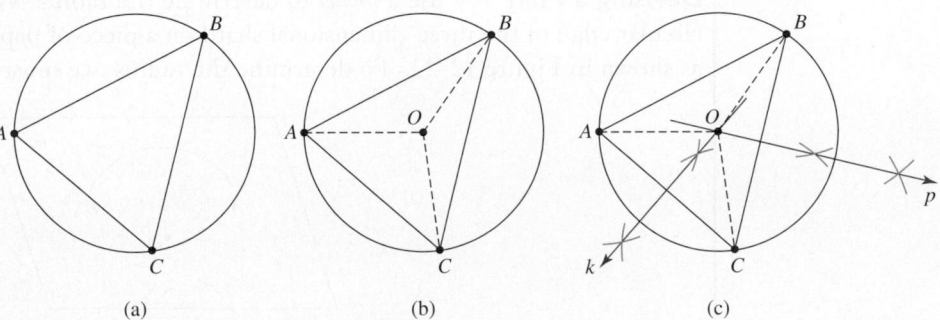

(a)   (b)   (c)

Figure 12-20

Suppose we know the location of center O of the circle, as in Figure 12-20(b). The properties of the center of a circle enable us to find its location. Because O is the center of the circle circumscribed about $\triangle ABC$, $OA = OC = OB$. The fact that $OA = OC$ implies that O is equidistant from the endpoints of segment $\overline{AC}$. Hence, by Theorem 12-3(a), O is on the perpendicular bisector, k, of $\overline{AC}$, as shown in Figure 12-20(c). Similarly, because $OC = OB$, O is on the perpendicular bisector, p, of $\overline{BC}$. Because O is on k and on p, it is the point of intersection of the two perpendicular bisectors. Thus, given $\triangle ABC$, we can construct the center of the circumscribed circle by constructing perpendicular bisectors of any two sides of the triangle. The point where the perpendicular bisectors intersect is the **circumcenter**. Thus, the required circle is the circle with the center at O and radius OA (the **circumradius**). The construction is shown in Figure 12-20(c).

Notice that because $OA = OB$, O is also on the perpendicular bisector of $\overline{AB}$ and therefore the three perpendicular bisectors of a triangle are **concurrent**, that is intersect in a single point.

 NOW TRY THIS 12-3

Construct an obtuse triangle, and then use the above reasoning to find the center of the circumcircle and construct the circle.

Problem Solving   **Archaeological Find**

At the site of an ancient settlement, archaeologists found a fragment of a saucer, as shown in Figure 12-21. To restore the saucer, the archaeologists needed to determine the radius of the original saucer. How can they do this?

Figure 12-21

Understanding the Problem   The border of the shard shown in Figure 12-21 is part of a circle. To reconstruct the saucer, we want to determine the center and the radius of the circle of which the shard is a part.

Devising a Plan   We use a *model* to determine the radius. We begin tracing an outline of the circular edge of the three-dimensional shard on a piece of paper. The result is an arc of a circle, as shown in Figure 12-22. To determine the radius, we must find the center O. We know that

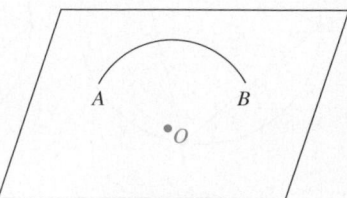

Figure 12-22

by connecting O to points of the circle, we obtain congruent radii. Also, each pair of radii determines an isosceles triangle. Consider points A, B, and O. Triangle ABO is isosceles and O is on the perpendicular bisector of $\overline{AB}$.

Carrying Out the Plan To find a line containing point O, construct a perpendicular bisector of $\overline{AB}$, as in Figure 12-23(a). Similarly, any other segment (for example, $\overline{AC}$) with endpoints on $\widehat{AB}$ has a perpendicular bisector containing O, as in Figure 12-23(b). The point of intersection of the two perpendicular bisectors is point O. To complete the problem, measure the length of either $\overline{OB}$ or $\overline{OA}$.

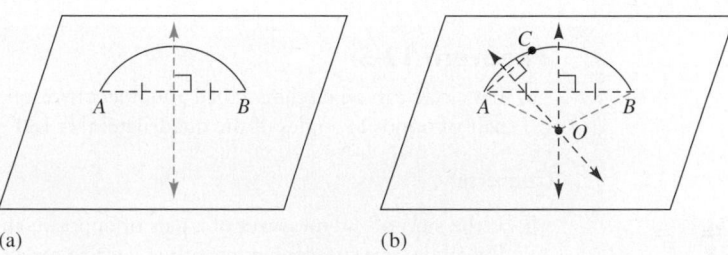

(a)   (b)

Figure 12-23

Looking Back Alternatively we can use the fact that a circle has infinitely many lines of symmetry, each containing the center of the circle. Thus we can find two lines of symmetry (by folding the arc onto itself and folding again a part of the arc onto itself) and the center of the circle is the point of intersection of the two lines. A related problem is, What would happen if the piece of pottery had been part of a sphere? Would the same ideas still work?

Circle Circumscribed About Some Quadrilaterals

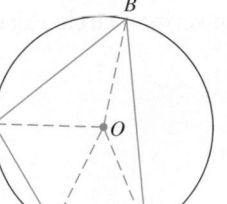

Figure 12-24

If a circle with center O is circumscribed about a quadrilateral $ABCD$ as in Figure 12-24, then the center O is equidistant from the endpoints of each of the sides and hence is on each of the four perpendicular bisectors of the sides. However, unlike in a triangle, the perpendicular bisectors of the sides of an arbitrary quadrilateral are not always concurrent (intersect in a single point) and hence it is not always possible to construct a circle circumscribed around every quadrilateral. For example, the four perpendicular bisectors of the sides of a parallelogram that is not a rectangle are not concurrent (Why?). If, on the other hand, the perpendicular bisectors of the sides of a quadrilateral are concurrent at some point, then that point is the center of the circle that circumscribes the quadrilateral. The radius of the circle is the distance from the point to any of the vertices.

 NOW TRY THIS 12-4

Construct the following quadrilaterals and the perpendicular bisectors of their sides. Based on your constructions why there is no circle that circumscribes the quadrilaterals?

a. A parallelogram that is not a rectangle
b. A non-isosceles trapezoid

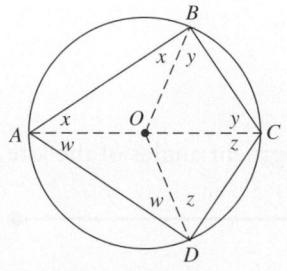

Figure 12-25

We next investigate a different criterion involving the angles of a convex quadrilateral for the quadrilateral to have a circumcircle. Suppose quadrilateral $ABCD$ has a circumcircle with center O as in Figure 12-25.

To find a relation among the angles of the quadrilateral, the measures of congruent base angles of the isosceles triangles AOB, BOC, COD, and DOC have been marked by x, y, z, and w, respectively. Notice that the sum of the measures of the opposite angles of the quadrilateral at B and D is

$$m(\angle B) + m(D) = x + y + w + z.$$

However, the sum of the measures of all the interior angles of any quadrilateral is 360°. Thus,

$$(x + y) + (y + z) + (z + w) + (w + x) = 360°,$$
$$2x + 2y + 2z + 2w = 360°,$$
$$x + y + z + w = 180°$$

Thus we have proved part (a) of Theorem 12-5 for the case when the center O is in the interior of the quadrilateral (if the center O of the circle is in the exterior of the quadrilateral or on a side, the result still holds and the proofs are similar).

Theorem 12-5

   a. If a circle can be circumscribed about a convex quadrilateral, then the sum of the measures of a pair of opposite angles of the quadrilateral is 180°; that is, the angles are supplementary.

Conversely,

   b. If the sum of the measures of a pair of opposite angles of a convex quadrilateral is 180° (that is, the angles are supplementary), then a circle can be circumscribed about the quadrilateral.

Notice that one pair of opposite angles of a quadrilateral is supplementary if, and only if, the other pair is supplementary (Why?).

EXAMPLE 12-4

Use Theorem 12-5 to find for which convex kites it is possible to construct a circumcircle.

Solution $ABCD$ is a convex kite as in Figure 12-26. By constructing the diagonal $\overline{AC}$ we obtain two isosceles triangles. Because the angles opposite congruent sides in each of the isosceles triangles are congruent, we see that $\angle A \cong \angle C$. By Theorem 12-5, a circumscribed circle exists if, and only if,

$m(\angle A) + m(\angle C) = 180°$. Because
$m(\angle A) = m(\angle C)$, we have
$m(\angle A) + m(\angle A) = 180°$ and hence
$m(\angle A) = 90°$.

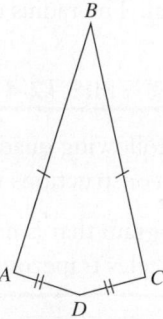

Figure 12-26

Thus, a circle can be circumscribed about a kite if, and only if, the congruent angles of the kite are right angles.

Assessment 12-1A

1. Given three points in the plane, is it always possible to find a point equidistant from the three points?
2. For what kind of triangles will the perpendicular bisectors of the sides intersect in the exterior of the triangle?
3. **a.** Use any tool to draw triangle ABC in which BC is greater than AC. Measure the angles opposite $\overline{BC}$ and $\overline{AC}$. Compare the angle measures. Repeat for different triangle. What did you find?
   b. Based on your finding in (a), make a conjecture concerning the lengths of sides and the measures of angles of a triangle.
4. In the following figure, the congruent sides and congruent right angles are marked. Prove that $ABCD$ is a parallelogram.

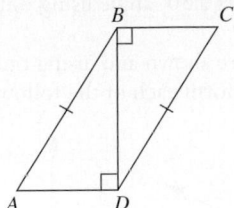

5. Use any tools to construct each of the following, if possible:
   a. A segment congruent to $\overline{AB}$ and an angle congruent to $\angle CAB$

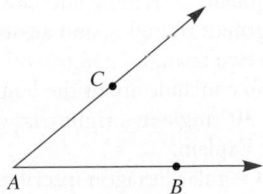

   b. A triangle with sides of lengths 2 cm, 3 cm, and 4 cm
   c. A triangle with sides of lengths 4 cm, 3 cm, and 5 cm (What kind of triangle is it?)
   d. A triangle with sides 4 cm, 5 cm, and 10 cm
   e. An equilateral triangle with sides 5 cm
   f. A triangle with sides 6 cm and 7 cm and an included angle of measure 75°
   g. A triangle with sides 6 cm and 7 cm and a nonincluded angle of measure 40°
   h. A triangle with sides 6 cm and 6 cm and a nonincluded angle of measure 40°
   i. A right triangle with legs 4 cm and 8 cm (the legs include the right angle)
   j. For each of the conditions in problem 5(b) through (i), does the given information determine a unique triangle? Explain why or why not.

6. Use the fact that the perpendicular bisector of the base of an isosceles triangle is also the angle bisector of the opposite angle to construct a 45° angle.
7. Refer to the figure shown and, using only a compass and a straightedge, perform each of the following:

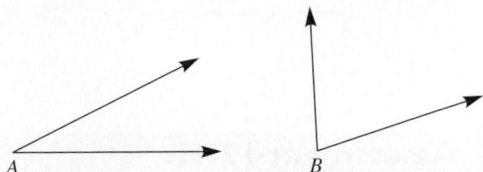

   a. Construct $\angle C$ so that $m(\angle C) = m(\angle A) + m(\angle B)$.
   b. Using the angles in (a), construct $\angle C$ so that $m(\angle C) = m(\angle B) - m(\angle A)$.
8. Is it possible to construct a right triangle with all sides congruent? Justify your answer.
9. Use the figure below to answer the following.
   a. Construct a circle with center O and two perpendicular diameters $\overline{AC}$ and $\overline{BD}$ of the circle. Why are the endpoints of the diameter vertices of a square inscribed in the circle?
   b. Construct the perpendicular bisector of $\overline{BC}$ and explain why $\overline{BE}$ is a side of a regular octagon inscribed in the circle.
   c. Use part (b) to construct the octagon.

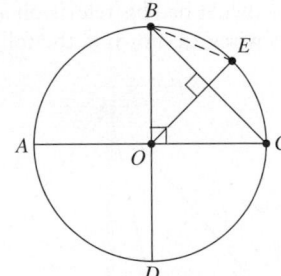

10. If you have a paper cutout of a square, how could you use only paperfolding to find the center of the circumscribing circle? What would be the radius of the circle?
11. Draw a line m and mark two points A and B on the same side of the line. Then use only a compass and straightedge to construct point C on m so that $AC = BC$. Describe how C can be found.
12. Draw a segment. Then use any instruments to construct a square whose side is congruent to the given segment. Then, using only a compass and a straightedge, construct the circle that circumscribes the square.

13. For which of the following figures is it possible to find a circle that circumscribes the figure? If it is possible to find such a circle, draw the figure and construct the circumscribing circle.
   a. A right triangle
   b. A regular hexagon
   c. A parallelogram that is not a rectangle
14. Using only a compass and a straightedge, perform each of the following:
   a. Construct an equilateral triangle with the following side $\overline{AB}$:

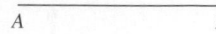

   A   B

   b. Construct a 60° angle.
   c. Construct a 120° angle.
15. Use the fact that an isosceles triangle is congruent to itself to answer the following.
   a. Write a congruence correspondence between an isosceles $\triangle ABC$ and itself when $AC = BC$.
   b. Use your answer in (a) to show that the base angles in an isosceles triangle are congruent.
16. Given 3 points in the plane is it always possible to find a fourth point that is the same distance from each of the 3 points? Justify your answer.

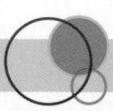

Assessment 12-1B

1. Given four points in the plane, no three of which are collinear, is it always possible to find a point in the same plane equidistant from the four points? Explain why or why not.
2. Is it possible to find in the plane four points equidistant from each other? Explain why or why not.
3. a. Use any tools to construct a triangle with one of the angles greater than another. Measure the sides opposite these angles. Repeat for different triangles.
   b. Based on your finding in part (a), make a conjecture concerning the measure of angles in a triangle and the lengths of sides.
   c. Based on your conjecture in part (b), explain why the hypotenuse is the longest side in a right triangle. Explain your reasoning.
4. A rural homeowner had his television antenna held in place by three guy wires, as shown in the following figure.

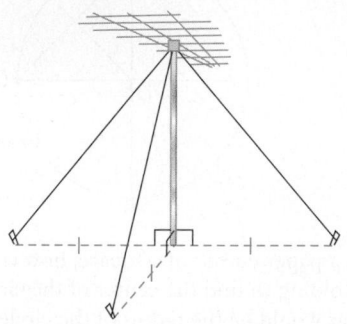

   a. If the stakes are on level ground and the distances from the stakes to the base of the antenna are the same, what is true about the lengths of the wires? Why?
   b. If the stakes are not on level ground yet are the same distance from the base of the antenna, explain whether you can make the same conclusion regarding the lengths of the wires.
5. For each of the following, determine whether the given conditions are sufficient to prove that $\triangle PQR \cong \triangle MNO$. Justify your answers.
   a. $\overline{PQ} \cong \overline{MN}, \overline{PR} \cong \overline{MO}, \angle P \cong \angle M$
   b. $\overline{PQ} \cong \overline{MN}, \overline{PR} \cong \overline{MO}, \overline{QR} \cong \overline{NO}$
   c. $\overline{PQ} \cong \overline{MN}, \overline{PR} \cong \overline{MO}, \angle Q \cong \angle N$
6. Use the fact that the perpendicular bisector of the base of an isosceles triangle is also the angle bisector of the opposite angle, to construct a 30° angle using only a compass and straightedge.
7. Refer to the figure shown and, using only a compass and a straightedge, perform each of the following:

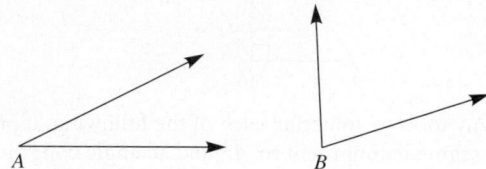

   A   B

   a. Construct $\angle C$ so that $m(\angle C) = 2m(\angle A) + m(\angle B)$.
   b. Construct $\angle C$ so that $m(\angle C) = 2m(\angle B) - m(\angle A)$.
8. Construct an equilateral triangle and one of its altitudes to obtain two congruent triangles, and answer the following:
   a. Why are the two triangles congruent?
   b. What can you conclude about the length of the leg opposite the 30° angle in a right triangle in relation to its hypotenuse? Explain.
9. a. $ABCDEF$ is a regular hexagon inscribed in a circle of radius r. The perpendicular bisector of $\overline{BC}$ intersects the arc $\overset{\frown}{BC}$ at G. Explain why $\overline{BG}$ is a side of a regular dodecagon (12-gon) inscribed in the circle.
   b. Prove that $\triangle BOC$ is equilateral.
   c. Construct a regular dodecagon.

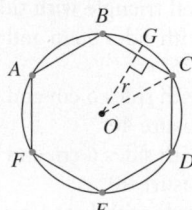

10. Let $ABCD$ be a square with diagonals $\overline{AC}$ and $\overline{BD}$ intersecting in point F, as shown in the figure.

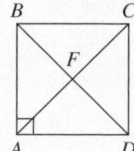

a. What is the relationship between point *F* and the diagonals $\overline{BD}$ and $\overline{AC}$? Why?

b. What are the measures of angles *BFA* and *AFD*? Why?

11. Using only a compass and a straightedge describe how to construct the center of a given circle if the center is not marked.

12. Use any instruments to construct a rectangle that is not a square. Then, using only a compass and a straightedge, construct the circle that circumscribes the rectangle.

13. For which of the following figures is it possible to find a circle that circumscribes the figure? If it is possible to find such a circle, draw the figure using any tools and construct the circumscribing circle.

   a. A rhombus
   b. A regular decagon

14. Use only a compass and straightedge and Theorem 12-14(b) in part (b) to construct the following:

   a. An isosceles triangle in which one of the angles measures 120°.
   b. A 150° angle.

15. a. Write all possible true congruence correspondences between an equilateral triangle and itself.

   b. Write all possible true congruence correspondences between a rectangle and itself.

16. a. Construct any right triangle *ABC*.

   b. Construct the circle that circumscribes the triangle in part (a) and conjecture the location of the center.

   c. Based on your conjecture in part (b) find the radius of the circle in terms of the hypotenuse.

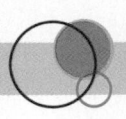

Mathematical Connections 12-1

Communication

1. In a circle with center *A* and radius *AB*, let *P* be a point of $\overline{AB}$ that is not an endpoint. Explain whether or not *P* is on the circle.

2. Write an argument to convince the class that Theorem 12-4 is true.

3. In the following kite, congruent segments are shown with tick marks.

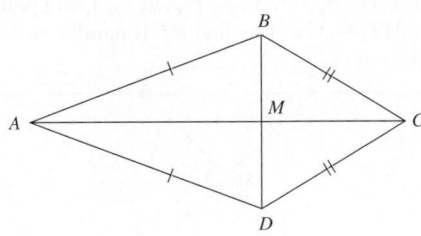

   a. Argue that the diagonal $\overline{AC}$ bisects ∠*A* and ∠*C*.
   b. Let *M* be the point on which the diagonals of kite *ABCD* intersect. Measure ∠*AMD* and make a conjecture concerning an angle formed by the diagonals of a kite. Justify your conjecture.
   c. Show that $\overline{BM} \cong \overline{MD}$.

4. If the kite in exercise 3 were concave, do the same answers to parts (a) through (c) hold? Justify your answer.

5. Are all rectangles whose diagonals are 19 in. long congruent? Justify your answer.

6. In view of the fact that each of the interior angles of a regular hexagon measures 120°, is it possible to construct an irregular hexagon with all the interior angles measuring 120°? Justify your answer.

7. Explain why the quadrilateral *ABCD* is a kite.

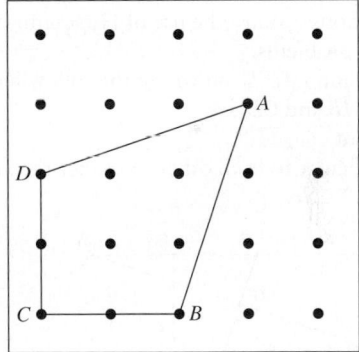

8. To construct the perpendicular to a line *ℓ* through a point *P* on the line, use a drafting triangle to align one of the legs of the triangle with the line, as shown in the figure. Next move the triangle along the leg until the other leg passes through the given point, *P*. Then answer the following questions. Will the construction described above work if the point *P* is not on line *ℓ*? Why or why not?

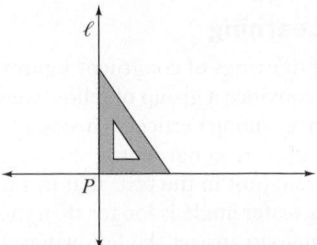

9. In the following drawing, a compass is used to draw circle O. A point P is marked on the circle. Using the compass with the same setting that was used to draw the circle and using point P as the center, draw an arc that intersects the original circle in two points A and B. Repeat the process using points A and B on the circle. Continue this process.

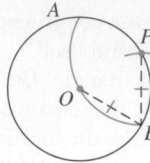

   a. What do you observe?
   b. If you connected the points on the circle in order with line segments, starting at P and going in a clockwise fashion around the circle, what figure would you draw? Why?

Open-Ended

10. a. Find at least five examples of congruent objects.
    b. Find at least five examples of similar objects that are not congruent.
11. Design a quilt pattern that involves rep-tiles or find a pattern and describe the rep-tiles involved.
12. Search the web or other sources for at least three problems whose solutions require the use of Hypotenuse-Leg theorem. Record the problems.
13. Use $\triangle AFB$ and $\triangle CED$ shown on the following dot paper to verify that $\overline{AB}$ and $\overline{CD}$ are
    a. congruent
    b. perpendicular to each other (consider the angles in $\triangle BGC$).

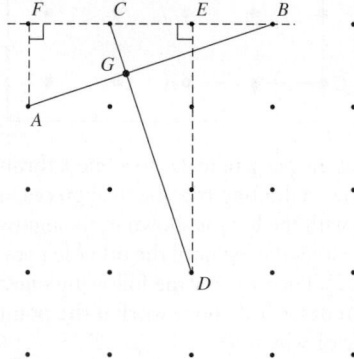

Cooperative Learning

14. Make a set of drawings of congruent figures and write an argument to convince a group of fellow students that the correspondence among vertices is necessary for determining that two triangles are congruent.
15. Use the theorem (not in the text) that in a triangle the side opposite the greater angle is longer than the side opposite the smaller angle to answer the following:
    a. Explain why the hypotenuse of a right triangle is longer than a leg of the triangle.
    b. Define the distance between a point not on a line and the line. Use part (a) to show that it is the length of the shortest segment among all the segments connecting the given point to points on the line.

Questions from the Classroom

16. On a test, a student wrote $AB \cong CD$ instead of $\overline{AB} \cong \overline{CD}$. Is this answer correct? Why?
17. A student asks if there are any constructions that cannot be done with a compass and a straightedge. How do you answer?
18. A student asks for a mathematical definition of congruence that holds for all figures. How do you respond?
19. A student constructed a line ℓ and three congruent segments $\overline{AD}$, $\overline{DE}$, and $\overline{EB}$ on the line. Then she constructed point C so that $AC = BC$. She claimed that by drawing $\overrightarrow{CD}$ and $\overrightarrow{CE}$, as shown in the following figure, she has trisected (divided into three congruent angles) $\angle ACB$. What is your response?

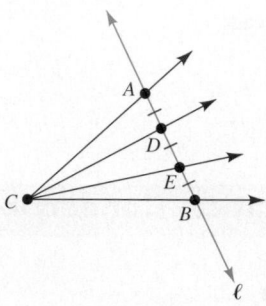

20. Claire claims that she discovered the following procedure for constructing a line through a given point P parallel to a given line ℓ. She would like to know how to prove that her construction is correct. How do you respond?

    Claire: I choose any point Q on line ℓ so that $\overline{PQ}$ is not perpendicular to ℓ and construct the perpendicular bisector of $\overline{PQ}$, intersecting ℓ at S and $\overline{PQ}$, at M. I find point T on $\overrightarrow{SM}$ such that $MT = MS$. The line PT is parallel to ℓ.

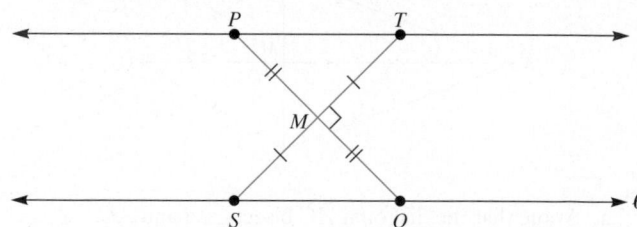

21. Zara claims that in spite of the fact that every triangle is congruent to itself, the statement $\triangle ABC \cong \triangle BAC$ is generally false. However, if $AC = BC$, the statement is true. Is she correct? If so, why?
22. A wooden gate is illustrated in the following figure. A student asks for the purpose of the diagonal boards on the gate. How do you respond?

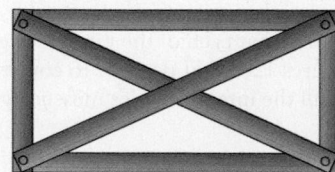

GSP/GeoGebra Activities

23. Use *GSP/GeoGebra* Lab 4 activities 1–7 to investigate congruence and noncongruence of triangles.

24. **a.** Use GSP Lab 1 activities 10 and 11 as well as Lab 2 activity 1 or Geogebra Lab 3 activities 1–4 to investigate measuring segments and angles.
    b. Use GSP or Geogebra to display a grid (choose *graph-show grid* in GSP or View > Grid in Geogebra). Then construct the triangles in problem 13 above and measure some of their sides and angles as well as ∠*CGB* to verify the claims in parts (a) and (b) of the problem.
25. Use GSP or Geogebra Lab 5 activities 1–4 to investigate the construction of a circle circumscribed about a triangle.
26. Use GSP or Geogebra activities 1–3 to investigate circles circumscribing rectangles and quadrilaterals.

Trends in Mathematics and Science Study (TIMSS) Questions

In square *EFGH*, which of these is FALSE?
  a. △*EIF* and △*EIH* are congruent.
  b. △*GHI* and △*GHF* are congruent.
  c. △*EFH* and △*EGH* are congruent.
  d. △*EIF* and △*GIH* are congruent.

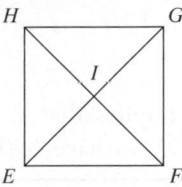

TIMSS, Grade 8, 2003

ABCD is a trapezoid.

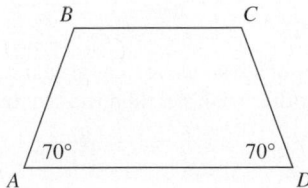

Another trapezoid, *GHIJ* (not shown), is congruent (the same size and shape as) to *ABCD*. Angles *G* and *J* each measure 70°. Which of these could be true?
  a. *GH* = *AB*
  b. Angle *H* is a right angle.
  c. All sides of *GHIJ* are the same length.
  d. The perimeter of *GHIJ* is 3 times the perimeter of *ABCD*.
  e. The area of *GHIJ* is less than the area of *ABCD*.

TIMSS, Grade 8, 2003

National Assessment of Educational Progress (NAEP) Question

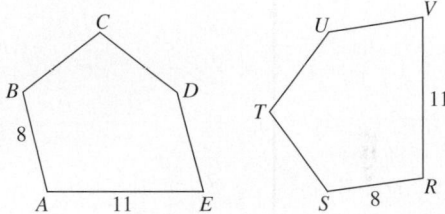

In the figure above, polygons *ABCDE* and *RSTUV* are congruent. Which side must have the same length as side *BC*?
  a. $\overline{CD}$
  b. $\overline{DE}$
  c. $\overline{ST}$
  d. $\overline{TU}$
  e. $\overline{UV}$

NAEP, Grade 8, 2009

12-2 Additional Congruence Properties

Angle, Side, Angle Property (ASA)

Triangles can be determined to be congruent by SSS and SAS. Can a triangle be constructed congruent to a given triangle by using two angles and a side? Figure 12-27 shows the construction of a triangle *A'B'C'* such that $\overline{A'C'} \cong \overline{AC}$, ∠*A'* ≅ ∠*A*, and ∠*C'* ≅ ∠*C*. It seems that △*A'B'C'* ≅ △*ABC*. This construction illustrates the **Angle, Side, Angle (ASA)** property of congruence.

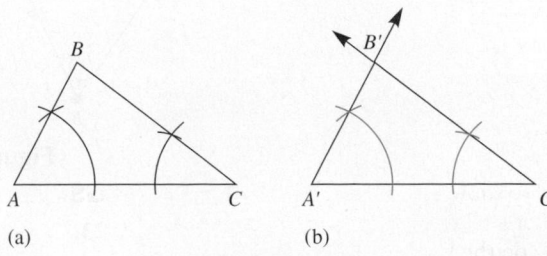

(a)   (b)

Figure 12-27

Theorem 12-6:   Angle, Side, Angle (ASA)

If two angles and the included side of one triangle are congruent to two angles and the included side of another triangle, respectively, then the triangles are congruent.

EXAMPLE 12-5

In Figure 12-28, $\triangle ABC$ and $\triangle DEF$ have two pairs of angles congruent and a pair of sides congruent. Show that the triangles are congruent.

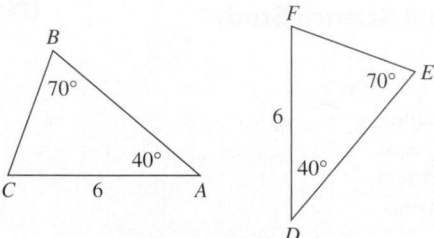

Figure 12-28

Solution Notice that $\angle A \cong \angle D$ and $\angle B \cong \angle E$, which implies that $\angle C \cong \angle F$ because the measure of each is $180° - (70 + 40)°$. Then, since $\overline{AC} \cong \overline{DF}$, we have $\triangle ABC \cong \triangle DEF$ by ASA.

Example 12-5 is a special case of the following theorem (which can be justified in a similar way).

Theorem 12-7:   Angle, Angle, Side (AAS)

If two angles and a side opposite one of these two angles of a triangle are congruent to the two corresponding angles and the corresponding side in another triangle, then the two triangles are congruent.

Notice in Theorem 12-7 the emphasis that the side in one triangle must correspond to the side in the other triangle. If that is not the case, two angles and a side do not ensure congruence, as shown in Figure 12-29, where a side and two angles in $\triangle ABC$ are congruent to a side and two angles in $\triangle DEF$. In that figure, $\overline{AB} \cong \overline{EF}$, $\angle A \cong \angle D$, and $\angle E \cong \angle B$, but the triangles are not congruent. Side $\overline{AB}$ is not opposite one of the angles mentioned. Consequently, the condition AAS does not ensure congruence unless the side is opposite one of the two angles.

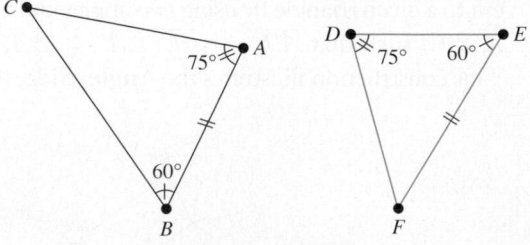

Figure 12-29

EXAMPLE 12-6

Show that the triangles in each part of Figure 12-30 are congruent.

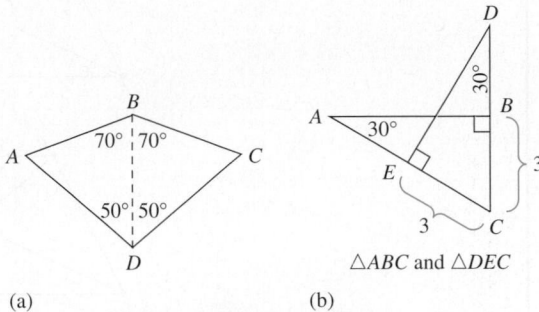

(a)   (b)

$\triangle ABC$ and $\triangle DEC$

Figure 12-30

Solution

a. $\angle ABD \cong \angle CBD$, $\overline{BD} \cong \overline{BD}$, and $\angle ADB \cong \angle CDB$. Consequently, by ASA, $\triangle ABD \cong \triangle CBD$.

b. Since $\angle C \cong \angle C$, $\angle ABC \cong \angle DEC$, and $\overline{EC} \cong \overline{BC}$, by AAS, $\triangle ABC \cong \triangle DEC$.

The ASA congruency condition can be used to prove that in a parallelogram (a quadrilateral with each pair of opposite sides parallel) opposite sides and opposite angles are congruent. In Figure 12-31 $ABCD$ is a parallelogram. Because a diagonal divides the parallelogram into two triangles, we construct $\overline{AC}$ as shown in Figure 12-31. Now $\triangle ABC \cong \triangle CDA$ by ASA because $\overline{AC} \cong \overline{CA}$ and the pairs of equally marked angles are congruent as they are alternate angles created by pairs of parallel sides and the transversal $\overleftrightarrow{AC}$. Therefore $\overline{AB} \cong \overline{CD}$ and $\overline{BC} \cong \overline{DA}$, by CPCTC. From the same congruent triangles we also get that $\angle B \cong \angle D$. By constructing the diagonal $\overline{BD}$ instead of $\overline{AC}$ we can show that $\angle BAD \cong \angle DCB$.

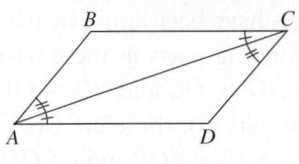

Figure 12-31

Thus we have proved the following:

Theorem 12-8

The opposite sides and the opposite angles of a parallelogram are congruent.

EXAMPLE 12-7

a. Use the definition of a parallelogram and Theorem 12-8 to prove that the diagonals of a parallelogram bisect each other; that is, in Figure 12-32(a), show that $AO = OC$ and $BO = OD$.

b. Draw a line through the point O where the diagonals of a parallelogram intersect, as in Figure 12-32(b). The line intersects the opposite sides of the parallelogram at points P and Q. Prove that $\overline{OP} \cong \overline{OQ}$.

c. In Figure 12-32(c) prove that $\overline{SQ} \cong \overline{PT}$.

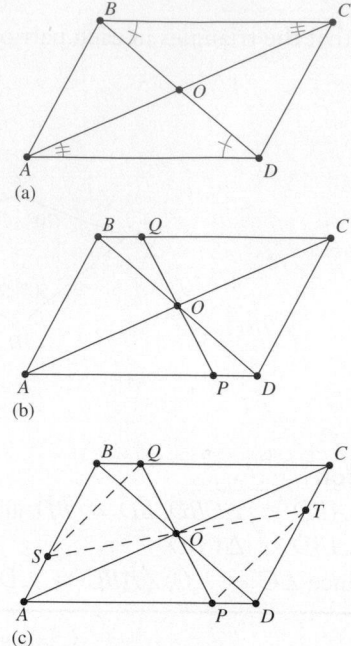

Figure 12-32

Solution

a. In Figure 12-32(a), we show that $\overline{AO} \cong \overline{OC}$ and $\overline{BO} \cong \overline{OD}$. These segments appear in two pairs of triangles that seem to be congruent. One pair is $\triangle AOD$ and $\triangle COB$. To prove these triangles congruent, we use Theorem 12-7 to conclude that $\overline{AD} \cong \overline{CB}$. We have no information about the other corresponding sides; however, we know that $\overline{BC} \parallel \overline{AD}$. Consequently, $\angle CBO \cong \angle ADO$, as these are alternate interior angles formed by the parallel lines and the transversal $\overleftrightarrow{BD}$. Similarly, using the transversal $\overleftrightarrow{AC}$, we have $\angle OAD \cong \angle OCB$ (congruent angles have been similarly marked in Figure 12-32(a)). Now $\triangle AOD \cong \triangle COB$ by ASA. Corresponding parts in these triangles are congruent so $\overline{AO} \cong \overline{OC}$ and $\overline{BO} \cong \overline{OD}$, and therefore $AO = OC$ and $BO = OD$.

b. As in part (a), there are two pairs of triangles that have segments $\overline{OQ}$ and $\overline{OP}$ as sides; one such pair is $\triangle AOP$ and $\triangle COQ$. In these triangles, $\overline{AO} \cong \overline{CO}$ (as was proved in part (a)). As in part (a), $\angle OAP \cong \angle OCQ$. Also, $\angle AOP \cong \angle COQ$ because these are vertical angles. Thus, $\triangle AOP \cong \triangle COQ$ and $\overline{OP} \cong \overline{OQ}$ because these are corresponding parts in these congruent triangles.

c. $\overline{OS} \cong \overline{OT}$ and $\overline{OQ} \cong \overline{OP}$ (using the results of part (b)). $\angle SOQ \cong \angle TOP$ because these are vertical angles. These imply $\triangle SQO \cong \triangle TPO$ by SAS. Consequently $\overline{SQ} \cong \overline{TP}$ by CPCTC.

NOW TRY THIS 12-5

If the diagonals of a quadrilateral bisect each other, must the quadrilateral be a parallelogram? Explain why or why not.

Using properties of congruent triangles, we can deduce various properties of quadrilaterals. Table 12-1 summarizes the definitions and lists some properties of six quadrilaterals, not all of which have been proved. These and other properties of quadrilaterals are further investigated in Assessment 12-2.

Table 12-1

Quadrilateral and Its Definition	Theorems about the Quadrilateral
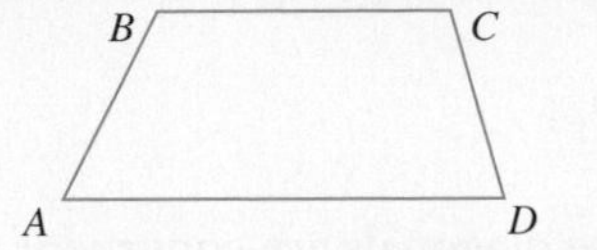 *Trapezoid:* A quadrilateral with at least one pair of parallel sides	Consecutive angles between parallel sides are supplementary.
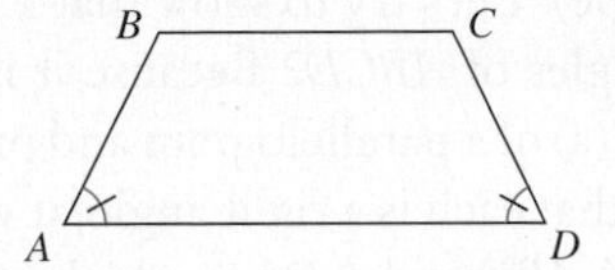 *Isosceles trapezoid:* A trapezoid with a pair of congruent base angles	**a.** Each pair of base angles are congruent. **b.** A pair of opposite sides are congruent. **c.** If a trapezoid has congruent diagonals it is isosceles.
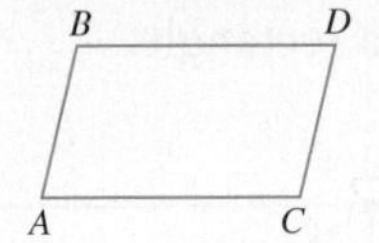 *Parallelogram:* A quadrilateral in which each pair of opposite sides is parallel	**a.** A parallelogram is a trapezoid in which each pair of opposite sides are parallel. **b.** Opposite sides are congruent. **c.** Opposite angles are congruent. **d.** Diagonals bisect each other. **e.** A quadrilateral in which the diagonals bisect each other is a parallelogram.
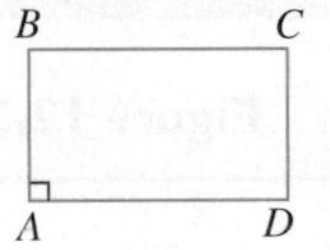 *Rectangle:* A parallelogram with a right angle	**a.** A rectangle has all the properties of a parallelogram. **b.** All the angles of a rectangle are right angles. **c.** A quadrilateral in which all the angles are right angles is a rectangle. **d.** The diagonals of a rectangle are congruent and bisect each other. **e.** A quadrilateral in which the diagonals are congruent and bisect each other is a rectangle.
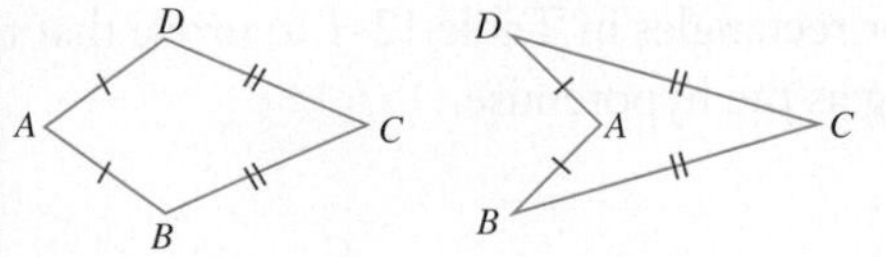 convex kite concave kite *Kite:* A quadrilateral with two adjacent sides congruent and the other two sides also congruent	**a.** Lines containing the diagonals are perpendicular to each other. **b.** A line containing one diagonal ($\overline{AC}$ in the figure) is a bisector of the other diagonal. **c.** One diagonal ($\overline{AC}$ in the figure) bisects nonconsecutive angles. **d.** A quadrilateral in which the line containing one diagonal is the perpendicular bisector of the other diagonal is a kite.
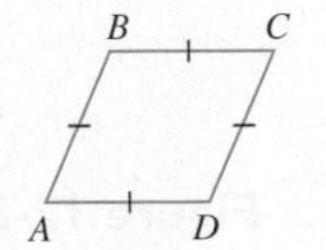 *Rhombus:* A parallelogram with two adjacent sides congruent	**a.** A rhombus has all the properties of a parallelogram and a kite. **b.** A quadrilateral in which all the sides are congruent is a rhombus. **c.** The diagonals of a rhombus are perpendicular to and bisect each other. Each diagonal bisects opposite angles. **d.** A quadrilateral in which the diagonals are perpendicular to and bisect each other is a rhombus.
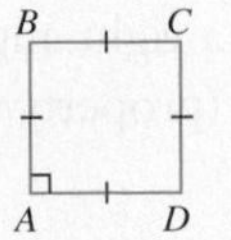 *Square:* A rectangle with two adjacent sides congruent	**a.** A square has all the properties of a parallelogram, a rectangle, and a rhombus. **b.** A rhombus with a right angle is a square.

The proofs of some of the theorems listed in Table 12-1 will be explored in the Assessments; meanwhile, in the following examples we assume some theorems from the table and prove other theorems.

EXAMPLE 12-8

Prove that a parallelogram whose diagonals are congruent is a rectangle.

Solution   We need to prove that one of the angles of the parallelogram in Figure 12-33 is a right angle (see definition of a rectangle). Let's try to show that $\angle BAD$ in Figure 12-33 is a right angle. What do we know about the angles of $ABCD$? Because it is a parallelogram, $\angle BAD$ and $\angle CDA$ are supplementary (see property (a) of a parallelogram and preperty of a trapezoid in Table 12-1 or Example 11-11). Thus, to show that each is a right angle, it will suffice to prove that the angles are congruent. These angles are in $\triangle ABD$ and $\triangle DCA$, which are congruent by SSS since $\overline{AB} \cong \overline{DC}$ (in a parallelogram opposite sides are congruent), $\overline{AD} \cong \overline{DA}$, and $\overline{BD} \cong \overline{CA}$ (given). Therefore, $\angle BAD \cong \angle CDA$ by CPCTC. Since these angles are supplementary and congruent, each must be a right angle. Therefore, $ABCD$ is a rectangle.

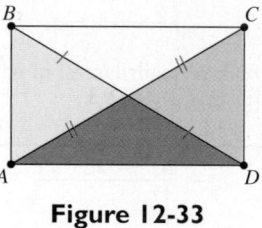

Figure 12-33

EXAMPLE 12-9

In Figure 12-34(a), $\triangle ABC$ is a right triangle, and $\overline{CD}$ is a **median** (segment connecting a vertex to the midpoint of the opposite side) to the hypotenuse $\overline{AB}$. Use property (e) for parallelograms in Table 12-1 and property (d) for rectangles in Table 12-1 to prove that the median to the hypotenuse in a right triangle is half as long as the hypotenuse.

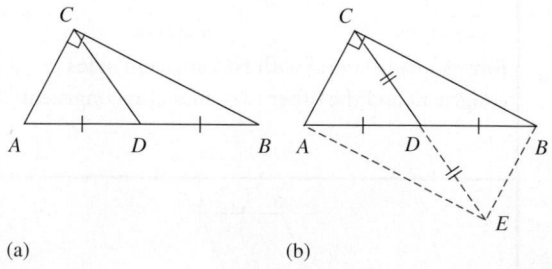

(a)   (b)

Figure 12-34

Solution   We are given that $\angle C$ is a right angle and D is the midpoint of $\overline{AB}$. We need to prove that $2CD = AB$. In Figure 12-34(b), we extend $\overline{CD}$ by its length to obtain $CE = 2CD$. Because D is the midpoint of $\overline{AB}$ and also the midpoint of $\overline{CE}$ (by our construction), the diagonals $\overline{AB}$ and $\overline{CE}$ of quadrilateral $ACBE$ bisect each other. Thus $ACBE$ is a parallelogram (property (e) for a parallelogram in Table 12-1). We also know that $\angle C$ is a right angle (given). Thus, $ACBE$ is a rectangle. Because the diagonals of a rectangle are congruent (property (d) for rectangles in Table 12-1), $CE = AB$ and therefore $2CD = AB$, or $CD = \frac{1}{2}AB$.

Congruence of Quadrilaterals

What minimum conditions determine congruence of quadrilaterals? For example, does the SSS condition for triangles extend to the SSSS condition for quadrilaterals? Figure 12-35 shows a parallelogram $ABCD$ with hinges at all the vertices and vertices A and D fixed so that their position cannot change. Dragging side $\overline{BC}$ to the right gives a new parallelogram AB_1C_1D with the same sides lengths but different angles, and therefore infinitely many noncongruent parallelograms. Thus knowing four sides does not determine a unique quadrilateral and therefore there is no SSSS congruency condition for quadrilaterals.

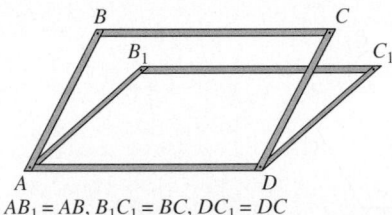

$AB_1 = AB, B_1C_1 = BC, DC_1 = DC$

Figure 12-35

One way to determine a quadrilateral is to give directions for drawing it. Start with a side and tell by what angle to turn at the end of each side to draw the next side (the turn is by an exterior angle). Thus, SASAS seems to be a valid congruence condition for quadrilaterals (this condition can be proved by dividing a quadrilateral into triangles and using what we know about congruence of triangles).

Assessment 12-2A

1. Use any tools to construct each of the following, if possible:
   a. A triangle with angles measuring 60° and 70° and an included side of 8 in.
   b. A triangle with angles measuring 60° and 70° and a nonincluded side of 8 cm on a side of the 60° angle
   c. A triangle with angles measuring 30°, 70°, and 80°

2. For each of the conditions in problem 1(a) through (c), is it possible to construct two noncongruent triangles? Explain why or why not.

3. For each of the following, determine whether the given conditions are sufficient to prove that $\triangle PQR \cong \triangle MNO$. Justify your answers.
   a. $\angle Q \cong \angle N, \angle P \cong \angle M, \overline{PQ} \cong \overline{MN}$
   b. $\angle R \cong \angle O, \angle P \cong \angle M, \overline{QR} \cong \overline{NO}$

4. A parallel ruler, shown as follows, can be used to draw parallel lines. The distance between the parallel segments $\overline{AB}$ and $\overline{DC}$ can vary. The ruler is constructed so that the distance between A and B equals the distance between D and C. The distance between A and C is the same as the distance between B and D. Explain why $\overline{AB}$ and $\overline{DC}$ are always parallel.

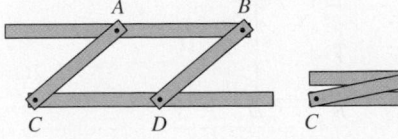
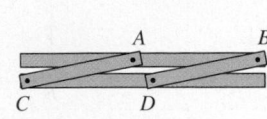

5. In each of the following, identify the word in the set {*parallelogram, rectangle, rhombus, trapezoid, kite,* or *square*} needed to make the resulting sentence true. If none of the words makes the sentence true, answer "none".
   a. A quadrilateral is a _____ if, and only if, its diagonals bisect each other.
   b. A quadrilateral is a _____ if, and only if, its diagonals are congruent.
   c. A quadrilateral is a _____ if, and only if, its diagonals are perpendicular.

6. Create several trapezoids that have a pair of opposite nonparallel sides congruent. Measure all angles and make a conjecture about the relationships among pairs of angles.

7. Classify each of the following statements as true or false. If the statement is false, provide a counterexample.
   a. The diagonals of a square are perpendicular bisectors of each other.
   b. If all sides of a quadrilateral are congruent, the quadrilateral is a rhombus.
   c. If a rhombus is a square, it must also be a rectangle.
   d. A trapezoid is a parallelogram.

8. a. Construct quadrilaterals having exactly one, two, or four right angles.
   b. Can a quadrilateral have exactly three right angles? Why?
   c. Can a parallelogram have exactly two right angles? Why?

9. Quadrilateral $ABCD$ pictured is a kite. Prove that $\overleftrightarrow{BD} \perp \overleftrightarrow{AC}$.

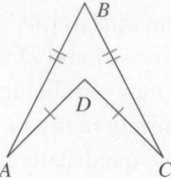

10. Classify each of the following figures formed from regular hexagon $ABCDEF$.

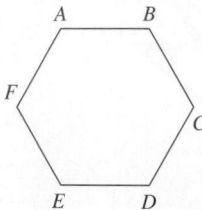

   a. Quadrilateral $ABDE$
   b. Quadrilateral $ABEF$

11. How many different rotational symmetries does a regular hexagon have? Why?

12. The game of Triominoes has equilateral-triangular playing pieces with numbers at each vertex, shown as follows:

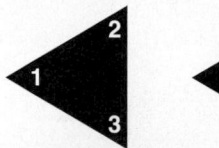 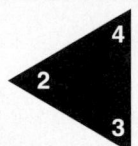

   If two pieces are placed together as shown in the following figure, explain what type of quadrilateral is formed:

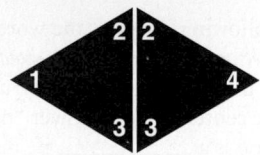

13. A **sector** of a circle is a section shaped of a piece of pie, bounded by two radii and an arc. What is a minimal set of conditions for determining that two sectors of the same circle are congruent?

14. In the rectangle $ABCD$ shown, X and Y are midpoints of the given sides and $DP = AQ$.

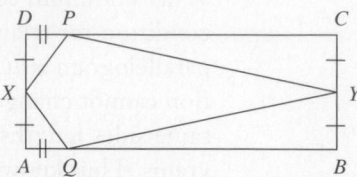

   a. What type of quadrilateral is $PYQX$? Prove your answer.
   b. If points P and Q are moved at a constant rate and in the same direction along $\overline{DC}$ and $\overline{AB}$, respectively, does this change your answer in part (a)? Why or why not?

15. Draw two quadrilaterals such that two angles and an included side in one quadrilateral are congruent, respectively, to two angles and an included side in the other quadrilateral but, the quadrilaterals are not congruent.

16. Using a straightedge and a compass, construct any convex kite. Then construct a second convex kite that is not congruent to the first but whose sides are congruent to the corresponding sides of the first kite.

17. What type of figure is formed by joining the midpoints of a rectangle (see the following figure)? Justify your answer.

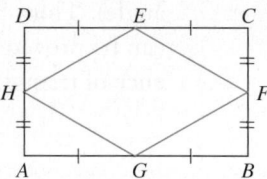

18. What minimum information is sufficient to determine congruency for each of the following?
   a. Two squares
   b. Two rectangles
   c. Two parallelograms

19. Prove each of the following:
   a. A quadrilateral with a pair of opposite sides parallel and congruent is a parallelogram.
   b. A quadrilateral whose diagonals bisect each other is a parallelogram.

20. Describe a set of minimal conditions to determine whether two regular polygons are congruent.

Assessment 12-2B

1. Use any tools to construct each of the following, if possible:
   a. A right triangle with one acute angle measuring 75° and a leg of 5 cm on a side of the 75° angle
   b. A triangle with angles measuring 30°, 60°, and 90°

2. For each of the conditions in problem 1(a) and (b), is it possible to construct two noncongruent triangles? Explain why or why not.

3. For each of the following, determine whether the given conditions are sufficient to prove that $\triangle PQR \cong \triangle MNO$. Justify your answers.

   a. $\overline{PQ} \cong \overline{MN}, \overline{PR} \cong \overline{MO}, \angle N \cong \angle Q$
   b. $\angle P \cong \angle M, \angle Q \cong \angle N, \angle R \cong \angle O$

4. k and ℓ are two lines with A and C on k and B and D on ℓ. If $AB = CD$ and $\overline{AB}$ and $\overline{CD}$ are perpendicular to line k, why must lines k and ℓ be parallel?

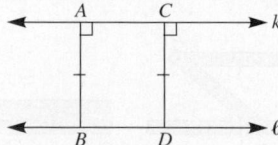

5. In each of the following, choose one word, *parallelogram, rectangle, rhombus, trapezoid, kite,* or *square,* so that the resulting sentence is true. If none of the words makes the sentence true, answer "none."
   a. A quadrilateral is a _____ if, and only if, its diagonals are congruent and bisect each other.
   b. A quadrilateral is a _____ if, and only if, its diagonals are perpendicular and bisect each other.
   c. A quadrilateral is a _____ if, and only if, its diagonals are congruent and perpendicular and they bisect each other.
   d. A quadrilateral is a _____ if, and only if, a pair of opposite sides is parallel and congruent.

6. *ABCD* in the following figure is a trapezoid in which $\overline{AD} \parallel \overline{BC}$ and $AD > BC$. Prove that if $AB = CD$ then $\angle A \cong \angle D$. (*Hint:* Through *C* draw a line parallel to $\overline{AB}$.)

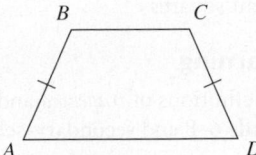

7. Classify each of the following statements as true or false. If the statement is false, provide a counterexample.
   a. No rectangle is a rhombus.
   b. No trapezoid is a square.
   c. Some squares are trapezoids.
   d. A rhombus is a parallelogram.
   e. A square is a rhombus.

8. a. Construct a pentagon with all sides congruent and exactly two right angles.
   b. Can a quadrilateral have four acute angles? Why?
   c. Can a kite have exactly two right angles? Why?

9. Are $\angle BAD$ and $\angle BCD$ congruent? Explain why or why not.

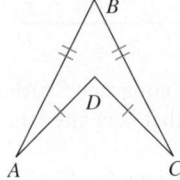

10. Classify each of the figures formed from regular hexagon *ABCDEF*.

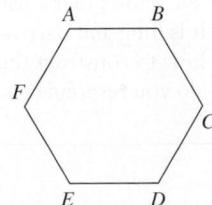

a. Quadrilateral *ABCD*
b. Quadrilateral *ABCE*

11. Use properties of a rectangle to show that a rectangle has a point symmetry (a 180° turn symmetry about some point).

12. Cut two congruent obtuse scalene triangles out of cardboard or paper and assemble them to form a parallelogram. Explain why the new figure is a parallelogram.

13. a. Use the definition of a *sector* of a circle in exercise 13 of Assessment 12-2A to identify two pairs of congruent sectors created by two diameters.
    b. Why do two diameters create pairs of congruent sectors?

14. a. Use only a compass and a straightedge to construct a kite.
    b. Use any tools to construct a rectangle so that the vertices of the kite from part (a) are on the sides of the rectangle.

15. Draw two noncongruent quadrilaterals such that two 45° angles and an included side in one quadrilateral are congruent, respectively, to two angles and an included side in the other quadrilateral.

16. a. Prove that a quadrilateral whose diagonals are congruent and bisect each other is a rectangle.
    b. Use part (a) and only a compass and straightedge to construct any rectangle.
    c. Construct another rectangle not congruent to the rectangle in part (b) but whose diagonals are congruent to the diagonals of the rectangle in part (b). Why are the rectangles not congruent?

17. a. What type of figure is formed by joining the midpoints of the sides of a parallelogram?
    b. Justify your answer to part (a).

18. What minimum information is sufficient to determine congruency for each of the following?
    a. Two rectangles
    b. Two kites
    c. Two isosceles triangles

19. Suppose polygon *ABCD* shown in the following figure is any parallelogram:

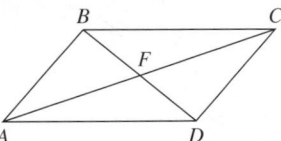

Use congruent triangles to justify each of the following:
a. $\angle A \cong \angle C$ and $\angle B \cong \angle D$ (opposite angles are congruent).
b. $\overline{BC} \cong \overline{AD}$ and $\overline{AB} \cong \overline{CD}$ (opposite sides are congruent).
c. $\overline{BF} \cong \overline{DF}$ and $\overline{AF} \cong \overline{CF}$ (the diagonals bisect each other).
d. $\angle DAB$ and $\angle ABC$ are supplementary.

20. Describe a set of minimal conditions to determine whether two rhombuses are congruent.

Mathematical Connections 12-2

Communication

1. Stan is standing on the bank of a river wearing a baseball cap. Standing erect and looking directly at the other bank, he pulls the bill of his cap down until it just obscures his vision of the opposite bank. He then turns around, being careful not to disturb the cap, and picks out a spot that is just obscured by the bill of his cap. He then paces off the distance to this spot and claims that the distance across the river is approximately equal to the distance he paced. Is Stan's claim true? Why?

2. Most ironing boards are collapsible for storage and can be adjusted to fit the height of the person using them. The surface of the board, though, remains parallel to the floor regardless of the height. Explain how to construct the legs of an ironing board to ensure that the surface is always parallel to the floor.

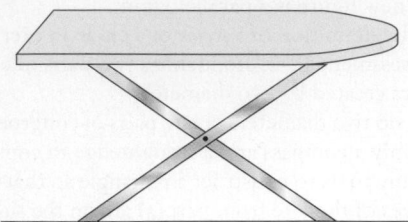

3. Using the fact that the perpendicular bisector of the base of an isosceles triangle is also the angle bisector to explain how to construct the angle bisector of a given angle using only a straightedge and a compass.

4. The marked angles and a side in the two triangles are congruent. Can you conclude that the triangles are congruent? Why or why not?

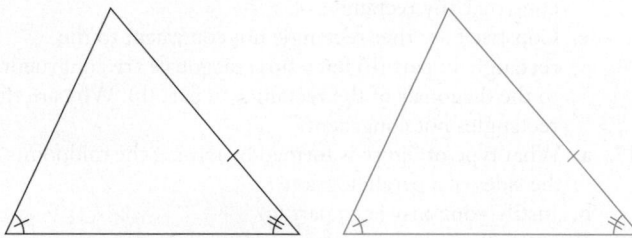

5. In the following figure, point P is on the angle bisector of $\angle A$. The distance from P to one of the sides of the angle is PB and to the other side it is PC. It seems that $PB = PC$.
   a. Pick another point on the angle bisector and construct the distances from that point to the sides of the angle. What seems to be true about the distances?

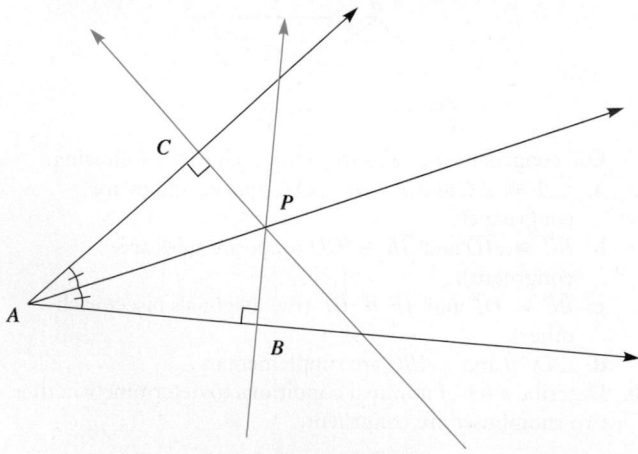

   b. State a conjecture about the points on the angle bisector of an angle by completing the following sentence:

   The distances from every point on the angle bisector of an angle …

   c. Prove your conjecture in part (b).

   d. State and prove the converse of your statement in part (b).

6. By definition, two quadrilaterals are congruent if there is a one-to-one correspondence between the vertices such that corresponding angles are congruent and corresponding sides are congruent. Thus, there are eight relations—four between sides and four between angles. If the following number of these conditions hold, will the quadrilaterals be congruent?
   a. 7
   b. 4

Open-Ended

7. On a 4 by 4 geoboard or dot paper (with 16 dots), create and answer questions about the following:
   a. Congruent triangles
   b. Isosceles trapezoids
   c. Noncongruent squares

Cooperative Learning

8. a. Record the definitions of *trapezoid* and *kite* given in different Grade 6–8 and secondary-school geometry textbooks.
   b. Compare the definitions found with those in this text and with those other groups found.
   c. Defend the use of one definition over another.

Questions from the Classroom

9. A student claims that polygon $ABCD$ in the following drawing is a parallelogram if $\angle 1 \cong \angle 2$. Is he correct? Why?

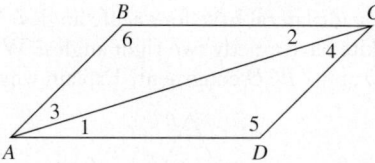

10. A student asks why "congruent" rather than "equal" is used to discuss triangles that have the same size and shape. What do you say?

11. A student asks why triangles often appear in structures such as bridges, building frames, and trusses. How do you respond?

12. A student says that she knows that a parallelogram has two altitudes but when it is long and narrow like the one shown, she does not know how to construct the altitude to the shorter sides. How do you respond?

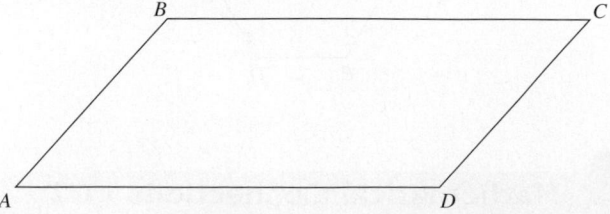

13. A student claims that if a line ℓ is drawn through vertex A of $\triangle ABC$ parallel to $\overline{BC}$ and a new point A' is chosen anywhere on ℓ, then the height of $\triangle A'BC$ to side $\overline{BC}$ stays the same.

He would like to know whether this is true and if so, why. How do you respond? *

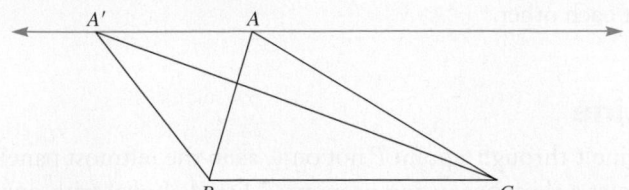

GSP/GeoGebra Activities

14. Use GSP Lab 1 activities 6–10 and Lab 2 activity 1 or Geogebra Lab 2 activity 2 and Lab 3 activities 1–4 to show the following using inductive reasoning.
   a. The opposite sides and opposite angles of a parallelogram are congruent.
   b. The diagonals of a parallelogram bisect each other.
   c. The diagonals of a rhombus are perpendicular bisectors of each other.
   d. If one pair of sides of a quadrilateral is congruent and parallel, then the quadrilateral is a parallelogram. Also prove this statement.

Review Problems

15. In the following regular pentagon, use the existing vertices to find all the triangles congruent to △ABC. Show that the triangles actually are congruent.

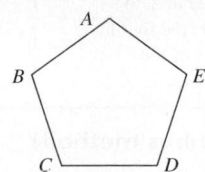

16. If possible, construct a triangle that has the three segments a, b, and c shown here as its sides; if not possible, explain why not.

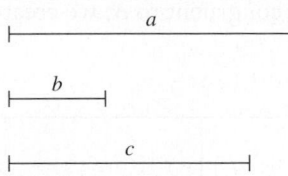

17. Construct an equilateral triangle whose sides are congruent to the following segment:

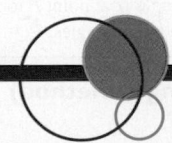

18. For each of the following pairs of triangles, determine whether the given conditions are sufficient to show that the triangles are congruent. If the triangles are congruent, tell which property can be used to verify this fact.

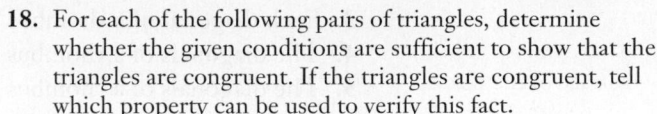

a.   b.

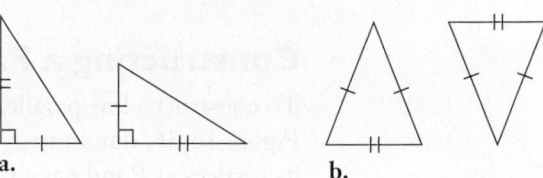

c.

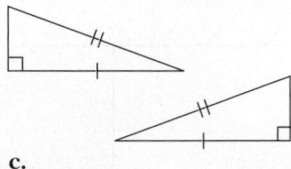

Trends in Mathematics and Science Study (TIMSS) Questions

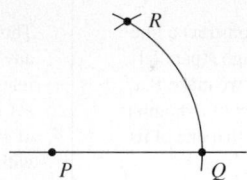

In the figure above, an arc of a circle with center P has been drawn to cut the line at Q. Then an arc with the same radius and center Q was drawn to cut the first arc at R. What would be the size of angle PRQ?
   a. 30°
   b. 45°
   c. 60°
   d. 75°

TIMSS, Grade 8, 2003

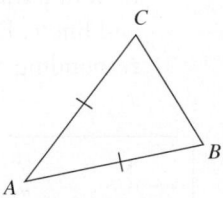

The triangle ABC has AB = AC.
Draw a line to divide triangle ABC into two congruent triangles.

TIMSS, Grade 8, 2003

12-3 Additional Constructions

We use the definition of a rhombus and the following properties (also listed in Table 12-1) to accomplish basic compass-and-straightedge constructions:

   1. A rhombus is a parallelogram in which all the sides are congruent.
   2. A quadrilateral in which all the sides are congruent is a rhombus.

3. Each diagonal of a rhombus bisects the opposite angles.
4. The diagonals of a rhombus are perpendicular.
5. The diagonals of a rhombus bisect each other.

Constructing a Parallel Line

To construct a line parallel to a given line ℓ through a point P not on ℓ, as in the leftmost panel of Figure 12-36, our strategy is to construct a rhombus (using property 2 listed above) with one of its vertices at P and one of its sides on line ℓ. Because the opposite sides of a rhombus are parallel, one of the sides through P will be parallel to ℓ. This construction is shown in Figure 12-36.

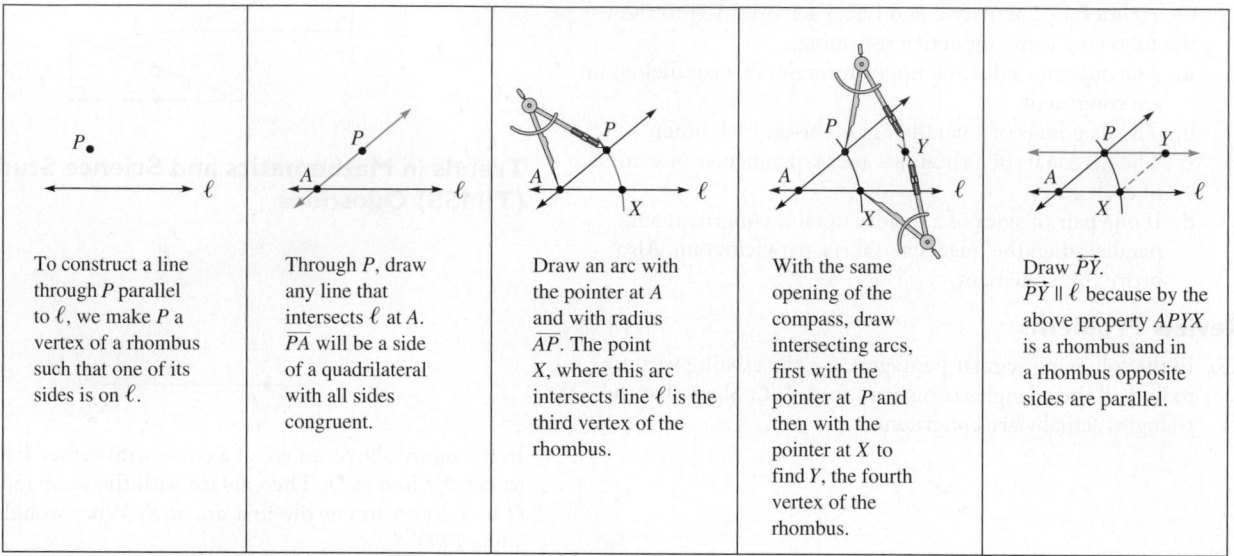

To construct a line through P parallel to ℓ, we make P a vertex of a rhombus such that one of its sides is on ℓ.

Through P, draw any line that intersects ℓ at A. $\overline{PA}$ will be a side of a quadrilateral with all sides congruent.

Draw an arc with the pointer at A and with radius $\overline{AP}$. The point X, where this arc intersects line ℓ is the third vertex of the rhombus.

With the same opening of the compass, draw intersecting arcs, first with the pointer at P and then with the pointer at X to find Y, the fourth vertex of the rhombus.

Draw $\overrightarrow{PY}$. $\overrightarrow{PY} \parallel \ell$ because by the above property $APYX$ is a rhombus and in a rhombus opposite sides are parallel.

Figure 12-36 Constructing parallel lines (rhombus method)

Figure 12-37 shows another way to construct parallel lines. If congruent corresponding angles are formed by a transversal cutting two lines, then the lines are parallel. Thus, the first step is to draw a transversal through P that intersects ℓ. The angle marked α is formed by the transversal and line ℓ. By constructing an angle with a vertex at P congruent to α, we create congruent corresponding angles; therefore, $\overleftrightarrow{PQ} \parallel \ell$.

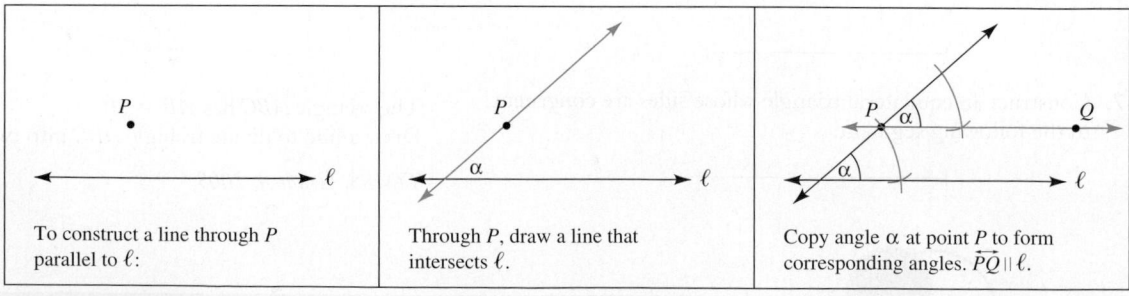

To construct a line through P parallel to ℓ:

Through P, draw a line that intersects ℓ.

Copy angle α at point P to form corresponding angles. $\overrightarrow{PQ} \parallel \ell$.

Figure 12-37 Constructing parallel lines (corresponding angle method)

 **NOW TRY THIS 12-6**

Parallel lines are frequently constructed using either a ruler and one triangle or two triangles. If a ruler and a triangle are used, the ruler is left fixed and the triangle is slid so that one side of the triangle touches the ruler at all times. In Figure 12-38, the hypotenuses of the right triangles are all parallel (also the legs not on the ruler are all parallel).

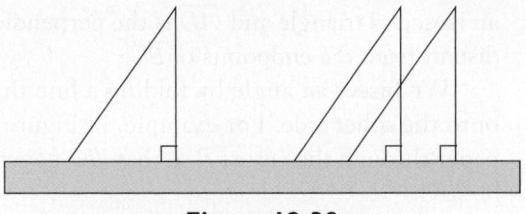

Figure 12-38

a. How can this method be used to construct a line through a given point parallel to a given line?

b. Why are the hypotenuses parallel?

Paperfolding can be used to construct parallel lines. For example, in Figure 12-39(a) to construct a line *m* parallel to line *p* through point *Q*, we can fold a perpendicular to line *p* so that the fold line does not contain point *Q*, as shown in Figure 12-39(b). Then by marking the image of point *Q* and connecting point *Q* and its image, *Q'*, we have $\overleftrightarrow{QQ'}$ parallel to line *p* (why?), as in Figure 12-39(c).

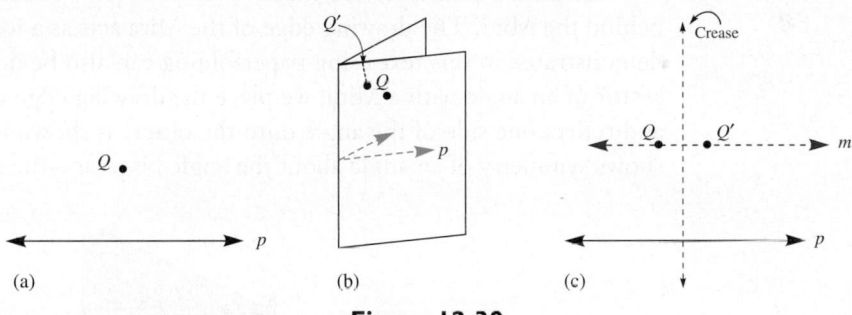

(a)   (b)   (c)

Figure 12-39

Constructing Angle Bisectors

Another construction based on a property of a rhombus is a construction of an *angle bisector*, (a ray that separates an angle into two congruent angles). The diagonal of a rhombus with vertex *A* bisects $\angle A$, as shown in Figure 12-40.

Another way to bisect an angle can be devised by using the fact that a perpendicular bisector of the base of an isosceles triangle contains the vertex and is the angle bisector of the opposite angle. Actually, the steps of the construction can be the same as in Figure 12-40, where $\triangle BAC$ is

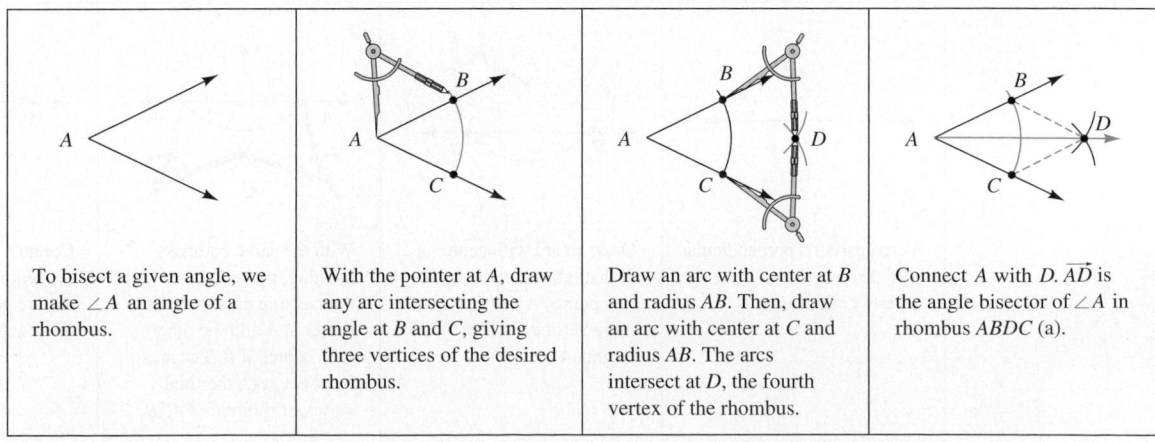

To bisect a given angle, we make $\angle A$ an angle of a rhombus.	With the pointer at *A*, draw any arc intersecting the angle at *B* and *C*, giving three vertices of the desired rhombus.	Draw an arc with center at *B* and radius *AB*. Then, draw an arc with center at *C* and radius *AB*. The arcs intersect at *D*, the fourth vertex of the rhombus.	Connect *A* with *D*. $\overrightarrow{AD}$ is the angle bisector of $\angle A$ in rhombus *ABDC* (a).

Figure 12-40 Bisecting an angle (rhombus method)

an isosceles triangle and $\overleftrightarrow{AD}$ is the perpendicular bisector of $\overline{BC}$, because both A and D are equidistant from the endpoints of $\overline{BC}$.

We bisect an angle by folding a line through the vertex so that one side of the angle folds onto the other side. For example, in Figure 12-41 we bisect $\angle ABC$ by folding and creasing the paper through the vertex B so that $\overrightarrow{BC}$ coincides with $\overrightarrow{BA}$.

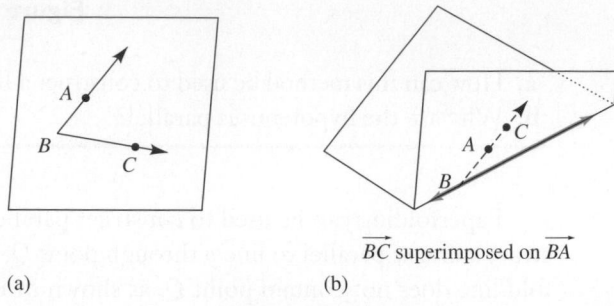

$\overrightarrow{BC}$ superimposed on $\overrightarrow{BA}$

(a)   (b)

Figure 12-41

A Mira is a plastic device that acts as a reflector so that the image of an object can be seen behind the Mira. The drawing edge of the Mira acts as a folding line on paper. Any construction demonstrated in this text using paperfolding can also be done with a Mira. To construct the bisector of an angle with a Mira, we place the drawing edge of the Mira on the vertex of the angle and reflect one side of the angle onto the other, as shown in Figure 12-42. Notice that the Mira shows symmetry of an angle about the angle bisector—the reflection line.

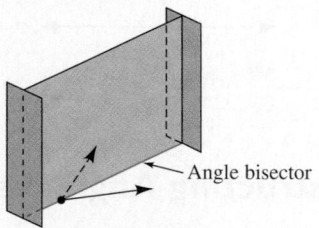

Angle bisector

Figure 12-42

Constructing Perpendicular Lines

To construct a line through P perpendicular to line ℓ, where P is not a point on ℓ, as in Figure 12-43, recall that the diagonals of a rhombus are perpendicular to each other. If we construct a rhombus

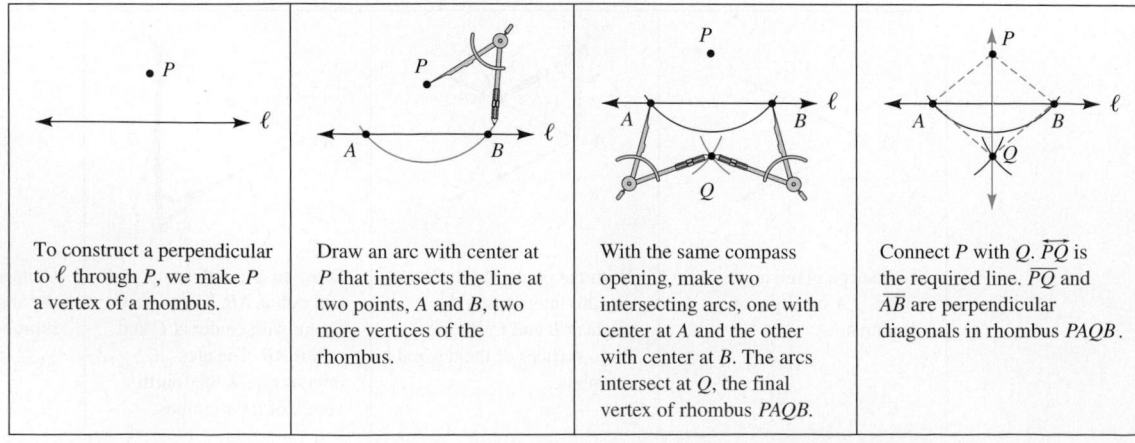

To construct a perpendicular to ℓ through P, we make P a vertex of a rhombus.	Draw an arc with center at P that intersects the line at two points, A and B, two more vertices of the rhombus.	With the same compass opening, make two intersecting arcs, one with center at A and the other with center at B. The arcs intersect at Q, the final vertex of rhombus $PAQB$.	Connect P with Q. $\overleftrightarrow{PQ}$ is the required line. $\overleftrightarrow{PQ}$ and $\overline{AB}$ are perpendicular diagonals in rhombus $PAQB$.

Figure 12-43 Constructing a perpendicular to a line from a point not on the line

with a vertex at P and a diagonal on ℓ, as in Figure 12-43, the segment connecting the fourth vertex Q to P is perpendicular to ℓ.

In Section 12-1 we saw how to construct the perpendicular bisector of a segment using a property of a perpendicular bisector stated in Theorem 12-3. Here we show how a property of a rhombus can also be used for constructing the perpendicular bisector of a segment.

To construct the perpendicular bisector of a line segment, as in Figure 12-44, we use the fact that the diagonals of a rhombus are perpendicular bisectors of each other.

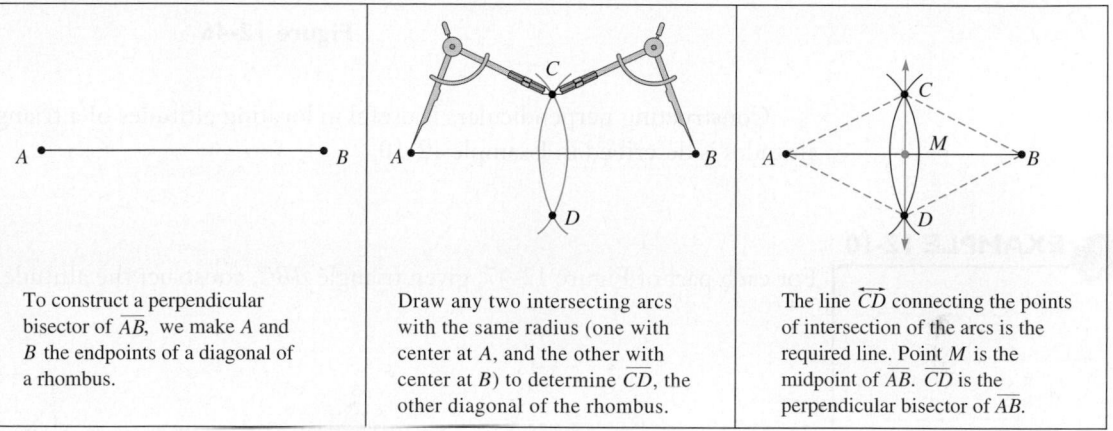

To construct a perpendicular bisector of $\overline{AB}$, we make A and B the endpoints of a diagonal of a rhombus.

Draw any two intersecting arcs with the same radius (one with center at A, and the other with center at B) to determine $\overline{CD}$, the other diagonal of the rhombus.

The line $\overleftrightarrow{CD}$ connecting the points of intersection of the arcs is the required line. Point M is the midpoint of $\overline{AB}$. $\overleftrightarrow{CD}$ is the perpendicular bisector of $\overline{AB}$.

Figure 12-44 Bisecting a line segment

Constructing a perpendicular to a line ℓ at a point M on ℓ is based on the fact that the diagonals of a rhombus are perpendicular bisectors of each other. Observe in Figure 12-44 that $\overleftrightarrow{CD}$ is a perpendicular to $\overline{AB}$ through M. Thus we construct a rhombus whose diagonals intersect at point M, as in Figure 12-45.

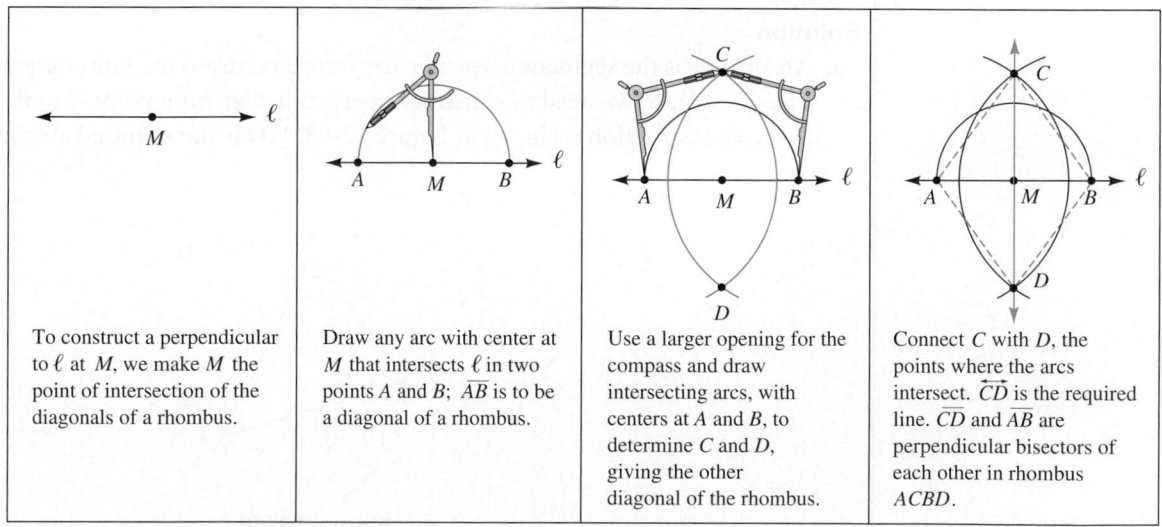

To construct a perpendicular to ℓ at M, we make M the point of intersection of the diagonals of a rhombus.

Draw any arc with center at M that intersects ℓ in two points A and B; $\overline{AB}$ is to be a diagonal of a rhombus.

Use a larger opening for the compass and draw intersecting arcs, with centers at A and B, to determine C and D, giving the other diagonal of the rhombus.

Connect C with D, the points where the arcs intersect. $\overleftrightarrow{CD}$ is the required line. $\overleftrightarrow{CD}$ and $\overline{AB}$ are perpendicular bisectors of each other in rhombus $ACBD$.

Figure 12-45 Constructing a perpendicular to a line from a point on the line

Perpendicularity constructions can also be completed by means of paperfolding or by using a Mira. To use paperfolding to construct a perpendicular to a given line ℓ at a point P on the line, we fold the line onto itself, as shown in Figure 12-46(a). The fold line is perpendicular to ℓ. To perform the construction with a Mira, we place the Mira with the drawing edge on P, as shown in Figure 12-46(b), so that ℓ is reflected onto itself. The line along the drawing edge is the required perpendicular.

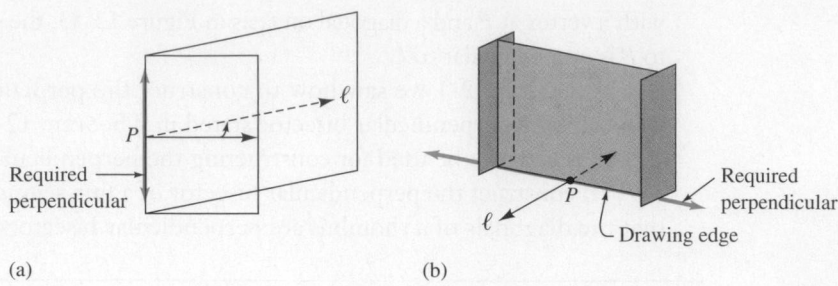

(a)   (b)

Figure 12-46

Constructing perpendiculars is useful in locating altitudes of a triangle. The construction of altitudes is described in Example 12-10.

EXAMPLE 12-10

For each part of Figure 12-47, given triangle ABC, construct the altitude from vertex A.

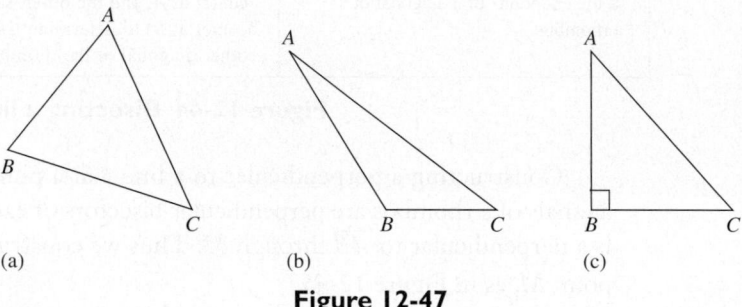

(a)   (b)   (c)

Figure 12-47

Solution

a. An altitude is the segment perpendicular from a vertex to the line containing the opposite side of a triangle, so we need to construct a perpendicular from point A to the line containing $\overline{BC}$. Such a construction is shown in Figure 12-48. $\overline{AD}$ is the required altitude.

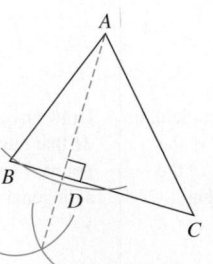

Figure 12-48

b. The construction of the altitude from vertex A is shown in Figure 12-49. Notice that the required altitude $\overline{AD}$ does not intersect the interior of $\triangle ABC$.

c. Triangle ABC is a right triangle. The altitude from vertex A is the side $\overline{AB}$. No construction is required.

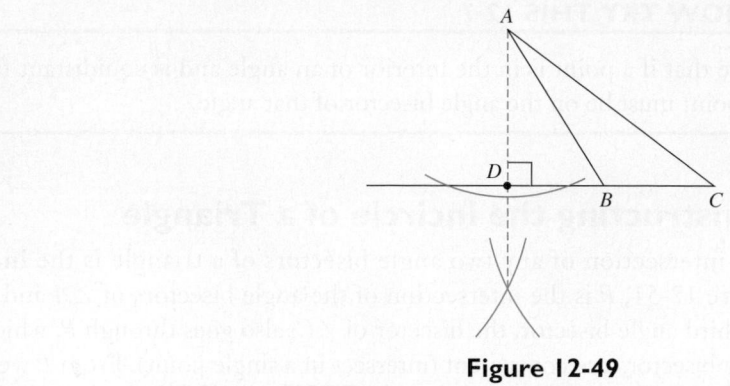

Figure 12-49

Properties of Angle Bisectors

To investigate properties of an angle bisector, recall that we defined the *distance from a point to a line* as *the length of the perpendicular segment from the point to the line*. Consider the angle bisector in Figure 12-50. It seems that any point P on the angle bisector is equidistant from the sides of the angle; that is, $PD = PE$.

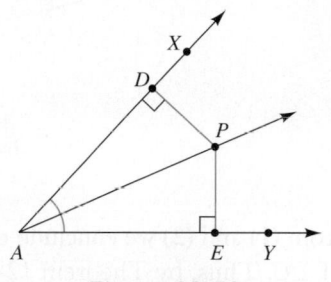

Figure 12-50

To justify this, we find two congruent triangles that have $\overline{PD}$ and $\overline{PE}$ as corresponding sides. The only triangles pictured are $\triangle ADP$ and $\triangle AEP$. Because $\overleftrightarrow{AP}$ is the angle bisector, $\angle DAP \cong \angle EAP$. Also, $\angle PDA$ and $\angle PEA$ are right angles and are thus congruent. $\overline{AP}$ is congruent to itself, so $\triangle PDA \cong \triangle PEA$ by AAS. Thus, $\overline{PD} \cong \overline{PE}$ because they are corresponding parts of congruent triangles PDA and PEA. Consequently, we have proved the first part of the following theorem.

Theorem 12-9

a. Any point P on an angle bisector is equidistant from the sides of the angle.

b. Any point in the interior of an angle that is equidistant from the sides of the angle is on the angle bisector of the angle.

Both parts of the theorem can be stated as: A point is on an angle bisector of an angle if, and only if, it is equidistant from the sides of the angle. Notice that the angle bisector is on the line of symmetry of the angle.

Notice that we have proved only part (a) of Theorem 12-9. Now Try This 12-9 investigates the proof of part (b).

NOW TRY THIS 12-7

Prove that if a point is in the interior of an angle and is equidistant from the sides of the angle, the point must be on the angle bisector of that angle.

Constructing the Incircle of a Triangle

The intersection of any two angle bisectors of a triangle is the **incenter** of the triangle. In Figure 12-51, P is the intersection of the angle bisectors of $\angle A$ and of $\angle B$. We will show that the third angle bisector, the bisector of $\angle C$, also goes through P, which will imply that the three angle bisectors are concurrent (intersect in a single point). From P we construct the perpendiculars to the three sides of the triangle, $\overline{PN}$, $\overline{PK}$, and $\overline{PM}$. Because P is on the angle bisector of $\angle A$, from Theorem 12-9(a) we know that

  1. $PN = PK$

Similarly, because P is on the angle bisector of $\angle B$, we know that

  2. $PK = PM$

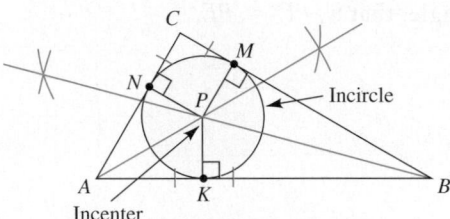

Figure 12-51

Now from (1) and (2) we conclude that $PN = PM$. But this means that P is equidistant from the sides of $\angle C$. Thus, by Theorem 12-9(b), P is on the angle bisector of $\angle C$. We have proved the following theorem.

> **Theorem 12-10**
>
> The angle bisectors of a triangle are concurrent and the three distances from the point of intersection to the sides are equal.

If in Figure 12-51 we draw a circle with the center P and radius PM (or PK or PN), the circle seems to fit exactly inside the triangle. In fact each line segment "touches" the circle at one point only. The sides of the triangle are **tangent** to the circle. *A line is tangent to a circle if it intersects the circle in one, and only one, point.* It is possible to show that a line perpendicular to a radius at the endpoint of the radius that is not the center is tangent to the circle. Thus, the sides of $\triangle ABC$ in Figure 12-51 are tangent to the circle. Such a circle is **inscribed** in the triangle. The center of the inscribed circle, the **incircle,** is the **incenter** of the triangle.

To find the radius of the inscribed circle, it is sufficient to construct from the incenter just one perpendicular (to one of the sides). The point of intersection of the perpendicular with the corresponding side when connected with P determines the radius.

NOW TRY THIS 12-8

Use any tools to investigate whether a circle can be inscribed in a square.

Assessment 12-3A

1. Draw a line ℓ and point P not on ℓ. Use a compass and a straightedge to construct a line m through P parallel to ℓ, using each of the following:
   a. Alternate interior angles
   b. A rhombus

2. Construction companies avoid vandalism at night by hanging expensive pieces of equipment from the boom of a crane, as shown in the following figure:

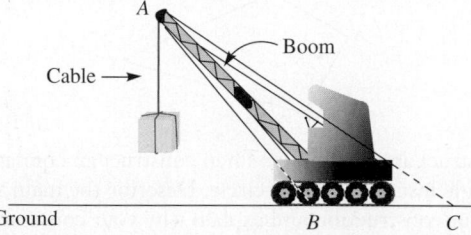

   a. If you consider a triangle with two vertices A and B as marked in the figure and the intersection of a line through the cable holding the equipment and the ground as the third vertex, what type of triangle is formed?

   b. If you consider the triangle formed by points A, B, and C, describe where the altitude containing vertex A of the triangle is.

3. Construct the perpendicular bisectors of each of the sides of the following triangles. Use any desired method.

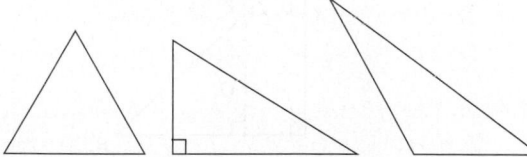

   a. Make a conjecture about the perpendicular bisectors of the sides of an acute triangle.
   b. Make a conjecture about the perpendicular bisectors of the sides of a right triangle.
   c. Make a conjecture about the perpendicular bisectors of the sides of an obtuse triangle.
   d. For each of the three triangles, construct the circle that circumscribes the triangle.

4. a. Given the following triangle ABC, construct a point P that is equidistant from the three vertices of the triangle and the incircle for the triangle. Explain why your construction is correct.

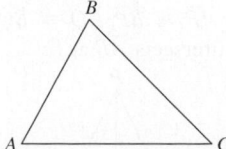

   b. Repeat (a) for the following obtuse triangle:

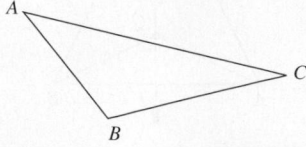

5. For which of the following figures is it possible to construct a circle that is inscribed in the figure (each side of the figure must be tangent to the circle)? If it is possible to find such a circle, draw the figure and construct the inscribed circle using only a straightedge and a compass; if not, explain how you decided that it is impossible to find an inscribed circle.
   a. A rectangle
   b. A rhombus
   c. A regular hexagon

6. Given $\overline{AB}$ in the following figure, construct a square with $\overline{AB}$ as a side:

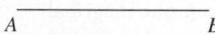

7. a. Given A, B, and C as vertices, use a compass and a straightedge to construct a parallelogram $ABCD$:

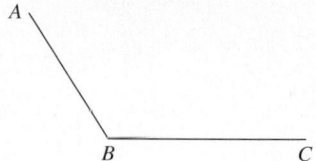

   b. Suppose you are "charged" 10¢ each time you use your straightedge to draw a line segment and 10¢ each time you use your compass to draw an arc. What is the cheapest way to construct the parallelogram in part (a).

8. In right triangle ABC, point O is the incenter, $CD = 2''$, $AC = 8''$, $AB = 10''$. Find OD.

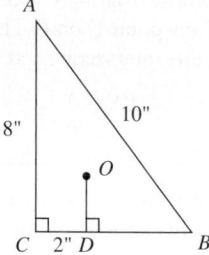

9. Describe how to construct the incircle of a regular pentagon.

10. In the parallelogram $ABCD$ shown, find EF.

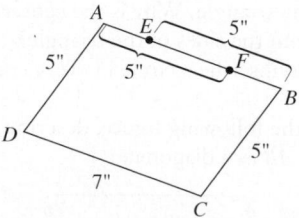

11. a. Show that any parallelogram can be dissected (cut apart) to form two non-isosceles trapezoids.
    b. Prove that any two congruent trapezoids can always be put together to form a parallelogram.

12. Some car jacks are constructed like a collapsed rhombus. When used to raise a car, the rhombus is easily seen.

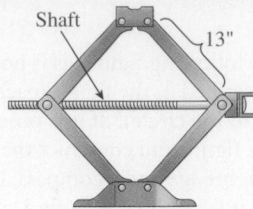

Shaft

13"

If the sides of the rhombus are 13″, what is the maximum length of the shaft?

13. Using any tools determine if it is possible to construct each of the following. If the construction is not possible, explain why.
   a. A square, given one side
   b. A rectangle, given one diagonal
   c. A triangle with two obtuse angles
   d. A parallelogram with exactly three right angles

14. Using only a compass and a straightedge, construct angles with each of the following measures:
   a. 30°
   b. 45°
   c. 75°

15. Construct $\overline{AB}$ close to the bottom of a blank page. Use a compass and a straightedge to construct the perpendicular bisector of $\overline{AB}$. You are not allowed to put any marks below $\overline{AB}$.

16. Draw a convex quadrilateral similar to the one shown below and construct each of the following points, if possible. If it is not possible, explain why. Describe each construction in words and explain why it produces the required point.
   a. The point that is equidistant from $\overline{AB}$ and $\overline{AD}$ and also from points B and C
   b. The point that is equidistant from $\overline{AB}$, $\overline{AD}$, and $\overline{BC}$

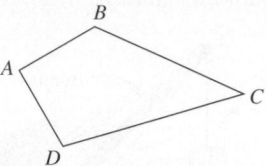

17. Construct any size circle. Then construct an equilateral triangle inscribed in that circle. Describe the main steps in your construction and explain why your construction produces an equilateral triangle.

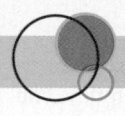

Assessment 12-3B

1. Draw a figure like the one below, then describe how to use a compass and a straightedge to construct line m through P parallel to ℓ, using each of the following:
   a. Perpendicular lines
   b. A quadrilateral whose diagonals bisect each other. (*Hint*: Connect P with any point Q on ℓ. Through the midpoint M of $\overline{PQ}$ draw any line intersecting ℓ at Q'.)

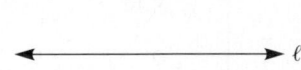

$P \bullet$

$\longleftrightarrow \ell$

2. Construct an obtuse triangle and the three altitudes. Mark the point where the lines that contain the altitudes are concurrent.

3. Construct an obtuse triangle and the perpendicular bisectors of two of its sides. Then construct the circle that circumscribes the triangle.

4. Draw a large obtuse triangle and construct the circle inscribed in the triangle. Why is the center of the circle equidistant from the sides of the triangle?

5. Is it possible to inscribe a circle in every convex kite? Why or why not?

6. Given $\overline{AB}$ in the following figure, describe how to construct a square with $\overline{AB}$ as a diagonal:

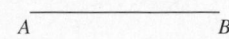

$A \qquad\qquad B$

7. Suppose you are "charged" 10¢ each time you use your straightedge to draw a line segment and 10¢ each time you use your compass to draw an arc. Determine the cheapest way to construct an equilateral triangle.

8. In the isosceles right triangle shown, O is the incenter; the radius of the inscribed circle is r. Find AX.

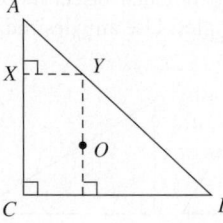

9. Describe how to construct the incircle of a regular decagon.

10. In the parallelogram shown, find x in terms of a and b.

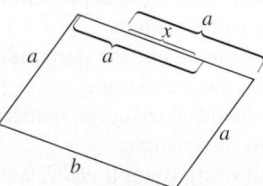

11. Explain why any rectangle can be dissected into two congruent trapezoids in many different ways.

12. In the following concave quadrilateral $APBQ$, $\overline{PQ}$ and $\overline{AB}$ are the diagonals; $\overline{AP} \cong \overline{BP}$, $\overline{AQ} \cong \overline{BQ}$, and $\overline{PQ}$ has been extended until it intersects $\overline{AB}$ at C:

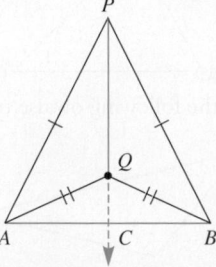

a. Make a conjecture concerning $\overline{PQ}$ and $\overline{AB}$.

b. Justify your conjecture in (a). *

c. Make conjectures concerning the relationships between $\overrightarrow{PQ}$ and $\angle APB$ and between $\overrightarrow{QC}$ and $\angle AQB$.

d. Justify your conjectures in (c).

13. Using any tools, determine if it is possible to construct a unique figure for each of the following. If the construction is not possible, explain why.

a. A square, given one diagonal

b. A parallelogram, given two of its adjacent sides

c. A rhombus, given its diagonals

d. A parallelogram given a side and all the angles

14. Using only a compass and a straightedge, construct angles with each of the following measures:

a. 15°

b. 105°

c. 120°

15. Draw a line ℓ and a point P not on the line and use the sliding triangle method described in Figure 12-38 to construct a perpendicular to ℓ through P using a straightedge and a right triangle.

16. Draw a convex quadrilateral similar to the one shown below and construct each of the following points, if possible. If it is not possible, explain why. Describe each construction in words and explain why it produces the required point.

a. The point that is equidistant from *A*, *B*, and *C*

b. The point that is equidistant from *A*, *B*, *C*, and *D*

c. The point that is equidistant from the four sides

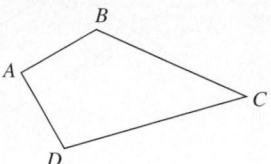

17. Construct a circle and an equilateral triangle for which the circle is the incircle. Describe the main steps in your construction and explain why your construction produces an equilateral triangle.

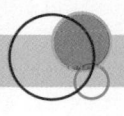

Mathematical Connections 12-3

Communication

1. Given an angle and a roll of tape, describe how you might construct the bisector of the angle.

2. Write a letter from you, a curriculum developer, to parents explaining whether or not the geometry curriculum in Grades 5–8 should include construction problems that use only a compass and straightedge.

3. Patty paper constructions are accomplished using the waxed paper that is sometimes applied between hamburger patties by commercial meat companies. Research patty paper constructions and organize a presentation on them.

4. If a line is tangent to a circle at point *P*, and *A* is any point on the line, then $\overline{AP}$ is a *tangent segment*. Use this definition to prove that the two tangent segments from a point *A* outside the circle to the circle are congruent. That is, prove that $\overline{AP} \cong \overline{AQ}$.

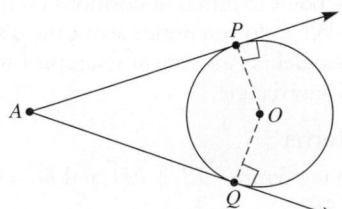

5. In the following figure a circle is inscribed in a quadrilateral. By the result of exercise 4, the lengths of congruent tangent segments have been designated by the same lowercase letter (*a*, *b*, *c*, or *d* in the figure).

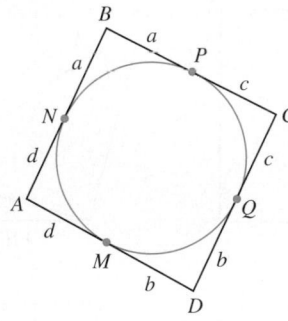

a. Prove that $AD + BC = AB + CD$.

b. Based on part (a) complete the following statement:

If a circle can be inscribed in a quadrilateral then _____.

c. Use part (b) to describe all the rectangles in which it is impossible to inscribe a circle. Justify your answer.

6. *ABCD* is a parallelogram. The diagonals divide the parallelogram into four triangles with incenters at *E*, *F*, *G*, and *H*. Prove that *EFGH* is a rhombus.

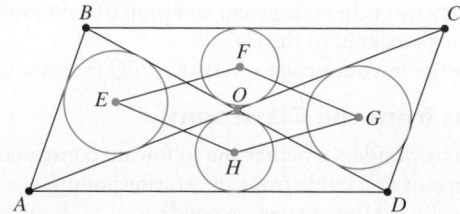

7. If two pieces of tape of the same width cross each other, what type of parallelogram is *ABCD*? Justify your answer.

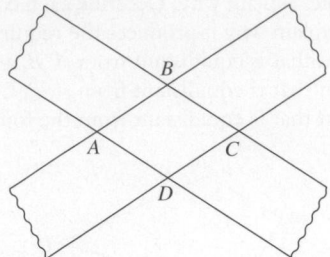

Open-Ended

8. Given a circle and using any tools construct a pentagon for which the circle is an incircle. Describe the main steps of your construction.
9. Search the web for problems related to the material in this section that involve a square. Describe two problems from your search that you find interesting.

Cooperative Learning

10. In your group, perform the paperfolding construction shown and answer the questions that follow. Let *P* and *Q* be two opposite vertices of a rectangular piece of paper, as shown in (i). Fold *P* onto *Q* so that a crease is formed. This results in (ii), where *A* and *B* are the endpoints of the crease. Next crease again so that point *A* folds onto point *B*. This results in (iii). Next, unfold to obtain two creases, as in (iv).

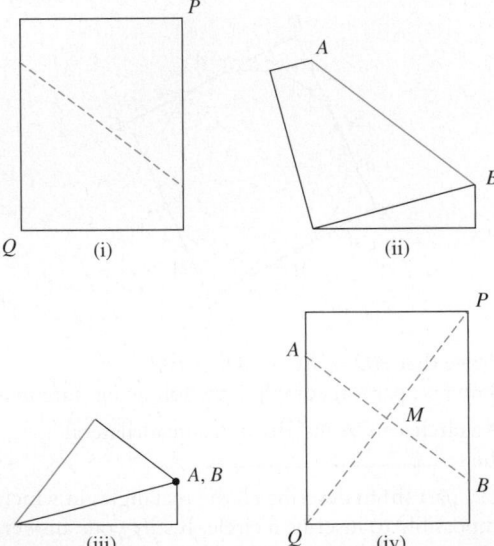

a. Individually write an explanation of why every point on $\overline{AB}$ is equidistant from the endpoints of $\overline{PQ}$. Compare the explanations in your group and prepare one explanation to be presented to the class.
b. Discuss in your group whether *APBQ* is a rhombus.

Questions from the Classroom

11. A student wonders whether the following construction of a tangent to a circle from an exterior point *P* is a valid construction. How do you respond?

Rotate the ruler about point *P* counterclockwise until it just touches the circle as shown in the figure below:

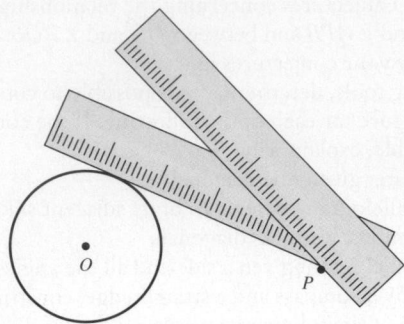

12. A student wants to know how to bisect a given angle drawn on a large sheet of paper using only an unmarked ruler. How do you respond?
13. Gail is asked to draw a triangle and to construct its incircle. She finds the point *O* where two angle bisectors intersect and point *D* where the angle bisector $\overleftrightarrow{BO}$ intersects the opposite side and draws the circle with center *O* and radius *OD*. The teacher marks the construction "wrong." Gail wants to know why. How do you respond?

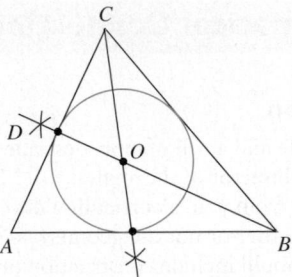

GSP/GeoGebra Activities

14. Use inductive reasoning to answer the following.
    a. Draw all the altitudes of a triangle. Make conjectures about the lines containing the altitudes of each of the following types of triangles: acute, right, and obtuse.
    b. Draw an angle and its angle bisector. Pick a point on the angle bisector and calculate the distance from the point to each side of the angle.
    c. Move the point to different positions on the angle bisector. What do you notice about the distances?
15. Use GSP/GeoGebra Assessment Exercise 1 to inscribe a circle in a given triangle.

Review Problems

16. In the following figure, $\overleftrightarrow{AB} \parallel \overleftrightarrow{ED}$ and $\overline{BC} \cong \overline{CE}$. Explain why *DE* = *AB*.

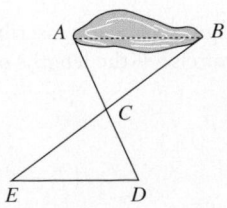

17. Draw △*ABC*. Then construct △*PQR* congruent to △*ABC* using each of the following combinations:
  a. Two sides of △*ABC* and an angle included between these sides
  b. The three sides of △*ABC*
  c. Two angles and a side included between these angles

18. In two right triangles, △*ABC* and △*DEF*, if ∠*A* and ∠*D* are congruent and $\overline{AC}$ and $\overline{DF}$ are congruent, what can you conclude about the two triangles. Why?

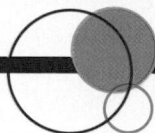

12-4 Similar Triangles and Other Similar Figures

When a germ is examined under a microscope or when a slide is projected on a screen, the shapes in each case usually remain the same, but the sizes are altered. Informally, such figures are **similar**. For example, see the student page and do Problem 3. The ratio of the corresponding side lengths is the **scale factor**. On the student page the scale factor in the top figures is $\frac{2}{3}$.

It seems that in any enlargement or reduction, the resulting figure is similar to the original; that is, the corresponding angle measures remain the same and the corresponding sides are proportional. In general, we have the following definition of similar triangles and similar polygons.

Definition of Similar Triangles

△*ABC* is similar to △*DEF*, written △*ABC* ~ △*DEF*, if, and only if, the corresponding interior angles are congruent and the lengths of the corresponding sides are proportional; that is, ∠*A* ≅ ∠*D*, ∠*B* ≅ ∠*E*, ∠*C* ≅ ∠*F*, and $\frac{AB}{DE} = \frac{AC}{DF} = \frac{BC}{EF}$.

Definition of Similar Polygons

Two polygons with the same number of vertices are similar if there is a one-to-one correspondence between the vertices of one and the vertices of the other such that the corresponding interior angles are congruent and corresponding sides are proportional.

In similarity as with congruence of triangles, we do not need all the conditions in the definition to conclude that the triangles are similar. In fact, the following conditions suffice to conclude that the triangles are similar: SSS, SAS and AA. We state these conditions in the following theorems, which could be proved in more advanced geometry texts.

Theorem 12-11: SSS Similarity for Triangles

If the lengths of corresponding sides of two triangles are proportional, then the triangles are similar.

Theorem 12-12: SAS Similarity for Triangles

If two sides are proportional to their corresponding sides and the included angles are congruent, then the triangles are similar.

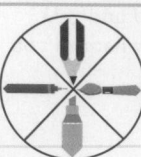

School Book Page SIMILAR FIGURES

Similar figures have the same shape, but not necessarily the same size. Corresponding angles of similar figures are congruent. Lengths of corresponding sides of similar figures are proportional.

EXAMPLE **Identifying Similar Figures**

② Show that the triangles are similar.

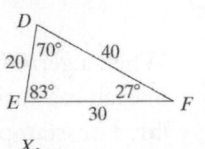

The measures of ∠D and ∠X are 70°.
The measures of ∠E and ∠Y are 83°.
The measures of ∠F and ∠Z are 27°.

$\frac{20}{30} = \frac{2}{3}$, $\frac{40}{60} = \frac{2}{3}$, and $\frac{30}{45} = \frac{2}{3}$

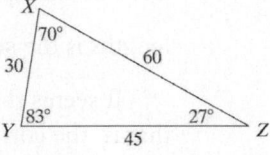

GO for Help
For help with checking proportions, go to Lesson 7-3, Example 1.

The measures of corresponding angles are equal. The lengths of corresponding sides form equal ratios. The triangles are similar.

✔ Quick Check

2. Is the triangle at the right similar to triangle *DEF* in Example 2? Explain your reasoning.

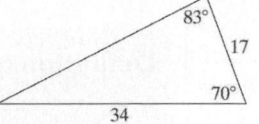

If the angles of two triangles are congruent, then the triangles are similar. The sides must also be congruent for the triangles to be congruent.

EXAMPLE **Application: Architecture**

③ Triangles *ABC* and *CDE* are similar. The measure of ∠A is 70°. The measure of ∠B is 55°. What is the measure of ∠E? Explain.

The measure of ∠ACB is 55°, since the sum of the angles in a triangle equals 180°. ∠ACB and ∠E are corresponding angles. So the measure of ∠E is 55°.

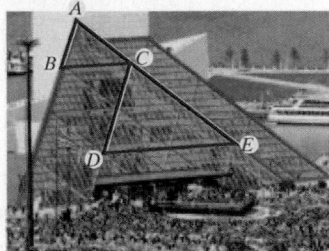

Online active math*

For: Congruence Activity
Use: Interactive Textbook, 8-6

✔ Quick Check

3. The measure of $\overline{AB}$ is 4 units, the measure of $\overline{BC}$ is 5 units, and the measure of $\overline{CD}$ is 12 units. Find the measure of $\overline{DE}$.

8-6 Congruent and Similar Figures   **393**

> **Theorem 12-13: Angle, Angle (AA) Similarity for Triangles**
> If two angles of one triangle are congruent, respectively, to two angles of a second triangle, then the triangles are similar.

We refer to these theorems as SSS similarity, SAS similarity, and AA similarity. Conditions like SSSS or AAAA are not valid for quadrilaterals (Why?).

In what follows we give examples in which the similarity conditions are used.

EXAMPLE 12-11

Given the pairs of similar triangles in Figure 12-52, find a one-to-one correspondence among the vertices of the triangles such that the corresponding angles are congruent. Then write the proportions for the corresponding sides that follow from the definition.

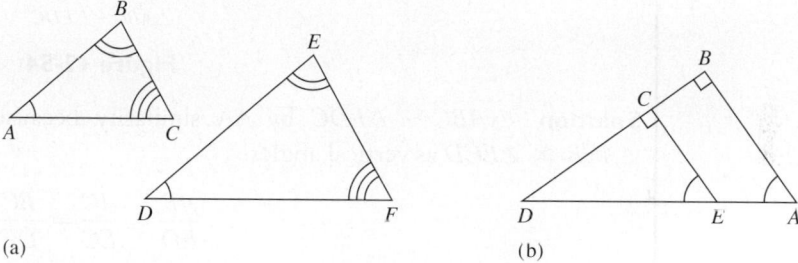

(a)   (b)

Figure 12-52

Solution

a. $\triangle ABC \sim \triangle DEF$

$$\frac{AB}{DE} = \frac{BC}{EF} = \frac{AC}{DF}$$

b. $\triangle ABD \sim \triangle ECD$

$$\frac{AB}{EC} = \frac{BD}{CD} = \frac{AD}{ED}$$

EXAMPLE 12-12

For each part of Figure 12-53, find a pair of similar triangles.

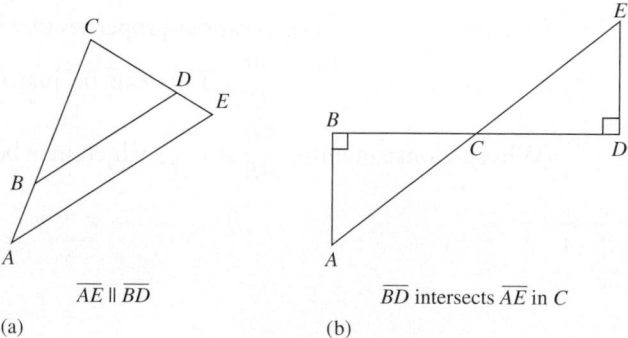

$\overline{AE} \parallel \overline{BD}$ $\overline{BD}$ intersects $\overline{AE}$ in C

(a)   (b)

Figure 12-53

Solution

a. Because $\overline{AE} \parallel \overline{BD}$, congruent corresponding angles are formed by a transversal cutting the parallel segments. Thus, $\angle CBD \cong \angle CAE$ and $\angle CDB \cong \angle CEA$. Thus $\triangle CBD \sim \triangle CAE$ by AA, similarity.

b. $\angle B \cong \angle D$ because both are right triangles. Also, $\angle ACB \cong \angle ECD$ because they are vertical angles. Thus, $\triangle ACB \sim \triangle ECD$ by AA similarity.

⬤ **NOW TRY THIS 12-9**

Are all right triangles in which the hypotenuse is twice as long as one of the legs similar? Explain your answer.

⬤ **EXAMPLE 12-13**

In Figure 12-54 show that $\triangle ABC \sim \triangle EDC$ and then solve for x.

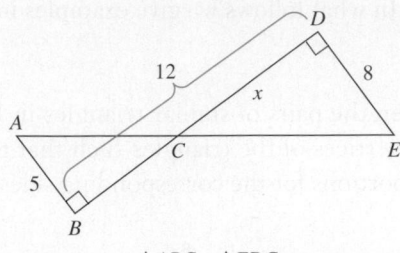

$$\triangle ABC \sim \triangle EDC$$

Figure 12-54

Solution $\triangle ABC \sim \triangle EDC$ by AA similarity because $\angle B$ and $\angle D$ are right angles and $\angle ACB \cong \angle ECD$ as vertical angles.

$$\frac{AB}{ED} = \frac{AC}{EC} = \frac{BC}{DC}$$

Now, $AB = 5, ED = 8$, and $CD = x$, so $BC = 12 - x$. Thus,

$$\frac{5}{8} = \frac{12 - x}{x}$$

$$5x = 8(12 - x)$$

$$5x = 96 - 8x$$

$$13x = 96$$

$$x = \frac{96}{13}$$

Properties of Proportion

Similar triangles give rise to various properties that involve proportions. For example, in Figure 12-55 if $\overline{BC} \| \overline{DE}$, then $\frac{AB}{BD} = \frac{AC}{CE}$. This can be justified as follows: $\overline{BC} \| \overline{DE}$, so $\triangle ADE \sim \triangle ABC$ (Why?). Consequently, $\frac{AD}{AB} = \frac{AE}{AC}$, which may be written as follows:

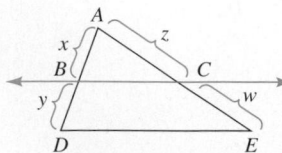

Figure 12-55

$$\frac{x + y}{x} = \frac{z + w}{z}$$

$$\frac{x}{x} + \frac{y}{x} = \frac{z}{z} + \frac{w}{z}$$

$$1 + \frac{y}{x} = 1 + \frac{w}{z}$$

$$\frac{y}{x} = \frac{w}{z}$$

$$\frac{x}{y} = \frac{z}{w}$$

This result is summarized in the following theorem.

> ### Theorem 12-14
> If a line parallel to one side of a triangle intersects the other sides, then it divides those sides into proportional segments.

The converse of Theorem 12-14 is also true; that is, if in Figure 12-55 we know that $\dfrac{AB}{BD} = \dfrac{AC}{CE}$, then we conclude that $\overline{BC} \parallel \overline{DE}$.

We can prove the converse using the SAS similarity condition. We can reverse the steps in the proof of Theorem 12-14. From $\dfrac{x}{y} = \dfrac{z}{w}$ we get $\dfrac{x+y}{x} = \dfrac{z+w}{z}$ and hence $\dfrac{AD}{AB} = \dfrac{AE}{AC}$. Using this proportion and the fact that $\triangle ABC$ and $\triangle ADE$ share $\angle A$, we conclude by SAS similarity that $\triangle ABC \sim \triangle ADE$. Consequently, $\angle ABC \cong \angle ADE$ and therefore $\overline{BC} \parallel \overline{DE}$. We summarize this result in the following theorem.

> ### Theorem 12-15
> If a line divides two sides of a triangle into proportional segments, then the line is parallel to the third side.

Similarly, if lines parallel to $\overline{DE}$ intersect $\triangle ADE$, as shown in Figure 12-56, so that $a = b = c = d$, it can be shown that $e = f = g = h$. This result is stated in the following theorem.

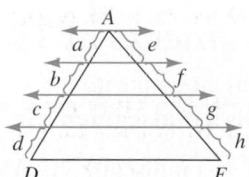

Figure 12-56

> ### Theorem 12-16
> If parallel lines cut off congruent segments on one transversal, then they cut off congruent segments on any transversal.

Theorem 12-16 can be used to divide a given segment into any number of congruent parts. For example, using only a compass and a straightedge, we can divide segment $\overline{AB}$ in Figure 12-57 into three congruent parts by making the construction resemble Figure 12-56.

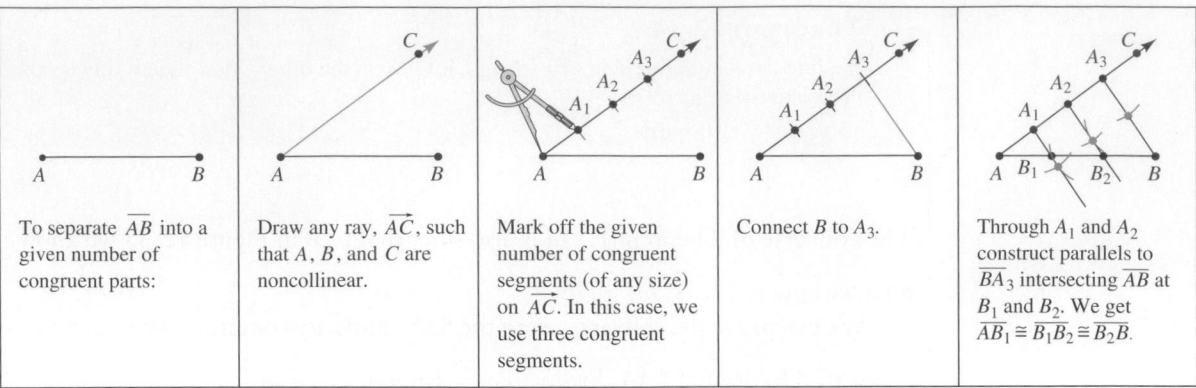

Figure 12-57 Separating a segment into congruent parts

It is only necessary to construct $\overline{A_2B_2}$. We can then use a compass to mark off point B_1 such that $B_1B_2 = BB_2$, as in Figure 12-58.

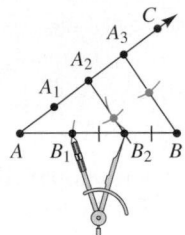

Figure 12-58

Midsegments of Triangles and Quadrilaterals

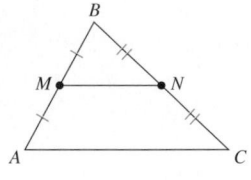

Figure 12-59

The segment connecting the midpoints of two sides of a triangle or two adjacent sides of a quadrilateral is a **midsegment**. In Figure 12-59, M and N are midpoints of $\overline{AB}$ and $\overline{BC}$, respectively, and $\overline{MN}$ is a midsegment. We have $\dfrac{BM}{AM} = \dfrac{BN}{CN} = 1$. Also by SAS similarity $\triangle MBN \sim \triangle ABC$. Therefore,

$$\frac{MN}{AC} = \frac{MB}{AB} = \frac{MB}{2MB} = \frac{1}{2}$$

Consequently, we have the following theorem.

> **Theorem 12-17:   The Midsegment Theorem**
> The midsegment joining the midpoints of two sides of a triangle is parallel to and is half as long as the third side..

The following theorem is an immediate consequence of Theorem 12-16.

> **Theorem 12-18**
> If a line bisects one side of a triangle and is parallel to a second side, then it bisects the third side and therefore is a midsegment.

EXAMPLE 12-14

In the quadrilateral *ABCD* in Figure 12-60, *M*, *N*, *P*, and *Q* are the midpoints of the sides.

a. What kind of quadrilateral is *MNPQ*?
b. For what kind of quadrilateral *ABCD* will *MNPQ* be a rhombus?

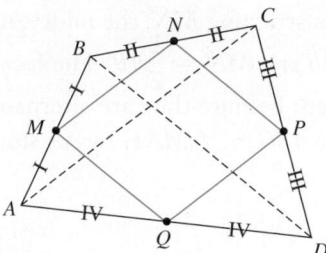
Figure 12-60

Solution

a. *MNPQ* appears to be a parallelogram. To prove this conjecture, observe that $\overline{NP}$ is a midsegment in $\triangle BCD$ and consequently $\overline{NP} \parallel \overline{BD}$ (and $NP = \frac{1}{2}BD$). Similarly, in $\triangle ABD$, $\overline{MQ}$ is a midsegment and therefore $\overline{MQ} \parallel \overline{BD}$. Consequently, $\overline{MQ} \parallel \overline{NP}$. In a similar way we could show that $\overline{MN} \parallel \overline{QP}$ (consider midsegments in $\triangle ABC$ and in $\triangle ADC$) and therefore by the definition of a parallelogram, *MNPQ* is a parallelogram.

b. Because $MQ = \frac{1}{2}BD$, $MN = \frac{1}{2}AC$ and *MNPQ* is a parallelogram, *MNPQ* will be a rhombus if, and only if, $MQ = MN$, or equivalently if, and only if, $BD = AC$. Thus *MNPQ* is a rhombus if, and only if, the diagonals of *ABCD* are congruent.

A **median** of a triangle is a segment connecting a vertex of the triangle to the midpoint of the opposite side. A triangle has three medians. A careful drawing of the three medians shown in Figure 12-61 suggests that the three medians are concurrent, which they are. The point of intersection, *G*, is the **center of gravity**, or the **centroid**, of the triangle.

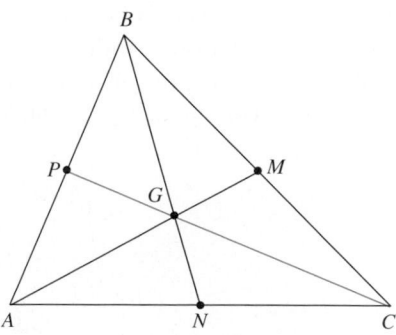
Figure 12-61

If a triangular piece of somewhat-thick, uniform material such as wood or metal is placed at its centroid on a sharp object like a pencil tip, it will balance. In the next example, we prove an interesting property of the centroid.

EXAMPLE 12-15

Show that the centroid of a triangle divides each median into two segments with lengths in the ratio 1:2.

Solution   In Figure 12-62, we have two medians of $\triangle ABC$ and their point of intersection G. We prove that $AG = 2MG$ and $BG = 2NG$. The fact that M and N are midpoints of two sides suggests constructing $\overline{MN}$, the midsegment. Theorem 12-17, the Midsegment Theorem tells us that $\overline{MN} \| \overline{AB}$ and $MN = \frac{1}{2}AB$. The fact that $\overline{MN} \| \overline{AB}$ implies that the comparably marked angles are congruent because they are alternate interior angles formed by a transversal of the parallel lines. Thus, $\triangle ABG \sim \triangle MNG$ by AA similarity. Using the fact that $AB = 2MN$ we have

$$\frac{BG}{NG} = \frac{AG}{MG} = \frac{AB}{MN} = 2.$$

Thus, $BG = 2GN$ and $AG = 2MG$.

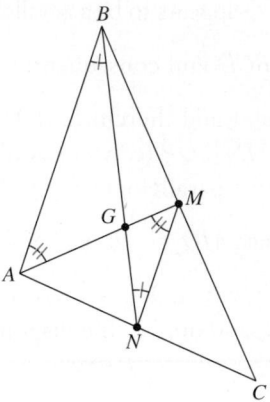

Figure 12-62

Indirect Measurements

Similar triangles have long been used to make indirect measurements. Thales of Miletus (ca. 600 BCE) is believed to have determined the height of the Great Pyramid of Egypt by using ratios involving shadows, similar to those pictured in Figure 12-63. The sun is so far away that it should make approximately congruent angles at B and B'. Because the angles at C and C' are right angles, $\triangle ABC \sim \triangle A'B'C'$. Hence,

$$\frac{AC}{A'C'} = \frac{BC}{B'C'}$$

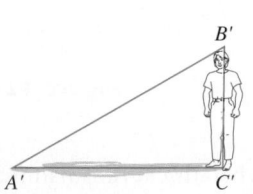
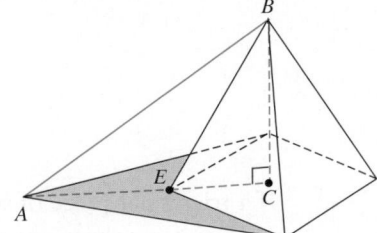

Figure 12-63

And because $AC = AE + EC$, the following proportion is obtained by substituting $AE + EC$ for AC in the above proportion:

$$\frac{AE + EC}{A'C'} = \frac{BC}{B'C'}$$

The person's height and shadow can be measured. Also, the length AE of the shadow of the pyramid can be measured, and EC is half the length of the diagonal of the base. Each term of the proportion except the height of the pyramid is then known. Thus, the height BC of the pyramid can be found by solving the proportion.

EXAMPLE 12-16

On a sunny day, a tall tree casts a 40-m shadow as in Figure 12-64. At the same time, a meterstick held vertically casts a 2.5-m shadow. How tall is the tree?

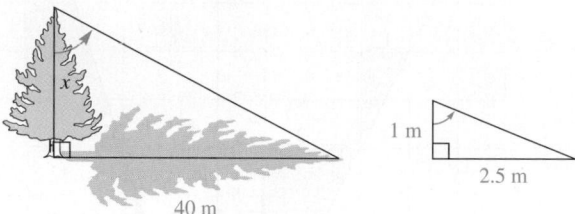

Figure 12-64

Solution In Figure 12-64, the triangles are similar by AA similarity because the tree and the stick both meet the ground at right angles and the angles formed by the sun's rays are congruent because the shadows are measured at the same time.

$$\frac{x}{40} = \frac{1}{2.5}$$
$$2.5x = 40$$
$$x = 16$$

The tree is 16 m tall.

BRAIN TEASER

Two neighbors, Smith and Wheeler, plan to erect flagpoles in their yards. Smith wants a 10-ft pole, and Wheeler wants a 15-ft pole. To keep the poles straight while the concrete bases harden, guy wires are to be tied from the tops of the flagpoles to a fence post on the property lines and to the bases of the flagpoles, as shown in Figure 12-65. How high should the fence post be and how far apart should they erect the flagpoles for this scheme to work?

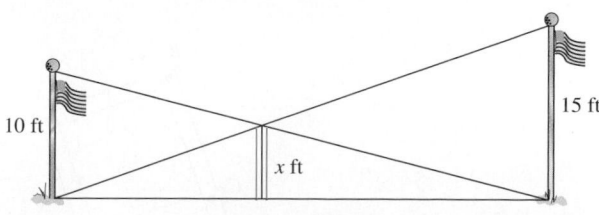

Figure 12-65

Using Similar Triangles to Determine Slope

In Chapter 8 we defined the slope of a line with equation $y = mx + b$ to be m or equivalently as $\dfrac{\text{Rise}}{\text{Run}}$.

In Figure 12-66, right triangles have been constructed and shaded on several lines. In each triangle, the horizontal side is the run, which is always positive, and the vertical side is the rise, which is positive if we go up toward the line and negative if we go down toward the line.

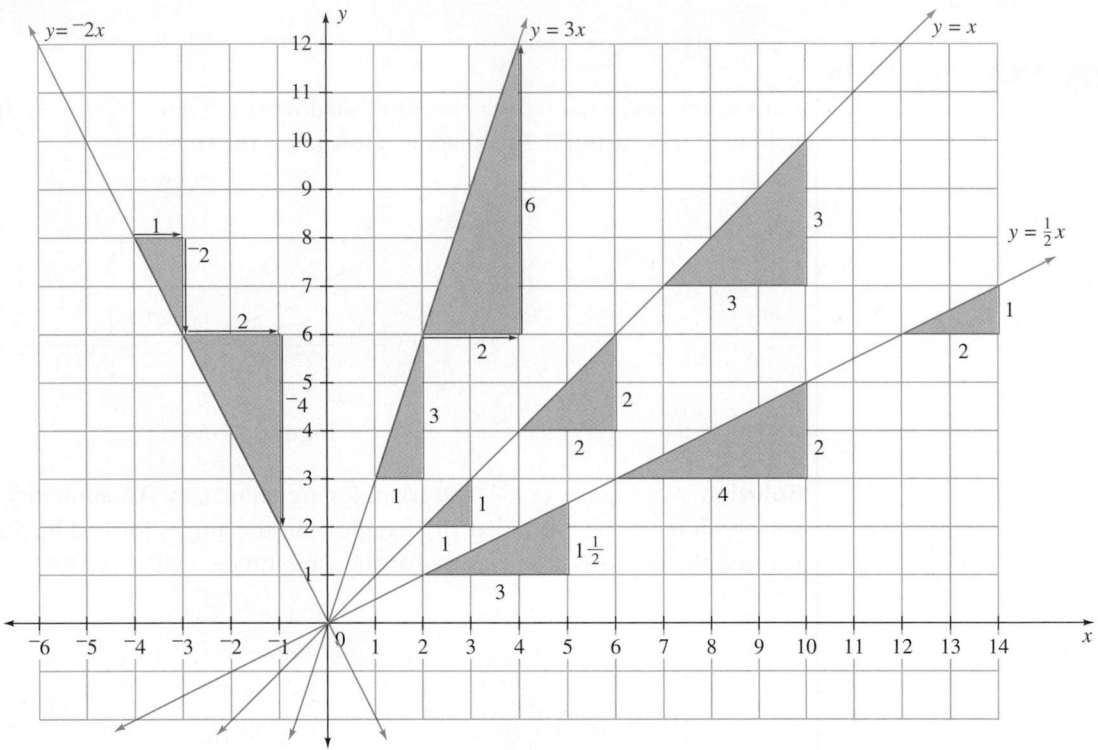

Figure 12-66

○ Historical Note

The device pictured in Figure 12-67 is a pantograph, invented in 1603 by the German theologian astronomer and geometer Christopher Scheiner (1575–1650). It was used to draw enlarged versions of figures. In Figure 12-67, the red dots represent either brads or nuts and bolts. The strips are made of lath or cardboard and are rigid. A pointer at D was used to trace along an original figure, which causes the pencil at F to draw an enlarged version of the figure.

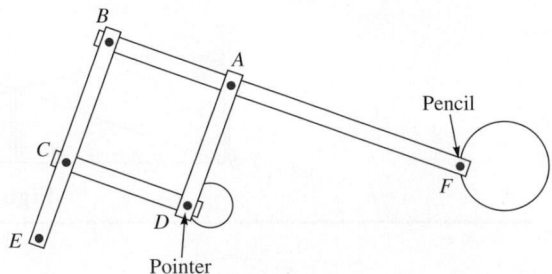

Figure 12-67

The slope of each line in Figure 12-66 can be calculated as the rise over the run in any of the shaded triangles, with hypotenuse along the given line. This is so because the right triangles with hypotenuses along the same line are similar. To test this fact, notice that

$$\text{for } y = \frac{1}{2}x, \quad m = \frac{\text{Rise}}{\text{Run}} = \frac{1\frac{1}{2}}{3} = \frac{2}{4} = \frac{1}{2}$$

$$\text{for } y = x, \quad m = \frac{\text{Rise}}{\text{Run}} = \frac{1}{1} = \frac{2}{2} = \frac{3}{3} = 1$$

$$\text{for } y = 3x, \quad m = \frac{3}{1} = \frac{6}{2} = 3$$

$$\text{for } y = {}^{-}2x, \quad m = \frac{{}^{-}4}{2} = \frac{{}^{-}2}{1} = {}^{-}2$$

Because for a given line all the right triangles with hypotenuses along the line and legs respectively parallel to the x- and y-axes are similar, the slope can be determined as $\dfrac{\text{Rise}}{\text{Run}}$ in any such triangle. Thus the slope formula given in Figure 12-68 gives the same value for any two points A and B on the same line.

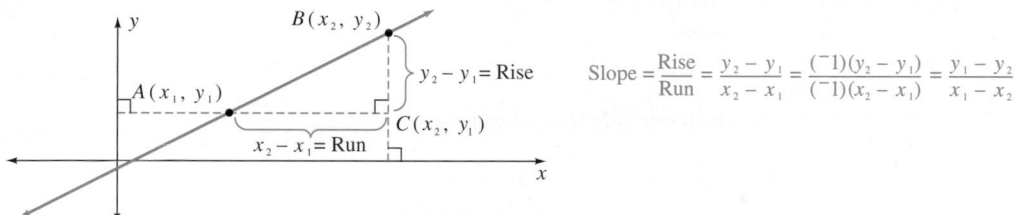

$$\text{Slope} = \frac{\text{Rise}}{\text{Run}} = \frac{y_2 - y_1}{x_2 - x_1} = \frac{({}^{-}1)(y_2 - y_1)}{({}^{-}1)(x_2 - x_1)} = \frac{y_1 - y_2}{x_1 - x_2}$$

Figure 12-68

Notice that in the slope formula we can use the differences of y-coordinates and the x-coordinates in any order as long as the order is the same in the numerator and the denominator. The following example illustrates the use of similar triangles to find the equation of a line.

EXAMPLE 12-17

Use similar triangles to find the equation of the line through the points $A(1, 2)$ and $B(3, 4)$.

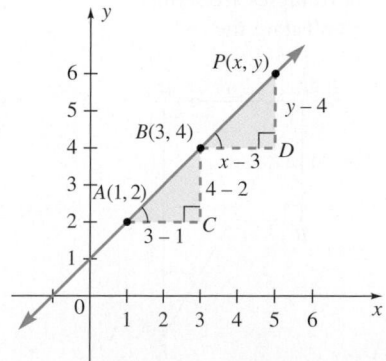

Figure 12-69

Solution Any point $P(x, y)$ different from B is on line $\overleftrightarrow{AB}$ in Figure 12-69 if, and only if, $\triangle ABC \sim \triangle BPD$ or equivalently if, and only if, $\dfrac{PD}{BD} = \dfrac{BC}{AC}$. That is, if, and only if, the slopes of $\overline{AB}$ and $\overline{BP}$ are equal (If $P = B$, the following equation involves $\dfrac{0}{0}$ which is not meaningful):

$$\frac{y - 4}{x - 3} = \frac{4 - 2}{3 - 1}$$
$$y - 4 = x - 3$$
$$y = x + 1$$

Notice that point B satisfies the equation $y - 4 = x - 3$ because we get $4 - 4 = 3 - 3$, a true statement. Thus all the points on the line satisfy the equation.

Assessment 12-4A

1. In rhombi $ABCD$ and $A_1B_1C_1D_1$, $\angle BAD \cong \angle B_1A_1D_1$. Are the rhombi similar? Why or why not?

2. Which of the following are always similar? Why?
   a. Any two equilateral triangles
   b. Any two squares
   c. Any two rectangles
   d. Any two rectangles in which one side is twice as long as the other.

3. Use grid paper to draw figures that have sides three times as large as the ones in the following figure and that are colored similarly.

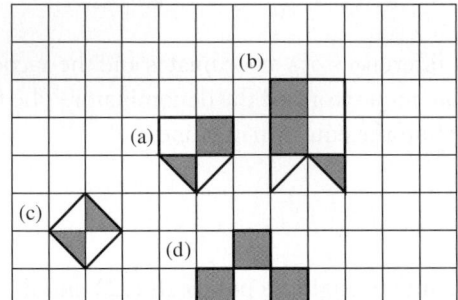

4. a. Which of the following pairs of triangles are similar? If they are similar, explain why and state the correspondence.

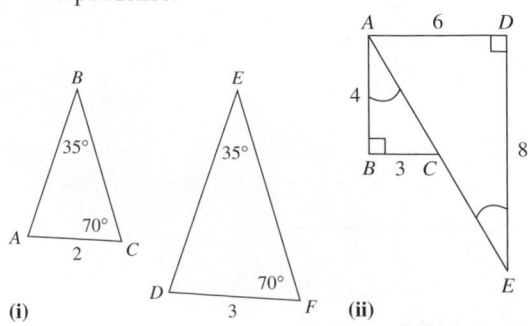

(i)   (ii)

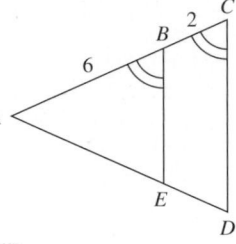

(iii)

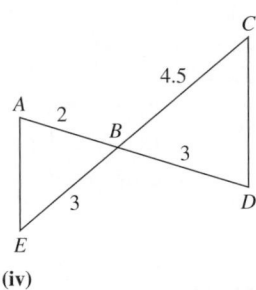

(iv)

   b. For each pair of similar triangles in part (a), find the scale factor of the sides of the triangles.

5. In the following figures, find the measure of the sides marked x.

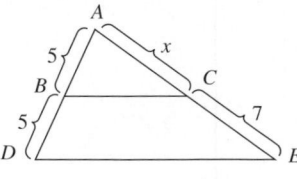

   a. $\triangle ABC \sim \triangle ADE$

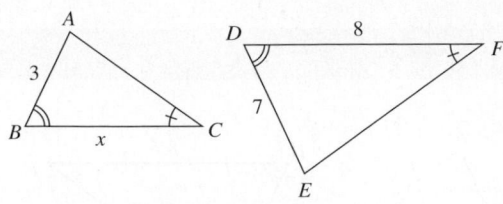

b. $\triangle ABC \sim \triangle EDF$

6. Draw a segment $\overline{AB}$ and then use a compass and a straightedge to separate $\overline{AB}$ into five congruent pieces.

7. Given segments of length a, b, and c as shown, construct a segment of length x so that $\dfrac{a}{b} = \dfrac{c}{x}$.

a _____

b _____

c _____

8. A photocopy of a polygon was reduced by 80% and then the copy was again reduced by 80%.
   a. Is the second photocopy similar to the original? Explain.
   b. What is the ratio of the corresponding sides of the second photocopy to the original?

9. Sketch with approximate measures two hexagons with corresponding sides proportional but so that they are not similar. (Consider a regular hexagon and a hexagon with all sides congruent but not all angles congruent.)

10. If you copy a page on a machine at 75%, you should get a similar copy of the page. What is the corresponding setting to obtain the original from the copy?

11. Samantha wants to know how far above the ground the top of a leaning flagpole is. At high noon, when the Sun is almost directly overhead, the shadow cast by the pole is 7 ft long. Samantha holds a plumb bob with a string 3 ft long up to the flagpole and determines that the point of the plumb bob touches the ground 13 in. from the base of the flagpole. How far above the ground is the top of the pole?

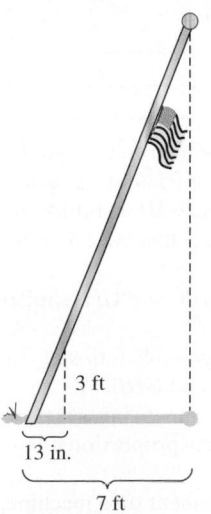

12. In the following figure, find the distance AB across the pond using the similar triangles shown:

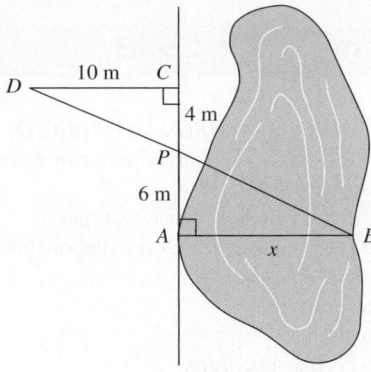

13. a. Examine several examples of similar polygons and make a conjecture concerning the ratio of their perimeters.
    b. Prove your conjecture in part (a).

14. The midpoints of the sides of a scalene triangle have been connected as shown.

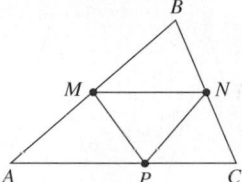

   a. Prove that the four smaller triangles are congruent.
   b. Is each of the smaller triangles similar to the original triangle? Why or why not?

15. The midpoints M, N, P, Q of the sides of a quadrilateral $ABCD$ have been connected and an interior quadrilateral is obtained. We have shown that $MNPQ$ is a parallelogram. What is the most you can say about the kind of parallelogram $MNPQ$ is if $ABCD$ is
   a. a rhombus?
   b. a kite?
   c. an isosceles trapezoid?
   d. a quadrilateral that is neither a rhombus nor a kite but whose diagonals are perpendicular to each other?

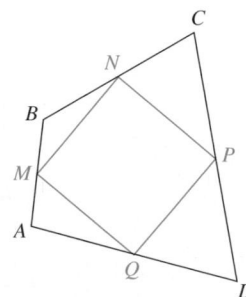

16. If you took cross sections of a typical ice cream cone parallel to the circular opening where the ice cream is usually placed, explain whether the cross sections would be similar.

Assessment 12-4B

1. Consider parallelograms $ABCD$ and $A_1B_1C_1D_1$ in which $\angle BAD \cong \angle B_1A_1D_1$. Explain whether or not the parallelograms are similar.

2. Which of the following are always similar?
   a. Any two rectangles in which the diagonal in one is twice as long as in the other.
   b. Any two rhombuses
   c. Any two circles
   d. Any two regular polygons
   e. Any two regular polygons with the same number of sides

3. On a piece of grid paper, draw similar but not congruent isosceles triangles.

4. Which of the following pairs of triangles are similar? If they are similar, explain why. Find the scale factor which is less than 1 and state the correspondence.

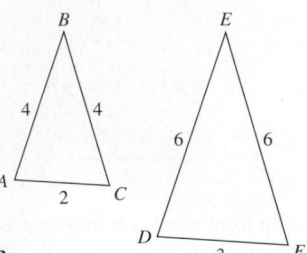

a.

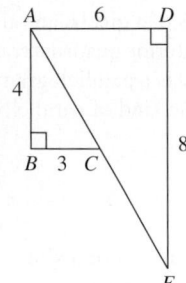

b.

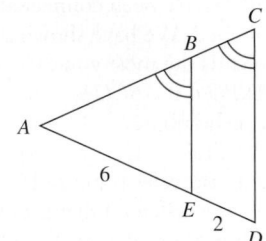

c.

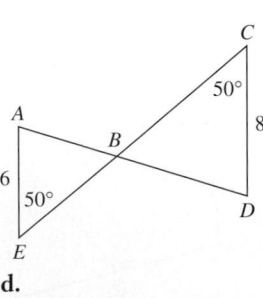

d.

5. Prove that in the following figures triangles in each part are similar and find the measure of the sides marked with x or y.

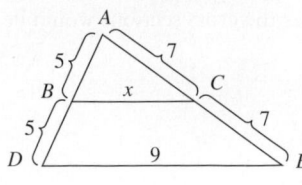

a.

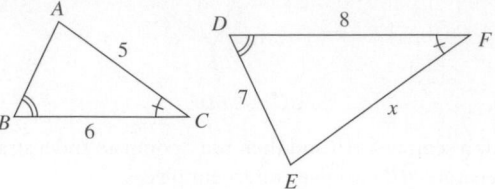

b.

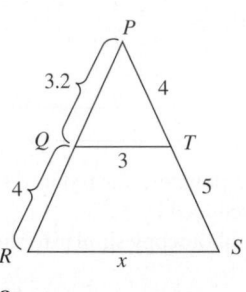

c.

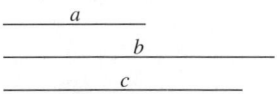

d.

6. Draw a segment $\overline{AB}$ and then use a compass and a straightedge to separate $\overline{AB}$ into three congruent pieces:

7. Given segments of length a, b, and c as shown, construct a segment of length x so that $\dfrac{a}{b} = \dfrac{x}{c}$.

$$\overline{a}$$
$$\overline{b}$$
$$\overline{c}$$

8. Triangle ABC is similar to triangle DEF with a side of triangle ABC that is 75% of its corresponding side in triangle DEF. Also, triangle GHI is similar to triangle DEF with a side of triangle GHI that is 32% of its corresponding side in triangle DEF.
   a. Are triangles ABC and GHI similar to each other? Why or why not?
   b. What are the possible ratios of corresponding sides of triangles ABC and GHI?

9. Sketch with approximate measures two pentagons with corresponding sides proportional but so that they are not similar.

10. If you copy a document on a machine, you should get a similar copy. Suppose you want to reduce a document to $\dfrac{1}{6}$ of its original size and you first reduce it by 25%. By what percent do you need to reduce the copy to obtain the desirable size?

11. To find the height of a tree, a group of Girl Scouts devised the following method. A girl walks toward the tree along its shadow until the shadow of the top of her head coincides with the shadow of the top of the tree. If the girl is 150 cm tall, her distance to the foot of the tree is 15 m, and the length of her shadow is 3 m, how tall is the tree?

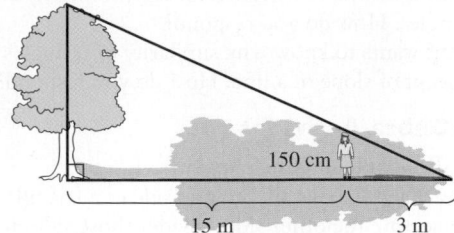

12. In the following figure, identify similar triangles and find DE and EA.

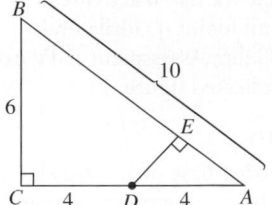

13. a. Examine several similar triangles and write a conjecture concerning the ratio of their corresponding heights.
    b. Prove your conjecture in part (a).

14. a. In the accompanying figure, $ABCD$ is a trapezoid. M is the midpoint of $\overline{AB}$. Through M, a line parallel to the bases has been drawn, intersecting $\overline{CD}$ at N. (i) Explain why N must be the midpoint of $\overline{CD}$ and (ii) express MN in terms of a and b, the lengths of the parallel sides of the trapezoid.

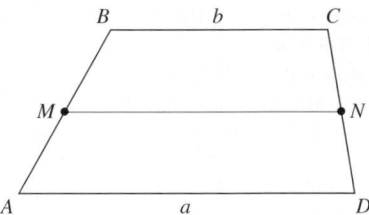

b. Denote MN by c. Use your answer to part (a) to show that b, c, and a form an arithmetic sequence.
c. In the trapezoid $ABCD$, the lengths of the bases are a and b as shown. Side $\overline{AB}$ has been divided into 9 congruent segments. Through the endpoints of the segments, lines parallel to $\overline{AD}$ have been drawn. In this way, 8 new segments connecting the sides $\overline{AB}$ and $\overline{CD}$ have been created. Show that the sequence of 10 terms, starting with b, proceeding with the lengths of the parallel segments, and ending with a, is an arithmetic sequence and find the sum of the sequence in terms of a and b.

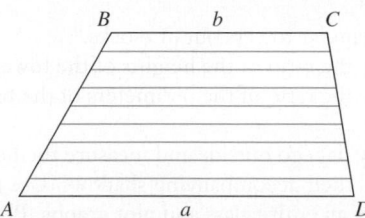

15. $ABCD$ is a convex quadrilateral and N, P, Q, M are the midpoints of its sides as shown. The diagonals $\overline{BD}$ and $\overline{AC}$ intersect at T, $\overline{NP}$ intersects $\overline{AC}$ at S, and $\overline{BD}$ intersects $\overline{MN}$ at V.
    a. Prove that $NSTV$ is a parallelogram
    b. Complete the following statements and prove them.
       i. $MNPQ$ is a rectangle if, and only if, $\overline{AC}$ and $\overline{BD}$ are _____.
       ii. $MNPQ$ is a square if, and only if, $\overline{AC}$ and $\overline{BD}$ are _____.

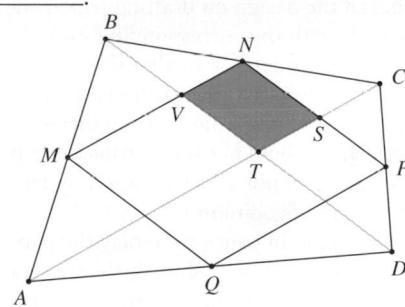

16. Are all semicircles (diameters not included) similar to each other?

Mathematical Connections 12-4

Communication

1. Do you think any two cubes are similar? Why or why not?
2. Architects frequently build three-dimensional scale drawings to construct scale models of projects. Are all the models similar to the finished products? Why or why not?
3. Assuming the lines on an ordinary piece of notebook paper are parallel and equidistant, describe a method for using the paper to divide a piece of licorice evenly among 2, 3, 4, or 5 children. Explain why it works.

4. In the following right triangle ABC, $\overline{CD} \perp \overline{AB}$:

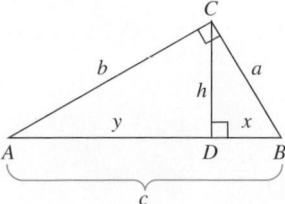

a. Find three pairs of similar triangles. Justify your answers.

b. Write the corresponding proportions for each set of similar triangles.

c. Use part (b) and the figure to show that $a^2 = xc$. Also argue that $b^2 = yc$.

d. Use part (c) to show that $a^2 + b^2 = c^2$. State this result in words using the legs and hypotenuse of a right triangle.

e. Use part (b) to show that $h^2 = xy$.

f. Show that x, h, and y form a geometric sequence.

Open-Ended

5. Build two similar towers out of blocks.
   a. What is the ratio of the heights of the towers?
   b. What is the ratio of the perimeters of the bases of the towers?

6. On a sunny day, go outside and measure the heights of objects and their accompanying shadows. Use the data gathered by an entire class and plot graphs. Plot all data points on the same graph, with shadow lengths on the horizontal axis and object heights on the vertical axis. What do you observe? Why?

Cooperative Learning

7. A building was to be built on a triangular piece of property. The architect was given the approximate measurements of the angles of the triangular lot as 54°, 39°, and 87° and the lengths of two of the sides as 100 m and 80 m. When the architect began the design on drafting paper, she drew a triangle to scale with the corresponding measures and found that the lot was considerably smaller than she had been led to believe. It appeared that the proposed building would not fit. The surveyor was called. He confirmed each of the measurements and could not see a problem with the size. Neither the architect nor surveyor could understand the reason for the other's opinion.
   a. Have one person in your group play the part of the architect and explain why she felt she was correct.
   b. Have one person in the group explain the reason for the miscommunication.
   c. Have the group suggest a way to provide an accurate description of the lot.

Questions from the Classroom

8. A student asks if, for the same n, all convex n-gons with all angles congruent are similar. How do you respond?

9. A student argues that in the following figure triangles ABC and DBE are similar because $\overline{DE}$ divides the sides proportionally. One side is divided in the ratio $\frac{2}{4}$, while the other is divided in the ratio $\frac{3}{6}$. How do you respond?

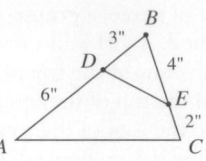

10. A student asks whether there is an ASA similarity condition for triangles. How do you respond?

11. A student wants to know why similarity of triangles explains the concept of slope of a line. How do you respond?

GSP/GeoGebra Activities

12. Use GSP/GeoGebra Lab 7 activities 1–6 to:
    a. show that a line parallel to one side of a triangle and intersecting the other sides divides those sides into proportional segments.
    b. investigate properties of a midsegment of a triangle.

13. Use GSP/GeoGebra Lab 8 activities 1–9 to investigate properties of a midpoint quadrilateral.

14. Use GSP/GeoGebra Assessment A-IV exercise 12 to construct the indicated figure.

Review Problems

15. Given the following base of an isosceles triangle and the altitude to that base, construct the triangle:

                     Base

                     Altitude

16. Given the following length of a side of an equilateral triangle, construct an altitude of the equilateral triangle.

17. Write a paragraph describing how to construct an isosceles right triangle when given the length of the hypotenuse of the 45°-45°-90° triangle.

18. Use a compass and a straightedge to draw a pair of obtuse vertical angles and the angle bisector of one of these angles. Extend the angle bisector. Does the extended angle bisector bisect the other vertical angle? Justify your answer.

19. Describe a minimal set of conditions that can be used to argue that two quadrilaterals are congruent.

Trends in Mathematics and Science Study (TIMSS) Question

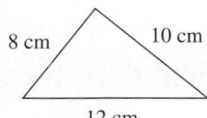

Which of the following triangles is similar to the triangle shown above?

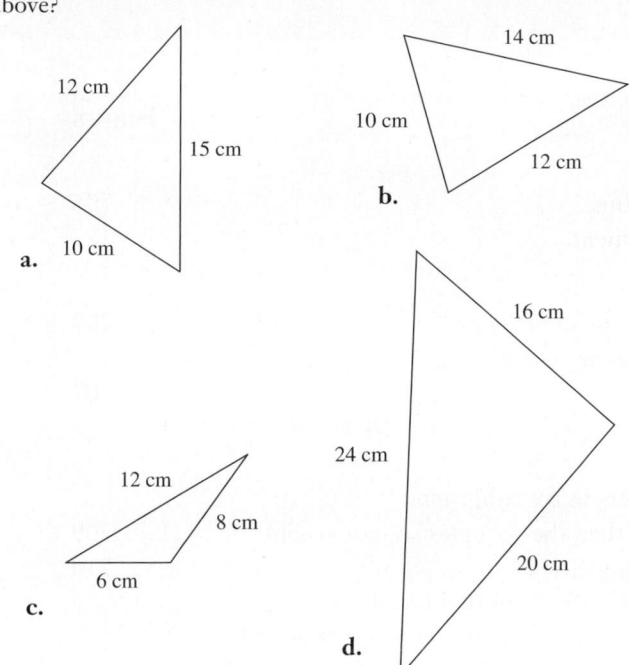

a.

b.

c.

d.

TIMSS, Grade 8, 2003

National Assessment of Educational Progress (NAEP) Question

The figure below shows two right angles. The length of *AE* is *x* and the length of *DE* is 40.

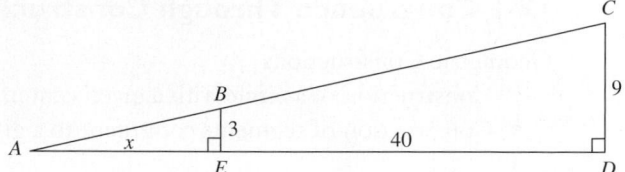

Show all of the steps that lead to finding the value of *x*. Your last step should give the value of *x*.

NAEP, Grade 8, 2007

○ **Historical Note** ─────────────────────────────────

In 1975, Benoit Mandelbrot invented the word *fractal* to describe certain irregular and fragmented shapes. These shapes are such that if one looks at a small part of the shape, the small part resembles the larger shape. These shapes are somewhat like rep-tiles, although the smaller structures of fractals are not necessarily identical to the larger structure. Fractal geometry was introduced for the purpose of modeling natural phenomena such as irregular coastlines, arteries and veins, the branching structure of plants, the thermal agitation of molecules in a fluid, and sponges. An example of a fractal—a computer-generated picture of the Mandelbrot set known as the "Tail of the Seahorse," appears below.

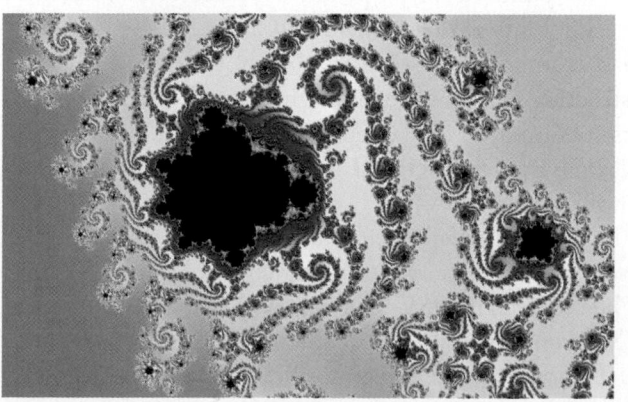

Hint for the Solution to the Preliminary Problem

$\triangle ABC$ on the map is similar to the triangle in reality.

Chapter 12 Summary

12-1 Congruence Through Constructions	**Pages**

Geometric Constructions
- Construction of a circle with a given center and radius.   706
- Construction of segments congruent to a given segment.

Congruence
- **Segments are congruent** if, and only if, they have the same length.   707
- **Angles are congruent** if, and only if, their measures are equal.
- **△ABC is congruent to △DEF**, written: 707
  △ABC ≅ △DEF if, and only if, ∠A ≅ ∠D, ∠B ≅ E, ∠C ≅ ∠F, $\overline{AB}$ ≅ $\overline{DE}$,
  $\overline{BC}$ ≅ $\overline{EF}$, and $\overline{AC}$ ≅ $\overline{DF}$
- **Side, Side, Side (SSS)**. If the three sides of one triangle are congruent,
  respectively, to the three sides of a second triangle, then the triangles are congruent.   709
- Construction of a triangle congruent to a given triangle.   710
- Given segments of lengths a, b, and c. A triangle with sides congruent to these
  segments can be constructed if, and only if, $a + b > c$, $a + c > b$, and $b + c > a$.
- **Theorem: Side, Angle, Side (SAS)**. If two sides and the included angle of one triangle
  are congruent to two sides and the included angle of another triangle, respectively,
  then the two triangles are congruent.   711–712
- **Theorem: Hypotenuse-Leg (HL)**. If the hypotenuse and a leg of one right triangle
  are congruent to the hypotenuse and a leg of another right triangle, then the triangles
  are congruent.   713
  - **Theorem:** For every isosceles triangle:
    a. the angles opposite the congruent sides are congruent (base angles of an
    isosceles triangle are congruent)   713
    b. the angle bisector of an angle formed by two congruent sides lies on the
    perpendicular bisector of the third side of the triangle.   713
  - **Theorem:**   714
    a. Any point equidistant from the endpoints of a segment is on the
    perpendicular bisector of the segment.
    b. Any point on the perpendicular bisector of a segment is equidistant from the
    endpoints of the segment.
- Construction of a perpendicular bisector of a segment can be accomplished by
  constructing two points equidistant from the endpoints of the segment.   715
- Construction of a circle circumscribed about a triangle.
  - A circle is **circumscribed** about a given △ABC if every vertex of the triangle is   715–716
    on the circle.
  - The point O, where the perpendicular bisectors intersect, is the **circumcenter**,
    the center of the circumscribing circle.
  - The required circle is the circle with center O and radius OA.
  - **Theorem:**   717–718
    a. If a circle can be circumscribed about a convex quadrilateral, then the sum of
    the measures of a pair of opposite angles of the quadrilateral is 180°, that is
    the angles are supplementary.
    b. If the sum of the measures of a pair of opposite angles of a convex quadrilateral
    is 180°, then a circle can be circumscribed about the quadrilateral.

12-2 Additional Congruence Properties

Theorem: Angle, Side, Angle (ASA) If two angles and the included side of one triangle are congruent to two angles and the included side of another triangle, respectively, then the triangles are congruent.	724
Theorem: Angle, Angle, Side (AAS) If two angles and a side opposite one of these two angles of a triangle are congruent to the two corresponding angles and the corresponding side in another triangle, then the two triangles are congruent.	724
Theorem: The opposite sides and the opposite angles of a parallelogram are congruent.	725

The various quadrilaterals have the following properties: 727
  • Trapezoid
     ○ Consecutive angles between parallel sides are supplementary.
  • Isosceles Trapezoid
     ○ Each pair of base angles are congruent.
     ○ A pair of opposite sides are congruent.
     ○ If a trapezoid has congruent diagonals it is isosceles.
  • Parallelogram
     ○ A parallelogram has all the properties of a trapezoid.
     ○ Opposite sides are congruent.
     ○ Opposite angles are congruent.
     ○ Diagonals bisect each other.
     ○ A quadrilateral in which the diagonals bisect each other is a parallelogram.
  • Rectangle
     ○ A rectangle has all the properties of a parallelogram.
     ○ All the angles of a rectangle are right angles.
     ○ A quadrilateral in which all the angles are right angles is a rectangle.
     ○ The diagonals of a rectangle are congruent and bisect each other.
     ○ A quadrilateral in which the diagonals are congruent and bisect each other is a rectangle.
  • Kite
     ○ Lines containing the diagonals are perpendicular to each other.
     ○ A line containing one diagonal is a bisector of the other diagonal.
     ○ One diagonal bisects nonconsecutive angles.
  • Rhombus
     ○ A rhombus has all the properties of a parallelogram and a kite.
     ○ A quadrilateral in which all the sides are congruent is a rhombus.
     ○ The diagonals of a rhombus are perpendicular to and bisect each other.
     ○ Each diagonal bisects opposite angles.
  • Square
     ○ A square has all the properties of a parallelogram, a rectangle, and a rhombus.

SASAS is one possible congruence condition for quadrilaterals.

12.3 Additional Constructions

Construction of a parallel line to a given line ℓ through a point P can be accomplished by any of the following:	734
• Constructing a rhombus with one side on ℓ.	
• Constructing a transversal and congruent corresponding angles.	734

Constructing an angle bisector 735
- The construction can be accomplished making the given angle the angle of a rhombus.

Constructing perpendicular lines 736
- Constructing a line perpendicular to ℓ through point P not on ℓ can be accomplished by making P a vertex of a rhombus and connecting it to the opposite vertex.
- Bisecting a line segment can be achieved by constructing the perpendicular bisector of the segment. 737
- Constructing a perpendicular to ℓ at a point M on ℓ can be accomplished by making M the point of intersection of the diagonals of a rhombus. 737

Properties of Angle Bisectors
- **Theorem:**
  a. Any point P on an angle bisector is equidistant from the sides of the angle. 739
  b. Any point in the interior of an angle, that is equidistant from the sides of an angle is on the angle bisector of the angle.
- The intersection of any two angle bisectors of a triangle is the **incenter.** 740
  o **Theorem:** The angle bisectors of a triangle are concurrent and the three distances from the point of intersection to the sides are equal.

12-4 Similar Triangles and Other Similar Figures

Two polygons with the same number of vertices are **similar** if there is a one-to-one correspondence between the vertices of one and the vertices of the other such that the corresponding interior angles are congruent and corresponding sides are proportional. 745

$\triangle ABC$ is **similar to** $\triangle DEF$, written $\triangle ABC \sim \triangle DEF$, if, and only if,

$\angle A \cong \angle D, \angle B \cong \angle E, \angle C \cong \angle F,$ and $\dfrac{AB}{DE} = \dfrac{AC}{DF} = \dfrac{BC}{EF}.$

- **Theorem: SSS Similarity for Triangles** 744
  If corresponding sides of two triangles are proportional, then the triangles are similar.
- **Theorem: SAS Similarity for Triangles** 744
  Given two triangles, if two sides are proportional and the included angles are congruent, then the triangles are similar.
- **Theorem: Angle, Angle (AA) Similarity for Triangles** 746
  If two angles of one triangle are congruent, respectively, to two angles of a second triangle, then the triangles are similar.

Properties of Proportion 748
- **Theorem:** If a line parallel to one side of a triangle intersects the other sides, then it divides those sides into proportional segments. 748
- **Theorem:** If a line divides two sides of a triangle into proportional segments, then the line is parallel to the third side. 748
- **Theorem:** If parallel lines cut off congruent segments on one transversal, then they cut off congruent segments on any transversal. 748
  o The above theorem can be used to divide a given segment into any number of congruent parts.

The segment connecting the midpoints of two sides of a triangle or the midpoints of two opposite sides of a quadrilateral is a **midsegment**.	749
• **Theorem: The Midsegment Theorem**	749
The midsegment joining the midpoints of two sides of a triangle is parallel to and is half as long as the third side.	749
• **Theorem:** If a line bisects one side of a triangle and is parallel to a second side, then it bisects the third side and therefore is a midsegment.	749
A **median** of a triangle is a segment connecting a vertex of the triangle to the midpoint of the opposite side. The point of intersection of any two medians is the **centroid**.	750
Similar triangles can be used to make indirect measurements.	751–754
Similar triangles can be used to define the slope of a line.	753–754

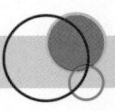

Chapter 12 Review

1. Each of the following figures contains at least one pair of congruent triangles. Identify them and tell why they are congruent.

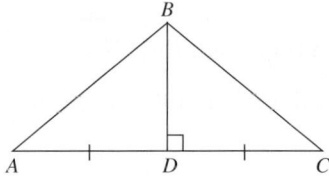

a.

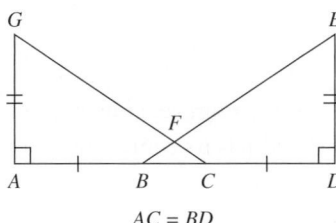

$AC = BD$

b.

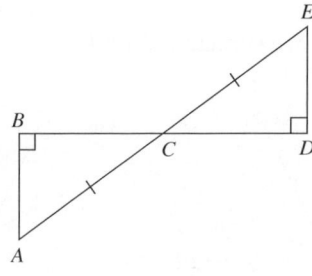

c.

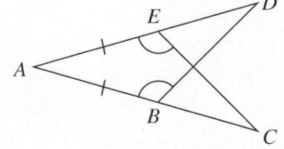

d.

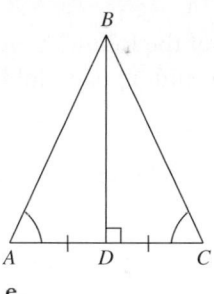

e.

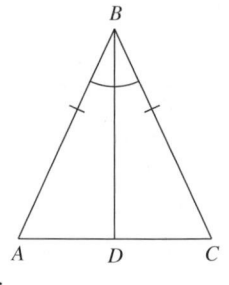

f.

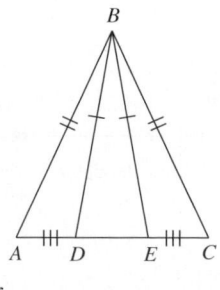

g.

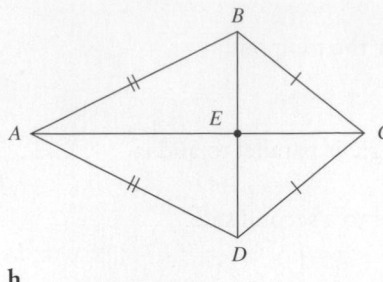

h.

2. In the following figure, *ABCD* is a square and $\overline{DE} \cong \overline{BF}$. What kind of figure is *AECF*? Justify your answer.

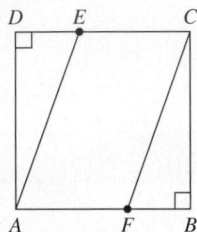

3. Construct each of the following by (1) using a compass and straightedge and (2) paperfolding:

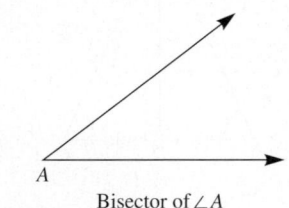

Bisector of ∠*A*

a.

Perpendicular to ℓ at *B*

b.

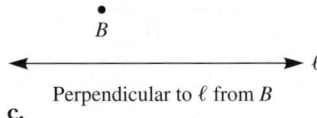

Perpendicular to ℓ from *B*

c.

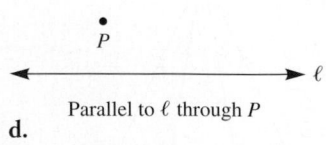

Parallel to ℓ through *P*

d.

4. For each of the following pairs of similar triangles, find the missing measures:

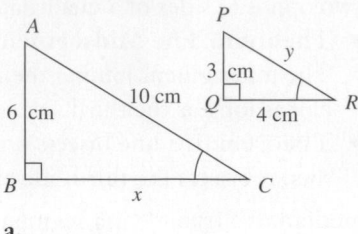

a.

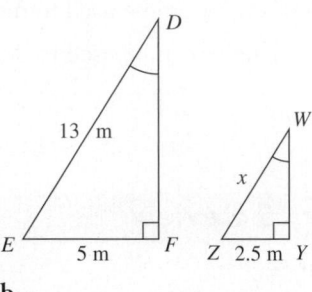

b.

5. Divide the following segment into five congruent parts:

—————————————

6. If *ABCD* is a trapezoid, $\overline{EF} \| \overline{AD}$, and $\overline{AC}$ is a diagonal, then what is the relationship between $\dfrac{a}{b}$ and $\dfrac{c}{d}$?

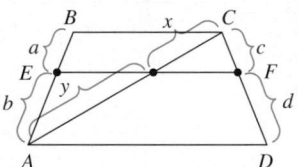

7. Given the following figure, construct a circle that contains *A* and *B* and has its center on ℓ:

8. For each of the following figures, show that appropriate triangles are similar and find *x*:

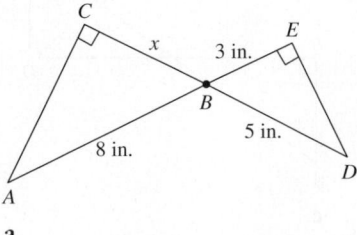

a.

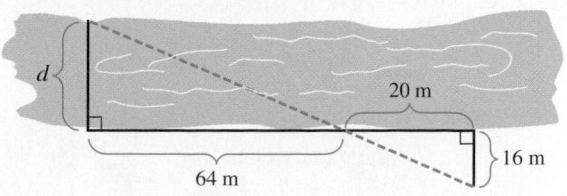

12. Find the distance d across the river sketched.

13. Describe how any parallelogram can be dissected and reassembled to form a rectangle.

14. A regular hexagon $ABCDEF$ can be divided into two congruent trapezoids by drawing $\overline{AD}$.
  a. What is the relation of $\overline{AD}$ to the circumcircle of the regular hexagon?
  b. Prove that $\angle ABD$ is a right angle.

15. Is the following statement always true or false? If false, what condition on the diagonals could we add to make a true statement?

  A quadrilateral whose diagonals are congruent and perpendicular is a square.

16. For each of the following find the equation of the line determined by the given pair of points.
  a. $(2, {}^-3)$ and $({}^-1, 1)$
  b. $({}^-3, 0)$ and $(3, 2)$

17. Why are two congruent triangles ABC and $A'B'C'$ similar?

18. $ABCD$ is a trapezoid with $\overline{BC} \| \overline{AD}$. Points M and N are midpoints of the diagonals. Line $\overleftrightarrow{CN}$ intersects $\overline{AD}$ at E.
  a. Prove that $\triangle BCN \cong \triangle DEN$
  b. Prove that $\overline{MN}$ is a midsegment of $\triangle ACE$.
  c. If $AD = a$, $BC = b$, express MN in terms of a and b. Justify your answers.

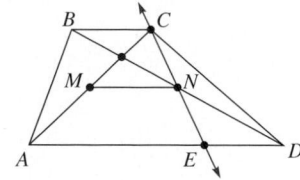

Given: $\angle ABC \cong \angle ADE$

b.

9. A person 2 m tall casts a shadow 1 m long when a building has a 6 m shadow. How high is the building?

10. a. Which of the following polygons can be inscribed in a circle? Assume that all sides of each polygon are congruent and that all the angles of polygons (ii) and (iii) are congruent.

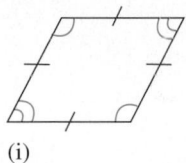 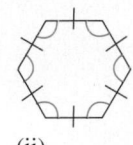 

  (i)   (ii)   (iii)

  b. Based on your answer in (a), make a conjecture about what kinds of polygons can be inscribed in a circle.

11. Determine the vertical height of the playground slide shown in the following figure:

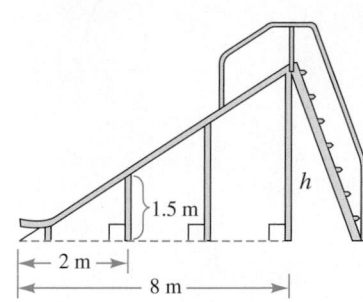

Congruence and Similarity with Transformations

Preliminary Problem

On a 1-m equilateral-triangle "pool table," a ball is hit at a 60° angle along a segment parallel to a side of the table and 10 cm from one pocket as shown. Find the length of the path as it travels around the table until it returns to where it started or falls into one of the pockets.

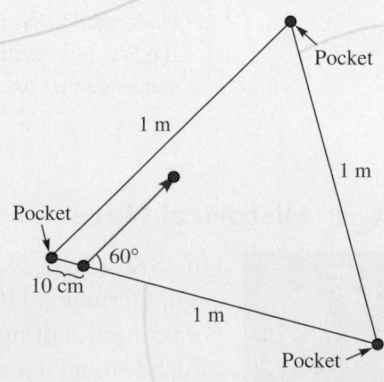

If needed, see Hint on page 822.

Euclid envisioned moving one geometric figure in a plane and placing it on top of another to determine whether the two figures were congruent. Intuitively, we know this can be done by making a tracing of one figure, then sliding, turning, flipping the tracing, or using some combination of these motions, to place it atop the other figure. Additionally, similarities (and/or scaling) can be accomplished through a combination of the motions and dilations (size-altering transformations introduced in this chapter).

In grades 3–5 *Focal Points*, students should:

- Predict and describe the results of sliding, flipping, and turning two-dimensional shapes. (p. 31)

- Describe a motion or a series of motions that will show that two shapes are congruent. (p. 32)

In grades 6–8 *Focal Points*, students should:

- Describe sizes, positions, and orientations of shapes under informal transformations such as flips, turns, slides, and scaling. (p. 37)

- Examine the congruence, similarity, and line or rotational symmetry of objects using transformations. (p. 37)

This chapter contains sections on motions and tessellations, with connections between symmetry and the motions.

In *PSSM*, we find the following:

Young children come to school with intuitions about how shapes can be moved. Students can explore motions such as slides, flips, and turns by using mirrors, paper folding, and tracing. Later, their knowledge about transformations should become more formal and systematic. In grades 3–5 students can investigate the effects of transformations and begin to describe them in mathematical terms. Using dynamic geometry software, they can begin to learn the attributes needed to define a transformation. In the middle grades, students should learn to understand what it means for a transformation to preserve distance, as translations, rotations, and reflections do…. (p. 43)

Additionally *PSSM* lists the following expectations for all students in prekindergarten through grade 2:

- recognize and apply slides, flips, and turns;

- recognize and create shapes that have symmetry. (p. 96)

In grade 8 *Common Core Standards*, students are expected to:

- Verify experimentally the properties of rotations, reflections, and translations:

  ○ Lines are taken to lines, and line segments to line segments of the same length.

  ○ Angles are taken to angles of the same measure.

  ○ Parallel lines are taken to parallel lines.

- Understand that a two-dimensional figure is congruent to another if the second can be obtained from the first by a sequence of rotations, reflections, and translations; given two congruent figures, describe a sequence that exhibits the congruence between them.

 Historical Note

 In 1872, at age 23, Felix Klein (1849–1925) was appointed to a chair at the University of Erlangen, Germany. His inaugural address, the *Erlanger Programm*, described geometry as the study of properties of figures that do not change under a particular set of transformations. Specifically, Euclidean geometry was described as the study of such properties of figures as area and length, which remain unchanged under a set of transformations called *isometries*.

- Describe the effect of dilations, translations, rotations, and reflections on two-dimensional figures using coordinates.

- Understand that a two-dimensional figure is similar to another if the second can be obtained from the first by a sequence of rotations, reflections, translations, and dilations; given two similar two-dimensional figures, describe a sequence that exhibits the similarity between them. (pp. 55–56)

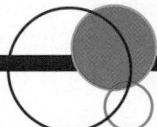

13-1 Translations and Rotations

Think about a rug that completely covers a floor. If a long pin were stuck through the rug, it pairs a point on the rug with a corresponding point on the floor below. There is a one-to-one correspondence between the points of the rug and the points of the floor. If the rug were flipped over, and placed back on the floor so that it still covered the floor completely, then another one-to-one correspondence could be established between the points of the rug and the points of the floor. Similarly, a one-to-one correspondence between a plane and itself can be established so that there is a **transformation** from the plane to itself with each point from the original plane having an **image** in the "transformed" plane, and vice versa. Any transformation of the plane that preserves distance is an **isometry** (or **motion**). Preserving distance means that given any two points P and Q in the original plane, the distance PQ between them is the same as the distance $P'Q'$ between the matching points P' and Q' in the transformed plane. Isometries also preserve geometric shapes and relations including parallelism and perpendicularity. In this section, we examine translations and rotations.

Translations

Figure 13-1(a) shows a two-dimensional representation of a child moving down a slide without twisting or turning. This type of motion is a **translation**, or **slide**. In Figure 13-1(a), the child (**preimage** or **original**) at the top of the slide moves a certain distance in a certain direction, indicated by a **slide arrow**, along a **slide line** to obtain the **image** at the bottom of the slide. In Figure 13-1(b) the translation is determined by the **slide arrow**, or **vector**, from M to N. The vector determines the image of any point in a plane in the following way: The image of a point A in the plane is the point A'

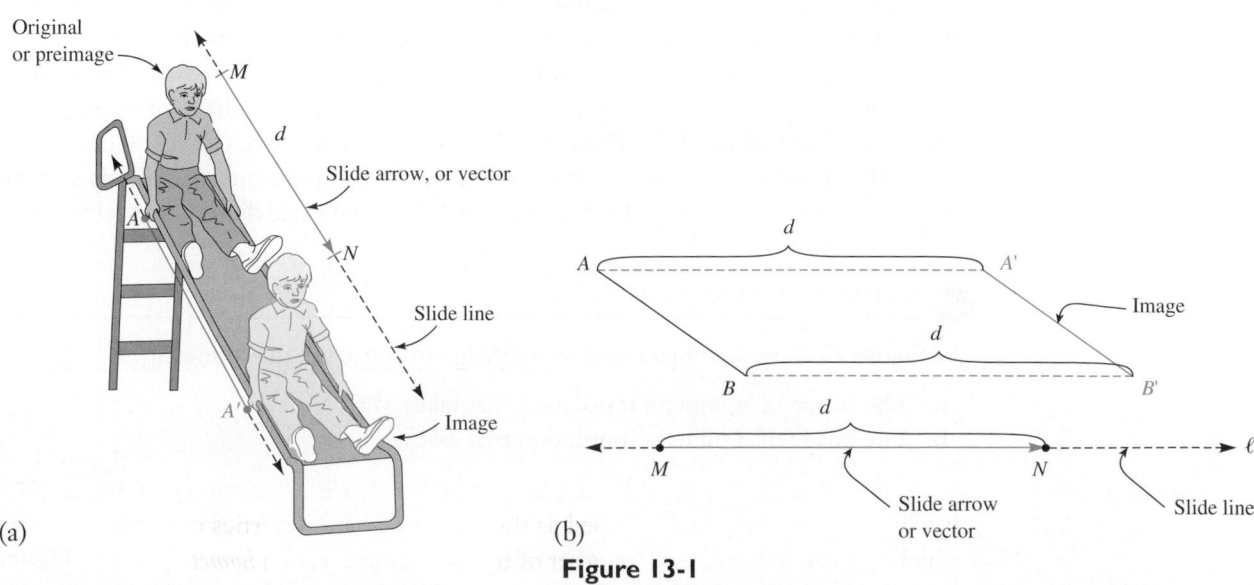

(a)   (b)

Figure 13-1

obtained by sliding A along a line parallel to slide line $\overleftrightarrow{MN}$ in the direction from M to N by the distance MN indicated by the slide arrow. (MN is also denoted by d in Figure 13-1(b).) Hence $AA' = MN = d$ and $\overleftrightarrow{MN}$ is parallel to $\overrightarrow{AA'}$. Under the translation, a figure changes neither its shape nor size. In fact, a translation preserves both length and angle size, and thus congruence of figures.

In Figure 13-1(b) $\overline{AB}$ is slid, or translated, by the slide arrow from M to N along slide line ℓ making $\overline{AB} \cong \overline{A'B'}$ and $AA' = BB' = MN = d$. Thus, quadrilateral $AA'B'B$ is a parallelogram, and $\overline{AB} \parallel \overline{A'B'}$. This illustrates that under a translation the image of a segment is a congruent segment parallel to the original (preimage).

A definition of a translation follows:

> ### Definition of a Translation
>
> A **translation** is a motion (or transformation) of a plane that moves every point of the plane a specified distance in a specified direction along a straight line.

Constructions of Translations

Using tracing paper is a natural way to construct a "motion." The image of a figure under a translation constructed with tracing paper is seen in Figure 13-2.

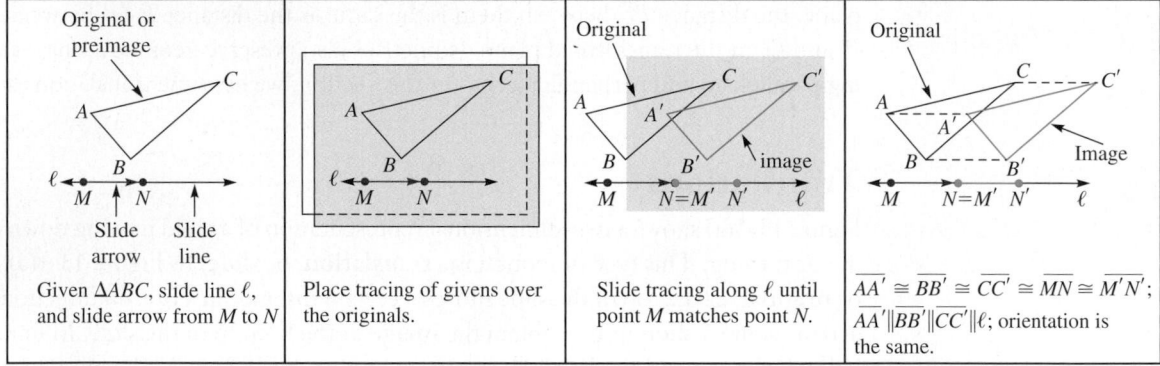

Figure 13-2 Construction of translation with tracing paper

Traditionally, most constructions in geometry have been accomplished with a compass and a straightedge. We leave the traditional construction as an activity in Now Try This 13-1.

To construct the image of an object under a translation, we first need to know how to construct the image A' of a single point A. In Figure 13-2, by definition of translation $MN = AA'$ and $\overleftrightarrow{MN}$ is parallel to $\overrightarrow{AA'}$. Thus, $MAA'N$ is a parallelogram. (Why?) Consequently, to construct the translation image A' of a point A with a compass and straightedge, we need only to construct a parallelogram $MAA'N$ so that $\overrightarrow{AA'}$ is in the same direction as $\overrightarrow{MN}$.

NOW TRY THIS 13-1

In Figure 13-3 use a compass and straightedge to construct the following:

a. The image of A under a translation that takes M to N.
b. The image of A under a translation that takes N to M.

Figure 13-3

We saw that the image of a segment under a translation was a parallel segment. Thus, to find the image of a triangle under a translation, it suffices to find the images of the three vertices of the triangle and to connect these images with segments to form the triangle's image.

It is possible to use a geoboard, dot paper, or a grid to find an image of a segment. The following example uses dot paper.

EXAMPLE 13-1

Find the image of $\overline{AB}$ under the translation from X to X' pictured on the dot paper in Figure 13-4.

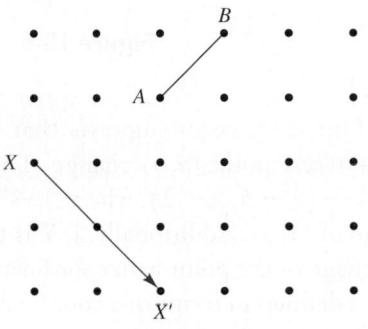

Figure 13-4

Solution   We want the image of $\overline{AB}$ to be parallel to $\overline{AB}$ so that $ABB'A'$ is a parallelogram. X', the image of X under the translation, could be obtained by shifting X 2 units vertically down and then 2 units horizontally to the right, as shown in Figure 13-5. This shifting determines the slide arrow from X to X'. Therefore the image of points A and B on the dot paper can be obtained by shifting each point first down 2 units and then 2 units to the right. Thus, $\overline{A'B'}$, the image of $\overline{AB}$, is found as in Figure 13-5.

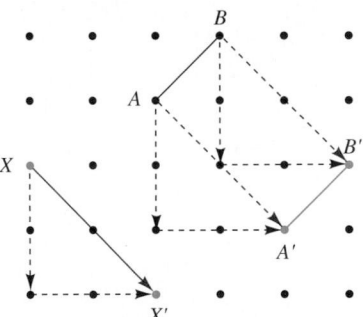

Figure 13-5

Coordinate Representation of Translations

In many applications of mathematics and as recommended in the grade 8 *Common Core Standards* (p. 56), it is necessary to use translations in a coordinate system. In Figure 13-6, $\triangle A'B'C'$ is the image of $\triangle ABC$ under the translation defined by the slide arrow from O to O', where O is the origin and O' has coordinates $(5, ^-2)$. Point O' is the image of point O under the given translation. The point $O'(5, ^-2)$ can be obtained by moving O horizontally to the right 5 units and then 2 units down. As each point in triangle ABC is translated in the direction from O to O' by the distance OO', we obtain the image of any point by moving it horizontally to the right 5 units and then vertically 2 units down. In Figure 13-6 the images A', B', and C' for points A, B, and C are shown. Table 13-1 shows how the coordinates of the image vertices A', B', and C' in Figure 13-6 are obtained from the coordinates of A, B, and C.

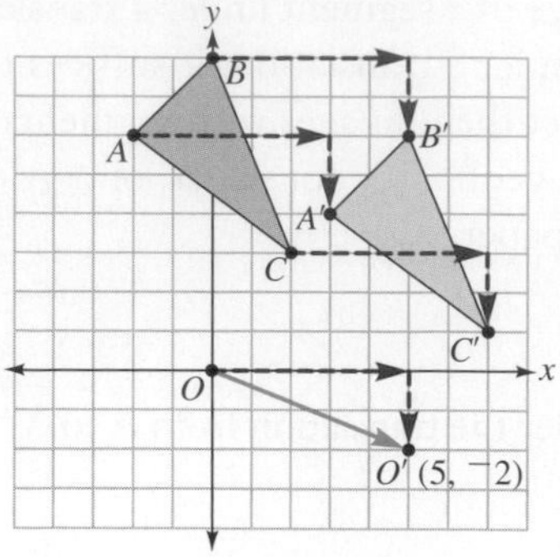

Table 13-1

	Image Point
Point (x, y)	$(x + 5, y - 2)$
$A(^-2, 6)$	$A'(3, 4)$
$B(0, 8)$	$B'(5, 6)$
$C(2, 3)$	$C'(7, 1)$

Figure 13-6

This discussion suggests that a translation is described by showing how the coordinates of any point (x, y) change. The translation in Table 13-1 can be written symbolically as $(x, y) \rightarrow (x + 5, y - 2)$, where "$\rightarrow$" denotes "moves to." In this notation, $(x + 5, y - 2)$ is the image of (x, y). Additionally, if T is the translation function with the plane as the domain, then the image of the point with coordinates (x, y) is $T(x, y) = (x + 5, y - 2)$. This idea of a translation is defined in terms of a coordinate system in the following:

Definition of the Translation in a Coordinate System

A **translation** is a function from the plane to the plane such that the image of the point with coordinates (x, y) is $(x + a, y + b)$, where a and b are real numbers.

EXAMPLE 13-2

Find the coordinates of the image of the vertices of quadrilateral $ABCD$ in Figure 13-7 under the translations in parts (a) through (c). Draw the image in each case.

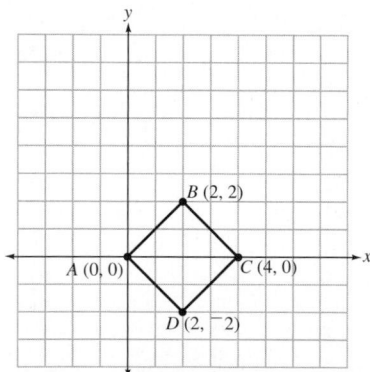

Figure 13-7

  a. $(x, y) \rightarrow (x - 2, y + 4)$
  b. A translation determined by the slide arrow from $A(0, 0)$ to $A'(^-2, 4)$
  c. A translation determined by the slide arrow from $S(4, ^-3)$ to $S'(2, 1)$

Solution
  a. Because $(x, y) \rightarrow (x - 2, y + 4)$, the images A', B', C', and D' of the corresponding points A, B, C, and D can be found as follows:

$$A(0, 0) \rightarrow A'(0 - 2, 0 + 4), \text{ or } A'(^-2, 4)$$
$$B(2, 2) \rightarrow B'(2 - 2, 2 + 4), \text{ or } B'(0, 6)$$
$$C(4, 0) \rightarrow C'(4 - 2, 0 + 4), \text{ or } C'(2, 4)$$
$$D(2, ^-2) \rightarrow D'(2 - 2, ^-2 + 4), \text{ or } D'(0, 2)$$

The square *ABCD* and its image *A'B'C'D'* are shown in Figure 13-8.

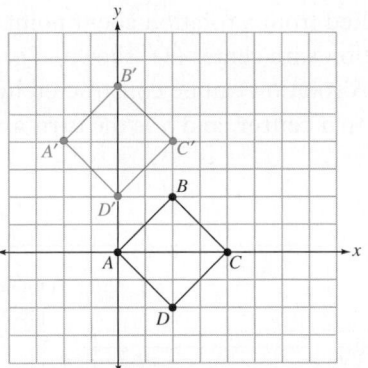

Figure 13-8

b. This is the same translation as in part (a) and hence the image of *ABCD* is *A'B'C'D'* as in part (a).

c. The translation from $S(4, {}^-3)$ to $S'(2, 1)$ moves $S(4, {}^-3)$ 2 units left to $(2, {}^-3)$ and then vertically up 4 units to $S'(2, 1)$. Thus, any point (x, y) moves to a point with coordinates $(x - 2, y + 4)$. This is the same translation as in part (a), so the image of *ABCD* is *A'B'C'D'*.

Translations are useful in determining frieze patterns such as those appearing in wallpaper designs. An example is seen in Figure 13-9.

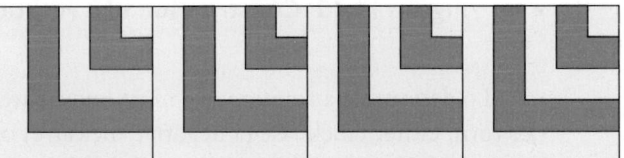

Figure 13-9

 NOW TRY THIS 13-2

In the wallpaper pattern of Figure 13-10, describe a translation that could be used to construct the wallpaper.

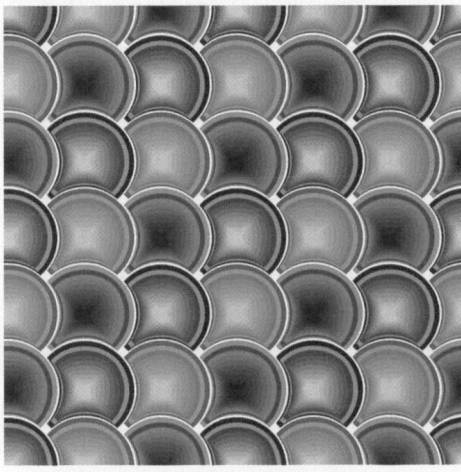

Figure 13-10

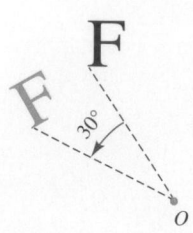

Rotations

A **rotation**, or **turn**, is another kind of isometry. Figure 13-11 illustrates congruent figures that resulted from a rotation about point O. The image of the letter **F** under a 30° counterclockwise rotation with center O is shown in green.

A rotation can be constructed by using tracing paper, as in Figure 13-12, where point O is the **turn center**, and α is the **turn angle**.

Figure 13-11

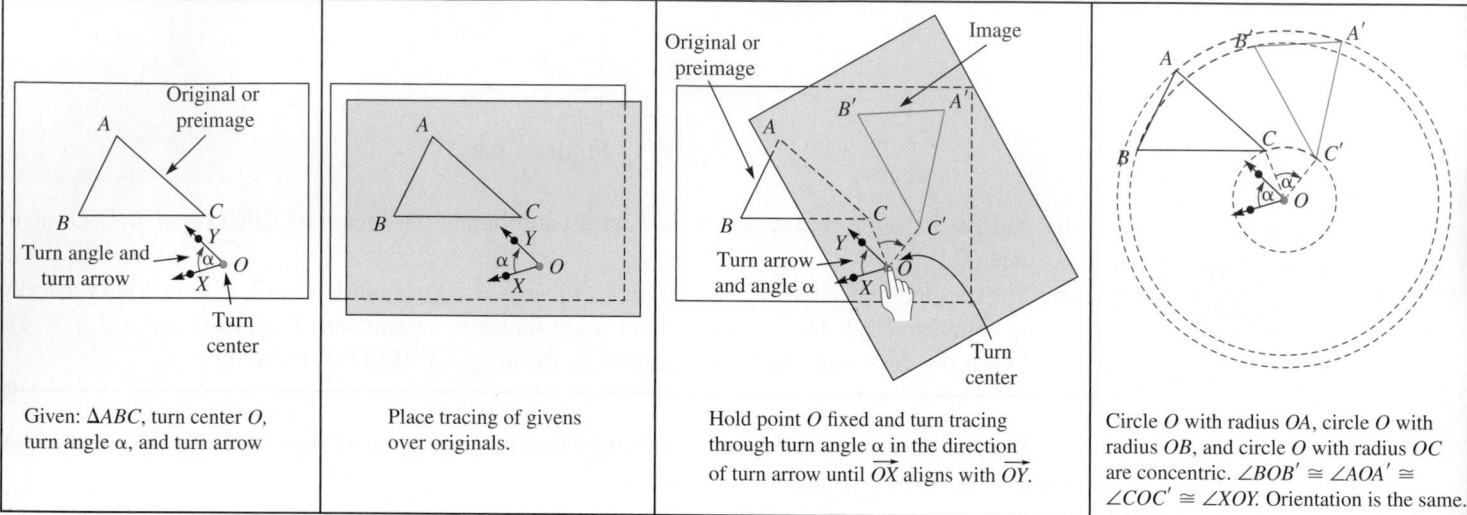

| Given: $\triangle ABC$, turn center O, turn angle α, and turn arrow | Place tracing of givens over originals. | Hold point O fixed and turn tracing through turn angle α in the direction of turn arrow until $\overrightarrow{OX}$ aligns with $\overrightarrow{OY}$. | Circle O with radius OA, circle O with radius OB, and circle O with radius OC are concentric. $\angle BOB' \cong \angle AOA' \cong \angle COC' \cong \angle XOY$. Orientation is the same. |

Figure 13-12 Construction of a rotation using tracing paper

To determine a rotation, we must know three pieces of data: the turn center; the direction of the turn, either clockwise (a negative measure) or counterclockwise (a positive measure); and the amount of the turn. The amount and the direction of the turn can be illustrated by a **turn arrow**, or they can be specified in numbers of degrees.

This discussion leads to the following definition.

Definition of Rotation

A **rotation** is a transformation of the plane determined by holding one point—the center—fixed and rotating the plane about this point by a certain amount in a certain direction (a certain number of degrees either clockwise or counterclockwise).

To construct an image of a figure under a rotation as in Figure 13-12, observe that every point on the tracing paper construction except the center moves along a circle. Also the angle formed by any point (not the turn center), the center of the rotation O, and the image of the point is the angle of the turn. (Why?) With this in mind, we construct the image of a point under a rotation in Now Try This 13-3.

 NOW TRY THIS 13-3

Use a compass and a straightedge to construct the image of point P under a rotation with center O through the angle and in the direction given in Figure 13-13. (*Hint*: Construct an isosceles

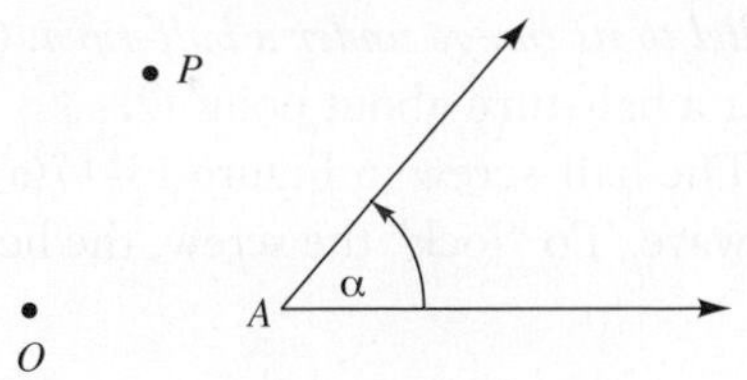

• *P*

• *O*

A α

Figure 13-13

triangle *BAC* with *B* on one side of the given angle and *C* on the other side so that *AB* = *AC* = *OP*. Then construct △*POP'* congruent to △*BAC*.)

It can be proven that a rotation is an isometry, so the image of a figure under a rotation is congruent to the original figure. It can be shown that *under any isometry, the image of a line is a line, the image of a circle is a circle, and the images of parallel lines are parallel lines*.

For certain angles, like 90°, rotations may be constructed on a geoboard or dot paper, as demonstrated in Example 13-3.

EXAMPLE 13-3

In Figure 13-14, find the image of △*ABC* under the rotation shown with center *O*.

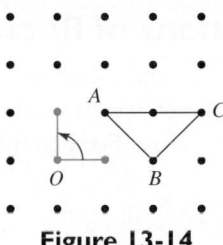

Figure 13-14

Solution △*A'B'C'*, the image of △*ABC*, is shown in Figure 13-15. The image of *A* is *A'* because ∠*A'OA* is a right angle (Why?) and *OA* = *OA'*. Similarly, *B'* is the image of *B*. To find the location of *C'*, we use the fact that rotation is an isometry and hence △*A'B'C'* ≅ △*ABC*. Thus, ∠*B* ≅ ∠*B'*. The location of point *C'* shown makes ∠*B* ≅ ∠*B'* and *C'B'* = *CB* (Why?). The direction of the rotation is counterclockwise as specified.

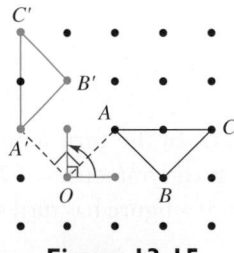

Figure 13-15

A rotation of 360° about a point moves any point and hence any figure onto itself. Such a transformation is an **identity transformation**. Any point may be the center of such a rotation.

A rotation of 180° about a point is a **half-turn**. Because a half-turn is a rotation, it has all the properties of rotations. Additionally, as suggested in Figure 13-16(b), *a line (or line segment) is*

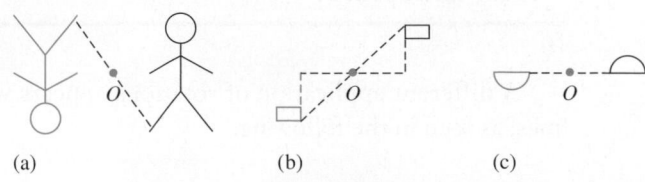

(a)   (b)   (c)

Figure 13-16

parallel to its image under a half-turn. (Why?) Figure 13-16 shows some shapes and their images under a half-turn about point *O*.

The half-screw in Figure 13-17(a) has a head that shows a practical example of a half-turn in hardware. To "lock" the screw, the head is turned 180° as shown in Figure 13-17(b).

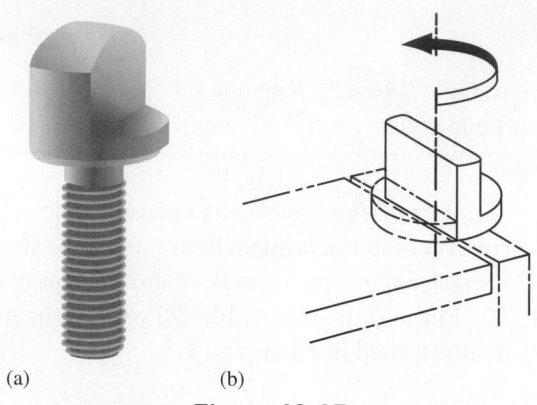

(a)   (b)

Figure 13-17

Applications of Rotations

One immediate application of a rotation is its use in turn symmetries. One way to describe a figure with turn symmetry is to see if there is a center and turn angle(s) that will take the original figure to itself as an image. For example, Figure 13-18 shows a star that has several turn symmetries.

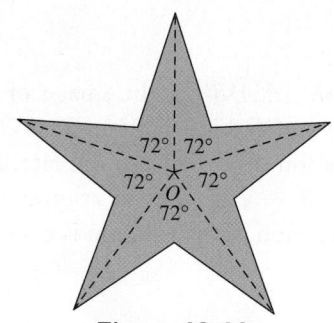

Figure 13-18

The center, *O*, of the turn symmetries is the center of the circle that circumscribes the star. As marked, the turn symmetries of 72°, 144°, 216°, and 288° are turn angles of rotations centered at *O*. Observe that if a figure has turn symmetry of 72°, it also has turn symmetries of multiples of 72°.

> **NOW TRY THIS 13-4**
>
> **a.** Draw a figure with point symmetry (see Chapter 11).
> **b.** Explain whether the figure drawn in part (a) is its own image under a half-turn.
> **c.** Given any line ℓ and a point *P* on the line, what is the image of ℓ under a half-turn about *P*?
> **d.** Given any line ℓ and a point *P* not on ℓ, describe how to construct the image of ℓ through a half-turn about *P*.

A different application of rotations appears when we consider the slopes of perpendicular lines, as seen in the following.

Slopes of Perpendicular Lines via Rotations

Transformations can be used to investigate various mathematical relationships. For example, they can be used to determine the relationship between the slopes of two perpendicular lines, neither of which is vertical.

We first consider a special case in which the lines go through the origin. Suppose the two perpendicular lines are as pictured in Figure 13-19 with one point, B, of ℓ_1 having coordinates (a, b). The coordinates of point B determine the coordinates of point $A(a, 0)$, and triangle OAB is formed. If the plane containing the lines is rotated $90°$ counterclockwise about center O, then $\ell_2 = \ell'_1$, the image of ℓ_1 is ℓ_2. The image of triangle OAB is triangle $OA'B'$. Because the rotation takes a triangle to a congruent triangle, $\triangle OAB \cong \triangle OA'B'$. Thus, A' has coordinates $(0, a)$ and B' has coordinates $(^-b, a)$. (Why?)

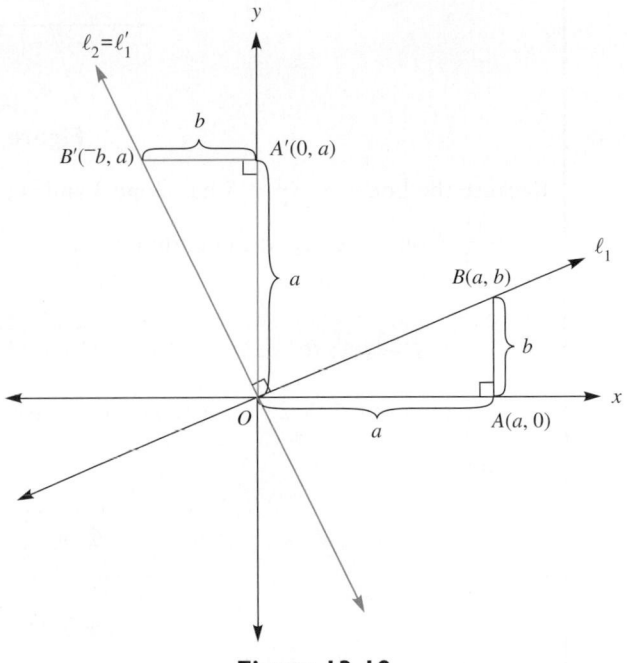

Figure 13-19

Now consider the slopes of ℓ_1 and ℓ_2. The slope of ℓ_1 is b/a, and the slope of ℓ_2 is $a/(^-b)$. The slopes are negative reciprocals of each other and their product is $^-1$.

If two nonvertical perpendicular lines ℓ and m, do not intersect at the origin, then two lines parallel to ℓ and m can be found that do pass through the origin. Because parallel lines have equal slopes, the relationship between the slopes, m_1 and m_2, of the perpendicular lines is the same as the relationship between the slopes of the perpendicular lines through the origin; that is, $m_1 m_2 = ^-1$.

It is also possible to prove the converse statement; that is, if the slopes of two lines satisfy the condition $m_1 m_2 = ^-1$, then the lines are perpendicular. We summarize these results in the following theorem.

Theorem 13-1:   Slopes of Perpendicular Lines

Two nonvertical lines, are perpendicular if, and only if, their slopes m_1 and m_2 satisfy the condition $m_1 m_2 = ^-1$. Every vertical line has no slope but is perpendicular to every line with slope 0.

EXAMPLE 13-4

Find the equation of line ℓ through point $(^-1, 2)$ and perpendicular to the line with equation $y = 3x + 5$.

Solution   If m is the slope of ℓ, as in Figure 13-20, then the equation of line ℓ is $y = mx + b$.

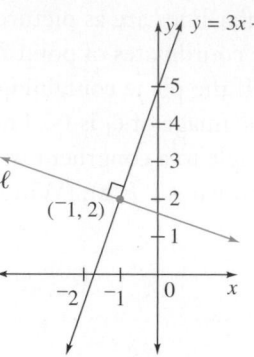

Figure 13-20

Because the line $y = 3x + 5$ has slope 3 and is perpendicular to ℓ, we have $3m = {}^-1$; therefore, $m = \dfrac{^-1}{3}$. Consequently, the equation of ℓ is

$$y = \frac{^-1}{3}x + b.$$

Because the point $(^-1, 2)$ is on ℓ, we can substitute $x = {}^-1, y = 2$ in $y = \dfrac{^-1}{3}x + b$ and solve for b as follows:

$$2 = \frac{^-1}{3}(^-1) + b$$

$$\frac{5}{3} = b.$$

Consequently, the equation of ℓ is

$$y = \frac{^-1}{3}x + \frac{5}{3}.$$

Assessment 13-1A

1. For each of the following, find the image of the given quadrilateral under a translation from A to B:

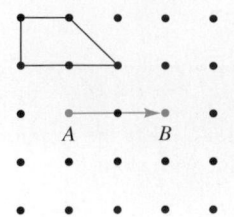

a.

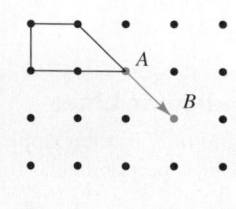

b.

2. Construct the image of $\overline{BC}$ under the translation pictured in the figure by using the following:
   a. Tracing paper
   b. Compass and straightedge

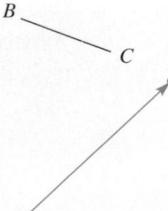

3. Find the coordinates of the image for each of the following points under the translation defined by $(x, y) \rightarrow (x + 3, y - 4)$:
   a. $(0, 0)$ b. $(^-3, 4)$ c. $(^-6, ^-9)$

4. Find the coordinates of the points whose images under the translation $(x, y) \rightarrow (x - 3, y + 4)$ are the following:
   a. $(0, 0)$ b. $(^-3, 4)$ c. $(^-6, ^-9)$

5. Consider the translation $(x, y) \rightarrow (x + 3, y - 4)$. In each of the following, draw the image of the figure under the translation and find the coordinates of the images of the labeled points:

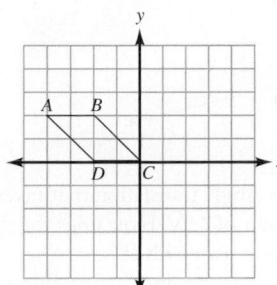 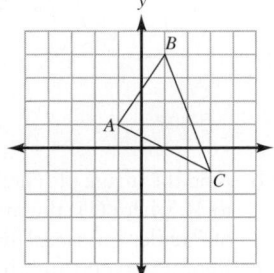

    a.   b.

6. Consider the translation $(x, y) \rightarrow (x + 3, y - 4)$. In the following, draw the figure whose image is shown:

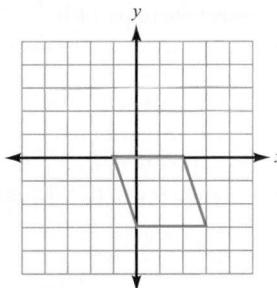

7. Find the image of the following quadrilateral in a 90° counterclockwise rotation about O:

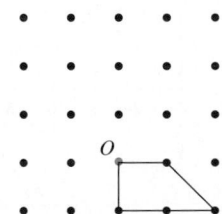

8. If ℓ is a line whose equation is $y = 2x - 1$, find the equation of the image of ℓ under each of the following translations:
   a. $(x, y) \rightarrow (x, y - 2)$
   b. $(x, y) \rightarrow (x + 3, y)$

9. If $y = ^-2x + 3$ is the image of line k under the translation $(x, y) \rightarrow (x + 3, y - 2)$, find the equation of k.

10. Use a compass and a straightedge to find the image of line ℓ under a half-turn about point O as shown.

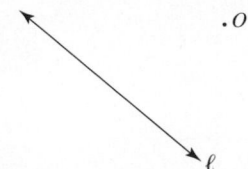

11. The image of $\overline{AB}$ under a rotation is given in the following figure. Find $\overline{AB}$.

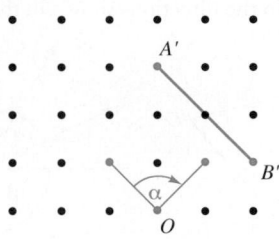

A rotation with
center O through α

12. The image of **NOON** is still **NOON** after a certain half-turn. List some other words that have the same property. What letters can such words contain?

13. a. Draw a line ℓ and any two points A and B so that $\overline{AB}$ is parallel to line ℓ. Find the image of ℓ under a translation from A to B.
   b. Draw a line ℓ and any two points A and B so that $\overline{AB}$ is not parallel to ℓ. Construct ℓ', the image of ℓ under the translation from A to B.
   c. How are ℓ and ℓ' in part (b) related? Why?
   d. What is the image of $\angle ABC$ under 10 successive rotations about B if $m(\angle ABC) = 36°$.

14. The images of any point under a rotation by certain angles can be found with only a compass and straightedge (without the use of a protractor). Construct P', the image of P when it is rotated about O, as shown in the figure, for angles with the following measures and direction:
   a. 45° counterclockwise
   b. 60° clockwise

                                • P

                    •O

15. For each of the following points, find the coordinates of the image point under a half-turn about the origin:
   a. $(4, 0)$
   b. $(2, 4)$
   c. $(^-2, ^-4)$
   d. (a, b)

16. In the following figure, find the image of the figure under a half-turn about O:

17. Draw any line and label it ℓ. Use tracing paper to find ℓ', the image of ℓ under each of the following rotations. In each case, describe how ℓ' is related to ℓ.
   a. Half-turn about point O on ℓ
   b. A 90° turn counterclockwise about point O not on ℓ

18. a. Find the final image of $\triangle ABC$ by performing two rotations in succession each with center O, one by angle α and one by angle β, in the directions shown in the figure.

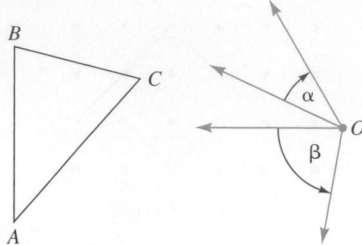

   b. Is the order of the rotations important?

   c. Could the result have been accomplished in one rotation?

19. Use a drawing similar to Figure 13-19 to find

   a. The images of the following points under a 90° rotation counterclockwise about the origin:

      i. $(2, 3)$ **ii.** $(^-1, 2)$

      iii. (m, n) in terms of m and n

   b. Show that under a half-turn with the origin as center, the image of a point with coordinates (a, b) has coordinates $(^-a, ^-b)$.

   c. Use what you found in part (a) to get the image of $P(a, b)$ under rotation clockwise by 90° about the origin. (*Hint:* Rotate first as in part (a), then apply a half-turn about the origin.)

20. Find the equation of the line through $(1, 0)$ and perpendicular to

   a. $y = x + 2$

   b. $y = ^-2x + 3$

   c. $x = ^-4$

   d. $y = ^-3$

21. The following figure is a rhombus with sides of length a and one of the vertices at (h, k):

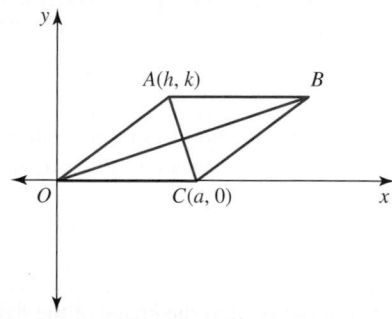

a. Find the coordinates of B in terms of a, h, and k.

b. Explain why $h^2 + k^2 = a^2$.

c. Prove that the diagonals of a rhombus are perpendicular to each other.

d. Describe the turn symmetries of the rhombus.

22. a. Draw two points O and M and a point P. Construct P', the image of P under a half-turn about O and then P'', the image of P' under a half-turn about M.

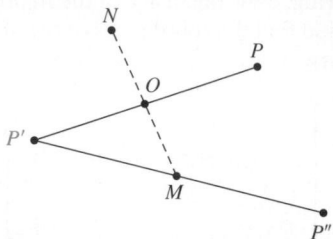

   b. Find the image of P under a single translation that takes N to M, where O is the midpoint of $\overline{MN}$. How does the image of P under this translation compare with P''?

   c. Repeat parts (a) and (b) for a new point Q (first find Q', the image of Q under a half-turn about O and then Q'', the image of Q' under a half-turn about M).

   d. Conjecture what single transformation will have the same effect on any point in the plane as a half-turn about O followed by a half-turn about M.

   e. Justify your conjecture in part (d).

23. How is the image of a line related to the original line under a half-turn? Explain your reasoning.

24. Name four geometric figures that can be their own images under a half-turn and one that cannot.

25. In the drawing below, points A, B, and C are given. Find all locations of point D such that a parallelogram is formed with the four points.

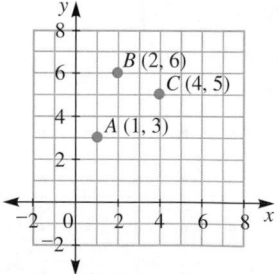

Assessment 13-1B

1. Find the figure whose image is given in each of the following under a translation from X to X':

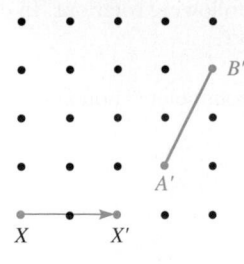

a.

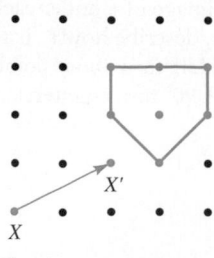

b.

2. Construct $\overline{BC}$ whose image $\overline{B'C'}$ under the translation is pictured in the figure by using the following:

   a. Tracing paper

   b. Compass and straightedge

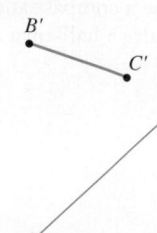

3. Find the coordinates of the image for each of the following points under the translation defined by $(x, y) \rightarrow (x + 3, y - 4)$:
   a. $(7, 14)$
   b. $(^-3, ^-5)$
   c. (h, k)

4. Find the coordinates of the points whose images under the translation $(x, y) \rightarrow (x - 3, y + 4)$ are the following:
   a. $(7, 14)$
   b. $(^-7, ^-10)$
   c. (h, k)

5. Consider the translation $(x, y) \rightarrow (x + 3, y - 4)$. In each of the following, draw the image of the figure under the translation, and find the coordinates of the images of the labeled points:

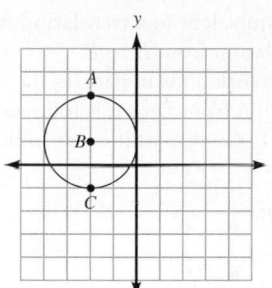
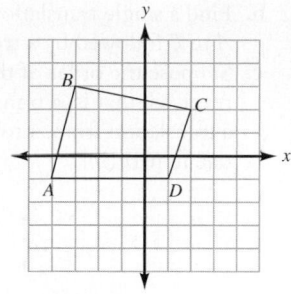

   a. b.

6. Consider the translation $(x, y) \rightarrow (x + 3, y - 4)$. In the following, draw the figure whose image is shown:

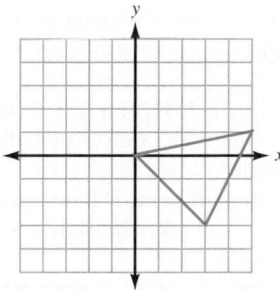

7. Find the figure whose image was found through a 90° clockwise rotation about point O and is shown below.

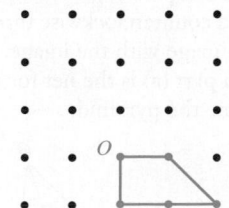

8. If ℓ is a line whose equation is $y = 2x - 1$, find the equation of the image of ℓ under each of the following translations:
   a. $(x, y) \rightarrow (x - 3, y + 2)$
   b. $(x, y) \rightarrow (x - 5, y - 4)$

9. If P' is the image of point P under a half-turn about O, what can be said about points P', P, and O? Why?

10. Use a compass and a straightedge to find the preimage of ℓ under a half-turn about point O as shown.

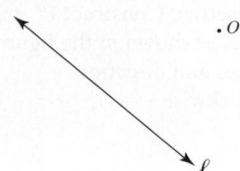

11. The image of $\overline{AB}$ under a rotation is given in the following figure. Find $\overline{AB}$.

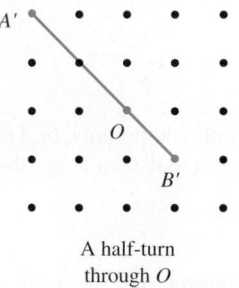

A half-turn
through O

12. The image of **MOW** is still **MOW** after a certain half-turn. Explain whether **MOM** has this property.

13. a. Refer to the following figure and use paper folding or any other method to show that if P' is the image of P under rotation about point O by a given angle, then O is on the perpendicular bisector of $\overline{PP'}$.

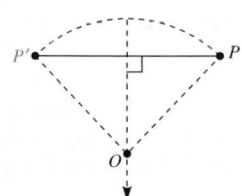

   b. $\triangle A'B'C'$ shown in the following figure was obtained by rotating $\triangle ABC$ about a certain point O. Explain how to find the point O and the angle of rotation.

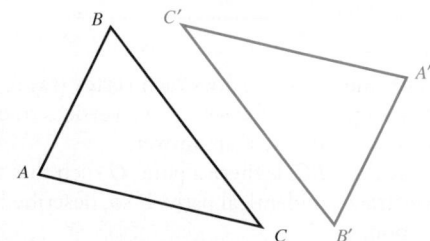

   c. Triangles ABC and $A'B'C'$ shown are congruent. Trace them and explain why it is impossible to find a rotation under which $\triangle A'B'C'$ is the image of $\triangle ABC$.

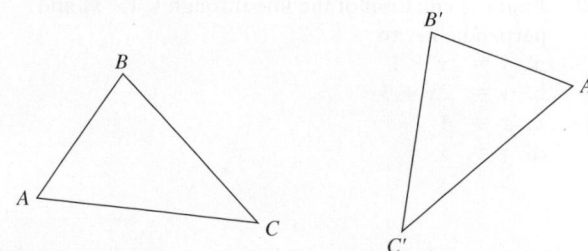

14. The images of any point under a rotation by certain angles can be found with only a compass and straightedge (without the use of a protractor). Construct P', the image of P when it is rotated about O, as shown in the figure, for angles with the following measures and direction:
    a. 30° counterclockwise
    b. 30° clockwise

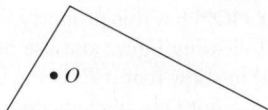

15. For each of the following points, find the coordinates of the image point under a half-turn about the origin:
    a. $(0, 3)$
    b. $(^-2, 5)$
    c. $(^-a, ^-b)$
16. In the following figure, find the image of the figure under a half-turn about O:

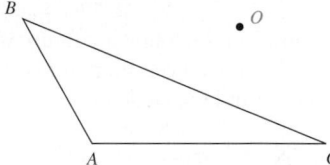

17. Draw any line and label it ℓ. Use tracing paper to find ℓ', the image of ℓ under each of the following rotations. In each case, describe in words how ℓ' is related to ℓ.
    a. Half-turn about point O, not on ℓ
    b. A 60° turn counterclockwise about point O, not on ℓ
18. When $\triangle ABC$ in the following figure is rotated about point O by 360°, each of the vertices traces a path:

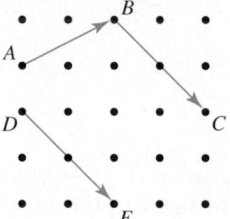

    a. What geometric figure does each vertex trace?
    b. Identify all points O for which two vertices trace an identical path. Justify your answer.
    c. Given any $\triangle ABC$, is there a point O such that the three vertices trace an identical path? If so, describe how to find such a point.
19. Find the equation of the image of the line $y = 3x - 1$ under the following transformations:
    a. Half-turn about the origin
    b. A 90° counterclockwise rotation about the origin
20. Find the equation of the line through $(^-1, ^-3)$ and perpendicular to
    a. $y = 2x + 1$
    b. $y = ^-2x + 3$
    c. $x = ^-4$
    d. $y = ^-3$

21. The following figure shows $\overline{AB}$, one side of a rhombus. Describe how to use a half-turn about O to find the coordinates of the other two vertices of the rhombus.

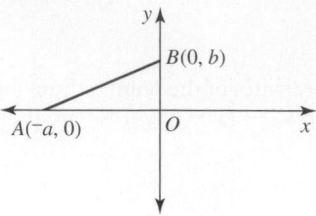

22. a. A translation from A to B is followed by a translation from B to C, as shown. Show that the same result can be accomplished by a single translation. What is that translation?
    b. Find a single translation equivalent to a translation from A to B followed by a translation from D to E.
    c. Suppose the order of the translations in part (b) is reversed; that is, a translation from D to E is followed by a translation from A to B. Is the result different from the one in part (b)?

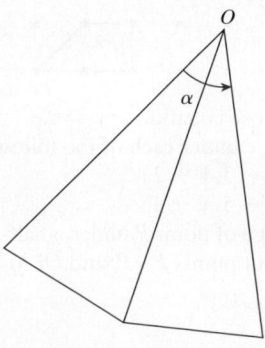

23. If $\overline{AB}$ is rotated 90° clockwise about point O, explain whether its image is perpendicular to $\overline{AB}$, regardless of the location of point O.
24. a. Points $A(1, 2)$, $B(^-2, 3)$, and $C(3, 8)$ are given. Find all possible coordinates of point D such that the four points form a rectangle.
    b. For each location of point D found in part (a), find a translation that will take $\overline{AB}$ to the opposite side of the rectangle formed.
25. Suppose one of the vertices of a square whose diagonals intersect at the origin has coordinates $(^-2, 5)$. Find the coordinates of the other three vertices.
26. The preimage below is part of a net for a pyramid. The preimage is rotated counterclockwise through α with center O.
    a. Sketch the preimage with the image.
    b. If the sketch in part (a) is the net for the lateral surface of a pyramid, name the pyramid.

Mathematical Connections 13-1

Communication

1. If we are given two congruent nonparallel segments, is it always possible to find a rotation so that the image of one segment will be the other segment? Explain why or why not.

2. **a.** If you rotate an object 180° clockwise or counterclockwise, using the same center, is the image the same in both cases? Explain.
   b. Answer part (a) if you rotate the object 360°.

3. Given any point with coordinates (x, y), what are the coordinates of the image of the point under a half-turn about the origin? Why?

4. For each of the following figures, trace the figure on tracing paper, rotate the tracing by 180° about the given point O, sketch the image, and then make a conjecture about the kind of figure that is formed by the union of the original figure and its image. In each case, explain why you think your conjecture is true.

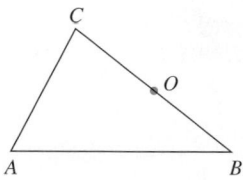

O is the midpoint of $\overline{BC}$.

a.

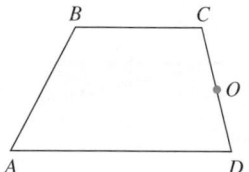

ABCD is a trapezoid and O is the midpoint of $\overline{CD}$.

b.

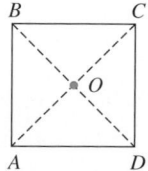

ABCD is a square and O is the intersection of its diagonals.

c.

Open-Ended

5. A drawing of a cube, shown in the following figure, can be created by drawing a square $ABCD$, finding its image under a translation defined by the slide arrow from A to A' so that

$AA' = AB$, and connecting the points A, B, C, and D with their corresponding images.

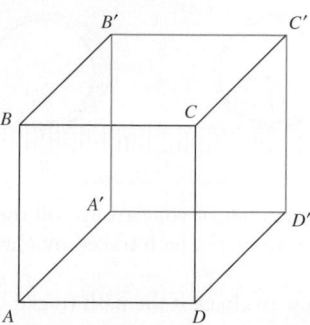

Draw several other perspective geometric figures using translations. In each case, name the figures and indicate the slide arrow that defines the translation.

6. Wall stenciling has been used to obtain an effect similar to that of wallpapering. The stencil pattern in the following figure can be used to create a border on a wall:

Measure the length of a wall of a room and design your own stencil pattern to create a border. Cut the pattern from a sheet of plastic or cardboard. Define the translation that will accomplish creating an appropriate border for the wall.

7. The following pattern can be created by rotating figure A about O clockwise by the indicated angle, then rotating the image B about O by the same angle, and then rotating the image C about O by the same angle, and so on until one of the images coincides with the original figure A:

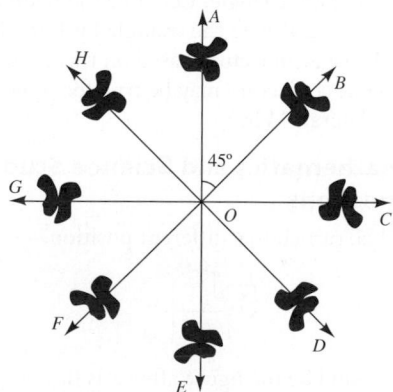

Make several designs with different numbers of congruent figures around a circle in which the image of each figure under the same rotation is the next figure and so that one of the images coincides with the original figure.

Cooperative Learning

8. Mark a point A on a sheet of paper and set a straightedge through A as shown in the following figure. Find a circular

shape (a jar lid is a good choice) and mark point *P* on the edge of the shape. Place the shape on the straightedge so that *P* coincides with *A*. Consider the path traced by *P* as the circle rolls so that its edge stays in contact with the straightedge all the time and until *P* comes in contact with the straightedge again at point *B*. (The drawing is not to scale.)

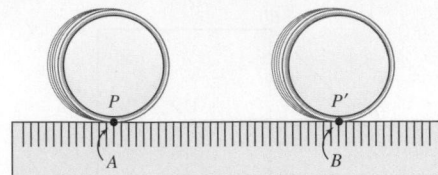

a. Have one member of your group roll the circular shape and another draw the path traced by *P* as accurately as possible.
b. Discuss how to check if the path traced by point *P* is an arc of a circle.
c. Find the length of $\overline{AB}$.

Questions from the Classroom

9. A student asks if every translation on a grid can be accomplished by a translation along a vertical direction followed by a translation along a horizontal direction. How do you respond?
10. A student asks why, in a half-turn, the direction is never specified. How do you respond?

GSP/Geogebra Activities

11. Use *GSP* Lab 11 activities 6–8 to investigate the following properties of translations:
    • A figure and its image are congruent.
    • The image of a line is a line parallel to it.
12. Use *GSP*/Geogebra Lab 11 activities 9–11 to investigate properties of rotations.
13. Use *GSP* or Geogebra to draw an equilateral triangle and two altitudes of the triangle. Let *O* be the point at which the altitudes intersect. Rotate the triangle by 120° about *O* in any direction. Make a conjecture based on this experiment. Do you think your conjecture may be true for some triangles that are not equilateral? Why?

Trends in Mathematics and Science Study (TIMSS) Questions

This figure will be turned to a different position.

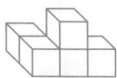

Which of these could be the figure after it is turned?

(a) (b) (c) (d)

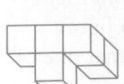 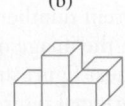 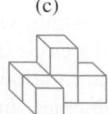 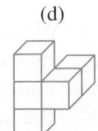

TIMSS, Grade 4, 2003

Rectangle *PQRS* can be rotated (turned) onto rectangle *UVST*.

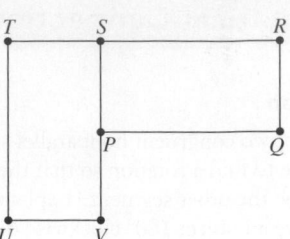

What point is the center of rotation?
a. *P*
b. *R*
c. *S*
d. *T*
e. *V*

TIMSS, Grade 8, 2003

A half-turn about point *P* in the plane is applied to the shaded figure.

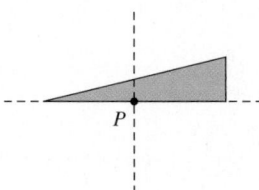

Which of the following shows the results of the half-turn?

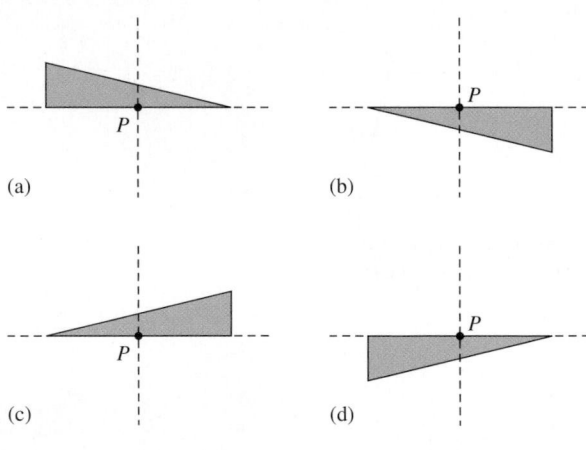

(a)   (b)

(c)   (d)

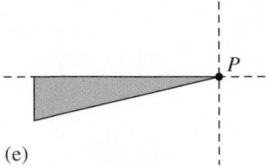

(e)

TIMSS, Grade 8, 2007

BRAIN TEASER

In Figure 13-21, a coin is shown above and touching another coin. Suppose the top coin is rotated around the circumference of the bottom coin without slipping until it rests directly below the bottom coin. Will the head be straight up or upside down?

Figure 13-21

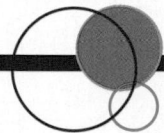

13-2 Reflections and Glide Reflections

Reflections

Another isometry is a **reflection**, or **flip**. One example of a reflection often encountered in our daily lives is a mirror image. Figure 13-22 shows a figure with its mirror image.

Mirror

Figure 13-22

Another reflection is shown in the following cartoon.

B.C.

by johnny hart

We obtain reflections in a line in various ways. Consider the half tree shown in Figure 13-23(a). Folding the paper along the **reflecting line** and drawing the image gives the **mirror image**, or **image**, of the half tree. In Figure 13-23(b), the paper is shown unfolded. Another way to simulate a reflection in a line involves using a Mira, as illustrated in Figure 13-23(c).

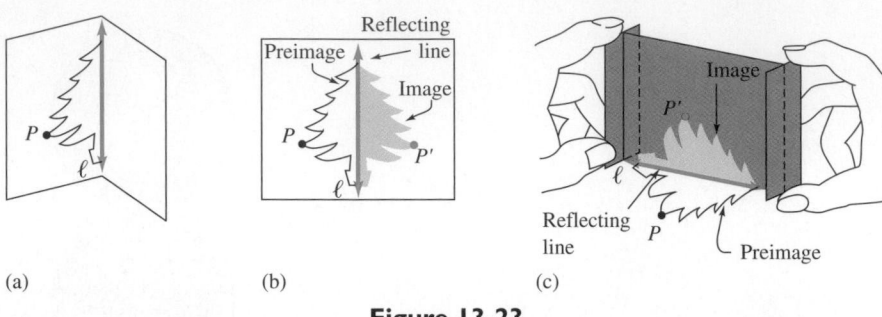

(a)   (b)   (c)

Figure 13-23

In Figure 13-24(a), the image of P under a reflection in line ℓ is P'. $\overline{PP'}$ is both perpendicular to and bisected by ℓ, or equivalently, ℓ is the perpendicular bisector of $\overline{PP'}$. In Figure 13-24(b), P is its own image under the reflection in line ℓ. If ℓ were a mirror, then P' would be the mirror image of P. This leads us to the following definition of a reflection.

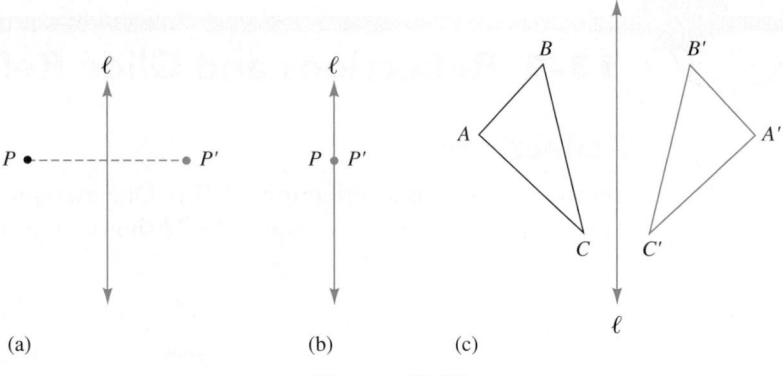

(a)   (b)   (c)

Figure 13-24

Definition of Reflection

A **reflection** in line ℓ is a transformation from the plane to the plane that pairs each point P with a point P' in such a way that ℓ is the perpendicular bisector of $\overline{PP'}$, as long as P is not on ℓ. If P is on ℓ, then $P = P'$.

In Figure 13-24(c), we see another property of a reflection. In the original triangle ABC, if we walk clockwise around the vertices, starting at vertex A, we see the vertices in the order A-B-C. However, in the reflection image of triangle ABC, if we start at A' (the image of A) and walk clockwise, we see the vertices in the following order: A'-C'-B'. Thus a reflection does something that neither a translation nor a rotation does; it reverses the **orientation** of the original figure.

There are many methods of constructing a reflection image. We already illustrated such constructions with paper folding and a Mira. Next, we illustrate the construction of the image of a figure under a reflection in a line with tracing paper in Figure 13-25.

E-manipulative Activity

Use *Transformations* to arrange figures and play a matching game involving rotations and reflections.

Constructing a Reflection by Using Tracing Paper

Aligning the reference point in Figure 13-25 ensures that no translating occurs along the reflecting line when the reflection is performed. If we wish the image to be on the paper with the original, we may indent the tracing paper or acetate sheet to mark the images of the original vertices.

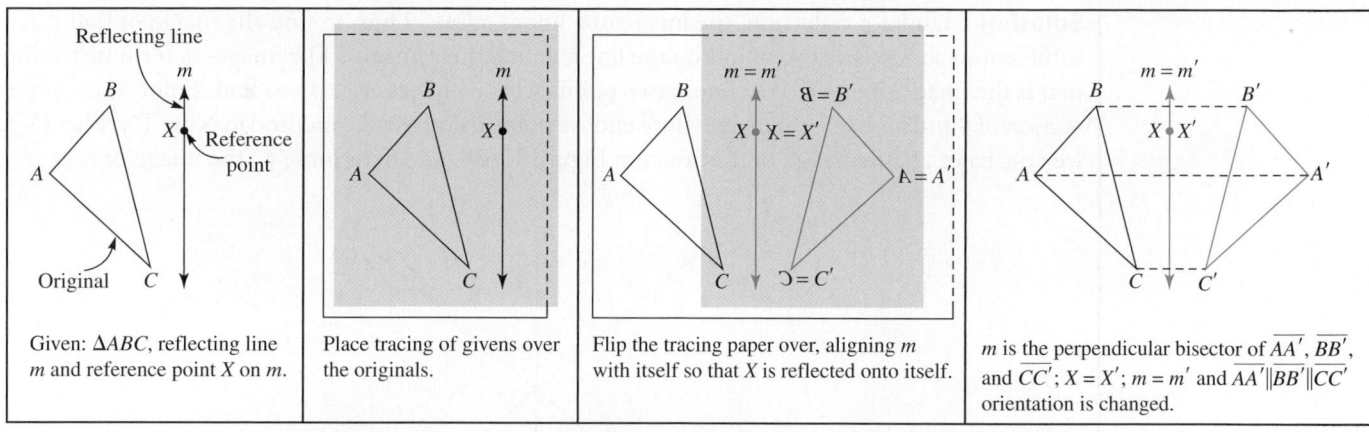

Figure 13-25 Construction of a reflection using tracing paper

As with other motions, a reflection may be constructed with a compass and straightedge as seen in Now Try This 13-5.

NOW TRY THIS 13-5

Use the definition of a reflection in a line and properties of a rhombus to construct the image P' of point P in Figure 13-26 under reflection in line m, using only a compass and straightedge.

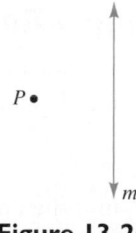

Figure 13-26

Some students have more difficulty applying motions to shapes, such as lines than to points. Example 13-5 deals with a reflection of a line.

EXAMPLE 13-5

Describe how to construct the image of line ℓ under a reflection in line m in Figure 13-27.

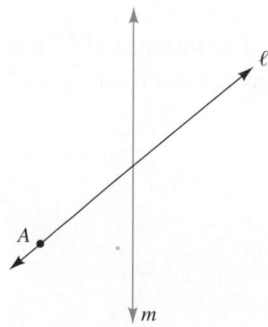

Figure 13-27

Solution   Under a reflection, the image of a line is a line. Thus, to find the image of line ℓ, it is sufficient to choose any two points on the line and find their images. The images determine the line that is the image of line ℓ. We choose two points whose images are easy to find. Point *X*, the intersection of ℓ and *m*, is its own image. If we choose point *A* and use the method in Now Try This 13-5, we construct *A′*, the image of *A*, shown in Figure 13-28, and determine ℓ′, the image of ℓ.

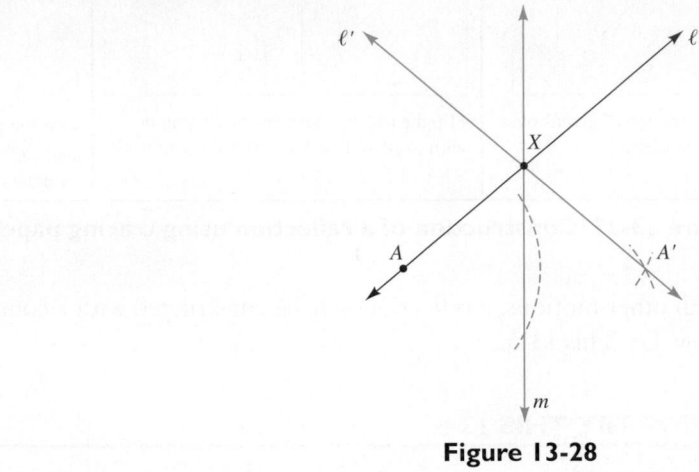

Figure 13-28

Constructing a Reflection on Dot Paper or a Geoboard

On dot paper or a geoboard, the images of figures under a reflection can sometimes be found by inspection, as seen in Example 13-6.

EXAMPLE 13-6

In Figure 13-29, find the image of △*ABC* under a reflection in line *m*.

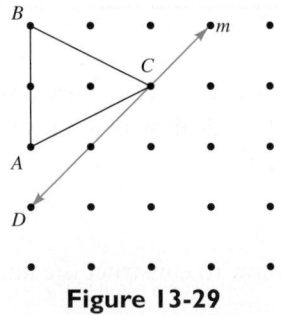

Figure 13-29

Solution   The image *A′B′C′* is given in Figure 13-30. Note that *C* is the image of itself and the images of the vertices *A* and *B* are *A′* and *B′* such that *m* is the perpendicular bisector of $\overline{AA'}$ and $\overline{BB'}$.

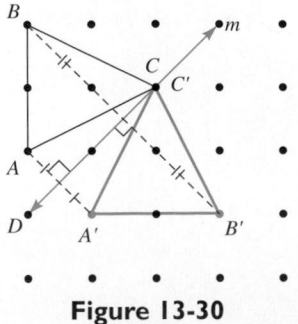

Figure 13-30

 **NOW TRY THIS 13-6**

Trace △*ABC* and △*A'B'C'* from Figure 13-30 on a sheet of paper (without tracing line *m*). Find the line of reflection using paper folding.

 Reflections, rotations, and translations appear in the elementary curriculum, as shown in the following partial student page.

School Book Page FLIP, TURN, AND SLIDE

Activity

How can I draw patterns?

a. This pattern was created with transformations starting with shape A. What transformation was used to move from shape A to shape B? shape B to shape C? and so on through shape I?

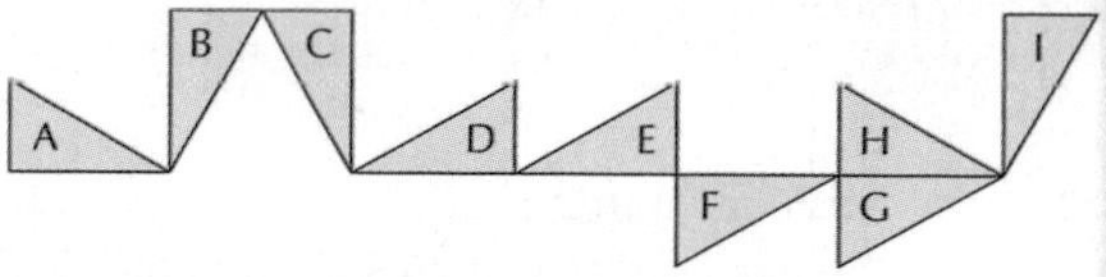

b. Work with a partner. Choose a polygon. Trace around it. Apply a transformation and trace around it again. Continue at least 8 more times. Show your partner your pattern and ask your partner to describe the transformations you used.

CHECK ✓

For another example, see Set 6-10 on p. 387.

Tell whether the figures in each pair are related by a slide, a flip, or a turn. If a turn, describe it.

1. 　　　　2. 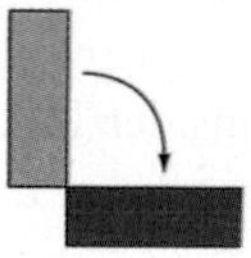　　　　3. 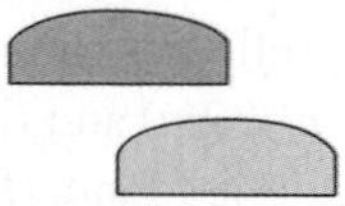　　　　4.

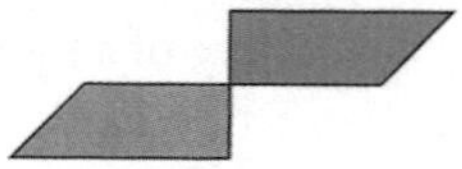

5. Reasoning Maria said the figures at the right are related by a 90° turn. Paul said they are related by a 270° turn. Who is right? Explain.

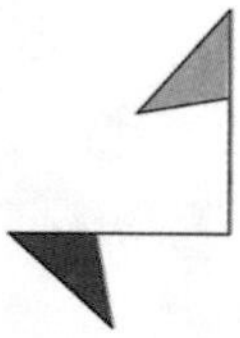

Section C Lesson 6-10　　365

Source: p. 365; From SCOTT FORESMAN-ADDISON WESLEY MATHEMATICS. Copyright © 2008 Pearson Education, Inc., or its affiliates. Used by permission. All Rights Reserved.

Reflections in a Coordinate System

For some reflecting lines, like the x-axis and y-axis and the line $y = x$, we can find the coordinates of the image, given the coordinates of the point. In Figure 13-31, the line $y = x$ bisects the angle between the x-axis and y-axis. The image of $A(1, 4)$ is the point $A'(4, 1)$. Also the image of $B(^-3, 0)$ is $B'(0, ^-3)$.

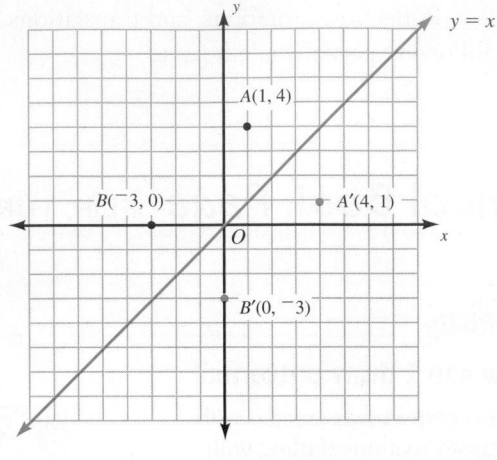

Figure 13-31

A reflection of a point with coordinates (a, b) in the line $y = x$ is generalized in Now Try This 13-7.

 **NOW TRY THIS 13-7**

Show that in general the image of $P(a, b)$ under reflection in line $y = x$ is the point $P'(b, a)$ and consequently that the reflection in the line $y = x$ interchanges the coordinates of the point.

Applications of Reflections

One natural application of reflections is in finding lines of symmetry of figures. A line of symmetry of a figure is a reflecting line that can be used to make a figure its own image under the reflection as in Figure 13-32(a) and (b). Consider Figure 13-32(a) where various lines of symmetry (or reflecting lines) can be found. (How many are there?)

(a)   (b)

Figure 13-32

Another application is seen when a ray of light bounces off a mirror or when a billiard ball bounces off the rail of a billiard table. The **angle of incidence**, the angle formed by the incoming ray in Figure 13-33 and a line perpendicular to the mirror, is congruent to the **angle of reflection**, the angle between the reflected ray and the line perpendicular to the mirror.

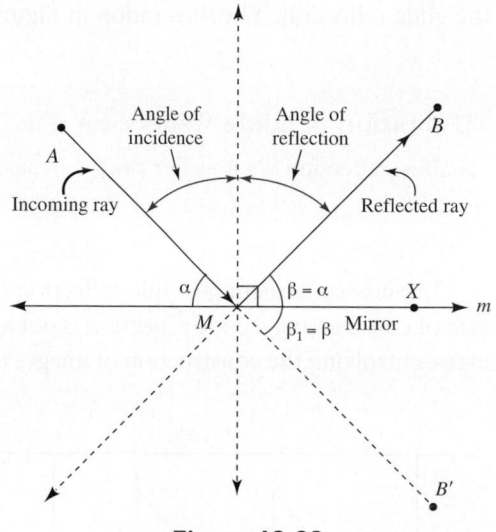

Figure 13-33

Because the angle of incidence is congruent to the angle of reflection, the respective complements of these angles must be congruent. If the measures of these complements are α and β, as indicated in Figure 13-33, then $\alpha = \beta$. Figure 13-33 shows B', the image of B under reflection in m, and hence $\angle XMB'$ is the image of $\angle XMB$. Notice that β, the measure of $\angle XMB$, must equal β_1, the measure of $\angle XMB'$, because reflection preserves angle measurement. Because $\alpha = \beta$ and $\beta = \beta_1$, we have $\alpha = \beta_1$. For that reason, points A, M, and B' are collinear (Why?). We can show that these facts imply *Fermat's Principle:* Light follows the path of shortest distance; that is, the path A-M-B that light travels is the shortest among all the paths connecting A with a point in the mirror to B.

NOW TRY THIS 13-8

Trace Figure 13-33 and mark on the mirror m a point P other than M. Prove that $AM + MB < AP + PB$ by showing that $MB = MB'$ and $BP = B'P$, and then by using the Triangle Inequality applied to $\triangle APB'$ and the fact that A, M, and B' are collinear.

Glide Reflections

Another basic isometry is a **glide reflection**. An example of a glide reflection is shown in the footprints of Figure 13-34. We consider the footprint labeled F_1 to have been translated to footprint F_2 and then

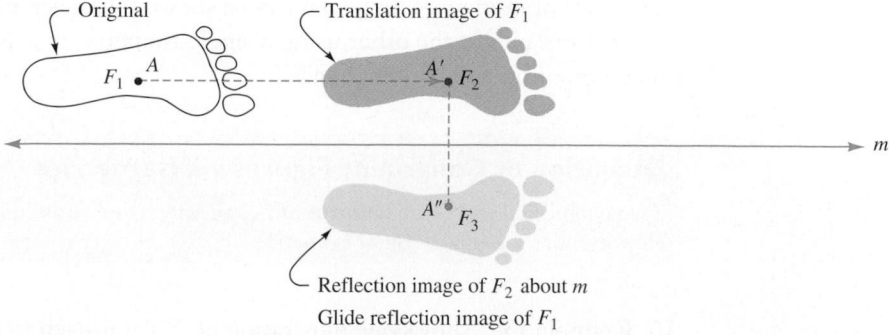

Figure 13-34

reflected over line m (parallel to the direction of the translation) to yield F_3, the image of F_2. F_3 is the final image of F_1. Note that point A is translated to find point A' and then point A' is reflected over line m (parallel to the direction of the translation) to obtain point A''. Thus, A'' is the image of A in the glide reflection. The illustration in Figure 13-34 leads us to the following definition.

Definition of Glide Reflection

A **glide reflection** is a transformation composed of a translation followed by a reflection in a line parallel to the slide arrow.

Because constructing a glide reflection involves constructing a translation and a reflection, the task of constructing a glide reflection is not a new problem. An example is seen in Figure 13-35. Exercises involving the construction of images of figures under glide reflections are left as assessments.

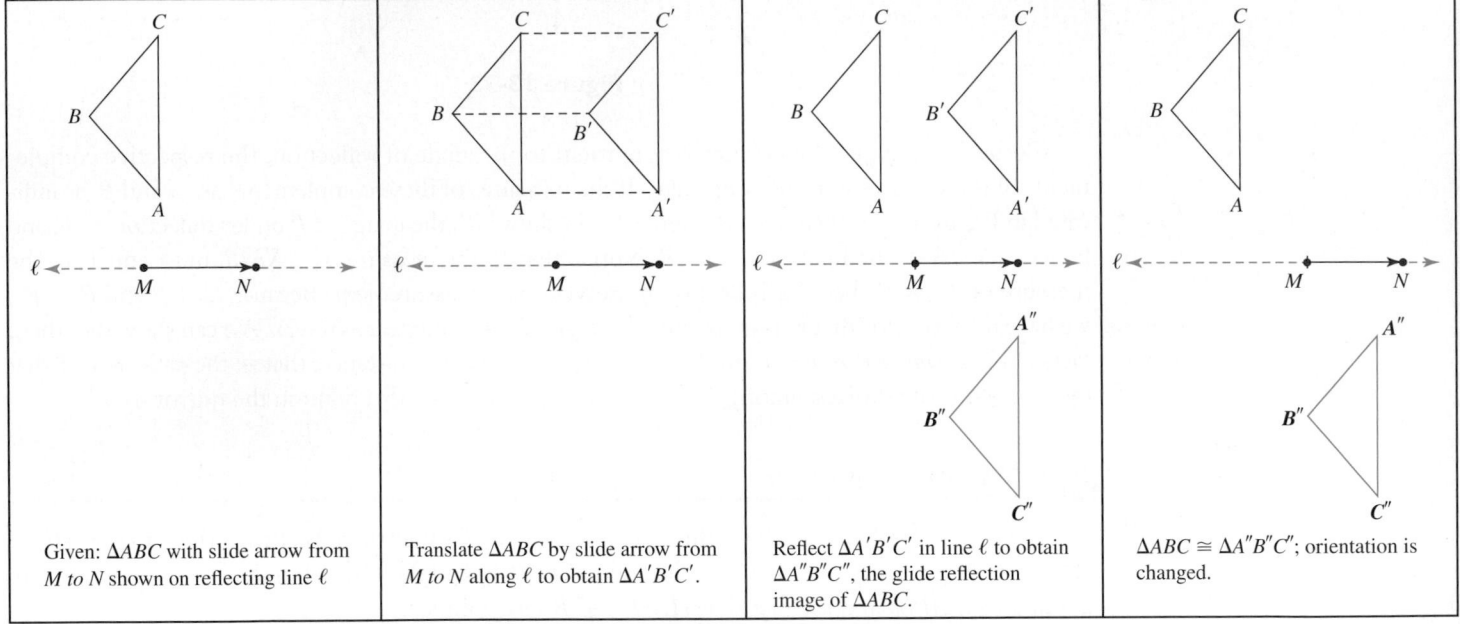

Figure 13-35 Construction of a glide reflection

Congruence via Isometries

We have seen that under an isometry, the image of a triangle is a congruent triangle. Also, given two congruent polygons, it is possible to show that one can be transformed to the other by using a sequence of isometries. In fact, it can be shown that given two congruent polygons, it is possible to transform one to the other using a single isometry, that is, one translation, one rotation, one reflection, or one glide reflection.

Definition of Congruent Figures via Isometries

Two geometric figures are **congruent** if, and only if, one is an image of the other under a single isometry or under a composition of isometries.

Example 13-7 shows one illustration of this approach to congruence.

EXAMPLE 13-7

ABCD in Figure 13-36 is a rectangle. Describe a sequence of isometries to show

 a. △*ADC* ≅ △*CBA*   **b.** △*ADC* ≅ △*BCD*   **c.** △*ADC* ≅ △*DAB*

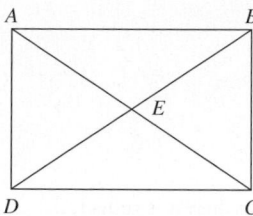

Figure 13-36

Solution

 a. A half-turn of △*ADC* with center *E* is one such transformation.
 b. A reflection in a line passing through *E* and parallel to $\overline{AD}$ is one such transformation.
 c. A reflection of △*ADC* in a line passing through *E* and parallel to $\overline{DC}$ is one such transformation.

Assessment 13-2A

1. Find the image of the given quadrilateral in a reflection in ℓ:

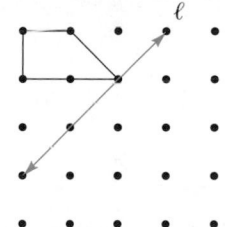

2. Determine which of the following figures have a reflecting line (or line of symmetry) such that the image of the figure under the reflection is the figure itself. In each case, find as many such reflecting lines as possible, sketching appropriate drawings.
 a. Circle
 b. Segment
 c. Line
 d. Square
 e. Scalene triangle
 f. Equilateral triangle
 g. Trapezoid whose base angles are not congruent
 h. Kite
 i. Regular hexagon
3. Determine the final result when △*ABC* is reflected in line ℓ and then the image is reflected again in ℓ.

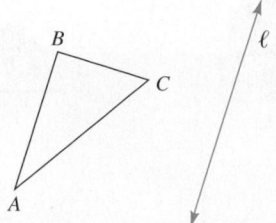

4. Draw a line and then draw a circle whose center is not on the line. Find the image of the circle under a reflection in the line.
5. **a.** Refer to the following figure and suppose lines ℓ and *m* are parallel and △*ABC* is reflected in ℓ to obtain △*A'B'C'* and then △*A'B'C'* is reflected in *m* to obtain △*A"B"C"*. Determine whether the same final image is obtained if △*ABC* is reflected first in *m* and then its image is reflected in ℓ.

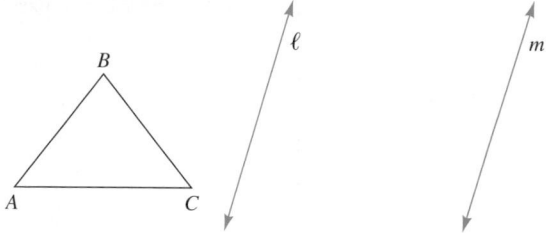

 b. Conjecture what single transformation will take △*ABC* directly to △*A"B"C"*. Check your conjecture using tracing paper.
6. Use a Mira, if available, to investigate exercise 5.
7. Given △*ABC* and its reflection image △*A'B'C'*, find the line of reflection.

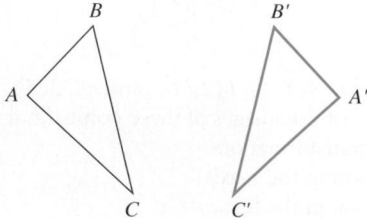

8. **a.** The word **TOT** is its own image when it is reflected through a vertical line through **O**, as shown in the

following figure. List some other words that are their own images when reflected similarly.

b. The image of **BOOK** is still **BOOK** when it is reflected through a horizontal line. List some other words that have the same property. Which uppercase letters can you use?

c. With an appropriate font, the image of 1881 is 1881 after reflection in either a horizontal or vertical line, as shown in the following figure. What other natural numbers less than 2000 have this property?

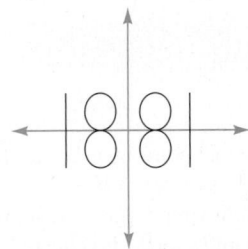

9. Find the equation of the image of the line with equation $y = 2x + 1$ when it is reflected in each of the following:
   a. x-axis   **b.** y-axis
   c. $y = x$

10. A glide reflection is determined by a translation parallel to a given line followed by a reflection in that line.
   a. Determine whether the same final image is obtained if the reflection is followed by the translation.
   b. Use your answer in (a) to determine whether the reflection and translation involved in the glide reflection commute.

11. Decide whether a reflection, a translation, a rotation, or a glide reflection will transform figure 1 into each of the other numbered figures (there may be more than one answer).

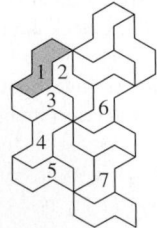

12. Given points $A(3, 4)$, $B(2, {}^-6)$, and $C({}^-2, 5)$, find the coordinates of the images of these points under each of the following transformations:
   a. Reflection in the x-axis
   b. Reflection in the line $y = x$

13. a. Conjecture what the image of a point with coordinates (x, y) will be under each of the transformations in exercise 12.
   b. Suppose a point P with coordinates (x, y) is reflected in the x-axis and then its image P' is reflected in the y-axis

to obtain P''. What are the coordinates of P'' in terms of x and y? Justify your answer.

14. Find the equations of the images of the following lines when reflected in the x-axis:
   a. $y = {}^-x + 3$
   b. $y = 0$

15. Find the equation of the images of the following lines when the reflection line is the y-axis.
   a. $y = {}^-x + 3$
   b. $y = 0$

16. a. The two circles "touch" each other at point P; that is, the tangent to one circle at P is also the tangent to the other circle. Find the line of reflection such that the image of the smaller circle will be in the interior of the larger circle, but still "touching" the larger circle.

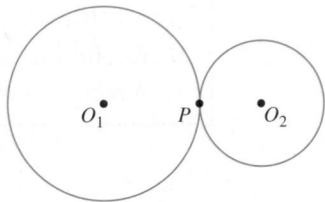

   b. Name a line of symmetry for the figure.

17. Two congruent circles with centers O_1 and O_2 intersect at points A and B.

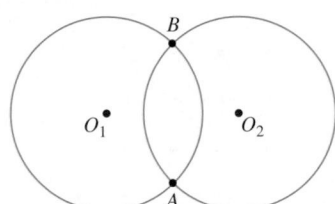

   a. In what line of reflection will the image of the circle with center O_1 be the circle with center O_2?
   b. Can the circle with center O_1 be transformed into the circle with center O_2 by a translation? If so, describe the translation.

18. Graph each of the following pairs of lines and for each construct the corresponding line of reflection so that the image of one line in the pair will be the second line. For each, identify the line of reflection.
   a. $y = {}^-x$ and $y = x$
   b. $y = 2x$ and $y = {}^-2x$

19. Two farm houses, H and T, are located away from a road, as shown. A telephone company wants to construct underground lines from point X at the edge of the road so that the sum of the lengths of the cables connecting the houses (H, and T) to X is as small as possible. Where should point X be located?

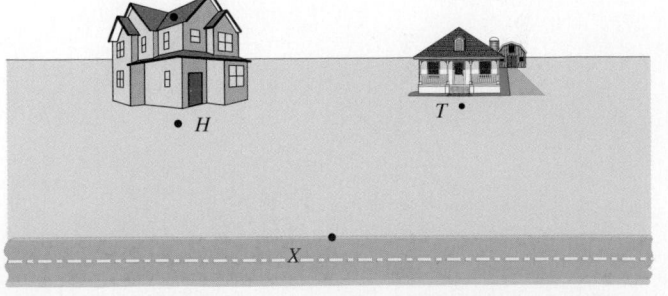

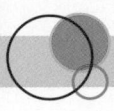

Assessment 13-2B

1. Find the image of the given quadrilateral in a reflection in ℓ:

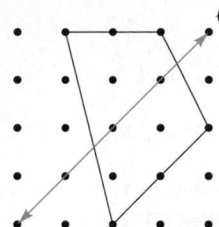

2. Determine which of the following figures have a reflecting line (or line of symmetry) such that the image of the figure under the reflection is the figure itself. In each case, find as many reflecting lines as possible, sketching appropriate drawings.
   a. Arc
   b. Ray
   c. Two perpendicular lines
   d. Rectangle
   e. Isosceles triangle
   f. Isosceles trapezoid
   g. Rhombus
   h. Regular n-gon

3. Determine the final result when △ABC is reflected in line ℓ and when the image is reflected again in ℓ.

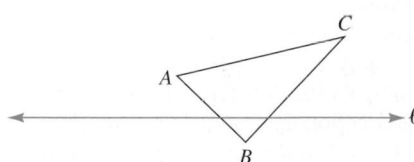

4. Sketch the image of the figure below through a reflection in line ℓ.

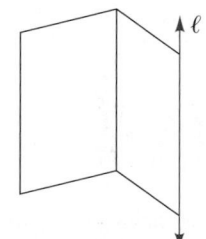

5. a. For the following figure, use any construction method to find the image of △ABC if △ABC is reflected in ℓ to obtain △A'B'C' and then △A'B'C' is reflected in m to obtain △A''B''C'' (ℓ and m intersect at O).

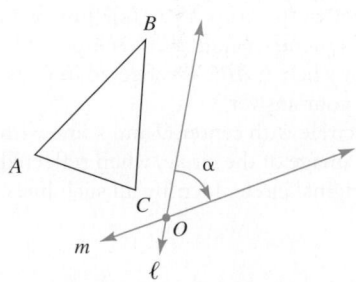

   b. Conjecture what single transformation will take △ABC directly to △A''B''C''. Check your conjecture using tracing paper.
   c. Answer the questions in (a) and in (b) for the case in which ℓ and m are perpendicular.

6. Use a Mira, if available, to investigate exercises 4 and 5.

7. a. Construct a "stylized" bow tie by using a reflection of the triangle below:

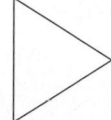

   b. Tell which reflecting line was used in part (a) and why.
   c. Is there more than one reflecting line that could have been used in part (a)? If so, where are others?

8. In a cartoon strip, a pet is named Otto.
   a. If Otto is spelled in all capital letters, draw a reflecting line to show that the name can be its own image in a reflection.
   b. Where is the reflecting line?

9. Describe the set of all lines that are their own images when reflected in each of the following:
   a. x-axis
   b. y-axis
   c. y = x

10. a. Copy the congruent triangles below on a piece of paper. Find as many reflecting lines as needed to take △XYZ to △X'Y'Z' using successive reflections.

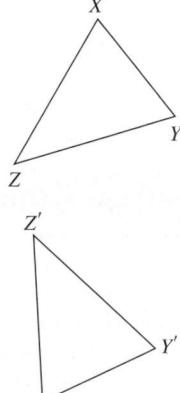

   b. Explain why the isometry in part (a) cannot be a translation, a rotation, or a reflection.

11. **a.** In the Pharlemina's Favorite quilt pattern below, describe a motion that will take part (a) green to part (b) blue.

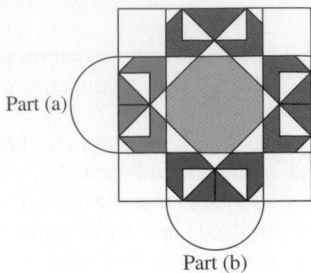

Part (a)

Part (b)

b. Follow the directions in (a) for the Dutchman's Puzzle quilt pattern shown

Part (a)   Part (b)

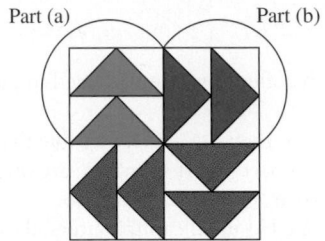

12. Given points $A(3, 4)$, $B(2, {}^-6)$, and $C({}^-2, 5)$, find the coordinates of the images of these points under each of the following transformations:
   a. Reflection in the y-axis
   b. Reflection in the line $y = {}^-x$

13. Conjecture what the image of a point with coordinates (x, y) will be under each of the transformations in exercise 12.

14. Find the equations of the images of the following lines when reflected in the x-axis:
   a. $y = 3x$
   b. $y = {}^-x$
   c. $x = 0$

15. Find the equation of the images of the lines in exercise 14 when the reflection line is the y-axis.

16. In which line will the two intersecting circles reflect onto themselves; that is, the image of the circles will be the same two circles?

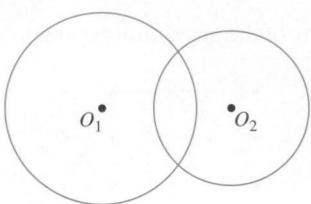

17. Construct a square and a circle of an appropriate size so that it will be possible to transform the circle by a reflection into the square such that the image of the circle will be tangent to all sides of the square. Identify the line of reflection.

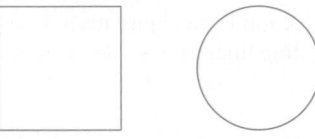

18. Graph each of the following pairs of lines and for each construct the corresponding line of reflection so that the image of one line in the pair will be the second line. For each, identify the line of reflection.
   a. $x = 0$ and $y = 0$
   b. $x = {}^-2$ and $x = 3$

19. A *fixed point* of a transformation is a point whose image is the point itself. List all the fixed points of the following transformations:
   a. Reflection in line ℓ
   b. Rotations by a given angle and direction about point O
   c. Translation with slide arrow from A to B; $A \neq B$
   d. Glide reflection determined by a translation with slide arrow from A to B followed by a reflection in line ℓ parallel to line AB; $A \neq B$

 **Mathematical Connections 13-2**

Communication

1. When a billiard ball bounces off a side of a pool table, assume the angle of incidence is congruent to the angle of reflection. In the following figure showing a scale drawing of a pool table, a cue ball is at point A. Show how a player should aim to hit two sides of the table and then the ball at B. Justify your solution.

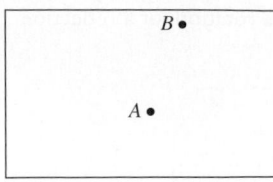

2. **a.** Draw an isosceles triangle ABC and then construct a line such that the image of $\triangle ABC$ when reflected in the line is $\triangle ABC$ though every point is not necessarily its own image. Explain why the line you constructed has the required property.
   b. For what kind of triangles is it possible to find more than one line with the property in (a)? Justify your answer.
   c. Given a scalene triangle ABC, is it possible to find a line ℓ such that when $\triangle ABC$ is reflected in ℓ, its image is itself? Explain your answer.
   d. Draw a circle with center O and a line with the property that the image of the circle, when reflected in the line, is the original circle. Identify all such lines. Justify your answer.

3. Use the following drawing to explain how a periscope works:

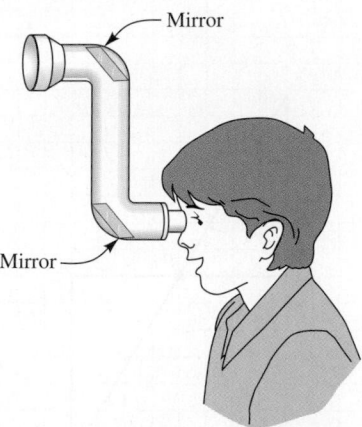

Mirror

Mirror

Open-Ended

4. In the following figure representing a miniature golf course hole, explain and justify the procedure showing how to aim the ball so that it gets in the hole if it is to bounce off
   a. one wall only.   **b.** two walls.

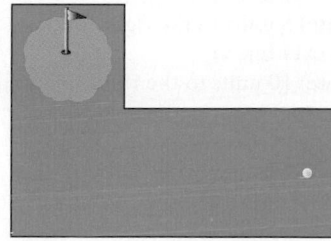

5. Design several wall stencil patterns using a reflection. In each case, explain how you would use the stencil in practice.
6. Design wall stencil patterns using a glide reflection.

Cooperative Learning

7. In the following figure representing a pool table, ball B is sent on a path that makes a 45° angle with the table wall, as shown. It bounces off the wall 5 times and returns to its original position.

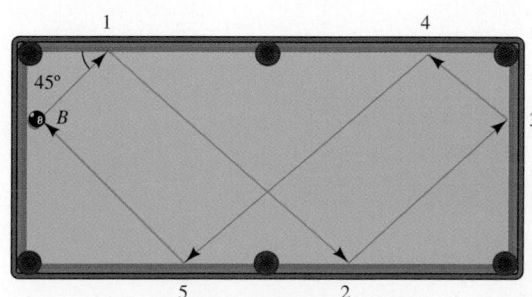

a. Have each member of your group use graph paper to construct rectangular models of different-sized pool tables. Simulate the experiment using any tools (such as a straightedge, compass, and protractor) by choosing different positions for ball B.

b. Share the results of your experiments with the rest of the group and together conjecture for which dimensions of the pool table and for what positions of B the experiment described in the problem will work.

Questions from the Classroom

8. A student asks, "If I have a point and its image, is that enough to determine whether the image was found using a translation, reflection, rotation, or glide reflection?" How do you respond?
9. Another student asks a question similar to exercise 8 but is concerned about a segment and its image. How do you respond to this student?
10. Sammi said that in the drawing below the image A' of point A under a reflection in line ℓ can also be found by a rotation through α shown with center O. Therefore, every reflection is actually a rotation. How do you respond?

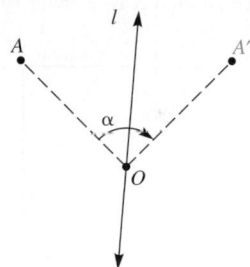

GSP/Geogebra Activities

11. Use *GSP/Geogebra* Lab 11 activities 1–5 to investigate properties of reflections.
12. Use *GSP* Lab 12, activity 1 to investigate glide reflections.
13. **a.** Use *GSP* or Geogebra to draw $\triangle ABC$ and three lines m, n, and p that intersect in a single point, as shown below. Reflect $\triangle ABC$ in line m to obtain its image $\triangle A'B'C'$. Then reflect $\triangle A'B'C'$ in line n to obtain $\triangle A''B''C''$. Finally, reflect $\triangle A''B''C''$ in line p to obtain the final image $\triangle A'''B'''C'''$.

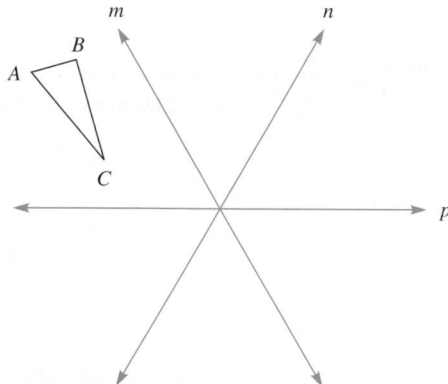

b. Find a single line q that could be used to reflect the original $\triangle ABC$ onto the final image.

Review Problems

14. Which single digits are their own images under a rotation by an angle whose measure is less than 360°?
15. What is the image of a point (a, b) under a half-turn about the origin?

16. Find all possible rotations that transform a circle onto itself.

17. Explain how an isometry can be used to construct a rectangle whose area is equal to that of the parallelogram *ABCD* in the following figure:

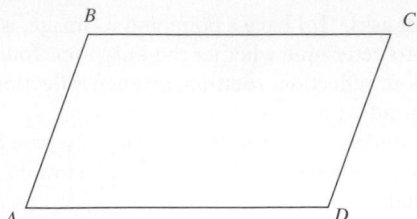

National Assessment of Educational Progress (NAEP) Questions

You will need the piece labeled *X* to answer this question.

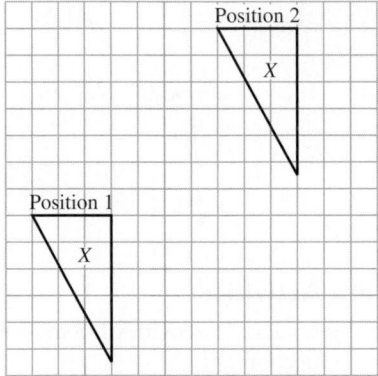

Which word best describe how to move the piece labeled *X* from position 1 to position 2?

a. Flip
b. Fold
c. Slide
d. Turn

NAEP, Grade 4, 2009

The figure below shows two triangles, labeled 1 and 2.

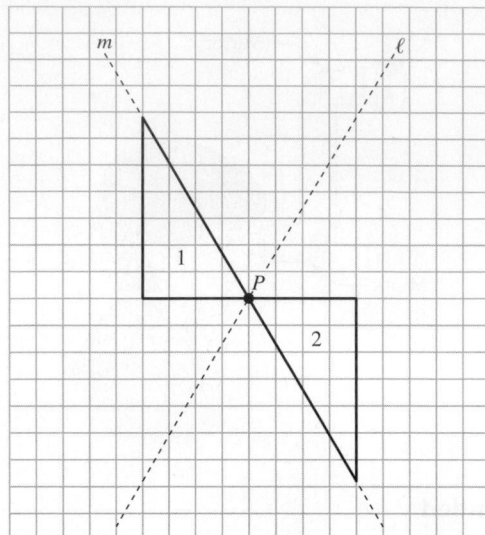

Which one of the following describes a way to move triangle 1 so that it completely covers triangle 2?

a. Turn (rotate) 180 degrees about point *P*.
b. Flip (reflect) over line *ℓ*.
c. Slide (translate) 5 units to the right followed by 8 units down.
d. Flip (reflect) over line *m*.
e. Slide (translate) 10 units to the right followed by 16 units down.

NAEP, Grade 8, 2009

BRAIN TEASER

Two cities are on opposite sides of a river, as shown in Figure 13-37.

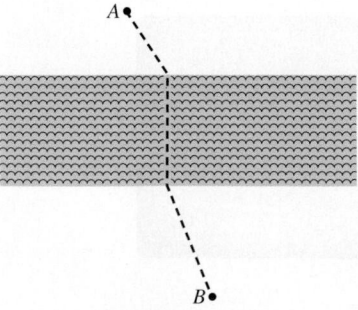

Figure 13-37

The cities' engineers want to build a bridge across the river that is perpendicular to the banks of the river and access roads to the bridge so that the distance between the cities is as short as possible. Where should the bridge and the roads be built?

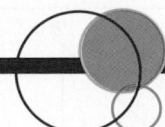

13-3 Dilations

Isometries preserve distance. Consequently, the image of a figure under an isometry is a figure congruent to the original. A different type of transformation happens when a slide of $\triangle ABC$ is projected on a screen to obtain $\triangle A'B'C'$. All objects on the slide are often enlarged on the screen by the same factor. Figure 13-38 shows another example of such a transformation, a **dilation** or a **size transformation**. (In *Focal Points*, the word *scaling* is used.) The point O is the *center* of the dilation and 2 is the *scale factor*. Points O, A, and A' are collinear and $OA' = 2OA$; also, O, C, and C' are collinear and $OC' = 2OC$. Similarly, O, B, and B' are collinear and $OB' = 2OB$. It can be shown that $\triangle A'B'C'$ is similar to $\triangle ABC$ and that each side of $\triangle A'B'C'$ is twice as long as the corresponding sides of $\triangle ABC$.

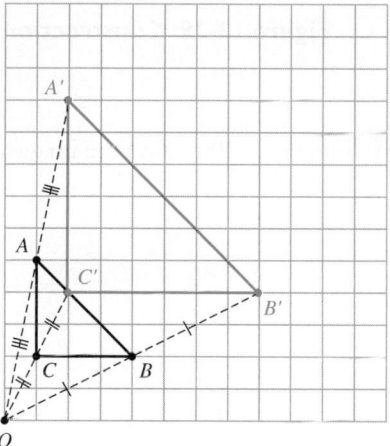

Figure 13-38

In general, we have the following definition and theorem.

Definition of Dilation

A **dilation** from the plane to the plane with center O and scale factor $r > 0$ is a transformation that assigns to each point A in the plane the point A' such that O, A, and A' are collinear, O is not between A and A', and $OA' = rOA$. O is its own image.

Theorem 13-2

Under a dilation, the image of a triangle is a similar triangle, and the image of a polygon is a similar polygon.

It is possible to define a dilation when the scale factor is negative except that O *must be between A and A'*.

Figure 13-39 shows a construction of a dilation with center O and scale factor of $\frac{1}{2}$. This construction refers to compass and straightedge constructions in Chapter 12.

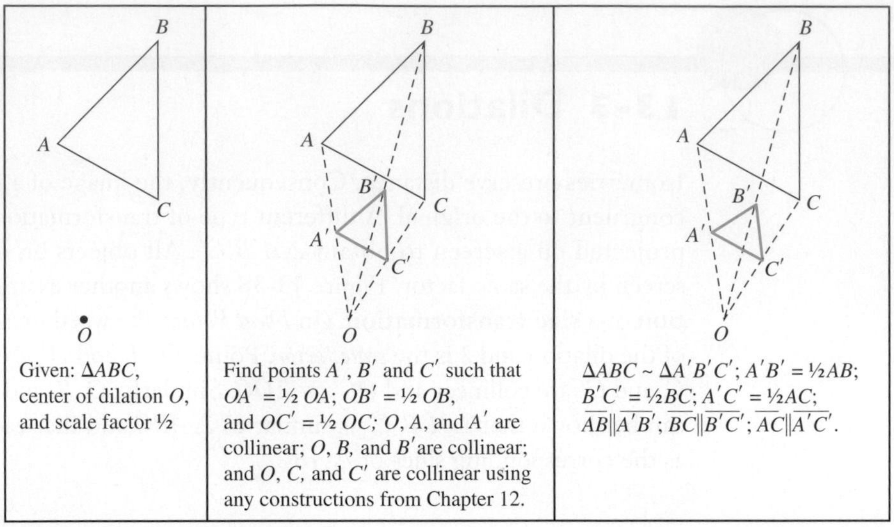

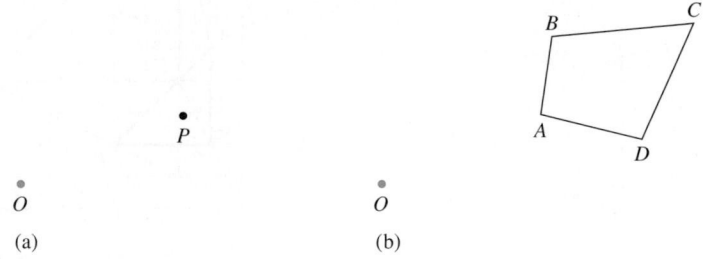

B
A
C
O
Given: ΔABC,
center of dilation O,
and scale factor ½
B
A
B′
C
A′
C′
O
Find points A', B' and C' such that
$OA' = \frac{1}{2} OA$; $OB' = \frac{1}{2} OB$,
and $OC' = \frac{1}{2} OC$; O, A, and A' are
collinear; O, B, and B' are collinear;
and O, C, and C' are collinear using
any constructions from Chapter 12.
B
A
B′
C
A′
C′
O
$\Delta ABC \sim \Delta A'B'C'$; $A'B' = \frac{1}{2}AB$;
$B'C' = \frac{1}{2}BC$; $A'C' = \frac{1}{2}AC$;
$\overline{AB} \parallel \overline{A'B'}$; $\overline{BC} \parallel \overline{B'C'}$; $\overline{AC} \parallel \overline{A'C'}$.

Figure 13-39 Construction of a dilation with center O and scale factor $\frac{1}{2}$

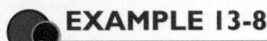

EXAMPLE 13-8

a. In Figure 13-40(a), find the image of point P under a dilation with center O and scale factor $\frac{2}{3}$.

P
O
(a)
B
C
A
D
O
(b)

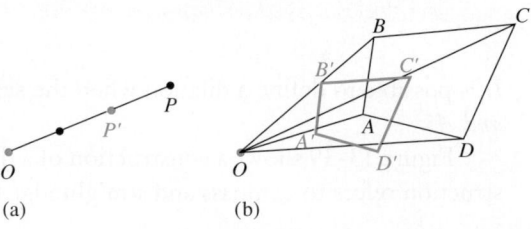

Figure 13-40

b. Find the image of the quadrilateral $ABCD$ in Figure 13-40(b) under the dilation with center O and scale factor $\frac{2}{3}$.

Solution

a. In Figure 13-41(a), we connect O with P and divide $\overline{OP}$ into three congruent parts. The point P' is the image of P because $OP' = \frac{2}{3}OP$.

b. We find the image of each of the vertices as in part (a) and connect the images to obtain the quadrilateral $A'B'C'D'$, shown in Figure 13-41(b).

P
P′
O
(a)
B
C
B′
C′
A
A′
D
D′
O
(b)

Figure 13-41

In Figure 13-41(b), the sides of the quadrilateral $A'B'C'D'$ are all parallel to the corresponding sides of the original quadrilateral, and the angles of the quadrilateral $A'B'C'D'$ are congruent to the corresponding angles of quadrilateral $ABCD$. Also, each side in the quadrilateral $A'B'C'D'$ is $\frac{2}{3}$ as long as the corresponding side of quadrilateral $ABCD$. These properties are true for any dilation and are summarized in the following theorem.

> **Theorem 13-3**
>
> A dilation with center O and scale factor $r > 0$ has the following properties:
>
> **a.** The image of a line segment is a line segment parallel to the original segment and r times as long.
> **b.** The image of an angle is an angle congruent to the original angle.

EXAMPLE 13-9

Show that $\triangle ABC$ in Figure 13-42 is the image of $\triangle ADE$ under a dilation. Identify the center of the dilation and the scale factor.

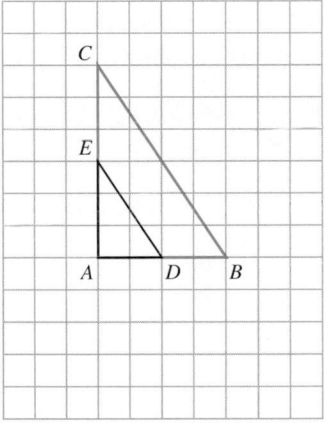

Figure 13-42

Solution Because $\dfrac{AB}{AD} = \dfrac{AC}{AE} = 2$, we choose A as the center of the dilation and 2 as the scale factor. Notice that under this transformation, the image of A is A itself. The image of D is B, and the image of E is C.

From Theorem 13-3, it follows that the image of a polygon under a dilation is a similar polygon. (Why?) However, for any two similar polygons it is not always possible to find a dilation such that the image of one polygon under the transformation is the other polygon. But, given two similar polygons, we can "move" one polygon to a place such that it will be the image of the other under a dilation. Example 13-10 shows such an instance.

EXAMPLE 13-10

Show that △ *ABC* in Figure 13-43 is the image of △ *APQ* under a succession of isometries with a dilation.

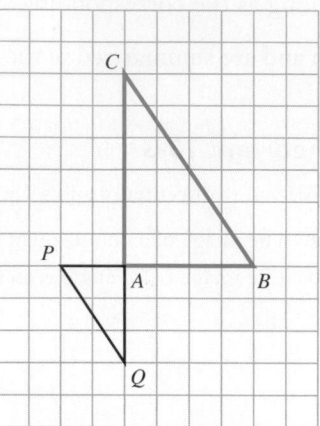

Figure 13-43

Solution   We use the strategy of *looking at a related problem*. In Example 13-9, the common vertex served as the center of the dilation. Here, we first rotate △*APQ* using a half-turn about *A* to obtain △ *AP′Q′*, as shown in Figure 13-44.

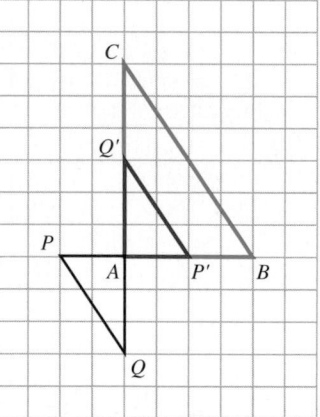

Figure 13-44

Now *C* is the image of *Q′* under a dilation with center at *A* and scale factor 2. Also, *B* is the image of *P′*, and *A* is the image of itself under this transformation. Thus, △ *ABC* can be obtained from △ *APQ* by first finding the image of △ *APQ* under a half-turn about *A* and then applying a dilation with center *A* and scale factor 2 to that image.

Earlier it was noted that a dilation could be defined using a negative scale factor. If that route had been taken, with *A* as the center, a dilation with scale factor ⁻2 could be used to achieve the desired result.

Examples 13-9 and 13-10 are a basis for a definition of similar figures.

> **Definition of Similar Figures via Isometries and Dilations**
>
> Two figures are **similar** if it is possible to transform one onto the other by an isometry (or a sequence of isometries) followed by a dilation.

Applications of Dilations

One way to make an object appear three-dimensional is to use a **perspective drawing**. For example, to make a letter appear three-dimensional we can use a dilation with an appropriate center O and a scale factor to create a three-dimensional effect, as shown in Figure 13-45 for the letter C.

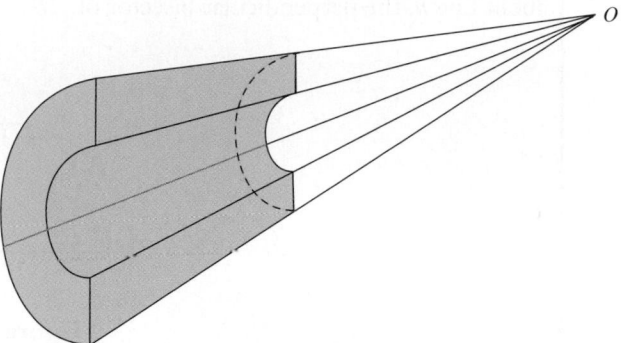

Figure 13-45

When a picture of an object is taken as with a box camera in Figure 13-46(a) with the pinhole lens at O, the object appears upside down on the negative. The picture of the object on the negative can be interpreted as an image under composition of a half-turn and a dilation. Figure 13-46(b) illustrates the image of an arrow from A to B under a composition of a half-turn followed by a dilation with scale factor $\frac{1}{2}$.

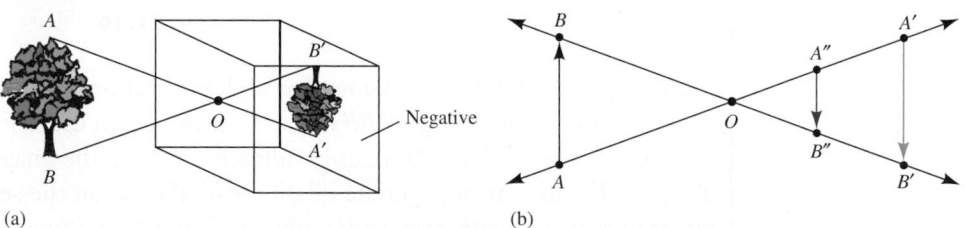

(a)   (b)

Figure 13-46

In Figure 13-46(b) the image A' of A under the half-turn with center O is found on the ray opposite $\overrightarrow{OA}$ so that $OA' = OA$. The point B', the image of B under the half-turn, is found similarly on the ray opposite $\overrightarrow{OB}$. The images of A' and B' under the dilation are A'' and B'', respectively. Consequently, the image of the arrow from A to B under the composition of the half-turn followed by the dilation is the arrow from A'' to B''.

A geometric construction application of dilations is shown in Example 13-11.

EXAMPLE 13-11

Inscribe a square in the semicircle and its given diameter in Figure 13-47 so that two of the vertices of the square are on the semicircle and two are on the diameter $\overline{AB}$.

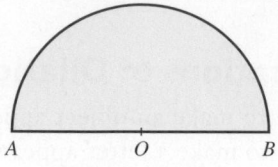

Figure 13-47

Solution The required square $XYZW$ is seen in Figure 13-48. It appears to be symmetrical about line n, the perpendicular bisector of $\overline{AB}$.

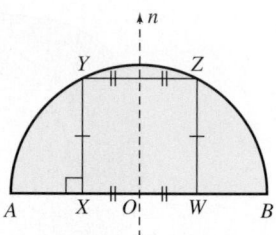

Figure 13-48

To solve the problem, we use the strategy of using a *related simpler problem*. It is easier to construct a square with two vertices on $\overline{AB}$ and the other two anywhere in the interior of the semicircular area as in Figure 13-49.

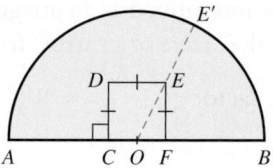

Figure 13-49

Such a square is $CDEF$. We know that under a dilation with center O with any scale factor, the image $C'D'E'F'$ of square $CDEF$ is also a square. From the definition of a dilation, we know that O, E, and E' are collinear. If we use a different dilation, the image E' of E will still be on $\overrightarrow{OE}$. The image of E under an appropriate dilation will also be on the semicircle. Thus we find the exact image of E by drawing a ray with endpoint O and containing E. The intersection of the ray and the semicircle is the desired E'. Similarly the image D' of D can be found to determine the desired square as seen in Figure 13-50. Next we drop perpendiculars from D' and E' to $\overline{AB}$ to determine C' and F'. $C'D'E'F'$ is the desired square.

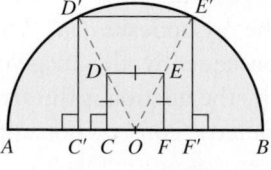

Figure 13-50

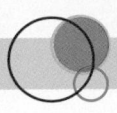

Assessment 13-3A

1. In the following figures, describe a sequence of isometries followed by a dilation so that the larger triangle is the final image of the smaller one:

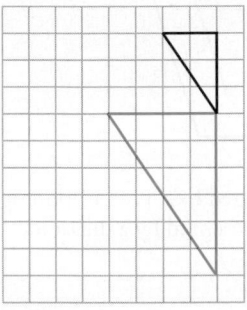

a.

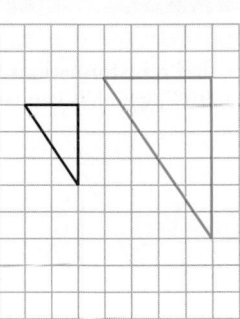

b.

2. In the following drawing, find the image of $\triangle ABC$ under the dilation with center O and scale factor $\dfrac{1}{2}$:

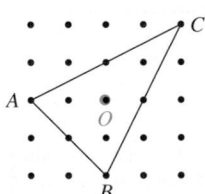

3. In each of the following drawings, find transformations that will take $\triangle ABC$ to its image, $\triangle A'B'C'$, which is similar:

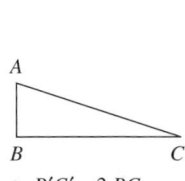
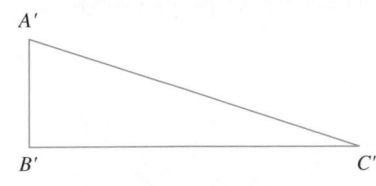

a. $B'C' = 2\,BC$

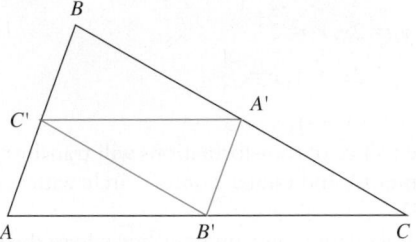

b. A', B', and C' are the respective midpoints of $\overline{BC}$, $\overline{AC}$, and $\overline{AB}$.

4. In the following figure, the smaller triangle is the image of the larger under a dilation centered at point O. Find the scale factor and the length of x and y as pictured.

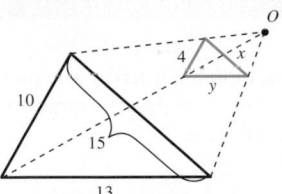

5. $\overline{A'B'}$ is the image of a candle $\overline{AB}$ produced by a box camera. Given the measurements in the figure, find the height of the candle.

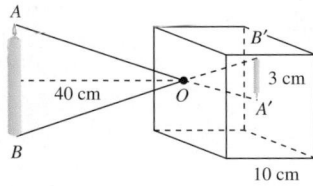

6. Find the coordinates of the images of $A(2, 3)$, and $B(^-2, 3)$ under the following transformations. Assume that all dilations are centered at the origin.
   a. A dilation with a scale factor 3 followed by a dilation with a scale factor 2
   b. A dilation with a scale factor 2 followed by a dilation with a scale factor 3

7. If a dilation with center O and scale factor r takes a quadrilateral $ABCD$ to $A'B'C'D'$, what dilation will take $A'B'C'D'$ back to $ABCD$?

8. a. Explain why in a coordinate system a dilation with center at the origin and scale factor $r > 0$ is given by $(x, y) \rightarrow (rx, ry)$.
   b. The transformation $(x, y) \rightarrow (^-2x, ^-2y)$ can be achieved by a dilation followed by an isometry. Find that dilation and the isometry.

c. Find the equations of the images of each of the following lines under the dilation in part (a) with the given scale factor r:

   i. $y = 2x, r = \dfrac{1}{2}$

   ii. $y = 2x, r = 2$

   iii. $y = 2x + 1, r = \dfrac{1}{3}$

   iv. $y = {}^-x - 1, r = 3$

9. What sequence of transformations will transform a circle with center O_1 and radius 2 onto a circle with center O_2 and radius 3?

10. Consider a dilation on a number line where the center of the transformation has 0 as its coordinate and the scale factor is 3. Describe the set of images of the points whose coordinates are integers.

11. In the following figure, the smaller pentagon is the image of the larger under a dilation centered at point O. Find the scale factor and the length of x and y as pictured.

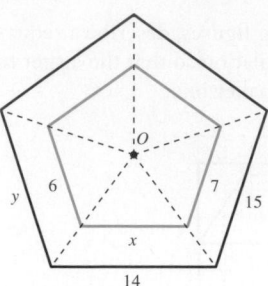

12. If a $2'' \times 3''$ photograph is enlarged to a $4'' \times 6''$ photograph, explain whether this can be represented by a dilation.

Assessment 13-3B

1. In the following figures, describe a sequence of isometries followed by a dilation so that the larger triangle is the final image of the smaller one:

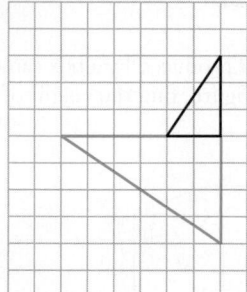

a.

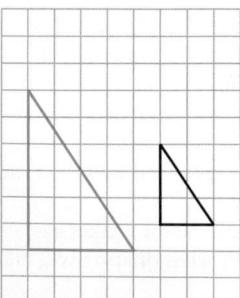

b.

2. Find the image of $\triangle ABC$ under the dilation with center O and scale factor $\dfrac{1}{2}$.

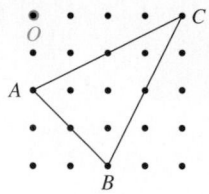

3. In each of the following drawings, find transformations that will take $\triangle ABC$ to its image, $\triangle A'B'C'$, which is similar:

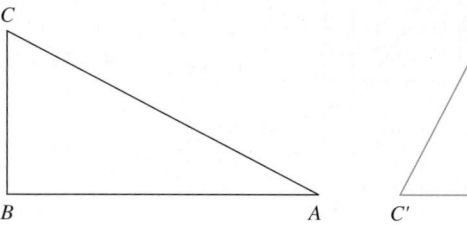

a. $AB = 2\,A'B'$

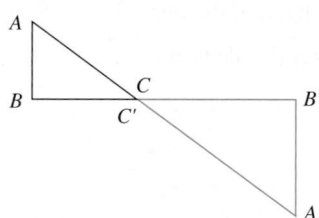

b. $\overline{AB} \parallel \overline{A'B'}$; points A, C, and A' are collinear; points B, C', and B' are collinear; $AB = \dfrac{2}{3}\,A'B'$

4. The following describes a dilation with center O and image of the segment of length 4 in blue. Find the scale factor and the lengths designated by x and y.

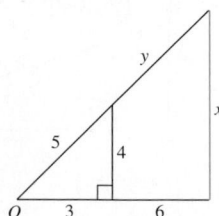

5. The following describes a dilation with center O and image in blue. Find the scale factor and the lengths designated by x and y.

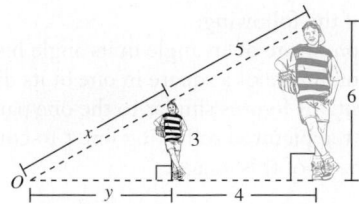

6. Find the coordinates of the images of $A(2, 3)$, and $B(^-2, 3)$ under the following transformations. Assume that all dilations are centered at the origin.
   a. A dilation with scale factor 2 followed by a translation with a slide arrow from $(2, 1)$ to $(3, 4)$
   b. A translation with a slide arrow from $(2, 1)$ to $(3, 4)$ followed by a dilation with scale factor 2
   c. What can you conclude from your answers to parts (a) and (b) concerning the order in which the transformations are performed?
7. What dilation will transform a circle with center at the origin and radius 4 onto a circle with the same center and radius 3?
8. What sequence of transformations will transform a circle with center O_1, and radius $\frac{1}{2}$ onto a circle with center O_2 and radius 3?
9. Sketch the image of the Octagon Quilt pattern with the center of the pattern as the center of a dilation with a scale factor of 2.

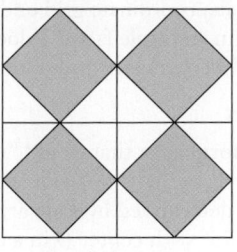

Octagon

10. Copy the following figure onto grid paper and determine the center and the scale factor of the dilation. Why is there only one possibility for the center?

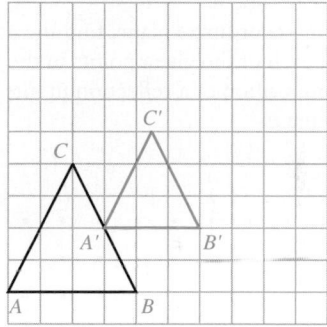

11. What dilation will undo a size transformation with center O and scale factor $\frac{3}{4}$?
12. Given any triangle ABC, construct a square $DEFG$ so that D is on $\overline{AB}$ and the vertices F and G are on $\overline{AC}$, and E is on $\overline{BC}$.

Mathematical Connections 13-3

Communication

1. Which of the following properties do not change under a dilation? Explain how you can be sure of your answers.
   a. Distance between points   b. Angle measure
   c. Parallelism; that is, if two lines are parallel to each other, then their images are parallel to each other.
2. Given two similar figures, explain how to tell if there is a dilation that transforms one of the figures onto the other.
3. a. Consider two consecutive dilations, each with center O and scale factors $\frac{1}{2}$ and $\frac{1}{3}$, respectively. Suppose the image of figure F under the first transformation is F' and the image of F' under the second transformation is F''. What single transformation will map F directly onto F''? Explain why.
   b. What would be the answer to (a) if the scale factors were r_1 and r_2?
4. a. Is the image of a circle with center O under a dilation with center O always a circle? Explain why or why not.
   b. Assume that under a dilation with scale factor r, the image of a segment of length a is a segment of length ra and answer part (a) of this question in case the center of the dilation is not at the center of the circle.
5. a. If a dilation had a scale factor of 1, describe the net result.
   b. Why could a dilation not have a scale factor of 0?

Open-Ended

6. Describe several real-life situations other than the ones discussed in this section in which dilations occur.
7. Use a sheet of graph paper with a coordinate system. Locate the origin as the center of a transformation with a scale factor of $^-1$ and find the image of some triangle located in the first quadrant. What other transformation studied has the same image as this transformation?

Cooperative Learning

8. Have members of your group draw several figures and find their images under a dilation with a scale factor of 3 and center of your choice.
   a. How does the perimeter of each image compare to the perimeter of the original figure? Compare your answers.
   b. Make a conjecture concerning the relationship between the perimeter of each image and the perimeter of the original figure under a dilation with a scale factor r.
   c. Discuss your findings and come up with a group conjecture.

Questions from the Classroom

9. A student asks, "If I have two triangles that are not similar, is it possible to transform one onto the other by a sequence of isometries followed by a dilation?" How do you respond?

10. A student says that a coordinate grid under a dilation with the center at the origin and scale factor 2 does not change the grid. The image is still a coordinate grid. How do you respond?

Review Problems

11. Describe a transformation that would "undo" each of the following:
    a. A translation determined by slide arrow from M to N
    b. A rotation of 75° with center O in a clockwise direction
    c. A rotation of 45° with center A in a counterclockwise direction.
    d. A glide reflection that is the composition of a reflection in line m and a translation that takes A to B where $\overrightarrow{AB} \parallel m$
    e. A reflection in line n

12. In the following coordinate plane, find the coordinates of the images of each of the given points in the transformation that is the composition of a reflection in line m followed by a reflection in line n.

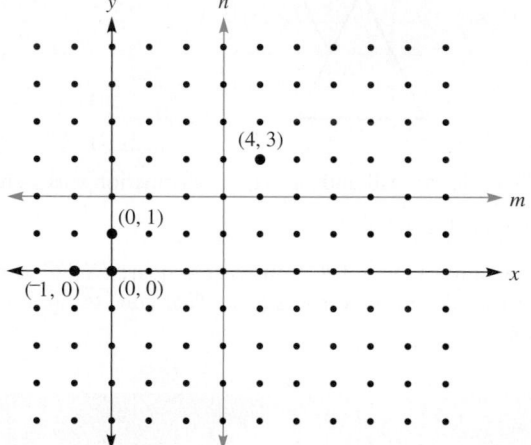

a. $(4, 3)$
b. $(0, 1)$
c. $(^-1, 0)$
d. $(0, 0)$

13. Find each of the following:
    a. Reflection image of an angle in its angle bisector
    b. Reflection image of a square in one of its diagonals

14. Zuni art contains figures similar to the one partially drawn below. Use the pictured reflecting line ℓ to complete an abstract version of this image.

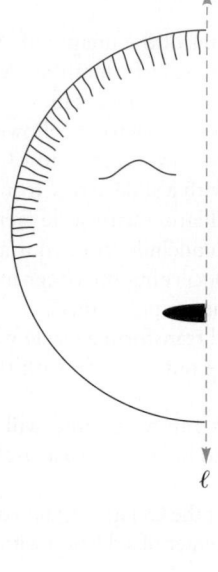

BRAIN TEASER Suppose you are looking in a mirror hung flat on a wall and your entire image goes from the top of the mirror to the bottom. How does the length of the part of your body that you see compare with the length of the mirror in Figure 13-51?

Figure 13-51

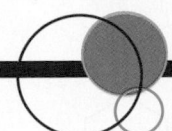

13-4 Tessellations of the Plane

In this section we use concepts from motion geometry to study *tessellations* of the plane. A **tessellation** of a plane is the filling of the plane with repetitions of congruent figures in such a way that no figures overlap and there are no gaps. (Similarly, one can tessellate space.) The tiling of a floor and various mosaics are examples of tessellations. Maurits C. Escher was a master of tessellations. Many of his drawings have fascinated mathematicians for decades. His *Study of Regular Division of the Plane with Reptiles* (pen, ink, and watercolor), 1939 is a tessellation of the plane by a lizardlike shape, as shown in Figure 13-52.

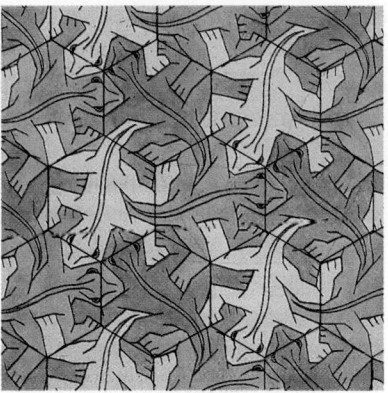

Study of Regular Division of the Plane with Reptiles (1939), M.C. Escher. Copyright © 2011 The M.C. Escher Company, Holland. All rights reserved. www.mcescher.com

Figure 13-52

At the heart of the tessellation in Figure 13-52, we see a regular hexagon. But perhaps the simplest tessellation of the plane can be achieved with squares. Figure 13-53 shows two different tessellations of the plane with squares.

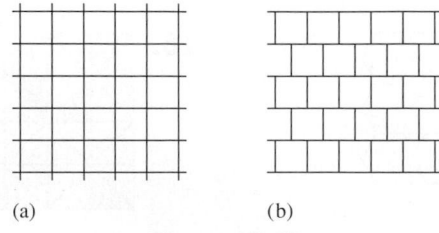

(a)   (b)

Figure 13-53

Tessellations with Regular Polygons

A **regular tessellation** constructed with a regular polygon is appealing and interesting because of its simplicity. Figure 13-54 shows portions of tessellations with equilateral triangles (a) and with regular hexagons (b).

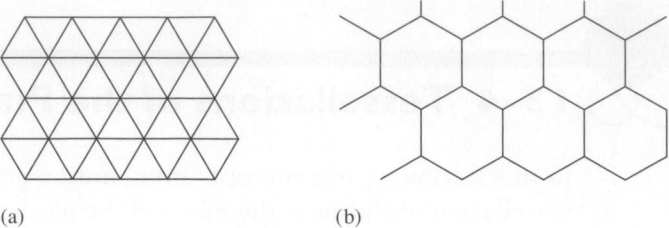

(a)   (b)

Figure 13-54

To determine other regular polygons that tessellate the plane, we investigate the possible size of the interior angle of a tessellating polygon. If n is the number of sides of a regular polygon, then because the sum of the measures of the exterior angles of the regular polygon is 360°, the measure of a single exterior angle of the polygon is $360°/n$. Hence, the measure of an interior angle is $180° - 360°/n$ because an exterior angle is the supplement of an interior angle. Table 13-2 gives some values of n, the type of regular polygon related to each, and the angle measure of an interior angle found by using the expression $180° - 360°/n$.

 The following section of a grade 4 student page introduces a connection between mathematics and art through tessellations.

 **School Book Page** MATH AND ART

 Math and Art

Tile patterns, called **tessellations,** can be drawn by sliding, flipping, and turning certain polygons. In each tessellation below, tell whether the red polygons are related by a slide, a flip, or a turn.

15. 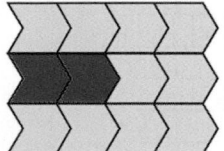   **16.** 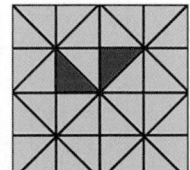   **17.**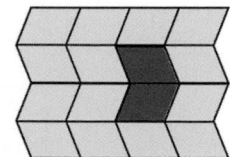

18. Draw your own tessellation. Tell what motions you used to create the pattern.

Table 13-2

Number of Sides (*n*)	Regular Polygon	Measure of Interior Angle
3	Triangle	60°
4	Square	90°
5	Pentagon	108°
6	Hexagon	120°
7	Heptagon	900/7°
8	Octagon	135°
9	Nonagon	140°
10	Decagon	144°

If a regular polygon tessellates the plane, the sum of the congruent angles of the polygons around every vertex must be 360°. If we divide 360° by each of the angle measures in the table, only 60°, 90°, and 120° divide 360°; hence of the listed regular polygons, only an equilateral triangle, a square, and a regular hexagon can tessellate the plane.

Can other regular polygons tessellate the plane? Notice that $\frac{360}{120} = 3$. Hence, 360 divided by a number greater than 120 is smaller than 3. However, the number of sides of a polygon cannot be less than 3. Because a polygon with more than six sides has an interior angle greater than 120°, it actually is not necessary to consider polygons with more than six sides. Consequently, no regular polygon with more than six sides can tessellate the plane.

Semiregular Tessellations

When more than one type of regular polygon is used and the arrangement of the polygons at each vertex is the same, the tessellation is **semiregular**. Figure 13-55 shows an example of a semiregular tessellation.

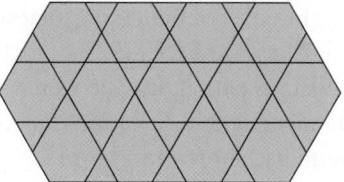

Figure 13-55

NOW TRY THIS 13-9

Create a semiregular tessellation consisting only of squares and equilateral triangles.

Tessellating with Other Shapes

Next, we consider tessellating the plane with arbitrary convex quadrilaterals. We investigate the problem, with the help of paper or cardboard quadrilaterals. Figure 13-56 shows an arbitrary convex quadrilateral and a way to tessellate the plane with the quadrilateral. Successive half-turns of the quadrilateral about the midpoints P, Q', and R'' of its sides produces four congruent quadrilaterals around a common vertex. Notice that the sum of the measures of the angles around vertex A is $a + b + c + d$. This is the sum of the measures of the interior angles of the quadrilateral, or 360°. Hence, four congruent quadrilaterals fit around vertex A. This process can be repeated so that four congruent quadrilaterals fit around each vertex of the original quadrilateral and its images. Thus, any convex quadrilateral tessellates a plane.

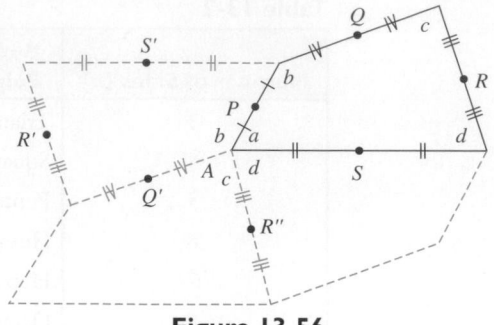

Figure 13-56

As we saw earlier in this section, a regular pentagon does not tessellate the plane. However, some nonregular pentagons do. One is shown in Figure 13-57, along with a tessellation of the plane by the pentagon.

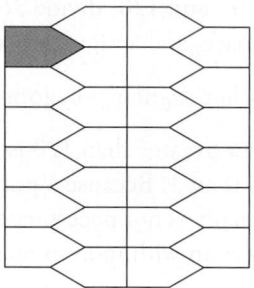

Figure 13-57

◯⊢ **Historical Note**

Marjorie Rice

Mathematicians thought they had solved which irregular pentagons tessellate when they had classified eight types of pentagons that would tessellate. They believed they had all of them. But then in 1975, Marjorie Rice (1923–), with no formal training in mathematics, discovered a ninth type of tessellating pentagon. She went on to discover four more by 1977. Her interest was piqued by reading an article in *Scientific American* by Martin Gardner (1914–2010). Two of the pentagons she found are shown in Figure 13-58. The problem of how many types of pentagons tessellate remains unsolved.

Type 9 discovered in February 1976

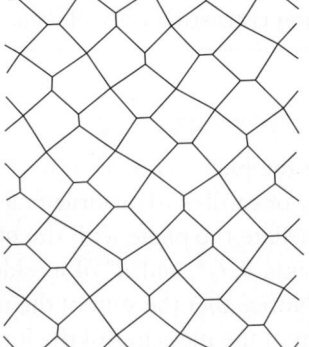

Type 13 discovered in December 1977

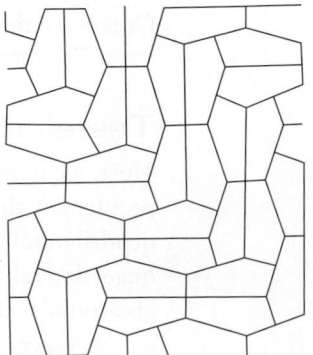

Figure 13-58

Creating Tessellations with Translations

What other types of designs can be made that tessellate a plane? The plane geometry and motions studied earlier provide some clues on how to design shapes that work. Consider the methods shown in Figure 13-59. In Figure 13-59(a), triangle *ABE* is removed from the left of parallelogram *ABCD* and slid to the right, forming the rectangle *BB'E'E* of Figure 13-59(b). This notion can be used to create a tessellating shape.

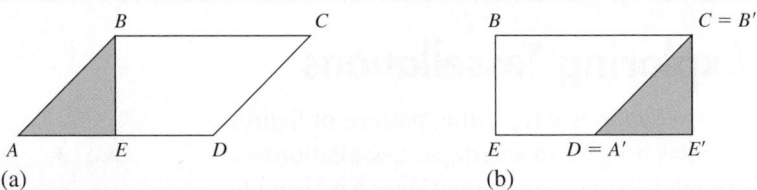

(a)   (b)

Figure 13-59

Consider any polygon known to tessellate a plane, such as rectangle *ABCD* in Figure 13-60(a). On the left side of the figure draw any shape in the interior of the rectangle, as in Figure 13-60(b). Cut this shape from the rectangle and slide it to the right by the slide that takes *A* to *B*, as shown in Figure 13-60(c). The resulting shape will tessellate the plane. (Why?)

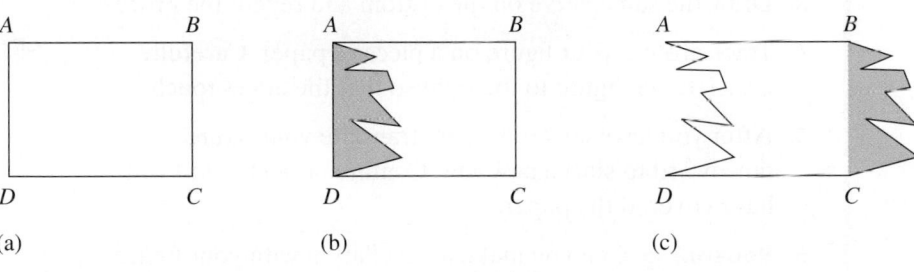

(a)   (b)   (c)

Figure 13-60

 The following partial student page shows how a square may be altered in two directions, shapes formed, and artistic details added to form "student art."

Creating Tessellations with Rotations

A second method of forming a tessellation involves a series of rotations of parts of a figure. In Figure 13-61(a), we start with an equilateral triangle *ABC*, choose the midpoint *O* of one side of the triangle, and cut out a shape, being careful not to cut away more than half of angle *B*.

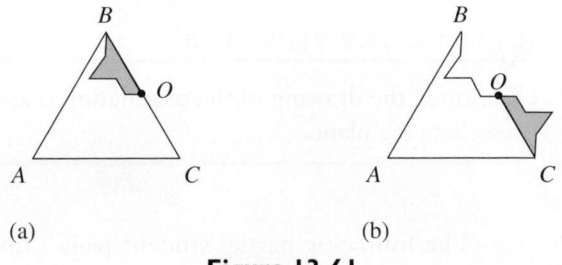

(a)   (b)

Figure 13-61

School Book Page EXPLORING TESSELLATIONS

Extension

For Use With Lesson 10-6

Exploring Tessellations

A *tessellation* is a repeating pattern of figures that has no gaps or overlaps. Tessellations are made using transformations. An example of a tessellation is shown at the right.

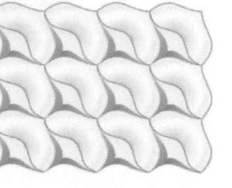

Use a square piece of cardboard.

1. Draw a curve from one vertex of the square to a neighboring vertex, as shown at the right.

2. Cut along the curve you drew. Translate the cutout piece to the opposite side of the square, and tape it down.

3. Draw the same curve on the bottom and repeat the process.

4. Trace around your figure on a piece of paper. Carefully translate the figure to the right so that the edges touch.

5. After you have covered a row, translate your figure downward to start a new row. Continue tracing until you have covered the paper.

6. **Reasoning** Can you make a tessellation with your figure using reflections? Explain.

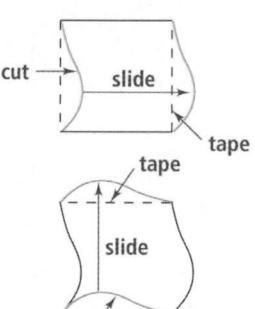

We then rotate the shape 180° clockwise around point O as in Figure 13-61(b). If we continue this process on the other two sides, then we obtain a shape that can be rotated around point A to tessellate the plane. Complete the tessellating shape and tessellate the plane with it in Now Try This 13-10.

 NOW TRY THIS 13-10

Continue the drawing of the tessellating shape in Figure 13-61. Cut out the shape and use it to tessellate the plane.

　　The following partial student page shows a grade 6 tessellation construction involving rotations.

School Book Page EXTENSIONS

PRACTICE

C Extensions

20. Follow these steps to change a square into a shape that tessellates the plane.

a. Cut out a square piece of paper about 8 cm on a side. Label the sides and vertices as shown.

b. Draw and cut any shape out of side 1. Do <u>not</u> cut off a corner.

c. Rotate the piece about point *A* and tape it to side 2. Be sure to tape the piece the same distance from point *A* as it was before you cut it.

d. Similarly, cut into side 3, rotate the piece about point *B,* and tape the piece to side 4.

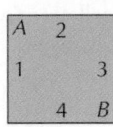 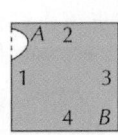 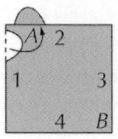 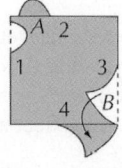

e. Repeat the process a few more times, cutting into any <u>unaltered</u> part of the square except a corner. Remember this plan:

- Rotate pieces from side 1 onto side 2, and from side 2 onto side 1. (Be sure to tape the piece the same distance from *A*.)

- Rotate pieces from side 3 onto side 4, and from side 4 onto side 3. (Be sure to tape the piece the same distance from *B*.)

f. Add artistic details and draw the tessellation.

Assessment 13-4A

1. On dot paper, draw a tessellation of the plane using the following figure:

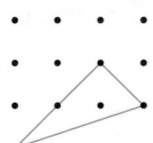

2. a. Tessellate the plane with the following quadrilateral:

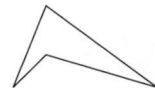

b. Is it possible to tessellate the plane with any quadrilateral? Why or why not?

3. On dot paper, use the following pentomino, to make a tessellation of the plane, if possible.

4. There are many ways to tessellate the plane by using combinations of regular polygons. An example follows.

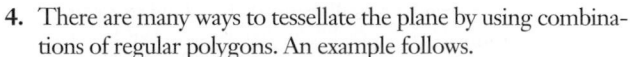
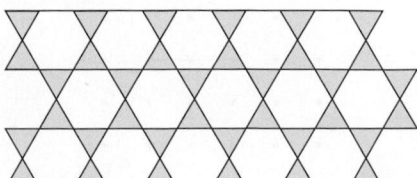

Produce other such tessellations using only equilateral triangles, squares, and regular hexagons.

5. The shaded figure in the Card Trick quilt pattern is formed from four ell-shaped figures. Explain whether the outlined figure will tessellate a plane.

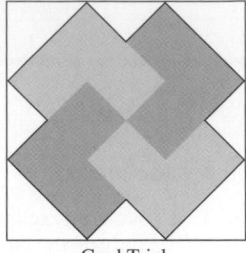

Card Trick

6. To determine whether a shape created using a glide reflection will tessellate the plane, complete the following:
   a. Start with a rectangle. Determine some shape that you might use with a slide to form a tessellating shape. Slide it as shown. Determine the horizontal line of symmetry of the rectangle, and reflect as shown.

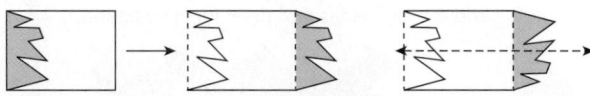

   b. Explain why the described series of motions is a glide reflection.
   c. Determine whether the final shape will tessellate the plane.
7. The **dual of a regular tessellation** is the tessellation obtained by connecting the centers of the polygons in the original tessellation if their sides intersect in more than one point. The dual of the tessellation of equilateral triangles is the tessellation of regular hexagons, shown in color in the following figure:

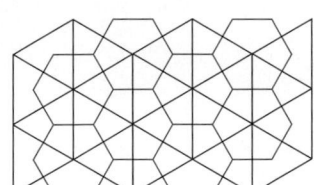

Describe and show the dual of each of the following:
   a. The regular tessellation of squares shown in Figure 13-53(a)
   b. The tessellation of squares in Figure 13-53(b)
   c. A tessellation of regular hexagons
8. A sidewalk is made of tiles of the type shown in the following figure:

Each tile is made of three regular hexagons with some sides removed. Draw a partial tessellation composed of seven such figures.
9. Which of the following tessellations are semiregular and which are not? Justify your answers.

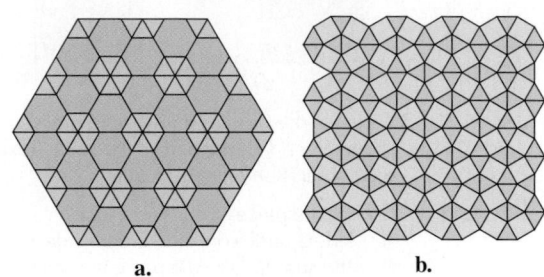

a.   b.

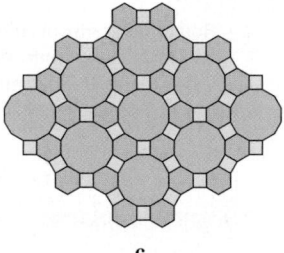

c.

10. A portion of a Pieced Star Quilt pattern is shown below. Will this hexagon tessellate a plane? Prove your answer.

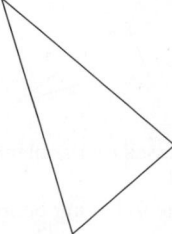

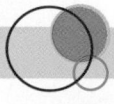 **Assessment 13-4B**

1. On dot paper, draw a tessellation of the plane using the following figure:

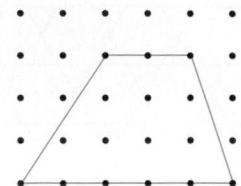

2. a. Tessellate the plane with the following triangle:

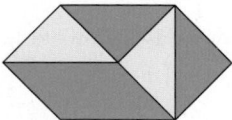

   b. Is it possible to tessellate a plane with any triangle? Why or why not?

3. On dot paper, use each of the following pentominoes, one at a time, to make a tessellation of the plane, if possible:

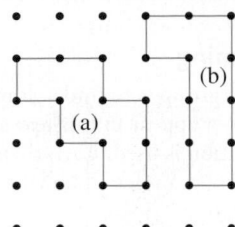

4. Produce a tessellation of a plane using a combination of regular octagons and squares.

5. A "bow-tie" figure is formed by right triangles (white) near the center of the Brown Goose quilt pattern. Explain whether this bow-tie figure will tessellate a plane.

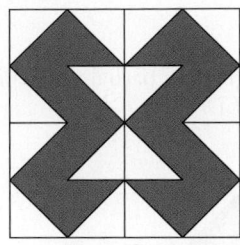

Brown Goose

6. a. Create a figure using a reflection that will tessellate a plane.
   b. Create a figure using a reflection that will not tessellate a plane.

7. The partial student page below shows "student work." Answer questions 17 and 18.

School Book Page MATH AND ART

Math and Art

The Dutch artist M. C. Escher (1898 to 1972) is renowned for his interesting tessellations. Inspired by Escher's tessellating shapes of fish and lizards, Jon and Miguel created the art below.

17. What transformation would move fish *A* to fish *B*?

18. What transformation would move lizard *C* to lizard *D*? lizard *C* to lizard *E*?

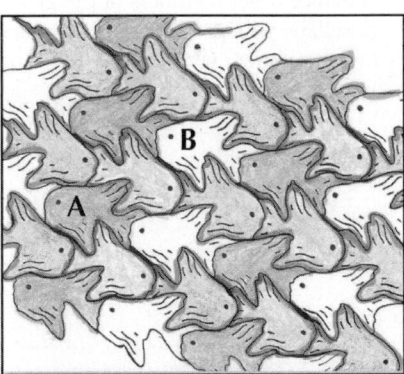

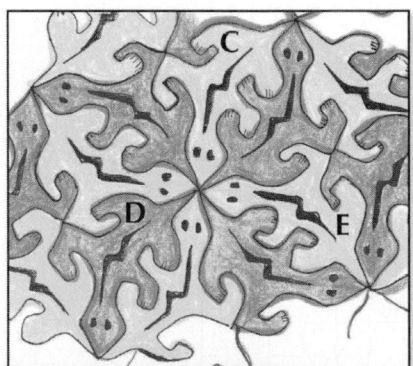

8. A concrete tile similar to that shown below is used to construct a sound barrier wall for an interstate highway in Washington, DC. Sketch a tessellation using the figure to show that a wall could be built of the tiles.

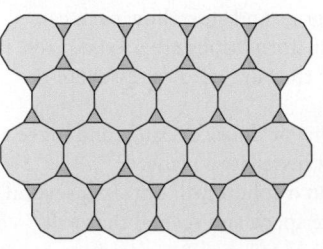

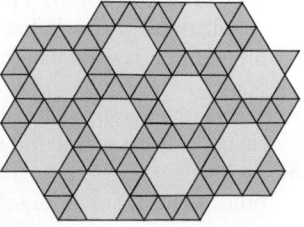

a.   b.

10. Explain whether the "arrow" below will tessellate a plane.

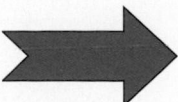

9. Which of the following tessellations are semiregular and which are not? Justify your answers.

Mathematical Connections 13-4

Communication

1. The following figure is a partial tessellation of the plane with the trapezoid *ABCD*:

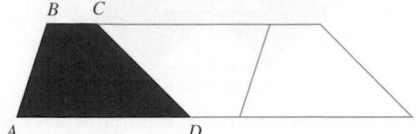

   a. Explain how the tessellation can be used to find a formula for the area of the trapezoid.
   b. Tessellate the plane with a triangle and show how the tessellation can be used to find the relationship between the length of the segment connecting the midpoints of the two sides of a triangle and the length of the third side (the Midsegment Theorem of Chapter 12).
2. Explain in your own words why only three types of regular polygons tessellate the plane.
3. The following figure shows how to tessellate the plane with irregular pentagons. Explain how the pentagons can be constructed.

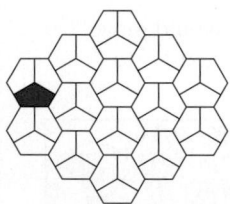

Open-Ended

4. There are endless numbers of figures that tessellate a plane. In the following drawing, the shaded figure is shown to tessellate the plane:

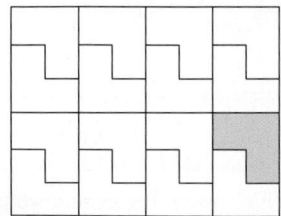

   Design several different polygons and show how each can tessellate the plane. What transformations are used in each of your designs? Explain how they are used to tessellate the plane.
5. Examine different quilt patterns or floor coverings and make a sketch of those you found that tessellate a plane.
6. A cube will tessellate space but a sphere will not. List several other solids that will tessellate space and several that will not.

Cooperative Learning

7. Each member of a group is to find a drawing by M. C. Escher that does not appear in this text and in which the concept of tessellation is used. Each then shows the other members of the group, in detail, how he or she thinks Escher created the tessellation.
8. a. Convince the members of your group that the following figure containing six equilateral triangles tessellates the plane:

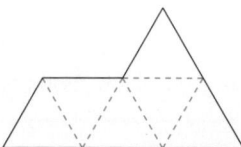

   b. As a group, find different figures that contain six equilateral triangles. How many such figures can you find? Discuss the meaning of "different."
   c. Find some of the figures in (b) that are rep-tiles. (A **rep-tile** is a figure whose copies can be used to form a larger figure similar to itself.) Convince other members of your group that your figures are rep-tiles and that they tessellate the plane.
9. Trace each of the following pentomino shapes and cut out several copies of each. Each member of a small group should pick a shape, decide if it tessellates the plane, and convince other members of the group that it does or does not tessellate. The group should report the answers with figures or an argument why a shape does not tessellate.

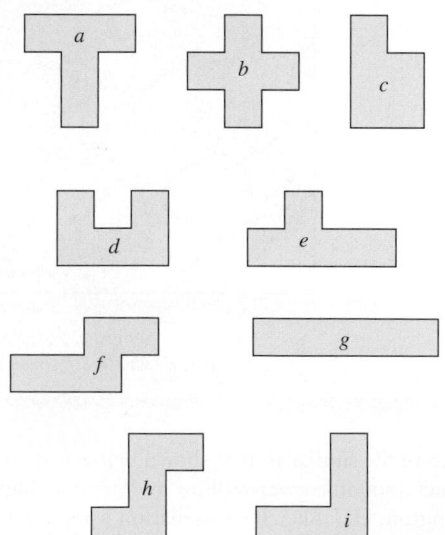

Questions from the Classroom

10. Isaiah says that the bricks used in a classroom wall construction form a tessellation. How do you respond?
11. A student asks if the image seen through a kaleidoscope tessellates a plane. How do you respond?

Review Questions

12. Find the image of the figure below in each of the following:
    a. A translation from M to N
    b. A 90° rotation counterclockwise through M

13. With the quilt patterns below, define isometries that will take each pattern to itself.

Dutchman's
Puzzle
a.

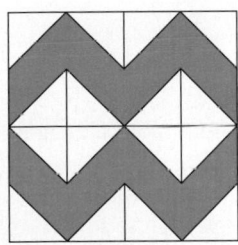

Ribbons
b.

14. Determine whether the figure below could be its own image under some isometry. If so, which one?

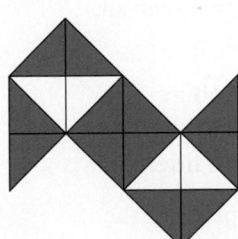

15. A quadrilateral has vertices at $(a, 0)$, $(0, a)$, $(^-a, 0)$, and $(0, ^-a)$, where $a > 0$.
    a. What is the quadrilateral?
    b. Find the image of each vertex under the reflection in the line $y = x$.

16. Prove that under a dilation with center O and scale factor r, where $0 < r < 1$, the image of $\overline{AB}$ is a parallel segment $\overline{A'B'}$.

17. What dilation, if any, allows a line with equation $y = kx$, $k > 0$, to be the image of the line with equation $y = x$?

National Assessment of Educational Progress (NAEP) Questions

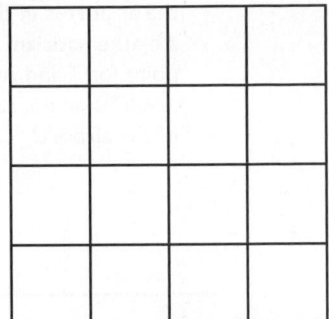

Identical puzzle pieces have been put together to form the large square shown above. Which of the following could be the shape of each puzzle piece?

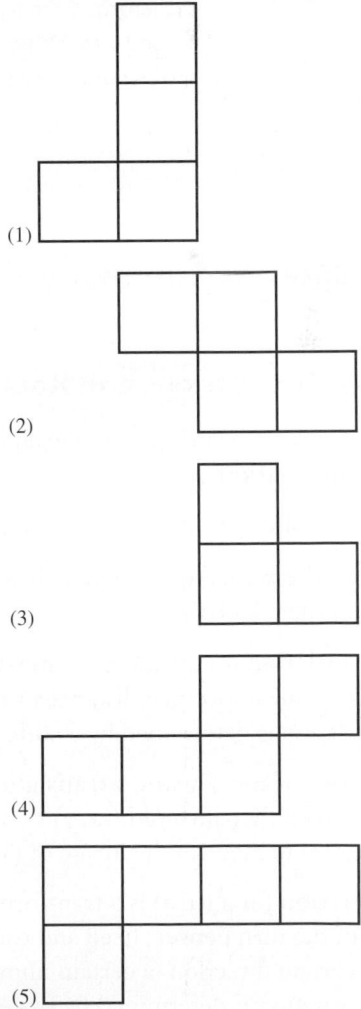

(1)

(2)

(3)

(4)

(5)

NAEP, 2009, Grade 8

An architect needs to determine the location of an airport T serving three cities A, B, and C as in Figure 13-62 so that the sum of the distances from the airport to the three cities is minimal and the airport is in the interior of triangle ABC. The architect's friend, a mathematician, suggests investigating the problem by picking any point for T and rotating $\triangle ATB$ clockwise about B by $60°$. Assuming $\triangle ABC$ is acute, can you help the architect to find the true location of the airport?

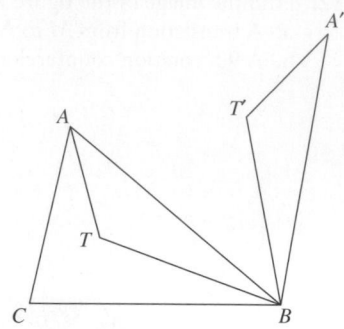

Figure 13-62

 **Hint for Solving the Preliminary Problem**

Consider the angles of incidence and reflection when the ball "bounces" off of the side. Then consider the length of the path from the start to the first wall and after the bounce the length of the path to the second wall. What is that sum? Then follow the ball along the rest of its path and decide if it hits the spot where it started.

Chapter 13 Summary

13-1 Translations and Rotations

	Pages
Any function from a plane to itself that is a one-to-one correspondence is a **transformation** of the plane.	771
Any transformation (or **motion**) that preserves length or distance is an **isometry.**	771
Isometries preserve geometric shapes and relations including congruence, parallelism, and perpendicularity.	771
A **translation** (or a **slide**) is a **transformation** of a plane that moves every point of the plane a specified distance in a specified direction along a straight line. The translation is determined by a **slide arrow**, or **vector**, and is along the **slide line**.	772
In terms of coordinates, a **translation** of the plane is a function from the plane to the plane such that any point (x, y) corresponds to the point $(x + a, y + b)$, where a and b are real numbers. The image of (x, y) is $(x + a, y + b)$.	774
A **rotation** (or a **turn**) is a transformation of the plane determined by holding one point, the turn **center**, fixed and rotating the plane about this point by a certain amount in a certain direction (a certain number of degrees either clockwise or counterclockwise). The rotation is determined by its center and a **turn angle.**	776
A **half-turn** is a rotation of 180 degrees about a center.	777
Theorem (Slopes of Perpendicular lines): Two nonvertical lines are perpendicular if, and only if, their slopes, m_1 and m_2, satisfy the condition that $m_1 m_2 = {}^-1$. Every vertical line has no slopes but is perpendicular to every line with slope 0.	779

Under any isometry, the image of a line is a line, the image of a circle is a circle, and the images of parallel lines are parallel lines. | 777

A figure has **turn symmetry** if there is a turn center and turn angle (whose measure is not a multiple of 360°) that will take the original to itself as an image. | 778

13-2 Reflections and Glide Reflections

A **reflection** in line ℓ is a transformation from the plane to the plane that pairs each point P with a point P' in such a way that ℓ is the perpendicular bisector of $\overline{PP'}$ as long as P is not on ℓ. If P is on ℓ, then $P = P'$. | 788

A reflection reverses **orientation** of the image of a figure. A translation and a rotation do not. | 788

A **glide reflection** is a transformation composed of a translation followed by a reflection in a line parallel to the slide arrow. | 794

Any two geometric figures are **congruent** if, and only if, one is an image of the other under a single isometry or under a composition of isometries. | 794

When a ray of light bounces off a mirror or when a billiard ball bounces off the rail of a billiard table, the **angle of incidence** is congruent to the **angle of reflection.** | 792–793

13-3 Dilations

A **dilation (size transformation)** from the plane to the plane with center O and scale factor $r > 0$ is a transformation that assigns to each point A in the plane the point A' such that O, A, and A' are collinear and $OA' = rOA$ such that O is not between A and A'. O is its own image. | 801

Theorem: Under a dilation, the image of a triangle is a similar triangle, and the image of a polygon is a similar polygon. | 801

Theorem: A dilation with center O and scale factor $r > 0$ has the following properties: | 803
  a. The image of a line segment is a line segment parallel to the original segment and r times as long.
  b. The image of an angle is an angle congruent to the original angle.

Two figures are **similar** if one is the image of the other by an isometry (or a sequence of isometries) followed by a dilation. | 805

13-4 Tessellations of the Plane

A **tessellation** of a plane is the filling of the plane with repetitions of congruent figures in such a way that no figures overlap and there are no gaps. | 811

A **regular tessellation** is a tessellation using a single regular polygon. | 811–812

When more than one type of regular polygon is used and the arrangement of the polygons at each vertex is the same, the tessellation is **semiregular.** | 813

Chapter 13 Review

1. Complete each of the following motions:

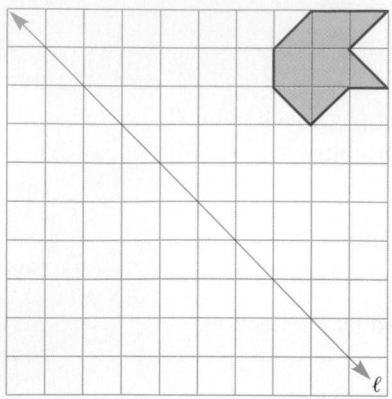

A reflection in ℓ

a.

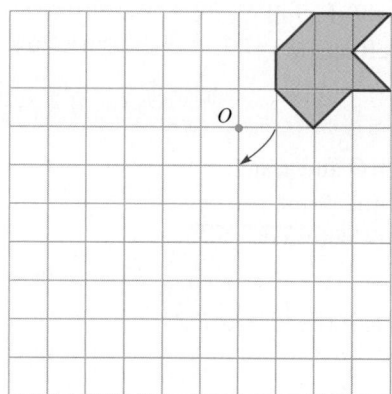

A rotation in O through the given arc

b.

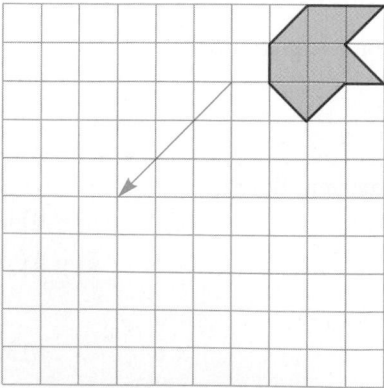

A translation, as pictured

c.

2. For each of the following figures, construct the image of $\triangle ABC$. (*Hint:* In each part, find the images of the vertices.)

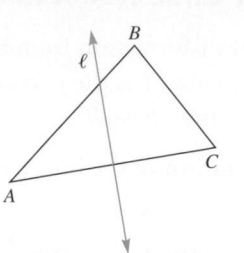

Through a reflection in ℓ

a.

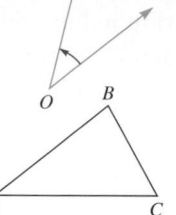

Through the given rotation in O

b.

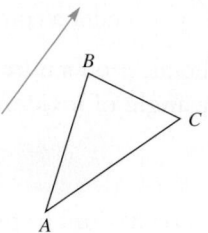

Through the translation arrow pictured

c.

3. Determine any reflections or rotations, that take the following figures to themselves:

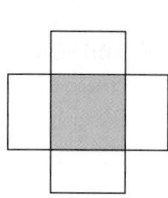

a.

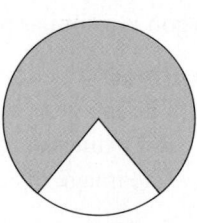

b.

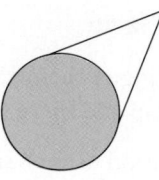

c.

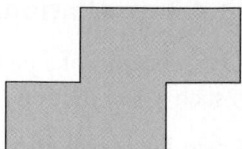

d.

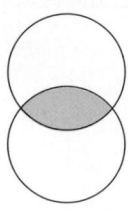

e.

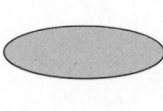

f.

4. In the following figure, △A'B'C' is the image of △ABC under a dilation.

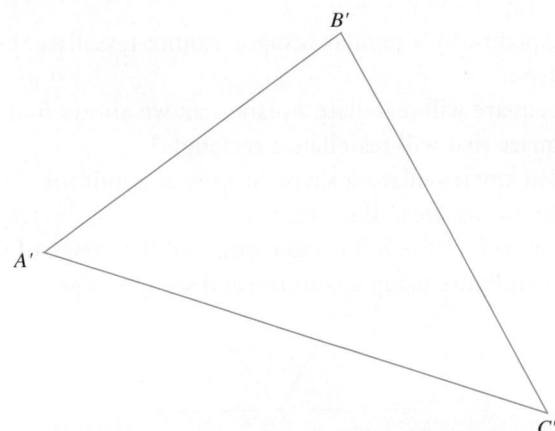

Locate points A, B, and C such that A' is the center of the dilation and $BC = \frac{1}{2}B'C'$.

5. Given that STAR in the figure shown is a parallelogram, describe a sequence of isometries to show the following:
   a. △STA ≅ △ARS
   b. △TSR ≅ △RAT

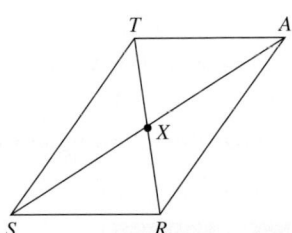

6. Given that BEAUTY in the figure shown is a regular hexagon, describe a sequence of isometries that will transform the following:
   a. BEAU onto AUTY
   b. BEAU onto YTUA

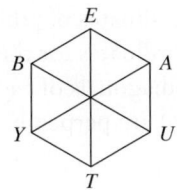

7. Given that △SNO ≅ △SWO in the following figure, describe one or more isometries that will transform △SNO onto △SWO.

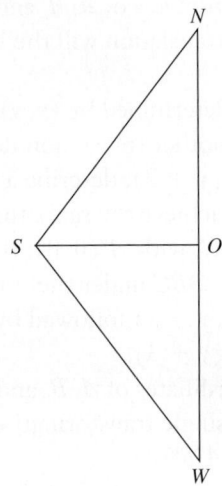

8. Show that △SER in the following figure is the image of △HOR under a succession of isometries with a dilation.

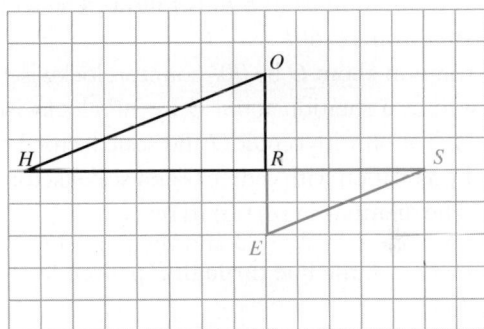

9. Show that △TAB in the following figure is the image of △PIG under a succession of isometries with a dilation.

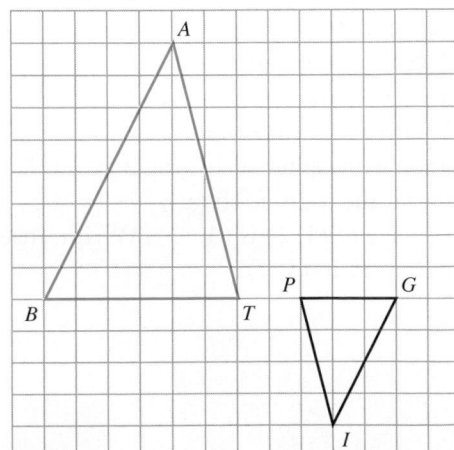

10. The triangle $A'B'C'$ with $A'(0, 7.91)$, $B'(^-5, ^-4.93)$, $C'(4.83, 0)$ is the image of triangle ABC under the translation $(x, y) \rightarrow (x + 3, y - 5)$.
   a. Find the coordinates of A, B, and C.
   b. Under what translation will the image of $\triangle A'B'C'$ be $\triangle ABC$?

11. If a translation determined by $(x, y) \rightarrow (x + 3, y - 2)$ is followed by another translation determined by $(x, y) \rightarrow (x - 3, y + 2)$, describe a single transformation that would achieve the same thing.

12. Suppose $\triangle A''B''C''$ with $A''(0, 0)$, $B''(1, 5)$, $C''(^-2, 7)$ is the image of $\triangle ABC$ under the translation $(x, y) \rightarrow (x + 2, y - 1)$ followed by the translation $(x, y) \rightarrow (x + 1, y + 3)$.
   a. Find the coordinates of A, B, and C.
   b. Under what single transformation will the image of $\triangle ABC$ be $\triangle A''B''C''$?

13. Write each of the following as a single transformation:
   a. i. A translation from A to B followed by a translation from B to C
      ii. A translation from B to C followed by a translation from A to B
   b. A rotation about O by 90° counterclockwise, followed by a rotation about O by 30° clockwise
   c. i. A dilation with center O and scale factor 3 followed by a dilation with center O and scale factor 2
      ii. The dilations in part (i) in reverse order

14. Given the line $y = 2x + 3$ and the point $P(^-1, 3)$, find the equation of the line through P perpendicular to the given line.

15. Find the equation of the image of the line $y = ^-x + 3$ under each of the following transformations:
   a. The translation $(x, y) \rightarrow (x + 2, y - 3)$
   b. Reflection in the x-axis
   c. Reflection in the y-axis
   d. Reflection in the line $y = x$
   e. Half-turn about the origin
   f. Dilation with center at the origin and scale factor 2

16. a. Consider the translation $(x, y) \rightarrow (x + 3, y - 5)$. The image of $(1, 2)$ is $(4, ^-3)$. What translation takes $(4, ^-3)$ to $(1, 2)$?

   b. What is the net result of following the translation $(x, y) \rightarrow (x + h, y + k)$ by the translation $(x, y) \rightarrow (x - h, y - k)$?

17. Explain why a regular octagon cannot tessellate the plane.

18. A square will tessellate a plane; can we always find a square that will tessellate a rectangle?

19. Can any tessellating shape be used as a unit for measuring area? If so, explain why.

20. For each of the following cases, find the image of the given figure using a compass and straightedge.

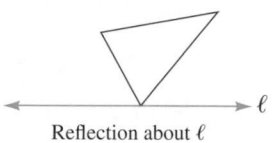
Reflection about ℓ
   a.

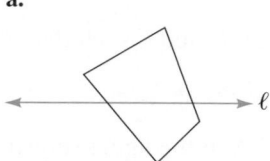
Reflection about ℓ
   b.

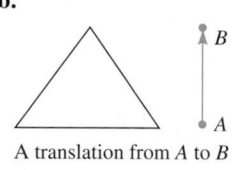
A translation from A to B
   c.

21. Determine whether the following shapes tessellate the plane.

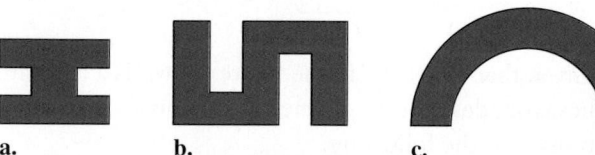
   a.   b.   c.

22. Three vertices of a rhombus are at $O(0, 0)$, $A(3, 4)$, $C(a, 0)$. Find the following:
   a. All possible coordinates of point C
   b. All possible coordinates for the fourth vertex B
   c. Verify that the diagonals of each of the rhombi you found in part (b) are perpendicular to each other.

Area, Pythagorean Theorem, and Volume

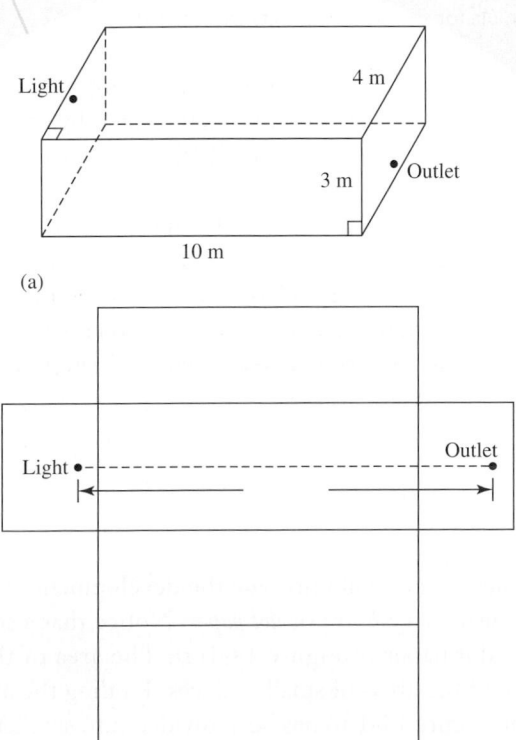

(a)

(b)

Preliminary Problem

A rectangular living room is 10 m long, 4 m wide, and 3 m high. A small light is fastened to the wall 30 cm from the ceiling and 2 m from each side wall. The only electrical outlet in the room is on the opposite wall, 30 cm from the floor and 2 m from each side wall. A drawing showing the room is given in part (a) of the figure on the left. An electrical cord connected directly from the light to the outlet would be hazardous to anyone in the room so the length of cord must be taped to the walls and/or ceiling, and/or floor. Al claims that the shortest distance for the path of the cord could be computed by going directly as shown in part (b) of the figure on the left. Betty claims that she can find a shorter path. Tell who is correct and why.

If needed, see Hint on page 913.

P*SSM* discusses the measurable attributes of objects as follows:

> A measurable attribute is a characteristic of an object that can be quantified. Line segments have length, plane regions have area, and physical objects have mass. As students progress through the curriculum from preschool through high school, the set of attributes they can measure should expand. . . . In grades 3–5, students should learn about area more thoroughly, as well as perimeter, volume, temperature, and angle measure. In these grades, they learn that measurements can be computed using formulas and need not always be taken directly with a measuring tool. Middle-grade students build on these earlier measurement experiences by continuing their study of perimeter, area, and volume and by beginning to explore derived measurements, such as speed. (p. 44)

Many concepts of measurement are more confusing for children than for adults because students lack everyday measurement experiences. In this chapter, we use both the English and the metric systems of measurement for length, area, volume, mass, and temperature with the philosophy that students should learn to think within a system. We develop formulas for the area of plane figures and for surface areas and volumes of three-dimensional figures. We use the concept of area in discussing the Pythagorean theorem.

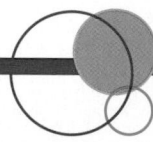

14-1   Areas of Polygons and Circles

In the grade 4 *Focal Points*, we find the following:

> Students recognize area as an attribute of two-dimensional regions. They learn that they can quantify area by finding the total number of same-sized units of area that cover the shape without gaps or overlaps. They understand that a square that is 1 unit on a side is the standard unit for measuring area. They select appropriate units, strategies (e.g., decomposing shapes), and tools for solving problems that involve estimating or measuring area. Students connect area measure to the area model that they have used to represent multiplication, and they use this connection to justify the formula for the area of a rectangle. (p. 16)

In this section, we quantify area using same-sized square units and we develop formulas for the area of various polygons. Technically, we find the areas of regions determined by shapes; for example, we use the vernacular "find the area of a triangle" instead of "find the area of the region determined by a triangle."

Area is measured using square units and the area of a region is the number of nonoverlapping square units that covers the region. A square measuring 1 ft on a side has an area of 1 square foot, denoted 1 ft^2. A square measuring 1 cm on a side has an area of 1 square centimeter, denoted 1 cm^2.

Students sometimes confuse an area of 5 cm^2 with the area of a square 5 cm on each side. The area of a square 5 cm on each side is $(5 \text{ cm})^2$, or 25 cm^2. Five squares each 1 cm by 1 cm have the area of 5 cm^2. Thus, $5 \text{ cm}^2 \neq (5 \text{ cm})^2$.

Areas on a Geoboard

Addition Method

In teaching the concept of area, intuitive activities should precede the development of formulas. Many such activities can be accomplished using a *geoboard* or *dot paper*. Notice that a square unit is defined in the upper left corner of the dot paper in Figure 14-1(a). The area of the shaded pentagon can be found by finding the sum of the areas of smaller pieces. Finding the area in this way uses the *addition method*. The region in Figure 14-1(b) has been divided into smaller pieces in Figure 14-1(c). What is the area of this shape?

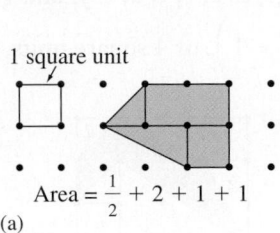

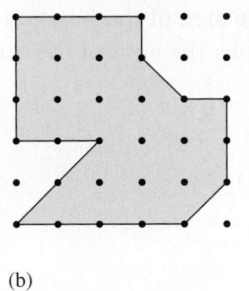

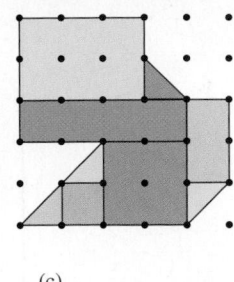

1 square unit

$$\text{Area} = \frac{1}{2} + 2 + 1 + 1$$

(a)

(b)

(c)

Figure 14-1

Rectangle Method

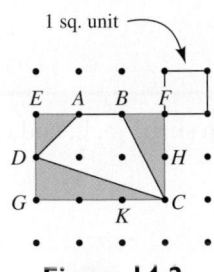

1 sq. unit

Figure 14-2

Another method of finding the area of shapes on the dot paper is the *rectangle method*. To find the area of quadrilateral *ABCD* in Figure 14-2, we construct the rectangle *EFCG* around the quadrilateral and then subtract the areas of the shaded triangles *EAD*, *BFC*, and *DGC*. The area of rectangle *EFCG* can be counted to be 6 square units. The area of $\triangle EAD$ is $\frac{1}{2}$ square unit, and the area of $\triangle BFC$ is half the area of rectangle *BFCK*, or $\frac{1}{2}$ of 2, or 1 square unit. Similarly, the area of $\triangle DGC$ is half the area of rectangle *DHCG*; that is, $\frac{1}{2} \cdot 3$, or $\frac{3}{2}$ square units. Consequently, the area of *ABCD* is $6 - \left(\frac{1}{2} + 1 + \frac{3}{2} \right)$, or 3 square units.

EXAMPLE 14-1

Using a geoboard, find the area of each of the shaded regions of Figure 14-3.

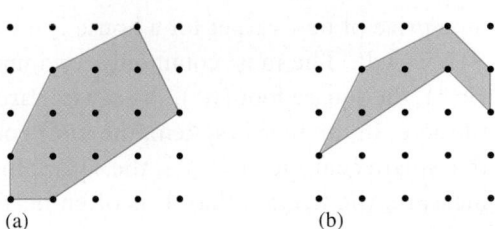

(a)   (b)

Figure 14-3

Solution

a. We construct a rectangle around the hexagon and then subtract the areas of regions *a*, *b*, *c*, *d*, and *e* from the area of this rectangle, as shown in Figure 14-4. Therefore, the area of the hexagon is $16 - (3 + 1 + 1 + 1 + 1)$, or 9 square units. The addition method could also be used in this problem.

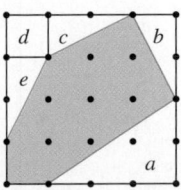

Figure 14-4

b. The area of the hexagon equals the area of the surrounding rectangle shown in Figure 14-5 minus the sum of the areas of figures *a*, *b*, *c*, *d*, *e*, *f*, and *g*. Thus the area of the hexagon is

$$12 - \left(3 + 1 + \frac{1}{2} + \frac{1}{2} + 1 + 1 + 1 \right), \text{ or 4 square units.}$$

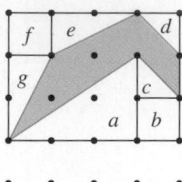

Figure 14-5

NOW TRY THIS 14-1

Find the area of the large double hexagon in Figure 14-6 using each given shape (a, b, and c) as a unit of area measure.

a. Triangles

b. Rhombuses

c. Trapezoids

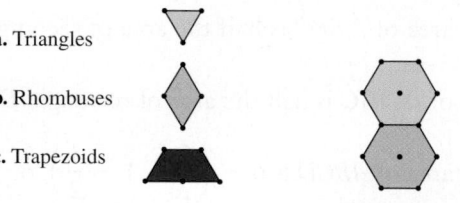

Figure 14-6

Converting Units of Area

The price of new carpet for a house is quoted in terms of dollars per square yard, for example, $12.50/yd^2. The most commonly used units of area in the English system are the square inch (in.^{2}), the square foot (ft^2), the square yard (yd^2), the square mile (mi^2), and, for land measure, the acre. In the metric system, the most commonly used units are the square millimeter (mm^2), the square centimeter (cm^2), the square meter (m^2), the square kilometer (km^2), and, for land measure, the hectare (ha). It is often necessary to convert from one area measure to another within a system.

To determine how many 1-cm squares are in 1 m^2, look at Figure 14-7(a). There are 100 cm in 1 m, so each side of the square meter has a measure of 100 cm. Thus, it takes 100 rows of 100 1-cm squares each to fill a square meter, that is, $100 \cdot 100$, or 10,000 1-cm squares. Because the area of each centimeter square is 1 cm $\cdot$ 1 cm, or 1 cm^2, there are 10,000 cm^2 in 1 m^2. In general, *the area A of a square that is s units on a side is s^2 square units*, as shown in Figure 14-7(b).

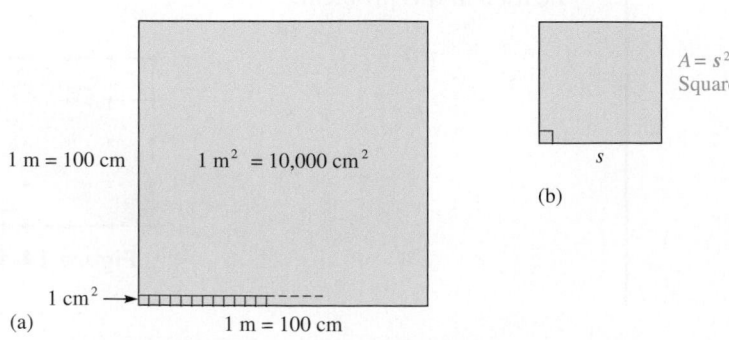

Figure 14-7

Other metric conversions of area measure can be developed similarly. For example, Figure 14-8(a) shows that $1 \text{ m}^2 = 10{,}000 \text{ cm}^2 = 1{,}000{,}000 \text{ mm}^2$. Likewise, Figure 14-8(b) shows that $1 \text{ m}^2 = 0.000001 \text{ km}^2$. Similarly, $1 \text{ cm}^2 = 100 \text{ mm}^2$ and $1 \text{ km}^2 = 1{,}000{,}000 \text{ m}^2$.

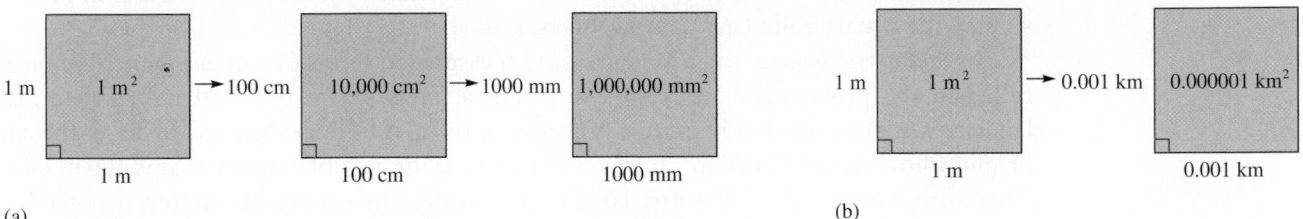

(a)   (b)

Figure 14-8

Table 14-1 shows symbols for metric units of area and their relationship to the square meter.

Table 14-1

Unit	Symbol	Relationship to Square Meter
square kilometer	km^2	1,000,000 m^2
*square hectometer	hm^2	10,000 m^2
*square dekameter	dam^2	100 m^2
square meter	**m^2**	**1** **m^2**
*square decimeter	dm^2	0.01 m^2
square centimeter	cm^2	0.0001 m^2
square millimeter	mm^2	0.000001 m^2

*Not commonly used

EXAMPLE 14-2

Complete each of the following:

a. $5 \text{ cm}^2 =$ _____ mm^2

b. $124{,}000{,}000 \text{ m}^2 =$ _____ km^2

Solution

a. $1 \text{ cm}^2 = 100 \text{ mm}^2$ implies $5 \text{ cm}^2 = 5 \cdot 1 \text{ cm}^2 = 5 \cdot 100 \text{ mm}^2 = 500 \text{ mm}^2$.

b. $1 \text{ m}^2 = 0.000001 \text{ km}^2$ implies $124{,}000{,}000 \text{ m}^2 = 124{,}000{,}000 \cdot 1 \text{ m}^2$
$= 124{,}000{,}000 \cdot 0.000001 \text{ km}^2 = 124 \text{ km}^2$.

Based on the relationship among units of length in the English system, it is possible to convert among English units of area. For example, because $1 \text{ yd} = 3 \text{ ft}$, it follows that $1 \text{ yd}^2 = (1 \text{ yd})^2 = 1 \text{ yd} \cdot 1 \text{ yd} = 3 \text{ ft} \cdot 3 \text{ ft} = 9 \text{ ft}^2$. Similarly, because $1 \text{ ft} = 12 \text{ in.}$, $1 \text{ ft}^2 = (1 \text{ ft})^2 = 1 \text{ ft} \cdot 1 \text{ ft} = 12 \text{ in.} \cdot 12 \text{ in.} = 144 \text{ in.}^2$ Table 14-2 summarizes various relationships among units of area in the English system.

Table 14-2

Unit of Area	Equivalent of Other Units
1 ft^2	$\frac{1}{9}$ yd^2, or 144 in.^{2}
1 yd^2	9 ft^2, or 1296 in.^{2}
1 mi^2	3,097,600 yd^2, or 27,878,400 ft^2

Land Measure

The concept of area is widely used in land measure. The common unit of land measure in the English system is the **acre**. Historically, an acre was the amount of land a man with one horse could plow in one day. There are 4840 yd^2 in 1 acre. For very large land measures in the English system, the **square mile** (mi^2), or 640 acres, is used.

    In the metric system, small land areas are measured in terms of a square unit 10 m on a side, called an **are** (pronounced "air") and denoted by **a**. Thus, 1 a $= 10$ m $\cdot 10$ m, or 100 m^2. Larger land areas are measured in **hectares**. A hectare is 100 a. A hectare, denoted by **ha**, is the amount of land whose area is 10,000 m^2. It follows that 1 ha is the area of a square that is 100 m on a side. Therefore, 1 ha $= 1$ hm^2. For very large land measures, the **square kilometer**, denoted by km^2, is used. One square kilometer is the area of a square with a side 1 km, or 1000 m, long. Land area measures are summarized in Table 14-3.

Table 14-3

Unit of Area	Equivalent of Other Units
1 a	100 m^2
1 ha	100 a, or 10,000 m^2, or 1 hm^2
1 km^2	1,000,000 m^2 or 100 ha
1 acre	4840 yd^2
1 mi^2	640 acres

EXAMPLE 14-3

a. A square field has a side of 400 m. Find the area of the field in hectares.
b. A square field has a side of 400 yd. Find the area of the field in acres.

Solution

a. $A = (400 \text{ m})^2 = 160{,}000 \text{ m}^2 = 160{,}000 \text{ m}^2 \left(\dfrac{1 \text{ ha}}{10{,}000 \text{ m}^2} \right) = 16 \text{ ha}$

b. $A = (400 \text{ yd})^2 = 160{,}000 \text{ yd}^2 = 160{,}000 \text{ yd}^2 \left(\dfrac{1 \text{ acre}}{4840 \text{ yd}^2} \right) \text{ acre} \approx 33.1 \text{ acres}$

Area of a Rectangle

In the grade 3 *Common Core Standards* students "find the area of a rectangle with whole-number side lengths by tiling it and show that the area is the same as would be found by multiplying the side lengths." (p. 25)

    Suppose the square in Figure 14-9(a) represents 1 square unit. Then, we can tile the rectangle *ABCD* as seen in Figure 14-9(b). It contains $3 \cdot 4$, or 12 square units.

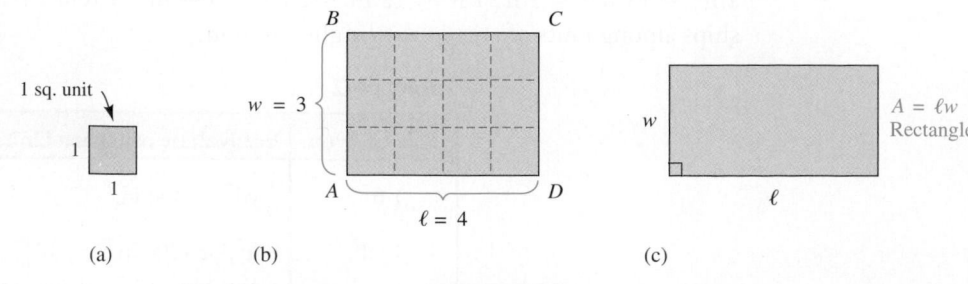

Figure 14-9

If the unit in Figure 14-9(a) is 1 cm², then the area of rectangle $ABCD$ is 12 cm². In general, the area A of any rectangle can be found by multiplying the lengths of two adjacent sides ℓ and w, or $A = \ell w$, as given in Figure 14-9(c).

EXAMPLE 14-4

Find the area of each rectangle in Figure 14-10.

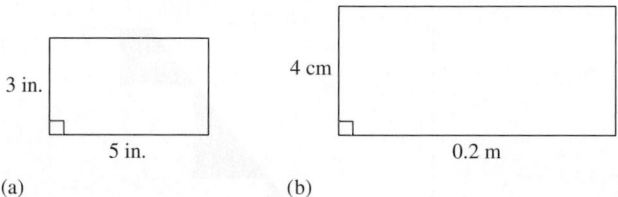

(a)   (b)

Figure 14-10

Solution

a. $A = (3 \text{ in.})(5 \text{ in.}) = 15 \text{ in.}^2$

b. First, write the lengths of the sides in the same unit of length. Because 0.2 m= 20 cm, $A = (4 \text{ cm})(20 \text{ cm}) = 80 \text{ cm}^2$. Alternatively, 4 cm = 0.04 m, so $A = (0.04 \text{ m})(0.2 \text{ m}) = 0.008 \text{ m}^2$.

NOW TRY THIS 14-2

Estimate the area in square centimeters of a dollar bill. Measure and calculate how close the estimate is to the actual area.

Area of a Parallelogram

The area of a parallelogram can be found by *reducing the problem to one that we already know how to solve*, in this case, finding the area of a rectangle. One important strategy for deriving the formula for the area of a parallelogram is **dissection**. In dissection, we cut a figure with unknown area into a number of pieces. By reassembling these pieces we obtain a figure whose area we know how to find.

Informally, to find the area of the parallelogram in Figure 14-11(a), we cut the shaded triangular piece of the parallelogram and translate it to obtain the rectangle in Figure 14-11(b). Because the shaded areas are congruent, the area of the parallelogram in Figure 14-11(a) is the same as the area of the rectangle in Figure 14-11(b); that is, *the area, A, of the parallelogram is the length of its base (b) times the corresponding height, h, that is, $A = b \cdot h$.*

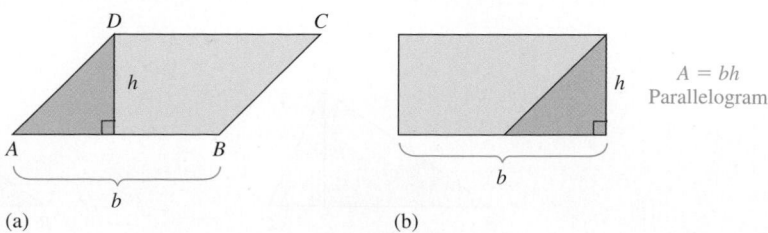

(a)   (b)

Figure 14-11

The technique above works for parallelograms like the one in Figure 14-11(a) but does not work for all parallelograms because sometimes, as in Figure 14-12(a), the height (*h*) from a vertex does not intersect the opposite base but intersects the line containing the base. If this happens we can use an adjacent side as a base to develop the general formula. The formula $A = bh$ is still valid even if the corresponding altitude does not intersect the base. One way to justify this is depicted in Figure 14-12(a) and Figure 14-12(b) where the white shapes in Figure 14-12(a) are translated to the let *b* units as shown in Figure 14-2b.

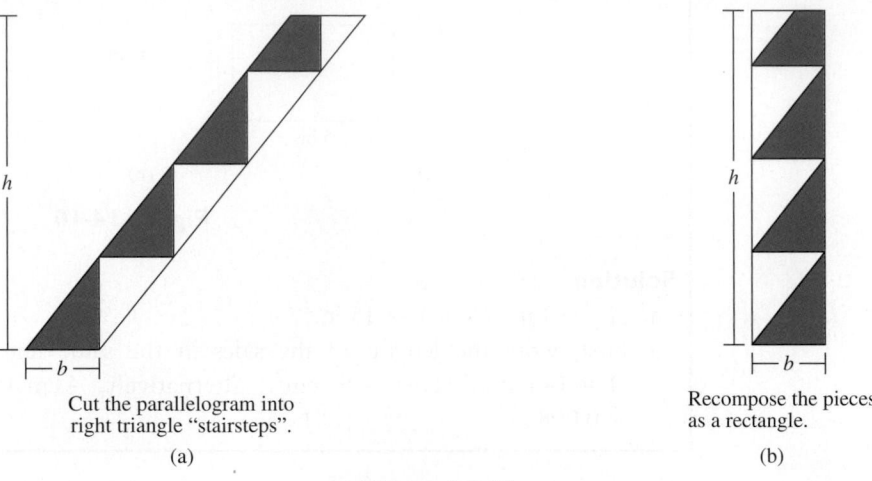

Cut the parallelogram into
right triangle "stairsteps".

(a)

Recompose the pieces
as a rectangle.

(b)

Figure 14-12

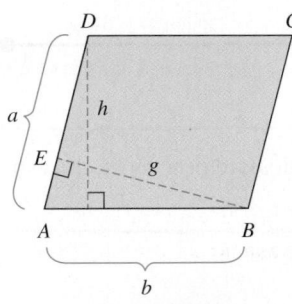

Figure 14-13

In general, any side of a parallelogram can be designated as a **base**. The **height** (*h*) (**altitude**) is the perpendicular distance between the bases and is always the length of a segment perpendicular to the lines containing the bases. A parallelogram has two heights. The area of parallelogram *ABCD* in Figure 14-13 is given by $A = bh$, that is, the length of the base times the corresponding height. Similarly, *EB*, or *g*, is the height that corresponds to the bases $\overline{AD}$ and $\overline{BC}$, each of which has measure *a*. Consequently, the area of the parallelogram *ABCD* is *ag*. Therefore, $A = ag = bh$.

Area of a Triangle

The formula for the area of a triangle can be derived from the formula for the area of a parallelogram. To explore this, suppose $\triangle BAC$ in Figure 14-14(a) has base of length *b* and height *h*. Let $\triangle ABC'$ be the image of $\triangle BAC$ when $\triangle BAC$ is rotated 180° about *M*, the midpoint of $\overline{AB}$, as in Figure 14-14(b). Proving that quadrilateral *BCAC'* is a parallelogram is left as an exercise. Parallelogram *BCAC'* has area *bh* and is constructed of congruent triangles *BAC* and *ABC'*. So the area of $\triangle ABC$ is $\frac{1}{2}bh$. In general, *the area of a triangle is equal to half the product of the length of a side and the height to that side.*

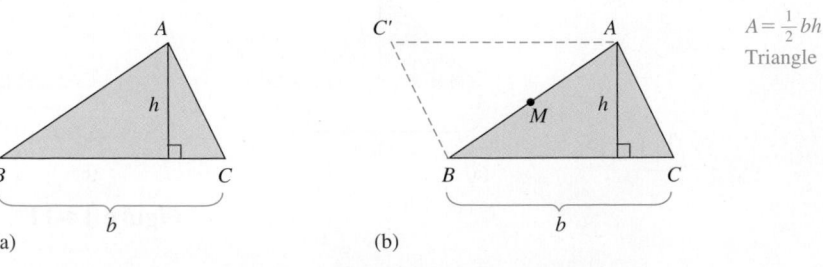

(a)

$A = \frac{1}{2}bh$
Triangle

(b)

Figure 14-14

In Figure 14-15, $\overline{BC}$ is a base of $\triangle ABC$, and the corresponding height h_1, or AE, is the distance from the opposite vertex A to the line containing $\overline{BC}$. Similarly, $\overline{AC}$ can be chosen as a base. Then h_2, or BG, the distance from the opposite vertex B to the line containing $\overline{AC}$, is the corresponding height. If $\overline{AB}$ is chosen as a base, then the corresponding height is h_3, or CF. Thus, the area of $\triangle ABC$ is

$$\text{Area}(\triangle ABC) = \frac{bh_1}{2} = \frac{ah_2}{2} = \frac{ch_3}{2}$$

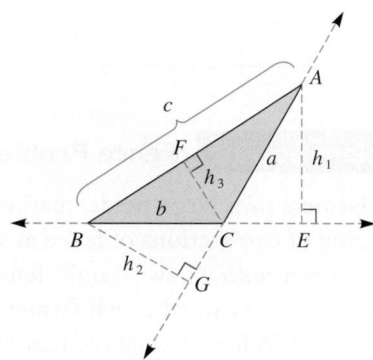

Figure 14-15

🔵 **NOW TRY THIS 14-3**

To derive the formula for the area of a triangle from the formula for the area of a rectangle, cut out any triangle as in Figure 14-16(a); fold to find the height h; and fold the altitude in half as shown in Figure 14-16(b). Next fold along the colored segments in the trapezoid in Figure 14-16(c) to obtain the rectangle in Figure 14-16(d).

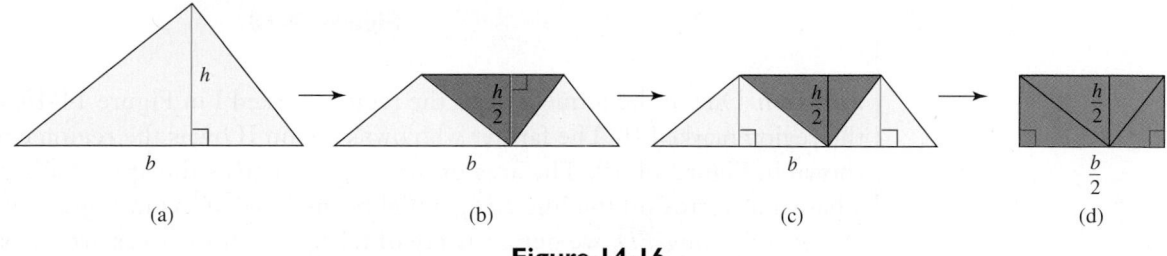

    (a)   (b)   (c)   (d)

Figure 14-16

  a. What is the area of the rectangle in Figure 14-16(d)?
  b. How can the formula for the area of a triangle be developed from your answer in part (a)?

NOW TRY THIS 14-4

In Figure 14-17, $\ell \parallel \overleftrightarrow{AB}$. How are the areas of $\triangle ABP$, $\triangle ABQ$, $\triangle ABR$, $\triangle ABS$, $\triangle ABT$, and $\triangle ABU$ related? Explain your answer.

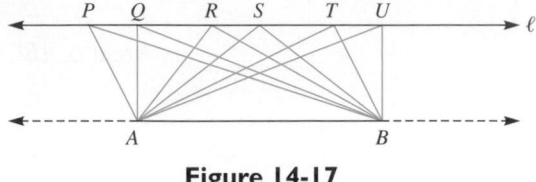

Figure 14-17

Problem Solving   **Fence Problem**

Two farmers own large fields, marked I and II in Figure 14-18, divided by a common border consisting of two sections of fence as shown in Figure 14-18. They want to replace the two sections of fence with a new straight fence so that the areas of the new regions are the same as the old areas. In other words, each farmer should have the same amount of land as before the border was changed. Where should the new "straight" fence be placed?

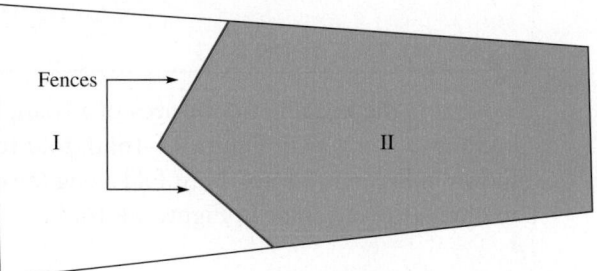

Figure 14-18

Solution One of the farmers owns the region marked I in Figure 14-18, and the other owns the region marked II. The farmer who owns region II owns the region enclosed by $\triangle ABC$ as shown in Figure 14-19. The area of that triangle equals the area of any triangle with $\overline{AC}$ as a base and vertex on the line ED parallel to the base $\overline{AC}$ as in Figure 14-19. Thus, if point B "moves" along $\overline{ED}$, we get a variety of triangles whose areas are the same as the area of $\triangle ABC$. If we choose $\triangle ADC$, the area of region II does not change and the new border $\overline{CD}$ is straight. Making the border become $\overline{AE}$ is another possibility.

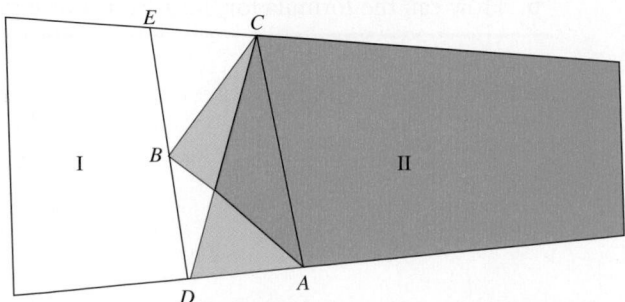

Figure 14-19

EXAMPLE 14-5

Find the areas of the figures in Figure 14-20. Assume the quadrilaterals *ABCD* in (a) and (b) are parallelograms. (Figures are not drawn to scale.)

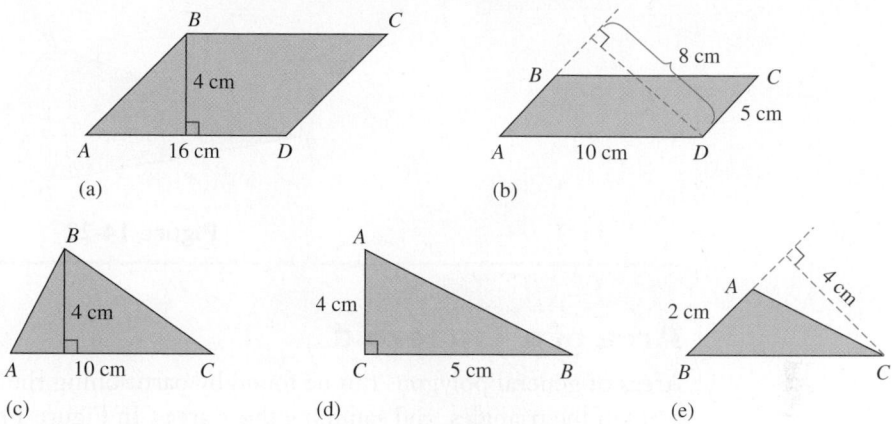

Figure 14-20

Solution

a. $A = bh = (16 \text{ cm})(4 \text{ cm}) = 64 \text{ cm}^2$

b. $A = bh = (5 \text{ cm})(8 \text{ cm}) = 40 \text{ cm}^2$

c. $A = \frac{1}{2}bh = \frac{1}{2}(10 \text{ cm})(4 \text{ cm}) = 20 \text{ cm}^2$

d. $A = \frac{1}{2}bh = \frac{1}{2}(5 \text{ cm})(4 \text{ cm}) = 10 \text{ cm}^2$

e. $A = \frac{1}{2}bh = \frac{1}{2}(2 \text{ cm})(4 \text{ cm}) = 4 \text{ cm}^2$

Area of a Kite

The area of a kite can be found by relating it to the area of a rectangle. Consider the kite shown in Figure 14-21(a), where d_1 and d_2 are the lengths of the diagonals. If the kite is dissected and re-assembled as shown in Figure 14-21(b), then a rectangle is formed with length d_1 and width $d_2/2$.

The rectangle in Figure 14-21(b) has area $(d_1 d_2)/2$, or $\frac{1}{2}(d_1 d_2)$. Therefore, the area of a kite is equal to half the product of the lengths of its diagonals. Partitioning the kite into triangles and finding the areas of the triangles can also be used to find the area of a kite.

Since rhombuses and squares are also kites, the area formula for kites developed above can be used to find the areas of rhombuses and squares.

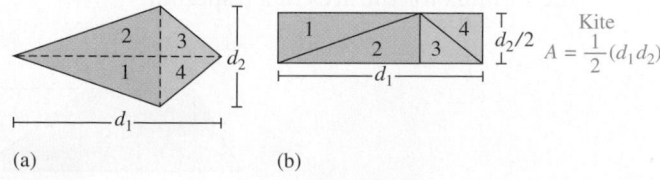

Figure 14-21

● NOW TRY THIS 14-5

We can find the area of a kite using the lengths of the two diagonals, as seen in Figure 14-21. Prove that the same formula works for any quadrilateral whose diagonals are perpendicular to each other, as seen in Figure 14-22.

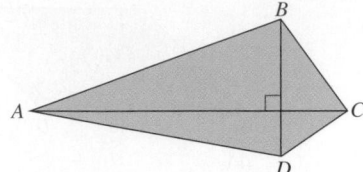

Figure 14-22

Area of a Trapezoid

Areas of general polygons can be found by partitioning the polygons into triangles, finding the areas of the triangles, and summing those areas. In Figure 14-23(a), trapezoid $ABCD$ has bases b_1 and b_2 and height h. By connecting points B and D, as in Figure 14-23(b), we create two triangles: one with base of length AB and height DE and the other with base of length CD and height BF. Because $\overline{DE} \cong \overline{BF}$, each has length h. Thus, the areas of triangles ADB and DCB are $\frac{1}{2}(b_1h)$ and $\frac{1}{2}(b_2h)$, respectively. Hence, the area of trapezoid $ABCD$ is $\frac{1}{2}(b_1h) + \frac{1}{2}(b_2h) = \frac{1}{2}h(b_1 + b_2)$; that is, *the area of a trapezoid is equal to half the height times the sum of the lengths of the bases.*

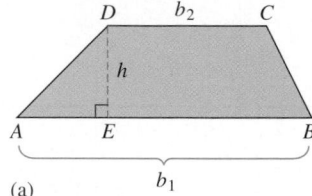
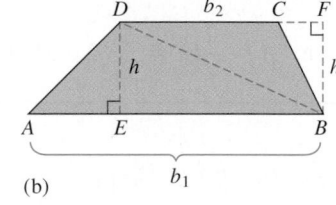

$$A = \tfrac{1}{2}h(b_1 + b_2)$$
Trapezoid

(a)   (b)

Figure 14-23

The formula for the area of a trapezoid can also be developed using the formula for the area of a parallelogram, as shown in Now Try This 14-6.

● NOW TRY THIS 14-6

Cut out a trapezoid $ABCD$ as shown in Figure 14-24. Copy and place as shown by the dashed trapezoid (rotate $ABCD$ 180° about the midpoint of $\overline{CB}$). Use the total figure obtained to derive the formula for the area of a trapezoid.

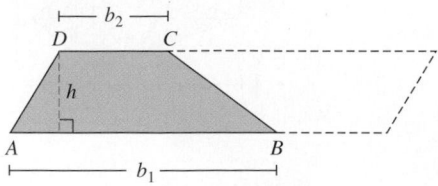

Figure 14-24

EXAMPLE 14-6

Find the areas of the trapezoids in Figure 14-25.

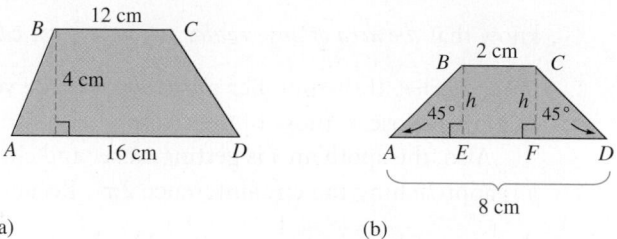

(a) (b)

Figure 14-25

Solution

a. $A = \frac{1}{2}h(b_1 + b_2) = \frac{1}{2}(4 \text{ cm})(12 \text{ cm} + 16 \text{ cm}) = 56 \text{ cm}^2$

b. To find the area of trapezoid $ABCD$, we use the strategy of *determining a subgoal* of finding the height, h. In Figure 14-25(b), $BE = CF = h$. Also, $\overline{BE}$ is a side of $\triangle ABE$, which has angles with measures of 45° and 90°. Consequently, the third angle in triangle ABE is $180° - (45° + 90°)$, or 45°. Therefore, $\triangle ABE$ is isosceles and $AE = BE = h$. Similarly, it follows that $FD = h$. Because $AD = 8 \text{ cm} = h + EF + h$, we could find h if we knew the value of EF. From Figure 14-25(b), $EF = BC = 2 \text{ cm}$ because $BCFE$ is a rectangle (Why?) and opposite sides of a rectangle are congruent. Now $h + EF + h = h + 2 + h = 8 \text{ cm}$. Thus, $h = 3 \text{ cm}$ and the area of the trapezoid is $A = \frac{1}{2}(3 \text{ cm})(2 \text{ cm} + 8 \text{ cm})$, or 15 cm^2.

Area of a Regular Polygon

The area of a triangle can be used to find the area of any regular polygon. This is illustrated *using a simpler case* strategy involving a regular hexagon in Figure 14-26(a). The hexagon can be separated into six congruent triangles with side s and height a as shown in Figure 14-26(a). The height of such a triangle of a regular polygon is the **apothem** and is denoted a. The area of each triangle is $\frac{1}{2}as$. Because 6 triangles make up the hexagon, the area of the hexagon is $6\left(\frac{1}{2}as\right)$, or $\frac{1}{2}a(6s)$. However, $6s$ is the perimeter, p, of the hexagon, so the area of the hexagon is $\frac{1}{2}ap$. The same process can be used to develop the formula for the area of any regular polygon; that is, *the area of any regular polygon is $\frac{1}{2}ap$, where a is the height of one of the triangles involved and p is the perimeter of the polygon*, as shown in Figure 14-26(b).

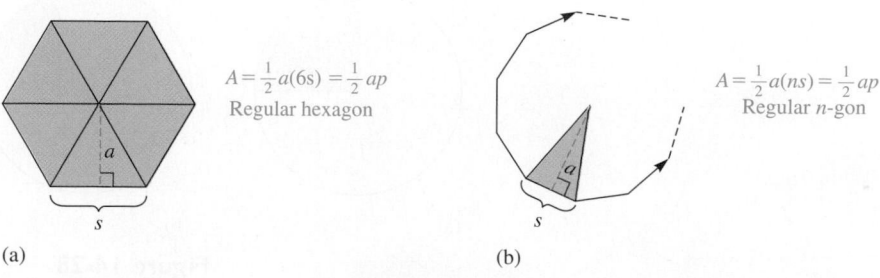

$A = \frac{1}{2}a(6s) = \frac{1}{2}ap$
Regular hexagon

$A = \frac{1}{2}a(ns) = \frac{1}{2}ap$
Regular n-gon

(a) (b)

Figure 14-26

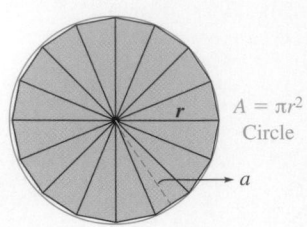

$A = \pi r^2$
Circle

Figure 14-27

Area of a Circle

We use the strategy of *examining a related problem* to find the area of a circle. The area of a regular polygon inscribed in a circle, as in Figure 14-27, approximates the area of the circle, and we know that *the area of any regular n-gon is* $\frac{1}{2}ap$, *where a is the height of a triangle of the n-gon and p is the perimeter*. If the number of sides n is made very large, then the perimeter and the area of the n-gon are close to those of the circle.

Also, the apothem a is getting closer and closer to the radius r of the circle, and the perimeter p is approaching the circumference $2\pi r$. Because the area of the circle is approximately equal to the area of the n-gon, $\frac{1}{2}ap \approx \frac{1}{2}r(2\pi r) = \pi r^2$. In fact, *the area of the circle is precisely* πr^2.

Another approach for leading students to discover the formula for finding the area of a circle is mentioned in the grade 7 *Focal Points*:

> Students see that the formula for the area of a circle is plausible by decomposing a circle into a number of wedges and rearranging them into a shape that approximates a parallelogram. (p. 19)

An example of this approach is seen in Now Try This 14-7 and the accompanying student page.

NOW TRY THIS 14-7

a. Draw a circle of radius 4 cm and divide it into eight equal-sized sectors.
b. Cut out the eight equal-sized sectors and rearrange the sectors to form a parallelogram-shaped figure such as the one on the student page.
c. The area of a parallelogram is found by multiplying the length of its base b by the height h, or $A = bh$. Use your figure to explain why this formula can be written as $A = \frac{1}{2}Cr$ to find the area of a circle where C is the circumference and r is the radius of the circle.

Area of a Sector

A **sector** of a circle is a piece-of-pie-shaped region of the circle determined by an actual angle of the circle. The area of a sector depends on the radius of the circle and the measure of the central angle determining the sector. If the angle has a measure of 90°, as in Figure 14-28(a), the area of the sector is one-fourth the area of the circle, or $\frac{90}{360}\pi r^2$. The area of a sector with central angle of 1° is $\frac{1}{360}$ of the area of the circle, and *a sector with central angle* $\theta°$ *has area* $\frac{\theta}{360}$ *of the area of the circle, or* $\frac{\theta}{360}(\pi r^2)$, as shown in Figure 14-28(b).

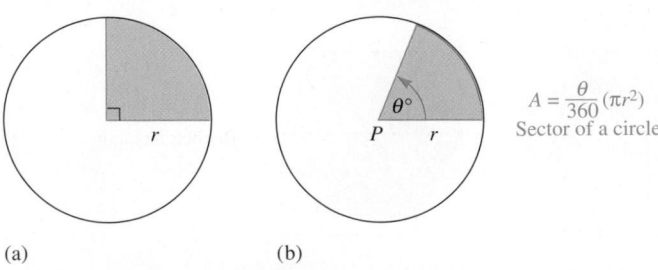

(a)   (b)

Figure 14-28

School Book Page AREA OF A CIRCLE

9-6 Area of a Circle

What You'll Learn
To find the area of a circle

Why Learn This?
To plant crops on a farm using a center-pivot irrigation system, farmers must calculate the area of a circle.

Suppose you cut a circle into equal-sized wedges. You can rearrange the wedges into a figure that resembles a parallelogram.

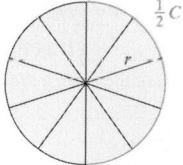

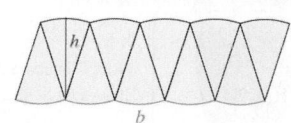

The base of the parallelogram is one half of the circumference of the circle, or πr. The height of the parallelogram is the same length as the circle's radius.

GO for Help

For help finding the area of a parallelogram, go to Lesson 9-4, Example 1.

$A = b \times h$ ← Use the formula for the area of a parallelogram.

$\quad = \pi r \times r$ ← Substitute πr for b and r for h.

$\quad = \pi r^2$ ← Simplify.

The calculations suggest a formula for the area of a circle.

KEY CONCEPTS Area of a Circle

$A = \pi r^2$

Finding the Areas of Other Shapes

Knowing how to find the area of simple figures, such as squares, rectangles, parallelograms, triangles, trapezoids, and circles can be used to find the area of irregular-shaped figures by separating them into familiar shapes. Examine the student page for two different methods of finding the area of an irregular shape.

 NOW TRY THIS 14-8

Find the area of the shape on the bottom of the student page.

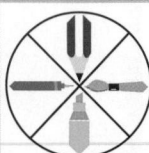

School Book Page MORE THAN ONE WAY

You can find the area of any figure by separating it into familiar figures.

● More Than One Way

Anna and Ryan are helping their friends build a large wooden deck. What is the area of the deck?

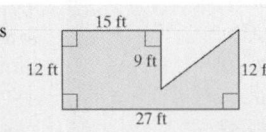

Anna's Method

I'll subtract the area of the triangle from the area of the rectangle.

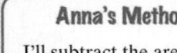

Area of the rectangle: Area of the triangle:

$A = bh$ $A = \frac{1}{2}bh$

$\quad = (27)(12) = 324$ $\quad = \frac{1}{2}(12)(9) = 54$

Now I'll subtract the area of the triangle from the area of the rectangle.

$A = 324 - 54 = 270$

The area of the deck is 270 ft^2.

Ryan's Method

I'll add the areas of the rectangle and the trapezoid.

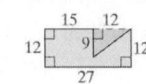

Area of the rectangle: Area of the trapezoid:

$A = bh$ $A = \frac{1}{2}h(b_1 + b_2)$

$\quad = (15)(12)$ $\quad = \frac{1}{2}(12)(3 + 12)$

$\quad = 180$ $\quad = 90$

Now I'll add the two areas together.

$A = 180 + 90 = 270$

The area of the deck is 270 ft^2.

Choose a Method

Find the area of the figure.

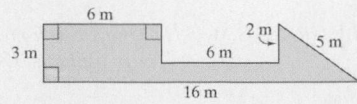

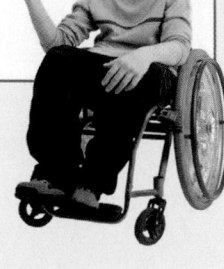

Assessment 14-1A

1. As an example of measuring area with a nonstandard measure, estimate the area of your desktop in terms of a piece of notebook paper as your unit of area. Then measure the area of your desktop with the paper and compare it to your estimate.

2. Choose the most appropriate metric units (cm^2, m^2, or km^2) and English units ($in.^2$, yd^2, or mi^2) for measuring each of the following:
   a. Area of a sheet of notebook paper
   b. Area of a quarter
   c. Area of a desktop
   d. Area of a classroom floor

3. Estimate and then measure each of the following using cm^2 or m^2.
   a. Area of a door
   b. Area of a desktop

4. Complete the following conversion table:

Item	m^2	cm^2	mm^2
a. Area of a sheet of paper		588	
b. Area of a cross section of a crayon			192
c. Area of a desktop	1.5		
d. Area of a dollar bill		100	
e. Area of a postage stamp		5	

5. Complete the following:
   a. $4000 \text{ ft}^2 = $ ———— yd^2
   b. $10^6 \text{ yd}^2 = $ ———— mi^2
   c. $10 \text{ mi}^2 = $ ———— acre
   d. $3 \text{ acres} = $ ———— ft^2

6. Find the areas of each of the following figures if the distance between two adjacent dots in a row or a column is 1 unit:

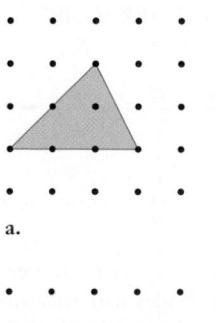

a.

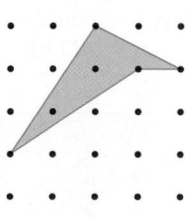

b.

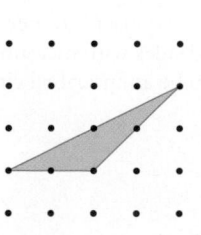

c.

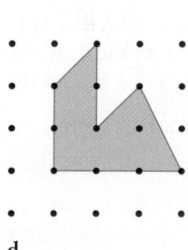

d.

7. Complete each of the following:
   a. A football field (with endzones included) is about 49 m × 100 m, or ____ m^2.
   b. About ____ a are in two football fields with endzones.
   c. About ____ ha are in two football fields with endzones.

8. Find the area of △ ABC in each of the following:

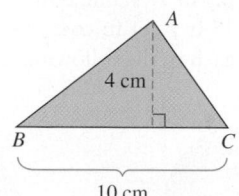

a.

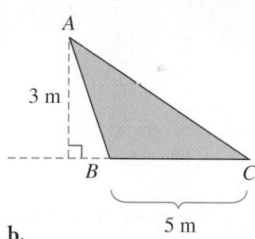

b.

9. If a triangle is inscribed in a circle so that one of the triangle's sides is a diameter of the circle, what is the greatest area that the triangle can have in terms of the radius, r, of the circle?

10. Two different squares have sides in the ratio $a : b$.
    a. Are the squares similar? Why?
    b. What is the ratio of the areas of the squares?

11. Find the area of each of the following quadrilaterals:

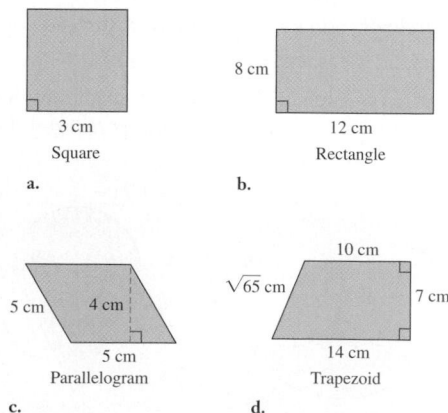

a. Square
b. Rectangle
c. Parallelogram
d. Trapezoid

12. a. A rectangular piece of land is 1300 m × 1500 m.
       i. What is the area in square kilometers?
       ii. What is the area in hectares?
    b. A rectangular piece of land is 1300 yd × 1500 yd.
       i. What is the area in square miles?
       ii. What is the area in acres?
    c. Explain which measuring system you would rather use to solve problems like those in (a) and (b).

13. For a parallelogram whose sides are 6 cm and 10 cm, which of the following is true?
  a. The data are insufficient to enable us to determine the area.
  b. The area has to be 60 cm².
  c. The area is greater than 60 cm².
  d. The area is less than 60 cm².

14. If the diagonals of a rhombus are 12 cm and 5 cm long, find the area of the rhombus.

15. Find the cost of carpeting the following rectangular rooms:
  a. Dimensions: 6.5 m × 4.5 m; cost = $13.85/m²
  b. Dimensions: 15 ft × 11 ft; cost = $30/yd²

16. Find the area of each of the following. Leave your answers in terms of π.

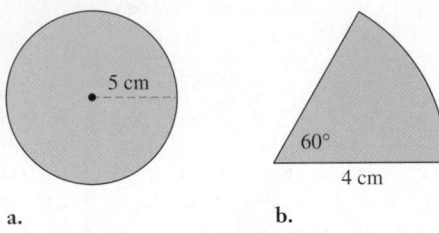

  a.   **b.**

17. Joe uses stick-on square carpet tiles to cover his 3 m × 4 m bathroom floor. If each tile is 10 cm on a side, how many tiles does he need?

18. Find the area of each of the following regular polygons:

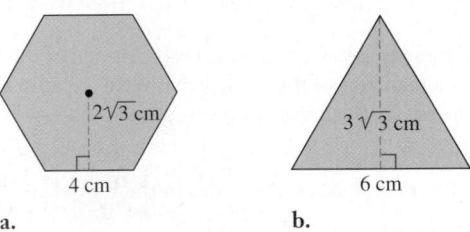

  a.   **b.**

19. a. If a circle has a circumference of 8π cm, what is its area?
  b. If a circle of radius r and a square with a side of length s have equal areas, express r in terms of s.

20. Find the area of each of the following shaded regions. Assume all arcs are circular with centers marked. Leave all answers in terms of π.

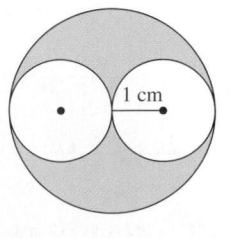 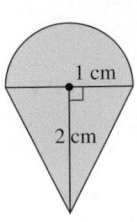

  a.   **b.**

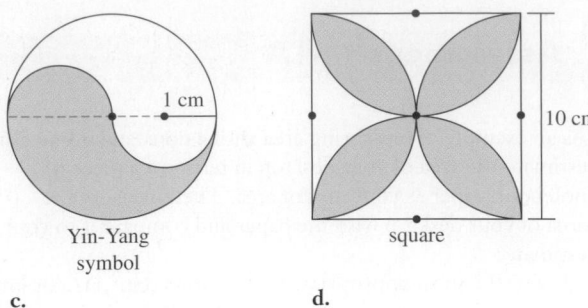

Yin-Yang symbol   square

  c.   **d.**

21. A circular flower bed is 6 m in diameter and has a circular sidewalk around it 1 m wide. Find the area of the sidewalk in square meters.

22. a. If the area of a square is 144 cm², what is its perimeter?
  b. If the perimeter of a square is 32 cm, what is its area?

23. a. What happens to the area of a square when the length of each side is doubled?
  b. If the ratio of the sides of two squares is 1 to 5, what is the ratio of their areas?

24. Find the shaded area in the following figure:

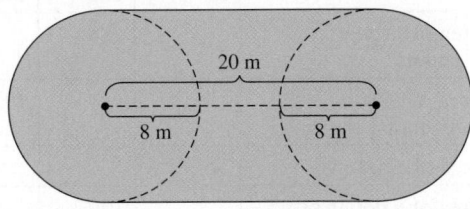

25. Complete and explain how to use geometric shapes to find an equivalent algebraic expression not involving parentheses for the following:

$a(b + c)$

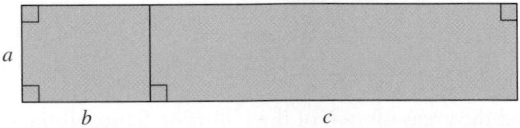

26. In the following figure, $\ell \parallel \overleftrightarrow{AB}$. If the area of $\triangle ABP$ is 10 cm², what are the areas of $\triangle ABQ$, $\triangle ABR$, $\triangle ABS$, $\triangle ABT$, and $\triangle ABU$? Explain your answers.

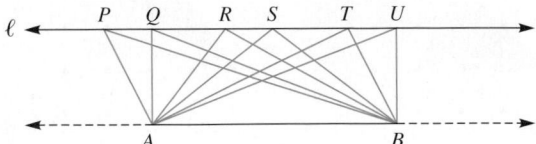

27. Rheba wanted a piece of red circular glass inset into her right triangular window. If the window had sides with measures of 3 ft, 4 ft, and 5 ft, and the glass was to be an inscribed circle as shown, what is the radius of the circular glass? (Areas may be used to solve this problem.)

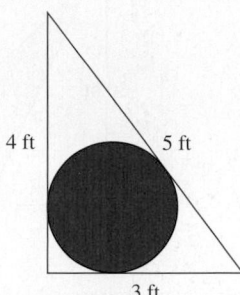

4 ft

5 ft

3 ft

28. Heron's formula can be used to find the area of a triangle if the lengths of the three sides are known. If the lengths of the three sides are a, b, and c units and the semiperimeter

$s = \dfrac{a + b + c}{2}$, then the area of the triangle is given by $\sqrt{s(s - a)(s - b)(s - c)}$. Use Heron's formula to find the areas of the right triangles with the sides given.
   a. 3 cm, 4 cm, 5 cm
   b. 5 cm, 12 cm, 13 cm

29. In the following figure, quadrilateral $ABCD$ is a parallelogram and P is any point on $\overline{AC}$. Prove that the area of $\triangle BCP$ is equal to the area of $\triangle DPC$.

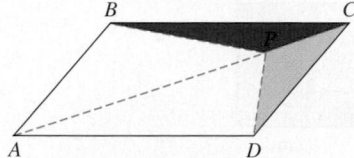

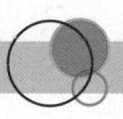

Assessment 14-1B

1. As an example of measuring area with a nonstandard measure, estimate the area of your desktop in terms of your hand as your unit of area. Then measure the area of your desktop with your hand and compare it to your estimate.
2. Choose the most appropriate metric units (cm², m², or km²) and English units (in.², yd², or mi²) for measuring each of the following:
   a. Area of a parallel parking space
   b. Area of an airport runway
   c. Area of your mathematics book cover
3. Estimate and then measure each of the following using cm², m², or km²:
   a. Area of a chair seat
   b. Area of a whiteboard or chalkboard
4. Complete the following conversion table:

	m²	cm²	mm²
a.	52		
b.			105
c.		86	
d.			10,000
e.	8.2		

5. Complete the following:
   a. 99 ft² = _____ yd²
   b. 10^6 yd² = _____ ft²
   c. 6.5 mi² = _____ acres
   d. 3 acres = _____ yd²

6. Find the areas of each of the following figures if the distance between two adjacent dots in a row or a column is 1 unit:

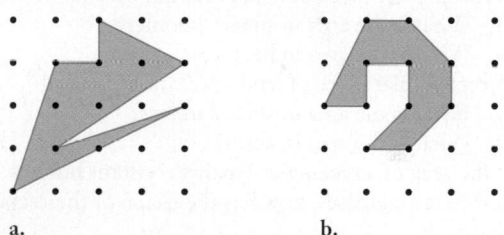

   a.   b.

7. If all vertices of a polygon are points on square-dot paper, the polygon is a **lattice polygon**. In 1899, G. Pick discovered a surprising theorem involving I, the number of dots *inside* the polygon, and B, the number of dots that lie *on* the polygon. The theorem states that the area of any lattice polygon is $I + \dfrac{1}{2}B - 1$. Check that this is true for the polygons in exercise 6.

8. Find the area of $\triangle ABC$ in each of the following. (Drawings are not to scale.)

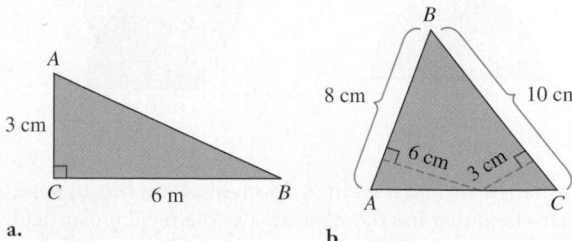

   a.   b.

9. If a triangle and a square have one side in common, where could the third vertex of the triangle lie if the area of the triangle is exactly equal to area of the square?

10. **a.** If triangle *ABC* is similar to triangle *DEF* and $\frac{AB}{DE} = \frac{2}{3}$, what is the ratio of the heights of the triangles?
    b. What is the ratio of the areas of the two triangles in part (a)?

11. Find the area of each of the following quadrilaterals:

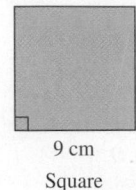

9 cm
Square

a.

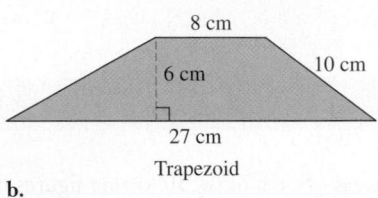

8 cm
6 cm
10 cm
27 cm
Trapezoid

b.

12. **a.** A rectangular piece of land is 1.2 km by 900 m.
    i. What is the area in square kilometers?
    ii. What is the area in hectares?
    b. A rectangular piece of land is 1.2 mi by 900 yd.
    i. What is the area in square miles?
    ii. What is the area in acres?

13. **a.** If the area of a rectangle remains constant but its perimeter increases, how has the shape of the rectangle changed?
    b. If the perimeter of a rectangle remains constant but its area increases, how does the shape of the rectangle change?

14. If the diagonals of a rhombus are 1.2 m and 40 cm long, find the area of the rhombus in square meters.

15. A rectangular plot of land is to be seeded with grass. If the plot is 22 m × 28 m and a 1 kg bag of seed is needed for 85 m² of land, how many bags of seed must you buy?

16. Find the area of each of the following. Leave your answers in terms of π.

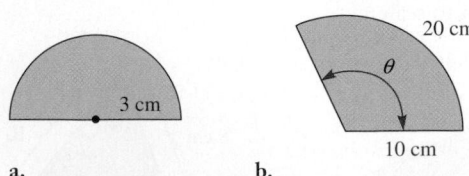

3 cm

20 cm
θ
10 cm

a.   **b.**

17. A rectangular field is 64 m × 25 m. Shawn wants to fence a square field that has the same area as the rectangular field. How long are the sides of the square field?

18. Find the area of the following figures.

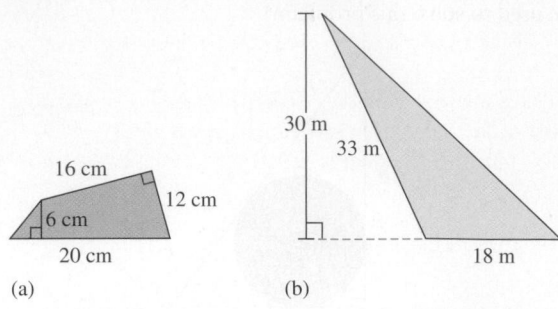

16 cm
12 cm
6 cm
20 cm
(a)

30 m
33 m
18 m
(b)

19. Suppose the largest square peg possible is placed in a circular hole as shown in the following figure and that the largest circular peg possible is placed in a square hole. In which case is there a smaller percentage of space wasted?

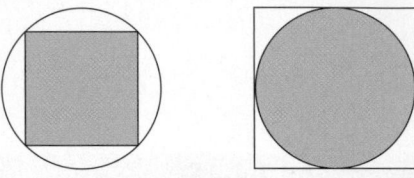

20. Find the area of each of the following shaded parts. Assume all arcs are circular. Leave all answers in terms of π.

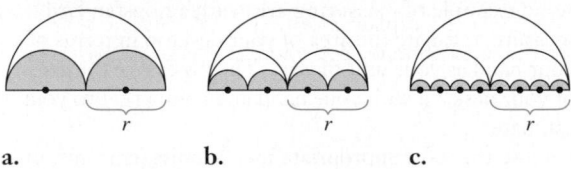

r   r   r
a.   **b.**   **c.**

21. For a dartboard (see the following figure), Joan is trying to determine how the area of the outside shaded region compares with the sum of the areas of the 3 inside shaded regions so that she can determine payoffs. How do they compare?

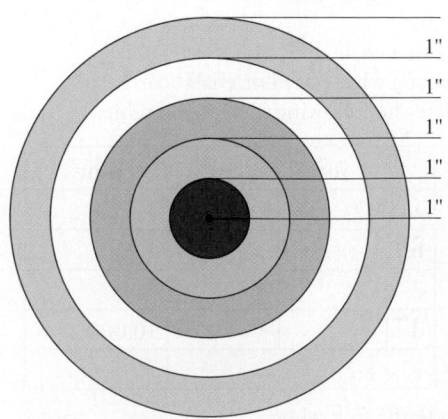

1"
1"
1"
1"
1"

22. If the area of a square is 169 in.², what is its perimeter?

23. **a.** What happens to the area of a circle if its diameter is doubled?

b. What happens to the area of a circle if its radius is increased by 10%?

c. What happens to the area of a circle if its circumference is tripled?

24. Quadrilateral *MATH* has been dissected into squares. The area of the red square is 64 square units and the area of the blue square is 81 square units. Determine the dimensions and the area of quadrilateral *MATH*.

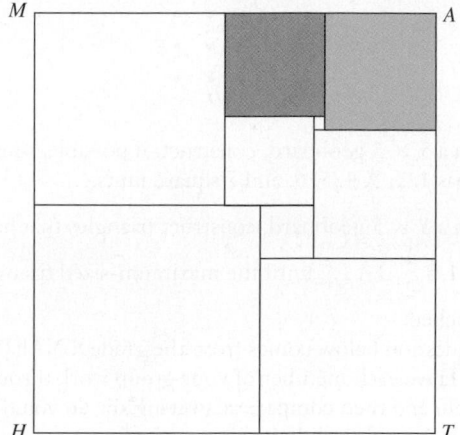

25. Complete and explain how to use geometric shapes to find an equivalent algebraic expression not involving parentheses for the following.
$(a + b)(c + d)$

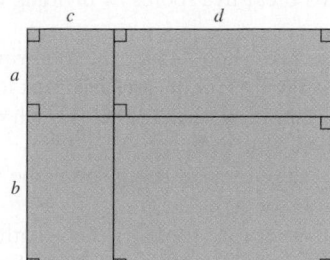

26. A store has wrapping paper on sale. One package is 3 rolls of $2\frac{1}{2}$ ft × 8 ft for $6.00. Another package is 5 rolls of $2\frac{1}{2}$ ft × 6 ft for $8.00. Which is the better buy per square foot?

27. Squares *A* and *B* are congruent. One vertex of *B* is at the center of square *A*. What is the ratio of the shaded area to the area of square *A*?

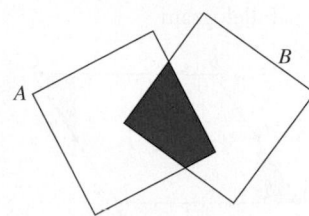

28. Given △*ABC* with parallel lines dividing $\overline{AB}$ into three congruent segments as shown, how does the area of △ *BDE* compare with the area of △*ABC*?

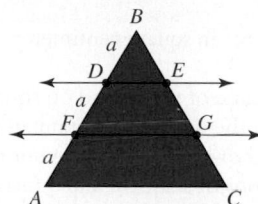

29. Use exercise 28 and compare the following areas:
a. Triangle *DBE* and trapezoid *DEGF*
b. Triangle *DBE* and trapezoid *FGCA*
c. Trapezoids *DEGF* and *FGCA*
d. Trapezoid *DEGF* and triangle *ABC*
e. Trapezoid *FGCA* and triangle *ABC*
f. Triangle *ABC* and trapezoid *DECA*

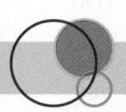

Mathematical Connections 14-1

Communication

1. Suppose a triangle has sides of lengths 6 in., 11 in., and 13 in. Explain how you can find the area of this triangle when the height is not given.

2. John claimed he had a garden twice as large as Al's rectangular-shaped garden that measured 15 ft by 30 ft. When they visited John's rectangular-shaped garden, they found it measured 18 ft by 50 ft. Al claimed that it could not be twice as large since neither the length nor the width were twice as large. Who was correct and why?

3. a. If a 10 in. (diameter) pizza costs $10, how much should a 20 in. pizza cost? Explain the assumptions made to obtain the answer.

b. If the ratio between the diameters of two pizzas is 1: *k*, what should the ratio be between the prices? Explain the assumptions made to obtain the answer.

4. a. Explain how the following drawing can be used to determine a formula for the area of △*ABC*:

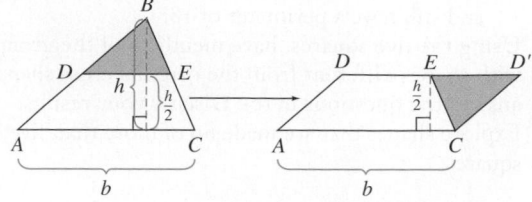

b. Use paper cutting to reassemble $\triangle ABC$ in (a) into parallelogram $ADD'C$.

5. The area of a parallelogram can be found by using the concept of a half-turn (a turn by 180°). Consider the parallelogram $ABCD$, and let M and N be the midpoints of $\overline{AB}$ and $\overline{CD}$, respectively. Rotate the shaded triangle with vertex M about M by 180° clockwise and rotate the shaded triangle with vertex N about N by 180° clockwise. What kind of figure do you obtain? Now complete the argument to find the area of the parallelogram.

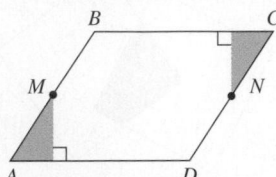

6. If the length of a rectangle is increased by 10% and the width of the rectangle is decreased by 10%, is the area changed? If so, does it increase or decrease and by what percent? Explain your Answer

Open-Ended

7. a. Estimate the area in square centimeters that your handprint will cover.

b. Trace the outline of your hand on square-centimeter grid paper and use the outline to obtain an estimate for the area. Explain how you arrived at your estimate.

8. a. Give dimensions of a square and a rectangle that have the same perimeter but such that the square has the greater area.

b. Give dimensions of a square and a rectangle that have equal area but such that the rectangle has greater perimeter.

Cooperative Learning

9. Use five 1×1 squares to build the cross shape shown and discuss the questions that follow.

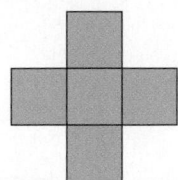

a. What is the area of this shape?

b. What is the perimeter of this shape?

c. Add squares to the shape so that each square added shares a complete edge with at least one other square.

   i. What is the minimum number of squares that can be added so that the shape has a perimeter of 18?

   ii. What is the maximum number?

   iii. What is the maximum area the new shape could have and still have a perimeter of 18?

d. Using the five squares, have members of the group start with shapes different from the original cross shape and answer the questions in (c). Discuss your results.

e. Explore shapes that are made up of more than five squares.

10. As a group work on the following activities.

a. On a 5×5 geoboard, make $\triangle DEF$ as shown below. Keep the rubber band around D and E fixed and move the vertex F to all the possible locations so that the triangles formed will have the same area as the area of $\triangle DEF$. How do the locations for the third vertex relate to $\overline{DE}$?

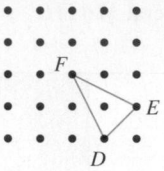

b. On a 5×5 geoboard, construct, if possible, squares of areas 1, 2, 3, 4, 5, 6, and 7 square units.

c. On a 5×5 geoboard, construct triangles that have areas $\frac{1}{2}$, 1, $1\frac{1}{2}$, 2, ..., until the maximum-sized triangle is reached.

11. The question below comes from the grade 8 NAEP test from 2009. Have each member of your group work through the problem and then compare answers. How do you think the eighth graders did on this question?

    The Morrisons are going to build a new one-story house. The floor of the house will be rectangular with a length of 30 feet and a width of 20 feet.

    The house will have a living room, a kitchen, two bedrooms, and a bathroom. In part (a) below create a floor plan that shows these five rooms by dividing the rectangle into rooms.

    Your floor plan should meet the following conditions.

- Each one of the five rooms must share at least one side with the rectangle in part (a); that is, each room must have at least one outside wall.

- The floor area of the bathroom should be 50 square feet.

- Each of the other four rooms (not the bathroom) should have a length of at least 10 feet and a width of at least 10 feet.

Be sure to label each room by name (living room, kitchen, bedroom, etc.) and include its length and width, in feet. (Do not draw any hallways on your floor plan.)

a. Draw your floor plan on the figure below. Remember to label your rooms by name and include the length and width in feet, for each room.

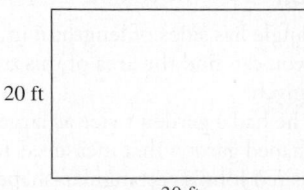

b. Complete the table below by filling in the floor area, in square feet, for each room in your floor plan.

Room	Floor Area (in square feet)
Living Room	
Kitchen	
Bedroom	
Bedroom	
Bathroom	
Total Floor Area	600

Questions from the Classroom

12. On a field trip, Glenda, a sixth-grade student, was looking at a huge dinosaur footprint and wondering about its area. Glenda said all you have to do is place a string around the border of the print and then take the string off and form it into a square and compute the area of the square. How would you help her?

13. A student asks, "Can I find the area of an angle?" How do you respond?

14. A student claims that because *are* and *hectare* are measures of area, we should say "square are" and "square hectare." How do you respond?

15. Larry and Gary are discussing whose garden has the most area to plant flowers. Larry claims that all they have to do is walk around the two gardens to get the perimeter and the one with the greatest perimeter has the greatest area. How would you help these students?

16. Jimmy claims that to find the area of a parallelogram he just has to multiply length times width. In the figure he multiplies $(25 \text{ in.})(20 \text{ in.}) = 500 \text{ in.}^2$? What would you tell him?

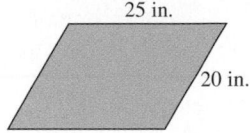

17. A student wants to know if it is possible to find the perimeters of the following figures if their areas are known. How do you respond?
   a. Square
   b. Rectangle
   c. Rhombus
   d. Circle

GSP/GeoGebra Activities

18. Use *GSP Lab 9* or *GeoGebra Lab 9* to investigate how to motivate the formulas for finding areas of rectangles, parallelograms, and trapezoids.

Review Problems

19. A glass table top is essentially square but has rounded circular corners. Find its perimeter.

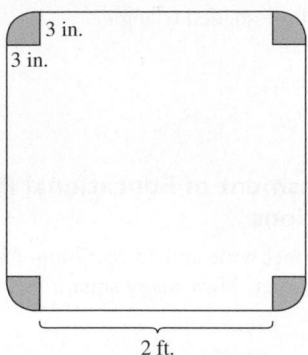

20. a. The earth has a circumference of approximately 39,750 km. With this circumference, what is its radius?
   b. Use the measurement in part (a) to determine the length of an arc from the North Pole to the equator.

21. Compare the perimeter of a regular hexagon to the circumference of its circumscribed circle.

Trends in Mathematics and Science Study (TIMSS) Questions

Jill had a rectangular piece of paper.

She cut her paper along the dotted line and made an L shape like this.

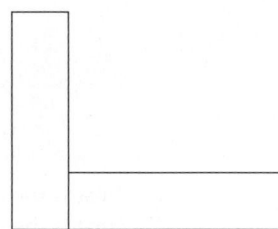

Which of these statements is true?
   a. The area of the L shape is greater than the area of the rectangle.
   b. The area of the L shape is equal to the area of the reactangle.
   c. The area of the L shape is less than the area of the rectangle.
   d. You cannot work out which area is greater without measuring.

TIMSS, Grade 4, 2007

The figure shows a shaded triangle inside a square.

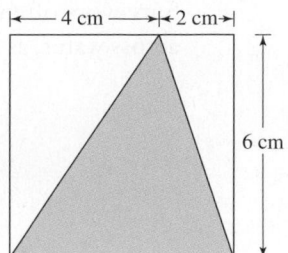

What is the area of the shaded triangle?

Answer: _____

TIMSS, Grade 8, 2007

National Assessment of Educational Progress (NAEP) Questions

Mark's room is 12 feet wide and 15 feet long. Mark wants to cover the floor with carpet. How many square feet of carpet does he need?

Answer: _____ square feet

The carpet costs $2.60 per square foot. How much will the carpet cost?

Answer: $_____

NAEP, Grade 4, 2007

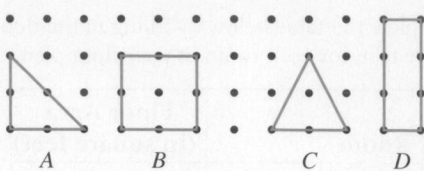

Which figure has the greatest area?
a. *A*   b. *B*   c. *C*   d. *D*

NAEP, Grade 4, 2009

How many square tiles, 5 inches on a side, does it take to cover a rectangular area that is 50 inches wide and 100 inches long?

NAEP, Grade 8, 2009

The rectangle in Figure 14-29(b) was apparently formed by cutting the square in Figure 14-29(a) along the dotted lines and reassembling the pieces as pictured.

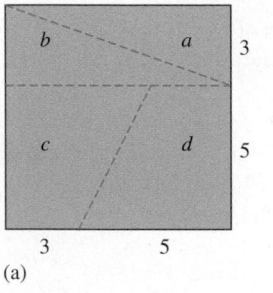

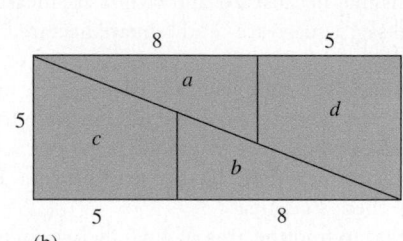

(a)   (b)

Figure 14-29

1. What is the area of the square in 14-29(a)?
2. What is the area of the rectangle in 14-29(b)?
3. How do you explain the discrepancy between the areas?

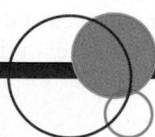

14-2 The Pythagorean Theorem, Distance Formula, and Equation of a Circle

Surveyors often have to calculate distances that cannot be measured directly, such as distances across water, as illustrated in Figure 14-30.

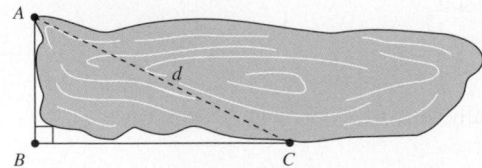

Figure 14-30

To measure the distance from point A to point C, they could use one of the most remarkable and useful theorems in mathematics: the Pythagorean theorem. This theorem is illustrated in Figure 14-31.

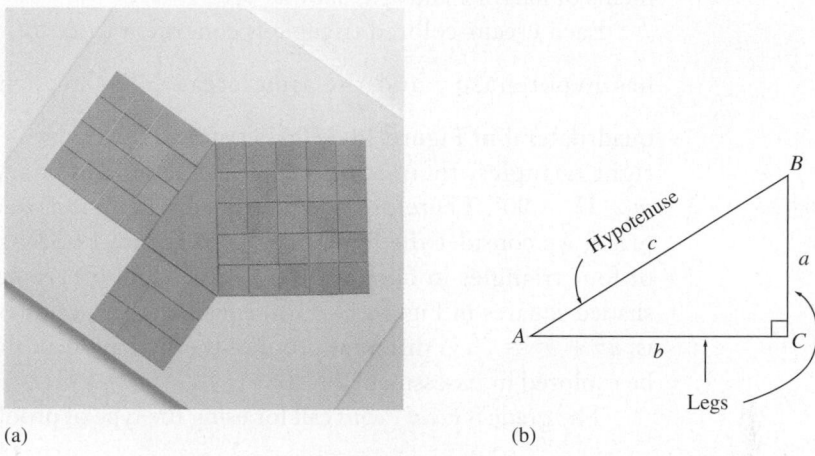

Figure 14-31

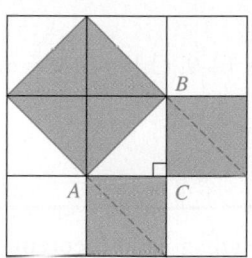

Figure 14-32

In the triangle in Figure 14-31(a) and in triangle ABC in Figure 14-31(b), the side opposite the right angle is the **hypotenuse**. The other two sides are **legs**. Interpreted in terms of area, the *Pythagorean theorem states that the area of a square with the hypotenuse of a right triangle as a side is equal to the sum of the areas of the squares with the legs as sides.*

Because the Pythagoreans affirmed geometric results on the basis of special cases, mathematical historians believe it is possible they may have discovered the theorem by looking at a floor tiling like the one illustrated in Figure 14-32. Each square can be divided by its diagonal into two congruent isosceles right triangles, so we see that the shaded square constructed with $\overline{AB}$ as a side consists of four triangles, each congruent to $\triangle ABC$. Similarly, each of the shaded squares with legs $\overline{BC}$ and $\overline{AC}$ as sides consists of two triangles congruent to $\triangle ABC$. Thus, the area of the larger square is equal to the sum of the areas of the two smaller squares. The theorem is true in general and is stated below.

Theorem 14-1: Pythagorean Theorem

If a right triangle has legs of lengths a and b and hypotenuse of length c, then $c^2 = a^2 + b^2$.

There are hundreds of known proofs for the Pythagorean theorem. The classic book *The Pythagorean Proposition*, by E. Loomis*, contains many of these proofs. Some proofs involve the strategy of *drawing diagrams* with a square area c^2 equal to the sum of the areas a^2 and b^2 of two other squares. One such proof is given in Figure 14-33. In Figure 14-33(a), the measures of the

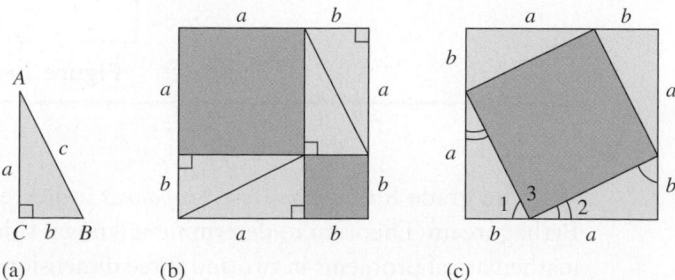

Figure 14-33

*Loomis, E. *The Pythagorean Proposition*. National Council of Teachers of Mathematics. Reston, UA: NCTM, 1976.

legs of a right triangle *ABC* are *a* and *b* and the measure of the hypotenuse is *c*. We draw a square with sides of length *a* + *b* and subdivide it, as shown in Figure 14-33(b). In Figure 14-33(c), another square with side of length *a* + *b* is drawn and each of its sides is divided into two segments of length *a* and *b*, as shown.

Each cream-colored triangle is congruent to $\triangle ABC$ (Why?). Consequently, each triangle has hypotenuse *c* and the same area, $\frac{1}{2}ab$. Thus, the length of each side of the blue quadrilateral in Figure 14-33(c) is *c* and so the figure is a rhombus. Because the triangles are right triangles, their acute angles are complementary. Hence $m(\angle 1) + m(\angle 2) = 90°$ so $m(\angle 3) = 90°$. Therefore, the blue quadrilateral is a square whose area is c^2. To complete the proof, we consider the four triangles in Figure 14-33(b) and (c). Because the areas of the sets of four triangles in both Figure 14-33(b) and (c) are equal, the sum of the areas of the two shaded squares in Figure 14-33(b) equals the area of the shaded square in Figure 14-33(c); that is, $a^2 + b^2 = c^2$. A different proof of the Pythagorean theorem just using Figure 14-32(c) will be explored in Assessment 14-2B.

The grade 8 *Focal Points* call for using the type of proof shown above, as seen in the following:

> Students explain why the Pythagorean theorem is valid by using a variety of methods—for example, by decomposing a square in two different ways. (p. 20)

Another example of a decomposition approach is given in Now Try This 14-9. Other proofs are explored in the assessment portion of this section.

NOW TRY THIS 14-9

In 1873, Henry Perigal, a London stockbroker, published a proof of the Pythagorean theorem. It is illustrated in Figure 14-34. Explain how this figure could be used to justify the theorem.

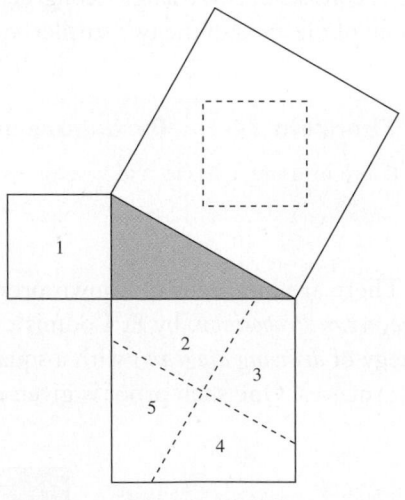

Figure 14-34

The grade 8 *Common Core Standards* indicate that students should be able to "apply the Pythagoream Theorem to determine unknown side lengths in right triangles in real-world and mathematical problems in two and three dimensions." (p. 56) Examples are given next.

EXAMPLE 14-7

a. For the drawing in Figure 14-35, find the value of x.

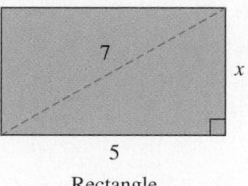

Rectangle

Figure 14-35

b. The size of a rectangular television screen is given as the length of the diagonal of the screen. If the length of the screen is 24 in. and the width is 18 in. as shown in Figure 14-36, what is the diagonal length?

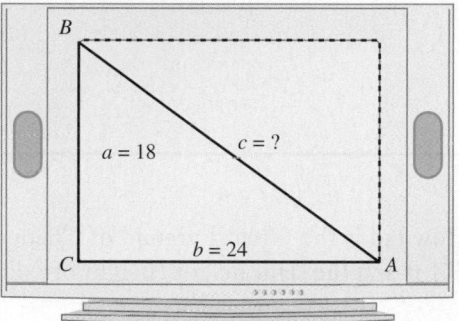

Figure 14-36

Solution

a. In the rectangle, the diagonal partitions the rectangle into two right triangles, each with length 5 units and width x units. Thus we have the following:

$$5^2 + x^2 = 7^2$$
$$25 + x^2 = 49$$
$$x^2 = 24$$
$$x = \sqrt{24}, \text{ or approximately } 4.9 \text{ units}$$

b. A right triangle is formed with the diagonal as the hypotenuse and the legs of measure 24 in. and 18 in. The Pythagorean theorem can be used to find the length of the diagonal.

$$c^2 = 18^2 + 24^2$$
$$c^2 = 324 + 576$$
$$c^2 = 900$$
$$c = 30$$

Because all the measurements are inches, the diagonal has length 30 in.

EXAMPLE 14-8

A pole $\overline{BD}$, 28 ft high, is perpendicular to the ground. Two wires $\overline{BC}$ and $\overline{BA}$, each 35 ft long, are attached to the top of the pole and to stakes A and C on the ground, as shown in Figure 14-37. If points A, D, and C are collinear, how far are the stakes A and C from each other?

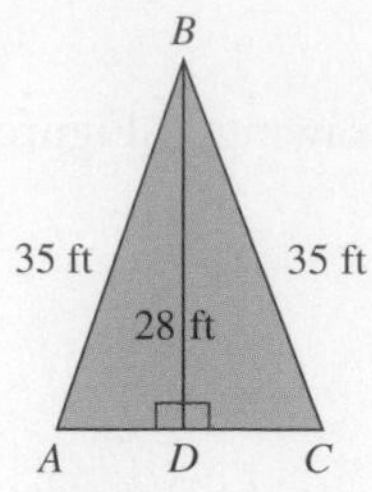

B

35 ft   35 ft

28 ft

A   D   C

Figure 14-37

Solution $\overline{AC}$ is not a side in any known right triangle, but we want to find AC. Because a point equidistant from the endpoints of a segment must be on the perpendicular bisector of the segment, it follows that $AD = DC$. Therefore, AC is twice as long as DC. Our *subgoal* is to find DC. We may find DC by applying the Pythagorean theorem in triangle BDC. This results in the following:

$$28^2 + (DC)^2 = 35^2$$
$$(DC)^2 = 35^2 - 28^2$$
$$DC = \sqrt{441}, \text{ or } 21 \text{ ft}$$
$$AC = 2DC = 42 \text{ ft}$$

EXAMPLE 14-9

How tall is the Great Pyramid of Cheops, a right regular square pyramid, if the base has a side 771 ft and the slant height (height of $\triangle EAB$) is 620 ft?

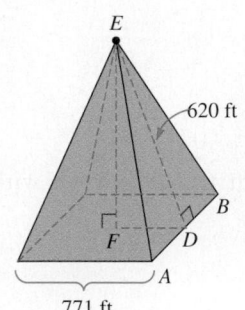

Figure 14-38

Solution   In Figure 14-38, $\overline{EF}$ is a leg of a right triangle formed by $\overline{FD}, \overline{EF}$, and $\overline{ED}$. Because the pyramid is a right regular pyramid, $\overline{EF}$ intersects the base at its center. Thus, $DF = \left(\dfrac{1}{2}\right) \cdot AB$, or $\left(\dfrac{1}{2}\right) \cdot 771$, or 385.5 ft. Now ED, the slant height, has length 620 ft, and we can apply the Pythagorean theorem as follows:

$$(EF)^2 + (DF)^2 = (ED)^2$$
$$(EF)^2 + (385.5)^2 = (620)^2$$
$$(EF)^2 = 235{,}789.75$$
$$EF \approx 485.6 \text{ ft}$$

Thus, the Great Pyramid is approximately 485.6 ft tall.

Special Right Triangles

An isosceles right triangle has two legs of equal length and two 45° angles. Any such triangle is a **45°-45°-90° right triangle**. Drawing a diagonal of a square forms two of these triangles, as shown in Figure 14-39(a).

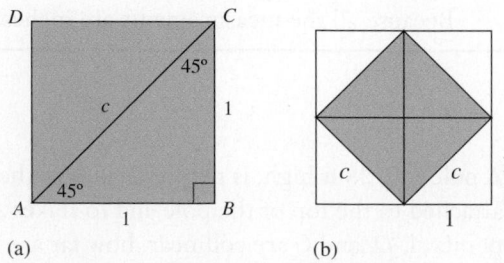

Figure 14-39

In Figure 14-39(b), we see several 45°-45°-90° triangles. Each side of the shaded square is the hypotenuse of a 45°-45°-90° triangle. The area of the shaded square is 2 square units. (Why?)

Therefore, $c^2 = 2$ and $c = \sqrt{2}$. Another way to see that $c = \sqrt{2}$ is to apply the Pythagorean theorem to one of the nonshaded triangles. Because $c^2 = 1^2 + 1^2 = 2$, then $c = \sqrt{2}$.

In one isosceles right triangle in Figure 14-39, each leg is 1 unit long and the hypotenuse is $\sqrt{2}$ units long. This is generalized when the isosceles right triangle has a leg of length a, as follows.

> **Theorem 14-2: 45°-45°-90° Triangle Relationships**
> In an isosceles right triangle with the length of each leg a, the hypotenuse has length $a\sqrt{2}$.

NOW TRY THIS 14-10

In the following cartoon, Jason runs a different pattern than he was told. Explain how he arrived at this pattern.

Foxtrot copyright © 1999 Bill Amend. Reprinted with permission of Universal Uclick. All rights reserved.

Figure 14-40(a) shows that in terms of area a 30°-60°-90° triangle is half of an equilateral triangle. When the equilateral triangle has side 2 units long, then in the 30°-60°-90° triangle, the leg opposite the 30° angle is 1 unit long and the leg opposite the 60° angle has a length of $\sqrt{3}$ units. (Why?)

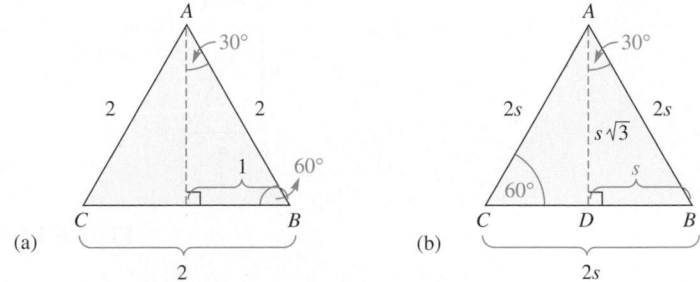

Figure 14-40

This example may also be generalized using the triangle in Figure 14-40(b). When the side of the equilateral triangle ABC is $2s$, then in triangle ABD, the side opposite the 30° angle, $\overline{BD}$, is s units long, and AD may be found using the Pythagorean theorem to have a length of $s\sqrt{3}$ units. This discussion is summarized in the following.

> ### Theorem 14-3: 30°-60°-90° Triangle Relationships
>
> In a 30°-60°-90° triangle, the length of the hypotenuse is twice the length of the leg opposite the 30° angle, and the leg length opposite the 60° angle is $\sqrt{3}$ times the length of the shorter leg.

Converse of the Pythagorean Theorem

The grade 8 *Common Core Standards* state that students should be able to "explain a proof of the Pythagoream Theorem and its converse." (p. 56)

The converse of the Pythagorean theorem is also true. It provided a useful way for early surveyors—in particular, the Egyptian rope stretchers—to determine right angles. Figure 14-41(a) shows a knotted rope with 12 equally spaced knots. Figure 14-41(b) shows how the rope might be held to form a triangle with sides of lengths 3, 4, and 5. The triangle formed is a right triangle and contains a 90° angle. Note that $5^2 = 3^2 + 4^2$.

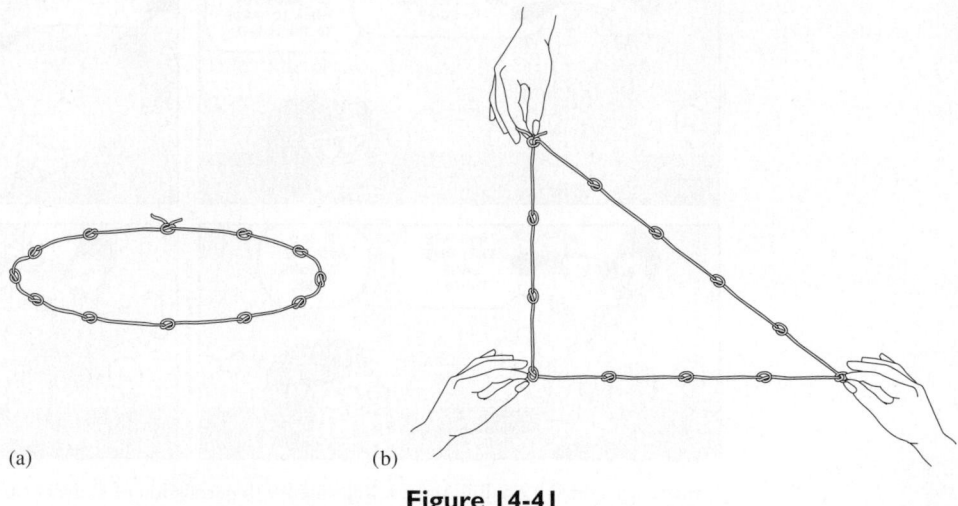

(a)   (b)

Figure 14-41

Given a triangle with sides of lengths a, b, and c such that $a^2 + b^2 = c^2$ as shown in Figure 14-42(a), must the triangle be a right triangle?

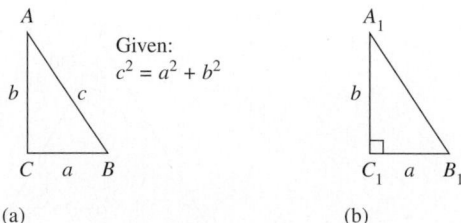

Given:
$c^2 = a^2 + b^2$

(a)   (b)

Figure 14-42

To investigate this question, we construct a right triangle with two sides a and b, as shown in Figure 14-42(b). By the Pythagorean theorem, $(A_1B_1)^2 = a^2 + b^2$. Therefore, $(A_1B_1)^2 = c^2$ and $A_1B_1 = c$. By the side-side-side (SSS) property, $\triangle ABC \cong \triangle A_1B_1C_1$ and hence $m(\angle C) = 90°$. Therefore, we have the following theorem, which is the converse of the Pythagorean theorem.

> ### Theorem 14-4:   Converse of the Pythagorean Theorem
>
> If $\triangle ABC$ is a triangle with sides of lengths a, b, and c such that $a^2 + b^2 = c^2$, then $\triangle ABC$ is a right triangle with the right angle opposite the side of length c.

EXAMPLE 14-10

Determine whether the following can be the lengths of the sides of a right triangle:

   a. 51, 68, 85   **b.** 2, 3, $\sqrt{13}$ **c.** 3, 4, 7

Solution

   a. $51^2 + 68^2 = 7225 = 85^2$, so 51, 68, and 85 can be the lengths of the sides of a right triangle.
   b. $2^2 + 3^2 = 4 + 9 = 13 = (\sqrt{13})^2$, so 2, 3, and $\sqrt{13}$ can be the lengths of the sides of a right triangle.
   c. $3^2 + 4^2 \neq 7^2$, so the measures cannot be the lengths of the sides of a right triangle. In fact, since $3 + 4 = 7$, segments with these lengths do not form a triangle.

NOW TRY THIS 14-11

   a. Draw three segments that can be used to form the sides of a right triangle.
   b. Multiply the lengths of the three segments in (a) by a fixed number and determine whether the resulting three lengths can be sides of a right triangle. Explain why or why not.
   c. Using three new numbers, repeat the experiment in (a) and (b). Form a conjecture based on your experiments.

The Distance Formula: An Application of the Pythagorean Theorem

The grade 8 *Common Core Standards* state that students should be able to "apply the Pythagorean Theorem to find the distance between two points in a coordinate system." (p. 56)

Given the coordinates of two points A and B, we can find the distance AB. We first consider the special case in which the two points are on one of the axes. For example, in Figure 14-43(a), $A(2, 0)$ and $B(5, 0)$ are on the x-axis. The distance between these two points is 3 units:

$$AB = OB - OA = 5 - 2 = 3$$

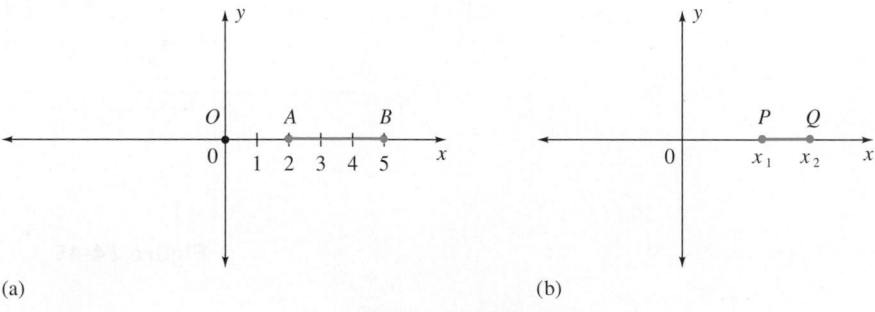

(a)   (b)

Figure 14-43

In general, if two points P and Q are on the x-axis, as in Figure 14-43(b), with x-coordinates x_1 and x_2, respectively, and $x_2 > x_1$, then $PQ = x_2 - x_1$. In fact, *the distance between two points on the x-axis is always the absolute value of the difference between the x-coordinates of the points.* (Why?) A similar result holds for any two points on the y-axis.

Figure 14-44 shows two points in the plane: $C(2, 5)$ and $D(6, 8)$. The distance between C and D can be found by using the strategy of *looking at a related problem*. We draw perpendiculars from the points to the x-axis and to the y-axis, as shown in Figure 14-44. The segments intersect

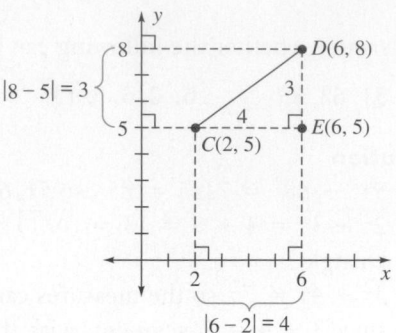

Figure 14-44

at point E forming right triangle CDE. The lengths of the legs of triangle CDE are found by using horizontal and vertical distances and properties of rectangles.

$$CE = |6 - 2| = 4$$
$$DE = |8 - 5| = 3$$

The distance between C and D can be found by applying the Pythagorean theorem to the triangle.

$$CD^2 = DE^2 + CE^2$$
$$= 3^2 + 4^2$$
$$= 25$$
$$CD = \sqrt{25}, \text{ or } 5$$

The method can be generalized to find a formula for the distance between any two points $A(x_1, y_1)$ and $B(x_2, y_2)$. Construct a right triangle with $\overline{AB}$ as its hypotenuse by drawing segments through A parallel to the x-axis and through B parallel to the y-axis, as shown in Figure 14-45. The lines containing the segments intersect at point C, forming right triangle ABC. Now, apply the Pythagorean theorem.

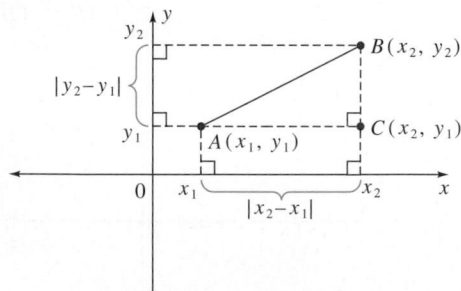

Figure 14-45

In Figure 14-45, we see that $AC = |x_2 - x_1|$ and $BC = |y_2 - y_1|$. By the Pythagorean theorem, $(AB)^2 = |x_2 - x_1|^2 + |y_2 - y_1|^2$, and consequently $AB = \sqrt{|x_2 - x_1|^2 + |y_2 - y_1|^2}$. Because $|x_2 - x_1|^2 = (x_2 - x_1)^2$ and $|y_2 - y_1|^2 = (y_2 - y_1)^2$, $AB = \sqrt{(x_2 - x_1)^2 + (y_2 - y_1)^2}$. This result is known as the **distance formula**.

Theorem 14-5:   Distance Formula

The distance between the points $A(x_1, y_1)$ and $B(x_2, y_2)$ is given by

$$AB = \sqrt{(x_2 - x_1)^2 + (y_2 - y_1)^2}.$$

NOW TRY THIS 14-12

Does it make any difference in the distance formula if $(x_1 - x_2)$ and $(y_1 - y_2)$ are used instead of $(x_2 - x_1)$ and $(y_2 - y_1)$, respectively.

EXAMPLE 14-11

  a. Show that $A(7, 4)$, $B(^-2, 1)$, and $C(10, ^-5)$ are the vertices of an isosceles triangle.
  b. Show that $\triangle ABC$ in (a) is a right triangle.

Solution
  a. Using the distance formula, we find the lengths of the sides.

$$AB = \sqrt{(^-2 - 7)^2 + (1 - 4)^2} = \sqrt{(^-9)^2 + (^-3)^2} = \sqrt{90} \text{ or } 3\sqrt{10}$$
$$BC = \sqrt{[10 - (^-2)]^2 + (^-5 - 1)^2} = \sqrt{12^2 + (^-6)^2} = \sqrt{180} \text{ or } 6\sqrt{5}$$
$$AC = \sqrt{(10 - 7)^2 + (^-5 - 4)^2} = \sqrt{3^2 + (^-9)^2} = \sqrt{90} \text{ or } 3\sqrt{10}$$

    Thus, $AB = AC$, and so the triangle is isosceles.
  b. Because $(\sqrt{90})^2 + (\sqrt{90})^2 = (\sqrt{180})^2$, $\triangle ABC$ is a right triangle with $\overline{BC}$ as hypotenuse and $\overline{AB}$ and $\overline{AC}$ as legs.

EXAMPLE 14-12

Determine whether the points $A(0, 5)$, $B(1, 2)$, and $C(2, ^-1)$ are collinear.

Solution   It is hard to tell by graphing the points whether the points are collinear (on the same line). If they are not collinear, they would be the vertices of a triangle, and hence $AB + BC$ would be greater than AC (triangle inequality). If $AB + BC = AC$, a triangle could not be formed and the points would be collinear. Using the distance formula, we find the lengths of the sides:

$$AB = \sqrt{(0 - 1)^2 + (5 - 2)^2} = \sqrt{1 + 9} = \sqrt{10}$$
$$BC = \sqrt{(2 - 1)^2 + (^-1 - 2)^2} = \sqrt{1 + 9} = \sqrt{10}$$
$$AC = \sqrt{(0 - 2)^2 + [5 - (^-1)]^2} = \sqrt{4 + 36} = \sqrt{40}$$

Thus, $AB + BC = 2\sqrt{10}$. Is this sum equal to $\sqrt{40}$? Because $\sqrt{40} = \sqrt{4 \cdot 10} = \sqrt{4}(\sqrt{10}) = 2\sqrt{10}$, we have $AB + BC = AC$, and consequently, A, B, and C are collinear.
    An alternative solution is to show that the slopes of each of the segments that can be formed are the same.

Using the Distance Formula to Develop the Equation of a Circle

Recall that a *circle* is the set of all points in a given plane equidistant from a given point, the center. Suppose that (x, y) is any point on a circle with the center at the origin, $(0, 0)$, as shown in Figure 14-46(a). Any point on the circle is the same distance from the center and this distance is the *radius, r*.

From the distance formula, we have $r = \sqrt{(x-0)^2 + (y-0)^2} = \sqrt{x^2 + y^2}$. An equivalent form can be obtained by squaring both sides, and we have $r^2 = x^2 + y^2$, which is the equation of a circle in standard form. We state this as Theorem 14-6.

Theorem 14-6:   Equation of the Circle with Center at the Origin

An equation of a circle with the center at the origin and radius r is $x^2 + y^2 = r^2$.

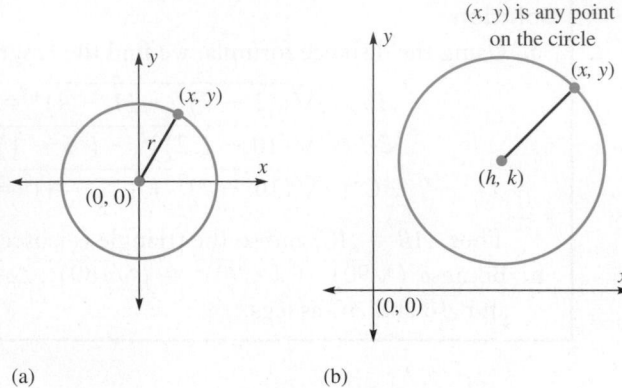

(a)   (b)

Figure 14-46

If the center is not at the origin, how would the equation change? Suppose the center is at (h, k) and r is the radius as shown in Figure 14-46(b). Then we can use the distance formula to obtain the following.

$$\sqrt{(x-h)^2 + (y-k)^2} = r \quad \text{or} \quad (x-h)^2 + (y-k)^2 = r^2$$

Thus, we have:

Theorem 14-7:   Equation of the Circle with Center at (h, k)

An equation of a circle with center (h, k) and radius r is $(x-h)^2 + (y-k)^2 = r^2$.

EXAMPLE 14-13

a. Find an equation of the circle with its center at $(2, {}^-5)$ and radius of 3.
b. Given the equation of a circle $(x-3)^2 + (y+4)^2 = 3$, find the radius and the center.

Solution

a. Let $(h, k) = (2, {}^-5)$ and $r = 3$; then using $(x-h)^2 + (y-k)^2 = r^2$ we have $(x-2)^2 + (y+5)^2 = 9$.

b. Rewriting $(x-3)^2 + (y+4)^2 = 3$ as $(x-3)^2 + [y - ({}^-4)]^2 = (\sqrt{3})^2$ and using $(x-h)^2 + (y-k)^2 = r^2$, we have $r^2 = (\sqrt{3})^2$ and $(h, k) = (3, {}^-4)$. Therefore, the radius is $\sqrt{3}$ and the center is at $(3, {}^-4)$.

BRAIN TEASER

A farmer has a square plot of land. An irrigation system can be installed with the option of one large circular sprinkler, as in Figure 14-47(a), or nine small sprinklers, as in Figure 14-47(b). The farmer wants to know which plan will provide water to the greater percentage of land in the field, regardless of the cost and the watering pattern. What advice would you give?

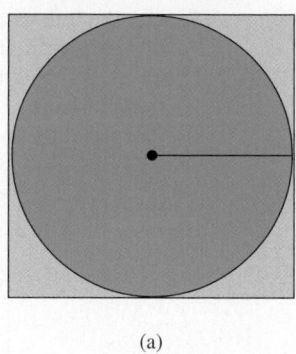
(a)

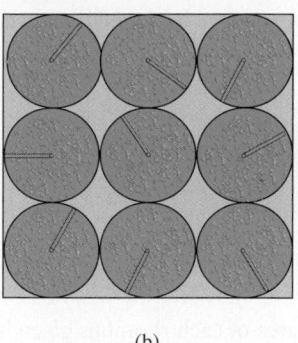
(b)

Figure 14-47

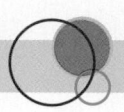

Assessment 14-2A

1. Find the length of the following segments. Assume that the horizontal and vertical distances between neighboring dots is 1 unit.

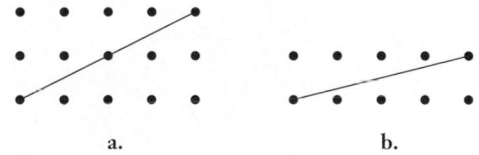

       a.   **b.**

2. Use the Pythagorean theorem to find x and y in each of the following:

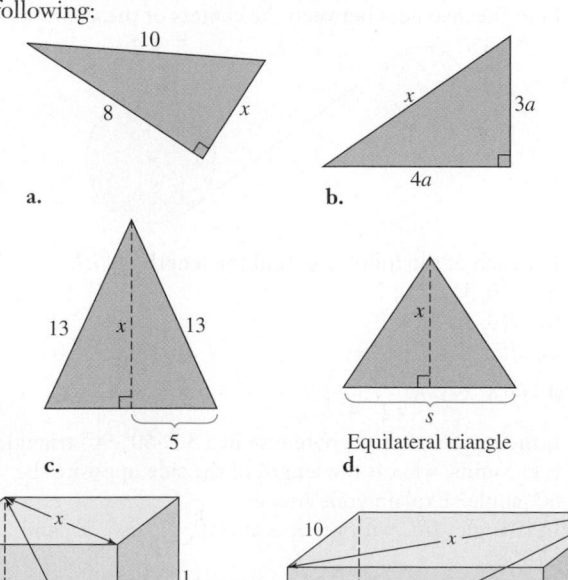

3. On a 5×5 geoboard, construct segments of the following lengths.
   a. $\sqrt{13}$
   b. $\sqrt{5}$

4. Find a square on a 4×4 geoboard that has an area of 2 square units and a perimeter of $4\sqrt{2}$ units.

5. What is the greatest perimeter of any isosceles triangle on a 5×5 geoboard?

6. If the hypotenuse of a right triangle is 30 cm long and one leg is twice as long as the other, how long are the legs of the triangle?

7. For each of the following, determine whether the given numbers represent lengths of sides of a right triangle:
   a. 10, 24, 16
   b. 16, 34, 30
   c. $\sqrt{2}, \sqrt{2}, 2$

8. What is the longest line segment that can be drawn in the interior of a right rectangular prism (like 2(f)) that is 12 cm wide, 15 cm long, and 9 cm high?

9. Starting from point A, a boat sails due south for 6 mi, then due east for 5 mi, and then due south for 4 mi. How far is the boat from A?

10. A 15 ft ladder is leaning against a wall and the wall is perpendicular to the ground. The base of the ladder is 3 ft from the wall. How high above the ground is the top of the ladder?

11. In the following figure, two poles are 25 m and 15 m high. A cable 14 m long joins the tops of the poles. Find the distance between the poles.

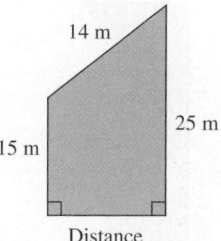

12. Find the area of each of the following figures:

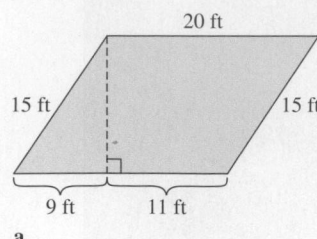

a.

b.

13. Find the area of each rhombus given below.

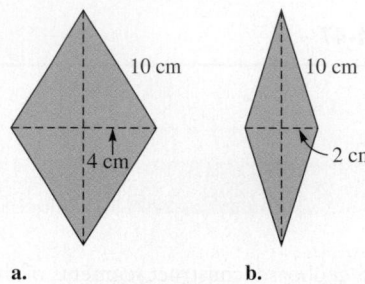

a.   b.

14. For each of the following, solve for the unknowns:

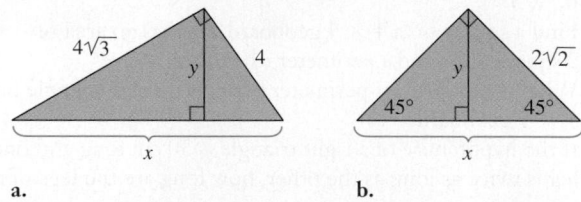

a.   b.

15. A company wants to lay a string of buoys across a lake. To find the length of the lake, they made the following measurements. What is the length of the lake?

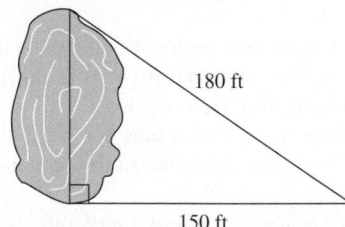

16. Before James Garfield was elected president of the United States, he discovered a proof of the Pythagorean theorem. He formed a trapezoid like the one that follows and found the area of the trapezoid in two ways. Can you discover his proof?

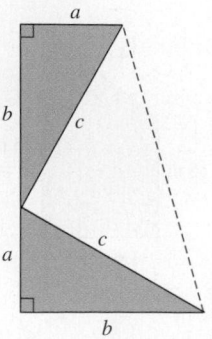

17. Use the following figure to prove the Pythagorean theorem by first proving that the quadrilateral with side c is a square. Then, compute the area of the square with side $a + b$ in two ways: (a) as $(a + b)^2$ and (b) as the sum of the areas of the four triangles and the square with side c.

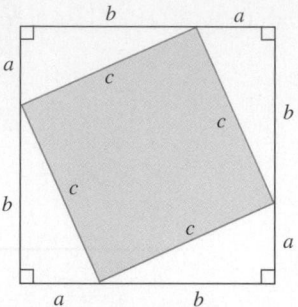

18. Equilateral triangles have been drawn on the sides of a right triangle as shown.

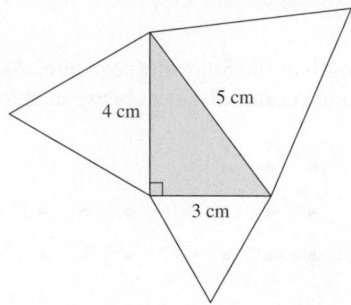

   a. Find the area of each equilateral triangle.
   b. How are the areas of the equilateral triangles related?

19. Find the distances between the centers of these circles.

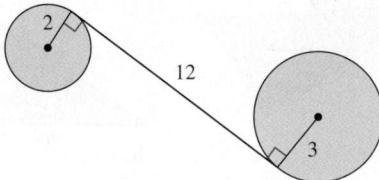

20. For each of the following, find the length of $\overline{AB}$:
   a. $A(0, 3), B(0, 7)$
   b. $A(0, 3), B(4, 0)$
   c. $A(^-1, 2), B(3, ^-4)$
   d. $A(4, ^-5), B\left(\dfrac{1}{2}, \dfrac{^-7}{4}\right)$

21. If the length of the hypotenuse in a 30°-60°-90° triangle is $c/2$ units, what is the length of the side opposite the 60° angle? Explain your answer.

22. In triangle ABC with vertices at $A(0, 0), B(6, 0)$, and $C(0, 8)$,
   a. write the equations of the lines that contain the altitudes.
   b. what are the coordinates of the intersection of the altitudes?

23. a. Find the possible coordinates of the third vertex of an isosceles right triangle that has endpoints of one leg with coordinates $(0, 0)$ and $(8, 8)$ and the hypotenuse lies on the x-axis.
   b. What are the lengths of the sides of the triangle in part (a)?

c. Show that the sides of the triangle satisfy the Pythagorean theorem.

24. Find an equation of the circle with the given center and radius:
  a. Center at $(^-3, 4)$, radius $= 4$
  b. Center at $(^-3, ^-2)$, radius $= \sqrt{2}$

25. Give the center of the circle and the radius in each of the following.
  a. $x^2 + y^2 = 16$
  b. $(x - 3)^2 + (y - 2)^2 = 100$
  c. $(x + 2)^2 + (y - 3)^2 = 5$
  d. $x^2 + (y + 3)^2 = 9$

26. A boat starts at point A, moves 3 km due north, then 2 km due east, then 1 km due south, and then 4 km due east to point B. Find the distance AB.

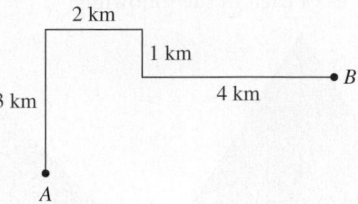

Assessment 14-2B

1. Find the length of the following segments. Assume that the horizontal and vertical distances between neighboring dots is 1 unit.

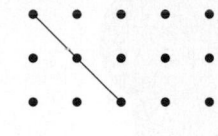

a.

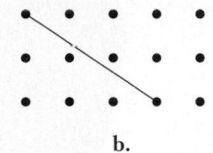

b.

2. Use the Pythagorean theorem to find x in each of the following:

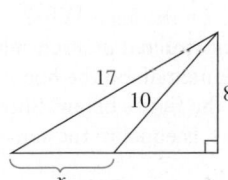

a.

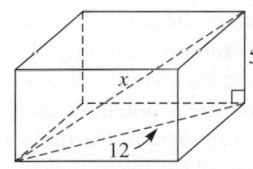

Right rectangular prism

b.

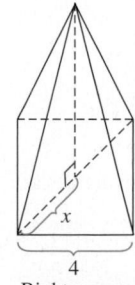

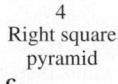

Right square pyramid

c.

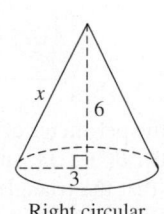

Right circular cone

d.

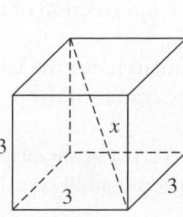

Cube

e.

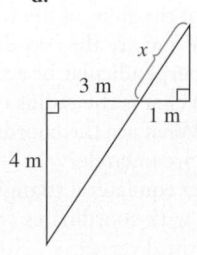

f.

3. On a 5 × 5 geoboard, construct segments of the following lengths.
  a. $\sqrt{10}$ **b.** $\sqrt{17}$

4. On a 4 × 4 geoboard, the greatest perimeter of any square is 12 units. Find a polygon on the geoboard with a greater perimeter. Prove your answer.

5. If possible, draw a square with the given number of square units on a 5 × 5 geoboard grid.
  a. 5
  b. 7
  c. 8
  d. 14
  e. 15

6. A door is 6 ft 6 in. tall and 36 in. wide. Can a piece of plywood that is 7 ft by 8 ft be carried through the door?

7. For each of the following, determine whether the given numbers represent lengths of sides of a right triangle:
  a. $\dfrac{3}{2}, \dfrac{4}{2}, \dfrac{5}{2}$
  b. $\sqrt{2}, \sqrt{3}, \sqrt{5}$
  c. 18, 24, 30

8. What is the longest piece of straight dry spaghetti that will fit in a cylindrical can that has a radius of 2 in. and height of 10 in.?

9. A CB radio station C is located 3 mi from the interstate highway h. The station has a range of 6.1 mi in all directions from the station. If the interstate is along a straight line, how many miles of highway are in the range of this station?

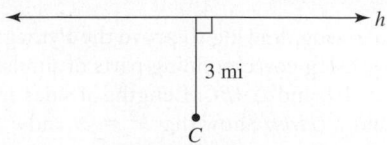

10. An access ramp enters a building 1 m above ground level and starts 3 m from the building. How long is the ramp?

11. To make a home plate for a neighborhood baseball park, we can cut the plate from a square, as shown in the following figure.

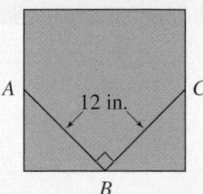

If A, B, and C are midpoints of the sides of the square, what is the length of the side of the square to the nearest tenth of an inch?

12. Find the area of each of the following:

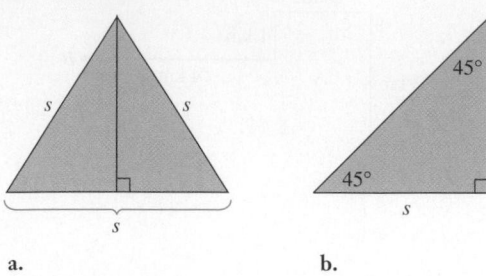

a.

b.

13. The following two rhombuses have perimeters that are equal. Use the properties of a rhombus to find the area of each rhombus.

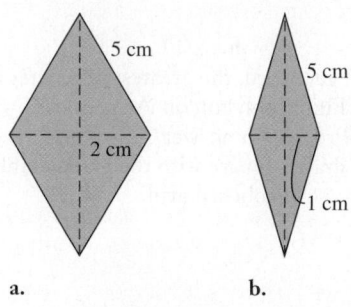

a.

b.

14. A builder needs to calculate the dimensions of a regular hexagonal window. Assuming the height CD of the window is 1.3 m, find the width AB (O is the midpoint of $\overline{AB}$) in the following figure:

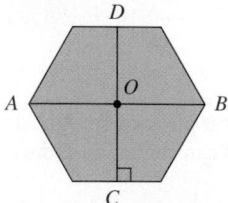

15. On a baseball field (square), if a player standing on third base throws on a straight line to a player on first base, how far is the ball thrown? (*Hint:* The distance from home plate to first base is 90 ft.)

16. Use the following drawing to prove the Pythagorean theorem by using corresponding parts of similar triangles $\triangle ACD$, $\triangle CBD$, and $\triangle ABC$. Lengths of sides are indicated by a, b, c, x, and y. (*Hint:* Show that $b^2 = cx$ and $a^2 = cy$.)

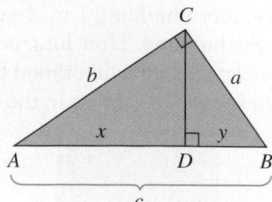

17. Show how the following figure could be used to prove the Pythagorean theorem:

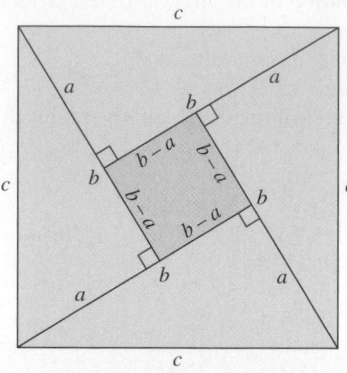

18. Construct semicircles on right triangle ABC with $\overline{AB}$, $\overline{BC}$, and $\overline{AC}$ as diameters, as shown below.

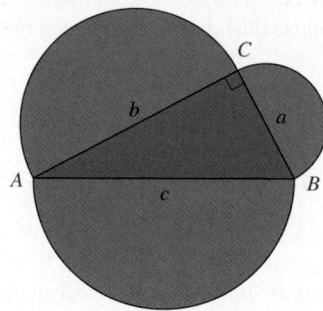

Is the area of the semicircle on the hypotenuse equal to the sum of the areas of the semicircles on the legs? Why?

19. In exercise 18, semicircles were constructed on each side of the right triangle ABC. Flip the semicircle on the hypotenuse over the hypotenuse, as shown in the figure below. Show that the area of the right triangle, t, is equal to the sum of the areas of the two crescents c_1 and c_2.

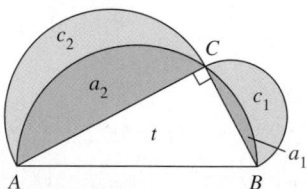

20. Find the perimeter of the triangle with vertices at $A(0, 0)$, $B(^-4, ^-3)$, and $C(^-5, 0)$.

21. Show that the triangle whose vertices are $A(^-2, ^-5)$, $B(1, ^-1)$, and $C(5, 2)$ is isosceles.

22. Triangle ABC has vertices at $A(0, 0)$, $B(6, 0)$, and $C(0, 8)$.
  a. Write the equations of the perpendicular bisectors of each of the sides of the triangle.
  b. What are the coordinates of the intersection of the perpendicular bisectors?
  c. What is the radius of the circumcircle for the triangle?
  d. What are the coordinates of the center of the circumcircle?

23. In an equilateral triangle with one vertex at the origin and one with coordinates $(8, 0)$, find the possible coordinates of the third vertex.

24. Find the equation of the circle with the given center and radius.
    a. Center at $(0, 0)$ and $r = \sqrt{5}$
    b. Center at $(^-6, ^-7)$ and $r = 6$
25. Give the center and radius for each circle.
    a. $(x - 3)^2 + (y + 2)^2 = 9$
    b. $3x^2 + 3y^2 = 9$
26. Find the area between two concentric (same center) circles whose equations are given below.

    $$(x + 4)^2 + (y - 8)^2 = 9$$

    and

    $$(x + 4)^2 + (y - 8)^2 = 64$$

27. The distance from point A to the center of a circle with radius r is d. Express the length of the tangent segment $\overline{AP}$ in terms of r and d.

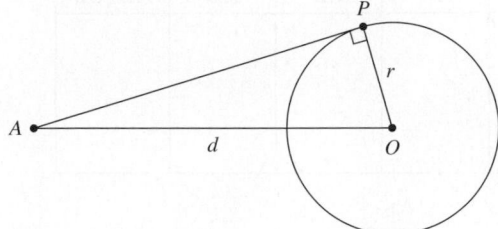

Mathematical Connections 14-2

Communication

1. Given the following square, describe how to use a compass and a straightedge to construct a square whose area is as follows:

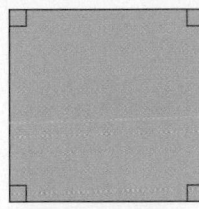

    a. Twice the area of the given square
    b. Half the area of the given square
2. Gail tried the Egyptian method of using a knotted rope to determine a right angle so that she could build a shed. She placed her knots so that each was 1 ft from the next. She stretched out her rope in the form of a triangle whose sides were of lengths 5, 12, and 13 ft. Did she have a right angle? Explain why or why not.
3. Explain how you would find the distance XY across the lake shown below, and then find XY.

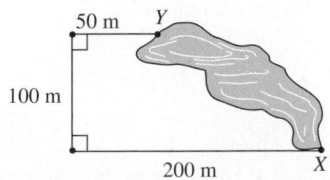

4. The sum of the squares of the lengths of all the sides of a right triangle is 200. Explain how to find the length of the hypotenuse and then find the length of the hypotenuse.
5. Explain how to construct a line segment of length $\sqrt{13}$ cm.

Open-Ended

6. Draw several kinds of triangles including a right triangle. Draw a square on each of the sides of the triangles. Compute the areas of the squares and use this information to investigate whether the Pythagorean theorem works for only right triangles. Use a geometry utility if available.
7. A **Pythagorean triple** is a sequence of three natural numbers a, b, and c that satisfy the relationship $a^2 + b^2 = c^2$. The least three numbers that form a Pythagorean triple are 3-4-5. Another triple is 5-12-13 because $5^2 + 12^2 = 13^2$.
    a. Find two other Pythagorean triples.
    b. Does doubling each number in a Pythagorean triple result in a new Pythagorean triple? Why or why not?
    c. Does adding a fixed number to each number in a Pythagorean triple result in a new Pythagorean triple? Why or why not?
    d. Suppose $a = 2uv$, $b = u^2 - v^2$, and $c = u^2 + v^2$, where u and v are natural numbers with $u > v$. Determine whether a-b-c is a Pythagorean triple.

Cooperative Learning

8. Have each person in the group use a 1-m string to make a different right triangle. Measure each side to the nearest centimeter. Use these measurements to see how closely the your measurements come to satisfying the Pythagorean theorem.

Questions from the Classroom

9. As part of the discussion of the Pythagorean theorem, squares were constructed on each side of a right triangle. A student asks, "If different similar figures are constructed on each side of the triangle, does the same type of relationship still hold?" How do you reply?
10. Amy says that the equation of the circle she just drew is $(x - 3)^2 + (y + 2)^2 = ^-16$. How do you respond?
11. One leg of a right triangle is 3 cm and the hypotenuse is 5 cm. Joni found the length of the other leg by evaluating $\sqrt{3^2 + 5^2}$. What error did she make?
12. June wants to know where all the points that satisfy the relation $x^2 + y^2 < 1$ are located. How do you respond?

GSP/GeoGebra Activities

13. Use *GSP Lab 10* or *GeoGebra Lab 13* to investigate the Pythagoream theorem.
14. Use *GSP/GeoGebra* to determine the relationship between the length of the hypotenuse of a 45°-45°-90° triangle and the length of a leg.
    a. Construct a 45°-45°-90° triangle, label the vertices as in the figure below, and measure the lengths of the sides.

Record the data for your triangle (triangle1) in the following table and compute the ratio:

	AC	CB	AB	AB/CB
Triangle 1				
Triangle 2				
Triangle 3				
Triangle 4				

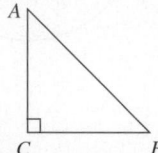

b. Repeat (a) for three other triangles.
c. Make a conjecture about the relationship between the length of the hypotenuse and the length of a leg for these triangles.
d. Given a 30°-60°-90° triangle, determine the relationships between the lengths of the hypotenuse and the shorter leg and the relationships between the lengths of the longer and shorter legs.

Review Problems

15. Arrange the following in decreasing order: 3.2 m², 322 cm², 0.032 km², 3020 mm².
16. Find the area of each of the following figures:

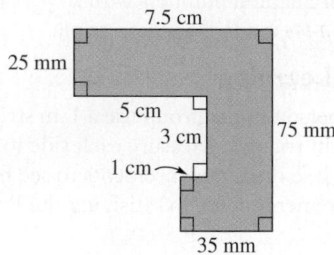

a.

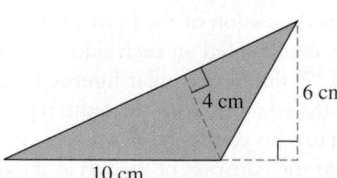

b.

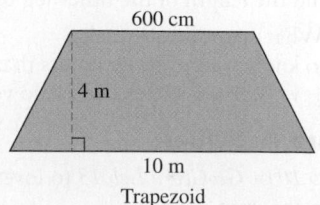

Trapezoid

c.

17. Complete the following table, which concerns circles:

	Radius	Diameter	Circumference	Area
a.	5 cm			
b.		24 cm		
c.				17π m²
d.			20π cm	

18. A 10-m wire is wrapped around a circular region. If the wire fits exactly, what is the area of the region?

National Assessment of Educational Progress (NAEP) Question

The endpoints of a line segment are the points with coordinates (2, 1) and (8, 9). What are the coordinates of the midpoint of this line segment?

a. $\left(2, 3\frac{1}{2}\right)$
b. (3, 4)
c. (5, 5)
d. $\left(4\frac{1}{2}, 5\frac{1}{2}\right)$
e. (10, 10)

NAEP Grade 8, 2005

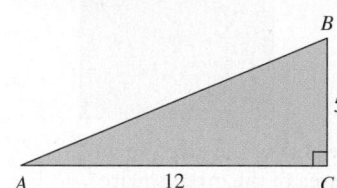

In the right triangle above, what is the length of *AB*?
a. 8.5
b. 12
c. 13
d. 17
e. 30

NAEP, Grade 8, 2009

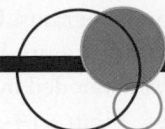

14-3 Geometry in Three Dimensions

Simple Closed Surfaces

In grade 5 *Focal Points* we find

> Students relate two-dimensional shapes to three-dimensional shapes and analyze properties of polyhedral solids, describing them by the number of edges, faces, or vertices as well as the types of faces. (p. 17)

A visit to the grocery store exposes us to many three-dimensional objects that have simple closed surfaces. Figure 14-48 shows some examples.

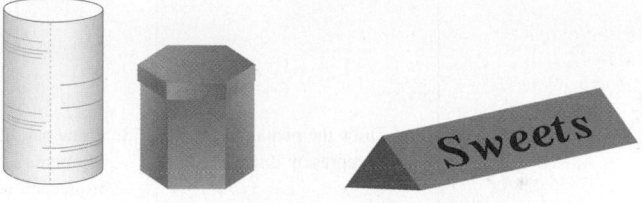

Figure 14-48

A **simple closed surface** has exactly one interior, has no holes, and is hollow. A **sphere** is the set of all points at a given distance from a given point, the **center**. The set of all points on a simple closed surface together with all interior points is a **solid**. Figure 14-49(a), (b), (c), and (d) are examples of simple closed surfaces; (e) and (f) are not. A **polyhedron** (*polyhedra* is the plural) is a simple closed surface made up of polygonal regions, or **faces**. The vertices of the polygonal regions are the **vertices** of the polyhedron, and the sides of each polygonal region are the **edges** of the polyhedron. Figure 14-49(a) and (b) are examples of polyhedra but (c), (d), (e), and (f) are not.

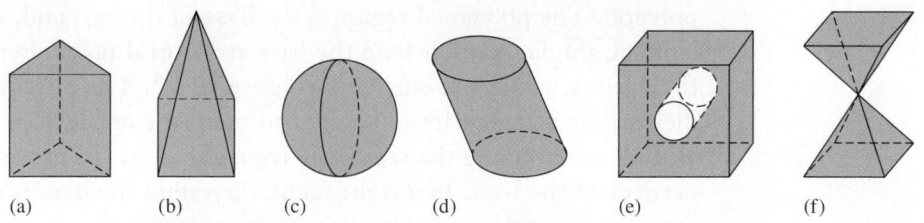

| (a) | (b) | (c) | (d) | (e) | (f) |

Figure 14-49

A **prism** is a polyhedron in which two congruent polygonal faces (bases) lie in parallel planes and the remaining faces are formed by the union of the line segments joining corrosponding points in the two bases. Figure 14-50 shows four different prisms. The shaded parallel faces of a prism are the **bases** of the prism. A prism usually is named after its bases, as the figure suggests.

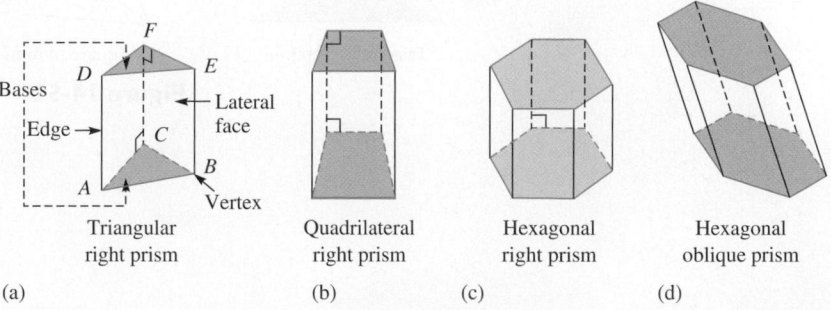

| Triangular right prism | Quadrilateral right prism | Hexagonal right prism | Hexagonal oblique prism |
| (a) | (b) | (c) | (d) |

Figure 14-50

The faces other than the bases are the **lateral faces** of a prism and are parallelograms. If the lateral faces of a prism are all bounded by rectangles, the prism is a **right prism**, as in Figure 14-50(a)–(c). Figure 14-50(d) is an **oblique prism** because some of its lateral faces are *not* bounded by rectangles.

Students often have difficulty drawing three-dimensional figures. Figure 14-51 gives an example of how to draw a right pentagonal prism.

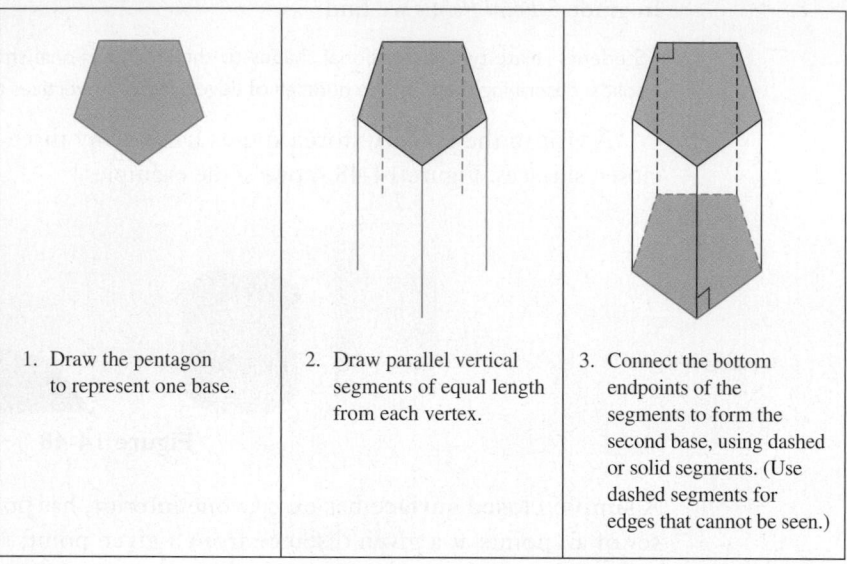

1. Draw the pentagon to represent one base.

2. Draw parallel vertical segments of equal length from each vertex.

3. Connect the bottom endpoints of the segments to form the second base, using dashed or solid segments. (Use dashed segments for edges that cannot be seen.)

Figure 14-51

A **pyramid** is a polyhedron determined by a polygon and a point not in the plane of the polygon. The pyramid consists of the triangular regions determined by the point and each pair of consecutive vertices of the polygon and the polygonal region determined by the polygon. The polygonal region is the **base** of the pyramid, and the point is the **apex**. As with a prism, the faces other than the base are **lateral faces**. Pyramids are classified according to their bases, which are shaded in Figure 14-52. The **altitude** of the pyramid is the perpendicular line segment from the apex to the plane of the base. A **right regular pyramid** has an altitude intersecting the regular polygonal base at the center of the circle passing through all vertices of the base. In a right regular pyramid the base is a regular polygon and the lateral faces are formed by congruent isosceles triangles.

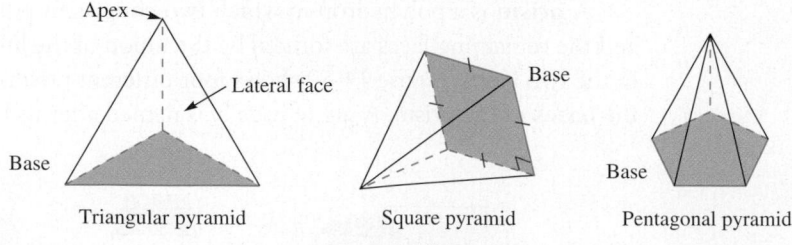

Apex

Lateral face

Base

Base

Triangular pyramid

Base

Square pyramid

Base

Pentagonal pyramid

Figure 14-52

To draw a pyramid, follow the steps in Figure 14-53.

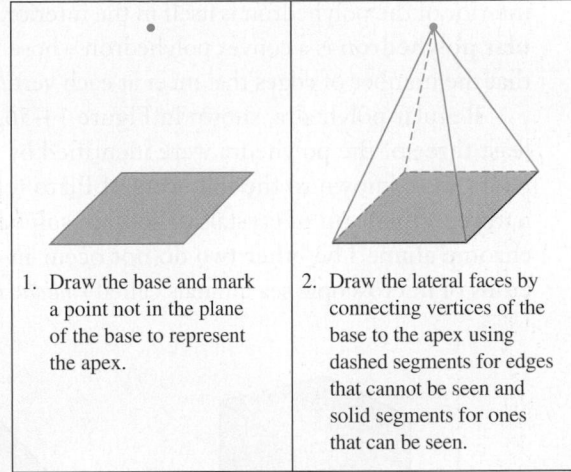

1. Draw the base and mark a point not in the plane of the base to represent the apex.

2. Draw the lateral faces by connecting vertices of the base to the apex using dashed segments for edges that cannot be seen and solid segments for ones that can be seen.

Figure 14-53

NOW TRY THIS 14-13

SHOE

Shoe-New Business copyright © 1994 MacNelly. Distributed by King Features Syndicate

Notice in the *SHOE* cartoon that the segments are solid for the top view of the Washington Monument. Find a photograph of the monument and draw other views with dashed segments for edges that cannot be seen.

If one or more corners of a polyhedron is removed by an intersecting plane or planes and the removed portion replaced by a polygond region, we say that the polyhedron is a **truncated polyhedron**. Figure 14-54 shows several truncated polyhedra.

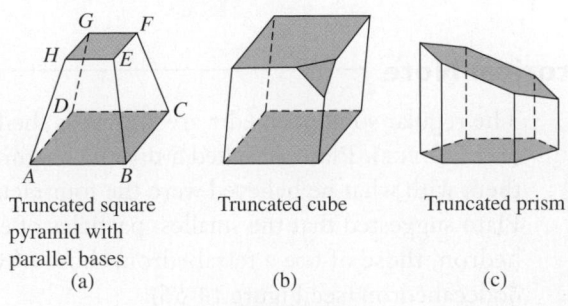

Truncated square pyramid with parallel bases
(a)

Truncated cube
(b)

Truncated prism
(c)

Figure 14-54

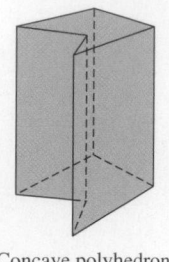

Concave polyhedron
Figure 14-55

Regular Polyhedra

A polyhedron is a **convex polyhedron** if, and only if, the segment connecting any two points in the interior of the polyhedron is itself in the interior. Figure 14-55 shows a concave polyhedron. A **regular polyhedron** is a convex polyhedron whose faces are congruent regular polygonal regions such that the number of edges that meet at each vertex is the same for all the vertices of the polyhedron.

Regular polyhedra, shown in Figure 14-56, have fascinated mathematicians for centuries. At least three of the polyhedra were identified by the Pythagoreans (ca. 500 BCE). Two other polyhedra were known to the followers of Plato (ca. 350 BCE). Three of the five polyhedra occur in nature in the form of crystals of sodium sulphantimoniate, sodium chloride (common salt), and chrome alum. The other two do not occur in crystalline form but have been observed as skeletons of microscopic sea animals called *radiolaria*.

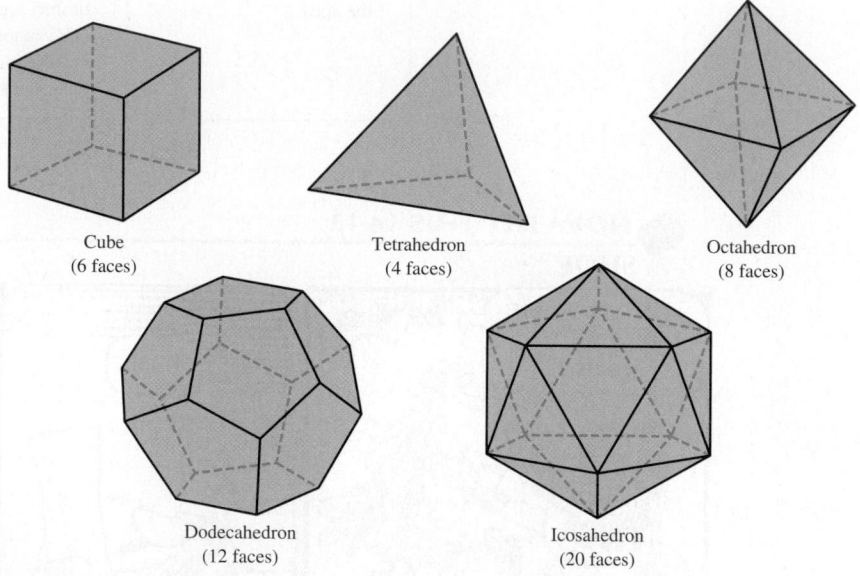

Cube
(6 faces)

Tetrahedron
(4 faces)

Octahedron
(8 faces)

Dodecahedron
(12 faces)

Icosahedron
(20 faces)

Figure 14-56

Problem Solving | Regular Polyhedra?

How many regular polyhedra are there?

Understanding the Problem Each face of a regular polyhedron is congruent to each of the other faces of that polyhedron, and each face is bounded by a regular polygon. We are to find the number of different regular polyhedra.

Devising a Plan The sum of the measures of all the angles of the faces at a vertex of a regular polyhedron must be less than 360°. We next examine the measures of the interior angles of regular polygons to determine which of the polygons could be faces of a regular polyhedron. Then we try to determine how many types of polyhedra there are.

Historical Note

The regular solid polyhedra are known as the **Platonic solids**, after the Greek philosopher Plato (ca. 350 BCE). Plato attached a mystical significance to the five regular polyhedra, associating them with what he believed were the four elements (earth, air, fire, and water) and the universe. Plato suggested that the smallest particles of earth have the form of a cube, those of air an octahedron, those of fire a tetrahedron, those of water an icosahedron, and those of the universe a dodecahedron (see Figure 14-56).

Carrying Out the Plan We determine the size of an angle of some regular polygons, as shown in Table 14-4. Could a regular heptagon be a face of a regular polyhedron? At least three figures must fit together at a vertex to make a polyhedron. (Why?) If three angles of a regular heptagon were together at one vertex, then the sum of the measures of these angles would be $\frac{3 \cdot 900°}{7}$, or $\frac{2700°}{7}$, which is greater than 360°. Similarly, more than three angles cannot be used at a vertex. Thus, a heptagon cannot be used to make a regular polyhedron.

Table 14-4

Polygon	Measure of an Interior Angle
Triangle	60°
Square	90°
Pentagon	108°
Hexagon	120°
Heptagon	$\left(\frac{900}{7}\right)°$

The measure of an interior angle of a regular polygon increases as the number of sides of the polygon increases. (Why?) Thus any polygon with more than six sides has an interior angle greater than 120°. So if three angles were to fit together at a vertex, the sum of the measures of the angles would be greater than 360°. This means that the only polygons that might be used to make regular polyhedra are equilateral triangles, squares, regular pentagons, and regular hexagons. Consider the possibilities given in Table 14-5.

Notice that we were not able to use six equilateral triangles to make a polyhedron because $6(60°) = 360°$ and the triangles would lie in a plane. Similarly, we could not use four squares or three hexagons. We also could not use more than three pentagons because if we did, the sum of

Table 14-5

Polygon	Measure of an Interior Angle	Number of Polygons at a Vertex	Sum of the Angles at the Vertex	Polyhedron Formed	Model
Triangle	60°	3	180°	Tetrahedron	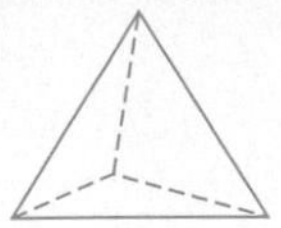
Triangle	60°	4	240°	Octahedron	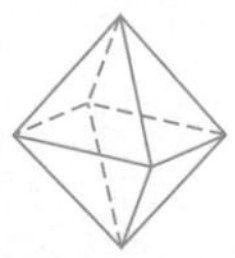
Triangle	60°	5	300°	Icosahedron	
Square	90°	3	270°	Cube	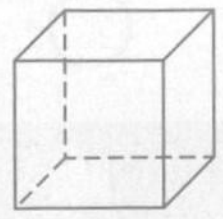
Pentagon	108°	3	324°	Dodecahedron	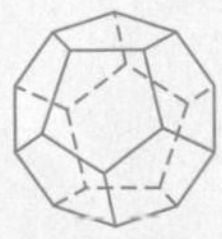

the measures of the angles would be more than 360°. Equilateral triangles, squares, and regular hexagons are used to tile floors.

Looking Back Interested readers may want to investigate **semiregular polyhedra**. These are also formed by using regular polygons as faces, but the regular polygons used need not have the same number of sides. In addition at each vertex semiregular polyhedra have the same number of faces of each kind. Two semiregular polyhedra are shown in Figure 14-57(a) and (b).

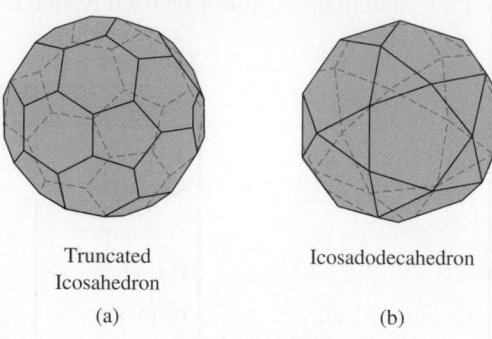

Truncated
Icosahedron

(a)

Icosadodecahedron

(b)

Figure 14-57

The patterns (nets) in Figure 14-58 can be used to construct the five regular polyhedra. A **net** is a pattern used to construct a polyhedron. It is left as an exercise to determine other nets for constructing the regular polyhedra.

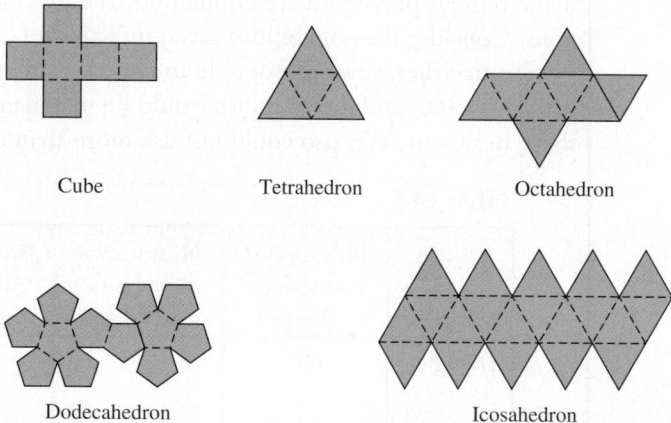

Cube

Tetrahedron

Octahedron

Dodecahedron

Icosahedron

Figure 14-58

NOW TRY THIS 14-14

A simple relationship among the number of faces, the number of edges, and the number of vertices of any polyhedron was discovered by the French mathematician and philosopher René Descartes (1596–1650) and rediscovered by the Swiss mathematician Leonhard Euler (1707–1783).

Historical Note

Leonhard Euler went blind in 1766 and for the remaining 17 years of his life continued to do mathematics by dictating to a secretary and by writing formulas in chalk on a slate for his secretary to copy down. He published 530 papers in his lifetime and left enough work to supply the *Proceedings of the St. Petersburg Academy* for the next 47 years.

Table 14-6 suggests a relationship among the numbers of vertices (V), edges (E), and faces (F). This relationship is known as Euler's formula: $V + F - E = 2$. Check that the relationship holds for the three truncated polyhedra in Figure 14-54.

Table 14-6

Name	V	F	E
Tetrahedron	4	4	6
Cube	8	6	12
Octahedron	6	8	12
Dodecahedron	20	12	30
Icosahedron	12	20	30

Cylinders and Cones

A *cylinder* is an example of a simple closed surface that is *not* a polyhedron. Cylinders are similar to prisms in that as the number of sides of a prism increases, the prism approaches the shape of a cylinder. Like prisms, cylinders have two congruent parallel bases. Consider two congruent simple closed curves that are contained in parallel planes. A **cylinder** is the union of the line segments joining the corresponding points on the simple closed curves and the interiors of the simple closed curves. The simple closed curves along with their interiors, are the **bases** of the cylinder, and the remaining points constitute the **lateral surface**.

If a base of a cylinder is a circular region, as in Figure 14-59(a) and (b), the cylinder is a **circular cylinder**. If a line segment connecting a point on one base to its corresponding point on the other base is perpendicular to the planes of the bases then the cylinder is a **right cylinder**. Circular cylinders that are not right cylinders are **oblique cylinders**. The cylinder in Figure 14-59(a) is a right cylinder; those in Figure 14-59(b) and (c) are oblique cylinders.

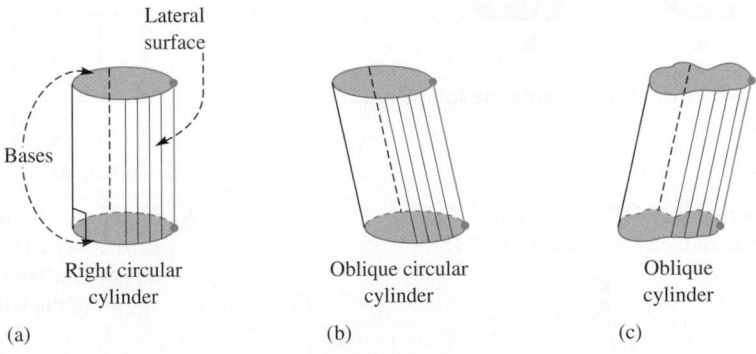

Figure 14-59

Suppose we have a simple closed curve, other than a polygon, in a plane and a point P not in the plane of the curve. The union of line segments connecting point P to each point of a simple closed curve, the simple closed curve, and the interior of the curve is a **cone**. Cones are pictured in Figure 14-60. Point P is the **apex** of the cone. The points of the cone not in the base constitute the *lateral surface of the cone*. A line segment from vertex P perpendicular to the plane of the base is the **altitude**. A **right circular cone**, such as the one in Figure 14-60(a), is a cone whose altitude intersects the base (a circular region) at the center of the circle. Figure 14-60(b) illustrates an **oblique circular cone**.

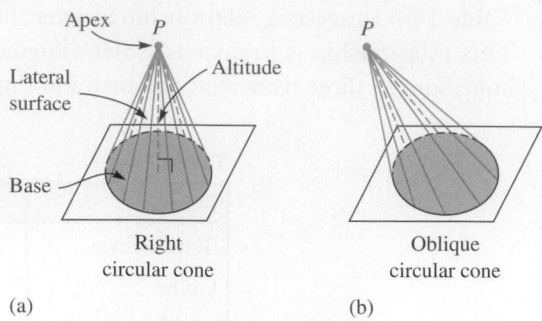

Apex

Lateral surface

Altitude

Base

Right circular cone

(a)

Oblique circular cone

(b)

Figure 14-60

Assessment 14-3A

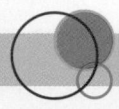

1. Identify each of the following polyhedra. If a polyhedron can be described in more than one way, give as many names as possible.

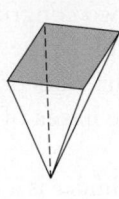 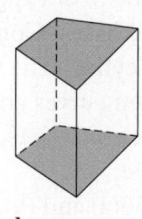 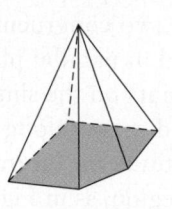

   a.   **b.**   **c.**

2. The following are pictures of stacks of solid cubes. In each case, determine the number of cubes in the stack and the number of faces that are glued together.

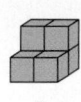 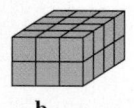

   a.   **b.**

3. Given the tetrahedron shown, name the following:
   a. Vertices
   b. Edges
   c. Faces
   d. Intersection of face *DRW* and edge $\overline{RA}$
   e. Intersection of face *DRW* and face *DAW*

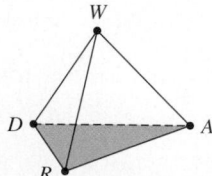

4. Determine the minimum number of faces possible for each of the following:
   a. Prism   **b.** Pyramid   **c.** Polyhedron

5. Classify each of the following as true or false:
   a. If the lateral faces of a prism are rectangles, it is a right prism.
   b. Every pyramid is a prism.
   c. Some pyramids are polyhedra.
   d. The bases of a prism lie in perpendicular planes.

6. If possible, sketch each of the following:
   a. An oblique square prism
   b. An oblique square pyramid

7. Name each polyhedron that can be constructed using the following nets.

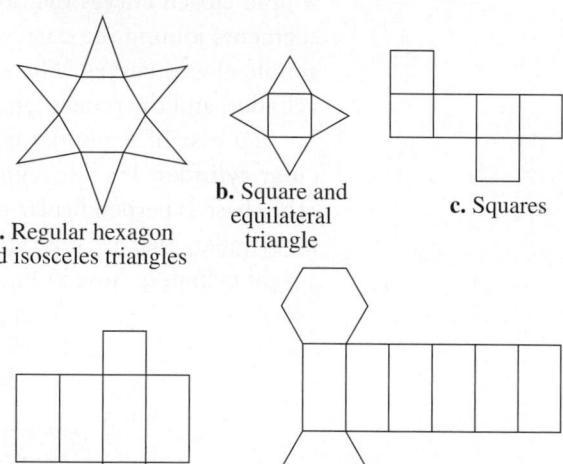

   a. Regular hexagon and isosceles triangles

   b. Square and equilateral triangle

   c. Squares

   d. Rectangles   **e.** Rectangles and regular hexagons

8. The figure on the left in each of the following represents a card attached to a wire, as shown. Match each figure on the left with what it would look like if you were to revolve it by spinning the wire between your fingers.

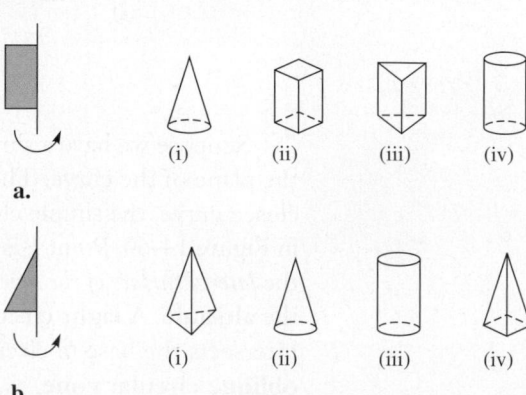

   (i)   (ii)   (iii)   (iv)

   a.

   (i)   (ii)   (iii)   (iv)

   b.

9. Which of the following three-dimensional figures could be used to make a shadow like in (a)? in (b)?

 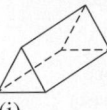 

a. (i) (ii) (iii) (iv)

 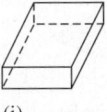 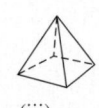 

b. (i) (ii) (iii) (iv)

10. Name the intersection of each of the following with the plane shown:

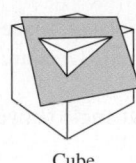

Cube

a.

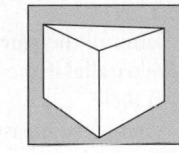

Remainder of unseen figure completes the cube

b.

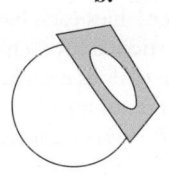

Sphere

c.

11. The right hexagonal prism shown has regular hexagons as bases.

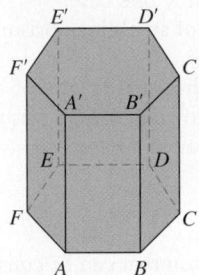

a. Name all the pairs of parallel lateral faces. (Faces are parallel if the planes containing the faces are parallel.)
b. What is the measure of the dihedral angle between two adjacent lateral faces? Why?

12. For each of the following figures, find $V + F - E$, where V, E, and F stand, respectively, for the number of vertices, edges, and faces:

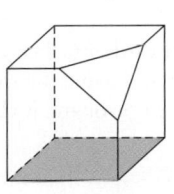 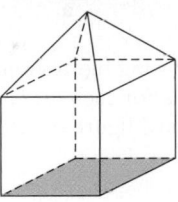

a. **b.**

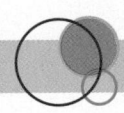

Assessment 14-3B

1. Identify each of the following:

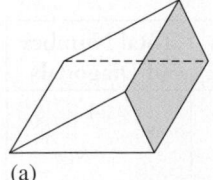 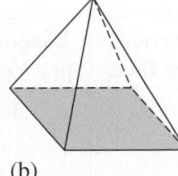 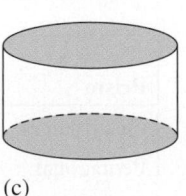

(a) (b) (c)

2. The following are pictures of solid cubes lying on a flat surface. In each case, determine the number of cubes in the stack and the number of faces that are glued together.

a. **b.**

3. Given the prism shown, name the following:

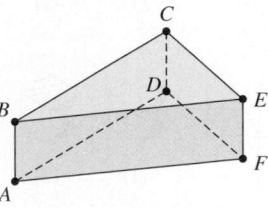

a. Vertices
b. Edges
c. Faces
d. Intersection of face BCE and edge $\overline{EF}$.
e. Intersection of face BCE and face ADF.

4. Determine the minimum number of edges possible for each of the following:
a. Prism **b.** Pyramid **c.** Polyhedron

5. Classify each of the following as true or false:
   a. The bases of all cones are circles.
   b. A cylinder has only one base.
   c. All lateral faces of an oblique prism are rectangular regions.
   d. All regular polyhedra are convex.

6. For each of the following, draw a prism and a pyramid that have the given region as a base:
   a. Triangle
   b. Pentagon
   c. Regular hexagon

7. Name the polyhedron that can be constructed using the following nets. Assume squares and equilateral triangles were used to form the nets.

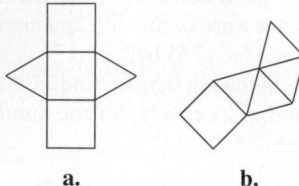

           a.   b.

8. The following figures represent cards attached to a wire, as shown. For each card, sketch and name the three-dimensional figure resulting from revolving it about the wire.

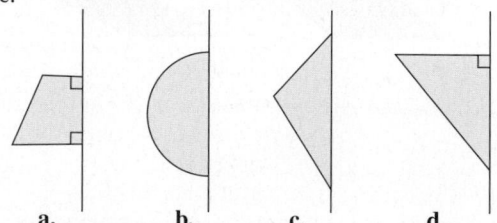

      a.   b.   c.   d.

9. On the left of each of the following figures is a net for a three-dimensional object. On the right are several objects. Which object will the net fold to make?

a.

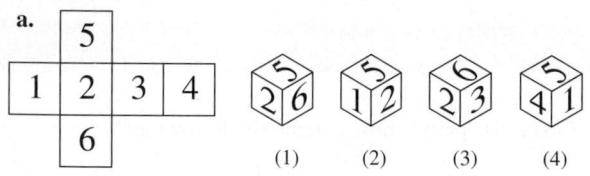

        (1)   (2)   (3)   (4)

b.

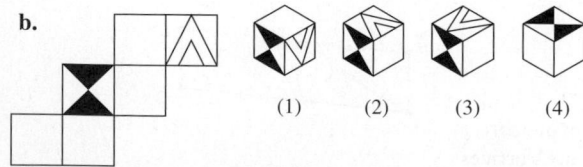

        (1)   (2)   (3)   (4)

10. Sketch the intersection of each of the following with the plane shown:

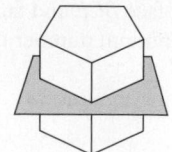

Right pentagonal prism

a.

Right circular cone
(plane parallel to base)

b.

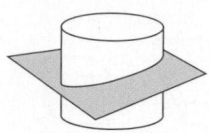

Right circular cylinder
(plane not parallel to base)

c.

11. Draw a right pentagonal prism with bases $ABCDE$ and $A'B'C'D'E'$.
   a. Name all the pairs of parallel lateral faces. (Faces are parallel if the planes containing the faces are parallel.)
   b. What is the measure of the dihedral angle between two adjacent lateral faces? Why?

12. Answer each of the following questions about a pyramid and a prism, each having an n-gon as a base:
   a. How many faces does each have?
   b. How many vertices does each have?
   c. How many edges does each have?
   d. Use your answers to (a), (b), and (c) to verify Euler's formula $V + F - E = 2$ for all pyramids and all prisms.

13. A diagonal of a prism is any segment determined by two vertices that do not lie in the same face, as shown in the following figure.

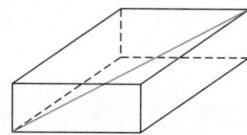

Complete the following table showing the total number of diagonals for various prisms:

Prism	Vertices per Base	Diagonals per Vertex	Total Number of Diagonals
Quadrilateral	4	1	4
Pentagonal	5		
Hexagonal			
Heptagonal			
Octagonal			
.			
.			
.			
n-gonal			

Mathematical Connections 14-3

Communication

1. How many possible pairs of bases does a rectangular prism have? Explain.
2. A circle can be approximated by a "many-sided" polygon. Use this notion to describe the relationship between each of the following:
   a. A pyramid and a cone
   b. A prism and a cylinder
3. Can either or both of the following be drawings of a quadrilateral pyramid? If yes, where would you be standing in each case? Explain why.

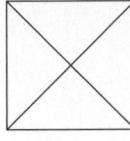   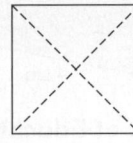

   a.   b.

4. When a plane intersects a three-dimensional object, a cross section is created. Sketch a cube and show how each of the following cross sections can be obtained. Explain your reasoning:
   a. An equilateral triangle
   b. A scalene triangle
   c. A rectangle
   d. A square
   e. A pentagon
   f. A hexagon
   g. A parallelogram that is not a rectangle
   h. A rhombus that is not square
5. The following is a picture of a right rectangular prism. M and N are points on two edges such that $\overline{MN} \| \overline{AD}$. Answer the following questions and explain your reasoning:

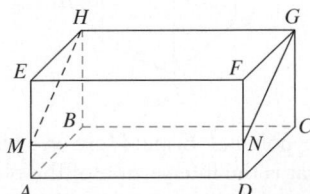

   a. Is $HGNM$ a rectangle or a parallelogram that is not a rectangle?
   b. If the vertex H is connected to each of the vertices A, B, C, and D, a rectangular pyramid is formed. Is it a right regular pyramid?
   c. A pyramid is formed by connecting the intersection of the diagonals $\overline{HF}$ and $\overline{EG}$ with the vertices of the base $ABCD$. Is it a right regular pyramid?

Open-Ended

6. When a box in the shape of a right rectangular prism, like the one in the following figure, is cut by a plane halfway between the opposite sides and parallel to these sides, that plane is a *plane of symmetry*. If a mirror is placed at the plane of symmetry, the reflection of the front part of the box will look just like the back part. Draw several space figures and find the number of planes of symmetry for each. Summarize your results in a table. Can you identify any figures with infinitely many planes of symmetry?

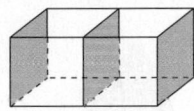

Cooperative Learning

7. In a two-person game, draw a three-dimensional figure without showing it to your partner. Tell your partner the shape of all possible cross sections of your figure sufficient to identify the figure. If your partner can identify your figure, 1 point is earned by your partner. If a figure is identified that has all the cross sections listed but that figure is not your figure, 2 points are earned by your partner. Each of you should take an equal number of turns.
8. Some of the following nets can be folded into cubes. Have each person in your group draw all the nets that can be folded into cubes. Share your findings with the group and decide how many different such nets there are. Discuss what "different" means in this case. Finally, compare your group's answers with those of other groups.

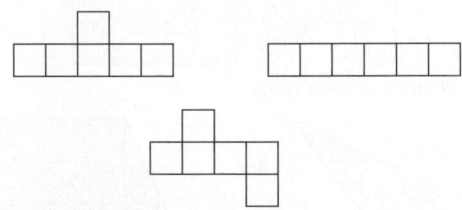

9. Choose different triples of points on the edges of each of the following figures. Then find the cross section of where a plane through each triple of points intersects the figure. (The

following figures show only a few examples.) List possible figures that can be obtained in this manner.

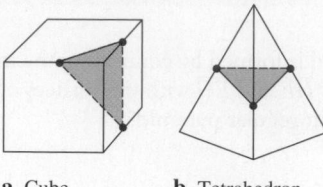

a. Cube   **b.** Tetrahedron

Questions from the Classroom

10. A student asks how to find the shortest path between two points A and B on two different faces which are neither the top or bottom of a right rectangular prism, without leaving the prism. How do you respond?

11. Jodi has a model of a tetrahedron (shown below) and would like to know how many different nets exist for the tetrahedron. How do you respond?

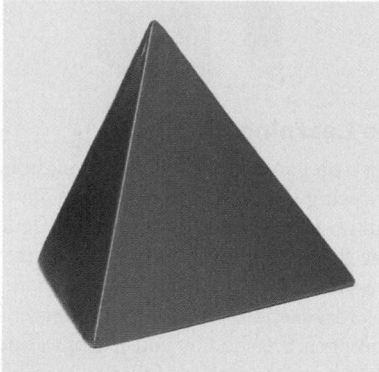

Review Problems

12. Find the area of the shape below. Each small square represents 1 square unit.

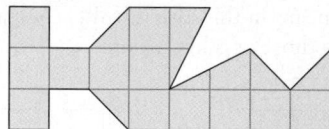

13. Find the area of each shape.

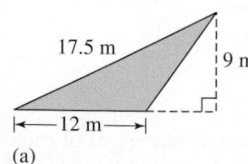
(a)

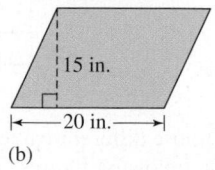
(b)

14. Find the measure of diagonal d in the figure below.

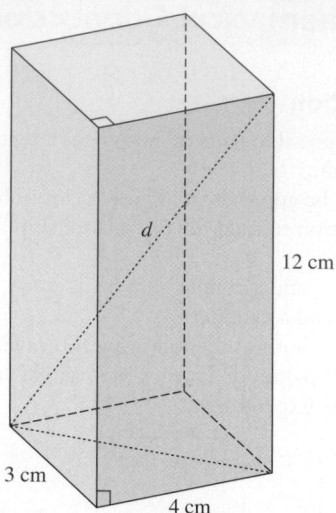

12 cm

3 cm

4 cm

National Assessment of Educational Progress (NAEP) Questions

Which of the following could NOT be folded into a cube?

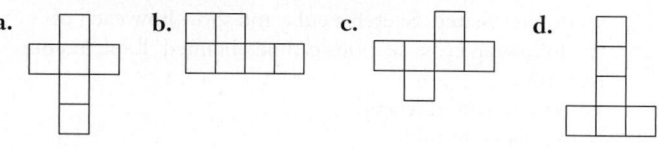

a.   **b.**   **c.**   **d.**

NAEP, Grade 8, 2003

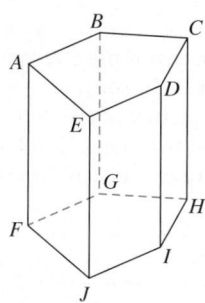

In the figure above, points A, E, and H are on a plane that intersects a right prism. What is the intersection of the plane with the right prism?

a. A line
b. A triangle
c. A quadrilateral
d. A pentagon
e. A hexagon

NAEP, Grade 12, 2003

BRAIN TEASER

A rectangular region can be rolled to form the lateral surface of a right circular cylinder. What shape of paper is needed to make an oblique circular cylinder?

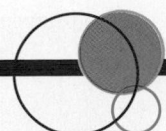

14-4 Surface Areas

Painting houses, buying roofing, seal-coating driveways, and buying carpet are among the common applications that involve computing areas. In many real-world problems, we must find the surface areas of such three-dimensional figures as prisms, cylinders, pyramids, cones, and spheres. Formulas for finding these areas are usually based on finding the area of two-dimensional pieces of the three-dimensional figures.

In the grade 5 *Focal Points* we find the following:

> They (students) decompose three-dimensional shapes and find surface areas and volumes of prisms. As they work with surface area, they find and justify relationships among the formulas for the areas of different polygons. They measure necessary attributes of shapes to use area formulas to solve problems. (p. 17)

Surface Area of Right Prisms

Consider the cereal box shown in Figure 14-61(a). Ignore the flaps for gluing the box together. To find the amount of cardboard necessary to make the box, we cut the box along the edges and make it lie flat, as shown in Figure 14-61(b). When we do this we obtain a *net* for the box. The box is composed of a series of rectangles. We find the area of each rectangle and sum those areas to find the surface area of the box.

(a)

(b)

Figure 14-61

NOW TRY THIS 14-15

a. Find the surface area of the box in Figure 14-61.
b. Could the box be made from a rectangular piece of cardboard 21 in. by 15 in.? If not, what size rectangle is needed and how would you do it?

A similar process can be used for many three-dimensional figures. For example, the surface area of the cube in Figure 14-62(a) is the sum of the areas of the faces of the cube. A net for a cube is shown in Figure 14-62(b). Because each of the six faces is a square of area 16 cm², the surface area is $6(16 \text{ cm}^2)$, or 96 cm². In general, for a cube whose edge has length e units as in Figure 14-62(c), the surface area is $6e^2$.

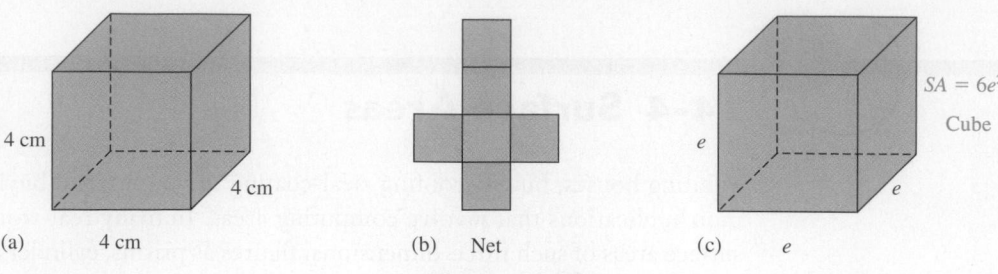

4 cm
4 cm
4 cm
(a)

(b)   Net

$SA = 6e^2$
Cube

e
e
(c) e

Figure 14-62

To find the surface area of a right prism, like the cereal box in Figure 14-61(a), we find the sum of the areas of the rectangles that make up the lateral faces and the areas of the top and bottom. The sum of the areas of the lateral faces is the **lateral surface area (LSA)**. The **surface area (SA)** is the sum of the lateral surface area and the area of the bases.

NOW TRY THIS 14-16

Figure 14-63 shows a right pentagonal prism with a net for the prism. If B stands for the area of each of the prism's bases, show that the surface area of the prism could be computed as $SA = ph + 2B$, where p is the perimeter of the base of the prism and h is the height. Does this formula hold for all right prisms? Why or why not?

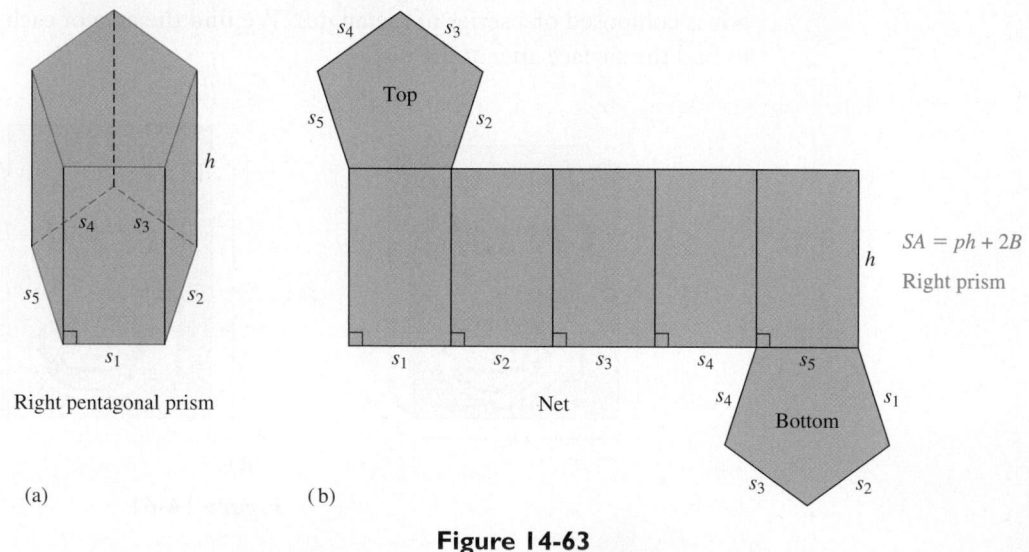

h
s_4 s_3
s_5 s_2
s_1
Right pentagonal prism

s_4 s_3
Top
s_5 s_2

s_1 s_2 s_3 s_4 s_5
Net

h
$SA = ph + 2B$
Right prism

s_4 s_1
Bottom
s_3 s_2

(a)   (b)

Figure 14-63

EXAMPLE 14-14

Find the surface area of each of the right prisms in Figure 14-64.

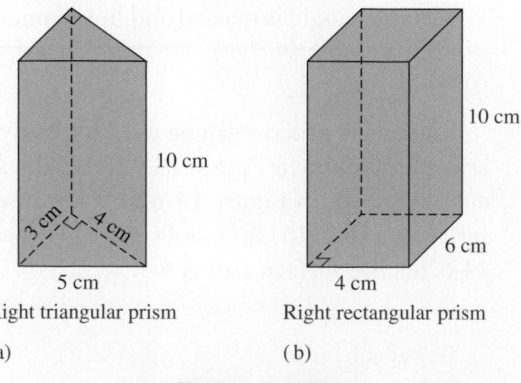

10 cm
3 cm   4 cm
5 cm
Right triangular prism
(a)

10 cm
6 cm
4 cm
Right rectangular prism
(b)

Figure 14-64

Solution

a. Each base is a right triangle. The area of the bases is $2\left(\dfrac{1}{2}(3 \cdot 4)\right)$, or 12 cm^2. The area of the three lateral faces is $4 \cdot 10 + 3 \cdot 10 + 5 \cdot 10$, or 120 cm^2. Thus, the surface area is 12 cm^2 + 120 cm^2, or 132 cm^2.

b. The area of the bases is $2(4 \cdot 6)$, or 48 cm^2. The lateral surface area is $2(10 \cdot 6) + 2(10 \cdot 4)$, or 200 cm^2. Thus, the surface area is 248 cm^2.

Note that in Example 14-14 we could have used the formula developed in Now Try This 14-16. For example, in part (a) $P = 3 + 4 + 5 = 12$ cm and the area of a base is 6 cm^2, so $SA = ph + 2B = 12 \cdot 10 + 12 = 120 + 12 = 132$ cm^2.

Surface Area of a Cylinder

As the number of sides of a right regular prism increases, as shown in Figure 14-65, the figure approaches the shape of a right circular cylinder.

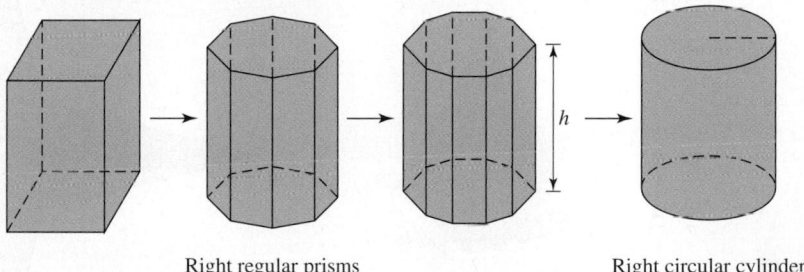

Right regular prisms   Right circular cylinder

Figure 14-65

To find the surface area of the right circular cylinder shown in Figure 14-66, we cut off the bases and slice the lateral surface open by cutting along any line perpendicular to the bases. Such a slice is shown as a dotted segment in Figure 14-66(a). Then we unroll the cylinder to form a rectangle, as shown in Figure 14-66(b). To find the total surface area, we find the area of the rectangle and the areas of the top and bottom circles. The length of the rectangle is the circumference of the circular base $2\pi r$, and its width is the height of the cylinder h. Hence, the area of the rectangle is $2\pi rh$. The area of each base is πr^2. *The surface area of a right circular cylinder is the sum of the areas of the two circular bases and the lateral surface area*; that is, $SA = 2\pi r^2 + 2\pi rh$.

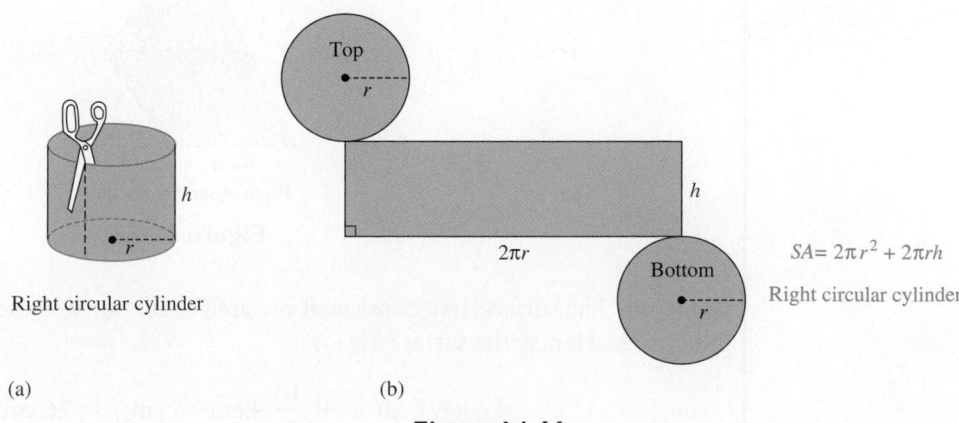

Right circular cylinder

$SA = 2\pi r^2 + 2\pi rh$

Right circular cylinder

(a)   (b)

Figure 14-66

Surface Area of a Pyramid

The surface area of a pyramid is the sum of the lateral surface area of the pyramid and the area of the base. A right regular pyramid is a pyramid such that the segments connecting the apex to each vertex of the base are congruent and the base is a regular polygon. The lateral faces of the right regular pyramid pictured in Figure 14-67(a) are congruent isosceles triangles. Each triangular face has an altitude of length ℓ, the *slant height*. Because the pyramid is right regular, each side of the base has the same length b. To find the lateral surface area of a right regular pyramid, we need to find the area of one face, $\frac{1}{2}b\ell$, and multiply it by n, the number of faces. Adding the lateral surface area $n\left(\frac{1}{2}b\ell\right)$ to the area of the base B gives the surface area, $SA = n\left(\frac{1}{2}b\ell\right) + B$. This formula can be simplified because nb is the perimeter, p, of the base. *Thus, the surface area of a right regular pyramid is given by $SA = \frac{1}{2}p\ell + B$, as shown in Figure 14-67(b).*

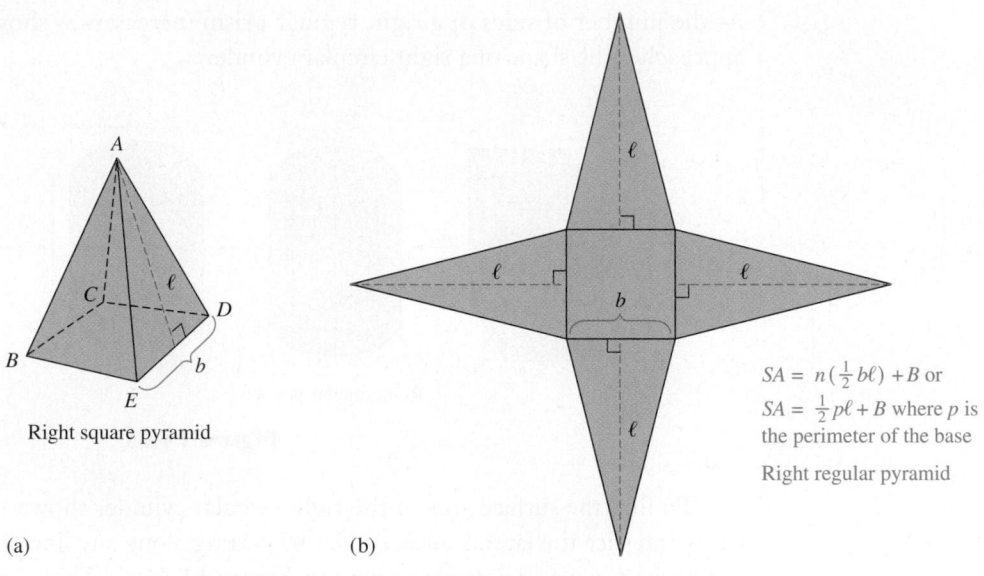

Right square pyramid

$SA = n\left(\frac{1}{2}b\ell\right) + B$ or
$SA = \frac{1}{2}p\ell + B$ where p is
the perimeter of the base

Right regular pyramid

(a)   (b)

Figure 14-67

EXAMPLE 14-15

Find the surface area of the right square pyramid in Figure 14-68.

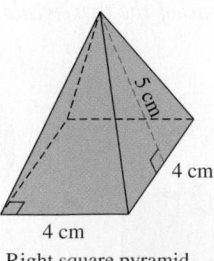

Right square pyramid

Figure 14-68

Solution The surface area consists of the area of the square base plus the area of the four triangular faces. Hence, the surface area is

$$4 \text{ cm} \cdot 4 \text{ cm} + 4\left(\frac{1}{2} \cdot 4 \text{ cm} \cdot 5 \text{ cm}\right) = 16 \text{ cm}^2 + 40 \text{ cm}^2$$

$$= 56 \text{ cm}^2.$$

EXAMPLE 14-16

The Great Pyramid of Cheops is a right square pyramid with a height of 148 m and a square base with perimeter of 940 m. The altitude of each triangular face is 189 m. The basic shape of the Transamerica Building in San Francisco is a right square pyramid that has a height of 260 m and a square base with a perimeter of 140 m. The altitude of each triangular face is 261 m. How do the lateral surface areas of the two structures compare?

Solution   The length of one side of the square base of the Great Pyramid is $\frac{940}{4}$, or 235 m.

Likewise, the length of one side of the square base of the Transamerica Building is $\frac{140}{4}$, or 35 m.

The lateral surface area (*LSA*) of the two are computed as follows:

$$(\text{Great Pyramid})\ LSA = 4\left(\frac{1}{2} \cdot 235 \cdot 189\right) = 88{,}830\ \text{m}^2$$

$$(\text{Transamerica})\ LSA = 4\left(\frac{1}{2} \cdot 35 \cdot 261\right) = 18{,}270\ \text{m}^2$$

Therefore, the lateral surface area of the Great Pyramid is approximately $\frac{88{,}830}{18{,}270}$ or 4.9 times as great as that of the Transamerica Building.

Surface Area of a Cone

As the number of sides of a right regular pyramid increases, as shown in Figure 14-69, the figure approaches the shape of a right circular cone.

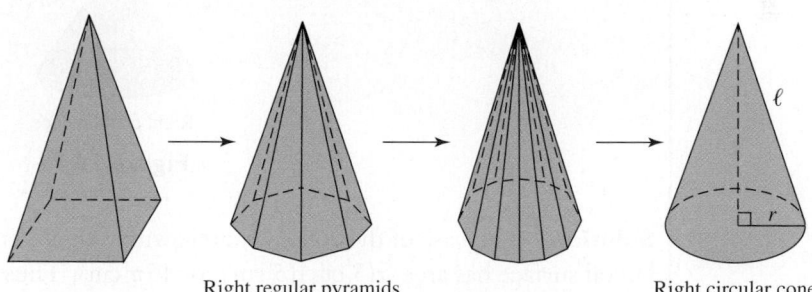

Right regular pyramids   Right circular cone

Figure 14-69

It is possible to find a formula for the surface area of a right circular cone by approximating the cone with a pyramid with many sides. We could inscribe in the circular base of the cone a regular polygon with many sides. The polygon can be used as the base of a right regular pyramid. The lateral surface area of the pyramid is close to the lateral surface area of the cone. The greater the number of faces of the pyramid, the closer the surface area of the pyramid is to that of the cone. In Figure 14-70(a), the lateral surface of the pyramid is $\frac{1}{2}ph$, where p is the perimeter of the base and h is the height of each triangle. With many sides in the pyramid, the perimeter of its base is close to the perimeter of the circle, $2\pi r$. The height of each triangle of the pyramid is close to the slant height ℓ, a segment that connects the apex of the cone with a point on the circular base. Consequently, it is reasonable that the lateral surface area of the cone becomes $\frac{1}{2} \cdot 2\pi r \cdot \ell$, or $\pi r\ell$. To find the total surface area of the cone, we add πr^2, the area of the base. Thus, *the surface area of a right circular cone is given by* $SA = \pi r^2 + \pi r\ell$, as shown in Figure 14-70(b).

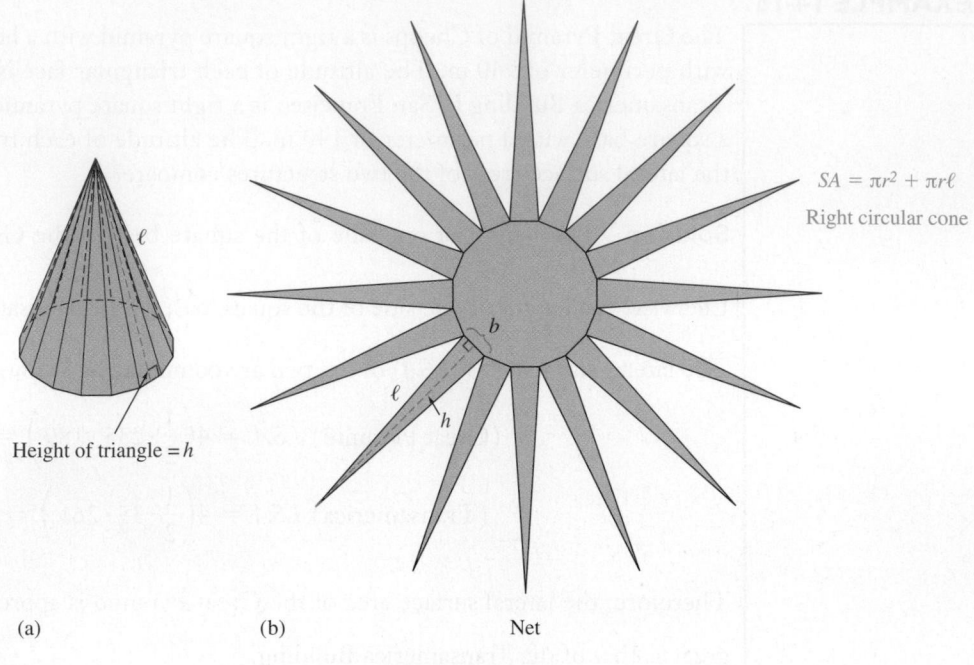

$SA = \pi r^2 + \pi r \ell$

Right circular cone

Height of triangle = h

(a)   (b)   Net

Figure 14-70

EXAMPLE 14-17

Find the surface area of the cone in Figure 14-71.

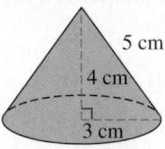

5 cm
4 cm
3 cm

Right circular cone

Figure 14-71

Solution   The base of the cone is a circle with radius 3 cm and area $\pi(3 \text{ cm})^2$, or $9\pi \text{ cm}^2$. The lateral surface has area $\pi(3 \text{ cm})(5 \text{ cm})$, or $15\pi \text{ cm}^2$. Thus we have the following surface area:

$$SA = \pi(3 \text{ cm})^2 + \pi(3 \text{ cm})(5 \text{ cm})$$
$$= 9\pi \text{ cm}^2 + 15\pi \text{ cm}^2$$
$$= 24\pi \text{ cm}^2$$

Surface Area of a Sphere

A **great circle** of a sphere is a circle on the sphere whose radius is equal to the radius of the sphere. A great circle is the intersection of a plane through the center of the sphere and the sphere. Because there are infinitely many different planes through the center, there are infinitely many great circles on a sphere. However, they are all congruent. Finding a formula for the surface area of a sphere is typically done using calculus, but it is not easy using only elementary mathematics. It can be shown that the surface area of a sphere is 4 times the area of a great circle of the sphere. Therefore, *the surface area of a sphere is given by $SA = 4\pi r^2$, as shown in Figure 14-72.

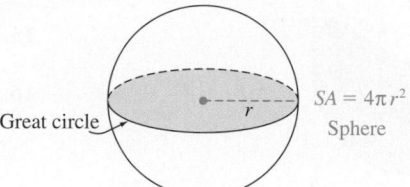

Great circle

$SA = 4\pi r^2$
Sphere

Figure 14-72

Assessment 14-4A

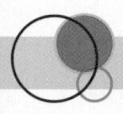

1. Which of these nets could be folded along the dotted segments to form a cube?

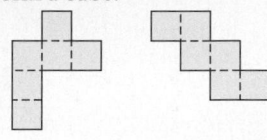

a.   b.

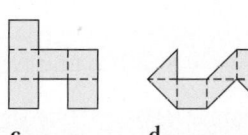

c.   d.

2. Find the surface area of each of the following:

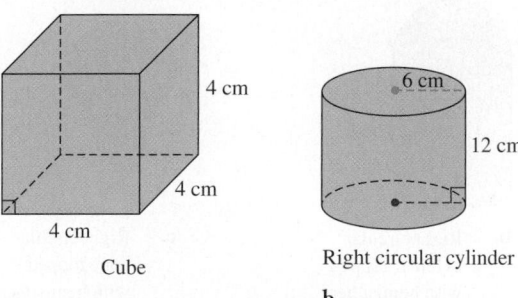

4 cm

4 cm

4 cm

Cube

a.

6 cm

12 cm

Right circular cylinder

b.

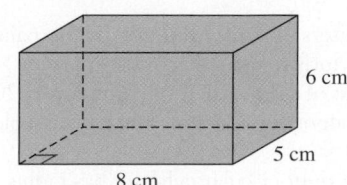

6 cm

5 cm

8 cm

Right rectangular prism

c.

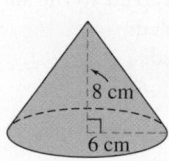

4 cm

Sphere

d.

8 cm

6 cm

Right circular cone

e.

3. How many liters of paint must you buy to paint the walls of a rectangular prism-shaped room that is 6 m × 4 m with a ceiling height of 2.5 m if 1 L of paint covers 20 m²? (Assume there are no doors or windows and paint comes in 1-L cans.)

4. The napkin ring pictured in the following figure is to be re-silvered. How many square millimeters must be covered?

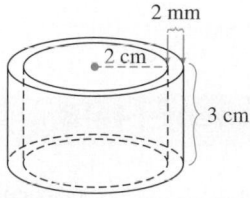

2 mm

2 cm

3 cm

5. Assume the radius of Earth is 6370 km and Earth is a sphere. What is its surface area?

6. Two cubes have sides of length 4 cm and 6 cm, respectively. What is the ratio of their surface areas?

7. The base of a right pyramid is a regular hexagon with sides of length 12 m. The altitude of the pyramid is 9 m. Find the exact total surface area of the pyramid.

8. A soup can has a $2\frac{5}{8}$ in. diameter and is 4 in. tall. What is the area of the paper that will be used to make the label for the can if the paper covers the entire lateral surface area?

9. The top of a right rectangular box has an area of 88 cm². The sides have areas 32 cm² and 44 cm². What are the dimensions of the box?

10. How does the lateral surface area of a right circular cone change if
    a. the slant height is tripled but the radius of the base remains the same?
    b. the radius of the base is tripled but the slant height remains the same?
    c. the slant height and the radius of the base are multiplied by 3?

11. Find the exact surface area of a right square pyramid if the area of the base is 100 cm² and the height of the pyramid is 20 cm.

12. The sector shown in the following figure is rolled into a cone so that the dashed edges just touch. Find the following:

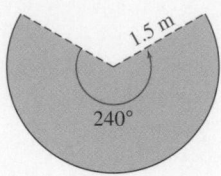

a. The lateral surface area of the cone
b. The total surface area of the cone

13. As seen in exercise 12, a sector of a circle can be used to construct a right circular cone. The length of the arc of the sector becomes the circumference of the circular base of the cone.
a. If the length of the arc is 6π, what is the radius of the base of the cone that can be constructed?
b. In part (a), the radius of the sector is 5 units; what is the slant height of the cone that can be constructed?
c. Using the information in parts (a) and (b), what is the height of the cone that can be constructed?
d. Using the information in parts (a)–(c), what is the angle measure for the original sector?

14. If the cardboard tube of a toilet paper roll has diameter of 2.5 in. and is 4 in. tall, what is the lateral surface area of the cardboard roll?
15. If two right circular cones are similar with radii of the bases in the ratio $1 : 2$, what is the ratio of their surface areas?
16. Water covers approximately 70% of the earth's surface. Assume the earth is a sphere with diameter about 13,000 km. What amount of the earth's surface is covered with water?
17. If two cubes have total surface areas of 64 in.^{2} and 36 in.^{2}, what is the ratio of the lengths of their edges?
18. The total surface area of a cube is 10,648 cm^2. What is the length of each of the following?
a. One of the edges
b. A diagonal that is not a diagonal of a face
19. Find the total surface area of the following stand, which was cut from a right circular cone:

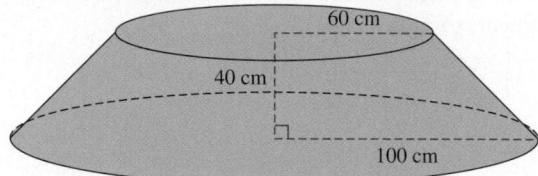

Assessment 14-4B

1. Which of the following nets could be folded to form a rectangular prism?

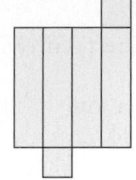   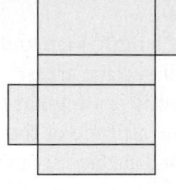   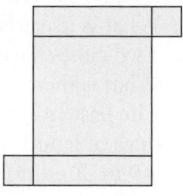

a.   **b.**   **c.**

2. Find the surface area of each of the following:

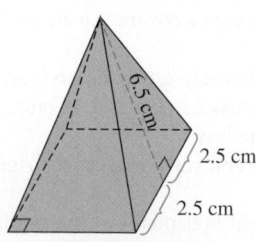

Right square pyramid

a.

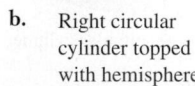

b. Right circular cylinder topped with hemisphere

c. Right circular cone topped with hemisphere

3. How many liters of paint must you buy to paint the walls of a rectangular prism-shaped room that is 8 m × 5 m with a ceiling height of 2.5 m if 1 L of paint covers 20 m^2? (Assume there are no doors or windows and paint is sold in 1-L cans.)
4. Suppose one right circular cylinder has radius 2 m and height 6 m and another has radius 6 m and height 2 m.
a. Which cylinder has the greater lateral surface area?
b. Which cylinder has the greater total surface area?
5. What happens to the surface area of a sphere if the radius is
a. doubled?
b. tripled?

6. How does the surface area of a right rectangular box (including top and bottom) change if
   a. each dimension is doubled?
   b. each dimension is tripled?
   c. each dimension is multiplied by a factor of k?

7. Approximately how much material is needed to make the tent illustrated in the following figure (both ends and the bottom as well as the sides should be included)?

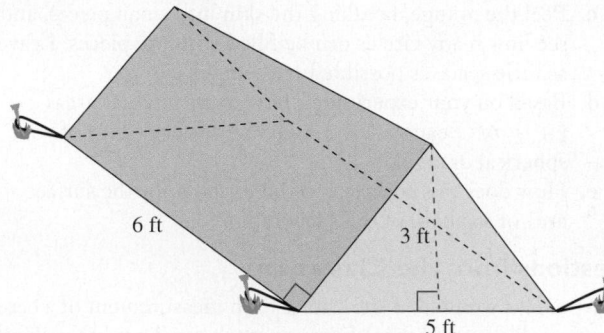

8. A square piece of paper 10 cm on a side is rolled to form the lateral surface area of a right circular cylinder and then a top and bottom are added. What is the surface area of the cylinder?

9. A structure is composed of unit cubes with at least one face of each cube connected to the face of another cube, as shown in the following figure.

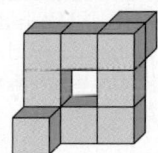

   a. If one cube is added, what is the maximum surface area the structure can have?
   b. If one cube is added, what is the minimum surface area the structure can have?
   c. Is it possible to design a structure so that one can add a cube and yet add nothing to the surface area of the structure? Explain your answer.

10. How does the lateral surface area of a right circular cylinder change if:
    a. the radius of the base is doubled?
    b. the height of the cylinder is doubled?

11. Find the surface area of a right square pyramid if the area of the base is 169 cm^2 and the slant height of the pyramid is 13 cm.

12. The region in each of the following figures revolves about the indicated axis. For each case, sketch the three-dimensional figure obtained and find its surface area.

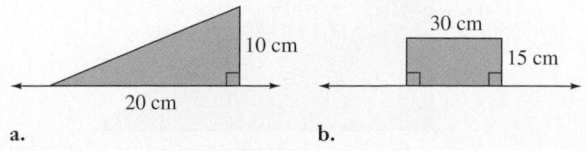

a.   b.

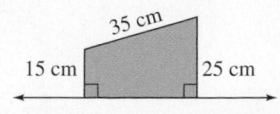

c.

13. If a right square pyramid and a right circular cone are inscribed in a cube as shown. Find the
    a. surface area of the pyramid.
    b. surface area of the cone.

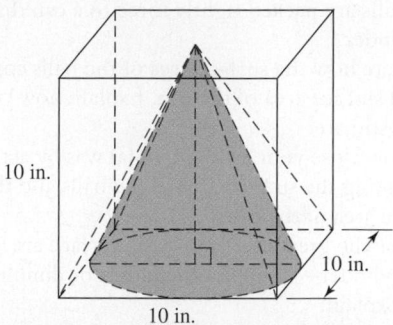

14. A gas storage tank is in the shape of a right circular cylinder that has a radius of the base of 3 ft and a height of 8 ft. The farmer wants to paint the tank including both bases but only has 1 gallon of paint. If 1 gallon of paint will cover 350 ft^2, will the farmer have enough paint to complete the job? Explain.

15. If two right circular cones are similar with radii of the bases in the ratio of 2:3, what is the ratio of their surface areas?

16. The diameter of Jupiter is about 11 times as great as the diameter of Earth. How do the surface areas compare?

17. If two cubes have edges of 2 ft and 4 ft, what is the ratio of their surface areas?

18. Find the surface area of the figure formed by removing a square section of the cube as shown below.

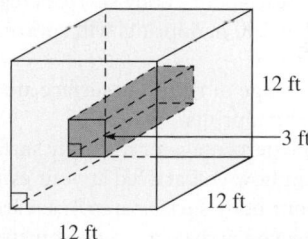

19. In the following figure, a right circular cylinder is inscribed in a right circular cone. Find the lateral surface area of the cylinder if the height of the cone is 40 cm, the height of the cylinder 30 cm, and the radius of the base of the cone is 25 cm.

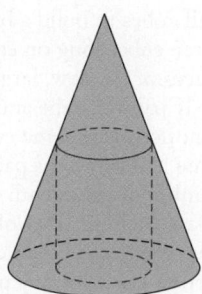

Mathematical Connections 14-4

Communication

1. Tennis balls are packed tightly three to a can that is shaped like a cylinder.
   a. Estimate how the surface area of the balls compares to the lateral surface area of the can. Explain how you arrived at your estimate.
   b. See how close your estimate in (a) was by actually computing the surface area of the balls and the lateral surface area of the can.

2. Which has the greater effect on the surface area of a right circular cylinder—doubling the radius or doubling the height? Explain.

3. In the drawing below, cube B was cut from a larger cube of surface area 216 cm² resulting in Figure A. The surface area of cube B is 24 cm². What effect did removing cube B from cube A have on the surface area? Explain.

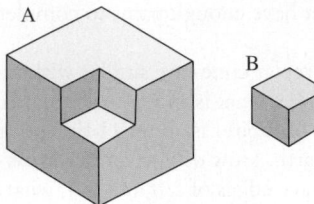

Open-Ended

4. One method of estimating body surface area in burn victims uses the fact that 100 handprints will approximately cover the whole body.
   a. What percentage of the body surface area is the surface area of two handprints?
   b. Estimate the percentage of the body surface area of one arm. Explain how you arrived at your estimate.
   c. Estimate your body surface area in square centimeters. Explain how you arrived at your estimate.
   d. Find the area of the flat part of your desk. How does the area of the desk compare with the surface area of your body?

5. Draw a net for a right prism that has surface area 80 cm².

Cooperative Learning

6. a. Shawn used small cubes to build a bigger cube that was solid and was three cubes long on each side. He then painted all the faces of the new, large cube red. He dropped the newly painted cube and all the little cubes came apart. He noticed that some cubes had only one face painted, some had two faces painted, and so on. Describe the number of cubes with 0, 1, 2, 3, 4, 5, or 6 faces painted. Have each member of the group choose a different number of sides and then combine your data to see if it makes sense. Look for any patterns that occur.
   b. What would the answers be if the large cube was four small cubes long on a side?
   c. Make a conjecture about how to count the cubes if the large cube were n small cubes long on a side.

7. As a group, work through the following activity to motivate the formula for the surface area of a sphere.
   a. Take an orange that is as close to spherical as you can find and determine its radius. (This will be approximate.)
   b. Take a compass and draw four disjoint circles that have the same radius as your orange.
   c. Peel the orange, breaking the skin into small pieces, and see how many circles can be filled with the pieces. Leave as little space as possible between pieces.
   d. Based on your experiment, how many circular areas $(A = \pi r^2)$ can be covered by the surface (skin) of your spherical orange?
   e. How does this compare to the formula for the surface area of a sphere of the same radius?

Questions from the Classroom

8. A student wonders if she triples each measurement of a cereal box, will she need three times as much cardboard to make the new box. How would you help her decide?

9. Jodi says that if you double the radius of a right circular cone and divide the slant height by 2, then the surface area of the cone stays the same since the 2s cancel each other out. How do you respond?

10. Abi used the formula $SA = \pi r(r + \ell)$, where r is the radius and ℓ is the slant height, to find the surface area of a right circular cone. Is she correct? What do you tell her?

11. Jan says that if you double each of the dimensions of a rectangular box, it will take twice as much wrapping paper to wrap it. How do you respond?

Review Problems

12. Complete each of the following:
   a. $10 \text{ m}^2 = \underline{\hspace{1cm}} \text{ cm}^2$
   b. $13{,}680 \text{ cm}^2 = \underline{\hspace{1cm}} \text{ m}^2$
   c. $5 \text{ cm}^2 = \underline{\hspace{1cm}} \text{ mm}^2$
   d. $2 \text{ km}^2 = \underline{\hspace{1cm}} \text{ m}^2$
   e. $10^6 \text{ m}^2 = \underline{\hspace{1cm}} \text{ km}^2$
   f. $10^{12} \text{ mm}^2 = \underline{\hspace{1cm}} \text{ m}^2$

13. The sides of a rectangle are 10 cm and 20 cm long. Find the length of a diagonal of the rectangle.

14. The length of the side of a rhombus is 30 cm. If the length of one diagonal is 40 cm, find the length of the other diagonal.

15. Find the perimeters and the areas of the following figures:

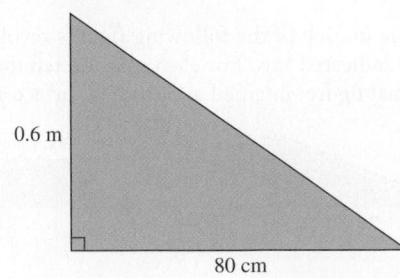

a.

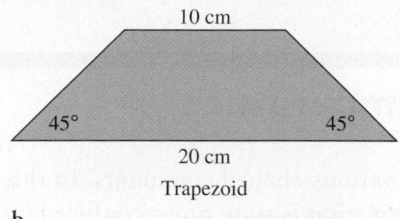

10 cm

45°   45°

20 cm

Trapezoid

b.

16. In the following, the length of the diagonal $\overline{AC}$ of rhombus *ABCD* is 40 cm; *AE* = 24 cm. Find the length of a side of the rhombus and the length of the diagonal $\overline{BD}$.

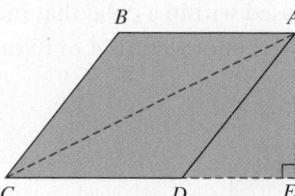

B   A

C   D   E

17. Find the length of the diagonal of a cube whose side is 1 unit long.

National Assessment of Educational Progress (NAEP) Question

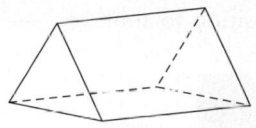

Which of the following can be folded to form the preceding prism?

a. 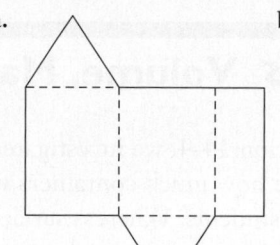   **b.**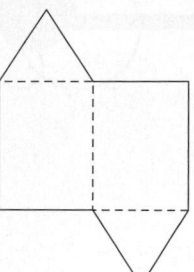

c. 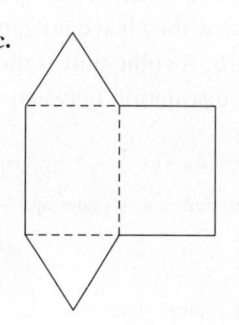   **d.**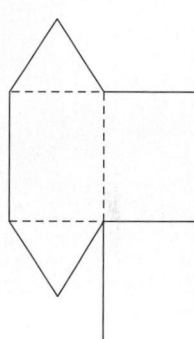

e.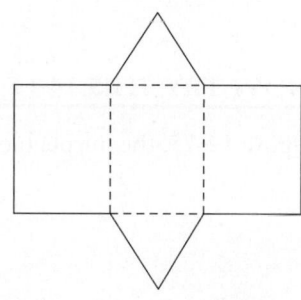

NAEP, Grade 8, 2005

BRAIN TEASER

A manufacturer of paper cups wants to produce paper cups in the form of truncated cones 16 cm high, with one circular base of radius 11 cm and the other of radius 7 cm, as shown in Figure 14-73. When the base of such a cup is removed and the cup is slit and flattened, the flattened region looks like a part of a circular ring. To design a pattern to make the cup, the manufacturer needs the data required to construct the flattened region. Find these data.

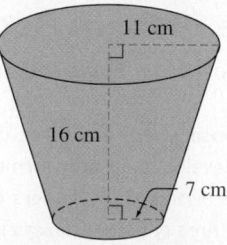

11 cm

16 cm

7 cm

Figure 14-73

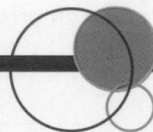

14-5 Volume, Mass, and Temperature

In Section 14-4, we investigated surface areas of various-shaped containers. In this section, we explore how much containers will hold. This distinction is sometimes confused by elementary school students. Whereas surface area is the number of square units covering a three-dimensional figure, volume describes how much space a three-dimensional figure contains. The unit of measure for volume must be a shape that tessellates space. Cubes tessellate space; that is, they can be stacked so that they leave no gaps and fill space. Standard units of volume are based on cubes and are *cubic units*. A cubic unit is the amount of space enclosed within a cube that measures 1 unit on a side. The distinction between surface area and volume is demonstrated in Figure 14-74.

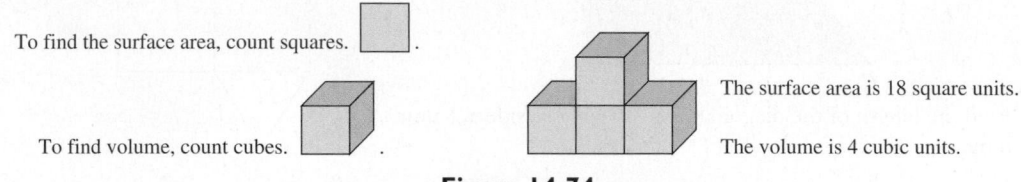

To find the surface area, count squares.

To find volume, count cubes.

The surface area is 18 square units.

The volume is 4 cubic units.

Figure 14-74

NOW TRY THIS 14-17

In Figure 14-75, the purple block is moved from one position to another.

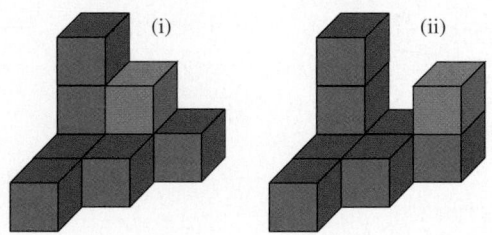

(i)　　　　(ii)

Figure 14-75

    a. How does the volume in (i) compare with the volume in (ii)?
    b. Do (i) and (ii) have the same surface area? If not, which has the greater surface area?
    c. Find the surface area of each figure.

Volume of Right Rectangular Prisms

In the grade 7 *Focal Points* we find the following:

> By decomposing two- and three-dimensional shapes into smaller, component shapes, students find surface areas and develop and justify formulas for the surface areas and volumes of prisms and cylinders. As students decompose prisms and cylinders by slicing them, they develop and understand formulas for their volumes (*Volume = Area of base × Height*). They apply these formulas in problem solving to determine volumes of prisms and cylinders. (p. 19)

The quote from *Focal Points* assumes the prisms and cylinders are right prisms and right circular cylinders. With this in mind, we investigate how to find the volume of a prism by using component shapes. In the grade 5 *Common Core Standards* we find that students:

> Find the volume of a right rectangular prism with whole-number side lengths by packing it with unit cubes, and show that the volume is the same as would be found by multiplying the edge lengths, equivalently by multiplying the height by the area of the base. (p. 37)

The volume of a right rectangular prism can be found by determining how many cubes are needed to build it as a solid. To find the volume, count how many cubes cover the base and then how many layers of these cubes are used to fill the prism. As shown in Figure 14-76(a), there are $8 \cdot 4$, or 32, cubes required to cover the base and there are five such layers. The volume of the rectangular prism is $(8 \cdot 4)5$, or 160 cubic units. For any right rectangular prism with dimensions ℓ, w, and h measured in the same linear units, the volume of the prism is given by the area of the base, ℓw, times the height, h, or $V = \ell wh$, as shown in Figure 14-76(b).

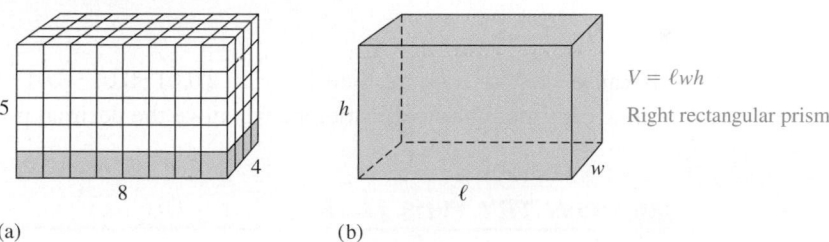

Figure 14-76

Converting Metric Measures of Volume

The most commonly used metric units of volume are the **cubic centimeter** and the **cubic meter**. A cubic centimeter is the volume of a cube whose length, width, and height are each 1 cm. One cubic centimeter is denoted 1 cm³. Similarly, a cubic meter is the volume of a cube whose length, width, and height are each 1 m. One cubic meter is denoted 1 m³. Other metric units of volume are symbolized similarly.

Figure 14-77 shows that since 1 dm = 10 cm, 1 dm³ = (10 cm)(10 cm)(10 cm) = 1000 cm³. Figure 14-78 shows that 1 m³ = 1,000,000 cm³ and that 1 dm³ = 0.001 m³. *Each metric unit of length is* 10 *times as great as the next smaller unit. Each metric unit of area is* 100 *times as great*

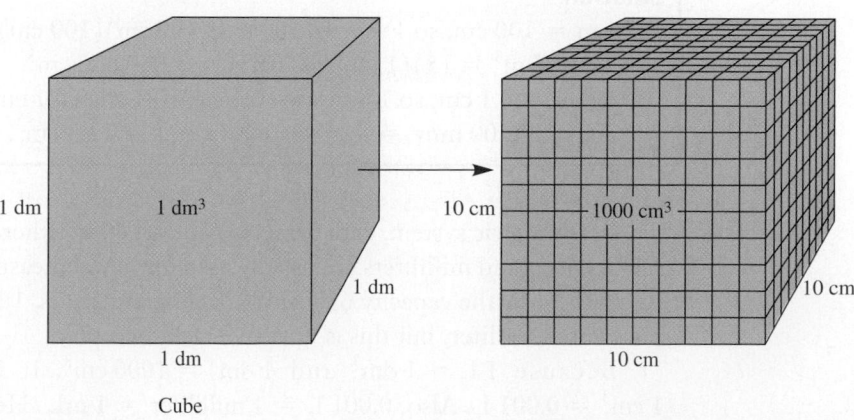

Figure 14-77

as the next smaller unit. Each metric unit of volume is 1000 times as great as the next smaller unit. For example:

$$1 \text{ cm} = 10 \text{ mm}$$
$$1 \text{ cm}^2 = 100 \text{ mm}^2$$
$$1 \text{ cm}^3 = 1000 \text{ mm}^3$$

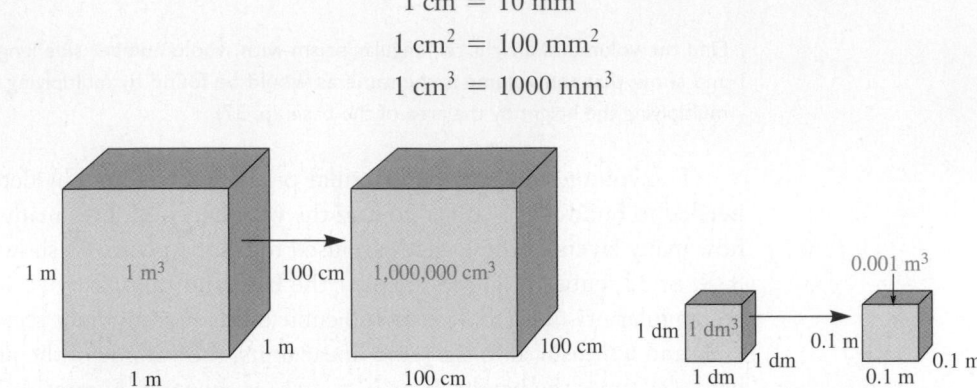

(a)   Cube   (b)   Cube

Figure 14-78

Because 1 cm = 0.01 m, then $1 \text{ cm}^3 = (0.01 \cdot 0.01 \cdot 0.01) \text{ m}^3$, or 0.000001 m^3. To convert from cubic centimeters to cubic meters, we move the decimal point six places to the left.

NOW TRY THIS 14-18

a. Describe how many places to move the decimal point and in what direction in a metric area conversion if you know how many places and the direction to move in the corresponding length conversion.

b. Describe how many places to move the decimal point and in what direction in a metric volume conversion if you know how many places and the direction to move in the corresponding length conversion.

EXAMPLE 14-18

Complete each of the following:

a. $5 \text{ m}^3 = $ _____ cm^3
b. $12,300 \text{ mm}^3 = $ _____ cm^3

Solution

a. $1 \text{ m} = 100 \text{ cm}$, so $1 \text{ m}^3 = (100 \text{ cm})(100 \text{ cm})(100 \text{ cm})$, or $1,000,000 \text{ cm}^3$.
   Thus, $5 \text{ m}^3 = (5)(1,000,000 \text{ cm}^3) = 5,000,000 \text{ cm}^3$.
b. $1 \text{ mm} = 0.1 \text{ cm}$, so $1 \text{ mm}^3 = (0.1 \text{ cm})(0.1 \text{ cm})(0.1 \text{ cm})$, or 0.001 cm^3.
   Thus, $12,300 \text{ mm}^3 = 12,300(0.001 \text{ cm}^3) = 12.3 \text{ cm}^3$.

In the metric system, cubic units may be used for either dry or liquid measure, although units such as liters and milliliters are usually used for liquid measures. By definition, a **liter**, symbolized L, equals, or is the capacity of, a cubic decimeter; that is, $1 \text{ L} = 1 \text{ dm}^3$. In the United States, L is the symbol for liter, but this is not universally accepted.

Because $1 \text{ L} = 1 \text{ dm}^3$ and $1 \text{ dm}^3 = 1000 \text{ cm}^3$, it follows that $1 \text{ L} = 1000 \text{ cm}^3$ and $1 \text{ cm}^3 = 0.001 \text{ L}$. Also, $0.001 \text{ L} = 1 \text{ milliliter} = 1 \text{ mL}$. Hence, $1 \text{ cm}^3 = 1 \text{ mL}$. These relationships are summarized in Figure 14-79 and Table 14-7.

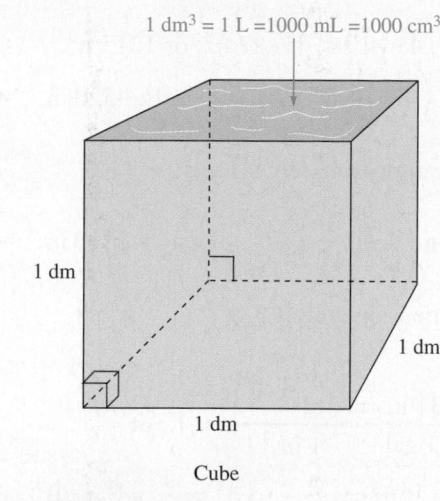

$1\ dm^3 = 1\ L = 1000\ mL = 1000\ cm^3$

$1\ cm^3 = 1\ mL$

1 dm

1 cm   1 cm
1 cm

1 dm

1 dm

Cube

Cube

Figure 14-79

Table 14-7

Unit	Symbol	Relation to Liter
kiloliter	kL	1000 L
*hectoliter	hL	100 L
*dekaliter	daL	10 L
liter	**L**	**1 L**
*deciliter	dL	0.1 L
centiliter	cL	0.01 L
milliliter	mL	0.001 L

*Not commonly used

EXAMPLE 14-19

Complete the following:

a. $27\ L = $ _____ mL
b. $362\ mL = $ _____ L
c. $3\ mL = $ _____ cm^3
d. $3\ m^3 = $ _____ L

Solution

a. $1\ L - 1000\ mL$, so $27\ L = 27 \cdot 1000\ mL = 27{,}000\ mL$. Alternatively,

$$27\ L = 27\ L \cdot \frac{1000\ mL}{1\ L} = 27{,}000\ mL.$$

b. $1\ mL = 0.001\ L$, so $362\ mL = 362(0.001\ L) = 0.362\ L$.
c. $1\ mL = 1\ cm^3$, so $3\ mL = 3\ cm^3$.
d. $1\ m^3 = 1000\ dm^3$ and $1\ dm^3 = 1\ L$, so $1\ m^3 = 1000\ L$ and $3\ m^3 = 3000\ L$.

Alternatively, $3\ m^3 = 3\ m^3 \cdot \dfrac{1000\ dm^3}{1\ m^3} \cdot \dfrac{1\ L}{1\ dm^3} = 3000\ L$.

Converting English Measures of Volume

Basic units of volume in the English system are the cubic foot ($1\ ft^3$), the cubic yard ($1\ yd^3$), and the cubic inch ($1\ in.^3$). In the United States, for liquid measure, 1 gallon (gal) $= 231\ in.^3$, which is about 3.8 L, and 1 quart (qt) $= \dfrac{1}{4}$ gal, or about $58\ in.^3$.

Relationships among the one-dimensional units enable us to convert from one unit of volume to another, as shown in the following example.

EXAMPLE 14-20

Convert each of the following, as indicated:

a. $45\ yd^3 = $ _____ ft^3
b. $4320\ in.^3 = $ _____ yd^3
c. $10\ gal = $ _____ ft^3
d. $3\ ft^3 = $ _____ yd^3

Solution

a. Because $1 \text{ yd}^3 = (3 \text{ ft})^3 = 27 \text{ ft}^3$, $45 \text{ yd}^3 = 45 \cdot 27 \text{ ft}^3$, or 1215 ft^3.

b. Because $1 \text{ in.} = \dfrac{1}{36} \text{ yd}$, $1 \text{ in.}^3 = \left(\dfrac{1}{36}\right)^3 \text{ yd}^3$. Consequently, $4320 \text{ in.}^3 =$

$4320\left(\dfrac{1}{36}\right)^3 \text{ yd}^3 \approx 0.0926 \text{ yd}^3$, or approximately 0.1 yd^3.

c. Because $1 \text{ gal} = 231 \text{ in.}^3$ and $1 \text{ in.}^3 = \left(\dfrac{1}{12}\right)^3 \text{ ft}^3$, $10 \text{ gal} = 2310 \text{ in.}^3 =$

$2310\left(\dfrac{1}{12}\right)^3 \text{ ft}^3 \approx 1.337 \text{ ft}^3$, or approximately 1.3 ft^3.

Alternatively, $10 \text{ gal} = 10 \text{ gal} \cdot \dfrac{231 \text{ in.}^3}{1 \text{ gal}} \cdot \dfrac{\left(\dfrac{1}{12}\right)^3 \text{ ft}^3}{1 \text{ in.}^3} \approx 1.3 \text{ ft}^3$

d. From (a), $1 \text{ ft}^3 = \dfrac{1}{27} \text{ yd}^3$. Hence, $3 \text{ ft}^3 = 3 \cdot \dfrac{1}{27} \text{ yd}^3 = \dfrac{1}{9} \text{ yd}^3 \approx 0.1 \text{ yd}^3$.

Volumes of Prisms and Cylinders

We have shown that the volume of a right rectangular prism, as shown in Figure 14-80, involves multiplying the area of the base times the height. If we denote the area of the base by B and the height by h, then $V = Bh$.

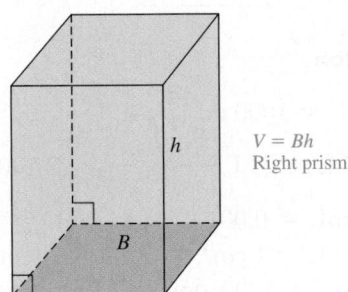

$V = Bh$
Right prism

Figure 14-80

Formulas for the volumes of many three-dimensional figures can be derived using the volume of a right prism. In Figure 14-81(a), a rectangular solid box has been sliced into thin layers. If the layers are shifted to form the solids in Figure 14-81(b) and (c), the volume of each of the three solids is the same as the volume of the original solid. This idea is the basis for **Cavalieri's Principle**.

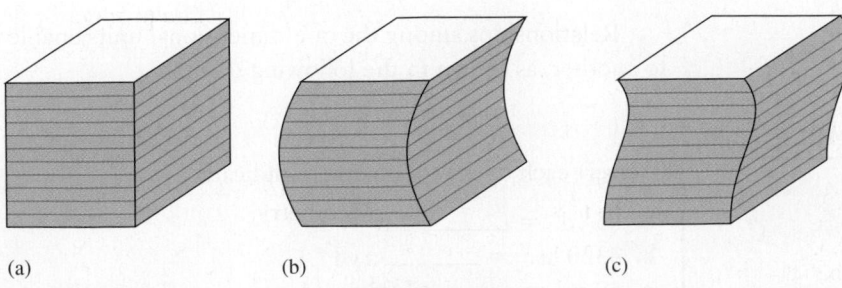

(a)   (b)   (c)

Figure 14-81

Cavalieri's Principle

Two solids each with a base in the same plane have equal volumes if every plane parallel to the bases intersects the solids in cross sections of equal area.

 NOW TRY THIS 14-19

a. The two right prisms in Figure 14-82 have the same height. How do their volumes compare? Explain why.

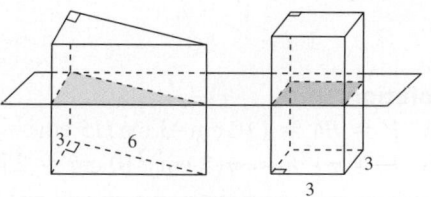

Figure 14-82

b. Consider the right prism and right circular cylinder in Figure 14-83(a) and (c) as stacks of papers. If the papers are shifted as shown in Figure 14-83(b) and (d), an oblique prism and an oblique cylinder, respectively, are formed.
   (i) Explain how the volume of the oblique prism is related to the volume of the right prism.
   (ii) Explain how the volume of the oblique cylinder is related to the volume of the right cylinder.

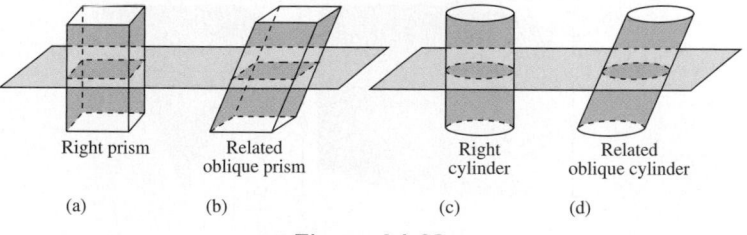

Right prism   Related   Right   Related
                   oblique prism   cylinder   oblique cylinder

(a)   (b)   (c)   (d)

Figure 14-83

Any oblique prism can be thought of as a stack of thin cards, all shaped like the base of the solid. The oblique stack can be straightened to form a right prism with the same volume. These prisms have the same height, h, and base area, B, so *the volume of any prism is given* by $V = Bh$.

The volume of a cylinder can be approximated using prisms with increasing numbers of sides in their bases. The volume of each prism is the product of the area of the base and the height. Similarly, the **volume, V, of a cylinder** is the product of the area of the base B and the height h. If the base is a circle of radius r, and the height of the cylinder is h, then $V = Bh = \pi r^2 h$.

 Historical Note

Bonaventura Cavalieri (1598–1647), an Italian mathematician and disciple of Galileo, contributed to the development of geometry, trigonometry, and algebra in the Renaissance. He became a Jesuit at an early age and later, after reading Euclid's *Elements*, was inspired to study mathematics. In 1629, Cavalieri became a professor at Bologna and held that post until his death. Cavalieri is best known for his principle concerning the volumes of solids.

EXAMPLE 14-21

Find the volume of each figure in Figure 14-84.

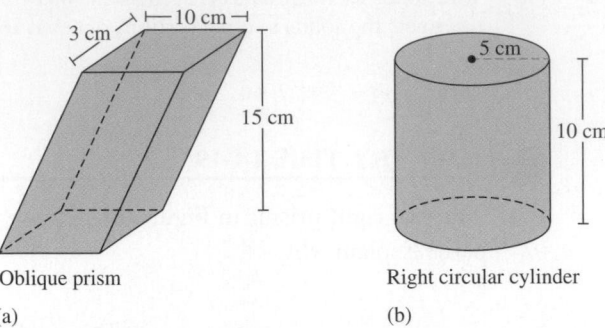

Oblique prism   Right circular cylinder
(a)   (b)

Figure 14-84

Solution

a. $V = Bh = (10 \text{ cm} \cdot 3 \text{ cm})15 \text{ cm} = 450 \text{ cm}^3$

b. $V = \pi r^2 h = \pi(5 \text{ cm})^2 10 \text{ cm} = 250\pi \text{ cm}^3$

Volumes of Pyramids and Cones

Figure 14-85(a) shows a right prism and a right pyramid with congruent bases and equal heights; h marks the perpendicular distance between the bases of the prism and cylinder and the perpendicular distance between the apex and bases in the pyramid and cone; it is the height or altitude in all.

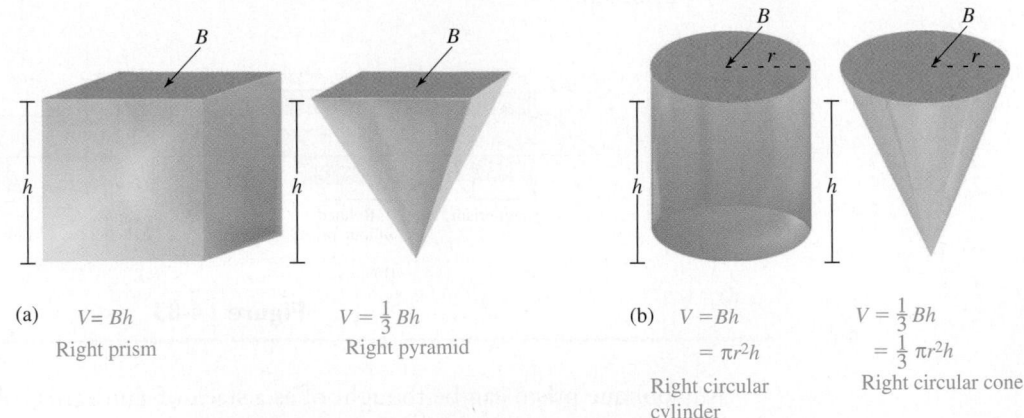

(a) $V = Bh$ $V = \frac{1}{3} Bh$ (b) $V = Bh$ $V = \frac{1}{3} Bh$
    Right prism   Right pyramid $= \pi r^2 h$ $= \frac{1}{3} \pi r^2 h$
                                                    Right circular   Right circular cone
                                                    cylinder

Figure 14-85

How are the volumes of the containers in Figure 14-85 related? Students may explore the relationship by filling the pyramid with water, sand, or rice and pouring the contents into the prism. They should find that it takes three full pyramids to fill the prism. Therefore the volume of the pyramid is equal to one-third the volume of the prism. This relationship between prisms and pyramids with congruent bases and heights, respectively, is true in general; that is, for a pyramid $V = (1/3)Bh$, where B is the area of the base and h is the height. The same relationship holds between the volume of a cone and the volume of a cylinder, where they have congruent bases and equal heights, as shown in Figure 14-85(b). Therefore, the volume of a right circular cone is given by $V = (1/3)Bh$, or $V = (1/3)\pi r^2 h$.

This relationship between the volume of a cone and the volume of a cylinder with congruent bases and congruent heights is explored on the student page on page 897. Answer parts A–D.

School Book Page CONES AND CYLINDERS

Problem 4.2 **Cones and Cylinders, Pyramids and Cubes**

- Roll a piece of stiff paper into a cone shape so that the tip touches the bottom of the cylinder you made in Problem 4.1.

- Tape the cone shape along the seam. Trim the cone so that it is the same height as the cylinder.

- Fill the cone to the top with sand or rice, and empty the contents into the cylinder. Repeat this as many times as needed to fill the cylinder completely.

A. What is the relationship between the volume of the cone and the volume of the cylinder?

B. Suppose a cylinder, a cone, and a sphere have the same radius and the same height. What is the relationship between the volumes of the three shapes?

C. Suppose a cone, a cylinder, and a sphere all have the same height, and that the cylinder has a volume of 64 cubic inches. How do you use the relationship in Question B to find

   1. the volume of a sphere whose radius is the same as the cylinder?

   2. the volume of a cone whose radius is the same as the cylinder?

D. Suppose the radius of a cylinder, a cone, and a sphere is 5 centimeters and the height of the cylinder and cone is 8 centimeters. Find the volume of the cylinder, cone, and sphere.

Investigation 4 Cones, Spheres, and Pyramids   **51**

Another way to determine the volume of a pyramid in terms of a prism is to start with a cube and draw three diagonals from one vertex to the non-adjacent vertices of one of the opposite faces as shown in figure 14-86(a). We can see that there are three pyramids formed inside the cube, as shown in Figure 14-86(b).

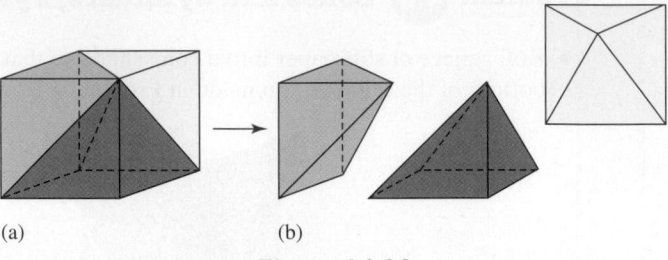

(a)   (b)

Figure 14-86

The three pyramids are identical in size and shape, do not overlap, and their union is the whole cube. Therefore, each of these pyramids has a volume one-third that of the cube. Once again we see that for a pyramid $V = (1/3)Bh$, where B is the area of the base and h is the height. This can be demonstrated by building three paper models of the pyramids and fitting them together into a prism. A net that can be enlarged and used for the construction is given in Figure 14-87.

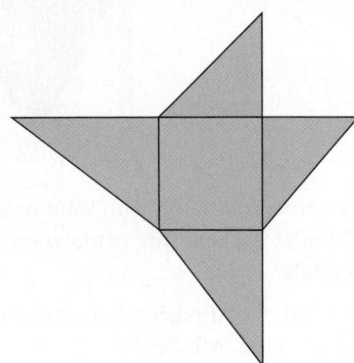

Figure 14-87

EXAMPLE 14-22

Find the volume of each figure in Figure 14-88.

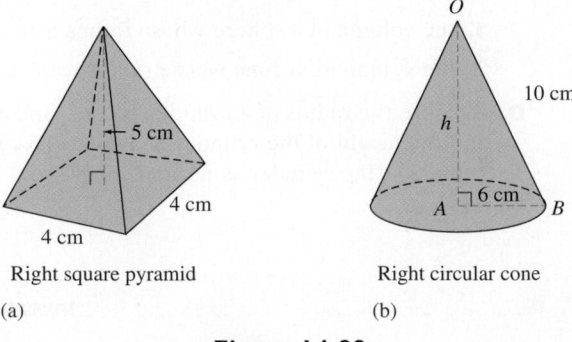

Right square pyramid

(a)

Right circular cone

(b)

Figure 14-88

Solution

a. The figure is a pyramid with a square base, whose area is $4 \text{ cm} \cdot 4 \text{ cm}$ and whose height is 5 cm. Hence, $V = \frac{1}{3} Bh = \frac{1}{3}(4 \text{ cm} \cdot 4 \text{ cm})(5 \text{ cm}) = \frac{80}{3} \text{ cm}^3$.

b. The base of the cone is a circle of radius 6 cm. Because the volume of the cone is given by $V = \frac{1}{3} \pi r^2 h$, we need to know the height. In the right triangle OAB, $OA = h$ and by the Pythagorean theorem, $h^2 + 6^2 = 10^2$. Hence, $h^2 = 100 - 36$, or 64, and $h = 8$ cm. Thus, $V = \frac{1}{3} \pi r^2 h = \frac{1}{3} \pi (6 \text{ cm})^2 (8 \text{ cm}) = 96\pi \text{ cm}^3$.

EXAMPLE 14-23

Figure 14-89 is a net for a pyramid. If each triangle is equilateral, find the volume of the pyramid.

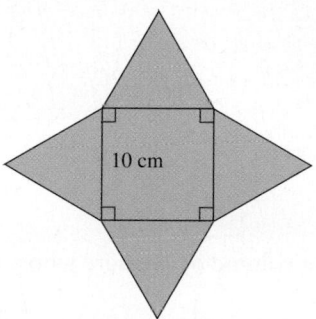

10 cm

Figure 14-89

Solution The pyramid obtained from the folded model is shown in Figure 14-90. The volume of the pyramid is $V = \frac{1}{3} Bh = \frac{1}{3} \cdot 10^2 h$. We must find h. Notice that h is a leg in the right triangle EOF, where F is the midpoint of $\overline{CB}$. We know that $OF = 5$ cm. If we knew EF, we could find h by applying the Pythagorean theorem to $\triangle EOF$. To find the length of $\overline{EF}$, notice that $\overline{EF}$ is a leg in the right triangle EBF. ($\overline{EF}$ is the perpendicular bisector of $\overline{BC}$ in the equilateral triangle BEC.) In the right triangle EBF, we have $(EB)^2 = (BF)^2 + (EF)^2$. Because $EB = 10$ cm and $BF = 5$ cm, it follows that $10^2 = 5^2 + (EF)^2$, or $EF = \sqrt{75}$ cm. In $\triangle EOF$, we have $h^2 + 5^2 = (EF)^2$, or $h^2 + 25 = 75$. Thus, $h = \sqrt{50}$ cm and $V \approx \frac{1}{3} \cdot 10^2 \cdot \sqrt{50} = \left(\frac{100}{3} \right) \cdot \sqrt{50}$ cm^3 or approximately 235.7 cm^3.

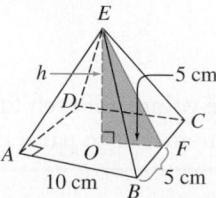

Figure 14-90

Volume of a Sphere

To find the volume of a sphere, imagine that a sphere is composed of a great number of right regular congruent pyramids with apexes at the center of the sphere and that the vertices of the base touch the sphere, as shown in Figure 14-91. If the pyramids have very small bases, then the height of each pyramid is nearly the radius r. Hence, the volume of each pyramid is $\frac{1}{3}Bh$ or $\frac{1}{3}Br$, where B is the area of the base. If there are n pyramids each with base area B, then the total volume of the pyramids is $V = \frac{1}{3}nBr$. Because nB is the total surface area of all the bases of the pyramids and because the sum of the areas of all the bases of the pyramids is very close to the surface area of the sphere, $4\pi r^2$, the volume of the sphere is given by $V = \frac{1}{3}(4\pi r^2)r = \frac{4}{3}\pi r^3$.

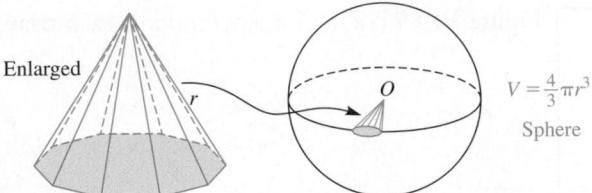

Figure 14-91

EXAMPLE 14-24

Find the volume of a sphere whose radius is 6 cm.

Solution $V = \frac{4}{3}\pi(6 \text{ cm})^3 = \frac{4}{3}\pi(216 \text{ cm}^3) = 288\pi \text{ cm}^3$

NOW TRY THIS 14-20

A cylinder, a cone, and a sphere have the same radius and same height, as shown in Figure 14-92.

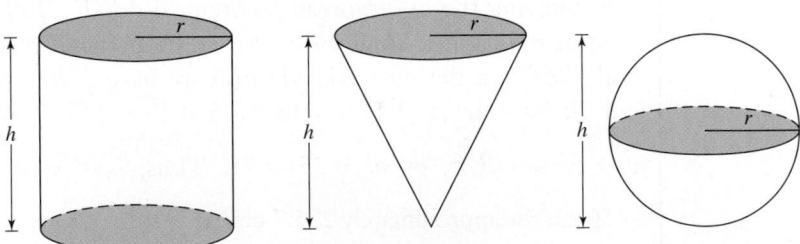

Figure 14-92

a. Find the volume of each figure in terms of r and h.
b. Use your answer to part (a) to show that the volumes are in the ratio 3:1:2.

Comparing Measurements of Similar Figures

Two planar figures are similar if they have the same shape but not necessarily the same size. As discussed in Chapter 12, the ratio of the corresponding side lengths is the *scale factor*, which we refer to as k. In Figure 14-93, $\triangle ABC \sim \triangle A_1B_1C_1$ with scale factor k; that is $\frac{a}{a_1} = \frac{b}{b_1} = \frac{c}{c_1} = k$.

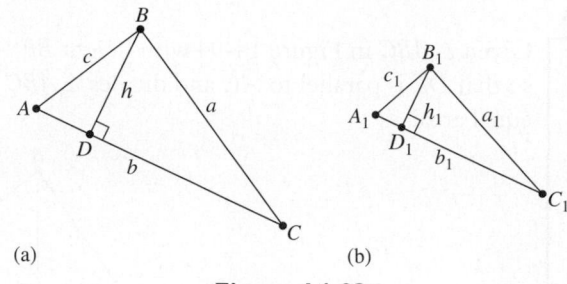

Figure 14-93

Also, because $\triangle ABD \sim \triangle A_1 B_1 D_1$, (Why?) we have $\dfrac{h}{h_l} = \dfrac{c}{c_i} = k$. Therefore the heights of the two similar triangles also have the same scale factor, k. This is true in general and is summarized in Theorem 14-6.

Theorem 14-6

The ratio of any *linear measurement* of two similar figures, such as, length, width, height, perimeter, diagonal, diameter, and slant height, has the same scale factor k.

Next we find the areas of the two similar triangles in Figure 14-93 and find the scale factor for the two areas.

$$\frac{Area(\triangle ABC)}{Area(\triangle A_1 B_1 C_1)} = \frac{(bh)/2}{(b_1 h_1)/2} = \frac{b}{b_1} \cdot \frac{h}{h_1} = k \cdot k = k^2$$

We have proved the following theorem.

Theorem 14-7

For similar triangles with scale factor k, the ratio of their areas is k^2.

Because any two similar polygons can be subdivided into nonoverlapping similar triangles, the analogous theorem for similar polygons holds.

Theorem 14-8

For similar polygons with scale factor k, the ratio of their areas is k^2.

EXAMPLE 14-25

Is it true that the viewing area of a 35-inch television screen is about twice the viewing area of a 25-inch television screen? Recall that television screens are measured diagonally.

Solution   Because we are dealing with areas of similar polygons, than the areas are in the ratio of $35^2 : 25^2$ or $1225 : 625$ or $1.96 : 1$. Therefore the viewing area of the 35-inch screen is almost twice that of a 25-inch screen.

EXAMPLE 14-26

Given $\triangle ABC$ in Figure 14-94 with height $BF = 10$ cm, how far above $\overline{AC}$ should $\overline{DE}$ be drawn so that $\overline{DE}$ is parallel to $\overline{AC}$ and divides $\triangle ABC$ into two regions (a triangle and a trapezoid) of equal area?

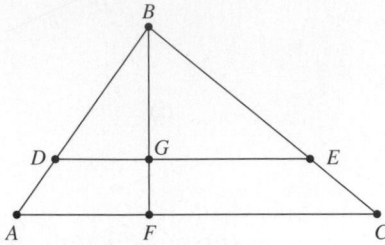

Figure 14-94

Solution   We need the area of $\triangle DBE$ to be half the area of $\triangle ABC$. When $\overline{DE}$ is drawn parallel to $\overline{AC}$ we have $\triangle DBE \sim \triangle ABC$. (Why?) We know that the ratio of areas of the triangles is the square of the scale factor for the corresponding sides or heights. Thus,

$$\left(\frac{BG}{BF}\right)^2 = \frac{1}{2}$$

$$\left(\frac{BG}{10}\right)^2 = \frac{1}{2}$$

$$\frac{BG}{10} = \sqrt{\frac{1}{2}} = \frac{1}{\sqrt{2}}$$

$$BG = \frac{10}{\sqrt{2}}$$

Consequently, $GF = 10 - BG = 10 - \dfrac{10}{\sqrt{2}}$ cm ≈ 2.93 cm. Therefore $\overline{DE}$ should be drawn $10 - \dfrac{10}{\sqrt{2}}$ cm ≈ 2.93 cm above $\overline{AC}$.

Next we investigate the ratio of the volumes of two three-dimensional figures. Consider the two similar rectangular prisms shown in Figure 14-95 with scale factor 3. They have the same shape but not the same size.

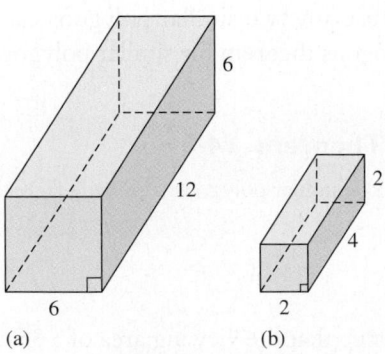

Figure 14-95

The surface areas and volumes of the two rectangular prisms in Figure 14-95 are:

$SA_{(a)} = 2(6 \cdot 6) + 2(6 \cdot 12) + 2(6 \cdot 12) = 360$ cm $SA_{(b)} = 2(2 \cdot 2) + 2(4 \cdot 2) + 2(4 \cdot 2) = 40$ cm

$V_{(a)} = Bh = 36 \cdot 12 = 432$ cm^2 $V_{(b)} = Bh = 4 \cdot 4 = 16$ cm^2

The ratios of the surface areas and the volumes for the two prisms are given below.

$$\frac{SA_{(a)}}{SA_{(b)}} = \frac{360}{40} = 9 \text{ or } 3^2 \qquad \frac{V_{(a)}}{V_{(b)}} = \frac{432}{16} = 27 = 3^3$$

This example suggests that if two similar prisms have scale factor k, the ratio of their surface areas is k^2, and the ratio of their volumes is k^3. This is true in general for similar three-dimensional figures and is summarized in Theorem 14-9.

> ### Theorem 14-9 Similarity Principle of Measurement:
> If the scale factor of two similar figures is k, then the ratio of their areas or surface areas is k^2, and the ratio of their volumes is k^3.

EXAMPLE 14-27

a. How does the surface area of a sphere 10 in. in diameter compare with the surface area of a sphere 5 in. in diameter?

b. How do the volumes of the spheres in part (a) compare?

Solution

a. Any two spheres are similar. The ratio of the diameters is $10:5$ or 2. By Theorem 14-9, the ratio of the surface areas of the spheres is 2^2 or 4. Therefore the 10-in. sphere has 4 times the surface area of the 5-in. sphere.

b. By Theorem 14-9 the ratio of the volumes is 2^3 or 8. Therefore the volume of the 10-in. sphere is 8 times that of the 5-in. sphere.

Problem Solving · Volume Comparisons: Cylinders and Boxes

A metal can manufacturer has a large quantity of rectangular metal sheets 20 cm × 30 cm. Without cutting the sheets, the manufacturer wants to make cylindrical pipes with circular cross sections from some of the sheets and box-shaped pipes with square cross sections from the other sheets. The volume of the box-shaped pipes is to be greater than the volume of the cylindrical pipes. Is this possible? If so, how would the pipes be made and what are their volumes?

Understanding the Problem We are to use 20 cm × 30 cm rectangular sheets of metal to make some cylindrical pipes as well as some box-shaped pipes with square cross sections that have a greater volume than do the cylindrical pipes. We need to determine if this is possible and if so, design the pipes and compute the volumes.

Figure 14-96 shows a sheet of metal and two sections of pipe made from it, one cylindrical and the other box-shaped. A model for such pipes can be designed from a piece of paper by bending it into a right circular cylinder or by folding it into a right rectangular prism, as shown in the figure.

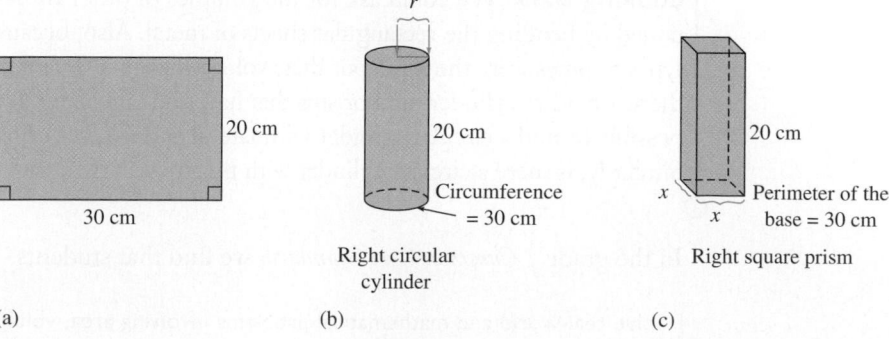

Figure 14-96

Devising a Plan If we compute the volume of the cylinder in Figure 14-96(b) and the volume of the prism in Figure 14-96(c), we can determine which is greater. If the prism has a greater volume, the solution of the problem will be complete. Otherwise, we look for other ways to design the pipes before concluding that a solution is impossible.

To compute the volume of the cylinder, we find the area of the base. The area of the circular base is πr^2. To find r, we note that the circumference of the circle, $2\pi r$, is 30 cm. Thus, $r = \dfrac{30}{2\pi}$, and the area of the circle is $\pi r^2 = \pi\left(\dfrac{30}{2\pi}\right)^2 = \dfrac{900\pi}{4\pi^2} = \dfrac{225}{\pi}$.

With the given information, we can also find the area of the base of the rectangular box. Because the perimeter of the base of the prism is $4x$, we have $4x = 30$, or $x = 7.5$ cm. Thus, the area of the square base is $x^2 = (7.5)^2$, or 56.25 cm^2.

Carrying Out the Plan Denoting the volume of the cylindrical pipe by V_1 and the volume of the box-shaped pipe by V_2, we have $V_1 = \left(\dfrac{225}{\pi}\right) \cdot 20$, or approximately 1432.4 cm^3. For the volume of the box-shaped pipe, we have $V_2 = 56.25 \cdot 20$, or 1125 cm^3. We see that in the first design for the pipes, the volume of the cylindrical pipe is greater than the volume of the box-shaped pipe. This is not the required outcome.

Rather than bend the rectangular sheet of metal along the 30 cm side, we could bend it along the 20 cm side to obtain either pipe, as shown in Figure 14-97. Denoting the radius of the cylindrical pipe by r, the side of the box-shaped pipe by y, and their volumes by V_3 and V_4, respectively, we have $V_3 = \pi r^2 \cdot 30 = \pi[20/(2\pi)]^2 30 = (10^2 \cdot 30)/\pi$, or approximately 954.9 cm^3. Also, $V_4 = y^2 30 = \left(\dfrac{20}{4}\right)^2 30 = 25 \cdot 30$, or 750 cm^3. Because $V_2 = 1125$ cm^3 and $V_3 \approx 945.9$ cm^3, we see that the volume of the box-shaped pipe with an altitude of 20 cm is greater than the volume of the cylindrical pipe with an altitude of 30 cm. We could order the volumes from greatest to least for the four shapes.

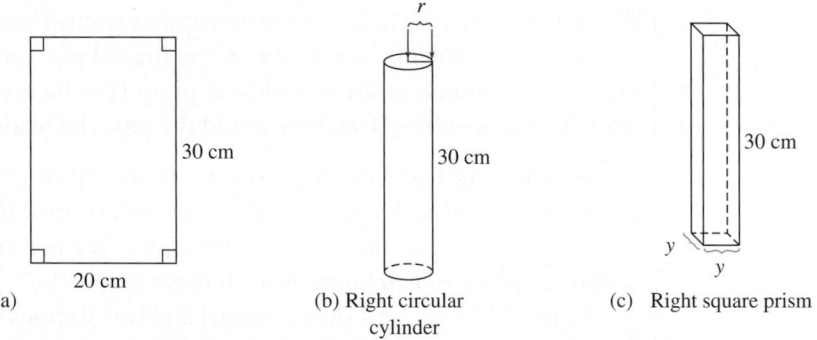

(a)   (b) Right circular cylinder   (c) Right square prism

Figure 14-97

Looking Back We could ask for the volumes of other three-dimensional objects that can be obtained by bending the rectangular sheets of metal. Also, because the lateral surface areas of the four types of pipes were the same but their volumes were different, we might want to investigate whether there are other cylinders and prisms that have the same lateral surface area and the same volume. Is it possible to find a circular cylinder with lateral surface area of 600 cm^2 and smallest possible volume? Similarly, is there a circular cylinder with the given surface area and greatest possible volume?

In the grade 7 *Common Core Standards* we find that students:

> Solve real-world and mathematical problems involving area, volume and surface area of two- and three-dimensional objects composed of triangles, quadrilaterals, polygons, cubes, and right prisms. (p. 50)

 On the partial student page that follows we see an example of the type of problem called for in the *Common Core Standards*. Work the problems involving the Hubble telescope described on the student page.

School Book Page PROBLEM SOLVING

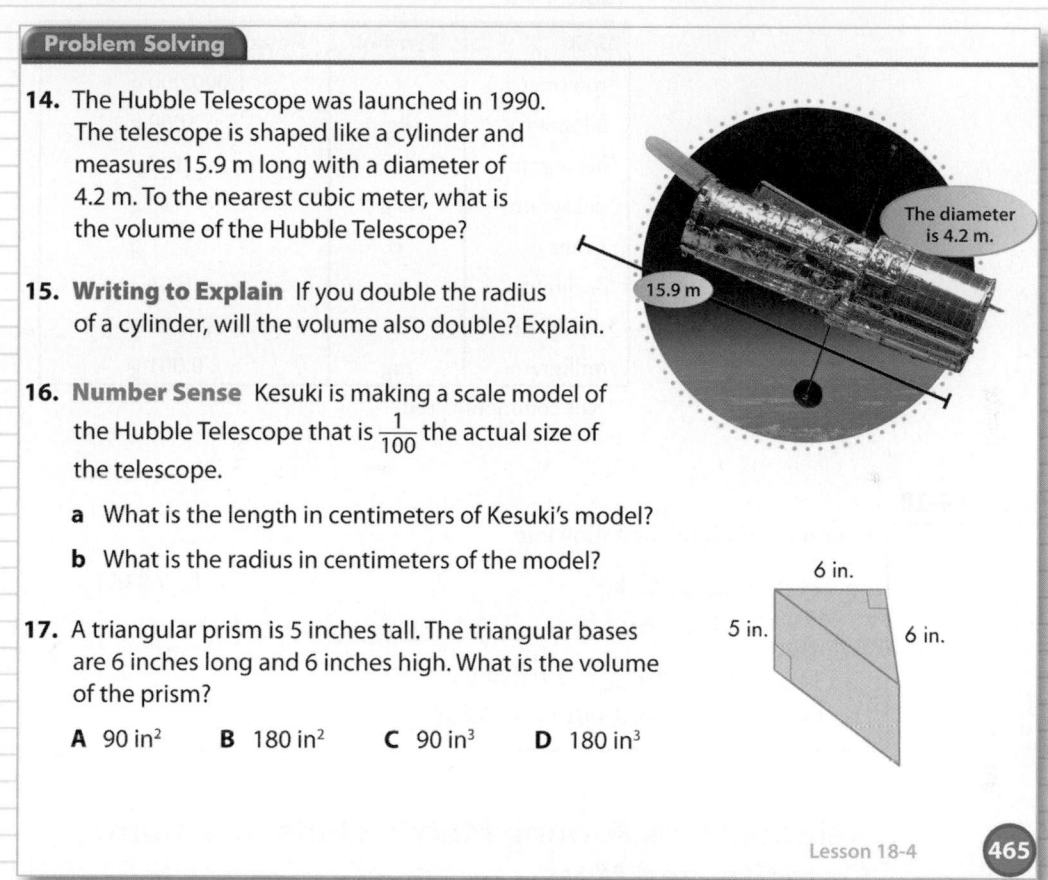

Problem Solving

14. The Hubble Telescope was launched in 1990. The telescope is shaped like a cylinder and measures 15.9 m long with a diameter of 4.2 m. To the nearest cubic meter, what is the volume of the Hubble Telescope?

15. Writing to Explain If you double the radius of a cylinder, will the volume also double? Explain.

16. Number Sense Kesuki is making a scale model of the Hubble Telescope that is $\frac{1}{100}$ the actual size of the telescope.

a What is the length in centimeters of Kesuki's model?

b What is the radius in centimeters of the model?

17. A triangular prism is 5 inches tall. The triangular bases are 6 inches long and 6 inches high. What is the volume of the prism?

A 90 in² **B** 180 in² **C** 90 in³ **D** 180 in³

The diameter is 4.2 m.

15.9 m

6 in.

5 in. 6 in.

Lesson 18-4 **465**

Mass

Three centuries ago, Isaac Newton pointed out that in everyday life the word *weight* is used for what is really mass. **Mass** is a quantity of matter, as opposed to **weight**, which is a force exerted by gravitational pull. When astronauts are in orbit above Earth, their weights have changed even though their masses remain the same. In common parlance on Earth, *weight* and *mass* are used interchangeably. In the English system, weight is measured in avoirdupois units such as tons, pounds, and ounces. One pound (lb) equals 16 ounces (oz) and 2000 lb equals 1 English ton.

In the metric system, a fundamental unit for mass is the **gram**, denoted g. An ordinary paper clip or a thumbtack each has a mass of about 1 g. As with other base metric units, prefixes are added to *gram* to obtain other units. For example, a kilogram (kg) is 1000 g. Two standard loaves of bread have a mass of about 1 kg. A person's mass is measured in kilograms. A newborn baby has a mass of about 3 kg. Another unit of mass is the *metric ton* (t), which is equal to 1000 kg. The metric ton is used to record the masses of objects such as cars and trucks. A small car has a mass of about 1 t. *Mega* (1,000,000) and *micro* (0.000001) are other prefixes used with gram.

Table 14-8 lists metric units of mass. Conversions that involve metric units of mass are handled in the same way as conversions that involve metric units of length.

Table 14-8

Unit	Symbol	Relationship to Gram
ton (metric)	t	1,000,000 g
kilogram	kg	1000 g
*hectogram	hg	100 g
*dekagram	dag	10 g
gram	**g**	**1 g**
*decigram	dg	0.1 g
*centigram	cg	0.01 g
milligram	mg	0.001 g

*Not commonly used

**EXAMPLE 14-28**

Complete each of the following:

a. $34 \text{ g} = $ _____ kg   **b.** $6836 \text{ kg} = $ _____ t

Solution

a. $34 \text{ g} = 34(0.001 \text{ kg}) = 0.034 \text{ kg}$
b. $6836 \text{ kg} = 6836(0.001 \text{ t}) = 6.836 \text{ t}$

Relationships Among Metric Units of Volume, Capacity, and Mass

The relationships among the units of volume, capacity, and mass in the metric system is illustrated in Figure 14-98. Does the English system have any relationships among units that are easy to remember?

1 cm³ (1 mL) of water   1 g

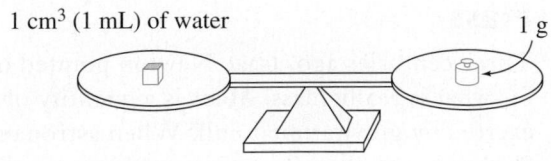

1 dm³ (1 L) of water

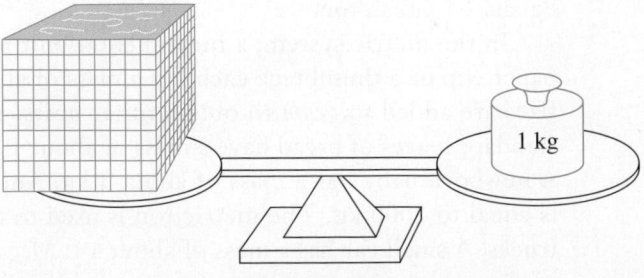

Figure 14-98

NOW TRY THIS 14-21

Find the following relationships:
  a. 1 cm^3 of water has a mass of 1 _____.
  b. 1 dm^3 of water has a mass of 1 _____.
  c. 1 L of water has a volume of 1 _____.
  d. 1 cm^3 of water has a capacity of 1 _____.
  e. 1 mL of water has a mass of 1 _____.
  f. 1 m^3 of water has a capacity of 1 _____.
  g. 1 m^3 of water has a mass of 1 _____.

EXAMPLE 14-29

A waterbed measures 180 cm × 210 cm × 20 cm.

  a. Approximately how many liters of water can it hold?
  b. What is its mass in kilograms when it is full of water?

Solution
  a. The volume of the waterbed (which is close to a rectangular prism) is approximated by multiplying the length ℓ times the width w times the height h.

$$V = \ell w h$$
$$= 180 \text{ cm} \cdot 210 \text{ cm} \cdot 20 \text{ cm}$$
$$= 756{,}000 \text{ cm}^3, \text{ or } 756{,}000 \text{ mL}$$

Because 1 mL = 0.001 L, the volume is 756 L.
  b. Because 1 L of water has a mass of 1 kg, 756 L of water has a mass of 756 kg, which is 0.756 t.

To see one advantage of the metric system, suppose the bed in Example 14-29 is 6 ft × 7 ft × 9 in. Try to approximate the volume in gallons and the weight of the water in pounds. In which system would you rather compute?

Temperature

For normal temperature measurements in the metric system, the base unit is the **degree Celsius**, named for Anders Celsius, a Swedish scientist. The Celsius scale has 100 equal divisions between 0 degrees Celsius (0°C), the freezing point of water, and 100 degrees Celsius (100°C), the boiling point of water, as seen in Figure 14-99. The **kelvin** (K) temperature scale is an extension of the degree Celsius scale down to *absolute zero*, a hypothetical temperature characterized by a complete absence of heat energy. The freezing point of water on this scale is 273.15 kelvins. In the English system, the Fahrenheit scale has 180 equal divisions between 32°F, the freezing point of water, and 212°F, the boiling point of water.

Figure 14-99 gives other temperature comparisons of the two scales and further illustrates the relationship between them. Because the Celsius scale has 100 divisions between the freezing point and the boiling point of water, whereas the Fahrenheit scale has 180 divisions, the relationship between the two scales is 100 to 180, or 5 to 9. For every 5 degrees on the Celsius scale, there are 9 degrees on the Fahrenheit scale, and for each degree on the Fahrenheit scale, there is $\frac{5}{9}$ degree on the Celsius scale. Because the ratio between the number of degrees above freezing on the Celsius scale and the number of degrees above freezing on the Fahrenheit scale remains the same and equals $\frac{5}{9}$, we may convert temperature from one system to the other.

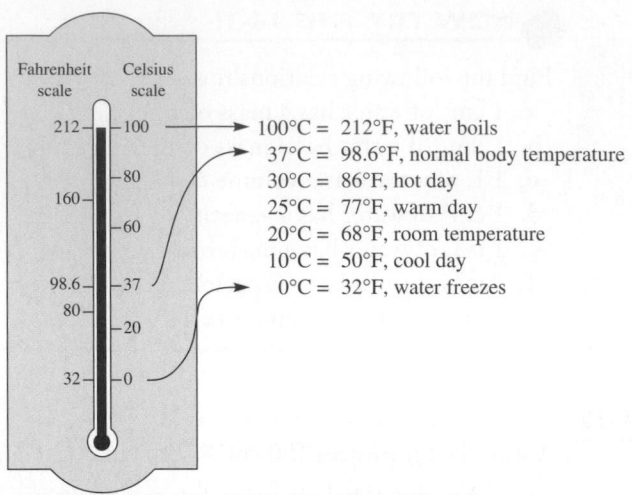

Figure 14-99

For example, suppose we want to convert 50° on the Fahrenheit scale to the corresponding number on the Celsius scale. On the Fahrenheit scale, 50° is 50 − 32, or 18°, above freezing, but on the Celsius scale, it is $\frac{5}{9} \cdot 18$, or 10°, above freezing. Because the freezing temperature on the Celsius scale is 0°, 10° above freezing is 10° Celsius. Thus, 50°F = 10°C. In general, F degrees is F − 32 above freezing on the Fahrenheit scale, but only $\frac{5}{9}(F - 32)$ above freezing on the Celsius scale. Thus, we have the relation $C = \frac{5}{9}(F - 32)$. If we solve the equation for F, we obtain $F = \frac{9}{5}C + 32$.

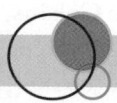 **NOW TRY THIS 14-22**

Does it ever happen that the temperature measured in Celsius degrees is the same if it is measured in Fahrenheit degrees? If so, when?

Assessment 14-5A

1. Complete each of the following:
   a. 8 m³ = _____ dm³
   b. 675,000 m³ = _____ km³
   c. 7000 mm³ = _____ cm³
   d. 400 in.³ = _____ yd³
   e. 0.2 ft³ = _____ in.³
2. If a faucet is dripping at the rate of 15 drops/min and there are 20 drops/mL, how many liters of water are wasted in a 30-day month?
3. The Great Pyramid of Cheops has a square base of 771 ft on a side and a height of 486 ft. How many rooms 35 ft × 20 ft × 8 ft would be needed to have a volume equivalent to that of the Great Pyramid?

4. Find the volume of each of the following:

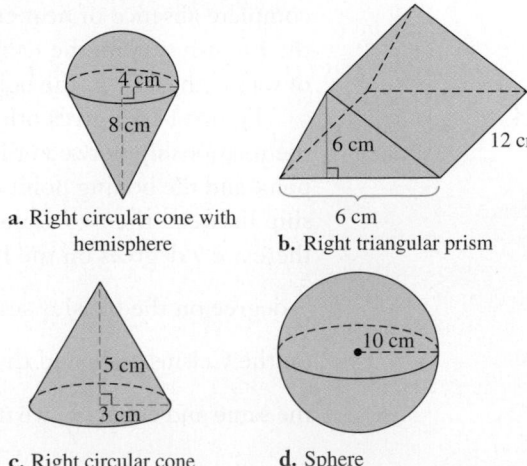

a. Right circular cone with hemisphere

b. Right triangular prism

c. Right circular cone

d. Sphere

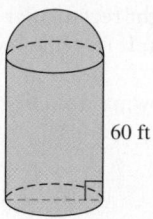

e. Right circular cylinder
with hemisphere

60 ft

20 ft

5. Complete the following table:

	a.	b.	c.	d.	e.	f.
cm³		500			750	4800
dm³	2					
L			1.5			
mL				5000		

6. Place a decimal point in each of the following to make it an accurate sentence:
   a. A paper cup holds about 2000 mL.
   b. A regular soft-drink bottle holds about 320 L.
   c. A quart milk container holds about 10 L.
   d. A teaspoonful of cough syrup is about 500 mL.
7. Two cubes have sides of lengths 4 cm and 6 cm, respectively. What is the ratio of their volumes?
8. What happens to the volume of a sphere if the radius is doubled?
9. Complete the following table for right rectangular prisms with the given dimensions:

	a.	b.	c.	d.
Length	20 cm	10 cm	2 dm	15 cm
Width	10 cm	2 dm	1 dm	2 dm
Height	10 cm	3 dm		
Volume (cm³)				
Volume (dm³)				7.5
Volume (L)			4	

10. Earth's diameter is approximately 4 times the Moon's and both bodies are spheres. What is the ratio of their volumes?
11. An Olympic-sized pool is in the shape of a right rectangular prism is 50 m × 25 m. If it is 2 m deep throughout, how many liters of water does it hold?
12. A standard drinking straw is 25 cm long and 4 mm in diameter. How much liquid can be held in the straw at one time?
13. **a.** What happens to the volume of an aquarium that is in the shape of a right rectangular prism if the length, width, and height are all doubled?
    b. What happens to the volume of the aquarium if all the measurements are tripled.
    c. When you multiply each linear dimension of the aquarium by a positive value n, what happens to the volume?

14. The Great Pyramid of Cheops is a right square pyramid with height of 148 m and a square base with a perimeter of 940 m. The Transamerica Building in San Francisco has the basic shape of a right square pyramid that has a square base with a perimeter of 140 m and a height of 260 m. Which one has the greater volume and by how many times as great?
15. A right circular cone-shaped paper water cup has a height of 8 cm and a radius of 4 cm. If the cup is filled with water to half its height, what portion of the volume of the cup is filled with water?
16. If each edge of a cube is increased by 30%, by what percent does the volume increase?
17. A tennis ball can in the shape of a right circular cylinder holds three tennis balls snugly. If the radius of a tennis ball is 3.5 cm, what percentage of the tennis ball can outside the tennis balls is occupied?
18. A box is packed with six soda cans, as in the following figure. What percentage of the volume of the interior of the box is not occupied by the cans?

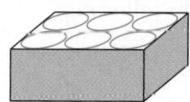

19. A right rectangular prism with base $ABCD$ at the bottom is shown in the following figure:

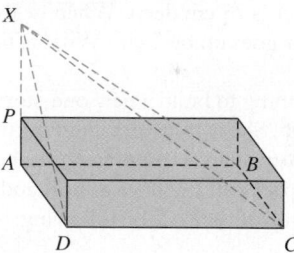

Suppose X is drawn so that $AX = 3AP$, where AP is the height of the prism, and X is connected to A, B, C, and D to form a pyramid. How do the volumes of the pyramid and the prism compare?
20. An engineer is to design a square-based pyramid whose volume is to be 100 m³.
    a. Find the dimensions (the length of a side of the square and the altitude) of one such pyramid.
    b. How many (noncongruent) such pyramids are possible? Why?
21. For each of the following, select the most appropriate metric unit of measure (gram, kilogram, or metric ton):
    a. Car
    b. Adult
    c. Can of frozen orange juice
    d. Elephant
22. For each of the following, choose the correct unit (milligrams, grams, or kilograms) to make each sentence reasonable:
    a. A staple has a mass of about 340 _____.
    b. A professional football player has a mass of about 110 _____.
    c. A vitamin tablet has a mass of about 1100 _____.

23. Complete each of the following:
    a. 15,000 g = _____ kg
    b. 0.036 kg = _____ g
    c. 4320 mg = _____ g
    d. 0.03 t = _____ kg
    e. 25 oz = _____ lb   (16 oz = 1 lb)
24. A paper dollar has a mass of approximately 1 g. Is it possible to lift $1,000,000 in the following denominations?
    a. $1 bills
    b. $10 bills
    c. $100 bills
    d. $1000 bills
    e. $10,000 bills

25. A fish tank, which is a right rectangular prism, is 40 cm × 20 cm × 20 cm. If it is filled with water, what is the mass of the water?
26. Convert each of the following from degrees Fahrenheit to the nearest integer degree Celsius:
    a. 10°F
    b. 30°F
    c. 212°F
27. Answer each of the following:
    a. The thermometer reads 20°C. Can you go snow skiing?
    b. Your body temperature is 39°C. Are you ill?
    c. The temperature reads 35°C. Should you go water skiing?
    d. It's 30°C in the room. Are you comfortable, hot, or cold?

Assessment 14-5B

1. Complete each of the following:
    a. 500 cm³ = _____ m³
    b. 3 m³ = _____ cm³
    c. 0.002 m³ = _____ cm³
    d. 25 yd³ = _____ ft³
    e. 1200 in.³ = _____ ft³
2. Jeremy has a fish tank that has a 40 cm by 70 cm rectangular base. The water is 25 cm deep. When he drops rocks into the tank, the water goes up by 2 cm. What is the volume in liters of the rocks?
3. Maggie is planning to build a new one-story house with floor area of 2000 ft². She is thinking about putting in a 9-ft ceiling instead of an 8-ft ceiling. If she does this, how many more cubic feet of space will she have to heat and cool?
4. Find the volume of each of the following:

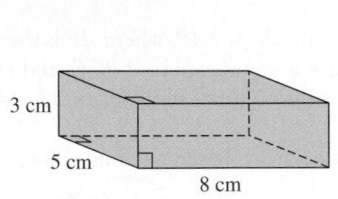

a. Right rectangular prism

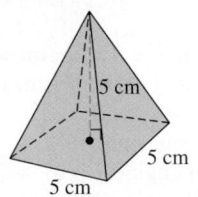

b. Right square pyramid

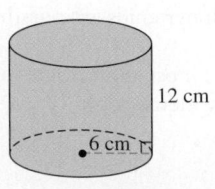

c. Right circular cylinder

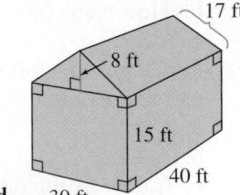

d.

e.

5. Complete the following chart:

	a.	b.	c.	d.	e.	f.
cm³		200			202	6500
dm³	6					
L				3		
mL			1200			

6. Place a decimal point in each of the following to make it an accurate sentence.
    a. A rectangular block that is 20 cm × 10 cm × 10 cm would displace 20,000 mL of water.
    b. A box with a volume of 5600 cm³ would hold 5600 L.
    c. Jerry used 20,000 L of water to fill his young son's wading pool.
    d. An eye dropper holds about 600 mL.
7. a. If two cubes have sides in the ratio 2:5, what is the ratio of their volumes?
    b. If two similar cones have heights in the ratio a : b, what is the ratio of their volumes?
8. What happens to the volume of a sphere if the radius is tripled?
9. Complete the following chart for right rectangular prisms with the given dimensions:

	a.	b.	c.	d.
Length	5 cm	8 cm	2 dm	15 cm
Width	10 cm	6 dm	1 dm	2 dm
Height	20 cm	4 dm		
Volume (cm³)				
Volume (dm³)				12
Volume (L)			10	

10. Two spherical cantaloupes of the same kind are sold at a fruit and vegetable stand. The circumference of one is 60 cm and that of the other is 50 cm. The larger melon is $1\frac{1}{2}$ times as expensive as the smaller. Which melon is the better buy and why?

11. A rectangular swimming pool with dimensions 10 m × 25 m is being built. The pool has a shallow end that is uniform in depth and a deep end that drops off as shown in the following figure. What is the volume of this pool in cubic meters?

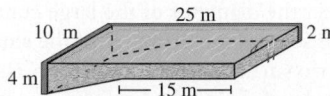

12. Determine how many liters a right circular cylindrical tank holds if it is 6 m long and 13 m in diameter.

13. Determine the volume of silver needed to make the napkin ring in the following figure out of solid silver. Give your answer in cubic millimeters.

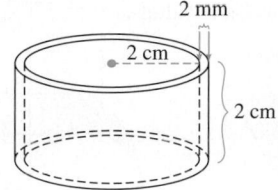

14. A theater decides to change the shape of its popcorn container from a regular box to a right regular pyramid, as shown in the following figure, and charge only half as much.

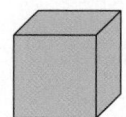 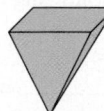

If the containers are the same height and the tops are the same size, is this a bargain for the customer? Explain.

15. A right circular cylindrical can is to hold approximately 1 L of water. What should be the height of the can if the radius is 12 cm?

16. One freezer measures 1.5 ft × 1.5 ft × 5 ft and sells for $350. Another freezer measures 2 ft × 2 ft × 4 ft and sells for $400. Which freezer is the better buy in terms of dollars per cubic foot?

17. A heavy metal sphere with radius 10 cm is dropped into a right circular cylinder with base radius of 10 cm. If the original cylinder has water in it that is 20 cm high, how high is the water after the sphere is placed in it?

18. A box contains a can, (right circular cylinder) as in the accompanying figure. What percentage of the volume of the box does the can occupy?

19. Half of the air is let out of a spherical balloon. If the balloon remains in the shape of a sphere, how does the radius of the smaller balloon compare to the original radius?

20. A square sheet of cardboard measuring y cm on a side is to be used to produce an open-top box when the maker cuts off a small square x cm × x cm from each corner and bends up the sides. Find the volume of the box if $y = 200$ and $x = 20$.

21. For each of the following, select the appropriate metric unit of measure (gram, kilogram, or metric ton):
    a. Jar of mustard
    b. Bag of peanuts
    c. Army tank
    d. Cat

22. For each of the following, choose the correct unit (milligrams, grams, or kilograms) to make each sentence reasonable:
    a. A dime has a mass of 2 _____ .
    b. The recipe said to add 4 _____ of salt.
    c. One strand of hair has a mass of 2 _____ .

23. Complete each of the following:
    a. 8000 kg = _____ t
    b. 72 g = _____ kg
    c. 5 kg 750 g = _____ g
    d. 2.6 lb = _____ oz (16 oz = 1 lb)
    e. 3.8 lb = _____ oz (16 oz = 1 lb)

24. A paper clip has a mass of about 1 g. Is it possible to lift:
    a. 1000 paper clips
    b. 100,000 paper clips

25. a. Rainfall is usually measured in linear measure. Suppose St. Louis received 2 cm of rain on a given day. If a certain lot in St. Louis has measure 1 ha, how many liters of rainfall fell on the lot?
    b. What is the mass of the water that fell on the lot?

26. Convert each of the following from degrees Fahrenheit to the nearest integer degree Celsius:
    a. 0°F   b. 100°F
    c. ⁻40°F

27. Answer each of the following:
    a. The thermometer reads 26°C. Will the outdoor ice rink be open?
    b. It is 40°C. Will you need a sweater at the outdoor concert?
    c. Your bath water is 16°C. Will you have a hot, warm, or chilly bath?

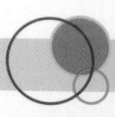

Mathematical Connections 14-5

Communication

1. **a.** Which will increase the volume of a right circular cylinder more: doubling its height or doubling its radius? Explain.
   b. Is your answer the same for a right circular cone? Why?
2. Explain how you would find the volume of an irregular shape.
3. Read the following problems (i) and (ii):
   i. A tank in the shape of a cube 5 ft 3 in. on a side is filled with water. Find the volume in cubic feet, the capacity in gallons, and the weight of the water in pounds.
   ii. A tank in the shape of a cube 2 m on a side is filled with water. Find the volume in cubic meters, the capacity in liters, and the mass of the water in kilograms.
   Discuss which problem is easier to work and why.
4. A furniture company gives an estimate for moving based upon the size of the rooms in an apartment. Write a rationale for why this is feasible. What assumptions are being made?

Open-Ended

5. A right circular cylinder has a 4-in. diameter, is 6 in. high, and is completely full of water. Design a right rectangular prism that will hold almost exactly the same amount of water.
6. Circular-shaped cookies are to be packaged 48 to a box. Each cookie is approximately 1 cm thick and has a diameter of 6 cm. Design a box that will hold this volume of cookies and has the least amount of surface area.
7. Design a cylinder that will hold 1 L of juice. Give the dimensions of your cylinder and tell why you designed the shape as you did.

Cooperative Learning

8. **a.** Find many different types of cans that are in the shape of a cylinder. Measure the height and diameter of each can.
   b. Compute the surface area and volume for each can.
   c. Based on the information collected, make a recommendation for designing an "ideal" can.

Questions from the Classroom

9. A student asks whether the volume of a prism can ever have the same numerical value as its surface area. How do you answer?
10. A student says the volume of the cube shown below is 5 cm.³ What is wrong with this interpretation?

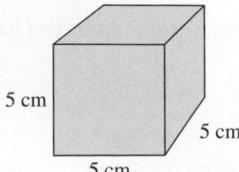

11. A student claims that it does not make any difference if his temperature is 2 degrees above normal Fahrenheit or 2 degrees above normal Celsius because in either case he is only 2 degrees above normal. How do you respond?

12. Andrea claims that if she doubles the length and width of the base of a rectangular prism and triples the height, she has increased the volume by a factor $2 \cdot 2 \cdot 3 = 12$. What would you tell her?
13. Jamie had $6.00 to spend on popcorn at a movie theater. She had to choose between buying two small right cylindrical containers at $3.00 each or one large cylindrical container for $6.00. She noticed that the containers were about the same height and that the diameter of the large container looked about twice as long as the diameter of the small container. She bought two small containers and then asked in math class the next day if she made the right choice to get the most popcorn for her money. How would you help her?
14. A student asks if all cubes are similar and all spheres are similar. How do you respond?

Review Problems

15. Find the perimeter and the area of the following figures. Leave answers as exact values.

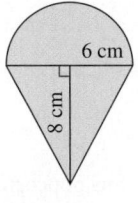

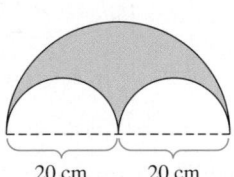

   a.   **b.** The shaded portion only

16. Complete the following:
   a. 350 mm = _____ cm
   b. 1600 cm² = _____ m²
   c. 0.4 m² = _____ mm²
   d. 5.2 cm² = _____ m²
17. Determine whether each of the following is a right triangle:

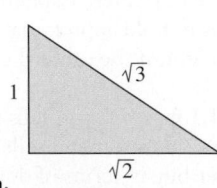

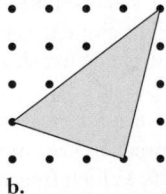

   a.   **b.**

18. Find the surface area of each of the following. Leave answers as exact values.

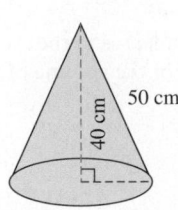

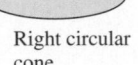

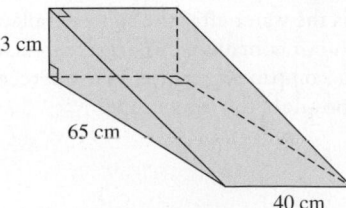

   a. Right circular cone   **b.** Right triangular prism

Trends in Mathematics and Science Study (TIMSS) Questions

All the small blocks are the same size. Which stack of blocks has a different volume from the others?

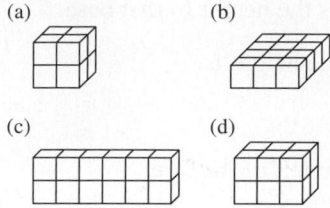

(a)   (b)

(c)   (d)

TIMSS, Grade 8, 2003

Oranges are packed in boxes. The average diameter of the oranges is 6 cm, and the boxes are 60 cm long, 36 cm wide, and 24 cm deep. Which of these is the BEST approximation of the number of oranges that can be packed in a box?

  a. 30   **b.** 240   **c.** 360   **d.** 1920

TIMSS, Grade 8, 2003

National Assessment of Educational Progress (NAEP) Question

How many 200-milliliter servings can be poured from a pitcher that contains 2 liters of juice?

  a. 20   **b** 15   **c.** 10   **d.** 5   **e.** 1

NAEP, Grade 8, 2007

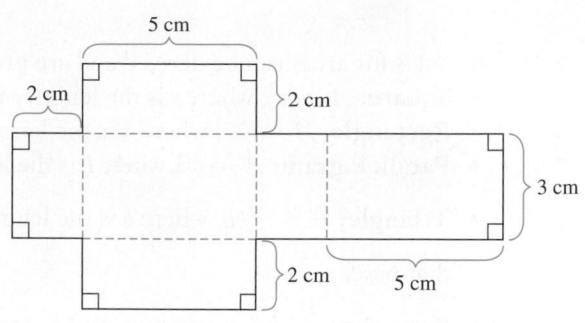

When the shape shown above is folded up, it will make a rectangular box. What is the volume of the box?

Answer: _____ cm^3

TIMSS, Grade 8, 2007

Hint for Solving the Preliminary Problem

Compute the distance using Al's method. Also, think of the room as a cardboard box that can be opened up as a net. Use the Pythagorean theorem to find the distance from the light to the outlet. Compare this distance to the distance computed using Al's method.

Chapter 14 Summary

14-1 Areas of Polygons and Circles	Pages
Areas of polygons can be found on a geoboard by using the **addition method** or the **rectangle method**.	828–829
The commonly used units of area in the metric system are the **square millimeter** (mm²), **square centimeter** (cm²), **square meter** (m²), and **square kilometer** (km²). Land is measured using the **are** (a) (100 m²) and the **hectare** (ha) (10,000 m²).	831–832
The commonly used units of area in the English system are the **square inch** (in.²), **square foot** (ft²), **square yard** (yd²), and the **square mile** (mi²). For land measure, the **acre** is used.	831–832

Formulas for areas can be derived and are given below.

- **Square:** $A = s^2$, where s is the length of a side. ……830
- **Rectangle:** $A = \ell w$, where ℓ is the length and w is the width. ……832
- **Parallelogram:** $A = bh$, where b is the length of the base and h is the height to that base. ……833
- **Triangle:** $A = \dfrac{1}{2}bh$, where b is the length of the base and h is the height to that base. ……834
- **Kite:** $A = \dfrac{1}{2}(d_1 d_2)$, where d_1 and d_2 are the lengths of the diagonals of the kite. ……837
- **Trapezoid:** $A = \dfrac{1}{2}h(b_1 + b_2)$, where b_1 and b_2 are the lengths of the bases and h is the height. ……838
- **Regular polygon:** $A = \dfrac{1}{2}ap$, where a is the apothem and p is the perimeter. ……839
- **Circle:** $A = \pi r^2$, where r is the radius. ……840
- **Sector:** $A = \theta \pi r^2 / 360$, where θ is the degree measure of the central angle forming the sector and r is the radius of the circle containing the sector. ……840

14-2 The Pythagorean Theorem, Distance Formula, and Equation of a Circle

Theorem: Pythagorean theorem: In any right triangle, the square of the length of the hypotenuse is equal to the sum of the squares of the lengths of the legs. ……851

Theorem: 45°-45°-90° triangle: The length of the hypotenuse of a 45°-45°-90° triangle is $\sqrt{2}$ times the length of a leg. ……854–855

Theorem: 30°-60°-90° triangle: The length of the hypotenuse in a 30°-60°-90° triangle is 2 times the length of the leg opposite the 30° angle, and the length of the leg opposite the 60° angle is $\sqrt{3}$ times the length of the short leg. ……855

Theorem: Converse of the Pythagorean theorem: If triangle ABC has sides of lengths a, b, and c such that $a^2 + b^2 = c^2$, then $\triangle ABC$ is a right triangle with the right angle opposite the side of length c. ……856

Theorem: Distance formula: The distance between the points $A(x_1, y_1)$ and $B(x_2, y_2)$ is given by $AB = \sqrt{(x_2 - x_1)^2 + (y_2 - y_1)^2}$. ……858

Theorem: Equation of the Circle with Center at the Origin: The equation of the circle with center at the origin and radius r is $x^2 + y^2 = r^2$. ……860

Theorem: Equation of the Circle with Center at (h, k): The equation of the circle with center (h, k) and radius r is $(x - h)^2 + (y - k)^2 = r^2$. ……860

14-3 Geometry in Three Dimensions

A **simple closed surface** has exactly one interior, has no holes, and is hollow. ……867

A **sphere** is the set of all points at a given distance from a given point, the **center**. ……867

A **polyhedron** is a simple closed surface made up of polygonal regions, or **faces**. The vertices of the polygonal regions are the **vertices** of the polyhedron, and the sides of each polygon are the **edges** of the polyhedron. ……867

A **prism** is a polyhedron in which two congruent polygonal faces (**bases**) lie in parallel planes and the remaining faces are formed by the union of the line segments joining corresponding points in the two bases. The parallel faces of a prism are the **bases** of the prism. The faces other than the bases are **lateral faces** and are parallelograms. ……867

A **right prism** is a prism in which the lateral faces are bounded by rectangles. If the lateral faces are parallelograms that are not rectangles then the prism is an **oblique prism**.	868
A **pyramid** is a polyhedron determined by a polygon and a point not in the plane of the polygon. The pyramid consists of the triangular region determined by the point and each pair of consecutive vertices of the polygon, and the polygonal region determined by the polygon. The polygonal region is the **base** and the point is the **apex**.	868
• A **right regular pyramid** is a pyramid that has a regular polygon as a base and has lateral faces that are congruent isosceles triangles.	868
A **truncated polyhedron** is a polyhedron with one or more corners removed by an intersecting plane or planes and the removed portion replaced by a polygonal region.	869
A polyhedron is a **convex polyhedron** if, and only if, the segment connecting any two points in the interior of the polyhedron is itself in the interior. If the polyhedron is not convex, it is **concave**.	870
A **regular polyhedron** is a convex polyhedron whose faces are congruent regular polygonal regions such that the number of edges that meet at each vertex is the same for all vertices of the polyhedron. There are only five regular polyhedra.	870
Euler's formula for polyhedra: $V + F - E = 2$, where V stands for the number of vertices, F for the number of faces, and E for the number of edges.	872–873
A **cylinder** is a simple closed surface that is the union of line segments joining corresponding points of congruent simple closed curves in parallel planes and the interiors of the simple closed curves. The simple closed curves along with their interiors are the **bases** of the cylinder and the remaining points constitute the **lateral surface** of the cylinder.	873
• If the base of a cylinder is a circular region, the cylinder is a **circular cylinder**.	873
• In a cylinder, if a line segment connecting a point on one base to its corresponding point on the other base is perpendicular to the planes of the bases, then it is a **right cylinder**. If a cylinder is not a right cylinder, it is an **oblique cylinder.**	873
A **cone** is the union of line segments connecting a point to a simple closed curve in a plane other than a polygon, the simple closed curve, and the interior of the curve. An **altitude** is a line segment from the apex of the cone perpendicular to the plane of the base.	873
• A **right circular cone** is a cone whose altitude intersects the base at the center of the circle.	873
• A circular cone that is not a right circular cone is an **oblique circular cone**.	873

14-4 Surface Areas

The sum of the areas of the lateral faces is the **lateral surface area**.	880
The **surface area** (*SA*) is the sum of the lateral surface area and the area of the base(s).	880
• **Cube:** $SA = 6e^2$, where e is the length of an edge.	880
• **Prism:** $SA = 2B + ph$, where B is the area of the base, p is the perimeter of the base, and h is the height of the prism.	880

- **Right circular cylinder:** $SA = 2\pi r^2 + 2\pi rh$, where r is the radius of the circular base and h is the height of the cylinder.   881
- **Right regular pyramid:** $SA = B + \frac{1}{2}p\ell$, where B is the area of the base, p is the perimeter of the base, and ℓ is the slant height.   882
- **Right circular cone:** $SA = \pi r^2 + \pi r\ell$, where r is the radius of the circular base and ℓ is the slant height.   883–884
- **Sphere:** $SA = 4\pi r^2$, where r is the radius of the sphere.   884–885

14-5 Volume, Mass, and Temperature

Formulas for finding the volumes of various shapes can be developed.
- **Right prism:** $V = \ell wh$ (or $V = Bh$), where ℓ is the length, w is the width, and h is the height.   894
- **Cube:** $V = e^3$, where e is an edge.   894
- **Circular cylinder:** $V = \pi r^2 h$, where r is the radius of the base and h is the height of the cylinder.   895
- **Pyramid:** $V = \frac{1}{3}Bh$, where B is the area of the base and h is the height of the pyramid.   896
- **Circular cone:** $V = \frac{1}{3}\pi r^2 h$, where r is the radius of the circular base and h is the height.   896
- **Sphere:** $V = \frac{4}{3}\pi r^3$, where r is the radius of the sphere.   900

Commonly used units for volume and capacity in the metric system are the **cubic meter** (m^3), **cubic decimeter** (dm^3), and **cubic centimeter** (cm^3).   891–892

$1\ dm^3 = 1\ L$ (liter) and $1\ cm^3 = 1\ mL$.   892–893

The base unit for capacity is the **liter**. Other units are shown below.   893

Unit	Symbol	Relation to Liter
kiloliter	kL	1000 L
*hectoliter	hL	100 L
*dekaliter	daL	10 L
liter	**L**	**1 L**
*deciliter	dL	0.1 L
centiliter	cL	0.01 L
milliliter	mL	0.001 L

*Not commonly used

Commonly used units for volume and capacity in the English system are the **cubic inch** ($in.^3$), **cubic foot** (ft^3), **cubic yard** (yd^3), and **gallon** (gal). $1\ gal = 231\ in^3$.   893

Commonly used units of mass are the **milligram** (mg), **gram** (g), **kilogram** (kg), and **metric ton** (t).   905–906

1 L of water has a mass of 1 kg and 1 mL of water has a mass of 1 g.   905–906

The base unit for mass is the **gram** other units are shown below.   905–906

Unit	Symbol	Relationship to Gram
ton (metric)	t	1,000,000 g
kilogram	kg	1000 g
*hectogram	hg	100 g
*dekagram	dag	10 g
gram	**g**	**1 g**
*decigram	dg	0.1 g
*centigram	cg	0.01 g
milligram	mg	0.001 g

*Not commonly used

In the metric system the unit commonly used for temperature is the **degree Celsius** (°C).
In the English system, the unit of temperature is the **degree Fahrenheit** (°F). The
scientific unit used is the **kelvin** (K).   907

- $C = \dfrac{5}{9}(F - 32)$ and $F = \dfrac{9}{5}C + 32$ 908

- Basic temperature reference points are the following: 908

    100°C—boiling point of water

    37°C—normal body temperature

    20°C—comfortable room temperature

    0°C—freezing point of water

Chapter 14 Review

1. Determine the area of the shaded region on each
   of the following geoboards if the unit of measure is
   1 cm².

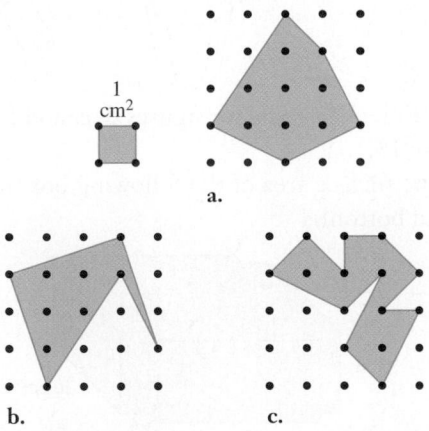

2. Explain how the formula for the area of a trapezoid
   can be found by using the following figures:

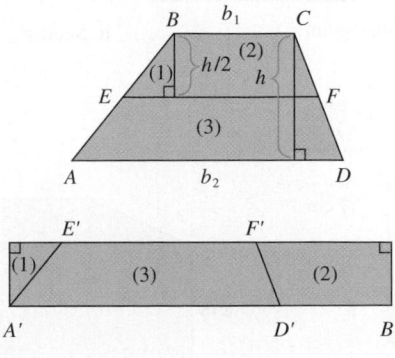

3. Lines *a*, *b*, and *c* are parallel to the line containing side $\overline{AB}$ of the triangles shown.

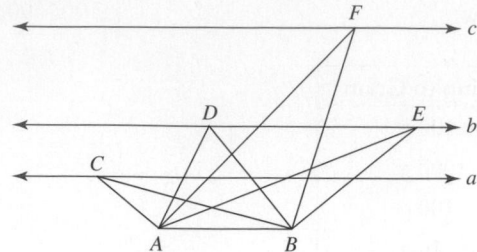

List the triangles in order of size of their areas from least to greatest. Explain why your order is correct.

4. Use the figure shown to find each of the following areas:
   a. The area of the regular hexagon
   b. The area of the circle

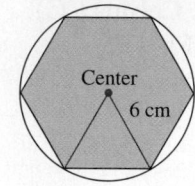

5. Find the area of each shaded region in the following figures.

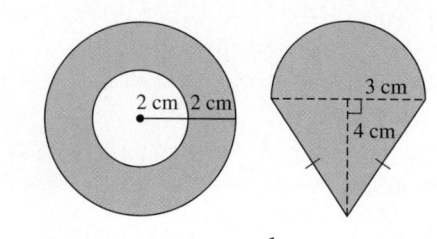

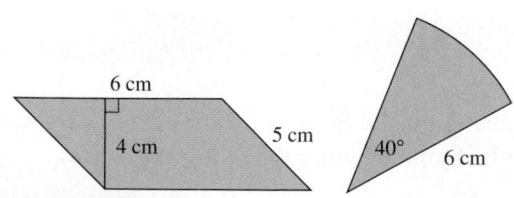

c. Parallelogram   d. Sector

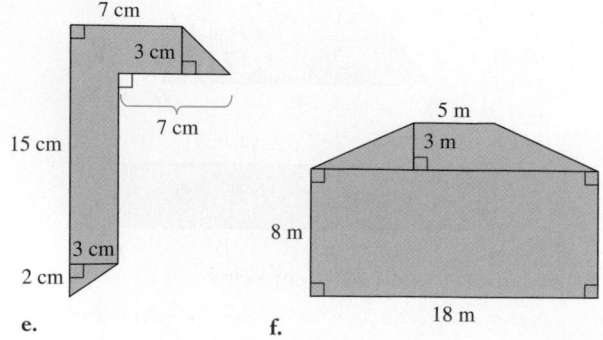

e.   f.

6. A baseball diamond is actually a square 90 ft on a side. What is the distance a catcher must throw from home plate to second base?

7. Find the length of segment *AG* in the spiral shown.

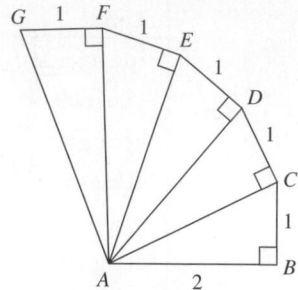

8. For each of the following, determine whether the measures represent sides of a right triangle.
   a. 5 cm, 12 cm, 13 cm
   b. 40 cm, 60 cm, 104 cm

9. If a pyramid has an octagon for a base, how many lateral faces does it have?

10. Carefully draw nets that can be folded into each of the following. Cut the nets out and fold them into the given polyhedra.
    a. Tetrahedron
    b. Square pyramid
    c. Right rectangular prism
    d. Right circular cylinder with a top and a bottom

11. Explain whether or not a cylinder has to have a circular base.

12. a. Draw as many different nets as possible that can be folded into an open cube (a cube without a top).
    b. How many different nets are possible in part (a)? Explain what "different" means to you.

13. In the following, draw in dotted lines for the unseen segments:

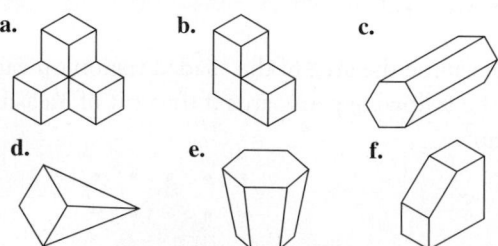

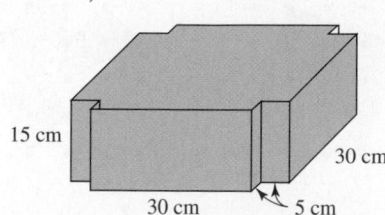

14. Verify Euler's formula for figures d, e, and f in exercise 13.

15. Find the surface area of the following box (include the top and bottom):

16. Find the surface area and volume of each of the following figures.

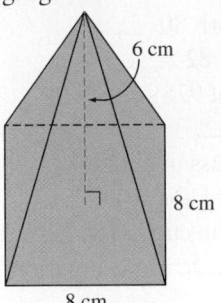

8 cm
6 cm
8 cm

Right square pyramid

a.

8 cm
6 cm

Right circular cone

b.

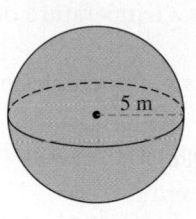

5 m

Sphere

c.

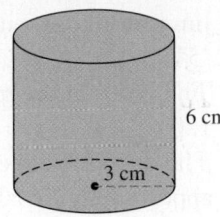

6 cm
3 cm

Right circular cylinder

d.

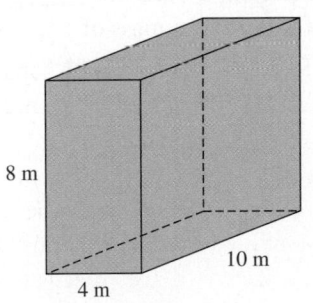

8 m
10 m
4 m

Right rectangular prism

e.

17. Find the lateral surface area of the following right circular cone.

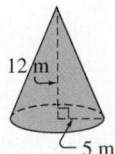

12 m
5 m

18. Doug's Dog Food Company wants to impress the public with the magnitude of the company's growth. Sales of Doug's Dog Food doubled from 2000 to 2008, so the company is displaying the following graph, which shows the radius of the base and the height of the 2008 can to be double those of the 2000 can. What does

the graph really show with respect to the company's growth? Explain your answer.

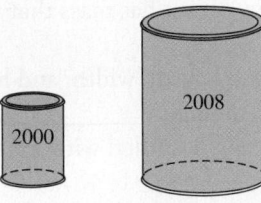

2000   2008

19. Find the area of the kite shown in the following figure.

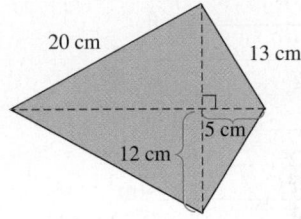

20 cm   13 cm
5 cm
12 cm

20. The diagonal of a rectangle has measure 1.3 m, and a side of the rectangle has measure 120 cm. Find the following:
   a. Perimeter of the rectangle
   b. Area of the rectangle

21. Find the area of a triangle that has sides of 3 m, 3 m, and 2 m.

22. A poster is to contain 0.25 m^2 of printed matter, with margins of 12 cm at top and bottom and 6 cm at each side. Find the width of the poster if its height is 74 cm.

23. A right circular cylinder of height 10 cm and a right circular cone share a circular base and have the same volume. What is the height of the cone?

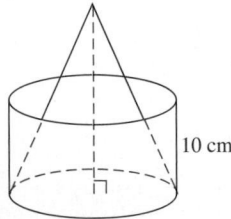

10 cm

24. On a 5 × 5 geoboard paper,
   a. draw a polygon whose perimeter is greater than 16.
   b. draw a polygon with least perimeter. What is the perimeter?
   c. draw a polygon with greatest area.

25. What is the length of the diagonal of a $8\frac{1}{2}$ × 11 in. piece of paper?

26. a. Find an equation of the circle with center at $(3, {}^-4)$ and a radius of 5.
   b. Given that the equation of a circle is $36 = (x - 5)^2 + (y + 3)^2$, find the radius and the center.

27. Complete each of the following with the appropriate metric unit.

  a. A very heavy object has mass that is measured in _____.

  b. A cube whose length, width, and height are each 1 cm has a volume of _____.

  c. If the cube in (b) is filled with water, the mass of the water is _____.

  d. $1 \text{ L} = $ _____ dm^3

  e. If a car uses 1 L of gas to go 12 km, the amount of gas needed to go 300 km is _____ L.

  f. $20 \text{ ha} = $ _____ a

  g. $51.8 \text{ L} = $ _____ cm^3

  h. $10 \text{ km}^2 = $ _____ m^2

  i. 50 L _____ mL

  j. $5830 \text{ mL} = $ _____ L

  k. $25 \text{ m}^3 = $ _____ dm^3

  l. $75 \text{ dm}^3 = $ _____ mL

  m. $52{,}813 \text{ g} = $ _____ kg

  n. 4800 kg _____ t

28. a. A tank that is a right rectangular prism is $1 \text{ m} \times 2 \text{ m} \times 3 \text{ m}$. If the tank is filled with water, what is the mass of the water in kilograms?

  b. Suppose the tank in (a) is exactly half full of water and then a heavy metal sphere of radius 30 cm is put into the tank. How high is the water now if the height of the tank is 3 m?

29. For each of the following, fill in the correct unit to make the sentence reasonable:

  a. Anna filled the gas tank with 80 ___.

  b. A man has a mass of about 82 ___.

  c. The textbook has a mass of 978 ___.

  d. A nickel has a mass of 5 ___.

  e. A typical adult cat has a mass of about 4 ___.

  f. A compact car has a mass of about 1.5 ___.

  g. The amount of coffee in the cup is 180 ____.

30. For each of the following, decide if the situation is likely or unlikely:

  a. Carrie's bath water has a temperature of 15°C.

  b. Anne found 26°C too warm and so lowered the thermostat to 21°C.

  c. Jim is drinking water that has a temperature of ⁻5°C.

  d. The water in the teakettle has a temperature of 120°C.

  e. The outside temperature dropped to 5°C, and ice appeared on the lake.

31. Complete each of the following:

  a. 2 dm^3 of water has a mass of _____ g.

  b. 1 L of water has a mass of _____ g.

  c. 3 cm^3 of water has a mass of _____ g.

  d. 4.2 mL of water has a mass of _____ kg.

  e. 0.2 L of water has a volume of _____ m^3.

Credits

Answers to Problems

Answers to Assessment A exercises, odd-numbered Mathematical Connections problems, Chapter Review problems, Now Try This problems, Brain Teasers, and Preliminary Problems appear in the Student Edition.

Chapter 1

Assessment 1-1A

1. (a) 4950 (b) 251,001 **2.** Building a staircase as seen in (a) gives a visual graphic of the sum $1 + 2 + \ldots + n$. Copying the staircase as in (b) and placing it as shown demonstrates that an array that is n units high and $n + 1$ units long is produced. There are $n(n + 1)$ units in (b) which is twice the number desired. So the sum $1 + 2 + \ldots + n$ must be $n(n + 1)/2$. Gauss's sum when $n = 100$ would be $100(100 + 1)/2$ or 5050. **3.** 10,248 **4.** 12 **5.** I was the only one going, so there is 1. **6.** 27 **7.** E is greater by 49. **8.** Dandy, Cory, Alabama, Bubba **9.** 12 **10.** $a = 42$; $b = 32$; $c = 37$; $d = 2$ **11.** 45 **12.** 9 **13.** If we choose Box B and pull out a fruit, we know exactly what the label should be. That box does not contain the apples and oranges but is either a box of only apples or of only oranges. If we pull out an orange, the box should be labeled Oranges. If we pull out an apple, the box should be labeled Apples. Once we label Box B correctly, we do know the labels for the other two boxes. **14.** $2.45 **15.** 23 rungs

Mathematical Connections 1-1

Communication

1. Answers vary. For example, problem-solving skills can help students to meet future challenges in work, life, and school. Problem-solving skills allow students to take on new tasks and problems and have the confidence to do so. If a first approach to a problem fails, good problem solvers can come up with alternative approaches. Much of the mathematics that students are taught is introduced through interesting problems. Students need to know how to problem solve to make progress on these problems and in turn learn the mathematics. **3.** (a) 28 (b) Answers vary depending on the strategies used. (c) Answers vary. (d) $\dfrac{n(n - 1)}{2}$

Open-Ended

5. Answers vary. For example, if the average reach of people in your group was 1.8 m, then it would take approximately 22,000,000 people.

Cooperative Learning

7. Answers vary. For example, it is in the last step where students examine whether the answer is reasonable and whether it checks given the original conditions in the problem. Many times students discover wrong answers at this point since they have never bothered to check whether the answer they arrived at makes sense. It is also at this step that students reflect on the mathematics used and determine whether there might be different ways of solving the problem. Also at this stage students reflect on any connections to other problems or generalizations. **9.** Answers vary.

For example, if these prime numbers are to be used in a magic square, then the sum in each of the three columns must be the same and must be a natural number. The natural number must be 1/3 of the sum of all nine numbers. However, $1 + 3 + 4 + 5 + 6 + 7 + 8 + 9 + 10 = 53$ and $53/3 = 17\dfrac{2}{3}$, which is not a natural number. Therefore these numbers cannot be used for a magic square.

Assessment 1-2A

1. (a) ▦▦▦▦ (b) ◁◁◁◁◁ (c) △

2. (a) 11, 13, 15; arithmetic (b) 250, 300, 350; arithmetic (c) 96, 192, 384; geometric (d) $10^6, 10^7, 10^8$; geometric (e) 33, 37, 41; arithmetic **3.** (a) $199; 2n - 1$ (b) $4950; 50(n - 1)$ (c) $3 \cdot 2^{99}; 3 \cdot 2^{n-1}$ (d) $10^{100}; 10^n$ (e) $405; 5 + 4n$ or $9 + 4(n - 1)$ **4.** 2, 7, 12 **5.** (a) 1331 (b) 729 (c) 2744 **6.** (a) 41 (b) $4n + 1$, or $5 + (n - 1)4$ (c) $12n + 4$ **7.** (a) 42 (b) $4n + 2$ or $6 + (n - 1)4$ **8.** 1200 students **9.** 23rd year **10.** (a) 3, 5, 9, 15, 23, 33 (b) 4, 6, 10, 16, 24, 34 (c) 15, 17, 21, 27, 35, 45 **11.** (a) 299, 447, 644 (b) 56, 72, 90 **12.** (a) 101 (b) 61 (c) 200 (d) 11 **13.** (a) 3, 6, 11, 18, 27 (b) 4, 9, 14, 19, 24 (c) 9, 99, 999, 9999, 99999 (d) 5, 8, 11, 14, 17 **14.** (a) Answers vary. For example, if $n = 5$, then

$\dfrac{5 + 5}{5} \neq 5 + 1$. (b) Answers vary. For example, if $n = 2$, then $(2 + 4)^2 \neq 2 + 16$. **15.** (a) 41 (b) $n^2 + (n - 1)$ (c) Yes, the 35th figure **16.** 12 **17.** (a) 1, 5, 9, 13, 17, 21, … (b) Answers vary. The sequences may be either 1, 5, 9, … or 5, 9, 13, …, depending on whether you start counting before or after the first cut. The total number of pieces after n cuts is $4n + 1$.

Mathematical Connections 1-2

Communication

1. (a) Yes. The difference between terms in the new sequence is the same as in the old sequence because a fixed number was added to each number in the sequence. (b) Yes. If the fixed number is k, the difference between terms of the second sequence is k times the difference between terms of the first sequence. (c) Yes. The difference of the new sequence is the sum of the original differences.

Open-Ended

3. Answers vary. For example, two more patterns follow:

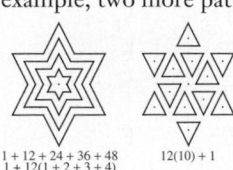

$1 + 12 + 24 + 36 + 48$
$1 + 12(1 + 2 + 3 + 4)$ $12(10) + 1$

5. Answers vary depending on sequence.

Cooperative Learning

7. (a) 81 **(b)** 40 **(c)** 3^{n-1}

Questions from the Classroom

9. Al and Betty should be told that both answers could be correct as long as each person's rule works for all the given terms. Al is thinking of a geometric sequence in which each term is multiplied by 2 to get the next term. His sequence is correct and he should be asked to explain his rule. Betty is thinking of a different sequence. She starts with 2, then adds 4 to get the 2nd term, then adds 6 to get the next term, then 8, then 10, then 12, and so on. Both students are correct because each rule works for all the given terms.

Review Problems

11. 90 **12.** 7 **13.** We need one 12-person tent and a combination of tents to hold 14 people. There are 10 ways: 662, 653, 6332, 62222, 5522, 5333, 53222, 33332, 332222, and 2222222.

Assessment 1-3A

1. (a) False statement **(b)** False statement **(c)** Not a statement **(d)** True statement **(e)** Not a statement **2.** Answers vary. **(a)** There exists a natural number n such that $n + 8 = 11$. **(b)** There exists a natural number n such that $n^2 = 4$. **(c)** For all natural numbers n, $n + 3 = 3 + n$. **(d)** For all natural numbers n, $5n + 4n = 9n$. **3.** Answers vary. **(a)** For every natural number, n, $n + 8 = 11$. **(b)** Every natural number n satisfies $n^2 = 4$. **(c)** There is no natural number n such that $n + 3 = 3 + n$. **(d)** There is no natural number n such that $5n + 4n = 9n$. **4. (a)** This book does not have 500 pages. **(b)** $3 \cdot 5 \neq 15$ **(c)** Some dogs do not have four legs. **(d)** No rectangles are squares. **(e)** All rectangles are squares. **(f)** Some dogs have fleas. **5. (a)** True **(b)** True

6. (a)

p	$\sim p$	$\sim(\sim p)$
T	F	T
F	T	F

(b)

p	$\sim p$	$p \vee \sim p$	$p \wedge \sim p$
T	F	T	F
F	T	T	F

(c) Yes **(d)** No **7. (a)** $q \wedge r$ **(b)** $r \vee \sim q$ **(c)** $\sim(q \wedge r)$ **(d)** $\sim q$ **8. (a)** False **(b)** True **(c)** False **(d)** False **(e)** False **9. (a)** False **(b)** True **(c)** False **(d)** False **(e)** True **10. (a)** No **(b)** No

11.

p	q	$\sim p$	$\sim p \wedge q$
T	T	F	F
T	F	F	F
F	T	T	T
F	F	T	F

12. (a) $p \rightarrow q$ **(b)** $\sim p \rightarrow q$ **(c)** $p \rightarrow \sim q$ **(d)** $p \rightarrow q$ **(e)** $\sim q \rightarrow \sim p$ **(f)** $q \leftrightarrow p$ **13. (a)** Converse: If $2x = 10$,

then $x = 5$. Inverse: If $x \neq 5$, then $2x \neq 10$. Contrapositive: If $2x \neq 10$, then $x \neq 5$. **(b)** Converse: If you do not like mathematics, then you do not like this book. Inverse: If you like this book, then you like mathematics. Contrapositive: If you like mathematics, then you like this book. **(c)** Converse: If you have cavities, then you do not use Ultra Brush toothpaste. Inverse: If you use Ultra Brush toothpaste, then you do not have cavities. Contrapositive: If you do not have cavities, then you use Ultra Brush toothpaste. **(d)** Converse: If your grades are high, then you are good at logic. Inverse: If you are not good at logic, then your grades are not high. Contrapositive: If your grades are not high, then you are not good at logic. **14.** Answers vary. If a number is not a multiple of 4, then it is not a multiple of 8. (Contrapositive) **15. (a)** Valid **(b)** Valid **(c)** Invalid **16.** Answers vary. **(a)** Some freshmen are intelligent. **(b)** If I study for the final, then I will look for a teaching job. **(c)** There exist triangles that are isosceles. **17. (a)** If a figure is a square, then it is a rectangle. **(b)** If a number is an integer, then it is a rational number. **(c)** If a polygon has exactly three sides, then it is a triangle. **18. (a)** $3 \cdot 2 \neq 6$ or $1 + 1 = 3$. **(b)** You cannot pay me now *and* you cannot pay me later.

Mathematical Connections 1-3

Communication

1. Commands and questions are not statements because they can't be classified as true or false. **3. (a)** A disjunction is in the form p or q and it is true if either p or q or both are true. The only time a disjunction is false is when both p and q are false. **(b)** An implication in the form if p, then q is false if, and only if, p is true and q is false. **5.** Dr. No is a male spy who is not poor and not tall. **7.** When a comma or a semicolon is used in an e-mail address (depending on the server), the logical meaning is "and" so that all addresses will receive the e-mail.

Cooperative Learning

9. Answers vary. Statements (a) and (b) are false. Statement (c) causes the problem. If statement (c) is also false, then that makes statement (c) a true assertion which is a contradiction. On the other hand, if statement (c) is true, then it wrongly asserts that it is false.

Questions from the Classroom

11. Consider the example.
Hypotheses: All teachers are over 6 ft tall.
               Kay is a teacher.
Conclusion: Kay is over 6 ft tall.

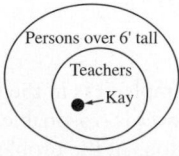

An Euler diagram can be drawn to show that all teachers belong to the set of people over 6 ft tall. Kay belongs to the set of teachers. Thus, the argument is valid even though the hypothesis is false.

Chapter Review

1. Monday **2.** Answers vary. One solution is to pair a number with its corresponding letter in the English alphabet giving the

message: SECRET CODES. **3.** The question is asked "What makes you come so soon?" when the scholar arrives two hours later than usual. This makes no sense *unless* the "ten o'clock" is at night. **4. (a)** 15, 21, 28; neither **(b)** 32, 27, 22; arithmetic **(c)** 400, 200, 100; geometric **(d)** 21, 34, 55; neither **(e)** 17, 20, 23; arithmetic **(f)** 256, 1024, 4096; geometric **(g)** 16, 20, 24; arithmetic **(h)** 125, 216, 343; neither
5. (a) $3n + 2$ if the sequence is arithmetic with difference of 3. **(b)** 3^n if the sequence is geometric with ratio 3. **(c)** $n^3 - 1$ if the sequence is based on the sequence of cubes, 1, 8, 27, 64, . . .
6. (a) 1, 4, 7, 10, 13 **(b)** 2, 6, 12, 20, 30 **(c)** 3, 7, 11, 15, 19 **7. (a)** 10,100 **(b)** 10,201 **8. (a)** False; for example, $3 + 3 = 6$ and 6 is not odd. **(b)** False; for example, 19 is odd and it ends in 9. **(c)** True; the sum of any two even numbers, $2m$ and $2n$, is even because $2m + 2n = 2(m + n)$
9.

16	3	2	13
5	10	11	8
9	6	7	12
4	15	14	1

10. 26 **11.** $2.00 **12.** 21 posts
13. 44, 000, 000 **14.** 20 **15.** 39 **16.** 235 **17.** 8 **18.** 4
19. 10 cm, 20 cm, 60 cm **20.** 12 **21.** Al-winter; Betty-fall; Carl-spring; Dan-summer **22.** Answers vary. For example, fill the 4-cup pot and empty it into the 7-cup pot. Repeat. There is now 1 cup in the 4-cup pot. Empty the 7-cup pot and pour the 1 cup into the 7-cup pot. Fill the 4-cup pot and empty it into the 7-cup pot. It will now contain 5 cups. **23. (a)** 30, 42, 56 **(b)** 10,100 **(c)** $n(n + 1)$
24. Answers vary. **(a)** 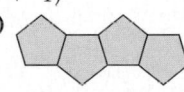 **(b)** 5, 8, 11, 14 **(c)** 302
(d) $3n + 2$ **25.** In statement (i) each and every student passed the final. In statement (ii) at least one student passed the final and possibly all the students passed. **26. (a)** Yes **(b)** Yes **(c)** No **(d)** Yes **27. (a)** No women smoke. **(b)** $3 + 5 \neq 8$ **(c)** Beethoven wrote some music that is not classical. **28.** Converse: If someone will faint, we will have a rock concert. Inverse: If we do not have a rock concert, then no one will faint. Contrapositive: If no one will faint, then we will not have a rock concert.
29.

p	q	$\sim p$	$\sim q$	$p \rightarrow \sim q$	$q \rightarrow \sim p$
T	T	F	F	F	F
T	F	F	T	T	T
F	T	T	F	T	T
F	F	T	T	T	T

Therefore, $p \rightarrow \sim q \equiv q \rightarrow \sim p$.
30. (a)

p	q	$\sim q$	$p \wedge \sim q$	$p \wedge q$	$(p \wedge \sim q) \vee (p \wedge q)$
T	T	F	F	T	T
T	F	T	T	F	T
F	T	F	F	F	F
F	F	T	F	F	F

(b)

p	q	$\sim p$	$p \vee q$	$(p \vee q) \wedge \sim p$	$[(p \vee q) \wedge \sim p] \rightarrow q$
T	T	F	T	F	T
T	F	F	T	F	T
F	T	T	T	T	T
F	F	T	F	F	T

31. Answers vary. **(a)** Joe Czernyu loves Mom and apple pie. **(b)** The structure of the Statue of Liberty will eventually rust. **(c)** Albertina passed Math 100.
32. Let the following letters represent the given sentences:
p: You are fair-skinned.
q: You will sunburn.
r: You do not go to the dance.
s: Your parents want to know why you didn't go to the dance.
Symbolically, $p \rightarrow q$, $q \rightarrow r$, $r \rightarrow s$. Using contrapositives we have: $\sim s \rightarrow \sim r$, $\sim r \rightarrow \sim q$, $\sim q \rightarrow \sim p$. By the chain rule, $\sim s \rightarrow \sim p$; that is, if your parents do not want to know why you didn't go to the dance, then you are not fair-skinned.
33. Valid, *modus tollens*

Answers to Now Try This

1-1. 11 pieces for 10 cuts; $(n + 1)$ pieces for *n* cuts **1-2. (a)** 2500 **(b)** 6960 **1-3.** 120 games **1-4.** 90 days **1-5.** Answers vary. For example, because each person owes $13, Al could pay $4.25 to Betty and $4.00 to Carl, Dani could pay $7.00 to Carl, and everyone would be even. **1-6.** 23 floors
1-7. Answers vary. For example,

$$\begin{array}{r} 132 \\ + 932 \\ \hline 1064 \end{array} \text{ or } \begin{array}{r} 173 \\ + 873 \\ \hline 1046 \end{array}$$

1-8. 83 **1-9. (a)** Answers vary. For example, the next three terms could be $\triangle, \triangle, \circ$. **(b)** The pattern could be one circle, two triangles, one circle, two triangles, and so on.
1-10. (a) Inductive reasoning **(b)** The next several numbers also work. **(c)** Yes, if $x = 11$, then $11^2 + 11 + 11$ is not prime because it is divisible by 11. **1-11. (a)** 4 **(b)** 7 **(c)** 12 **(d)** 20 **(e)** 33 **(f)** The sum of the first *n* Fibonacci numbers is one less than the Fibonacci number two numbers later in the sequence.
1-12. (a) 118,098 bacteria, $2 \cdot 3^n$ bacteria. **(b)** After 10 hours, there are $2 + 10 \cdot 3 = 32$ bacteria, and after *n* hours, there are $2 + n(3)$ bacteria. We can see that after only 10 hours geometric growth is much faster than arithmetic growth. In this case, 118,098 versus 32. This is true in general when $n > 1$.
1-13. (a) [grid figure] **(b)**

1	2	3	4
4	12	24	40

(c) 4 12 24 40 60 84 112
     ⌄ ⌄ ⌄ ⌄ ⌄ ⌄
      8 12 16 20 24 28
       ⌄ ⌄ ⌄ ⌄ ⌄
        4 4 4 4

(d) No, finding differences for the 100th term is very hard. It is easier to find a pattern involving the number of horizontal sticks and vertical sticks; that is, the 100th term is $= 101 \cdot 100 + 101 \cdot 100 = 20,200$.

1–14.

p	q	$\sim p$	$\sim q$	$p \vee q$	$\sim(p \vee q)$	$\sim p \wedge \sim q$
T	T	F	F	T	F	F
T	F	F	T	T	F	F
F	T	T	F	T	F	F
F	F	T	T	F	T	T

$\sim(p \vee q) \equiv \sim p \wedge \sim q$

1–15.

p	q	$p \to q$	$\sim(p \to q)$	$\sim q$	$p \wedge \sim q$
T	T	T	F	F	F
T	F	F	T	T	T
F	T	T	F	F	F
F	F	T	F	T	F

$\sim(p \to q) \equiv p \wedge \sim q$

1–16.

p	q	$p \to q$	$q \to p$	$(p \to q) \wedge (q \to p)$
T	T	T	T	T
T	F	F	T	F
F	T	T	F	F
F	F	T	T	T

Answers to Brain Teasers

Section 1-1

35 moves. This can be solved using the strategy of examining simpler cases and looking for a pattern. If one person is on each side, 3 moves are necessary. If two people are on each side, 8 moves are necessary. With 3 people on each side, 15 moves are necessary. If n people are on each side, $(n + 1)^2 - 1$ moves are required.

Section 1-1

The number of moves is $2^n - 1$ for n coins. This can be solved using the strategy of examining a simpler problem. If there is one coin, 1 move is necessary. If there are two coins, 3 moves are necessary. For three coins, the number of moves is 7. For four coins, the number of moves is 15.

At the rate of one move per second, it would take approximately 585 billion years to move 64 coins.

Section 1-2

3 1 2 2 11; the pattern counts the number of times a number occurs in the previous row. For example, to find the sixth row we examine the fifth row. There are three 1s, two 2s, and one 1, so the sixth row is 3 1 2 2 11. The pattern continues using this rule.

Answer to Preliminary Problem

If eggs are removed 2 at a time, there is 1 left, revealing us that the number of eggs must be an odd number. Also, the number could not be 1 because one cannot remove eggs 2 at a time if there is only 1 in the basket. So there must be at least 3 and the least number must be in arithmetic sequence 3, 5, 7, 9, 11, 13, 15, 17, 19, 21, 23, 25, 27, ...

Similarly we can deduce from the second statement that there are 2 eggs left when they are removed 3 at a time that the number of eggs must be in the pattern 5, 8, 11, 14, 17, 20, 23, 26, . . . and that there must be at least 5 eggs in the basket.

Finally, because there are 3 left when the eggs are removed 5 at a time, the number of eggs has to be in the pattern 8, 13, 18, 23, . . . so there must be at least 8 in the basket.

But more importantly for the problem, there is one number—23—that appears in all sequences and is the least number of eggs that satisfy the conditions. Thus, 23 is the minimum number of eggs that must be in the basket.

Chapter 2

Assessment 2-1A

1. (a) $\overline{\overline{\text{M}}}$CDXXIV; the double bar over M represents $1000 \cdot 1000 \cdot 1000$. **(b)** 46,032; the 4 in 46,032 represents 40,000 whereas the 4 in 4632 represents only 4000. **(c)** ❮ ▼▼; the space in the latter number indicates ❮ is multiplied by 60. **(d)** 𒐕∩I; the 𒐕 represents 1000 whereas 𝟫 represents only 100. **(e)** 👁 represents three groups of 20 plus zero 1s (or 60) and ☰ represents three 5s and three 1s (or 18). **2.** MCML; MCMXL-VIII **(b)** ❮❮ ❮▼▼; ❮❮ ❮ **(c)** 𒐕𝟫𝟫I; 𝟤𝟫∩∩∩∩∩∩∩∩∩∩IIIIIIIII **(d)** ☰;☰ **3.** 1922 **4. (a)** CXXI **(b)** XLII

5. (c) MCCXXIII; ⋮⋮ **6. (a)** Hundreds **(b)** Tens

7. (a) 3,004,005 **(b)** 20,001 **8.** 811 or 910 **9. (a)** 86 **(b)** 11 **10.** 2112_{four} **11. (a)** $(1, 10, 11, 100, 101, 110, 111, 1000, 1001, 1010, 1011, 1100, 1101, 1110, 1111)_{\text{two}}$ **(b)** $(1, 2, 3, 10, 11, 12, 13, 20, 21, 22, 23, 30, 31, 32, 33)_{\text{four}}$ **12.** 20 **13.** $2032_{\text{four}} = (2 \cdot 10^3 + 0 \cdot 10^2 + 3 \cdot 10^1 + 2 \cdot 1)_{\text{four}}$
$= 2 \cdot 4^3 + 0 \cdot 4^2 + 3 \cdot 4^1 + 2 \cdot 1$
$= 142$

14. (a) 111_{two} **(b)** EEE_{twelve} **15. (a)** ETE_{twelve}; $EE1_{\text{twelve}}$ **(b)** 11111_{two}; 100001_{two} **(c)** 554_{six}; 1000_{six} **16. (a)** There is no numeral 4 in base four. **(b)** There are no numerals 6 or 7 in base five. **17.** 7; 3 blocks, 1 flat, 1 long, 2 units

18.

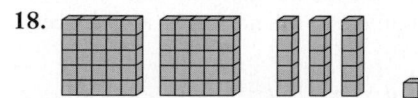

19. (a) 8 pennies can be traded for 1 nickel and 3 pennies. After the trade, we have 2 quarters, 10 nickels, and 3 pennies. 10 nickels can be traded for 2 quarters. After this trade, we have 4 quarters, 0 nickels, and 3 pennies. **(b)** Assume that you have 73 cents in any possible combination, for example, 10 nickels and 23 pennies. Because 23 pennies can be traded for 4 nickels and 3 pennies, we have 14 nickels and 3 pennies. 14 nickels can be traded for 2 quarters and 4 nickels. After the second trade, we should have 2 quarters, 4 nickels, and 3 pennies. Thus, $73 = 243_{\text{five}}$. **20. (a)** 3030_{three} **(b)** $EOTE_{\text{twelve}}$ **21. (a)** 10 flats = 1 block; 10 flats in base ten = 1000 **(b)** 20 flats = 1 block + 8 flats; 20 flats in base twelve = 1800_{twelve} **(c)** 10000_{two} **22. (a)** 3311_{five} **(b)** 1046_{twelve} **(c)** 100000_{two} **23. (a)** 117 **(b)** 45 **(c)** 1331 **24.** 1 prize of \$625, 2 prizes of \$125, and 1 of \$25 **25. (a)** 8 weeks, 2 days **(b)** 1 day, 5 hr **26. (a)** 6 **(b)** 1 **27. (a)** 30300_{five} **(b)** 208010_{twelve} **28. (a)** Answers vary. For example, subtract 2020. **(b)** Answers vary, for example, subtract 50.

Mathematical Connections 2-1

Communication

1. Answers vary. Ben is incorrect. Zero is a place holder in the Hindu-Arabic system. It is used to differentiate between numbers like 54 and 504. If zero were nothing, then we could eliminate it without changing our number system. **3. (a)** Answers vary. We name our numbers in groups of three digits. Thus the grouping in threes is natural. For example, 323 thousand, 243. It helps with readability. It has been proposed with the metric system to drop the commas and simply use spaces instead. **(b)** Answers vary. **5.** 4; 1, 2, 4, 8; 1, 2, 4, 8, 16

Cooperative Learning

7. A base-ten system uses place value and face value. The Roman system uses the additive, subtractive, and multiplicative properties and does not use place value.

Assessment 2-2A

1. (a) $\{m, a, t, h, e, i, c, s\}$ **(b)** $\{x \mid x$ is a natural number and $x > 20\}$ or $\{21, 22, 23, \ldots\}$ **2. (a)** $P = \{a, b, c, d\}$ **(b)** $\{1, 2\} \subset \{1, 2, 3, 4\}$ **(c)** $\{0, 1\} \not\subseteq \{1, 2, 3, 4\}$ **(d)** $0 \notin \{\ \}$ or $0 \notin \varnothing$ **3. (a)** Yes **(b)** Yes **(c)** No **4. (a)** 720 **(b)** $n(n - 1)(n - 2) \cdot \ldots \cdot 3 \cdot 2 \cdot 1 = n!$ **5. (a)** 24 **(b)** 6 **(c)** 12 **6.** $A = C, E = H, I = L$ **7. (a)** $1100 - 100$, or 1000 if arithmetic **(b)** 501 if arithmetic **(c)** 11 if geometric **(d)** 100 **(e)** 5 **8.** $\overline{A}$ is the set of all college students with at least one grade that is not an A; that is, those college students who do not have a straight-A average. **9. (a)** 7 **(b)** 0 **10. (a)** $n(D) = 5$ **(b)** $C = B$ **11. (a)** $2^5 = 32$ **(b)** $2^5 - 1 = 31$ **(c)** 8 **12.** 8 **13.** $B \subset A, C \subset A, C \subset B$ **14. (a)** $\notin$ **(b)** $\notin$ **(c)** $\in$ **(d)** $\in$ **15. (a)** $\not\subseteq$ **(b)** $\subseteq$ **(c)** $\not\subseteq$ **(d)** $\not\subseteq$ **16. (a)** Yes **(b)** No. A may equal B. **(c)** Yes **(d)** No. Consider $A = \{1\}$ and $B = \{1, 2\}$. **17. (a)** Let $A = \{1, 2, 3\}$ and $B = \{1, 2, 3, 4, \ldots, 100\}$. Since $A \subset B, n(A)$ is less than $n(B)$. So $3 < 100$. **(b)** Let $A = \varnothing$ and $B = \{1, 2, 3\}$. Since $A \subset B, n(A) = 0$ is less than $n(B) = 3$, so $0 < 3$. **18.** 35 **19.** 81

Mathematical Connections 2-2

Communication

1. A set is well defined if any object can be classified as belonging to the set or not belonging to the set. For example, the set of U.S. presidents is well defined but the set of rich U.S. presidents is not well defined because "rich" is a matter of opinion. **3.** Yes, $\varnothing \subset A$ for all non-empty sets A since A contains at least one element and $\varnothing$ contains none; also $\varnothing \subseteq A$. **5.** To show $A \not\subseteq B$, we must be able to find at least one element from set A that does not belong to set B. **7.** If A and B are finite sets, we say $n(A) \leq n(B)$ in case A is a subset (not necessarily proper) of B.

Open-Ended

9. (a) Answers vary. Let A be the set of all natural numbers not equal to 1 with N as the universal set. Then $\overline{A} = \{1\}$ is finite. **(b)** Answers vary. Let A be the set of even natural numbers with N as the universal set. $\overline{A}$ is the set of odd natural numbers and so is infinite.

Cooperative Learning

11. (a) There are $2^{64} \approx 1.84 \times 10^{19}$ subsets of $\{1, 2, 3, \ldots, 64\}$. If a computer can list one every millionth of a second, then it would take about

$$1.84 \times 10^{19} \times 0.000001 \text{ sec} \times \frac{1 \text{ yr}}{31,536,000 \text{ sec}} \approx 580,000 \text{ yr to}$$

list all the subsets. **(b)** There are $64 \cdot 63 \cdot 62 \cdot \ldots \cdot 2 \cdot 1 \approx 1.27 \times 10^{89}$ one-to-one correspondences between the two sets. So it would take about $1.27 \times 10^{89} \times 0.000001 \text{ sec} \times$ $\frac{1 \text{ year}}{31,536,000 \text{ sec}} \approx 4 \times 10^{75}$ yr to list all the one-to-one correspondences between the sets.

Questions from the Classroom

13. If $A \subseteq B$, then every element in A is an element of B. Since $B \subseteq C$ every element in B is an element of C. Thus every element in A is in C and $A \subseteq C$. **15.** The student is incorrect. Consider $A = \{1\}$ and $B = \{a\}$. Neither $A \subseteq B$ nor $B \subseteq A$ is true.

Review Problems

17. 1410 **19.** 10111_{twelve}

Assessment 2-3A

1. (a) A or C **(b)** N **(c)** $\varnothing$ **2. (a)** Yes **(b)** Yes **(c)** Yes **(d)** Yes **3. (a)** True **(b)** False. Let $A = \{a, b, c\}$ and $B = \{a, b\}$. Then $A - B = \{c\}$, but $B - A = \varnothing$. **(c)** False. Let $U = \{a, b, c\}, A = \{a\}$, and $B = \{b\}$. Then $A \cap B = \varnothing$ and $\overline{A \cap B} = U. \overline{A} = \{b, c\}; \overline{B} = \{a, c\}$ and $\overline{A} \cap \overline{B} = \{c\}$. $\overline{A \cap B} \neq \overline{A} \cap \overline{B}$. **(d)** False. Let $A = \{a, b\}; B = \{b\}$. $A \cup B = \{a, b\}; (A \cup B) - A = \varnothing \neq B$. **(e)** False. Let $A = \{1, 2, 3\}, B = \{3, 4, 5\}$. Then $(A - B) \cup A = \{1, 2, 3\}$, but $(A - B) \cup (B - A) = \{1, 2\} \cup \{4, 5\} = \{1, 2, 4, 5\}$. **4. (a)** $A \cap B = B$ **(b)** $A \cup B = A$

5. (a) 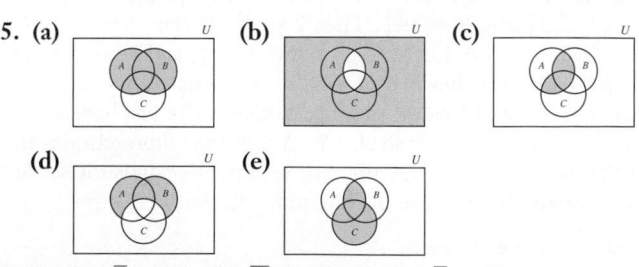 **(b)** **(c)**

(d) **(e)**

6. (a) $S \cup \overline{S} = U$ **(b)** $\overline{U} = \varnothing$ **(c)** $S \cap \overline{S} = \varnothing$ **(d)** $\varnothing \cap S = \varnothing$ **7. (a)** $A - B = A$ **(b)** A **(c)** $A - B = \varnothing$ **8.** Yes. By definition $A - B$ is the set of all elements in A that are not in B. If $A - B$ is the empty set, then this means that there are no elements in A that are not in B, which makes A a subset of B. **9.** Answers vary.

10.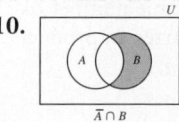

$\overline{A} \cap B$

11. (a) False

$A \cup (B \cap C)$ $\neq$ $(A \cup B) \cap C$

(b) False

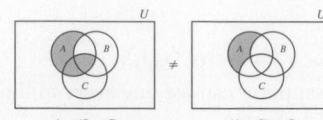

$A - (B - C)$ ≠ $(A - B) - C$

12. (a) $A \cap B \cap C \subseteq A \cap B$ **(b)** $A \cup B \subseteq A \cup B \cup C$
13. (a) (i) 5 **(ii)** 2 **(iii)** 2 **(iv)** 3   **(b) (i)** $n + m$ **(ii)** The smaller
of the two numbers m and n **(iii)** m **(iv)** n **14. (a)** Greatest is
15; least is 6.   **(b)** Greatest is 4; least is 0.   **15. (a)** The set of
college basketball players more than 200 cm tall   **(b)** The set of
humans who are not college students or who are college students
less than or equal to 200 cm tall   **(c)** The set of humans who are
college basketball players or who are college students taller than
200 cm   **(d)** The set of all humans who are not college basketball
players and who are not college students taller than 200 cm
(e) The set of all college students taller than 200 cm who are not
basketball players   **(f)** The set of all college basketball players
less than or equal to 200 cm tall   **16.** 18   **17.** 4
18. (a) 20 **(b)** 10 **(c)** 10   **19.** 3. Using the following Venn
diagram and the fact that the set of people who are O negative is
$100 - n(A \cup B \cup C)$, we see that the answer is 3.

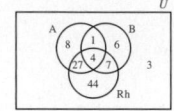

20. (a) False. Let $A = \{a, b, c\}$ and $B = \{1, 2, 3\}$.   **(b)** False.
Let $A = \{1, 2, 3\}$ and $B = \{1, 2, 3, 4\}$.   **(c)** True.   **21.** Steelers
versus Jets, Vikings versus Packers, Bills versus Redskins, Cowboys
versus Giants   **22. (a)** $A \times B = \{(x, a), (x, b), (x, c), (y, a),$
$(y, b), (y, c)\}$ **(b)** $B \times A = \{(a, x), (a, y), (b, x), (b, y),$
$(c, x), (c, y)\}$ **(c)** No   **23. (a)** $C = \{a\}, D = \{b, c, d, e\}$
(b) $C = \{1, 2\}, D = \{1, 2, 3\}$ **(c)** $C = D = \{0, 1\}$

Mathematical Connections 2-3

Communication

1. (a) Yes. $A \cap B \subseteq A \cup B$ **(b)** No. For example, let
$A = \{1, 2, 3\}$ and $B = \{4\}$. Then $2 \in A \cup B$, but
$2 \notin A \cap B$.   **3.** No. Let $A = \{1\}$ and $B = \{a\}$. Then
$A \times B = \{(1, a)\}$, but $B \times A = \{(a, 1)\}$. These are not
equal.   **5.** B could have 14 or more elements. B must have at
least one element more than A.   **7.** A highway intersection is the
part that highways have in common. Set intersection is the set of
elements that the sets have in common.   **9.** Answers vary

Questions from the Classroom

11. Elements are not repeated in sets. In the student question,
b and c are listed twice.   **13.** Even though the Cartesian product
of sets includes all pairings in which each element of the first set is
the first component in a pair with each element of the second set,
this is not necessarily a one-to-one correspondence. A one-to-one
correspondence implies that there must be the same number of
elements in each set. This is not the case in a Cartesian product.
For example, consider sets $A = \{1\}$ and $B = \{a, b\}$.

Review Problems

15. The number "two" exists in base two as 10_{two} but
there is no single digit to represent "two."   **17. (a)** Answers vary, for example, $\{x \mid 3 < x < 10,$
and $x \in N\}$ **19. (a)** 8 **(b)** 8 **(c)** 12 **(d)** 4
(e) 16, 16   **(f)** Every subset of A is a subset of B. The others can

be listed by adjoining the element 5 to each subset of A. So there
are twice as many subsets of B as subsets of A. A has 16 subsets
and B has 32 subsets.   **21.** Answers vary.   **23.** 60

Chapter Review

1. (a) tens   **(b)** thousands   **(c)** hundreds   **2. (a)** 400,044
(b) 117   **(c)** 1704   **(d)** 11   **(e)** 1448   **3. (a)** CMXCIX
(b) ∩∩∩∩∩∩∩∩||||||| **(c)** ÷   **(d)** 2341_{five} **(d)** 11011_{two} **4. (a)** 3^{17}
(b) 2^{21} **5.** 2020_{three} **6.** 1 block, 2 flats, 2 longs, 0 units
7. (a)

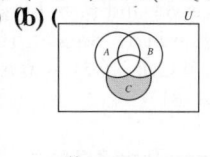

8. (a) 10^{10} **(b)** 5^8 **(c)** 2^{31} **9. (a)** 10,000,000,023
(b) $10,000,000,001_{two}$ **(c)** $10,000,000,001_{five}$ **(d)** 9,999,999,999
(e) $1,111,111,111_{two}$ **(f)** $EEEEE_{twelve}$ **10.** Answers vary. Selling pencils by the units, dozen, and gross is an example of the use
of base 12.   **11. (a)** The Egyptian system had seven symbols. It
was a *tally* system and a *grouping* system, and it used the *additive
property*. It did not have a symbol for zero but this was not very
important because it did not use place value.   **(b)** The Babylonian
system used only two symbols. It was a place value system (base 60)
and was additive within the positions. It lacked a symbol for zero
until around 300 BCE.   **(c)** The Roman system used seven symbols. It was additive, subtractive, and multiplicative. It did not have
a symbol for 0.   **(d)** The Hindu-Arabic system uses 10 symbols. It
uses place value and it has a symbol for 0.   **12. (a)** 1003_{five}
(b) 10000000_{two} **(c)** $T8_{twelve}$ **13. (a)** 10000001000_{two}
(b) $E0T018_{twelve}$ **14.** eight
15. 15   **16.** $\varnothing, \{m\}, \{a\}, \{t\}, \{b\}, \{m, a\}, \{m, t\},$
$\{m, b\}, \{a, t\},$
$\{a, b\}, \{t, b\}, \{m, a, t\}, \{m, a, b\}, \{m, t, b\}, \{a, t, b\}, \{m, a, t, b\}$
17. (a) $A \cup B = A$ **(b)** $C \cap D = \{l, e\}$
(c) $\overline{D} = \{u, n, i, v, r\}$ **(d)** $A \cap \overline{D} = \{r, v\}$
(e) $\overline{B \cup C} = \{s, v, u\}$ **(f)** $(B \cup C) \cap (D \cup C), a\}$
(g) $\{i, n\}$ **(h)** $\{e\}$ **(i) (b)** (
18. (a)

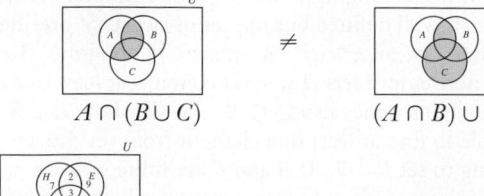

$A \cup B \cap C$

19. 5040   **20. (a)** Answers will vary.
$t \longleftrightarrow e; b \longleftrightarrow n e \longleftrightarrow d$ **(b)** 6   **21.** It is not true that
$A \cap (B \cup C) = (A \cap B) \cup C$ for all A, B, and C, as shown in the
following diagrams.

$A \cap (B \cup C)$ ≠ $(A \cap B) \cup C$

22.

23. Answers vary.   **(a)** $B \cup (C \cap A)$ **(b)** $B - C$ or $B \cap \overline{C}$
24. (a) False. Consider the sets $\{a\}$ and $\{2\}$.   **(b)** False. It is
not a proper subset of itself.   **(c)** False. Consider the sets $\{t, b, e\}$
and $\{e, n, d\}$. They have the same number of elements, but they
are not equal.   **(d)** False. This can be placed into one-to-one
correspondence with the set of natural numbers.   **(e)** False. The
set $\{5, 10, 15, 20, \dots\}$ is a proper subset of the natural numbers

and is equivalent to the natural numbers, since there is a one-to-one correspondence between the two sets. **(f)** False. Let $B = \{1, 2, 3\}$ and A be the set of natural numbers. **(g)** True **(h)** False. Let $A = \{1, 2, 3\}$ and $B = \{a, b, c\}$. **25. (a)** Because $A \cup B$ is the union of disjoint sets, $A - B$, $B - A$, and $A \cap B$, the equation is true. **(b)** True, because $A - B$ and B are disjoint as well as $B - A$ and A; also, $A \cup B = (A - B) \cup B = (B - A) \cup A$. **26. (a)** 17 **(b)** 34 **(c)** 0 **(d)** 17 **27.** 7 **28. (a)** False. $U = \{1, 2, 3, 4\}$, $A = \{1, 2\}$, $B = \{1, 2, 3\}$, $\overline{A} = \{3, 4\}$, $\overline{B} = \{4\}$, $\overline{A} \not\subseteq \overline{B}$. **(b)** True **(c)** True **(d)** True **(e)** True **(f)** True **29.** 24 **30. (a)** False, $U = \{1, 2, 3, 4\}$, $A = \{1, 2, 3\}$, $B = \{2, 3\}$, $C = \{2\}$, $A \cup B = A \cup C$, $B \neq C$. **(b)** False, $U = \{1, 2, 3, 4\}$, $A = \{1, 2, 3\}$ $B = \{1, 2, 4\}$, $A \cap B = A \cap C$, $B \neq C$ **31. (a)** Let $A = \{1, 2, 3, \ldots, 13\}$ and $B = \{1, 2, 3\}$; then B is a proper subset of A. Therefore, B has fewer elements than A, so $n(B)$ is less than $n(A)$; thus, $3 < 13$. **(b)** Let $A = \{1, 2, 3, \ldots, 12\}$ and $B = \{1, 2, 3, \ldots, 9\}$. B is a proper subset of A, so A has more elements than B; thus $n(A)$ is greater than $n(B)$, $12 > 9$. **32.** 12 outfits

Answers to Now Try This

2-1. (a) 3 blocks 12 flats 11 longs 17 units
= 3 blocks (1 block 2 flats) (10 longs 1 long) (10 units 7 units)
= 4 blocks 2 flats (1 flat 1 long) (1 long 7 units)
= 4 blocks 3 flats 2 longs 7 units
= 4327
(b) $3282 = 3 \cdot 10^3 + 2 \cdot 10^2 + 8 \cdot 10 + 2 \cdot 1$
2-2. (a) 𓏤𓇳𓏤𓏤𓏤𓏥 999∩∩||| **(b)** 203,034
(c) Answers vary. For example, writing large numbers is very cumbersome as the system is additive and does not use place value. Performing operations involving addition, subtraction, multiplication, and division is hard because of the way that numbers are represented. **2-3. (a)** ▼▼▼ <<▼▼▼▼▼ <<▼ **(b)** $2 \cdot 60^2 + 11 \cdot 60 + 1 = 7861$ **(c)** Answers vary. The Hindu-Arabic system has a symbol for 0 and this is very important in a system that uses place value. Because it uses base sixty, the Babylonian system requires the use of many symbols to write numbers such as 59.
2-4. (a) 249 **(b) (i)** MDCXXXIV **(ii)** $\overline{\text{V}}$CCLXXX

(iii) LXXXVIII **(c) (i)** ⦂≣ **(ii)** ⦂⦂
2-5.
```
5| 728
 5| 145    3
  5| 29     0
   5| 5      4
      1 → 0
```
Thus, the answer is 10403_{five}. **2-6. (a)** and **(b)** represent the correspondence:

$$
\begin{array}{ccc}
1 & \leftrightarrow & A \\
2 & \leftrightarrow & B \\
3 & \leftrightarrow & C \\
4 & \leftrightarrow & D
\end{array}
\quad \text{as} \quad
\begin{array}{c}
1\ 2\ 3\ 4 \\
A\ B\ C\ D
\end{array}
$$

The 24 one-to-one correspondences are

1 2 3 4	1 2 3 4	1 2 3 4	1 2 3 4
A B C D	B A C D	C A B D	D A B C
A B D C	B A D C	C A D B	D A C B
A C B D	B C A D	C B A D	D B A C
A C D B	B C D A	C B D A	D B C A
A D B C	B D A C	C D A B	D C A B
A D C B	B D C A	C D B A	D C B A.

(c) We notice that $24 = 4 \cdot 3 \cdot 2 = 4 \cdot 3 \cdot 2 \cdot 1$. We also notice that we had four choices for people to swim in lane 1. After making a choice, we see that we had three choices for the swimmer in lane 2, leaving us with two choices for lane 3 and, finally, one choice for lane 4. Extrapolating from this, we conjecture that there are
$$5 \cdot 4 \cdot 3 \cdot 2 \cdot 1 = 120$$
distinct one-to-one correspondences between a pair of five-element sets. **2-7.** $n(n - 1)(n - 2) \cdot \ldots \cdot 1 = n!$
2-8. (a) No. Two sets may be equivalent without being equal. To see this consider the following example:
$$A = \{a, b, c\}$$
$$B = \{1, 2, 3\}.$$
Then,
$$a \leftrightarrow 1$$
$$b \leftrightarrow 2$$
$$c \leftrightarrow 3$$
is a one-to-one correspondence between A and B, and therefore $A \sim B$. However, $A \neq B$. **(b)** Yes. If two sets A and B are equal, then they have exactly the same elements. Therefore there is a one-to-one correspondence between A and B given by pairing each element in A with that same element in B. **2-9.** The set of natural numbers is $N = \{1, 2, 3, 4, 5, \cdots \}$. If N were finite, then there would be some greatest element q in N. However, $q + 1$ is a natural number greater than q and would still be in the set N. Thus, there can be no greatest element in N and it cannot be finite. **2-10. (a)** Yes. By definition, $A \subseteq B$ means that every element of A is an element of B. Similarly, $A \subset B$ means that every element of A is an element of B but there exists an element in B that is not an element of A. Hence, if $A \subset B$, it is true that every element of A is in B. Consequently, $A \subseteq B$. Notice that if the more stringent condition $A \subset B$ is satisfied, then the weaker condition $A \subseteq B$ must also be satisfied. **(b)** No. Consider the following counterexample:
$$A = \{a, b, c\}$$
$$B = \{a, b, c\}$$
Then, $A \subseteq B$. Notice that $A \not\subset B$ since $A = B$.
(c) No. By definition, if the empty set were a proper subset of the empty set, then the empty set would have to contain an element that is not in the empty set. But the empty set has no elements. Therefore, $\emptyset \not\subset \emptyset$. **2-11. (a)** Assuming that a simple majority forms a winning coalition, we see that any subset consisting of three or more senators is a winning coalition. There are 16 such subsets. To see this, let $\{A, B, C, D, E\}$ be the set of five senators on the committee. Then the following are all possible winning coalitions:

$\{A, B, C\}$	$\{A, B, D\}$	$\{A, B, E\}$	$\{A, C, D\}$
$\{A, C, E\}$	$\{A, D, E\}$	$\{B, C, D\}$	$\{B, C, E\}$
$\{B, D, E\}$	$\{C, D, E\}$	$\{A, B, C, D\}$	$\{A, B, C, E\}$
$\{A, B, D, E\}$	$\{A, C, D, E\}$	$\{B, C, D, E\}$	$\{A, B, C, D, E\}$

From the list, we see that there are five subsets containing exactly four members. We also see that there are five senators on the committee. To understand why these numbers are the same, notice that creating a four-element subset is equivalent to deleting a single element from the total set. That is, we can give a one-to-one correspondence between the set of four-element subsets of $\{A, B, C, D, E\}$ and the set of senators by corresponding to each four-element subset the senator who is not in that subset, as shown:
$$
\begin{array}{ccc}
\{A, B, C, D\} & \leftrightarrow & E \\
\{A, B, C, E\} & \leftrightarrow & D \\
\{A, B, D, E\} & \leftrightarrow & C
\end{array}
$$

$$\{A, C, D, E\} \leftrightarrow B$$
$$\{B, C, D, E\} \leftrightarrow A$$

(b) We can give a one-to-one correspondence between the three-element subsets and the two-element subsets of $\{A, B, C, D, E\}$ by matching each three-element subset with the unique two-element subset that contains the senators on the committee but not in the subset; for example, $\{A, B, C\} \leftrightarrow \{D, E\}$. From part (a), we know that there are exactly 10 three-element subsets of the committee; hence, there must be 10 two-element subsets of the committee.

2-12. (a) 32 **(b)** 15 **(c)** $2^n - 1$ **(d)** 16

2-13. The formula is $n(A \cup B) = n(A) + n(B) - n(A \cap B)$. To justify this formula, notice that in $n(A \cup B)$, the elements of $A \cap B$ are counted only once. In $n(A) + n(B)$, the elements of $A \cap B$ are counted twice, once in A and once in B. Thus, subtracting $n(A \cap B)$ from $n(A) + n(B)$ makes the number equal to $n(A \cup B)$. For example, if $A = \{a, b, c\}$ and $B = \{c, d\}$, then $A \cup B = \{a, b, c, d\}$ and $n(A \cup B) = 4$. However, $n(A) + n(B) = 3 + 2 = 5$ since c is counted twice. Because $A \cap B = \{c\}$, $n(A \cap B) = 1$ and $n(A) + n(B) - n(A \cap B) = 4$.

2-14. It is always true that $A \cap (B \cap C) = (A \cap B) \cap C$. The following figure gives Venn diagrams of each side of this equation. Because the Venn diagrams result in the same set, the equation is always true.

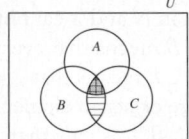 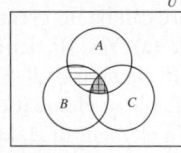

Similarly, it is always true that $A \cup (B \cup C) = (A \cup B) \cup C$. The following Venn diagrams justify this statement.

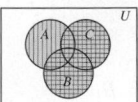 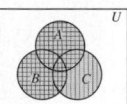

However, in general, $A - (B - C) \neq (A - B) - C$. To see this, consider the following counterexample:

$$A = \{1, 2, 3, 4, 5\}$$
$$B = \{1, 2, 3\}$$
$$C = \{3, 4\}$$

Then, $A - (B - C) = A - \{1, 2\} = \{3, 4, 5\}$, but $(A - B) - C = \{4, 5\} - C = \{5\}$. Thus for the preceding choice of A, B, and C, we have that $A - (B - C) \neq (A - B) - C$. **2-15.** The following Venn diagrams show $A \cup (B \cap C) = (A \cup B) \cap (A \cup C)$.

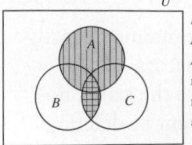 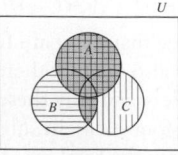

The property should be called the distributive property of set union over intersection.

2-16. 1. 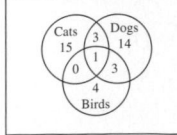 **2.** 4; 4 **3.** 0 **4.** 5 **5.** Answers vary

Answer to Brain Teaser

Section 2-2

Consider the set of students in Mr. Gonzales's class. This set has $2^{24} - 1$ non-empty subsets. The set of students in Ms. Chan's class has $2^{25} - 1$ non-empty subsets. Using the Fundamental Counting Principle, we find that $(2^{24} - 1) \cdot (2^{25} - 1)$, or approximately 563 trillion school committees could be formed to contain at least one student from each class. This number is greater than the population of the world, which is approximately 7 billion. Thus, Linda is right.

Answer to the Preliminary Problem

We can sort the information using a Venn diagram. Suppose M is the set of students taking mathematics, C is the set of students taking chemistry, and P is the set of students taking physics. We draw a general Venn diagram representing all possible intersections as shown in Figure (a).

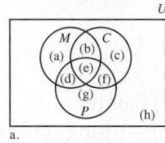 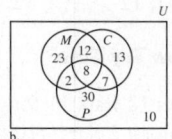

Note that there are 8 distinct regions formed and these are labeled (a)–(h) for reference. As we determine the number of students represented in each region, these numbers can be placed as shown in Figure (b). Because 10 students take none of the 3 subjects, we can enter 10 in region (h). Next because 8 students take all three subjects, we can place an 8 in region (e). Because 10 students take both mathematics and physics and we already have 8 students taking these two subjects, we need to place 2 students in region (d). In a similar manner we determine that there are 12 students in region (b) and 7 students in region (f).

Now we know a total of 45 students are taking mathematics and we already have $2 + 8 + 12 = 22$, so there must be 23 students in region (a). In a similar manner we can determine that there are 13 students in region (c) and 30 students in region (g).

The completed figure with all the data is shown in Figure (b). If we total the number of students in all 8 regions we see that there are 105 students. However, this contradicts John's data that reports that only 100 students were surveyed. So just on the basis of this report John was possibly not hired for the job.

Chapter 3

Assessment 3-1A

1. (a) true **(b)** false **(c)** true **2.** $n(A \cap B) = 2$ **3.** For example, let $A = \{1, 2\}$, $B = \{2, 3\}$; then $A \cup B = \{1, 2, 3\}$. Thus, $n(A) = 2$, $n(B) = 2$, $n(A \cup B) = 3$, but $n(A) + n(B) = 2 + 2 = 4 \neq n(A \cup B)$. **4. (a)** 3, 4, 5, 6 **(b)** 3

5. (a) yes **(b)** yes **(c)** yes **(d)** No; $3 + 5 \notin V$. **(e)** yes **(f)** No, $1 + 1$ not in the set **6.** Answers vary. For example, 4, 6, 8 and 5, 7, 9 **7. (a)** 1, commutative property of addition **(b)** 7, commutative property of addition **(c)** 0, additive identity **(d)** 7, associative and commutative properties of addition

8. (a) commutative property of addition **(b)** associative property of addition **(c)** commutative property of addition **(d)** identity property of addition **(e)** commutative property of

addition   **(f)** associative property of addition   **9.** No. If $k = 0$, we would have $k = 0 + k$, implying $k < k$ which is false.
10. (i) For any whole numbers a and b, $a < b$ if, and only if, there exists a natural number k such that $b - k = a$, or equivalently if, and only if, $b - a$ is a natural number.   **(ii)** For any whole numbers a and b, $a > b$ if, and only if, there exists a natural number k such that $a - k = b$, or equivalently if and only if $a - b$ is a natural number.   **11. (a)** 33, 38, 43   **(b)** 56, 49, 42
12. (a) 9   **(b)** 8   **(c)** 3 **(d)** 6 or 8   **(e)** 5   **(f)** 4 or 8   **(g)** 9
13. $9 + 8 + 7 + \cdots + 1$ or $\dfrac{10 \cdot 9}{2} = 45$
14. (a) Vera is the tallest and Kent is the shortest **(b)** Answers can vary, for example: Kent, 140 cm; Mischa, 142 cm; Sally, 148 cm; Vera, 152 cm   **15. (a)** $9 = 7 + x$ or $9 = x + 7$
(b) $x = 6 + 3$ or $x = 3 + 6$ **(c)** $9 = x + 2$ or $9 = 2 + x$
16. (a) $a \geq b$ **(b)** $b \geq c$ and $a \geq b - c$
17. (a) $x + (y + z) = (x + y) + z$; associative
$$= z + (x + y); \text{ commutative}$$
(b) $x + (y + z) = (y + z) + x$; commutative
$$= y + (z + x); \text{ associative}$$
$$= y + (x + z); \text{ commutative}$$
18. (a)   **(b)** The solution to $\square + 5 = 8$ is 3

        Take-away
        5 x's.
        $8 - 5 = 3$
   3 x's left

(c) difference is 3   **(d)** $8 - 5 \rightarrow 5$; $0\,1\,2\,3\,4\,5\,6\,7\,8$ **19. (a)** 4   **(b)** 8   **(c)** 5
(d) 9   **(e)** all whole numbers   **(f)** 11   **(g)** no solution   **(h)** all whole numbers   **20.** $a - (b + c)$

Mathematical Connections 3-1

Communication

1. No; sets A (students taking algebra) and B (students taking biology) do not have to be disjoint. For example, suppose there are 11 students taking both algebra and biology. This implies the following Venn diagram:

      A B
      11 (11) 19

From this Venn diagram we see that $n(A \cup B) = 41$ and there are not necessarily 52 students taking algebra or biology.
3. Answers vary. For example, **(i)** if $a > b$ then $a = b + k$ for some counting number k. Then $a + c = b + c + k$ and therefore $a + c > b + c$.   **(ii)** If Adam has a dollars and Bob has b dollars and Adam has more money than Bob and each gets additional c dollars, Adam will still have more than Bob.   **5.** It is very useful to have students learn more than one method to model addition and subtraction if the methods are to generalize to sets of numbers other than whole numbers. The missing-addend approach is useful for all sets of numbers in solving subtraction problems. The method of counting to compute additions is not effective with sets of fractions or real numbers.   **7. (a)** When you put 9 and 4 next to each other, it is equal to the same length as 13.
(b) If you put 9 and 4 together on top of 4 and 9 together, they both equal 13.   **(c)** Take the length 9 away from 13 and the length left is equal to 4.   **(d)** Take the length 4 away from 13 and the length left is equal to 9.   **9.** For 0 to be an identity for subtraction, the following would have to hold: For every whole number a, $a - 0 = a = 0 - a$. In this case, $a - 0 = a$ works but $0 - a = a$ is not true for all whole numbers.

Open-Ended

11. Answers vary. For example, let $A = \{a, b\}$ and $B = \{a, b, c, d\}$. Then $4 - 2 = n(B - A) = n(\{c, d\}) = 2$.

Cooperative Learning

13. (a) The table shows that if you add any two single-digit whole numbers, your answer is also a whole number.   **(b)** The table shows that if you add any two single-digit whole numbers, the order is not important; that is, if $a \in W$ and $b \in W$, $a + b = b + a$. Each row of answers has a corresponding column of identical answers.
(c) The first row and first column show that if you add any digit to the identity, 0, you get the identical digit back again.   **(d)** The properties reduce the number of facts to be remembered: for example, the 19 facts in the first row and column can be learned by just knowing that 0 is the additive identity. The commutative property also reduces the number of facts; for example, if you know $9 + 2$, then you know $2 + 9$.   **15.** Answers vary

Questions from the Classroom

17. Answers vary. For example, the take-away model or the missing-addend model do not work for all types of subtraction problems. The model that is appropriate in this case is the *comparison* model. The pencils Karly has can be matched to some of the pencils that Sam has and then the answer is the number of Sam's pencils left over.   **19.** Answers vary. For example, John is correct in that he can get the same answer by adding up or adding down. He is using the fact that if a, b, and c are whole numbers, then $(a + b) + c = (c + b) + a$. The commutative and associative properties can be used to justify the steps given below.
$$a + b + c = (a + b) + c = c + (a + b)$$
$$= c + (b + a) = (c + b) + a$$

Assessment 3-2A

1. (a)
```
  981
 +421
 1402
```
(b)
```
 2025
 1196
+3148
 6369
```
2.

3. (a) Answers vary one possibility:
```
  863
+ 752
 1615
```
(b) Answers vary one possibility:
```
  368
+ 257
  625
```
4.
```
  3428
+ 5631
  9059
```
5. (a)
```
  93      93 + 3      96
- 37  →  - (37 + 3) → - 40
              56
```
(b)
```
 321      321 + 2      323      323 + 60      383
- 38  →  - (38 + 2) → - 40  →  - (40 + 60) → - 100
                                              283
```
6. (a) (i)
```
  687
 +549
   16
   12
   11
 1236
```
(ii)
```
  359
 +673
   12
   12
    9
 1032
```

(b) The algorithm works because the placement of partial sums still accounts for place value. It is fairly easy in the example because only two digits are added at a time. This process can be adapted if more than two numbers are added.   **7.** Answers vary. For example, **(a)** The student added $8 + 5 = 13$ and wrote down 13 with no regrouping. He then added $2 + 7 = 9$ and wrote down 9.
(b) The student added $8 + 5 = 13$ and instead of writing down 3 and regrouping with the 1, he wrote down 1 and regrouped with the 3.   **(c)** The student only recorded the difference in the units $(9 - 5)$, the tens $(5 - 0)$, and the hundreds $(3 - 2)$. The student always subtracted the smaller number from the larger number ignoring what number was "on top" and which was "on the bottom."
(d) The student regrouped 3 hundreds as 2 hundreds and 10 tens but then did not regroup the 10 tens as $9 \cdot 10 + 15$.   **8.** No.
9. Step 1—place value
Step 2—commutative and associative properties of addition
Step 3—distributive property of multiplication over addition
Step 4—single-digit addition facts
Step 5—place value

10. (a)
$$68 + 23 = (6 \cdot 10 + 8) + (2 \cdot 10 + 3)$$
$$= (6 \cdot 10 + 2 \cdot 10) + (8 + 3)$$
$$= (6 + 2)10 + (8 + 3)$$
$$= 8 \cdot 10 + 11$$
$$= 8 \cdot 10 + (1 \cdot 10 + 1)$$
$$= (8 \cdot 10 + 1 \cdot 10) + 1$$
$$= (8 + 1) \cdot 10 + 1 = 9 \cdot 10 + 1 = 91$$

(b)
$$174 + 285 = (1 \cdot 100 + 7 \cdot 10 + 4) + (2 \cdot 100 + 8 \cdot 10 + 5)$$
$$= (1 \cdot 100 + 2 \cdot 100) + (7 \cdot 10 + 8 \cdot 10) + (4 + 5)$$
$$= (1 + 2) \cdot 100 + (7 + 8) \cdot 10 + (4 + 5)$$
$$= 3 \cdot 100 + 15 \cdot 10 + 9$$
$$= 3 \cdot 100 + (10 + 5) \cdot 10 + 9$$
$$= 3 \cdot 100 + 1 \cdot 100 + 5 \cdot 10 + 9$$
$$= (3 + 1) \cdot 100 + 5 \cdot 10 + 9$$
$$= 4 \cdot 100 + 5 \cdot 10 + 9$$
$$= 459$$

11. (a)

```
    4  3  5  8
 +  3  8  6  4
  0 /1 /1 /1
 /7 /1 /1 /2
    8  2  2  2
```

(b)

```
    4  9  2  3
 +  9  8  9  7
  1 /1 /1 /1
 /3 /7 /1 /0
 1  4  8  2  0
```

12. (a) 121_{five} **(b)** 20_{five} **(c)** 1010_{five} **(d)** 14_{five} **(e)** 1001_{two}
(f) 1010_{two}

13.

+	0	1	2	3	4	5	6	7
0	0	1	2	3	4	5	6	7
1	1	2	3	4	5	6	7	10
2	2	3	4	5	6	7	10	11
3	3	4	5	6	7	10	11	12
4	4	5	6	7	10	11	12	13
5	5	6	7	10	11	12	13	14
6	6	7	10	11	12	13	14	15
7	7	10	11	12	13	14	15	16

Base eight

(a) 474_{eight} **(b)** 667_{eight} **14. (a)** 9 hr 33 min 25 sec **(b)** 1 hr 39 min 40 sec **15. (a)** 34; 34; 34 **(b)** 34 **(c)** 34 **(d)** yes **(e)** yes

16. (a)

```
     1 1
     4 3 2
     9 7 6
   1 4 1
 + 1 4 1 8₆
   2 8 2 6
```

(b)

```
   3 3 2
   1 3 0
   2 2
   4 3 0
   2 0 3
   1 2 0
   3 1 0 five
```

17. There is no numeral 5 in base five: $22_{\text{five}} + 33_{\text{five}} = 110_{\text{five}}$.
18. (a) 1241_{five} **(b)** 101_{two} **(c)** TET_{twelve} **(d)** 4000_{five}
19. There is no numeral 5 in base five: $22_{\text{five}} + 33_{\text{five}} = 110_{\text{five}}$. **20. (a)** Answers vary. For example, place 5 in the middle and the following pairs along the same direction $(1, 9), (2, 8), (3, 7), (4, 6)$. **(b)** three $(5, 1, 9)$ **21.** The method produces a palindrome in each case: **(a)** 363 **(b)** 9339 **(c)** 5005.

Mathematical Connections 3-2

Communication

1. Answers vary. This approach emphasizes the meaning of place value of the digits and may be easier for young children than the standard algorithm. It can also serve as a transition to the standard algorithm.   **3.** Answers vary. For example, the "scratch marks" represent the normal "carries" and is just a way to keep track of them.

Open-Ended

5. Students are encouraged to use the bibliographies at the end of the chapters to seek mathematics education articles to answer this question.

Questions from the Classroom

7. Answers vary. For example, you could talk about how adding 1 and subtracting 1 has the effect of adding 0 to the problem, which produces an equivalent problem. The reason for changing the original problem to the new problem is that the numbers are easier to work with. You could tell her that her technique works well and discuss how it works.   **9.** Answers vary. For example, the teacher could ask Joe to perform the computation with base ten blocks.   **11.** Betsy is confused on what you should add to check a subtraction. She should try smaller numbers to get a feeling for what needs to be done—for example, $9 - 5 = 4$—to check we add $4 + 5$ to see that we get back to 9. Number lines or colored rods could be used to show that $9 - 5 = 4$ and that $4 + 5$ gets her back to the length of 9.   **13.** Answers vary. For example, if you have a dollars and you spend b dollars you will be left with $a - b$ dollars. However, if you have c dollars less and you spend c dollars less, you will be left with the same amount as before.

Review Problems

15. For example, $(2 + 3) + 4 = 2 + (3 + 4)$.

Assessment 3-3A

1. (a) 5 **(b)** 4 **(c)** any whole number **2. (a)** yes **(b)** yes
(c) yes **3. (a)** No, $2 + 3 = 5$. **(b)** yes
4. (a) $ac + ad + bc + bd$ **(b)** $\square \cdot \Delta + \square \cdot \bigcirc$

(c) $ab + ac - ac$ or ab **5. (a)** $(5 + 6) \cdot 3 = 33$
(b) No parentheses needed   **(c)** No parentheses needed
(d) $(9 + 6) \div 3 = 5$ **6. (a)** $y(x + y)$ **(b)** $x(y + 1)$
(c) $ab(a + b)$ **7. (a)** 6   **(b)** 0   **(c)** 4   **8.** 72   **9. (a)** $4 \cdot 2$
(b) $2 \cdot 4$ or $4 \cdot 2$ **10. (a)** associative property of multiplication   **(b)** commutative property of multiplication   **(c)** commutative property of multiplication   **(d)** identity property of
multiplication   **(e)** zero multiplication property   **(f)** distributive property of multiplication over addition   **11. (a)** closure
property of multiplication   **(b)** zero multiplication
property   **(c)** identity property of multiplication
12. (a) distributive property of multiplication over addition
(b) $32 \cdot 12 = 32(10 + 2)$
$$= 32 \cdot 10 + 32 \cdot 2$$
$$= 320 + 64$$
$$= 384$$
13. (a) $9(10 - 2) = 9 \cdot 10 - 9 \cdot 2 = 90 - 18 = 72$
(b) $20(8 - 3) = 20 \cdot 8 - 20 \cdot 3 = 160 - 60 = 100$
14. (a) $(a + b)^2 = (a + b)(a + b) = (a + b)a +$
$(a + b)b = a^2 + ba + ab + b^2 = a^2 + 2ab + b^2$ **(b)** The area
of the square with side $a + b$ can be expressed as $(a + b)(a + b)$
and also as the sum of the areas of four regions: two squares,
a^2 and b^2, and two rectangles ab and ba. Hence,
$$(a + b)^2 = a^2 + b^2 + ab + ba = a^2 + 2ab + b^2.$$

	a	b
a	a^2	ab
b	ba	b^2

$\}\,a + b$

15. The area of the complete square is $(a + b)^2$. The area of the
small square is $(a - b)^2$. The area of each of the $a \times b$ rectangles
is ab, so the area of the four rectangles is $4ab$. Therefore, the
area of the large square minus the small square is $(a + b)^2 -$
$(a - b)^2 = 4ab$.   **16. (a)** $(50 - 1)(50 + 1) = 50^2 - 1 = 2499$
(b) $(100 - 2)(100 + 2) = 10,000 - 4 = 9,996$
(c) $(20 + 6)(20 - 6) = 400 - 36 = 364$
(d) $(100 - 98) \cdot (100 + 98) = 2 \cdot 198 = 396$
17. (a) $(ab)c = c(ab)$ commutative property of multiplication
$\qquad\quad = (ca)b$ associative property of multiplication
(b) $(a + b)c = c(a + b)$ commutative property of
$\qquad\qquad\qquad\qquad\qquad$ multiplication
$\qquad\qquad = c(b + a)$ commutative property of addition
18. (a) $y(x - y)$ **(b)** $47(101 - 1)$ **(c)** $ab(b - a)$
19. (a) $40 = 8 \cdot 5$ **(b)** $326 = 2 \cdot x$ **20.** Let x be a number. Then the directions result in $(5x + 5) \div 5 - 1$ or
$5(x + 1) \div 5 - 1 = x + 1 - 1 = x$. Thus we always get back
the original number.   **21. (a)** $(8 \div 4) \div 2 \neq 8 \div (4 \div 2)$
(b) $8 \div (2 + 2) \neq (8 \div 2) + (8 \div 2)$ **22.** Suppose there
are two bags of marbles, with a marbles in one bag and b marbles
in the other. We want to distribute the marbles equally among c
students. Then the number of marbles that each student gets can
be found in two ways as follows: Putting all the marbles in one
bag, there are $a + b$ marbles and each student gets $(a + b) \div c$
marbles. Or divide the marbles in the first bag first, and then
divide the marbles in the second bag. In this way each student
would get $(a \div c) + (b \div c)$ marbles. Alternatively, $a \div c = x$
and $b \div c = y$. Then $a = cx$ and $b = cy$. Consequently,
$$a + b = cx + cy$$
$$= c(x + y)$$

Now by definition of division, $x + y = (a + b) \div c$. When you
substitute for x and y, the property follows.   **23. (a)** 4   **(b)** 3
(c) 2 **(d)** 95   **24.** A possible answer is given, resulting in $4 \cdot 3$, or
12 color schemes.

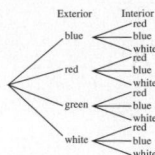

25. (a) 3   **(b)** 2   **(c)** 2   **(d)** 6   **(e)** 4   **26.** yes, 64
27. $(10 \cdot 12) \div 8$ or 15   **28. (a)** Subtract 18 from 45
(b) Divide 54 by 9   **(c)** Add 11 and 48   **(d)** Add 6 and 8
29. $\{4n + 1 \,|\, n \in W\} = \{1, 5, 9, 13, \ldots\}$ **30. (a)** Yes
(b) Yes   **(c)** Yes, a **(d)** Answers vary. For example,
$(a \odot b) \odot c = b \odot c = a, a \odot (b \odot c) = a \odot a = a$.

Mathematical Connections 3-3

Communication

1. Answers vary. For example, $ab = a(b - 1) + a$. If a and b are
both odd, then $a(b - 1)$ is even since $b - 1$ is even. When an odd
number is added to an even number the result is odd, and therefore $a(b - 1) + a$ is odd. Alternatively a number is odd if and
only if it has the form $2k + 1$ for some whole number k.
Hence the product of two odd numbers can be written as
$(2k + 1)(2m + 1) = 2(2km + k + m) + 1$, an odd number.
3. Answers vary. For example, you might think of $9 \cdot 7$ as $7 \cdot 9$
and see if that helps. You might think of $9 \cdot 7$ as $9 \cdot 6 + 9$ or
$54 + 9 = 63$ or $9 \cdot 7 = 9 \cdot 5 + 9 \cdot 2 = 45 + 18 = 63$. Also
$9 \cdot 7 = (10 - 1)7 = 70 - 7 = 63$ **5.** This is the case when x is
either 0 or 1.

Open-Ended

7. Answers vary. For example, a taxi driver charges $3 for en-
tering a cab and $2 per minute for 6 minutes. (The prices here
are unrealistic, but this is the type of problem students may
suggest.)

Cooperative Learning

9. (a) yes   **(b)** 18 and 19   **(c)** In general, even numbers appear to reach 1 quicker   **(d)** The process terminates for particular choices, but it is not known if it terminates for all numbers.

Questions from the Classroom

11. You could encourage Sue to substitute numbers for a and
b to see if her claim is true. For example, if $a = 2$ and $b = 4$,
then $3(2 \cdot 4) = 3 \cdot 8 = 24$ and $(3 \cdot 2)(3 \cdot 4) = 6 \cdot 12 = 72$.
Therefore, we have a counterexample to show Sue's claim is false.
At this point the correct associative and distributive properties could
be demonstrated.   **13.** The student probably points out that if
$a \in W, a \div 1 = a$. Since $1 \div a$ is not defined in the set of whole
numbers, $1 \div a \neq a$ and therefore 1 is not the identity for division.

Review Problems

15. No. For example, $5 - 2 \neq 2 - 5$.

Assessment 3-4A

1. (a)
$$
\begin{array}{r}
426 \\
\times\ 783 \\
\hline
1278 \\
3408 \\
2982 \\
\hline
333558
\end{array}
$$

(b)
$$
\begin{array}{r}
327 \\
\times\ 941 \\
\hline
327 \\
1308 \\
2943 \\
\hline
307707
\end{array}
$$

2. (a)

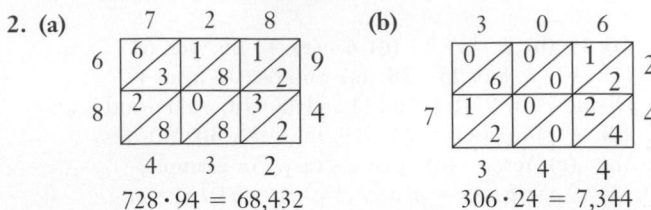

$728 \cdot 94 = 68{,}432$

(b)

$306 \cdot 24 = 7{,}344$

3. In millions of liters: 168,000; 40,000; 6400; 48,750; 11,500; 1500; 3600　**4. (a)** 5^{19}　**(b)** 6^{15}　**(c)** 10^{313}　**(d)** 10^{12} or $2^{12} \cdot 5^{12}$　**5. (a)** 2^{100} because $2^{80} + 2^{80} = 2^{80}(1 + 1) = 2^{80} \cdot 2 = 2^{81}$　**(b)** 2^{102} because $2^{101} = 2^{100} \cdot 2$ and $2^{102} = 2^{100} \cdot 4$
6. The following partial products, which are obtained through the distributive property of multiplication over addition, are shown in the model.

(a)
$$
\begin{array}{rl}
22 & \\
\times\ 13 & \\
\hline
6 & (3 \times 2) \\
60 & (3 \times 20) \\
20 & (10 \times 2) \\
200 & (10 \times 20) \\
\hline
286 &
\end{array}
$$

(b)
$$
\begin{array}{rl}
15 & \\
\times\ 21 & \\
\hline
5 & (1 \times 5) \\
10 & (1 \times 10) \\
100 & (20 \times 5) \\
200 & (20 \times 10) \\
\hline
315 &
\end{array}
$$

$15 \cdot 21 = 315$

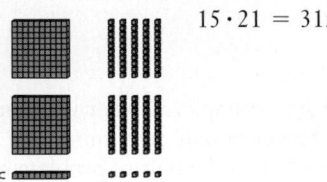

(c)

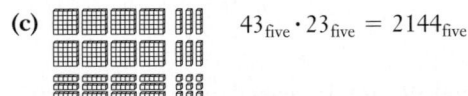

$43_{\text{five}} \cdot 23_{\text{five}} = 2144_{\text{five}}$

To find $43_{\text{five}} \cdot 23_{\text{five}}$, count the number of flats, longs, and ones; we have $4 \cdot 2$ flats, $2 \cdot 3 + 3 \cdot 4$ longs, and $3 \cdot 3$ units. Remembering that 5 units = 1 long, 5 longs = 1 flat, and 5 flats = 1 block, we have $43_{\text{five}} \cdot 23_{\text{five}} = 2$ blocks, 1 flat, 4 longs, and 4 units = 2144_{five}.　**7. (a)** 10010_{two}　**(b)** In base two, two is 10_{two}. We illustrate the property on a three-digit number in base two:

$$
\begin{aligned}
abc_{\text{two}} \cdot 10_{\text{two}} &= (a \cdot 10_{\text{two}}^2 + b \cdot 10_{\text{two}} + c_{\text{two}}) \cdot 10_{\text{two}} \\
&= a \cdot 10_{\text{two}}^3 + b \cdot 10_{\text{two}}^2 + c_{\text{two}} \cdot 10_{\text{two}} + 0 \\
&= abc0_{\text{two}}
\end{aligned}
$$

(c) When we multiply by four we multiply by 10_{two}^2, that is we multiply by 10_{two} and then again by 10_{two}. Each time we append a 0. Hence altogether we append 00.
(d)
$$
\begin{aligned}
110_{\text{two}} \cdot 11_{\text{two}} &= 110_{\text{two}} \cdot (10_{\text{two}} + 1_{\text{two}}) \\
&= 110_{\text{two}} \cdot 10_{\text{two}} + 110_{\text{two}} \\
&= 1100_{\text{two}} + 110_{\text{two}} \\
&= 10010_{\text{two}}
\end{aligned}
$$

8.
$$
\begin{array}{rcr}
\rightarrow & 17 \times & 63 \\
& 8 & 126 \\
& 4 & 252 \\
& 2 & 504 \\
\rightarrow & 1 & 1008,
\end{array}
$$
and $63 + 1008 = 1071$
9. (a) 1332 calories　**(b)** Jane, 330 more calories　**(c)** Maurice, 96 more calories　**10.** No　**11.** $375　**12. (a)** 77 remainder 7　**(b)** 8 remainder 10　**(c)** 10 remainder 91　**13. (a)** 3754　**(b)** 7345　**14.** 3
15. (a) Answers vary. For example,
$$
\begin{array}{cc}
32 & 23 \\
\times\ 69 & \times\ 96 \\
\hline
2208 & 2208
\end{array}
$$
(b) $(10a + b)(10c + d) = (10b + a)(10d + c)$ implies $100ac + 10bc + 10ad + bd = 100bd + 10ad + 10bc + ac$ or $99ac = 99bd$, which implies that $ac = bd$. Therefore this works whenever the product of the ones digits is the same as the product of the tens digits.　**16.** 1356, 2712, and 452　**17.** $142
18. (a) 5 was multiplied by 6 to obtain 30; the 3 was regrouped; then 3 was multiplied by 2 to obtain 6; the "regroup" was added to obtain 9, which was recorded.　**(b)** When 1 was brought down, the quotient of 0 was not recorded.　**19.** place value; distributive property of multiplication over addition; associative property of multiplication; definition of a^n; identity property for addition and zero multiplication property; place value　**20. (a)** $\{4n + 3 \mid n \in W\}$　**(b)** 3, 7, 11, 15, 19, ...
(c) It is an arithmetic sequence because we add 4 to each term to get the next term.

22. (a)

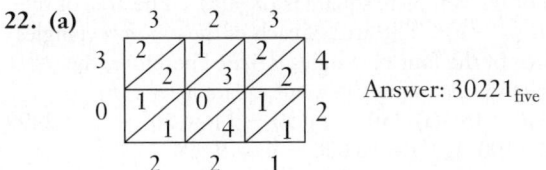

Answer: 30221_{five}

(b) $a = 5, b = 7$　**23. (a)**
$$
\begin{array}{r}
763 \\
\times\ 8 \\
\hline
6104
\end{array}
$$
(b)
$$
\begin{array}{r}
678 \\
\times\ 3 \\
\hline
2034
\end{array}
$$
24. (a) 233_{five}　**(b)** $4_{\text{five}}\ \text{R}1_{\text{five}}$　**(c)** 1513_{six}　**(d)** 31_{five}　**(e)** 110_{two}　**(f)** 1101110_{two}

Mathematical Connections 3-4

Communication

All the answers may vary.
1. Because $345 \cdot 678 = 345 \cdot (6 \cdot 10^2 + 7 \cdot 10 + 8)$, one explanation could be as follows: Using the distributive property of multiplication over addition, first multiply 345 by 6 and the result by 10^2; then multiply 345 by 7 and the result by 10; then multiply 345 by 8. Add all the numbers previously obtained.　**3.** The result is always 4. Let the original number be x. The calculation appears as follows:

$$[(2x)3 + 24] \div 6 - x = 4$$

5. *PSSM* suggests that the traditional algorithm for division be considered toward the end of grades 3–5 as one efficient way to calculate. Students should be able to justify the procedure. However, NCTM recommends that students consolidate and practice a small number of strategies for division and be able to use them fluently. Long division can be practiced by working on context problems and not necessarily just drill problems. If a student can demonstrate that he or she can handle division problems, there is no need to continue to practice them in excess. Division can then receive less emphasis.

Cooperative Learning

7. Arguments vary depending upon groups. Some will argue that addition should be followed by subtraction because they are inverses of each other. Others will argue that addition should be followed by multiplication because multiplication is repeated addition. This would also postpone subtraction until students are more ready for it.

Questions from the Classroom

9. Evidently the student does not understand the process of long division. The repeated subtraction method should help in understanding the mistake.

$$
\begin{array}{r}
6\overline{)36} \\
-6 \quad \text{1 six} \\
\hline
30 \\
-30 \quad \text{5 sixes} \\
\hline
\text{6 sixes}
\end{array}
$$

Instead of adding 1 and 5, the student wrote 15.   **11.** If the number has three digits with the unit digit 0, we have $ab0 \div 10 = ab$ since $ab \cdot 10 = ab0$. This will not work if the 0 digit is not the unit digit.   **13. (a)** The first equation is true because $39 + 41 = 39 + (1 + 40) = (39 + 1) + 40 = 40 + 40$. Now, $39 \cdot 41 = (40 - 1)(40 + 1) = 40^2 - 1^2$, and $40^2 - 1 \neq 40^2$.
(b) Yes, this pattern continues because the numbers being considered are in the form $(a - 1)(a + 1)$, which is equal to $a^2 - 1$.

Review Problems

15. (a) $(a + b + 2)x$ **(b)** $(3 + x)(a + b)$
17. (a) $36 = 4 \cdot 9$ **(b)** $112 = 2x$ **(c)** $48 = x \cdot 6$, or $48 = 6x$ **(d)** $x = 7 \cdot 17$

Assessment 3-5A

1. (a) 160 **(b)** 120 **2. (a)** $(9 \cdot 6) \cdot (2 \cdot 5) = 54 \cdot 10 = 540$
(b) $(8 \cdot 7) \cdot (25 \cdot 4) = 56 \cdot 100 = 5600$
3. (a) $(475 + 525) + 49 = 1049$ **(b)** $375 - 75 - 1 = 299$
4. (a) 605 **(b)** 963 **5. (a)** 36 **(b)** 120 **(c)** 46 **(d)** 97
6. (a) $28 + 2 = 30; 30 + 20 = 50; 50 + 3 = 53$; so the answer is $2 + 20 + 3 = 25$. **(b)** $47 + 3 = 50; 50 + 10 = 60; 60 + 3 = 63$; so the answer is $3 + 10 + 3 = 16$.
7. 496 mi **8. (a)** $86 + 37 = (80 + 30) + (7 + 6) = 123$ by adding the tens and the units separately. **(b)** $97 + 54 = 97 + 3 + 54 - 3 = 100 + 51 = 151$ by adding 3 to 97 and subtracting it from 54. **(c)** $230 + 60 + 70 + 44 + 40 + 6 = (230 + 70) + (60 + 40) + (44 + 6) = 450$ by using compatible numbers. **9. (a)** 5300 **(b)** 100,000
(c) 120,000 **(d)** 2330 **10. (a)** $900 \div 30 = 30$
(b) $25,000 - 20,000 = 5000$ **(c)** $30 \cdot 30 = 900$
(d) Answers vary $2000 + 3000 + 6000 + 1000 = 12,000$
11. Answers vary. For example, **(a)** $2 + 3 + 5 = 10$, so 10,000 is an initial estimate. $10,000 + 2000$ (adjustment) $= 12,000$ for the final estimate. **(b)** The sum of the front digits is 22, so $2200 + 270$ adjustment gives 2470. **12. (a)** The first set of numbers is not clustered. The second set is clustered about 500, so an estimate is 2500. **(b)** Estimates vary. **13. (a)** The

range is 600 $(20 \cdot 30)$ to 1200 $(30 \cdot 40)$ **(b)** The range is 700 $(100 + 600)$ to 900 $(200 + 700)$. **(c)** The range is 230 $(200 + 30)$ to 340 $(300 + 40)$. **14.** Answers vary. For example, $3300 - 100 - 300 - 400 - 500 = 2000$. **15.** For example, $35 \cdot 20 = 700$ seats or $40 \cdot 25 = 1000$ seats; 700 will be low and 1000 high. **16. (a)** The first product is greater because one of the factors is the same and the other factor is greater. **(b)** Same answers because 22 was divided by 2 to obtain 11 while 32 was multiplied by 2 to obtain 64. The result is multiplying the original computation by 2 and dividing by 2, which does not change it. **(c)** Same answers because the first number was multiplied by 3 and the second number was divided by 3, which does not change it. **17. (a)** false **(b)** false **(c)** false
(d) true **18.** The clustering strategy gives $6 \cdot 70,000$, or 420,000.
19. (a) high; $299 \cdot 3 < 300 \cdot 3$
(b) low; $6001 \div 299 > 6000 \div 300$
(c) low; $6000 \div 299 > 6000 \div 300$
(d) low; $10 \cdot 99$ is only 990
20. One possibility is that to find $(10x + 5)^2$ write $(10x + 5)^2 = 100x^2 + 50x + 50x + 25 = 100x^2 + 100x + 25 = 100x(x + 1) + 25$. For example, in 65^2 we take $6 \cdot 7 = 42$ and append 25 to obtain 4225.

Mathematical Connections 3-5

Communication

1. Answers vary. The front-end estimation is almost always less than the exact sum because each of the estimates is usually less than the actual values being added. The only case where this is not true is when each of the front-end numbers that is used is followed by zeroes. In this case, the estimate would equal the exact sum. **3.** The estimate is always too high. If we increase x by an amount p and decrease y by an amount q, then the estimate is $(x + p) - (y - q)$. This can be rewritten as $(x - y) + (p + q)$. This shows that the estimate is always $(p + q)$ greater than the difference $(x - y)$.

Open-Ended

5. Answers vary. For example an estimate is sufficient, when you are determining the amount of a tip for a waiter in a restaurant.

Cooperative Learning

7. Answers vary.

Questions from the Classroom

9. Molly has a good idea and almost has it right. Point out that she needs to add 2 at the end instead of subtract 2. The reasons for this should be explained. For example,

$$
\begin{aligned}
261 - 48 &= 261 - (50 - 2) \\
&= 261 - 50 + 2 \\
&= (261 - 50) + 2 \\
&= 211 + 2 \\
&= 213
\end{aligned}
$$

11. By using the calculator first, she is not learning the estimation skills that are so useful in making decisions. One important use of estimation is to determine whether an answer is reasonable. For example, if the problem is $492 \cdot 63$, by estimating, the student would know that the answer should be approximately 30,000. If she gets an answer such as 17,712 (492·36) or 59,346 (942·63), she would know that there was an error in computation. (Maybe she transposed two digits.)

Review Problems

13. (a)

```
   18)623
  − 180   ← 10 (18s)
    443
  − 180   ← 10 (18s)
    263
  − 180   ← 10 (18s)
     83
  −  18   ← 1 (18s)
     65
  −  18   ← 1 (18)
     47
  −  18   ← 1 (18)
     29
  −  18   ← 1 (18)
     11      34 (18s)
```

```
      34
  18)623
   −54
    83
   −72
    11
```

(b)

```
   21)493
  − 210   ← 10 (21s)
    283
  − 210   ← 10 (21s)
     73
  −  21   ← 1 (21)
     52
  −  21   ← 1 (21)
     31
  −  21   ← 1 (21)
     10      23 (21s)
```

```
      23
  21)493
   −42
    73
   −63
    10
```

(c)

```
   97)1000
  − 970   ← 10 (97s)
     30
```

```
      10
  97)1000
   −97
    30
   − 0
    30
```

Chapter Review

1. (a) Distributive property of multiplication over addition **(b)** Commutative property of addition **(c)** Identity property of multiplication **(d)** Distributive property of multiplication over addition **(e)** Commutative property of multiplication **(f)** Associative property of multiplication **2. (a)** $3 < 13$ because there exists a natural number, namely 10, such that $3 + 10 = 13$ **(b)** $12 > 9$ or $9 < 12$ because there exists a natural number, namely 3, such that $9 + 3 = 12$. **3. (a)** 15, 14, 13, 12, 11, or 10 **(b)** 10 **(c)** Any whole number **(d)** Whole numbers from 0 through 26 **4. (a)** $15a$ **(b)** $5x^2$ **(c)** $xa + xb + xy$ **(d)** $3x + 15 + xy + 5y$ or $(x + 5)(3 + y)$ **(e)** $x(3x + 1)$ **(f)** $x^3(2x^2 + 1)$ **5.** 40 cans **6.** 12 outfits **7.** 26 **8.** $6000 for 80 people is cheaper. **9.** $214 **10.** $400

11. (a) If n is the original number, then each of the following lines shows the result of performing the instruction:

$$n$$
$$n + 17$$
$$2(n + 17) = 2n + 34$$
$$2n + 30$$
$$4n + 60$$
$$4n + 80$$
$$n + 20$$
$$n$$

(b) Answers vary. For example, the next two lines could be subtract 65 and then divide by 4. **(c)** Answers vary. **12.** 1119 **13.** 60,074 **14. (a)** 5 remainder 243 **(b)** 91 remainder 10 **(c)** 120_{five} remainder 2_{five} **(d)** 11_{two} remainder 10_{two} **15. (a)** $912 \cdot 5 + 243 = 4803$ **(b)** $11 \cdot 91 + 10 = 1011$ **(c)** $23_{\text{five}} \cdot 120_{\text{five}} + 2_{\text{five}} = 3312_{\text{five}}$ **(d)** $11_{\text{two}} \cdot 11_{\text{two}} + 10_{\text{two}} = 1011_{\text{two}}$ **16. (a)** $(19 \cdot 194)10 = 36,860$ **(b)** $(379 \cdot 193)100 = 7,314,700$ **(c)** $481 \cdot 73 \cdot (8 \cdot 125) = (481 \cdot 73)1000 = 35,113,000$ **(d)** $374 \cdot 893 \cdot (200 \cdot 50) = (374 \cdot 893) \cdot 10,000 = 3,339,820,000$ **17.** $4380 **18.** 2600 cases **19.** $2.16 **20.** 36 bikes and 18 tricycles **21. (a)** 212_{five} **(b)** 101_{two} **(c)** 1442_{five} **(d)** 101101_{two}

22. $44_{\text{five}} \cdot 34_{\text{five}} = (4 \cdot 10_{\text{five}} + 4) \cdot 34_{\text{five}}$
$= 4 \cdot 34_{\text{five}} \cdot 10_{\text{five}} + 4 \cdot 34_{\text{five}}$
$= 3010_{\text{five}} + 301_{\text{five}}$
$= 3311_{\text{five}}$

23.
```
   4_five)434_five | 100
      400_five     |  4
       34
       31
        3
```

Thus $434_{\text{five}} = 104_{\text{five}} = 104_{\text{five}} \cdot 4 + 3$ **24. (a)** For example, $(26 + 24) + (37 − 7) = 50 + 30 = 80$ **(b)** For example, $(7 \cdot 9)(4 \cdot 25) = 63 \cdot 100 = 6300$ **25. (a)** 441 **(b)** 36 **(c)** 180 **(d)** 406 **26.** Answers vary. For example, (a) $2300 + 300$ (adjustment) $= 2600$, (b) 2600 **27.** $2400 \cdot 4 = 9600$ **28. (a)** $999 \cdot 47 + 47 = 47(999 + 1) = 47 \cdot 1000 = 47,000$ **(b)** $43 \cdot 59 + 41 \cdot 43 = 43 \cdot (59 + 41) = 43 \cdot 100 = 4300$ **(c)** $1003 \cdot 79 − 3 \cdot 79 = 79 \cdot (1003 − 3) = 79 \cdot 1000 = 79,000$ **(d)** $1001 \cdot 113 − 113 = 113 \cdot (1001 − 1) = 113 \cdot 1000 = 113,000$ **(e)** $101 \cdot 35 = (100 + 1)35 = 100 \cdot 35 + 1 \cdot 35 = 3500 + 35 = 3535$ **(f)** $98 \cdot 35 = (100 − 2)35 = 100 \cdot 35 − 2 \cdot 35 = 3500 − 70 = 3430$ **29. (a)** $3x^3 + 4x^2 + 7x + 8 + (5x^2 + 2x + 1) = 3x^3 + 9x^2 + 9x + 9$ **(b)** Answers vary. **(c)** Answers vary. For example, $34 \cdot 10^2 = (3 \cdot 10 + 4) \cdot 10^2 = 3 \cdot 10^3 + 4 \cdot 10^2 + 0 \cdot 10 + 0 = 3400$ and $(3x + 4) x^2 = 3x^3 + 4x^2$.

Answers to Now Try This

3-1. For example, if the sets of elements are $\{a, b, c\}$ and $\{a, d\}$, then the union of the sets of elements is $\{a, b, c, d\}$. The union has only 4 elements while the original sets have 3 and 2, respectively. The sum of 3 and 2 is 5, not 4, the number of elements in

the set union. **3-2.** Answers vary; for example, students are used to starting with 1 when they count, so they sometimes start with 1 on the number line. It should be pointed out that we are working with whole numbers and the first whole number is 0. Next, to represent 3 on a number line an arrow (vector) of length 3 units must be used. The length of the arrow in Figure 3-3 is 2 units.
3-3. (a) Closed; an even number plus an even number is always even. (b) Not closed; for example, $1 + 3 = 4$ and 4 is not an element of F. (c) Not closed; $2 + 2 = 4$ and 4 is not in G.
3-4. (a) Use disjoint sets A and B such that $n(A) = 3$ and $n(B) = 5$. Then $n(A) + n(B) = n(A \cup B)$, $n(B) + n(A) = n(B \cup A)$. Because $A \cup B = B \cup A$, $n(A) + n(B) = n(B) + n(A)$

(b)

3-5. Let $B \subseteq A$. If $n(A) = a$ and $n(B) = b$, then $a - b = n(A - B)$. For example we can model the subtraction $4 - 1 = 3$ as follows: $A = \{a, b, d, e\}$ and $B = \{a\}, A - B = \{b, d, e\}, n(A) = 4, n(B) = 1,$ $n(A - B) = 3$. **3-6.** (a) The set of whole numbers is not closed under subtraction; for example, $2 - 5$ is not a whole number. (b) Subtraction is not associative for whole numbers; for example, $9 - (7 - 2) \neq (9 - 7) - 2$. (c) Subtraction is not commutative for whole numbers; for example, $3 - 2 = 1$, but $2 - 3$ is not a whole number. (d) There is no identity for whole-number subtraction; for example, $5 - 0 = 5 \neq 0 - 5$.
3-7. (a) 5 (b) 7 (c) 10 (d) $a - b$
3-8. $182 + 61$ can be represented by:

1 block	8 flats	2 units
	6 flats	1 unit

1 block (10 flats + 4 flats) 3 units
or 1 block (1 block + 4 flats) 3 units
or 2 blocks   4 flats   3 units
or 243

3-9. (i) The method is valid because subtracting and adding the same number does not change the original sum.
(ii) $97 + 69 = (97 + 3) + (69 - 3)$
$= 100 + 66$
$= 166$
3-10. (a) 1000_{five} (b) 112_{five}
3-11.

+	0_{two}	1_{two}
0_{two}	0_{two}	1_{two}
1_{two}	1_{two}	10_{two}

(b) (i) $\begin{array}{r} 1111_{\text{two}} \\ + \ 111_{\text{two}} \\ \hline 10110_{\text{two}} \end{array}$ (ii) $\begin{array}{r} 1101_{\text{two}} \\ - \ 111_{\text{two}} \\ \hline 110_{\text{two}} \end{array}$

3-12. One possible explanation is as follows: The first shirt can be worn with each of the 5 pairs of pants for a total of 5 different outfits. Also, the second shirt can be worn with each of the 5 pairs of pants for a total of 5 outfits different from the previous outfits. Similarly, the 3rd, 4th, 5th, and 6th shirts can be combined with the 5 pairs of pants for new outfits. In this way, each of the 6 shirts can be used for 5 new outfits for a total $5 + 5 + 5 + 5 + 5 + 5 = 6 \cdot 5$ outfits. **3-13.** 91 members
3-14. (a) The set of whole numbers is not closed under

division; for example, $8 \div 5$ is not a whole number. Likewise, $8 \div 2 \neq 2 \div 8$ and $(8 \div 4) \div 2 \neq 8 \div (4 \div 2)$ show it is not commutative or associative. (b) 1 is not the identity for whole-number division because $n \div 1 = 1$ for all whole numbers n, but $1 \div n \neq n$ except when $n = 1$.
3-15. $10 \cdot 600 = 10^1(6 \cdot 10^2)$
$= (6 \cdot 10^2)10^1$
$= 6 \cdot (10^2 10^1)$
$= 6 \cdot 10^3$
$= 6 \cdot 10^3 + 0 \cdot 10^2 + 0 \cdot 10 + 0 \cdot 1$
$= 6000$
$20 \cdot 300 = (2 \cdot 10^1)(3 \cdot 10^2)$
$= (2 \cdot 3)(10^1 \cdot 10^2)$
$= 6(10^1 \cdot 10^2)$
$= 6 \cdot 10^3$
$= 6 \cdot 10^3 + 0 \cdot 10^2 + 0 \cdot 10 + 0 \cdot 1$
$= 6000$
3-16. $7 \cdot 4589 = 7(4 \cdot 10^3 + 5 \cdot 10^2 + 8 \cdot 10 + 9)$
$= (7 \cdot 4)10^3 + (7 \cdot 5)10^2 + (7 \cdot 8) \cdot 10 + 7 \cdot 9$
$= 28000 + 3500 + 560 + 63$
$= 32123$
3-17. Answers vary. For example, (a) $40 + 160 = 200$ and $29 + 31 = 60$ so the sum is 260. (b) $3679 - 400 = 3279$ and $3279 - 74 = 3205$. (c) $75 + 25 = 100$ and $100 + 3 = 103$. (d) $2500 - 500 = 2000$ and $2000 - 200 = 1800$. **3-18.** Answers vary. For example, (a) $4 \cdot 25 = 100$ and $32 \cdot 100 = 3200$. (b) $123 \cdot 3 = 100 \cdot 3 + 23 \cdot 3 = 300 + 69 = 369$. (c) $25 \cdot 35 = (30 - 5)(30 + 5) = 30^2 - 5^2 = 900 - 25 = 875$. (d) $5075 \div 25 = 5000 \div 25 + 75 \div 25 = 200 + 3 = 203$. **3-19.** Answers vary. For example, (a) To estimate $4525 \cdot 9$, we know $4525 \cdot 10 = 45,250$ and since we have only 9 sets of 4525 we can take away approximately 5000 from our estimate and we have 40,250. (b) To estimate $3625 \div 42$, we know the answer will be close to $3600 \div 40$, or 90.

Answers to Brain Teasers

Section 3-1

Answers vary. For example,

1	2	3
8	9	4
7	6	5

Section 3-2

The license plate number is 10968.

Section 3-4

1. (a) a computer (b) Answers vary. **2.** (a) When a person tells his or her age by listing cards, the person is giving the base two representation for his or her age. The number can then be determined by adding the numbers in the upper left-hand corners of the named cards. (b) Card F would have 32 in the upper left corner. Each of the numbers 1 through 63 could be written in base two so you can tell where to place them on the cards.

Section 3-5

There are 85 members of a PTA to contact using a phone chain. We are to assume that each call takes 30 sec and that everyone is at home and answers the phone. The key to solving this problem is to realize that when a caller calls one of his/her two people, the first person being called does not wait for the second person to be called before this first person can begin calling his/her two people. A strategy is to build a model of this phone tree and a table to keep track of the number of people contacted. A model for the first 3 min is given below. The table on the right of the model shows the number of *People called* in each 30-sec interval and it also shows the *Total number called*.

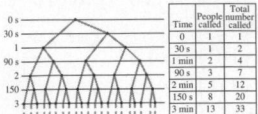

Time	People called	Total number called
0		
30 s	1	2
1 min	2	4
90 s	3	7
2 min	5	12
150 s	8	20
3 min	13	33

If we examine the numbers in the *People called* column. we notice a familiar sequence, that is, the *Fibonacci* sequence where each number in the sequence is the sum of the previous two numbers. We can see from the model that this will continue to happen because the number of dots in each segment is the sum of the dots in the previous two segments. Therefore, we need to continue the table to see at what time the total number of people called reaches or exceeds 85. Continuing the table, we see that there would be $8 + 13 = 21$ more people called at 210 sec, giving a total of $33 + 21 = 54$. For 4 min, there would be $13 + 21 = 34$ more people called giving a total of $54 + 34 = 88$. Therefore, we can call 85 people in 4 min.

We can extend the problem by trying different numbers of members of the PTA or generalizing the problem to finding the length of time required to call n people. We could also investigate how the Fibonacci sequence shows up in other contexts, such as pine cones or the breeding habits of some animals.

Answer to Preliminary Problem

Suppose there were c children and each took n cookies. Then $nc = 124 - 7$ or 117. We have:

$$nc = 3 \cdot 39 = 3 \cdot 3 \cdot 13 = 9 \cdot 13 = 3 \cdot 39.$$

Because the number of children was between 10 and 30, the only possibility is that there were 13 children and each took 9 cookies.

Chapter 4

Assessment 4-1A

1. (a) true **(b)** true **(c)** true **(d)** true **(e)** true **(f)** False; there is no value $c \in W$ such that $30c = 6$. **2. (a)** yes **(b)** no **(c)** yes **3. (a)** 2, 3, 4, 6, 11 **(b)** 2, 3, 6, 9 **(c)** 2, 3, 5, 6, 10 **4. (a)** No, $17|34,000$ and $17 \nmid 15$, so $17 \nmid 34,015$. **(b)** Yes, $17|34,000$ and $17|51$, so $17|34,051$. **(c)** No, $19|19,000$ and $19 \nmid 31$, so $19 \nmid 19,031$. **(d)** Yes, 5 is a factor of $2 \cdot 3 \cdot 5 \cdot 7$. **(e)** No, $5|2 \cdot 3 \cdot 5 \cdot 7$ and $5 \nmid 1$, so $5 \nmid (2 \cdot 3 \cdot 5 \cdot 7) + 1$. **5. (a)** True by Theorem 4-1 **(b)** True by Theorem 4-2(b) **(c)** none **(d)** True by Theorem 4-2(b) **(e)** True by Theorem 4-1 **6. (a)** False, $2|6$ but $(2 + 5) \nmid (6 + 5)$ **(b)** True. Because $b|a$, there is a c such that $a = bc$. Then $a^3 = b^3 c^3 = (bc^3)b^2$, which means $b^2|a^3$. **(c)** True. Because $b|a$, and $b|b$ imply $b|(a + b)$ **7. (a)** $210 = 7 \cdot 30$ **(b)** $19|1900$ and $19|38$ **(c)** $6|(2 \cdot 3) \cdot 2^2 \cdot 3 \cdot 17^4$ and $6|6 \cdot 2^2 \cdot 3 \cdot 17^4$ **(d)** $7|4200$ but $7 \nmid 22$ **8. (a)** true **(b)** false **(c)** false **9. (a)** 7 **(b)** 7

(c) 6 **10. (a)** Any digit 0 through 9 **(b)** 1; 4; 7 **(c)** 1; 3; 5; 7; 9 **(d)** 7 **(e)** 7 **11. (a)** True. Because $13|390000$ and $13|26$; 13 divides the sum. **(b)** True. Because $13|260,000$ and $13 \nmid 33$; 13 does not divide the sum. **(c)** True. Because $31|93$; $31|93 \cdot 93^{10}$. **(d)** True. Because $23|690,000$ and $23 \nmid 68$; $23 \nmid (690,000 + 68)$. **12.** 17 **13.** Each pencil costs 19¢. **14. (a)** $(7 + 2 + 4 + 2 + 8 + 1) + 5$. Since the sum in the parentheses is divisible by 3 the remainder is found when 5 is divided by 3. The remainder is 2. **(b)** A number in base ten like $abcd$ can be written as $(a \cdot 999 + b \cdot 99 + c \cdot 9) + (a + b + c + d)$. Because the number in the first parentheses is divisible by 3 it leaves remainder 0 upon division by 3. So the remainder when $abcd$ is divided by 3 is the same as when $a + b + c + d$ is divided by 3. **(c)** Yes, because $a \cdot 999 + b \cdot 99 + c \cdot 9$ also leaves remainder 0 upon division by 9 (see part (b)). **15. (a)** $12,343 + 4546 + 56 = 16,945$; $4 + 1 + 2 = 7$ has a remainder of 7 when divided by 9, as does $1 + 6 + 9 + 4 + 5$. **(b)** $987 + 456 + 8765 = 10,208$; $6 + 6 + 8 = 20$ has a remainder of 2 when divided by 9, as does $1 + 0 + 2 + 0 + 8 = 11$. **(c)** $10,034 + 3004 + 400 + 20 = 13,458$; $8 + 7 + 4 + 2 = 21$ has a remainder 3 when divided by 9, as does $1 + 3 + 4 + 5 + 8$. **(d)** $1003 - 46 = 957$; $4 - 1 = 3$ has a remainder of 3 when divided by 9, as does $9 + 5 + 7 = 21$. **(e)** $345 \cdot 56 = 19,320$. 345 has a remainder of 3 when divided by 9; 56 has a remainder of 2 when divided by 9; $3 \cdot 2 = 6$ has remainder 6 when divided by 9. $1 + 9 + 3 + 2 + 0 = 15$ has a remainder of 6 when divided by 9. **(f)** Answers vary. **16. (a)** No. For example, $11 \nmid 10,001$ **(b)** Yes. Because

$$abccba = (a \cdot 10^5 + a) + (b \cdot 10^4 + b \cdot 10) + c \cdot 10^3 + c \cdot 10^2$$
$$= a \cdot 100,001 + b \cdot 10,001 + c \cdot 1100$$
$$= 11(a \cdot 9091 + b \cdot 910 + c \cdot 100)$$

17. (a) 1, 3, and 7 divide n. 1 divides every number. Also, because $n = 21 \cdot d, d \in I$, then $n = (3 \cdot 7)d = 3(7d) = 7(3d)$, which implies that n is divisible by 3 and by 7. **18. (a)** Yes, if $5|x$ and $5|y$, then $5|(x + y)$. **(b)** Yes, if $5|y$, then $5|(x - y)$. **(c)** Yes, if $5|x$, then 5 divides all the multiples of x and in particular $5|xy$. **19. (a)** False; $2|4$, but $2 \nmid 1$ and $2 \nmid 3$. **(b)** False (same example as in (a)) **(c)** False; $12|72$, but $12 \nmid 8$ and $12 \nmid 9$. **(d)** true **(e)** true **20.** Let $n = a10^4 + b10^3 + c10^2 + d10 + e$.

$$a \cdot 10^4 = a \cdot (10,000) = a \cdot (9999 + 1) = a \cdot 9999 + a$$
$$b \cdot 10^3 = b \cdot (1000) = b \cdot (999 + 1) = b \cdot 999 + b$$
$$c \cdot 10^2 = c \cdot (100) = c \cdot (99 + 1) = c \cdot 99 + c$$
$$d \cdot 10 = d \cdot (10) = d \cdot (9 + 1) = d \cdot 9 + d$$

Thus, $n = (a \cdot 9999 + b \cdot 999 + c \cdot 99 + d \cdot 9) + (a + b + c + d + e)$. Because $9|9$; $9|99$; $9|999$; $9|9999$, it follows that if $9|(a + b + c + d + e)$ then $9|[(a \cdot 9999 + b \cdot 999 + c \cdot 99 + d \cdot 9) + (a + b + c + d + e)]$; that is, $9|n$. If, on the other hand, $9 \nmid (a + b + c + d + e)$, it follows that $9 \nmid n$.

Mathematical Connections 4-1

Communication

1. No, any amount of postage must be a multiple of 3 (being the sum of a multiple of 6 and a multiple of 9, both of which are multiples of 3). **3. (a)** Yes, $4|52,832$, so 4 divides any whole number times 52,832. Therefore, 4 divides $52,832 \cdot 324,518$, which is the area. **(b)** Yes, 2 is a factor of 52,834 and 2 is a factor of 324,514, so $2 \cdot 2$ or 4 is a factor of $52,834 \cdot 324,514$, which is the area. **5. (a)** No. If $10|x$, then $x = 10n = 5 \cdot 2n$ for some whole number n. Therefore x is divisible by 5. **(b)** Yes, all odd multiples of 5 are not divisible by 10 but are divisible by 5.   **7.** 243; yes; Consider any number n of the form $abcabc$. Then we have the following:

$$n = (a \cdot 10^5) + (b \cdot 10^4) + (c \cdot 10^3) + (a \cdot 10^2)$$
$$+ (b \cdot 10^1) + c$$
$$= a \cdot (10^5 + 10^2) + b \cdot (10^4 + 10^1) + c \cdot (10^3 + 1)$$
$$= a \cdot (100,000 + 100) + b \cdot (10,000 + 10) + c \cdot 1001$$
$$= a \cdot (1001 \cdot 100) + b \cdot (1001 \cdot 10) + c \cdot 1001$$
$$= (1001)[a \cdot 100 + b10 + c1]$$
$$= (7 \cdot 11 \cdot 13)[a \cdot 100 + b \cdot 10 + c \cdot 1]$$

Therefore, if you divide by 1001, the quotient is abc.
9. (a) Let $abcd$ be a number. Subtracting the last digit gives $abc0$. Thus, the result is divisible by 2, 5, and 10.   **(b)** If one subtracts cd from $abcd$, one obtains $ab00$. This number is divisible by 2, 4, 5, 10, 20, 25, 50, and 100.   **(c)** $abcd - (a + b + c + d) = a \cdot 10^3 + b \cdot 10^2 + c \cdot 10 + d - a - b - c - d = a \cdot 999 + b \cdot 99 + c \cdot 9 = 9(a \cdot 111 + b \cdot 11 + c)$. Thus, the result is divisible by 3 and 9.
(d) • Consider a four-digit number in base five, $abcd_{five}$. Then subtracting the last digit produces $abc0_{five} = 10_{five} \cdot abc_{five}$, which is divisible by 10_{five} or 5 in base ten.
• $abcd_{five} - cd_{five} = ab00_{five}$, which is divisible by 10_{five} and 100_{five} or 5 and 25 in base ten.
• $abcd_{five} - a - b - c - d = 444_{five} \cdot a + 44_{five} \cdot b + 4_{five} \cdot c = 4_{five}(111_{five} \cdot a + 11_{five} \cdot b + c)$, which is divisible by 2 and 4.

Open-Ended

11. (a) Answers vary. For example, upon inspection of the given numbers we notice that all the numbers are multiples of 3. Since 3 divides each number, then 3 divides the sum of any of the numbers. Because 100 is not divisible by 3, there is no winning combination of the given numbers that will sum to 100.   **(b)** Answers vary. For example, since many of these multiples of 3 sum to 99 $(33 + 66, 45 + 51 + 3, \text{etc.})$, the company could place at most 1000 cards with the number 1 on the card. This would ensure that there are at most 1000 winners. Other numbers could also be used.

Cooperative Learning

13. Answers vary.

Questions from the Classroom

15. Yes if $a \neq 0$; the student's conclusion is that $a|0$ and this is true because if $a \neq 0$, the equation $a \cdot k = 0$ has a unique solution, $k = 0$.   **17.** It has been shown that any four-digit number n can be written in the form $n = a \cdot 10^3 + b \cdot 10^2 + c \cdot 10 + d = (a \cdot 999 + b \cdot 99 + c9) + (a + b + c + d)$. The test for divisibility by some number g will depend on the sum of the digits $a + b + c + d$ if, and only if, $g|(a \cdot 999 + b \cdot 99 + c \cdot 9)$ regardless of the values of a, b, and c. Since the only numbers greater than 1 that divide 9, 99, and 999 are 3 and 9, the test for divisibility by dividing the sum of the digits by the number works only for 3 and 9. A similar argument works for any m-digit number.
19. It is true that a number is divisible by 21 if, and only if, it is divisible by 3 and by 7. However, the general statement is false. For example, 12 is divisible by 4 and by 6 but not by $4 \cdot 6$, or 24. One part of the statement is true, that is, "if a number is divisible by $a \cdot b$, then it is divisible by a and b." The statement "if a number is divisible by a and by b, it is divisible by ab" is true if a and b have no primes in common. This follows from the Fundamental Theorem of Arithmetic which will be introduced in Section 4-2.

1. 30   **2. (a)** prime   **(b)** not prime   **(c)** not prime   **(d)** prime   **(e)** prime   **(f)** prime   **(g)** prime   **(h)** not prime
3. (a)   504   **(b)**   2475

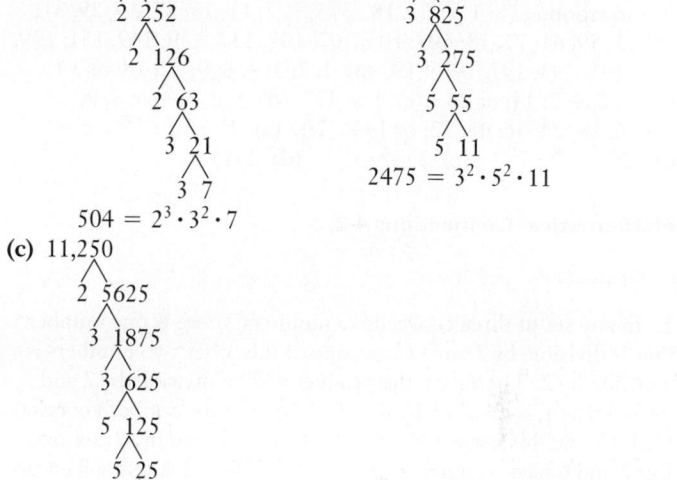

$504 = 2^3 \cdot 3^2 \cdot 7$

$2475 = 3^2 \cdot 5^2 \cdot 11$

(c) 11,250

$11,250 = 2 \cdot 3^2 \cdot 5^4$

4. (a)   210

(b) You could multiply $2 \cdot 3 \cdot 7 \cdot 5$.   **5.** 73   **6. (a)** $2^8 \cdot 3^4 \cdot 5^2 \cdot 7$
(b) $2^3 \cdot 5^2 \cdot 7^{20} \cdot 13$ **(c)** 251   **(d)** $7 \cdot 11 \cdot 13$ **7. (a)** 1 by 48; 2 by 24; 3 by 16; 4 by 12   **(b)** Only one, 1 by 47
8. LCM$(2, 3, 4, \ldots, 10) = 2520$ **9. (a)** The Fundamental Theorem of Arithmetic states that n can be written as a product of primes in one and only one way. Since $2|n$ and $3|n$ and 2 and 3 are both prime, they must be included in the unique factorization:

$$2 \cdot 3 \cdot p_1 \cdot \ldots \cdot p_m = n;$$
$$(2 \cdot 3)(p_1 \cdot p_2 \cdot \ldots \cdot p_m) = n; \text{ therefore}$$
$$6|n.$$

(b) Yes. If $a|n$, there exists a whole number c such that $ca = n$. If $b|n$ there exists a whole number d such that $db = n$. Therefore $(ca)(db) = n^2 \Rightarrow (dc)(ab) = n^2 \Rightarrow ab|n^2$.   **10.** 625   **11.** No, because n is not written as a product of primes. Notice that $2 \cdot 3 \cdot 5 \cdot 7 \cdot 11 \cdot 13$ is a product of primes, but when we add 1 to it, the primes 2, 3, 5, 7, 11, and 13 do not divide n. In fact $n = 30,031 = 59 \cdot 509$, which is its prime factorization.
12. No, because $8^z = 2^{3z}$ which is its prime factorization. Because of the uniqueness of prime factorization 2^{3z} cannot have a 5 to a non-zero power in its prime factorization.   **13. (a)** Every other number is greater than 2 and even and hence composite.
(b) The terms in the infinite sequence 111, 111111, 111111111, ... in which each term consists of $3n$ ones for $n = 1, 2, 3, \ldots$ are all divisible by 3.
14. $2^5 n = 2^5 \cdot 2 \cdot 3^5 \cdot 5^4 \cdot 7^3 \cdot 11^7$.
Hence $n = 2 \cdot 3^5 \cdot 5^4 \cdot 7^3 \cdot 11^7$
$$= 2 \cdot 3 \cdot 5 \cdot 7 \cdot 11^6 \cdot (3^4 \cdot 5^3 \cdot 7^2 \cdot 11)$$
Thus $2 \cdot 3 \cdot 5 \cdot 7 \cdot 11^6$ is a factor (divisor) of n.
15. Yes, because $7^5 \cdot 11^3 = 7 \cdot (7^4 \cdot 11^3)$ **16. (a)** $3 \cdot 5 \cdot 7 \cdot 11 \cdot 13$ is composite because it is divisible by 3, 5, 7, 11, and 13.

(b) $(3 \cdot 4 \cdot 5 \cdot 6 \cdot 7 \cdot 8) + 2 = 2[(3 \cdot 2 \cdot 5 \cdot 6 \cdot 7 \cdot 8) + 1]$ and so it is composite. **(c)** $(3 \cdot 5 \cdot 7 \cdot 11 \cdot 13) + 5 = 5[(3 \cdot 7 \cdot 11 \cdot 13) + 1]$ and so it is composite. **(d)** $10! + 7 = 7[(10 \cdot 9 \cdot 8 \cdot 6 \cdot 5 \cdot 4 \cdot 3 \cdot 2 \cdot 1) + 1]$ and so it is composite. **17.** $2^3 \cdot 3^2 \cdot 25^3$ is not a prime factorization because 25 is not prime. The prime factorization is $2^3 \cdot 3^2 \cdot 5^6$. **18.** 3, 5; 5, 7; 11, 13; 17, 19; 29, 31; 41, 43; 59, 61; 71, 73; 101, 103; 107, 109; 137, 139; 149, 151; 179, 181; 191, 193; 197, 199 **19. (a)** 1; 2; 3; 4; 6; 9; 12; 18; or 36 **(b)** 1; 2; 4; 7; 14; or 28 **(c)** 1 or 17 **(d)** 1; 2; 3; 4; 6; 8; 9; 12; 16; 18; 24; 36; 48; 72; or 144 **20. (a)** $2^{35} \cdot 3^{35} \cdot 7^{40}$ **(b)** $2^{200} \cdot 3^{40} \cdot 5^{200}$ **(c)** $2 \cdot 3^6 \cdot 5^{110}$ **(d)** 2311

Mathematical Connections 4-2

Communication

1. In any set of three consecutive numbers, there is one number that is divisible by 3 and at least one of the other two numbers is divisible by 2. Therefore, the product will be divisible by 2 and by 3 and so it is divisible by 6. **3.** No, they are not both correct. Using 3 and 4 is correct because they have no common divisors. But 2 and 6 have a common divisor of 2. Using this test will ensure only that the number is divisible by 6. **5.** Let $a = 2 \cdot 3 \cdot 5 \cdot 7$ and $b = 11 \cdot 13 \cdot 17 \cdot 19$. Then each prime p less than or equal to 19 appears in the prime factorization of a or of b but not in both. If p is in the prime factorization of a, then $p \mid a$ but $p \nmid b$ and hence $p \nmid (a + b)$. A similar argument holds if p is in the prime factorization of b. **7.** Suppose n is composite and d is its least positive divisor other than 1. We need to show that d is prime. If not, then some prime p less than d will divide d and hence will divide n, which contradicts the fact that d is the smallest divisor of n greater than 1.

Open-Ended

9. (a) Answers vary. **(i)** 25 **(ii)** 21 **(b)** 13, in the interval 100–199 **(c) (i)** 8 **(ii)** 7 **(d)** Answers vary.

Cooperative Learning

11. The students must have had the first 23 prime numbers for the number of tiles; that is, 2, 3, 5, 7, 11, 13, 17, 19, 23, 29, 31, 37, 41, 43, 47, 53, 59, 61, 67, 71, 73, 79, 83 tiles. Therefore, the number of tiles is the sum of the first 23 prime numbers, which is 874 tiles.

Questions from the Classroom

13. Bob has almost the right idea but it needs a little work. It should be pointed out that if a number is not divisible by 2 and 3, then it couldn't be divisible by 6, so there is no need to check 6. Also, if a number is not divisible by 2, then it can't be divisible by 4 or 8, so there is no need to check 4 and 8. Next, if a number is not divisible by 5, then it can't be divisible by 10, so there is no need to check 10. Thus we have cut Bob's list to 2, 3, and 5, which are all primes. The Sieve of Eratosthenes can be used to motivate these concepts. Next we need to explore what happens when the number to be checked is large and show that just checking 2, 3, and 5 is not adequate. For example, just checking for divisibility by 2, 3, and 5 is not enough to check whether 169 is prime. We can use the sieve to show that if we are trying to determine if a positive integer n is prime, we only have to check primes p such that $p^2 \le n$. **15.** The student is wrong. Only the perfect squares have an odd number of divisors. The perfect squares less than 1000

are $1^2, 2^2, 3^2, \ldots, 31^2$. Therefore, there are 31 perfect squares between 1 and 1000 and hence 31 numbers with an odd number of divisors. Consequently, there are $1000 - 31 = 969$ numbers between 1 and 1000 that have an even number of divisors. **17.** The student is right. In every sequence of six consecutive numbers greater than 3, only the numbers before and after a multiple of 6 can be prime. For example, consider the numbers 17, 18, 19, 20, 21, and 22. 18, 20 and 22 are even. 18 and 21 are multiples of 3. Only 17 and 19, the numbers before and after 18 (a multiple of 6), can be prime. All of the others are multiples of 2 or 3.

Review Problems

19. (a) 2; 3; 6 **(b)** 2; 3; 5; 6; 9, 10 **21.** Yes, among eight people, each would get $422.

Assessment 4-3A

1. (a) $D_{18} = \{1, 2, 3, 6, 9, 18\}$
$D_{10} = \{1, 2, 5, 10\}$
$GCD(18, 10) = 2$
$M_{18} = \{18, 36, 54, 72, 90, \ldots\}$
$M_{10} = \{10, 20, 30, 40, 50, 60, 70, 80, 90, \ldots\}$
$LCM(18, 10) = 90$
(b) $D_{24} = \{1, 2, 3, 4, 6, 8, 12, 24\}$
$D_{36} = \{1, 2, 3, 4, 6, 9, 12, 18, 36\}$
$GCD(24, 36) = 12$
$M_{24} = \{24, 48, 72, 96, 120, 144, 168, \ldots\}$
$M_{36} = \{36, 72, 108, 144, 180, \ldots\}$
$LCM(24, 36) = 72$
(c) $D_8 = \{1, 2, 4, 8\}$
$D_{24} = \{1, 2, 3, 4, 6, 8, 12, 24\}$
$D_{52} = \{1, 2, 4, 13, 26, 52\}$
$GCD(8, 24, 52) = 4$
$M_8 = \{8, 16, 24, 32, 40, 48, 56, 64, 72, 80, 88, 96, \ldots\}$
$M_{24} = \{24, 48, 72, 96, 120, 144, 168, 192, 216, 240, 264, 288, 312, \ldots\}$
$M_{52} = \{52, 104, 156, 208, 260, 312, \ldots\}$
$LCM(8, 24, 52) = 312$
(d) $D_7 = \{1, 7\}, D_9 = \{1, 9\}$
$GCD(7, 9) = 1$
$M_7 = \{7, 14, 21, 28, 35, 42, 49, 56, 63, \ldots\}$
$M_9 = \{9, 18, 27, 36, 45, 54, 63, \ldots\}$
$LCM(7, 9) = 63$
2. (a) $132 = 2^2 \cdot 3 \cdot 11$
$504 = 2^3 \cdot 3^2 \cdot 7$
$GCD(132, 504) = 2^2 \cdot 3 = 12$
$LCM(132, 504) = 2^3 \cdot 3^2 \cdot 7 \cdot 11 = 5544$
(b) $65 = 5 \cdot 13$
$1690 = 2 \cdot 5 \cdot 13^2$
$GCD(65, 1690) = 5 \cdot 13 = 65$
$LCM(65, 1690) = 2 \cdot 5 \cdot 13^2 = 1690$
(c) $96 = 2^5 \cdot 3$
$900 = 2^2 \cdot 3^2 \cdot 5^2$
$630 = 2 \cdot 3^2 \cdot 5 \cdot 7$
$GCD(96, 900, 630) = 2 \cdot 3 = 6$
$LCM(96, 900, 630) = 2^5 \cdot 3^2 \cdot 5^2 \cdot 7 = 50,400$
(d) $108 = 2^2 \cdot 3^3$
$360 = 2^3 \cdot 3^2 \cdot 5$
$GCD(108, 360) = 2^2 \cdot 3^2 = 36$
$LCM(108, 360) = 2^3 \cdot 3^3 \cdot 5 = 1080$

3. (a) $\text{GCD}(2924, 220) = \text{GCD}(220, 64) =$
$\text{GCD}(64, 28) = \text{GCD}(28, 8) = \text{GCD}(8, 4) = \text{GCD}(4, 0) = 4$
(b) $\text{GCD}(14595, 10856) = \text{GCD}(10856, 3739) =$
$\text{GCD}(3739, 3378) = \text{GCD}(3378, 361) = \text{GCD}(361, 129) =$
$\text{GCD}(129, 103) = \text{GCD}(103, 26) = \text{GCD}(26, 25) =$
$\text{GCD}(25, 1) = 1$ **4. (a)** 72 **(b)** 1440 **(c)** 630
(d) $9^{100} \cdot 25^{100}$ or $3^{200} \cdot 5^{200}$ or 15^{200} **5. (a)** $220 \cdot 2924/4$, or
160,820 **(b)** $14{,}595 \cdot 10{,}856/1$, or 158,443,320
6. (a) $\text{LCM}(15, 40, 60) = 120$ min $= 2$ hr, so the clocks alarm
again together at 8:00 A.M. **(b)** no **7.** $\text{GCD}(6, 10) = 2$,
$\text{LCM}(6, 10) = 30$ **8.** 24 **9.** 15 cookies **10.** 36 min
11. They should pass the starting point after $\text{LCM}(12, 18, 16) =$
144 min. **12. (a)** ab **(b)** $\text{GCD}(a, a) = a; \text{LCM}(a, a) = a$
(c) $\text{GCD}(a^2, a) = a, \text{LCM}(a^2, a) = a^2$ **(d)** $\text{GCD}(a, b) = a$;
$\text{LCM}(a, b) = b$ **13. (a)** True, if a and b are even, then
$\text{GCD}(a, b) \geq 2$. **(b)** True, $\text{GCD}(a, b) = 2$ implies that a and b
are even. **(c)** False, the GCD could be a multiple of 2; for example, $\text{GCD}(8, 12) = 4$. **14. (a)** 15 **(b)** 1 **15.** $4 = 2^2$. Since
97,219,988,751 is odd, it has no prime factors of 2. Consequently,
1 is their only common divisor and they are relatively prime.
16. The 60th caller **17.** 12 revolutions

18. (a)

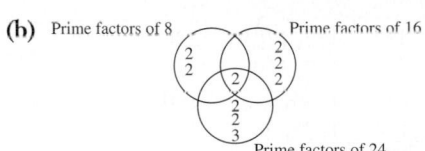

Prime factors of 10   Prime factors of 15

2 5 3
2

Prime factors of 60

(b) Prime factors of 8   Prime factors of 16

2 2 2 2
2 2
2
2
3

Prime factors of 24

19. All whole numbers between 1 and 49 except $1 \cdot 7, 2 \cdot 7, 3 \cdot 7,$
$4 \cdot 7, 5 \cdot 7, 6 \cdot 7,$ and 7^2. **20.** $1, 2, 2^2, 2^3, \ldots, 2^{20}$
21. (a) $6x^3 y^3 (2x + 3y)$ **(b)** $6x^2 y^2 z^2 (2x + 3y^2 z + 4x^2 yz^2)$
22. $5^9 = 1{,}953{,}125$ and $2^9 = 512$

Mathematical Connections 4-3

Communication

1. No, the set of common multiples is infinite and therefore there
could be no greatest common multiple. **3.** No. For example,
consider $\text{GCD}(2, 4, 10) = 2$. $\text{LCM}(2, 4, 10) = 20$, and the
$\text{GCD} \cdot \text{LCM} = 2 \cdot 20 = 40$, while $abc = 2 \cdot 4 \cdot 10 = 80$. **5.** No.
Let $a = 2 \cdot 3, b = 3 \cdot 5, c = 5 \cdot 7$. Then $\text{GCD}(a, b, c) = 1$, but
$\text{GCD}(a, b) = 3$ and $\text{GCD}(b, c) = 5$. **7.** The LCM equals the
product of the numbers if, and only if, the numbers have no prime
factors in common. Because $\text{GCD}(a, b) \cdot \text{LCM}(a, b) = ab$,
$\text{LCM}(a, b) = ab$ if, and only if, $\text{GCD}(a, b) = 1$; that is, a and b
have no prime factors in common.
9. Answers vary.
- $m = 6$ and $n = 9$; then $\text{GCD}(6, 9) = 3$ and
  $\text{LCM}(6, 9) = 18$. $\text{GCD}(15, 18) = 3$.
- $m = 7$ and $n = 11$; then $\text{GCD}(7, 11) = 1$ and
  $\text{LCM}(7, 11) = 77$. $\text{GCD}(18, 77) = 1$.
- $m = 8$ and $n = 16$; then $\text{GCD}(8, 16) = 8$ and
  $\text{LCM}(8, 16) = 16$. $\text{GCD}(24, 16) = 8$.

(The conjecture is true but examples don't prove it.)

Open-Ended

11. Answers vary. From the answer to problem 8 it follows that
$\text{LCM}(a, b) < ab$ if, and only if, $\text{GCD}(a, b) > 1$; that is, if, and
only if, a and b have at least one prime in common.

Cooperative Learning

13. (a) Answers vary. **(b)** Answers vary.

Questions from the Classroom

15. The student is partially correct. If a and b are distinct
natural numbers, then the student is correct. By definition,
$a \leq \text{LCM}(a, b); b \leq \text{LCM}(a, b)$. Also $\text{GCD}(a, b) \leq a$
and $\text{GCD}(a, b) \leq b$. Hence, $\text{GCD}(a, b) \leq \text{LCM}(a, b)$. However, the equality holds if $a = b$.

Review Problems

17. $x = 15{,}625; y = 64$ **19.** No, 3 divides the sum of the digits
and therefore 3 divides 3111. **21.** 27,720

Chapter 4 Review

1. (a) false **(b)** false **(c)** true **(d)** False; 12, for example
(e) False; 9, for example **2. (a)** False; $7 | 7$ and $7 \nmid 3$, yet $7 | 3 \cdot 7$.
(b) False; $3 \nmid (3 + 4)$, but $3 | 3$ and $3 \nmid 4$. **(c)** true **(d)** true
(e) False; $4 \nmid 2$ and $4 \nmid 22$, but $4 \nmid 44$. **3. (a)** Divisible by 2, 3,
4, 5, 6, 8, 9, 11 **(b)** Divisible by 3, 11 **4.** If 10,007 is prime,
$17 \nmid 10{,}007$. We know $17 | 17$, so $17 \nmid (10{,}007 + 17)$ by
Theorem 4-2(b). **5. (a)** 87$\underline{2}$4; 87$\underline{5}$4; 87$\underline{8}$4 **(b)** 41,856; 44,856;
47,856 **(c)** 87,1 74; 87,4 64; 87,$\underline{7}$54 **6. (a)** The student's
claim is true. Examples vary. **(b)** Let n be a whole number.
Then $n + (n + 1) + (n + 2) + (n + 3) + (n + 4) =$
$5n + 10 = 5(n + 2)$. Thus, the sum is divisible by 5.
7. (a) composite **(b)** prime **8.** Check for divisibility by 3
and 8, $24 | 4152$. **9.** No, they are the same if the numbers are
equal. **10.** $\text{LCM}(a, b, c) = \text{LCM}(m, c)$, where $m = \text{LCM}(a, b)$,
$\text{LCM}(a, b) = \dfrac{ab}{\text{GCD}(a, b)}$, and $\text{LCM}(m, c) = \dfrac{mc}{\text{GCD}(m, c)}$.
Each of these GCDs can be found using the Euclidean algorithm.
11. The number is not divisible by 2, 3, 5, 7, 11, and 13 because each of these primes divides one product in the sum
$2 \cdot 3 \cdot 5 \cdot 7 + 11 \cdot 13$ but not the other. The student
checked that $17 \nmid 353$ and because $19^2 = 361 > 353$,
no other primes need to be tested. **12. (a)** 4 **(b)** 73
13. (a) $2^4 \cdot 5^3 \cdot 7^4 \cdot 13 \cdot 29$ **(b)** 77,562 **14.** Answers vary. For
example, 16. To obtain five divisors, we raise a prime (2) to the
$(5 - 1)$ power. **15.** 1, 2, 3, 4, 6, 8, 9, 12, 16, 18, 24, 36, 48, 72,
144 **16. (a)** $2^2 \cdot 43$ **(b)** $2^5 \cdot 3^2$ **(c)** $2^2 \cdot 5 \cdot 13$ **(d)** $3 \cdot 37$
17. The LCM of all positive integers less than or equal to 12 is
$2^3 \cdot 3^2 \cdot 5 \cdot 7 \cdot 11$ or 27720. **18.** \$0.31 **19.** 9:30 A.M. **20.** We
know that the $\text{GCD}(a, b) \cdot \text{LCM}(a, b) = ab$. Because
$\text{GCD}(a, b) = 1$, then $\text{LCM}(a, b) = ab$. **21.** 5 packages
22. 15 min **23.** 71 lattes. Because $9869 = 71 \cdot 139$ and 71 as well
as 139 are prime, she sold 71 lattes at \$1.39 each. **24. (a)** $2^{10} \cdot 3^{10}$
(b) $(2 \cdot 17)^n = 2^n \cdot 17^n$ **(c)** 97^4 since 97 is prime
(d) $(2^3)^4 \cdot (2 \cdot 3)^3 \cdot (2 \cdot 13)^2$
$= 2^{12} \cdot 2^3 \cdot 3^3 \cdot 2^2 \cdot 13^2$
$= 2^{17} \cdot 3^3 \cdot 13^2$
(e) $2^3 \cdot 3^2 (1 + 2 \cdot 3 \cdot 7) = 2^3 \cdot 3^2 \cdot 43$
(f) $2^4 \cdot 5^6 (3 \cdot 5 + 1) = 2^8 \cdot 5^6$

25. By the division algorithm every prime number greater than 3 can be written as $12q + r$, where $r = 1, 5, 7,$ or 11, since for other values of r, $12q + r$ is composite as 12 and r will share a common factor, making it not prime.

26. $n = a \cdot 10^2 + b \cdot 10 + c$
$n = a \cdot (99 + 1) + b \cdot (9 + 1) + c$
$n = 99a + 9b + a + b + c$

Since $9 \mid 99a$ and $9 \mid 9b$, then $9 \mid [99a + 9b + (a + b + c)]$ if, and only if, $9 \mid (a + b + c)$.   **27.** We first show that among any three consecutive odd whole numbers, there is always one that is divisible by 3. For that purpose, suppose that the first whole number in the triplet is not divisible by 3. Then by the division algorithm, that whole number can be written in the form $3n + 1$ or $3n + 2$ for some whole number n. Then the three consecutive odd whole numbers are $3n + 1, 3n + 3, 3n + 5$ or $3n + 2, 3n + 4, 3n + 6$. In the first triplet, $3n + 3$ is divisible by 3, and in the second, $3n + 6$ is divisible by 3. This implies that if the first odd whole number is greater than 3 and not divisible by 3, then the second or the third must be divisible by 3, and hence cannot be prime.   **28. (a)** If the least divisor greater than 1 were composite it could not be the least because a prime factor of the divisor would divide the number.

(b) If $d \mid n$ then $n = kd$ where $k \mid n$ and $k = \dfrac{n}{d}$.

Answers to Now Try This

4-1. If $5 \nmid a$ and $5 \nmid b$, then $5 \nmid (a + b)$ is not always true. For example, $5 \nmid 8$ and $5 \nmid 12$. But $5 \mid 8 + 12$. The statement is sometimes true for example $5 \nmid 7$ and $5 \nmid 12$ and $5 \nmid (7 + 12)$.   **4-2.** Yes, it is true. If $3 \mid x$, then 3 divides any whole number times x and in particular $3 \mid xy$.   **4-3.** $1 + 2 + 5 + 0 + 6 + 5 = 19$, so we must find numbers x and y such that $9 \mid [19 + (x + y)]$. Any two numbers that sum to 8 or 17 will satisfy this. Therefore, the blanks could be filled with 8 and 9, or 9 and 8, or the pairs $(8, 0), (0, 8), (1, 7), (7, 1), (2, 6), (6, 2), (3, 5), (5, 3), (4, 4)$.   **4-4. (a)** Answers vary. For example, only square numbers are listed in column 3; 2 is the only even number that will ever be in column 2, and column 2 contains prime numbers. The powers of 2 appear in successive columns.   **(b)** There will never be other entries in column 1 because 1 is the only number with one factor. Other numbers have at least the number itself and 1.   **(c)** 49, 121, 169   **(d)** 64   **(e)** The square numbers have an odd number of factors. Factors occur in pairs; for example, for 16 we have 1 and 16, 2 and 8, and 4 and 4. When we list the factors, we list only the distinct factors, so 4 is not listed twice, thereby making the number of factors of 16 an odd number. Similar reasoning holds for all square numbers.   **4-5. (a)** When you obtain a whole-number answer, it means the number you started with is divisible by the number you divided by.   **(b)** 2261 is divisible by both 17 and 19 and therefore the choice of color is not determined uniquely.   **4-6. (a)** 1; 2; 3; 6; 9   **(b)** 1; 2; 3; 4; 6; 8   **(c)** Only white rods can be used to form one-color trains for prime numbers if two or more rods must be used.   **(d)** The number must have at least 8 factors: 1; 2; 3; 5; 6; 10; 15; 30.   **4-7. (a)** No, because the multiples of 2 have 2 as a factor   **(b)** The multiples of 3: $\{3, 6, 9, 12, 15, \ldots\}$ **(c)** The multiples of 5: $\{5, 10, 15, 20, \ldots\}$ **(d)** The multiples of 7: $\{7, 14, 21, \ldots\}$ **(e)** We have to check only divisibility by 2, 3, 5, and 7.   **4-8.** The 1, 2, 3, and 6 rods can all be used to build both the 24 and 30 train. The greatest of these is 6, so $\mathrm{GCD}(24, 30) = 6$.   **4-9. (a)** The leftmost area is the factors of 24 that are not factors of 40. The center or intersection area

is the factors of both 24 and 40. The rightmost area is the factors of 40 that are not factors of 24.   **(b)** 8

(c)   Factors of 36   Factors of 44

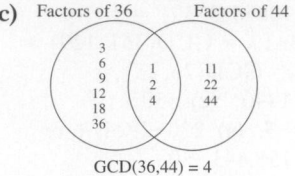

GCD(36,44) = 4

4-10.   Multiples of 8   Multiples of 12

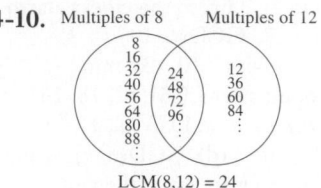

LCM(8,12) = 24

Answers to Brain Teasers

Section 4-1

The part of the explanation that is incorrect is the division by $(e - a - d)$, which is equal to 0. Division by 0 is impossible.

Section 4-2

Assuming that the ages are whole numbers, we list the decompositions of 2450 into three factors each followed by the sum of the three factors.

1, 1, 2450	2452	1, 2, 1225	1228
1, 5, 490	496	1, 7, 350	358
1, 10, 245	256	1, 14, 175	190
1, 35, 70	106	1, 25, 98	124
1, 49, 50	100	2, 5, 245	252
2, 25, 49	76	2, 7, 175	184
2, 35, 35	72	5, 14, 35	54
5, 10, 49	**64**	5, 5, 98	108
5, 7, 70	82	**7, 7, 50**	**64**
7, 10, 35	52		
7, 14, 25	46		

The only sum of three factors that occurs more than once is indicated in bold. If all sums were different, Natasha would have known which one it is, because she was told that the "right" sum was twice her age. Because she needed more information, we can conclude that her age is 32 and the ages of Jody's friends are 5, 10, and 49 or 7, 7, and 50. That Natasha determined the answer after Jody's reply that she is at least one year younger than the oldest of the three friends eliminates 5, 10, 49 among these ages. If Jody's age had been 48 or less, Natasha would still need more information. Hence, the friend's ages must be 7, 7, and 50 and Jody is 49 years old.

Section 4-3

If n is the width of the rectangle and m is the length of the rectangle, then the number of squares the diagonal crosses is $(n + m) - \mathrm{GCD}(n, m)$ or $(n + m) - 1$.

Answer to the Preliminary Problem

If Jacob has n marbles, then $n + 1$ is a multiple of 4, of 5, and of 7. The $\mathrm{LCM}(4, 5, 7) = 4 \cdot 5 \cdot 7 = 140$. Thus $n + 1$ is a multiple of 140. Since we want the smallest such whole number n, we conclude that $n + 1 = 140$ and $n = 139$. Consequently the least number of marbles Jacob has is 139.

Chapter 5

Assessment 5-1A

1. (a) ‾2 **(b)** 5 **(c)** ‾m **(d)** 0 **(e)** m **(f)** ‾$a − b$ or ‾$(a + b)$
or ‾a + ‾b **2. (a)** 2 **(b)** m **(c)** 0 **3. (a)** 5 **(b)** 10
(c) ‾5 **(d)** ‾5

4. (a)

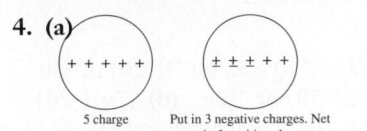

5 charge   Put in 3 negative charges. Net
result: 2 positive charges; answer: 2

(b)

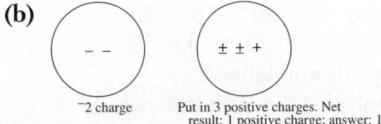

‾2 charge   Put in 3 positive charges. Net
result: 1 positive charge; answer: 1

(c)

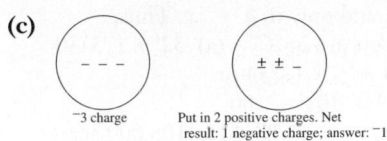

‾3 charge   Put in 2 positive charges. Net
result: 1 negative charge; answer: ‾1

(d)

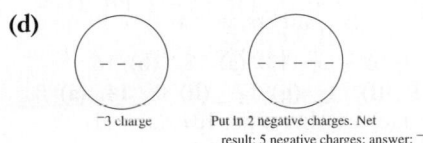

‾3 charge   Put in 2 negative charges. Net
result: 5 negative charges; answer: ‾5

5. (a)

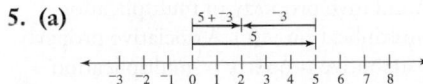

(b)

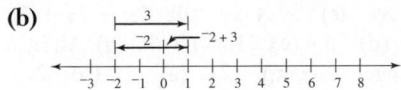

(c)

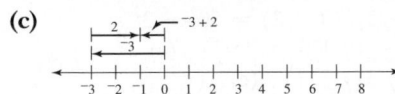

(d)

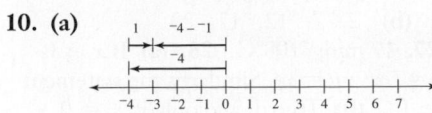

6. (a) $3 + ‾(‾2) = 5$ **(b)** ‾$3 + ‾2 = ‾5$
(c) ‾$3 + ‾(‾2) = ‾1$ **7. (a)** $3 − (‾2) = x$ if, and only if,
$3 = ‾2 + x$. Thus, $x = 5$. **(b)** ‾$3 − 2 = x$ if, and only if,
‾$3 = 2 + x$. Therefore, $x = ‾5$. **(c)** ‾$3 − (‾2) = x$ if, and only
if, ‾$3 = ‾2 + x$. Therefore, $x = ‾1$. **8. (a)** ‾$17 + 10 = ‾7$
giving a net change of ‾7 points **(b)** ‾$10 + 8 = ‾2$ giving a new temperature of ‾2°C **(c)** $5000 + (‾100)$, or
4900 ft **9. (a)** ‾$45 + ‾55 + ‾165 + ‾35 + ‾100 +$
$75 + 25 + 400$

10. (a)

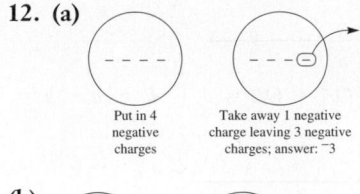

(b)

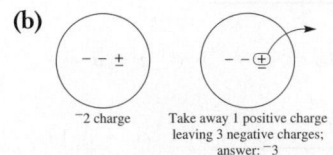

11. (a) ‾$4 − 2 = ‾6$; ‾$4 − 1 = ‾5$; ‾$4 − 0 = ‾4$; ‾$4 − ‾1 = ‾3$
(b) $3 − 1 = 2$; $2 − 1 = 1$; $1 − 1 = 0$; $0 − 1 = ‾1$;
‾$1 − 1 = ‾2$; ‾$2 − 1 = ‾3$

12. (a)

Put in 4   Take away 1 negative
negative   charge leaving 3 negative
charges   charges; answer: ‾3

(b)

‾2 charge   Take away 1 positive charge
leaving 3 negative charges;
answer: ‾3

13. (a) ‾9 **(b)** 3 **(c)** 1 **14. (a)** $55 − 60$; $55 + ‾60$; ‾5°F
(b) $200 − 220$; $200 + ‾220$; ‾20 **15. (a)** $3 + ‾(2 − 4x) =$
$3 + (‾2 + ‾[‾(4x)]) = (3 + ‾2) + 4x = 1 + 4x$
(b) ‾$(‾x + ‾y) = x + ‾(‾x) + ‾(‾y) = x + x + y = 2x + y$
16. The equation holds if, and only if, $c = 0$
(a and b can be any integers). Justification: It can be shown
that $a − (b − c) = a − b + c$ for all integers a, b,
and c. Thus, the original equation holds if, and only if,
$(a − b) + ‾c = (a − b) + c$, which in turn holds if, and only if,
‾$c = c$. This last equation is true if, and only if, $c = 0$.
17. (a) I **(b)** W **(c)** $I − \{0\}$ **(d)** $\varnothing$ **(e)** $\varnothing$
18. Answers may vary.

‾3	4	‾1
2	0	‾2
1	‾4	3

19. (a) 0 **(b)** ‾101 **(c)** 1 **(d)** $a − 1$ **(e)** ‾4 **20. (a)** All
negative integers **(b)** All positive integers **(c)** All integers less
than ‾1 **(d)** 2 or ‾2 **21. (a)** 9 **(b)** 2 **(c)** 0 and 2
22. (a) 0 or 12 **(b)** ‾8 or 8 **(c)** Every integer satisfies this
equation. **23. (a)** 89 **(b)** 19 **24. (a)** $d = ‾3$, next terms:
‾12, ‾15 **(b)** $d = ‾y$, next terms: $x − 2y$, $x − 3y$ **25. (a)** True
(b) True **(c)** True **26. (a)** ‾4 **(b)** 3 **(c)** ‾5 **27. (a)** ‾18
(b) ‾6 **(c)** 22 **(d)** ‾18 **(e)** 23

Mathematical Connections 5-1

Communication

1. He could have driven 12 mi in either direction from
milepost 68. Therefore, his location could be either at
the $68 − 12 = 56$ milepost or at the $68 + 12 = 80$ milepost. **3.** Two numbers are additive inverses if, and only if,
their sum is 0. We have

$$(b − a) + (a − b) = b + ‾a + a + ‾b$$
$$= b + 0 + ‾b$$
$$= b + ‾b$$
$$= 0$$

5. Answers vary. Students may substitute different integers for a
and b, plot them on a number line, and show that in each case the
distance between the points is $|a − b|$. They should give at least
three examples: When both integers are positive, one is positive

and the other negative, and both are negative. A more general approach is as follows:

If $a > b > 0$, then $AB = OA - OB = a - b = |a - b|$.

If $a > 0$ and $b < 0$, then $AB = OA + OB = a + {}^-b = a - b = |a - b|$.

If $a < b < 0$, then $AB = OA - OB = {}^-a - {}^-b = {}^-a + b = b - a = |a - b|$.

Open-Ended

7. Answers vary. For example, suppose Alice was figuring how much money she spent yesterday. She had one bill for $50, another bill for $85 arrived in the mail, and then she returned some books for a $30 credit. **9. (a)** 4 **(b)** Yes. Let x be our number and let a be the answer. Then,

$$
\begin{aligned}
{}^-[(x - 10) + {}^-3] &= a \\
{}^-[(x + {}^-10) + {}^-3] &= a \\
{}^-({}^-x + 10 + {}^-3) &= a \\
{}^-({}^-x + 7) &= a \\
x + {}^-7 &= a
\end{aligned}
$$

From here, $x = a + 7$. **(c)** Answers vary.

Cooperative Learning

11. Answers vary.

Questions from the Classroom

13. The algorithm is correct, and the student should be congratulated for finding it. One way to encourage such creative behavior is to name and refer to the procedure after the student who invented it; for example, "David's subtraction method." In fourth grade, the technique can be explained by using a money model. Suppose you have $4 in one checking account and $80 in another, for a total of $84. You spent $27 by withdrawing $7 from the first account and $20 from the second. The first checking account is overdrawn by $3; that is, the balance is ${}^-\$3$. The balance in the second account is $60. After transferring $3 from the second account to the first, the balance in the first account is $0 and in the second $57; that is, the total balance is $57. The algorithm will always work. **15.** The picture is supposed to illustrate that an integer and its opposite are mirror images of each other. Because a could be negative, the picture is correct. For example, possible values for a and ${}^-a$ are $a = {}^-1, {}^-a = 1$, and $a = {}^-7, {}^-a = 7$. At this point, the teacher could remind the students that the "−" sign in ${}^-a$ *does not* mean that ${}^-a$ is negative. If a is positive, ${}^-a$ is negative, but if a is negative, ${}^-a$ is positive.

Assessment 5-2A

1. $3({}^-1) = {}^-3; 2({}^-1) = {}^-2; 1({}^-1) = {}^-1; 0({}^-1) = 0;$
$({}^-1)({}^-1) = 1$, by continuing the pattern of an arithmetic sequence with fixed difference 1.

2.

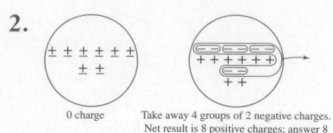

0 charge   Take away 4 groups of 2 negative charges.
Net result is 8 positive charges; answer 8

3.

$$2({}^-4) = {}^-8$$

4. (a) $({}^-3)({}^-3) = 9$ **(b)** $({}^-5)2 = {}^-10$ **5. (a)** $4({}^-20)$ or ${}^-80$
(b) $({}^-4)({}^-20)$ or 80 more **(c)** $n({}^-20)$ or ${}^-20n$ **(d)** $({}^-n)({}^-20)$
or $20n$ more **6. (a)** 5 **(b)** ${}^-11$ **(c)** Undefined; since $0k = {}^-5$
has no integer solution **7. (a)** ${}^-10$ **(b)** ${}^-10$ **(c)** Not defined
(d) Not defined **(e)** 1 **8. (a)** ${}^-30; {}^-30 \div {}^-6 = 5;$
${}^-30 \div 5 = {}^-6$ **(b)** 20; $20 \div {}^-5 = {}^-4; 20 \div {}^-4 = {}^-5$
(c) $0; 0 \div {}^-3 = 0$; division by 0 is not defined.
9. (a) $(4x) \div 4 = a$ if, and only if, $4x = 4a$ if, and only
if, $a = x$. Thus $(4x) \div 4 = x$. **(b)** ${}^-xy \div y = a$ if,
and only if, ${}^-xy = ya = ay$ if, and only if, $a = {}^-x$. Thus,
${}^-xy \div y = {}^-x$. **10.** All answers are in °C. **(a)** $32 + ({}^-3)30$
or $32 - 3 \cdot 30 = {}^-58$ **(b)** $0 + ({}^-4)({}^-25)$ or
$4 \cdot 25 = 100$ **(c)** ${}^-20 + ({}^-4)({}^-30) = 100$
(d) $25 + 3({}^-20) = 25 - 3 \cdot 20 = {}^-35$°C **11.** 108,000 acres
12. (a) ${}^-1({}^-5 + {}^-2) = {}^-1({}^-7) = 7;$ $({}^-1)({}^-5) + ({}^-1)({}^-2) =$
$5 + 2 = 7$ **(b)** ${}^-3({}^-3 + 2) = {}^-3({}^-1) = 3;$
$({}^-3)({}^-3) + ({}^-3)(2) = 9 + {}^-6 = 3$ **13. (a)** ${}^-8$ **(b)** 16
(c) ${}^-1000$ **(d)** 81 **(e)** 1 **(f)** ${}^-1$ **(g)** 12 **(h)** 0 **14. (a)** 0
(b) 0 **(c)** 9 **15. (b)** and **(c)** are always positive, **(a)** is always negative. **(d)** and **(e)** are neither **16. (b)** = **(c)**;
(d) = **(e)** **17. (a)** Commutative property of multiplication
(b) Closure property of multiplication **(c)** Associative property
of multiplication **(d)** Distributive property of multiplication
over addition **18. (a)** xy **(b)** $2xy$ **(c)** $3x - y$ **(d)** ${}^-x$
19. (a) ${}^-2x + 2y$ **(b)** $x^2 - xy$ **(c)** ${}^-x^2 + xy$ **(d)** ${}^-2x - 2y + 2z$
20. (a) ${}^-2$ **(b)** 2 **(c)** 0 **(d)** ${}^-6$ **(e)** ${}^-36$ **(f)** 6 **(g)** All integers except 0 **(h)** All integers except 0 **21. (a)** ${}^-5$ **(b)** ${}^-2$
(c) No solutions **(d)** ${}^-2$ **(e)** ${}^-2$ or 2 **(f)** ${}^-2$ or 4 **(g)** ${}^-1$
(h) 1 or ${}^-3$ **22. (a)** $(50 + 2)(50 - 2) = 50^2 - 2^2 =$
$2500 - 4 = 2496$ **(b)** $25 - 10,000 = {}^-9975$ **(c)** $x^2 - y^2$
23. (a) $(3 + 5)x$ or $8x$ **(b)** $x(y + 1)$ **(c)** $x(x + y)$
(d) $x(3y + 2 - z)$ **(e)** $a[b(c + 1) - 1]$ **(f)** $(4 + a)(4 - a)$
(g) $(2x + 5y)(2x - 5y)$
24. (a) $(a - b)^2 = (a + {}^-b)(a + {}^-b)$
$$
\begin{aligned}
&= a(a + {}^-b) + {}^-b(a + {}^-b) \\
&= a^2 + a({}^-b) + ({}^-b)a + ({}^-b)({}^-b) \\
&= a^2 + {}^-(ab) + {}^-(ba) + b^2 \\
&= a^2 - 2ab + b^2
\end{aligned}
$$
(b) (i) $98^2 = (100 - 2)^2 = 100^2 - 2(200) + 2^2 =$
$10,000 - 400 + 4 = 9604$ **(ii)** $99^2 = (100 - 1)^2 =$
$100^2 - 2(100) + 1^2 = 10,000 - 200 + 1 = 9801$
(iii) $997^2 = (1000 - 3)^2 = 1000^2 - 2(3000) + 3^2 =$
$1,000,000 - 6000 + 9 = 994,009$ **25. (a)** $8, 11, d = 3$, nth
term is $3n - 13$. **(b)** ${}^-128, {}^-256, r = 2$, nth term is ${}^-2^n$.
(c) $2^7, {}^-2^8, r = {}^-2$, nth term is $2({}^-2)^{n-1}$ or ${}^-({}^-2)^n$.
26. (a) ${}^-9, {}^-6, {}^-1, 6, 15$ **(b)** ${}^-2, {}^-7, {}^-12, {}^-17, {}^-22$
(c) ${}^-3, 3, {}^-9, 15, {}^-33$ **27.** 17 min; ${}^-108$°C **28. (a)** If $x \geq 0$
and $y \leq 0$, then ${}^-|x| \, |y| = {}^-x({}^-y) = xy$. Similarly, the statement
is true for $x \leq 0$ and $y \geq 0$. **(b)** True if, and only if, $x = 0$.
If $x \neq 0$, ${}^-x^2$ is negative and x^2 is positive and hence the
expressions cannot be equal.

Mathematical Connections 5-2

Communication

1. No; it is not of the form $(a - b)(a + b)$.
3. One does not normally replace a variable a with ab. However, $(^-1)(ab) = ^-(ab)$ is true by Theorem 5-6 for all integers and ab is an integer.
$(^-1)(ab) = (^-1 \cdot a)b$ by Associative Property of multiplication.
$(^-1 \cdot a)b = (^-a)$ by Theorem 5-6.
Therefore, $(^-a)b = ^-(ab)$ by property of equality. The argument is valid.
5. Answers vary.   **(a)** The argument is logical if we assume that $(^-1)(^-1)$ is either 1 or $^-1$. That assumption, however, has a flaw; $(^-1)(^-1)$ could be an integer other than 1 or $^-1$.   **(b)** Yes; for example, if we assume that $(^-a)(^-b)$ is either $^-(ab)$ or ab, we could show in a similar way that $^-(ab)$ causes a contradiction.

Open-Ended

7. Answers vary. For example, if the student answered only one problem and missed it, he or she would score $^-1$. If the student answered 5 correct and missed more than 20, he or she would receive a negative score. Any values of x and y such that $4x < y$ would result in a negative score, where x is the number of correct and y is the number of incorrect answers.   **9.** Answers vary.

Cooperative Learning

11. (a) Answers vary.   **(b)** Answers vary.

Questions from the Classroom

13. The student is correct that a debt of $5 is greater than a debt of $2. However, what this means is that on a number line $^-5$ is farther to the left than is $^-2$. The fact that $^-5$ is farther to the left than $^-2$ on a number line implies that $^-5 < ^-2$.   **15.** The procedure can be justified as follows. Since for all integers c, $^-c = (^-1)c$, the effect of performing the opposite of an algebraic expression is the same as multiplying the expression by $^-1$. However, in the expression $x - (2x - 3)$, the "$-$" is used to denote subtraction, not simply finding the opposite. If the expression is first rewritten as $x + ^-(2x + ^-3)$, then it is the case that $^-(2x + ^-3) = ^-1(2x + ^-3)$, or $^-2x + 3$. Now the expression can be rewritten as $x + ^-2x + 3$, which a student might obtain from the father's rule.

Review Problems

17.

19. (a) 14   **(b)** 21   **(c)** $^-4$ **(d)** 22

Chapter Review

1. (a) $^-3$ **(b)** a **(c)** $^-1$ **(d)** $^-x - y$ **(e)** $x - y$ **(f)** $x + y$
(g) 32   **(h)** 32   **2. (a)** $^-7$ **(b)** 8   **(c)** 8   **(d)** 0   **(e)** 8   **(f)** 15
3. (a) 3   **(b)** $^-5$ **(c)** Any integer except 0   **(d)** No integer will work.   **(e)** $^-41$ **(f)** Any integer   **4.** $2(^-3) = ^-6$;
$1(^-3) = ^-3; 0(^-3) = 0$; if the pattern continues, then $^-1(^-3) = 3$;
$^-2(^-3) = 6$.   **5. (a)** $10 - 5 = 5$ **(b)** $1 - (^-2) = 3$ **6. (a)** ^-x
(b) $y - x$ or $^-x + y$ **(c)** $3x - 1$ **(d)** $2x^2$ **(e)** 0
(f) $^-x^2 - 6x - 9$

(g) $4 - x^2$ **7. (a)** ^-2x **(b)** $x(x + 1)$ **(c)** $(x - 6)(x + 6)$
(d) $(9y^2 + 4x^2)(3y - 2x)(3y + 2x)$ **(e)** $5(1 + x)$ **(f)** $(x - y)x$
8. (a) False; it is not positive for $x = 0$.   **(b)** False, if one value is positive and one is negative.   **(c)** False; let $a = 2, b = ^-5$
(d) True   **9. (a)** $1 \div 2 \neq 2 \div 1$ **(b)** $3 - (4 - 5) \neq (3 - 4) - 5$
(c) $1 \div 2$ is not an integer.   **(d)** $8 \div (4 - 2) \neq (8 \div 4) -$
$(8 \div 2)$ **10. (a)** $^-10$ **(b)** $^-2^{99}$ **(c)** 2^{89} **(d)** 0   **(e)** 3 or $^-3$
(f) $x \leq 0$; that is, $0, ^-1, ^-2, ^-3, \ldots$ **(g)** $x \geq 4$ or $x \leq ^-4$; that is,
$\{\ldots ^-6, ^-5, ^-4\} \cup \{4, 5, 6, 7, \ldots\}$ **(h)** $x = 11$ or $^-9$
11. (a) $^-1, 1, ^-1, 1, ^-1, 1$ **(b)** $^-2, 4, ^-8, 16, ^-32, 64$
(c) $^-5, ^-8, ^-11, ^-14, ^-17, ^-20$ **12. (a)** Geometric, ratio $^-1$
(b) Geometric, ratio $^-2$ **(c)** Arithmetic, $d = ^-3$
13. 26   **14.** Oregon, Nevada, Iowa, California, Arizona, Ohio, West Virginia, Alabama   **15. (a)** $^-22$ **(b)** 7080 ft   **(c)** Answers may vary, but the expectation is that the elevation would be quite high (Mauna Kea).   **16.** Answers vary. One difference is $^-274$ ft.   **(b)** Answers vary.   **17.** Answers may vary but the depth is 3963 m below sea level or $^-3963$ m.
18. (a) degree Celsius $\approx$ kelvin $- 273$ **(b)** 373 kelvin
(c) 233 kelvin   **19. (a)** Answers may vary. For example, number engaged is positive, number of casualties is negative.   **(b)** $^-51,112$ or 51,112 casualties   **20.** $^-2$ could mean two floors below ground level.

Answers to Now Try This

5-1. (a) The sum of two negative integers is always negative. Consider the number line model for examples.
(b) It can be either depending on the values of the numbers.
(c)

5-2. (a) Since $x \leq 0$, $|x| = ^-x$ and $|x| + x = ^-x + x = 0$.
(b) Since $x \leq 0$, $^-|x| + x = ^-(^-x) + x = 2x$.
(c) Since $x \geq 0$, $^-|x| + x = ^-x + x = 0$.   **5-3. (a)** Answers vary. For example, a mail carrier brings you three letters, one with a check for $23 and the other two with bills for $13 and $12, respectively. Are you richer or poorer and by how much?
(b) Answers will vary. For example, a mail carrier brings you one letter with a check for $18 and takes away a bill for $37 that was intended for someone else. Are you richer or poorer and by how much?   **5-4. (a)** Yes, because $a - b = a + ^-b$ and the sum of two integers is an integer.   **(b)** None of the properties holds for integers because $a - b \neq b - a$ (if $a \neq b$) and $(a - b) - c \neq a - (b - c)$ (if $c \neq 0$). There is no single integer i such that for all integers $a - i = a$ and $i - a = a$ (the first equation implies that $i = 0$ but 0 does not satisfy the second equation).   **5-5.** The entries in column A are 5, 4, 3, 2, 1, 0, $^-1, ^-2, \ldots$. The entries in column B are the same. The entries in column C are 25, 16, 9, 4, 1, 0, 1, 4, 9, $\ldots$. The spreadsheet suggests that a positive integer times itself is positive, zero times itself is zero, and a negative integer times itself is a positive integer.
5-6. (a) $101 \cdot 99 = (100 + 1)(100 - 1) = 100^2 - 1^2 = 10,000 - 1 = 9999$ **(b)** $22 \cdot 18 = (20 + 2)(20 - 2) = 20^2 - 2^2 = 400 - 4 = 396$ **(c)** $24 \cdot 36 = (30 - 6)(30 + 6) = 30^2 - 6^2 = 900 - 36 = 864$
(d) $998 \cdot 1002 = (1000 - 2)(1000 + 2) = 1000^2 - 2^2 = 1,000,000 - 4 = 999,996$ **5-7.** $a \div 0 = x$, if, and only if, $0 \cdot x = a$ and x is unique. Because $0x = 0$ for all integers x, the equation has no solution if $a \neq 0$. If $a = 0$, then for all integers x,

$0x = 0$. Because the solution is not unique, $0 \div 0$ is not defined. **5-8.** $GCD(206, 23) = GCD(206 - 9 \cdot 23, 23)$ because if $d|206$ and $d|23$, then $d|206$ and $d|9 \cdot 23$. Also, if $d|206$ and $d|9 \cdot 23$, then $d|(206 - 9 \cdot 23)$. Because $206 - 9 \cdot 23$ is a negative number ($^-1$) then d is greater than any other divisor of both $206 - 9 \cdot 23$ and 23. With $d|^-1$ and $d|23$ and $d > 0$, then d is $GCD(^-1, 23)$. Thus, $d = 1$ because 1 is the only positive divisor of $^-1$.

Answers to Brain Teasers

Section 5-1
$123 - 45 - 67 + 89 = 100$

Section 5-2
Answers may vary. $1 = 4^4/4^4$; $2 = (4 \cdot 4)/(4 + 4)$;
$3 = 4 - (4/4)^4$; $4 = [(4 - 4)/4] + 4$; $5 = 4 + 4^{(4-4)}$;
$6 = 4 + [(4 + 4)/4]$; $7 = (44/4) - 4$; $8 = [(4 + 4)/4]4$;
$9 = 4 + 4 + 4/4$; $10 = (44 - 4)/4$

Answer to the Preliminary Problem

A table showing combinations of acceptable sets of numbers appears below.

Multiples of 6	Multiples of 5	Total Barrels	Total Weight
$8 \cdot 6 = 48$	$1 \cdot 5 = 5$	9	53
$3 \cdot 6 = 18$	$7 \cdot 5 = 35$	10	53

Multiples of $^-6$	Multiples of 5	Total Barrels	Total Weight
$2 \cdot ^-6 = ^-12$	$13 \cdot 5 = 65$	15	53
$13 \cdot ^-6 = ^-78$	$5 \cdot 5 = 25$	18	$^-53$

Multiples of 6	Multiples of $^-5$	Total Barrels	Total Weight
$2 \cdot 6 = 12$	$13 \cdot ^-5 = ^-65$	18	$^-53$
$13 \cdot 6 = 78$	$5 \cdot ^-5 = ^-25$	18	53

Multiples of $^-6$	Multiples of $^-5$	Total Barrels	Total Weight
$8 \cdot ^-6 = ^-48$	$1 \cdot ^-5 = ^-5$	9	$^-53$
$3 \cdot ^-6 = ^-18$	$7 \cdot ^-5 = ^-35$	10	$^-53$

Thus there are 8 possible ways that the barrels could have been labeled as seen above.

Chapter 6

Assessment 6-1A

1. (a) The solution to $8x = 7$ is $\frac{7}{8}$. **(b)** Jane ate $\frac{7}{8}$ of the pizza. **(c)** The ratio of boys to girls is 7 to 8.
2. (a) $\frac{1}{6}$ **(b)** $\frac{1}{4}$ **(c)** $\frac{2}{6}$ or $\frac{1}{3}$ **(d)** $\frac{7}{12}$ **3. (a)** $\frac{2}{3}$ **(b)** $\frac{4}{6}$ or $\frac{2}{3}$

(c) $\frac{6}{9}$ or $\frac{2}{3}$ **(d)** $\frac{8}{12}$ or $\frac{2}{3}$. The diagram illustrates the Fundamental Law of Fractions. **4. (a)** No, the parts do not have equal areas. **(b)** yes **(c)** yes **5. (a)** **(b)**

(c) **6. (a)** $\frac{9}{24}$ or $\frac{3}{8}$ **(b)** $\frac{12}{24}$ or $\frac{1}{2}$ **(c)** $\frac{4}{24}$ or $\frac{1}{6}$

(d) $\frac{8}{24}$ or $\frac{1}{3}$ **7. (a)** $\frac{4}{18}, \frac{6}{27}, \frac{8}{36}$ **(b)** $\frac{^-4}{10}, \frac{2}{^-5}, \frac{^-10}{25}$ **(c)** $\frac{0}{1}, \frac{0}{2}, \frac{0}{4}$

(d) $\frac{2a}{4}, \frac{3a}{6}, \frac{4a}{8}$ **8. (a)** $\frac{52}{31}$ **(b)** $\frac{3}{5}$ **(c)** $\frac{^-5}{7}$ **9. (a)** undefined
(b) undefined **(c)** 0 **(d)** cannot be simplified **(e)** cannot be simplified **10. (a)** $\frac{a-b}{3}, a \neq ^-b$ **(b)** $\frac{2x}{9y}, x, y \neq 0$

11. (a) equal **(b)** equal **12. (a)** not equal **(b)** not equal

13. Answers vary. **14.** $\frac{36}{48}$ **15.**

16. (a) $\frac{32}{3}$ **(b)** $^-36$ **17. (a)** $>$ **(b)** $>$ **(c)** $<$ **18. (a)** $\frac{11}{13}, \frac{11}{16}, \frac{11}{22}$

(b) $\frac{^-1}{5}, \frac{^-19}{36}, \frac{^-17}{30}$ **19.** The nth term of the sequence is $\frac{n}{n+2}$.

Hence, the $(n + 1)$st term is $\frac{n+1}{n+3}$. We need to show that

$\frac{n+1}{n+3} > \frac{n}{n+2}$ for $n \geq 1$. By Theorem 6-3 this is true if, and only if, $(n + 1)(n + 2) > n(n + 3)$, which is equivalent to $n^2 + 2n + n + 2 > n^2 + 3n$, or $2 > 0$. Because the last statement is true, the statement $\frac{n+1}{n+3} > \frac{n}{n+2}$ is true. Another approach is to use Theorem 6-5 and notice that each term of the sequence starting from the second can be obtained by adding the numerators and the denominators of the neighboring terms. **20.** Answers may vary. The following are possible answers: **(a)** $\frac{10}{21}, \frac{11}{21}$

(b) $\frac{^-22}{27}, \frac{^-23}{27}$ **21.** 456 mi **22. (a)** $\frac{6}{16}$, or $\frac{3}{8}$; $\frac{6}{32,000}$,

or $\frac{3}{16,000}$ **(b)** $\frac{10}{100}$, or $\frac{1}{10}$ **(c)** $\frac{15}{60}$ or $\frac{1}{4}$ **(d)** $\frac{8}{24}$, or $\frac{1}{3}$

23. $\frac{1}{6}$ **24.** Impossible to determine partly because

$\frac{20}{25} = \frac{24}{30} = \frac{4}{5}$, the same fraction of students passed in each class, but the actual scores in one class could have been higher than in the other. More students *passed* Ms. Price's test because there were more students. **25.** Bren's class

Mathematical Connections 6-1

Communication

1. Answers vary; for example, $\frac{3}{4}$ of a quantity is found by dividing that quantity by 4 and then taking 3 of the parts. Hence, $\frac{3}{4}$ of 4

is 3. On the other hand, four $\frac{3}{4}$ is 12 quarters, or 3.   **3.** The new fraction is equal to $\frac{1}{2}$. The principle can be generalized as follows:

$$\frac{a}{b} = \frac{ar_1}{br_1} = \frac{ar_2}{br_2} = \frac{ar_3}{br_3} = \ldots = \frac{ar_n}{br_n}$$
$$= \frac{a(r_1 + r_2 + r_3 + \ldots + r_n)}{b(r_1 + r_2 + r_3 + \ldots + r_n)} = \frac{a}{b}$$

5. Answers vary.

Cooperative Learning

7. (a) Area of $a = \frac{1}{4}$ **(b)** Area of $a = 1$

Area of $b = \frac{1}{4}$ Area of $b = 1$

Area of $c = \frac{1}{16}$ Area of $c = \frac{1}{4}$

Area of $d = \frac{1}{8}$ Area of $d = \frac{1}{2}$

Area of $e = \frac{1}{16}$ Area of $e = \frac{1}{4}$

Area of $f = \frac{1}{8}$ Area of $f = \frac{1}{2}$

Area of $g = \frac{1}{8}$ Area of $g = \frac{1}{2}$

Questions from the Classroom

9. The first student's approach is correct. What the second student has done is to treat the problem as if it had been $\left(\frac{1}{5}\right)\left(\frac{5}{3}\right) = \frac{1}{3}$, when in reality, the problem is $\frac{15}{53}$. One cancels factors, not digits.   **11.** The student is applying incorrectly

Theorem 6-1, which is true only for products. The student can be shown that $\frac{9}{5} = \frac{6+3}{2+3} \neq \frac{6}{2} = 3$.   **13.** Answers vary; for example, she could show $\frac{3}{4}$ of the faces by picking 3 of the 4 large faces and 3 of the 4 small faces.   **15. (a)** Suppose the rational numbers are $\frac{2}{16}$ and $\frac{1}{4}$. In this case, $\frac{1}{4} > \frac{2}{16}$ and Iris is incorrect.   **(b)** Suppose the rational numbers are $\frac{2}{3}$ and $\frac{1}{2}$. In this case, $\frac{2}{3} > \frac{1}{2}$. Shirley is incorrect.   **17.** Daryl is not correct because the whole is not divided into 3 equal-size pieces.   **19.** If you write each fraction over a common denominator, it is easier to compare them. Thus we have $\frac{3}{8} = \frac{9}{24}$ and $\frac{2}{3} = \frac{16}{24}$. Because $\frac{16}{24} > \frac{9}{24}$, then $\frac{2}{3} > \frac{3}{8}$.

Assessment 6-2A

1. (a) $\frac{7}{6}$ or $1\frac{1}{6}$ **(b)** $\frac{-4}{12}$ or $\frac{-1}{3}$ **(c)** $\frac{5y-3x}{xy}$

(d) $\frac{-3y + 5x + 14y^2}{2x^2y^2}$ **(e)** $\frac{71}{24}$ or $2\frac{23}{24}$ **(f)** $\frac{-23}{3}$ or $-7\frac{2}{3}$

2. (a) $18\frac{2}{3}$ **(b)** $-2\frac{93}{100}$ **3. (a)** $\frac{27}{4}$ **(b)** $\frac{-29}{8}$ **4. (a)** $\frac{1}{3}$, high

(b) $\frac{1}{6}$, low   **(c)** $\frac{3}{4}$, low   **(d)** $\frac{1}{2}$, high   **5.** $8\frac{4}{12} = 8\frac{1}{3}$

6. (a) Beavers   **(b)** Ducks   **(c)** Bears

7.   About 0   About $\frac{1}{2}$ About 1   **8. (a)** $\frac{1}{2}$, high   **(b)** 0, low
$\frac{1}{10}$ $\frac{1}{100}$ $\frac{4}{7}$ $\frac{8}{12}$ $\frac{2}{5}$ $\frac{9}{18}$ $\frac{7}{8}$ $\frac{13}{10}$

(c) $\frac{3}{4}$, high   **(d)** 1, high   **9. (a)** 2   **(b)** $\frac{3}{4}$ **10. (a)** $\frac{1}{4}$

(b) 0   **11. (a)** A   **(b)** H   **(c)** T   **(d)** H   **12.** Approximately, $4 \cdot 3 = 12$.   **13.** $\frac{1}{4}$ **14.** $6\frac{7}{12}$ yd   **15.** $2\frac{5}{6}$ yd   **16. (a)** $\frac{1}{2} + \frac{3}{4} \in Q$

(b) $\frac{1}{2} + \frac{3}{4} = \frac{3}{4} + \frac{1}{2}$ **(c)** $\left(\frac{1}{2} + \frac{1}{3}\right) + \frac{1}{4} = \frac{1}{2} + \left(\frac{1}{3} + \frac{1}{4}\right)$

17. (a) $\frac{3}{2}, \frac{7}{4}, 2$ **(b)** $\frac{6}{7}, \frac{7}{8}, \frac{8}{9}$, not arithmetic, $\frac{2}{3} - \frac{1}{2} \neq \frac{3}{4} - \frac{2}{3}$

18. $1, \frac{7}{6}, \frac{8}{6}, \frac{9}{6}, \frac{10}{6}, \frac{11}{6}, 2$

19. (a) (i) $\frac{1}{4} + \frac{1}{3 \cdot 4} = \frac{1}{4} + \frac{1}{12} = \frac{16}{48} = \frac{1}{3}$

(ii) $\frac{1}{5} + \frac{1}{4 \cdot 5} = \frac{1}{5} + \frac{1}{20} = \frac{25}{100} = \frac{1}{4}$

(iii) $\frac{1}{6} + \frac{1}{5 \cdot 6} = \frac{1}{6} + \frac{1}{30} = \frac{36}{180} = \frac{1}{5}$

(b) $\frac{1}{n} = \frac{1}{n+1} + \frac{1}{n(n+1)}$ **20. (a)** $\frac{5}{6}$ **(b)** $\frac{21}{6}$ or $3\frac{1}{2}$

21. (a) $\frac{3x^2 + y^3}{x^2y^2}$ **(b)** $\frac{az - by}{xy^2z}$ **(c)** $\frac{2ab - b^2}{a^2 - b^2}$ **22.** $\frac{1}{2}$ mi

23. No, you need $\frac{1}{4}$ c more of milk.

Mathematical Connections 6-2

Communication

1. Answers vary. For example, $\frac{1}{3} + \frac{1}{4}$ or $\frac{7}{12}$ does not represent the amount received since the fractions did not come from the same size "whole."   **3. (a)** No, but it is easier to reduce if we do choose a least common denominator.   **(b)** No. For example, $\frac{1}{3} + \frac{1}{6} = \frac{3}{6}$, which is not in simplest form.   **5.** No, Kara spent $\frac{1}{3}$ of $\frac{1}{2}$ of her allowance on Sunday, so she has $\frac{1}{2} - \frac{1}{3} \cdot \frac{1}{2} = \frac{3}{6} - \frac{1}{6} = \frac{2}{6}$, or $\frac{1}{3}$, of her allowance left.   **7. (a)** Yes. If a, b, c, and d are integers, $b \neq 0$, $d \neq 0$, then $\frac{a}{b} - \frac{c}{d} = \frac{ad - bc}{bd}$ is a rational number.   **(b)** No; for example, $\frac{1}{2} - \frac{1}{4} \neq \frac{1}{4} - \frac{1}{2}$.   **(c)** No; for example, $\frac{1}{2} - \left(\frac{1}{4} - \frac{1}{8}\right) \neq \left(\frac{1}{2} - \frac{1}{4}\right) - \frac{1}{8}$.   **(d)** No. If there is an identity for subtraction, it must be 0, since only for 0 does

$\frac{a}{b} - 0 = \frac{a}{b}$. However, in general, $0 - \frac{a}{b} \neq \frac{a}{b} - 0$, and hence there is no identity.   **(e)** No; since there is no identity, an inverse cannot be defined.

Open-Ended

9. Answers vary. For example, Mike ate a third of a pizza, leaving $\frac{2}{3}$ of the pizza for Ann. Ann gave $\frac{1}{4}$ of the entire pizza to her little brother. What fraction of the pizza was left for Ann?

11. (a) (i) $\frac{1}{3} = \frac{1}{4} + \frac{1}{12}; \frac{1}{3} = \frac{1}{6} + \frac{1}{6}$ **(ii)** $\frac{1}{7} = \frac{1}{8} + \frac{1}{56};$
$\frac{1}{7} = \frac{1}{14} + \frac{1}{14}$ **(b)** $\frac{1}{n} - \frac{1}{n+1} = \frac{(n+1) - n}{n(n+1)} = \frac{1}{n(n+1)}$
(c) Rewriting part (b) as a sum gives $\frac{1}{n} = \frac{1}{n+1} + \frac{1}{n(n+1)}$.
(d) $\frac{1}{17} = \frac{1}{18} + \frac{1}{17(18)} = \frac{1}{18} + \frac{1}{306}$. Also,
$\frac{1}{17} = \frac{1}{34} + \frac{1}{34}$.

Questions from the Classroom

13. Kendra's picture shows that $\frac{1}{3}$ of the 3-square whole combined with $\frac{3}{4}$ of the 4-square whole gives $\frac{4}{7}$ of a 7-square whole. To do what Kendra is trying to do, she would need to use the same size whole rather than 3 different wholes. When $\frac{1}{3} + \frac{3}{4}$ are added, the same whole must be used.   **15.** It might be easier but she would be incorrect. Using Sally's approach $\frac{1}{2} + \frac{1}{2}$ would result in $\frac{1+1}{2+2}$ or $\frac{1}{2}$ and not 1.

Review Problems

17. (a) $\frac{2}{3}$ **(b)** $\frac{13}{17}$ **(c)** $\frac{25}{49}$ **(d)** $\frac{a}{1}$ **(e)** simplified **(f)** $a + b$

Assessment 6-3A

1. (a) $\frac{1}{4} \cdot \frac{1}{3} = \frac{1}{12}$ **(b)** $\frac{2}{4} \cdot \frac{3}{5} = \frac{6}{20}$ **2. (a)**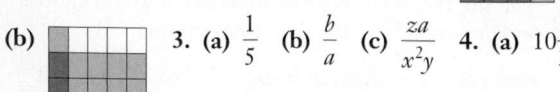

(b)   **3. (a)** $\frac{1}{5}$ **(b)** $\frac{b}{a}$ **(c)** $\frac{za}{x^2 y}$ **4. (a)** $10\frac{1}{2}$

(b) $8\frac{1}{3}$ **5. (a)** $^-3$ **(b)** $\frac{3}{10}$ **(c)** $\frac{y}{x}$ **(d)** $\frac{^-1}{7}$ **6. (a)** $\frac{21}{12}$ or $\frac{7}{4}$

(b) $\frac{6}{4}$ or $\frac{3}{2}$ **(c)** $\frac{^-1}{8}$ **(d)** $\frac{3}{2}$ **7.** Answers vary. For example,
(a) $6 \div 2 \neq 2 \div 6$ **(b)** $(8 \div 4) \div 2 \neq 8 \div (4 \div 2)$ **8. (a)** 26
(b) 29 **(c)** 92 **(d)** 18 **9. (a)** 20 **(b)** 16 **(c)** 1
10. (a) less than 1 **(b)** less than 1 **(c)** greater than 2
11. (c) **12.** 9600 students **13.** $\frac{1}{6}$ **14.** \$240.00 **15.** \$225.00
16. 32 marbles **17. (a)** $\frac{1}{3^{13}}$ **(b)** 3^{13} **(c)** 5^{11} **(d)** 5^{19}

(e) $\frac{1}{(^-5)^2}$ or $\frac{1}{5^2}$ **(f)** a^5 **18. (a)** $\left(\frac{1}{2}\right)^{10}$ **(b)** $\left(\frac{1}{2}\right)^3$

(c) $\left(\frac{2}{3}\right)^9$ **(d)** 1 **19. (a)** False; $2^3 \cdot 2^4 \neq (2 \cdot 2)^{3+4}$. **(b)** False;
$2^3 \cdot 2^2 \neq (2 \cdot 2)^{3 \cdot 2}$. **(c)** False; $2^3 \cdot 2^3 \neq (2 \cdot 2)^{2 \cdot 3}$. **(d)** True,
since $ab \neq 0$. **(e)** False; $(2 + 3)^2 \neq 2^2 + 3^2$. **(f)** False;
$(2 + 3)^{-2} \neq \frac{1}{2^2} + \frac{1}{3^2}$. **20. (a)** 5 **(b)** 6 or $^-6$ **(c)** $^-2$ **(d)** $^-4$
21. (a) $x \leq 2$ **(b)** $x < 2$ **(c)** $x \geq 2$ **(d)** $x \geq 1$
22. (a) $\left(\frac{1}{2}\right)^3$ **(b)** $\left(\frac{3}{4}\right)^8$ **(c)** $\left(\frac{4}{3}\right)^{10}$ **(d)** $\left(\frac{4}{5}\right)^{10}$
23. (a) and **(b)** $2S = 1 + \frac{1}{2} + \frac{1}{2^2} + \ldots + \frac{1}{2^{63}}$

$$2S - S = 1 + \left(\frac{1}{2} + \frac{1}{2^2} + \ldots + \frac{1}{2^{63}}\right)$$
$$- \left(\frac{1}{2} + \frac{1}{2^2} + \ldots + \frac{1}{2^{63}}\right) - \frac{1}{2^{64}}$$
$$= 1 - \frac{1}{2^{64}}$$

Note that $2S = 1 + S - \frac{1}{64}$.

Hence, $2S - S = 1 + S - \frac{1}{2^{64}} - S = 1 - \frac{1}{2^{64}}$. **(c)** $1 - \frac{1}{2^n}$
24. (a) $\frac{3}{2}, \frac{3}{4}, \frac{3}{8}, \frac{3}{16}, \frac{3}{32}$ **(b)** The common ratio is $\frac{1}{2}$.
(c) $n \geq 10$ **25. (a)** 32^{50}, since $32^{50} = (2^5)^{50} = 2^{250}$
and $4^{100} = (2^2)^{100} = 2^{200}$ **(b)** $(^-3)^{-75}$, since
$(^-27)^{-15} = (^-3)^{-45} = \frac{^-1}{3^{45}} < \frac{^-1}{3^{75}}$ **26.** Assume $x < y$,
i.e., $y - x > 0$. Then $\frac{x+y}{2} - x = \frac{x+y-2x}{2} = y - x > 0$.
Hence, $x < \frac{x+y}{2}$. Similarly $\frac{x+y}{2} < y$.

Mathematical Connections 6-3

Communication

1. Answers vary. For example, $\frac{1}{2}$ of a number x is equivalent to
$\frac{1}{2} \cdot x$ or $\frac{x}{2}$, whereas dividing a number x by $\frac{1}{2}$ is equivalent to
$\frac{x}{\frac{1}{2}} = \frac{x}{1} \cdot \frac{2}{1} = \frac{2x}{1} = 2x$. Therefore, they are not the same.

3. Never less than n. $0 < \frac{a}{b} < 1$ implies $0 < 1 < \frac{b}{a}$.
The last inequality implies $0 < n < n\left(\frac{b}{a}\right)$. Therefore
$n \div \left(\frac{a}{b}\right) = n\left(\frac{b}{a}\right) > n$. **5.** The second number must be the
reciprocal of the first and must be a positive number less than 1.

Open-Ended

7. (a) Answers vary. For example, if you have a board $1\frac{3}{4}$ yd long,
how many $\frac{1}{2}$ yd lengths can you divide it into? **(b)** Answers vary.

Cooperative Learning

9. The answers depend on the size of the bricks and the size of the joints. In all likelihood, the measurements will be made in fractions of inches for the size of the joints. The size of the bricks may be done in inches. An alternative is to measure in centimeters. All measurements are approximate and some rounding or estimation may occur.

Questions from the Classroom

11. Answers vary. For example, Carl is not correct since 0 is a rational number and the inverse of $\frac{0}{1}$ does not exist.   **13.** Joel is wrong.

$$2\frac{3}{5} \cdot 3\frac{4}{5} = \left(2 + \frac{3}{5}\right)\left(3 + \frac{4}{5}\right) = \frac{247}{25} \neq \left(2 + \frac{4}{5}\right)\left(3 + \frac{3}{5}\right) = \frac{252}{25}$$

15. The student is generalizing the distributive property of multiplication over addition to the distributive property of multiplication over multiplication. The latter does not hold.   **17.** The student is wrong unless $n = 0$ or $p = m$. The Fundamental Law of Fractions holds only for multiplication. For example,

$\frac{7 + 3}{2 + 3} \neq \frac{7}{2}$ **19.** Answers vary. For example, suppose we have $4\frac{2}{3}$ c of flour, and $\frac{1}{3}$ of a cup is required to make a bun. Then $4\frac{2}{3} \div \frac{1}{3}$ is the number of buns that can be made with $4\frac{2}{3}$ c of flour. In 1 c of flour we have 3 one-thirds of a cup. In 4 c, we have $4 \cdot 3$, or 12 one-thirds. Also, in $\frac{2}{3}$ of a cup we have 2 one-thirds of a cup of flour, so all together we have 14 one-thirds of a cup of flour and hence 14 buns. A picture of cups and of a bun can be very helpful.

Review Problems

21. (a) $\frac{25}{16}$ or $1\frac{9}{16}$ **(b)** $\frac{25}{18}$ or $1\frac{7}{18}$ **(c)** $\frac{5}{216}$ **(d)** $\frac{259}{30}$ or $8\frac{19}{30}$ **(e)** $\frac{37}{24}$ or $1\frac{13}{24}$ **(f)** $\frac{-39}{4}$ or $-9\frac{3}{4}$

Assessment 6-4A

1. (a) $5:21$ **(b)** $21:5$ **(c)** $21:26$ **(d)** Answers vary. For example, minor.   **2. (a)** 30   **(b)** $-3\frac{1}{3}$ **(c)** $23\frac{1}{3}$ **(d)** $10\frac{1}{2}$ **3. (a)** $2:5$. Because the ratio is $2:3$, there are $2x$ boys and $3x$ girls; hence, the ratio of boys to all students is $\frac{2x}{2x + 3x} = \frac{2}{5}$.

(b) $m:(m + n)$ **(c)** $3:2$ **4.** 36 lb   **5.** 12 grapefruits for $1.80   **6.** 270 mi   **7.** 64 pages   **8. (a)** 42, 56   **(b)** 24 and 32 (or -24 and -32)   **9.** $16,400.00; $24,600.00; $41,000.00   **10.** $77.00 and $99.00   **11.** 135   **12. (a)** $\frac{1}{6}$ **(b)** $\frac{1}{1}$ **(c)** $\frac{7}{12}$ **13.** Answers vary. $\frac{36 \text{ oz}}{12 ¢} = \frac{48 \text{ oz}}{16 ¢}; \frac{12 ¢}{16 ¢} = \frac{36 \text{ oz}}{48 \text{ oz}}, \frac{16 ¢}{12 ¢} = \frac{48 \text{ oz}}{36 \text{ oz}}$ **14. (a)** $\frac{5}{7}$

(b) 6 ft   **15. (a)** 27   **(b)** 20   **16.** approximately 34 cm   **17.** 312 lb   **18. (a)** $\frac{2}{3}$ tsp mustard seeds, 1 c scallions, $2\frac{1}{6}$ c beans   **(b)** $\frac{2}{3}$ tsp mustard seeds, 2 c tomato sauce, $2\frac{1}{6}$ c beans   **(c)** $\frac{7}{13}$ tsp mustard seeds, $1\frac{8}{13}$ c tomato sauce, $\frac{21}{26}$ c scallions   **19.** 15.12 Ω **20.** about

74.6 cm   **21.** The ratio between the mass of the gold in the ring and the mass of the ring is $18/24$. If x is the number of ounces of pure gold in the ring which weighs 0.4 oz, we have $18/24 = x/0.4$. Hence, $x = (18 \cdot 0.4)/24$, or 0.3 oz. Consequently, the price of the gold in the ring is $0.3 \cdot \$1800.00$ or $540.00.   **22. (a)** $320.00

(b) 8   **23. (a)** $1:2$ **(b)** Let $\frac{a}{b} = \frac{c}{d} = \frac{e}{f} = r$.

Then, $a = br$
$c = dr$
$e = fr$.

So, $a + c + e = br + dr + fr$
$a + c + e = r(b + d + f)$
$\frac{a + c + e}{b + d + f} = r$.

Mathematical Connections 6-4

Communication

1. (a) $40/700$, or $4/70$, or $2/35$ **(b)** 525 cm   **(c)** For the first set, $\frac{\text{footprint length}}{\text{thighbone length}} = \frac{40}{100} = \frac{20}{50}$; that is, a 50-cm thighbone would correspond to a 20-cm footprint. Thus, it is not likely that the 50-cm thighbone is from the animal that left the 30-cm footprint. $\left(\text{Notice that } \frac{20}{50} \neq \frac{30}{50}.\right)$

3. If $\frac{a}{b} = \frac{c}{d}$, then $ad = bc$.

$ad - bd = bc - bd$
$(a - b)d = b(c - d)$
$\frac{d}{c - d} = \frac{b}{a - b}$

5. No, the dimensions don't vary proportionally because $4/6 \neq 5/7$ and $4/6 \neq 8/10$, and $5/7 \neq 8/10$. This can be seen by cross multiplication or just by reducing all three fractions to simplest terms—$\frac{2}{3}, \frac{5}{7}$, and $\frac{4}{5}$—and noticing that no two are equal.

7. Answers vary; for example, let m be the number of adult men living in the condo and w be the number of adult women. The number of married men is equal to the number of married women, so $\frac{2}{3}m = \frac{3}{4}w$. The ratio of married people to the total adult population is

$$\frac{\frac{2m}{3} + \frac{3w}{4}}{m + w} = \frac{\frac{2m}{3} + \frac{2m}{3}}{m + w} = \frac{\frac{4m}{3}}{m + w}.$$

Because $\frac{2m}{3} = \frac{3w}{4}, w = \frac{8m}{9}$. Thus,

$$\frac{\frac{4m}{3}}{m + w} = \frac{\frac{4m}{3}}{m + \frac{8m}{9}} = \frac{\frac{4m}{3}}{\frac{17m}{9}} = \frac{12}{17},$$

which is the desired ratio.

Open-Ended

9. Answers vary. For example, ratios and proportions are seen in census reports, economic data about joblessness, and baseball statistics.   **11.** Answers vary. For example, the rectangle that is most pleasing to the eye has its sides in the golden ratio, which is approximately 1.618:1.

13. Mary has not paid attention to the corresponding parts of the proportion. If she looks at cross multiplication, it is clear that $12 \cdot 60 \neq 1 \cdot 5$; or if she reduces $5/60$ to $1/12$, it is clear that $12/1 \neq 1/12$. She needs to compare the appropriate measures as in $12/1 = 60/5$ or $1/12 = 5/60$ or $12/60 = 1/5$.

15. The way Al set up the proportion assumes the tree is 15 ft tall. The correct proportion comparing the object to its shadow is

$$\frac{5 \text{ ft}}{\frac{3}{2} \text{ ft}} = \frac{x \text{ ft}}{15 \text{ ft}}, \quad \text{so } x = 50 \text{ ft.}$$

The key is to have the same units for shadows and the same units for heights. **17.** Collect data on the ratio of arm length to nose length by measuring the arms and noses of other students in her class. Find the average ratio, then use it to find the expected length of the nose of the Statue of Liberty.

Chapter Review

1. (a) ▭▭▭▭▭ **(b)** ▭▭▭ **(c)**

2. Answers may vary. For example, $\frac{10}{12}, \frac{15}{18}, \frac{20}{24}$. **3. (a)** $\frac{6}{7}$

(b) $\frac{ax}{b}$ **(c)** $\frac{0}{1}$ **(d)** $\frac{5}{9}$ **(e)** b **(f)** $\frac{2}{27}$ **(g)** Cannot be further simplified. **(h)** Cannot be further simplified. **4. (a)** = **(b)** > **(c)** > **(d)** < **5. (a)** $^-3, \frac{1}{3}$ **(b)** $^-3\frac{1}{7}, \frac{7}{22}$ **(c)** $\frac{^-5}{6}, \frac{6}{5}$

(d) $\frac{3}{4}, \frac{^-4}{3}$ **6.** $^-2\frac{1}{3}, ^-1\frac{7}{8}, 0, \left(\frac{71}{140}\right)^{300}, \frac{69}{140}, \frac{1}{2}, \frac{71}{140},$

$\left(\frac{74}{73}\right)^{300}$ **7.** Yes. By the definition of multiplication and the commutative and associative laws of multiplication, we can do the following:

$$\frac{4}{5} \cdot \frac{7}{8} \cdot \frac{5}{14} = \frac{4 \cdot 7 \cdot 5}{5 \cdot 8 \cdot 14}$$
$$= \frac{4 \cdot 7 \cdot 5}{8 \cdot 14 \cdot 5}$$
$$= \frac{4}{8} \cdot \frac{7}{14} \cdot \frac{5}{5}$$

8. (a) 24, because $\frac{1}{3}(8 \cdot 9)$ is equal to $\left(\frac{1}{3} \cdot 9\right) \cdot 8 = 3 \cdot 8 = 24$.

(b) 66, because $36 \cdot 1\frac{5}{6}$ is equal to $36 \cdot \frac{11}{6} = 6 \cdot 11 = 66$.

9. (a) 17 pieces **(b)** $\frac{11}{6}$ yd **10. (a)** 15 **(b)** 15

(c) 4 **11.** Answers vary; see Section 6–3. **12.** Answers vary.

13. $\frac{76}{100}, \frac{78}{100}$, but answers may vary. **14.** [5][0][4][7][9][2][×]

[2][3][1/x][=] **15.** $333\frac{1}{3}$ calories **16.** 752 times

17. $\frac{240}{1000} = \frac{6}{25}$ **18.** It is not reasonable to say that the university

won $\frac{3}{4} + \frac{5}{8}$, or $\frac{11}{8}$, of its basketball games. The correct fraction cannot be determined without additional information but it is between $\frac{5}{8}$ and $\frac{3}{4}$. **19.** $16\frac{2}{3}$ ft **20.** You should show him that the given fraction could be written as an integer over an integer. In this case, the result is $\frac{8}{9}$. **21.** 112 bags

22. $\frac{4}{15}$ **23.** $\frac{^-12}{10}$ is greater than $\frac{^-11}{9}$ because $^-12 \cdot 9 > ^-11 \cdot 10$. Alternatively, $\frac{^-12}{10} - \frac{^-11}{9} = \frac{^-108}{90} - \frac{^-110}{90} = \frac{2}{90}$ which is positive; therefore $\frac{12}{10} > \frac{^-11}{9}$. **24. (a)** 3 **(b)** 3 **25. (a)** $\frac{5}{4}$, or $1\frac{1}{4}$

(b) $\frac{19}{6}$, or $3\frac{1}{6}$ **(c)** -100 **(d)** $\frac{13}{3}$, or $4\frac{1}{3}$ **26. (a)** $\frac{a^3}{x^7}$ **(b)** $\frac{y^8}{x^{10}}$

27. (a) $\frac{3ax+b}{x^2y^2}$ **(b)** $\frac{15-2y^2}{3xy^2}$ **(c)** $\frac{a-bx^2y}{x^3y^2z}$ **(d)** $\frac{31}{2^3 3^3}$, or $\frac{31}{216}$

28. Answers vary. For example, the problem is to find how many $\frac{1}{2}$-yd pieces of ribbon there are in $1\frac{3}{4}$ yd. There are 3 pieces of length $\frac{1}{2}$ yd with $\frac{1}{4}$ yd left over. This $\frac{1}{4}$ yd is $\frac{1}{2}$ of a $\frac{1}{2}$-yd piece. Therefore, there are 3 pieces of $\frac{1}{2}$-yd ribbon and 1 piece that is $\frac{1}{2}$ of the $\frac{1}{2}$-yd piece or $3\frac{1}{2}$ of the $\frac{1}{2}$-yd pieces. Thus "3 pieces and $\frac{1}{4}$ yard left" and "$3\frac{1}{2}$ pieces" are correct answers. **29.** $\frac{2ab}{ab+bc+ac}$

30. (a) $17:30$ **(b)** $17:13$ **(c)** $13:17$ **31.** 64 fl oz for $3.60 **32.** No, $18/6 \neq 12/3$. **33. (a)** $16\frac{4}{5}$ **(b)** $192\frac{1}{2}$

(c) 1 **34.** $8\frac{1}{4}$ oranges, 44 grapes **35.** 7.5 m **36.** The ratio of hydrogen to the total is $1:9$. Therefore, $\frac{1}{9} = \frac{x}{16}$ implies $x = 1\frac{7}{9}$ oz. **37.** No, the ratio depends on how many chips came from each plant. **38.** $1:r^2$ **39. (a)** $18:7$ **(b)** $18:25$ **40. (a)** $1:5$ **(b)** $8:15$ **41.** 9

Answers to Now Try This

6-1. A Venn diagram depicting the relationship among natural numbers, whole numbers, integers, and rational numbers follows:

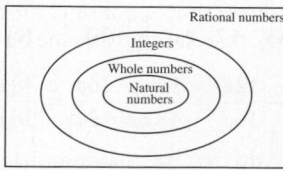

6-2. (a) One quarter of 100¢ is 25¢, one quarter of 60 min is 15 min. **(b)** Fractions are defined in relation to a whole. In this case, there are 2 wholes. In Figure 6-2(a), $\frac{1}{3}$ of a large circular whole is shaded. In Figure 6-2(b), $\frac{1}{2}$ of a smaller square whole is shaded. To show $\frac{1}{3} > \frac{1}{2}$, the fractions would have to be associated with the same whole, for example, by comparing $\frac{1}{3}$ of the circle with $\frac{1}{2}$ of the circle. It is true that the area shaded in the circle is larger than the area shaded in the square, but this does not show $\frac{1}{3} > \frac{1}{2}$.

6-3.

$$\begin{array}{c} \leftarrow\!\!\!+\!\!+\!\!+\!\!+\!\!+\!\!+\!\!+\!\!+\!\!+\!\!+\!\!+\!\!+\!\!+\!\!+\!\!+\!\!+\!\!+\!\!+\!\!\rightarrow \\ {\scriptstyle -2\frac{7}{4} \quad -1 \quad -\frac{1}{2} \quad 0 \quad \frac{1}{2} \quad 1 \quad \frac{3}{2} \quad 2} \end{array}$$

6-4. If n were zero, then bn would be zero and division by zero is undefined.   **6-5.** If a, b, and c are integers and $b > 0$, then $\frac{a}{b} > \frac{c}{b}$ if, and only if, $a > c$. The question is to investigate whether the theorem is true if $b < 0$. Consider $2 > 1$ and $^-1 < 0$. Now $2/(^-1) < 1/(^-1)$, which contradicts an expanded theorem when $b < 0$.   **6-6.** $\frac{^-6}{11}, \frac{^-4}{9}, \frac{^-3}{8}, \frac{9}{16}, \frac{5}{8}, \frac{2}{3}, \frac{3}{4}$ **6-7.** If $a > 0$ and $c > b > 0$, then $\frac{a}{b} > \frac{a}{c}$.   **6-8.** Consider two rational numbers $\frac{a}{b}$ and $\frac{c}{d}$, where $\frac{a}{b} < \frac{c}{d}$. By the denseness property of rational numbers we can find a rational number x_1, between the two fractions. Since $\frac{a}{b} < x_1$, there is a rational number x_2 between $\frac{a}{b}$ and x_1. We next can find a rational number x_3 between $\frac{a}{b}$ and x_2 and so on. This process can be repeated indefinitely and hence we obtain infinitely many rational numbers $x_1, x_2, x_3, \ldots$ between $\frac{a}{b}$ and $\frac{c}{d}$.

6-9. Because $\frac{a}{b} < \frac{c}{d}$ with $b > 0$ and $d > 0$, Theorem 6-3 implies that $ad < bc$. Now $\frac{a}{b} < \frac{a+c}{b+d}$ if $a(b+d) < b(a+c)$, which is equivalent to $ab + ad < ba + bc$ or $ad < bc$. Thus $\frac{a}{b} < \frac{a+c}{b+d}$. Similarly we can show that $\frac{a+c}{b+d} < \frac{c}{d}$.   **6-10.** $2\frac{3}{4} + 5\frac{3}{8} = 2 + \frac{3}{4} + 5 + \frac{3}{8} = 7 + \frac{6}{8} + \frac{3}{8} = 7 + \frac{9}{8} = 7 + 1 + \frac{1}{8} = 8\frac{1}{8}$.

6-11. (a) $8\frac{1}{4} + 6\frac{1}{2} = \frac{33}{4} + \frac{13}{2} = \frac{33}{4} + \frac{26}{4} = \frac{59}{4} = 14\frac{3}{4}$

(b) $5\frac{5}{6}$ **6-12.** $\frac{3}{4}$ is greater than $\frac{1}{2}$; $\frac{1}{2} + \frac{1}{2} = 1$, so $\frac{3}{4} + \frac{1}{2} > 1$. $\frac{4}{6}$ is less than 1, so it cannot be the correct answer for $\frac{3}{4} + \frac{1}{2}$.

6-13. Answers vary. For example, consider the following. If Caleb has $10.00, how many chocolate bars can he buy if **(a)** the price of one bar is $2.00? **(b)** the price of one bar is $\$\frac{1}{2}$? For (a) the answer is $10 \div 2$ or 5. For (b) the answer is $10 \div \frac{1}{2}$, which is the same as finding the number of $\frac{1}{2}$s in 10. Since there are two halves in 1, in 10 there are 20. Hence Caleb can buy 20 bars.

6-14. $\frac{a}{b} \div \frac{c}{d} = \dfrac{\frac{a}{b}}{\frac{c}{d}} = \dfrac{\frac{a}{b} \cdot \frac{d}{c}}{\frac{c}{d} \cdot \frac{d}{c}} = \dfrac{\frac{a}{b} \cdot \frac{d}{c}}{1} = \frac{a}{b} \cdot \frac{d}{c}$

6-15. $\frac{a \div c}{b \div d} = \dfrac{\frac{a}{c}}{\frac{b}{d}} = \dfrac{\frac{a}{c} \cdot \frac{d}{b}}{\frac{b}{d} \cdot \frac{d}{b}} = \dfrac{\frac{a}{c} \cdot \frac{d}{b}}{1} = \frac{ad}{bc} = \frac{a}{b} \div \frac{c}{d}$

6-16. (a) 20 lawns **(b)** $30.00, the strategy is to find the cost of one movie, $6.00, and then multiply by 5.

Answers to Brain Teasers

Section 6-1

At 6:00, the hands on the clock form a straight line, but the second hand is on 12. After that, the hands form a straight line approximately every 1 hr, 5 min, and 27 seconds. So after 6:00 the minute and hour hands form a straight line at: 7:05:27, 8:10:55, 9:16:22, 10:21:49, 11:27:16, 12:32:44, 1:38:11, 2:43:38, 3:49:05, 4:54:33.

Section 6-3

The prince started with 46 bags of gold. Observe that after crossing each bridge, the prince was left with half the bags he had previously minus one additional bag of gold. To determine the number he had prior to crossing the bridge, we can use the inverse operations; that is, add 1 and multiply by 2. The prince had one bag left after crossing the fourth bridge. He must have had two before he gave the guard the extra bag. Finally, he must have had four bags before he gave the guard at the fourth bridge any bags. The entire procedure is summarized in the following table.

Bridge	Bags After Crossing	Bags Before Guard Given Extra	Bags Prior to Crossing
Fourth	1	2	4
Third	4	5	10
Second	10	11	22
First	22	23	46

Section 6-4

No. The will is impossible because the fractions of cats to be shared do not add up to the whole units of cats. $\frac{1}{2}x + \frac{1}{3}x + \frac{1}{9}x = \frac{17}{18}x$, but the sum should be $1x$, or $\frac{18}{18}x$.

Answer to Preliminary Problem

The original ratio of female dolls (F) to male dolls (M) is 3:1; that is, $\frac{F}{M} = \frac{3}{1}$. This implies that $F = 3M$. If 30 of the female dolls are sold then the number of female dolls is now $F - 30$. The new ratio of female dolls to male dolls is $\frac{F - 30}{M} = \frac{1}{2}$. Because $F = 3M$ we have $\frac{3M - 30}{M} = \frac{1}{2}$. Solving we have $6M - 60 = M$ or $M = 12$. Therefore the store now has 12 male dolls.

Chapter 7

Assessment 7-1A

1. (a) $0 \cdot 10^0 + 0 \cdot 10^{-1} + 2 \cdot 10^{-2} + 3 \cdot 10^{-3}$
(b) $2 \cdot 10^2 + 0 \cdot 10^1 + 6 \cdot 10^0 + 0 \cdot 10^{-1} + 6 \cdot 10^{-2}$
(c) $3 \cdot 10^2 + 1 \cdot 10^1 + 2 \cdot 10^0 + 0 \cdot 10^{-1} + 1 \cdot 10^{-2} +$
$0 \cdot 10^{-3} + 3 \cdot 10^{-4}$ (d) $0 \cdot 10^0 + 0 \cdot 10^{-1} + 0 \cdot 10^{-2} + 0 \cdot 10^{-3} +$
$1 \cdot 10^{-4} + 3 \cdot 10^{-5} + 2 \cdot 10^{-6}$ 2. (a) 4356.78   (b) 4000.608
(c) 40,000.03   (d) 0.2004007   3. (a) 536.0076   (b) 3.008
(c) 0.000436   (d) 5,000,000.2   4. (a) Thirty-four hundredths
(b) Twenty and thirty-four hundredths   (c) Two and thirty-four
thousandths   (d) Thirty-four millionths
5. (a) $\dfrac{436}{1000} = \dfrac{109}{250}$ (b) $\dfrac{2516}{100} = \dfrac{629}{25}$ (c) $\dfrac{^-316,027}{1000}$
(d) $\dfrac{281,902}{10,000} = \dfrac{140,951}{5000}$ (e) $\dfrac{^-43}{10}$ (f) $\dfrac{^-6201}{100}$ 6. (a) Yes
(b) Yes   (c) Yes   (d) Yes   (e) No   (f) Yes   7. (a) 0.8
(b) 3.05   (c) 0.5   (d) 0.03125   (e) No   (f) 0.2128
8. Nonterminating; the fraction $\dfrac{7}{60}$ does not terminate when
written as a decimal. The denominator has a factor of 3 when in
simplest form.   9. Answers may vary. Some of the numbers that
are composed of whole-number powers of 2 and 5 are 1, 5, 10,
25, and 50. These all divide 100 and can be written as a two-digit
decimal between 0 and 1, but there are others.   10. (a) 13.492,
13.49199, 13.4919, 13.49183   (b) $^-$1.4053, $^-$1.45, $^-$1.453, $^-$1.493
11. (a) 0.014   (b) 365.24

12.
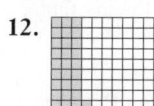

There are 32 of 100 squares shaded, representing $\dfrac{32}{100}$ of the whole
grid, or 0.32 of the grid.   13. 0.84   14. For example, 8.3401
15. (a) One method is to append enough zeros to the right of one of
the numbers to make the numbers have the same number of decimal
places and then to append a single digit, say 5, to the smaller decimal
(e. g. 1.237005 is between 1.237 and 1.24589).   (b) The process in
part (a) can be repeated over and over with the number found and
the original greater number. For the example in part (a) we would
get the infinite sequence 1.235005, 1.2350055, 1.23500555, . . .
16. A meaning could be as follows: $3 \cdot 6^0 + 1 \cdot 6^{-1} + 4 \cdot 6^{-2} + 5 \cdot 6^{-3}$
in base ten. In base six, 10_{six} would be used instead of 6.
17. Rhonda, Martha, Kathy, Molly, Emily   18. $\dfrac{3}{32} = 0.09375$

Mathematical Connections 7-1

Communication

1. (a) $3^① \, 2^② \, 5^③ \, 6^④$ (b) $0^① \, 0^② \, 3^③ \, 2^④$
3. One day cannot be expressed as a terminating decimal because
$\dfrac{1}{365}$ cannot be written as a terminating decimal.   5. The true
meaning would be $\dfrac{5}{100}¢$, or $0.0005, per copy and not the real in-
tended price of 5¢ per copy.   7. Answers vary. For example, a fraction can be written as a terminating decimal if it can be written as a

fraction with a denominator that is a power of 10. The denominator
can be written as a power of 10 if it contains only factors of 2 and 5.
Other factors may appear in the denominator if the fraction
is not in simplest form. For example, in $\dfrac{28}{35}$ the denominator of 35 has
a factor of 7, but in its simplest form, $\dfrac{4}{5}$, there is no factor of 7.

Open-Ended

9. Answers vary, but many countries will be similar to ours.

Cooperative Learning

11. Answers vary depending on the objects chosen. For example,
if a 5×5 flat is used to represent 1, then a 1×5 long would represent 0.1, and a 1×1 unit would represent 0.01.

Questions from the Classroom

13. The student is mistaken. $0.86 = \dfrac{86}{100}; 0.9 = \dfrac{90}{100}$ and
$\dfrac{90}{100} > \dfrac{86}{100}$, so $0.9 > 0.86$.   15. $0.304 = \dfrac{304}{100}, 0.34 = 0.340 =$
$\dfrac{340}{100}$. Because $340 \neq 304$, the decimals are not equal.

Assessment 7-2A

1. $231.24

2. (a)

8.2	1.9	6.4
3.7	5.5	7.3
4.6	9.1	2.8

(b) (i) Yes (ii) 19.05   3. $4.50   4. Approximately 6391 cm^3
(rounded to the nearest cm^3)   5. 73.005   6. 26.84 in Canadian
dollars   7. (a) about $5.80 to heat the house for 1 day
(b) 199 hr (rounded)   8. approximately 8.64 liters
9. (a) 5.4, 6.3, 7.2   (b) 1.3, 1.5, 1.7   10. 1.12464   11. 0.2222
could be written as the sum of a geometric sequence as follows:
$\dfrac{2}{10} + \left(\dfrac{2}{10}\right)\left(\dfrac{1}{10}\right) + \left(\dfrac{2}{10}\right)\left(\dfrac{1}{10}\right)^2 + \left(\dfrac{2}{10}\right)\left(\dfrac{1}{10}\right)^3$ or as
$0.2 + (0.2)(0.1) + (0.2)(0.1)^2 + (0.2)(0.1)^3$.

12.
13. No, the bank is over $7.74.   14. (a) 0.0000000032
(b) 3,200,000,000   (c) 0.42   (d) 620,000   15. (a) $1.27 \cdot 10^7$ m
(b) $4.486 \cdot 10^9$ km   (c) $5 \cdot 10^7$ cans   16. (a) $^-$2.78
(b) 3.5   17. (a) 0.000000000753 g   (b) 298,000 km/sec
(c) 778,570,000 km   18. (a) $4.8 \cdot 10^{28}$ (b) $4 \cdot 10^7$ (c) $2 \cdot 10^2$
19. (a) 200   (b) 200   (c) 204   (d) 203.7   (e) 203.65
20. 20.25 mpg   21. $55 + 5 + 18, 78$ dollars   22. Estimates
vary; exact answers follow. (a) 122.06   (b) 57.31   (c) 25.40
(d) 136.15   23. $2.3 \cdot 1 = 2.3; 8.7 \cdot 9 = 78.3$ 24. 49,736.5281
25. Answers vary. For example, $40 \cdot \$8 = \$320; 40\left(\dfrac{1}{4}\right) = \$10$;
so her salary is $330.   26. (a) 12   (b) 0.6   (c) $2b$ (d) $2b$

27. (i), (ii), and (iv) have equal quotients.   **28. (a)** $1.5^2; 2.5^2;$ $4.3 + 0.25 = 3.5^2; 5.4 + 0.25 = 4.5^2$ **(b)** $n(n-1) + 0.25 = (n-0.5)^2$

Mathematical Connections 7-2

Communication

1. Suppose a deposit is recorded with $10.00 over the true amount and a check is recorded with $10.00 more than the true amount. The ultimate balance is correct.   **3.** Answers vary. For example, many of the estimation techniques that work for whole-number division also work for decimal division. The long division algorithm is more efficient when good estimates are used. Also, estimates are important to determine whether an answer obtained by long division is reasonable. Estimation techniques can also be used to place the decimal point in the quotient when decimals are divided.   **5.** Answers vary. For example, lining up the decimal points and using place value.

Open-Ended

7. Answers vary depending on the articles chosen.   **9.** Answers may vary.

Cooperative Learning

11. Answers vary.

Questions from the Classroom

13. It is evident what happens when 0.5 is raised to large powers. Because $\left(\frac{1}{2}\right)^{10} = \frac{1}{1024}$ and $\frac{1}{2^{20}} = \left[\left(\frac{1}{2}\right)^{10}\right]^2 = \frac{1}{1,048,576}$, therefore $\frac{1}{2}$ raised to a positive integer gets quickly close to 0. In fact, any number between 0 and 1 when raised to a sufficiently large exponent will get as close to 0 as we wish. Using the $\boxed{x^2}$ key repeatedly, we get: 0.998001, 0.996005996, 0.99202794407, 0.98411944182, 0.96849107576, . . ., 0.3589714782, which are approximate values of $0.999^2, 0.999^4, 0.999^8, 0.999^{16}, \ldots, 0.999^{1024}$. We see that the 10th term in the sequence is less than 0.5 and hence further squaring should quickly result in numbers closer and closer to 0.

Review Problems

15. $14.0479 = 1 \cdot 10^1 + 4 \cdot 10^0 + 0 \cdot 10^{-1} + 4 \cdot 10^{-2} + 7 \cdot 10^{-3} + 9 \cdot 10^{-4}$ **17.** Yes; for example, $\frac{13}{26} = \frac{1}{2} = 0.5$

Assessment 7-3A

1. (a) $0.\overline{4}$ **(b)** $0.\overline{285714}$ **(c)** $0.\overline{27}$ **(d)** $0.0\overline{6}$ **(e)** $0.02\overline{6}$ **(f)** $0.\overline{01}$ **(g)** $0.8\overline{3}$ **(h)** $0.\overline{076923}$ **(i)** $0.0\overline{47619}$ **(j)** $0.\overline{157894736842105263}$ **2. (a)** $\frac{4}{9}$ **(b)** $\frac{61}{99}$ **(c)** $\frac{461}{330}$ **(d)** $\frac{5}{9}$ **(e)** $\frac{-211}{90}$ **(f)** $\frac{-2}{90}$ **3.** $0.01\overline{6}$ hr
4. $-1.454 > -1.45\overline{4} > -1.\overline{454} > -1.4\overline{54} = -1.\overline{45}$
5. Answers vary. $\frac{6}{7}, \frac{7}{8}, \frac{8}{9}$, or $0.\overline{857142}, 0.875, 0.\overline{8}$
6. $0.\overline{308641975}$ **7.** The repeating decimal part is determined by the 7 in the denominator of $\frac{22}{7}$. The denominator, 7 has prime

factors other than 2 or 5.   **8.** $0.\overline{446355}$; six digits   **9.** Yes. Zeros can be repeated or 9s could be repeated.   **10. (a)** Answers vary. For example, 3.221, 3.2211, 3.22111.   **(b)** Answers vary. For example, 462.2425, 462.2426, 462.2427.   **11.** $0.47\overline{2}$
12. (a) Answers vary. For example, 0.751, 0.752, 0.753.
(b) Answers vary. For example, 0.334, 0.335, 0.336.   **13. (a)** 8
(b) 7   **14. (a) (i)** $\frac{1}{9}$ **(ii)** $\frac{1}{99}$ **(iii)** $\frac{1}{999}$ **(b)** $\frac{1}{9999}$ **(c)** $0.0\overline{1}$
15. (a) 0.1 **(b)** 0.4 **(c)** 10 **16. (a)** $\frac{2}{9}$ **(b)** $\frac{3}{9}$ or $\frac{1}{3}$ **(c)** 10
17. (a) $\frac{5}{99}$ **(b)** $\frac{3}{999}$ or $\frac{1}{333}$ **18.** 0.775 **19. (a)** $\frac{3}{10}$ **(b)** $\frac{2009}{990}$
20. 12; to make the repeating parts match   **21. (a)** $\frac{-7}{3}$ or $2.\overline{3}$
(b) Same as (a)   **(c)** $\frac{7}{3}$ or $2.\overline{3}$

Mathematical Connections 7-3

Communication

1. (a) Mathematically the cost is $66\frac{2}{3}$ ¢, but realistically, the cost is 67¢.   **(b)** See part (a).   **(c)** The cost is rounded up. **(d)** Most cash registers do not allow repeating decimals, so grocery stores do not use them.   **(e)** See part (d).   **3.** Answers vary. For example, it is easier to compute addition of fractions when there is a common denominator, as in $\frac{1}{7} + \frac{5}{7}$. When nonrepeating decimals are involved, it becomes hard to do additions because of lining up the place value. When denominators are different and terminating decimals are obtained, it is easier to use decimals, as in $\frac{2}{5} + \frac{1}{4} = 0.40 + 0.25 = 0.65$.

Open-Ended

5. (a) Any integer n can be written as a decimal by appending .0 to the right.   **(b) (i)** $0.\overline{6}$ **(ii)** 1 or $1.\overline{0}$; 1 is the simplest form because it requires fewer symbols.   **(iii)** $1.\overline{06}$ **(c)** It would have to be adapted to allow multiplication from the left.   **(d)** A repeating decimal can be written as a rational number in the form $\frac{a}{b}$, where $b \neq 0$. The fractions can be multiplied and the product can be converted to a decimal.   **7.** Most would prefer $\frac{7}{3}$ as a solution. However depends on the context. For example when melons are priced at 3 for $7, we might prefer to know that they are $2.333 . . . each and hence actually $2.34 each.

Questions from the Classroom

9. Answers may vary. A calculator is a tool that may aid in the computation of decimals. It can help by furnishing estimates for answers. This has to do with the value of the calculator as a tool. The value of repeating decimals is a different question. One major white-collar crime dealt with rounding off fractions of pennies and depositing that money in a bank account. These fractions could have been the result of using repeating decimals and rounding. Depending upon the context, the result of not using such decimals could lead to serious errors.   **11.** Yes, because every rational number can be written as a repeating decimal. For example, $0.45 = 0.44\overline{9}$.

13. $22,761.95   **15.** 0.077. The rule says that the decimal point should be placed four digits from the right of the product of 22 and 35. This product is 770, and therefore the answer is 0.0770, which equals 0.77. Because 0.077 and 0.0770 are equivalent, the rule still works.

Assessment 7-4A

1. (a) 789% (b) 19,310% (c) $83\frac{1}{3}$% (d) 12.5% (e) 62.5%
(f) 80% **2.** (a) 0.16 (b) 0.002 (c) $0.13\overline{6}$ (d) $0.00\overline{3}$
3. (a) 4 (b) 2 (c) 25 (d) 200 (e) 12.5 **4.** (a) 2.04
(b) 50% (c) 60 (d) 3.43 **5.** (a) $\frac{5x}{100}$ (b) $10a$ **6.** 63 boxes
7. $25,500 **8.** (a) Bill sold 221. (b) Joe sold 90%. (c) Ron started with 265. **9.** 20% **10.** approximately 46% decrease
11. $22.40 **12.** $336 **13.** 35% **14.** 1200 employees
15. $\frac{325}{500}; \frac{325}{500} = \frac{650}{1000} = \frac{65}{100} = 65\%$, while $\frac{600}{1000} = \frac{60}{100} = 60\%$.
16. (a) $76; $76 (b) They cost the same. **17.** $440
18. (a) $3.30 (b) $24.00 (c) $1.90 (d) $24.50 **19.** Apprentice makes $700. Journeyman makes $1400. Master makes $2100.
20. (a) 4% (b) (i) 44 (ii) 8.8% **21.** $82,644.63
22. (a) The answer is essentially true. Spending at that rate will amount to $2.57 trillion. (b) Approximately 0.12% **23.** No, 56% of the salary is more than twice 25% of the salary, but $950 is less than twice $500. **24.** (a) 3 (b) 0.003% **25.** (a) $4.50
(b) 50% (c) 100% **26.** 550 students **27.** *Interest Rate per Period*
(c) $\frac{10}{12}$% or $\frac{5}{6}$% or $0.8\overline{3}$% (d) $\frac{12}{365}$% **28.** $3675.00 **29.** $43,059 **30.** $12,905.80 **31.** Approximately $2.53 **32.** Approximately 14.03% **33.** Geometric sequence **34.** $3483.81
35. The Pay More Bank offers a better rate. **36.** No, now she owes $2539.47.

Mathematical Connections 7-4

Communication

1. Answers vary. For example, 10% of 850 is 85 and 1% of 850 is 8.5, so 11% of 850 is 93.5. **3.** It means that not only did you meet 100% of your savings goal, you surpassed it by 25%. If your savings goal was $100, then you saved $100 plus an extra $25. **5.** (a) Greater, because if 25% of x is 55, then x must be 4 times as great as 55, or 220. (b) Less, because if 150% of x is 55, then x is only $\frac{2}{3}$ of 55, or $36\frac{2}{3}$. **7.** The *whole* in each part is different, so 50% of the greater quantity is greater than 50% of the smaller quantity. To be equal, we would have to have the same size whole to begin with. **9.** Let x be the amount invested. The first stock will be worth $(1.15x)0.85$ after 2 yr. The second stock will be worth $(0.85x)1.15$. Because each of these equals $(1.15 \cdot 0.85)x$ (commutative and associative properties of multiplication) the investments are equally good. **11.** If we want our money doubled, then $2 = 1(1 + 0.02)^n$, or $2 = 1.02^n$, and we are looking for n. By trial and error, we can find that $1.02^{35} \approx 2$, so it would take about 35 yr to double an investment at 2% compounded annually.

Open-Ended

13. Answers vary. **15.** Answers vary.
(a) 115 is 37% of what number? (b) a is p% of what number?
17. Answers vary.

Cooperative Learning

19. Answers vary.

Questions from the Classroom

21. $3\frac{1}{4}\% = 3\% + \frac{1}{4}\% = \frac{3}{100} + \frac{\left(\frac{1}{4}\right)}{100} = 0.03 + 0.0025 =$
0.0325. Knowing that $\frac{1}{4} = 0.25$, the student incorrectly wrote
$\frac{1}{4}\% = 0.25$. **23.** By definition, $p\% = \frac{p}{100}$, where p is any real
number. Hence, $0.01\% = \frac{0.01}{100}$. Because $\frac{0.01}{100} \neq 0.01$, the
student is wrong. **25.** Because the formula for compound interest is $A = P(1 + r)^n$ and $P = d$ dollars and $r = 1$, then we have $A = d(1 + 1)^n = d \cdot 2^n$. Thus, the student is correct.

Review Problems

27. (a) $\frac{418}{25}$ (b) $\frac{3}{1000}$ (c) $\frac{-507}{100}$ (d) $\frac{123}{1000}$ **29.** $\frac{3}{12,500}$
31. (a) 208,000 (b) 0.00038

Chapter Review

1. (a) $A = 0.02, B = 0.05, C = 0.11$
(b) ◄——┼┼┼┼┼┼┼┼┼┼┼┼┼┼┼┼┼┼——►   **2.** (a) $\frac{32012}{1000}$ or $\frac{8003}{250}$
    0   0.1   0.2
    A B D C   E
(b) $\frac{103}{100,000}$ **3.** A fraction in simplest form, $\frac{a}{b}$, can be written
as a terminating decimal if, and only if, the prime factorization of the denominator contains no primes other than 2 or 5.
4. 8 shelves **5.** (a) $0.\overline{571428}$ (b) 0.125 (c) $0.\overline{6}$ (d) 0.625
6. (a) $\frac{7}{25}$ (b) $\frac{-607}{100}$ (c) $\frac{1}{3}$ (d) $\frac{94}{45}$ **7.** (a) 307.63 (b) 307.6
(c) 308 (d) 300 **8.** (a) $4.26 \cdot 10^5$ (b) $3.24 \cdot 10^{-4}$
(c) $2.37 \cdot 10^{-6}$ (d) $-3.25 \cdot 10^{-1}$ **9.** $1.45\overline{19}, 1.4\overline{519}, 1.45\overline{19},$
$1.4519, 0.13\overline{401}, -0.134, -0.13\overline{401}$. **10.** (a) $1.78341156 \cdot 10^6$
(b) $3.47 \cdot 10^{-6}$ (c) $4.93 \cdot 10^9$ (d) $2.94 \cdot 10^{17}$ (e) $4.7 \cdot 10^{35}$
(f) $1.536 \cdot 10^{-6}$ **11.** (a) Answers vary. For example: 0.105, 0.104, 0.103, 0.102, 0.101 (b) Answers may vary; 0.0005, 0.001, 0.002, 0.004 (c) 0.15, 0.175, 0.1875, 0.19375. **12.** (a) 25%
(b) 192 (c) $56.\overline{6}$ (d) 20% **13.** (a) 12.5% (b) 7.5%
(c) 627% (d) 1.23% (e) 150% **14.** (a) 0.60 (b) $0.00\overline{6}$
(c) 1 **15.** $9280 **16.** $3.\overline{3}$% **17.** 88.6% **18.** $5750
19. It makes no difference, the discount is always 31.6%. **20.** $80
21. Approximately 31% **22.** Answers vary. For example, if the dress was originally priced at $100, then 60% off would result in a sale price of $40. Then the 40% coupon would give a final price of $24. This model could be applied to the actual list price of the dress. **23.** $15,000 **24.** Approximately $15,110.69

Answers to Now Try This

7-1. (a) $\frac{1}{10}$ or 0.1 (b) $\frac{1}{100}$ or 0.01

(c)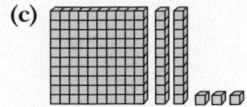

7-2. The cartoon reports the cost of tomatoes is .99¢, which is less than 1¢. This mistake is commonly made in stores when the money amount is less than 1 dollar; for example, an item is marked for .50¢ when it really should be $0.50 or 50¢. Note that the clerk did not raise the price.   **7-3. (a)** 2.5

(b) $\dfrac{3}{8} = \dfrac{3 \cdot 5^3}{2^3 \cdot 5^3} = \dfrac{375}{1000} = 0.375$

(c) $\dfrac{3}{20} = \dfrac{3 \cdot 5}{2^2 \cdot 5 \cdot 5} = \dfrac{15}{2^2 \cdot 5^2} = \dfrac{15}{100} = 0.15$

7-4. (a) $3.6 \cdot 1000 = 3.6 \cdot 10^3 = \left(3 + \dfrac{6}{10}\right)10^3 = 3 \cdot 10^3 +$

$\dfrac{6}{10} \cdot 10^3 = 3 \cdot 10^3 + 6 \cdot 10^2 = 3 \cdot 10^3 + 6 \cdot 10^2 + 0 \cdot 10^1 +$

$0 \cdot 1 = 3600.$ Thus, we see that multiplication by 1000 results in moving the decimal point three places to the right. **(b)** In general, multiplication by 10^n, where n is a positive integer, results in moving the decimal point n places to the right.   **7-5.** $1.19/32 oz is about $0.037/oz, while $1.43/48 oz is about $0.030/oz, so the 48-oz jar is a better buy.   **7-6.** Answers vary. For example, using the front digits the first estimate is $2 + $0 + $6 + $4 + $5 = $17. Next, we adjust the estimate. Because $0.89 + $0.13 is about $1.00 and $0.75 + $0.05 is $0.80 and $0.80 + $0.39 is about $1.20, the adjustment is $2.20 and the estimate is $19.20.   **7-7. (a)** The answer is rounded to 0.0000014.

(b) The answer is rounded to 0.00000136.   **7-8. (a)** $\dfrac{1}{9} = 0.\overline{1}$

(b) (i) $\dfrac{2}{9} = 2(0.\overline{1}) = 0.\overline{2}$ **(ii)** $\dfrac{3}{9} = 3(0.\overline{1}) = 0.\overline{3}$

(iii) $\dfrac{5}{9} = 5(0.\overline{1}) = 0.\overline{5}$ **(iv)** $\dfrac{8}{9} = 8(0.\overline{1}) = 0.\overline{8}$

7-9. $0.\overline{9} = 1.$ If $x = 0.\overline{9}$ then $10x = 9.\overline{9}, 9x = 9, x = 1$

7-10. Answers may vary. One possible answer is: $0.3\overline{55}; 0.36\overline{5}$

7-11. (a) $\dfrac{11}{20}, 55\%$ **(b)** $\dfrac{3}{5}, 60\%$ **(c)** $\dfrac{13}{25}, 52\%$ **(d)** $\dfrac{1}{5}, 20\%$

7-12. (a) Answers vary. For example, most calculators will convert the decimal form of a number to a percent by moving the decimal point two places to the left. Other calculators actually place a $\boxed{\%}$ symbol in the display when the % key is pushed.   **(b)** $33.\overline{3}\%$ **7-13.** Olives—$37\dfrac{1}{2}\%$, 3 slices; Plain—$12\dfrac{1}{2}\%$, 1 slice; Remainder—50%, 4 slices

Answer to Brain Teaser

Section 7-4

Let C = amount of crust, P = amount of pie, and R = amount of crust removed. $C = 25\%$ of $P = 0.25P$ and $C - R = 0.2$ $(P - R)$. Therefore, $R = 0.25C$ and so the amount of crust should be reduced by 25%.

Answer to the Preliminary Problem

If the wholesale price of the item is p, then the retail price is 130% of p, or $1.3p$. The price after the 30% discount is 70% of $1.3p$,

or $0.7 \cdot 1.3p = 0.91p$. Since $0.91p < p$, the sale price is less than the wholesale price. More concretely, if the wholesale price of the item is $100, the marked-up price is $130 and the price after the discount is $91. Therefore the store is losing on the item.

Chapter 8

Assessment 8-1A

1. Answers vary. One answer is $0.232233222333\ldots$
2. (a) $\sqrt{6}$ **(b)** 2   **(c)** 3   **3.** $0.\overline{9}, 0.9\overline{98}, \sqrt{0.98}, 0.\overline{98}, 0.9\overline{88},$ $0.9, 0.\overline{898}$ **4. (a)** Yes   **(b)** No   **(c)** No   **(d)** Yes   **(e)** Yes
(f) Yes   **5. (a)** 15   **(b)** 13   **(c)** Impossible   **(d)** 25
6. (a) 2.65   **(b)** 0.11   **7. (a)** False; $\sqrt{2} + 0 = \sqrt{2}$ **(b)** False; $^-\sqrt{2} + \sqrt{2} = 0$ **(c)** False; $(\sqrt{2})(\sqrt{2}) = 2$ **(d)** True; $\sqrt{2} - \sqrt{2} = 0$ **8. (a)** Answers vary; for example, $\sqrt{2}, \sqrt{3},$ and $\sqrt{5}$.   **(b)** Answers vary. For example, assume the following patterns continue in each number: $0.54544544454444\ldots,$ $0.545444544445\ldots,$ and $0.545444545545554\ldots$ **(c)** Answers vary. For example, assume the following patterns continue in each number: $0.51511511151111\ldots, 0.52522522252222\ldots,$ $0.53533533353333\ldots$ **9.** Answers may vary. For example, between any two rational numbers we could find three irrational numbers. Because there are infinitely many disjoint intervals bounded by rational numbers, there must be infinitely many irrational numbers.   **10. (a)** R **(b)** $\varnothing$ **(c)** Q **(d)** $\varnothing$
(e) R **(f)** R **11. (a)** Q, R **(b)** N, I, Q, R **(c)** R, S
(d) I, Q, R **(e)** Q, R **12. (a)** N, I, Q, R **(b)** Q, R **(c)** R, S
(d) None   **(e)** Q, R **13. (a)** 64   **(b)** None   **(c)** $^-64$
(d) None   **(e)** All real numbers greater than zero   **(f)** None

14. (a) $6\sqrt{5}$ **(b)** $11\sqrt{3}$ **(c)** $6\sqrt{7}$ **15. (a)** $^-3\sqrt[3]{2}$
(b) $2\sqrt[5]{3}$ **(c)** $5\sqrt[3]{2}$ **(d)** $^-3$ **16. (a)** $5, 5\sqrt[3]{2}, 5\sqrt[3]{4}, 10$

(b) Answers may vary. $2, 2\sqrt[4]{\dfrac{1}{2}}, 2\sqrt[4]{\dfrac{1}{4}}, 2\sqrt[4]{\dfrac{1}{8}}, 1$ **17.** 6.4 ft

18. (a) 2^{10} **(b)** 2^{11} **(c)** 2^{12} **19. (a)** 4   **(b)** $\dfrac{3}{2}$ **(c)** $\dfrac{^-4}{7}$

(d) $\dfrac{5}{6}$ **(e)** 12   **20. (a)** Rational   **(b)** Rational   **21.** Between

18 and 19   **22.** Answers may vary. For example, $\dfrac{^-2}{\sqrt{3}} > \dfrac{^-3}{\sqrt{5}}$ if,

and only if, $^-2\sqrt{5} > ^-3\sqrt{3}$ if, and only if, $3\sqrt{3} > 2\sqrt{5}$ if, and

only if, $1.5\sqrt{3} > \sqrt{5}$ if, and only if, $(1.5)(1.73) > 2.23$. The last

is true so $\dfrac{^-2}{\sqrt{3}} > \dfrac{^-3}{\sqrt{5}}$. A calculator was not needed to consider

the estimations given.

Mathematical Connections 8-1

Communication

1. The mathematician proba bly meant that there are more irrational numbers than rational numbers. Georg Cantor proved this.
3. False: $\sqrt{64 + 36} \neq \sqrt{64} + \sqrt{36}$ **5.** No; $\sqrt{13}$ is an irrational number. So when it is expressed as a decimal, it is nonterminating and nonrepeating.

Open-Ended

7. Answers vary. For example: **(a)** $1, 2, 3, 4, 5, \ldots$
(b) $\sqrt{2}, \sqrt{3}, \sqrt{5}, \sqrt{7}, \ldots$ or $\pi, \pi + 1, \pi + 2, \pi + 3, \ldots$

Cooperative Learning

9. (a) When a number between 0 and 1 is raised to larger and larger exponents, the results approach 0.
(b) Answers vary.
(c) Answers vary.

Questions from the Classroom

11. To be a rational number a number must be able to be written in the form $\dfrac{a}{b}$ with $b \neq 0$ and both a and b must be integers. $\sqrt{2}$ is not an integer. **13.** The principal square root of 25 is 5 and not $^-5$. While it is true that $(^-5)^2 = 25$, $^-5$ is not the principal square root of 25; the roots don't have to be equal. **15.** The student is incorrect. One example in carpentry is to find the hypotenuse of a right triangle. Likely the result will be estimated, but if the lengths of the shorter sides are 1 in. and 1 in., the length of the hypotenuse is $\sqrt{2}$ in.

Assessment 8-2A

1. (a) $10 + 2d$ **(b)** $2n - 10$ **(c)** $10n^2$ **(d)** $n^2 - 2n$
2. (a) $\dfrac{7(n + \sqrt{3}) - 14}{7} - n$ **(b)** $\sqrt{3} - 2$ **3. (a)** $2(n + 1)$
(b) $(n + 2)^2 - 2(n + 1)$ **4. (a)** $20 + 25h$ dollars **(b)** $175d$ cents
(c) $3x + 3$ **(d)** $q2^n$ **(e)** $(40 - 3t)°F$ **(f)** $4s + 15,000$ dollars
(g) $3x + 6$ **(h)** $3m$ **5.** $S = 20P$ **6.** $g = 5 + b$ **7.** $6n + 4$
8. (a) $P = 8t$ dollars **(b)** $P = 15 + 10(t - 1)$ dollars for $t \geq 1$ **9.** $5x + 1300$ dollars **10.** Answers in dollars:
(a) Youngest x, eldest $3x$, middle $\dfrac{3x}{2}$ **(b)** Middle y, eldest $2y$, youngest $\dfrac{2y}{3}$ **(c)** Eldest z, youngest $\dfrac{z}{3}$, middle $\dfrac{z}{2}$
11. (a) (i) $^-399$; **(ii)** $^-3 + (n - 1)(^-4)$, or $^-4n + 1$
(b) (i) $1 + 99\sqrt{2}$; **(ii)** $1 + (n - 1)\sqrt{2}$ **(c) (i)** $\pi + 198.5$;
(ii) $\pi + 0.5 + (n - 1)2$, or $2n + \pi - 1.5$ **12. (a) (i)** 384;
(ii) $3(\sqrt{2})^{n-1}$ **(b) (i)** $1/\pi^{13}$; **(ii)** $\pi\left(\dfrac{1}{\pi}\right)^{n-1}$ or $\left(\dfrac{1}{\pi}\right)^{n-2}$
(c) (i) $128 + 128\sqrt{2}$; **(ii)** $(1 + \sqrt{2})(\sqrt{2})^{n-1}$ **13.** $14 - 9\sqrt{2}$
14. 128 **15. (a)** $1, 1, 2, 3, 5, 8, 13$ **(b)** 4 **(c)** 7 **(d)** 12
(e) 20 **(f)** 33 **(g)** Answers may vary; the sum of the first n Fibonacci numbers is 1 less than the Fibonacci number two terms later in the sequence. **(h)** $F_1 + F_2 + F_3 + \ldots + F_n = F_{n+2} - 1$

Mathematical Connections 8-2

Communication

1. Both are correct if they define x. For the first student, x is the first of the three consecutive natural numbers. The second chose x to be the second of the three consecutive natural numbers.

Open-Ended

3. Answers vary.

Cooperative Learning

5. Answers vary.

Questions from the Classroom

7. The student thinks about the distributive property of multiplication over addition and thinks that a similar property of multiplication over multiplication is true. A counterexample shows that this is wrong. **9.** The student is right. The first statement is the commutative property of set union, and the second of set intersection.

Review Problems

11. $\sqrt[3]{729} = 9$ **13.** 1

Assessment 8-3A

1. If $\triangle + \square = 12$, then $\bigcirc + \bigcirc + 12 = 18$ and $\bigcirc = 3$.
Then $\square + \bigcirc + \bigcirc = \square + 6 = 10$, so $\square = 4$. If $\square = 4$, then $4 + \triangle = 12$ and $\triangle = 8$. Therefore, $\bigcirc = 3, \square = 4$, and $\triangle = 8$. **2. (a)** $21 + \sqrt{3}$ **(b)** $45/2$, or $22\frac{1}{2}$ **(c)** $^-1$
(d) 3 **(e)** 3 **3.** 22 **4.** 524 student tickets
5. Let x be the amount the youngest receives. Then $x + 3x + x + 14,000 = 486,000$, or $5x = 472,000$. Youngest received \$94,400.00; oldest \$283,200.00; middle \$108,400.00. **6.** 41, 41, and 38 in. **7.** Let x be the number of nickels. Then $67 - x$ is the number of dimes. So $10(67 - x) + 5x = 420, x = 50$. Thus, the box contains 50 nickels and 17 dimes. **8.** Ricardo 4, Miriam 14 **9.** 625 **10.** 350 yd by 175 yd **11.** 148 and 151

Mathematical Connections 8-3

Communication

1. Both are correct. For the first student, x is the first of the three consecutive whole numbers. The second chose x to be the second of the three consecutive whole numbers.

Open-Ended

3. Answers vary. For example:
(a) $(x + 1)(x + 2) = x^2 + 3x + 2$ **(b)** $2x + 3 = 2x - 1$
(c) $3x + 1 = 2x + 1$

Cooperative Learning

5. Answers vary.

Questions from the Classroom

7. Equations can be set up with one unknown as follows. If Jillian delivers x papers, then Abby delivers $2x$ and Brandy $2x + 50$. Thus, $x + 2x + 2x + 50 = 550, 5x = 500$, and hence $x = 100$. Jillian delivers 100 papers, Abby 200, and Brandy 250.
9. He misused the equality sign. What he wrote implies that $4x + 5 = 40$, which is not true.

Review Problems

11. $n - 4 + n - 2 + n + n + 2 + n + 4 = 5n$
13. (a) $P = \dfrac{[60 + 5(t - 1)]t}{2}$ dollars
(b) $15d$ dollars **15.** $\sqrt{5}/32$

Assessment 8-4A

1. (a) Double the input number.   **(b)** Add 6 to the input number.   **2. (a)** This is not a function, since the input 1 is paired with two outputs (*a* and *d*).   **(b)** This is a function.
3. (a) Answers will vary. For example:

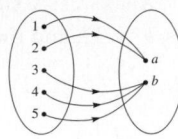

(b) 30

4. (a)

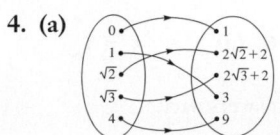

(b) $\{(0, 1), (1, 3), (\sqrt{2}, 2\sqrt{2} + 1), (\sqrt{3}, 2\sqrt{3} + 1), (4, 9)\}$

(c)

x	$f(x)$
0	1
1	3
$\sqrt{2}$	$2\sqrt{2} + 1$
$\sqrt{3}$	$2\sqrt{3} + 1$
4	9

(d)

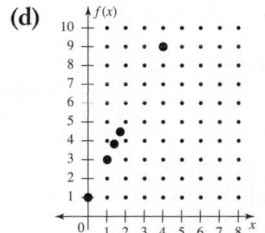

5. (a) This is a function.   **(b)** This is a function.

6. (a) (i)

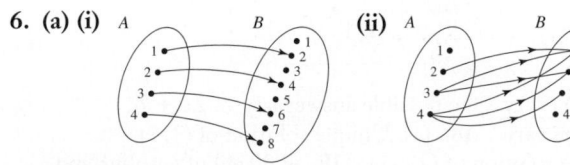

(b) Part (i) is a function from *A* to *B*. For each element in *A*, there is a unique element in *B*. The range of the function is $\{2, 4, 6, 8\}$.

7.

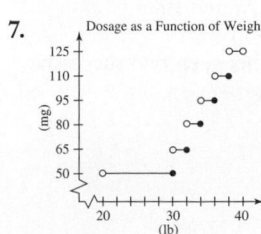

8. Answers vary. **(a)** $L(n) = 2n + (n - 1)$, or $3n - 1$
(b) $L(n) = n^2 + 1$

9. (a)

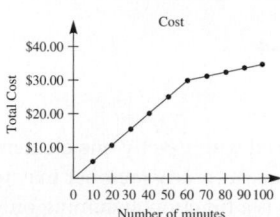

Notice that because we can't really depict 100 dots on the graphs a dot is drawn only for a multiple of 10 minutes.   **(b)** That the company charges for each part of a minute at the rate of $0.50 per minute   **(c)** The two segments represent different charges per minute. The one representing the higher cost is steeper.
(d) $C(t) = 0.50t$ if $0 \le t \le 60$, $C(t) = 30 + 0.10(t - 60)$ if $t > 60$.   **10. (a)** $5n - 2$ **(b)** 3^n **11. (a)** 30   **(b)** 65
(c) $7\sqrt{7} - 5$ **(d)** $^-5$ **12. (a)** The first three are.   **(b)** Only (ii) and (iii)   **(c)** The first three are.   **13. (a)** 16; 16; 16; $4\sqrt{5}$
(b) $\{(1, 9), (2, 8), (3, 7), (4, 6), (5, 5), (6, 4), (7, 3), (8, 2), (9, 1)\}$ **(c)** The domain is $R^+ \times R^+$, and the range is R^+.
14. (a) 50 cars   **(b)** Between 6:00 A.M. and 6:30 A.M.   **(c)** 0
(d) between 8:30 A.M. and 9 A.M., by 100 cars   **(e)** Segments are used because the data are continuous rather than discrete. For example, there are a number of cars at 5:20 A.M.
15. (a) $H(2) = 192$; $H(6) = 192$; $H(3) = 240$; $H(5) = 240$. Some of the heights correspond to the ball going up, some to the ball coming down.

(b)

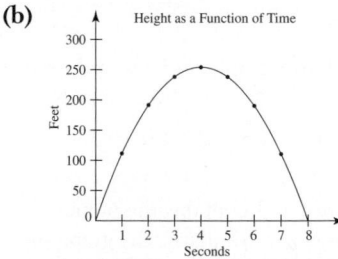

At $t = 4$ sec, the ball's height is $H(4) = 256$ ft above the ground.   **(c)** 8 sec   **(d)** $0 \le t \le 8$ **(e)** $0 \le H(t) \le 256$
16. (a) (i) 40   **(ii)** 49   **(b) (i)** $S(n) = 2n(n + 1)$
(ii) $S(n) = (n + 1)^2 + (n + 2)n$ or $2n^2 + 4n + 1$
17. (a) $\dfrac{n(n + 1)}{2}$ **(b)** 4^{n-1} **18.** The converse is false.

For example, the set of ordered pairs $\{(1, 2), (1, 3)\}$ does not represent a function, since the element 1 is paired with two different second components.   **19.** Only (b) does not represent a function; if *x*, the input, is any given real number, then *y* is not unique, as it could be any real number *y* such that $y > x - 2$.   **20.** Only (b) is not. For $x = 1$ there are many values of *y*.   **21. (a)** Boys: B, H; girls: $A, C, D, G, I, \mathcal{J}, E, F$

(b) $\{(A, B), (A, C), (A, D), (C, A), (C, B), (C, D), (D, A),$
$(D, B), (D, C), (F, G), (G, F), (I, \mathcal{J}), (\mathcal{J}, I), (E, H)\}$
(c) No **22. (a)** Yes **(b)** No **23. (a)** None **(b)** Reflexive, symmetric, and transitive (and so an equivalence relation)
(c) Reflexive, symmetric, and transitive (and so an equivalence relation) **(d)** Reflexive, symmetric, and transitive (and so an equivalence relation) **(e)** Symmetric **(f)** Reflexive and symmetric
(g) Transitive

Mathematical Connections 8-4

Communication

1. Yes; each element of A is paired with exactly one element of B.
3. (a) This is not a function, since a faculty member may teach more than one class. **(b)** This is a function (assuming only one teacher per class). **(c)** This is not a function, since not every senator is paired with a committee. (Not every senator is the chairperson of a committee.)

Open-Ended

5. Answers will vary. **7.** Answers will vary. **9.** Answers will vary.

Cooperative Learning

11. (a) Answers will vary. **(b)** Answers will vary. **(c)** Answers will vary. **(d)** Answers will vary. **(e)** Answers will vary.
(f) Answers will vary. **(g)** Answers will vary. **(h)** Answers will vary. **(i)** Answers will vary.

Questions from the Classroom

13. It is a function. Each input has exactly one output. **15.** If the domain does not include the numbers between two inputs, the points corresponding to the two inputs cannot be connected.

Review Problems

17. 4 hr **19.** Answers will vary. For example, $\sqrt{3} + {}^-\sqrt{3} = 0$, a rational number.

Assessment 8-5A

1. (a) and **(b)** The graph of $y = mx + 3$ contains the point $(0, 3)$ and is parallel to the line $y = mx$. Similarly, the graph of $y = mx - 3$ contains the point $(0, {}^-3)$ and is parallel to $y = mx$.

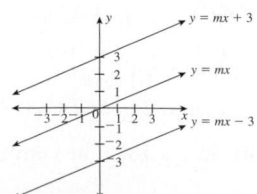

2. (a)

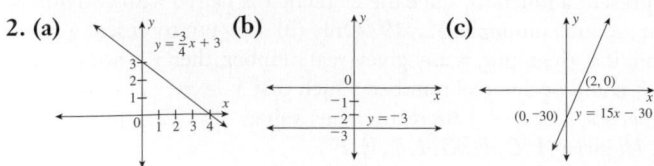

3.

	x-intercept	y-intercept
(a)	4	3
(b)	None	$^-3$
(c)	2	$^-30$

4. (a) Using $(0, 32)$ and $(100, 212)$, the slope is $(212 - 32)/(100 - 0) = \dfrac{9}{5}$. So $F = \left(\dfrac{9}{5}\right)C + b$. Substitute in the point $(0, 32)$. Thus, $b = 32$, and the equation is $F = \left(\dfrac{9}{5}\right)C + 32$. **(b)** $C = \dfrac{5}{9}(F - 32)$
(c) $({}^-40°, {}^-40°)$ **5. (a)** $y = \left(\dfrac{1}{3}\right)x$; slope $\dfrac{1}{3}$, y-intercept 0
(b) $y = {}^-x + 3$; slope $^-1$, y-intercept 3
(c) $y = \left(\dfrac{1}{3}\right)x$; slope $\dfrac{1}{3}$, y-intercept 0 **6. (a)** $y = {}^-x - 1$
(b) $y = \left(\dfrac{1}{2}\right)x$ **(c)** $y = x - \dfrac{1}{2}$ **7.** Answers vary.
8. (a) $x = {}^-2$; y is any real number. **(b)** $x > 0$ and $y < 0$; x and y are real numbers. **9.** perimeter $= 12$ units; area $= 8$ sq units.
10. (a) $x = 3$ **(b)** $y = 5$ **11. (a)** $\dfrac{1}{3}$ **(b)** 0 **(c)** 1 if $a \neq b$
12. (a) $y = \dfrac{1}{3}x + \dfrac{5}{3}$ **(b)** $y = 2$ **(c)** $y = x$ **13.** Answers vary, depending on estimates from the fitted line; for example:
(a) From the fitted line, estimate point coordinates of $(50, 8)$ and $(60, 18)$; the slope is $\dfrac{10}{10} = 1$. Use the point $(50, 8)$ and substitute $T = 50$ and $C = 8$ into $C = 1T + b$ (i.e., an equation of the form $y = mx + b$), so $8 = 1(50) + b$; $b = {}^-42$; the equation is then $C = T - 42$. **(b)** 48 chirps in 15 sec **(c)** $N = 4T - 168$

14. (a) $(3, 0)$ or $({}^-3, 0)$ **(b)** $x = 3$ or $x = {}^-3$ **15.** $y = 4$

16. (a)

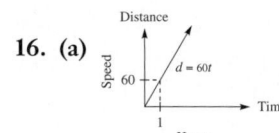

(b) 60 **17. (a) (i)** The plotted points are shown in the figure.

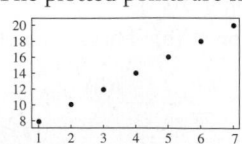

(ii) Answers vary. One possible answer is $y = 2x + 6$.
(b) Answers vary. **18. (a)** Unique solution of $(2, 5)$
(b) Unique solution of $(1, {}^-5)$ **19.** 4000 gal of gasoline and 1000 gal of kerosene **20.** 17 quarters, 10 dimes

Mathematical Connections 8-5

Communication

1. No (unless $r = 1$), because the slopes between two successive points are not the same.

Open-Ended

3. Answers vary.

Cooperative Learning

5. Answers vary.

Questions from the Classroom

7. The student is correct. Suppose there is a set of points with no association.   **9.** A line with undefined slope is perpendicular to the *x*-axis. Two lines perpendicular to another line are parallel.

Review Problems

11. Approximately 2.45   **13. (a)** $2\sqrt{7}$ **(b)** $7 - \sqrt{7}$ **(c)** $^-5\sqrt{7}$
15. (a) $\{x \mid x \in R \text{ and } x \geq ^-1\}$ **(b)** $\{x \mid x \in R \text{ and } x \leq 0\}$

Chapter Review

1. (a) Irrational   **(b)** Irrational   **(c)** Rational   **(d)** Rational
(e) Irrational   **2. (a)** $11\sqrt{2}$ **(b)** $12\sqrt{2}$
(c) $6\sqrt{5}$ **(d)** $3\sqrt[3]{6}$ **3. (a)** No; $\sqrt{2} + (^-\sqrt{2})$ is rational.
(b) No; see (a).   **(c)** No; $\sqrt{2} \cdot \sqrt{2}$ is rational.   **(d)** No; $\sqrt{2}/\sqrt{2}$
is rational.   **4.** 4.796   **5.** Approximately 1.26   **6. (a)** $5\sqrt{2}$
or $^-5\sqrt{2}$ **(b)** $\dfrac{1}{\sqrt[4]{4}}, \dfrac{1}{\sqrt[4]{16}}, \dfrac{1}{\sqrt[4]{64}}$ or $\dfrac{^-1}{\sqrt[4]{4}}, \dfrac{1}{\sqrt[4]{16}}, \dfrac{^-1}{\sqrt[4]{64}}$ **7.** $\sqrt{7}$
8. $1 + (n - 1)(\sqrt{2} - 1)$ **9.** $S = 13P$ **10.** There are
103 times as many girls as boys.   **11.** $f = 3y$ **12.** $10S - 10n$
13. 26   **14. (a)** If *n* is the original number, then each of the following lines shows the result of performing the instruction:

$$n$$
$$n + 17$$
$$2(n + 17) = 2n + 34$$
$$2n + 30$$
$$4n + 60$$
$$4n + 80$$
$$n + 20$$
$$n$$

(b) Answers will vary. For example, the next two lines could be subtract 65 and then divide by 4.   **15. (a)** $\sqrt{10} + 2$ **(b)** 29   **(c)** 3
(d) no solution   **(e)** Every real number is a solution.   **16.** Paige
111, Jordan 222, and Mike 666   **17.** Science books 17 days, other
books 3 days   **18.** Rashid 50, Dahlia 150, Jacobo 300
19. (a) Function   **(b)** Not a function   **(c)** Function
20. (a) Range $= \{3, 4, 5, 6\}$ **(b)** Range $= \{14, 29, 44, 59\}$
(c) Range $= \{0, 1, 4, 9, 16\}$ **(d)** Range $= \{5, 9, 15\}$
21. (a) This is not a function, since one student can have
two majors.   **(b)** This is a function. The range is the subset
of the natural numbers that includes the number of pages in
each book in the library.   **(c)** This is a function. The range is
$\{6, 8, 10, 12, \ldots\}$.   **(d)** This is a function. The range is $\{0, 1\}$.
(e) This is a function. The range is *N*.
22. (a) $C(x) = 200 + 55(x - 1)$

(b)
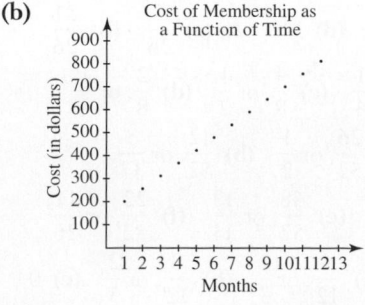

(c) In the ninth month, the cost exceeds \$600.   **(d)** In the
107th month   **23.** 5   **24. (a)** Yes, each input has exactly one
output.   **(b)** No, for $x = 4$, there are two values for *y*.
(c) No, for $x = 5$, there are two values for *y*.   **25. (a)** 14, 18, 22,

26 (b)
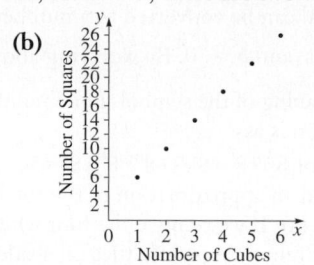

(c) $y = 4x + 2$ **(d)** The graph consists of points on the line
$y = 4x + 2$ for $x = 1, 2, 3, 4, \ldots$. The graph does not contain
all the points on the line, and hence is not a straight line. The
points lie along the line but the line contains more points.
26. (a)
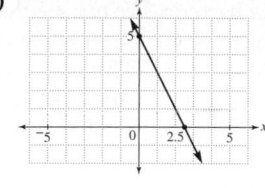
(b)
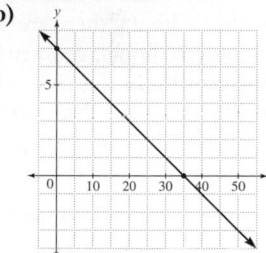

(c)
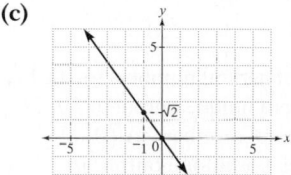

27. $\left(\dfrac{13}{3}, \dfrac{5}{3}\right)$ or $(4.\overline{3}, 1.\overline{6})$ **28. (a)** $\dfrac{^-11}{3}$ **(b)** 36   **(c)** $^-1$

(d) 3 or $^-3$ **(e)** $\dfrac{7}{2}$ or $\dfrac{^-2}{3}$ **(f)** $^-2$ or 3   **(g)** $\dfrac{1}{2}$ or 1

29. $\dfrac{1}{\pi}$ **30.** No; cash registers do not handle irrational numbers.

Answers to Now Try This

8-1. (a) $\sqrt{13} \approx 3.6056$ **(b)** $Guess2 = \dfrac{\dfrac{n}{Guess1} + Guess1}{2}$

8-2. The fraction $\sqrt{\dfrac{1}{5}}$ can be written as $\dfrac{1}{\sqrt{5}}$.   **(a)** Using the
student text, $\dfrac{1}{\sqrt{5}}$ is not a fraction. Fractions are depicted as rational numbers only.   **(b)** With the Billstein, Libeskind, Lott
text, $\dfrac{1}{\sqrt{5}}$ is a fraction because numerators and denominators could be real numbers.   **8-3. (a)** Yes, 1   **(b)** Using
the normal mathematics order of operations, the answer is
no.   **(c)** $^-593$ **(d)** 1   **(e)** The answer is no. The exponentiations and multiplications must be done before subtractions and
additions.   **8-4. (a)** The approach works as shown:

$$\sqrt{\sqrt{\sqrt{a}}} = \left(\left(a^{1/2}\right)^{1/2}\right)^{1/2} = \left(a^{1/4}\right)^{1/2}$$
$$= a^{1/8}$$
$$= \sqrt[8]{a}$$

(b) For $n = 2^k$, where k is a positive integer. As shown in part (a), by repeatedly applying the square-root function to a we get $\sqrt[2^k]{a}$.
8-5. For a number like $8^{0.\overline{3}}$, one meaning is $8^{1/3}$ or $\sqrt[3]{8}$. Any repeating or terminating decimal can be converted to a number in $\frac{a}{b}$ form where a, b are integers, and $b \neq 0$. By worrying about signs, one can determine the meaning of the symbol. For a number like $8^{0.101001\cdots}$, one interpretation is as
$8^{\left(\frac{1}{10} + \frac{0}{100} + \frac{1}{1000} + \frac{0}{10000} + \frac{0}{100000} + \frac{1}{1000000} + \cdots\right)}$ or $8^{\frac{1}{10}} \cdot 8^0 \cdot 8^{\frac{1}{1000}} \cdot 8^0 \cdot 8^0 \cdot 8^{\frac{1}{1000000}} \cdots$.
Each part could be evaluated and an approximation of the total value could be obtained. **8-6. (a)** If we remove the four white corner tiles, then the number of remaining white tiles on a side is the same as the number of shaded tiles, i.e., n. On four sides we have $4n$. Adding the four corner tiles, we get $4n + 4$.
(b) (i) 206 **(ii)** $2n + 6$ **8-7. (a)** After 10 hours, there are $2 \cdot 3^{10} = 118,098$ bacteria, and after n hours, there are $2 \cdot 3^n$ bacteria. **(b)** After 10 hours, there are $2 + 10 \cdot 3 = 32$ bacteria, and after n hours, there are $2 + 3n$ bacteria. We can see that after only 10 hours, geometric growth is much faster than arithmetic growth. In this case, 118,098 versus 32. This is true in general when $r > 1$. **8-8. (a)** $\square = 3, \triangle = 9$ **(b)** $\square = 4, \triangle = 2$
8-9. If Bob delivered b papers then Abby delivered $3b$ papers and Connie $3b + 13$ papers.

$$b + 3b + 3b + 13 = 496$$
$$b = 69$$
$$a = 3b = 207$$
$$c = 3b + 13 = 220$$

8-10. (a) It is a function from the set of natural numbers to $\{0, 1\}$, because for each natural number there is a unique output in $\{0, 1\}$. **(b)** It is a function from the set of natural numbers to $\{0, 1\}$, because for each natural number there is a unique output in $\{0, 1\}$. **8-11.** Answers vary. The graphs are parallel lines with the same slope. **8-12. (a)** Girls: A, C, D, F, G, I; boys: B, J **(b)** E and H **8-13. (a)** If $m = 0$, the line is the x-axis. **(b)** As m increases from 0, the slope is positive and the graph of the line becomes steeper as m becomes larger. **(c)** As m decreases from 0, the slope is negative and the graph of the line becomes steeper as the absolute value of m becomes larger. **8-14.** If there were no restrictions on the domain, a_n is replaced by y, and n is replaced by x, the equation becomes $y = a_1 + (x - 1)d$ or $y = dx + (a_1 - d)$. The slope of this line is d and the y-intercept is $a_1 - d$. **8-15. (a)** All the points on a horizontal line have the same y-coordinate. Thus, two points on a horizontal line have
the form (x_1, y_1) and (x_2, y_2), where $y_2 = y_1$. The slope is

$\frac{y_2 - y_1}{x_2 - x_1} = \frac{0}{x_2 - x_1} = 0$. **(b)** For any vertical line, $x_2 = x_1$ and therefore $x_2 - x_1 = 0$. If we attempted to find the slope, we would have to divide by 0, which is impossible. Hence, the slope of a vertical line is not defined. **8-16.** The solution to $\begin{cases} 2x + y = 60 \\ x + 2y = 75 \end{cases}$ can be found by substitution.
If $x - 2y = 75$, then $x = 75 + {}^-2y$. Substituting, we have

$$2(75 + {}^-2y) + y = 60$$
$$150 + {}^-4y + y = 60$$
$${}^-3y = {}^-90$$
$$y = 30$$
$$x = 15$$

and $2 \cdot 15 + y = 60$ implies $y = 30$. Thus a shirt costs $15 and a sweater costs $30. **8-17. (a)** The graphs are lines that intersect at $x = 4$ and $y = 3$. **(b)** The lines are parallel and not equal and hence the system has no solution. Algebraic approach: Assuming that there is a solution x and y, we multiply the first equation by 2 and add it side-by-side to the second. Therefore, x and y must satisfy $0 \cdot x + 0 \cdot y = 5$. However, no x and y satisfies this equation (a solution would imply $0 = 5$). This contradicts our assumption that the original system has a solution. **(c)** The lines are parallel and not equal. The algebraic approach is similar to part (b).

Answers to Brain Teasers

Section 8-1

March 14 is 3/14, the first three digits of π. Many schools also have a special event at 1:59 P.M. to celebrate the first six digits of π(3.14159).

One fallacy in the "proof" is in the line,

  "But also $\sqrt{{}^-1}\sqrt{{}^-1} = {}^-1$."

This is not true in the set of real numbers because the $\sqrt{{}^-1}$ does not exist. Thus, that sentence is meaningless.

Section 8-5

The solutions are all $({}^-1, 2)$.

Answer to the Preliminary Problem

Half of the length of the expanded rail is $(1/2)(1\ \text{mi} + 2\ \text{ft})$ and because 1 mi = 5280 ft, this half length becomes $(1/2)(5280 + 2)$ or 2641 ft. The 1-mi of non-expanded rail is $(1/2)(5280\ \text{ft})$ or 2640 ft. These two lengths are the hypotenuse and a leg of a right triangle where the second leg is the height of the buckle. Using the Pythagorean Theorem, we have the following:

$$2640^2 + \text{height}^2 = 2641^2$$
$$\text{height}^2 = 2641^2 - 2640^2$$
$$\text{height}^2 = 5281$$
$$\text{height} \approx 72.67\ \text{ft}$$

Thus, the 2-ft expansion in the rail would cause an almost 73-ft buckle, and there could be a major issue for trains with a buckle of this height.

Chapter 9

Assessment 9-1A

1. (a) No **(b)** Yes **(c)** Yes **(d)** No **2. (a)** $\frac{5}{26}$ **(b)** $\frac{21}{26}$
(c) $\frac{11}{26}$ **3. (a)** $\frac{3}{8}$ **(b)** $\frac{2}{8}$, or $\frac{1}{4}$ **(c)** $\frac{4}{8}$, or $\frac{1}{2}$ **(d)** $\frac{2}{8}$, or $\frac{1}{4}$
(e) 0 **(f)** $\frac{3}{8}$ **(g)** $\frac{1}{8}$ **4. (a)** $\frac{26}{52}$, or $\frac{1}{2}$ **(b)** $\frac{12}{52}$, or $\frac{3}{13}$
(c) $\frac{28}{52}$, or $\frac{7}{13}$ **(d)** $\frac{4}{52}$, or $\frac{1}{13}$ **(e)** $\frac{48}{52}$, or $\frac{12}{13}$ **(f)** $\frac{22}{52}$, or $\frac{11}{26}$
(g) $\frac{3}{52}$ **(h)** $\frac{30}{52}$, or $\frac{15}{26}$ **5. (a)** $\frac{4}{12}$, or $\frac{1}{3}$ **(b)** $\frac{8}{12}$, or $\frac{2}{3}$ **(c)** 0

(d) $\frac{6}{12}$, or $\frac{1}{2}$ (e) 1 **6.** (a) $\frac{1}{6}$ (b) $\frac{4}{6}$, or $\frac{2}{3}$ **7.** 70%,

$P(\text{No Rain}) = 1 - P(\text{Rain}) = 1 - 0.30 = 0.70$ **8.** (a) $\frac{18}{38}$, or

$\frac{9}{19}$ (b) $\frac{2}{38}$, or $\frac{1}{19}$ (c) $\frac{26}{38}$, or $\frac{13}{19}$ (d) $\frac{20}{38}$, or $\frac{10}{19}$

9. 10 times **10.** $\frac{350}{1380}$, or $\frac{35}{138}$ **11.** 0.7 **12.** (a) $\frac{45}{80}$, or $\frac{9}{16}$

(b) $\frac{10}{80}$, or $\frac{1}{8}$ (c) $\frac{60}{80}$, or $\frac{3}{4}$ (d) $\frac{30}{80}$, or $\frac{3}{8}$ **13.** (a) 22 (b) 10

14. No, she is not correct. The probability of the event of drawing

a white ball from box #1 is $\frac{3}{4}$, or $\frac{6}{8}$. The probability of the event of

drawing a white ball from box #2 is $\frac{5}{8}$. Because $\frac{6}{8} > \frac{5}{8}$, the probability

of drawing a white ball is greater for box #1. **15.** The probability

is $\frac{1}{2}$ because it is a fair coin. The coin has no memory, so the

probability of the event of a head on the 16th toss is the same as

the probability of the event of a head on any toss regardless of this

history. **16.** (a) $\frac{2}{7}$ (b) $\frac{5}{7}$

Mathematical Connections 9-1

Communication

1. All probabilities are less than or equal to 1 and greater
than or equal to 0. If events are mutually exclusive, then
$P(A \cup B) = P(A) + P(B)$. Therefore, if A and B are mutually
exclusive, $P(A \cup B) = P(A) + P(B) = 0.8 + 0.9 = 1.7$, which
is impossible. Therefore, events A and B are not mutually exclusive. **3.** Joe's conjecture is incorrect. Each of the numbers

1 through 4 has approximate probability $\frac{1}{4}$ of occurring because each

angle where the arrow is located seems to measure 90 degrees and
the spinner has the same chance of landing in any of the regions.

Open-Ended

5. Answers vary depending on the book selected. **7.** Answers
vary. For example, event A is an impossible event such as rolling a
10 on a single roll of a standard die. Event B has low probability,
such as the chance of rain being 20%. Event C is around 0.5, so
this might be something like obtaining 5 heads when tossing a fair
coin 10 times. Event D has a high probability of happening, but it
is not certain (for example, tossing a number less than 6 on a toss
of a standard die). Event E has probability 1, so it has to happen
(for example, tossing either a head or a tail on the toss of a fair
coin). Event F has probability greater than 1, but this cannot happen, so no event is possible.

Cooperative Learning

9. (a) Answers vary. (b) The person who receives 4 times the
value of the number on the die wins when 1, 2, and 3 are tossed.
The person who receives the square of the number showing wins
when a 5 or a 6 is tossed. When a 4 is tossed, the game is a draw.
The person who receives 4 times the value of the die has a greater
chance of winning, and therefore the game is not fair.

11. A fair coin has "no memory." Hence, the probability of a tail

on each toss is $\frac{1}{2}$ regardless of how many tails appeared in previous

tosses. **13.** Tossing three heads on the first three tosses of a coin
does not imply the coin is unfair. Only when a fair coin is tossed a
much greater number of times can we expect to get approximately
equal numbers of tails and heads. The probability of the event of

three heads in three tosses is $\frac{1}{8}$. **15.** The probability of an event

is a ratio and does not necessarily reflect the number of elements
in the event or in the sample space. For example, if $n(S) = 20$
and $n(A) = 12$, then

$P(A) = \frac{12}{20}$, which could also be reported as $P(A) = \frac{3}{5}$.

Assessment 9-2A

1. (a)

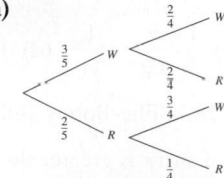

The probability of the event that two balls are of different colors

is $\frac{3}{5} \cdot \frac{2}{4} + \frac{2}{5} \cdot \frac{3}{4} = \frac{3}{5}$.

(b) $\frac{12}{25}$

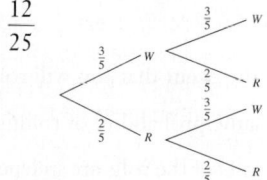

2. (a) $\frac{1}{216}$ (b) $\frac{1}{120}$ **3.** $\frac{1}{30}$ **4.** (a) $\frac{64}{75}$ (b) $\frac{11}{75}$ **5.** $\frac{1}{16}$

6. (a) $\frac{4}{10}$, or $\frac{2}{5}$ (b) Yes, now $P(\text{Even}) = \frac{18}{20} = \frac{9}{10}$ **7.** (a) $\frac{1}{320}$

(b) $\frac{63}{4000}$ (c) 0 (d) $\frac{171}{320}$ **8.** (a) The first spinner. If you
choose the first spinner, you win if, and only if, the spinning
combinations are as follows:

Outcome on spinner A	Outcome on spinner B
4	3
6	3 or 5
8	3 or 5

The probability of this happening is $\frac{1}{3} \cdot \frac{1}{3} + \frac{1}{3} \cdot \frac{2}{3} + \frac{1}{3} \cdot \frac{2}{3} = \frac{5}{9}$.

If you choose spinner B, the probability of winning is only $\frac{4}{9}$.

(b) Answers vary. One choice is as follows:

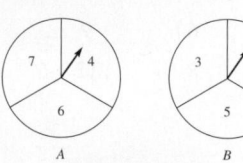

If you choose A, the probability of winning is $\frac{5}{9}$. If you choose B,

the probability of winning is only $\frac{4}{9}$. **9.** $\frac{1}{32}$ **10. (a)** $\frac{16}{81}$ **(b)** $\frac{8}{27}$

11. (a) $\frac{1}{25}$ **(b)** $\frac{8}{25}$ **(c)** $\frac{16}{25}$ **12.** $\frac{1}{12}$ **13.** $\frac{2}{28}$, or $\frac{1}{14}$

14. (a)

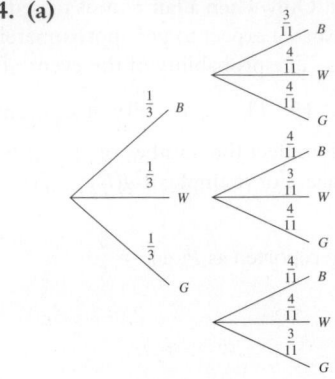

(b) $\frac{1}{3} \cdot \frac{3}{11} + \frac{1}{3} \cdot \frac{3}{11} + \frac{1}{3} \cdot \frac{3}{11}$, or $\frac{3}{11}$ **(c)** $\frac{1}{3} \cdot \frac{3}{11} = \frac{1}{11}$ **(d)** 1

15. (a) $\frac{11}{16}$ **(b)** $\frac{11}{16}$ **16.** 4 **17.** $\frac{7}{12}$ **18.** Billie-Bobby-Billie because the probability of winning two in a row is greater this way. Note, it does not say win two out of three.

Mathematical Connections 9-2

Communication

1. If the die is fair, the probability of the event that Jim will roll a 3 on the next roll is $\frac{1}{6}$, which is the same probability of rolling any one of the other numbers. The results of the rolls are independent. **3.** Put the red ball into one of the boxes and the three white balls into the other box. In this case the probability of the event of getting the red ball is $\frac{1}{2} \cdot 1 + \frac{1}{2} \cdot 0$, or $\frac{1}{2}$. In the three other arrangements, the probability of the event of getting the red ball is smaller.

Open-Ended

5. One possibility is that one die has the numbers 0, 0, 0, 3, 3, 3, and the other die has the numbers 1, 2, 3, 7, 8, 9. The probability of each outcome is $\frac{3}{36}$, or $\frac{1}{12}$.

Cooperative Learning

7. (a) Answers vary. **(b)** Answers vary.
(c)

×	1	2	3	4
1	1	2	3	4
2	2	4	6	8
3	3	6	9	12
4	4	8	12	16

(d) Yes, each player has probability $\frac{1}{2}$ of winning. **(e)** Answers vary. For example, one player wins with products, 1, 2, 3, 4, 5, 6, 8, 10 and the other player wins with products 9, 12, 15, 16, 18, 20, 24, 25, 30, 36. Then each player had probability $\frac{1}{2}$ of winning.

9. (a) The game is not fair. You should choose spinner A.
(b) The game is not fair. You should choose spinner C. It has a winning probability of $\frac{35}{99}$.

Questions from the Classroom

11. *Independence* refers to whether one event affects another. A second outcome does not depend upon the first. For example, if you get a "head" on the first flip of a coin, the probability of the event of getting a "head" on the second flip is not changed. The term *mutually exclusive* is used to describe one event that can occur in more than one way. For example, when you flip a coin, there are two possible outcomes that cannot occur at the same time. **13.** One multiplies probabilities when events are independent. Adding probabilities occurs when separate outcomes are parts of the desired event. **15.** Alva's reasoning is not correct. If she chooses bag A, her chances of getting a red ball are $\frac{1}{5}$. If she chooses bag B, her chances of getting a red ball are $\frac{3}{16}$. Because $\frac{1}{5} > \frac{3}{16}$, she should choose bag A.

Review Problems

17. (a) $\frac{1}{30}$ **(b)** 0 **(c)** $\frac{19}{30}$

Assessment 9-3A

1. Answers vary; likely this would not work because the outcomes are not equally likely. **2.** Answers vary. For example, let the digits 1 and 2 represent Diamonds, digits 3 and 4 represent Hearts, digits 5 and 6 represent Spades, and 7 and 8 represent Clubs. If 0 or 9 appear, then disregard these digits. Read the random digits in pairs to simulate two draws. **3. (a)** Let 1, 2, 3, 4, 5, and 6 represent the numbers on the die and ignore the numbers 0, 7, 8, 9. **(b)** Number the persons 01, 02, 03, . . . , 18, 19, 20. Go to the random-digit table and mark off groups of two. The three persons chosen are the first three whose numbers appear. Skip over pairs of digits that exceed 20 or represent a person already chosen. **(c)** Represent red by the numbers 0, 1, 2, 3, 4; green by the numbers 5, 6, 7; yellow by the number 8; and white by the number 9. **4.** Mark off 30 blocks of three digits in a random-digit table. Disregard 000 or any numbers that are greater than 500. Also disregard duplicates. These are the numbers of the 30 students who will be chosen for the trip. **5.** To simulate Monday, let the digits 1 through 8 represent rain and 0 and 9 represent no rain. If rain occurred on Monday, repeat the same process for Tuesday. If it did not rain on Monday, let the digits 1 through 7 represent rain and 0, 8, and 9 represent dry. Repeat a similar process for the rest of the week. **6.** Answers vary. For example, mark off blocks of two digits and let the digits 00, 01, 02, . . . , 13, 14 represent contracting the disease and 15 to 99 represent no disease. Mark off blocks of six digits to represent the three children. If at least one of the numbers of successive pairs within your 6-digit block representing the family is in the range 00 to 14, then this represents a child in the three-child family having strep.   **7.** $\frac{3}{10}$ **8.** 1200 fish   **9. (a)** 7   **(b)** Answers vary. For example, let the digits 0, 1, 2, 3, 4 represent a victory by team A and the digits 5, 6, 7, 8, 9 represent a victory by team B. Then go to the random-digit table, pick a starting place, and count the number of digits (games) it takes for one of the teams to win. Repeat this experiment a number of times for four games and for seven games and use the definition of probability to compute the experimental probability.   **10.** One way to simulate the problem is to use a random-digit table. Because Carmen's probability of making any basket is 80%, we could use the occurrence of a 0, 1, 2, 3, 4, 5, 6, or 7 to simulate making the basket and the occurrence of an 8 or a 9 to simulate missing the basket. Another way to simulate the problem is to construct a spinner with 80% of the spinner devoted to making a basket and 20% of the spinner devoted to missing the basket. This could be done by constructing the spinner with 80% of the 360 degrees (that is, 288 degrees) devoted to making the basket and 72 degrees devoted to missing the basket.

We can compute the theoretical probability for this experiment by using a *tree diagram*, as shown here. Thus, theoretical probabilities for the number of points scored in 25 attempts can be computed. These theoretical estimates are given in the following table. Compare these results with the experimental probability obtained by simulation.

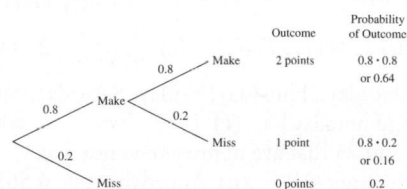

Number of Points	Expected Number of Times Points Are Scored in 25 Attempts
0	5
1	4
2	16

Mathematical Connections 9-3

Communication

1. Answers will vary. One possibility is to see Kanold, C. "Teaching Probability Theory Modeling Real Problems." *Mathematics Teacher* 85 (April 1994): 232–235.   **(a)** should be about 2.08 ≈ 2   **(b)** 1:1

Open-Ended

3. Answers vary; one could use a random-digit table with blocks of two digits, or a spinner could be designed with 12 sections representing the different months. The spinner could be spun 5 times to represent the birthdays of five people. We could then keep track of how many times at least two people have the same birthday. The theoretical probability is approximately 0.618.   **5.** Let the 10 ducks be represented by the digits 0, 1, 2,

3, . . . , 8, 9. Then pick a starting point in the table and mark off 10 digits to simulate which ducks the hunters shoot at. Count how many of the digits 0 through 9 are not in the 10 digits; this represents the ducks that escaped. Do this experiment many times and take the average to determine an answer. See how close your simulation comes to 3.5 ducks.

Cooperative Learning

7. (a)–(b) Answers vary.   **(c)** 15   **(d)** $\left(\frac{1}{2}\right)^{10} = \frac{1}{1024}$

9. Pick a starting spot in the table and count the number of digits it takes before all the numbers 1 through 9 are obtained. Ignore 0. Repeat this experiment many times and find the average number of boxes.

Questions from the Classroom

11. Maximilian could be correct, but students should be asked to explain how their simulations were designed and carried out.

Review Problems

13. (a) $\frac{1}{4}$ **(b)** $\frac{1}{52}$ **(c)** $\frac{48}{52}$, or $\frac{12}{13}$ **(d)** $\frac{3}{4}$ **(e)** $\frac{1}{2}$ **(f)** $\frac{1}{52}$ **(g)** $\frac{16}{52}$, or $\frac{4}{13}$ **(h)** 1   **15.** $\frac{8}{27}$

Assessment 9-4A

1. (a) 12 to 40, or 3 to 10   **(b)** 40 to 12, or 10 to 3   **2.** 30 to 6, or 5 to 1   **3.** 15 to 1   **4.** $\frac{5}{8}$ **5.** 73:27   **6.** 5:3   **7.** 3:5

8. $\frac{2}{11}$ **9.** 1:1   **10.** 5:88   **11.** 9:1   **12.** 1:35   **13. (a)** $\frac{1}{2}$

(b) $\frac{1}{4}$ **14.** 8.4¢   **15.** $10,000   **16.** No, the expected value is $^-$$1.00.

Mathematical Connections 9-4

Communication

1. Odds are given from probabilities. The *odds in favor* of an event, E, are given by $\frac{P(E)}{P(\overline{E})}$. The *odds against* an event are given by $\frac{P(\overline{E})}{P(E)}$.

Open-Ended

3. Answers vary. For example, a person pays $1.00 to play the game; he/she wins $2.00 if coins match and nothing if they don't match   **5.** Answers vary. One possibility is a game involving a spinner with five sectors of equal area numbered 1 through 5. You pay $2.00 for playing the game. If the spinner lands on regions 1 or 2, you win $5.00. The expected winnings in this game are $5 \cdot \frac{2}{5} - 2$, or 0.

Questions from the Classroom

7. Suppose the odds in favor of winning a game are 1:2. Even when the outcomes are equally likely, it does not mean that out of every three games the player will win one game. It only means that if a large number of games are played, the ratio between the number of wins and the number of losses is close to $\frac{1}{2}$. However, the answer is partially correct because while the ratio of wins to losses is $\frac{P(W)}{1 - P(W)} = \frac{a}{b}$, $P(Win) = \frac{a}{a + b}$. **9.** The probability is $\frac{1}{3}$ that an individual does not vote. Maria expects $48 - \frac{1}{3} \cdot 48$, or 32 to vote. Since she is guaranteed 24 votes and $24 > (51\%)(32)$, she could be confident of winning.

Another way to consider the situation is to consider that 32 are expected to vote, BUT perhaps $\frac{1}{3}$ of 24, or 8, that she expected to vote for her might not. In that situation, she is guaranteed $24 - 8$, or 16, votes but needs $51\%(32)$, or 17. She is missing 1 vote. Her ground is certainly less safe in this scenario.

Review Problems

11. The blue section must have angle measure 300 degrees; the red has angle measure 60 degrees.

Assessment 9-5A

1. 224 **2.** 10,000 **3.** 180 **4. (a)** true **(b)** false **(c)** false **5. (a)** 40,320 **(b)** 19,958,400 **6.** 15

7. (a) 24,360 **(b)** 4060 **8.** $\frac{1}{120}$ **9.** 45 **10.** 1260

11. 8 people **12.** 2,598,960 different 5-card hands (order within the hand is not important.) **13.** $\frac{1}{25,827,165}$

14. (a) $\frac{{}_{12}C_4}{{}_{22}C_4}$, or $\frac{9}{133}$ **(b)** $\frac{{}_{10}C_4}{{}_{22}C_4}$, or $\frac{6}{209}$

(c) $1 - \frac{{}_{10}C_4}{{}_{22}C_4}$, or $\frac{203}{209}$ **15. (a)** $({}_{20}C_2 \cdot {}_{21}C_4 \cdot {}_4C_2) \div {}_{45}C_8 \approx 0.032$

(b) $\frac{{}_{25}C_8}{{}_{45}C_8} \approx 0.005$ **(c)** $1 - \frac{{}_{25}C_8}{{}_{45}C_8} \approx 0.995$ **(d)** $\frac{{}_{20}C_8}{{}_{45}C_8} \approx 5.84 \cdot 10^{-4}$

16. $\frac{1}{120}$ **17. (a)** 10^4, or 10,000 **(b)** 5040 **(c)** $\frac{625}{10,000}$, or $\frac{1}{16}$

18. ${}_6C_4$ or 15 **19.** 56 choices

Mathematical Connections 9-5

Communication

1. Answers vary. For example, the Fundamental Counting Principle (FCP) says that to find the number of ways of making several decisions in a row, multiply the number of choices that can be made for each decision. The FCP can be used to find the number of permutations. A permutation is an arrangement of things in a definite order. A combination is a selection of things in which the order is not important. We could find the number of combinations by using the FCP and then dividing by the number of ways in which the things chosen for the permutation can be arranged. **3. (a)** $8! \cdot 3!$. If the family is considered a unit and each of the remaining people also a unit, we have 8 units. There are 8! ways to arrange the 8 units. For each of the 8! ways, the family unit can be arranged in 3! ways and hence the number of seating arrangements is $8! \cdot 3!$ or 241,920. **(b)** $10! - 8! \cdot 3!$ or 3,386,880.

We need to subtract the answer in part (a) from the number of all possible seating arrangements.

Open-Ended

5. (a) 10^6, or 1,000,000 **(b)** Answers vary. For example, you would first find the population of California and then experiment with using letters in the license plates. This would help because the choice is for 26 letters in a slot rather than 10 numbers.

Questions from the Classroom

7. The student is confused about choosing four objects none at a time. There is one way to choose no objects. Therefore, we say ${}_4P_0 = 1$. **9.** If the principle is used properly, one can compute the number of permutations. That does not imply that they are not needed. It is a concept that comes up a lot, so having a name for it and formulas to help with computations are useful.

Review Problems

11. $\frac{3}{36}$, or $\frac{1}{12}$ **13. (a)** No **(b)** The expected payoff is $\frac{1}{38} \cdot \$36.00$, or approximately 95¢. Therefore, you can expect to lose on the average about 5¢ a game if you play it a large number of times.

Chapter Review

1. (a) $\{HHH, HHT, HTH, HTT, TTT, TTH, THH, THT\}$
(b) $\{HHH, HHT, HTH, THH\}$ **(c)** $\frac{4}{8}$, or $\frac{1}{2}$ **2. (a)** {Monday, Tuesday, Wednesday, Thursday, Friday, Saturday, Sunday}
(b) {Tuesday, Thursday} **(c)** $\frac{2}{7}$ **3.** There are 800 blue ones, 125 red ones, and 75 that are neither blue nor red.
4. (a) Approximately 0.497 **(b)** Approximately 0.503
(c) About 1.01 to 1 **5. (a)** $\frac{5}{12}$ **(b)** $\frac{9}{12}$, or $\frac{3}{4}$ **(c)** $\frac{5}{12}$
(d) $\frac{9}{12}$, or $\frac{3}{4}$ **(e)** 0 **(f)** 1 **6. (a)** $\frac{13}{52}$, or $\frac{1}{4}$ **(b)** $\frac{1}{52}$ **(c)** $\frac{22}{52}$, or $\frac{11}{26}$
(d) $\frac{48}{52}$, or $\frac{12}{13}$ **7. (a)** $\frac{64}{729}$ **(b)** $\frac{24}{504}$, or $\frac{1}{21}$ **8.** $\frac{6}{25}$
9. $\frac{14}{80}$, or $\frac{7}{40}$ **10.** $\frac{7}{45}$ **11.** 4 to 48, or 1 to 12 **12.** 3 to 3, or 1 to 1 **13.** $\frac{3}{8}$ **14.** \$0.30 **15.** $33\frac{1}{3}$¢, so if 34¢ is charged the game is not fair. **16.** The expected value is \$1.50 . The expected net winnings are $^-\$0.50$. **17.** 900 **18.** 120 **19.** 5040
20. $\frac{2}{20}$, or $\frac{1}{10}$ **21. (a)** $5 \cdot 4 \cdot 3$, or 60 **(b)** $\frac{3}{60}$, or $\frac{1}{20}$ **(c)** $\frac{1}{60}$
22. $\frac{15}{36}$, or $\frac{5}{12}$ **23.** $\frac{2}{5}$ **24.** $7! = 5040$ **25.** 0.027 **26.** $\frac{63}{80}$
27. Answers vary. For example, **(a)** Randomly select digits 0–5 from a random-digit table; discard digits 6–9. **(b)** Mark off two-digit numbers in a random digit table. Discard any other than 01–12. Select the months represented in the first three non-repeated numbers. **(c)** Let random digits 0–2 represent red; digits 3–5 represent white; digits 6–8 represent blue; discard any 90. **28.** These events are not equally likely. If each probablity

is computed, we have $P(3H) = \frac{1}{8}$, $P(2H) = \frac{3}{8}$, $P(1H) = \frac{3}{8}$, and

$P(0H) = \frac{1}{8}$. **29. (a)** $\frac{1}{8}$ **(b)** $\frac{1}{4}$ **(c)** $\frac{1}{16}$ **30.** $\frac{8}{20}$, or $\frac{2}{5}$ **31.** $\frac{47}{85}$

Answers to Now Try This

9-1. (a) 1 **(b)** 1 **(c)** Yes, they always sum to 1. 1 is the sum of the probabilities of all the different elements in any sample space.
(d) $\frac{1}{4}$, $\frac{90°}{360°} = \frac{1}{4}$ **9-2. 1. (a)** $P(\text{both balls are red}) = \frac{1}{9}$

(b) $P(\text{no ball is red}) = \frac{4}{9}$ **(c)** $P(\text{at least one ball is red}) = \frac{5}{9}$

(d) $P(\text{at most one ball is red}) = \frac{8}{9}$

(e) $P(\text{both balls are the same color}) = \frac{3}{9} = \frac{1}{3}$

9-3. (a) $P(\text{rain both days}) = (0.3)(0.6) = 0.18$

(b) $P(\text{no rain either day}) = (0.7)(0.4) = 0.28$
(c) $P(\text{rain exactly one day}) = (0.3)(0.4) + (0.7)(0.6) = 0.12 + 0.42 = 0.54$ **(d)** $P(\text{rain at least one day}) = 1 - 0.28 = 0.72$
(e) No. If conditions are favorable for rain tonight, they are more likely to be favorable for rain tomorrow. **9-5. (a)** yes
(b) Answers vary. With replacement: any game with the same number of white and colored marbles. Without replacement: 3 colored and 6 white or 3 white and 6 colored, 6 colored and 10 white or 6 white and 10 colored marbles. **(c)** Without replacement: the white and colored marbles need to be two consecutive triangular numbers, that is, $1 + 2 + 3 + \ldots + n$ and $1 + 2 + 3 + \ldots + n + n + 1$ (this is easier to discover and verify using combinations introduced in Section 9–5).

9-6. (a) Asnwers vary. **(b)** $\frac{3}{8}$ **(c)** No, simulations will not always result in the same probability as the theoretical probability. However, if the experiment is repeated a greater number of times, the simulated probability should approach the theoretical probability. **9-7. 1.** Let the digits 0, 1, 2, and 3 represent type A blood donors. Go to a random-digit table and group the numbers in pairs. Any pair with exactly one of the digits 0, 1, 2, or 3 represents exactly one of the donors in two having type A blood. Use this simulation to calculate the probability. **2.** Answers vary. **9-8. (b)** There is one way to toss a head and one way of not tossing a head, so the odds in favor are $1:1$. **(c)** There are 4 ways to draw an ace and 48 ways of not drawing an ace, so the odds in favor are $4:48$, or $1:12$. **(d)** There are 13 wasy of drawing a heart and 39 ways of not drawing a heart, so the odds in favor are $13:39$, or $1:3$ **9-9. (a)** $n(n-1)$, $n(n-1)(n-2)$, $n(n-1)(n-2)(n-3)$ **(b)** $n(n-1)(n-2) \cdot \ldots \cdot (n-r+1)$ **(c)** $4 \cdot 3 \cdot 2 = 24$ ways to choose a president, vice president, and secretary **9-10.** Answers vary. **(a)** We get an error message because 100! and 98! are too large for the calculator to handle. **(b)** $\frac{100!}{98!} = \frac{98! \cdot 99 \cdot 100}{98!} = 9900$

Answers to Brain Teasers

Section 9-2

Page 507 These dice are truly remarkable in that no matter which die the other player chooses, you can always choose one

that will beat it $\frac{2}{3}$ of the time. Therefore, the strategy for playing this game is to go second and make your choice of die based on the information in the following table:

First Person's Choice	Second Person's Choice	Probability of Second Choice Winning
A	D	$\frac{2}{3}$
B	A	$\frac{2}{3}$
C	B	$\frac{2}{3}$
D	C	$\frac{2}{3}$

Section 9-4

Page 521 In order to have the same number of heads, Al and Betty must both toss 0 heads or they must both toss 1 head. The probability that Al tosses 0 heads is $\frac{1}{2}$, and the probability he tosses 1 head is $\frac{1}{2}$. For Betty, the probability that she tosses 0 heads is the same as tossing 2 tails, which is $\frac{1}{4}$. The probability that she tosses 1 head is equal to $P(HT) + P(TII) = \frac{1}{4} + \frac{1}{4} = \frac{1}{2}$. Next, the probability of Al tossing 0 heads and Betty tossing 0 heads is $\frac{1}{2} \cdot \frac{1}{4} = \frac{1}{8}$. The probability of Al tossing 1 head and Betty tossing 1 head is $\frac{1}{2} \cdot \frac{1}{2} = \frac{1}{4}$. Therefore, the probability that they toss the same number of heads is $\frac{1}{8} + \frac{1}{4} = \frac{3}{8}$.

Page 523 We first find the probability that all 40 children have different birthdays and then subtract the result from 1. We get

$$1 - \left(\frac{365-1}{365}\right)\left(\frac{365-2}{365}\right) \cdots \left(\frac{365-39}{365}\right) \approx 0.89$$

Thus, the probability that the friend wins the bet is approximately 0.89, or 89%. For 50 people, the probability rises to approximately 97%.

A different approach to solving the problem is to think about it as an occupancy problem where the names of 40 children are randomly put in 365 boxes, one box for each day of the year. The sample space S is the set of all permutations of 365 possible birthdays (repetition allowed). Thus, S has 365^{40} elements. The probability is

$$\frac{365^{40} - 365 \cdot 364 \cdot \ldots \cdot (365-39)}{365^{40}}$$

This expression can of course be written in the form found in the first approach.

Section 9-5

(a) The probability of a successful flight with two engines is 0.9999. **(b)** The probability of a successful flight with four engines is 0.99999603.

Answer to the Preliminary Problem

Using the hint, $\dfrac{n + 3}{n + 10} = 75\%$. Solving for n, we have:

$$n + 3 = (n + 10)\,0.75$$
$$n + 3 = 0.75n + 7.5$$
$$0.25n = 4.5$$
$$n = 18$$

Therefore, 18 blue marbles can be added to the bag. The probability can be checked to see if $(3 + 18)/(18 + 10) = 75\%$.

Chapter 10

Assessment 10-1A

1. Answers will vary. Among the questions that the class will have to determine are the following:
 • Do you count the houses all around the block or only on the side where your house is?
 • Do you count the houses across the street from you? Are they considered on your block?
 • What happens if you are in a new part of town where the blocks aren't developed yet?
Data to be collected will be determined by the question asked but likely will be a frequency count in the second grade. The frequency count could be shown in a bar graph and interpretations could be made about the graph. 2. Answers will vary. First and foremost will be the question of what it means to be active. Other questions to consider might include the following:
 • Does this mean that they do the same type of activities but one group does it more than the other?
 • Does this mean that one group sleeps more than the other? In middle school, there will be choices of both adults and sixth-grade students. Depending on understanding, a randomization might be made but likely not on a big scale. Convenient samples might be a class and their adult caregivers where answers to the question chosen could be obtained. If randomization is done, then it will likely be confined to the school itself. One unanswered question is, "What is an adult?" Is this term understood in the same way by all who do the study?
3. Answers will vary. You might ask if the students want to have the water sampled at different times of the day; if they want to invite an "unbiased" expert to taste the water on the different floors and see if the experts can discern differences; if they want to determine if the county health department tests the water of the school to have a baseline. A major question to be answered is, "What is meant by 'better'?" Is it taste? Is it chemical content? Is it as simple as warmness and coolness of the water? 4. The sample in (a) is more likely to be random and is more likely to provide better input to answer the question. The sample in (b) would most likely sample students at only one grade level and they would be likely to have similar tastes. 5. (a) fair (b) biased 6. Answers will vary. (a) Elementary students might blindfold the adults and simply hand them two different cans to see if the adults could tell the difference. (b) Middle grade students might pour the liquids in unmarked cups, use a series of taste tests to eliminate random guessing, and so on. 7. A major question is, "What does it mean to visit a country?" Does this mean that your airline landed in the country; that you spent a night there; that you

spent an extended vacation there? If a country has now separated into different countries as the USSR did, do you count each part that was visited when the country was the USSR, or do you not count a visit to the new countries at all because your visit was before they became separate countries? 8. A strong criticism of the prediction is that the sample of owners who had telephones and automobiles was not likely to be representative of voters in 1936. 9. This type of representation is likely not to be indicative of the class at all. Representative samples of comments would be more "fair" and would give a better picture of the class. 10. Choosing names from a hat is the more random process and is more likely to give a representative selection than is choosing the first 50 people in a line. 11. Second-grade students are likely to make many observations which cannot be determined from the graph but from their knowledge of each other. An example is "Alfie was the only one to wear dress shoes today." Reasonable interpretations are that on Tuesday the most popular shoes worn are tennis shoes and Crocs. A different interpretation is that on Tuesdays the most popular shoes are those with softer soles.

Mathematical Connections 10-1

Communication

1. Answers will vary. The answer should depend on the grade level identified for the problem.

Open-Ended

3. Answers will vary according to grade level and sophistication.

Cooperative Learning

5. Answers will vary.

Questions from the Classroom

7. Answers will vary. For example, the parents might appeal to statistical reason and talk about a stampede of people independent of the event at which it occurred. They might also talk about the mix of animals and people in this "stampede," which could be different from the examples that are cited. And finally, they, could also talk to her about how the evidence that she has is not generalizable to other events.

Assessment 10-2A

1. (a) Tuesday
(b) Approximately 1500 (c) 3

2.

Glasses of Lemonade Sold	
Monday	🥤🥤
Tuesday	🥤🥤
Wednesday	🥤🥤🥤
Thursday	🥤
Friday	

🥤 represents 10 glasses

3.

4. (a) 72, 74, 81, 81, 82, 85, 87, 88, 92, 94, 97, 98, 103, 123, 125
(b) 72 lb (c) 125 lb

5.

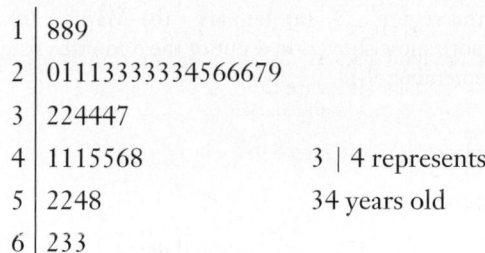

Weights of Students in East Junior
High Algebra I Class

6. Answers vary.   **(a)** A dot plot will have two columns of *x*'s, one for Heads and one for Tails.   **(b)** A bar graph will have two bars, one for Heads and one for Tails. The vertical axis will be partitioned to show frequencies of each.   **7. (a)** November, 30 cm   **(b)** 50 cm   **8. (a)** Ages of HKM Employees

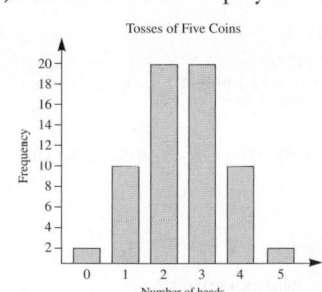

1	889
2	01113333334566679
3	224447
4	1115568
5	2248
6	233

3 | 4 represents
34 years old

(b) There are more employees in their 40s.   **(c)** 20   **(d)** 17.5%

9.

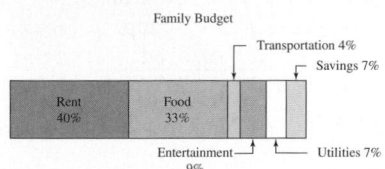

Tosses of Five Coins

10. Percentages are approximate.

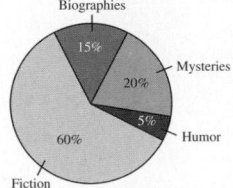

Family Budget

Transportation 4%
Savings 7%
Rent 40%
Food 33%
Entertainment 9%
Utilities 7%

11. Favorite Types of Books

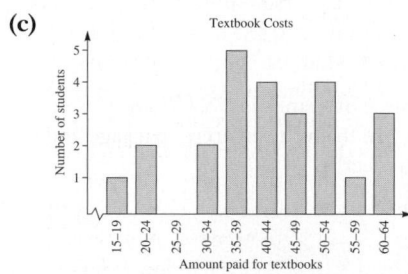

Biographies 15%
Mysteries 20%
Humor 5%
Fiction 60%

12. 12.5%   **13.** 19% of 21 people is 4 people. 14% of 21 people is 3 people. 52% of 21 people is 11 people. Each of the 5% sectors possibly represents 1 person. There are only 21 people total.   **14. (a)** Answers will vary. A graph of this type might be suspect. To have a survey where there are more people in their 70s–90s might mean the people are in a home for the aging. Also to have more people in their 90s than 80s reading might make one suspicious. Without more information, one cannot

tell.   **(b)** From this particular survey, one would suspect that people in their 70s read the most, **but** the graph only depicts the number of readers of mysteries.

15. (a)   Fall Textbook Costs

1	6
2	33
3	0357799
4	0122589
5	00138
6	022

2 | 3 represents $23

(b)   Fall Textbook Costs

Classes	Tally	Frequency
$15–19	|	1
$20–24	||	2
$25–29		0
$30–34	||	2
$35–39	⩂⩂	5
$40–44	||||	4
$45–49	|||	3
$50–54	||||	4
$55–59	|	1
$60–64	|||	3
		25

(c)

Textbook Costs

16. (a) women   **(b)** Approximately 1 yr   **(c)** About 5 yr.
17. Measures are approximate: savings 0.8 cm; rent 2.4 cm; food 0.9 cm; auto payment 2.2 cm; tuition 1.7 cm.
18. (a) Connecticut   **(b)** South Dakota   **(c)** Around $42,000
(d) Approximately 73%

Mathematical Connections 10-2

Communication

1. (a) Answers vary. For example, in a pictograph, we can observe change over time and make comparisons between similar situations. In a circle graph, it is hard to make these observations.
(b) Answers vary. For example, circle graphs are used when comparing parts to a whole. Circle graphs allow for visual comparisons of fractional parts. Bar graphs can't handle these comparisons

as well. **(c)** Answers vary. For example, in a stem and leaf plot, we can see the same shape and information as in a histogram but no data points are lost in the display as they may be in a histogram, especially when intervals are used. **3.** Answers vary. For example, the percents are of a whole and the circle represents the complete whole, or 100%. Because of rounding, it may happen that the sum of the percents is close to 100% but not exactly 100% **5. (a)** Answers vary. For example, a different graph may be more appropriate since we have continuous data changing over time. A line graph is given though students see this in the next section.

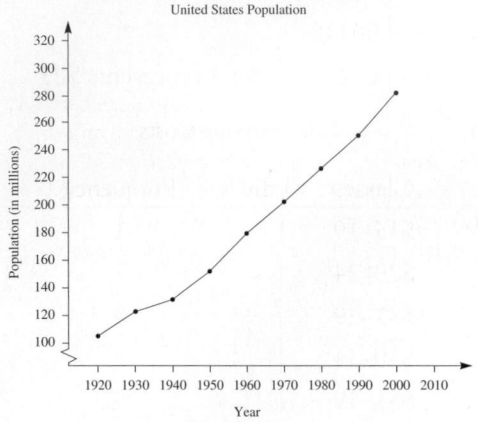

(b) The data fall into distinct categories and are not continuous, so we use a bar graph.

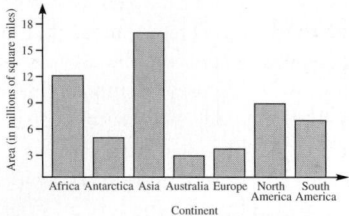

Open Ended

7. Answers vary; a good source of graphs is *USA Today*.
9. (a) Note: The data here are from the cigarette package and does not depict *all* Canadian deaths.

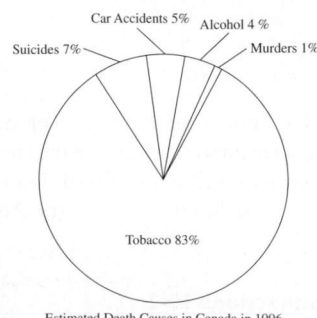

(b) Answers vary.

Cooperative Learning

11. Answers vary. Expectations of word length on future pages depend on the page chosen.

Questions from the Classroom

13. Ailene's graph is acceptable though it may be difficult for some to read and needs a title.

Review Problems

15. Technique (ii) would provide better feedback. The first technique involves self-selection bias while the second technique generates a random sample of the population.

Assessment 10-3A

1. (a) Approximately $8400 **(b)** $14,000 **(c)** Approximately $7000 **(d)** Right after 2 yr **2. (a)** About 1802 **(b)** About 13,000,000 **(c)** 1810–1820 because the slope of the line is steeper in this region. **3. (a)** January **(b)** March **(c)** 2011, they sold more snow shovels in 4 out of the 5 months listed.
4. 8-D -September, 9-D
5. (a)

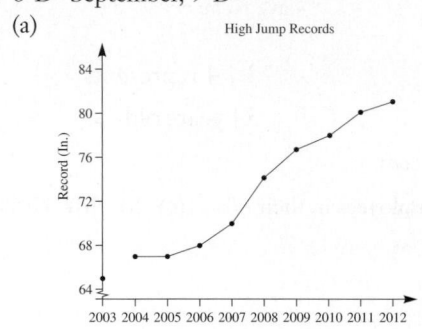

6. (a) Negative **(b)** Approximately 10 **(c)** About 22 yr old
7. (a), (b) Arithmetic Sequence

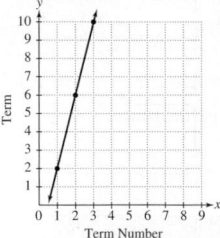

(c) $y = 4x - 2$ **8. (a) (i)** Negative **(ii)** Answers will vary.

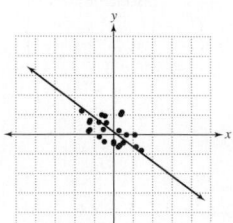

(b) No association **9. (a)** $y = \dfrac{3}{8}x - \dfrac{3}{10}$ **(b)** $y = \dfrac{1}{2}x + \dfrac{3}{2}$

(c) $y = \dfrac{3}{4}x + \dfrac{1}{2}$ **10.** Answers vary. For example, 2804 in.

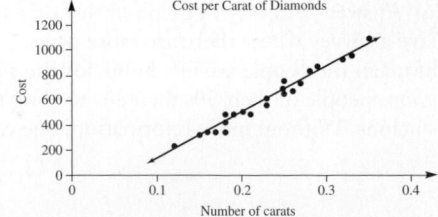

Using ordered data and drawing a trend line, an estimate for 0.5 carat is about $1700, but it is unwise to estimate outside the given range of data.   **11. (a)** ⁻1   **(b)** 47   **(c)** 5   **(d)** 20
12. Negative   **13.** The data has no apparent linear association or the data could be constant.

Mathematical Connections 10-3

Communications

1. Answers will vary, but if association is positive, the slope of the trend line is positive; if association is negative, so is the slope of the trend line.

Cooperative Learning

3. A line graph would be a good way to depict data but all data have to be using the same distance measure. Thus before a graph can be drawn, some measures will have to be converted.

Open-Ended

5. Answers vary.

Questions from the Classroom

7. Merle has a valid point, but we need to understand the graph construction.

Review Problems

9. If the miscellaneous category were only 10%, the stacked bar graph follows:

Smith Expenses

| Misc. 10% |
| Gas 13% |
| Utilities 5% |
| Food 20% |
| Rent 32% |
| Taxes 20% |

Assessment 10-4A

1. (a) Mean = 6.625, median = 7.5, mode = 8.
(b) Mean 13.$\overline{4}$, median = 12, mode = 12.   **(c)** Mean ≈ 19.9, median = 18, modes = 18 and 22.   **2.** Answers vary. For example, 1, 2, 3, 4, 5, 6, 6.   **3. (a) (i)** 70   **(ii)** 70   **(iii)** 60 and 80
(b) 1500   **4.** 78.$\overline{3}$ **5.** 62.5   **6.** $\dfrac{100m + 50n}{m + n}$ **7.** Approximately 2.59   **8.** Approximately 215 lb   **9. (a)** $41,275
(b) $38,000   **(c)** $38,000   **10. (a)** $132,000   **(b)** $12,143.75
(c) Approximately $21,756.60   **(d)** $13,500   **11. (a)** Answers vary. The median of $38,000 reported with the interquartile range of $13,500 would be one appropriate set of choices.   **(b)** Based on the answer in (a), we know that at least 50% of the salaries are at least within $13,500 of $38,000.   **12.** $19\dfrac{2}{3}$ mpg   **13.** 58 years old

14. 91   **15. (a)** A—$25, B—$50   **(b)** B   **(c)** $80 at B
(d) Answers vary. There is more variation at Theater B; also there are higher prices at Theater B.
16.

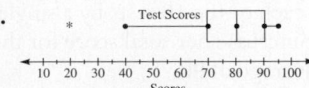

17. (a)

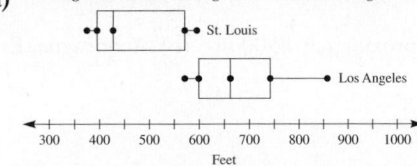

(b) Answers vary. For example, in the buildings listed, Los Angeles has the tallest buildings, and about 75% of the tallest buildings in Los Angeles are taller than the tallest listed building in St. Louis.
18. $s ≈ 7.3$ cm   **19. (a)** Approximately 76.8   **(b)** 76   **(c)** 71
(d) 14   **(e)** Approximately 156.8   **(f)** Approximately 12.5
(g) Approximately 9.04   **20.** Between 60.5 in. and 70.5 in.
21. 16%   **22. (a)** 65   **(b)** 53   **(c)** 77   **(d)** 65   **23.** 0.68
24. 90   **25.** 8400   **26.** 60th percentile   **27.** Jill would have the higher standing since Jack could be at the 65th percentile.

Mathematical Connections 10-4

Communication

1. Mode, answers vary. If you considered the most common hat size, that would be the mode. Both the mean and median might be non-whole numbers which would not be used for hat sizes.
3. The government probably uses the mean of data collected over a period of years.   **5.** mean   **(a)** For example, 10, 30, 70, and 90.
(b) Choose four numbers whose mean is 50.   **(c)** The mean of the new numbers is 50.   **7.** No. To find the average speed we divide the distance traveled by the time it takes to drive it. The first part of the trip took $\dfrac{5}{30}$, or $\dfrac{1}{6}$, of an hour. The second part of the trip took $\dfrac{5}{50}$, or $\dfrac{1}{10}$, of an hour. Therefore, to find the average speed we compute $\dfrac{10}{\dfrac{1}{6} + \dfrac{1}{10}}$ to obtain 37.5 mph.   **9.** Answers vary. A set of data that contains wait time might be appropriate *if* the times were determined during peak periods of the events. In that case, the mean with mean absolute deviation and standard deviation would be appropriate for this numerical data.

Open-Ended

11. Answers vary depending on student choice. One choice might be a box plot with two sets of data depicted so that comparisons could be seen easily.

Questions from the Classroom

13. The new mean is $\dfrac{9(10,000) + 20,000}{10} = 11,000.$
Consequently, the new mean has increased by $1000. The median and mode may change in special cases.   **15.** Since the median is 90, at least half of the class had grades of 90 or more. Since Tom scored 80, he did not do better than half of the class.   **17.** If the mean is less than the median, then one can be certain that there

were more scores above the mean than below it. The low scores tend to be further from the mean than the high scores.
19. Mel did not really miss the cutoff by a single point. She would have had to increase her score on each of the 10 tests by a single point to reach an average of 90 or increase her total score for the 10 tests by 10 points to reach an average of 90.

Review Problems

21. (a) Everest, approximately 8500 m **(b)** Aconcagua, Everest, McKinley

Assessment 10-5A

1. Answers vary in all parts of this question. **(a)** A question to ask is how quiet is a glider. **(b)** One question is how many motorcycles were sold in what years. **(c)** 11% is not much more than 10% with the given base for the percentage. **(d)** Any time there is a percentage involved, one should ask "percentage of what?" Also, how does one measure freshness on a numerical scale? **2.** Answers vary. One possibility is that the temperature is always 25°C or close to it. **3.** It could very well be that most of the pickups sold in the last 10 yr were actually sold during the last 2 yr. In such a case, most of the pickups have been on the road for only 2 yr, and therefore the given information might imply, but would not substantiate, that the average life of a pickup is around 10 yr. **4.** The horizontal axis does not have uniformly sized intervals and both the horizontal axis and the graph are not labeled. **5.** The three-dimensional drawing distorts the graph. The result of doubling the radius and the height of the can is to increase the volume by a factor of 8. **6.** One would need more information; for example, do men drive more than women? **7.** There were more scores above the mean than below, but the mean was affected more by low scores. **8. (a)** False; prices vary only by $30. **(b)** False; the bar has 4 times the area but this is not true of prices. **(c)** True **9. (a)** This bar graph could have perhaps 20 accidents at the point where the scale starts. Then 38 in 2012 would appear to be over four times the 24 of 2008, when in fact it is only 58% higher. **(b)** A bar graph with a vertical scale going to 100 would minimize the effect. **10.** Answers vary, but one such would be 5, 5, 5, 5, 5, 5, 100, 100. The mean would be 28.75 and the median 5. Neither is representative. **11.** A median might be misleading, depending on the number of data points given. Also the mean would not be sufficient. A report of the mean, median, and standard deviation would be the most helpful of all "averages" studied. **12.** Answers vary. A sample size is needed. Then, one way to pick a random sample of adults in the town is to use the telephone book. This method will not list all adults in the town, but this is probably the most accessible set of data. Then choose the adults at random using a random digit table. **13.** The student is incorrect and, in addition, is misusing percentages. She mixes percentages of "effectiveness" with percentages of times taking the drug. **14.** Answers vary. **(a)** You probably would not have a representative sample. This is a *convenience sample*. **(b)** You probably would not. The students eating in the cafeteria could likely be a biased sample. **(c)** Assign students numbers from 1 to *n*, where there are *n* students. Choose your sample using random-number selection methods. Make sure your sample is big enough. **15.** The data are indicators of passenger complaints and not an overall airline rating. Larger airlines may have more complaints. The numbers are not percentages. The number of complaints

might depend on how many flights each airline has per day. **17.** Only part (b) might not be helpful.

Mathematical Connections 10-5

Review Problems

1. (a) About 74.17 **(b)** 75 **(c)** 65 **(d)** About 237.97 **(e)** About 15.43 **(f)** 13 **3.** 76.6 **5.** Women's Olympic 100 m Swim Times 1960–2008

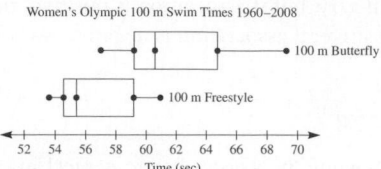

From examining the box plot, we can see that the times on the 100 m butterfly are much greater (relatively speaking) than the times on the 100 m freestyle.

Chapter Review

1. If the average is 2.41 children, then the mean is being used. If the average is 2.5, then the mean or the median might have been used. **2.** 12 **3. (a)** Mean = 30, median = 30, mode = 10. **(b)** Mean = 5, median = 5, modes = 3, 5, 6. **4. (a)** Range = 50, variance ≈ 371.4, standard deviation ≈ 19.3, mean absolute deviation ≈ 17.14, IQR = 40. **(b)** Range = 8, variance = 5.2, standard deviation = 2.28, mean absolute deviation = 1.8, IQR = 3.

5. (a)

Ms Rider's Class
Masses in Kilograms

(b)

Ms Rider's Class
Masses in Kilograms

3	99
4	001122223345678999 4|0 represents 40 kg

(c)

Ms Rider's Class
Masses in Kilograms

Mass	Tally	Frequency
39	||	2
40	||	2
41	||	2
42	||||	4
43	||	2
44	|	1
45	|	1
46	|	1
47	|	1
48	|	1
49	|||	$\frac{3}{20}$

(d)

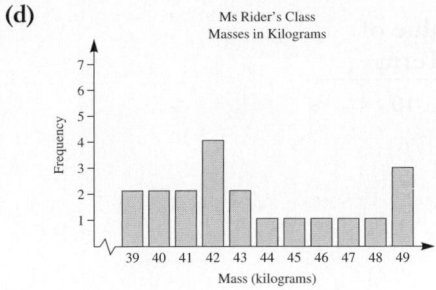

Ms Rider's Class
Masses in Kilograms

6. (a) Test Grades

Classes	Tally	Frequency
61–70	|||||	6
71–80	|||||||||	11
81–90	|||||||	7
91–100	|||||	6
		30

(b)

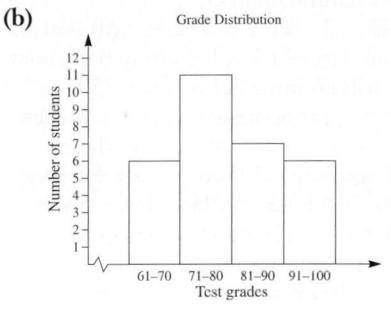

Grade Distribution

7.

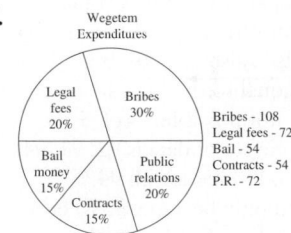

Wegetem
Expenditures

Bribes - 108
Legal fees - 72
Bail - 54
Contracts - 54
P.R. - 72

8. The widths of the bars are not uniform.   **9.** $2840
10.

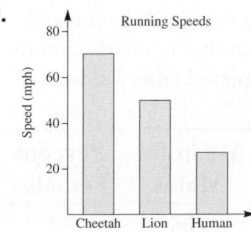

Running Speeds

11. (a) Life Expectancies of Males and Females

Females		Males
	67	1446
	68	28
	69	156
	70	0049
	71	02235578
	72	01145
	73	1689

Females		Males
7	74	123599
9310	75	14
86	76	
88532	77	
9999854332211	78	
99655443210	79	
42	80	

7 | 74 | represents | 67 | 1 represents
74.7 years old   67.1 years old
(b)

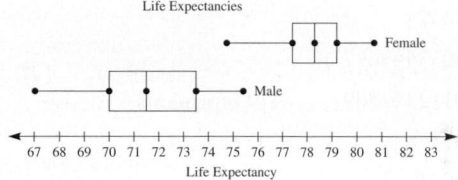

Life Expectancies

(c) 75% of the women's data is significantly above that of men; women are expected to live longer.   **12.** Larry was correct because his average was $3.2\overline{6}$, while Marc's was $2.7\overline{3}$.   **13. (a)** 360
(b) None   **(c)** 350   **(d)** $s \approx 108.2$ **(e)** 320   **(f)** 200
(g) About 11,711.11   **(h)** $84.\overline{4}$ **14. (a)** 67 mph
(b) $Q_3 = 74$, $Q_1 = 64$

(c)

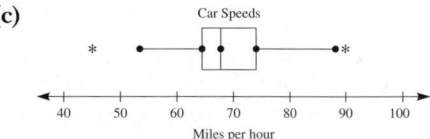

Car Speeds

(d) 30%   **(e)** About 50% of people checked drove between 65 and 75 mph.   **(f)** 64; 74; 67   **15. (a)** Positive   **(b)** 170 lb
(c) 67 in.   **(d)** 64 in.   **(e)** 50 lb   **16. (a)** Collette is more consistent. There is less deviation from the mean in her scoring.
(b) Collette: 18–30; Rudy: 10–38   **(c)** Collette scored more than might have been expected.   **17.** 2.5%   **18. (a)** The score for each   **(b)** Same as (a)   **(c)** same as (a)   **(d)** 0   **(e)** 0   **19.** The bars would be approximately the same height in a vertical bar graph.   **20. (a)** 19 lb   **(b)** It is nearly impossible for this to happen so the probability is near 0.   **(c)** Since the company claims a person can lose up to 6 lbs, any average loss from 0 to 6 is legitimate.   **21.** The bar graph is more appropriate. Line graphs are used to show change over time. The line graph indicates there are values between the colors, which is not true.   **22.** No reportable single number would be meaningful for such a pool.   **23.** Answers vary. A reasonable way to make the claim is to pull boxes at random after they are packed and weigh the contents. If more than 98% of the boxes' contents have weights within 1 g of 138 g, the claim is reasonable. The choices of 98% and 1 g are arbitrary but reasonable choices to test this claim.   **24. (a)** Answers vary. Perhaps to be with those of comparable age a typical senior might choose the first.   **(b)** The first   **25.** Answers vary. The weights are hard to read and the three dimensions distort the data. The graphs compare volume not area. Also, the data starts at 110 with no squiggle to show break.   **26.** Answers vary.   **(a)** One way would be to leave the television on, even if no one is watching.   **(b)** They show very popular shows during "ratings sweeps" periods.   **27.** Answers vary. For example, graphs may show area or volume instead of relative size; another is to select a horizontal

baseline that will support the point being made. **28. (a)** About 3 billion **(b)** About 6 billion **(c)** The population doubled. **(d)** About 9.3 billion **(e)** About 55% **29. (a)** 25 **(b)** 475 **(c)** 0.16 **30.** 475 **31. (a)** 525 **(b)** 600 **(c)** 675 **32.** The percentages in the circle graph sum to 193% instead of 100%.

Answers to Now Try This

10-1. Quick Check: 2a-biased, 2b-fair. Check your understanding: 1. C; 2. A; 3. B; 4. population is seventh-graders and sample is 20 students at the soccer game 5. was not 6. biased **10-2.** Ages of Presidents at Death

4	69
5	36778
6	003344567778
7	0112347889
8	01358
9	0033

4|9 represents 49 years old

10-3. It appears that the fifth-period class did better. We can see that there are more scores grouped toward the bottom, which is where the higher grades are located.

10-4. (a) 2.

Number of Students in Math Classes

3.

Age	Tally	Frequency
20–29	IIII III	8
30–39	IIII	4
40–49	III	3
50–59	II	2
60–69	II	2
70–79	I	1

Ages of Employees

(b) Yes, the interval 0–2 should be represented on the graph or a break shown as in part (a) above. **10-5.** The implication is that the data are continuous. **10-6.** Answers will vary with the number of terms.

(a)

# of Term	Value of Term
1	⁻10
2	⁻6
3	⁻2
4	2

(b)

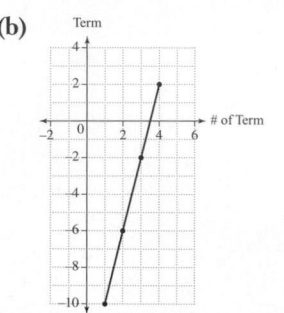

(c) $y = 4x - 14$, where x is a natural number.

10-7. (a) $y = 212.5x$ **(b)** 425 cal **(c)** $y = 112.5x$ **(d)** 450 cal **(e)** About 1190 cal **(f)** From 0 to 1 hr it will go from 0 to about 112 then from 1 hr to 4 hr it will go from 112 to about 750.

10-8. (a) A histogram is more appropriate to compare numbers of data that are grouped into numerical intervals. With the line graph there appears to be a frequency for *Times per month* such as 4.5, etc. **(b)** Connecting the dots is visually okay, but is mathematically meaningless. **(c) (i)** A circle graph is appropriate for showing the division of a whole into parts. **(ii)** A line graph is appropriate for showing how data values change over time.

10-9. (a) Answers vary. For example, all the puppies could weigh 7 lb. Then $(6 \cdot 7)/6 = 7$ lb. Another possibility is 4, 6, 6, 7, 8, 9, 9. The mean for these weights is also 7. **(b)** If all the scores were equal, then the mean is equal to the high score and also the low score. Otherwise, this is not possible. **(c)** 5

10-10. Answers vary. For example, one set of data is 92, 94, 94, 94, 96, in which the mean, median, and mode are all 94.

10-11. (a) The average of 2.58 could only be a mean. To be a mode it would have to be a whole number. To be a median it would have to be a whole number or a whole number plus $\frac{1}{2}$. **(b)** The average of 2.5 could be a mean or a median. It could not be a mode since the mode for the number of children must be a whole number. **10-12.** The completed table follows:

	Overall Mean	Mean for Females	Mean for Males	Percent Females
Ramirez	218	230	205	52
Jonsey	221	224	199	88

The percentage of males in Ms. Jonsey's class is 100% − 88%, or 12%. Also the mean for Ms. Jonsey's class is determined as follows:

$$(88\%)(224) + 12\%(\text{Mean for males}) = 221$$

Solving, we find that the mean for males in Ms. Jonsey's class is 199. For the entry for Mr. Ramirez's class, we know that the percentages of males and females must sum to 100%. If F is the percentage of females, then the percentage of males in that class must be $1.00 - F$. Further, we know that the overall mean is determined as follows:

$$(F)(230) + (1.00 - F)(205) = 218$$

Solving for F, we find that $F = 52\%$.

Finally, we observe that the means for both females and males were higher in Mr. Ramirez's class, but the overall mean is higher in Ms. Jonsey's class.   **10-13.** One possible set is 0, 0, 40, 40, 40, 70, 70, 80, 90, 90, 100.   **10-14. (a)** For the compact car data, 48 is an outlier. There are no outliers for the midsize car. For the SUV cars scores of 27, 28, and 30 are outliers.

(b)

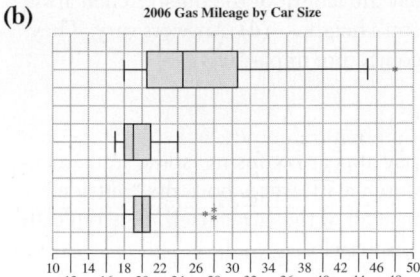

2006 Gas Mileage by Car Size

10-15. Because there is no base for the percentages given, arguments are difficult to make. Assuming the base is the set of all parents, consider the following: **(a)** The argument cannot be made because we do not know even if the other 82% answered this part of the survey.   **(b)** The argument cannot be made. The statement may be the result of considering test/learning to be two equal parts that make 18%. Hence, half of 18% is 9%.   **(c)** The argument cannot be made; there is no cause-effect relationship shown.   **(d)** This argument cannot be made. Possibly it could be made for 11%.   **(e)** The argument cannot be made. The indicated 9% does not tell us about the other 91% of parents.

Answers to Brain Teasers

Page 604

We know the sum of the ages of the first seven guests divided by 7 is 21. Therefore, the sum of the ages is $7 \cdot 21 = 147$. If we add another age of 29 to the sum, we have $147 + 29 = 176$. If we divide by 8, we see the mean is now 22. When another 29-year-old enters the room, the sum of the ages increases to $176 + 29 = 205$ and we now have nine people. If we let the honoree's age be represented by x, then the sum of the ages of the ten people now present is $205 + x$. If we divide this sum by 10, then we would obtain the mean of 27. If we solve the equation for x, we will have the honoree's age.

$$\frac{205 + x}{10} = 27$$
$$205 + x = 270$$
$$x = 65$$

Therefore, the honoree is 65 years old.

Page 622

The sum of Ouida's current test scores is 492. We divide by 6 to find that the average (mean) is 82. If she replaces the highest and

lowest scores with one score, she will have only five scores. If the final exam score is represented by x, the sum of the five scores is $326 + x$. If we solve the equation, we will have the needed score.

$$\frac{326 + x}{5} = 82$$
$$326 + x = 410$$
$$x = 84$$

Ouida needs to earn 84 points on the final exam to maintain her average.

Answer to the Preliminary Problem

The first person has the calculator and picks a large random number. The first person then records the number, adds his salary to it, and passes the calculator to the second person. That person adds his salary and passes it to the next person. This continues for all five people. When the calculator is returned to the first person, he subtracts the random number that was used and divides by 5 to obtain the mean salary.

Chapter 11

Assessment 11-1A

1. (a) 12 ways   **(b)** 60 ways   **2. (a)** Answers may vary. For example: $\overrightarrow{BC}$ and $\overleftrightarrow{DH}$ or $\overrightarrow{AE}$ and $\overrightarrow{BD}$ **(b)** Parallel   **(c)** No   **(d)** Empty set   **(e)** In plane FHG, consider $\overrightarrow{FK}$ (not shown) perpendicular to $\overleftrightarrow{FH}$. In plane BDH we have $\overleftrightarrow{FB} \perp \overleftrightarrow{FH}$. Because $\overleftrightarrow{FB}$ is perpendicular to the floor, $\overleftrightarrow{FB} \perp \overleftrightarrow{FK}$ and hence the dihedral angle between the planes measures 90°. Thus, the planes are perpendicular.   **3. (a)** $\varnothing$ **(b)** $\{C\}$ **(c)** $\{A\}$ **(d)** Answers vary; for example, $\overleftrightarrow{AC}$ and $\overleftrightarrow{BE}$ **(e)** $\overleftrightarrow{AC}$ and $\overleftrightarrow{DE}$ or $\overleftrightarrow{AD}$ and $\overleftrightarrow{CE}$ **(f)** Plane BCD or plane BEA **4.** 14   **5.** Answers vary.   **6. (a)** 110°   **(b)** 40°   **(c)** 20°   **(d)** 130°   **7. (a) (i)** 41°31'10"   **(ii)** 11°44'   **(b) (i)** 54'   **(ii)** 15°7'48"   **8. (a) (i)** 90°   **(ii)** 12°30'   **(iii)** 205°   **(b)** 52°30'   **9. (a)** $m(\angle AOB) = 22.5°$, $m(\angle COD) = 67.5°$ **(b)** $m(\angle BOC) = 31.25°$, $m(\angle AOB) = 58.75°$ **(c)** $3x + 3y = 180$, $x + y = 60$, $m(\angle BOC) = 60°$. If we change the position of $\overrightarrow{OE}$, $3x$ and $3y$ will have different values and hence x and y will have different values, but $3x + 3y$ remains 180° and therefore $x + y = 60° = m(\angle BOC)$.   **10. (a)** 6   **(b)** 8   **(c)** $2(n - 1)$ **11. (a)**

		0	1	2	3	4	5
Number of Lines	2			Not possible	Not possible	Not possible	Not possible
	3					Not possible	Not possible
	4				Not possible		
	5				Not possible	Not possible	
	6				Not possible	Not possible	Not possible	

Number of Intersection Points

(b) $n(n - 1)/2$ **12.** Answers vary.   **13.** Answers vary.   **14.**   **15. (a)** 70°   **(b)** 110°   **(c)** 70°   **(d)** 110°   **(e)** 180°

Mathematical Connections 11-1

Communication

1. (a) Answers vary. **(b)** The fire is located near the intersection of the two bearing lines. **(c)** Answers vary. **3.** Yes. Through P in plane α, draw a line k perpendicular to $\overleftrightarrow{AB}$. By definition, since n is perpendicular to α it is perpendicular to every line in α through P. Thus n and k are perpendicular, which implies that planes α and β are perpendicular.

Cooperative Learning

5. (a) Answers vary. An angle of 20° can be drawn by tracing a 50° angle and a 30° angle, as shown in the following figure. Another 20° angle adjacent to the first 20° angle can be drawn in a similar way, thus creating a 40° angle.

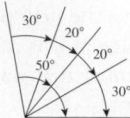

(b) Yes; answers vary. **(c)** All positive multiples of 10°; once a 10° angle is determined using the 40° angle from (a) and the given 30° angle, the other multiples can be determined.

Open-Ended

7. Answers depend on students.

Questions from the Classroom

9. It is a matter of definition. Two distinct lines are parallel if they do not intersect and are in the same plane. Lines that do not intersect and do not lie in a single plane are skew lines. **11.** Extending the rays does not change the angle measure or the angle's opening. The angle measure is a measure of the opening not length. **13.** The carpenter can check if the dihedral angle is 90° by placing a well-constructed box on the floor and moving it as close to the wall as possible. If an entire face of the box touches the wall, then the wall is perpendicular to the floor.

Assessment 11-2A

1. (a) 1 cm **(b)** 1 cm **(c)** 8 cm **(d)** 0.5 cm **(e)** 0.7 cm
(f) 5 cm **(g)** 7.3 cm **(h)** 5.2 cm **2.** Answers vary. **3. (a)** $2\frac{7}{9}$ yd
(b) 14,400 in. **(c)** 100 yd **(d)** 31 ft **4.** Answers vary.
5. (a) Approximately 75 mm **(b)** Approximately 7.5 cm
6. (a) Centimeters **(b)** Centimeters or millimeters **(c)** Centimeters or meters **7. (a)** Inches **(b)** Inches **(c)** Feet or inches
8. (a) 0.35; 350 **(b)** 163; 1630 **(c)** 0.035; 3.5 **(d)** 0.1; 10
(e) 200; 2000 **9. (a)** 13.50 **(b)** 0.770 **(c)** 10.0 **(d)** 15.5
10. 6 m; 5218 mm; 245 cm; 700 mm; 91 mm; 8 cm **11.** Answers vary. **12. (a)** π **(b)** $\frac{\pi}{2}$ units **13. (a)** Approximately 7 cm
(b) Approximately 11 cm **14. (a)** 1 cm **(b)** 0.262 km
(c) 3000 m **(d)** 0.03 m **15. (a)** $AB + BC > AC$
(b) $BC + CA > AB$ **(c)** $AB + CA > BC$ **16. (a)** Yes **(b)** No
17. (a) Approximately 32.1 in. **(b)** Approximately 45.6 in.
18. (a) 6 cm **(b)** $\frac{3}{\pi}$ m **19. (a)** 6π cm **(b)** 4 cm

20. The circumference will double. **21. (a)** 3096 km/hr
(b) 1032 m/sec **(c)** Mach 4.04 or about Mach 4 **22. (a)** 0.5 m
(b) 0.05 cm **(c)** 0.005 m or 5 mm

Mathematical Connections 11-2

Communication

1. Answers vary, but the accomplishment was not trivial.
3. The outer curve has a greater radius and a correspondingly greater distance (that is, arc length) to run. To compensate for the extra distance, the runner in the outer lane is given an apparent head start.

Cooperative Learning

5. (a) Answers vary. **(b)** Answers vary. **(c)** Answers vary. Students should conjecture that the length of the outstretched arms is about the same as a person's height. **(d)** Answers vary. They should all agree with the conjecture in part (c).

Questions from the Classroom

7. Though the question may appear confusing, some students may think that the circumference measures both the "outside" and "inside" of the curve instead of the curve itself. One typically discuses the length of the semicircle.

Review Problems

9. Yes; see the drawing.

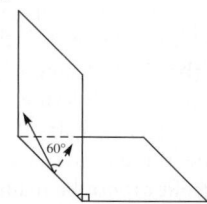

11. Perpendicular lines are intersecting lines, and two intersecting lines determine a plane.

Assessment 11-3A

1. (a) 1, 4, 6, 7, 8 **(b)** 1, 6, 7, 8 **(c)** 6, 7 **(d)** 1, 8 **2.** 8
3. A concave polygon **4. (a)** and **(c)** are convex, since no matter which two points in the interior of the figure are chosen, the entire segment connecting the points lies within the figure; (b) and (d) are concave since two points within the figure can be chosen such that the segment formed by the two points does not lie entirely within the figure. **5.** Squares **6.** Drawings vary, but (c) and (d) are impossible because each angle of an equilateral triangle has measure 60°. **7. (a)** 5 **(b)** 35 **(c)** 170 **(d)** $\frac{n(n-3)}{2}$

8. (a)   **(b)**

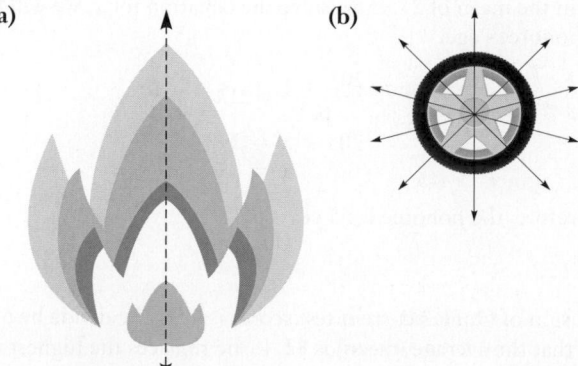

9. (a) Isosceles and equilateral **(b)** Isosceles **(c)** Scalene
10. (a) Has line symmetry, turn symmetry and point symmetry
(b) Has line symmetry only **11.** Answers vary.

12.

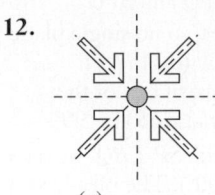   

(a)   (b)

13.

 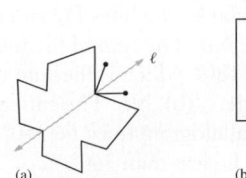

(a)   (b)

Mathematical Connections 11-3

Communication

1. (a) Answers vary. An $8\frac{1}{2} \times 11$-in. piece of rectangular paper requires only one fold. **(b)** Folding the square along one diagonal and then along the other, 4 congruent angles are formed. Each is $\frac{360°}{4}$ or $90°$. The paperfolding implies that the diagonals bisect each other. Unfolding and then folding the square so that opposite sides fall onto each other shows that the diagonals are congruent.

Cooperative Learning

3. Answers vary. **5. (a)** Reuleaux triangles aren't truly triangles. They are curves of constant width. **(b)** Reuleaux triangles have equal-sized arcs, whereas equilateral triangles have sides of the same length.

Questions from the Classroom

7. Cut the cylinder open along a line perpendicular to the base and through point B. The shortest path between A and B is the segment connecting A and B. Now fold the rectangle back into the cylinder; the path will appear on the cylinder. **9.** A square is a particular kind of rectangle in which all sides have equal measures. All squares are rectangles.

GSP/GeoGebra Activities

11. See online appendix.

Review Problems

13. $\varnothing$; 1 point; 2 points; ray **15. (a)** 0.007 m **(b)** $\frac{17}{36}$ yd
(c) 400 cm **(d)** 61.2 in.

Assessment 11-4A

1. 6 **2. (a)** $60°$ **(b)** $45°$ **(c)** $60°$ **(d)** $60°$ **3. (a)** Yes. A pair of corresponding angles are $50°$ each. **(b)** Yes. A pair of corresponding angles are $70°$ each. **(c)** Yes. A pair of alternate interior angles are $40°$ each. **(d)** Yes. A pair of corresponding angles are $90°$ each. **(e)** Yes. A pair of alternate exterior angles

are congruent making a pair of alternate interior angles congruent. **(f)** One way is to extend $\overline{AC}$ to intersect n at D. Because x is the measure of an exterior angle in triangle BCD, $x = z + w$ (x is a supplement of $\angle BCD$ and so is $z + w$). But $x = y + z$ so $z + w = z + y$ implying $w = y$. Because w and y are measures of congruent alternate interior angles, $m \parallel n$.

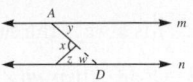

4. $70°$ and $20°$ **5.** 20 **6.** $360°$
7. $110°$, $114°$, $118°$, $122°$, $126°$, $130°$ **8. (a)** $70°$
(b) $70°$ **(c)** $65°$ **(d)** $45°$ **9. (a)** $x = 40°$ **(b)** $x = 18°$
(c) $x = 120°$ **10. (a)** $60°$ **(b)** $90°$ **11.** $360°$ **12.** $111°$
13. $m(\angle 1) = 60°$; $m(\angle 2) = 30°$; $m(\angle 3) = 110°$
14. (a) $100°$ **(b)** The measure of the exterior angle equals the sum of the measures of the remote interior angles. Since $\angle 1$ and $\angle ACB$ are supplementary, then $m(\angle 1) + m(\angle ACB) = 180°$. We also know that the sum of the measures of the angles of a triangle equals $180°$; that is, $m(\angle A) + m(\angle B) + m(\angle C) = 180°$. So $m(\angle 1) = m(\angle A) + m(\angle B)$.

15. $x = \dfrac{180°}{13}$ or $x \approx 13.846°$.

$y = \dfrac{360°}{13}$ or $y \approx 27.692°$.

$z = \dfrac{720°}{13}$ or $z \approx 55.385°$.

$w = \dfrac{1080°}{13}$ or $w \approx 83.077°$.

Mathematical Connections 11-4

Communication

1. (a) No. Two or more obtuse angles will produce a sum of more than $180°$. **(b)** Yes. For example, each angle may have measure $60°$. **(c)** No. The sum of the measures of the three angles would be more than $180°$. **(d)** No. It may have an obtuse or right angle as well. **3. (a)** Five triangles will be constructed in which the sum of the angles of each triangle is $180°$. The sum of the measures of the angles of all the triangles equals $5(180°)$, from which we subtract $360°$ (the sum of all the measures of the angles of the triangles with vertex P). Thus, $5(180°) - 360° = 540°$. **(b)** Here, n triangles are constructed, so we have $n(180°)$ from which we subtract $360°$ (the sum of all the measures of the angles of the triangles with vertex P). Thus, we obtain $n(180°) - 360°$ or $(n - 2)180°$. **5. (a)** Divide the quadrilateral into two triangles: $\triangle ABC$ and $\triangle ACD$. The sum of the measures of the interior angles of each triangle is $180°$, so the sum of the measures of the interior angles of the quadrilateral is $2(180°)$, or $360°$. **(b)** This is also true for nonconvex polygons. **(c)** Any concave pentagon can be divided into three triangles, and a concave hexagon can be divided into four triangles. In general, any concave n-gon can be divided into $n - 2$ triangles, so that the sum of the measures of the angles of the triangles is the sum of the measures of the interior angles of the n-gon. **7.** Yes. Because the sum of the measures of the marked angles is $360°$ and opposite angles are congruent we get $2m(\angle 2) + 2m(\angle 1) = 360°$ or $m(\angle 2) + m(\angle 1) = 180°$. From problem 3(a) of Assessment 11-4B, we get $a \parallel b$.

Open-Ended

9. Answers vary; you should get a multiple of 360°.

Cooperative Learning

11. (a) If $m(\angle A) = \alpha$ and $m(\angle B) = \beta$, then $m(\angle D) = 180 - \left(\dfrac{\alpha}{2} + \dfrac{\beta}{2}\right)$.   **(b)** If $m(\angle C)$ is always the same, no matter what the measures of $\angle A$ and $\angle B$ are, then $m(\angle D)$ is always the same.   **(c)** Answers vary. The following is a possible solution. From (a):

$$m(\angle D) = 180 - \left(\dfrac{\alpha + \beta}{2}\right) = \dfrac{360 - (\alpha + \beta)}{2}.$$ Because

$\alpha + \beta = 180 - m(\angle C)$, we get the following:

$$m(\angle D) = \dfrac{360 - (180 - m(\angle C))}{2}$$

$$= \dfrac{360 - 180 + m(\angle C)}{2}$$

$$= 90 + \dfrac{1}{2}m(\angle C)$$

Because the answer for $m(\angle D)$ depends only on $m(\angle C)$, the conjecture is justified.

Questions from the Classroom

13. The only possible figures are acute triangles. To see why, let an *n*-gon ($n \ge 4$) have the property that each angle is acute. Then all the exterior angles would be obtuse. A quadrilateral would have more than $90 \cdot 4$, or 360°, as the sum of the measures of the exterior angles and for $n > 4$ the polygon would have more than $90 \cdot 4$ or 360° as the sum of the measures of its exterior angles, which is not possible.   **15.** Rory is correct if only one exterior angle at a vertex is considered but incorrect if all are. With a diagram and to marking the angle measures is being added, Rory should be convinced that an error was made.

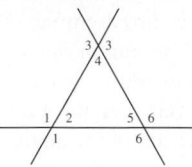

In the figure above, we have
$m(\angle 6) = m(\angle 2) + m(\angle 4)$
$m(\angle 1) = m(\angle 4) + m(\angle 5)$
$m(\angle 3) = m(\angle 2) + m(\angle 5)$
$m(\angle 6) + m(\angle 1) + m(\angle 3) = m(\angle 2) + m(\angle 4) + m(\angle 4)$
$\qquad\qquad\qquad\qquad\qquad + m(\angle 5) + m(\angle 2) + m(\angle 5)$
$\qquad\qquad\qquad\qquad = 2 \cdot 180 = 360$

Review Problems

17. (a) Two sets of parallel sides—*A, B, C, D, E, F, G*; one set of parallel sides—*A, B, C, D, E, F, G, I,* and *J* (at least one set of parallel sides); no parallel sides: *H*   **(b)** Four right angles—*D, F, G*; exactly two right angles—*I*   **(c)** Four congruent sides—*B, C, F, G*; at least two pairs of congruent sides *A, B, C, D, E, F, G, H*; at least one pair of congruent sides—*A, B, C, D, E, F, G, H, I*; no sides congruent—*I*   **19.** Answers vary.   **21. (a)** Rotational symmetries of 90°, 180°, 270° about the center of the large square   **(b)** 180° rotational symmetry about the center of the large square

Chapter Review

1. (a) $\overrightarrow{AB}$, $\overrightarrow{BC}$, and $\overrightarrow{AC}$ **(b)** $\overrightarrow{BA}$ and $\overrightarrow{BC}$ **(c)** $\overline{AB}$ **(d)** $\overline{AB}$ **2. (a)** $\overleftrightarrow{PQ}$ and $\overleftrightarrow{AB}$ **(b)** Planes *APQ* and *BPQ*   **(c)** $\overrightarrow{AQ}$ **(d)** No; $\overleftrightarrow{PQ}$ and $\overleftrightarrow{AB}$ are skew lines, so no single plane contains them.   **3. (a)** 132°11′   **(b)** 56°28′   **(c)** 132°21′   **(d)** 45′0″   **(e)** 35°8′35″   **4. (a)** The measure of one of the dihedral angles formed by planes α and γ is $m(\angle APS) = 90°$ because α and γ intersect in $\overleftrightarrow{PQ}$, $\overline{AP} \perp \overleftrightarrow{PQ}$ and $\overline{SP} \perp \overleftrightarrow{PQ}$ (since the dihedral angle *S–AB–Q* measures 90°). The measure of one of the dihedral angles formed by planes β and γ is $m(\angle BPQ) = 90°$ because line *PQ* is perpendicular to line *AB* (given).   **(b)** Line *AB* is perpendicular to the lines *PQ* and *PS* in plane γ.   **5.** Answers vary.   **6. (a)** No. The sum of the measures of two obtuse angles is greater than 180°, which is the sum of the measures of the angles of any triangle.   **(b)** No. The sum of the measures of the four angles in a parallelogram must be 360°. If all the angles are acute, the sum would be less than 360°.   **7.** 18°, 36°, 126°   **8.** 48°   **9.** 90 sides   **10.** $m(\angle 3) = m(\angle 4) = 45°$ **11. (a)** 60°   **(b)** 120°   **(c)** 120°   **12.** 55°   **13. (a)** 40°   **(b)** 55°   **14. (a)** $x = 50°$ and $y = 60°$ **(b)** $x = 83°$ **15.** $m(\angle 1) = m(\angle 2) = 65°; m(\angle 3) = m(\angle 4) = 115°; m(\angle 5) = 70°$.   **16. (a)** 540°   **(b)** From the sum of the measures of all the interior angles of the *n* triangles we need to subtract the measures of the angles with vertex at *P*, that is, 360°. We get $n180° - 360°$ or $(n - 2)180°$.   **(c)** Answers vary. One way is to connect *B* with *E* and *A* with *F*. We get two quadrilaterals and a triangle. Thus, the sum of all the measures of the interior angles of the polygon is $2 \cdot 360° + 180°$, or 900°.   **17. (a)** Alternate interior angles are congruent.   **(b)** Corresponding angles are congruent.   **(c)** $m(\angle B) + m(\angle C) + m(\angle BAC) = m(\angle BAD) + m(\angle DAE) + m(\angle BAC) = 180°$ **18.** If $a = b$, or $b = c$, or $a = c$, the proof is automatic. If none of $a = b$, or $b = c$, or $a = c$ are true, proceed as follows: Draw any line *t* intersecting the three lines. $a \parallel b$ implies $\alpha = \beta$. Next $b \parallel c$ implies $\beta = \gamma$. Consequently, $\alpha = \gamma$, which implies $a \parallel b$.

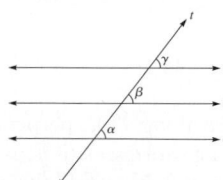

19. Out of context of a geometry course, assuming that a rectangle has four right angles and that a diagonal divides it into two triangles with the same angles, Wally's claim is correct. As explained in the problem, the sum of the measures of the interior angles in each right triangle is 180°. Therefore, the sum of the measures of the angles in $\triangle ACD$ and $\triangle BCD$ is $2 \cdot 180°$, or 360°. In this sum all the angles of the original triangle are included as well as the two right angles at *D*. Hence the sum of the measures of the angles in the original triangle is $360° - 2 \cdot 90°$, or 180°. On the other hand, to prove that a rectangle has four right angles requires the use of the property that if lines are parallel and cut by a transversal then corresponding angles are congruent. (This last property is also required to prove that the sum of the measures of the interior angles in a triangle is 180°.) Thus Wally's reasoning is incomplete.

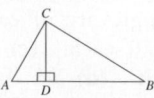

20. One approach is to draw through A a line c parallel to a and then prove that it is parallel to b. If $c \parallel a$, then $\alpha_1 = z$, as these are corresponding angles. Since $x = \alpha_1 + \beta_1$ and also $x = \alpha_1 + \beta_1 = y + z$, we get $y = \beta_1$, which implies that $c \parallel b$.

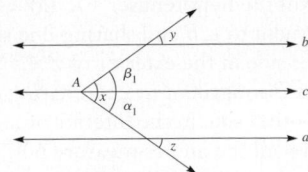

21. (a) $\angle 3$ and $\angle 4$ are supplements of congruent angles. (b) $m(\angle 3) > m(\angle 4)$. Reason: $m(\angle 3) = 180° - m(\angle 1)$ and $m(\angle 4) = 180° - m(\angle 2)$. Because $m(\angle 1) < m(\angle 2)$, $^-m(\angle 1) > {}^-m(\angle 2)$. After adding $180°$ to both sides of the last inequality, the result follows. **22.** (a) $30°$ (b) $60°$ (c) $240°$ **23.** Answers vary. **24.** (a) $_{10}C_3 = \dfrac{10 \cdot 9 \cdot 8}{3!} = 120$. The number of triangles equals the number of ways of choosing 3 points out of 10 points where order is not important.
(b) $_nC_3 = \dfrac{n(n-1)(n-2)}{6}$ **25.** (a) $\dfrac{360}{20}$, or 18 (b) Does not exist; $25 \nmid 360$. (c) Does not exist; the sum is always $360°$. (d) Does not exist; the equation $\dfrac{n(n-3)}{2} = 4860$, or $n(n-3) = 9720$, has no solution in integers because if $n = 100$ then $100 \cdot 97 = 9700 < 9720$ and if $n = 101$ then $101 \cdot 98 = 9898 > 9720$. **26.** (a) 4 (b) 1 (c) 1 (d) None (e) 2 (f) 2 **27.** (a) Line and turn (b) Line, turn, and point (c) Line **28.** Answers vary. **29.** (a) $60°$ (b) $30°$ (c) $30°$ (d) $60°$ (e) $60°$ (f) $120°$ **30.** (a) $16\dfrac{2}{3}$ yd (b) $\dfrac{947}{1760}$, or approximately 0.538 m. (c) 3960 ft (d) $9\dfrac{25}{36}$ or approximately 9.694 yd (e) 5000 m (f) 1.65 m (g) 520 mm (h) 0.125 km **31.** (a) Not possible, $p > q + r$ (b) Not possible, $p = q + r$

32. 340 cm **33.** $\dfrac{3}{2\pi}$ m **34.** $\dfrac{400°}{\pi}$ **35.** She is incorrect. The circumference of a circle with radius 6 cm is 12π cm.

Answers to Now Try This

11-1. (a) They do not have to be. Consider any point D such that $AD + DB > AB$. Point A, B, and D cannot be collinear. (b) No, the points must be collinear. (c) (i) $\overleftrightarrow{AB}$ (ii) $\overline{AB}$ (iii) $\overleftrightarrow{AB}$ (iv) $\overrightarrow{AB}$ (v) $\overleftrightarrow{AB}$ (vi) $\overleftrightarrow{AB}$ **11-2.** $n + 1$ **11-3.** The problem gives the number of lines that can be drawn through n points. This problem is a model for the number of handshakes that take place if everyone shakes hands with everyone else. Everyone shakes hands with everyone except himself. Because we don't want to count one handshake twice, we need to divide by 2. For n people there will be $\dfrac{n(n-1)}{2}$ handshakes. For 20 people there will be $\dfrac{20 \cdot 19}{2} = 190$ handshakes. **11-4.** $\dfrac{1}{2}n^2 + \dfrac{1}{2}n + 1$ is a correct conjecture. **11-5.** (a) No. If skew lines had a point in common there would be a single plane that contains the lines which would contradict the definition of skew lines. (b) Skew lines cannot be parallel. By definition, parallel lines are in the same

plane. Skew lines are not. (c) Lines n and o are parallel; lines n, m, and ℓ are skew as are lines o and m.
11-6. $8.42° = 8° + 0.42(60') = 8°25.2' = 8°25' + (0.2)(60'') = 8°25'12''$ **11-7.** Answers vary. **11-8.** (a) By folding a crease onto itself, we have created two angles that are both supplementary and congruent. They must have measure $90°$, making the creases perpendicular. (b) Construction (c) One obtains a $45°$ angle by bisecting a $90°$ angle. The $135°$ angle is $90° + 45°$. The $22°30'$ angle is obtained by bisecting the $45°$ angle. (d) The least angle measures $90°$. (e) The least angle measures $45°$.
(f) The least angle measures $\dfrac{360°}{2^n}$. **11-9.** (a) It is possible for a line intersecting a plane to be perpendicular to only one line in the plane. Drawings will vary. (b) It is not possible for a line intersecting a plane to be perpendicular to two distinct lines and not be perpendicular to the plane. (c) Yes. If a line intersects a plane in point P and is perpendicular to two lines in the plane through P, then it is perpendicular to every line in the plane through P. (d) ℓ is perpendicular to α. **11-10.** (a)–(e) Answers vary. (f) One would expect measures to be different because of different hand sizes. **11-11.** (a) Decidollar (b) Centidollar (c) Dekadollar (d) Hectodollar (e) Kilodollar **11-12.** (a) 115 cm (b) 55 cm **11-13.** Answers vary but the average should be approximately 3.1. The line should have an equation approximately $y = 3.1x$. The slope is 3.1. **11-14.** Answers vary. For example, the Greek letter α. **11-15.** One approach to this problem is to start shading the area surrounding point X. If we stay between the lines, we should be able to decide whether the shaded area is inside or outside the curve. The shaded part of the following figure indicates that point X is located outside the curve.

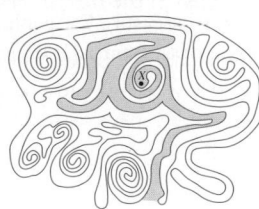

11-16. Answers vary.
(a)   (b)

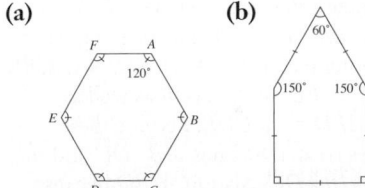

11-17. (a) The statement fits the hierarchy. (b) (1) True (2) True (3) True (4) True (5) True (6) True (7) True (8) Flase. **11-18.** Answers vary, but a square will do. **11-19.** A square is a quadrilateral with exactly four lines of symmetry: the two lines containing the diagonals and the two lines containing the midpoints of the opposite sides. **11-20.** (a) A 180-sided regular polygon has the desired symmetry. (b) A square has turn symmetries of $90°$, $180°$, and $270°$. **11-21.** (a) The sum of the measures of the interior and exterior angles is $n180°$. Therefore the sum of the measures of the interior angles is $n180° - 360°$ or $(n-2)180°$. (b) $\dfrac{(n-2) \cdot 180°}{n}$ or $180° - \dfrac{360°}{n}$

11-22. (a) 720° **(b)** 36°

11-23. $V = 60, E = 90, F = 32$ and
$$V - E + F = 60 - 90 + 32$$
$$= 2$$

Answer to the Preliminary Problem

If we think of AB as the length of one side of the triangle, $BC + CD$ as the sum of the lengths of two sides of the square used, $DE + EF + FG$ as the sum of the lengths of three sides of the pentagon used and continue in this manner, we find that we are looking at the sum $1 + 2 + 3 + 4 + 5 + \ldots + 14$. This is the sum of a finite arithmetic sequence. Thus,

$$1 + 2 + 3 + 4 + 5 + \ldots + 14 = \frac{14(14 + 1)}{2} = 105. \text{ Thus,}$$

the length of the path is 105 m.

Answers to Brain Teasers

Section 11-2

We could walk under the wire because it will be over 3 m above the Earth.

Section 11-4

The problem can be solved by walking around the star, but a simpler approach is to notice that the vertices of the star can be obtained by extending pairs of sides of a regular 12-gon. Choosing any side and a fifth side of the 12-gon, we obtain an isosceles triangle whose base is the longest diagonal of the 12-gon and whose base angles can be shown to measure 75° each. Thus, the measure of a marked interior angle of the star is $180° - 2 \cdot 75°$ or 30°.

Chapter 12

Assessment 12-1A

1. It is not possible if the 3 points are collinear. If the 3 points are not collinear then the point is the center of the circumcircle. **2.** For obtuse triangles **3. (a)** $m(\angle A) > m(\angle B)$ **(b)** The side of greater length is opposite the angle of greater measure. **4.** $\triangle ADB \cong \triangle CBD$ by *HL* because $\overline{AB} \cong \overline{CD}, \overline{DB} \cong \overline{BD}$, and $\angle ADB$ as well as $\angle CBD$ are right angles. Thus $\angle ABD \cong \angle CDB$. Because these angles are alternate interior angles created by lines $\overleftrightarrow{AB}, \overleftrightarrow{DC}$ and the transversal $\overleftrightarrow{BD}$, it follows that $\overline{AB} \parallel \overline{DC}$. Also $\overline{BC} \parallel \overline{AD}$ because $\angle ADB \cong \angle CBD$ (alternate interior angles). **5. (a)** Construct a triangle congruent to $\triangle CAB$ or an isoceles triangle congruent to any isoceles triangle with vertex angle at A. **(b)** First construct a segment AB of length 4 (or 2 or 3) then find point C as the intersection of circles centered at A and B with respective radii of 2 cm and 3 cm. **(c)** A scalene right triangle **(d)** Not possible **(e)** Proceed as in part (b). **(f)** One way is to use a protractor to construct the 75° angle and then mark the 6 cm and 7 cm sides on the sides of the angle. **(g)** Use a protractor to construct the 40° angle and mark its vertex as A. Then on one of the sides mark point B such that $AB = 7$ cm. Then let C be the intersection of the other side of the angle with a circle centered at B with radius 6 cm. Two triangles are possible. **(h)** Either proceed as in part (g) or construct a triangle with two sides of 6 cm each and included angle of 100°. **(i)** Construct a right angle and then mark on its

sides legs of 4 cm and 8 cm. **(j)** (b) Yes, by SSS, (c) Yes, by SSS, (d) There is no triangle because $4 + 5 < 10$ (e) Yes, by SSS, (f) Yes, by SAS (g) The triangle is not unique (h) Yes, by SAS (i) Yes, by SAS **6.** Construct a right isosceles triangle and the perpendicular bisector of the hypotenuse. **7. (a)** Construct an angle with vertex A congruent to $\angle B$ and sharing one side with a side of $\angle A$ and the other side in the exterior of $\angle A$. **(b)** Construct an angle with vertex B congruent to $\angle A$, sharing one side with a side of $\angle B$ and the other side in the interior of $\angle B$. **8.** No, if a triangle is equilateral all the angles measure 60°. **9. (a)** Construction. The triangles ABO, BCO, CDO, and DAO are congruent (SAS) isosceles right triangles. Therefore the congruent angles in each triangle measure 45°. Consequently all the angles in $ABCD$ are 90° and all the sides are congruent. **(b)** Because the arcs $\overarc{BE}$ and $\overarc{EC}$ each measures 45°, the chords BE and EC are congruent. Bisecting each of the right central angles we get the vertices of the regular octagon. **(c)** Answers vary. Construct F such that $CF = BE$. Then extend $\overline{OE}$ to intersect the circle at G. Next extend $\overline{OF}$ to intersect the circle at H. Now connect E, C, F, D, G, A, H, B to obtain a regular octagon. **10.** Answers vary. Fold the squares in half so that one pair of opposite sides fall on each other. Unfold and repeat for the other pair of opposite sides. Where the two folds intersect is the required center. The segment connecting the center to one of the vertices of the square is the radius. **11. (a)** 6 **(b)** 24 **(c)** Point C can be found as the intersection of the line and the perpendicular bisector of $\overline{AB}$. **12.** The center of the circle is the intersection of the perpendicular bisectors of any two adjacent sides of the square. The radius of the circle is the distance from the center to any of the vertices of the square. Alternatively the center can be found as the intersection of the diagonals. **13. (a)** Yes **(b)** Yes **(c)** No **14. (a)** Construct point C as the intersection of two circles with centers at A and B and radius AB. **(b)** Any of the angles of the triangle in part (a) will measure 60°. **(c)** Any of the exterior angles of the triangle in part (a) will measure 120°. **15. (a)** $\triangle CAB \cong \triangle CBA$ **(b)** From part (a) $\angle A \cong \angle B$ by CPCTC. **16.** No. This is possible if, and only if, the three points are not colinear.

Mathematical Connections 12-1

Communication

1. P is not on the circle because it doesn't fit the definition. **3. (a)** $\triangle ABC \cong \triangle ADC$ by SSS. Hence, $\angle BAC \cong \angle DAC$ and $\angle BCM \cong \angle DCM$ by CPCTC. Therefore, $\overleftrightarrow{AC}$ bisects $\angle A$ and $\angle C$. **(b)** The angles formed are right angles. By part (a), $\angle BAM \cong \angle DAM$. Hence, $\triangle ABM \cong \triangle ADM$ by SAS. $\angle BMA \cong \angle DMA$ by CPCTC. Since $\angle BMA$ and $\angle DMA$ are adjacent congruent angles, each must be a right angle. Since vertical angles formed are congruent, all four angles formed by the diagonals are right angles. **(c)** By part (a) and SAS, $\triangle BAM \cong \triangle DAM$. Hence, $\overline{BM} \cong \overline{MD}$ by CPCTC **5.** No. If the diagonals of a rectangle are perpendicular it is a square. If they are not perpendicular the rectangle is not a square. In each case the diagonals can be 19 in. long. **7. (a)** $\overline{CB} \cong \overline{CD}$ since each is 2 units long. $\overline{AB} \cong \overline{AD}$ because each can be viewed as a hypotenuse of two congruent right triangles with legs 1 and 3 units long. **9. (a)** Following the directions in the problem after points A and B are constructed, we construct points C and D so that $AC = BD = r$ where r is the radius of the circle. We next construct point E so that $CE = r$. If we now draw an arc with center D and radius r we find that the arc intersects the circle at the existing point E. Thus only six points are determined.

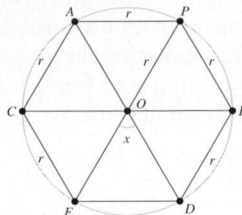

(b) The figure is a regular hexagon. To see why notice that the fact that the arc with center D and radius r intersects the circle at the existing point E is an experimental observation. However, rather than drawing that arc we can just connect E with D and prove that $ED = r$. From the above construction it follows that the five triangles: $\triangle OBP$, $\triangle OPA$, $\triangle OAC$, $\triangle OCE$, and $\triangle OBD$ are equilateral and therefore all their angles measure 60°. Consequently $x = m(\angle EOD) = 360° - 5 \cdot 60° = 60°$. Because $\triangle OED$ is isosceles, $m(\angle OED) = m(\angle ODE)$ and therefore each of these angles measures $\dfrac{180° - 60°}{2}$ or 60°. Thus $\triangle EOD$ is equilateral and therefore $ED = r$. Because the angles at the vertices of the hexagon are each 60° + 60°, or 120°, we obtain a regular hexagon.

Open-Ended

11. Answers may vary. One possible pattern is the stamp-block quilt which is a quilt constructed entirely of squares.
13. (a) $\triangle AFB \cong \triangle CED$ by SAS because $\overline{AF} \cong \overline{CE}, \overline{FB} \cong \overline{ED}$, and $\angle F \cong \angle E$ (each is a right angle). Hence, $\overline{AB} \cong \overline{CD}$ (corresponding parts in congruent triangles).   **(b)** Because $\triangle AFB$ is a right triangle, we have $m(\angle A) + m(\angle B) = 90°$. The congruence of triangles in part (a) implies that $\angle A \cong \angle C$. Hence, $m(\angle C) + m(\angle B) = 90°$. This implies that $\angle G$ in $\triangle BCG$ is a right angle and hence that $\overline{AB}$ and $\overline{CD}$ are perpendicular.

Cooperative Learning

15. (a) The right angle is the greatest angle in a right triangle.
(b) The distance is the length of the perpendicular segment from the point to the line. Any other segment will be a hypotenuse of a right triangle in which the perpendicular segment is a leg and hence shorter than the hypotenuse.

Questions from the Classroom

17. Some of the constructions that cannot be done using a compass and straightedge are angle trisection, duplication of a cube, and squaring the circle. Given any arbitrary angle, it is impossible in general with only a compass and straightedge to find two rays that divide the angle into three congruent angles. Some angles, but not all, can be trisected with straightedge and compass. For example, a right angle can be trisected. The duplication of a cube involves constructing the edge of a cube whose volume is twice the volume of a given cube. Squaring a circle involves constructing a square that has the same area as a given circle. For over 2000 years, mathematicians tried to perform these three constructions. In the nineteenth century, it was finally proved that these constructions cannot be done with straightedge and compass alone.   **19.** Using GSP or GeoGebra one could measure the

angles and find that they are not congruent. It is also possible to prove that if the three angles were congruent it would follow that $\angle EDC$ is obtuse, which is impossible because $\triangle CDE$ can be proved to be isosceles. For a detailed discussion of the trisection problem, see *The Trisection Problem* by Robert Yates (Washington, D.C.: NCTM Publications, 1971).   **21.** She is correct, if $AC = BC$ all corresponding parts are congruent.

GSP/GeoGebra Activities

See online appendix.

Assessment 12-2A

1. Constructions   **2. (a)** No; by ASA, the triangle is unique.
(b) No; by AAS, the triangle is unique. **(c)** Yes, the sides can be of any length.   **3. (a)** Yes; ASA   **(b)** Yes; AAS   **4.** When the parallel ruler is open at any setting, the distance $BC = BC$; it is given that $AB = DC$ and $AC = BD$, so $\triangle ABC \cong \triangle DCB$ by SSS. Hence, $\angle ABC \cong \angle DCB$ by CPCTC. Because these are alternate interior angles formed by lines $\overleftrightarrow{AB}$ and $\overleftrightarrow{CD}$ with transversal line $\overleftrightarrow{BC}$, then $\overline{AB} \parallel \overline{DC}$.   **5. (a)** Parallelogram   **(b)** None   **(c)** None
6. An isosceles trapezoid is formed and the angles formed with each base are congruent. The congruent angles are sometimes referred to as the base angles of the isosceles trapezoid. Other pairs of angles are supplementary.   **7. (a)** True   **(b)** True   **(c)** True
(d) False. A trapezoid may have only one pair of opposite sides parallel.   **8. (a)** Answers may vary.   **(b)** No. If the quadrilateral has three right angles, then the fourth must also be a right angle because the sum of the measures of the four angles is 360°.   **(c)** No; any parallelogram with a pair of right angles must have right angles as its other pair and hence it must be a rectangle.   **9.** Because $AB = BC$, B is equidistant from A and C. By Theorem 12-4, point B is on the perpendicular bisector of $\overline{AC}$. Similarly D is on the perpendicular bisector of $\overline{AC}$. Because two points determine a unique line, $\overleftrightarrow{BD}$ is the perpendicular bisector of $\overline{AC}$.
10. (a) Rectangle   **(b)** Isosceles trapezoid   **11.** When O, the center of the circumcircle is connected with the 6 vertices, 6 equilateral triangles result. A rotation about O by any multiple of 60° will map the vertices on themselves and hence there are 5 rotational symmetries by 60°, $2 \cdot 60°$, $3 \cdot 60°$, $4 \cdot 60°$, $5 \cdot 60°$.   **12.** A rhombus, because all the sides are congruent   **13.** Either the arcs or the central angles must have the same measure (radii are the same since the sectors are part of the same circle). An alternative condition is that the segments joining endpoints of each arc are congruent.   **14. (a)** Kite. $\overline{DX} \cong \overline{AX}$; $\angle D$ and $\angle A$ are right angles; $\overline{DP} \cong \overline{AQ}$. Thus, $\triangle PDX \cong \triangle QAX$ by SAS. Hence, $\overline{PX} \cong \overline{QX}$ by CPCTC. $\overline{DC} \cong \overline{BA}$ as opposite sides of a rectangle. $\overline{PD} \cong \overline{QA}$. Hence, $\overline{CP} \cong \overline{BQ}$. $\overline{CY} \cong \overline{BY}$, given; and $\angle C$ and $\angle B$ are right angles. Therefore, $\triangle CYP \cong \triangle BYQ$ by SAS, so $\overline{QY} \cong \overline{PY}$ and $PXQY$ is a kite.   **(b)** The answer does not change. However, when P and Q are midpoints of $\overline{DC}$ and $\overline{AB}$, respectively, $PXQY$ is a rhombus.   **15.** Make one of the quadrilaterals a square and the other a rectangle.   **16.** Construct the first kite by constructing a segment to become a diagonal and then construct two isosceles triangles with the segment as a common base. Construct the second kite starting with a segment not congruent to the first and construct two isosceles triangles with that segment as their common base but the sides congruent to the corresponding sides of the isosceles triangles in the first construction.   **17.** Rhombus; use SAS to prove that

$\triangle ECF \cong \triangle GBF \cong \triangle EDH \cong \triangle GAH.$ **18. (a)** The lengths of one side of each square must be equal. **(b)** The lengths of the two perpendicular sides of the rectangles must be equal. **(c)** Answers may vary; one solution is that two adjacent sides must have equal lengths and the included angle of one must be congruent to the other. **19. (a)** If in the accompanying figure $\overline{AD} \parallel \overline{BC}$ and $\overline{AD} \cong \overline{BC}$ then the pairs of alternate interior angles are congruent. That is, $\angle OBC \cong \angle ADO$ and $\angle BCO \cong \angle DAO$. Thus, $\triangle CBO \cong \triangle ADO$ by ASA. Consequently, $\overline{BO} \cong \overline{OD}$ and $\overline{AO} \cong \overline{OC}$ (by CPCT). Now $\triangle ABO \cong \triangle DCO$ by SAS as the angles at O are vertical. Hence, $\angle BAO \cong \angle DCO$. These angles are alternate interior angles created by $\overleftrightarrow{AB}, \overleftrightarrow{CD}$, and the transversal $\overleftrightarrow{AC}$. Thus $\overline{AB} \parallel \overline{CD}$ and by definition $ABCD$ is a parallelogram ($\overline{BC} \parallel \overline{AD}$ is given).

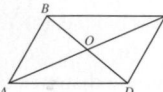

(b) In this figure $AO = OC$ and $BO = OD$. In addition $\angle BOC \cong \angle DOA$ are vertical angles. Thus by SAS, $\triangle BOC \cong \triangle DOA$. Consequently $\angle ADO \cong \angle CBO$. Because these are alternate interior angles created by lines $\overleftrightarrow{BC}, \overleftrightarrow{AD}$ and transversal $\overleftrightarrow{BD}, \overline{BC} \parallel \overline{AD}$. From the congruence of the triangles it also follows that $\overline{BC} \cong \overline{DA}$. Thus $ABCD$ is a quadrilateral with pair of opposite sides parallel and congruent and therefore a parallelogram. **20.** Answers may vary; for example, the polygons must have the same number of sides with one pair congruent (all regular polygons with the same number of sides are similar, so if they have the same number of sides with one pair congruent, they are congruent).

Mathematical Connections 12-2

Communication

1. The triangle formed by Stan's head, Stan's feet, and the opposite bank is congruent to the triangle formed by Stan's head, Stan's feet, and the spot just obscured by the bill of his cap. These triangles are congruent by ASA since the angle at Stan's feet is 90° in both triangles, Stan's height is the same in both triangles, and the angle formed by the bill of his cap is the same in both triangles. The distance across the river is approximately equal to the distance he paced off, since these distances are corresponding parts of congruent triangles. **3.** Given $\angle BAC$, construct a circle with center A and any radius. Let D and E be the points where the circle intersects the sides of the angle. Next construct the perpendicular bisector of $\overline{DE}$. **5. (a)** The distances are equal. **(b)** The distances from every point on an angle bisector of the angle to the sides of the angle are equal. **(c)** $\triangle APC \cong \triangle APB$ by AAS; hence $\overline{PC} \cong \overline{PB}$. **(d)** If a point P is equidistant from the sides of an angle, then it is on the angle bisector of the angle. To prove this statement we assume that $PC = PB$ and prove that $\overleftrightarrow{AP}$ bisects $\angle A$. We have $\triangle APC \cong \triangle APB$ by Hypotenuse-Leg congruency condition. Thus $\angle CAP \cong \angle PAB$ as these are corresponding angles in the congruent triangles.

Open-Ended

7. Answers vary. Possible questions and answers concerning figures with vertices on the nails (or dots of the dot paper) are: **(a)** How many noncongruent right isosceles triangles are possible? Answer: 6 **(b)** How many noncongruent isosceles trapezoids that are not rectangles are there? Answer: 4 **(c)** How many noncongruent squares are there? Answer: 5

9. The student is wrong. $\angle 1 \cong \angle 2$ implies that $\overline{AD}$ and $\overline{BC}$ are parallel, but does not imply that the other two sides are parallel. **11.** The SSS congruency condition assures that triangles are rigid and hence contribute to supporting the structure. **13.** This is true because the distance from A or from A' to the line BC is the distance between the two parallel lines.

GSP/GeoGebra Activities

See online appendix.

Review Problems

15. The triangles that are congruent to triangle ABC are triangles BCD, CDE, DEA, and EAB, as well as triangles CBA, DCB, EDC, AED, and BAE. They are all congruent by SAS. **17.** Constructions

Assessment 12-3A

1. (a) Construction **(b)** Construction **2. (a)** A right triangle **(b)** The altitude is the extension of the cable from vertex A to the ground. **3.** Construction **(a)** The perpendicular bisectors of the sides of an acute triangle meet inside the triangle. **(b)** The perpendicular bisectors of the sides of a right triangle meet at the midpoint of the hypotenuse. **(c)** The perpendicular bisectors of the sides of an obtuse triangle meet outside the triangle. **(d)** Construction **4. (a)** This point is equidistant from all vertices because it is on all three perpendicular bisectors. Being at the intersection of two of the perpendicular bisectors forces the point to be equidistant from all three vertices. **(b)** Same as part (a) **5. (a)** If the rectangle is not a square, it is impossible to construct an inscribed circle. The angle bisectors of a rectangle do not intersect in a single point. **(b)** Possible. The center of the circle is the intersection of the diagonals (which are also the angle bisectors of the vertices) and the radius of the circle is the distance from the center to any of the sides. **(c)** Possible. The intersection of the three longest diagonals is the center of the circle. **6.** Construction **7. (a)** Construction **(b)** 40¢ **8.** 4″ **9.** The center is where the angle bisectors of the angles meet (the incenter). The radius is the distance from the incenter to a side. **10.** $x = 3''$ **11.** Answers vary **(a)** If the parallelogram is not a rectangle, cut along an altitude. If the parallelogram is a rectangle, cut along any line through the point where the diagonals meet and such that the line is not a diagonal and is not parallel to any side. **(b)** Make a copy of the given trapezoid and put it upside down next to $\overline{CD}$ as shown. More precisely, extend $\overline{BC}$ so that $CE = a$ and extend $\overline{AD}$ so that $\overline{DF} = b$. Because $\overline{BE} \parallel \overline{AF}$ and $BE = AF$ (the length of each is $a + b$), $ABEF$ is a parallelogram.

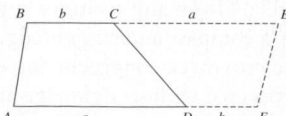

12. As close to 26 in. as the jack can close **13. (a)** Possible **(b)** There are infinitely many noncongruent rectangles. The endpoints of two segments bisecting each other and congruent to the given diagonal determine a rectangle, but since the segments may intersect at any angle, there are infinitely many such rectangles. **(c)** Not possible because the sum of the measurements of the angles would be greater than 180°. **(d)** There is no unique

parallelogram because the fourth angle must also be a right angle.
14. (a) Construct an equilateral triangle and bisect one of its
angles.   **(b)** Bisect a 30° angle, then add 30° and 15° angles
or bisect a right angle.   **(c)** Add 60° and 15° angles or 45° and
30° angles.   **15.** Construction   **16. (a)** The point is determined
by the intersection of the angle bisector of $\angle A$ and the perpendicular bisector of $\overline{BC}$. Because the point is on the angle bisector
of $\angle BAD$, it is equidistant from its sides. Because it is on the
perpendicular bisector of $\overline{BC}$, it is equidistant from B and C.
(b) The point is determined by the intersection of the angle
bisector of $\angle A$ and $\angle B$.   **17.** Answers vary. One way is to construct six 60° central angles with common vertex at the center
O. Then construct three 120° central angles with vertex O. The
sides of these angles intersect the circle at points A, B, and C,
which are vertices of an equilateral triangle. $\triangle ABC$ is equilateral
because $\triangle AOB$, $\triangle BOC$, and $\triangle AOC$ are congruent (by SAS).
Another approach is to construct first a regular inscribed hexagon
(see Assessment 12-1B problem 9). Then choose any vertex of the
hexagon and two other nonconsecutive vertices.

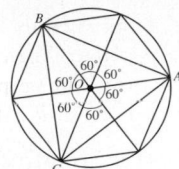

Mathematical Connections 12-3

Communication

1. Given $\angle BAC$, place one strip of tape so that an edge of the
tape is along $\overline{AB}$ and another strip of the tape so that one of its
edges is on $\overline{AC}$, as shown. Two edges of the strips of tape intersect in the interior of the angle at D. Connect A with D. $\overline{AD}$ is
the angle bisector. Because the diagonals of a rhombus bisect its
angles, this contruction can be justified by showing that $AEDF$ is
a rhombus. (E and F are the points of intersection of the tops of
the tape pieces and the opposite sides.) $AEDF$ is a parallelogram.
It remains to be shown that $\overline{AF} \cong \overline{AE}$. For that purpose, we
show that $\triangle FAG \cong \triangle EAH$. We have $\overline{FG} \cong \overline{EH}$ because the two
strips of tape have the same width. $\angle A \cong \angle A$, and the angles at
H and G are right angles. Thus, the triangles are congruent by
AAS.   **3.** Answers vary.   **5. (a)** Referring to the figure in the
problem: $AD + BC = (b + d) + (a + c) = a + b + c + d$
$AB + CD = (a + d) + (b + c) = a + b + c + d$ Hence,
$AD + BC = AB + CD$.   **(b)** The sum of the lengths in one
pair of opposite sides equals the sum of the lengths of the other
pair.   **(c)** If the sides of the rectangle have lengths a and b and a
circle can be inscribed in the rectangle then by part (a) $2a = 2b$ or
$a = b$. Thus if, the rectangle is not a square, a circle cannot be inscribed in the rectangle.   **7.** Rhombus. For proof see the solution
to problem 1 above.

Open-Ended

9. Answers vary.

Questions from the Classroom

11. The construction is not valid if the only tools allowed are
straightedge and compass. The procedure is sometimes referred to
as eyeballing, where we depend on our sight to judge if the ruler
touches the circle only at one point.   **13.** $\overline{OD}$ is the radius if, and

only if, it is perpendicular to $\overline{AC}$. If $AB \neq BC$, then $\overline{BD}$ is not perpendicular to $\overline{AC}$.   **13.** $\overline{OD}$ is the radius if, and only if, it is perpendicular to $\overline{AC}$. If $AB \neq BC$, then $\overline{BD}$ is not perpendicular ot $\overline{AC}$.

GSP/GeoGebra Activities

15. See online appendix.

Review Problems

17. construction

Assessment 12-4A

1. It is given that $\angle BAD \cong \angle B_1A_1D_1$. Because the pairs $\angle ABC$
and $\angle BAD$ as well as $\angle A_1B_1C_1$ and $\angle B_1A_1D_1$ are supplementary, it
follows that $\angle ABC \cong \angle A_1B_1C_1$. Thus the rhombi have congruent
corresponding angles. Because in a rhombus adjacent sides are
congruent, $\dfrac{AD}{AB} = 1 = \dfrac{A_1D_1}{A_1B_1}$. This implies $\dfrac{AD}{A_1D_1} = \dfrac{AB}{A_1B_1}$
and therefore ratios of corresponding sides are proportional.
2. (a) Yes; AAA   **(b)** Yes; sides are proportional and angles are
congruent.   **(c)** No. Two rectangles that are not squares are
similar if, and only if, the ratio of the longer side to the shorter
is the same for both.   **(d)** Always similar because the ratio of
the longer side equals the ratio of the corresponding shorter sides
and the angles are congruent.   **3.** Make all dimensions 3 times as
long; that is, in part (c) each side would be 3 diagonal units long.
One possible solution set is shown here:

(a)   **(b)**

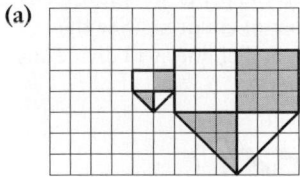   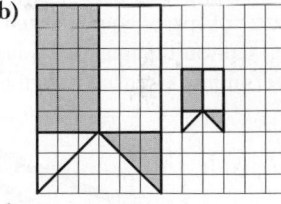

(c)   **(d)**

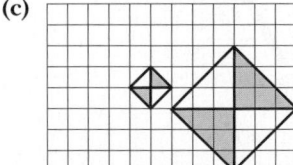   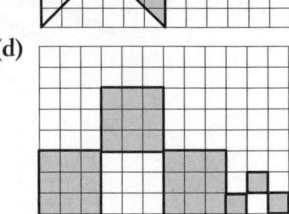

4. (a) (i) $\triangle ABC \sim \triangle DEF$ (by AA) **(ii)** $\triangle ABC \sim \triangle EDA$
(by SAS) (by AA)   **(iii)** $\triangle ACD \sim \triangle ABE$ (by AA)
(iv) $\triangle ABE \sim \triangle DBC$ by SAS similarity condition since $\dfrac{2}{3} = \dfrac{3}{4.5}$
and the vertical angles at 3 are congruent.   **(b) (i)** $\dfrac{2}{3}$ **(ii)** $\dfrac{1}{2}$ **(iii)** $\dfrac{4}{3}$
(iv) $\dfrac{2}{3}$ or the reciprocals   **5. (a)** 7   **(b)** $\dfrac{24}{7}$ **6.** Construction

7. Construction as follows:

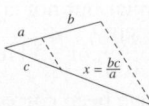

8. (a) Yes, because the angles stay the same and the ratios of corresponding sides are proportional.   **(b)** 0.64   **9.** Sketch
10. $133\frac{1}{3}\%$ but most copy machines will not allow this setting.

11. About 233 in. **12.** 15 m **13. (a)** The ratio of the perimeters is the same as the ratio of the sides. **(b)** If a, b, c, d are the sides of one quadrilateral and a_1, b_1, c_1, d_1 the corresponding sides of a similar quadrilateral then $\dfrac{a}{a_1} = \dfrac{b}{b_1} = \dfrac{c}{c_1} = \dfrac{d}{d_1} = r$ (the scale factor)

$$\text{Now,} \quad \frac{a + b + c + d}{a_1 + b_1 + c_1 + d_1} = \frac{a_1 r + b_1 r + c_1 r + d_1 r}{a_1 + b_1 + c_1 + d_1}$$
$$= \frac{(a_1 + b_1 + c_1 + d_1)r}{a_1 + b_1 + c_1 + d_1} = r.$$

Hence the ratio of the perimeters is r. An analogous proof works for any two similar n-gons. **14. (a)** From the Midsegment Theorem we know that each midsegment is half the measure of the opposite side of $\triangle ABC$. Thus the sides of each of the smaller triangles are half as long as the corresponding sides of $\triangle ABC$. Hence by SSS they are congruent to each other. **(b)** Yes; the ratio of the corresponding sides is $\dfrac{1}{2}$ and therefore by SSS the triangles are similar. **15. (a)** Rectangle **(b)** Rectangle **(c)** Rhombus **(d)** Rectangle **16.** They are circles, so have exactly the same shape, and hence are similar.

Mathematical Connections 12-4

Communication

1. Any two cubes are similar because all the faces are squares and the dihedral angles between two adjacent faces measure $90°$. **3.** Lay the licorice diagonally on the paper so that it spans a number of spaces equal to the number of children. (See the figure.) Cut on the lines. Equidistant parallel lines will divide any transversal into congruent segments.

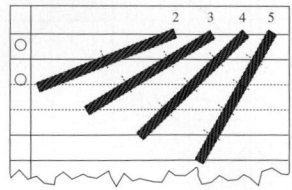

Open-Ended

5. (a) The answers may vary depending upon the construction. **(b)** The perimeters of the bases should be in the same ratio as the heights in part (a).

Cooperative Learning

7. (a) Answers vary. **(b)** The following are two different-size triangles with the given data. The given measures of the angles was only approximate; such triangles with these exact measures do not exist. The triangles are similar but not congruent. (The ratio of the corresponding sides is $\dfrac{80}{100}$, or $\dfrac{4}{5}$.) Hence, the surveyor and the architect could both have been correct in their conclusions.

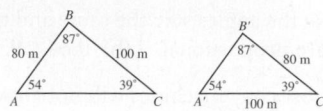

(c) All answers vary.

9. The student is incorrect. The corresponding sides are not proportional. **11.** Slope of a nonvertical line was defined as the slope or $\dfrac{\text{rise}}{\text{run}}$ of a segment determined by any two points on the line. To show that this is a valid definition we need to show that for every two points we pick on the line, the slope of the corresponding segment is the same. This is the case because the right triangles shown in the figure are similar.

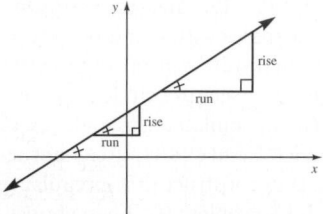

GSP/GeoGebra Activities

13. See online appendix.

Review Problems

15. Start with the given base and construct the perpendicular bisector of the base. The vertex of the required triangle must be on that perpendicular bisector. Starting at the point where the perpendicular bisector intersects the base, mark on the perpendicular bisector a segment congruent to the given altitude. The endpoint of the segment not on the base is the vertex of the required isosceles triangle. **17.** Answers may vary. Students may suggest that angles of measure $45°$ be constructed with the endpoints of the hypotenuse as vertices of the $45°$ angles and the hypotenuse as one of the sides of the angles. Both angles need to be constructed on the same side of the hypotenuse. **19.** SASAS or ASASA

Chapter Review

1. (a) $\triangle ADB \cong \triangle CDB$ by SAS **(b)** $\triangle GAC \cong \triangle EDB$ by SAS **(c)** $\triangle ABC \cong \triangle EDC$ by AAS **(d)** $\triangle BAD \cong \triangle EAC$ by ASA **(e)** $\triangle ABD \cong \triangle CBD$ by ASA or SAS **(f)** $\triangle ABD \cong \triangle CBD$ by SAS **(g)** $\triangle ABD \cong \triangle CBE$ by SSS; $\triangle ABE \cong \triangle CBD$ by SSS **(h)** $\triangle ABC \cong \triangle ADC$ by SSS; $\triangle ABE \cong \triangle ADE$ by SSS or SAS; $\triangle EBC \cong \triangle EDC$ by SSS or SAS **2.** Parallelogram; $\triangle ADE \cong \triangle CBF$ by SAS, so $\angle DEA \cong \angle CFB$; $\angle DEA \cong \angle EAF$ (alternate interior angles between the parallels $\overline{DC}$ and $\overline{AB}$ and the transversal $\overline{AE}$), so $\angle EAF \cong \angle CFB$; and therefore $\overline{AE} \parallel \overline{FC}$. $\overline{EC} \parallel \overline{AF}$ (parallel sides of the square); two pairs of parallel opposite sides implies a parallelogram. **3.** Constructions **4. (a)** $x = 8$ cm $y = 5$ cm **(b)** $x = 6.5$ m **5.** Construction **6.** $\dfrac{a}{b} = \dfrac{c}{d}$ **7.** Find point O, the intersection of the perpendicular bisector of $\overline{AB}$ and line ℓ. The required circle has center O and radius OA. **8. (a)** $\triangle ACB \sim \triangle DEB$ by AA similarity, $x = \dfrac{24}{5}$ in. **(b) (i)** $\triangle AED \sim \triangle ACB$ by AA **(ii)** $y = \dfrac{4'}{3}$ **(iii)** $x = \dfrac{55'}{6}$ **9.** 12 m **10. (a)** Polygons (ii) and (iii) **(b)** Any convex regular polygon can be inscribed in a circle. **11.** $h = 6$ m **12.** $d = \dfrac{256}{5}$ m **13.** Slice along a pair of parallel sides for which a perpendicular slice exists. Slide the shaded triangle to the right.

14. **(a)** AD is a diameter. **(b)** $\overline{AD}$ and $\overline{BE}$ are diameters that bisect each other (at the center). Hence $ABDE$ is a rectangle and so $\angle ABD$ is a right angle. **15.** False. If in addition the diagonals bisect each other, then the quadrilateral is a square. If not, it is not a square. **16.** **(a)** $y = \dfrac{-4}{3}x - \dfrac{1}{3}$ **(b)** $y = \dfrac{1}{3}x + 1$

17. Congruent triangles have two corresponding angles that are congruent. This makes them similar by AA.
18. **(a)** $\triangle BCN \cong \triangle DEN$ by ASA because $BN = ND$ (given), $\angle CBN \cong \angle EDN$ are alternate interior angles created by the parallels $\overline{BC}$ and $\overline{AD}$ and transversal BD, and the angles with vertex N are vertical angles. **(b)** From the congruence in part (a) it follows that $CN = EN$, and $AM = CM$ (given). **(c)** Because $\overline{MN}$ is a midsegment in $\triangle ACE$, $MN = \frac{1}{2}AE$. Now $AE = AD - ED = AD - BC$ because $ED = BC$ (from congruence in part (a)). Thus $AE = a - b$ and therefore $MN = \frac{1}{2}AE = \frac{1}{2}(a - b)$.

Answers to Now Try This

12-1.

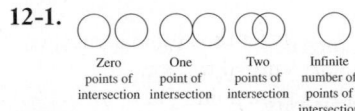

| Zero points of intersection | One point of intersection | Two points of intersection | Infinite number of points of intersection |

12-2. Construct an arc with center at P and any radius. Mark A and B as the points of intersection of the arc with ℓ. Now construct the perpendicular bisector of $\overline{AB}$. **12-3.** Construction.
12-4. (a) and (b) the opposite angles are not supplementary.
12-5. Yes. If $ABCD$ is a quadrilateral whose diagonals intersect at M, then congruent vertical angles with vertex at M are formed. Because the diagonals bisect each other, it follows from SAS that $\triangle AMD \cong \triangle CMB$. Hence, $\angle MAD \cong \angle MCB$ and therefore $\overline{BC}$ is parallel to $\overline{AD}$. In a similar way, we can show that $\overline{AB}$ is parallel to $\overline{DC}$ and hence $ABCD$ is a parallelogram. **12-6.** **(a)** One way to accomplish the construction is to place the hypotenuse of the triangle on the given line and to place the ruler so that one of the legs of the triangle will be on the ruler. Now, keeping the ruler fixed, slide the triangle so that the side of the triangle on the ruler touches the ruler all the time. Slide the triangle on the ruler until the given point is on the hypotenuse. The line containing the hypotenuse is parallel to the given line. **(b)** The angles firmed by the hypotenuses and the ruler are congruent. Because they are corresponding angles the hypotenuses are parallel. **12-7.** In Figure 12-50 if $PD = PE$ then by the Hypotenuse-Leg congruence condition (Theorem 12-2), $\triangle APD \cong \triangle APE$ and hence $\angle PAD \cong \angle PAE$. **12-8.** Construct the diagonals of the square. Use the intersection of the diagonals as the center of the inscribed circle. **12-9.** Using the result (in Example 12-9) that the median to the hypotenuse is half as long as the hypotenuse, we can prove that one of the angles in all such triangles is 60°. Thus by AA the triangles are similar.

Answer to the Preliminary Problem

Because $\triangle ABC$ on the map is similar to the corresponding triangle in reality we have the following proportions:

$$\frac{BC}{10.1} = \frac{512}{4.2}, \quad BC = \frac{10.1 \cdot 512}{4.2} \approx 1{,}231.24 \text{ mi.}$$

$$\frac{AC}{12.2} = \frac{512}{4.2}, \quad AC = \frac{12.2 \cdot 512}{4.2} \approx 1{.}487.2 \text{ mi.}$$

Chapter 13

Assessment 13-1A

1. **(a)**   **(b)**

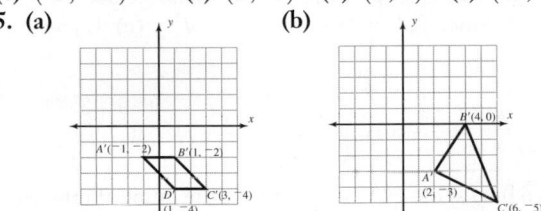

2. **(a)** Construct as suggested by the following: Trace $\overline{BC}$ and the line containing the slide arrow on the tracing paper and label the trace of B as B' and the trace of C as C'. Mark on the original paper and on the tracing paper the initial point of the arrow by P and the head of the arrow by Q. Slide the tracing paper along the line $\overleftrightarrow{PQ}$ so that P will fall on Q. The segment $\overline{B'C'}$ overlays the image of $\overline{BC}$ under the translation. **(b)** Construct a parallelogram with $\overline{BC}$ and $\overline{B'C'}$ as opposite sides. **3.** **(a)** $(3, {}^-4)$ **(b)** $(0, 0)$
(c) $({}^-3, {}^-13)$ **4.** **(a)** $(3, {}^-4)$ **(b)** $(0, 0)$ **(c)** $({}^-3, {}^-13)$
5. **(a)**   **(b)**

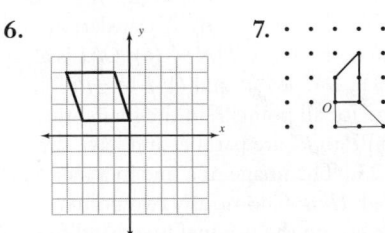

6.   **7.**

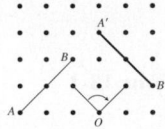

8. **(a)** $y = 2x - 3$ **(b)** $y = 2x - 7$ **9.** $y = {}^-2x - 1$
10. Answers vary. Choose points P and Q on ℓ. Construct P', not P, on $\overleftrightarrow{OP}$ such that $OP = OP'$. Similarly find Q'. $\overleftrightarrow{P'Q'}$ is the desired image. **11.** Reverse the rotation (to the counterclockwise direction) to locate $\overline{AB}$, that is, the preimage.

12. Answers may vary, but H, I, N, O, S, X, or Z could appear in such rotational words. Examples include SOS. Variations could use M and W in rotational images, for example, MOW.
13. **(a)** The image is the line ℓ itself. **(b)** Construction
(c) ℓ and ℓ' are parallel. If P and Q are any points on ℓ and P' and Q' their respective images, then from the definition of a translation $\overline{PP'}$ and $\overline{QQ'}$ are parallel and congruent to $\overline{AB}$. Hence, $\overline{PP'}$ and $\overline{QQ'}$ are parallel and congruent. Thus, $PP'Q'Q$ is a parallelogram and therefore $\ell' \parallel \ell$. **(d)** The image is $\angle ABC$ itself. **14.** *Hint:*

An angle whose measure is 45° can be constructed by bisecting a right angle. An angle whose measure is 60° can be constructed by first constructing an equilateral triangle.   **15. (a)** $(^-4, 0)$ **(b)** $(^-2, ^-4)$ **(c)** $(2, 4)$ **(d)** $(^-a, ^-b)$ **16.** *Hint*: Find the images of the vertices.

17. (a) $\ell' = \ell$ **(b)** $\ell' \perp \ell$ **18. (a)** First rotate $\triangle ABC$ by angle α to obtain $\triangle A'B'C'$, and then rotate $\triangle A'B'C'$ by angle β to obtain $\triangle A''B''C''$.   **(b)** No   **(c)** Yes, by rotation about O by angle $|\alpha - \beta|$ in the direction of the larger of α and β
19. (a) (i) $(^-3, 2)$ **(ii)** $(^-2, ^-1)$ **(iii)** $(^-n, m)$ **(b)** This could be demonstrated with graph paper. The image of the point with coordinates (a, b) under a half-turn with the origin as center must be on a circle with center at $(0, 0)$ having equation $x^2 + y^2 = a^2 + b^2$ and being the other endpoint of the diameter having (a, b) as one endpoint. The only point satisfying these conditions is $(^-a, ^-b)$.   **(c)** $(b, ^-a)$ **20. (a)** $y = ^-x + 1$
(b) $y = \frac{1}{2}x - \frac{1}{2}$ **(c)** $y = 0$ **(d)** $x = 1$ **21. (a)** The image of $A(b, k)$ under the translation from O to C is $B(b + a, k + 0)$ or $B(b + a, k)$.   **(b)** Using the distance formula, we get $OA^2 = b^2 + k^2$. Since $OA = OC$, $b^2 + k^2 = a^2$.   **(c)** Using part (b), the product of the slopes of $\overline{OB}$ and $\overline{AC}$ is

$$\frac{k}{b + a} \cdot \frac{k}{b - a} = \frac{k^2}{b^2 - a^2} = \frac{k^2}{^-k^2} = -1.$$

(d) There is a turn symmetry of 180° about the point of intersection of the two diagonals.   **22. (a)** Construction   **(b)** It is the same.   **(c)** The images of Q are the same.   **(d)** A translation taking N to M (where O is the midpoint of $\overline{MN}$)   **(e)** $\overline{OM}$ is a midsegment in $\triangle PP'P''$. Hence, $\overline{OM} \parallel \overline{PP''}$ and $OM = \frac{1}{2}PP''$. Because O and M are the same for all points P in the plane, the vectors from N to M and from P to P'' are parallel and have the same length and direction.   **23.** The image of a line in a half-turn is a parallel to the original. *Hint:* Choose any two points on the original if the center is not on the original line. Find the images of these points. Show that the triangles formed using the points and the center are congruent making angles congruent which determines parallel lines.   **24.** Answers vary. Four that can be their own images are a circle, a square, a regular octagon, and a regular hexagon. An equilateral triangle cannot be its own image.   **25.** The possibilities are points with coordinates $(5, 8)$, $(^-1, 4)$, and $(3, 2)$.

Mathematical Connections 13-1

Communication

1. Yes. If the congruent segments are $\overline{AB}$ and $\overline{CD}$, connect A with C and B with D, as shown. The perpendicular bisectors of $\overline{AC}$ and $\overline{BD}$ intersect at O. (If they do not intersect, connect A with D and B with C). The point O is the center of the required rotation. The angle of rotation is $\angle AOC$ in the direction from A to C. The image of A will be C because the rotation is by $\angle AOC$. To show that under this rotation the image of $\overline{AB}$ is $\overline{CD}$, we need only show that $\angle AOC \cong \angle BOD$. This can be checked experimentally by performing the rotation or by measuring $\angle AOC$ and $\angle BOD$.

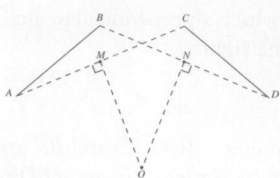

3. The coordinates of the image are $(^-x, ^-y)$. The image of a point P through a half-turn about the origin O is on $\overrightarrow{PO}$ so that O is the midpoint of $\overline{PP'}$; $(0, 0)$ are the coordinates of O, the midpoint of all segments with endpoints with coordinates (x, y) and $(^-x, ^-y)$.

Open-Ended

5. Answers vary.   **7.** Answers vary. If the rotation is by 30°, there will be 360/30, or 12 images.

Questions from the Classroom

9. The answer is yes, the student is correct because the two described components could be constructed as components of the given translation. An analogy is the vertical and horizontal component of a vector.

GSP/Geogebra Activities

11. (6) Construction   **(7) (a)** Congruent   **(b)** Yes   **(c)** No
(8) (a) The segments are congruent.   **(b)** Measurements
(c) Yes   **13.** Construction; answers vary.

Assessment 13-2A

1. Locate the images of vertices directly across (perpendicular to) ℓ on the geoboard.

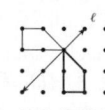

2. Reflecting lines are described for each.   **(a)** All diameters (infinitely many)   **(b)** Perpendicular bisector of the segment or the line containing the segment   **(c)** Any line perpendicular to the given line or the line itself   **(d)** Perpendicular bisectors of the sides and lines containing the diagonals   **(e)** None
(f) Perpendicular bisectors of each side   **(g)** None   **(h)** If not a rhombus, the line containing the diagonal determined by vertices of the noncongruent angles; if a rhombus then both diagonals are lines of symmetry   **(i)** Perpendicular bisectors of parallel sides and three diameters determined by vertices on the circumscribed circle   **3.** The original figure   **4.** Construction   **5. (a)** The final images are congruent but in different locations, and hence not the same.   **(b)** A translation by a slide arrow from P to R is determined as follows: Let P be any point on ℓ and Q on m such that $\overline{PQ} \perp \ell$. Point R is on $\overrightarrow{PQ}$ such that $PQ = QR$.
6. Constructions vary, but final image is a translation of the original
7. The line of reflection is the perpendicular bisector of $\overline{BB'}$.
8. (a) Examples include MOM, WOW, TOOT, and HAH.
(b) Examples include BOX, HIKE, CODE, etc. B, C, D, E, H, I, K (depending on font), O, and X may be used.   **(c)** 1, 8, 11, 88, 101, 111, 181, 808, 818, 888, 1001, 1111, 1881   **9. (a)** $y = ^-2x - 1$

(b) $y = {}^-2x + 1$ **(c)** $y = \frac{1}{2}x - \frac{1}{2}$ **10. (a)** The images
are the same.   **(b)** Yes   **11.** None of the images has a reverse orientation, so there are no reflections or glide reflections involved. Thus,

1 to 2 is a rotation.
1 to 3 is a rotation.
1 to 4 is a translation down.
1 to 5 is a rotation (with an exterior point as the center of rotation).
1 to 6 is a translation.
1 to 7 is a translation.

12. (a) $A'(3, {}^-4); B'(2, 6); C'({}^-2, {}^-5)$ **(b)** $A'(4, 3),$
$B'({}^-6, 2), C'(5, {}^-2)$ **13. (a) (i)** $(x, {}^-y)$ **(ii)** (y, x)
(b) $({}^-x, {}^-y)$; the following figure:

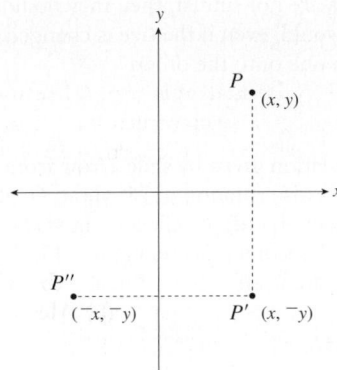

14. (a) $y = x - 3$ **(b)** $y = 0$ **15. (a)** $y = x + 3$ **(b)** $y = 0$
16. (a) The line through P perpendicular to $\overleftrightarrow{O_1O_2}$ or the perpendicular bisector of QQ', where Q and Q' are the points of intersection other than P of $\overleftrightarrow{O_1O_2}$ with circle O_2 and circle O_1, respectively.   **(b)** $\overleftrightarrow{O_1O_2}$ **17. (a)** $\overleftrightarrow{AB}$ **(b)** yes; a translation taking O_1 to O_2 **18. (a)** $x = 0$ or $y = 0$ **(b)** $x = 0$ or $y = 0$ **19.** Let H represent one house, X the roadside spot, and T the other house, and T' (the reflection of T in the road r). The intersection of $\overline{HT'}$ and r determines the point on the road at which X should be placed.

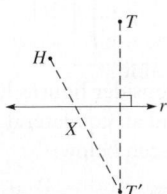

Mathematical Connections 13-2

Communication

1. Answers may vary. Find A', the image of A under reflection in $\overline{EH}$, and B', the image of B under reflection in $\overline{GH}$. Mark the intersections of $\overline{A'B'}$ with $\overline{EH}$ and $\overline{GH}$, respectively, by C and D. The player should aim the ball at A toward the point C. The ball will hit D, bounce off, and hit B. To justify the answer, we need to show that the path A–C–D–B is such that $\angle 1 \cong \angle 3$ and $\angle 4 \cong \angle 6$. Notice that $\angle 1 \cong \angle 2$ and $\angle 6 \cong \angle 5$ because the image of an angle under reflection is congruent to the original angle.

Also, $\angle 2 \cong \angle 3$ and $\angle 4 \cong \angle 5$, as each pair constitutes vertical angles. Consequently, $\angle 1 \cong \angle 3$ and $\angle 4 \cong \angle 6$.

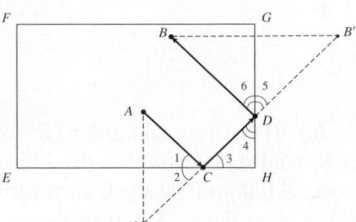

3. The angle of incidence is the same as the angle of reflection. With the mirrors tilted 45°, the object's image reflects to 90° down the tube and then 90° to the eyepiece. The two reflections "counteract" each other, leaving the image upright.

Open-Ended

5. Answers vary.

Cooperative Learning

7. (a) Constructions will vary.   **(b)** The experiment will work for rectangular tables in which the length is twice the width and for any position of B.

Questions from the Classroom

9. Having only a segment and its image is not enough to determine the transformation. It requires three noncollinear points. For a further examination of this problem, see *Transformational Geometry*, by Richard Brown (Palo Alto, CA: Dale Seymour Publications, 1989).

GSP/Geogebra Activities

11. (5) (a) The measure of an angle and the measure of its reflected angle are equal.   **(b)** The length of a segment and the length of its reflected segment are equal.   **(c)** A polygon and its reflection are congruent.   **(d)** The orientations of an object and its reflection are opposite.   **(e)** The reflecting line contains the midpoint of each segment joining a point and its reflection.
13. (a) Construction   **(b)** The line desired, q, is such that the angle formed by lines m and n in that order is the same as the angle formed by the lines q and p in that order.

Review Problems

15. $({}^-a, {}^-b)$ **17.** Construct $\overline{BE}$ perpendicular to $\overline{AD}$, as shown. Next translate $\triangle ABE$ by the slide arrow from B to C. The image of $\triangle ABE$ is $\triangle DCE'$. The rectangle $BCE'E$ is the required rectangle.

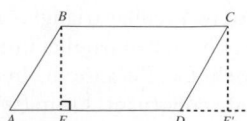

Assessment 13-3A

1. Answers vary.   **(a)** Slide the smaller triangle down 3 units (translation). Then complete a dilation with scale factor 2 using the top-right vertex as the center. **(b)** Slide the smaller triangle right 5, and up 1. Then complete the dilation as in (a).

2.

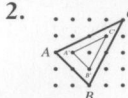

3. Answers vary. **(a)** Translation taking B to B' followed by a dilation with center B' (and scale factor 2) **(b)** Dilation with scale factor $1/2$ and center A followed by a half-turn with the midpoint of $\overline{C'B'}$ as center. **4.** $x = 6; y = 5.2$; scale factor $2/5$ **5.** 12 cm **6. (a)** $(12, 18), (^{-}12, 18)$ **(b)** same as in (a) **7.** Dilation with center O and scale factor $1/r$ **8. (a)** Answers vary. A possible explanation follows:

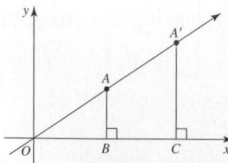

Let $A'(x', y')$ be the image of $A(x, y)$ under the dilation. From the definition of a dilation, we have $\dfrac{OA'}{OA} = r$. Since $\triangle OA'C \cong \triangle OAB$, $\dfrac{OC}{OB} = r$, which implies that $\dfrac{x'}{x} = r$ and hence $x' = rx$. Similarly, $y' = ry$. **(b)** Answers vary; the dilation with center at the origin and scale factor 2 followed by a half-turn about the origin. **(c) (i)** $y = 2x$ **(ii)** $y = 2x$ **(iii)** $y = 2x + \dfrac{1}{3}$ **(iv)** $y = ^{-}x - 3$ **9.** Answers vary; a translation taking O_1 to O_2 followed by a dilation with center at O_2 and scale factor $\dfrac{3}{2}$. **10.** The set of images of the set of integers on a number line is the set of points where coordinates are multiples of 3; that is, $\{\ldots, ^{-}6, ^{-}3, 0, 3, 6, 9, \ldots\}$. **11.** $x = 98/15$; $y = 90/7$; scale factor $= 7/15$ **12.** The enlargement of a $2'' \times 3''$ photograph to a $4'' \times 6''$ photograph can be achieved by a dilation with any center.

Mathematical Connections 13-3

Communication

1. (a) It does change. For example, consider the segment whose endpoints are $(0, 0)$ and $(1, 1)$ which has length $\sqrt{2}$. Under the dilation with center at $(0, 0)$ and scale factor 2, the image of the segment is a segment whose endpoints are $(0, 0)$ and $(2, 2)$. That segment has length $2\sqrt{2}$. **(b)** It does not change. Under a dilation, the image of a triangle is a similar triangle and the corresponding angles of two similar triangles are congruent. **(c)** It does not change. Given two parallel lines, draw a transversal that intersects each line. Because the lines are parallel, the corresponding angles are congruent. From (b), the images of the angles will also be congruent and hence the image lines will be parallel. **3. (a)** A single dilation with center O and scale factor $\dfrac{1}{2} \cdot \dfrac{1}{3}$ or $\dfrac{1}{6}$; let P be any point and P' its image under the first dilation and P'' the image of P' under the second dilation. Then $\dfrac{OP'}{OP} = \dfrac{1}{2}$ and $\dfrac{OP''}{OP'} = \dfrac{1}{3}$. Consequently, $\left(\dfrac{OP'}{OP}\right) \cdot \left(\dfrac{OP''}{OP'}\right) = \dfrac{1}{2} \cdot \dfrac{1}{3}$,

or $\dfrac{OP''}{OP} = \dfrac{1}{6}$. Thus, P' can be obtained from P by a dilation with center O and scale factor $\dfrac{1}{2} \cdot \dfrac{1}{3}$ or $\dfrac{1}{6}$. **(b)** the dilation with center O and scale factor $r_1 r_2$ **5. (a)** The net result is the identity transformation. Every point is its own image. **(b)** A scale factor of 0 maps every point in the plane to the center of the dilations. Here lines are no longer mapped to parallel lines but to a single point. This transformation is not a dilation.

Open-Ended

7. The result is equivalent to a half-turn in the origin.

Questions from the Classroom

9. If two triangles are not similar, then they do not have the same shape and never would, even if the size is changed. It is not possible to transform one onto the other.

Review Problems

11. (a) The translation given by slide arrow from N to M **(b)** A counterclockwise rotation of 75° about O **(c)** A clockwise rotation of 45° about A **(d)** A reflection in m and translation from B to A **(e)** A second reflection in n. **13. (a)** The angle itself **(b)** the square itself

Assessment 13-4A

1. Answers vary.

2. (a) *Hint:* Perform half-turns about the midpoints of all sides. **(b)** Yes. Successive 180° turns of a quadrilateral about the midpoints of its sides will produce four congruent quadrilaterals around a common vertex, with each of the quadrilateral's angles being represented at each vertex. **3.** Answers vary.

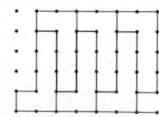

4. Answers vary. *Hint:* Consider figures like a pentagon formed by combining a square and an equilateral triangle. **5.** The shape will t essellate a plane as seen below:

6. (a) Construction **(b)** It combines a translation and a reflection. **(c)** It will tessellate the plane. **7. (a)** The dual is another tessellation of squares (congruent to those given). **(b)** A tessellation

of isosceles triangles   **(c)** The tessellation of equilateral triangles; it is illustrated in the statement of the problem.
8.

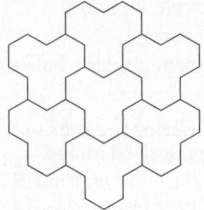

9. (a) Is not semiregular; **(b)** and **(c)** are semiregular.   **10.** The hexagon will tessellate the plane as shown:

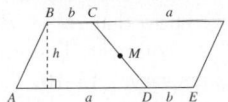

Mathematical Connections 13-4

Communication

1. (a) The image of *ABCD* under a half-turn about *M* (the midpoint of *CD*) is the trapezoid *FEDC*. Because the trapezoids are congruent, *ABFE* is a parallelogram. The area of the parallelogram is $AE \cdot h$ or $(a + b)h$. The parallelogram is the union of two nonoverlapping congruent trapezoids. The area of each trapezoid is $(a + b)(h/2)$.

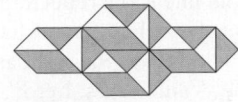

(b) To tessellate the plane with △*ABC*, we find △*A'CB*—the image of △*ABC* under a half-turn about *M*, the midpoint of $\overline{BC}$. If *N* is the midpoint of $\overline{AB}$, then the image of *N* is *N'*. It can be shown that *N, M,* and *N'* are collinear and hence that *ANN'C* is a parallelogram. Because the image of $\overline{NM}$ is $\overline{MN'}$, it follows that $NM = MN'$. Hence, $NM = \frac{1}{2}NN' = \frac{1}{2}AC$. Thus, $NM = \frac{1}{2}AC$. Also $\overline{NM} \parallel \overline{AC}$.   **3.** First tessellate the plane with regular hexagons and then divide each hexagon into congruent pentagons by three segments connecting the center of a hexagon to the midpoint of every second side. The segments inside one hexagon should not have points in common with segments inside a neighboring hexagon. Actually, one could draw the *Y*-shape segments in one hexagon and then translate the shape in all directions by a translation, taking the center of one hexagon to the center of a neighboring one.

Open-Ended

5. Answers vary.

Cooperative Learning

7. Answers vary.   **9.** All will tessellate. Answers vary.

Questions from the Classroom

11. Answers vary depending on the kaleidoscope.

13. (a) 90°, 180°, 270° rotation about center   **(b)** 180° rotation about the center (half-turn in that center); reflections across vertical and horizontal lines through the center   **15. (a)** It is a square.   **(b)** The respective images are $(0, a), (a, 0), (0, \bar{a}), (\bar{a}, 0)$.   **17.** Any dilation with center on the line $y = x$ with scale factor 1.

Chapter Review

1. (a) 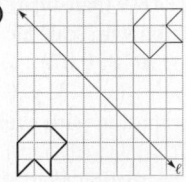   **(b)** 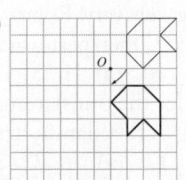   **(c)**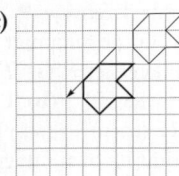

2. *Hint:* In each part, find the images of the vertices.
3. (a) 4 reflections through dashed lines as shown; also rotations of 90°, 180°, and 270° about a center at the intersection of the dashed lines.

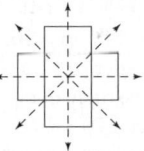

(b) one reflection in the dashed line shown

(c) one reflection in the dashed line shown

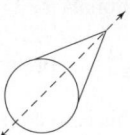

(d) a half-turn about center *O*

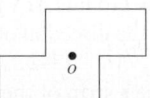

(e) two reflections in dashed lines; a half-turn about center *O*

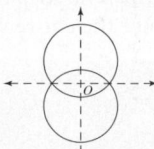

(f) two reflections in dashed lines; a half-turn about center *O*

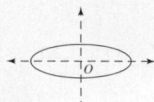

4. $A = A'$, B is the midpoint of $\overline{A'B'}$, and C is the midpoint of $\overline{A'C'}$. **5.** In each case, half-turn about X **6. (a)** Clockwise rotation by $120°$ about the center of the hexagon **(b)** A reflection in the perpendicular bisector of $\overline{BY}$ **7.** Reflection in $\overleftrightarrow{SO}$
8. Answers vary. Let $\triangle H'O'R'$ be the image of $\triangle HOR$ under a half-turn about R. Then $\triangle SER$ is the image of $\triangle H'O'R'$ under a dilation with center R and scale factor $\dfrac{2}{3}$. Thus, $\triangle SER$ is the image of $\triangle HOR$ under the half-turn about R followed by the dilation described above. **9.** Answers vary. Rotate $\triangle PIG$ $180°$ (half-turn) about the midpoint of $\overline{PT}$, then perform a dilation with scale factor 2 and center $P'(=T)$. **10. (a)** $A(^-3, 12.91)$, $B(^-8, 0.07)$, $C(1.83, 5)$ **(b)** Under the translation $(x, y) \rightarrow (x - 3, y + 5)$
11. $(x, y) \rightarrow (x, y)$. It is the identity transformation with every point its own image. **12. (a)** $A(^-3, ^-2)$, $B(^-2, 3)$, $C(^-5, 5)$
(b) $(x, y) \rightarrow (x + 3, y + 2)$ **13. (a) (i)** A translation from A to C
(ii) Same as (i) **(b)** A rotation about O by $60°$ counterclockwise.
(c) (i) A dilation with center O and scale factor 6
(ii) Same as (i) **14.** $y = -\dfrac{1}{2}x + \dfrac{5}{2}$ **15. (a)** $y = {^-}x + 2$
(b) $y = x - 3$ **(c)** $y = x + 3$ **(d)** $y = {^-}x + 3$
(e) $y = {^-}x - 3$ **(f)** $y = {^-}x + 6$ **16. (a)** $(x, y) \rightarrow$
$(x - 3, y + 5)$ **(b)** The identity transformation $(x, y) \rightarrow (x, y)$
17. An interior angle of a regular octagon has measure $135°$. Two together have measure $270°$; three together have measure more than $360°$. Octagons cannot fit together at a point without overlapping or leaving a gap. **18.** No **19.** Yes, but unless the shape fits exactly into the object being measured, it may be difficult. **20.** Constructions. *Hint:* Find images of the vertices.
21. (a) It will tessellate. **(b)** It will tessellate. **(c)** It will not tessellate. **22. (a)** $C(5, 0)$, $(^-5, 0)$, $(6, 0)$, or $\left(\dfrac{25}{6}, 0\right)$
(b) $B(8, 4)$, $(^-2, 4)$, $(3, ^-4)$, or $\left(\dfrac{^-7}{6}, 4\right)$ **(c)** The product of the slopes of respective diagonals is $^-1$.

Answers to Now Try This

13-1. (a) One way to do this is to connect A and M. Draw a line parallel to line MN through A; find a point A' such that $MN = AA'$ and $\overrightarrow{AA'}$ is the same direction as $\overrightarrow{MN}$. **(b)** Connect A and N. Draw a line parallel to line MN through A; find a point A' such that $MN = AA'$ and the direction of $\overrightarrow{AA'}$ must be the same as that of the direction of $\overrightarrow{NM}$. **13-2.** A translation along line ℓ that takes M to N produce a strip of the wallpaper. The heavily outlined symbol is the object to be translated. The translation that takes N to M yields the strip. The strip may then be translated back, up, or down to produce the entire wallpaper.

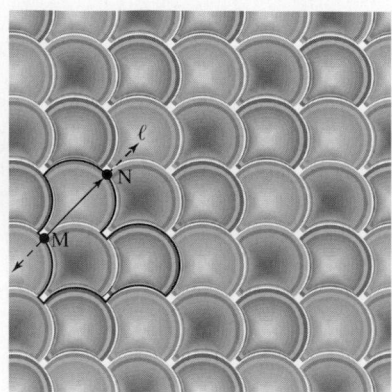

13-3. Draw a circle with center O and radius OP. Next draw the same size circle with A as center. Use a compass to measure the arc of circle A cut off by the angle. On circle O, mark the image P' of P. **13-4. (a)** Answers will vary.
(b) A figure with point symmetry is its own image under a half-turn.
(c) ℓ is its own image ℓ'. **(d)** Connect P to any point X on ℓ. Find the point X' on $\overrightarrow{XP}$ (other than X) such that $\overline{XP} \cong \overline{XP}$. Construct ℓ' through X' parellel to ℓ. ℓ' is the desired image.
13-5. *Hint:* Construct a perpendicular from P to line m. Find P' so that line m is the perpendicular bisector of $\overline{PP'}$. Points P and P' are the endpoints of a diagonal of a rhombus. The other diagonal lies along line m. **13-6.** Find the line such that the figure folds onto the image. The fold line is the reflecting line. **13-7.** The accompanying figure suggests the reason why the image of $P(a, b)$ is $P'(b, a)$. The line $y = x$ bisects the right angle in the first quadrant formed by the x- and y-axes. Let P' be the reflection image of P in the line $y = x$. Hence $\triangle OPB \cong \triangle OP'B'$. Thus $OB = OB' = b$ and $PB = P'B' = a$. The coordinates of P' are (b, a).

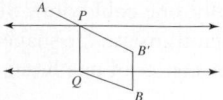

13-8. $MB = MB'$ and $BP = B'P$. Points A, P, and B' make a triangle. Therefore, they are not all on the same line and by the triangle inequality we know that $AP + PB'$ must be longer than AB'. A, M, and B' are collinear. Therefore, $AM + MB < AP + PB$. **13-9.** Answers vary. **13-10.** Continue to create the shapes and rotate them to form the tessellation.

Answers to Brain Teasers

Section 13-1

It will be as it is now; that is, upside down.

Section 13-2

Translate B toward A in the direction perpendicular to the banks of the river a distance equal to the width of the river. Connect A with the image B'. The point P where $\overline{AB'}$ intersects the far bank is the point where the bridge should be built.

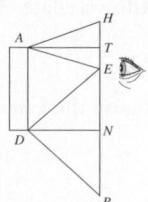

Section 13-3

In the drawing below, $\triangle AHT \cong \triangle AET$ and $\triangle DEN \cong \triangle DBN$. (Why?) Thus, $HT = ET$, $EN = BN$, and $HT + TE + EN + NB = 2(TE + EN)$. The image length HB is twice the length of the mirror because $TN = AD$.

Section 13-4

In figure (a) $\triangle BT'A'$ is the image of $\triangle BTA$ under rotation clockwise by $60°$ about B. Because $\angle TBT'$ is $60°$ and $BT' = BT$, $\triangle TBT'$ must be equilateral (the base angles at T and T' are congruent, their sum is $120°$ and hence each is $60°$). Thus

$$CT + BT + AT = CT + TT' + T'A'$$

Consequently, for each point T in the interior of the triangle the sum of the distances to the three cities is the length of the path C-T-T'-A'. The shortest path is along a straight line connecting C to A' (see figure (b)). The location of the required point T for which the sum of the distances is minimal is on line CA' and can be determined by the fact that $\triangle TBT'$ is equilateral and thus $\angle BTA'$ is $60°$. Hence we need only construct through B a line that makes $60°$ angle with line CA' (see figure (b)). For that purpose, we construct an equilateral triangle DEF with vertex D anywhere on line DA' so that F is on line CA'. Through B we draw a line parallel to line DE (see figure (b)). The point of intersection of that line with line CA' is the required point T.

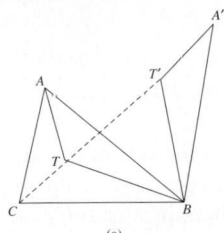

(a)

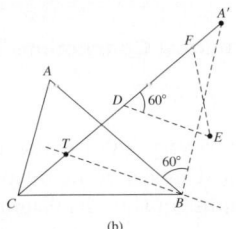

(b)

Solution to the Preliminary Problem

As the path is traced, the length of $XY + YZ$ is the same as the length AC. We know this because of $60°$ angles formed as the ball bounces at Y making $\angle BYZ$ have a measure of $60°$. (*Hint:* Remember congruent angles of incidence and angles of reflection.) Thus $AXYZ$ is a parallelogram making $\overline{AX} \cong \overline{ZY}$ and $\overline{XY} \cong \overline{AZ}$. Thus $XY + YZ = AX + AZ$.

Also $\triangle BYZ$ is equilateral so $\overline{BZ} \cong \overline{ZY}$. Thus $XY + YZ = AZ + ZB$ or AB. Hence $XY + YZ = 1$ m. Similarly $ZW + WP = BC = 1$ m, and $PQ + QX = AC = 1$ m. Thus the length of the path is $3(1$ m$) = 3$ m. The perimeter of $\triangle ABC$ is also 3 m, so the ball travels a length equal to the perimeter. **(b)** Answers vary; the dilation with center at the origin and scale factor 2 followed by a half-turn in the origin

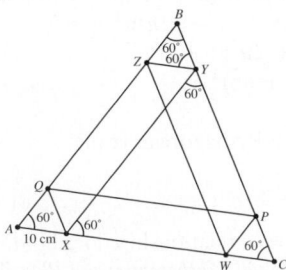

Chapter 14

Assessment 14-1A

1. Answers vary. **2. (a)** cm^2, in.^{2} **(b)** cm^2 or mm^2, in.^{2} **(c)** m^2 or cm^2, yd^2 or in.^{2} **(d)** m^2, yd^2 **3.** Answers vary. For example, **(a)** 1.6 m^2 **(b)** 2500 cm^2. **4. (a)** 0.0588 m^2,

58,800 mm^2 **(b)** 0.000192 m^2, 1.92 cm^2 **(c)** 15,000 cm^2, 1,500,000 mm^2 **(d)** 0.01 m^2, 10,000 mm^2 **(e)** 0.0005 m^2, 500 mm^2 **5. (a)** $444\frac{4}{9}$ **(b)** approx. 0.32 **(c)** 6400 **(d)** 130,680 **6. (a)** 3 sq. units **(b)** 3 sq. units **(c)** 2 sq. units **(d)** 5 sq. units **7. (a)** 4900 **(b)** 98 **(c)** 0.98 **8. (a)** 20 cm^2 **(b)** 7.5 m^2 **9.** r^2 **10. (a)** Yes. All squares are similar. **(b)** $a^2 : b^2$ **11. (a)** 9 cm^2 **(b)** 96 cm^2 **(c)** 20 cm^2 **(d)** 84 cm^2 **12. (a) (i)** 1.95 km^2 **(ii)** 195 ha **(b) (i)** Approx. 0.63 mi^2. **(ii)** Approx. 403 acres **(c)** Answers vary. For example, the metric system is easier because you only have to move the decimal point to convert units within the system. **13. (a)** True **(b)** False **(c)** False, the maximum area is 60 cm^2. **(d)** Don't know, since the height is unknown. The area is less than *or equal to* 60 cm^2. **14.** 30 cm^2 **15. (a)** \$405.11 **(b)** \$550 **16. (a)** 25π cm^2 **(b)** $(8/3)\pi$ cm^2 **17.** 1200 tiles **18. (a)** $24\sqrt{3}$ cm^2 **(b)** $9\sqrt{3}$ cm^2 **19. (a)** 16π cm^2 **(b)** $r = \dfrac{s}{\sqrt{\pi}}$ **20. (a)** 2π cm^2 **(b)** $\left(\dfrac{\pi}{2} + 2\right)$ cm^2 **(c)** 2π cm^2 **(d)** $(50\pi - 100)$ cm^2 **21.** 7π m^2 **22. (a)** 48 cm **(b)** 64 cm^2 **23. (a)** The area is quadrupled. **(b)** 1:25 **24.** $(320 + 64\pi)$ m^2

25.

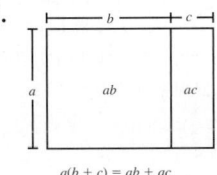

$a(b + c) = ab + ac$

26. The area of each triangle is 10 cm^2 because the base of each triangle is $\overline{AB}$ and the height of each triangle is the perpendicular distance between the two lines. Because each triangle has the same base and height, the areas of the triangles are the same. **27.** 1 ft **28. (a)** 6 cm^2 **(b)** 30 cm^2 **29.** Draw altitudes $\overline{BE}$ and $\overline{DF}$ of triangles BCP and DCP, respectively. $\triangle ABE \cong \triangle CDF$ by AAS. Thus, $\overline{BE} \cong \overline{DF}$. Because $\overline{CP}$ is a base of $\triangle BCP$ and $\triangle DCP$, and because the heights are the same, the areas must be equal.

Mathematical Connections 14-1

Communication

1. Answers vary. The triangle may be drawn on graph paper and estimated. Students may look up Heron of Alexandria to find the formula $A = \sqrt{s(s - a)(s - b)(s - c)}$, where s is one-half of the perimeter of the triangle with sides a, b, and c. In this case, the area is $\sqrt{15(15 - 6)(15 - 11)(15 - 13)}$, or $\sqrt{1080}$, approximately 32.9 in.^{2} **3. (a)** The area of the 10-in. pizza is 25π in.^{2} The area of the 20-in. pizza is 100π in.^{2}. Because the area of the 20-in. pizza is 4 times as great, this pizza might cost 4 times as much, or \$40. However, this is not the case because other factors are considered rather than just the area of the pizza. **(b)** If the price is based only on the area of pizza, then the ratio between prices should be $1:k^2$. **5.** After the rotations, a rectangle is formed. The area of the rectangle is length times width, which in the case of the parallelogram is the same as the base times the height.

Open-Ended

7. (a) Answers vary depending on the size of the hand. **(b)** Answers vary. For example, many people will trace their hands on square-centimeter paper and then count the number of squares that are entirely contained in the outline. Next, they count the number of squares that are partially contained in the hand outline and multiply this number by $1/2$. The final estimate is the sum of the two numbers.

Cooperative Learning

9. (a) 5 sq. units **(b)** 12 units **(c) (i)** 3 **(ii)** 15 **(iii)** 20 **(d)** Answers vary depending on chosen shape. **(e)** Answers vary depending on chosen shape. **11.** Answers vary.

Questions from the Classroom

13. No. An angle is a union of two rays. The student probably means the area of the interior of an angle. However, because the area of the interior of an angle is infinite, it does not have a measurable area. **15.** The area of a garden does not depend on the perimeter of the garden. For example, a garden that is 2 m by 6 m has a perimeter of 16 m and an area of 12 m². A garden that is 4 m by 4 m also has a perimeter of 16 m but has an area of 16 m². **17. (a)** Yes, because if $A = s^2$, then $s = \sqrt{A}$ and $p = 4s = 4\sqrt{A}$. **(b)** No, for example, if $A = 20$ then the sides could be 4 and 5 and then $p = 18$ or the sides could be 2 and 10 and $p = 24$. **(c)** No, for example, if $A = 20$ then the side could be 5, the height 4 and $p = 4 \cdot 5 = 20$ or the side could be 10, then height 2 and $p = 4 \cdot 10 = 40$. **(d)** Yes, if you know the area, then you know the radius and $C = 2\pi\left(\sqrt{\dfrac{A}{\pi}}\right)$.

Review problems

19. $\left(8 + \dfrac{\pi}{2}\right)$ ft, or approximately 9.57 ft **21.** $(2\pi - 6)r$, or approximately $0.28r$ is the difference.

Assessment 14-2A

1. (a) $\sqrt{20} \approx 4.5$ **(b)** $\sqrt{17} \approx 4.1$ **2. (a)** 6 **(b)** $5a$ **(c)** 12 **(d)** $\dfrac{s}{2}\sqrt{3}$ **(e)** $x = \sqrt{2}, y = \sqrt{3}$ **(f)** $x = \sqrt{325}, y = \sqrt{374}$

3. (a) Make a right triangle with legs of length 2 and 3, and then the hypotenuse has length $\sqrt{13}$. **(b)** Make a right triangle with legs of length 1 and 2, and then the hypotenuse has length $\sqrt{5}$. **4.** Answers vary, for example,

5. $8 + 4\sqrt{2} \approx 13.66$ units **6.** $6\sqrt{5}$ cm, $12\sqrt{5}$ cm **7. (a)** no **(b)** yes **(c)** yes **8.** $\sqrt{450}$ cm, or $15\sqrt{2}$ cm **9.** $\sqrt{125}$ mi, or about 11.2 mi **10.** $6\sqrt{6}$, or about 14.7 ft **11.** $4\sqrt{6}$ m or approx 9.8 m **12. (a)** 240 ft² **(b)** Approx 793.7 ft² **13. (a)** $8\sqrt{84}$ **(b)** $4\sqrt{96}$ **14. (a)** $x = 8, y = 2\sqrt{3}$ **(b)** $x = 4, y = 2$ **15.** $\sqrt{9900}$ ft or approximately 99.5 ft **16.** The area of the trapezoid is equal to the sum of the areas of the three triangles. Thus,

$$\frac{1}{2}(a + b)(a + b) = \left(\frac{1}{2}\right)ab + \left(\frac{1}{2}\right)ab + \left(\frac{1}{2}\right)c^2$$

$$\frac{1}{2}(a^2 + 2ab + b^2) = ab + \left(\frac{1}{2}\right)c^2$$

$$\frac{a^2}{2} + ab + \frac{b^2}{2} = ab + \frac{c^2}{2}$$

Subtracting ab from both sides and multiplying both sides by 2, we have $a^2 + b^2 = c^2$. The reader should also verify that the angle formed by the two sides of length c has measure 90°. **17.** The area of the large square is equal to the sum of the areas of the small squares and the areas of the four right triangles. Therefore, $(a + b)^2 = c^2 + 4(ab/2)$. Thus, $a^2 + 2ab + b^2 = c^2 + 2ab$, which in turn implies that $c^2 = a^2 + b^2$. **18. (a)** $\dfrac{9\sqrt{3}}{4}, \dfrac{16\sqrt{3}}{4}, \dfrac{25\sqrt{3}}{4}$ **(b)** The area of the largest triangle is equal to the sum of the areas of the other two triangles. **19.** 13 **20. (a)** 4 **(b)** 5 **(c)** $2\sqrt{13}$ **(d)** $\dfrac{\sqrt{365}}{4}$, or approx. 4.78 **21.** $\dfrac{c\sqrt{3}}{4}$ **22. (a)** $x = 0$; $y = 0; y = \dfrac{3}{4}x$ **(b)** $(0, 0)$ **23. (a)** $(16, 0)$ **(b)** $8\sqrt{2}, 8\sqrt{2}$, and 16 **(c)** $(8\sqrt{2})^2 + (8\sqrt{2})^2 = 128 + 128 = 256 = 16^2$ **24. (a)** $(x + 3)^2 + (y - 4)^2 = 16$ **(b)** $(x + 3)^2 + (y + 2)^2 = 2$ **25. (a)** $(0, 0), 4$ **(b)** $(3, 2), 10$ **(c)** $(^-2, 3), \sqrt{5}$ **(d)** $(0, ^-3), 3$ **26.** $\sqrt{40}$, or approx. 6.3 km

Mathematical Connections 14-2

Communication

1. Answers vary. **(a)** Let the length of the side of the square be s. Draw the diagonal of the square and make the new square have side lengths equal to the diagonal. Then the area is $(\sqrt{2}s)^2 = 2s^2$. **(b)** Make the lengths of the sides of the new square $1/2$ the length of the diagonal, or $(\sqrt{2}s/2)$; then the area of the new square is $(\sqrt{2}s/2)^2 = s^2/2$. **3.** Draw a perpendicular segment from Y to the segment representing 200 m. Draw segment $\overline{XY}$. A right triangle is formed with side lengths 100 m and 150 m. Using the Pythagorean theorem, we find that $XY = \sqrt{100^2 + 150^2} = \sqrt{32,500} = 180.3$ m. **5.** Because $13 = 2^2 + 3^2$. We need to construct a right triangle with legs 2 cm and 3 cm.

Open-Ended

7. (a) Answers vary. For example, 6-8-10, and 12-16-20. **(b)** Yes, we know that if a-b-c is a Pythagorean triple, then $a^2 + b^2 = c^2$. This implies that $4(a^2 + b^2) = 4c^2$, and that $(2a)^2 + (2b)^2 = (2c)^2$. **(c)** No. For example, consider adding 2 to 3-4-5, which results in 5-6-7, which is not a Pythagorean triple. **(d)** $a^2 + b^2 = (2uv)^2 + (u^2 - v^2)^2$
$$= 4u^2v^2 + u^4 - 2u^2v^2 + v^4$$
$$= u^4 + 2u^2v^2 + v^4$$
$$= (u^2 + v^2)^2$$
$$= c^2$$

Therefore $a \cdot b \cdot c$ is a Pythagorean triple.

Questions from the Classroom

9. Yes, the same type of relationship does hold. For a proof and discussion, see G. Polya, *Mathematics and Plausible Reasoning*, Vol. 1 (Princeton, N.J.: Princeton University Press, 1954, pp. 15–17). Several examples were given in the Assessment sections. **11.** If x is the length of the missing leg, then $3^2 + x^2 = 5^2$. Therefore, $x^2 = 5^2 - 3^2$ and $x = \sqrt{5^2 - 3^2}$, not $\sqrt{3^2 + 5^2}$.

Gsp/GeoGebra Activities

13. See Appendix answers.

Review Problems

15. 0.032 km², 322 cm², 3.2 m², 3020 mm². **17. (a)** 10 cm, 10π cm, 25π cm², **(b)** 12 cm, 24π cm, 144π cm², **(c)** $\sqrt{17}$ m, $2\sqrt{17}$ m, $2\pi\sqrt{17}$ m **(d)** 10 cm, 20 cm, 100π cm²

Assessment 14-3A

1. (a) quadrilateral pyramid **(b)** quadrilateral prism; possibly a trapezoid prism **(c)** pentagonal pyramid **2. (a)** 6 cubes; 14 faces glued. **(b)** 24 cubes; 92 faces glued **3. (a)** *A, D, R, W* **(b)** $\overline{AR}, \overline{RD}, \overline{AD}, \overline{AW}, \overline{WR}, \overline{WD}$ **(c)** $\triangle ARD, \triangle AWD$, $\triangle AWR, \triangle WDR$ **(d)** R **(e)** $\overline{DW}$ **4. (a)** 5 **(b)** 4 **(c)** 4 **5. (a)** True **(b)** False **(c)** True **(d)** False **6.** All are possible. **7. (a)** right regular hexagonal pyramid **(b)** right square pyramid **(c)** cube **(d)** right square prism **(e)** right hexagonal prism **8. (a)** iv **(b)** ii **9. (a)** i, ii, and iii **(b)** ii, iii, and iv **10. (a)** **(b)** **(c)**

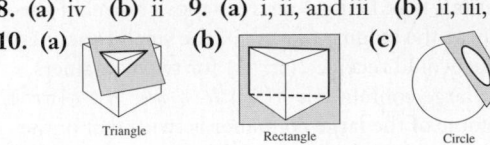

Triangle   Rectangle   Circle

11. (a) Three pairs of faces determined by three pairs of opposite sides of a hexagonal base **(b)** 120° since $\angle FAB$ is an angle whose measure is the measure of the dihedral angle formed by two adjacent faces. **12. (a)** $10 + 7 - 15 = 2$ **(b)** $9 + 9 - 16 = 2$

Mathematical Connections 14-3

Communication

1. Three. Each pair of parallel faces could be considered bases. **3.** Both could be drawings of a quadrilateral pyramid. In **(a)**, we are directly above the pyramid, and in **(b)**, we are directly below the pyramid. **5. (a)** A rectangle. $\overline{AD}$ is perpendicular to $\overline{AB}$ and to $\overline{AE}$ and therefore perpendicular to the plane *AEHB*. Since $\overline{MN}\|\overline{AD}$, $\overline{MN}$ is perpendicular to *AEHB*. Thus $\overline{MN}$ is perpendicular to every line through M in that plane. So $\overline{MN} \perp \overline{MH}$ and $\angle NMH$ is a right angle. Similarly the other angles of *MHGN* are right angles. **(b)** No, the edges of the pyramid are not the same length. **(c)** Yes, all the edges are the same length.

Cooperative Learning

7. Answers vary. **9. (a)** Parallelogram, rectangle, square, scalene triangle, isosceles triangle, equilateral triangle, pentagon, hexagon, a square and its interior, trapezoid that is not a parallelogram and a parallelogram that is neither a rhombus nor a rectangle **(b)** Triangle, quadrilateral, line segment triangle and its interior

Questions from the Classroom

11. There are two different nets shown below (assuming that the faces are indistinguishable).

Review Problems

13. (a) 54 m² **(b)** 300 in.²

Assessment 14-4A

1. (a) No **(b)** Yes **(c)** No **(d)** Yes **2. (a)** 96 cm² **(b)** 216π cm² **(c)** 236 cm² **(d)** 64π cm² **(e)** 96π cm² **3.** 2.5 L so buy 3 L **4.** 2688π mm² **5.** 162,307,600π km² **6.** 4 : 9 **7.** $(108\sqrt{21} + 216\sqrt{3})$ m² **8.** 10.5π in.² or approx. 33 in.² **9.** $\ell = 11$ cm, $w = 8$ cm, $h = 4$ cm **10. (a)** The lateral surface area is multiplied by 3. **(b)** The lateral surface area is multiplied by 3. **(c)** The lateral area is multiplied by 9. **11.** $(100 + 100\sqrt{17})$ cm² **12. (a)** 1.5π m² **(b)** 2.5π m² **13. (a)** 3 units **(b)** 5 units **(c)** 4 units **(d)** 216° **14.** 10π in.² **15.** 1 : 4 **16.** 118,300,000 π km² **17.** 4 : 3 **18. (a)** $\sqrt{\frac{10648}{6}}$ or approx. 42 cm **(b)** $\sqrt{5324}$ cm or approximately 72.97 cm **19.** $(6400\pi\sqrt{2} + 13,600\pi)$ cm²

Mathematical Connections 14-4

Communication

1. (a) Estimates vary. **(b)** The height of the can is $3d$ or $6r$, where d is the diameter of the ball. The circumference of the can is $2\pi r$. Therefore, the *LSA* is $(2\pi r)6r = 12\pi r^2$. The surface area of three balls is $3(4\pi r^2) = 12\pi r^2$. Therefore, the surface area of the three balls is the same as the lateral surface area of the can. **3.** No effect as 12 cm² were lost and 12 cm² were added.

Open-Ended

5. Answers vary. For example, draw a net for a prism that is 4 cm × 4 cm × 3 cm.

Cooperative Learning

7. The peeling of an orange can be used to demonstrate the formula for the surface area of a sphere, $SA = 4\pi r^2$.

Questions from the Classroom

9. Jodi is incorrect. Because the surface area of a cone is $SA = \pi r^2 + \pi r\ell$, if she doubles the radius and halves the slant height she will have $SA = \pi(2r)^2 + \pi(2r)\frac{\ell}{2}$, which reduces to $SA = 4\pi r^2 + \pi r\ell$, which is not the same as the original surface area. **11.** Jan is incorrect. You could point out that if you double each side of a rectangle, you increase the area by a factor of 4. For example, $A_1 = \ell \cdot w$ and $A_2 = (2\ell)(2w) = 4(\ell \cdot w) = 4A_1$. You might also just look at a cube with side s. Then $SA = 6s^2$. If you double each side, then you have $SA = 6(2s)^2 = 6 \cdot 4s^2 = 4(6s^2)$, which is 4 times as great as the original surface area.

Review Problems

13. $10\sqrt{5}$ cm **15. (a)** 240 cm; 2400 cm² **(b)** $(10\sqrt{2} + 30)$ cm, 75 cm² **17.** $\sqrt{3}$

Assessment 14-5A

1. (a) 8000 **(b)** 0.000675 **(c)** 7 **(d)** $\frac{25}{2916}$ or approx. 0.00857 **(e)** 345.6 **2.** 32.4 L **3.** Approx. 17,193 rooms

4. (a) $\left(\dfrac{256}{3}\right)\pi$ cm^3 **(b)** 216 cm^3 **(c)** 15π cm^3
(d) $\left(\dfrac{4000}{3}\right)\pi$ cm^3 **(e)** $\left(\dfrac{20,000}{3}\right)\pi$ ft^3
5.

	a.	b.	c.	d.	e.	f.
cm^3	2000	500	1500	5000	750	4800
dm^3	2	0.5	1.5	5	0.750	4.8
L	2	0.5	1.5	5	0.750	4.8
mL	2000	500	1500	5000	750	4800

6. (a) 200.0 **(b)** 0.320 **(c)** 1.0 **(d)** 5.00 **7.** 8 : 27
8. It is multiplied by 8.
9.

	a.	b.	c.	d.
Length	20 cm	10 cm	2 dm	15 cm
Width	10 cm	2 dm	1 dm	2 dm
Height	10 cm	3 dm	2 dm	2.5 dm
Volume (cm^3)	2000	6000	4000	7500
Volume (dm^3)	2	6	4	7.5
Volume (L)	2	6	4	7.5

10. 64 to 1 **11.** 2,500,000 L **12.** π mL **13. (a)** It is multiplied by 8. **(b)** It is multiplied by 27. **(c)** It is multiplied by n^3.
14. The Great Pyramid has the greater volume. It is approx. 25.7 times as great. **15.** 1/8 of the cone is filled.

16. It is multiplied by 2.197; 119.7% increase. **17.** $66\dfrac{2}{3}\%$ occupied by balls, so $33\dfrac{1}{3}\%$ occupied outside the balls.

18. $1 - \dfrac{\pi}{4}$ or about 21.5% **19.** They are equal. **20. (a)** Answers vary. For example, a square base with sides 5 m and a height of 12 m. **(b)** Infinitely many. Because $V = 100 = (1/3)a^2h$, where a is a side of the square base, then $300 = a^2h$. This equation has infinitely many solutions. **21. (a)** Metric ton **(b)** Kilogram **(c)** Gram **(d)** Metric ton **22. (a)** milligrams **(b)** kilograms **(c)** milligrams **23. (a)** 15 **(b)** 36 **(c)** 4.320 **(d)** 30 **(e)** 1.5625 **24. (a)** No **(b)** Possibly **(c)** Yes **(d)** Yes **(e)** Yes **25.** 16 kg **26. (a)** ⁻12°C **(b)** −1°C **(c)** 100°C **27. (a)** No **(b)** Yes **(c)** Yes **(d)** Hot

Mathematical Connections 14-5

Communication

1. (a) Doubling the height will only double the volume. Doubling the radius will multiply the volume by 4. This happens because the value of the radius is squared after it is doubled. **(b)** Yes
3. (i) The volume is approx. 145 ft^3, the capacity is approx. 1082 gal, and the weight of the water is approx. 9034 lb.
(ii) The volume is 8 m^3, the capacity is 8000 L, and the mass is 8000 kg. The metric problem is much easier to work because the conversions are much easier. They just involve moving the decimal point. The relationships among length, volume, capacity, and mass are much easier than in the English system.

Open-Ended

5. Answers vary but should have volumes close to 24π or 75.4 in.^{3}; for example, a rectangular prism that is 6 in. long, 4 in. wide, and 3.14 in. high. **7.** Answers vary, but the volume must be 1000 cm^3. Some students will worry about shelf space while others will worry about what shape is easiest to hold.

Questions from the Classroom

9. Consider a cube with side 6 cm. We need $s^3 = 6s^2$. This happens when $s = 6$. The volume is 216 cm^3 and its surface area is 216 cm^2. **11.** For each degree change in Celsius, there is a 9/5 degree change in Fahrenheit. When a person's temperature is 2 degrees above normal Celsius, it is (9/5)2 or 3.6 degrees above normal Fahrenheit. Therefore, being 2 degrees above normal Celsius is more serious than being 2 degrees above normal Fahrenheit. **13.** Suppose the volume of each of the small containers is $\pi r^2 h$. Then Jamie would receive $2(\pi r^2 h)$ for two containers. The volume of the large container is $\pi(2r)^2 h = \pi 4r^2 h = 4\pi r^2 h$. Therefore, the volume of the large container is twice that of the two small containers combined.

Review Problems

15. (a) $(20 + 6\pi)$ cm; $(48 + 18\pi)$ cm^2 **(b)** 40π cm; 100π cm^2
17. (a) yes **(b)** no

Chapter Review

1. (a) $8\dfrac{1}{2}$ cm^2 **(b)** $6\dfrac{1}{2}$ cm^2 **(c)** 7 cm^2 **2.** The pieces of the trapezoid are rearranged to form a rectangle with width $h/2$ and length $(b_2 + b_1)$. The area is $A = h/2(b_2 + b_1)$, which is the area of the initial trapezoid. **3.** Area($\triangle ABC$) < Area($\triangle ABD$) = Area($\triangle ABE$) < Area($\triangle ABF$). All the triangles have the same base, so the ordering is just by height and $\triangle ABD$ and $\triangle ABE$ have the same height. **4. (a)** $54\sqrt{3}$ cm^2 **(b)** 36π cm^2 **5. (a)** 12π cm^2 **(b)** $(12 + 4.5\pi)$ cm^2 **(c)** 24 cm^2 **(d)** 4π cm^2 **(e)** 64.5 cm^2 **(f)** 178.5 m^2 **6.** $\sqrt{16,200} = 90\sqrt{2}$ ft or approx. 127.3 ft **7.** 3
8. (a) yes **(b)** no **9.** 8 **10.** Answers vary. **11.** No; see definition. **12. (a)** Answers vary. **(b)** Answers vary.
13. Answers vary. **14.** $V + F - E = 2$ in (d) $5 + 5 - 8 = 2$; in (e) $12 + 8 - 18 = 2$; in (f) $10 + 7 - 15 = 2$ **15.** 5400 cm^2
16. (a) $SA = 32(2 + \sqrt{13})$ cm^2: $V = 128$ cm^3
(b) $SA = 96\pi$ cm^2; $V = 96\pi$ cm^3
(c) $SA = 100\pi$ m^2; $V = (500\pi)/3$ m^3 **(d)** $SA = 54\pi$ cm^2; $V = 54\pi$ cm^3 **(e)** $SA = 304$ m^2; $V = 320$ m^3 **17.** 65π m^2
18. The graph on the right has 8 times the volume of the figure on the left, rather than double as it should be. **19.** 252 cm^2
20. (a) 340 cm **(b)** 6000 cm^2 **21.** $2\sqrt{2}$ m^2 **22.** 62 cm
23. 30 cm
24. (a) Answers vary. **(b)** Answers vary. **(c)**

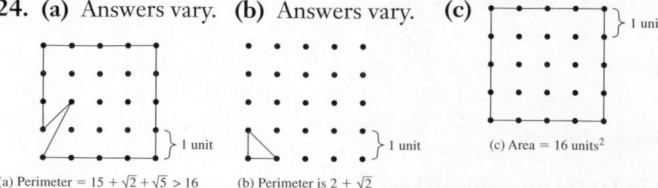

(a) Perimeter = $15 + \sqrt{2} + \sqrt{5} > 16$ (b) Perimeter is $2 + \sqrt{2}$ (c) Area = 16 units^2

25. $\sqrt{193.25}$ or approximately 13.9 in.
26. (a) $(x - 3)^2 + (y + 4)^2 = 25$ **(b)** $6, (5, ^-3)$
27. (a) metric tons **(b)** 1 cm^3 **(c)** 1 g **(d)** 1 **(e)** 25
(f) 2000 **(g)** 51,800 **(h)** 10,000,000 **(i)** 50,000 **(j)** 5.830
(k) 25,000 **(l)** 75,000 **(m)** 52.813 **(n)** 4.8 **28. (a)** 6000 kg
(b) Approx. 1.5565 m **29. (a)** L **(b)** kg **(c)** g **(d)** g **(e)** kg
(f) t **(g)** mL **30. (a)** Unlikely **(b)** Likely **(c)** Unlikely
(d) Unlikely **(e)** Unlikely **31. (a)** 2000 **(b)** 1000 **(c)** 3
(d) 0.0042 **(e)** 0.0002

Answers to Now Try This

14-1. (a) 12 **(b)** 6 **(c)** 4 **14-2.** Approximately 100 cm^2
14-3. (a) $\frac{b}{2} \cdot \frac{h}{2} = \frac{bh}{4}$ **(b)** The triangle was twice as large as the folded rectangle, so multiply by 2. Thus, the area of the triangle is $2 \cdot \frac{bh}{4} = \frac{1}{2}bh$. **14-4.** The areas of all the triangles are equal because the base of each triangle is the same and all have the same height. **14-5.** Let h_1 be the height of $\triangle ABC$ and h_2 be the height of $\triangle ADC$. Then the area of $\triangle ABC$ is $\frac{1}{2}(AC)h_1$ and the area of $\triangle ADC$ is $\frac{1}{2}(AC)h_2$. Therefore, the area of quadrilateral $ABCD$ is $\frac{1}{2}(AC)h_1 + \frac{1}{2}(AC)h_2 = \frac{1}{2}(AC)(h_1 + h_2)$. AC is the measure of one diagonal and $(h_1 + h_2)$ is the measure of the other diagonal, so the formula works.

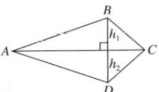

14-6. The new figure is a parallelogram with base $(b_1 + b_2)$ and height h, where b_1 and b_2 are the bases of the original trapezoid. The area of the parallelogram is $A = h(b_1 + b_2)$. Because this is twice the area of the original trapezoid, we divide by 2 to obtain $A = \frac{h(b_1 + b_2)}{2}$, which is the formula for the area of the original trapezoid. **14-7. (a)** Construction **(b)** Construction **(c)** The length is approximately equal to $\pi \cdot$ radius, half of the circumference. The height is approximately equal to the radius. Therefore $A = \frac{1}{2}Cr = \pi r^2$. As the circle is cut into more and more sectors and put back together, the shape becomes more and more like a parallelogram. **14-8.** 30 m^2 **14-9.** The square on one leg that is labeled 1 could be cut off and placed in the dashed space on the square on the hypotenuse. Then pieces 2, 3, 4, and 5 could be cut off and placed around piece 1 so that the square on the hypotenuse is filled exactly with the five pieces. This shows that the sum of the areas of the squares on the two legs of a right triangle is equal to the area of the square on the hypotenuse. **14-10.** Jason ran the hypotenuse of a right triangle with legs each 10 yd long. **14-11. (a)** You could build the triangle and then measure the angles to see whether there was a right angle. If the angle is a right angle, then the triangle is a right triangle. You could measure the three sides and use the Pythagorean theorem and its converse to see whether a right triangle is formed. **(b)** If the three lengths of a right triangle are multiplied by a fixed number, then the resulting lengths determine a right triangle; for example, if the right triangle lengths are 3-4-5, and the fixed number is 5, then 15-20-25 is a right triangle. **(c)** If the

three lengths of a right triangle are multiplied by a fixed number, then the resulting numbers determine a right triangle. **14-12.** It makes no difference in the distance formula if $(x_1 - x_2)$ and $(y_1 - y_2)$ are used instead of $(x_2 - x_1)$ and $(y_2 - y_1)$, respectively. Because both quantities in the formula are squared, the result is the same whether the difference is positive or negative. **14-13.** Answers vary. **14-14.** $V + F - E = 2$; **(a)** $8 + 6 - 12 = 2$; $10 + 7 - 15 = 2$; $12 + 8 - 18 = 2$ **14-15. (a)** $SA = 2 \cdot (5/2) \cdot 8 + 2 \cdot (8 \cdot 11) + 2 \cdot (5/2) \cdot 11 = 271$ in.^{2} **(b)** No, the rectangle would have to be 21 in. by 16 in. **14-16.** Because we want the surface area of a right prism, we must include the top and bottom, so we need $2B$ (where B is the area of the base, which is the same as the area of the top) in the formula $SA = ph + 2B$. From the net, we see that the lateral surface area opens up into a rectangle that has width equal to the height, h. The length of the rectangle is equal to the sum of the lengths of the sides of the base, which is the perimeter of the base. Therefore, the area of the rectangle (lateral surface area of the prism) is $A = \ell w = ph$. Hence, the surface area for any right prism is given by $SA = ph + 2B$. **14-17. (a)** Both figures have a volume of 9 cubic units. **(b)** No, the second figure has the greater surface area. **(c)** The first figure has surface area 34 square units and the second figure has surface area 36 square units.
14-18. (a) Move the decimal point twice as many places as in a linear conversion and in the same direction. For example, the area of a square that is 1 m on each side is 1 m^2; 1 m $=$ 10 dm; 1 m$^2 =$ 100 dm^2. **(b)** Move the decimal point 3 times as many places as in a linear conversion and in the same direction. 1 m$^3 =$ 1000 dm^3. **14-19. (a)** The two figures have bases in the same plane and the figures have the same height. Figure 14-82 shows that if a plane parallel to the base is passed through the figures, then equal areas are obtained. By Cavalieri's Principle, these two figures have equal volumes. **(b) (i)** By Cavalieri's Principle, the volumes are the same. **(ii)** By Cavalieri's Principle, the volumes are the same. **14-20. (a)** We know that the height of each figure is $2r$ because the height of the sphere is $2r$. The volume of the cylinder is $\pi r^2 2r = 2\pi r^3$. The volume of the cone is $\frac{1}{3}\pi r^2 \cdot 2r = \frac{2}{3}\pi r^3$. The volume of the sphere is $\frac{4}{3}\pi r^3$.
(b) Using a common denominator of 3, the three formulas are $\frac{6}{3}\pi r^3, \frac{2}{3}\pi r^3$, and $\frac{4}{3}\pi r^3$. The ratio is 6:2:4, which simplifies to 3:1:2.
14-21. (a) g **(b)** kg **(c)** dm^3 **(d)** mL **(e)** g **(f)** kL
(g) t **14-22.** Yes, when it is $^-40$°C it is $^-40$°F.

Answers to Brain Teasers

Section 14-1

1. 64 square units **2.** 65 square units **3.** Although the pieces look like they should fit together, they do not really fit. To see this, assume the pieces do fit. We then obtain the following figure:

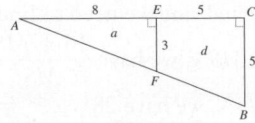

Since $\triangle AEF \sim \triangle ACB$, we have $\frac{8}{13} = \frac{3}{5}$, which is a contradiction.

This implies that pieces like those in the figure cannot fit together to form a triangle. In order for the pieces to fit together, the measure of $\overline{EF}$ must satisfy $\dfrac{8}{13} = \dfrac{EF}{5}$; hence, $EF = \dfrac{40}{13} = 3\dfrac{1}{13}$. Since $3\dfrac{1}{13}$ is close to 3, the discrepancy is so small that the pieces appear to fit.

Section 14-2

Let s be the length of the side of the square field. Then the area of the field is s^2 and the area covered by the large sprinkler is $\left(\dfrac{1}{2}s\right)^2 \pi$. The area of the field not covered by the sprinkler is

$$s^2 - \left(\frac{1}{2}s\right)^2 \pi = s^2 - \frac{1}{4}s^2\pi$$

Consider a square enclosing each small circle. The length of the side of each small square is $\dfrac{1}{3}s$. The radius of each small circle is $\dfrac{1}{6}s$. The area of each small square that is not covered by the small circle is

$$\left(\frac{1}{3}s\right)^2 - \left(\frac{1}{6}s\right)^2 \pi = \frac{1}{9}s^2 - \frac{1}{36}s^2\pi$$

Multiply this by 9 to find the total area of the field that is not covered:

$$9\left(\frac{1}{9}s^2 - \frac{1}{36}s^2\pi\right) = s^2 - \frac{1}{4}s^2\pi$$

which is the same as the area not covered by the large sprinkler.

Therefore, both sprinkler systems cover the same percentage of the square field and it does not matter which system is used if the only selection criterion is the amount of land covered by the system.

A related problem is to change the number of circles contained in the square. Will the percentage always be the same? Changing the shape of the field can also vary the problem.

Section 14-3

See "Making a Better Beer Glass" by A. Hoffer in the May 1982 issue of *Mathematics Teacher*, pp. 378–379, to see the shape of the glass.

Section 14-4

The cone and the flattened region obtained by slitting the cone along a slant height are shown.

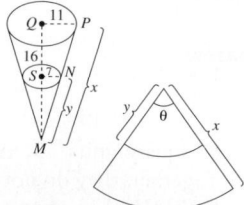

To construct the flattened ring we need to find x, y, and θ.

Because $\triangle MQP \sim \triangle MSN$, we have $\dfrac{16 + MS}{MS} = \dfrac{11}{7}$. Hence, $MS = 28$ cm. In $\triangle MSN$, we have $28^2 + 7^2 = y^2$, or $y \approx 28.86$. In $\triangle PQM$: $x^2 = 11^2 + 44^2$, or $x \approx 45.35$ cm. To find θ, we roll the sector with radius y and central angle θ into the cone whose base

is 7 cm and whose slant height is y. Hence, $2\pi y \cdot \dfrac{\theta}{360} = 2\pi \cdot 7$, or $\theta \approx \dfrac{7 \cdot 360}{28.86} \approx 87°19'$.

Answer to the Preliminary Problem

If we compute the distance using Al's method as shown in figure (a) at the right, the distance is

$$d = 10\text{ m} + 2.7\text{ m} + 0.3\text{ m} = 13\text{ m}$$

Using the net in figure (b) below and the Pythagorean theorem we compute the distance, d, as:

$$d^2 = 7^2 + 10.6^2$$
$$d^2 = 49 + 112.36$$
$$d^2 = 161.36$$
$$d = \sqrt{161.36} \approx 12.70\text{ m}$$

This distance is shorter than 13 m so Betty is correct.

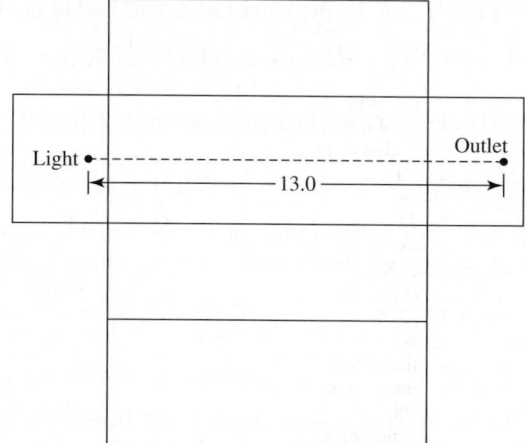

(a)

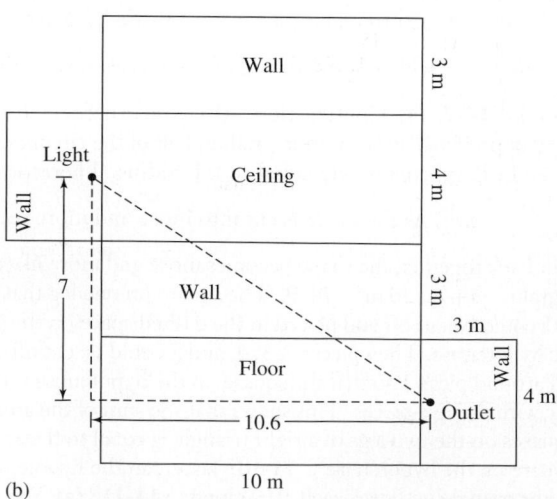

(b)

Note that for rooms of other dimensions with different placements of the light and outlet, Al's method may yield a shorter distance.

Index

A

AA. *See* Angle, angle (AA) similarity for triangles
AAS. *See* Angle, angle, side (AAS) property
Abscissa (*x*-coordinate), 430
Absolute deviation, 585
Absolute value, 227
Absolute zero, 907
Acre, 830, 832
Acute angle, 641, 642
Acute triangle, 668
Addends, 99
Adding up, 161
Addition
  base-five, 120–121
  in bases other than ten, 120–121
  charged-field model for, 224
  chip model for, 224
  of decimals, 342–343, 344
  defined for addition of integers, 228–229
  definition of subtraction of rational numbers
     in terms of, 284
  estimation, 162–164
  of integers, 224–231
  mental mathematics, 160
  of mixed numbers, 281, 282
  number-line model for, 224–226
  pattern model for, 226
  of rational numbers
    with like denominators, 277–278
    with unlike denominators, 278–279
  of real numbers, 398
  relation to multiplication of whole numbers, 128
  relation to subtraction of whole numbers,
    105, 106
  symbol for, 100
  of two-digit numbers, 115
  of whole numbers, 98–101, 108
Addition algorithms, 113–117
  lattice, 117
  left-to-right, 116
  scratch, 117
Addition and subtraction properties of equations,
  416
Addition facts, 104–105
Addition method, for area, 828–829
Addition properties
  additive inverse property, 230
    of rational numbers, 281, 283
    of real numbers, 398
  associative property of addition
    of integers, 230
    of real numbers, 398
    of whole numbers, 102, 103
  closure property of addition
    of integers, 230
    of real numbers, 398
    of whole numbers, 101
  commutative (order) property of addition
    of integers, 230
    of real numbers, 398
    of whole numbers, 101–102, 103
  of equality, 416
    for integers, 230–231
    for rational numbers, 283
    of real numbers, 416

identity (zero) property of addition
  of integers, 230
  of real numbers, 398
  of whole numbers, 102, 103
  for rational numbers, 281–283
  whole-number, 101–104
Additive identity, 230
Additive inverse, 230, 235
Additive inverse property
  of integers, 230
  of rational numbers, 281, 283
  of real numbers, 398
Additive property
  in Egyptian numeration system, 56
  in Roman numeration system, 58
Additive relationship, 314
Adjacent angles, 638
Adleman, Leonard, 201
Algebra
  Boolean, 37
  equations, 416–425
  fractions in, 267, 285
  functions, 425–445
  history of, 404
  missing-addend model and, 106
  overview, 404–405
  variables, 405–415
  whole-number addition and subtraction, 106
  whole-number multiplication and division, 126
Algebraic expressions, 405, 406
  written as fraction, 267
Algebraic thinking, 404, 411–412
Algorithms
  addition, 113–117
  cashier's, 161
  defined, 113
  division, 135–136, 149–151
    of rational numbers, 299–300
    of rational numbers, alternate, 300–301
  equal-additions, 120
  Euclidean, 208–209
  expanded, 113, 114
  multiplication, 145–149
  standard, 113, 114, 116
  subtraction, 117–119
  trading off, 159
al-Khowarizmi, Mohammed ibn Musa, 404
  See also al-Khwarizmi, Abu.
al-Khwarizmi, Abu, 113
  See also al-Khowarizmi, Mohammed ibn Musa.
all, 35
Alternate exterior angles, 682
Alternate interior angles, 682
Altitude
  of cone, 873
  of parallelogram, 834
  of pyramid, 868
  of triangles, 714–715, 738
Amount, 377
and
  biconditionals and, 40
  in compound statement, 36
  for intersection of set, 81
Angle, Angle, Side (AAS) property, 724
Angle, Angle (AA) similarity for triangles, 747

Angle, Side, Angle property (ASA), 723–728
Angle bisectors, 713
  constructing, 735–736
  properties of, 739–740
Angles, 638, 680–695
  acute, 641, 642
  adjacent, 638
  alternate exterior, 682
  alternate interior, 682
  base angles of an isosceles triangle, 714
  central, of circle, 643
  complementary, 681
  congruent, 667, 704, 705, 710–711
    constructing, 710–711
  congruent segments and, 666–667
  dihedral, 645–646
  exterior, 666, 682
  of incidence, 792–793
  included, 711
  interior, 638, 666, 682
  measurement of, 639–641, 642
  obtuse, 641, 642
  parallel lines and, 682
  planar, 642
  of polygon, 666
  of reflection, 793
  reflex, 639–640
  right, 641, 642
  size transformations and, 803
  sum of measures of
    exterior angles of convex *n*-gon, 686–689
    interior angles of convex polygon with *n*
     sides, 685–686
    of triangle, 683–685
  supplementary, 641, 681
  symbol for, 639
  transversal, 681
  turn, 673, 776
  types of, 641–642
  vertical, 680–681
  walking around stars, 688–689
Annual percentage yield (APY), 379
Annuities upon Lives (De Moivre), 603
Apex
  of cone, 873
  of pyramid, 868
Apothem, 839
Applications
  involving dilations, 805–806
  involving equations, 420–422
  involving normal curves, 602–604
  involving percent, 371–375
  involving reflections, 792–793
  involving rotations, 778–780
APY. *See* Annual percentage yield (APY)
Arc, 642–643
  length of, 660
  major, 643
  minor, 643
Archimedes, 396, 658
Area. *See also* Surface area (SA)
  of a circle, 840, 841
  converting units of, 830–831
  on geoboard, 828–830

Glossary of Symbols

Symbol	Meaning
$=$	is equal to
a_n	nth term of a sequence
a^n	a to the nth power
S_n	sum of the first n terms of a sequence
$p \vee q$	p or q
$\sim p$	negation of p or not p
$p \wedge q$	p and q
$p \equiv q$	p is logically equivalent to q
$p \rightarrow q$	p implies q
$p \leftrightarrow q$	p implies q and q implies p; or p if, and only if, q
4_{five}	4 base five
$E2T_{\text{twelve}}$	base ten meaning $11 \cdot 12^2 + 2 \cdot 12^1 + 10 \cdot 1$
$\{a, b, c\}$	set containing elements a, b, and c
$\{x \mid \ldots\}$	set builder notaion
$\in$	is an element of
$\notin$	is not an element of
$\varnothing$ or $\{\ \}$	empty set
$\subset$	is a proper subset of
$\subseteq$	is a subset of
$A \cup B$	union of sets A and B
$A \cap B$	intersection of sets A and B
U	universal set
$\overline{A}$	the complement of set A
$B - A$	set difference or complement of A relative to B
$A \sim B$	A is equivalent to B
$A \times B$	Cartesian product of sets A and B
$n(S)$	cardinal number of set S
$>$	is greater than
$<$	is less than
$\geq$	is greater than or equal to
$\leq$	is less than or equal to
$f(x)$	f of x, or output of f at x
(a, b)	ordered pair
$g \circ f$	composition of f with g
$(g \circ f)(x)$	$g(f(x))$

Symbol	Meaning
^{-}a	opposite of a or the additive inverse of a
$\lvert a \rvert$	absolute value of a
$a \mid b$	a divides b
$a \nmid b$	a does not divide b
$\sqrt{\ }$	principal square root
GCD	greater common divisor
LCM	least common multiple
$\dfrac{a}{b}$	fraction "a over b" or ratio, $a \div b$ with $b \neq 0$
$5\dfrac{3}{4}$	mixed number $5 + \dfrac{3}{4}$
a^0	$1, a \neq 0$
a^{-n}	$\dfrac{1}{a^n}, a \neq 0$
$\doteq$	is approximately equal
$\approx$	is approximately equal
$0.\overline{18}$	repeating decimal $0.18181818\ldots$
$\sqrt[n]{\ }$	the nth root
$a^{\frac{1}{n}}$	a to the $\dfrac{1}{n}$ power; or nth root of a
$a^{\frac{m}{n}}$	$\sqrt[n]{a^m}$
$\%$	percent
$P(E)$	probability of an event E
$P(\overline{A})$	probability of the complement of A
$n!$	n factorial which is equal to $n(n - 1)(n - 2) \cdot \ldots \cdot 3 \cdot 2 \cdot 1$ for $n \geq 1$
$0!$	zero factorial which equals 1
$_nP_r$	number of permutations of n objects chosen r at a time
$_nC_r$	number of combinations of n objects chosen r at a time
$P(B \mid A)$	conditional probability that event B occurs given that event A has occurred
E	mathematical expectation
$\overline{x}$	the arithmetic mean
MAD	mean absolute deviation